MW00352105

Structural Analysis

SOLID MECHANICS AND ITS APPLICATIONS
Volume 163

Series Editor: G.M.L. GLADWELL
Department of Civil Engineering
University of Waterloo
Waterloo, Ontario, Canada N2L 3GI

Aims and Scope of the Series

The fundamental questions arising in mechanics are: *Why?*, *How?*, and *How much?*
The aim of this series is to provide lucid accounts written by authoritative research-
ers giving vision and insight in answering these questions on the subject of mechanics as it relates to solids.

The scope of the series covers the entire spectrum of solid mechanics. Thus it includes the foundation of mechanics; variational formulations; computational mechanics; statics, kinematics and dynamics of rigid and elastic bodies: vibrations of
solids and structures; dynamical systems and chaos; the theories of elasticity, plasticity and viscoelasticity; composite materials; rods, beams, shells and membranes;
structural control and stability; soils, rocks and geomechanics; fracture; tribology;
experimental mechanics; biomechanics and machine design.

The median level of presentation is the first year graduate student. Some texts are
monographs defining the current state of the field; others are accessible to final year
undergraduates; but essentially the emphasis is on readability and clarity.

For other titles published in this series, go to
www.springer.com/series/6557

O.A. Bauchau ● J.I. Craig

Structural Analysis

With Applications to Aerospace Structures

 Springer

O.A. Bauchau
School of Aerospace Engineering
Georgia Institute of Technology
Atlanta, Georgia
USA

J.I. Craig
School of Aerospace Engineering
Georgia Institute of Technology
Atlanta, Georgia
USA

ISBN 978-90-481-2515-9   e-ISBN 978-90-481-2516-6
Springer Dordrecht Heidelberg London New York

Library of Congress Control Number: 2009932893

Printed on acid-free paper

Springer is part of Springer Science+Business Media (www.springer.com)

To our wives, Yi-Ling and Nancy, and our families

Preface

Engineered structures are almost as old as human civilization and undoubtedly began with rudimentary tools and the first dwellings outside caves. Great progress has been made over thousands of years, and our world is now filled with engineered structures from nano-scale machines to soaring buildings. Aerospace structures ranging from fragile human-powered aircraft to sleek jets and thundering rockets are, in our opinion, among the most challenging and creative examples of these efforts.

The study of mechanics and structural analysis has been an important area of engineering over the past 300 years, and some of the greatest minds have contributed to its development. Newton formulated the most basic principles of equilibrium in the 17th century, but fundamental contributions have continued well into the 20th century. Today, structural analysis is generally considered to be a mature field with well-established principles and practical tools for analysis and design. A key reason for this is, without doubt, the emergence of the finite element method and its widespread application in all areas of structural engineering. As a result, much of today's emphasis in the field is no longer on structural analysis, but instead is on the use of new materials and design synthesis.

The field of aerospace structural analysis began with the first attempts to build flying machines, but even today, it is a much smaller and narrower field treated in far fewer textbooks as compared to the fields of structural analysis in civil and mechanical engineering. Engineering students have access to several excellent texts such as those by Donaldson [1] and Megson [2], but many other notable textbooks are now out of print.

This textbook has emerged over the past two decades from our efforts to teach core courses in advanced structural analysis to undergraduate and graduate students in aerospace engineering. By the time students enroll in the undergraduate course, they have studied statics and covered introductory mechanics of deformable bodies dealing primarily with beam bending. These introductory courses are taught using texts devoted largely to applications in civil and mechanical engineering, leaving our students with little appreciation for some of the unique and challenging features of aerospace structures, which often involve thin-walled structures made of fiber-reinforced composite materials. In addition, while in widespread use in industry and the subject of numerous specialized textbooks, the finite element method is only slowly finding its way into general structural analysis texts as older applied methods and special analysis techniques are phased out.

The book is divided into four parts. The first part deals with basic tools and concepts that provide the foundation for the other three parts. It begins with an introduction to the equations of linear elasticity, which underlie all of structural analysis. A second chapter presents the constitutive laws for homogeneous, isotropic and linearly elastic material but also includes an introduction to anisotropic materials and particularly to transversely isotropic materials that are typical of layered composites. The first part concludes with chapter 4, which defines isostatic and hyperstatic problems and introduces the fundamental solution procedures of structural analysis: the displacement method and the force method.

Part 2 develops Euler-Bernoulli beam theory with emphasis on the treatment of beams presenting general cross-sectional configurations. Torsion of circular cross-sections is discussed next, along with Saint-Venant torsion theory for bars of arbitrary shape. A lengthy chapter is devoted to thin-walled beams typical of those used in aerospace structures. Coupled bending-twisting and nonuniform torsion problems are also addressed.

Part 3 introduces the two fundamental principles of virtual work that are the basis for the powerful and versatile energy methods. They provide tools to treat more realistic and complex problems in an efficient manner. A key topic in Part 3 is the development of methods to obtain approximate solution for complex problems. First, the Rayleigh-Ritz method is introduced in a cursory manner; next, applications of the weak statement of equilibrium and of energy principles are presented in a more formal manner; finally, the finite element method applied to trusses and beams is presented. Part 3 concludes with a formal introduction of variational methods and general statements of the energy principles introduced earlier in more applied contexts.

Part 4 covers a selection of advanced topics of particular relevance to aerospace structural analysis. These include introductions to plasticity and thermal stresses, buckling of beams, shear deformations in beams and Kirchhoff plate theory.

In our experience, engineering students generally grasp concepts more quickly when presented first with practical examples, which then lead to broader generalizations. Consequently, most concepts are first introduced by means of simple examples; more formal and abstract statements are presented later, when the student has a better grasp of the significance of the concepts. Furthermore, each chapter provides numerous examples to demonstrate the application of the theory to practical problems. Some of the examples are re-examined in successive chapters to illustrate alternative or more versatile solution methods. Step-by-step descriptions of important solution procedures are provided.

As often as possible, the analysis of structural problems is approached in a unified manner. First, kinematic assumptions are presented that describe the structure's displacement field in an approximate manner; next, the strain field is evaluated based on the strain-displacement relationships; finally, the constitutive laws lead to the stress field for which equilibrium equations are then established. In our experience, this approach reduces the confusion that students often face when presented with developments that don't seem to follow any obvious direction or strategy but yet, inevitably lead to the expected solution.

The topics covered in parts 1 and 2 along with chapters 9 and 10 from part 3 form the basis for a four semester-hour course in advanced aerospace structural analysis taught to junior and senior undergraduate students. An introductory graduate level course covers part 2 and selected chapters in parts 3 and 4, but only after a brief review of the material in part 1. A second graduate level course focusing on variational end energy methods covers part 3 and selected chapters in part 4. A number of homework problems are included throughout these chapters. Some are straightforward applications of simple concepts, others are small projects that require the use of computers and mathematical software, and others involve conceptual questions that are more appropriate for quizzes and exams.

A thorough study of differential calculus including a basic treatment of ordinary and partial differential equations is a prerequisite. Additional topics from linear algebra and differential geometry are needed, and these are reviewed in an appendix.

Notation is a challenging issue in structural analysis. Given the limitations of the Latin and Greek alphabets, the same symbols are sometimes used for different purposes, but mostly in different contexts. Consequently, no attempt has been made to provide a comprehensive list of symbols, which would lead to even more confusion. Also, in mechanics and structural analysis, sign conventions present a major hurdle for all students. To ease this problem, easy to remember sign conventions are used systematically. Stresses and force resultants are positive on positive faces when acting along positive coordinate directions. Moments and torques are positive on positive faces when acting about positive coordinate directions using the right-hand rule.

In a few instances, new or less familiar terms have been chosen because of their importance in aerospace structural analysis. For instance, the terms "isostatic" and "hyperstatic" structures are used to describe statically determinate and indeterminate structures, respectively, because these terms concisely define concepts that often puzzle and confuse students. Beam bending stiffnesses are indicated with the symbol "H" rather than the more common "EI." When dealing exclusively with homogeneous material, notation "EI" is easy to understand, but in presence of heterogeneous composite materials, encapsulating the spatially varying elasticity modulus in the definition of the bending stiffness is a more rational approach.

It is traditional to use a bold typeface to represent vectors, arrays, and matrices, but this is very difficult to reproduce in handwriting, whether in a lecture or in personal notes. Instead, we have adopted a notation that is more suitable for handwritten notes. Vectors and arrays are denoted using an underline, such as $\underline{u}$ or $\underline{F}$. Unit vectors are used frequently and are assigned a special notation using a single overbar, such as $\bar{\imath}_1$, which denotes the first Cartesian coordinate axis. We also use the overbar to denote non-dimensional scalar quantities, $i.e.$, $\bar{k}$ is a non-dimensional stiffness coefficient. This is inconsistent, but the two uses are in such different contexts that it should not lead to confusion. Matrices are indicated using a double-underline, $i.e.$, $\underline{\underline{C}}$ indicates a matrix of M rows and N columns.

Finally, we are indebted to the many students at Georgia Tech who have given us helpful and constructive feedback over the past decade as we developed the course notes that are the predecessor of this book. We have tried to constructively utilize their initial confusion and probing questions to clarify and refine the treatment of important but confusing topics. We are also grateful for the many discussions and valuable feedback from our colleagues, Profs. Erian Armanios, Sathya Hanagud, Dewey Hodges, George Kardomateas, Massimo Ruzzene, and Virgil Smith, several of whom have used our notes for teaching advanced aerospace structural analysis here at Georgia Tech.

Atlanta, Georgia,   *Olivier Bauchau*
July 2009   *James Craig*

Contents

Part II   Beams and thin-wall structures

Part III   Energy and variational methods

Part I

Basic tools and concepts

1

Basic equations of linear elasticity

Structural analysis is concerned with the evaluation of deformations and stresses arising within a solid object under the action of applied loads. If time is not explicitly considered as an independent variable, the analysis is said to be static; otherwise it is referred to as structural dynamic analysis, or simply structural dynamics. Under the assumption of small deformations and linearly elastic material behavior, three-dimensional formulations result in a set of fifteen linear first order partial differential equations involving the displacement field (three components), the stress field (six components) and the strain field (six components). This chapter presents the derivation of these governing equations. In many applications, this complex problem can be reduced to simpler, two-dimensional formulations called plane stress and plane strain problems.

For most situations, it is not possible to develop analytical solutions of these equations. Consequently, structural analysis is concerned with the analysis of *structural components*, such as bars, beams, plates, or shells, which will be addressed in subsequent chapters. In each case, assumptions are made about the behavior of these structural components, which considerably simplify the analysis process. For instance, given a suitable set of assumptions, the analysis of bar and beam problems reduces to the solution of one-dimensional equations for which analytical solutions are easily obtained.

1.1 The concept of stress

1.1.1 The state of stress at a point

The state of stress in a solid body is a measure of the intensity of forces acting within the solid. It can be visualized by cutting the solid by a plane normal to unit vector, $\bar{n}$, to create two free bodies which reveal the forces acting on the exposed surfaces. From basic statics, it is well-known that the distribution of forces and moments that will appear on the surface of the cut can be represented by an *equipollent* force, $\underline{F}$, acting at a point of the surface and a couple, $\underline{M}$. Newton's 3rd law also requires a force and couple of equal magnitudes and opposite directions to act on the two surfaces created by the cut through the solid, as depicted in fig. 1.1. (See appendix A for a description of the vector, array and matrix notations used in this text.)

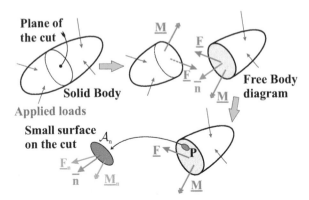

Fig. 1.1. A solid body cut by a plane to isolate a free body.

Consider now a small surface of area $\mathcal{A}_n$ located at point **P** on the surface generated by the cut in the solid. The forces and moments acting on this surface are equipollent to a force, $\underline{F}_n$, and couple, $\underline{M}_n$; note that these resultants are, in general, different, in both magnitude and orientation, from the corresponding resultants acting on the entire surface of the cut, as shown in fig. 1.1. Let the small surface be smaller and smaller until it becomes an element of infinitesimal area $\mathrm{d}\mathcal{A}_n \to 0$. As the surface shrinks to a differential size, the force and couple acting on the element keep decreasing in magnitude and changing in orientation whereas the normal to the surface remains the unit vector $\bar{n}$ of constant direction in space. This limiting process gives rise to the concept of *stress vector*, which is defined as

$$\underline{\tau}_n = \lim_{\mathrm{d}\mathcal{A}_n \to 0} \left(\frac{F_n}{\mathrm{d}\mathcal{A}_n} \right). \tag{1.1}$$

The existence of the stress vector, *i.e.*, the existence of the limit in eq. (1.1), is a *fundamental assumption of continuum mechanics*. In this limiting process, it is assumed that the couple, $\underline{M}_n$, becomes smaller and smaller and, in the limit, $\underline{M}_n \to 0$ as $\mathrm{d}\mathcal{A}_n \to 0$; this is also an assumption of continuum mechanics which seems to be reasonable because in the limiting process, both forces and moment arms become increasingly small. Forces decrease because the area they act on decreases and moment arms decrease because the dimensions of the surface decrease. At the limit, the couple is the product of a differential element of force by a differential element of moment arm, giving rise to a negligible, second order differential quantity.

In conclusion, whereas an equipollent couple might act on the entire surface of the cut, the equipollent couple is assumed to vanish on a differential element of area of the same cut. The total force acting on a differential element of area, $\mathrm{d}\mathcal{A}_n$, is

$$\underline{F}_n = \mathrm{d}\mathcal{A}_n \, \underline{\tau}_n. \tag{1.2}$$

Clearly, the stress vector has units of force per unit area. In the SI system, this is measured in Newton per square meters, or Pascals (Pa).

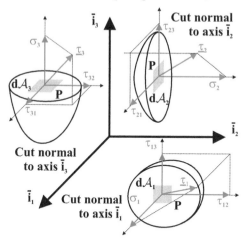

Fig. 1.2. A rigid body cut at point P by three planes orthogonal to the Cartesian axes.

During the limiting process described in the previous paragraph, the surface orientation, as defined by the normal to the surface, is kept constant in space. Had a different normal been selected, a different stress vector would have been obtained.

To illustrate this point, consider a solid body and a coordinate system, $\mathcal{I}$, consisting of three mutually orthogonal unit vectors, $\mathcal{I} = (\bar{\imath}_1, \bar{\imath}_2, \bar{\imath}_3)$, as shown in fig. 1.2. First, the solid is cut at point $\mathbf{P}$ by a plane normal to axis $\bar{\imath}_1$; on the surface of the cut, at point $\mathbf{P}$, a differential element of surface with an area $\mathrm{d}\mathcal{A}_1$ is defined and let $\underline{\tau}_1$ be the stress vector acting on this face. Next, the solid is cut at the same point by a plane normal to axis $\bar{\imath}_2$; at point $\mathbf{P}$, let $\underline{\tau}_2$ be the stress vector acting on the differential element of surface with an area $\mathrm{d}\mathcal{A}_2$. Finally, the process is repeated a third time for a plane normal to axis $\bar{\imath}_3$; at point $\mathbf{P}$, the stress vector $\underline{\tau}_3$ is acting on the differential element of surface with an area $\mathrm{d}\mathcal{A}_3$. Clearly, three stress vectors, $\underline{\tau}_1$, $\underline{\tau}_2$, and $\underline{\tau}_3$ are acting at the same point $\mathbf{P}$, but on three mutually orthogonal faces normal to axes $\bar{\imath}_1$, $\bar{\imath}_2$, and $\bar{\imath}_3$, respectively. Because these three stress vectors are acting on three faces with different orientations, there is no reason to believe that those stress vectors should be identical.

To further understand the state of stress at point $\mathbf{P}$, the components of each stress vectors acting on the three faces are defined

$$\underline{\tau}_1 = \sigma_1 \bar{\imath}_1 + \tau_{12} \bar{\imath}_2 + \tau_{13} \bar{\imath}_3, \tag{1.3a}$$

$$\underline{\tau}_2 = \tau_{21} \bar{\imath}_1 + \sigma_2 \bar{\imath}_2 + \tau_{23} \bar{\imath}_3, \tag{1.3b}$$

$$\underline{\tau}_3 = \tau_{31} \bar{\imath}_1 + \tau_{32} \bar{\imath}_2 + \sigma_3 \bar{\imath}_3. \tag{1.3c}$$

The stress components σ_1, σ_2, and σ_3 are called *direct*, or *normal stresses*; they act on faces normal to axes $\bar{\imath}_1$, $\bar{\imath}_2$, and $\bar{\imath}_3$, respectively, in directions along axes $\bar{\imath}_1$, $\bar{\imath}_2$, and $\bar{\imath}_3$, respectively. The stress components τ_{12} and τ_{13} are called *shearing* or *shear stresses*; both act on the face normal to axis $\bar{\imath}_1$, in directions of axes $\bar{\imath}_2$ and $\bar{\imath}_3$, respectively. Similarly, stress components τ_{21} and τ_{23} both act on the face normal to axis $\bar{\imath}_2$, in directions of axes $\bar{\imath}_1$ and $\bar{\imath}_3$, respectively. Finally, stress components τ_{31} and τ_{32} both act on the face normal to axis $\bar{\imath}_3$, in directions along axes $\bar{\imath}_1$ and $\bar{\imath}_2$, respectively. The various stress components appearing in eq. (1.3) are referred to as the *engineering stress components*. The units of stress components are identical to those of the stress vector, force per unit area, or Pascal.

The stress components represented in fig. 1.2 are all defined as positive. Furthermore, the three faces depicted in this figure are positive faces. A face is *positive* when the outward normal to the face, *i.e.*, the normal pointing away from the body, is in the same direction as the axis to which the face is normal; a face is *negative* when its outward normal is pointing in the direction opposite to the axis to which the face is normal. The positive directions of stress components acting on negative faces are the opposite of those for stress components acting on positive faces. This sign convention is illustrated in fig. 1.3, which

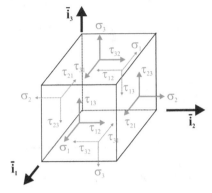

Fig. 1.3. Sign conventions for the stress components acting on a differential volume element. All stress components shown here are positive.

shows positive stress components acting on the six faces of a cube of differential size. Positive stress components are shown in solid lines on the three positive faces of the cube; positive stress components are shown in dotted lines on the three negative (hidden) faces of the cube.

Taken together, the direct stress components σ_1, σ_2, and σ_3 and the shear stress components, τ_{12} and τ_{13}, τ_{21} and τ_{23}, and τ_{31} and τ_{32}, fully characterize the state of stress at point **P**. It will be shown in a later section that if the stress components acting on three orthogonal faces are known, it is possible to compute the stress components acting at the same point, on a face of arbitrary orientation. This discussion underlines the fact that the state of stress at a point is a complex concept: its complete definition requires the knowledge of nine stress components acting on three mutually orthogonal faces.

This should be contrasted with the concept of force. A force is vector quantity that is characterized by its magnitude and orientation. Alternatively, a force can be defined by the three components of the force vector in a given coordinate system. The definition of a force thus requires three quantities, whereas the definition of the stress state requires nine quantities.

A force is a vector, which is referred to as a *first order tensor*, whereas a state of stress is a *second order tensor*. Several quantities commonly used in solid mechanics are also second order tensors: the strain tensor, the bending stiffnesses of a beam, and the mass moments of inertia of a solid object. The first two of these quantities will be introduced in later sections and chapters. Much like the case for vectors, all second order tensors will be shown to possess certain common characteristics.

1.1.2 Volume equilibrium equations

In general, the state of stress varies throughout a solid body, and hence, stresses acting on two parallel faces located a small distance apart are not equal. Consider, for instance, the two opposite faces of a differential volume element that are normal to axis $\bar{\imath}_2$, as shown in fig. 1.4. The axial stress component on the negative face at coordinate x_2 is σ_2, but the stress components on the positive face at coordinate $x_2 + dx_2$ will be slightly different and written as $\sigma_2(x_2 + dx_2)$. If $\sigma_2(x_2)$ is an analytic function, it is then possible to express $\sigma_2(x_2 + dx_2)$ in terms of $\sigma_2(x_2)$ using a Taylor series expansion to find

$$\sigma_2(x_2 + dx_2) = \sigma_2(x_2) + \left.\frac{\partial \sigma_2}{\partial x_2}\right|_{x_2} dx_2 + \ldots \text{higher order terms in } dx_2.$$

This expansion is a fundamental step in the derivation of the differential equations governing the behavior of a continuum. The stress component on the positive face at coordinate $x_2 + dx_2$ can be written as $\sigma_2(x_2 + dx_2) \approx \sigma_2 + (\partial \sigma_2/\partial x_2)dx_2$. The same Taylor series expansion technique can be applied to all other direct and shear stress components.

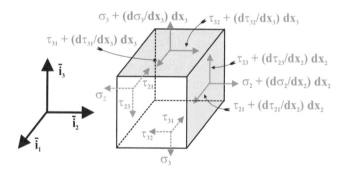

Fig. 1.4. Stress components acting on a differential element of volume. For clarity of the figure, the stress components acting on the faces normal to $\bar{\imath}_1$ are not shown.

Consider now the differential element of volume depicted in fig. 1.4. It is subjected to stress components acting on its six external faces and to body forces per unit volume, represented by a vector $\underline{b}$ acting at its centroid. These body forces could be gravity forces, inertial forces, or forces of an electric or magnetic origin; the components of this body force vector resolved in coordinate system $\mathcal{I} = (\bar{\imath}_1, \bar{\imath}_2, \bar{\imath}_3)$ as $\underline{b} = b_1\bar{\imath}_1 + b_2\bar{\imath}_2 + b_3\bar{\imath}_3$. The units of the force vector are force per unit volume or Newton per cubic meter.

Force equilibrium

According to Newton's law, static equilibrium requires the sum of all the forces acting on this differential element to vanish. Considering all the forces acting along the direction of axis $\bar{\imath}_1$, the equilibrium condition is

$$
-\sigma_1 \, \mathrm{d}x_2\mathrm{d}x_3 + \left(\sigma_1 + \frac{\partial\sigma_1}{\partial x_1}\mathrm{d}x_1\right)\mathrm{d}x_2\mathrm{d}x_3
$$
$$
-\tau_{21} \, \mathrm{d}x_1\mathrm{d}x_3 + \left(\tau_{21} + \frac{\partial\tau_{21}}{\partial x_2}\mathrm{d}x_2\right)\mathrm{d}x_1\mathrm{d}x_3
$$
$$
-\tau_{31} \, \mathrm{d}x_1\mathrm{d}x_2 + \left(\tau_{31} + \frac{\partial\tau_{31}}{\partial x_3}\mathrm{d}x_3\right)\mathrm{d}x_1\mathrm{d}x_2 + b_1 \, \mathrm{d}x_1\mathrm{d}x_2\mathrm{d}x_3 = 0.
$$

This equation states an equilibrium of forces, and therefore the stress components must be multiplied by the area of the surface on which they act to yield the corresponding force. Similarly, the component of the body force per unit volume of the body is multiplied by the volume of the differential element, $\mathrm{d}x_1\mathrm{d}x_2\mathrm{d}x_3$, to give the body force acting on the element. After simplification, this equilibrium condition becomes

$$
\left[\frac{\partial\sigma_1}{\partial x_1} + \frac{\partial\tau_{21}}{\partial x_2} + \frac{\partial\tau_{31}}{\partial x_3} + b_1\right]\mathrm{d}x_1\mathrm{d}x_2\mathrm{d}x_3 = 0.
$$

This equation is satisfied when the expression in brackets vanishes, and this yields the equilibrium equation in the direction of axis $\bar{\imath}_1$

$$
\frac{\partial\sigma_1}{\partial x_1} + \frac{\partial\tau_{21}}{\partial x_2} + \frac{\partial\tau_{31}}{\partial x_3} + b_1 = 0.
$$

For the same reasons, forces along axes $\bar{\imath}_2$ and $\bar{\imath}_3$ must vanish as well, and a similar reasoning yields the following three equilibrium equations

$$
\frac{\partial\sigma_1}{\partial x_1} + \frac{\partial\tau_{21}}{\partial x_2} + \frac{\partial\tau_{31}}{\partial x_3} + b_1 = 0, \tag{1.4a}
$$
$$
\frac{\partial\tau_{12}}{\partial x_1} + \frac{\partial\sigma_2}{\partial x_2} + \frac{\partial\tau_{32}}{\partial x_3} + b_2 = 0, \tag{1.4b}
$$
$$
\frac{\partial\tau_{13}}{\partial x_1} + \frac{\partial\tau_{23}}{\partial x_2} + \frac{\partial\sigma_3}{\partial x_3} + b_3 = 0, \tag{1.4c}
$$

which must be satisfied at all points inside the body.

The equilibrium conditions implied by Newton's law, eqs. (1.4), have been written by considering an differential element *of the undeformed body*. Of course, when forces are applied, the body deforms and so does every single differential element. Strictly speaking, equilibrium should be enforced *on the deformed configuration of the body*, rather than its undeformed configuration. Indeed, stresses are only present when external forces are applied and the body is deformed. When no forces are applied, the body is undeformed, but stresses all vanish.

Unfortunately, it is difficult to write equilibrium conditions on the deformed configuration of the body because this configuration is unknown; indeed, the goal of the theory of elasticity is to predict the deformation of elastic bodies under load. It is a basic assumption of the *linear theory of elasticity* developed here that the displacements of the body under the applied loads are very small, and hence, the difference between the deformed and undeformed configurations of the body is very small. Under this assumption, it is justified to impose equilibrium conditions to the undeformed configuration of the body, because it is nearly identical to its deformed configuration.

Moment equilibrium

To satisfy all equilibrium requirements, the sum of all the moments acting on the differential element of volume depicted in fig. 1.4 must also vanish. Consider first the moment equilibrium about axis $\bar{\imath}_1$. The contributions of the direct stresses and of the body forces can be eliminated by choosing an axis passing through the center of the differential element. The resulting moment equilibrium equation is

$$
\tau_{23}\, \mathrm{d}x_1\mathrm{d}x_3\frac{\mathrm{d}x_2}{2} + \left(\tau_{23} + \frac{\partial\tau_{23}}{\partial x_2}\mathrm{d}x_2\right)\mathrm{d}x_1\mathrm{d}x_3\frac{\mathrm{d}x_2}{2}
$$
$$
-\tau_{32}\, \mathrm{d}x_1\mathrm{d}x_2\frac{\mathrm{d}x_3}{2} - \left(\tau_{32} + \frac{\partial\tau_{32}}{\partial x_3}\mathrm{d}x_3\right)\mathrm{d}x_1\mathrm{d}x_2\frac{\mathrm{d}x_3}{2}
$$
$$
= \left[\tau_{23} - \tau_{32} + \frac{\partial\tau_{23}}{\partial x_2}\frac{\mathrm{d}x_2}{2} - \frac{\partial\tau_{32}}{\partial x_3}\frac{\mathrm{d}x_3}{2}\right]\mathrm{d}x_1\mathrm{d}x_2\mathrm{d}x_3 = 0.
$$

The bracketed expression must vanish and after neglecting higher order terms, this reduces to the following equilibrium condition

$$
\tau_{23} - \tau_{32} = 0.
$$

Enforcing the vanishing of the sum of the moments about axes $\bar{\imath}_2$ and $\bar{\imath}_3$ leads to similar equations,

$$
\tau_{23} = \tau_{32}, \ \tau_{13} = \tau_{31}, \ \tau_{12} = \tau_{21}. \quad (1.5)
$$

The implication of these equalities is summarized by the principle of reciprocity of shear stresses, which is illustrated in fig. 1.5.

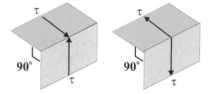

Fig. 1.5. Reciprocity of the shearing stresses acting on two orthogonal faces.

Principle 1 (Principle of reciprocity of shear stresses) *Shear stresses acting in the direction normal to the common edge of two orthogonal faces must be equal in magnitude and be simultaneously oriented toward or away from the common edge.*

Another implication of the reciprocity of the shear stresses is that of the nine components of stresses, six only are independent. It is common practice to arrange the stress tensor components in a 3×3 matrix format

$$
\begin{bmatrix} \sigma_1 & \tau_{12} & \tau_{13} \\ \tau_{12} & \sigma_2 & \tau_{23} \\ \tau_{13} & \tau_{23} & \sigma_3 \end{bmatrix}. \quad (1.6)
$$

The principle of reciprocity implies the *symmetry of the stress tensor*.

1.1.3 Surface equilibrium equations

At the outer surface of the body, the stresses acting inside the body must be in equilibrium with the externally applied *surface tractions*. Surface tractions are represented by a stress vector, $\underline{t}$, that can be resolved in reference frame $\mathcal{I} = (\bar{\imath}_1, \bar{\imath}_2, \bar{\imath}_3)$ as $\underline{t} = t_1 \bar{\imath}_1 + t_2 \bar{\imath}_2 + t_3 \bar{\imath}_3$. Figure 1.6 shows a free body in the form of a differential tetrahedron bounded by three negative faces cut through the body in directions normal to axes $\bar{\imath}_1$, $\bar{\imath}_2$, and $\bar{\imath}_3$, and by a fourth face, **ABC**, of area $\mathrm{d}\mathcal{A}_n$, which is a differential element of the outer surface of the body. The unit normal to this element of area is denoted $\bar{n}$, and its components in coordinate system $\mathcal{I}$ are $\bar{n} = n_1 \bar{\imath}_1 + n_2 \bar{\imath}_2 + n_3 \bar{\imath}_3$. Note that n_1, n_2, and n_3 are the cosines of the angle between $\bar{n}$ and $\bar{\imath}_1$, $\bar{n}$ and $\bar{\imath}_2$, and $\bar{n}$ and $\bar{\imath}_3$, respectively, also called the *direction cosines* of $\bar{n}$: $n_1 = \bar{n} \cdot \bar{\imath}_1 = \cos(\bar{n}, \bar{\imath}_1)$, $n_2 = \bar{n} \cdot \bar{\imath}_2 = \cos(\bar{n}, \bar{\imath}_2)$, and $n_3 = \bar{n} \cdot \bar{\imath}_3 = \cos(\bar{n}, \bar{\imath}_3)$.

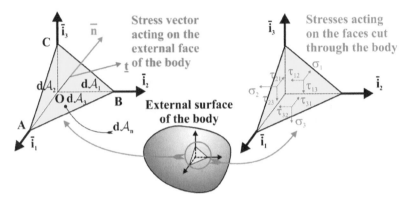

Fig. 1.6. A tetrahedron with one face along the outer surface of the body.

Equilibrium of forces acting along axis $\bar{\imath}_1$ implies

$$t_1 \mathrm{d}\mathcal{A}_n = \sigma_1 \mathrm{d}\mathcal{A}_1 + \tau_{21} \mathrm{d}\mathcal{A}_2 + \tau_{31} \mathrm{d}\mathcal{A}_3 - b_1 \frac{\mathrm{d}x_1 \mathrm{d}x_2 \mathrm{d}x_3}{6}, \qquad (1.7)$$

where $\mathrm{d}\mathcal{A}_1$, $\mathrm{d}\mathcal{A}_2$, and $\mathrm{d}\mathcal{A}_3$ are the areas of triangles **OBC**, **OAC** and **OAB**, respectively, and the last term represents the body force times the volume of the tetrahedron. The areas of the three faces normal to the axes are found by projecting face **ABC** onto planes normal to the axes using the direction cosines to find

$$\mathrm{d}\mathcal{A}_1 = n_1 \mathrm{d}\mathcal{A}_n, \quad \mathrm{d}\mathcal{A}_2 = n_2 \mathrm{d}\mathcal{A}_n, \quad \text{and} \quad \mathrm{d}\mathcal{A}_3 = n_3 \mathrm{d}\mathcal{A}_n. \qquad (1.8)$$

Dividing eq. (1.7) by $\mathrm{d}\mathcal{A}_n$ then yields the first component of the surface traction vector

$$t_1 = \sigma_1 n_1 + \tau_{21} n_2 + \tau_{31} n_3,$$

where the body force term vanishes because it is a higher order differential term. The same procedure can be followed to express equilibrium conditions along the directions of axes $\bar{\imath}_2$ and $\bar{\imath}_3$. The three components of the surface traction vector then become

$$t_1 = \sigma_1 \, n_1 + \tau_{12} \, n_2 + \tau_{13} \, n_3, \tag{1.9a}$$

$$t_2 = \tau_{12} \, n_1 + \sigma_2 \, n_2 + \tau_{23} \, n_3, \tag{1.9b}$$

$$t_3 = \tau_{31} \, n_1 + \tau_{32} \, n_2 + \sigma_3 \, n_3. \tag{1.9c}$$

A body is said to be in equilibrium if eqs. (1.4) are satisfied at all points inside the body, and eqs. (1.9) are satisfied at all points of its external surface.

1.2 Analysis of the state of stress at a point

The state of stress at a point is characterized in the previous section by the normal and shear stress components acting on the faces of a differential element of volume cut from the solid. The faces of this cube are cut normal to the axes of a Cartesian reference frame $\mathcal{I} = (\bar{\imath}_1, \bar{\imath}_2, \bar{\imath}_3)$, and the stress vector acting on these faces are resolved along the same axes. Clearly, another face at an arbitrary orientation with respect to these axes can be selected. In section 1.2.1, it will be shown that the stresses acting on this face can be related to the stresses acting on the faces normal to axes $\bar{\imath}_1, \bar{\imath}_2$, and $\bar{\imath}_3$. This important result implies that once the stress components are known on three mutually orthogonal faces at a point, they are known on *any* face passing through that point. Hence, the state of stress at a point is fully defined once the stress components acting on three mutually orthogonal faces at a point are known.

1.2.1 Stress components acting on an arbitrary face

To establish relationships between stresses, it is necessary to consider force or moment equilibrium due to these stresses, and this must be done with reference to a specific free body diagram. Figure 1.7 shows a specific free body constructed from a tetrahedron defined by three faces cut normal to axes $\bar{\imath}_1, \bar{\imath}_2$, and $\bar{\imath}_3$, and a fourth face normal to unit vector $\bar{n} = n_1 \, \bar{\imath}_1 + n_2 \, \bar{\imath}_1 + n_3 \, \bar{\imath}_3$, of arbitrary orientation. This tetrahedron is known as *Cauchy's tetrahedron*. The components, n_1, n_2, and n_3, of this unit vector are the *direction cosines* of unit vector $\bar{n}$, *i.e.*, the cosines of the angles between $\bar{n}$ and $\bar{\imath}_1$, $\bar{n}$ and $\bar{\imath}_2$, and $\bar{n}$ and $\bar{\imath}_3$, respectively.

Figure 1.7 shows the stress components acting on faces **COB**, **AOC** and **AOB**, of area $\mathrm{d}\mathcal{A}_1$, $\mathrm{d}\mathcal{A}_2$, and $\mathrm{d}\mathcal{A}_3$, respectively; the stress vector, τ_n, acts on face **ABC** of area $\mathrm{d}\mathcal{A}_n$. The body force vector, $\underline{b}$, is also acting on this tetrahedron. Equilibrium of forces acting on tetrahedron **OABC** requires

$$\underline{\tau}_1 \mathrm{d}\mathcal{A}_1 + \underline{\tau}_2 \mathrm{d}\mathcal{A}_2 + \underline{\tau}_3 \mathrm{d}\mathcal{A}_3 = \underline{\tau}_n \mathrm{d}\mathcal{A}_n + \underline{b} \, \mathrm{d}\mathcal{V},$$

where $\underline{\tau}_1, \underline{\tau}_2$ and $\underline{\tau}_3$ are the stress vectors acting on the faces normal to axes $\bar{\imath}_1, \bar{\imath}_2$, and $\bar{\imath}_3$, respectively, and $\mathrm{d}\mathcal{V}$ is the volume of the tetrahedron.

Dividing this equilibrium equation by $\mathrm{d}\mathcal{A}_n$ and using eq. (1.8) gives the stress vector acting of the inclined face as

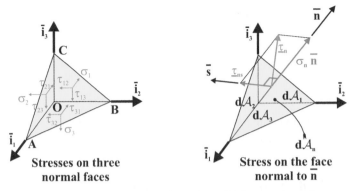

Stresses on three normal faces   **Stress on the face normal to $\bar{n}$**

Fig. 1.7. Differential tetrahedron element with one face, **ABC**, normal to unit vector $\underline{n}$ and the other three faces normal to axes $\bar{\imath}_1, \bar{\imath}_2,$ and $\bar{\imath}_3$, respectively.

$$\underline{T}_n = \underline{T}_1 n_1 + \underline{T}_2 n_2 + \underline{T}_3 n_3 - \underline{b}\, dV/d\mathcal{A}_n$$

The body force term is multiplied by a higher order term, $dV/d\mathcal{A}_n$, which can be neglected in the equilibrium condition. Expanding the three stress vectors in terms on the stress components then yields

$$\underline{T}_n = (\sigma_1 \bar{\imath}_1 + \tau_{12}\bar{\imath}_2 + \tau_{13}\bar{\imath}_3)\, n_1 + (\tau_{21}\bar{\imath}_1 + \sigma_2\bar{\imath}_2 + \tau_{23}\bar{\imath}_3)\, n_2 + (\tau_{31}\bar{\imath}_1 + \tau_{32}\bar{\imath}_2 + \sigma_3\bar{\imath}_3)\, n_3. \tag{1.10}$$

To determine the direct stress, σ_n, acting on face **ABC**, it is necessary to project this vector equation in the direction of unit vector $\bar{n}$. This can be achieved by taking the dot product of the stress vector by unit vector $\bar{n}$ to find

$$\bar{n} \cdot \underline{T}_n = \bar{n}\cdot [(\sigma_1 \bar{\imath}_1 + \tau_{12}\bar{\imath}_2 + \tau_{13}\bar{\imath}_3)\, n_1 + (\tau_{21}\bar{\imath}_1 + \sigma_2\bar{\imath}_2 + \tau_{23}\bar{\imath}_3)\, n_2 +$$
$$(\tau_{31}\bar{\imath}_1 + \tau_{32}\bar{\imath}_2 + \sigma_3\bar{\imath}_3)\, n_3].$$

Because $\bar{n} = n_1\,\bar{\imath}_1 + n_2\,\bar{\imath}_1 + n_3\,\bar{\imath}_3$, this yields

$$\sigma_n = (\sigma_1 n_1 + \tau_{12}n_2 + \tau_{13}n_3)\, n_1 + (\tau_{21}n_1 + \sigma_2 n_2 + \tau_{23}n_3)\, n_2$$
$$+ (\tau_{31}n_1 + \tau_{32}n_2 + \sigma_3 n_3)\, n_3,$$

and finally, after minor a rearrangement of terms,

$$\sigma_n = \sigma_1 n_1^2 + \sigma_2 n_2^2 + \sigma_3 n_3^2 + 2\tau_{23}n_2 n_3 + 2\tau_{13}n_1 n_3 + 2\tau_{12}n_1 n_2. \tag{1.11}$$

The stress components acting in the plane of face **ABC** can be evaluated in a similar manner by projecting eq. (1.10) along a unit vector in the plane of face **ABC**. Consider a unit vector, $\bar{s} = s_1\,\bar{\imath}_1 + s_2\,\bar{\imath}_1 + s_3\,\bar{\imath}_3$, normal to $\bar{n}$, $i.e.$, such that $\bar{n}\cdot\bar{s} = 0$. The shear stress component acting on face **ABC** in the direction of unit vector $\bar{s}$ is denoted τ_{ns} and is obtained by projecting eq. (1.10) along vector $\bar{s}$ to find

$$\tau_{ns} = (\sigma_1 s_1 + \tau_{12}s_2 + \tau_{13}s_3)\, n_1 + (\tau_{21}s_1 + \sigma_2 s_2 + \tau_{23}s_3)\, n_2$$
$$+ (\tau_{31}s_1 + \tau_{32}s_2 + \sigma_3 s_3)\, n_3,$$

and finally, after minor a rearrangement of terms,

$$\tau_{ns} = \sigma_1 n_1 s_1 + \sigma_2 n_2 s_2 + \sigma_3 n_3 s_3 + \tau_{12}(n_2 s_1 + n_1 s_2)$$
$$+ \tau_{13}(n_1 s_3 + n_3 s_1) + \tau_{23}(n_2 s_3 + n_3 s_2). \tag{1.12}$$

Equations. (1.11) and (1.12) express an important result of continuum mechanics. They imply that once the stress components acting on three mutually orthogonal faces are known, the stress components on a face of arbitrary orientation can be readily computed. To evaluate the direct stress component acting on an arbitrary face, all that is required are the direction cosines of the normal to the face. Evaluation the shear stress component acting on the same face requires, in addition, the direction cosines of the direction of the shear stress component in that face.

Consider the following question: how much information is required to fully define the state of stress at point $\mathbf{P}$ of a solid? Clearly, the body can be cut at this point by a plane of arbitrary orientation. The stress vector acting on this face gives information about the state of stress at point $\mathbf{P}$. The stress vector acting on a face with another orientation would give additional information about the state of stress at the same point. If additional faces are considered, each new stress vector provides additional information. This reasoning would seem to imply that the complete knowledge of the state of stress at a point requires an infinite amount of information, specifically, the stress vectors acting on *all* the possible faces passing through point $\mathbf{P}$. Equations. (1.11) and (1.12), however, demonstrate the fallacy of this reasoning: once the stress vectors acting on three mutually orthogonal faces are known, the stress vector acting on *any* other face can be readily predicted. In conclusion, complete definition of the state of stress at a point only requires knowledge of the stress vectors, or equivalently of the stress tensor components, acting on three mutually orthogonal faces.

1.2.2 Principal stresses

As discussed in the previous section, eqs. (1.11) and (1.12) enable the computation of the stress components acting on a face of arbitrary orientation, based on the knowledge of the stress components acting on three mutually orthogonal faces. As illustrated in fig. 1.7, the stress vector acting on a face of arbitrary orientation has, in general, a component $\sigma_n \, \bar{n}$, acting in the direction normal to the face, and a component $\tau_{ns} \, \bar{s}$, acting within the plane of the face.

This discussion raises the following question: is there a face orientation for which the stress vector is exactly normal to the face? In other words, does a particular orientation, $\bar{n}$, exist for which the stress vector acting on this face *consists solely* of $\underline{\tau}_n = \sigma_p \, \bar{n}$, where σ_p is the yet unknown magnitude of this direct stress component? Introducing this expression into eq. (1.10) results in

$$\sigma_p \bar{n} = (\sigma_1 \bar{\imath}_1 + \tau_{12} \bar{\imath}_2 + \tau_{13} \bar{\imath}_3) \, n_1 + (\tau_{21} \bar{\imath}_1 + \sigma_2 \bar{\imath}_2 + \tau_{23} \bar{\imath}_3) \, n_2 + (\tau_{31} \bar{\imath}_1 + \tau_{32} \bar{\imath}_2 + \sigma_3 \bar{\imath}_3) \, n_3.$$

This equation alone does not allow the determination of both σ_p and of unit vector $\bar{n}$. Projecting this vector relationship along axes $\bar{\imath}_1$, $\bar{\imath}_2$, and $\bar{\imath}_3$ leads to the following three scalar equations

$$(\sigma_1 - \sigma_p)\, n_1 + \tau_{12}\, n_2 + \tau_{13}\, n_3 = 0,$$
$$\tau_{12}\, n_1 + (\sigma_2 - \sigma_p)\, n_2 + \tau_{23}\, n_3 = 0,$$
$$\tau_{13}\, n_1 + \tau_{23}\, n_2 + (\sigma_3 - \sigma_p)\, n_3 = 0,$$

respectively. The unknowns of the problem are the direction cosines, n_1, n_2, and n_3 that define the orientation of the face on which shear stresses vanish, and the magnitude, σ_p, of the direct stress component acting on this face.

These equations are recast as a homogeneous system of linear equations for the unknown direction cosines

$$\begin{bmatrix} \sigma_1 - \sigma_p & \tau_{12} & \tau_{13} \\ \tau_{12} & \sigma_2 - \sigma_p & \tau_{23} \\ \tau_{13} & \tau_{23} & \sigma_3 - \sigma_p \end{bmatrix} \begin{Bmatrix} n_1 \\ n_2 \\ n_3 \end{Bmatrix} = 0. \tag{1.13}$$

Since this is a homogeneous system of equations, the trivial solution, $n_1 = n_2 = n_3 = 0$, is, in general, the solution of this system. When the determinant of the system vanishes, however, non-trivial solutions will exist. The vanishing of the determinant of the system leads to the cubic equation for the magnitude of the direct stress

$$\sigma_p^3 - I_1 \sigma_p^2 + I_2 \sigma_p - I_3 = 0, \tag{1.14}$$

where the quantities I_1, I_2, and I_3 are defined as

$$I_1 = \sigma_1 + \sigma_2 + \sigma_3, \tag{1.15a}$$
$$I_2 = \sigma_1 \sigma_2 + \sigma_2 \sigma_3 + \sigma_3 \sigma_1 - \tau_{12}^2 - \tau_{13}^2 - \tau_{23}^2, \tag{1.15b}$$
$$I_3 = \sigma_1 \sigma_2 \sigma_3 - \sigma_1 \tau_{23}^2 - \sigma_2 \tau_{13}^2 - \sigma_3 \tau_{12}^2 + 2\tau_{12}\tau_{13}\tau_{23}, \tag{1.15c}$$

are called the three *stress invariants*.

The solutions of eq. (1.14) are called the *principal stresses*. Since this is a cubic equation, three solutions exist, denoted σ_{p1}, σ_{p2}, and σ_{p3}. For each of these three solutions, the matrix of the system of equations defined by eq. (1.13) has a zero determinant, and a non-trivial solution exists for the directions cosines that now define the direction of a face on which the shear stresses vanish. This direction is called a *principal stress direction*. Because the equations to be solved are homogeneous, their solution will include an arbitrary constant, which can be determined by enforcing the normality condition for unit vector $\bar{n}$, $n_1^2 + n_2^2 + n_3^2 = 1$.

This solution process can be repeated for each of the three principal stresses. This will result in three different principal stress directions. It can be shown that these three directions are mutually orthogonal.

1.2.3 Rotation of stresses

In the previous sections, free body diagrams are formed with faces cut in directions normal to axes of the orthonormal basis $\mathcal{I} = (\bar{\imath}_1, \bar{\imath}_2, \bar{\imath}_3)$, and the stress vectors are resolved into stress components along the same directions. The orientation of this basis is entirely arbitrary: basis $\mathcal{I}^* = (\bar{\imath}_1^*, \bar{\imath}_2^*, \bar{\imath}_3^*)$ could also have been selected, and an analysis identical to that of the previous sections would have led to the definition of normal stresses σ_1^*, σ_2^*, σ_3^*, and shear stresses τ_{23}^*, τ_{13}^*, τ_{12}^*. A typical equilibrium equation at a point of the body would be written as

$$\frac{\partial \sigma_1^*}{\partial x_1^*} + \frac{\partial \tau_{21}^*}{\partial x_2^*} + \frac{\partial \tau_{31}^*}{\partial x_3^*} + b_1^* = 0, \tag{1.16}$$

where the notation $(.)^*$ is used to indicate the components of the corresponding quantity resolved in basis $\mathcal{I}^*$. A typical surface traction would be defined as

$$t_1^* = n_1^* \, \sigma_1^* + n_2^* \, \tau_{21}^* + n_3^* \, \tau_{31}^*. \tag{1.17}$$

Although expressed in different reference frames, eqs. (1.4) and (1.16), or (1.9) and (1.17) express the same equilibrium conditions for the body. Two orthonormal bases, $\mathcal{I}$ and $\mathcal{I}^*$, are involved in this problem. The orientation of basis $\mathcal{I}^*$ relative to basis $\mathcal{I}$ is discussed in section A.3.1 and leads to the definition of the matrix of direction cosines, or rotation matrix, $\underline{\underline{R}}$, given by eq. (A.36).

Consider the stress component σ_1^*: it represents the magnitude of the direct stress component acting on the face normal to axis $\bar{\imath}_1^*$. Equation (1.11) can now be used to express this stress component in terms of the stress components resolved in axis system $\mathcal{I}$ to find

$$\sigma_1^* = \sigma_1 \ell_1^2 + \sigma_2 \ell_2^2 + \sigma_3 \ell_3^2 + 2\tau_{23}\ell_2\ell_3 + 2\tau_{13}\ell_1\ell_3 + 2\tau_{12}\ell_1\ell_2, \tag{1.18}$$

where ℓ_1, ℓ_2, and ℓ_3, are the direction cosines of unit vector $\bar{\imath}_1^*$. Similar equations can be derived to express the stress components σ_2^* and σ_3^* in terms of the stress components resolved in axis system $\mathcal{I}$. For σ_2^*, the direction cosines ℓ_1, ℓ_2, and ℓ_3 appearing in eq. (1.18) are replaced by direction cosines m_1, m_2, and m_3, respectively, whereas direction cosines n_1, n_2, and n_3 will appear in the expression for σ_3^*. Coordinate rotations are defined in appendix A.3.

The shear stress components follow from eq. (1.12) as

$$\begin{aligned} \tau_{12}^* = {}&\sigma_1\ell_1 m_1 + \sigma_2\ell_2 m_2 + \sigma_3\ell_3 m_3 + \tau_{12}(\ell_2 m_1 + \ell_1 m_2) \\ &+ \tau_{13}(\ell_1 m_3 + \ell_3 m_1) + \tau_{23}(\ell_2 m_3 + \ell_3 m_2). \end{aligned} \tag{1.19}$$

Here again, similar relationships can be derived for the remaining shear stress components, τ_{13}^* and τ_{23}^*, through appropriate cyclic permutation of the indices.

All these relationships can be combined into the following compact matrix equation

$$\begin{bmatrix} \sigma_1^* & \tau_{12}^* & \tau_{13}^* \\ \tau_{21}^* & \sigma_2^* & \tau_{23}^* \\ \tau_{31}^* & \tau_{32}^* & \sigma_3^* \end{bmatrix} = \underline{\underline{R}}^T \begin{bmatrix} \sigma_1 & \tau_{12} & \tau_{13} \\ \tau_{12} & \sigma_2 & \tau_{23} \\ \tau_{13} & \tau_{23} & \sigma_3 \end{bmatrix} \underline{\underline{R}}, \tag{1.20}$$

where $\underline{\underline{R}}$ is the rotation matrix defined by eq. (A.36). This equation concisely encapsulates the relationship between the stress components resolved in two different coordinate systems, and it can be used to compute the stress components resolved in basis $\mathcal{I}^*$ in terms of the stress components resolved in basis $\mathcal{I}$.

Finally, since the principal stresses at a point are independent of the particular coordinate system used to define the stress state, the coefficients of the cubic equation that determines the principal stresses, eq. (1.14), must be invariant with respect to reference frames. This is the very reason why quantities I_1, I_2, and I_3 defined by eq. (1.15) are called the *stress invariants*. The word "invariant" refers to the fact that these quantities are *invariant with respect to a change of coordinate system*. Let $\mathcal{I}^*$ and $\mathcal{I}$ be two different orthonormal bases,

$$I_1 = \sigma_1^* + \sigma_2^* + \sigma_3^* = \sigma_1 + \sigma_2 + \sigma_3, \tag{1.21a}$$

$$I_2 = \sigma_1^* \sigma_2^* + \sigma_2^* \sigma_3^* + \sigma_3^* \sigma_1^* - \tau_{12}^{*2} - \tau_{13}^{*2} - \tau_{23}^{*2}$$
$$= \sigma_1 \sigma_2 + \sigma_2 \sigma_3 + \sigma_3 \sigma_1 - \tau_{12}^2 - \tau_{13}^2 - \tau_{23}^2, \tag{1.21b}$$

$$I_3 = \sigma_1^* \sigma_2^* \sigma_3^* - \sigma_1^* \tau_{23}^{*2} - \sigma_2^* \tau_{13}^{*2} - \sigma_3^* \tau_{12}^{*2} + 2\tau_{12}^* \tau_{13}^* \tau_{23}^*$$
$$= \sigma_1 \sigma_2 \sigma_3 - \sigma_1 \tau_{23}^2 - \sigma_2 \tau_{13}^2 - \sigma_3 \tau_{12}^2 + 2\tau_{12} \tau_{13} \tau_{23}. \tag{1.21c}$$

Tedious algebra using eqs. (1.20) to write the stress components resolved in basis $\mathcal{I}^*$ in terms of the stresses components resolved in basis $\mathcal{I}$ will reveal that the above relationships are correct.

Example 1.1. *Computing principal stresses*

Consider the following stress tensor

$$\underline{\underline{S}} = \begin{bmatrix} -5 & -4 & 0 \\ -4 & 1 & 0 \\ 0 & 0 & 1 \end{bmatrix}.$$

Compute the principal stresses and the principal stress directions. The stress invariants defined by eq. (1.15) are computed as $I_1 = -3$, $I_2 = -25$ and $I_3 = -21$. The principal stress equation, eq. (1.14), now becomes

$$\sigma_p^3 + 3\sigma_p^2 - 25\sigma_p + 21 = (\sigma_p - 1)(\sigma_p^2 + 4\sigma_p - 21) = 0,$$

The solutions of this cubic equations yield the principal stresses as $\sigma_{p1} = 3$, $\sigma_{p2} = 1$ and $\sigma_{p3} = -7$.

Next, the principal direction associated with $\sigma_{p1} = 3$ is computed. The homogeneous system defined by eq. (1.13) becomes

$$\begin{bmatrix} -8 & -4 & 0 \\ -4 & -2 & 0 \\ 0 & 0 & -2 \end{bmatrix} \begin{Bmatrix} n_1 \\ n_2 \\ n_3 \end{Bmatrix} = 0.$$

The determinant of this system vanishes because the first two equations are a multiple of each other. The first equation yields $n_1 = \alpha$ and $n_2 = -2\alpha$, where α is an arbitrary constant, whereas the third equation gives $n_3 = 0$. Since the principal direction must be unit vector, $n_1^2 + n_2^2 + n_3^2 = 1$, or $5\alpha^2 = 1$; finally $n_1 = 1/\sqrt{5}$, $n_2 = -2/\sqrt{5}$ and $n_3 = 0$. Proceeding in a similar manner for the other two principal stresses, the three principal directions are found to be

$$\bar{n}_1 = \frac{1}{\sqrt{5}}\left\{\begin{matrix} 1 \\ -2 \\ 0 \end{matrix}\right\} ; \quad \bar{n}_2 = \left\{\begin{matrix} 0 \\ 0 \\ 1 \end{matrix}\right\} ; \quad \bar{n}_3 = \frac{1}{\sqrt{5}}\left\{\begin{matrix} -2 \\ -1 \\ 0 \end{matrix}\right\}.$$

It is easily verified that the principal directions are orthogonal to each other; indeed, $\bar{n}_1 \cdot \bar{n}_2 = \bar{n}_2 \cdot \bar{n}_3 = \bar{n}_3 \cdot \bar{n}_1 = 0$.

Example 1.2. Principal stresses as an eigenproblem

Consider the following stress tensor

$$\underline{\underline{S}} = \begin{bmatrix} 5.0 & 2.5 & -1.3 \\ 2.5 & 7.8 & -3.4 \\ -1.3 & -3.4 & -4.5 \end{bmatrix}.$$

Compute the principal stresses and the principal stress directions. Rather than following the procedure described in the previous examples, the homogeneous system of linear equations, eq. (1.13), that govern the problem is recast as

$$\begin{bmatrix} \sigma_1 & \tau_{12} & \tau_{13} \\ \tau_{12} & \sigma_2 & \tau_{23} \\ \tau_{13} & \tau_{23} & \sigma_3 \end{bmatrix} \left\{\begin{matrix} n_1 \\ n_2 \\ n_3 \end{matrix}\right\} = \sigma_p \left\{\begin{matrix} n_1 \\ n_2 \\ n_3 \end{matrix}\right\}. \tag{1.22}$$

In this form, it becomes clear (see appendix A.2.4) that the determination of the principal stresses and principal stress directions is equivalent to the determination of the three eigenvalues, σ_{p1}, σ_{p2} and σ_{p3}, of the stress tensor, and determination of the corresponding three eigenvectors, $\bar{n}_1$, $\bar{n}_2$, and $\bar{n}_3$. Using a standard linear algebra software package, the three eigenpairs of the above stress tensor are found to be

$$\sigma_{p1} = -5.4180, \; \bar{n}_1 = \left\{\begin{matrix} -0.064 \\ -0.237 \\ -0.969 \end{matrix}\right\} ; \quad \sigma_{p2} = 3.5693, \; \bar{n}_2 = \left\{\begin{matrix} 0.879 \\ -0.473 \\ 0.058 \end{matrix}\right\} ;$$

$$\sigma_{p3} = 10.1487, \; \bar{n}_3 = \left\{\begin{matrix} 0.472 \\ 0.849 \\ -0.239 \end{matrix}\right\}.$$

Here again, it is easily verified that the principal directions are orthogonal to each other by computing $\bar{n}_i \cdot \bar{n}_j$ for any combination of i and j. This can be represented in a more compact way by creating a matrix, denoted $\underline{\underline{P}}$, that is constructed by arranging the principal stress direction vectors as the columns

$$\underline{\underline{P}} = [\bar{n}_1, \bar{n}_2, \bar{n}_3] = \begin{bmatrix} -0.0640 & 0.8791 & 0.4723 \\ -0.2372 & -0.4731 & 0.8485 \\ -0.9693 & 0.0577 & -0.2388 \end{bmatrix}.$$

Because the principal directions are mutually orthogonal unit vectors, this matrix is orthogonal, that is: $\underline{\underline{P}}^T\underline{\underline{P}} = \underline{\underline{I}}$, where $\underline{\underline{I}}$ is the 3×3 identity matrix. Furthermore, since matrix $\underline{\underline{P}}$ stores the eigenvectors of the stress tensor $\underline{\underline{S}}$, it follows that the transformation $\underline{\underline{P}}^T\underline{\underline{S}}\,\underline{\underline{P}}$ will diagonalize the stress tensor. That is,

$$\underline{P}^T \underline{S} \, \underline{P} = \begin{bmatrix} \sigma_{p1} & 0 & 0 \\ 0 & \sigma_{p2} & 0 \\ 0 & 0 & \sigma_{p3} \end{bmatrix} = \begin{bmatrix} -5.4180 & 0 & 0 \\ 0 & 3.5693 & 0 \\ 0 & 0 & 10.1487 \end{bmatrix},$$

and this can easily be verified by direct computation.

Example 1.3. Stresses acting on the octahedral face

Figure 1.8 shows a tetrahedron cut along three faces normal to the principal stress directions defined by axes $\bar{\imath}_1^*$, $\bar{\imath}_2^*$ and $\bar{\imath}_3^*$. The three mutually orthogonal edges of the tetrahedron each are of unit length. The fourth face of the tetrahedron is the *octahedral face* which is, by definition, the face that is equally inclined with respect to the principal stress directions. The normal to the octahedral face is $\bar{n}^T = \{1, 1, 1\} / \sqrt{3}$, *i.e.*, the direction cosines of this unit vector are $1/\sqrt{3}$ with respect to each of the three principal stress directions. Find the stress components acting on the octahedral face.

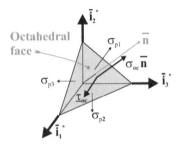

Fig. 1.8. The octahedral face.

By definition, the principal stress directions are such that on the corresponding faces, the shear stresses vanish. Hence, fig. 1.8 shows only the principal stress acting on each face. The stress vector acting on the octahedral face can be resolved into the octahedral direct stress vector, $\sigma_{oc} \, \bar{n}$, acting in the direction normal to the octahedral face, and octahedral shear stress vector, $\underline{\tau}_{oc}$, acting in the plane of the octahedral face. Using eq. (1.11), the magnitude of the direct octahedral stress is

$$\sigma_{oc} = \sigma_{p1} \left(\frac{1}{\sqrt{3}}\right)^2 + \sigma_{p2} \left(\frac{1}{\sqrt{3}}\right)^2 + \sigma_{p3} \left(\frac{1}{\sqrt{3}}\right)^2 = \frac{\sigma_{p1} + \sigma_{p2} + \sigma_{p3}}{3}. \quad (1.23)$$

The direct stress acting on the octahedral face is the *average of the principal stresses*.

The equilibrium condition for the tetrahedron in fig. 1.8 is now

$$\frac{1}{2}\sigma_{p1} \, \bar{\imath}_1^* + \frac{1}{2}\sigma_{p2} \, \bar{\imath}_2^* + \frac{1}{2}\sigma_{p3} \, \bar{\imath}_3^* = \frac{\sqrt{3}}{2}(\sigma_{oc} \, \bar{n} + \underline{\tau}_{oc}), \quad (1.24)$$

where the factor of 1/2 represents the area of each of the three faces normal to the principal axes directions and $\sqrt{3}/2$ the area of the octahedral face which is an equilateral triangle with sides of length $\sqrt{2}$. The octahedral shear stress vector now becomes

$$\sqrt{3}\,\underline{\tau}_{oc} = (\sigma_{p1} - \sigma_{oc})\bar{\imath}_1^* + (\sigma_{p2} - \sigma_{oc})\bar{\imath}_2^* + (\sigma_{p3} - \sigma_{oc})\bar{\imath}_3^*.$$

The magnitude of the octahedral shear stress, $\tau_{oc} = \|\underline{\tau}_{oc}\|$, is

$$\tau_{oc} = \frac{1}{\sqrt{3}}\left[(\sigma_{p1}^2 + \sigma_{p2}^2 + \sigma_{p3}^2) - \frac{1}{3}(\sigma_{p1} + \sigma_{p2} + \sigma_{p3})^2\right]^{1/2}. \qquad (1.25)$$

The first two invariants of the stress state, see eqs. (1.21a) and (1.21b), are easily expressed in terms of principal stresses as $I_1 = \sigma_{p1} + \sigma_{p2} + \sigma_{p3}$ and $I_2 = \sigma_{p1}\sigma_{p2} + \sigma_{p2}\sigma_{p3} + \sigma_{p3}\sigma_{p1}$. The octahedral stresses are now expressed in terms of these invariants as

$$\sigma_{oc} = \frac{I_1}{3}, \quad \tau_{oc} = \frac{\sqrt{2}}{3}\sqrt{I_1^2 - 3I_2}.$$

1.2.4 Problems

Problem 1.1. Stresses on an inclined face
Consider the tetrahedron shown in fig. 1.7. A set of three mutually orthogonal unit vectors will be defined: $\bar{\ell}$ is a unit vector parallel to vector **AB**, $\bar{m}$ is such that $\bar{m} = \bar{n} \times \bar{\ell}$, and $\bar{n}$ is the normal to face **ABC**. Let the stress vector acting on face **ABC** be resolved along these axes, i.e., let $\underline{\tau}_n = \tau_{n\ell}\,\bar{\ell} + \tau_{nm}\,\bar{m} + \sigma_n\,\bar{n}$. (1) Find the stress components, $\tau_{n\ell}$, τ_{nm} and σ_n, in terms of the stress components acting on the faces normal to axes $\bar{\imath}_1$, $\bar{\imath}_2$, and $\bar{\imath}_3$.

Problem 1.2. Principal stresses
Given a state of stress defined by: σ_1=200 MPa, σ_2=300 MPa, $\sigma_3 = -100$ MPa, $\tau_{12} = 50$ MPa, $\tau_{13} = -80$ MPa and $\tau_{23} = 100$ MPa, (1) Determine the principal stresses. (2) Determine the principal stress directions. Note: you should consider using a software package to handle the computations.

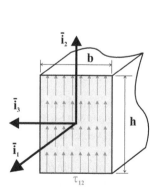

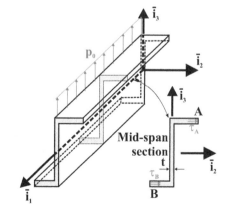

Fig. 1.9. Uniform distribution of shear stresses over the cross-section of a beam.

Fig. 1.10. Shear stresses at points **A** and **B** on cross-section.

Problem 1.3. Shear stress distribution over the cross-section of a beam
Figure 1.9 depicts a beam with a rectangular cross-section of a width b and height h. This beam is subjected to a vertical shear force, V_2, and the resulting shear stress distribution is assumed to be uniformly distributed over the cross-section, *i.e.*, $\tau_{12} = V_3/(bh)$. *(1)* Is this assumption reasonable? Explain your answer.

Problem 1.4. Shear stresses in a "Z" section
Figure 1.10 depicts a cantilevered beam with a "Z" cross-section subjected to a distributed transverse load p_0. Due to this loading, direct and shear stresses will develop in the beam. *(1)* Evaluate the shear stresses, denoted τ_A and τ_B, acting in the plane of the beam's mid-span cross-section at points **A** and **B**, respectively. Explain your answer.

1.3 The state of plane stress

A particular state of stress of great practical importance is the *plane state of stress*. In this case, all stress components acting along the direction of axis $\bar{\imath}_3$ are assumed to vanish, or to be negligible compared to the stress components acting in the other two directions. The only non-vanishing stress components are σ_1, σ_2, and τ_{12}, and furthermore, these stress components are assumed to be independent of x_3. This state of stress occurs, for instance, in a very thin plate or sheet subjected to loads applied in its own plane. This type of situation is illustrated by the thin sheet shown in fig. 1.11. For the plane stress state, the two flat surfaces of the thin sheet must be stress free.

1.3.1 Equilibrium equations

The equations of equilibrium derived for the general, three-dimensional case, see eq. (1.4), considerably simplify in the plane stress case. The equation in the $\bar{\imath}_3$ direction is satisfied, and the remaining two equations reduce to

$$\frac{\partial \sigma_1}{\partial x_1} + \frac{\partial \tau_{21}}{\partial x_2} + b_1 = 0; \qquad \frac{\partial \tau_{12}}{\partial x_1} + \frac{\partial \sigma_2}{\partial x_2} + b_2 = 0. \tag{1.26}$$

Similar simplifications take place for the definition of surface tractions in eq. (1.9),

$$t_1 = n_1 \sigma_1 + n_2 \tau_{21}; \qquad t_2 = n_1 \tau_{12} + n_2 \sigma_2. \tag{1.27}$$

For this two-dimensional problem, the boundary of the thin sheet on which externally applied stresses and forces may act is the thin edge defined by the curve $\mathcal{C}$ as shown in fig. 1.11. The outer normal to this curve is the unit vector $\bar{n} = n_1 \bar{\imath}_1 + n_2 \bar{\imath}_2$ and the tangent direction is the unit vector $\bar{s} = s_1 \bar{\imath}_1 + s_2 \bar{\imath}_2$. If θ is the angle between the normal and axis $\bar{\imath}_1$, it follows that $n_1 = \cos\theta$, $n_2 = \sin\theta$, $n_3 = 0$ and $s_1 = -\sin\theta$, $s_2 = \cos\theta$, $s_3 = 0$. The surface traction component in the direction of vector $\bar{n}$ then follows from eq. (1.11) as

$$t_n = \cos^2\theta\, \sigma_1 + \sin^2\theta\, \sigma_2 + 2\sin\theta\, \cos\theta\, \tau_{12}, \tag{1.28}$$

and eq. (1.12) yields the surface traction component in the direction of the tangent $\bar{s}$ to curve $\mathcal{C}$ as

$$t_s = \sin\theta\cos\theta(\sigma_2 - \sigma_1) + (\cos^2\theta - \sin^2\theta)\,\tau_{12}. \tag{1.29}$$

Thus, for plane stress problems, the equilibrium equations, eq. (1.26), must be satisfied at all points within the body, and along curve $\mathcal{C}$, the surface equilibrium equations, eq. (1.27), or eqs. (1.28) and (1.29), must be satisfied.

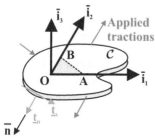

Fig. 1.11. Plane stress problem in thin sheet with in-plane tractions.

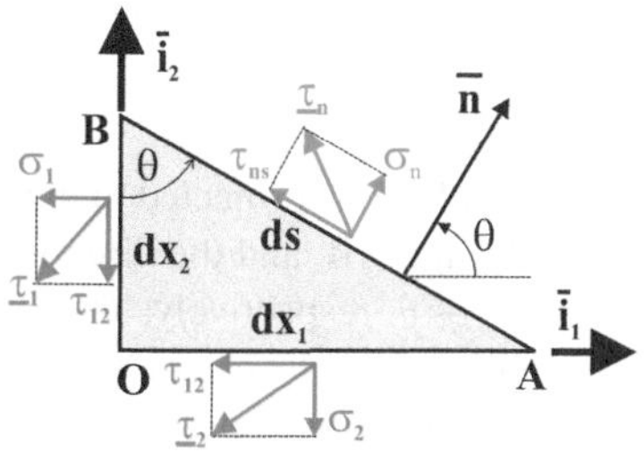

Fig. 1.12. Differential element with a face at an angle θ.

1.3.2 Stresses acting on an arbitrary face within the sheet

Figure 1.12 shows a free body **OAB** taken from within the thin sheet in fig. 1.11. It is a differential triangle with two sides cut normal to axes $\bar{\imath}_1$ and $\bar{\imath}_2$, and the third side cut normal to a unit vector, $\bar{n} = n_1\,\bar{\imath}_1 + n_2\,\bar{\imath}_2$, at an arbitrary orientation angle θ with respect to axis $\bar{\imath}_1$. Clearly, $n_1 = \cos\theta$ and $n_2 = \sin\theta$.

Triangle **OAB** is the two-dimensional version of Cauchy's tetrahedron presented in section 1.2.1 and depicted in fig. 1.7. Hence, the results derived in section 1.2 are directly applicable to the present case. Figure 1.12 shows the stress components acting on sides **OA** and **OB**, of length dx_1, and dx_2, respectively. On side **AB**, of length ds, the stress vector τ_n is acting. Finally, the body force vector, $\underline{b}$, is also acting on this triangle. For convenience, the thickness of the body in the direction of axis $\bar{\imath}_3$ is taken to be unity.

Equilibrium of forces acting on triangle **OAB** can be expressed by multiplying each of the stress vectors by the area over which they acts, *i.e.*, the length times the unit thickness, and this yields

$$\underline{\tau}_2 dx_1 + \underline{\tau}_1 dx_2 = \underline{\tau}_n ds + \underline{b}\,dx_1 dx_2/2,$$

where $\underline{\tau}_1$ and $\underline{\tau}_2$ are the stress vectors acting on the faces normal to axes $\bar{\imath}_1$ and $\bar{\imath}_2$, respectively. Dividing this equilibrium equation by ds gives the stress vector acting on the inclined face as

$$\underline{\mathcal{T}}_n = \underline{\mathcal{T}}_1 n_1 + \underline{\mathcal{T}}_2 n_2 - \underline{b}\, \mathrm{d}x_1 \mathrm{d}x_2/2\mathrm{d}s$$

The body force term is multiplied by a higher order differential term, which can neglected. Expanding the stress vectors in terms of the stress components then yields

$$\underline{\mathcal{T}}_n = (\sigma_1 \bar{\imath}_1 + \tau_{12} \bar{\imath}_2)\,\cos\theta + (\tau_{21} \bar{\imath}_1 + \sigma_2 \bar{\imath}_2)\,\sin\theta. \tag{1.30}$$

The three-dimensional equivalent of this relationship is given by eq. (1.10).

Projecting this vector equation in the direction of unit vector $\bar{n}$ yields the direct stress component, σ_n, acting on this face as $\sigma_n = (\sigma_1 \cos\theta + \tau_{12}\sin\theta)\,\cos\theta + (\tau_{21}\cos\theta + \sigma_2 \sin\theta)\,\sin\theta$, or after rearrangement,

$$\sigma_n = \sigma_1\,\cos^2\theta + \sigma_2\,\sin^2\theta + 2\tau_{12}\,\cos\theta\sin\theta. \tag{1.31}$$

Next, eq. (1.30) is projected in the direction normal to unit vector $\bar{n}$. This is in the direction of edge **AB**, and the direction cosines of this vector with axes $\bar{\imath}_1$ and $\bar{\imath}_2$ are $-\sin\theta$ and $\cos\theta$, respectively. The shear stress component, τ_{ns}, acting on side **AB** then becomes $\tau_{ns} = (-\sigma_1\sin\theta + \tau_{12}\cos\theta)\,\cos\theta + (-\tau_{21}\sin\theta + \sigma_2\cos\theta)\,\sin\theta$ which, after rearrangement, becomes

$$\tau_{ns} = -\sigma_1\,\cos\theta\sin\theta + \sigma_2\,\sin\theta\cos\theta + \tau_{12}(\cos^2\theta - \sin^2\theta). \tag{1.32}$$

Equations (1.31) and (1.32) could have been directly derived from their three-dimensional equivalent, eqs. (1.11) and (1.12), respectively, by noting that for the plane stress case, $n_1 = \cos\theta$, $n_2 = \sin\theta$, $n_3 = 0$ and $s_1 = -\sin\theta$, $s_2 = \cos\theta$, $s_3 = 0$.

These important results show that knowledge of the stress components σ_1, σ_2, and τ_{12} on two orthogonal faces allows computation of the stress components acting on a face with an arbitrary orientation. In other words, the knowledge of the stress components on two orthogonal faces fully defines the state of stress at a point.

1.3.3 Principal stresses

Principal stresses and their directions can also be determined for plane stress situations. It is a straightforward process to simply write eqs. (1.13), (1.14) and (1.15) with $\sigma_3 = \tau_{23} = \tau_{13} = 0$. This yields a vanishing principal stress along axis $\bar{\imath}_3$ and a quadratic equation for the remaining two principal stresses, which must lie in plane $(\bar{\imath}_1, \bar{\imath}_2)$. The computational procedure is otherwise unchanged.

It is more interesting, however, to consider eq. (1.31) as defining the direct stress, σ_n, acting on side **AB** of triangle **OAB**, see fig. 1.12. The magnitude of this direct stress is a function of θ, the orientation angle of this face. The particular orientation, θ_p, that maximizes (or minimizes) the magnitude of this stress component is determined by requiring the vanishing of the derivative of σ_n with respect to angle θ, to find

$$\frac{\mathrm{d}\sigma_n}{\mathrm{d}\theta} = -2\sigma_1\cos\theta_p\sin\theta_p + 2\sigma_2\cos\theta_p\sin\theta_p + 2\tau_{12}(\cos^2\theta_p - \sin^2\theta_p) = 0.$$

Using the elementary double-angle trigonometric identities, the orientation of the side that gives the extreme direct stress is found to be

$$\tan 2\theta_p = \frac{2\tau_{12}}{\sigma_1 - \sigma_2}. \tag{1.33}$$

This equation possesses two solutions θ_p and $\theta_p + \pi/2$ corresponding to two mutually orthogonal principal stress directions. The maximum axial stress is found along one direction, and the minimum along the other.

To determine these axes unambiguously, it is convenient to develop separate equations for both $\sin 2\theta_p$ and $\cos 2\alpha_p$ as follows. If eq. (1.33) is rewritten as

$$\tan 2\theta_p = \frac{2\tau_{12}}{\sigma_1 - \sigma_2} = \frac{\sin 2\theta_p}{\cos 2\theta_p},$$

it is then possible to identify $\sin 2\theta_p = \tau_{12}/\Delta$ and $\cos 2\theta_p = (\sigma_1 - \sigma_2)/2\Delta$, where Δ is determined by the following trigonometric identity, $\sin^2 2\theta_p + \cos^2 2\theta_p = 1$, to find

$$\Delta = \left[\left(\frac{\sigma_1 - \sigma_2}{2} \right)^2 + (\tau_{12})^2 \right]^{1/2}.$$

Thus, the sine and cosine of angle $2\theta_p$ can be expressed as follows

$$\sin 2\theta_p = \frac{\tau_{12}}{\Delta}, \quad \cos 2\theta_p = \frac{\sigma_1 - \sigma_2}{2\Delta}, \tag{1.34}$$

where

$$\Delta = \sqrt{\left(\frac{\sigma_1 - \sigma_2}{2} \right)^2 + \tau_{12}^2}. \tag{1.35}$$

This result is equivalent to eq. (1.33), but it gives a unique solution for θ_p because both the sine and cosine of the angle are known. The maximum and minimum axial stresses, denoted σ_{p1} and σ_{p2}, respectively, act in the directions θ_p and $\theta_p + \pi/2$, respectively. These maximum and minimum axial stresses, called the *principal stresses*, are evaluated by introducing eq. (1.34) into eq. (1.31) to find

$$\sigma_{p1} = \frac{\sigma_1 + \sigma_2}{2} + \Delta; \quad \sigma_{p2} = \frac{\sigma_1 + \sigma_2}{2} - \Delta. \tag{1.36}$$

The principal stresses are maximum and minimum values of the axial stress in an algebraic sense. Note that it is possible, however, to have $|\sigma_{p2}| > |\sigma_{p1}|$.

The shear stress acting on the faces normal to the principal stress directions vanishes, as expected. This can be verified by introducing eq. (1.34) into eq. (1.32)

$$\tau_{ns} = -\frac{\sigma_1 - \sigma_2}{2} \sin 2\theta_p + \tau_{12} \cos 2\theta_p = -\frac{\sigma_1 - \sigma_2}{2} \frac{\tau_{12}}{\Delta} + \tau_{12} \frac{\sigma_1 - \sigma_2}{2\Delta} = 0.$$

It is also interesting to find the orientation of the faces leading to the maximum value of the shear stress. Indeed, in view of eq. (1.32), the shear stress is also a function of the face orientation angle. The orientation, θ_s, of the face on which the maximum (or minimum) shear stress acts satisfies the following extremal condition

$$\frac{\mathrm{d}\tau_{ns}}{\mathrm{d}\theta} = -\frac{\sigma_1 - \sigma_2}{2} 2\cos 2\theta_s - \tau_{12} 2\sin 2\theta_s = 0, \tag{1.37}$$

or

$$\tan 2\theta_s = -\frac{\sigma_1 - \sigma_2}{2\tau_{12}} = -\frac{1}{\tan 2\theta_p}, \tag{1.38}$$

where the last equality follows from eq. (1.34). Here again, this equation presents two solutions, θ_s and $\theta_s + \pi/2$, corresponding to two mutually orthogonal faces. To define these orientations unequivocally, separate definitions of the sine and cosines of angle $2\theta_s$ are given as follows

$$\sin 2\theta_s = -\frac{\sigma_1 - \sigma_2}{2\Delta}; \quad \cos 2\theta_s = \frac{\tau_{12}}{\Delta}, \tag{1.39}$$

where Δ is again given by eq. (1.35).

The maximum shear stress acting on these faces results from introducing eq. (1.39) into eq. (1.32) to find

$$\tau_{\max} = \Delta = \frac{\sigma_{p1} - \sigma_{p2}}{2}. \tag{1.40}$$

Since $\tan 2\theta_s = -1/\tan 2\theta_p$, trigonometric identities reveal that

$$\theta_s = \theta_p - \frac{\pi}{4}. \tag{1.41}$$

This means that the faces on which the maximum shear stresses occur are inclined at a 45° angle with respect to the principal stress directions.

The axial stresses acting on these faces are found by introducing eq. (1.39) into eq. (1.31) and using the first stress invariant property to find

$$\sigma_{1s} = \sigma_{2s} = \frac{\sigma_1 + \sigma_2}{2} = \frac{\sigma_{p1} + \sigma_{p2}}{2}. \tag{1.42}$$

1.3.4 Rotation of stresses

In the previous sections, faces are cut in planes normal to the two axes of an orthonormal basis $\mathcal{I} = (\bar{\imath}_1, \bar{\imath}_2)$, and the stress vectors are resolved into stress components along the same directions. It is clear that the orientation of this basis is entirely arbitrary: an orthonormal basis $\mathcal{I}^* = (\bar{\imath}_1^*, \bar{\imath}_2^*)$ could have been selected, and an analysis identical to that of the previous sections would have led to the definition of axial stresses σ_1^* and σ_2^*, and shear stress τ_{12}^*. A typical equilibrium equation at a point of the body would be written as

$$\frac{\partial \sigma_1^*}{\partial x_1^*} + \frac{\partial \tau_{21}^*}{\partial x_2^*} + b_1^* = 0; \tag{1.43}$$

where the notation $(\cdot)^*$ is used to indicate the components of the corresponding quantity resolved in $\mathcal{I}^*$. A typical surface traction is be defined as

$$t_1^* = n_1^* \, \sigma_1^* + n_2^* \, \tau_{21}^*. \tag{1.44}$$

Although expressed in different reference frames, eqs. (1.26) and (1.43), or (1.27) and (1.44) express the same equilibrium conditions for the body. The problem at hand involves two distinct orthonormal bases, $\mathcal{I}$ and $\mathcal{I}^*$, and the relationship between these two basis is developed in appendix A.3.3.

Consider the stress component σ_1^*: it represents the magnitude of the direct stress component acting on the face normal to axis $\bar{\imath}_1^*$. Let θ be the angle between unit vector $\bar{\imath}_1^*$ and axis $\bar{\imath}_1$. Equation (1.31) can now be used to express the stress component σ_1^* in terms of the stress components resolved in axis system $\mathcal{I}$ to find

$$\sigma_1^* = \sigma_1 \cos^2 \theta + \sigma_2 \sin^2 \theta + 2\tau_{12} \sin \theta \cos \theta. \tag{1.45}$$

A similar equation can be derived to express σ_2^* in terms of the stress components resolved in axis system $\mathcal{I}$ by replacing angle θ by $\theta + \pi/2$ in the above equation; $\theta + \pi/2$ is the angle between unit vector $\bar{\imath}_2^*$ and axis $\bar{\imath}_1$.

Finally, the shear stress component can be computed from eq. (1.32) as

$$\tau_{12}^* = -\sigma_1 \sin \theta \cos \theta + \sigma_2 \sin \theta \cos \theta + \tau_{12}(\cos^2 \theta - \sin^2 \theta). \tag{1.46}$$

These results can be combined into a compact matrix form as

$$\begin{Bmatrix} \sigma_1^* \\ \sigma_2^* \\ \tau_{12}^* \end{Bmatrix} = \begin{bmatrix} \cos^2 \theta & \sin^2 \theta & 2 \sin \theta \cos \theta \\ \sin^2 \theta & \cos^2 \theta & -2 \sin \theta \cos \theta \\ -\sin \theta \cos \theta & \sin \theta \cos \theta & \cos^2 \theta - \sin^2 \theta \end{bmatrix} \begin{Bmatrix} \sigma_1 \\ \sigma_2 \\ \tau_{12} \end{Bmatrix}. \tag{1.47}$$

This relationship can be easily inverted by recognizing that the inverse transformation is obtained simply by replacing θ by $-\theta$ to find

$$\begin{Bmatrix} \sigma_1 \\ \sigma_2 \\ \tau_{12} \end{Bmatrix} = \begin{bmatrix} \cos^2 \theta & \sin^2 \theta & -2 \sin \theta \cos \theta \\ \sin^2 \theta & \cos^2 \theta & 2 \sin \theta \cos \theta \\ \sin \theta \cos \theta & -\sin \theta \cos \theta & \cos^2 \theta - \sin^2 \theta \end{bmatrix} \begin{Bmatrix} \sigma_1^* \\ \sigma_2^* \\ \tau_{12}^* \end{Bmatrix}. \tag{1.48}$$

With the help of double-angle trigonometric identities, the transformation rules for stress components, eq. (1.47), can also be written in the following useful form

$$\sigma_1^* = \frac{\sigma_1 + \sigma_2}{2} + \frac{\sigma_1 - \sigma_2}{2} \cos 2\theta + \tau_{12} \sin 2\theta, \tag{1.49a}$$

$$\sigma_2^* = \frac{\sigma_1 + \sigma_2}{2} - \frac{\sigma_1 - \sigma_2}{2} \cos 2\theta - \tau_{12} \sin 2\theta, \tag{1.49b}$$

$$\tau_{12}^* = \qquad - \frac{\sigma_1 - \sigma_2}{2} \sin 2\theta + \tau_{12} \cos 2\theta. \tag{1.49c}$$

These important results show that knowledge of the stress components σ_1, σ_2, and τ_{12} on two orthogonal faces allows computation of the stress components acting on a face with an arbitrary orientation. In other words, the knowledge of the stress components on two orthogonal faces fully defines the state of stress at a point.

1.3.5 Special states of stress

Two plane stress states are of particular interest. One is called the *hydrostatic stress state* and the other is called the *pure shear state*. A third special state of plane stress is the stress developed in a thin-walled cylindrical pressure vessel.

Hydrostatic stress state. A stress state of practical importance is the *hydrostatic state of stress*. In this case, the principal stresses are equal, *i.e.*, $\sigma_{p1} = \sigma_{p2} = p$, where p is the *hydrostatic pressure*. It follows from eq. (1.49) that the stresses acting on a face with any arbitrary orientation are

$$\sigma_1 = \sigma_2 = p, \quad \tau_{12} = 0. \tag{1.50}$$

Pure shear state. A stress state of great practical importance is the state of *pure shear* characterized by principal stresses of equal magnitude but opposite signs, *i.e.*, $\sigma_{p2} = -\sigma_{p1}$, as depicted in fig. 1.13. Equations (1.45) and (1.46) then reveal the direct and shear stresses, respectively, acting on a face inclined at a $45°$ angle with respect to the principal stress directions as

$$\tau_{12}^* = -\sigma_{p1}; \quad \sigma_1^* = \sigma_2^* = 0. \tag{1.51}$$

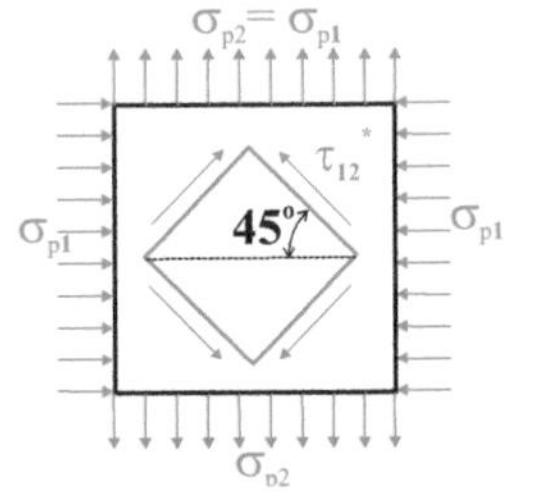

Fig. 1.13. A differential plane stress element in pure shear.

On faces oriented at $45°$ angles with respect to the principal stress directions, the direct stresses vanish and the shear has a maximum value, equal in magnitude to the common magnitudes of the two principal stresses.

Stress state in thin-walled pressure vessels. The stress state in the walls of thin-walled tanks, called pressure vessels, of certain shapes consists of two in-plane normal stresses and an in-plane shear stress. Although the pressure vessel may be subjected to a large internal pressure that will produce a pressure loading on the interior wall in the transverse direction, the magnitude of this stress often is orders of magnitude smaller than the in-plane stress components and is therefore usually neglected. The spherical pressure vessel and a long cylindrical pressure vessel (ignoring the effect of the ends) are two useful examples.

A thin-walled ($t \ll R$) cylindrical pressure vessel subjected to an internal pressure, p_i, is depicted in fig. 1.14, where it is assumed that the only stresses are the two in-plane stress components, σ_a in the axial direction, and σ_h in the circumferential or "hoop" direction, and possibly a shear stress, τ_{ah}. In the central portion of the cylinder, it is possible to create the simple free body shown in the figure, which will allow direct calculation of these stresses. From axial force equilibrium, it follows that $\sigma_a \pi R t = p_i \pi R^2 / 2$, and hence, $\sigma_a = p_i R / 2t$. Equilibrium in the tangential (hoop) direction implies $2\sigma_h b t = p_i 2Rb$, and hence, $\sigma_h = p_i R / t$. Finally, it should be clear that $\tau_{ah} = 0$ for this axis orientation.

It is left as an exercise to show that by a similar free body analysis of a spherical thin-walled pressure vessel, $\sigma_a = \sigma_h = p_i R / t$ in any direction and the shear stress

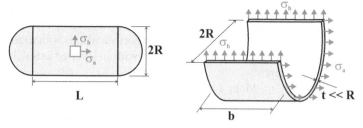

Fig. 1.14. Long, thin-walled cylindrical pressure vessel (left) and free body diagram (right) used to calculate in-plane stresses σ_h and σ_a.

vanishes. This is a special case of two-dimensional hydrostatic stress. A more formal analysis of pressure vessels is presented in section 4.4.

1.3.6 Mohr's circle for plane stress

Equation (1.49) expresses the direct and shear stresses acting on a face oriented at an arbitrary angle θ with respect the axis $\bar{\imath}_1$, but the presence of trigonometric functions involving the angle 2θ makes it difficult to give a simple, geometric interpretation of these formulæ. A useful geometric interpretation, however, called *Mohr's circle*, can be developed. Let the state of stress at a point be defined by its principal stresses, σ_{p1} and σ_{p2}. Equation (1.49) then implies that the stresses acting on a face oriented at an angle θ with respect to the principal stress directions can be written as

$$\sigma^* = \sigma_a + R \cos 2\theta; \quad \tau^* = -R \sin 2\theta, \tag{1.52}$$

where $\sigma_a = (\sigma_{p1} + \sigma_{p2})/2$ and $R = (\sigma_{p1} - \sigma_{p2})/2$. With this notation and the help of basic trigonometric identities, eq. (1.52) becomes

$$(\sigma^* - \sigma_a)^2 + (\tau^*)^2 = R^2. \tag{1.53}$$

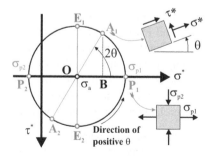

Fig. 1.15. Mohr's circle for visualizing plane stress state.

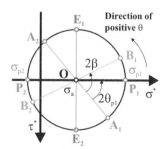

Fig. 1.16. Mohr's circle construction procedure.

This equation clearly represents the *equation of a circle*, known as *Mohr's circle* in which σ^* is plotted along the horizontal axis, τ^* is plotted along the vertical axis, and the circle is centered at a coordinate σ_a on the horizontal axis with a radius of R, as depicted in fig. 1.15[1]. The reason for plotting τ^* with an inverted axis will become clear in the next paragraphs.

Consider point $\mathbf{A_1}$ on Mohr's circle such that segment $\mathbf{OA_1}$ makes an angle 2θ with the horizontal. The coordinates of this point are $\sigma^* = \sigma_a + R\cos 2\theta$ and $\tau^* = -R\sin 2\theta$; hence, in view of eq. (1.52), the coordinates of point $\mathbf{A_1}$ represent the state of stress on a face oriented at an angle θ. In fact, each point on Mohr's circle represents the state of stress acting on a face at a specific orientation.

An **important sign convention** must be defined: on Mohr's circle, a positive angle θ is measured in the **counterclockwise direction**, see fig. 1.15, to match the positive direction of angle θ that identifies the orientation of a face in fig. 1.12. Given the sign convention for angle θ, the shear stress must be positive downward on the ordinate of Mohr's circle depicted in fig. 1.15.[2]

The following observations are made.

- At point $\mathbf{P_1}$, the stress state is $\sigma^* = \sigma_{p1}$ and $\tau^* = 0$; this corresponds, as expected, to the stress components acting in the principal stress direction. Similar results are found at point $\mathbf{P_2}$ which represents the stress components acting in the second principal direction.
- At point $\mathbf{E_1}$, associated with an angle $\theta = \pi/4$, the stress components are $\tau^*_{\max} = R = (\sigma_{p1} - \sigma_{p2})/2$ and $\sigma^* = \sigma_a = (\sigma_{p1} + \sigma_{p2})/2$. These results are identical to those expressed by eqs. (1.40) and (1.42), respectively. In the graphical representation of stress states given by Mohr's circle, it becomes obvious that the maximum shear stress is found on faces oriented at $\pm 45°$ angles with respect to the principal stress directions, and this is defined by points $\mathbf{E_1}$ and $\mathbf{E_2}$ in fig. 1.15.
- Points $\mathbf{A_1}$ and $\mathbf{A_2}$ represent the stress components acting on two faces oriented $90°$ apart. The shear stresses acting on those two faces are equal in magnitude and of opposite sign, as required by the principle of reciprocity of shear stresses illustrated in fig. 1.5. The direct stresses correspond to stresses σ_1^* and σ_2^* in eqs. (1.49).

In the above discussion, Mohr's circle is constructed based on the knowledge of the principal stresses represented by points $\mathbf{P_1}$ and $\mathbf{P_2}$ in figs. 1.15 and 1.16. In practice, it is often the case that the state of stress at a point is defined by known stress components σ_1, σ_2 and τ_{12}. These three stress components define two diametrically opposed points, $\mathbf{A_1}$ and $\mathbf{A_2}$, on Mohr's circle depicted in fig. 1.16. Once this circle is constructed with the help of the procedure described below, the stress components acting on any face rotated by an angle β in a counterclockwise direction, represented

---

[1] A Mohr's circle representation that describes the rotation of a three-dimensional second order tensor can be constructed but it involves three interdependent circles and is quite tedious to construct and to use.

[2] An equivalent construction of Mohr's circle has the shear stress positive upwards along the ordinate, but angle θ is then positive in the clockwise direction.

by points $\mathbf{B_1}$ and $\mathbf{B_2}$ in fig. 1.16, can be directly obtained from simple geometric constructions.

1. Draw a first point, identified as point $\mathbf{A_1}$, at coordinates (σ_1, τ_{12}). This represents the direct and shear stresses acting on one face of the solid.
2. Draw a second point, identified as point $\mathbf{A_2}$, at coordinates $(\sigma_2, -\tau_{12})$. This represents the direct and shear stresses acting on a face of the solid at a $90°$ angle counterclockwise with respect to the first face. Since the two faces are $90°$ apart, these two points must define diametrically opposite points on Mohr's circle.
3. Draw a straight line segment joining points $\mathbf{A_1}$ and $\mathbf{A_2}$; the intersection of this segment with the horizontal axis defines the center of the Mohr's circle of diameter $\mathbf{A_1 O A_2}$ at point $\mathbf{O}$. Points $\mathbf{A_1}$ and $\mathbf{A_2}$ represent the stress components on two orthogonal faces, that is, on faces of relative orientation $\theta = 90°$, since the angle between segments $\mathbf{OA_1}$ and $\mathbf{OA_2}$ is $2\theta = 180°$.
4. Once Mohr's circle is drawn, the stress state on faces at any orientation angle can be computed. For instance, the stress components acting on a face oriented at an angle β from the face on which stress components σ_1 and τ_{12} act can be computed by constructing a new diameter $\mathbf{B_1 O B_2}$ rotated 2β degrees from the reference diameter $\mathbf{A_1 O A_2}$. The coordinates of point $\mathbf{B_1}$ yield the new stress components.

Mohr's circle displays in a graphical manner many important features characterizing the state of stress at a point.

1. The principal stresses, σ_{p1} and σ_{p2}, shown in figs. 1.15 and 1.16, are represented by the points $\mathbf{P_1}$ and $\mathbf{P_2}$ at the intersection of Mohr's circle with the horizontal axis. Clearly, these points define the orientation of the faces on which the direct stresses take on maximum and minimum values and for which the shear stress vanishes.
2. The faces on which the maximum shear stresses occurs are represented by the points at the intersection of Mohr's circle with a vertical line passing through its center. It is clear that the magnitude of the maximum shear stress equals the radius of Mohr's circle: $\tau_{max} = (\sigma_{p1} - \sigma_{p2})/2$, see eq. (1.40). The angle between the principal stress directions and those of the face of maximum shear is $45°$, because the angle $\mathbf{P_1 O E_1}$ is $90°$, see eq. (1.41). Finally, the direct stresses acting on the faces of maximum shear equal the average of the principal stresses, $\sigma_{1s} = \sigma_{2s} = (\sigma_{p1} + \sigma_{p2})/2$, see eq. (1.42).
3. The stress components acting on two mutually orthogonal faces are represented by two diametrically opposite points on Mohr's circle. Since the center of the circle is on the horizontal axis, the shear stresses on those two faces are equal in magnitude and opposite in sign, as required by the principle of reciprocity of shear stresses illustrated in fig. 1.5.
4. Finally, note that *all the points on Mohr's circle represent the same state of stress* at one point of the solid. Of course, this state of stress is represented by stress components that depend on the orientation of the face on which they act.

Mohr's circle is a graphical representation of all the possible stress components corresponding to a single state of stress.

1.3.7 Lamé's ellipse

Lamé's ellipse provides an elegant geometric interpretation of the state of stress at a point. Consider a material in a plane state of stress and let $\underline{\tau}_n$ be the stress vector acting on the face with a unit normal $\bar{n}$ at an angle θ with respect to axes $\bar{\imath}_1^*$, as depicted in fig. 1.17. As angle θ varies, the tip of the stress vector, $\underline{\tau}_n$, draws an ellipse, called *Lamé's ellipse*, with its center at **O** and its semi-axes given by the absolute value of the principal stresses, $|\sigma_{p1}|$ and $|\sigma_{p2}|$, respectively. The minor and major axes of the ellipse are aligned with the principal stress directions so that axes $\bar{\imath}_1^*$ and $\bar{\imath}_2^*$ are the principal stress directions.

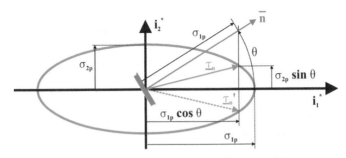

Fig. 1.17. Lamé's ellipse. Stress vector $\underline{\tau}_n$ corresponds to positive principal stresses whereas stress vector $\underline{\tau}_n'$ corresponds to $\sigma_{p1} > 0$ and $\sigma_{p2} < 0$.

To prove the stated claim that the locus of the tip of the stress vector draws the ellipse shown in fig. 1.17, the stress vector acting on the face at an angle θ with respect to axis $\bar{\imath}_1^*$ can be expressed with the help of eq. (1.30) as $\underline{\tau}_n = \sigma_{1p} \cos\theta\, \bar{\imath}_1^* + \sigma_{2p} \sin\theta\, \bar{\imath}_2^*$, where it is noted that $\sigma_1 = \sigma_{1p}$, $\sigma_2 = \sigma_{2p}$, and $\tau_{12} = 0$ because the selected axis system coincides with the principal stress directions. Let x_1 and x_2 be the coordinates of the tip of the stress vector, hence, $\underline{\tau}_n = x_1\, \bar{\imath}_1^* + x_2\, \bar{\imath}_2^*$. It then follows that $x_1 = \sigma_{1p} \cos\theta$ and $x_2 = \sigma_{2p} \sin\theta$, and elimination of the angle θ using the elementary trigonometric identity leads to

$$\left(\frac{x_1}{\sigma_{1p}}\right)^2 + \left(\frac{x_2}{\sigma_{2p}}\right)^2 = 1. \tag{1.54}$$

This is the equation of an ellipse with semi-axes equal to $|\sigma_{p1}|$ and $|\sigma_{p2}|$, respectively, proving the stated claim.

As the orientation of the face changes, the tip of the stress vector sweeps around Lamé's ellipse. Note that while the shape of the ellipse is not affected by the sign of the principal stresses, the orientation of the stress vector does depend on their sign. For instance, the stress vector $\underline{\tau}_n$ shown in fig. 1.17 corresponds to $\sigma_{p1} > 0$ and

$\sigma_{p2} > 0$; for the case where $\sigma_{p1} > 0$ and $\sigma_{p2} < 0$, however, the stress vector acting on the same face is now represented by vector $\underline{\tau}'_n$.

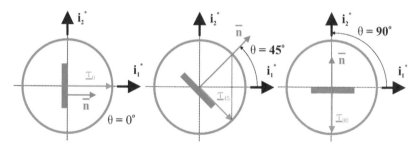

Fig. 1.18. Lamé's ellipse for the case of pure shear; the three figures illustrate the stress vectors acting on faces at 0, 45, and 90 degrees with respect to axis $\overline{\imath}^*_1$.

An interesting case is provided by the pure shear state of stress discussed in section 1.3.5. This is defined by the principal stresses $\sigma_{1p} = \tau$ and $\sigma_{2p} = -\tau$: the principal stresses are equal in magnitude but of opposite sign. Since the two semi-axes of Lamé's ellipse are equal, it becomes the circle depicted in fig. 1.18, and hence, the norm of the stress vector remains constant as the face on which it act rotates. When the face is oriented at a 45 degree angle, the stress vector acts at a -45 degree angle with respect to axis $\overline{\imath}^*_1$ and the face is subjected to only a shear stress, as expected. Finally, note that while the face rotates counterclockwise, the stress vector describes Lamé's ellipse in the clockwise direction.

1.3.8 Problems

Problem 1.5. Stress states on two sets of faces
The plane stress state at a point is known and characterized by the following stress components: $\sigma_1 = 250$ MPa, $\sigma_2 = 250$ MPa, and $\tau_{12} = 0$ MPa in a coordinate system $\mathcal{I} = (\overline{\imath}_1, \overline{\imath}_2)$. Find the stress components σ^*_1, σ^*_2, and τ^*_{12} in a coordinate system $\mathcal{I}^* = (\overline{\imath}^*_1, \overline{\imath}^*_2)$, where $\overline{\imath}^*_1$ is at a 25 degree angle with respect to $\overline{\imath}_1$.

Problem 1.6. Stress invariants for plane stress state
The stress invariants defined in eq. (1.15) for three-dimensional problems. *(1)* Show that for plane stress problems, the following two quantities are invariants

$$I_1 = \sigma_1 + \sigma_2; \quad I_2 = \sigma_1 \sigma_2 - \tau^2_{12}. \tag{1.55}$$

(2) Prove your claim of invariance by showing that these quantities are identical when computed in terms of the principal stresses and in terms of stresses acting on a face at an arbitrary orientation.

Problem 1.7. Stress rotation formulæ in matrix form
Show that the plane stress stress rotation formulae given by eq. (1.47) can be recast in the following compact matrix form

$$\begin{bmatrix} \sigma_1^* & \tau_{12}^* \\ \tau_{12}^* & \sigma_2^* \end{bmatrix} = \begin{bmatrix} \cos\theta & \sin\theta \\ -\sin\theta & \cos\theta \end{bmatrix} \begin{bmatrix} \sigma_1 & \tau_{12} \\ \tau_{12} & \sigma_2 \end{bmatrix} \begin{bmatrix} \cos\theta & -\sin\theta \\ \sin\theta & \cos\theta \end{bmatrix}.$$

Problem 1.8. Mohr's circle

Draw Mohr's circle for the state of stress defined by $\sigma_1 = 80$ MPa, $\sigma_2 = -20$ MPa and $\tau_{12} = 40$ MPa. Using this circle, (1) calculate the stress on axes rotated 60 degrees counterclockwise from the reference axes, and (2) determine the principal stresses and the corresponding directions. Do these results agree with the results in section 1.3.3?

Problem 1.9. Mohr's circle for the state of pure shear

Draw Mohr's circle for the state of pure shear defined in section 1.3.5. Show how eq. (1.51) can be readily derived from Mohr's circle.

Problem 1.10. Mohr's circle for the hydrostatic state of stress

Draw Mohr's circle for the state of hydrostatic stress defined in section 1.3.5. Show how eq. (1.50) can be readily derived from Mohr's circle.

Problem 1.11. Stresses in a pressure vessel

A cylindrical pressure vessel of radius R and thickness t is subjected to an internal pressure p_i. At any point in the cylindrical portion of vessel wall, two stress components are acting: the hoop stress, $\sigma_h = Rp_i/t$ and the axial stress, $\sigma_a = Rp_i/(2t)$. The radial stress, acting in the direction perpendicular to the wall, is very small, $\sigma_r \approx 0$. The pressure vessel features a weld line at a 45 degree angle with respect to the axis of the cylinder, as shown in fig. 1.19. (1) Find the direct stress acting in the direction perpendicular to the weld line. (2) Find the shear stress acting along the weld line.

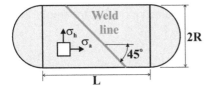

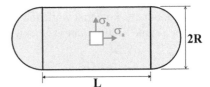

Fig. 1.19. Pressure vessel with a weld line.   **Fig. 1.20.** Stresses acting in a pressure vessel.

Problem 1.12. Maximum stresses in a pressure vessel

Figure 1.20 shows a cylindrical pressure vessel of radius R and thickness t subjected to an internal pressure p_i. At any point in the cylindrical portion of vessel wall, two stress components are acting: the hoop stress, $\sigma_h = Rp_i/t$ and the axial stress, $\sigma_a = Rp_i/(2t)$. The radial stress, acting in the direction perpendicular to the wall, is very small, $\sigma_r \approx 0$. (1) Find the orientation of the face on which the maximum direct stress is acting. What is the value of the maximum direct stress? (2) Find the orientation of the face on which the maximum shear stress is acting acting. What is the value of the maximum shear stress?

Problem 1.13. Stresses in a composite material layer

A layer of unidirectional composite material is subjected to a state of stress $\sigma_1 = 245$ MPa, $\sigma_2 = -175$ MPa, and $\tau_{12} = 95$ MPa. As depicted in fig. 1.21, the fibers in the unidirectional composite material layer run at an angle $\theta = 25$ degrees with respect to axis $\bar{\imath}_1$. (1) Find the direct stress acting in the direction of the fiber. (2) Find the direct stress acting in the direction perpendicular to the fiber.

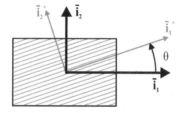

Fig. 1.21. Layer of unidirectional composite material with fiber direction.

1.4 The concept of strain

The state of strain at a point is a characterization of the deformation in the neighborhood of a material point in a solid. The description of the state of strain at a point is a great deal more complicated than that of the stress state, and the presence of nonlinear terms is much more obvious. The state of strain is concerned with the deformation of a solid in the neighborhood of a given point, say point **P**, located by a position vector $\underline{r} = x_1 \bar{\imath}_1 + x_2 \bar{\imath}_2 + x_3 \bar{\imath}_3$, as depicted in fig. 1.22.

To visualize this deformation, a small rectangular parallelepiped **PQRST** of differential size dx_1 by dx_2 by dx_3 is cut in the neighborhood of point **P**. The *reference configuration* is the configuration of the solid in its undeformed state. Under the action of applied loads, the body deforms and assumes a new configuration, called the *deformed configuration*. All the material particles that formed the rectangular parallelepiped **PQRST** in the reference configuration now form the parallelepiped **PQRST** in the deformed configuration. The state of strain at a point characterizes the deformation of the parallelepiped without any consideration for the loads that created the deformation.

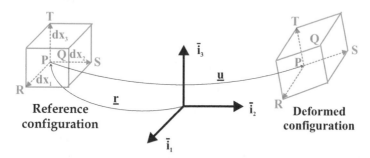

Fig. 1.22. The neighborhood of point **P** in the reference and deformed configurations.

While position vector, $\underline{r}$, locates material point **P**, the *displacement vector, $\underline{u}$*, is a measure of how much a material point moves from the reference to the deformed configuration. The components of the displacement vector resolved in coordinate system $\mathcal{I} = (\bar{\imath}_1, \bar{\imath}_2, \bar{\imath}_3)$ can be expressed as

$$\underline{u}(x_1, x_2, x_3) = u_1(x_1, x_2, x_3)\,\bar{\imath}_1 + u_2(x_1, x_2, x_3)\,\bar{\imath}_2 + u_3(x_1, x_2, x_3)\,\bar{\imath}_3. \quad (1.56)$$

This displacement field describes the displacement of a point at position (x_1, x_2, x_3) within the solid and consists of two parts: a rigid body motion and a *deformation* or *straining* of the solid. The rigid body motion itself consists of two parts: a rigid body translation and a rigid body rotation. By definition, a rigid body motion does not produce strain in the body. Consequently, the strain-displacement equations must extract from the displacement field the information that describes only the deformation of the body while ignoring its rigid body motion.

1.4.1 The state of strain at a point

A *material line* is the ensemble of material particles that form a straight line in the reference configuration of the body. For instance, segments **PR**, **PS** and **PT** of the reference configuration are material lines. Due to the deformation of the body, all the material particles forming material line **PR** will move to segment **PR** in the deformed configuration. Due to the differential nature of this segment, it can be assumed to remain straight in the deformed configuration.

When comparing segment **PR** in the reference and deformed configurations, the motion consists of two parts: a change in orientation and a change in length. Clearly, the change in length is a deformation or stretching of the material line. Similarly, segments **PR** and **PS** form a rectangle in the reference configuration, but they form a parallelogram in the deformed configuration. The angular distortion of the rectangle into a parallelogram represents a deformation of the body. Stretching of a material line and angular distortion between two material lines will be selected as measures of the state of strain at a point.

The stretching or *relative elongations* of material lines **PR**, **PS** and **PT** will be denoted as ϵ_1, ϵ_2 and ϵ_3, respectively. The *angular distortions* between segments **PS** and **PT**, **PR** and **PT**, and **PR** and **PS** will be denoted γ_{23}, γ_{13}, and γ_{12}, respectively.

Relative elongations or extensional strains

The relative elongation, ϵ_1, of material line **PR** is defined as

$$\epsilon_1 = \frac{\|\mathbf{PR}\|_{\text{def}} - \|\mathbf{PR}\|_{\text{ref}}}{\|\mathbf{PR}\|_{\text{ref}}}, \tag{1.57}$$

where the subscripts $(\cdot)_{\text{ref}}$ and $(\cdot)_{\text{def}}$ are used to indicate the reference and deformed configurations, respectively, and $\|\cdot\|$ means magnitude of a vector. The relative elongation is a non-dimensional quantity. The length of the material line in the reference configuration is

$$\|\mathbf{PR}\|_{\text{ref}} = \|\mathrm{d}x_1 \, \bar{\imath}_1\| = \mathrm{d}x_1, \tag{1.58}$$

whereas in the deformed configuration, it is

$$\|\mathbf{PR}\|_{\text{def}} = \|dx_1 \bar{\imath}_1 + \underline{u}(x_1 + dx_1) - \underline{u}(x_1)\|$$

$$= \|dx_1 \bar{\imath}_1 + \bar{u}(x_1) + \frac{\partial \underline{u}}{\partial x_1} dx_1 - \underline{u}(x_1)\| = \|dx_1 \bar{\imath}_1 + \frac{\partial \underline{u}}{\partial x_1} dx_1\|$$

$$= \|\bar{\imath}_1 dx_1 + \left(\frac{\partial u_1}{\partial x_1} \bar{\imath}_1 + \frac{\partial u_2}{\partial x_1} \bar{\imath}_2 + \frac{\partial u_3}{\partial x_1} \bar{\imath}_3 \right) dx_1\| \qquad (1.59)$$

$$= \sqrt{1 + 2\frac{\partial u_1}{\partial x_1} + \left(\frac{\partial u_1}{\partial x_1} \right)^2 + \left(\frac{\partial u_2}{\partial x_1} \right)^2 + \left(\frac{\partial u_3}{\partial x_1} \right)^2} \, dx_1,$$

where the higher order differential terms in the Taylor series expansion of the displacement field are neglected. The relative elongation now becomes

$$\epsilon_1 = \sqrt{1 + 2\frac{\partial u_1}{\partial x_1} + \left(\frac{\partial u_1}{\partial x_1} \right)^2 + \left(\frac{\partial u_2}{\partial x_1} \right)^2 + \left(\frac{\partial u_3}{\partial x_1} \right)^2} - 1. \qquad (1.60)$$

A fundamental assumption of linear elasticity is that all displacement components remain very small so that *all second order terms can be neglected*. This can be stated as requiring

$$\left| \frac{\partial u_1}{\partial x_1} \right| \ll 1, \quad \left| \frac{\partial u_2}{\partial x_1} \right| \ll 1, \quad \left| \frac{\partial u_3}{\partial x_1} \right| \ll 1, \quad \left| \frac{\partial u_1}{\partial x_2} \right| \ll 1, \quad \text{etc.} \qquad (1.61)$$

With these assumptions and by making use of the binomial expansion[3], the expression for the relative elongation given in eq. (1.60) reduces to

$$\epsilon_1 \approx 1 + \frac{\partial u_1}{\partial x_1} - 1 = \frac{\partial u_1}{\partial x_1}. \qquad (1.62)$$

A similar reasoning applied to material lines **PS** and **PT** yields expressions for the three components of relative elongation

$$\epsilon_1 = \frac{\partial u_1}{\partial x_1}, \quad \epsilon_2 = \frac{\partial u_2}{\partial x_2}, \quad \epsilon_3 = \frac{\partial u_3}{\partial x_3}. \qquad (1.63)$$

Angular distortions or shear strains

The angular distortion, γ_{23}, between two material lines **PT** and **PS** is defined as the change of the initially right angle

$$\gamma_{23} = \langle TPS \rangle_{\text{ref}} - \langle TPS \rangle_{\text{def}} = \frac{\pi}{2} - \langle TPS \rangle_{\text{def}}, \qquad (1.64)$$

where the notation $\langle TPS \rangle$ is used to indicate the angle between segments **PT** and **PS**. Both relative elongation and angular distortion are non-dimensional quantities. To eliminate the difference between the two angles, basic properties of the sine function are used: the sine of the angular distortion becomes

---

[3] When $|a| \ll 1$, it is possible to expand $(1 \pm a)^n \approx 1 \pm na$.

$$\sin \gamma_{23} = \sin \left(\frac{\pi}{2} - \langle TPS \rangle_{\text{def}} \right) = \cos \langle TPS \rangle_{\text{def}}. \tag{1.65}$$

The cosine of the angle between the two material lines is computed from the law of cosines applied to triangle TPS in the deformed configuration

$$\|\mathbf{TS}\|_{\text{def}}^2 = \|\mathbf{PT}\|_{\text{def}}^2 + \|\mathbf{PS}\|_{\text{def}}^2 - 2\cos\langle TPS \rangle_{\text{def}}\|\mathbf{PT}\|_{\text{def}}\|\mathbf{PS}\|_{\text{def}}. \tag{1.66}$$

The angular distortion thus becomes

$$\gamma_{23} = \arcsin \frac{\|\mathbf{PT}\|_{\text{def}}^2 + \|\mathbf{PS}\|_{\text{def}}^2 - \|\mathbf{TS}\|_{\text{def}}^2}{2\|\mathbf{PT}\|_{\text{def}}\|\mathbf{PS}\|_{\text{def}}}. \tag{1.67}$$

The same procedure as used above in determining ϵ_1 can be used to compute $\|\mathbf{PR}\|_{\text{def}}$ and $\|\mathbf{PS}\|_{\text{def}}$ but since the present computations are a bit more tedious, it will be convenient to introduce two temporary vectors, $\underline{A}$ an $\underline{B}$, defined as follows

$$\mathbf{PT}_{\text{def}} = \left(\bar{\imath}_3 + \frac{\partial \underline{u}}{\partial x_3} \right) dx_3 = \underline{A}, \quad \mathbf{PS}_{\text{def}} = \left(\bar{\imath}_2 + \frac{\partial \underline{u}}{\partial x_2} \right) dx_2 = \underline{B},$$

and hence

$$\mathbf{TS}_{\text{def}} = \mathbf{PS}_{\text{def}} - \mathbf{PT}_{\text{def}} = \underline{B} - \underline{A}.$$

With the help of this notation, the numerator, N, of eq. (1.67) becomes

$$N = \underline{A} \cdot \underline{A} + \underline{B} \cdot \underline{B} - (\underline{B} - \underline{A}) \cdot (\underline{B} - \underline{A}) = 2\underline{A} \cdot \underline{B}$$
$$= 2 \left(\bar{\imath}_2 + \frac{\partial \underline{u}}{\partial x_2} \right) \left(\bar{\imath}_3 + \frac{\partial \underline{u}}{\partial x_3} \right) dx_2 dx_3 = 2 \left(\frac{\partial u_2}{\partial x_3} + \frac{\partial u_3}{\partial x_2} + \frac{\partial \underline{u}}{\partial x_2} \frac{\partial \underline{u}}{\partial x_3} \right) dx_2 dx_3$$
$$= 2 \left(\frac{\partial u_2}{\partial x_3} + \frac{\partial u_3}{\partial x_2} + \frac{\partial u_1}{\partial x_2} \frac{\partial u_1}{\partial x_3} + \frac{\partial u_2}{\partial x_2} \frac{\partial u_2}{\partial x_3} + \frac{\partial u_3}{\partial x_2} \frac{\partial u_3}{\partial x_3} \right) dx_2 dx_3. \tag{1.68}$$

The denominator, D, can be expressed in the same manner to find

$$D = 2\|\underline{A}\|\,\|\underline{B}\| = 2\sqrt{\underline{A} \cdot \underline{A}}\,\sqrt{\underline{B} \cdot \underline{B}}$$
$$= 2 dx_2 dx_3 \sqrt{1 + 2\frac{\partial u_2}{\partial x_2} + \left(\frac{\partial u_1}{\partial x_2} \right)^2 + \left(\frac{\partial u_2}{\partial x_2} \right)^2 + \left(\frac{\partial u_3}{\partial x_2} \right)^2} \tag{1.69}$$
$$\sqrt{1 + 2\frac{\partial u_3}{\partial x_3} + \left(\frac{\partial u_1}{\partial x_3} \right)^2 + \left(\frac{\partial u_2}{\partial x_3} \right)^2 + \left(\frac{\partial u_3}{\partial x_3} \right)^2}.$$

Finally, these results can be combined in eq. (1.67) to yield the rather cumbersome expression $\gamma_{23} = \arcsin N/D$. With the help of the small displacement assumption, see eq. (1.61), the numerator simplifies to $N \approx 2(\partial u_2/\partial x_3 + \partial u_3/\partial x_2)\, dx_2 dx_3$, whereas the denominator reduces to $D \approx 2(1 + \partial u_2/\partial x_2 + \partial u_3/\partial x_3)\, dx_2 dx_3$. With these simplifications, the shearing strain component becomes

$$\gamma_{23} \approx \frac{\partial u_2/\partial x_3 + \partial u_3/\partial x_2}{1 + \partial u_2/\partial x_2 + \partial u_3/\partial x_3} \approx \frac{\partial u_2}{\partial x_3} + \frac{\partial u_3}{\partial x_2}. \tag{1.70}$$

A similar reasoning applies to the other material lines to yield the three angular distortions or shear strains as

$$\gamma_{23} = \frac{\partial u_2}{\partial x_3} + \frac{\partial u_3}{\partial x_2}, \quad \gamma_{13} = \frac{\partial u_1}{\partial x_3} + \frac{\partial u_3}{\partial x_1}, \quad \gamma_{12} = \frac{\partial u_1}{\partial x_2} + \frac{\partial u_2}{\partial x_1}. \tag{1.71}$$

Summary

The relative elongations, eqs. (1.63), and angular distortions, eqs. (1.71) characterize the state of deformation at a point. The relative elongations are also called *direct strains* or *axial strains*, whereas the angular distortions are called *shearing strains* or simply *shear strains*. It is important to note that the *strain-displacement relationships*, eqs. (1.63) and (1.71) are obtained under the small displacement assumption defined in eq. (1.61). If the displacements become large, expressions (1.60) and (1.67) should be used instead. It is clear that the small displacement assumption implies that all strain components also remain very small, *i.e.*,

$$|\epsilon_1| \ll 1, \ |\epsilon_2| \ll 1, \ |\epsilon_3| \ll 1, \ |\gamma_{23}| \ll 1, \ |\gamma_{13}| \ll 1, \ |\gamma_{12}| \ll 1. \tag{1.72}$$

Rigid body rotation

In general, the motion of a solid body can be decomposed into a rigid body motion and straining or deformation. The previous sections are focused on the deformation of the solid, but the rigid body motion can also be extracted from the displacement field. The components of the rotation vector associated with the displacement field are

$$\omega_1 = \frac{1}{2} \left(\frac{\partial u_3}{\partial x_2} - \frac{\partial u_2}{\partial x_3} \right), \tag{1.73a}$$

$$\omega_2 = \frac{1}{2} \left(\frac{\partial u_1}{\partial x_3} - \frac{\partial u_3}{\partial x_1} \right), \tag{1.73b}$$

$$\omega_3 = \frac{1}{2} \left(\frac{\partial u_2}{\partial x_1} - \frac{\partial u_1}{\partial x_2} \right). \tag{1.73c}$$

Each components of the rotation vector $\underline{\omega}^T = \{\omega_1, \omega_2, \omega_3\}$ represent the rotation of the solid about axes $\bar{\imath}_1$, $\bar{\imath}_2$, and $\bar{\imath}_3$, respectively.

1.4.2 The volumetric strain

Consider the block of material defined by the three segments **PR**, **PS**, and **PT**. The volume of this block in the reference configuration is $dx_1 dx_2 dx_3$. After deformation, this volume becomes

$$\mathcal{V} \approx (1+\epsilon_1)(1+\epsilon_2)(1+\epsilon_3) \, dx_1 dx_2 dx_3 \approx (1+\epsilon_1+\epsilon_2+\epsilon_3) \, dx_1 dx_2 dx_3, \tag{1.74}$$

where the higher order strain quantities are neglected in view of eq. (1.72). The relative change in volume is now

$$e = \frac{(1 + \epsilon_1 + \epsilon_2 + \epsilon_3)\, dx_1 dx_2 dx_3 - dx_1 dx_2 dx_3}{dx_1 dx_2 dx_3} = \epsilon_1 + \epsilon_2 + \epsilon_3. \qquad (1.75)$$

The quantity e is known as the *volumetric strain* and measures the relative change in volume of the material.

1.5 Analysis of the state of strain at a point

The state of strain at a point is characterized in the previous section by the relative elongations of three material lines and their relative angular distortions. The orientations of three material lines are selected parallel to the axes of the Cartesian reference frame $\mathcal{I} = (\bar{\imath}_1, \bar{\imath}_2, \bar{\imath}_3)$. It is clear that the orientation of this reference frame is entirely arbitrary: a reference frame $\mathcal{I}^* = (\bar{\imath}_1^*, \bar{\imath}_2^*, \bar{\imath}_3^*)$ could have been selected, and an analysis identical to that of the previous section would have led to the definition of relative elongations

$$\epsilon_1^* = \frac{\partial u_1^*}{\partial x_1^*}, \quad \epsilon_2^* = \frac{\partial u_2^*}{\partial x_2^*}, \quad \epsilon_3^* = \frac{\partial u_3^*}{\partial x_3^*}, \qquad (1.76)$$

and angular distortions

$$\gamma_{23}^* = \frac{\partial u_2^*}{\partial x_3^*} + \frac{\partial u_3^*}{\partial x_2^*}, \quad \gamma_{13}^* = \frac{\partial u_1^*}{\partial x_3^*} + \frac{\partial u_3^*}{\partial x_1^*}, \quad \gamma_{12}^* = \frac{\partial u_1^*}{\partial x_2^*} + \frac{\partial u_2^*}{\partial x_2^*}. \qquad (1.77)$$

Although expressed in different reference frames, the strain displacements equations, eq. (1.63) and (1.71), or (1.76) and (1.77) both characterize the state of deformation at a point of the body. Therefore, the strain components resolved in the two reference frames should be closely related. Because strain components are purely geometric in nature, it should not be unexpected that the relationship between the strain components resolved in two different coordinate systems is also purely geometric in nature.

1.5.1 Rotation of strains

In this section, the strain components resolved two different bases, $\mathcal{I}$ and $\mathcal{I}^*$, will be related to each other. The orientation of basis $\mathcal{I}^*$ relative to basis $\mathcal{I}$ is discussed in appendix A.3.1 and leads to the definition of the matrix of direction cosines, or rotation matrix, $\underline{\underline{R}}$, given by eq. (A.36).

With the help of the chain rule for derivatives, the first component of strain given by eq. (1.76) becomes

$$\epsilon_1^* = \frac{\partial u_1^*}{\partial x_1^*} = \frac{\partial u_1^*}{\partial x_1}\frac{\partial x_1}{\partial x_1^*} + \frac{\partial u_1^*}{\partial x_2}\frac{\partial x_2}{\partial x_1^*} + \frac{\partial u_1^*}{\partial x_3}\frac{\partial x_3}{\partial x_1^*} = \frac{\partial u_1^*}{\partial x_1}\ell_1 + \frac{\partial u_1^*}{\partial x_2}\ell_2 + \frac{\partial u_1^*}{\partial x_3}\ell_3, \quad (1.78)$$

where eq. (A.39) is used to express the derivatives of x_1, x_2, and x_3 with respect to x_1^*.

Next the displacement component u_1^* is expressed in terms of the displacement components in coordinate system $\mathcal{I}$ with the help of eq. (A.39), to find

$$
\epsilon_1^* = \ell_1 \frac{\partial}{\partial x_1} (\ell_1 u_1 + \ell_2 u_2 + \ell_3 u_3) + \ell_2 \frac{\partial}{\partial x_2} (\ell_1 u_1 + \ell_2 u_2 + \ell_3 u_3)
$$
$$
+ \ell_3 \frac{\partial}{\partial x_3} (\ell_1 u_1 + \ell_2 u_2 + \ell_3 u_3).
$$
(1.79)

The last step is to use the strain-displacement relationships, eqs. (1.63) and (1.71), to find

$$
\epsilon_1^* = \epsilon_1 \ell_1^2 + \epsilon_2 \ell_2^2 + \epsilon_3 \ell_3^2 + \gamma_{12} \ell_1 \ell_2 + \gamma_{13} \ell_1 \ell_3 + \gamma_{23} \ell_2 \ell_3.
$$

A similar analysis for the other direct strain components results in the following expressions for the extensional strain in system $\mathcal{I}^*$,

$$
\epsilon_1^* = \epsilon_1 \ell_1^2 + \epsilon_2 \ell_2^2 + \epsilon_3 \ell_3^2 + 2\frac{\gamma_{12}}{2} \ell_1 \ell_2 + 2\frac{\gamma_{13}}{2} \ell_1 \ell_3 + 2\frac{\gamma_{23}}{2} \ell_2 \ell_3,
$$
(1.80a)
$$
\epsilon_2^* = \epsilon_1 m_1^2 + \epsilon_2 m_2^2 + \epsilon_3 m_3^2 + 2\frac{\gamma_{12}}{2} m_1 m_2 + 2\frac{\gamma_{13}}{2} m_1 m_3 + 2\frac{\gamma_{23}}{2} m_2 m_3,
$$
(1.80b)
$$
\epsilon_3^* = \epsilon_1 n_1^2 + \epsilon_2 n_2^2 + \epsilon_3 n_3^2 + 2\frac{\gamma_{12}}{2} n_1 n_2 + 2\frac{\gamma_{13}}{2} n_1 n_3 + 2\frac{\gamma_{23}}{2} n_2 n_3.
$$
(1.80c)

Proceeding in a similar manner yields the shear strain components expressed in basis $\mathcal{I}^*$

$$
\frac{\gamma_{12}^*}{2} = \epsilon_1 \ell_1 m_1 + \epsilon_2 \ell_2 m_2 + \epsilon_3 \ell_3 m_3 + \frac{\gamma_{12}}{2} (\ell_1 m_2 + \ell_2 m_1)
$$
$$
+ \frac{\gamma_{13}}{2} (\ell_1 m_3 + \ell_3 m_1) + \frac{\gamma_{23}}{2} (\ell_2 m_3 + \ell_3 m_2),
$$
(1.81a)
$$
\frac{\gamma_{13}^*}{2} = \epsilon_1 \ell_1 n_1 + \epsilon_2 \ell_2 n_2 + \epsilon_3 \ell_3 n_3 + \frac{\gamma_{12}}{2} (\ell_1 n_2 + \ell_2 n_1)
$$
$$
+ \frac{\gamma_{13}}{2} (\ell_1 n_3 + \ell_3 n_1) + \frac{\gamma_{23}}{2} (\ell_2 n_3 + \ell_3 n_2),
$$
(1.81b)
$$
\frac{\gamma_{23}^*}{2} = \epsilon_1 m_1 n_1 + \epsilon_2 m_2 n_2 + \epsilon_3 m_3 n_3 + \frac{\gamma_{12}}{2} (m_1 n_2 + m_2 n_1)
$$
$$
+ \frac{\gamma_{13}}{2} (m_1 n_3 + m_3 n_1) + \frac{\gamma_{23}}{2} (m_2 n_3 + m_3 n_2).
$$
(1.81c)

Expressions (1.80) and (1.81) are quite tedious, but the permutations of indices are readily observed. In these equations, it should be noted that the shear strain components are divided by a factor of 2. In this form, eqs. (1.80) and (1.81) become similar to eqs. (1.11) and (1.12), respectively; the axial strain take the place of the axial stresses and the shear strain that of the shear stresses.

The shearing strain components γ_{23}, γ_{13} and γ_{12} are called the *engineering shear strain components*, whereas the *tensor shear strain components*, ϵ_{23}, ϵ_{13} and ϵ_{12}, are defined as

$$
\epsilon_{23} = \frac{\gamma_{23}}{2}, \quad \epsilon_{13} = \frac{\gamma_{13}}{2}, \quad \epsilon_{12} = \frac{\gamma_{12}}{2}.
$$
(1.82)

When using the tensor strain components, the strain rotation expressions, eqs. (1.80) and (1.81), can be written in a compact matrix form as

$$
\begin{bmatrix} \epsilon_1^* & \epsilon_{12}^* & \epsilon_{13}^* \\ \epsilon_{12}^* & \epsilon_2^* & \epsilon_{23}^* \\ \epsilon_{13}^* & \epsilon_{23}^* & \epsilon_3^* \end{bmatrix} = \underline{\underline{R}}^T \begin{bmatrix} \epsilon_1 & \epsilon_{12} & \epsilon_{13} \\ \epsilon_{12} & \epsilon_2 & \epsilon_{23} \\ \epsilon_{13} & \epsilon_{23} & \epsilon_3 \end{bmatrix} \underline{\underline{R}}, \tag{1.83}
$$

where $\underline{\underline{R}}$ is the rotation matrix defined by eq. (A.36).

Comparing this result with eq. (1.20) for the rotation of stress components, it becomes clear that the transformation equations for these second order tensors are identical. Equation (1.20) expresses the transformation rules for the components of the second order stress tensor, whereas eq. (1.83) expresses the same rule for the second order strain tensor. In fact, a second order tensor is defined as a mathematical entity whose components measured in two different coordinate systems transform according to the ruled expressed by eqs. (1.20) or (1.83).

1.5.2 Principal strains

Because it has been established that stress and strain components are the components of the second order stress and strain tensors, respectively, it should not be unexpected that the concept of principal stresses, discussed in section 1.2.2 for the stress tensor, has its equivalent when it comes to the strain tensor.

To introduce the concept of principal strains, the following question is asked: is there a coordinate system $\mathcal{I}^*$ for which the shear strains vanish? If such a coordinate system exists, eq. (1.83) implies that

$$
\begin{bmatrix} \epsilon_1^* & 0 & 0 \\ 0 & \epsilon_2^* & 0 \\ 0 & 0 & \epsilon_3^* \end{bmatrix} = \underline{\underline{R}}^T \begin{bmatrix} \epsilon_1 & \epsilon_{12} & \epsilon_{13} \\ \epsilon_{12} & \epsilon_2 & \epsilon_{23} \\ \epsilon_{13} & \epsilon_{23} & \epsilon_3 \end{bmatrix} \underline{\underline{R}},
$$

where $\epsilon_1^* = \epsilon_{p1}$, $\epsilon_2^* = \epsilon_{p2}$, and $\epsilon_3^* = \epsilon_{p3}$ are the principal strains. By pre-multiplying by $\underline{\underline{R}}$ and reversing the equality, this equation can be written in the following form

$$
\begin{bmatrix} \epsilon_1 & \epsilon_{12} & \epsilon_{13} \\ \epsilon_{12} & \epsilon_2 & \epsilon_{23} \\ \epsilon_{13} & \epsilon_{23} & \epsilon_3 \end{bmatrix} \underline{\underline{R}} = \underline{\underline{R}} \begin{bmatrix} \epsilon_{p1} & 0 & 0 \\ 0 & \epsilon_{p2} & 0 \\ 0 & 0 & \epsilon_{p3} \end{bmatrix},
$$

where the orthogonality of the direction cosine matrix, eq. (A.37), is used. It follows that the principal strains, ϵ_{p1}, ϵ_{p2} and ϵ_{p3}, are the solutions of three systems of three equations of the form

$$
\begin{bmatrix} \epsilon_1 & \epsilon_{12} & \epsilon_{13} \\ \epsilon_{12} & \epsilon_2 & \epsilon_{23} \\ \epsilon_{13} & \epsilon_{23} & \epsilon_3 \end{bmatrix} \begin{Bmatrix} n_1 \\ n_2 \\ n_3 \end{Bmatrix} = \epsilon_p \begin{Bmatrix} n_1 \\ n_2 \\ n_3 \end{Bmatrix},
$$

where ϵ_p represents each of the three principal strains and n_1, n_2, and n_3 the principal strain directions. These equations can be recast as a homogeneous system of linear equations

$$
\begin{bmatrix} \epsilon_1 - \epsilon_p & \epsilon_{12} & \epsilon_{13} \\ \epsilon_{12} & \epsilon_2 - \epsilon_p & \epsilon_{23} \\ \epsilon_{13} & \epsilon_{23} & \epsilon_3 - \epsilon_p \end{bmatrix} \begin{Bmatrix} n_1 \\ n_2 \\ n_3 \end{Bmatrix} = 0. \tag{1.84}
$$

Clearly, this homogeneous system is equivalent to system (1.13) that defines the principal stresses.

Since this is a homogeneous system of equations, the trivial solution, $n_1 = n_2 = n_3 = 0$, is, in general, the solution of this system. When the determinant of the system vanishes, however, non-trivial solutions will exist. The vanishing of the determinant of the system leads to a cubic equation for the magnitude of the principal strains given by

$$\epsilon_p^3 - I_1\epsilon_p^2 + I_2\epsilon_p - I_3 = 0, \tag{1.85}$$

where the quantities, I_1, I_2, and I_3, defined as

$$I_1 = \epsilon_1 + \epsilon_2 + \epsilon_3, \tag{1.86a}$$

$$I_2 = \epsilon_1\epsilon_2 + \epsilon_2\epsilon_3 + \epsilon_3\epsilon_1 - \epsilon_{12}^2 - \epsilon_{13}^2 - \epsilon_{23}^2, \tag{1.86b}$$

$$I_3 = \epsilon_1\epsilon_2\epsilon_3 - \epsilon_1\epsilon_{23}^2 - \epsilon_2\epsilon_{13}^2 - \epsilon_3\epsilon_{12}^2 + 2\epsilon_{12}\epsilon_{13}\epsilon_{23}, \tag{1.86c}$$

are called the three *strain invariants*.

The solutions of eq. (1.85) are called the *principal strains*. Because this is a cubic equation, there will be three solutions, denoted ϵ_{p1}, ϵ_{p2}, and ϵ_{p2}. For each of these three solutions, the matrix of the system of equations defined by eq. (1.84) has a zero determinant, and a non-trivial solution exists for the directions cosines that now define the direction for which the shear strains vanish. Such direction is called a *principal strain direction*. Because the equations to be solved are homogeneous, their solution will include an arbitrary constant which can be determined by enforcing the normality condition for unit vector $\bar{n}$, $n_1^2 + n_2^2 + n_3^2 = 1$.

Since there exist three principal strains, three principal strain directions must also exist. It can be shown that these three directions are mutually orthogonal.

1.6 The state of plane strain

A particular state of strain of great practical importance is the *plane state of strain*. In this case, the displacement component along the direction of axis $\bar{\imath}_3$ is assumed to vanish, or to be negligible compared to the displacement components in the other two directions. This means that the only non-vanishing strain components are ϵ_1, ϵ_2, and γ_{12}, and furthermore, these strain components are assumed to be independent of x_3.

Unlike the plane state of stress considered in section 1.3, plane strain problems are not characterized by having one dimension much thinner than the others. Instead, displacement in one direction is zero. An example of a plane strain problem is that of a very long buried pipe aligned with the $\bar{\imath}_3$ direction. Such a problem is clearly three-dimensional in its overall geometry, but if the displacement along the direction of axis $\bar{\imath}_3$ is small or negligible, the pipe is in a plane state of strain.

1.6.1 Strain-displacement relations for plane strain

If the material is in a plane state of strain, *i.e.*, if $u_3 = 0$ and $\partial/\partial x_3 = 0$, eqs (1.63) and (1.71) reduce to

$$\epsilon_1 = \frac{\partial u_1}{\partial x_1}, \quad \epsilon_2 = \frac{\partial u_2}{\partial x_2}, \quad \gamma_{12} = \frac{\partial u_1}{\partial x_2} + \frac{\partial u_2}{\partial x_1}. \tag{1.87}$$

1.6.2 Rotation of strains

Next, the strain components measured two different orthonormal bases, $\mathcal{I}$ and $\mathcal{I}^*$, will be related to each other. Since this problem involves two distinct orthonormal bases, the relationship between these two basis, as explored in appendix A.3.3, is relevant to this development.

With the help of the chain rule for derivatives, the first component of strain given by eq. (1.76) becomes

$$\epsilon_1^* = \frac{\partial u_1^*}{\partial x_1^*} = \frac{\partial u_1^*}{\partial x_1}\frac{\partial x_1}{\partial x_1^*} + \frac{\partial u_1^*}{\partial x_2}\frac{\partial x_2}{\partial x_1^*} = \frac{\partial u_1^*}{\partial x_1}\cos\theta + \frac{\partial u_1^*}{\partial x_2}\sin\theta,$$

where eq. (A.43) is used to express the derivatives of x_1 and x_2 with respect to x_1^*. Next, the displacement component u_1^* is expressed in terms of the displacement components resolved in coordinate system $\mathcal{I}$ with the help of eq. (A.43) to yield

$$\epsilon_1^* = \cos\theta\frac{\partial}{\partial x_1}\left(u_1\cos\theta + u_2\sin\theta\right) + \sin\theta\frac{\partial}{\partial x_2}\left(u_1\cos\theta + u_2\sin\theta\right). \tag{1.88}$$

The last step is to use the strain-displacement relationships, eqs. (1.87), to find

$$\epsilon_1^* = \cos^2\theta\,\epsilon_1 + \sin^2\theta\,\epsilon_2 + \sin\theta\cos\theta\,\gamma_{12}. \tag{1.89}$$

Proceeding in a similar manner yields the shear strain components in the $\mathcal{I}^*$ coordinate system

$$\frac{\gamma_{12}^*}{2} = -\epsilon_1\cos\theta\sin\theta + \epsilon_2\sin\theta\cos\theta + \frac{\gamma_{12}}{2}(\cos^2\theta - \sin^2\theta). \tag{1.90}$$

Here again, it is convenient to use the tensor component of shearing strain, $\epsilon_{12} = \gamma_{12}/2$, see eq. (1.82).

These results can be written in a matrix form as

$$\left\{\begin{array}{c} \epsilon_1^* \\ \epsilon_2^* \\ \epsilon_{12}^* \end{array}\right\} = \left[\begin{array}{ccc} \cos^2\theta & \sin^2\theta & 2\sin\theta\cos\theta \\ \sin^2\theta & \cos^2\theta & -2\sin\theta\cos\theta \\ -\sin\theta\cos\theta & \sin\theta\cos\theta & \cos^2\theta - \sin^2\theta \end{array}\right] \left\{\begin{array}{c} \epsilon_1 \\ \epsilon_2 \\ \epsilon_{12} \end{array}\right\}. \tag{1.91}$$

This relationship can be readily inverted by recognizing that the inverse transformation is obtained by replacing θ by $-\theta$ to find

$$\left\{\begin{array}{c} \epsilon_1 \\ \epsilon_2 \\ \epsilon_{12} \end{array}\right\} = \left[\begin{array}{ccc} \cos^2\theta & \sin^2\theta & -2\sin\theta\cos\theta \\ \sin^2\theta & \cos^2\theta & 2\sin\theta\cos\theta \\ \sin\theta\cos\theta & -\sin\theta\cos\theta & \cos^2\theta - \sin^2\theta \end{array}\right] \left\{\begin{array}{c} \epsilon_1^* \\ \epsilon_2^* \\ \epsilon_{12}^* \end{array}\right\}. \tag{1.92}$$

Note that these transformation formulæ are identical to those derived for the stress tensor, see eqs. (1.47) and (1.48), respectively.

With the help of double-angle trigonometric identities, the transformation equations for tensor strain components, eq. (1.91), can also be written as

$$\epsilon_1^* = \frac{\epsilon_1 + \epsilon_2}{2} + \frac{\epsilon_1 - \epsilon_2}{2} \cos 2\theta + \epsilon_{12} \sin 2\theta, \tag{1.93a}$$

$$\epsilon_2^* = \frac{\epsilon_1 + \epsilon_2}{2} - \frac{\epsilon_1 - \epsilon_2}{2} \cos 2\theta - \epsilon_{12} \sin 2\theta, \tag{1.93b}$$

$$\epsilon_{12}^* = \qquad - \frac{\epsilon_1 - \epsilon_2}{2} \sin 2\theta + \epsilon_{12} \cos 2\theta. \tag{1.93c}$$

While use of the strain tensor components, ϵ_{ij}, renders the treatment of stress and strain component rotation formulæ identical, it is customary to use the engineering shear strain components, γ_{ij}, instead of their tensor counterparts, and hence, the previous equations become

$$\epsilon_1^* = \frac{\epsilon_1 + \epsilon_2}{2} + \frac{\epsilon_1 - \epsilon_2}{2} \cos 2\theta + \frac{\gamma_{12}}{2} \sin 2\theta, \tag{1.94a}$$

$$\epsilon_2^* = \frac{\epsilon_1 + \epsilon_2}{2} - \frac{\epsilon_1 - \epsilon_2}{2} \cos 2\theta - \frac{\gamma_{12}}{2} \sin 2\theta, \tag{1.94b}$$

$$\gamma_{12}^* = \qquad - (\epsilon_1 - \epsilon_2) \sin 2\theta + \gamma_{12} \cos 2\theta. \tag{1.94c}$$

This important result shows that knowledge of the plane strain components ϵ_1, ϵ_2, and γ_{12} in two orthogonal directions allows the computation of the strain components in an arbitrary orientation. In other words, the knowledge of the plane strain components in two orthogonal directions fully defines the state of strain at that point.

1.6.3 Principal strains

The relative elongation in an arbitrary direction, θ, can be computed with the help of eq. (1.94). The orientation, θ_p, in which the maximum (or minimum) elongation occurs is determined by requiring the derivative of ϵ_1^* with respect to θ to vanish, and this yields

$$\frac{d\epsilon_1^*}{d\theta} = -\frac{\epsilon_1 - \epsilon_2}{2} 2 \sin 2\theta_p + \frac{\gamma_{12}}{2} 2 \cos 2\theta_p = 0. \tag{1.95}$$

This can be solved for $2\theta_p$ to find the orientation of extreme elongation as

$$\tan 2\theta_p = \frac{\gamma_{12}/2}{(\epsilon_1 - \epsilon_2)/2}, \tag{1.96}$$

where the factor 2 has not been canceled out in order to retain the similarity with eq. (1.33) if τ_{12} is replaced with $\gamma_{12}/2$ and σ_1 and σ_2 with ϵ_1 and ϵ_2, respectively. This equation presents two solutions, θ_{p1} and $\theta_{p2} = \theta_{p1} + \pi/2$, corresponding to two mutually orthogonal principal strain directions. The maximum axial strain is found along one direction, and the minimum is found along the other.

To define these orientations unequivocally, it is convenient to separately define the sine and cosines of angle $2\theta_p$ as follows

$$\sin 2\theta_p = \frac{\gamma_{12}}{2\Delta}, \quad \cos 2\theta_p = \frac{\epsilon_1 - \epsilon_2}{2\Delta}, \tag{1.97}$$

where

$$\Delta = \sqrt{\left(\frac{\epsilon_1 - \epsilon_2}{2}\right)^2 + \left(\frac{\gamma_{12}}{2}\right)^2}. \tag{1.98}$$

This result is equivalent to eq. (1.96), but it gives a unique solution for θ_p because both the sine and cosine of the angle are known. The maximum and minimum axial strains, denoted ϵ_{p1} and ϵ_{p2}, respectively, act in the directions θ_{p1} and $\theta_{p2} = \theta_{p1} + \pi/2$, respectively. These maximum and minimum axial strains, called the *principal strains*, are evaluated by introducing eq. (1.97) into eq. (1.94) to find

$$\epsilon_{p1} = \frac{\epsilon_1 + \epsilon_2}{2} + \Delta; \quad \epsilon_{p2} = \frac{\epsilon_1 + \epsilon_2}{2} - \Delta. \tag{1.99}$$

Finally, the shear strain in the principal directions vanishes as can be verified by introducing eq. (1.97) into eq. (1.94).

The development of the equations for the state of strain at a point yield equations that are very similar to those developed in section 1.2.2 for the state of stress at a point. In particular, the transformation equations are similar in form (identical if the strain tensor components, $\epsilon_{12} = \gamma_{12}/2$, are used to define the shear strain) and lead in both cases to the existence of principal stresses and principal strains. The orientations of the principal stresses and principal strains are not necessarily identical.

1.6.4 Mohr's circle for plane strain

Equations (1.94) express the direct and shear strains along an arbitrary direction defined by angle θ with respect the axis $\bar{\imath}_1$, but the presence of trigonometric functions involving the angle 2θ makes it difficult to give a geometric interpretation of these formulae. Let the state of strain at a point be defined by its principal strains, ϵ_{p1} and ϵ_{p2}; eq. (1.94) then implies that the strains along a direction defined by angle θ with respect to the principal strain directions can be written as

$$\epsilon^* = \epsilon_a + R \cos 2\theta, \quad \frac{\gamma^*}{2} = -R \sin 2\theta, \tag{1.100}$$

where $\epsilon_a = (\epsilon_{p1} + \epsilon_{p2})/2$, and $R = (\epsilon_{p1} - \epsilon_{p2})/2$. With this notation and the help of trigonometric identities, eq. (1.100) becomes

$$(\epsilon^* - \epsilon_a)^2 + \left(\frac{\gamma^*}{2}\right)^2 = R^2. \tag{1.101}$$

This equation represents the equation of a circle which is known as *Mohr's circle*. When ϵ^* is plotted along the horizontal axis and $\gamma^*/2$ along the vertical axis, the center of the circle is at a coordinate ϵ_a on the horizontal axis, and the radius of the circle is R, as depicted in fig. 1.23.

Consider now point $\mathbf{A}_1$ on Mohr's circle such that segment $\mathbf{OA}_1$ makes an angle 2θ with the horizontal. The coordinates of this point are $\epsilon^* = \epsilon_a + R\cos 2\theta$ and $\gamma^*/2 = -R\sin 2\theta$; hence, in view of eq. (1.100), the coordinates of point $\mathbf{A}_1$ represent the strain components along a direction defined by angle θ. In fact, each point on Mohr's circle represents the strain components along a specific orientation.

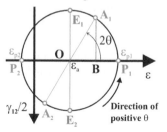

Fig. 1.23. Mohr's circle for visualizing plane strain state.

An **important sign convention** must be defined: on Mohr's circle, a positive angle θ is measured in the **counterclockwise direction**, see fig. 1.23, to match the positive direction of angle θ that identifies the orientation of a face in fig. 1.12. Given the sign convention for angle θ, the shear strain must be positive downward on the ordinate of Mohr's circle depicted in fig. 1.23.[4]

All the developments presented in section 1.3.6 for visualizing a plane state of stress using Mohr's circle also apply to the present problem of visualizing the plane state of strain, provided however, that the strain tensor is used. This means that $\gamma_{12}/2$ must be plotted on the vertical axis.

1.7 Measurement of strains

The goal of the theory of elasticity is to predict the state of stress at any point of an elastic body, given the applied loading. Such predictions must be validated by measuring the state of stress at specific points of a body, then comparing these measurements with the corresponding predictions. Unfortunately, no practical experimental device can measure stresses directly. An indirect measurement must therefore be made by first measuring the state of strain, then computing the corresponding state of stress using the constitutive laws for the material.

Strain gauges

Measurement of the state of strain itself is not an entirely straightforward process. First, it is relatively difficult to measure the strain state at an interior point of a solid body, so most measurement methods focus on measuring strains on the body's external surface. As noted in previous sections, the two-dimensional strain state is characterized by both direct and shear components. Measurements of the very small angular changes associated with shear strains are difficult to perform, but measurements of extensional strains on a surface are surprisingly easy to acquire.

The relative elongation at the surface of a body can be measured with the help of what are called electrical resistance strain gauges, or more simply *strain gauges*. This device consists of a very thin electric wire, or an etched foil pattern, which is glued to the surface of the solid. When the solid experiences an extensional strain, this strain is transferred through the glue to the gauge, hence increasing the length

---

[4] An equivalent construction of Mohr's circle has the shear strain positive upwards along the ordinate, but angle θ is then positive in the clockwise direction.

of the wire. In turn, the wire's cross-section is reduced by Poisson's effect, thereby slightly increasing its electrical resistance. The reverse happens for compressional strains, and the electrical resistance is slightly reduced in this case.

An accurate electrical measurement of this resistance change, using a Wheatstone bridge circuit for instance, yields an estimate of the length change, which in turn, allows an accurate estimate of the relative elongation and finally, the extensional strain in the direction of the wire. Because strain quantities are very small, strain measurements are often labeled in *micro-strains*, which indicates a relative elongation of μ m/m = 10^{-6} m/m. Because strains are non-dimensional quantities, the units employed for measurement of elongation can be any units of length.

Chevron strain gauge

Figure 1.24 shows the external surface of a body with two strain gauges forming at a 90 degree angle with respect to each other; this configuration is sometimes called *chevron strain gauge*. This device is of finite size, and hence, the two extensional strain measurements are not made exactly at the same point, but if the chevron strain gauge is very small and the strain gradients are small compared to its size, it can be assumed that the two gauges experience the same strain state.

Let e_{+45} and e_{-45} be the experimentally measured relative elongations in the two gauge directions. The two gauges of the chevron are oriented at $\pm$ 45 degrees with respect to a triad, $\mathcal{I} = (\bar{\imath}_1, \bar{\imath}_2)$, as shown in fig. 1.24. The state of strain at that point is defined by the three strain components, ϵ_1, ϵ_2, and γ_{12}, resolved in triad $\mathcal{I}$. With the help of eq. (1.94a), these measurements can be expressed as follows

$$e_{+45} = \frac{\epsilon_1 + \epsilon_2}{2} + \frac{\epsilon_1 - \epsilon_2}{2}\cos(2 \times \ 45°) + \frac{\gamma_{12}}{2}\sin(2 \times \ 45°) = \frac{\epsilon_1 + \epsilon_2}{2} + \frac{\gamma_{12}}{2},$$

$$e_{-45} = \frac{\epsilon_1 + \epsilon_2}{2} + \frac{\epsilon_1 - \epsilon_2}{2}\cos(2 \times 135°) + \frac{\gamma_{12}}{2}\sin(2 \times 135°) = \frac{\epsilon_1 + \epsilon_2}{2} - \frac{\gamma_{12}}{2}.$$

Clearly, the two measurements, e_{+45} and e_{-45}, are not sufficient to determine the strain state at the chevron's location. Indeed, three measurements would be required to determine the three strain components, ϵ_1, ϵ_2, and γ_{12}. It is possible, however, to unequivocally determine the shear strain by subtracting the above equations from each other to find

$$\gamma_{12} = e_{+45} - e_{-45}. \tag{1.102}$$

Adding the two equations yields $\epsilon_1 + \epsilon_2 = e_{+45} + e_{-45}$, but the two normal strain components, ϵ_1 and ϵ_2, cannot be determined individually.

The complete state of strain at the surface of the body is specified by three independent quantities, *i.e.*, either two extensional and a shear strain, or two principal strains and a principal direction. These can be computed from the measurement of relative elongation in three distinct directions on the surface.

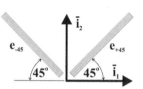

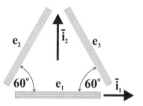

Fig. 1.24. Two strain gauges at the surface of a solid.

Fig. 1.25. Three strain gauges forming a rosette at the surface of a solid.

Strain gauge rosette

The experimental determination of the strain state at the surface of a body requires three independent measurements. One approach is to locate three strain gauges forming an equilateral triangle at the external surface of a body, as depicted in fig. 1.25. This type of device is commonly referred to as a *strain gauge rosette*; the configuration shown in the figure is often called a "delta rosette." Once again, this rosette is of finite size, and hence, the three extensional strain measurements are not made exactly at the same point, but if the rosette is very small and the strain gradients are small compared to the size of the rosette, it can be assumed that the three gauges experience the same strain state.

Let e_1, e_2, and e_3 be the experimentally measured relative elongations in the three gauge directions. With the help of eq. (1.94a), these measurements can be related to the strain components measured in triad $\mathcal{I} = (\bar{\imath}_1, \bar{\imath}_2)$ as follows

$$e_1 = \frac{\epsilon_1 + \epsilon_2}{2} + \frac{\epsilon_1 - \epsilon_2}{2},$$

$$e_2 = \frac{\epsilon_1 + \epsilon_2}{2} + \frac{\epsilon_1 - \epsilon_2}{2} \cos(+2 \times 60°) + \frac{\gamma_{12}}{2} \sin(+2 \times 60°),$$

$$e_3 = \frac{\epsilon_1 + \epsilon_2}{2} + \frac{\epsilon_1 - \epsilon_2}{2} \cos(-2 \times 60°) + \frac{\gamma_{12}}{2} \sin(-2 \times 60°).$$

These relationships can be inverted to yield the strain components in terms of the measured axial strains

$$\epsilon_1 = e_1, \quad \epsilon_2 = \frac{2}{3}\left(e_2 + e_3 - \frac{e_1}{2}\right), \quad \gamma_{12} = \frac{2}{\sqrt{3}}\left(e_2 - e_3\right). \tag{1.103}$$

The principal strain directions then follow from (1.97)

$$\sin 2\theta_p = \frac{e_2 - e_3}{\sqrt{3}\Delta}, \quad \cos 2\theta_p = \frac{2e_1 - e_2 - e_3}{3\Delta}, \tag{1.104}$$

and the principal strains are

$$\epsilon_{p1} = \bar{e} + \Delta, \quad \epsilon_{p2} = \bar{e} - \Delta, \tag{1.105}$$

where $\bar{e} = (e_1 + e_2 + e_3)/3$ and $\Delta = 2/3\sqrt{e_1^2 + e_2^2 + e_3^2 - e_2 e_3 - e_1 e_3 - e_1 e_2}$.

Various commonly used strain gauge arrangements are depicted in fig. 1.26. Note that a complete evaluation of the state of strain requires the knowledge of three strain components, and thus requires three independent measurements in three distinct directions. Combinations (a) and (c) of fig. 1.26 provide three independent measurements from which the strain state can be evaluated using a similar approach to that developed above for the delta strain gauge rosette shown in fig. 1.25.

Combinations (B) and (D) allow four independent measurements to be made to provide enough information in the event when one of the gauges is damaged. If the four gauges are properly working, the redundant information can be used to compensate for experimental errors, as illustrated in example 1.4 for the T-Delta rosette.

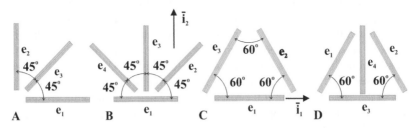

Fig. 1.26. Various commonly used strain gauge arrangements.

Example 1.4. Data reduction for the T-Delta rosette

Consider the T-Delta rosette shown in fig. 1.26 D. Given the output of the four gauges, e_1, e_2, e_3 and e_4, find the state of strain at the location of the rosette. First, the four measurements are expressed in terms of the three strain components with the help of eq. (1.94a) to find

$$e_1 = \frac{\epsilon_1 + \epsilon_2}{2} + \frac{\epsilon_1 - \epsilon_2}{2}\cos 120 + \frac{\gamma_{12}}{2}\sin 120 = \frac{\epsilon_1 + \epsilon_2}{2} - \frac{\epsilon_1 - \epsilon_2}{4} + \frac{\sqrt{3}}{4}\gamma_{12},$$

$$e_2 = \frac{\epsilon_1 + \epsilon_2}{2} + \frac{\epsilon_1 - \epsilon_2}{2}\cos 240 + \frac{\gamma_{12}}{2}\sin 240 = \frac{\epsilon_1 + \epsilon_2}{2} - \frac{\epsilon_1 - \epsilon_2}{4} - \frac{\sqrt{3}}{4}\gamma_{12},$$

$$e_3 = \frac{\epsilon_1 + \epsilon_2}{2} + \frac{\epsilon_1 - \epsilon_2}{2} = \epsilon_1,$$

$$e_4 = \frac{\epsilon_1 + \epsilon_2}{2} + \frac{\epsilon_1 - \epsilon_2}{2}\cos 180 + \frac{\gamma_{12}}{2}\sin 180 = \epsilon_2.$$

These relationships form a set of four equations for three unknowns, the strain components ϵ_1, ϵ_2, and γ_{12}, which can be written in a compact matrix form as

$$\begin{bmatrix} 1/4 & 3/4 & \sqrt{3}/4 \\ 1/4 & 3/4 & -\sqrt{3}/4 \\ 1 & 0 & 0 \\ 0 & 1 & 0 \end{bmatrix} \begin{Bmatrix} \epsilon_1 \\ \epsilon_2 \\ \gamma_{12} \end{Bmatrix} = \begin{Bmatrix} e_1 \\ e_2 \\ e_3 \\ e_4 \end{Bmatrix}.$$

These equations form an over-determined set of equations to evaluate the three components of strain. Since the strain measurement are likely to involve experimental errors, it seems appropriate to solve the over-determined system in a least squares sense, as explained in appendix A.2.10. For this problem, the least-squares solution given by eq. (A.33) becomes

$$\frac{1}{8}\begin{bmatrix} 9 & 3 & 0 \\ 3 & 17 & 0 \\ 0 & 0 & 3 \end{bmatrix}\begin{Bmatrix} \epsilon_1 \\ \epsilon_2 \\ \gamma_{12} \end{Bmatrix} = \begin{Bmatrix} (e_1+e_2)/4+e_3 \\ 3(e_1+e_2)/4+e_4 \\ \sqrt{3}\,(e_1-e_2)/4 \end{Bmatrix}.$$

The solution of this 3×3 linear system then yields the desired strain components as

$$\epsilon_1 = \frac{2e_1+2e_2+17e_3-3e_4}{18}; \ \epsilon_2 = \frac{6e_1+6e_2-3e_3+9e_4}{18}; \ \gamma_{12} = \frac{2(e_1-e_2)}{\sqrt{3}}.$$

1.7.1 Problems

Problem 1.14. Data reduction for the delta rosette
Consider the delta rosette shown in fig. 1.26 C. The measured data are $e_1 = 410\mu$, $e_2 = -290\mu$, and $e_3 = 610\mu$. *(1)* Find the state of strain at this location. *(2)* Draw Mohr's circle for this state of strain. *(3)* Find the orientation of the principal strain directions, and *(4)* find the principal strains. Use a software package to carry out these calculations.

Problem 1.15. Data reduction for the rectangular rosette
Consider the rectangular rosette shown in fig. 1.26 A. The measured data are $e_1 = -510\mu$, $e_2 = 780\mu$, $e_3 = 340\mu$. *(1)* Develop expressions similar to eq. (1.103) for the state of strain with respect to a surface axis system aligned with gauges #1 and #2. *(2)* Find the state of strain at this location for the given data. *(3)* Draw Mohr's circle for this state of strain. *(4)* Find the orientation of the principal strain directions, and *(5)* the principal strains.

Problem 1.16. Data reduction for the T-V rosette
Consider the T-V rosette shown in fig. 1.26 B. The measured data is $e_1 = 910\mu$, $e_2 = 990\mu$, $e_3 = 310\mu$ and $e_4 = 190\mu$. Use a least square approach to solve this problem. *(1)* Find the state of strain at this location. *(2)* Draw Mohr's circle for this state of strain. *(3)* Find the orientation of the principal strain directions, and *(4)* the principal strains.

Problem 1.17. Correlating rosette strain measurements
Consider the strain gauge arrangements shown in fig. 1.26 B. If the strain measurements e_1, e_2 and e_3 are given find the strain e_4.

Problem 1.18. Correlating rosette strain measurements
Consider the strain gauge arrangements shown in fig. 1.26 D. If the strain measurements e_1, e_2 and e_4 are given, find the strain e_3.

Problem 1.19. Misaligned Delta rosette
The delta rosette depicted in fig. 1.27 has been improperly installed on a solid: instead of aligning the rosette with axes $\bar{\imath}_1$ and $\bar{\imath}_2$, as desired, the gage was installed at an angle θ with respect to the desired directions. This implies that the gauge measurements will be e_1^*, e_2^* and e_3^*, instead of the desired e_1, e_2 and e_3. Since the misalignment is unintentional, the experimentalist will use the measurements, e_1^*, e_2^* and e_3^*, *as if they were* e_1, e_2 and e_3, respectively.

In other words, he will use the measurements e_1^*, e_2^* and e_3^* to extract the strain state, thinking that $\theta = 0$. *(1)* If the strain state is $\epsilon_1 = 1245\mu$, $\epsilon_2 = -780\mu$ and $\gamma_{12} = 675\mu$, determine the state of strain that the experimentalist will erroneously extract, denoted $\hat{\epsilon}_1$, $\hat{\epsilon}_2$ and $\hat{\gamma}_{12}$, as function of the misalignment angle. *(2)* On one graph, plot the relative errors $(\hat{\epsilon}_1 - \epsilon_1)/\epsilon_1$, $(\hat{\epsilon}_2 - \epsilon_2)/\epsilon_2$ and $(\hat{\gamma}_{12} - \gamma_{12})/\gamma_{12}$, as functions of $\theta \in [-10, 10]$ degrees.

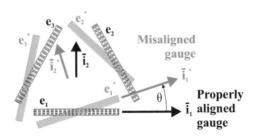

Fig. 1.27. Delta rosette with an angular misalignment of θ.

Problem 1.20. Transverse shear strain in beams
In beam theory, it is assumed that the planar cross-section of the beam remains planar and remains perpendicular to the axis of the beam as it bends. This implies that two material lines, the axis of the beam and a material line in the plane of the cross-section, remain perpendicular to each other. In view of this assumption, what is the transverse shear strain along the axis of the beam?

1.8 Strain compatibility equations

The displacement field uniquely defines the deformation of a solid body. Six strain components, however, are defined to characterize the state of deformation at a point. Hence, the strain components are not independent and must satisfy a set of relationships called the *strain compatibility equations*. Consider the following derivatives of the shear strain components

$$\frac{\partial^2 \gamma_{23}}{\partial x_2 \partial x_3} = \frac{\partial^2}{\partial x_2 \partial x_3}\left(\frac{\partial u_2}{\partial x_3} + \frac{\partial u_3}{\partial x_2}\right) = \frac{\partial^3 u_2}{\partial x_2 \partial x_3^2} + \frac{\partial^3 u_3}{\partial^2 x_2 \partial x_3} = \frac{\partial^2 \epsilon_2}{\partial x_3^2} + \frac{\partial^2 \epsilon_3}{\partial x_2^2}.$$

This implies that the shear and axial strain components are not independent. Consider now a different set of derivatives

$$\frac{\partial^2 \epsilon_1}{\partial x_2 \partial x_3} = \frac{\partial^3 u_1}{\partial x_1 \partial x_2 \partial x_3}, \quad \frac{\partial \gamma_{23}}{\partial x_1} = \frac{\partial^2 u_2}{\partial x_1 \partial x_3} + \frac{\partial^2 u_3}{\partial x_1 \partial x_2};$$

$$\frac{\partial \gamma_{13}}{\partial x_2} = \frac{\partial^2 u_1}{\partial x_2 \partial x_3} + \frac{\partial^2 u_3}{\partial x_1 \partial x_2}, \quad \frac{\partial \gamma_{12}}{\partial x_3} = \frac{\partial^2 u_1}{\partial x_2 \partial x_3} + \frac{\partial^2 u_2}{\partial x_1 \partial x_3},$$

which imply

$$2\frac{\partial^2 \epsilon_1}{\partial x_2 \partial x_3} = \frac{\partial}{\partial x_1}\left(-\frac{\partial \gamma_{23}}{\partial x_1} + \frac{\partial \gamma_{13}}{\partial x_2} + \frac{\partial \gamma_{12}}{\partial x_3}\right).$$

This is another relationship between the shear and axial strain components.

Similar relationships can be obtained through cyclical permutations of the indices to yield *Saint-Venant's strain compatibility equations*

$$\frac{\partial^2 \gamma_{23}}{\partial x_2 \partial x_3} = \frac{\partial^2 \epsilon_2}{\partial x_3^2} + \frac{\partial^2 \epsilon_3}{\partial x_2^2}, \tag{1.106a}$$

$$\frac{\partial^2 \gamma_{13}}{\partial x_1 \partial x_3} = \frac{\partial^2 \epsilon_1}{\partial x_3^2} + \frac{\partial^2 \epsilon_3}{\partial x_1^2}, \tag{1.106b}$$

$$\frac{\partial^2 \gamma_{12}}{\partial x_1 \partial x_2} = \frac{\partial^2 \epsilon_1}{\partial x_2^2} + \frac{\partial^2 \epsilon_2}{\partial x_1^2}, \tag{1.106c}$$

$$2\frac{\partial^2 \epsilon_1}{\partial x_2 \partial x_3} = \frac{\partial}{\partial x_1}\left(-\frac{\partial \gamma_{23}}{\partial x_1} + \frac{\partial \gamma_{13}}{\partial x_2} + \frac{\partial \gamma_{12}}{\partial x_3}\right), \tag{1.106d}$$

$$2\frac{\partial^2 \epsilon_2}{\partial x_1 \partial x_3} = \frac{\partial}{\partial x_2}\left(+\frac{\partial \gamma_{23}}{\partial x_1} - \frac{\partial \gamma_{13}}{\partial x_2} + \frac{\partial \gamma_{12}}{\partial x_3}\right), \tag{1.106e}$$

$$2\frac{\partial^2 \epsilon_3}{\partial x_1 \partial x_2} = \frac{\partial}{\partial x_3}\left(+\frac{\partial \gamma_{23}}{\partial x_1} + \frac{\partial \gamma_{13}}{\partial x_2} - \frac{\partial \gamma_{12}}{\partial x_3}\right). \tag{1.106f}$$

Some reflection is needed to fully understand the need for stating the compatibility equations. Clearly, if the state of deformation is defined by the three components of the displacement vector, *i.e.*, if the displacement field is given, it is a simple matter to compute the six strain components using eqs. (1.63) and (1.71). The inverse problem, however, is not so simple: if the state of deformation is defined by six components of strain, *i.e.*, given the strain field, it is not obvious to determine the displacement components that give rise to this strain field. Indeed, the six strain components are generated based on three displacement components only. Furthermore, some strain states could possibly be associated with displacement fields that include discontinuities or jumps corresponding to gaps or tears in the continuous body. In summary, if the six components of the strain field are derived from the three components of the displacement field, they are not independent and must satisfy Saint-Venant's strain compatibility equations. Three only of the six compatibility equations are independent.

2

Constitutive behavior of materials

The solution of elasticity problems requires three types of relationships. First, the equilibrium equations discussed in section 1.1.2, second, the strain-displacement relationships of section 1.4.1. Finally, the stress and strain fields must be related through a set of *constitutive laws*. These constitutive laws characterize the mechanical behavior of the material and consist of a set of mathematical idealizations of their observed behavior.

Homogeneity and isotropy

Constitutive laws for homogeneous, isotropic materials will be presented first. A *homogeneous material* is a material for which the physical properties are *identical at each point* within the sample. An *isotropic material* is a material for which the physical properties are *identical in all directions*. A sample of mild steel or aluminum can usually be assumed to be both homogeneous and isotropic.

Many engineering materials, however, are neither homogeneous nor isotropic. Consider a composite material consisting of long fibers aligned along a single direction and embedded in a matrix material. Such material is not homogeneous: the properties of the fibers are, in general, very different from those of the matrix material. Furthermore, it is not isotropic: if loading is applied along the fibers, the response of the material is likely to be very different from that observed when the loading is applied in a direction transverse to the fiber orientation. Such a material is referred to as being heterogeneous and anisotropic and will be examined in the second half of this chapter.

The assumptions of homogeneity or isotropy are *scale dependent*. For instance, it seems reasonable to consider a sample of aluminum to be both homogeneous and isotropic. Of course, at the atomic level, aluminum is neither homogeneous nor isotropic. Hence, assumptions of homogeneity and isotropy only hold for samples containing a very large number of atoms.

For high temperature turbine blade applications, either poly-crystalline or single crystal materials might be used. For single crystal materials, the atoms are arranged to form regular lattice structures that create a clearly defined orientation in the material. In such a case, a sample containing a large number of atoms could be assumed to be homogeneous but anisotropic, because the response of the material will be different when stresses are applied in different directions with respect to the lattice directions.

For poly-crystalline material samples containing a large number of crystals, the material could be considered homogeneous, but if the crystals are generally oriented in a specific direction, the material will be anisotropic. This can be the case for forged metals where the forging process aligns the crystals. For poly-crystalline material samples containing a large number of crystals arranged at random orientations with respect to each other, the material can be considered both homogeneous and isotropic, and this is often the case for common structural metals such as steel and aluminum.

For composite material reinforced with long fibers all aligned in the same direction, the material is clearly anisotropic because the fiber direction defines a preferential direction for the material. For samples containing just a few fibers, the material is not homogeneous, whereas for samples containing a very large number of fibers it is a reasonable assumption to consider the material to be homogeneous.

Material testing

At present, no first-principles based models models accurately describe the constitutive properties of structural materials. Most practical constitutive models are based on empirical data, and various types of constitutive laws have been proposed to represent the many types of experimentally observed material behaviors.

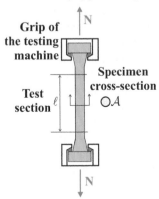

Fig. 2.1. Homogeneous bar loaded by a single stress component σ_1

If the deformation of the body remains very small, however, the stress-strain relationship can often be assumed to be linear. This widely used approximation, in which stress is proportional to strain, will be discussed in section 2.1.1. As the magnitude of the deformation increases, the stress-strain relationship can no longer be assumed to remain linear.

The stress-strain relationship for large deformations has distinctly different characteristics depending on whether the material is *ductile* or *brittle*. Constitutive relationships for ductile materials are presented in section 2.1.4 and relationships for brittle materials are presented in section 2.1.5.

Typically, material behavior is characterized by carrying out a tensile test similar to that sketched in fig. 2.1, in which a bar of circular cross-sectional area, A, is loaded in a testing machine that applies an axial force, N, to the test specimen. The test section is a representative portion of the test specimen of length, ℓ, located at a sufficient distance away from the grips of the testing machine to avoid the end effects they generate. The grips of the testing machine move slowly, applying an increasing load to the specimen. During the test, the extensional strain in the sample is computed by dividing the change in length of the test section by its original length, $\epsilon_1 = \Delta\ell/\ell$. The stress in the sample is computed by dividing the applied load by the sample cross-sectional area, $\sigma_1 = N/\mathcal{A}$. The results of the test are presented in the form of a stress-strain diagram: the strain is plotted along the abscissa, the stress along the ordinate.

2.1 Constitutive laws for isotropic materials

2.1.1 Homogeneous, isotropic, linearly elastic materials

For specimens undergoing small deformations, the stress-strain diagram often exhibits a linear behavior. Although this is a very crude approximation to the behavior of actual materials, it is a convenient assumption that is often used for preliminary evaluation. A linear relationship between stress and strain can be expressed as

$$\sigma_1 = E\,\epsilon_1, \tag{2.1}$$

where the coefficient of proportionality, E, is called *Young's modulus* or *modulus of elasticity*. Since strains are non-dimensional quantities, this coefficient has the same units as stress quantities, *i.e.*, Pa. This linear relationship is known as *Hooke's law*.

The elongation of a bar in the direction of the applied stress is accompanied by a lateral contraction that is also proportional to the applied stress. The resulting deformations for this uniaxial state of stress can therefore be described by the following strains

$$\epsilon_1 = \frac{1}{E}\,\sigma_1, \quad \epsilon_2 = -\frac{\nu}{E}\,\sigma_1, \quad \epsilon_3 = -\frac{\nu}{E}\,\sigma_1, \tag{2.2}$$

where ν is called *Poisson's ratio* and is a non-dimensional constant.

If a stress component, σ_2, is applied to the same material, similar deformations will result

$$\epsilon_1 = -\frac{\nu}{E}\,\sigma_2, \quad \epsilon_2 = \frac{1}{E}\,\sigma_2, \quad \epsilon_3 = -\frac{\nu}{E}\,\sigma_2. \tag{2.3}$$

Note that the assumption of material isotropy implies identical values of Young's modulus and Poisson's ratio in eq. (2.2) and (2.3). Similar relationships hold for an applied stress, σ_3.

Generalized Hooke's law

When the three stress components are applied simultaneously, the resulting deformation is the sum of the deformations obtained for each stress component applied individually because of the assumed linear behavior of the material. This results in the *generalized Hooke's law* for extensional strains

$$\epsilon_1 = \frac{1}{E}\left[\sigma_1 - \nu(\sigma_2 + \sigma_3)\right], \tag{2.4a}$$

$$\epsilon_2 = \frac{1}{E}\left[\sigma_2 - \nu(\sigma_1 + \sigma_3)\right], \tag{2.4b}$$

$$\epsilon_3 = \frac{1}{E}\left[\sigma_3 - \nu(\sigma_1 + \sigma_2)\right]. \tag{2.4c}$$

The extensional strains depend only on the direct stresses and not on the shear stresses. This is a key characteristic of isotropic materials, which does not hold for anisotropic materials.

Shear stress-shear strain relationships

The relationship between the shear strains and the shear stresses is bit more complicated to deduce, but it is revealed by the following reasoning. Consider the state of pure shear in a plane stress state described in section 1.3.5, which is characterized by two principal stresses that are equal and opposite in magnitude and with the third principal stress equal to zero. Assume that the principal stresses are $\sigma_{p2} = -\sigma_{p1}$, $\sigma_{p3} = 0$. The corresponding extensional strain components then follow from the generalized Hooke's law eq. (2.4a) and (2.4b) while the shear strain must be zero in the principal axes

$$\epsilon_1 = \frac{1+\nu}{E}\sigma_{p1}, \quad \epsilon_2 = -\frac{1+\nu}{E}\sigma_{p1}, \quad \gamma_{12} = 0. \tag{2.5}$$

In the analysis of the pure shear stress state, the state of stress on faces oriented at a 45° angle with respect to the principal stress directions is shown to take on an extreme value given by

$$\tau_{s12}^* = \sigma_{p2} = -\sigma_{p1}, \quad \sigma_{s1}^* = \sigma_{s2}^* = 0 \tag{2.6}$$

where the asterisk and subscript "s" are used to designate this special rotated axis system with maximum shear stresses. The strains in this rotated axis system are readily obtained from eq. (1.94), with $\theta_s = 45°$,

$$\gamma_{s12}^* = -(\epsilon_1 - \epsilon_2) = -\frac{2(1+\nu)}{E}\sigma_{p1}; \quad \epsilon_{s1}^* = \epsilon_{s2}^* = 0. \tag{2.7}$$

The relationship between τ_{s12}^* and γ_{s12}^* is then obtained by comparing eq. (2.6) and eq. (2.7) above to find $\gamma_{s12}^* = -2(1+\nu)\sigma_{p1}/E = 2(1+\nu)\tau_{s12}^*/E$, or $\tau_{s12}^* = G\,\gamma_{s12}^*$, where

$$G = \frac{E}{2(1+\nu)} \tag{2.8}$$

is defined as the *shear modulus*.

The above reasoning can be repeated for a state of pure shear in the other two orthogonal planes leading to similar results for the other shear stresses and strains, and this can be summarized by the *generalized Hooke's law for shear strains*

$$\gamma_{23} = \tau_{23}/G, \quad \gamma_{13} = \tau_{13}/G, \quad \gamma_{12} = \tau_{12}/G. \tag{2.9}$$

Here again, the shear modulus is the same in all directions due to the assumed isotropy of the material.

The shear strain-shear stress relationships, eq. (2.9), are established for the case of pure shear. They remain valid, however, for more complex stress states involving axial stresses, because, in view of eq. (2.4), axial stresses create no shear strains. Similarly, the generalized Hooke's law, eq. (2.4), is established when only axial stresses are applied. They do remain valid for more complex stress states involving shear stresses because, in view of eq. (2.9), shear stresses create no axial strains.

Matrix form of the constitutive laws

The constitutive laws, eqs. (2.4) and (2.9), are often called the generalized Hooke's laws. They can be expressed in a compact matrix form as

$$\underline{\epsilon} = \underline{\underline{S}} \, \underline{\sigma}, \tag{2.10}$$

where $\underline{\epsilon}$ and $\underline{\sigma}$ are the strain and stress arrays, respectively, and store the six strain and stress components, respectively,

$$\underline{\epsilon} = \left\{ \epsilon_1, \epsilon_2, \epsilon_3, \gamma_{23}, \gamma_{13}, \gamma_{12} \right\}^T, \tag{2.11a}$$

$$\underline{\sigma} = \left\{ \sigma_1, \sigma_2, \sigma_3, \tau_{23}, \tau_{13}, \tau_{12} \right\}^T, \tag{2.11b}$$

and the 6×6 *material compliance matrix*, $\underline{\underline{S}}$, is defined as

$$\underline{\underline{S}} = \frac{1}{E} \begin{bmatrix} 1 & -\nu & -\nu & 0 & 0 & 0 \\ -\nu & 1 & -\nu & 0 & 0 & 0 \\ -\nu & -\nu & 1 & 0 & 0 & 0 \\ 0 & 0 & 0 & 2(1+\nu) & 0 & 0 \\ 0 & 0 & 0 & 0 & 2(1+\nu) & 0 \\ 0 & 0 & 0 & 0 & 0 & 2(1+\nu) \end{bmatrix}. \tag{2.12}$$

The upper left 3×3 partition of the compliance matrix represents generalized Hooke's law, eq. (2.4), whereas the lower right 3×3 partition represents the shear stress-shear strain relationships, eq. (2.9). The vanishing of the upper right and lower left partitions stems from the absence of coupling between axial stresses and shear strains, and shear stresses and axial strains, respectively.

In summary, a homogeneous, linearly elastic, isotropic material is characterized by the constitutive laws given by eqs. (2.4) and (2.9) or combined as eq. (2.10). Only two material parameters are involved in these laws, Young's modulus, E, and Poisson's ratio, ν. The shear modulus G can be evaluated from eq. (2.8).

The constitutive laws are often presented in the compliance form of eq. (2.10), *i.e.*, strains are expressed as a function of stress. A straightforward algebraic process, however, yields the stiffness form of the same constitutive laws, where stresses are expressed as a function of strains,

$$\underline{\sigma} = \underline{\underline{C}}\,\underline{\epsilon}, \qquad (2.13)$$

where the 6×6 *material stiffness matrix*, $\underline{\underline{C}}$, is defined as

$$\underline{\underline{C}} = \frac{E}{(1+\nu)(1-2\nu)} \begin{bmatrix} 1-\nu & \nu & \nu & 0 & 0 & 0 \\ \nu & 1-\nu & \nu & 0 & 0 & 0 \\ \nu & \nu & 1-\nu & 0 & 0 & 0 \\ 0 & 0 & 0 & \frac{1-2\nu}{2} & 0 & 0 \\ 0 & 0 & 0 & 0 & \frac{1-2\nu}{2} & 0 \\ 0 & 0 & 0 & 0 & 0 & \frac{1-2\nu}{2} \end{bmatrix}. \qquad (2.14)$$

Plane stress state

The state of plane stress is studied in section 1.3. It will be convenient to define stress and strain arrays that include only the relevant components of stress and strain,

$$\underline{\epsilon} = \{\epsilon_1, \epsilon_2, \gamma_{12}\}^T, \qquad (2.15a)$$

$$\underline{\sigma} = \{\sigma_1, \sigma_2, \tau_{12}\}^T. \qquad (2.15b)$$

For the state of plane stress, $\sigma_3 = \tau_{13} = \tau_{23} = 0$, and the stiffness matrix reduces to a 3×3 matrix,

$$\underline{\underline{C}} = \frac{E}{(1-\nu^2)} \begin{bmatrix} 1 & \nu & 0 \\ \nu & 1 & 0 \\ 0 & 0 & \frac{1-\nu}{2} \end{bmatrix}. \qquad (2.16)$$

The constitutive laws for plane stress then become $\underline{\sigma} = \underline{\underline{C}}\,\underline{\epsilon}$, where stress and strain arrays are defined by eqs. (2.15), and the stiffness matrix by eq. (2.16). Note that due to Poisson's ratio effect, the strain component ϵ_3 does not vanish, $\epsilon_3 = -\nu(\sigma_1+\sigma_2)$.

Plane strain state

For the plane strain case, $\epsilon_3 = \gamma_{13} = \gamma_{23} = 0$, the stiffness matrix again reduces to a 3×3 matrix, but now different from eq. (2.16),

$$\underline{\underline{C}} = \frac{E}{(1+\nu)(1-2\nu)} \begin{bmatrix} 1-\nu & \nu & 0 \\ \nu & 1-\nu & 0 \\ 0 & 0 & \frac{1-2\nu}{2} \end{bmatrix}. \qquad (2.17)$$

The constitutive laws for plane stress then become $\underline{\sigma} = \underline{\underline{C}}\,\underline{\epsilon}$, where stress and strain arrays are defined by eqs. (2.15), and the stiffness matrix by eq. (2.17). Note that the stress component, σ_3, does not vanish due to Poisson's ratio effect, $\sigma_3 = \nu E(\epsilon_1 + \epsilon_2)/[(1+\nu)(1-2\nu)]$.

The bulk modulus

The volumetric strain is readily evaluated with the help of eq. (1.75)

$$e = \epsilon_1 + \epsilon_2 + \epsilon_3 = \frac{1 - 2\nu}{E}(\sigma_1 + \sigma_2 + \sigma_3) = \frac{1 - 2\nu}{E} I_1, \qquad (2.18)$$

where I_1 is the first stress invariant defined by eq. (1.15a).

In the special case of an applied hydrostatic pressure, $\sigma_1 = \sigma_2 = \sigma_3 = p$, a linear relationship is found between the applied pressure and the resulting volumetric strain

$$p = \kappa e, \qquad (2.19)$$

where

$$\kappa = \frac{E}{3(1 - 2\nu)}, \qquad (2.20)$$

is known is the *bulk modulus* of the material. When Poisson's ratio approaches a value of $1/2$, the bulk modulus approaches infinity, implying the vanishing of the volumetric strain under an applied pressure. Such a material is called an *incompressible material*. Many types of rubber materials are nearly incompressible, and metals undergoing plastic deformations are often assumed to be nearly incompressible.

2.1.2 Thermal effects

When a sample of a material is heated, its dimensions will change. Under a change in temperature, homogeneous isotropic materials will expand equally in all directions, generating *thermal strains*, $\epsilon^t = f(\Delta T)$, where $f(\Delta T)$ is a function of the change in temperature ΔT. The volume of most materials increases when they are subjected to increased temperatures, whereas temperature decreases generally cause the material to shrink. There are, however, notable exceptions. For example, the transition from water to ice under decreasing temperature is accompanied by a volume increase.

For moderate temperature changes, it is often adequate to assume that $f(\Delta T)$ is a linear function of the temperature change, *i.e.*, $f(\Delta T) = \alpha \Delta T$, where α is the *coefficient of thermal expansion*, a positive number if the material expands under increased temperature. The thermal strain now becomes

$$\epsilon^t = \alpha \Delta T. \qquad (2.21)$$

Two important aspects of thermal deformations must be emphasized. First, thermal strains are purely extensional: temperature changes do not induce shear strains. Second, thermal strains do not generate internal stresses, in contrast with mechanical strains that are related to internal stresses through the material constitutive law. An unconfined material sample subjected to a temperature change simply expands, but remains unstressed.

For homogeneous isotropic materials, the total strain is the sum of the thermal and mechanical strains. Thermal strains are the consequence of temperature changes, whereas mechanical strains result from the application of stresses. The total strains are the superposition of the mechanical strains, given by eq. (2.4), and their thermal counterparts, given by eq. (2.21),

$$\epsilon_1 = \frac{1}{E}[\sigma_1 - \nu(\sigma_2 + \sigma_3)] + \alpha\Delta T; \tag{2.22a}$$

$$\epsilon_2 = \frac{1}{E}[\sigma_2 - \nu(\sigma_1 + \sigma_3)] + \alpha\Delta T; \tag{2.22b}$$

$$\epsilon_3 = \frac{1}{E}[\sigma_3 - \nu(\sigma_1 + \sigma_2)] + \alpha\Delta T. \tag{2.22c}$$

Because temperature changes induce no shear strains, the shear stress-shear strain relationships given by eq. (2.9) remain unchanged.

When dealing with constrained material samples, temperature changes will indirectly generate stresses in the material. For example, consider a bar constrained at its two ends by rigid walls that prevent any extension of the bar. When subjected to a temperature change, ΔT, the bar tries to expand in all directions, but the rigid walls prevent expansion of the bar along its axis, $\bar{\imath}_1$. The stress components in the transverse direction, σ_2 and σ_3, must vanish because the bar is free to expand in those directions, whereas the axial strain, ϵ_1, must vanish, due to the presence of the rigid walls. Eq. (2.22a) then implies

$$\epsilon_1 = \frac{1}{E}[\sigma_1] + \alpha\Delta T = 0,$$

and hence, $\sigma_1 = -E\alpha\Delta T$; the temperature change induces a compressive stress in the bar. Such stresses are called *thermal stresses*. If same the bar is allowed to freely expand, *i.e.*, if the end walls are removed, axial equilibrium of the bar implies $\sigma_1 = 0$ and eq. (2.22a) then yields $\epsilon_1 = \alpha\Delta T$: the temperature change induces thermal strains, but no thermal stresses.

Example 2.1. Material sample confined in a rigid circular cylinder
Consider a sample of linearly elastic, isotropic material confined in an infinitely rigid circular cylinder and subjected to an applied stress σ_3, as depicted in fig. 2.2. Because the circular cylinder cannot deform in the directions perpendicular to the applied stress direction, the corresponding strain components must vanish, $\epsilon_1 = \epsilon_2 = 0$. The first two equations of the generalized Hooke's laws, eqs. (2.4a) and (2.4b), then yield $\sigma_1 = \sigma_2 = \nu/(1-\nu)\,\sigma_3$. Introducing these results into the last of the generalized Hooke's laws, eq. (2.4c), leads to

$$\epsilon_3 = \frac{(1+\nu)(1-2\nu)}{E(1-\nu)}\,\sigma_3.$$

The apparent modulus of elasticity of the sample is defined as $E_a = \sigma_3/\epsilon_3$, and

$$E_a = \frac{(1-\nu)}{(1+\nu)(1-2\nu)}\,E.$$

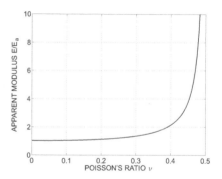

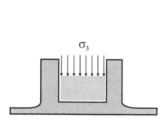

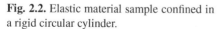

Fig. 2.2. Elastic material sample confined in a rigid circular cylinder.

Fig. 2.3. Normalized apparent modulus of elasticity versus Poisson's ratio.

As Poisson's ratio approaches $1/2$, the normalized apparent modulus of elasticity E_a/E increases rapidly, as shown in fig. 2.3. For $\nu = 0.45$, $E_a/E = 3.79$, i.e., the apparent modulus of the sample is 3.79 times that of the material.

Example 2.2. State of strain at the outer surface of a body
An experimentalist has measured the state of strain, ϵ_1, ϵ_2, and γ_{12}, at the outer surface of a three-dimensional body made of a homogeneous, isotropic, linearly elastic material. Axes $\bar{\imath}_1$ and $\bar{\imath}_2$ define the plane tangent to the outer surface of the body, and axis $\bar{\imath}_3$ is normal to this outer surface. Find the strain components ϵ_3, γ_{13} and γ_{23}.

Since the outer surface of the body is stress free, equilibrium requires $\sigma_3 = \tau_{13} = \tau_{23} = 0$. Hooke's law for shear components, see eq. (2.9), then readily implies that $\gamma_{13} = \gamma_{23} = 0$. The determination of the last strain component is more arduous. For this particular situation, generalized Hooke's laws, eqs. (2.4), become

$$\epsilon_1 = \frac{1}{E}\left[\sigma_1 - \nu\sigma_2\right], \quad \epsilon_2 = \frac{1}{E}\left[\sigma_2 - \nu\sigma_1\right], \quad \epsilon_3 = \frac{1}{E}\left[-\nu(\sigma_1 + \sigma_2)\right],$$

since $\sigma_3 = 0$. Adding together the first two equations yields $\epsilon_1 + \epsilon_2 = (1 - \nu)(\sigma_1 + \sigma_2)/E$. Introducing this result in the last equation then yields

$$\epsilon_3 = -\frac{\nu}{1 - \nu}(\epsilon_1 + \epsilon_2).$$

Typically, the three strain components at the outer surface of the body, ϵ_1, ϵ_2, and γ_{12}, are measured with the help of strain gauges. The determination of the remaining strain components is based on the equilibrium conditions at the surface of the body and on the constitutive laws, in this case Hooke's law.

2.1.3 Problems

Problem 2.1. Stresses expressed in terms of strains
It is sometimes necessary invert Hooke's law to express the stress in terms of the strain components. *(1)* Based on eqs. (2.4) and (2.9) prove the following relationships

$$\sigma_1 = \frac{E}{(1+\nu)(1-2\nu)}\left[(1-\nu)\epsilon_1 + \nu\epsilon_2 + \nu\epsilon_3\right], \qquad (2.23a)$$

$$\sigma_2 = \frac{E}{(1+\nu)(1-2\nu)}\left[\nu\epsilon_1 + (1-\nu)\epsilon_2 + \nu\epsilon_3\right], \qquad (2.23b)$$

$$\sigma_3 = \frac{E}{(1+\nu)(1-2\nu)}\left[\nu\epsilon_1 + \nu\epsilon_2 + (1-\nu)\epsilon_3\right], \qquad (2.23c)$$

and

$$\tau_{12} = G\gamma_{12}, \quad \tau_{23} = G\gamma_{23}, \quad \tau_{13} = G\gamma_{13}. \qquad (2.24)$$

Note: do not simply expand eq. (2.13) for your answer.

Problem 2.2. Independent coefficients for linearly elastic, isotropic materials

For a linearly elastic, isotropic material, the constitutive laws involve three parameters: Young's modulus, E, Poisson's ratio, ν, and the shear modulus, G. *(1)* Are these three coefficients independent of each other? *(2)* If not, give the equations that relate them.

Problem 2.3. Constitutive laws for stress and strain invariants

Let I_1^ϵ be the first invariant of the strain tensor, as defined by eq. (1.86), and I_1^σ be the first invariant of the stress tensor, as defined by eq. (1.15). *(1)* Find the constitutive law relating these two invariants if the material obeys the generalized Hooke's law.

Problem 2.4. Relationship between the principal stress and strain axes orientations

Prove that the principal stress and principal strain directions are always coincident at any point of a three-dimensional body made of a homogeneous, isotropic, linearly elastic material.

Problem 2.5. Stress data reduction for a strain gauge rosette

Consider the strain gauge rosette depicted in fig. 1.26A, bonded to the external surface of a body made of a homogeneous, isotropic, linearly elastic material. The following strains have been measured: $e_1 = 3657\mu$, $e_2 = -1245\mu$, $e_3 = 956\mu$. *(1)* Find the strain state at this point. *(2)* Find the principal strains and the principal strain directions at this point. *(3)* Sketch the rosette and superpose on this sketch the principal strain directions. *(4)* compute the state of stress at this point. *(5)* Find the principal stresses and the principal stress directions at this point. *(6)* Find the relationship between the principal strain and stress directions. For this material, $E = 73$ GPa and $\nu = 0.3$

Problem 2.6. Data reduction for the "stress gauge"

A "fish-bone" strain gauge has the configuration shown in fig. 2.4. The various sub-gauges, inclined at angles $+\alpha$ and $-\alpha$ with respect to the gauge direction, measure strains along those two directions, denoted ϵ_α and $\epsilon_{-\alpha}$, respectively. The sub-gauges are electrically connected in such a way that a single measurement is made, $e = \epsilon_\alpha + \epsilon_{-\alpha}$. The fish-bone gauge, also known as a "stress gauge," is intended to measure the stress, σ, along the direction of the gauge, *independently of any other stress components acting at that location. (1)* Find the value of angle α for which the gauge measurement, e, becomes independent of the other stress components. *(2)* Find the relationship between the measurement and the stress, σ, in the gauge direction.

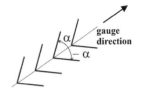

Fig. 2.4. Configuration of the "fish-bone" gauge.

2.1.4 Ductile materials

The linearly elastic behavior described in the previous section is a highly idealized behavior. In general, materials will present a nonlinear relationship between stress and strain.

Figure 2.5 shows a typical stress-strain diagram for a ductile material such as mild steel. From point **O** to point **A**, the material behaves in a linear manner, and this can be described by Hooke's law. In this region, the slope of the stress-strain diagram is constant and its value equals Young's modulus, E. If the loading is released, the specimen will come back to its original configuration without sustaining any permanent deformations, and it is referred to as being "elastic."

Beyond point **A**, the behavior is no longer proportional (linear), and hence, this point is called the *limit of proportionality*. The corresponding stress level is denoted σ_e, see fig. 2.5. The material may continue to be elastic beyond point **A**, but at some point it will begin to deform plastically, and when the load is removed, a permanent deformation will remain. The stress at which this occurs is called the *yield stress*, σ_y. More often than not, especially for mild steels, little difference exists between the limit of proportionality and the yield stress, and so σ_e and σ_y are often used interchangeably.

Beyond point **B**, the material undergoes extensive deformation at a nearly constant stress level, denoted by σ_p. From point **B** to point **C**, the material is undergoing a *plastic flow* under nearly constant stress level. Figure 2.5 shows that the strain over this region amounts to about 5% (*i.e.*, $\epsilon_1 = 0.05$), but for highly ductile steels and other materials, this may amount to more than 10%.

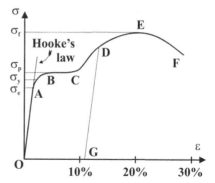

Fig. 2.5. Stress-strain diagram for a ductile material such as mild steel.

Beyond point **C**, an increasing stress level is required to continue deforming the material. The stress level increases up to point **E**, where the maximum stress level, denoted by σ_f, is reached.

Past this point, the cross-sectional area of the specimen decreases significantly at a particular location along the test section: this phenomenon is called *"necking"* of the specimen. Because the stress level is determined by dividing the applied load by the original cross-sectional area, the stress level will seem to decrease beyond point **E**, but if the stress level is computed by dividing the applied load by the reduced cross-section area of the specimen at the location where necking occurs, this *true stress* level sill continue to increase past point **E**.

With most experimental testing equipment, a controlled load (rather than a controlled deformation) is applied, and hence, point **F** is never recorded. Instead, once point **E** is reached, necking develops and the specimen breaks almost immediately afterwards. Consequently, the stress at point **E** is called the *failure stress* and designated by σ_f. Only when controlled extension is applied to the specimen is it possible to follow the behavior of the specimen from point **E** to point **F**.

Clearly, ductile materials undergo very large deformations before failure, corresponding to the portion of the stress-strain curve from point **B** to **E** in fig. 2.5. Experiments show that if the specimen is unloaded at a point between **B** and **E**, for example at point **D**, the stress-strain relationship will follow curve **DG**, parallel to **AO**, and while unloading, the material behaves elastically, although a permanent deformation of magnitude **OG** will remain after all loading is removed. If the specimen is reloaded, the stress-strain relationship will follow curve **GD**, and if additional loading is applied, it will follow curve **DEF**, as if the prior unloading had not taken place. The reloading curve **GD** is linear and reaches a higher stress level at point **D** before yield occurs and plastic deformation begins again. This increase in the yield stress is called *strain hardening*[1].

The discussion presented in the previous paragraphs is focused on diagrams of axial stress versus axial strain obtained from a tensile test as depicted in fig. 2.1. It is not unexpected that material behavior under shear exhibits nonlinear characteristics of a nature that is similar to that observed under tension. Figure 2.6 shows a typical shear stress-shear strain diagram for a ductile material such as mild steel. Here again, upon unloading, the material tends to behave in a linear manner, although a permanent deformation of magnitude **OG** will be remain after unloading.

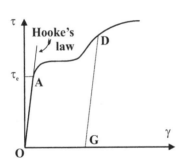

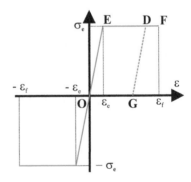

Fig. 2.6. Shear stress-shear strain diagram for a ductile material.

Fig. 2.7. Stress-strain diagram for an elastic-perfectly plastic material.

It is sometimes convenient to idealize the stress-strain diagram of ductile materials as presenting an initial elastic regime, followed by a perfectly plastic regime. This idealization is depicted in fig. 2.7. For a strain range $-\epsilon_e \leq \epsilon \leq \epsilon_e$, the material is linearly elastic, but for strain level outside this range, the material is *perfectly plastic*, that is, the material flows under a constant stress level, σ_e which is also the yield stress σ_y. This highly idealized material behavior is called *elastic-perfectly*

---

[1] Strain hardening is particularly noticeable in annealed copper such as might be encountered in new copper tubing. After the tube is initially bent, it requires a considerably greater effort to begin bending again or to try to reverse the initial bend.

plastic material behavior. Failure occurs when the strain reaches the level ϵ_f. Such a constitutive model is a good first approximation to the behavior of a ductile metallic material such as mild steel or annealed aluminum. Of course, if the material is unloaded at point **D**, the unloading curve follows segment **DG**, parallel to **OE**.

Other ductile materials such as aluminum and copper do not exhibit a plastic flow regime like the portion of the stress-strain diagram from point **B** to **C** in fig. 2.5 observed for mild steel. Figure 2.8 shows a representative stress-strain diagram for aluminum. For such materials, no pronounced limit of proportionality is present nor is the yield stress (elastic limit) evident. Instead, it is convenient to define the yield stress, denoted σ_y, as the stress level for which the specimen will exhibit a small permanent residual strain upon unloading. For aluminum, this residual strain is specified as 0.2% or $\epsilon = 0.002$. The yield stress can be determined from the stress-strain diagram by constructing a straight line parallel to the initial linearly elastic portion of the curve at a 0.2% offset and recording the stress at the intersection with the stress-strain curve, as illustrated in fig. 2.8.

2.1.5 Brittle materials

Ductile materials are characterized by stress-strain diagrams such as those presented in figs. 2.5 and 2.8: large deformations occur when stress levels greater than that corresponding to the elastic limit of the material are applied. For brittle materials, very little deformation is observed beyond the elastic limit. Typically, failure occurs abruptly at strain levels much smaller than those observed for ductile materials. Figure 2.9 shows a stress-strain diagram typical of that observed for brittle materials such as glass, concrete, stone, wood, unidirectional composites or ceramic materials.

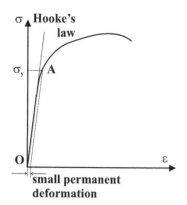
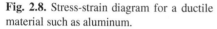
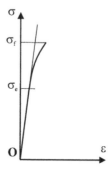

Fig. 2.8. Stress-strain diagram for a ductile material such as aluminum.

Fig. 2.9. Stress-strain diagram for a brittle material.

2.2 Allowable stress

A central problem of structural analysis is to determine the optimal configuration of a structure subjected to specific loads. The design will be influenced by many factors associated with various structural characteristics, such as those listed below.

1. *The strength of the structure.* When the local stress in the structure exceeds a specific value, the material will break or sustain permanent damage such as cracks or plastic deformations.
2. *The elastic deformation of the structure* under load. Even when subjected to small loads, a structure can present undesirable levels of elastic deformation. For example, the elastic deflection of a part may lead to interference with other parts in a structural assembly.
3. *The dynamics characteristics of the structure.* If the structure is subjected to dynamic loads, the time history of its response becomes important. More often than not, its natural frequencies must be carefully placed to avoid resonance. For aerospace structures, aeroelastic phenomena such as flutter will put stringent requirements on the torsional natural frequencies of wings and fuselages.
4. *The stability characteristics of the structure.* When parts of the structures are subjected to compressive loads, the equilibrium configuration can become unstable, resulting in buckling. During level flight, the upper skin of a wing of an aircraft is subjected to compressive loads. Wing design is significantly affected by buckling considerations.
5. *The time dependent deformations of the structure associated with creep* of the constitutive materials. Creep considerations play an important role in aircraft turbine engine design, because they are subjected to high temperatures.

The strength of a structure is the focus of the present section, although a good design must incorporate all the above characteristics. A structure is said to fail if it breaks, collapses, or develops significant permanent damage. Clearly, the applied service loads must be less, and often much less, than those corresponding to failure. The main reason for decreasing service loads is due to the numerous uncertainties about the problem. Among these are

1. The *actual magnitude of the applied service loads is not accurately known.* In an aircraft, maneuver loads or loads associated with a rough landing conditions cannot be precisely evaluated. Accidental overloads might also take place during flight or ground operations of the aircraft.
2. The *strength of materials presents statistical characteristics.* Measurements of the strengths of two nominally identical samples of aluminum will be different due to material inhomogeneities, processing difference, and experimental errors.
3. *Manufacturing variability* also plays an important role. For instance, machining fittings of complex shapes is a delicate operation. Dimensional tolerances might vary from part to part; the strength of the resulting material might not be equal to that measured in laboratory samples, and quality control sometimes fails to detect some types of defects in manufactured parts.

4. The strength of the material might decrease in time due to *corrosion, wear,* or the presence of a *chemically aggressive environment* such as salt water, fuels or solvents.
5. Finally, if failure is predicted based on the computed value of internal stresses in the structure, *these predicted stresses might be very different from their actual value*, because simplifying assumptions are used to predict these stresses.

Consequently, service loads must be limited to a conservative level, and as the uncertainty about the problem increases, so must the level of conservatism in the design. It is common practice to account for all these uncertainties by defining a *load factor*

$$\text{load factor} = \frac{\text{failure load}}{\text{service load}}, \qquad (2.25)$$

where *failure load* is the load at which the part fails and *service load* is the maximum load that is expected in normal service. Of course, the load factor should be larger than unity, and it is sometimes as large as 10. Engineering judgment must be carefully exercised in choosing this load factor. If a low value is selected, the likelihood of accidental failure will increase, whereas for high values, the design might be too expensive or too heavy for its intended purpose.

The load factor might be viewed as a *factor of safety* with respect to failure: limiting the service loads to a fraction of the failure loads implies a safe operation of the structure. Using the load factor as a factor of safety is not always practical because the failure load is often unknown. Indeed, it is not practical, nor cost effective to test all structures to failure to determine the failure load. A more common approach is to compute the local stresses induced by the applied loads and limit the these local stresses to an allowable level. This can be written as

$$\text{allowable stress} = \frac{\text{yield stress}}{\text{safety factor}}, \quad \text{or} \quad \sigma_{\text{allow}} = \frac{\sigma_y}{n}, \qquad (2.26)$$

where σ_{allow} is the allowable stress, σ_y the yield stress of the material, and n the factor of safety. This definition is adequate for ductile materials described in section 2.1.4. Once the yield stress is reached, permanent deformation occurs. On the other hand, for brittle materials such as those discussed in section 2.1.5, the following definition of the allowable stress is more appropriate

$$\text{allowable stress} = \frac{\text{ultimate stress}}{\text{safety factor}}, \quad \text{or} \quad \sigma_{\text{allow}} = \frac{\sigma_f}{n}, \qquad (2.27)$$

where the ultimate stress is the failure stress for the material.

In summary, the stress level, σ, that a structure is subjected to during service should be smaller than the allowable stress, leading to the following strength criterion

$$|\sigma| \leq \sigma_{\text{allow}}. \qquad (2.28)$$

For some materials, the allowable stress in tension and in compression are different, and the actual stress level should then be compared to the appropriate allowable stress.

2.3 Yielding under combined loading

The concept of allowable stress discussed in the previous section is focused on the highly idealized case where a structural component is subjected to a *single stress component*. The yield criterion is then simply expressed in terms of the single stress component as $\sigma \leq \sigma_y$. As depicted in fig. 1.3, a differential element of material can be subjected to a number of stress components simultaneously. The question is now: what is the proper yield criterion to be used when *multiple stress components are acting simultaneously?* Consider an aircraft propeller connected to a homogeneous, circular shaft. The engine applies a torque to the shaft resulting in shear stresses, τ, throughout the shaft. On the other hand, the propeller creates a thrust that generates uniform axial stresses, σ, over the cross-section. If the torque acts alone, the yield criterion is $\tau < \tau_y$; if the axial force acts alone, the corresponding criterion is $\sigma < \sigma_y$. In the actual structure, both stress components are acting simultaneously, and it is natural to ask: what is the proper criterion to apply?

The yield criteria to be presented in this section are applicable to isotropic, homogeneous material subjected to a general three-dimensional state of stress. Because the material is isotropic, the direction of application of the stress is irrelevant. If the material is subjected to a single stress component, it should yield under the same stress level regardless of the direction in which this stress component is applied. In contrast, if the material is anisotropic, the direction of application of stress becomes important.

For isotropic materials, there is no directional dependency of the yield criterion, even when subjected to a combined state of stress. An arbitrary state of stress can be represented by the six stress components defining the stress tensor at that point, for example, see eq. (1.3). Alternatively, the state of stress can be represented by the three principal stresses, σ_{p1}, σ_{p2}, and σ_{p3} and the three orientations defining the faces on which these principal stresses act, see section 1.2.2. If the yield criterion should not depend on directional information because of material isotropy, it is clear that *only the magnitudes of the principal stress* should appear in its expression.

In addition, empirical evidence indicates that hydrostatic stress does not cause yielding. This implies that changes in the state of stress in which the three principal stresses are increased equally will not result in yielding. Other empirical evidence also suggests that yielding is directly related to the maximum shear stress in the material which, in turn, is directly proportional to the differences between the principal stresses.

Two specific criteria will be presented here, Tresca's criterion, see section 2.3.1, and von Mises' criterion, see section 2.3.2. Both compute an equivalent maximum shear stress intensity but yield slightly different results for some cases. A more detailed discussion of yield criteria can be found in section 13.1.

2.3.1 Tresca's criterion

Tresca's yield criterion is expressed in terms of the following three principal stress inequalities

$$|\sigma_{p1} - \sigma_{p2}| \leq \sigma_y, \quad |\sigma_{p2} - \sigma_{p3}| \leq \sigma_y, \quad |\sigma_{p3} - \sigma_{p1}| \leq \sigma_y, \qquad (2.29)$$

where σ_y is the yield stress observed in a uniaxial test such as that described in fig. 2.5. The material operates in the linearly elastic range when the stress state it is subjected to satisfies the three inequalities expressed by eq. (2.29). Conversely, yielding develops whenever any one of these conditions is violated. Tresca's criterion clearly meets the two conditions stated above: it depends only on the principal stresses, and a hydrostatic state of stress will not produce yielding.

Tresca's criterion can be interpreted in the following manner. Let $\bar{\imath}_1^*$, $\bar{\imath}_2^*$, and $\bar{\imath}_3^*$ be the principal stress directions. Consider now a rotation of magnitude θ about axis $\bar{\imath}_3^*$. The shear stress on this face inclined with respect to the principal stress directions is then given by eq. (1.49), where $\sigma_1 = \sigma_{p1}$, $\sigma_2 = \sigma_{p2}$ and $\tau_{12} = 0$, to yield $\tau_{12} = -(\sigma_{p1} - \sigma_{p2})/2 \sin 2\theta$. Clearly, the maximum shear stress is found on a face inclined at an angle $\theta = 45$ degrees and gives $\tau_{12max} = |\sigma_{p1} - \sigma_{p2}|/2$. Similar arguments for rotations about axes $\bar{\imath}_2^*$ and $\bar{\imath}_1^*$ lead to $\tau_{13max} = |\sigma_{p1} - \sigma_{p3}|/2$, and $\tau_{23max} = |\sigma_{p2} - \sigma_{p3}|/2$, respectively. Tresca's criterion is now recast as $\tau_{23max} \leq \sigma_y/2$, $\tau_{13max} \leq \sigma_y/2$ and $\tau_{12max} \leq \sigma_y/2$.

By denoting $\tau_{max} = \max(\tau_{23max}, \tau_{13max}, \tau_{12max})$, Tresca's criterion can be expressed by a single condition, $\tau_{max} \leq \sigma_y/2$: *the material reaches the yield condition when the maximum shear stress equals half the yield stress under a uniaxial stress state.* This physical interpretation of Tresca's criterion helps explain why it is sometimes called the *maximum shear stress criterion*. Tresca's criterion is now applied to a few combined loading cases of practical interest.

Uniaxial stress state

First, consider the case of a material subjected to an uniaxial state of stress, σ_{p1}, $\sigma_{p2} = \sigma_{p3} = 0$. The sole non-vanishing principal stress is σ_{p1}, and Tresca's yield criterion reduces to $\sigma_{p1} \leq \sigma_y$. This result is identical to the yield criterion discussed in section 2.2, as expected.

Plane state of stress

Consider a material under a plane state of stress, as defined in section 1.3. If σ_1, σ_2, and τ_{12} are the stress components in an arbitrary coordinate system, the principal stresses are readily found as

$$\sigma_{p1}, \sigma_{p2} = \frac{\sigma_1 + \sigma_2}{2} \pm \sqrt{\left(\frac{\sigma_1 - \sigma_2}{2}\right)^2 + \tau_{12}^2}, \quad \sigma_{p3} = 0. \qquad (2.30)$$

Tresca's criterion now implies the following three conditions

$$2\sqrt{\left(\frac{\sigma_1 - \sigma_2}{2}\right)^2 + \tau_{12}^2} \leq \sigma_y, \quad \left|\frac{\sigma_1 + \sigma_2}{2} \pm \sqrt{\left(\frac{\sigma_1 - \sigma_2}{2}\right)^2 + \tau_{12}^2}\right| \leq \sigma_y.$$

$$(2.31)$$

Pure shear state

The state of pure shear is a special case of a plane stress state where $\sigma_1 = \sigma_2 = 0$ and $\tau_{12} = \tau$. The only remaining condition of Tresca's criterion, eq. (2.31), is $\tau \leq \sigma_y/2$. According to Tresca's criterion, the *shear stress level at which the material yields in a pure shear state is one half the level observed under uniaxial stress state.*

2.3.2 Von Mises' criterion

Von Mises' yield criterion is expressed by the following inequality

$$\sigma_{eq} = \frac{1}{\sqrt{2}}\sqrt{[(\sigma_{p1} - \sigma_{p2})^2 + (\sigma_{p2} - \sigma_{p3})^2 + (\sigma_{p3} - \sigma_{p1})^2]} \leq \sigma_y, \qquad (2.32)$$

where the first equality defines the *equivalent stress*, σ_{eq}. Von Mises' criterion states that *the yield condition is reached under the combined loading, when the equivalent stress, σ_{eq}, reaches the yield stress for a uniaxial stress state, σ_y.* Von Mises' criterion clearly meets the requirement stated above: it only depends on the principal stresses and a hydrostatic state of stress will not produce yielding.

The physical nature of this equivalent stress is better understood by considering the octahedral face discussed in example 1.3 on page 18. The magnitude of the shear stress acting on this octahedral face is given by eq. (1.25), and simple algebra then reveals

$$3\tau_{oc}^2 = (\sigma_{p1}^2 + \sigma_{p2}^2 + \sigma_{p3}^2) - \frac{1}{3}(\sigma_{p1} + \sigma_{p2} + \sigma_{p3})^2 = \frac{2}{3}\sigma_{eq}^2. \qquad (2.33)$$

This result implies that the equivalent stress is proportional the octahedral shear stress: $\sigma_{eq} = 3/\sqrt{2}\,\tau_{oc}$. Von Mises' criterion can now be restated as: *the yield condition is reached under combined loading when the octahedral shear stress reaches $3/\sqrt{2}$ of the yield stress for a uniaxial stress state, σ_y.*

When applying von Mises' criterion, the first step is to compute the equivalent stress defined by eq. (2.32). Given a loading state defined by the direct stress components σ_1, σ_2, and σ_3 and shear stress components τ_{23}, τ_{13}, and τ_{12}, it is necessary to first compute the principal stresses, σ_{p1}, σ_{p2}, and σ_{p3} using the procedure described in section 1.2.2. This laborious procedure can be bypassed by noticing that the first two invariants of the stress tensor, see eq. (1.15a) and (1.15b), can be written as $I_1 = \sigma_{p1} + \sigma_{p2} + \sigma_{p3}$ and $I_2 = \sigma_{p1}\sigma_{p2} + \sigma_{p2}\sigma_{p3} + \sigma_{p3}\sigma_{p1}$, respectively, because the shear stresses vanish on the faces normal to the principal stress directions. The following algebraic manipulations show that the equivalent stress is readily expressed in terms of these two invariant as

$$\begin{aligned}
\sigma_{eq}^2 &= [(\sigma_{p1} - \sigma_{p2})^2 + (\sigma_{p2} - \sigma_{p3})^2 + (\sigma_{p3} - \sigma_{p1})^2]/2 \\
&= (\sigma_{p1}^2 + \sigma_{p2}^2 + \sigma_{p3}^2) - (\sigma_{p1}\sigma_{p2} + \sigma_{p2}\sigma_{p3} + \sigma_{p3}\sigma_{p1}) \qquad (2.34) \\
&= (\sigma_{p1} + \sigma_{p2} + \sigma_{p3})^2 - 3(\sigma_{p1}\sigma_{p2} + \sigma_{p2}\sigma_{p3} + \sigma_{p3}\sigma_{p1}) = I_1^2 - 3I_2.
\end{aligned}$$

If the first two stress invariants are now expressed in terms of the stress components in an arbitrarily oriented axis system using eqs. (1.15a) and (1.15b), von Mises' yield criterion can then be written as

$$\sigma_{eq} = \sqrt{\sigma_1^2 + \sigma_2^2 + \sigma_3^2 - \sigma_2\sigma_3 - \sigma_3\sigma_1 - \sigma_1\sigma_2 + 3(\tau_{23}^2 + \tau_{13}^2 + \tau_{12}^2)} \leq \sigma_y. \tag{2.35}$$

This criterion is now applied to several combined loading cases of practical interest.

Uniaxial stress state

First, consider the case of a material subjected to an uniaxial state of stress, σ_{p1}, $\sigma_{p2} = \sigma_{p3} = 0$. The sole non-vanishing principal stress is σ_{p1}, and von Mises's yield criterion reduces to $\sigma_{p1} \leq \sigma_y$. This result is identical to the yield criterion discussed in section 2.2, as expected.

Plane state of stress

Consider a material under a plane state of stress as defined in section 1.3. If σ_1, σ_2 and τ_{12} are the stress components in an arbitrary coordinate system, the equivalent stress, eq. (2.35), now reduces to

$$\sigma_{eq} = \sqrt{\sigma_1^2 + \sigma_2^2 - \sigma_1\sigma_2 + 3\tau_{12}^2} \leq \sigma_y. \tag{2.36}$$

Pure shear state

The state of pure shear is a special case of plane stress where $\sigma_1 = \sigma_2 = 0$ and $\tau_{12} = \tau$. Von Mises' criterion, eq. (2.36), reduces to $\tau \leq \sigma_y/\sqrt{3}$. According to von Mises' criterion, the shear stress level at which the material yields in a pure shear state is $1/\sqrt{3} \approx 0.577$, i.e., about 60% of the level observed under uniaxial stress state. Experimentation shows that this prediction is slightly more accurate than that of Tresca's criterion. This and computational simplicity are the reasons why von Mises' criterion is more widely used than Tresca's.

2.3.3 Comparing Tresca's and von Mises' criteria

A useful geometric interpretation of Tresca's and von Mises' criteria can be obtained by considering a plane stress problem for which $\sigma_{p3} = 0$. In the stress space of the two remaining principal stresses, σ_{p1} and σ_{p2}, Tresca's criterion, see eq. (2.29), reduce to three inequalities

$$|\frac{\sigma_{p1}}{\sigma_y}| \leq 1, \quad |\frac{\sigma_{p2}}{\sigma_y}| \leq 1, \quad |\frac{\sigma_{p2}}{\sigma_y} - \frac{\sigma_{p1}}{\sigma_y}| \leq 1.$$

When taken as the limiting equalities, these three equations define the six straight line segments depicted in fig. 2.10. In the construction of this graph, the principal stresses are normalized by the yield stress. Safe stress levels correspond to stress states falling within the irregular hexagon enclosed by the six dashed line segments. For this plane stress state, the yield envelope is therefore the hexagon shown in dashed lines in fig. 2.10.

For the same stress states, von Mises' criterion, see eq. (2.32), becomes the oblique ellipse defined by

$$\left(\frac{\sigma_{p1}}{\sigma_y}\right)^2 + \left(\frac{\sigma_{p2}}{\sigma_y}\right)^2 - \left(\frac{\sigma_{p1}}{\sigma_y}\right)\left(\frac{\sigma_{p2}}{\sigma_y}\right) = 1.$$

Safe stress levels correspond to stress states falling within the ellipse shown in fig. 2.10 which forms the *yield envelope*.

At the six vertices of the hexagon, the yield conditions predicted by the two criteria are identical. For all other stress conditions, Tresca's criterion is slightly more conservative. In most experimental studies,

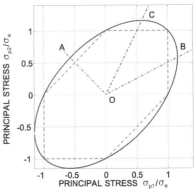

Fig. 2.10. Comparison of Tresca's and von Mises' criteria for a plane stress case.

yielding is observed to occur at points falling between these two criteria. As a purely practical matter, von Mises' criterion is often preferred because of its relatively simpler representation as a single analytical expression in contrast with the three separate inequalities that must be evaluated for Tresca's criterion.

When a set of loads is applied to a structure, it is natural to assume that they are all increased proportionally. Consequently, the components of the stress state, and therefore the principal stresses, increase proportionally as well. Three special stress states will be contrasted. In all three cases, the principal stresses are assumed to remain proportional as the load is applied, and hence, a single stress parameter, denoted σ, will be used to describe the loading for each case. The three stress cases are: *(1)* $\sigma_{p1} = -\sigma_{p2} = \sigma$, *(2)* $\sigma_{p1} = 2\sigma_{p2} = \sigma$, and *(3)* $\sigma_{p2} = 2\sigma_{p1} = \sigma$, and these correspond to the three radial lines **OA**, **OB**, and **OC**, respectively, shown in fig. 2.10. Table 2.1 shows a quantitative comparison of the cases. These three loading cases give the maximum discrepancy in the predictions of the two criteria. For all other loading configurations, the prediction differ by less than 15%.

Table 2.1. Comparison of the Tresca and von Mises yield criteria.

Stress state	Radial line in fig. 2.10	Tresca's yield stress	von Mises' yield stress	Percent difference
$\sigma_{p1} = -\sigma_{p2} = \sigma$	**OA**	$\sigma_y/2$	$\sigma_y/\sqrt{3}$	15.5%
$\sigma_{p1} = 2\sigma_{p2} = \sigma$	**OB**	σ_y	$2\sigma_y/\sqrt{3}$	15.5%
$\sigma_{p2} = 2\sigma_{p1} = \sigma$	**OC**	σ_y	$2\sigma_y/\sqrt{3}$	15.5%

2.3.4 Problems

Problem 2.7. Yield criterion for a confined cylindrical sample

Consider a sample of homogeneous, isotropic material of Poisson's ratio ν and yield stress σ_y confined in a rigid cylinder, as depicted in fig. 2.2. A single stress component is applied to the material and it is assumed that there is no friction between the sample and the enclosure. *(1)* Find the stress level σ_3 for which the sample will yield as a function of σ_y and ν if the material obeys Tresca's criterion. Plot your results. *(2)* Find the stress level σ_3 for which the sample will yield as a function of σ_y and ν if the material obeys von Mises' criterion. Plot your results. Use a range of Poisson's ratios $\nu \in [0, 0.5]$.

Problem 2.8. Yield criterion for a pressure vessel

A cylindrical pressure vessel of radius R and thickness t is subjected to an internal pressure p_i, as shown in fig. 1.20. At any point in the cylindrical portion of vessel wall, two stress components are acting: the hoop stress, $\sigma_h = Rp_i/t$ and the axial stress, $\sigma_a = Rp_i/(2t)$. The radial stress, acting in the direction perpendicular to the wall, is very small, $\sigma_r \approx 0$. The yield stress for the material is σ_y. *(1)* If the material is assumed to follow von Mises' criterion, find the maximum internal pressure the vessel can carry. *(2)* If the material is assumed to follow Tresca's criterion, find the maximum internal pressure the vessel can carry.

2.4 Material selection for structural performance

An important phase of structural design is the selection of a specific material. Table 2.2 lists the physical properties of three commonly used metals: aluminum, titanium, and steel. This table lists their respective ultimate stress, modulus of elasticity, and density. Table 2.3 lists the corresponding properties for a number of fibers.

Table 2.2 shows that the ultimate stress and modulus of elasticity of steel are far superior to those of titanium or aluminum. Why then is steel not always preferred, since it is far stronger and stiffer? A second look at table 2.2 shows that while steel is far stronger and stiffer, it is also far heavier that the other two metals. In a weight sensitive design, a compromise must be made between these conflicting characteristics. The same observations can be made when comparing the properties of fibers, as listed in table 2.3.

It is important to compare the performance of these various materials for specific structural applications. Three categories of structural design situations will be investigated, namely *strength design*, *stiffness design*, and *buckling design*. A *performance index* of the material will be derived in each case.

Table 2.2. Physical properties of a few metals.

	Ultimate stress [MPa]	Modulus of elasticity [GPa]	Density [kg/m^3]
Aluminum	620	73	2700
Titanium	1900	115	4700
Steel	4100	210	7700

Table 2.3. Physical properties of a few fibers.

	Ultimate stress [MPa]	Modulus of elasticity [GPa]	Density [kg/m³]
E-Glass	3400	72	2550
S-Glass	4800	86	2500
Carbon	1700	190	1410
Boron	3400	400	2570
Graphite	1700	250	1410

2.4.1 Strength design

Consider a sheet of material of length L, width b, and thickness t, subjected to a tension load P, as depicted in the left portion of fig. 2.11. Assuming the stress distribution to be uniform over the sheet's cross-section, the total load the material can carry is $P_{max} = \sigma_{ult} bt$, where σ_{ult} is the ultimate allowable stress for the material.

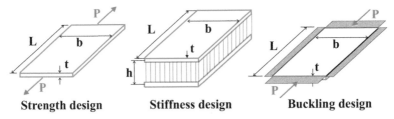

Strength design   **Stiffness design**   **Buckling design**

Fig. 2.11. Three types of design situation.

The total mass, M, of the structure is $M = \rho\, btL$, where ρ is the material density. Eliminating the sheet thickness between these two equations yields

$$P_{max} = \frac{M}{L}\frac{\sigma_{ult}}{\rho}. \tag{2.37}$$

For a given mass and geometry of the structure, the maximum load it can carry is

$$P_{max} \propto \frac{\sigma_{ult}}{\rho}. \tag{2.38}$$

The desired material performance index for strength design is σ_{ult}/ρ, and it is proportional to the maximum load that can be carried by a structure of given geometry and mass.

2.4.2 Stiffness design

In many instances, the stiffness of a structure is specified, but more often than not, it is the natural frequency of the structure that must be maximized. Consider the cantilevered, thin-walled beam of length L consisting of two thin skins of width b

and thickness t separated by a distance h as shown in the middle portion of fig. 2.11. Under certain conditions, the natural frequency of this structure is

$$\omega \propto \frac{1}{L^2} \left[\frac{H^c_{22}}{m} \right]^{1/2} \tag{2.39}$$

where H^c_{22} is the bending stiffness, and m the mass per unit span of the beam; these quantities are readily found as $H^c_{22} = Ebth^2/2[1 + 1/3(t/h)^2]$ and $m = 2\rho bt$, respectively. For a thin-walled structure $t/h \ll 1$, and the natural frequency becomes

$$\omega \propto \frac{h}{L^2} \left[\frac{E}{\rho} \right]^{1/2} \tag{2.40}$$

For a given configuration of the structure, h and L are given quantities, and the desired material performance index for stiffness design is $\sqrt{E/\rho}$, and it is proportional to the natural frequency of a structure of a given geometry and mass.

2.4.3 Buckling design

The right portion of fig. 2.11 shows a thin plate of length L, width b, and thickness t. The plate is supported around all its edges, and subjected to an in-plane compressive load P. The critical value of the load that will cause the plate to buckle is

$$P_{cr} \propto \frac{Et^3}{b}. \tag{2.41}$$

This formula will be derived in section 16.7. The total mass of the structure is $M = \rho\, btL$; eliminating the thickness of the plate from then yields

$$P_{cr} \propto \frac{M^3}{b^4 L^3} \frac{E}{\rho^3}. \tag{2.42}$$

For a given mass and geometry of the structure, the desired performance index is E/ρ^3, and it is proportional to the maximum compressive load that can be carried by a structure of given geometry and mass.

Table 2.4 lists the performance indices σ_{ult}/ρ, $\sqrt{E/\rho}$, and E/ρ^3 for strength, stiffness, and buckling designs, respectively. Table 2.5 lists the corresponding quantities for a few fibers.

Consider first the data of table 2.4. Steel is clearly the best material for strength design. When it comes to stiffness design, however, the three metals perform about equally well, with only a slight disadvantage for titanium. Finally, comparing the strength and buckling designs, the ranking of the materials is now reversed: aluminum performs far better than steel and titanium in buckling design.

The same observations can be made about the fibers for which data is listed in table 2.5. In a strength design, S-glass out performs the other fibers. The situation is reversed for stiffness and buckling designs. It is clear that the third power of the density in the denominator of the buckling design performance index makes lighter materials perform well in buckling sensitive designs.

It is not possible to directly compare the materials in tables 2.4 and 2.5. Indeed, the various metals can be used as structural materials, whereas the fibers cannot be used, as such, as structural materials. It is clear, however, that the remarkably high performance indices of these fibers justifies a closer look at their potential use in structural applications.

Table 2.4. Structural design performance indices for a few metals.

Performance index	Strength design σ_{ult}/ρ [10^3 m^2/sec^2]	Stiffness design $\sqrt{E/\rho}$ [10^3 m/sec]	Buckling design E/ρ^3 [m^8/(kg^{2}sec^2)]
Aluminum	230	5.2	3.7
Titanium	405	4.9	1.1
Steel	530	5.2	0.46

Table 2.5. Structural design performance indices for a few fibers.

Performance Index	Strength design σ_{ult}/ρ [10^3 m^2/sec^2]	Stiffness design $\sqrt{E/\rho}$ [10^3 m/sec]	Buckling design E/ρ^3 [m^8/(kg^2 sec^2)]
E-Glass	1330	5.3	4.3
S-Glass	1920	5.9	5.5
Carbon	1200	11.6	68
Boron	1320	12.5	23
Graphite	1200	13.3	89

2.5 Composite materials

2.5.1 Basic characteristics

Advanced composite materials for structural applications are made by embedding in a matrix material fibers that are all aligned in a single direction. A number of polymeric materials can be used as matrix materials. Thermoset materials such as epoxy have been extensively used as matrices for composite materials. The mechanical properties of epoxy are

$$\sigma_{ult}^{tens} = 50 \text{ MPa}, \quad \sigma_{ult}^{comp} = 140 \text{ MPa}, \tag{2.43}$$

for the ultimate allowable stress in tension and compression, respectively. The modulus of elasticity, and the density of the material are

$$E = 3.5 \text{ GPa}, \quad \rho = 1300 \text{ kg/m}^3. \tag{2.44}$$

A very crude way of approximating the strength of a composite material consisting of fibers all aligned in a single direction embedded in a matrix is to use a *rule of mixture*

$$S_c = V_f S_f + V_m S_m, \tag{2.45}$$

where S_c, S_f, and S_m are the strength of the composite, fiber, and matrix materials, respectively; and V_f and V_m the volume fractions of fiber, and matrix materials, respectively. If the material contains no voids $V_f + V_m = 1$.

Consider a composite material consisting of graphite fibers ($V_f = 0.6$), embedded in an epoxy matrix ($V_m = 0.4$). The strength of the composite can be estimated using eq. (2.45) and the data of table 2.3

$$S_c = 1700 \times 0.6 + 50 \times 0.4 = 1020 + 20 = 1040 \text{ MPa}. \tag{2.46}$$

Clearly, the matrix material contributes very little to the strength of the composite.

The stiffness of the composite can also be crudely estimated from the following reasoning. Assume the various phases of the material to be perfectly bonded together,

$$\epsilon_m = \epsilon_f = \epsilon_c, \tag{2.47}$$

where ϵ_m, ϵ_f, and ϵ_c are the strains in the matrix, fiber, and composite materials, respectively. If a sheet of this material is subjected to a tensile load P, the average stress in the composite, σ_c, can be defined as follows

$$P = A_c \sigma_c = A_f \sigma_f + A_m \sigma_m, \tag{2.48}$$

where σ_f and σ_m are the stresses in the fiber and matrix materials, respectively, and A_c, A_f, and A_m are the cross-sectional areas of composite, fiber, and matrix materials, respectively. Dividing eq. (2.48) by A_c yields

$$\sigma_c = \frac{A_f}{A_c}\sigma_f + \frac{A_m}{A_c}\sigma_m = V_f \sigma_f + V_m \sigma_m. \tag{2.49}$$

If both fiber and matrix materials are assumed to be linearly elastic, isotropic materials, the following constitutive laws adequately describe their behavior

$$\sigma_f = E_f \epsilon_f, \quad \sigma_m = E_m \epsilon_m, \tag{2.50}$$

where E_f and E_m are the moduli of elasticity for the fiber and matrix materials, respectively. Similarly, the modulus of elasticity E_c for the composite is defined as

$$\sigma_c = E_c \epsilon_c. \tag{2.51}$$

Introducing eqs. (2.50) and (2.51) into eq. (2.49), and taking into account the assumed equality of the strain in the various materials, eq. (2.47), yields

$$E_c = V_f E_f + V_m E_m. \tag{2.52}$$

For the graphite epoxy material considered above, the composite modulus of elasticity can be estimated as

$$E_c = 250 \times 0.6 + 3.5 \times 0.4 = 150 + 1.4 \approx 150 \, \text{GPa}. \qquad (2.53)$$

Here again, the intrinsic stiffness of the matrix material contributes little to the stiffness of the composite.

The above discussion clearly shows that a significant fraction of the high strength and stiffness of the fibers is directly inherited by the composite. The matrix material, however, contributes little to the strength and stiffness of the composite. This observation prompts the following question: what is the role of the matrix material in a composite? The matrix is needed to keep all the fibers together, and to provide an adequate surface finish. A less obvious role of the matrix is to diffuse the stresses among the otherwise isolated fibers. This aspect is explored in the next section.

2.5.2 Stress diffusion in composites

Consider a lamina consisting of fibers all aligned in a single direction embedded in a matrix material. The lamina is subjected to a far-field stress, σ_0. If all the fibers are continuous, it is easy to see how the entire load will be carried by the fibers only, with no significant contribution of the matrix. In practical situations, however, numerous broken fibers will be present in the lamina.

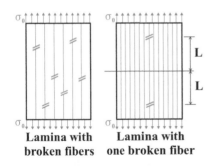

Lamina with broken fibers   Lamina with one broken fiber

Fig. 2.12. Lamina with a broken fiber.

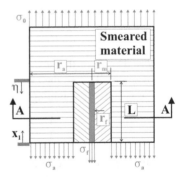

Fig. 2.13. Stress diffusion problem.

Figure 2.12 shows the geometry of the problem, including a single broken fiber of length $2L$. At the two ends of the fiber, the stress in the fiber must vanish. Nevertheless, the matrix material adjacent to the broken fiber will transfer stress from the surrounding material to the broken fiber. This stress diffusion process is a very important phenomenon because it allows *all fibers*, including broken fibers, to carry the applied load.

A simplified model of this phenomenon is depicted in the cross-section shown in fig. 2.13. It consists of a cylindrical fiber of radius r_f, surrounded by circular sleeve of matrix material of outer radius r_m, itself surrounded by a circular sleeve of composite material of outer radius r_a. The following assumptions will be made. *(1)* The matrix material carries shear stresses only. This assumption can be justified by the fact that the stiffness of the fiber is far greater than that of the matrix, and hence, the axial stress it carries is far greater than that carried by the matrix. *(2)* The axial stress in the fiber is uniformly distributed over its cross-section. *(3)* The properties of the composite material surrounding the core fiber/matrix cylinder can *smeared, i.e.*, the existence of individual fibers can be ignored, except for the specific broken fiber at the heart of the model. *(4)* The various phases of the model are perfectly bonded together.

Due to the symmetry of the problem, the displacements at $x_1 = 0$ are set to zero. Figure 2.14 shows the displacements $u_f(x_1)$, and $u_a(x_1)$ of the fiber, and composite, respectively, at section A-A in fig. 2.13. The stress-displacement relationships for the various constituents of the model are

$$\epsilon_f = \frac{du_f}{dx_1}, \tag{2.54a}$$

$$\epsilon_a = \frac{du_a}{dx_1}, \tag{2.54b}$$

$$\gamma_m = \frac{u_a - u_f}{r_m - r_f}, \tag{2.54c}$$

where ϵ_f, ϵ_a, and γ_m are the axial strains in the fiber and composite, and the shear strain in the matrix, respectively.

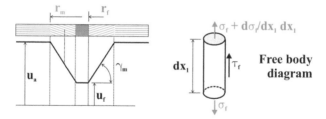

Fig. 2.14. Displacement definition and free body diagram of a differential element of fiber.

A free body diagram of a differential element of the fiber is shown in fig. 2.14. A summation of the forces along the axis of the fiber yields

$$\frac{d\sigma_f}{dx_1} + \frac{2}{r_f}\tau_m = 0, \tag{2.55}$$

where $\sigma_f(x_1)$ is the uniform axial stress in the fiber, and $\tau_m(x_1)$ the shear stress in the matrix. On the other hand, the free body diagram of the entire model depicted in fig. 2.13 yields an overall equilibrium equation

$$\sigma_a = \frac{\sigma_0}{1 - r_m^2/r_a^2} - \frac{r_f^2}{r_a^2}\frac{\sigma_f}{1 - r_m^2/r_a^2} \approx \sigma_0. \tag{2.56}$$

It is clear that the fiber has a much smaller radius than the overall composite, *i.e.*, $r_f/r_a \ll 1$, and the second term of this equation become negligible. Furthermore, $r_m/r_a \ll 1$, *i.e.*, $1 - (r_m/r_a)^2 \approx 1$.

If E_f, E_a, and G_m are the moduli of elasticity of the fiber, and composite, and the shearing modulus of the matrix, respectively, the constitutive laws for the various constituents of the model are

$$\sigma_f = E_f \epsilon_f, \tag{2.57a}$$

$$\sigma_a = E_a \epsilon_a, \tag{2.57b}$$

$$\tau_m = G_m \gamma_m. \tag{2.57c}$$

Introducing the matrix material constitutive law, eq. (2.57c), and the definition of the shear strain, eq. (2.54c), into the fiber equilibrium, eq. (2.55), yields

$$\frac{d\sigma_f}{dx_1} + \frac{2G_m}{r_f(r_m - r_f)}(u_a - u_f) = 0.$$

Taking a derivative of this equation with respect to x_1, introducing the definition of the fiber and composite strains, eqs. (2.54a) and (2.54b), respectively, and using the constitutive laws, eqs. (2.57a) and (2.57b) leads to

$$\frac{d^2\sigma_f}{dx_1^2} + \frac{2G_m}{r_f(r_m - r_f)}\left(\frac{\sigma_a}{E_a} - \frac{\sigma_f}{E_f}\right) = 0.$$

Finally, the stress in the composite, σ_a, is eliminated by means of the overall equilibrium eq. (2.56) to find the governing equation for the fiber stress

$$\frac{d^2\sigma_f}{dx_1^2} - \frac{2}{r_f(r_m - r_f)}\frac{G_m}{E_f}\sigma_f = -\frac{2}{r_f(r_m - r_f)}\frac{G_m}{E_f}\frac{E_f}{E_a}\sigma_0.$$

As shown in fig. 2.13, the non-dimensional variable $\eta = (L - x_1)/(2r_f)$ measures the distance from the fiber break divided by the fiber diameter. The governing equation for the fiber stress becomes

$$\sigma_f'' - \lambda^2 \sigma_f = -\lambda^2 \frac{E_f}{E_a}\sigma_0,$$

where the notation $(.)'$ is used to denote a derivative with respect to η; and $\lambda^2 = 8(G_m/E_f)(r_f/r_m)/(1 - r_f/r_m)$. The volume fraction of the material is $V_f = (\pi r_f^2)/(\pi r_m^2) = (r_f/r_m)^2$. Furthermore, the rule of mixture for the modulus of elasticity, eq. (2.52), yields $E_f/E_a = E_f/(V_f E_f + V_m E_m) \approx E_f/(V_f E_f) = 1/V_f$, where the fact that $E_m \ll E_f$ is taken into account. The governing equation finally can be recast as

$$\sigma_f'' - \lambda^2 \sigma_f = -\lambda^2 \frac{\sigma_0}{V_f}, \tag{2.58}$$

where

$$\lambda^2 = 8\frac{G_m}{E_f}\frac{\sqrt{V_f}}{1 - \sqrt{V_f}}. \tag{2.59}$$

The boundary conditions are $\sigma_f = 0$ at the broken end of the fiber, i.e., at $\eta = 0$. At $\eta = L/2r_f$, the symmetry of the problem requires $\sigma_f' = 0$. The solution of eq. (2.58) subjected to these boundary conditions is

$$\frac{\sigma_f}{\sigma_0} = \frac{1}{V_f} \left(1 - \frac{\cosh \lambda (L/2r_f - \eta)}{\cosh(\lambda L/2r_f)} \right) \approx \frac{1}{V_f} \left(1 - e^{-\lambda \eta} \right). \qquad (2.60)$$

To illustrate the distribution of stress near a fiber break, three material systems will be considered. Table 2.6 lists the relevant parameters for the three material systems: boron, graphite, and kevlar fibers in an epoxy matrix with a shearing modulus of 1.35 GPa.

Table 2.6. Physical properties of three material systems.

Material system	Volume fraction	E_f [GPa]	λ, Eq. (2.59)	δ/d_f, Eq. (2.62)
Boron/Epoxy	0.5	400	0.255	11
Graphite/Epoxy	0.6	250	0.385	7.3
Kevlar/Epoxy	0.6	130	0.534	5.3

The fiber stress at a large distance from the fiber break can be obtained from eq. (2.49): $\sigma_0 = V_f \sigma_{f\infty} + (1 - V_f)\sigma_{m\infty} \approx V_f \sigma_{f\infty}$. The stress distribution, eq. (2.60), then becomes

$$\frac{\sigma_f}{\sigma_{f\infty}} = 1 - e^{-\lambda \eta}, \qquad (2.61)$$

where the notation $(.)_\infty$ is used to denote the value of the corresponding quantity at a large distance from the fiber break. The non-dimensional parameter, λ, characterizes the fiber axial stress distribution near the fiber break, which is plotted in fig. 2.15 for the three material systems. At $\eta = 0$, which corresponds to the fiber break, the fiber axial stress vanishes. The fiber axial stress grows rapidly to its far field value $\sigma_{f\infty}$.

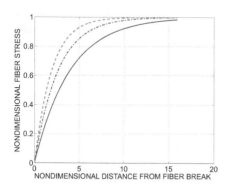

Fig. 2.15. Distribution of fiber axial stress near a fiber break for three material systems.

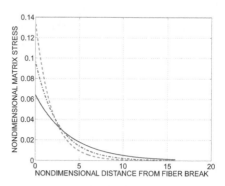

Fig. 2.16. Distribution of matrix shear stress near a fiber break for three material systems.

It is convenient to define the fiber *ineffective length* δ as the distance it takes for the fiber stress to reach 95% of its far field value, *i.e.*, $0.95 = 1 - \exp(-\lambda \delta/d_f)$, where d_f is the fiber diameter. Solving this equation yields the ineffective length as

$$\frac{\delta}{d_f} \approx \left[\frac{E_f}{G_m} \frac{1 - \sqrt{V_f}}{\sqrt{V_f}} \right]^{1/2}. \tag{2.62}$$

The ineffective length can be thought of as the length of fiber, near a fiber break, that does not carry axial stress at full capacity. Table 2.6 lists the ineffective length for the three material systems. It appears that 5.3 fiber diameters away from a break, the Kevlar fiber is already carrying 95% of its far field stress. This means that the matrix material transfers the load from the surrounding material to the broken fiber *very rapidly*. This mechanism is called the *shear lag* mechanism because the shear stress in the matrix is effectively transferring the load to the fiber. The shear stress in the matrix can be readily evaluated from the fiber equilibrium equation (2.55) as

$$\frac{\tau_m}{\sigma_{f\infty}} = \frac{\lambda}{4} e^{-\lambda\eta}. \tag{2.63}$$

Figure 2.16 shows the distribution of shear stress in the matrix near a fiber break, for the three material systems. The shear stress is maximum near the fiber break, then decays very rapidly.

An important role of the matrix material now becomes apparent in light of the above analysis. Near a fiber break, the matrix material transfers stresses from the surrounding material to the broken fiber. The shearing of the matrix near the fiber break is the mechanism that allows this stress transfer to occur. This mechanism is very efficient: for the material systems described above, the broken fiber is fully loaded within about ten fiber diameters from the fiber break. The zone affected by the fiber break is about 2δ in length (δ on each side of the break). For a graphite fiber with a diameter of 10 microns, the zone affected by a fiber break is therefore only about 200 microns in length.

Another way of looking at this fact is to say that a fiber is continuous or *infinitely long*, if its total length is much larger, say 100 times larger, than its ineffective length. Hence, a 10 micron diameter graphite fiber can be considered continuous or infinitely long when its length is greater than $100 \times 100 \, 10^{-6} = 10$ mm. From a load carrying stand point, a 10 millimeter long graphite fiber can be considered continuous or infinitely long.

2.6 Constitutive laws for anisotropic materials

Section 2.1 focuses on the constitutive behavior of isotropic materials. Due to the growing importance of composite materials, the linearly elastic behavior of anisotropic materials will be addressed here. The physical properties of anisotropic materials are directional, *i.e.*, the physical response of the material depends on the direction in which it is acted upon.

Consider, as an example, the stiffness of the unidirectional composite material described in section 2.5: in the fiber direction the stiffness of the composite is dominated by the high stiffness of the fiber, see eq. (2.52). In the direction transverse to the fiber, however, the stiffness of the composite is dominated by that of the matrix material, which is far small than that of the fiber. This contrasts with isotropic materials for which the mechanical response is identical in all directions.

The straining of the material will be measured by the engineering strain components which are stored in array ϵ, defined by eq. (2.11a). Similarly, the state of stress in the material is measured by the engineering stress components stored in array σ, defined by eq. (2.11b). A linearly elastic, anisotropic material is characterized a linear relationship between the stress and strain measures,

$$\underline{\sigma} = \underline{\underline{C}}\,\underline{\epsilon}; \quad \underline{\epsilon} = \underline{\underline{S}}\,\underline{\sigma}, \tag{2.64}$$

where $\underline{\underline{C}}$ is the 6×6 *stiffness matrix* and $\underline{\underline{S}}$ the 6×6 *compliance matrix*. These two matrices are the inverse of each other, *i.e.*,

$$\underline{\underline{S}} = \underline{\underline{C}}^{-1}. \tag{2.65}$$

The strain energy, A, stored in a differential element of the material is

$$A = \frac{1}{2}\,\underline{\epsilon}^T\underline{\sigma} = \frac{1}{2}\,\underline{\epsilon}^T\underline{\underline{C}}\,\underline{\epsilon} = \frac{1}{2}\,\underline{\sigma}^T\underline{\underline{S}}\,\underline{\sigma}. \tag{2.66}$$

The stored strain energy is a positive quantity for whatever deformation or stress state the material is subjected to. This implies that both stiffness and compliance matrices are symmetric and definite positive.

In general, the 6×6 stiffness matrix has $6 \times 6 = 36$ independent coefficients. The symmetry requirement, however, reduces the number of independent coefficients to 21. The stress-strain relationship, eq. (2.64), written in expanded form is

$$
\begin{Bmatrix} \sigma_1 \\ \sigma_2 \\ \sigma_3 \\ \tau_{23} \\ \tau_{13} \\ \tau_{12} \end{Bmatrix} =
\begin{bmatrix}
C_{11} & C_{12} & C_{13} & C_{14} & C_{15} & C_{16} \\
 & C_{22} & C_{23} & C_{24} & C_{25} & C_{26} \\
 & & C_{33} & C_{34} & C_{35} & C_{36} \\
 & & & C_{44} & C_{45} & C_{46} \\
 & & & & C_{55} & C_{56} \\
sym & & & & & C_{66}
\end{bmatrix}
\begin{Bmatrix} \epsilon_1 \\ \epsilon_2 \\ \epsilon_3 \\ \gamma_{23} \\ \gamma_{13} \\ \gamma_{12} \end{Bmatrix}, \tag{2.67}
$$

where the entries in the lower triangular part of the stiffness matrix are equal to the corresponding upper triangular entries. The 21 constants, C_{ij}, characterize the behavior of the material. Each constant must be determined experimentally. A material characterized by relationship (2.67) is called an *anisotropic* or *triclinic* material.

Materials sometimes possess a plane of symmetry. Let plane $(\bar{\imath}_1, \bar{\imath}_2)$ be a plane of symmetry of the material. The stress-strain relationship reduces to

$$
\begin{Bmatrix} \sigma_1 \\ \sigma_2 \\ \sigma_3 \\ \tau_{23} \\ \tau_{13} \\ \tau_{12} \end{Bmatrix} =
\begin{bmatrix}
C_{11} & C_{12} & C_{13} & 0 & 0 & C_{16} \\
 & C_{22} & C_{23} & 0 & 0 & C_{26} \\
 & & C_{33} & 0 & 0 & C_{36} \\
 & & & C_{44} & C_{45} & 0 \\
 & & & & C_{55} & 0 \\
sym & & & & & C_{66}
\end{bmatrix}
\begin{Bmatrix} \epsilon_1 \\ \epsilon_2 \\ \epsilon_3 \\ \gamma_{23} \\ \gamma_{13} \\ \gamma_{12} \end{Bmatrix}. \tag{2.68}
$$

The stiffness coefficient, C_{14}, must vanish, because, if it does not vanish, an axial strain, ϵ_1, would give rise to a shear stress, τ_{23}. The presence of a shear stress, τ_{23}, however, would violate the symmetry of the response, which is a natural consequence of the material symmetry. A systematic application of this symmetry argument shows that the 8 coefficients indicated as "0" in eq. (2.68) must vanish, leaving $21 - 8 = 13$ independent coefficients. This type of material is called a *monoclinic* material.

Some materials show a higher level of symmetry characterized by two mutually orthogonal planes of symmetry; for instance, let planes $(\bar{\imath}_1, \bar{\imath}_2)$ and $(\bar{\imath}_2, \bar{\imath}_3)$ be planes of symmetry. The stress-strain relationships then reduces to

$$
\begin{Bmatrix} \sigma_1 \\ \sigma_2 \\ \sigma_3 \\ \tau_{23} \\ \tau_{13} \\ \tau_{12} \end{Bmatrix} = \begin{bmatrix} C_{11} & C_{12} & C_{13} & 0 & 0 & 0 \\ & C_{22} & C_{23} & 0 & 0 & 0 \\ & & C_{33} & 0 & 0 & 0 \\ & & & C_{44} & 0 & 0 \\ & & & & C_{55} & 0 \\ sym & & & & & C_{66} \end{bmatrix} \begin{Bmatrix} \epsilon_1 \\ \epsilon_2 \\ \epsilon_3 \\ \gamma_{23} \\ \gamma_{13} \\ \gamma_{12} \end{Bmatrix}. \tag{2.69}
$$

Here again symmetry arguments can be used to prove that the 12 coefficients indicated as "0" in the above matrix must vanish, leaving $21 - 12 = 9$ independent coefficients. This type of material is called an *orthotropic* material.

A case of particular importance to the study of laminated composite materials is that of materials presenting two orthogonal planes of symmetry, and one plane of isotropy. Let planes $(\bar{\imath}_1, \bar{\imath}_2)$ and $(\bar{\imath}_2, \bar{\imath}_3)$ be the mutually orthogonal planes of symmetry, and let the material be isotropic in plane $(\bar{\imath}_2, \bar{\imath}_3)$. This means, for instance, that the coefficients C_{12} and C_{13} should be identical due to the isotropic response of the material in plane $(\bar{\imath}_2, \bar{\imath}_3)$. The stress-strain relationships now reduce to

$$
\begin{Bmatrix} \sigma_1 \\ \sigma_2 \\ \sigma_3 \\ \tau_{23} \\ \tau_{13} \\ \tau_{12} \end{Bmatrix} = \begin{bmatrix} C_{11} & C_{12} & C_{12} & 0 & 0 & 0 \\ & C_{22} & C_{23} & 0 & 0 & 0 \\ & & C_{22} & 0 & 0 & 0 \\ & & & \frac{C_{22}-C_{23}}{2} & 0 & 0 \\ & & & & C_{55} & 0 \\ sym & & & & & C_{55} \end{bmatrix} \begin{Bmatrix} \epsilon_1 \\ \epsilon_2 \\ \epsilon_3 \\ \gamma_{23} \\ \gamma_{13} \\ \gamma_{12} \end{Bmatrix}. \tag{2.70}
$$

Only five constants remain for this material called *transversely isotropic*.

Finally, an *isotropic* material is characterized by an identical response in all directions, leading to the following stress-strain relationship

$$
\begin{Bmatrix} \sigma_1 \\ \sigma_2 \\ \sigma_3 \\ \tau_{23} \\ \tau_{13} \\ \tau_{12} \end{Bmatrix} = \begin{bmatrix} C_{11} & C_{12} & C_{12} & 0 & 0 & 0 \\ & C_{11} & C_{12} & 0 & 0 & 0 \\ & & C_{11} & 0 & 0 & 0 \\ & & & \frac{C_{11}-C_{12}}{2} & 0 & 0 \\ & & & & \frac{C_{11}-C_{12}}{2} & 0 \\ sym & & & & & \frac{C_{11}-C_{12}}{2} \end{bmatrix} \begin{Bmatrix} \epsilon_1 \\ \epsilon_2 \\ \epsilon_3 \\ \gamma_{23} \\ \gamma_{13} \\ \gamma_{12} \end{Bmatrix}. \tag{2.71}
$$

Two independent constants only are left for this type of material. Relations (2.67) to (2.71) give the structure of the stiffness matrix, $\underline{\underline{C}}$, for various types of materials. The compliance matrix, $\underline{\underline{S}}$ can be obtained by inversion, see eq. (2.65).

While the actual structure of the stiffness matrix is obtained based on energy and symmetry arguments, the physical interpretation of the various terms appearing in this matrix is not clear. For example, isotropic materials are shown to be characterized by two independent coefficients, C_{11} and C_{12}; in practice, isotropic materials are generally characterized by their Young's modulus and Poisson's ratio, which have a clear physical interpretation. These constants are called the *engineering constants* because they can be readily measured experimentally. For the various types of materials, the stiffness and compliance terms can be expressed in terms of the engineering constants. The following section discusses the experimental determination and the physical interpretation of the engineering constants for a lamina made of unidirectional fibers embedded in a matrix material.

2.6.1 Constitutive laws for a lamina in the fiber aligned triad

Consider a thin sheet of composite material made of unidirectional fibers embedded in a matrix. Let axis $\bar{\imath}_1^*$ be oriented along the fiber direction, $\bar{\imath}_2^*$ in the transverse direction, and $\bar{\imath}_3^*$ is perpendicular to the plane of the thin sheet. Triad $\mathcal{I}^* = (\bar{\imath}_1^*, \bar{\imath}_2^*, \bar{\imath}_3^*)$ is called the fiber aligned triad and the superscript $(\cdot)^*$ will be used to indicate quantities measured in this triad.

If the diameter of the fiber is small compared to the thickness of the sheet, the material can be assumed to be a homogeneous, transversely isotropic material. The existence of individual fibers can be ignored: fibers and matrix materials are *smeared* into an equivalent, homogeneous, anisotropic material. For a linearly elastic, transversely isotropic material the constitutive laws reduce to eq. (2.70).

It will be assumed that the thin sheet of material is in a plane stress state, see section 1.3, *i.e.*, $\sigma_3^* \approx \tau_{13}^* \approx \tau_{23}^* \approx 0$. The constitutive laws expressed in compliance form are written in the following form

$$\left\{ \begin{array}{c} \epsilon_1^* \\ \epsilon_2^* \\ \gamma_{12}^* \end{array} \right\} = \left[\begin{array}{ccc} 1/E_1^* & -\nu_{21}^*/E_2^* & 0 \\ -\nu_{12}^*/E_1^* & 1/E_2^* & 0 \\ 0 & 0 & 1/G_{12}^* \end{array} \right] \left\{ \begin{array}{c} \sigma_1^* \\ \sigma_2^* \\ \tau_{12}^* \end{array} \right\}. \tag{2.72}$$

The compliance matrix is expressed in terms of four constants, E_1^*, E_2^*, ν_{12}^*, and G_{12}^*, which are called the *engineering constants*. Note that the compliance matrix must be symmetric, thus $\nu_{12}^*/E_1^* = \nu_{21}^*/E_2^*$. This means that although five constants appear in the expression of the compliance matrix, one of them say ν_{21}^*, can be computed from the other, and hence, is not an independent quantity.

The engineering constants can be readily measured experimentally. Consider a simple test where the composite is subjected to a known stress in the fiber direction only, σ_1^*, *i.e.*, $\sigma_2^* = \tau_{12}^* = 0$, as depicted in the left part of fig. 2.17. The first equation of (2.72) now reduces to $\epsilon_1^* = \sigma_1^*/E_1^*$. The strain in the fiber direction, ϵ_1^*, can be measured as a function of the applied stress, σ_1^*, by means of a strain gauge, and the modulus of elasticity is then computed as $E_1^* = \sigma_1^*/\epsilon_1^*$. Clearly, E_1^* is the modulus of elasticity of the material in the fiber direction.

The second equation of (2.72) becomes $\epsilon_2^* = -\nu_{12}\sigma_1^*/E_1^*$. The strain in the direction transverse to the fiber, ϵ_2^*, can also be measured by means of a strain gauge. Poisson's ratio now becomes $\nu_{12}^* = -E_1^*\epsilon_2^*/\sigma_1^*$.

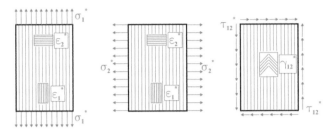

Fig. 2.17. Three simple tests for the determination of the engineering constants.

Consider next a second test where the composite material is subjected to a known stress in the direction transverse to the fiber, σ_2^*, i.e., $\sigma_1^* = \tau_{12}^* = 0$, as depicted in the middle portion of fig. 2.17. Using the same approach as before, a measurement of the transverse strain, ϵ_2^*, as a function of the transverse stress, σ_2^*, will then yield E_2^*, the modulus of elasticity of the material in the direction transverse to the fiber. An additional measurement of the strain in the fiber direction, ϵ_1^*, will yield ν_{21}^*. The symmetry of the compliance matrix can be verified experimentally by checking that the various measured quantities satisfy the symmetry condition $\nu_{12}^*/E_1^* = \nu_{21}^*/E_2^*$, within the expected experimental errors.

Finally, in the last test, the composite material is subjected to a known shear stress, τ_{12}^*, only, i.e., $\sigma_1^* = \sigma_2^* = 0$, as depicted in right portion of fig. 2.17. The last equation of (2.72) reduces to $\gamma_{12}^* = \tau_{12}^*/G_{12}^*$. A measurement of the shear strain, γ_{12}^*, then allows the evaluation of the shearing modulus, $G_{12}^* = \tau_{12}^*/\gamma_{12}^*$.

The stiffness matrix is obtained by inverting eq. (2.72) to find

$$\left\{\begin{array}{c} \sigma_1^* \\ \sigma_2^* \\ \tau_{12}^* \end{array}\right\} = \left[\begin{array}{ccc} \dfrac{E_1^*}{1 - \nu_{12}^{*2}E_2^*/E_1^*} & \dfrac{\nu_{12}^*E_2^*}{1 - \nu_{12}^{*2}E_2^*/E_1^*} & 0 \\ \dfrac{\nu_{12}^*E_2^*}{1 - \nu_{12}^{*2}E_2^*/E_1^*} & \dfrac{E_2^*}{1 - \nu_{12}^{*2}E_2^*/E_1^*} & 0 \\ 0 & 0 & G_{12}^* \end{array}\right] \left\{\begin{array}{c} \epsilon_1^* \\ \epsilon_2^* \\ \gamma_{12}^* \end{array}\right\}. \tag{2.73}$$

To simplify the writing of the above relationships, the following stress and strain arrays are introduced

$$\underline{\sigma}^* = \left\{\sigma_1^*, \sigma_2^*, \tau_{12}^*\right\}^T, \quad \underline{\varepsilon}^* = \left\{\epsilon_1^*, \epsilon_2^*, \gamma_{12}^*\right\}^T. \tag{2.74}$$

The constitutive laws, eqs. (2.73) and (2.72), are written in compact form as

$$\underline{\sigma}^* = \underline{\underline{C}}^*\underline{\varepsilon}^*, \quad \text{and} \quad \underline{\varepsilon}^* = \underline{\underline{S}}^*\underline{\sigma}^*, \tag{2.75}$$

respectively. The stiffness and compliance matrices are then

$$\underline{\underline{C}}^* = \begin{bmatrix} \dfrac{E_1^*}{1-\nu_{12}^{*2}E_2^*/E_1^*} & \dfrac{\nu_{12}^*E_2^*}{1-\nu_{12}^{*2}E_2^*/E_1^*} & 0 \\ \dfrac{\nu_{12}^*E_2^*}{1-\nu_{12}^{*2}E_2^*/E_1^*} & \dfrac{E_2^*}{1-\nu_{12}^{*2}E_2^*/E_1^*} & 0 \\ 0 & 0 & G_{12}^* \end{bmatrix} = \begin{bmatrix} C_{11}^* & C_{12}^* & 0 \\ C_{12}^* & C_{22}^* & 0 \\ 0 & 0 & C_{66}^* \end{bmatrix}, \qquad (2.76)$$

and

$$\underline{\underline{S}}^* = \begin{bmatrix} 1/E_1^* & -\nu_{21}^*/E_2^* & 0 \\ -\nu_{12}^*/E_1^* & 1/E_2^* & 0 \\ 0 & 0 & 1/G_{12}^* \end{bmatrix} = \begin{bmatrix} S_{11}^* & S_{12}^* & 0 \\ S_{12}^* & S_{22}^* & 0 \\ 0 & 0 & S_{66}^* \end{bmatrix}, \qquad (2.77)$$

respectively.

The engineering constant for lamina made of a few different type of materials are listed in table 2.7. This table lists the volume fraction V_f, engineering constants E_1^*, E_2^*, ν_{12}^*, and G_{12}^*, as well as the density of the various lamina.

Table 2.7. Engineering constants for lamina made of different materials.

Material system	V_f	E_1^* [GPa]	E_2^* [GPa]	ν_{12}^*	G_{12}^* [GPa]	density [kg/m³]
Graphite/Epoxy (T300/5208)	0.70	180.	10.	0.28	7.0	1600
Graphite/Epoxy (AS/3501)	0.66	138.	9.	0.30	7.0	1600
Boron/Epoxy (T300/5208)	0.50	204.	18.	0.23	5.6	2000
Scotchply (1002)	0.45	39.	8.	0.26	4.0	1800
Kevlar 49	0.60	76.	5.5	0.34	2.3	1460

2.6.2 Constitutive laws for a lamina in an arbitrary triad

In the previous section, the constitutive laws for a lamina made of a transversely isotropic material are discussed. The stresses and strains are measured in the fiber aligned triad, $\mathcal{I}^*$. In many cases, however, the constitutive laws for the lamina are required for a direction that might not coincide with that of the fibers. Figure 2.18 shows a transversely isotropic lamina with a reference triad, $\mathcal{I} = (\bar{\imath}_1, \bar{\imath}_2, \bar{\imath}_3)$ and the fiber

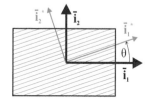

Fig. 2.18. Definition of two axis systems for a lamina.

aligned triad, $\mathcal{I}^*$. The fibers run at an angle θ with respect to a reference triad; angle θ is counted positive in the counterclockwise direction. Let $\underline{\sigma}$, and $\underline{\varepsilon}$ be the arrays of in-plane stresses and strains, respectively, measured in the reference triad $\mathcal{I}$. The lamina constitutive laws, measured in triad $\mathcal{I}$, now become

$$\underline{\sigma} = \underline{\underline{C}}\,\underline{\varepsilon}. \qquad (2.78)$$

Stiffness matrix $\underline{\underline{C}}$ could be obtained experimentally by performing a series a tests on the lamina, applying a stress along axis $\bar{\imath}_1$ first, then along axis $\bar{\imath}_2$, as described in the previous section. Although conceptually feasible, this approach is not practical because a series of tests would have to be performed each time the constitutive laws are desired for a specific angle θ. A better approach would be to relate the stiffness properties at an angle θ to those measured in the fiber direction. This can be readily achieved with the help of the formulae for computing stresses and strains in a rotated axis system.

Rotation of the stiffness matrix

All the elements required to relate the constitutive laws in the two triads are now in place. The constitutive laws for a lamina expressed in the fiber aligned triad, $\underline{\sigma}^* = \underline{\underline{C}}^* \underline{\varepsilon}^*$, are the starting point of the development. Introducing the rotation formulae for stresses, eq. (1.47), and strain, eq. (1.91), yields

$$
\begin{bmatrix} m^2 & n^2 & 2mn \\ n^2 & m^2 & -2mn \\ -mn & mn & m^2 - n^2 \end{bmatrix} \begin{Bmatrix} \sigma_1 \\ \sigma_2 \\ \tau_{12} \end{Bmatrix} = \underline{\underline{C}}^* \begin{bmatrix} m^2 & n^2 & mn \\ n^2 & m^2 & -mn \\ -2mn & 2mn & m^2 - n^2 \end{bmatrix} \begin{Bmatrix} \epsilon_1 \\ \epsilon_2 \\ \gamma_{12} \end{Bmatrix},
$$

where $m = \cos\theta$ and $n = \sin\theta$. Multiplying from the left by the inverse of the rotation matrix for stresses results in

$$
\begin{Bmatrix} \sigma_1 \\ \sigma_2 \\ \tau_{12} \end{Bmatrix} = \begin{bmatrix} m^2 & n^2 & -2mn \\ n^2 & m^2 & 2mn \\ mn & -mn & m^2 - n^2 \end{bmatrix} \underline{\underline{C}}^* \begin{bmatrix} m^2 & n^2 & mn \\ n^2 & m^2 & -mn \\ -2mn & 2mn & m^2 - n^2 \end{bmatrix} \begin{Bmatrix} \epsilon_1 \\ \epsilon_2 \\ \gamma_{12} \end{Bmatrix}.
$$

Comparing this relationship to (2.78) then leads to

$$
\underline{\underline{C}} = \begin{bmatrix} m^2 & n^2 & -2mn \\ n^2 & m^2 & 2mn \\ mn & -mn & m^2 - n^2 \end{bmatrix} \underline{\underline{C}}^* \begin{bmatrix} m^2 & n^2 & 2mn \\ n^2 & m^2 & -2mn \\ -mn & mn & m^2 - n^2 \end{bmatrix}. \tag{2.79}
$$

Performing this triple matrix multiplication yields the various terms of the stiffness matrix

$$
C = \begin{bmatrix} C_{11} & C_{12} & C_{16} \\ C_{12} & C_{22} & C_{26} \\ C_{16} & C_{26} & C_{66} \end{bmatrix}, \tag{2.80}
$$

where $C_{11} = m^4 C_{11}^* + n^4 C_{22}^* + 2m^2 n^2 C_{12}^* + 4m^2 n^2 C_{66}^*$ and similar expressions hold for the other entries. In view of the complexity of this result that involves powers of trigonometric functions, an alternative expression can be derived based on well-known trigonometric identities to find

$$
\begin{aligned}
C_{11} &= \alpha_1 + \alpha_2 + & \alpha_3 \cos 2\theta & + \alpha_4 \cos 4\theta, \\
C_{22} &= \alpha_1 + \alpha_2 - & \alpha_3 \cos 2\theta & + \alpha_4 \cos 4\theta, \\
C_{12} &= \alpha_1 - \alpha_2 & & - \alpha_4 \cos 4\theta, \\
C_{66} &= \phantom{\alpha_1 -{}} \alpha_2 & & - \alpha_4 \cos 4\theta, \\
C_{16} &= & (\alpha_3/2) \sin 2\theta & + \alpha_4 \sin 4\theta, \\
C_{26} &= & (\alpha_3/2) \sin 2\theta & - \alpha_4 \sin 4\theta,
\end{aligned} \tag{2.81}
$$

where the four material invariants, α_1, α_2, α_3, and α_4, are defined as

$$\alpha_1 = \frac{E_1^* + E_2^* + 2\nu_{12}^* E_2^*}{4\alpha_0}, \qquad \alpha_2 = \frac{E_1^* + E_2^* - 2\nu_{12}^* E_2^*}{8\alpha_0} + \frac{G_{12}^*}{2}, \qquad (2.82a)$$

$$\alpha_3 = \frac{E_1^* - E_2^*}{2\alpha_0}, \qquad \alpha_4 = \frac{E_1^* + E_2^* - 2\nu_{12}^* E_2^*}{8\alpha_0} - \frac{G_{12}^*}{2}, \qquad (2.82b)$$

and where $\alpha_0 = 1 - \nu_{12}^{*2} E_2^* / E_1^*$.

This relationship is written in a more compact manner by defining the following matrix

$$\underline{\underline{\chi}}(\theta) = \begin{bmatrix} 1 & 1 & \cos 2\theta & \cos 4\theta \\ 1 & 1 & -\cos 2\theta & \cos 4\theta \\ 1 & -1 & 0 & -\cos 4\theta \\ 0 & 1 & 0 & -\cos 4\theta \\ 0 & 0 & \frac{1}{2}\sin 2\theta & \sin 4\theta \\ 0 & 0 & \frac{1}{2}\sin 2\theta & -\sin 4\theta \end{bmatrix}, \qquad (2.83)$$

which is a function of the lamina orientation angle only. Next, the array of stiffness component is defined as

$$\underline{C} = \left\{ C_{11}, C_{22}, C_{12}, C_{66}, C_{16}, C_{26} \right\}^T, \qquad (2.84)$$

and finally, the array of material invariants

$$\underline{\alpha} = \left\{ \alpha_1, \alpha_2, \alpha_3, \alpha_4 \right\}^T. \qquad (2.85)$$

With these notations, the entries of the stiffness matrix for a lamina with a fiber orientation angle θ can be written as

$$\underline{C}(\theta) = \underline{\underline{\chi}}(\theta)\underline{\alpha}. \qquad (2.86)$$

In summary, the stiffness matrix for a lamina can be obtained as follows.

1. Determine the engineering constants, E_1^*, E_2^*, ν_{12}^*, and G_{12}^* by performing a series of test on the lamina, as discussed in section 2.6.1.
2. Compute the stiffness matrix, $\underline{\underline{C}}^*$, in the fiber aligned triad, see eq. (2.76).
3. Compute the material invariants, α_1, α_2, α_3, and α_4, with the help of eqs.(2.82). Set up the array of material invariants defined by eq. (2.85).
4. Set up matrix $\underline{\underline{\chi}}(\theta)$ given by eq. (2.83) for the desired fiber orientation angle, θ, and evaluate the components of the stiffness matrix using eq. (2.86).

The material invariants for lamina made of a few different type of materials are listed in table 2.8. This table lists the material invariants computed from eqs. (2.82) based on the data of table 2.7.

Figure 2.19 shows the stiffness components, C_{11} and C_{22}, as a function of the lamina angle, θ, for the Graphite/Epoxy T300/5208 material system. Note the rapid decline of the stiffness coefficient, C_{11}, when the lamina angle moves away from 0 degrees. This sharp decline is due to the high directionality of the lamina stiffness

Table 2.8. Material invariants for lamina made of different materials.

Material system	α_1 [GPa]	α_2 [GPa]	α_3 [GPa]	α_4 [GPa]
Graphite/Epoxy (T300/5208)	49.11	26.65	85.37	19.65
Graphite/Epoxy (AS/3501)	38.32	21.30	64.88	14.30
Boron/Epoxy (T300/5208)	57.84	29.64	93.44	24.04
Scotchply (1002)	12.97	7.43	15.72	3.43
Kevlar 49	21.49	10.95	35.55	8.65

properties. The shearing stiffness component, C_{66}, shown in fig. 2.20, drastically increases when the lamina angle is 45 degrees. This can be explained as follows: a state of pure shear, see section 1.3.5, is equivalent to stresses in tension and compression acting at 45 and 135 degree angles, respectively. Theses stresses are now aligned with the fiber direction, which presents very high stiffness.

The coupling stiffness terms, C_{16} and C_{26}, do not vanish. These terms express a coupling between extension and shearing of the lamina. In contrast, the stiffness matrix, $\underline{C}^*$, expressed in the fiber aligned triad, has vanishing terms in the corresponding entries. Indeed, when the loading is applied along the fiber direction, which is the intersection of two planes of symmetry, the response of the system must be symmetric, precluding extension-shear coupling. When the loading is no longer aligned with the intersection of the two planes of symmetry, a coupled response of the lamina is intuitively expected. Figure 2.21 shows the stiffness components, C_{16} and C_{26}, as a function of lamina angle θ.

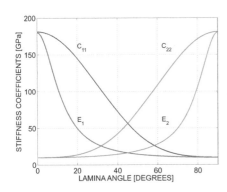

Fig. 2.19. Variation of the stiffness coefficients, C_{11} and C_{22}, and the engineering constants, E_1 and E_2, as a function of θ.

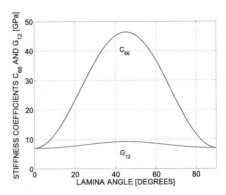

Fig. 2.20. Variation of the stiffness coefficient, C_{66}, and engineering constant, G_{12} as a function of θ.

Rotation of the compliance matrix

The lamina constitutive laws can be expressed in the stiffness form, eq. (2.78), or in the compliance form as

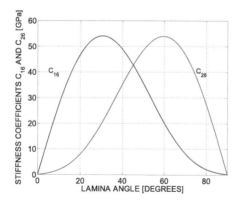

Fig. 2.21. Variation of the coupling stiffness coefficients, C_{16} and C_{26}, with the lamina angle θ.

$$\underline{\varepsilon} = \underline{\underline{S}}\,\underline{\sigma}, \tag{2.87}$$

where $\underline{\underline{S}} = \underline{\underline{C}}^{-1}$ is the compliance matrix measured in the arbitrary triad. Of course, the compliance matrix can be obtained by inverting the stiffness matrix, as indicated by eq. (2.65), but a direct determination is also possible. Introducing the stress rotation formula, eq. (1.47), and strain rotation formula, eq. (1.91), into the constitutive laws, eq. (2.75), and identifying the result with eq. (2.87) yields

$$\underline{\underline{S}} = \begin{bmatrix} m^2 & n^2 & -mn \\ n^2 & m^2 & mn \\ 2mn & -2mn & m^2 - n^2 \end{bmatrix} \underline{\underline{S}}^* \begin{bmatrix} m^2 & n^2 & 2mn \\ n^2 & m^2 & -2mn \\ -mn & mn & m^2 - n^2 \end{bmatrix}. \tag{2.88}$$

Performing this triple matrix multiplication yields the terms of the compliance matrix

$$\underline{\underline{S}} = \begin{bmatrix} S_{11} & S_{12} & S_{16} \\ S_{12} & S_{22} & S_{26} \\ S_{16} & S_{26} & S_{66} \end{bmatrix} = \begin{bmatrix} 1/E_1 & -\nu_{21}/E_2 & \nu_{61}/G_{12} \\ -\nu_{12}/E_1 & 1/E_2 & \nu_{62}/G_{12} \\ \nu_{16}/E_1 & \nu_{26}/E_2 & 1/G_{12} \end{bmatrix}, \tag{2.89}$$

where E_1, E_2, ν_{12}, G_{12}, ν_{16}, and ν_{26}, define the engineering constants in the arbitrary triad. Due to the symmetry of the compliance matrix, the following relationships hold $\nu_{12}/E_1 = \nu_{21}/E_2, \nu_{16}/E_1 = \nu_{61}/G_{12}$, and $\nu_{26}/E_2 = \nu_{62}/G_{12}$. The first entry of the compliance matrix is $S_{11} = m^4 S_{11}^* + n^4 S_{22}^* + 2m^2 n^2 S_{12}^* + m^2 n^2 S_{66}^*$, and similar expressions can be obtained for the other entries. In view of the complexity of this result that involves powers of trigonometric functions, an alternative expression is derived based on well-known trigonometric identities to find

$$\begin{aligned} S_{11} &= \beta_1 + \beta_2 + \beta_3 \cos 2\theta + \beta_4 \cos 4\theta, \\ S_{22} &= \beta_1 + \beta_2 - \beta_3 \cos 2\theta + \beta_4 \cos 4\theta, \\ S_{12} &= \beta_1 - \beta_2 \qquad\qquad - \beta_4 \cos 4\theta, \\ S_{66} &= \quad 4\beta_2 \qquad\qquad - 4\beta_4 \cos 4\theta, \\ S_{16} &= \qquad\qquad \beta_3 \sin 2\theta + 2\beta_4 \sin 4\theta, \\ S_{26} &= \qquad\qquad \beta_3 \sin 2\theta - 2\beta_4 \sin 4\theta, \end{aligned} \tag{2.90}$$

where the material invariants, β_1, β_2, β_3, and β_4, are defined as

$$\beta_1 = \frac{1}{4}\left(\frac{1}{E_1^*} + \frac{1}{E_2^*} - \frac{2\nu_{12}^*}{E_1^*}\right), \quad \beta_2 = \frac{1}{8}\left(\frac{1}{E_1^*} + \frac{1}{E_2^*} + \frac{2\nu_{12}^*}{E_1^*}\right) + \frac{1}{8G_{12}^*}, \quad (2.91a)$$

$$\beta_3 = \frac{1}{2}\left(\frac{1}{E_1^*} - \frac{1}{E_2^*}\right), \qquad \beta_4 = \frac{1}{8}\left(\frac{1}{E_1^*} + \frac{1}{E_2^*} + \frac{2\nu_{12}^*}{E_1^*}\right) - \frac{1}{8G_{12}^*}. \quad (2.91b)$$

Explicit expression for the engineering constants can be obtained from eq. (2.89)

$$E_1 = 1/\left(\beta_1 + \beta_2 + \beta_3 \cos 2\theta + \beta_4 \cos 4\theta\right), \qquad (2.92a)$$

$$E_2 = 1/\left(\beta_1 + \beta_2 - \beta_3 \cos 2\theta + \beta_4 \cos 4\theta\right), \qquad (2.92b)$$

$$\nu_{12} = -\left(\beta_1 - \beta_2 - \beta_4 \cos 4\theta\right)/\left(\beta_1 + \beta_2 + \beta_3 \cos 2\theta + \beta_4 \cos 4\theta\right), \qquad (2.92c)$$

$$G_{12} = 1/\left(4\beta_2 - 4\beta_4 \cos 4\theta\right), \qquad (2.92d)$$

$$\nu_{16} = \left(\beta_3 \sin 2\theta + 2\beta_4 \sin 4\theta\right)/\left(\beta_1 + \beta_2 + \beta_3 \cos 2\theta + \beta_4 \cos 4\theta\right), \qquad (2.92e)$$

$$\nu_{26} = \left(\beta_3 \sin 2\theta - 2\beta_4 \sin 4\theta\right)/\left(\beta_1 + \beta_2 + \beta_3 \cos 2\theta + \beta_4 \cos 4\theta\right). \qquad (2.92f)$$

These engineering constants can also be measured experimentally by performing the various tests depicted in fig. 2.22. The tests are similar to those discussed in section 2.6.1, except for the fact that stresses are now applied at an angle θ with respect to the fibers.

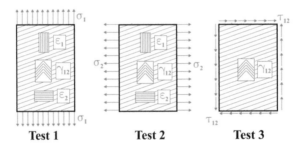

Fig. 2.22. Three simple tests for the determination of the engineering constants.

Figure 2.19 shows the variation of the modulus of elasticity, E_1, as a function of the lamina angle θ. Note the precipitous drop in the modulus of elasticity when the lamina angle moves away from 0 degrees. This drop is much more pronounced than that of the stiffness coefficient, C_{11}.

It is important to understand the difference between the stiffness coefficient, C_{11}, and the engineering constant, E_1. Mathematically, these two quantities clearly are different: $E_1 = 1/S_{11}$ but $1/S_{11} \neq C_{11}$ because the inverse of a matrix is not simply the inverse of its terms. This difference is easily understood in physical terms by looking at the tests that would allow the measurement of these quantities. Figure 2.22 shows the tests to be performed to measure the engineering constants, and fig. 2.23 shows the corresponding tests to be performed to measure the stiffness coefficients.

Focusing on the first test depicted in fig. 2.22, a single stress component, σ_1, is applied, *i.e.*, $\sigma_2 = \tau_{12} = 0$. A complex state of strain results that involves ϵ_1, ϵ_2, and γ_{12}. The measurement of the strain component, ϵ_1, yields E_1 from the first eq. (2.87), the measurement of ϵ_2 yields ν_{12} from the second eq. (2.87), and the measurement of γ_{12} yields ν_{16} from the last eq. (2.87).

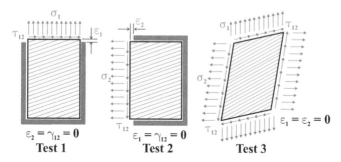

Fig. 2.23. Three simple tests for the determination of the stiffness coefficients.

In the first test depicted in fig. 2.23, a single strain component, ϵ_1, is applied, *i.e.*, $\epsilon_2 = \gamma_{12} = 0$. A complex state of stress results that involves stress components σ_1, σ_2, and τ_{12}. Measurements of these stresses would yield the stiffness coefficients, C_{11}, C_{12}, and C_{16}, from the first, second, and last equation of eqs. (2.78), respectively. Although conceptually simple, the tests depicted in fig. 2.23 are very difficult to perform in practice. For the first test, the test specimen would have to be constrained to prevent any deformations except for strain component ϵ_1, and the resulting stresses components would then need to be measured. Furthermore, friction between the sample and the side restraints should be completely eliminated. Clearly, such test is difficult to perform in practice.

Considering the first test in fig. 2.23 it is clear that the stiffness coefficient, C_{11}, reflects the stiffness of the material when it is constrained, *i.e.*, when $\epsilon_2 = \gamma_{12} = 0$. The effect of these constraints is to considerably stiffen the response of the material. At a 20 degree lamina angle, the stiffness coefficient C_{11} is about 130 GPa, see fig. 2.19, whereas the engineering constant E_1 is only about 50 GPa. The effect of constraining the material is clearly very important.

A similar effect is observed in fig. 2.19, which compares the stiffness coefficient, C_{66}, and the shearing modulus, G_{12}. The stiffness coefficient increases considerably, whereas the shearing modulus rises very modestly. Both quantities, however, reach their maximum values for a 45 degree lamina angle.

Figure 2.24 shows the Poisson's ratios, ν_{12}, and ν_{21}. Poisson's ratio ν_{12} has a value of 0.28 at a 0 degree lamina angle, but a value of about 0.02 only for a 90 degree lamina angle. For most metals, Poisson's ratio is about 0.3; with composite materials, a much wider range of value is observed. Finally, fig. 2.24 also shows the variation of the engineering constants ν_{16}, ν_{61}, ν_{26}, and ν_{62} as a function the lamina angle.

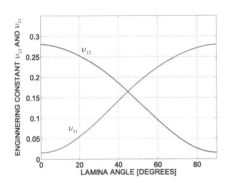

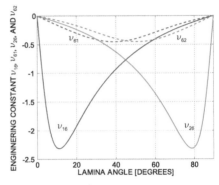

Fig. 2.24. Variation of engineering constants ν_{12} and ν_{21} with lamina angle θ.

Fig. 2.25. Variation of engineering constants ν_{16}, ν_{61}, ν_{26}, and ν_{62} with lamina angle θ.

2.7 Strength of a transversely isotropic lamina

The constitutive laws for a linearly elastic, transversely isotropic material are investigated in section 2.6.2. The equations developed in that section express a linear relationship between stress and strain, but provide no information about the strength of the material.

2.7.1 Strength of a lamina under simple loading conditions

The strength of a lamina made of transversely isotropic material can be experimentally determined by performing a series of simple tests. In practical applications, this lamina will be under plane state of stress. Consider a first test where the lamina is subjected to a single tensile stress, σ_1^*, applied in the fiber direction, *i.e.*, $\sigma_2^* = \tau_{12}^* = 0$, as depicted in fig. 2.26. As the applied stress increases, a point is reached where the material fails. Let σ_{1t}^{*f} be the stress level at which failure occurs. The same test could be repeated for a compressive stress σ_1^*, and let σ_{1c}^{*f} be the absolute value of the compressive stress at failure. There is no reason to believe that σ_{1t}^{*f} and σ_{1c}^{*f} are, in general, equal. Therefore, the subscripts $(.)_t$ and $(.)_c$ will be used to distinguish the tensile and compressive failure stresses, respectively.

In a second test, depicted in fig. 2.26, the lamina is subjected to a single tensile stress, σ_2^*, applied in the direction transverse to the fiber, *i.e.*, $\sigma_1^* = \tau_{12}^* = 0$. The applied stress level that corresponds to failure of the lamina is denoted σ_{2t}^{*f}, and let σ_{2c}^{*f} be the absolute value of the compressive stress that corresponds to failure. Figure 2.26 also shows the third test to be performed in which the lamina is subjected to a shear stress, τ_{12}^*, whereas $\sigma_1^* = \sigma_2^* = 0$. Let τ_{12}^{*f} denote the level of applied shear stress that corresponds to failure. Clearly, the failure level in shear does not depend on the sign of the shear stress.

Although conceptually simple, the above tests can be very difficult to perform in practice. Care must be taken in the tensile tests to reinforce the ends of the test

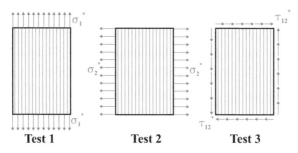

Fig. 2.26. Three tests for the determination of the strength of a lamina

specimens that fit into the grips of the testing machine to avoid premature failure near the grips. Furthermore, the specimen must be long enough to ensure that the test section is free of end effects. The test setup to measure compressive strength is far more complex because buckling of the specimen must be prevented. This can be achieved by providing lateral support of the test sample. Performing the shear test is also very complex. Subjecting a flat specimen to a state of pure shear is very difficult to achieve experimentally. Of course, a tubular specimen can be used, but at a far greater cost. Table 2.9 lists the typical failure stress levels for lamina made of different materials.

Table 2.9. Typical failure stresses for lamina made of different materials.

Material system	σ_{1t}^{*f} [MPa]	σ_{1c}^{*f} [MPa]	σ_{2t}^{*f} [MPa]	σ_{1c}^{*f} [MPa]	τ_{12}^{*f} [MPa]
Graphite/Epoxy (T300/5208)	1500	1500	40	240	68
Graphite/Epoxy (AS/3501)	1450	1450	52	205	93
Boron/Epoxy (T300/5208)	1260	2500	61	202	67
Scotchply (1002)	1060	610	31	118	72
Kevlar 49	1400	235	12	53	34

2.7.2 Strength of a lamina under combined loading conditions

In practical design situations, the lamina might be subjected to several stress components simultaneously. Consider, for instance, a lamina subjected to stresses along both the fiber direction and the transverse direction. Figure 2.27 shows the corresponding stress space and the failure stress levels σ_{1t}^{*f}, σ_{1c}^{*f}, σ_{2t}^{*f}, and σ_{2c}^{*f} which correspond to the various failure stress levels measured in the tests described previously. Assume that equal stresses are applied in both directions simultaneously, *i.e.*, $\sigma_1^* = \sigma_2^*$. These stress states form a 45 degree line in the stress space. As the applied stresses increase, failure will occur at a certain level. Of course, the applied stresses σ_1^* and σ_2^* could be applied in any proportion, corresponding to various radial lines emanating from the origin of the stress space. A different failure level will correspond to each radial line.

To cover all possible combinations, the failure envelope, depicted in fig. 2.27, should be known. All stress states within the failure envelope correspond to stress levels the material can sustain without failing, whereas the stress states outside the failure envelope result in failure.

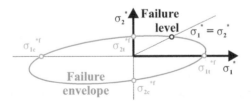

Fig. 2.27. Stress space for a lamina in biaxial stress state.

Clearly, the failure envelope could be obtained experimentally by performing a large number of tests with various combinations of applied stress components, σ_1^*, σ_2^*, and τ_{12}^*. This approach is not practical because it would require an overwhelming amount of testing to determine the failure envelope. A more desirable approach would be to determine the failure envelope based on the knowledge of a few failure stress levels such as $\sigma_{1t}^{*f}, \sigma_{1c}^{*f}, \sigma_{2t}^{*f}, \sigma_{1c}^{*f}$, and τ_{12}^{*f}. This can be achieved by means of a failure criterion that predicts failure under combined loads. Although many different failure criteria have been proposed, none is fully satisfactory, in the sense that their predictions are not always in very good agreement with the experimentally measured failure stresses. They are, however, widely used in preliminary design.

It is important to note that when designing with composite materials, the failure mode is often as important as the failure stress. Indeed, consider the case of a lamina subjected to a load transverse to the fibers: the lamina will fail at a very low stress level which is indicative of the low load carrying capability of the matrix material. On the other hand, if the same lamina is subjected to a stress aligned with the fibers, it will fail at a far higher stress level which reflects the high strength of the fiber. The failure modes in the two cases are quite different: matrix failure for the former, fiber failure for the latter. Failure of the matrix due to a transverse load does not substantially decrease the ability of the lamina to continue to carry high loads in the fiber direction, whereas with fiber failure, load carrying capability is completely lost. Clearly, a matrix failure is not always a catastrophic event in contrast to fiber failure, which completely eliminates any load carrying capability.

2.7.3 The Tsai-Wu failure criterion

A commonly used failure criterion is the Tsai-Wu failure criterion. This criterion states that the failure condition is reached when the combined applied stresses satisfy the following equality

$$F_{11}^*\sigma_1^{*2} + 2F_{12}^*\sigma_1^*\sigma_2^* + F_{22}^*\sigma_2^{*2} + F_{66}^*\tau_{12}^{*2} + F_1^*\sigma_1^* + F_2^*\sigma_2^* = 1, \qquad (2.93)$$

where the coefficients F_{11}^*, F_{12}^*, F_{22}^*, F_{66}^*, F_1^*, and F_2^* must be determined experimentally. Note that the stress components appearing in the criterion are expressed in the fiber aligned triad. Consider first the test described earlier where a single stress component σ_1^* is applied. At failure in tension and in compression, the above equality must be satisfied, implying

$$F_{11}^*\sigma_{1t}^{*f2} + F_1^*\sigma_{1t}^{*f} = 1, \quad F_{11}^*\sigma_{1c}^{*f2} - F_1^*\sigma_{1c}^{*f} = 1.$$

The second test involves stress component σ_2^* only and yields

$$F_{22}^*\sigma_{2t}^{*f2} + F_2^*\sigma_{2t}^{*f} = 1, \quad F_{22}^*\sigma_{2c}^{*f2} - F_2^*\sigma_{2c}^{*f} = 1.$$

Finally, the last test involves τ_{12}^* only and implies $F_{66}^*\tau_{12}^{*f2} = 1$. These five equations can be solved for five of the coefficients appearing in eq. (2.93) to find

$$F_{11}^* = \frac{1}{\sigma_{1t}^{*f}\sigma_{1c}^{*f}}, \quad F_{22}^* = \frac{1}{\sigma_{2t}^{*f}\sigma_{2c}^{*f}}, \quad F_{66}^* = \frac{1}{\tau_{12}^{*f2}};$$

$$F_1^* = \frac{\sigma_{1c}^{*f} - \sigma_{1t}^{*f}}{\sigma_{1t}^{*f}\sigma_{1c}^{*f}}, \quad F_2^* = \frac{\sigma_{2c}^{*f} - \sigma_{2t}^{*f}}{\sigma_{2t}^{*f}\sigma_{2c}^{*f}}.$$

These results are introduced in the initial statement of the failure criterion, eq. (2.93), to yield

$$\bar{\sigma}_{11}^{*2} + 2\bar{F}_{12}^*\bar{\sigma}_{11}^*\bar{\sigma}_{22}^* + \bar{\sigma}_{22}^{*2} + \bar{\tau}_{12}^{*2} + \bar{F}_1^*\bar{\sigma}_{11}^* + \bar{F}_2^*\bar{\sigma}_{22}^* = 1, \qquad (2.94)$$

where the following non-dimensional stress components are defined,

$$\bar{\sigma}_{11}^* = \frac{\sigma_1^*}{\sqrt{\sigma_{1t}^{*f}\sigma_{1c}^{*f}}}; \quad \bar{\sigma}_{22}^* = \frac{\sigma_2^*}{\sqrt{\sigma_{2t}^{*f}\sigma_{2c}^{*f}}}; \quad \bar{\tau}_{12}^* = \frac{\tau_{12}^*}{\tau_{12}^{*f}}; \qquad (2.95)$$

as well as the following non-dimensional coefficients:

$$\bar{F}_1^* = \frac{\sigma_{1c}^{*f} - \sigma_{1t}^{*f}}{\sqrt{\sigma_{1t}^{*f}\sigma_{1c}^{*f}}}; \quad \bar{F}_2^* = \frac{\sigma_{2c}^{*f} - \sigma_{2t}^{*f}}{\sqrt{\sigma_{2t}^{*f}\sigma_{2c}^{*f}}}. \qquad (2.96)$$

Coefficient $\bar{F}_{12}^*$ is as yet undetermined. Clearly, an additional test involving a biaxial state of applied stress (i.e., a test where both σ_1^* and σ_2^* are applied simultaneously) is required to determine this coefficient. Because such a biaxial test is very difficult to perform, coefficient $\bar{F}_{12}^*$ is often selected by fitting the prediction of the criterion to available experimental data. $\bar{F}_{12}^* = -1/2$ has been found to provide the best fit. The final statement of the Tsai-Wu criterion becomes

$$\bar{\sigma}_{11}^{*2} - \bar{\sigma}_{11}^*\bar{\sigma}_{22}^* + \bar{\sigma}_{22}^{*2} + \bar{\tau}_{12}^{*2} + \bar{F}_1^*\bar{\sigma}_{11}^* + \bar{F}_2^*\bar{\sigma}_{22}^* = 1. \qquad (2.97)$$

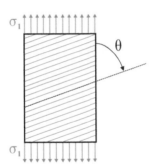

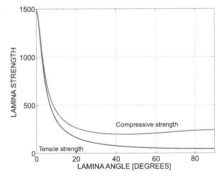

Fig. 2.28. Strength test for a lamina at an angle θ.

Fig. 2.29. Variation of the tensile and compressive failure loads with lamina angle θ.

Example 2.3. Tsai-Wu failure criterion for uniaxial stress

As an example of application of this criterion consider the simple test shown in fig. 2.28. A single stress component, σ_1, is applied to a lamina with fibers running at an angle θ. The stress rotation formula (1.47) yields the applied stresses in the fiber aligned triad as $\sigma_1^* = \sigma_1 \cos^2\theta$, $\sigma_2^* = \sigma_1 \sin^2\theta$, and $\tau_{12}^* = -\sigma_1 \cos\theta \sin\theta$.

The level of applied stress that corresponds to failure satisfies the failure criterion 2.97, *i.e.*,

$$\sigma_1^2 \left[\frac{\cos^4\theta}{\sigma_{1t}^{*f}\sigma_{1c}^{*f}} - \frac{\sin^2\theta \cos^2\theta}{\sqrt{\sigma_{1t}^{*f}\sigma_{1c}^{*f}\sigma_{2t}^{*f}\sigma_{2c}^{*f}}} + \frac{\sin^4\theta}{\sigma_{2t}^{*f}\sigma_{2c}^{*f}} + \frac{\sin^2\theta \cos^2\theta}{\tau_{12}^{*f2}} \right]$$
$$+ \sigma_1 \left[\frac{\bar{F}_1^* \cos^2\theta}{\sqrt{\sigma_{1t}^{*f}\sigma_{1c}^{*f}}} + \frac{\bar{F}_2^* \sin^2\theta}{\sqrt{\sigma_{2t}^{*f}\sigma_{2c}^{*f}}} \right] - 1 = 0.$$

This second order equation can be solved to find the failure load. The two solutions correspond to the failure loads in tension and compression. Figure 2.29 shows the absolute value of these failure loads as a function of the lamina angle θ for the Graphite/Epoxy materials (T300/5208) whose properties are given in table 2.9. Note the precipitous drop in strength as the lamina angle moves away from 0 degrees.

2.7.4 The reserve factor

The concept of reserve factor is often used in stress computations. The reserve factor, R, is defined as the factor by which the applied stress can be multiplied to reach failure, *i.e.*,

$$\sigma_{\text{fail}} = R\,\sigma_{\text{appl}}. \tag{2.98}$$

From this definition it follows that:

- $R = 1$ means that the applied stresses causes failure;

- $R > 1$ means that the applied stresses level is safe, *i.e.*, it is below the failure level. A reserve factor of two means that the applied stresses can be doubled before failure occurs;
- $R < 1$ means that the applied stresses is above the failure stress.

Let σ_1^*, σ_2^*, and τ_{12}^* be the stresses applied to a lamina. By definition of the reserve factor, it follows that $R\sigma_1^*$, $R\sigma_2^*$, $R\tau_{12}^*$ is the stress level that will cause failure. Assuming failure can be predicted by the Tsai-Wu failure criterion, eq. (2.97), the failure condition can be written as

$$
\frac{(R\sigma_1^*)^2}{\sigma_{1t}^{*f}\sigma_{1c}^{*f}} - \frac{(R\sigma_1^*)(R\sigma_2^*)}{\sqrt{\sigma_{1t}^{*f}\sigma_{1c}^{*f}\sigma_{2t}^{*f}\sigma_{2c}^{*f}}} + \frac{(R\sigma_2^*)^2}{\sigma_{2t}^{*f}\sigma_{2c}^{*f}} + \frac{(R\tau_{12}^*)^2}{\tau_{12}^{*f2}}
$$

$$
+ \bar{F}_1^* \frac{R\sigma_1^*}{\sqrt{\sigma_{1t}^{*f}\sigma_{1c}^{*f}}} + \bar{F}_2^* \frac{R\sigma_2^*}{\sqrt{\sigma_{2t}^{*f}\sigma_{2c}^{*f}}} - 1 = 0.
$$

Introducing the non-dimensional stresses, eq. (2.95), and regrouping the powers of R yields the following quadratic equation for the reserve factor

$$
\left(\bar{\sigma}_{11}^{*2} - \bar{\sigma}_{11}^*\bar{\sigma}_{22}^* + \bar{\sigma}_{22}^{*2} + \bar{\tau}_{12}^{*2}\right) R^2 + \left(\bar{F}_1^*\bar{\sigma}_{11}^* + \bar{F}_2^*\bar{\sigma}_{22}^*\right) R - 1 = 0. \tag{2.99}
$$

This quadratic equation has two roots, R_1 and R_2, which are positive and negative, respectively. The positive root gives the failure stress level, and the negative root gives the failure stress level when the sign of the applied stresses is reversed. In general, $|R_1| \neq |R_2|$ since the failure stress level in tension and compression are different.

3

Linear elasticity solutions

The equations of linear elasticity are derived in chapters 1 and 2, and can be divided into three groups: the equilibrium equations, the strain displacement equations, and the constitutive laws. Figure 3.1 shows these three groups of equations in a block diagram.

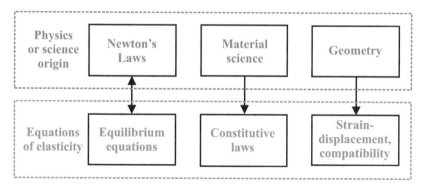

Fig. 3.1. The elasticity equations separated into three groups.

The *equilibrium equations* express the equilibrium conditions for a differential element of the body in terms of the stress field. These equilibrium conditions are a direct consequence Newton's laws applied to a differential element of the deformable body. They consists of the three partial differential equations of equilibrium, eqs. (1.4).

The *strain-displacement equations*, also called the *kinematic equations*, describe the deformation of the body without reference to the forces that create the deformation. The strain components are defined based on a purely kinematic description of the deformed and undeformed configurations of the solid. The strain-displacement equations consists of the six partial differential equations relating the strain components to the displacement components, eqs. (1.63) and (1.71).

The *constitutive laws* describe the behavior of materials under load. More specifically, they take the form of relationships linking the stress and strain components at a point. Constitutive laws are rooted in material science and express an approximation to the observed behavior of actual materials. For Hooke's law, they consists of six algebraic equations, eqs. (2.4) and (2.9).

A total of 15 equations of linear elasticity are obtained. Given the proper boundary conditions, these 15 equations can be solved to obtain the following 15 unknowns: the three components of the displacement vector, the six components of the strain tensor, and the six components of the stress tensor.

In addition, the six partial differential strain compatibility equations, eqs. (1.106), impose certain continuity conditions on the displacement components that may arise from a state of strain. While these compatibility equations are not part of the basic 15 equations of elasticity, their use may be a critical element of any solution procedure. In this chapter, solutions of this set of equations will be presented for very simple problems. Indeed, exact solutions for realistic problems are very difficult to obtain in general.

3.1 Solution procedures

The linear equations of elasticity form a set of coupled partial differential equations that are elegantly simple but like most partial differential equations, are often quite difficult to solve for realistic problems. Considerable simplification can be achieved when the general, three-dimensional formulation is reduced to a two-dimensional formulation by assuming the problem to be either plane stress or plane strain, as discussed in sections 1.3 or 1.6, respectively. Further simplification can be achieved for problems presenting specific symmetries. For example, the governing equations for two-dimensional problems featuring cylindrical symmetry reduce to ordinary differential equations. It is often necessary, however, to reformulate the elasticity equations in cylindrical or spherical coordinates to take advantage of specific symmetries or easily impose boundary conditions.

Three approaches are available for the solution of elasticity problems.

1. *Displacement formulations:* the objective is to derive three equations for the three unknown displacement components.
2. *Stress formulations:* the objective is to solve for the state of stress in the body. This means that six equations are required for the six stress components.
3. *Semi-inverse approaches:* assumptions are made to solve the problem for a subset of the variables. With that solution at hand, the remaining equations of the problem are solved. If all equations can be exactly satisfied, an exact solution is obtained and the initial assumptions are validated.

For all three approaches, dimensional reduction is often performed first. Under specific conditions, the initial three-dimensional problem can be reduced to a two- and sometimes one-dimensional problem, considerably easing the solution process. Examples of these various approaches are given in the following sections.

3.1.1 Displacement formulation

A formulation leading to equations involving only the displacement components, u_1, u_2, and u_3, is readily developed based on the following procedure.

1. Substitute the stress-strain equations (2.4) and (2.9) into the three equilibrium equations (1.4) to obtain three equations expressed in terms of strain components.
2. Substitute the strain-displacement equations (1.63) and (1.71) into these equations to obtain a set of three equilibrium equations expressed in terms of the displacement components, u_1, u_2, and u_3 alone.

These equations are generally referred to as *Navier's equations*. Given appropriate boundary conditions expressed in terms of displacement components, solution of Navier's equations yields the unknown displacement field throughout the body. While specification of displacement boundary conditions is straightforward, the specification of traction boundary conditions in terms of displacements often lead to complicated formulations. It is left as an exercise to show that Navier's equations are

$$\frac{E}{2(1+\nu)(1-2\nu)}\frac{\partial e}{\partial x_1} + G\,\nabla^2 u_1 + b_1 = 0 \tag{3.1a}$$

$$\frac{E}{2(1+\nu)(1-2\nu)}\frac{\partial e}{\partial x_2} + G\,\nabla^2 u_2 + b_2 = 0 \tag{3.1b}$$

$$\frac{E}{2(1+\nu)(1-2\nu)}\frac{\partial e}{\partial x_3} + G\,\nabla^2 u_3 + b_3 = 0, \tag{3.1c}$$

where e is the volumetric strain defined by eq. (1.75). The differential operator, ∇^2, called the *Laplacian*, is defined as

$$\nabla^2 = \frac{\partial^2}{\partial x_1^2} + \frac{\partial^2}{\partial x_2^2} + \frac{\partial^2}{\partial x_3^2}. \tag{3.2}$$

If body forces are constant throughout the body, taking a derivative with respect to x_1, x_2, and x_3 of eqs. (3.1a), (3.1b) and (3.1c), respectively, and summing up the resulting equations leads to

$$\frac{\partial e}{\partial x_1^2} + \frac{\partial e}{\partial x_2^2} + \frac{\partial e}{\partial x_3^2} = \nabla^2 e = 0. \tag{3.3}$$

Thus, for constant body forces, the volumetric strain satisfies the homogeneous Laplace's equation.

3.1.2 Stress formulation

It is a much more difficult task to formulate elasticity equations in terms of the stress components. The three equilibrium equations alone are not sufficient to determine the six unknown stress components. In this case, the compatibility equations must be included to insure that stress components correspond to a deformation state that is continuous and sufficiently smooth. The formulation is quite tedious but can be accomplished by following the steps.

1. Substitute the stress-strain equations (2.4) and (2.9) into the six compatibility equations (1.106).
2. Further simplify these six equations into three equations for the normal stresses and three equations for the shear stresses.

The resulting equations are called *Beltrami-Michell's equations*, which can be written as

$$\nabla^2\sigma_1 + \frac{1}{1+\nu}\frac{\partial^2 I_1}{\partial x_1^2} + \frac{\nu}{1-\nu}\left(\frac{\partial b_1}{\partial x_1} + \frac{\partial b_2}{\partial x_2} + \frac{\partial b_3}{\partial x_3}\right) + 2\frac{\partial b_1}{\partial x_1} = 0, \qquad (3.4a)$$

$$\nabla^2\sigma_2 + \frac{1}{1+\nu}\frac{\partial^2 I_1}{\partial x_2^2} + \frac{\nu}{1-\nu}\left(\frac{\partial b_1}{\partial x_1} + \frac{\partial b_2}{\partial x_2} + \frac{\partial b_3}{\partial x_3}\right) + 2\frac{\partial b_1}{\partial x_2} = 0, \qquad (3.4b)$$

$$\nabla^2\sigma_3 + \frac{1}{1+\nu}\frac{\partial^2 I_1}{\partial x_3^2} + \frac{\nu}{1-\nu}\left(\frac{\partial b_1}{\partial x_1} + \frac{\partial b_2}{\partial x_2} + \frac{\partial b_3}{\partial x_3}\right) + 2\frac{\partial b_1}{\partial x_3} = 0, \qquad (3.4c)$$

$$\nabla^2\tau_{12} + \frac{1}{1+\nu}\frac{\partial^2 I_1}{\partial x_1\partial x_2} + \left(\frac{\partial b_1}{\partial x_2} + \frac{\partial b_2}{\partial x_1}\right) = 0, \qquad (3.4d)$$

$$\nabla^2\tau_{23} + \frac{1}{1+\nu}\frac{\partial^2 I_1}{\partial x_2\partial x_3} + \left(\frac{\partial b_2}{\partial x_3} + \frac{\partial b_3}{\partial x_2}\right) = 0, \qquad (3.4e)$$

$$\nabla^2\tau_{31} + \frac{1}{1+\nu}\frac{\partial^2 I_1}{\partial x_3\partial x_1} + \left(\frac{\partial b_1}{\partial x_3} + \frac{\partial b_3}{\partial x_1}\right) = 0, \qquad (3.4f)$$

where I_1 is the first stress invariant given by eq. (1.15a). These equations, along with appropriate stress boundary conditions, can be solved for the stress state within the body. Solutions to all but the simplest problems are extremely difficult to construct. Moreover, many problems of practical interest involve boundary conditions expressed in terms of displacement components over parts of the body and in terms of stress components over other portions of the body; this leads to so called "mixed boundary value problems," which are very difficult to handle for all but the simplest problems.

If body forces are constant throughout the body, summing up eqs. (3.4a) to (3.4c) leads to $\nabla^2 I_1 = 0$, *i.e.*, the first stress invariant satisfies the homogeneous Laplace's equation. Introducing eq. (2.18), this becomes $E/(1 - 2\nu)\,\nabla^2 e = 0$ and finally $\nabla^2 e = 0$, a result that is obtained in the previous section, see eq. (3.3).

3.1.3 Solutions to elasticity problems

Solutions to practical problems in three dimensions are very difficult to achieve for all but the simplest geometries. This is largely due to the large number of partial differential equations in the governing equations of linear elasticity and the fact that solutions to partial differential equations involve arbitrary functions (rather than the much simpler arbitrary constants that occur in solutions to ordinary differential equations). The choice of such functions generally depends critically on the particular geometry of the problem under consideration. In this section, one single problem will be treated to illustrate the difficulty of the solution procedure.

Example 3.1. Rectangular bar hanging under its own weight

To illustrate the solution problem for a simple practical problem, the stress and displacement distributions in a prismatic bar hanging vertically under its own weight will be evaluated. For simplicity, consider a prismatic bar of length L, with a rectangular cross-section of width b and thickness t, hanging vertically under the action of gravity as shown in fig. 3.2. The cross-sectional dimensions of the bar are assumed to be far smaller than its length, $i.e.$, $b/L \ll 1$ and $t/L \ll 1$.

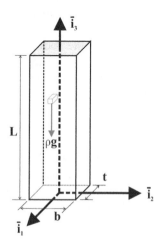

Fig. 3.2. Prismatic bar hanging under its own weight.

While a number of approaches to this problem are possible, perhaps the simplest is to seek a solution for the stress field. This is a natural choice because all the sides of the bar are stress-free, except for the top surface where it is attached to the support. Expressing these stress-free boundary conditions is relatively easy in a stress based formulation. While the six Beltrami-Michell equations, eqs. (3.4), could be used as the starting point of this development, it is easier to adopt a semi-inverse method, in which simplifying assumptions are made prior to solving the governing equations.

Because the cross-sectional dimensions of the bar are small compared to its length, it seems reasonable to assume that, (1) all transverse stress components vanish, and (2) the axial stress, σ_3, is solely a function of the variable x_3. With these simplifications, the three equilibrium equations reduce to the single equation, $d\sigma_3/dx_3 + b_3 = 0$. The applied load per unit volume of the bar is $b_3 = -\rho g$, where ρ is the material mass density and g the gravitational constant; this equation is integrated to find the axial stress component, σ_3, as $\sigma_3 = \int -b_3 \, dx_3 = \rho g x_3 + C$, where C is an integration constant.

The stress boundary conditions can be expressed using eq. (1.9), but since all surfaces are perpendicular to one of the coordinate axes, it follows that $\sigma_1 = \tau_{12} = \tau_{13} = 0$ on faces normal to axis $\bar{\imath}_1$, $\sigma_2 = \tau_{21} = \tau_{23} = 0$ on faces normal to $\bar{\imath}_2$, and $\sigma_3 = \tau_{31} = \tau_{32} = 0$ on the lower face. The assumed stress state satisfies all these boundary conditions except for the condition that $\sigma_3 = 0$ on the lower face. Imposing this condition on the stress field yields $C = 0$ and hence,

$$\sigma_3 = \rho g x_3. \tag{3.5}$$

All all other stress components vanish. This solution implies that the stress on the upper surface, at $x_3 = L$, is $\sigma_3 = \rho g L$. The net force on this area is the integral of the stress over the cross-section, which is equal to $\rho g L b t$, the total weight of the bar, as expected from elementary statics.

Now that the stress field throughout the body has been established, the corresponding displacement field must be evaluated. The first step in determining the displacement components is to express the strains in terms of the stresses using the constitutive equations (2.4) and (2.8) to find

$$\epsilon_1 = -\frac{\nu \rho g x_3}{E}, \quad \epsilon_2 = -\frac{\nu \rho g x_3}{E}, \quad \epsilon_3 = \frac{\rho g x_3}{E}, \quad \gamma_{12} = \gamma_{13} = \gamma_{23} = 0. \tag{3.6}$$

The shear strain components vanish, while the direct strain components are linear functions of x_3, hence, all six compatibility equations (1.106) are satisfied.

To determine the displacements, it is necessary to integrate the strain-displacement equations (1.63) and (1.71) which can be stated as follows

$$\frac{\partial u_1}{\partial x_1} = -\frac{\nu \rho g}{E} x_3, \tag{3.7a}$$

$$\frac{\partial u_2}{\partial x_2} = -\frac{\nu \rho g}{E} x_3, \tag{3.7b}$$

$$\frac{\partial u_3}{\partial x_3} = \frac{\rho g}{E} x_3, \tag{3.7c}$$

$$\frac{\partial u_1}{\partial x_2} + \frac{\partial u_2}{\partial x_1} = 0, \tag{3.7d}$$

$$\frac{\partial u_1}{\partial x_3} + \frac{\partial u_3}{\partial x_1} = 0, \tag{3.7e}$$

$$\frac{\partial u_2}{\partial x_3} + \frac{\partial u_3}{\partial x_2} = 0. \tag{3.7f}$$

Integration of these partial differential equations to determine the displacement field turns out to be a bit more challenging than it appears. Integrating eq. (3.7c) yields the third displacement component as

$$u_3 = \frac{\rho g}{2E} x_3^2 + f_1(x_1, x_2), \tag{3.8}$$

where the constant of partial integration is a function, $f_1(x_1, x_2)$, rather than simply a constant, as would be the case for ordinary differential equations. This result can now be substituted into equations (3.7e) and (3.7f) to find $\partial u_1/\partial x_3 = -\partial f_1/\partial x_1$ and $\partial u_2/\partial x_3 = -\partial f_1/\partial x_2$. These equations can be integrated to yield

$$u_1 = -\frac{\partial f_1}{\partial x_1}x_3 + f_2(x_1, x_2), \quad u_2 = -\frac{\partial f_1}{\partial x_2}x_3 + f_3(x_1, x_2), \tag{3.9}$$

where $f_2(x_1, x_2)$ and $f_3(x_1, x_2)$ are arbitrary functions arising from the integration. While eqs. (3.7c), (3.7e) and (3.7f) have been used already, the above displacements can be substituted into eqs. (3.7a) and (3.7b) to find $-(\partial^2 f_1/\partial x_1^2)\, x_3 + (\partial f_2/\partial x_1) = -(\nu\rho g/E)\, x_3$ and $-(\partial^2 f_1/\partial x_2^2)\, x_3 + (\partial f_3/\partial x_2) = -(\nu\rho g/E)\, x_3$, which can be rearranged into a more useful form as

$$\left(\frac{\partial^2 f_1}{\partial x_1^2} - \frac{\nu\rho g}{E}\right) x_3 = \frac{\partial f_2}{\partial x_1}, \quad \left(\frac{\partial^2 f_1}{\partial x_2^2} - \frac{\nu\rho g}{E}\right) x_3 = \frac{\partial f_3}{\partial x_2}. \tag{3.10}$$

These results must be carefully examined: functions $f_1(x_1, x_2)$, $f_2(x_1, x_2)$, and $f_3(x_1, x_2)$ are all three independent of x_3. Because the above equations must hold for any value of x_3, the expressions in parentheses, which depend only on x_1 and x_2, must vanish, as must the righthand sides of the equations, implying that

$$\frac{\partial^2 f_1}{\partial x_1^2} = \frac{\nu\rho g}{E}, \tag{3.11a}$$

$$\frac{\partial^2 f_1}{\partial x_2^2} = \frac{\nu\rho g}{E}, \tag{3.11b}$$

$$\frac{\partial f_2}{\partial x_1} = 0, \tag{3.11c}$$

$$\frac{\partial f_3}{\partial x_2} = 0. \tag{3.11d}$$

These expressions are still insufficient to determine the functions $f_1(x_1, x_2)$, $f_2(x_1, x_2)$ and $f_3(x_1, x_2)$, but eq. (3.7d) has not yet been used. Substituting u_1 and u_2 from eq. (3.9) into eq. (3.7d) yields

$$-2\frac{\partial^2 f_1}{\partial x_1 \partial x_2}x_3 + \frac{\partial f_3}{\partial x_1} + \frac{\partial f_2}{\partial x_2} = 0.$$

The reasoning used earlier applies here again: because the above equation must hold for any value of x_3 and because f_2 and f_3 are functions of x_1 and x_2 only, both the coefficient of x_3 and the independent term must vanish, leading to

$$\frac{\partial^2 f_1}{\partial x_1 \partial x_2} = 0 \tag{3.12a}$$

$$\frac{\partial f_3}{\partial x_1} + \frac{\partial f_2}{\partial x_2} = 0. \tag{3.12b}$$

Equations (3.11) and (3.12) now constitute a set of equations that can be solved for the unknown functions f_1, f_2 and f_3. Equations (3.11c) and (3.11d) can be integrated to yield $f_2 = C_1 a_1(x_2) + C_2$ and $f_3 = C_3 a_2(x_1) + C_4$, where $a_1(x_2)$ and $a_2(x_1)$ are arbitrary functions and C_1, C_2, C_3 and C_4 arbitrary constants. Substituting these into eq. (3.12b) results in

$$C_3 \frac{da_2(x_1)}{dx_1} + C_1 \frac{da_1(x_2)}{dx_2} = 0,$$

where the functional dependence is explicitly shown and the partial derivatives become regular derivatives. Inspection of this result reveals that the only possible solution is $a_1 = x_2$, $a_2 = x_1$ and $C_3 = -C_1$, leading to

$$f_2 = C_1 x_2 + C_2 \quad \text{and} \quad f_3 = -C_1 x_1 + C_4. \tag{3.13}$$

Next, eqs. (3.11a) and (3.11b) can be integrated to yield two different expressions for f_1: $f_1 = (\nu \rho g / 2E)\, x_1^2 + f_4(x_2)\, x_1 + C_5$ and $f_1 = (\nu \rho g / 2E)\, x_2^2 + f_5(x_1)\, x_2 + C_6$. Equation (3.12a) now implies $(\partial^2 f_1)/(\partial x_1 \partial x_2) = df_4/dx_2 = 0$ and $(\partial^2 f_1)/(\partial x_1 \partial x_2) = df_5/dx_1 = 0$, and hence, $f_4 = C_7$ and $f_5 = C_8$. Finally, it is possible to combine these results into a single expression for f_1

$$f_1 = \frac{\nu \rho g}{2E}(x_1^2 + x_2^2) + C_7 x_1 + C_8 x_2 + C_9, \tag{3.14}$$

where the C_7, C_8 and C_9 are arbitrary constants. The functions expressed in eqs. (3.13) and (3.14) can now be substituted into eqs. (3.8) and (3.9) to yield solutions for the displacement components

$$u_1 = \qquad -\frac{\nu \rho g}{E} x_1 x_3 \qquad - C_7 x_3 + C_1 x_2 + C_3,$$

$$u_2 = \qquad -\frac{\nu \rho g}{E} x_2 x_3 \qquad - C_8 x_3 - C_1 x_1 + C_4, \tag{3.15}$$

$$u_3 = \frac{\rho g}{2E} x_3^2 + \frac{\nu \rho g}{2E}(x_1^2 + x_2^2) + C_7 x_1 + C_8 x_2 + C_9.$$

At this point, the only remaining task is to determine the integration constants appearing in the displacement field. Two requirements must be met: the bar undergoes no rigid body translation and no rigid body rotation. The simplest way to impose these conditions is to enforce the vanishing of displacements and rotations at the center of the upper surface along which the bar is attached: vanishing of the displacements implies $u_1(0, 0, L) = u_2(0, 0, L) = u_3(0, 0, L) = 0$, whereas vanishing of the rotations leads to $\omega_1(0, 0, L) = \omega_2(0, 0, L) = \omega_3(0, 0, L) = 0$, see eqs. (1.73). Application of these boundary conditions to the displacement field given by eq. (3.15) is left as an exercise; the final expression for the displacement field is

$$u_1 = -\frac{\nu \rho g}{E} x_1 x_3, \quad u_2 = -\frac{\nu \rho g}{E} x_2 x_3, \quad u_3 = \frac{\rho g}{2E}\left[x_3^2 - L^2 + \nu(x_1^2 + x_2^2)\right]. \tag{3.16}$$

Equations (3.5) and (3.16) describe the state of stress and displacement, respectively, inside the prismatic bar hanging vertically under its own weight. A number of features of this solution are worth examining in more detail. The stress field consists of a single component, σ_3, which linearly increases from the lower to the upper end of the bar, as expected from basic statics requirements. The displacement solution is a bit more complex but quite revealing. The vertical displacement of the lower surface of the bar, i.e., at $x_3 = 0$, is given by

$$u_3(x_3 = 0) = -\frac{\rho g}{2E}\left[L^2 - \nu(x_1^2 + x_2^2)\right].$$

Figure 3.3 shows this distribution of non-dimensional displacement, $u_3/(\rho g/2EL^2)$, over the cross-section of the bar. The vertical displacement at the centerline, i.e., at $x_1 = x_2 = 0$, is that which would be obtained from a one dimensional analysis ignoring the finite dimension of the cross-section. The vertical displacement away from the centerline is reduced by a factor proportional to Poisson's ratio and the square of the distance from the centerline, resulting in a spherical shape for the deflected surface; the central portion of the bar deflects more than the outer regions. The vertical displacement of the upper surface vanishes only at the centerline, as required by the imposed boundary conditions, but is otherwise parabolic. These results are consistent with the stress-free boundary conditions assumed at the lower surface, but had the upper surface been assumed to remain planar, a completely different solution would have resulted. This behavior is perhaps easier to visualize if one imagines the bar to be made of a very soft material like gelatin; in this case, the parabolic displacement of the bar's cross-section becomes more intuitive.

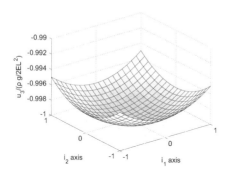

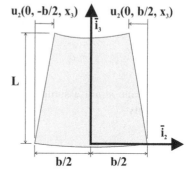

Fig. 3.3. Vertical displacement component, u_3, of lower surface of the prismatic bar.

Fig. 3.4. Lateral displacement component, u_2, of the left and right sides of prismatic bar.

The displacements of the sides of the bar reveal additional details of the deformation behavior. Figure 3.4 shows a greatly exaggerated plot of the shape of a section of the bar taken through the centerline and perpendicular to axis $\bar{i}_1$. As indicated by eq. (3.16), the sides of the bar taper inwards for increasing values of x_3 so that the transverse dimensions of the upper cross-section are smaller than those of the lower. This behavior is due to Poisson's effect, and the presence of Poisson's ratio, ν, in the equations for u_1 and u_2 clearly indicates the origin of this phenomenon.

3.2 Plane strain problems

The assumption of plane strain state introduced in section 1.6 reduces three-dimensional problems to two-dimensional problems and results in considerable simplification of the governing equations. In plane strain problems, the displacements, body forces and changes in properties are assumed to vanish along a preferential direction; it is always possible to select axis $\bar{\imath}_3$ to coincide with that preferential direction. Problems meeting these conditions are not necessarily two-dimensional in appearance, such as a thin sheet or a flat plate, but instead, experience no deformation in one direction. For example, the cross-section of a buried pipe or a cross-section of a long dam could be modeled as plane strain problems under the assumption that there is no displacement in the axial direction.

For plane strain states, the strain-displacement equations, eqs. (1.63) and (1.71), reduce to

$$\epsilon_1 = \frac{\partial u_1}{\partial x_1}, \quad \epsilon_2 = \frac{\partial u_2}{\partial x_2}, \quad \gamma_{12} = \frac{\partial u_1}{\partial x_2} + \frac{\partial u_2}{\partial x_1}, \tag{3.17}$$

while the axial and transverse shear strain components vanish, $\epsilon_3 = \gamma_{13} = \gamma_{23} = 0$. Similarly, the equilibrium equations, eqs. (1.4), reduce to

$$\frac{\partial \sigma_1}{\partial x_1} + \frac{\partial \tau_{21}}{\partial x_2} + b_1 = 0, \quad \frac{\partial \tau_{12}}{\partial x_1} + \frac{\partial \sigma_2}{\partial x_2} + b_2 = 0. \tag{3.18}$$

The transverse shear stress components vanish, $\tau_{13} = \tau_{23} = 0$, while the axial stress does not due to Poisson's effect, $\sigma_3 = \nu(\sigma_1 + \sigma_2)$. If the material is assumed to obey Hooke's law, the vanishing of the axial and transverse shear strain components results in the following reduced constitutive laws

$$\epsilon_1 = \frac{1+\nu}{E} \left[(1-\nu)\sigma_1 - \nu\sigma_2\right], \quad \epsilon_2 = \frac{1+\nu}{E} \left[(1-\nu)\sigma_2 - \nu\sigma_1\right], \quad \gamma_{12} = \frac{\tau_{12}}{G}. \tag{3.19}$$

Under plane strain assumptions, Navier's equations, eqs. (3.1), reduce to two equations only,

$$\frac{E}{2(1+\nu)(1-2\nu)} \frac{\partial e}{\partial x_1} + G\, \nabla^2 u_1 + b_1 = 0, \tag{3.20a}$$

$$\frac{E}{2(1+\nu)(1-2\nu)} \frac{\partial e}{\partial x_2} + G\, \nabla^2 u_2 + b_2 = 0, \tag{3.20b}$$

where the volumetric strain, see eq. (1.75), reduces to $e = \epsilon_1 + \epsilon_2$. The differential operator ∇^2 is now the *two-dimensional Laplacian*

$$\nabla^2 = \frac{\partial^2}{\partial x_1^2} + \frac{\partial^2}{\partial x_2^2}. \tag{3.21}$$

Taking derivatives with respect to x_1 and x_2 of eqs. (3.20a) and (3.20b), respectively, and summing up the resulting equations leads to

$$\frac{2(1-\nu)G}{1-2\nu}\nabla^2 e = -\left(\frac{\partial b_1}{\partial x_1} + \frac{\partial b_2}{\partial x_2}\right). \tag{3.22}$$

The constitutive law for the volumetric strain, given by eq. (2.18), reduces to $e = (1 - 2\nu)(1 + \nu)(\sigma_1 + \sigma_2)/E$ for plane strain state. It then follows that

$$\nabla^2(\sigma_1 + \sigma_2) = -\frac{1}{1-\nu}\left(\frac{\partial b_1}{\partial x_1} + \frac{\partial b_2}{\partial x_2}\right). \tag{3.23}$$

Unfortunately, this equation alone is insufficient to determine the three stress components, σ_1, σ_2 and τ_{12}. To overcome this problem, a novel approach, first proposed by Airy, is introduced. It is assumed that the body forces, b_1 and b_2, are *conservative forces*, i.e., they can be derived from a potential: $b_1 = -\partial V/\partial x_1$ and $b_2 = -\partial V/\partial x_2$, where $V(x_1, x_2)$ is the *potential of the body forces*. Next, the stress field is written in terms of *Airy's stress function*, $\phi(x_1, x_2)$, as

$$\sigma_1 = \frac{\partial^2 \phi}{\partial x_2^2} + V, \quad \sigma_2 = \frac{\partial^2 \phi}{\partial x_1^2} + V, \quad \tau_{12} = -\frac{\partial^2 \phi}{\partial x_1 \partial x_2}. \tag{3.24}$$

The stress field written in terms of Airy's stress function *automatically satisfies the equilibrium equations of the problem*, as can be verified by introducing eqs. (3.24) into eqs. (3.18). This is the very reason why Airy's stress function is introduced in the first place: instead of working with three stress components, σ_1, σ_2 and τ_{12}, a single unknown, the stress function, ϕ, remains. Furthermore, the stress field derived from Airy's stress function through eqs. (3.24) automatically satisfies equilibrium conditions.

Introducing the stress components expressed in terms of Airy's stress function into equilibrium equation (3.23) yields a single equation for the stress function

$$\frac{\partial^4 \phi}{\partial x_1^4} + 2\frac{\partial^4 \phi}{\partial x_1^2 \partial x_2^2} + \frac{\partial^4 \phi}{\partial x_2^4} = \nabla^4 \phi = -\frac{1-2\nu}{1-\nu}\nabla^2 V. \tag{3.25}$$

This is a nonhomogeneous, two-dimensional *bi-harmonic partial differential equation*. When the body forces vanish or are harmonic function, i.e., when $\nabla^2 V = 0$, the governing equation becomes the homogeneous bi-harmonic equation. The bi-harmonic equation has been extensively studied and a number of solution procedures are available.

3.3 Plane stress problems

The assumption of plane stress state introduced in section 1.3 reduces three-dimensional problems to two-dimensional problems and results in considerable simplification of the governing equations. In plane stress problems, the stress components and body forces are assumed to vanish along a preferential direction. It is always possible to select axis $\bar{\imath}_3$ to coincide with that preferential direction; hence $\sigma_3 = \tau_{13} = \tau_{23} = 0$ and $b_3 = 0$. Next, it is assumed that the response of the solid does not vary along axis $\bar{i}_3$, leading to further simplification of the governing equations. This latter assumption is realistic for bodies in the form of thin sheets loaded by forces acting in the plane of the sheet.

For plane stress states, the equilibrium equations are identical to those for plane strain states, eqs. (3.18). If the material is assumed to obey Hooke's law, the vanishing of the axial and shear stress components leads to the following reduced constitutive laws

$$\epsilon_1 = \frac{1}{E}(\sigma_1 - \nu\sigma_2), \quad \epsilon_2 = \frac{1}{E}(\sigma_2 - \nu\sigma_1), \quad \gamma_{12} = \frac{1}{G}\tau_{12}. \qquad (3.26)$$

The inverse relationships are

$$\sigma_1 = \frac{E}{1-\nu^2}(\epsilon_1 + \nu\epsilon_2), \quad \sigma_2 = \frac{E}{1-\nu^2}(\epsilon_2 + \nu\epsilon_1), \quad \tau_{12} = G\,\gamma_{12}, \qquad (3.27)$$

Finally, the strain along axis $\bar{i}_3$ is $\epsilon_3 = -\nu(\sigma_1 + \sigma_2)/E$. Although Hooke's law is used for both plane strain and plane stress problems, the reduced constitutive law differ for the two cases, see eqs (3.19) and (3.26), respectively.

It is convenient here again to use Airy's stress function to satisfy equilibrium conditions and substitute the stress components expressed in terms of the stress function, eq. (3.24), into the constitutive equations to obtain the following expressions for the strain components

$$\epsilon_1 = \frac{1}{E}\left[\frac{\partial^2\phi}{\partial x_2^2} - \nu\frac{\partial^2\phi}{\partial x_1^2} + (1-\nu)V\right], \quad \epsilon_2 = \frac{1}{E}\left[\frac{\partial^2\phi}{\partial x_1^2} - \nu\frac{\partial^2\phi}{\partial x_2^2} + (1-\nu)V\right]$$

$$\epsilon_3 = -\frac{\nu}{E}\left[\frac{\partial^2\phi}{\partial x_2^2} + \frac{\partial^2\phi}{\partial x_1^2} + 2V\right], \quad \gamma_{12} = -\frac{1}{G}\frac{\partial^2\phi}{\partial x_1\partial x_2}.$$

Of course, the transverse shear strain components vanish, $\gamma_{13} = \gamma_{23} = 0$.

These strain components can be substituted into the strain compatibility equations (1.106c), (1.106b), (1.106a) and (1.106f) to obtain

$$\nabla^4\phi = -(1-\nu)\nabla^2 V, \qquad (3.28a)$$

$$\frac{\partial^2\epsilon_3}{\partial x_1^2} = \frac{\partial^4\phi}{\partial x_1^4} + \frac{\partial^4\phi}{\partial x_1^2\partial x_2^2} + 2\frac{\partial^2 V}{\partial x_1^2} = 0, \qquad (3.28b)$$

$$\frac{\partial^2\epsilon_3}{\partial x_2^2} = \frac{\partial^4\phi}{\partial x_2^4} + \frac{\partial^4\phi}{\partial x_1^2\partial x_2^2} + 2\frac{\partial^2 V}{\partial x_2^2} = 0, \qquad (3.28c)$$

$$\frac{\partial^2\epsilon_3}{\partial x_1\partial x_2} = \frac{\partial^4\phi}{\partial x_1^3\partial x_2} + \frac{\partial^4\phi}{\partial x_1\partial x_2^3} + 2\frac{\partial^2 V}{\partial x_1\partial x_2} = 0, \qquad (3.28d)$$

respectively, while the last two compatibility equations, eqs. (1.106d) and (1.106e), are automatically satisfied. This appears to be a complicated situation with four equations to define the stress function. It can be shown, however, that failure to satisfy the last three equations, eqs. (3.28b) to (3.28d), does not lead to large errors. Hence, a single equation for Airy's stress function remains

$$\nabla^4\phi = -(1-\nu)\nabla^2 V. \tag{3.29}$$

When the body forces vanish or are harmonic function, *i.e.*, when $\nabla^2 V = 0$, the governing equation becomes the homogeneous bi-harmonic equation, as is the case for the plane strain state.

In conclusion, both plane strain and plane stress states lead to nonhomogeneous bi-harmonic equations, eqs. (3.25) and (3.29), respectively. The two equations present only slight differences in their nonhomogeneous parts. For plane strain and plane stress problems, boundary conditions will differ considerably and the constitutive relationships are also different; hence, identical solutions of the two problems should not be expected. Nonetheless, the wealth of knowledge about how to solve bi-harmonic equations is useful for both types of problems.

3.4 Plane strain and plane stress in polar coordinates

A number of practical plane strain or plane stress problems present circular boundaries or cylindrical symmetry. Examples include such problems as thick-walled tubes subjected to torsion or internal pressure, thin sheets with circular holes, curved beams and many others.

To formulate these types of problems, the governing equations of elasticity must be recast in a polar (or cylindrical) coordinate system. While this can be accomplished by re-examination of differential volume and area elements defined in the cylindrical coordinate system, the equations can also be obtained from those derived in Cartesian coordinates through appropriate transformations. To this end, consider the coordinate system $(\bar{\imath}_1, \bar{\imath}_2)$ that forms the basis of a Cartesian system and the unit vectors of the polar system, $(\bar{\imath}_r, \bar{\imath}_\theta)$, as depicted fig. 3.5.

Polar coordinates are expressed in terms of their Cartesian counterparts through the following well-known relationships

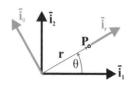

$$r = \sqrt{x_1^2 + x_2^2}, \quad \theta = \arctan\frac{x_2}{x_1}, \tag{3.30}$$

where r is the radial coordinate and θ is the angular coordinate, while the inverse transformation is readily obtained as

Fig. 3.5. Coordinate rotation from Cartesian into Polar.

$$x_1 = r\,\cos\theta, \quad x_2 = r\,\sin\theta. \tag{3.31}$$

Transformations of the displacement components expressed in the two coordinates systems are particular cases of the transformations expressed by eqs. (A.43), recast as

$$\begin{Bmatrix} u_r \\ u_\theta \end{Bmatrix} = \begin{bmatrix} \cos\theta & \sin\theta \\ -\sin\theta & \cos\theta \end{bmatrix} \begin{Bmatrix} u_1 \\ u_2 \end{Bmatrix}, \quad \begin{Bmatrix} u_1 \\ u_2 \end{Bmatrix} = \begin{bmatrix} \cos\theta & -\sin\theta \\ \sin\theta & \cos\theta \end{bmatrix} \begin{Bmatrix} u_r \\ u_\theta \end{Bmatrix}. \tag{3.32}$$

It will also be necessary to express the transformations of partial derivatives with respect to both coordinates system. The chain rule for derivatives implies that $\partial/\partial x_1 = (\partial/\partial r)(\partial r/\partial x_1) + (\partial/\partial\theta)(\partial\theta/\partial x_1)$, with a similar expression for the partial derivative with respect to x_2. It then follows that

$$
\left\{\begin{array}{c}\dfrac{\partial}{\partial x_1}\\[2mm]\dfrac{\partial}{\partial x_2}\end{array}\right\} = \begin{bmatrix}\cos\theta & -\sin\theta\\ \sin\theta & \cos\theta\end{bmatrix}\left\{\begin{array}{c}\dfrac{\partial}{\partial r}\\[2mm]\dfrac{1}{r}\dfrac{\partial}{\partial\theta}\end{array}\right\}. \tag{3.33}
$$

The derivatives of polar coordinates with respect to their Cartesian counterparts are easily developed from eq. (3.30) to find

$$
\frac{\partial r}{\partial x_1} = \frac{x_1}{r} = \cos\theta, \quad \frac{\partial r}{\partial x_2} = \frac{x_2}{r} = \sin\theta, \quad \frac{\partial\theta}{\partial x_1} = -\frac{\sin\theta}{r}, \quad \frac{\partial\theta}{\partial x_2} = \frac{\cos\theta}{r}. \tag{3.34}
$$

Next, the strain components expressed in the two coordinate systems will be related to each other using the general two-dimensional strain rotation expressions given by eqs. (1.91). The radial strain component, ϵ_r, becomes

$$
\epsilon_r = \epsilon_1 \cos^2\theta + \epsilon_2 \sin^2\theta + \gamma_{12} \sin\theta\cos\theta, \tag{3.35}
$$

where the Cartesian strain components, ϵ_1, ϵ_2, and γ_{12}, are computed by means of the strain-displacement equations, eqs. (1.63) and (1.71), to find

$$
\epsilon_1 = \frac{\partial u_1}{\partial x_1} = \left(\cos\theta\frac{\partial}{\partial r} - \frac{\sin\theta}{r}\frac{\partial}{\partial\theta}\right)(u_r\cos\theta - u_\theta\sin\theta),
$$

$$
\epsilon_2 = \frac{\partial u_2}{\partial x_2} = \left(\sin\theta\frac{\partial}{\partial r} + \frac{\cos\theta}{r}\frac{\partial}{\partial\theta}\right)(u_r\sin\theta + u_\theta\cos\theta),
$$

$$
\gamma_{12} = \frac{\partial u_1}{\partial x_2} + \frac{\partial u_2}{\partial x_1} = \left(\sin\theta\frac{\partial}{\partial r} + \frac{\cos\theta}{r}\frac{\partial}{\partial\theta}\right)(u_r\cos\theta - u_\theta\sin\theta)
$$
$$
+ \left(\cos\theta\frac{\partial}{\partial r} - \frac{\sin\theta}{r}\frac{\partial}{\partial\theta}\right)(u_r\sin\theta + u_\theta\cos\theta).
$$

Note that the partial derivatives and displacement components are evaluated with the help of eqs. (3.33) and (3.32), respectively. Finally, these strain components are substituted into eq. (3.35) to find, after considerable algebraic manipulation,

$$
\epsilon_r = (\cos^4\theta + 2\sin^2\theta\cos^2\theta + \sin^4\theta)\frac{\partial u_r}{\partial r} = \frac{\partial u_r}{\partial r}. \tag{3.36}
$$

A similar procedure can be followed to derive the three components of strain in the polar coordinate system as

$$
\epsilon_r = \frac{\partial u_r}{\partial r}, \tag{3.37a}
$$

$$
\epsilon_\theta = \frac{1}{r}\frac{\partial u_\theta}{\partial\theta} + \frac{u_r}{r}, \tag{3.37b}
$$

$$
\gamma_{r\theta} = \frac{\partial u_\theta}{\partial r} + \frac{1}{r}\frac{\partial u_r}{\partial\theta} - \frac{u_\theta}{r}. \tag{3.37c}
$$

A similar development can be carried out to express stress components in polar coordinates in terms of Airy's stress function. The two-dimensional stress component transformation equations (1.47) are used to express the radial stress component, σ_r, in terms of its Cartesian counterparts to find

$$\sigma_r = \sigma_1 \cos^2 \theta + \sigma_2 \sin^2 \theta + 2\tau_{12} \sin \theta \cos \theta$$
$$= \frac{\partial^2 \phi}{\partial x_2^2} \cos^2 \theta + \frac{\partial^2 \phi}{\partial x_1^2} \sin^2 \theta - 2 \frac{\partial^2 \phi}{\partial x_1 \partial x_2} \sin \theta \cos \theta,$$

where the Cartesian stress components are expressed in terms of Airy's stress function using eq. (3.24) and body force terms are neglected. The final step is to use eq. (3.33) to express the derivatives with respect to Cartesian coordinates in terms of derivatives with respect to polar coordinates. Tedious algebra then yields

$$\sigma_r = \frac{1}{r} \frac{\partial \phi}{\partial r} + \frac{1}{r^2} \frac{\partial^2 \phi}{\partial \theta^2}.$$

The same procedure can be used to obtain expressions for the remaining stress components in polar coordinates, leading to

$$\sigma_r = \frac{1}{r} \frac{\partial \phi}{\partial r} + \frac{1}{r^2} \frac{\partial^2 \phi}{\partial \theta^2}, \tag{3.38a}$$

$$\sigma_\theta = \frac{\partial^2 \phi}{\partial r^2}, \tag{3.38b}$$

$$\tau_{r\theta} = \frac{1}{r^2} \frac{\partial \phi}{\partial \theta} - \frac{1}{r} \frac{\partial^2 \phi}{\partial r \partial \theta}. \tag{3.38c}$$

To obtain a complete set of governing equations, it is also necessary to express the two equilibrium equations in polar coordinates. Figure 3.6 shows a differential element of area in polar coordinates with normal and shear stresses acting on each of its four faces. Since the stress state is assumed to vary smoothly, stress components on opposite faces of the differential element are expanded in Taylor series, using the first term of the series only.

The first equilibrium equation is obtained by projecting all forces along axis $\bar{\imath}_r$. Forces are obtained by multi-

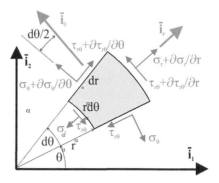

Fig. 3.6. Stresses acting on a differential area defined in polar coordinates.

plying the the stress components by the area on which they act, and a unit thickness of the volume element is assumed. Note that in view of the shape of the element, the circumferential stress component, σ_θ, contributes to the axial equilibrium equation because this components acts in a direction that forms an angle $d\theta/2$ with axis $\bar{\imath}_\theta$.

The second equilibrium equation is obtained by projecting forces along axis $\bar{\imath}_\theta$. It is left as an exercise to show that the resulting equilibrium equations are

$$\frac{\partial \sigma_r}{\partial r} + \frac{1}{r}\frac{\partial \tau_{r\theta}}{\partial \theta} + \frac{\sigma_r - \sigma_\theta}{r} = 0, \tag{3.39a}$$

$$\frac{1}{r}\frac{\partial \sigma_\theta}{\partial \theta} + \frac{\partial \tau_{r\theta}}{\partial r} + 2\frac{\tau_{r\theta}}{r} = 0. \tag{3.39b}$$

Finally, since the bi-harmonic equation governs both plane strain and plane stress problems, see eqs. (3.25) and (3.29), respectively, it is necessary to develop an expression for the Laplacian, ∇^2, in polar coordinates. This task is achieved by using eq. (3.33), which relates the derivatives with respect to Cartesian coordinates to derivatives with respect to polar coordinates, to find

$$\begin{aligned}\nabla^2 &= \frac{\partial^2}{\partial x_1^2} + \frac{\partial^2}{\partial x_2^2} = \left(\cos\theta\frac{\partial}{\partial r} - \frac{\sin\theta}{r}\frac{\partial}{\partial \theta}\right)^2 + \left(\sin\theta\frac{\partial}{\partial r} + \frac{\cos\theta}{r}\frac{\partial}{\partial \theta}\right)^2 \\ &= \frac{\partial^2}{\partial r^2} + \frac{1}{r}\frac{\partial}{\partial r} + \frac{1}{r^2}\frac{\partial^2}{\partial \theta^2}.\end{aligned} \tag{3.40}$$

The bi-harmonic operator then becomes

$$\nabla^4\phi = \left(\frac{\partial^2}{\partial r^2} + \frac{1}{r}\frac{\partial}{\partial r} + \frac{1}{r^2}\frac{\partial^2}{\partial \theta^2}\right)\left(\frac{\partial^2\phi}{\partial r^2} + \frac{1}{r}\frac{\partial\phi}{\partial r} + \frac{1}{r^2}\frac{\partial^2\phi}{\partial \theta^2}\right). \tag{3.41}$$

In the next section, several example problems will be solved to illustrate the use of polar coordinates for problems with cylindrical geometry.

3.5 Problem featuring cylindrical symmetry

Problems featuring cylindrical symmetry, that is, problems for which it is possible to assume that $\partial/\partial\theta = 0$, represent an important class of problems for which solutions are easily obtained because the process developed in the previous sections leads to ordinary, rather than partial differential equations. Such problems are also called *axisymmetric problems*, and the relationship between polar stress components and Airy's stress function, see eq. (3.38), reduces to

$$\sigma_r = \frac{1}{r}\frac{\partial\phi}{\partial r}, \quad \sigma_\theta = \frac{\partial^2\phi}{\partial r^2}, \quad \text{and} \quad \tau_{r\theta} = 0. \tag{3.42}$$

In the absence of body forces, the governing equation for both plane strain and plane stress problems becomes the bi-harmonic equation, see eqs. (3.25) and (3.29), respectively. In view of eq. (3.41), the governing equation becomes

$$\nabla^4\phi = \frac{d^4\phi}{dr^4} + \frac{2}{r}\frac{d^3\phi}{dr^3} - \frac{1}{r^2}\frac{d^2\phi}{dr^2} + \frac{1}{r^3}\frac{d\phi}{dr} = 0. \tag{3.43}$$

This is now an ordinary differential equation called the *Euler-Cauchy differential equation*. It can be transformed into an ordinary differential equation with constant coefficients through the following change of variables: $r = e^\xi$. Using the chain rule for derivatives, $d\phi/dr = (d\phi/d\xi)(d\xi/dr) = e^{-\xi} d\phi/d\xi$. Equation (3.43) then becomes

$$\frac{d^4\phi}{d\xi^4} - 4\frac{d^3\phi}{d\xi^3} + 4\frac{d^2\phi}{d\xi^2} = 0. \tag{3.44}$$

The solution to this equation is in the form $\phi = e^{z\xi}$, where z is a constant. This leads to the characteristic equation, $z^4 - 4z^3 + 4z^2 = z^2(z-2)^2 = 0$, with solutions $z = 0, 0, 2, 2$. In view of the repeated roots, the solution can then be written as $\phi(\xi) = C_1 + C_2\xi + C_3 e^{2\xi} + C_4\xi e^{2\xi}$ in terms of ξ, and finally, in terms of r as

$$\phi(r) = C_1 + C_2 \ln r + C_3 r^2 + C_4 r^2 \ln r,$$

where C_1, C_2, C_3, and C_4 are integration constants. In view of eq. (3.42), the stress components now become

$$\sigma_r = \frac{1}{r}\frac{d\phi}{dr} = \frac{C_2}{r^2} + 2C_3 + C_4(1 + 2\ln r), \quad \sigma_\theta = \frac{d^2\phi}{dr^2} = -\frac{C_2}{r^2} + 2C_3 + C_4(3 + 2\ln r). \tag{3.45}$$

Of course, the shear stress still vanishes, *i.e.*, $\tau_{r\theta} = 0$.

The determination of the integration constants and of the displacement field depends on the nature of the problem and the boundary conditions. The examples below illustrate the solution process.

Example 3.2. Thick-walled tube in plane strain state

Figure 3.7 shows a thick-walled cylinder of inner and outer radii, R_i and R_e, respectively, and subjected to internal and external pressures, p_i and p_e, respectively. Determine the stress and displacement distributions through the thickness of the tube.

The problem clearly presents cylindrical symmetry and hence, eq. (3.45) defines the stress state in the tube. In this example, the tube is assumed to be in a state of plane strain, *i.e.*, the axial strain component vanishes.

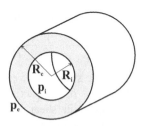

Fig. 3.7. Thick-walled tube subjected to internal and external pressures.

The applied pressures translate into boundary conditions at the inner and outer surfaces of the tube, $\sigma_r(r = R_e) = -p_e$ and $\sigma_r(r = R_i) = -p_i$. No boundary condition exists for the circumferential stress component, σ_θ, since this stress does not act on any of the boundaries of the system. Using eqs. (3.45), these boundary conditions become

$$-p_e = \frac{C_2}{R_e^2} + 2C_3 + C_4(1 + 2\ln R_e), \quad -p_i = \frac{C_2}{R_i^2} + 2C_3 + C_4(1 + 2\ln R_i). \quad (3.46)$$

The solution process now seems to have reached an impasse: three unknown coefficients, C_2, C_3 and C_4, must be evaluated to determine the stress components, but only two boundary conditions, eqs. (3.46), are available. To obtain the missing condition, the other fields of the problem, the strain and displacement fields, must be evaluated. First, the strain components are expressed in terms of their stress counterparts with the help of the constitutive laws, eqs. (3.19), to find

$$\epsilon_r = \frac{1 - \nu^2}{E}\sigma_r - \frac{\nu(1 + \nu)}{E}\sigma_\theta = C_a\sigma_r - C_b\sigma_\theta$$
$$= C_a\left[\frac{C_2}{r^2} + 2C_3 + C_4(1 + 2\ln r)\right] - C_b\left[-\frac{C_2}{r^2} + 2C_3 + C_4(3 + 2\ln r)\right],$$

$$\epsilon_\theta = \frac{1 - \nu^2}{E}\sigma_\theta - \frac{\nu(1 + \nu)}{E}\sigma_r = C_a\sigma_\theta - C_b\sigma_r$$
$$= C_a\left[-\frac{C_2}{r^2} + 2C_3 + C_4(3 + 2\ln r)\right] - C_b\left[\frac{C_2}{r^2} + 2C_3 + C_4(1 + 2\ln r)\right],$$

where $C_a = (1 - \nu^2)/E$ and $C_b = \nu(1 + \nu)/E$.

For problems presenting cylindrical symmetry, the strain-displacement equations, eqs. (3.37), reduce to $\epsilon_r = du_r/dr$ and $\epsilon_\theta = u_r/r$. Eliminating the radial displacement components from these two equations yields the strain compatibility condition: $\epsilon_r - \epsilon_\theta = r\, d\epsilon_\theta/dr$. Introducing the strain components computed above yields the following condition: $4C_aC_4 = 0$, and finally, $C_4 = 0$.

Equations (3.46) now involve only two unknown coefficients, which are easily found as $C_2 = -R_i^2 R_e^2 (p_i - p_e)/(R_e^2 - R_i^2)$ and $C_3 = (R_i^2 p_i - R_e^2 p_e)/2(R_e^2 - R_i^2)$, leading to the following expressions for the two stress components

$$\sigma_r(r) = \frac{R_i^2 p_i - R_e^2 p_e}{R_e^2 - R_i^2} - \frac{1}{r^2}\frac{(p_i - p_e)R_i^2 R_e^2}{R_e^2 - R_i^2},$$
$$\sigma_\theta(r) = \frac{R_i^2 p_i - R_e^2 p_e}{R_e^2 - R_i^2} + \frac{1}{r^2}\frac{(p_i - p_e)R_i^2 R_e^2}{R_e^2 - R_i^2}. \quad (3.47)$$

In view of the assumption of plane strain state, $\epsilon_3 = 0$ and $u_3 = 0$ and $\sigma_3 = \nu(\sigma_r + \sigma_\theta) = 2\nu(p_i R_i^2 - p_e R_e^2)/(R_e^2 - R_i^2)$: the axial stress component is constant through the thickness of the pipe. Since the shear stress components vanish, stress components σ_r, σ_θ, and σ_3 are, in fact, the principal stresses. Von Mises' equivalent stress is then readily obtained from eq. (2.32) as $2\sigma_{eq}^2 = (\sigma_r - \sigma_\theta)^2 + (\sigma_\theta - \sigma_3)^2 + (\sigma_3 - \sigma_r)^2$.

Figures 3.8 (a), (b) and (c) show the non-dimensional radial stress, σ_r/p_i, hoop stress, σ_θ/p_i, and Von Mises' equivalent stress, σ_e/p_i, respectively, when the cylinder is subjected to an internal pressure, p_i, i.e., when $p_e = 0$. Results are presented for three different ratios of the outer to inner radii ($\bar{R} = R_e/R_i = 1.5, 2.0$ and 3.0). The radial stress is compressive through the thickness of the cylinder and vanishes at the outer radial location, whereas the hoop stress is tensile. The maximum stress component is the hoop stress at $r = R_i$. A similar behavior is observed for the various values of $\bar{R}$. Clearly, von Mises' equivalent stress peaks at the inner radial location, *i.e.*, yield will initiate at the inside surface of the thick cylinder.

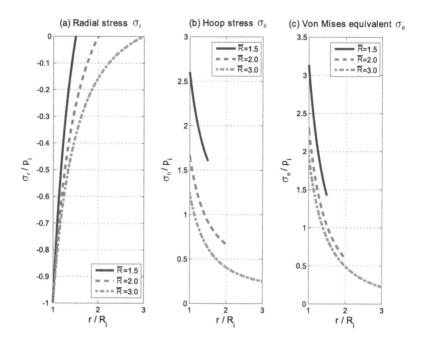

Fig. 3.8. Plots of the non-dimensional radial stress, σ_r/p_i, hoop stress, σ_θ/p_i, and Von Mises' equivalent stress, σ_e/p_i, for three different thickness ratios $\bar{R} = 1.5, 2.0, 3.0$.

The radial and hoop strain components are readily obtained from the constitutive laws, and finally, the radial displacement field is obtained as $u_r = r\epsilon_\theta$, leading to

$$u_r(r) = \frac{(1+\nu)(1-2\nu)}{E} \frac{R_i^2 p_i - R_e^2 p_e}{R_e^2 - R_i^2} r + \frac{1+\nu}{E} \frac{(p_i - p_e)R_i^2 R_e^2}{R_e^2 - R_i^2} \frac{1}{r}. \tag{3.48}$$

Figures 3.9 (a), (b) and (c) show the non-dimensional radial strain, $E\epsilon_r/p_i$, hoop strain, $E\epsilon_\theta/p_i$, and radial displacement, $Eu_r/(R_i p_i)$, respectively, when the cylinder is subjected to an internal pressure, p_i, *i.e.*, when $p_e = 0$. Results are presented for three different ratios of the outer to inner radii ($\bar{R} = R_e/R_i = 1.5, 2.0$ and 3.0).

Example 3.3. Thick-walled tube in plane stress state
Figure 3.7 shows a thick-walled cylinder of inner and outer radii, R_i and R_e, respectively, and subjected to internal and external pressures, p_i and p_e, respectively. In

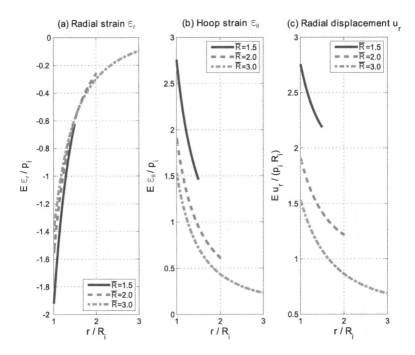

Fig. 3.9. Plots of the non-dimensional radial strain, $E\epsilon_r/p_i$, hoop strain $E\epsilon_\theta/p_i$, and radial displacement, $Eu_r/(R_ip_i)$, for three different thickness ratios $\bar{R} = 1.5, 2.0, 3.0$.

this example, the cylinder is assumed to be in a *state of plane stress*, in contrast with the plane strain assumption of example 3.2. Determine the stress and displacement distributions through the thickness of the tube.

The approach followed in the previous example could be used again here but with the constitutive laws associated with the plane stress state rather than those corresponding to the plane strain state. Instead of using Airy's stress function to satisfy the stress equilibrium and the compatibility equations, a displacement approach is used in this example.

It is assumed here that the cylinder is closed at both ends; hence, it is subjected to an axial load, $\pi R_i^2 p_i - \pi R_e^2 p_e$, which is assumed to be uniformly distributed over the cross-section of the tube, $\pi R_e^2 - \pi R_i^2$, leading to an axial stress

$$\sigma_3 = \frac{R_i^2 p_i - R_e^2 p_e}{R_e^2 - R_i^2}. \tag{3.49}$$

The constitutive laws for the material are given by Hooke's law, eqs. (2.4a) and (2.4b), as $E\epsilon_r = \sigma_r - \nu(\sigma_\theta + \sigma_3)$ and $E\epsilon_\theta = \sigma_\theta - \nu(\sigma_r + \sigma_3)$, respectively. Once recast in a matrix form, these relationships are readily inverted to find

$$\left\{ \begin{matrix} \sigma_r \\ \sigma_\theta \end{matrix} \right\} = \frac{E}{1-\nu^2} \begin{bmatrix} 1 & \nu \\ \nu & 1 \end{bmatrix} \left\{ \begin{matrix} \epsilon_r + \nu\sigma_3/E \\ \epsilon_\theta + \nu\sigma_3/E \end{matrix} \right\}. \tag{3.50}$$

Next, the radial and circumferential strain components are expressed in terms of the radial displacement component with the help of eqs. (3.37a) and (3.37b) to find

$$\sigma_r - \frac{\nu\sigma_3}{1-\nu} = \frac{E}{1-\nu^2}(\epsilon_r + \nu\epsilon_\theta) = \frac{E}{1-\nu^2}\left(\frac{du_r}{dr} + \frac{\nu u_r}{r}\right), \tag{3.51a}$$

$$\sigma_\theta - \frac{\nu\sigma_3}{1-\nu} = \frac{E}{1-\nu^2}(\epsilon_\theta + \nu\epsilon_r) = \frac{E}{1-\nu^2}\left(\frac{u_r}{r} + \nu\frac{du_r}{dr}\right). \tag{3.51b}$$

Note that the circumferential displacement component, u_θ, vanishes for this problem featuring cylindrical symmetry.

Finally, the radial and circumferential stress components are introduced into the radial equilibrium equation (3.39a) to obtain a single equation for the radial displacement component

$$\frac{d^2 u_r}{dr^2} + \frac{1}{r}\frac{du_r}{dr} - \frac{u_r}{r^2} = 0. \tag{3.52}$$

This is now an ordinary differential equation, similar to the Euler-Cauchy differential equation defined in eq. (3.43). It is, in fact, Navier's equation for this problem, and it could have been obtained by expressing eqs. (3.1) in polar coordinates, then imposing the cylindrical symmetry requirements.

Using the variable transformation $r = e^\xi$ and proceeding as before yields the displacement field as

$$u_r = C_1 r + C_2/r, \tag{3.53}$$

where C_1 and C_2 are two integration constants. The stress field, eqs. (3.51), becomes

$$\sigma_r - \frac{\nu\sigma_3}{1-\nu} = \frac{E}{1-\nu^2}\left[(1+\nu)C_1 - (1-\nu)\frac{C_2}{r^2}\right], \tag{3.54a}$$

$$\sigma_\theta - \frac{\nu\sigma_3}{1-\nu} = \frac{E}{1-\nu^2}\left[(1+\nu)C_1 + (1-\nu)\frac{C_2}{r^2}\right]. \tag{3.54b}$$

The integration constants are evaluated with the help of the boundary conditions at the inner and outer surfaces of the tube, $\sigma_r(r = R_e) = -p_e$ and $\sigma_r(r = R_i) = -p_i$, to find

$$\frac{EC_1}{1-\nu} = \frac{R_i^2 p_i - R_e^2 p_e}{R_e^2 - R_i^2} - \frac{\nu\sigma_3}{1-\nu}, \quad \text{and} \quad \frac{EC_2}{1+\nu} = \frac{(p_i - p_e)R_i^2 R_e^2}{R_e^2 - R_i^2}. \tag{3.55}$$

Introducing these constants into eqs. (3.54) yields the stress field as

$$\begin{aligned}
\sigma_r(r) &= \frac{R_i^2 p_i - R_e^2 p_e}{R_e^2 - R_i^2} - \frac{1}{r^2}\frac{(p_i - p_e)R_i^2 R_e^2}{R_e^2 - R_i^2}, \\
\sigma_\theta(r) &= \frac{R_i^2 p_i - R_e^2 p_e}{R_e^2 - R_i^2} + \frac{1}{r^2}\frac{(p_i - p_e)R_i^2 R_e^2}{R_e^2 - R_i^2}.
\end{aligned} \tag{3.56}$$

It is interesting to note that this stress field is identical to that found for the plane strain case, see eq. (3.47).

Note, however, that the axial displacements are different. Introducing the integration constants, eqs. (3.55), and the axial stress field, eq. (3.49), into eq. (3.53), yields

$$u_r = \frac{1 - 2\nu}{E} \frac{R_i^2 p_i - R_e^2 p_e}{R_e^2 - R_i^2} r + \frac{1 + \nu}{E} \frac{(p_i - p_e) R_i^2 R_e^2}{R_e^2 - R_i^2} \frac{1}{r}. \tag{3.57}$$

This expression should be compared with the corresponding displacement field for the plane strain case, see eq. (3.48).

If no end caps are present, the cylinder is not pressurized and $\sigma_3 = 0$. The analysis presented in this example remains valid, and the stress distributions are still given by eq. (3.56) and the integration constants by eq. (3.55) with $\sigma_3 = 0$. Finally, the axial displacement field becomes

$$u_r = \frac{1 - \nu}{E} \frac{R_i^2 p_i - R_e^2 p_e}{R_e^2 - R_i^2} r + \frac{1 + \nu}{E} \frac{(p_i - p_e) R_i^2 R_e^2}{R_e^2 - R_i^2} \frac{1}{r}. \tag{3.58}$$

Example 3.4. Thin-walled tube in plane stress state
Consider the thin-walled tube of mean radius R_m and thickness t subjected to an internal pressure p_i, as depicted in fig. 3.10. This problem is the limiting case of example 3.3, where $R_i = R_m - t/2$ and $R_e = R_m + t/2$, with $t/R_m \ll 1$.

Due of the internal pressure, a hoop force N acts in the tube. The free body diagram of a unit length of the upper part of the tube shown in fig. 3.10 yields the following equilibrium equation for the forces acting in the vertical direction

$$2N = \int_0^\pi p_i R_m \sin\theta \, d\theta = 2p_i R_m, \tag{3.59}$$

or $N = p_i R_m$. For thin-walled tubes, it is reasonable to assume that the hoop stresses are uniformly distributed through the thickness of the wall, leading to $N = t\sigma_\theta$, where σ_θ the hoop stress. It then follows that

$$\sigma_\theta = \frac{R_m p_i}{t}. \tag{3.60}$$

It is easy to show that this hoop stress is the average of the distribution predicted by the more detailed solution derived in example 3.3 for a thick tube under the same conditions. Indeed, the average of the circumferential stress given in eq. (3.56) is

$$\bar{\sigma}_\theta = \frac{1}{t} \int_{R_i}^{R_e} \left[\frac{R_i^2 p_i - R_e^2 p_e}{R_e^2 - R_i^2} + \frac{1}{r^2} \frac{(p_i - p_e) R_i^2 R_e^2}{R_e^2 - R_i^2} \right] dr = \frac{R_i p_i}{t}. \tag{3.61}$$

Since $R_m \approx R_i$ for thin-walled tubes, the two results are equivalent.

The hoop strain, ϵ_θ, is easily obtained as $\epsilon_\theta = \sigma_\theta/E = (R_m p_i)/(tE)$ and the radius of the ring increases by an amount $u_r = (R_m^2 p_i)/(Et)$, the radial displacement of the tube. Here again, this result can be checked by averaging the radial displacement distribution found earlier, see eq. (3.57), to find

$$\bar{u}_r = \frac{1}{t} \int_{R_i}^{R_e} u_r \, dr = \frac{1-\nu}{Et} \frac{R_i^2 p_i}{2} + \frac{1+\nu}{Et} \frac{R_i^2 R_e^2 p_i}{R_e^2 - R_i^2} \ln \frac{R_e}{R_i}$$

$$\approx \frac{1-\nu}{Et} \frac{R_m^2 p_i}{2} + \frac{1+\nu}{Et} \frac{R_m^2 p_i}{2} = \frac{R_m^2 p_i}{Et}.$$

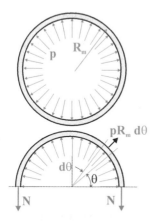

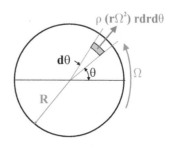

Fig. 3.10. Thin ring under internal pressure.   **Fig. 3.11.** Turbine disk rotating at high angular velocity Ω.

Example 3.5. *Turbine disk at high angular velocity*

Consider a homogeneous turbine disk of radius R rotating at high angular velocity Ω, as depicted in fig. 3.11. Due to the rotational speed of the turbine disk, each point on the disk is subjected to a centrifugal force $\rho(r\Omega^2)\, rdrd\theta$, where ρ is the material mass density, $-r\Omega^2$ the centripetal acceleration of the mass point, and $rdrd\theta$ the element of area on which the centrifugal force acts. Clearly, this centrifugal force acts in the radial direction of the polar coordinate system, and hence, the radial equilibrium equation, eq. (3.39a), must be modified to include a body force term,

$$\sigma_r - \sigma_\theta + r\frac{d\sigma_r}{dr} + \rho\Omega^2 r^2 = 0. \tag{3.62}$$

The disk is assumed to be in a state of plane stress, *i.e.*, $\sigma_3 = 0$, and the stresses can then be expressed in terms of the displacement field as

$$\sigma_r = \frac{E}{1-\nu^2}\left(\frac{du_r}{dr} + \frac{\nu u_r}{r}\right), \quad \sigma_\theta = \frac{E}{1-\nu^2}\left(\frac{u_r}{r} + \nu\frac{du_r}{dr}\right). \tag{3.63}$$

These expressions should be compared with eqs. (3.51).

Introducing the stress components, eq. (3.63), into the equilibrium equation of the problem, eq. (3.62), leads the governing equation for the radial displacement component

$$\frac{d^2 u_r}{dr^2} + \frac{1}{r}\frac{du_r}{dr} - \frac{u_r}{r^2} + (1 - \nu^2)\frac{\rho\Omega^2 r}{E} = 0. \tag{3.64}$$

The solution of this equation is

$$u_r = C_1 r + \frac{C_2}{r} - (1 - \nu^2)\frac{\rho\Omega^2}{E}\frac{r^3}{8}, \tag{3.65}$$

where the first two terms represent the solution of the homogeneous equation and the last term is the particular solution associated with the nonhomogeneous term in the equation. The displacement at the center of the disk, *i.e.*, at $r = 0$, must remain finite, and hence, $C_2 = 0$. The remaining integration constant, C_1, is determined by the boundary condition $\sigma_r(r = 0) = 0$ to give

$$C_1 = \frac{3 + \nu}{8(1 + \nu)}(1 - \nu^2)\frac{\rho\Omega^2}{E}R^2. \tag{3.66}$$

The stress field then becomes

$$\frac{\sigma_r}{\rho R^2 \Omega^2} = \frac{3 + \nu}{8}(1 - \bar{r}^2), \quad \frac{\sigma_\theta}{\rho R^2 \Omega^2} = \frac{3 + \nu}{8}\left(1 - \frac{1 + 3\nu}{3 + \nu}\bar{r}^2\right), \tag{3.67}$$

where $\bar{r} = r/R$. Note that the maximum stresses are found at the center of the disk, where $\sigma_r/(\rho R^2 \Omega^2) = \sigma_\theta/(\rho R^2 \Omega^2) = (3 + \nu)/8$. According to von Mises' criterion, the equivalent stress at that point becomes $\sigma_{eq}/(\rho R^2 \Omega^2) = (3 + \nu)/8$. The disk yields when $\sigma_{eq} = \sigma_y$, where σ_y is the material yield stress. The maximum speed at which the disk can rotate before centrifugal forces induce yielding is then

$$\Omega = \sqrt{\frac{8\sigma_y}{(3 + \nu)\rho R^2}}. \tag{3.68}$$

Example 3.6. Thin sheet with hole under uniaxial stress
Consider a thin sheet of material featuring a small hole. This problem is an idealization of a frequently encountered situation in aircraft structures. For instance, holes or cutout are common occurrences in aircraft skins to make a place for bolts, rivets, windows or access covers; similarly, bulkheads may have many holes that are passageways for cables, wires, or hydraulic lines. If the thin sheet is subjected to in-plane loading, a plane stress distribution will develop in the skin. Intuitively, the presence of the hole will increase the stress level in the sheet as compared to the stress level in the absence of a hole. The hole is said to be a *stress riser or stress concentrator*. This example will evaluate the stress distribution around the hole to identify the maximum stress level. The ratio of this maximum stress level to that observed in the absence of a hole is called the *stress concentration factor*.

Figure 3.12 shows the configuration considered here. A square plate of side dimension b presents a central circular hole of radius R_i, such that $R_i/b \ll 1$. The sheet is subjected to a far field unidirectional stress σ_a. A Cartesian coordinate system is selected with its origin at the center of the hole and axis $\bar{\imath}_2$ is aligned with the direction of the applied stress, σ_a. Since the hole is circular, it is natural to also make use of a polar coordinate system with its origin at the center of the hole; as shown in fig. 3.12, angle θ is measured from axis $\bar{\imath}_1$. Clearly, this problem does not present the cylindrical symmetry of the previous examples. It will be shown, however, that the problem can be treated as the superposition of two simpler problems: an axisymmetric and a non-axisymmetric problem.

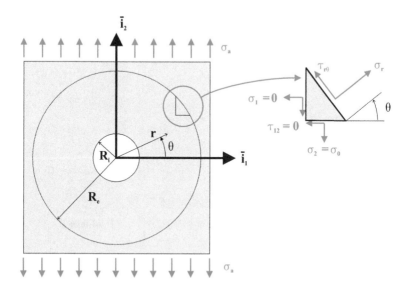

Fig. 3.12. Thin sheet with central hole of radius R_i subjected to uniaxial stress $\sigma_2 = \sigma_a$; also shown is the far field circle at $r = R_e$ where boundary conditions are applied.

The sheet is in a state of plane stress, and in the absence of body forces, the governing equation for Airy's stress function is the homogeneous form of the bi-harmonic partial differential equation (3.29). The boundary conditions around the edge of the hole are easily expressed in polar coordinates: both radial and shear stress components must vanish, $\sigma_r(r = R_i) = 0$ and $\tau_{r\theta}(r = R_i) = 0$. Because the circumferential stress, σ_θ, is not exposed around the inner edge of the circle, no condition is imposed on this stress component.

To avoid specifying boundary conditions in the Cartesian coordinate system, the far field stress σ_a is assumed to act on a circle of radius $R_e \gg R_i$; this assumption is consistent with the fact that the dimensions of the plate are much larger than the radius of the hole, $b \gg R_i$, as stated before. This implies $\sigma_1(r = R_e) = 0$, $\sigma_2(r = R_e) = \sigma_a$ and $\tau_{12}(r = R_e) = 0$.

These boundary conditions are stated in an awkward manner: stress components in a Cartesian system, σ_1, σ_2 and τ_{12}, are given at locations specified by polar coordinates $r = R_e$ and angle θ is arbitrary. To resolve this discrepancy, the stress components in the Cartesian system are transformed to their polar counterparts using the formulas for the rotation of stress components, eqs. (1.49a) and (1.49c), to find

$$\sigma_r(r = R_e, \theta) = \frac{\sigma_1 + \sigma_2}{2} + \frac{\sigma_1 - \sigma_2}{2} \cos 2\theta + \tau_{12} \sin 2\theta = \frac{\sigma_a}{2} - \frac{\sigma_a}{2} \cos 2\theta,$$

$$\tau_{r\theta}(r = R_e, \theta) = \qquad\qquad -\frac{\sigma_1 - \sigma_2}{2} \sin 2\theta + \tau_{12} \cos 2\theta = \frac{\sigma_a}{2} \sin 2\theta.$$

$$(3.69)$$

These equations could also be obtained directly from a Mohr's circle visualization, or they could be developed directly by expressing the equilibrium conditions of the triangular differential element depicted in the right portion of fig. 3.12.

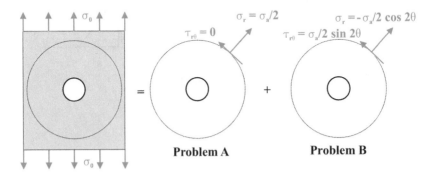

Fig. 3.13. The solution of the desired problem is found by superposing the solutions of two simpler problems: an axisymmetric problem, denoted "problem A" and a non-axisymmetric problem, denoted "problem B".

It now becomes possible to split the original problem into two simpler problems, both expressed in terms of polar coordinates as illustrated in fig. 3.13.

1. The axisymmetric problem, denoted Problem A, is subjected to the following boundary conditions: $\sigma_r = \tau_{r\theta} = 0$ around the edge of the hole, *i.e.*, at $r = R_i$, and $\sigma_r = \sigma_a/2$ and $\tau_{r\theta} = 0$ around the far field circular boundary, *i.e.*, at $r = R_e$. This problem is axisymmetric because the geometry of the problem presents cylindrical symmetry and the boundary conditions are independent of θ.

2. The non-axisymmetric problem, denoted Problem B, is subjected to the following boundary conditions $\sigma_r = \tau_{r\theta} = 0$ around the edge of the hole, *i.e.*, for $r = R_i$, and $\sigma_r = -\sigma_a/2 \cos 2\theta$ and $\tau_{r\theta} = \sigma_a/2 \sin 2\theta$ around the far field circular boundary, *i.e.*, for $r = R_e$. This problem is not axisymmetric because while the geometry of the problem does present cylindrical symmetry, the boundary conditions do depend on θ.

The solution to Problem A is developed in example 3.3. It consists of a thin cylinder subjected to an external pressure, $p_e = -\sigma_a/2$. The stress field is readily obtained by introducing $p_i = 0$ and $p_e = -\sigma_a/2$ eqs. (3.47) to find

$$\sigma_r^A = \frac{\sigma_a}{2} \left[\frac{R_e^2}{R_e^2 - R_i^2} - \frac{R_i^2 R_e^2}{R_e^2 - R_i^2} \frac{1}{r^2} \right], \quad \sigma_\theta^A = \frac{\sigma_a}{2} \left[\frac{R_e^2}{R_e^2 - R_i^2} + \frac{R_i^2 R_e^2}{R_e^2 - R_i^2} \frac{1}{r^2} \right].$$
(3.70)

The solution of Problem B is more difficult, and requires the solution of the homogeneous bi-harmonic equation in polar coordinates given by eq. (3.41). Since the bi-harmonic operator only contains even derivatives with respect to θ, an approach based on separation of variables seems appropriate. A solution of the following form is proposed

$$\phi(r, \theta) = \eta(r) \cos 2\theta.$$
(3.71)

Substituting this assumed solution into the homogeneous bi-harmonic equation in polar coordinates, eq. (3.41), leads to the following equation for Airy's stress function

$$\nabla^4 \phi = \left(\frac{d^4 \eta}{dr^4} + \frac{2}{r} \frac{d^3 \eta}{dr^3} - \frac{9}{r^2} \frac{d^2 \eta}{dr^2} + \frac{9}{r^3} \frac{d\eta}{dr} \right) \cos 2\theta = 0.$$

Because this expression must be valid for *all values* θ, the term in parentheses must vanish, and hence,

$$\frac{d^4 \eta}{dr^4} + \frac{2}{r} \frac{d^3 \eta}{dr^3} - \frac{9}{r^2} \frac{d^2 \eta}{dr^2} + \frac{9}{r^3} \frac{d\eta}{dr} = 0.$$

This is another instance of the Euler-Cauchy differential equation first encountered in section 3.5. Using the same procedure as before, the following solution is found: $\eta(r) = C_1 + C_2 r^2 + C_3 r^4 + C_4/r^2$. Airy's stress function now becomes

$$\phi(r, \theta) = \left[C_1 + C_2 r^2 + C_3 r^4 + \frac{C_4}{r^2} \right] \cos 2\theta.$$

Next, the stress field is obtained by introducing the stress function into eqs.(3.38) to find the stress components as

$$\sigma_r = \frac{1}{r} \frac{\partial \phi}{\partial r} + \frac{1}{r^2} \frac{\partial^2 \phi}{\partial \theta^2} = -\left[2C_2 + \frac{4C_1}{r^2} + \frac{6C_4}{r^4} \right] \cos 2\theta,$$

$$\sigma_\theta = \frac{\partial^2 \phi}{\partial r^2} = \left[2C_2 r + 12C_3 r^2 + \frac{6C_4}{r^4} \right] \cos 2\theta,$$

$$\tau_{r\theta} = \frac{1}{r^2} \frac{\partial \phi}{\partial \theta} - \frac{1}{r} \frac{\partial^2 \phi}{\partial r \partial \theta} = \left[2C_2 + 6C_3 r^2 - \frac{2C_1}{r^2} - \frac{6C_4}{r^4} \right] \sin 2\theta.$$

The boundary conditions, $\sigma_r = 0$ and $\tau_{r\theta} = 0$ at $r = R_i$, yield the following two equations

$$\sigma_r(r = R_i) = -\left[\frac{4}{R_i^2} C_1 + 2C_2 + \frac{6C_4}{R_i^4} \right] \cos 2\theta = 0,$$

$$\tau_{r\theta}(r = R_i) = \left[-\frac{2}{R_i^2} C_1 + 2C_2 + 6R_i^2 C_3 - \frac{6C_4}{R_i^4} \right] \cos 2\theta = 0,$$

whereas the boundary conditions, $\sigma_r = -\sigma_a/2 \cos 2\theta$ and $\tau_{r\theta} = \sigma_a/2 \sin 2\theta$ at $r = R_e$, lead to

$$\sigma_r(r = R_e) = -\left[\frac{4}{R_e^2}C_1 + 2C_2 + \frac{6C_4}{R_e^4}\right]\cos 2\theta = -\frac{\sigma_a}{2}\cos 2\theta,$$

$$\tau_{r\theta}(r = R_e) = \left[-\frac{2}{R_e^2}C_1 + 2C_2 + 6R_e^2C_3 - \frac{6C_4}{R_e^4}\right]\sin 2\theta = \frac{\sigma_a}{2}\sin 2\theta.$$

These four algebraic equations are used to determine the four integration constants, C_1, C_2, C_3 and C_4. This task is more easily achieved by recasting the equations in a matrix form as

$$\begin{bmatrix} -4/R_i^2 & -2 & 0 & -6/R_i^4 \\ -2/R_i^2 & 2 & 6R_i^2 & -6/R_i^4 \\ 4/R_e^2 & 2 & 0 & 6/R_e^4 \\ -2/R_e^2 & 2 & 6R_e^2 & -6/R_e^4 \end{bmatrix} \begin{Bmatrix} C_1 \\ C_2 \\ C_3 \\ C_4 \end{Bmatrix} = \begin{Bmatrix} 0 \\ 0 \\ \sigma_a/2 \\ \sigma_a/2 \end{Bmatrix}. \tag{3.72}$$

Note that the canceling of the trigonometric function of angle θ indicates that the assumed form of Airy's stress function, eq. (3.71), is able to satisfy all the boundary conditions for the particular problem. The solution to this set of algebraic equations is readily accomplished, but results are long and tedious expressions.

Since the interest is not in solutions for finite values of the outer radius, R_e, it is easier to immediately consider the situation where $R_e \to \infty$, or more specifically, where $1/R_e \to 0$. Applying this to eq. (3.72) results in

$$\begin{bmatrix} -4/R_i^2 & -2 & 0 & -6/R_i^4 \\ -2/R_i^2 & 2 & 6R_i^2 & -6/R_i^4 \\ 0 & 2 & 0 & 0 \\ 0 & 0 & 6 & 0 \end{bmatrix} \begin{Bmatrix} C_1 \\ C_2 \\ C_3 \\ C_4 \end{Bmatrix} = \begin{Bmatrix} 0 \\ 0 \\ \sigma_a/2 \\ 0 \end{Bmatrix}.$$

The last equation is divided by R_e^2 before taking the limit to insure that the third term remained finite. These equations can be solved to yield

$$C_1 = -\frac{R_i^2}{2}\sigma_a, \quad C_2 = \frac{\sigma_a}{4}, \quad C_3 = 0, \quad C_4 = \frac{R_i^4}{4}\sigma_a.$$

The solutions to Problem A and Problem B can now be combined to yield the complete solution for the state of stress around the circular hole of radius R_i

$$\sigma_r(r, \theta) = \frac{\sigma_a}{2}\left[\left(1 - \frac{R_i^2}{r^2}\right) + \left(-1 + 4\frac{R_i^2}{r^2} - 3\frac{R_i^4}{r^4}\right)\cos 2\theta\right],$$

$$\sigma_\theta(r, \theta) = \frac{\sigma_a}{2}\left[\left(1 + \frac{R_i^2}{r^2}\right) + \left(1 + 3\frac{R_i^4}{r^4}\right)\cos 2\theta\right],$$

$$\tau_{r\theta}(r, \theta) = \frac{\sigma_a}{2}\left[1 + 2\frac{R_i^2}{r^2} - 3\frac{R_i^4}{r^4}\right]\sin 2\theta.$$

These results show that the stress components decrease in the inverse proportion of the square of the distance from the center of the hole. As expected, at a large distance from the hole, the far field uniaxial stress state is recovered, $\sigma_r = \sigma_a/2\,(1 - \cos 2\theta)$,

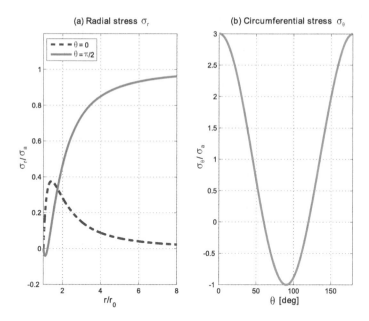

Fig. 3.14. Plots of stress state around circular hole in a thin sheet: (a) σ_r/σ_a along radii at $\theta = 0$ and $\theta = \pi/2$, and (b) σ_θ/σ_a around inside edge of hole.

$\sigma_\theta = \sigma_a (1 + \cos 2\theta)$, and $\tau_{r\theta} = \sigma_a/2 \sin 2\theta$, which in the Cartesian coordinate system, corresponds to $\sigma_2 = \sigma_a$ and $\sigma_1 = \tau_{12} = 0$. Figure 3.14(a) shows that the radial stress component, σ_r, rapidly approaches its asymptotic values of zero and σ_a, along two radial lines corresponding to $\theta = 0$ and 90 degrees, respectively.

Around the edge of the hole, *i.e.*, for $r = R_i$, the radial and shear stress components vanish, as required, but the circumferential stress does not: $\sigma_\theta(\theta) = \sigma_a(1 + 2 \cos 2\theta)$. Figure 3.14(b) shows the distribution of this hoop stress around the hole; note the peak values of $3\sigma_a$ at $\theta = 0$ or π, and of $-\sigma_a$ at $\theta = \pi/2$ or $3\pi/2$. The distribution of hoop stress over the other half of the hole, *i.e.*, for $\pi \leq \theta \leq 2\pi$, is the mirror image of that on the upper half of the hole.

Several important conclusions can be drawn from this example. The most significant is that the presence of a circular hole in a thin sheet under a uniaxial state of stress causes the appearance of a peak circumferential stress at the edge of the hole. This stress component peaks at a level that is 3 times as large as that of the applied stress, *i.e.*, the hole creates a *stress concentration factor* of 3. If the sheet is designed based on a simple yield criterion, $\sigma_{\max} < \sigma_y$, where σ_y is the yield stress for the material, the presence of the hole reduces the load carrying capacity of the sheet by a factor of three. The stress concentration factor is *independent of the hole size*; the above analysis just requires the hole diameter to be much smaller than the dimensions of the sheet. Consequently, no matter how small the hole is, the load carrying capability of the panel is reduced by a factor of three. In practice, because the hoop stress peaks in a relatively small region, the material will locally yield, and the load carrying capacity of the sheet will not be reduced as dramatically. If the panel is subjected to cyclic loads, however, cracks are likely to develop in the high stress area, possibly reducing the life of the component significantly.

The disturbance in the far field stress caused by the presence of the hole quickly decays away from the center of the hole, as illustrated in fig. 3.14. This means that the presence of the hole in "felt" only in a small area. Finally, it is interesting to note that the hoop stress is actually negative, $\sigma_\theta = -\sigma_a$, in the area around $\theta = \pm\pi/2$, that is, in the regions above and below the hole along axis $\bar{\imath}_2$. Consequently, secondary attachments might be made in this area without causing further problems.

The solution presented above is readily generalized to the case where $\sigma_1 = \sigma_b$ and $\sigma_2 = 0$, simply by replacing σ_a by σ_b and θ by $\theta + \pi/2$ in the above solution. Indeed, in the above solution, the applied loading direction is arbitrarily selected to coincide with that of axis $\bar{\imath}_2$. The solution for a sheet subjected to the biaxial state of stress $\sigma_1 = \sigma_b$ and $\sigma_2 = \sigma_a$ would then be obtained by superposing the two solutions. For example, if the far field stress is the pure shear stress state, the solution is obtained by setting $\sigma_1 = -\tau_0$ and $\sigma_2 = \tau_0$.

Example 3.7. Reinforced hole in a thin panel
In example 3.6, the presence of a hole in a thin panel is shown to cause a considerable disturbance in the stress field in the panel, and a stress concentration factor appears around the edge of the hole. In this example, the following question is raised: is it possible to eliminate this stress concentration by reinforcing the edge of the hole? Figure 3.15 shows the configuration to be investigated: the panel features a hole of radius R_i, but this time, a circular ring of cross-sectional area $\mathcal{A}$ and thickness t reinforces the hole. The circular ring that reinforces the hole is called a "boss."

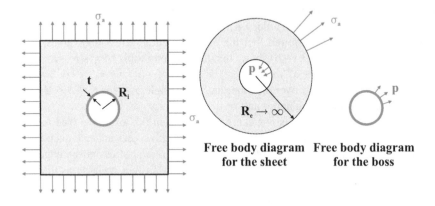

Free body diagram **Free body diagram**
for the sheet **for the boss**

Fig. 3.15. A thin panel with a hole subjected to a biaxial state of stress.

The panel is subjected to a biaxial state of stress, $\sigma_1 = \sigma_2 = \sigma_a$. Furthermore, the dimensions of the panel are assumed to be much larger than the radius of the hole. Consequently, the square panel can be replaced by a circular panel of radius $R_e \rightarrow \infty$, subjected to an external pressure, $p_e = -\sigma_a$. Finally, the boss fits into the circular hole, and hence, the boss and panel must interact through an unknown pressure p. Figure 3.15 shows the free body diagram of the panel and boss, separately. The circular panel is subjected to an external pressure, $p_e = -\sigma_a$, and an internal pressure, $p_i = -p$, of unknown magnitude. On the other hand, the boss is a thin ring subjected to an internal pressure of magnitude p. The magnitude of this unknown pressure will be found by imposing displacement compatibility: the radial displacements of the boss and hole in which it fits must match.

Because the circular panel is in a state of plane stress, the results developed in example 3.3 do apply. In particular, the stress field in the panel is given by eqs. (3.56), and hence $\sigma_r = -p_e - (p_i - p_e)/\bar{r}^2$ and $\sigma_\theta = -p_e + (p_i - p_e)/\bar{r}^2$, where $\bar{r} = r/R_i$. In this case, $p_e = -\sigma_a$ and $p_i = -p$, leading to the following stress field in the panel

$$\sigma_r = \sigma_a - \frac{\sigma_a - p}{\bar{r}^2}, \qquad \sigma_\theta = \sigma_a + \frac{\sigma_a - p}{\bar{r}^2}. \tag{3.73}$$

Next, the radial displacement distribution follows from eq. (3.58) as $Eu_r/R_i = -(1-\nu)p_e\bar{r} + (1+\nu)(p_i - p_e)/\bar{r}$. Because, $p_e = -\sigma_a$ and $p_i = -p$, the radial displacement of the edge of the hole becomes

$$u_r(r = R_i) = \frac{R_i}{E}[(1-\nu)\sigma_a + (1+\nu)(\sigma_a - p)] = \frac{R_i}{E}[2\sigma_a - (1+\nu)p].$$

On the other hand, the radial displacement of the boss is evaluated with the help of eq. (3.62) to find $u_r = (R_i^2 p)/(Et)$. If w is the width of the boss, its cross-sectional area is then $\mathcal{A} = wt$, and the radial displacement becomes $u_r = (R_i^2 wp)/(E\mathcal{A})$. Compatibility requires the radial displacement of the hole in the sheet to be identical to that of the boss, i.e., $R_i[2\sigma_a - (1+\nu)p]/E = (R_i^2 wp)/(E\mathcal{A})$. This condition yields the interface pressure between the boss and sheet as

$$p = \frac{2\sigma_a}{(1+\nu) + R_i w/\mathcal{A}}. \tag{3.74}$$

The stress field in the panel is evaluated by introducing the value of this pressure into eqs. (3.73).

It is now possible to answer the question raised at the beginning of this example: is it possible to eliminate this stress concentration by reinforcing the edge of the hole with the boss? A cursory examination of eq. (3.73) reveals that if $p = \sigma_a$, the stress components in the panel are $\sigma_r = \sigma_\theta = \sigma_a$, i.e., the stress field is *identical as that in the panel without a hole*. If $p = \sigma_a$, eq. (3.74) then implies

$$\mathcal{A} = \frac{wR_i}{1-\nu}. \tag{3.75}$$

In other words, if the cross-sectional area of the boss is given by the above relationship, the stress field in the panel is undisturbed by the presence of the hole: the panel "does not see" or "does not feel" the presence of the hole. A similar technique is used in aircraft fuselages: a boss is placed around the windows of the fuselage so as to leave the stress field undisturbed to the largest possible extent. Of course, since the fuselage is subjected to a variety of loading conditions, the boss minimizes the effect of the window on the fuselage stress distribution without completely eliminating it.

Example 3.8. Thin-walled spherical pressure vessel

The reasoning developed in example 3.4 can readily be extended to the situation of a thin-walled sphere of radius R and thickness t subjected to an internal pressure p, as shown in fig. 3.16. This type of configuration is representative of spherical pressure vessels.

First, the sphere is cut by a horizontal plane passing through its center, to reveal the free body diagram shown in the figure. Due to the symmetry of the problem, the pressure acting on the upper half of the sphere will be equilibrated by a hoop stress, σ_h, which is uniformly distributed around the circle at the intersection of the sphere with the plane of the cut. The total upward force generated by the pressure, $\pi R^2 p$, is equilibrated by the downward force generated by the distributed hoop stress, $2\pi R t \sigma_h$, assumed to be uniformly distributed through the thickness of the wall. This yields the following result

$$\sigma_h = \frac{pR}{2t}. \tag{3.76}$$

Note that the hoop stress is half of that in a pressurized tube of equal radius and thickness, see eq. (3.76).

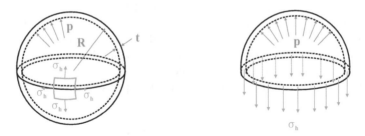

Fig. 3.16. Thin sphere under internal pressure.

Of course, in view of the spherical symmetry of the problem, the orientation of the plane of the cut is arbitrary. Hence, the hoop stress derived above is acting on a face with an arbitrary orientation. As shown in fig. 3.16, the stresses acting on an arbitrary differential element cut from the thin-walled sphere are σ_h in two orthogonal directions. Because the shear stress component vanishes, these are the principal stresses, and hence, $\sigma_{p1} = \sigma_{p2} = \sigma_h$. Note that Mohr's circle then reduces to a single point at ordinate σ_h.

For a linearly elastic material, the hoop strain, ϵ_h, is obtained from Hooke's law, eq. (2.4), as

$$\epsilon_1 = \epsilon_2 = \epsilon_h = \frac{1-\nu}{2} \frac{R}{t} \frac{p}{E}. \tag{3.77}$$

The deformation is identical in all directions, due to the spherical symmetry of the problem. Since the shear strain components vanish, the principal strains are $\epsilon_{p1} = \epsilon_{p2} = \epsilon_h$. The radius of the sphere increases by an amount $\Delta R = (1 - \nu)(pR^2)/(2Et)$.

If the wall thickness is very much smaller than the radius of the sphere, its curvature become unimportant. Hence, it is possible to look at the sphere as a thin, flat sheet of material subjected to a biaxial state of stress where $\sigma_a = \sigma_h$, as depicted in fig. 3.15.

Pressure vessels must often be drilled to install manifolds that monitor the internal pressure or to let the pressurized gas or fluid in and out of the vessel. Such situation is identical to that discussed in example 3.7: a thin panel under a biaxial state of stress featuring a circular hole. To minimize the effect of the hole on the stress distribution in the pressure vessel, it is common to reinforce the hole with a circular ring, as discussed in the example. For the optimum boss design given by eq. (3.75), the stress distribution in the spherical pressure vessel will remain undisturbed by the presence of a circular hole.

3.5.1 Problems

Problem 3.1. Navier's equations
Develop the three Navier equations following the procedure described in section 3.1.1.

Problem 3.2. A solution to Navier's equations
In principle, Navier's equations should allow solution for the unknown displacements within a solid body from which the stresses can be computed using the strain-displacement and stress-strain equations. However, they are not as useful as expected because it is very difficult to express the necessary displacement boundary conditions for most practical problems. Nonetheless a few solutions can be illustrated. Consider a problem with body forces given by: $b_1 = -6Gx_2x_3$, $b_2 = 2Gx_3x_1$, and $b_3 = 10Gx_1x_2$, and assume displacements given by $u_1 = C_1x_1^2x_2x_3$, $u_2 = C_2x_1x_2^2x_3$, and $u_3 = C_3x_1x_2x_3^2$. Also assume $G = E/2(1 + \nu)$ and $\nu = 1/4$. Determine the constants, C_1, C_2, and C_3 which allow satisfaction of the Navier equations. Hint: you will eventually need to solve 3 simultaneous equations.

Problem 3.3. Equilibrium equations in polar coordinates
Derive the plane stress equilibrium equations (one equation in the r and a second in the θ directions). Figure 3.6 provides the appropriate free body diagram. Make sure when you write a force equilibrium equation that you multiply all stresses by appropriate areas (assume the material has a unit thickness). You will need to account for the slight difference $(d\theta)$ in the direction of on opposite sides of the element when writing the equilibrium equations in both the r and θ directions. You will also need to use Taylor Series to express the differential changes in σ_r and σ_θ in the same manner as is done for rectangular differential areas.

Problem 3.4. Strain compatibility equations in polar coordinates
For plane stress problems presenting cylindrical symmetry, the strain-displacement equations expressed in polar coordinates are: $\epsilon_r = du_r/dr$, $\epsilon_\theta = u_r/r$, and $\gamma_{r\theta} = 0$. (1) How many strain compatibility equations exist for this problem? (2) Derive the strain compatibility equations, if any.

Problem 3.5. Thick-walled cylinder under internal pressure

Consider a thick-walled cylinder of internal and external radii R_i and R_e, respectively, in a state of plane strain subjected to an internal pressure p_i. *(1)* Plot the non-dimensional radial stress, σ_r/p_i, distribution through the thickness of the cylinder. *(2)* Plot the distribution of non-dimensional circumferential stress, σ_θ/p_i. *(3)* Plot the distribution of von Mises' equivalent stress, σ_e/p_i. *(4)* If the yield stress for the material is σ_y, plot the maximum internal pressure the thick-walled cylinder can carry as a function of $\rho = R_e/R_i$. What is the maximum pressure p_i/σ_y that can be carried by a very thick cylinder? *(5)* Plot the distribution of non-dimensional radial strain, $E\epsilon_r/p_i$. *(6)* Plot the distribution of non-dimensional circumferential strain, $E\epsilon_\theta/p_i$. *(7)* Plot the distribution of non-dimensional radial displacement, $Eu_r/(R_ip_i)$. Present all your results for $\rho = 1.5, 2.0$ and 3.0; use the radial coordinate $\bar{r} = r/R_i$.

Problem 3.6. Thick-walled cylinder under external pressure

Consider a thick-walled cylinder of internal and external radii R_i and R_e, respectively, in a state of plane strain subjected to an external pressure p_e. *(1)* Plot the non-dimensional radial stress, σ_r/p_e, distribution through the thickness of the cylinder. *(2)* Plot the distribution of non-dimensional circumferential stress, σ_θ/p_e. *(3)* Plot the distribution of von Mises' equivalent stress, σ_e/p_e. *(4)* If the yield stress for the material is σ_y, plot the maximum external pressure the thick-walled cylinder can carry as a function of $\rho = R_e/R_i$. What is the maximum pressure p_e/σ_y that can be carried by a very thick cylinder? *(5)* Plot the distribution of non-dimensional radial strain, $E\epsilon_r/p_e$. *(6)* Plot the distribution of non-dimensional circumferential strain, $E\epsilon_\theta/p_e$. *(7)* Plot the distribution of non-dimensional radial displacement, $Eu_r/(R_ip_e)$. Present all your results for $\rho = 1.5, 2.0$ and 3.0; use the radial coordinate $\bar{r} = r/R_i$.

Problem 3.7. Thick-walled cylinder under internal pressure

Consider a thick-walled cylinder of internal and external radii R_i and R_e, respectively, in a state of plane stress subjected to an internal pressure p_i. *(1)* Plot the non-dimensional radial stress, σ_r/p_i, distribution through the thickness of the cylinder. *(2)* Plot the distribution of non-dimensional circumferential stress, σ_θ/p_i. *(3)* Plot the distribution of von Mises' equivalent stress, σ_e/p_i. *(4)* If the yield stress for the material is σ_y, plot the maximum internal pressure the thick-walled cylinder can carry as a function of $\rho = R_e/R_i$. What is the maximum pressure p_i/σ_y that can be carried by a very thick cylinder? *(5)* Plot the distribution of non-dimensional radial strain, $E\epsilon_r/p_i$. *(6)* Plot the distribution of non-dimensional circumferential strain, $E\epsilon_\theta/p_i$. *(7)* Plot the distribution of non-dimensional radial displacement, $Eu_r/(R_ip_i)$. Present all your results for $\rho = 1.5, 2.0$ and 3.0; use the radial coordinate $\bar{r} = r/R_i$.

Problem 3.8. Thick-walled cylinder under external pressure

Consider a thick-walled cylinder of internal and external radii R_i and R_e, respectively, in a state of plane stress subjected to an external pressure p_e. *(1)* Plot the non-dimensional radial stress, σ_r/p_e, distribution through the thickness of the cylinder. *(2)* Plot the distribution of non-dimensional circumferential stress, σ_θ/p_e. *(3)* Plot the distribution of von Mises' equivalent stress, σ_e/p_e. *(4)* If the yield stress for the material is σ_y, plot the maximum external pressure the thick-walled cylinder can carry as a function of $\rho = R_e/R_i$. What is the maximum pressure p_e/σ_y that can be carried by a very thick cylinder? *(5)* Plot the distribution of non-dimensional radial strain, $E\epsilon_r/p_e$. *(6)* Plot the distribution of non-dimensional circumferential strain, $E\epsilon_\theta/p_e$. *(7)* Plot the distribution of non-dimensional radial displacement, $Eu_r/(R_ip_e)$. Present all your results for $\rho = 1.5, 2.0$ and 3.0; use the radial coordinate $\bar{r} = r/R_i$.

Problem 3.9. Disk rotating at high speed

A disk of mass density ρ, and inner and outer radii denoted a and b, respectively, is spinning about a fixed point at an angular velocity Ω. *(1)* Plot the distribution of non-dimensional radial stress, $\sigma_r/(\rho a^2 \Omega^2)$, through the thickness of the disk. *(2)* Plot the distribution of non-dimensional circumferential stress, $\sigma_\theta/(\rho a^2 \Omega^2)$. *(3)* Plot the distribution of non-dimensional von Mises' equivalent stress, $\sigma_e/(\rho a^2 \Omega^2)$. Present your stress distributions for $b/a = 1.5, 2.0$ and 3.0, as a function of $\bar{r} = r/a$. *(4)* First, let the inner radius, a, be fixed. Plot the maximum allowable non-dimensional angular speed, $\Omega_{max}\sqrt{\rho a^2/\sigma_y}$ as a function of $b/a \in [1.0, 10.0]$, *i.e.*, as the outer radius of the cylinder increases. Use von Mises' criterion to predict yielding, σ_y denotes the yield stress. *(5)* Next, let the outer radius, b, be fixed. Plot the maximum allowable non-dimensional angular speed, $\Omega_{max}\sqrt{\rho b^2/\sigma_y}$ as a function of $a/b \in [0.0, 1.0]$, *i.e.*, as the inner radius of the cylinder decreases. Comment on the significance of these last two results. Hint: the boundary conditions of the problem are $\sigma_r(r = a) = 0$ and $\sigma_r(r = b) = 0$.

Problem 3.10. Two cylinder assembly

Figure 3.17 shows two cylinders that have been assembled by a process called "shrink-fitting." The inner cylinder has nominal internal and external radii of a and b, respectively, whereas the corresponding quantities for the external cylinder are b and c, respectively. Assume that the unconstrained external radius of the inner cylinder *exceeds* the initially unconstrained internal radius of the external cylinder by an amount δ, where $\delta \ll b$. The two components are assembled by first heating the outer cylinder so that it expands, slipping the outer cylinder

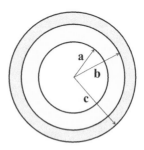

Fig. 3.17. Two concentric cylinder assembled by heat treatment.

over the inner, then letting the two components cool down. *(1)* Find the pressure, p, acting between the two cylinder after cool down. *(2)* Find the common radial displacement of the two cylinder at their interface. Hint: draw a free body diagram of the two cylinders separately. The internal cylinder is acted upon by an external pressure, p, whereas the external cylinder carries an internal pressure, p. This pressure can be found by imposing the compatibility of radial displacement at the interface between the cylinders.

Problem 3.11. Von Mises' equivalent stress around a hole in thin sheet

Consider a thin panel with a central circular hole of radius R_i subjected to a far field biaxial state of stress $\sigma_1 = \sigma_b$ and $\sigma_2 = \sigma_a$. *(1)* Evaluate the stress field in the panel. *(2)* Evaluate the non-dimensional Von Mises' equivalent stress σ_{eq}/σ_a, where σ_{eq} is defined by eq. (2.36). *(3)* Plot the distribution of the equivalent stress for $1 \le \bar{r} \le 5$, where $\bar{r} = r/R_i$, and $0 \le \theta \le 2\pi$. Plot your results for $\sigma_b/\sigma_a = -1.0$, *i.e.*, when the panel is in a state of pure shear, and for $\sigma_b/\sigma_a = 1.0$. *(4)* What are the stress concentration factors in each case? Note: use a software package to generate the three dimensional plots.

4

Engineering structural analysis

Solutions of the fifteen governing equations of linear elasticity are not easy to develop for practical problems. Chapter 3 outlined the complexity of the problem and presented solutions to a few of the simpler practical problems that can be easily treated. These equations define what is commonly called the *linearly elastic theory of solid mechanics* or more simply *linear elasticity* theory. Engineers and mathematicians have studied these equations for more than two centuries, and their efforts to develop solutions have led to broad areas of applied mathematics. The range of useful analytical solutions, however, still remains quite limited, and the problems for which solutions are available are usually very simple.

To analyze practical structures that are generally more complicated, it is almost always necessary to make judicious simplifications that reduce the governing equations to a form that can be solved with modest effort. This approach is widely referred to as *engineering structural analysis* or more simply *structural analysis*. These efforts have produced a rich collection of solutions to practical problems, and structural analysis is an important part of many areas of engineering. The chapters that follow treat the subject of structural analysis, but the developments are based on the fundamental theory of solid mechanics presented in the first three chapters. This chapter introduces basic solution processes, and subsequent chapters extend them to a range of useful structural elements.

4.1 Solution approaches

One of the most direct ways to simplify solid mechanics problems is to reduce their dimensionality. The plane stress and plane strain assumptions presented in sections 1.3 and 1.6, respectively, reduce three-dimensional problems to two-dimensional problem. In some cases, a problem can be further simplified to a one-dimensional form. For example, plane stress problems presenting cylindrical symmetry involve stress and strain fields that are functions of only the radial variable when polar coordinates are used to formulate the problem.

In addition to the simplifications mentioned above, various procedures are available to solve the resulting governing equations. Depending on the problem, different solution procedures may require vastly different analytical skills and/or computational efforts.

In general, the objective of structural analysis is to determine the stress and deformation fields that arise from applied loads. Once appropriate simplifications have been made, two approaches to the solution of the problem are possible.

1. In the first approach, a solution for the stress field is developed based on the equilibrium equations of the problem (and possibly using the compatibility equations). Next, the strain field is obtained from the stress field with the help of the constitutive laws. Finally, the strain-displacement equations are integrated to obtain the displacement field. As illustrated in example 3.1, this last step is often very tedious, even for the simplest problem. In addition, because three displacement components must be determined from the six components of the strain field, it is often necessary to invoke the auxiliary compatibility equations, eqs. 1.106. Note that the solution process sequentially moves through the three groups of equations of elasticity.
2. In the second approach, the solution process invokes the three groups of equations of elasticity in the reverse order. First, a set of purely kinematic assumptions are formulated. Typically, the displacement field of the structure under load is assumed. Next, the strain-displacement equations are used to evaluate the strain field, and the constitutive laws then yield the corresponding stress field. Finally, substitution of the stress components in the equilibrium equations leads to a complete solution of the problem.

To illustrate these two solution approaches, this chapter examines the simplest, one-dimensional problems involving a single, direct stress component that can be either tensile or compressive. A slender, homogeneous prismatic bar subjected only to axial loads is a structural component that meets these conditions. The analysis of this type of components and the associated solution procedures are described in the following sections for a variety of such structures. In the process, the two fundamental solution procedures described above are examined in more detail, and solutions are developed for a number of practical cases.

4.2 Bar under constant axial force

Figure 4.1 depicts an idealized problem consisting of an infinitely long, homogeneous bar with constant properties along its span and subjected to end loads P. The first step to the development of an approximate model for this structural component is to describe its kinematic behavior, *i.e.*, to describe how the component deforms under load. Since the axial load and physical properties are constant along the span, the local deformation of the bar must be identical at all points along its span.

Consider now an initially plane cross-section, S, at a point along the span of the bar as shown in fig. 4.1. All the material particles that form cross-section S before deformation will form a new section, S', after deformation. The symmetry of the problem requires the two semi-infinite halves of the deformed bar to be identical, and therefore, the deformed section, S', must remain *planar and normal to the axis of the bar*.

Fig. 4.1. Infinitely long bar under end loads.

For the more realistic problem of a bar of finite length, the above conclusions still hold, except for the portions of the bar near the end points where complex stress distributions may arise. For instance, if the bar is held in the grips of a testing machine, complex stress and displacement fields will develop under the grips. Very different stress and displacement fields will develop in a bar that is loaded by a pin passing through a hole drilled in the bar. In both cases, however, displacements and stresses will eventually become uniform through the cross-section, at large distances from these end zones. The solution developed here is not valid in these end zones, but it does apply in the portions of the bar that are a good distance from these end zones, as implied by Saint-Venant's principle, principle 2 on page 169.

Consider again cross-section nm shown in fig. 4.1. Since the cross-sections of the bar must remain planar, the axial deformation must be identical at all points of the section, and the axial strain, ϵ_1, will also be uniform over the cross-section. Clearly, from the basic definition of extensional strain, it follows that $\epsilon_1 = e/L$, where L is the length of the bar unaffected by the end regions, and e its change in length resulting from the applied load.

If the bar is slender, it is reasonable to assume that the direct stress components in the transverse direction, σ_2 and σ_3, are much smaller than the component, σ_1, aligned with the applied load. This means that $\sigma_2 \approx 0$ and $\sigma_3 \approx 0$. Finally, if the load is not excessive, stress and strain components remain proportional to each other. Hooke's law then applies and eq. (2.1) reduces to $\sigma_1 = E\epsilon_1$.

Since the axial stress component, σ_1, is assumed to be uniformly distributed over the cross-section, equilibrium of the section then requires that

$$\sigma_1 = \frac{P}{\mathcal{A}}, \tag{4.1}$$

where $\mathcal{A}$ is the cross-sectional area of the bar. The elongation of the bar resulting from the application of the load is now easily found as

$$e = \epsilon_1 L = \frac{\sigma_1 L}{E} = \frac{PL}{E\mathcal{A}}. \tag{4.2}$$

The above results are valid for both tensile and compressive load. However, in the case of compressive loads, the equilibrium configuration of the bar might become unstable as the load increases, leading to lateral buckling of the bar; this subject is treated in chapter 14.

Equation (4.2) shows that the elongation of the bar is proportional to the applied load; this can be emphasized by recasting the equation as $e = P/k$, where k is the *axial stiffness of the bar* given by

$$k = \frac{E\mathcal{A}}{L}. \tag{4.3}$$

Under an axial force, the bar behaves like a simple spring of constant stiffness, k, subjected to the same load.

One of the most common structural components that can be modeled as a bar under an axial force is a bar in a truss structure. A truss is a two or three-dimensional structure consisting of slender bars pinned at their ends to joints, which allow only axial forces to be transmitted into each member. In chapter 5, the simple model developed here will be extended to treat a broader class of slender bar problems featuring anisotropic materials, nonuniform cross-sections, and subjected to distributed axial loads varying along the bar's span.

The solution approaches outlined in section 4.1 will now be illustrated for axially loaded uniform bars in several examples.

Example 4.1. Series connection of axially loaded bars

The simplest example of bars subjected to axial forces is a series of bars connected in a straight line and subjected to axial forces applied at the bar ends. Figure 4.2 depicts a configuration featuring two bars connected in series; the left bar is clamped at point **A**, whereas the second bar is loaded by force P at point **C**. An axial load, $3P$, is applied at the junction point **B** between the two bars.

For this problem, the axial force equilibrium conditions can be written for each joint as shown in fig. 4.2. It then follows from equilibrium equations at points **B** and **C** that

$$F_{AB} = 4P, \quad F_{BC} = P, \tag{4.4}$$

where F_{AB} and F_{BC} are the axial forces in bar **AB** and **BC**, respectively. The sign convention used here and consistently throughout this book is that a tensile force in the bar is positive; this is the same convention used for the direct stress components.

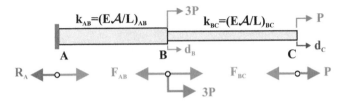

Fig. 4.2. Two bars connected in series and subjected to two loads.

Next, the constitutive law, eq. (4.2), is used to find the extension of each bar as

$$e_{AB} = \frac{4P}{k_{AB}} \quad \text{and} \quad e_{BC} = \frac{P}{k_{BC}}$$

where $k_{AB} = (EA/L)_{AB}$ and $k_{BC} = (EA/L)_{BC}$ are the axial stiffnesses of bars **AB** and **BC**, respectively. The notation $(EA/L)_{AB}$ is used as a shorthand notation for the more cumbersome $E_{AB}A_{AB}/L_{AB}$, where L_{AB}, A_{AB} and E_{AB} denote the length, cross-sectional area, and Young's modulus, respectively, for bar **AB**. Similar conventions are used for bar **BC**. Finally, the overall extension of the bar, which is the displacement of point **C**, is found from the compatibility condition, $d_C = e_{AB} + e_{BC}$, to yield

$$d_C = e_{AB} + e_{BC} = \left(\frac{4}{k_{AB}} + \frac{1}{k_{BC}} \right) P.$$

This is a particularly simple example not because two bars only are present, but rather because the forces in the bars and the reaction force at point **A** can be found from equilibrium considerations alone. The deflections then follow immediately from the force-deformation equations.

Example 4.2. Series connection of axially loaded bars (displacement approach)
Consider now the situation shown in fig. 4.3, which is similar to that depicted in fig. 4.2, except that both ends of the system, at points **A** and **C**, are now fixed and only the load applied at point **B** remains. In this case, the problem involves two reactions forces, R_A and R_C, and two bar forces, F_{AB} and F_{BC}, for a total of four unknowns. On the other hand, only three equations of equilibrium can be written, one at each of the three joints: $R_A = F_{AB}$, $F_{BC} - F_{AB} + 3P = 0$, and $R_C = F_{BC}$.

In contrast with the previous example, the equilibrium equations are no longer sufficient to determine the bar forces. Such problems are known as *hyperstatic* systems, or "statically indeterminate," or "statically redundant" systems in contrast with *isostatic* or "statically determinate" systems, such as that presented in example 4.1.

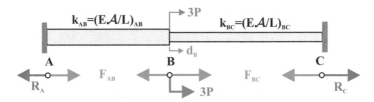

Fig. 4.3. Two bars connected in series with ends fixed.

To find the solution of this problem, deformations must also be considered. The constitutive laws of the system can be expressed as $e_{AB} = F_{AB}/k_{AB}$ and $e_{BC} = F_{BC}/k_{BC}$ for bars **AB** and **BC**, respectively. Introducing these results into the equilibrium equation for point **B** yields

$$k_{AB}\, e_{AB} - k_{BC}\, e_{BC} = 3P. \tag{4.5}$$

Finally, the kinematics of the system are used to express bar extensions in terms of the displacements of points **B** and **C** as $d_B = e_{AB}$ and $d_C = e_{AB} + e_{BC}$, respectively. The displacement at point **C**, however, must vanish because this point is clamped, $d_C = 0$, which implies $e_{AB} = -e_{BC}$ and $d_B = e_{AB} = -e_{BC}$.

Introducing these results into eq. (4.5) then yields a single equation for the unknown displacement at point **B**, $(k_{AB} + k_{BC}) d_B = 3P$. This is the equilibrium equation of the problem written in terms of the unknown displacement, d_B. This equation can be solved for the displacement, d_B, and the bar elongations can then be computed as $e_{AB} = -e_{BC} = d_B = 3P/(k_{AB} + k_{BC})$. Back substitution yields the forces in the bars

$$F_{AB} = k_{AB}\, e_{AB} = \frac{3k_{AB}P}{k_{AB} + k_{BC}}, \quad F_{BC} = -k_{BC}\, e_{BC} = -\frac{3k_{BC}P}{k_{AB} + k_{BC}}. \quad (4.6)$$

It is interesting to compare these internal forces with those obtained for the isostatic problem in example 4.2, see eq. (4.4). In the solution of the isostatic problem, the internal forces only depend on the externally applied loads, whereas in the solution of the hyperstatic problem, the internal forces depend on the applied loads, as expected, but also on the stiffness of the structure: indeed, the stiffnesses of the bars, k_{AB} and k_{BC}, appear in the final answer.

Example 4.3. Series connection of axially loaded bars (force approach)

The problem presented in the previous example, see fig. 4.3, will be analyzed again, but a different solution procedure will be followed. As noted previously, the problem involves two reactions forces, R_A and R_C, and two bar forces, F_{AB} and F_{BC}, for a total of four unknowns. Only three equations of equilibrium can be written, one at each of the three joints: $R_A = F_{AB}$, $F_{BC} - F_{AB} + 3P = 0$, and $R_C = F_{BC}$.

If any one of the four internal forces is known, the three others can be directly determined from the equilibrium equations. For instance, if F_{AB} is known, all other internal forces can be readily computed. More formally, the force in bar **AB**, denoted R, is assumed to be known. The three equilibrium equations then yield $F_{BC} = R - 3P$, $R_A = R$, and $R_C = R - 3P$.

The next step is to substitute these forces into the constitutive equations to determine the system deformation, *i.e.*, the bar extensions, as

$$e_{AB} = \frac{F_{AB}}{k_{AB}} = \frac{R}{k_{AB}}, \quad e_{BC} = \frac{F_{BC}}{k_{BC}} = \frac{R - 3P}{k_{BC}}.$$

Next, the strain-displacement equations express the relationship between the system deformations and the displacements of points **A**, **B**, and **C**. Figure 4.3 shows that $d_A = d_C = 0$ and $d_B = e_{AB}$, but the compatibility of deformation between the fixed points **A** and **C** also requires $e_{AB} + e_{BC} = 0$. This compatibility condition provides the necessary equation to solve for R,

$$e_{AB} + e_{BC} = \frac{R}{k_{AB}} + \frac{R - 3P}{k_{BC}} = 0, \quad \text{or} \quad R = \frac{3k_{AB}}{k_{AB} + k_{BC}}\, P. \quad (4.7)$$

Finally, the equilibrium equations yield $F_{AB} = R = 3P\, k_{AB}/(k_{AB} + k_{BC})$ and $F_{BC} = R - 3P = -3P\, k_{BC}/(k_{AB} + k_{BC})$. The displacement of point **B** becomes

$d_B = e_{AB} = F_{AB}/k_{AB}$. The solution is identical to that found in the previous example using the displacement approach.

In the force method, the determination of the unknown force, R, is based on the enforcement of compatibility conditions for system deformations. While this process is carried out here in abstract, mathematical terms, a physical description of the procedure is often helpful in formulating the solution.

In the first step, the system is assumed to be "cut" at a location that reveals the unknown internal force, R. Since this force acts in bar **AB**, the cut is made at an arbitrary point along this bar, for instance at point **A**, as depicted in fig. 4.4.

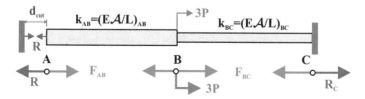

Fig. 4.4. Two bars connected in series with ends fixed and a cut at point **A**.

In the second step, under the action of the externally applied loads, a relative displacement of the two sides of the cut, denoted d_{cut}, will develop. Of course, in the real system this cut does not exist, *i.e.*, $d_{cut} = 0$. It is convenient to think of force R as an externally applied load, as illustrated in fig. 4.4. The extensions of the two bars can be written in terms of the forces as

$$e_{AB} = \frac{F_{AB}}{k_{AB}} = \frac{R}{k_{AB}}, \quad e_{BC} = \frac{F_{BC}}{k_{BC}} = \frac{R - 3P}{k_{BC}}.$$

In the third step, the compatibility condition is enforced. The displacement at the cut is the sum of the elongations of the two bars, $d_{cut} = e_{AB} + e_{BC}$. In this example, the displacement is positive if the two sides of the cut overlap and negative when a gap forms between the two sides of the cut. In the actual system, the cut is not present and the relative displacement at the cut must vanish: $d_{cut} = 0$. This condition leads to

$$d_{cut} = e_{AB} + e_{BC} = R/k_{AB} + (R - 3P)/k_{BC} = 0.$$

This equation expresses the displacement compatibility at the cut, and it is written in terms of forces and flexibilities (*i.e.*, the inverse of stiffnesses). The equation can be solved for the unknown force, R, as

$$R = \frac{3/k_{BC}}{1/k_{AB} + 1/k_{BC}} P = \frac{3k_{AB}}{k_{AB} + k_{BC}} P.$$

It then follows that $F_{AB} = R = 3P \, k_{AB}/(k_{AB} + k_{BC})$ and $F_{BC} = R - 3P = -3P \, k_{BC}/(k_{AB} + k_{BC})$, and finally, $d_B = e_{AB} = F_{AB}/k_{AB}$.

4.3 Hyperstatic systems

The examples treated in the previous section reveal fundamental differences between two types of systems that are commonly encountered in structural analysis. For some systems, the number of equations of equilibrium is *equal* to the total number of unknown internal forces. Internal forces include reaction forces and forces acting in the members of the system. Such systems are called *statically determinate* or *isostatic* systems. The term "isostatic," where "iso" means "the same," refers to the fact that the number of equilibrium equations is *the same* as the number of force unknowns. For isostatic problems, the unknown forces can be determined from the equations of equilibrium alone, *without* using the strain-displacement equations or constitutive laws. This is a special situation since, in general, the solution of elasticity problems requires the simultaneous solution of the three fundamental groups of equations: the equilibrium, strain-displacement, and constitutive equations. A very simple isostatic system is treated in example 4.1.

For other systems, the total number of unknown internal force and reactions is *larger than* the number of equilibrium equations. Such systems are called *statically indeterminate* or *hyperstatic* systems. The term "hyperstatic," where "hyper" means "larger," refers to the fact the number of force unknowns is *larger than the number of equilibrium equations*. In this case, the equilibrium equations are not sufficient to determine the internal forces in the system. The equilibrium equations by themselves present an infinite number of solutions.

The *degree of redundancy*, N_R, of a system is defined as the number of unknown internal forces minus the number of equations of equilibrium. For instance, the problem presented in example 4.2 features four unknown internal forces and three equations of equilibrium. Hence, its degree of redundancy is $N_R = 4 - 3 = 1$; the system is referred to as having *a single degree of redundancy* or being *hyperstatic of order 1*. The treatment of hyperstatic systems will require the simultaneous solution of the three fundamental groups of equations to evaluate all the unknown quantities of the problem.

The difference between iso- and hyperstatic systems might appear to be rather technical at first, but it is, in fact, very fundamental. A few of the key differences are discussed in the following paragraphs.

First, the solution procedure for the two types of systems is different. For isostatic systems, the equations of equilibrium are written first, then immediately solved for the unknown internal forces. Indeed, no other equations are needed to evaluate these forces. It is only when evaluating deformations and displacements that the constitutive laws and then the strain-displacement equations must be invoked. In contrast, the solution process for hyperstatic problems is somewhat more complex. The equilibrium equations cannot be solved independently of the other two sets of equations of elasticity, the strain-displacement equations and the constitutive laws. Clearly, hyperstatic problems are inherently more difficult to solve because the three sets of equations of elasticity shown in fig. 3.1 are now coupled.

Two main approaches are available for the solution of these coupled equations: the displacement method and the force method, which are presented in examples 4.2

and 4.3, respectively. These two solution procedures will be more formally developed in the next section.

A second difference is observed in the nature of the solution for the unknown internal forces. Compare the expressions given in eqs. (4.4) and (4.6) for the internal forces of an isostatic and a hyperstatic problem, respectively. For isostatic systems, internal forces can be expressed in terms of the externally applied loads, whereas for hyperstatic systems, internal forces depend on the applied loads, as expected, but also on the stiffness of the structure because the bar stiffnesses, k_{AB} and k_{BC}, appear in the final answer. This difference reflects the fact that the solution process for hyperstatic systems requires the use of the material constitutive laws. Consequently, material stiffness characteristics, such as the Young's modulus of the material, explicitly appear in the expressions for the internal forces. In other words, the internal force distribution in hyperstatic systems depends on the stiffness characteristics of the structure, whereas for isostatic systems, this distribution is independent of structural stiffnesses.

The third difference is best explained by considering once again the iso- and hyperstatic systems treated in examples 4.2 and 4.3, respectively. The hyperstatic system features two load paths: one load path, bar **AB**, carries a portion of the applied load to the ground, *i.e.*, to a fixed support while the other load path, bar **BC**, carries the remaining portion of the applied load to the other support. This system is said to present "dual load paths," see fig. 4.3. This contrasts with the isostatic problem that features a single load path: the applied loads are carried back to the single support at point **A** through the serially-connected bars **AB** and **BC**, see fig. 4.2. In the hyperstatic system, the equilibrium equations are not sufficient to determine how much of the load will be carried by load path **AB** and how much will be carried by load path **BC**. In fact, the applied load is split between the two load paths according to their relative stiffnesses, $F_{AB}/F_{BC} = -k_{AB}/k_{BC}$, where the minus sign reflects the sign conventions for the bar internal forces. The stiffer load path will carry more load than the more compliant one.

Systems with multiple load paths are inherently more damage tolerant than systems with a single load path. Indeed, if bar **AB** fails, the single load path system can no longer carry any load, whereas the dual load path system might still be able to carry the applied load, assuming that bar **BC** is designed to safely carry the entire load in the event of a failure of the other bar.

4.3.1 Solution procedures

Two general approaches are available for the solution of hyperstatic systems. The first approach is illustrated in example 4.2 and involves the following steps. First, write the equilibrium equations of the system. Second, use the constitutive laws to express internal forces in terms of member deformations. Third, use the strain-displacement equations to express system deformations in terms of displacements.

At this point, all the equations of elasticity have been written: the rest of the procedure manipulates these equations to obtain the solution of the problem. The deformations written in terms displacements are introduced in the constitutive laws to find the internal forces in terms of displacements, and finally, these internal forces are introduced into the equilibrium equations to yield the equations of equilibrium expressed in terms of displacements. Solution of these equilibrium equations then yield the displacements of the system. Deformations then follow by back substituting the displacements in the strain-displacement equations; finally, the internal forces are obtained from the constitutive laws by back substitution of the deformations. This solution approach is called the *displacement method* or the *stiffness method,* because the governing equations are equilibrium equations written in terms of unknown displacements and component stiffnesses.

The second approach is illustrated in example 4.3 and involves the following steps. First, write the equilibrium equations of the system. Next, determine the system degree of redundancy, N_R, which equals the number of unknown internal forces minus the number of equations of equilibrium. The system is now "cut" at N_R locations. At each of the N_R cuts, a *redundant force* is assumed to act, and a single relative displacement is defined to measure the relative displacement across the cut.

With the addition of these N_R cuts and the specification of the N_R redundant forces, the *originally hyperstatic system is transformed into an isostatic system* for which the internal forces can be determined in terms of the applied loads and the N_R redundant forces from the equilibrium equations alone, *i.e.,* the redundant forces are treated as externally applied loads. Next, the relative displacements at the N_R cuts are determined by first invoking the constitutive laws to yields system deformations in terms of the applied loads and the N_R redundant forces. Finally, the strain displacement equations can be used to find the relative displacements at the N_R cuts. The original hyperstatic system, however, cannot develop these relative displacements because it has no cuts. These compatibility requirements impose the vanishing of the relative displacements at the cuts, and this leads to a set of N_R equations for the N_R redundant forces. This approach is called the *force method* or the *flexibility method* because the governing equations express compatibility requirements in terms of the redundant forces and component flexibilities.

The displacement and force methods are general solution procedures that can be used to solve a wide range of hyperstatic problems. Hence, it is useful to formally describe these procedures in details. For clarity and simplicity, each step of the procedures is explained in terms of the structural components and variables encountered in the analysis of axially loaded bars. In later chapters, the same methods will be generalized for application to other, more complex structural components and systems.

4.3.2 The displacement or stiffness method

The displacement method focuses on expressing the governing equilibrium equations in terms of displacements, and the resulting equations are solved for these displacements. The forces and moments in the system are then computed from the displacements using the force-deformation relationships. This can be formalized in the following steps.

1. Write the *equilibrium equations of the system*. Equilibrium conditions express the vanishing of the sum of the forces and moments acting on the system. This step typically involves construction of free body diagrams of the various sub-components of the system, and then formulation of the equilibrium conditions.
2. Use the *constitutive laws* to express internal forces in terms of member deformations or strains.
3. Use the *strain-displacement equations* to express system deformations in terms of displacements. At this point the three groups of equations of elasticity have been utilized. The total number of unknowns of the problem should be equal to the total number of equations.
4. *Introduce the deformations-displacements equations* derived in step 3 *into the constitutive laws* derived in step 2 to find the internal forces in terms of displacements.
5. *Introduce the internal forces* derived in step 4 *into the equilibrium equations* derived in step 1 to yield the equations of equilibrium expressed in terms of displacements.
6. *Solve the equilibrium equations* derived in step 5 to find the displacements of the system.
7. *Find system deformations* by back-substituting the displacements into the strain-displacement equations derived in step 3.
8. *Find system internal forces* by back-substituting the deformations into the constitutive laws derived in step 2.

The displacement method focuses first on determining the displacement of the system, and system deformations are then obtained by back substituting displacements into the strain-displacement equations. Finally, the internal forces follow from back-substitution of deformations into the constitutive laws. The number of displacement variable is exactly equal to the number of equilibrium equations. All equilibrium equations will involve one or more displacement variables, and hence, the solution for the displacements in step 6 typically requires the solution of a set of linear equations. If this system of equations is large, computational tools will ease the solution process.

Example 4.4. Hyperstatic three-bar truss. Displacement method solution
The three-bar truss depicted in fig. 4.5 is a very simple system of axially loaded bars that exhibits all the characteristics of hyperstatic systems. The system is subjected to a vertical load P applied at point **O**, where the three bars are pinned together. The three bars will be identified by the points at which they are pinned to the ground, denoted points **A**, **B**, and **C**. $\mathcal{A}_A$, $\mathcal{A}_B$ and $\mathcal{A}_C$ are the cross-sectional areas of bars **A**, **B**, and **C**, respectively, and E_A, E_B and E_C denote their respective Young's moduli.

The truss features geometric and material symmetry about vertical axis **OB**: the cross-sectional areas of bars **A** and **C** are equal, $\mathcal{A}_A = \mathcal{A}_C$, and so are their Young's moduli, $E_A = E_C$. Consequently, the forces acting in bars **A** and **C**, denoted F_A and F_C, respectively, are also equal, $F_A = F_C$. The vertical displacement of point **O** is denoted Δ, and the displacement method focuses on determining this displacement

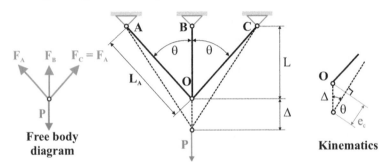

Fig. 4.5. Three bar truss.

first. Due to the symmetry of the problem, the horizontal displacement component vanishes.

Step 1 of the displacement method described in section 4.3.2 is to derive the equation of equilibrium of the problem. The free body diagram drawn in fig. 4.5 yields

$$F_B + 2F_A \cos \theta = P. \tag{4.8}$$

Clearly, the two unknown forces, F_A and F_B, cannot be determined from this single equilibrium equation: this is a hyperstatic system of order 1.

Step 2 invokes the constitutive laws to relate the forces in the bars to the corresponding bar deformations as follows

$$e_A = e_C = \frac{F_A L_A}{(EA)_A} = \frac{F_A L}{(EA)_A \cos \theta}, \quad e_B = \frac{F_B L}{(EA)_B}, \tag{4.9}$$

where e_A, e_B, and e_C are the elongations of the three bars.

Step 3 deals with the last set of equations of elasticity, the strain-displacement equations. Relating the vertical displacement, Δ, of point **O** to the elongations of the bars is a difficult task if Δ is arbitrarily large; for small displacement, however, *i.e.*, when $\Delta \ll L$, angle θ changes little during deformation, and the kinematics diagram in fig. 4.5 shows that $e_C \approx \Delta \cos \theta$. It follows that

$$e_A = e_C = \Delta \cos \theta, \quad e_B = \Delta. \tag{4.10}$$

All equations of elasticity have now been utilized for this problem. Step 4 is a purely algebraic step combining eqs. (4.9) and (4.10) to express the internal forces in terms of displacements to find

$$\frac{F_A}{(EA)_B} = \frac{F_C}{(EA)_B} = \frac{\Delta}{L} \bar{k}_A \cos^2 \theta, \quad \frac{F_B}{(EA)_B} = \frac{\Delta}{L}, \tag{4.11}$$

where $\bar{k}_A = (EA)_A/(EA)_B$ is the non-dimensional stiffness of bar **A**.

Step 5 is another purely algebraic step combining eqs. (4.8) and (4.11) to express the single equilibrium condition of the problem in terms of the single displacement component, Δ, to find

$$\frac{\Delta}{L} + 2\frac{\Delta}{L}\bar{k}_A \cos^3\theta = \frac{P}{(E\mathcal{A})_B}.$$

Step 6 solves this linear equation for the single displacement component, Δ, to find

$$\frac{\Delta}{L} = \frac{1}{1 + 2\bar{k}_A \cos^3\theta}\frac{P}{(E\mathcal{A})_B}. \tag{4.12}$$

This relationship can be written as $\Delta = P/k$, where k is the equivalent vertical stiffness of the three-bar truss, $k = \left[(E\mathcal{A})_B + 2(E\mathcal{A})_A \cos^3\theta\right]/L$.

In step 7, the deformations of the structure are recovered by introducing the displacement given by eq. (4.12) into the strain-displacement equations, eqs. (4.10), to find the elongations as

$$\frac{e_A}{L} = \frac{e_C}{L} = \frac{\cos\theta}{1 + 2\bar{k}_A \cos^3\theta}\frac{P}{(E\mathcal{A})_B}, \quad \frac{e_B}{L} = \frac{1}{1 + 2\bar{k}_A \cos^3\theta}\frac{P}{(E\mathcal{A})_B}. \tag{4.13}$$

The final step of the displacement method, step 8, recovers the forces in the bars by introducing the elongations, given by eq. (4.13) into the constitutive laws, eq. (4.9), to find

$$\frac{F_A}{P} = \frac{F_C}{P} = \frac{\bar{k}_A \cos^2\theta}{1 + 2\bar{k}_A \cos^3\theta}, \quad \frac{F_B}{P} = \frac{1}{1 + 2\bar{k}_A \cos^3\theta}. \tag{4.14}$$

Note that the internal forces in the bars depend on the stiffnesses of the system, $\bar{k}_A = (E\mathcal{A})_A/(E\mathcal{A})_B$. In fact, the ratio of the forces in bars **A** and **B** is $F_A/F_B = \bar{k}_A \cos^2\theta$, i.e., the ratio of the forces in the two bars is in proportion to the ratio of their stiffnesses.

Example 4.5. Rigid plate suspended by four elastic cables: displacement method
The hyperstatic system depicted in fig. 4.6 is more complicated than the previous example, but the same displacement method can be applied. In this example, a rigid square plate of side dimension ℓ is supported by four identical elastic cables of length h, cross-sectional area $\mathcal{A}$, and Young's modulus E. A vertical load P is applied to the rigid plate at point **K** located by coordinates x_{1p} and x_{2p} as indicated in the figure. Find the elongations and forces in the four cables.

The complication in this example arises from the kinematics. Because the plate is assumed to be perfectly rigid, it is easy to understand that the vertical displacements of points **A**, **B**, **C**, and **D**, denoted Δ_A, Δ_B, Δ_C, and Δ_D, respectively, are not independent. Indeed, any three points uniquely define a plane. For example, the displacements of points **A**, **B**, and **C** uniquely defined the configuration of the plate, and the displacement of the fourth point, **D**, follows. For a square plate it is easy to show that $\Delta_A + \Delta_B = \Delta_C + \Delta_D$ is the condition that ensures the infinite rigidity of the plate.

Step 1 of the displacement method described in section 4.3.2 is to derive the equations of equilibrium of the problem from the free body diagram shown in fig. 4.6

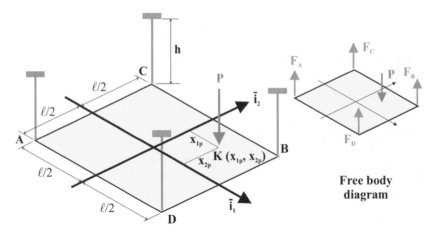

Fig. 4.6. A rigid plate supported by four identical elastic cables.

$$F_A + F_B + F_C + F_D = P, \tag{4.15a}$$

$$-F_A + F_B + F_C - F_D = \frac{2x_{2p}}{\ell}\, P, \tag{4.15b}$$

$$-F_A + F_B - F_C + F_D = \frac{2x_{1p}}{\ell}\, P, \tag{4.15c}$$

where the first equation corresponds to the equilibrium of forces in the vertical direction, and the next two equations are moment equilibrium equations about axes $\bar{\imath}_1$ and $\bar{\imath}_2$, respectively. Clearly, the four unknown forces, F_A, F_B, F_C, and F_D, cannot be determined from these three equilibrium equations. This is therefore a hyperstatic system of order 1.

Step 2 invokes the constitutive laws to relate the forces in the cables to the corresponding system deformations as follows

$$e_A = \frac{F_A h}{E\mathcal{A}}, \quad e_B = \frac{F_B h}{E\mathcal{A}}, \quad e_C = \frac{F_C h}{E\mathcal{A}}, \quad e_D = \frac{F_D h}{E\mathcal{A}}, \tag{4.16}$$

where e_A, e_B, e_C, and e_D are the elongations of the four cables. Step 3 deals with the strain-displacement equations which are particularly simple in this case:

$$e_A = \Delta_A, \quad e_B = \Delta_B, \quad e_C = \Delta_C, \quad e_D = \Delta_D. \tag{4.17}$$

All equations of elasticity have now been utilized for this problem.

Step 4 is a purely algebraic step combining eqs. (4.16) and (4.17) to express the internal forces in terms of displacements to find

$$F_A = \frac{E\mathcal{A}}{h}\Delta_A, \quad F_B = \frac{E\mathcal{A}}{h}\Delta_B, \quad F_C = \frac{E\mathcal{A}}{h}\Delta_C, \quad F_D = \frac{E\mathcal{A}}{h}\Delta_D. \tag{4.18}$$

Step 5 is another purely algebraic step combining eqs. (4.15) and (4.18) to express the equilibrium conditions of the problem in terms of the unknown displacements to yield the first three equations below,

$$\Delta_A + \Delta_B + \Delta_C + \Delta_D = \frac{Ph}{E\mathcal{A}}, \tag{4.19a}$$

$$-\Delta_A + \Delta_B + \Delta_C - \Delta_D = \frac{2x_{2p}}{\ell}\frac{Ph}{E\mathcal{A}}, \tag{4.19b}$$

$$-\Delta_A + \Delta_B - \Delta_C + \Delta_D = \frac{2x_{1p}}{\ell}\frac{Ph}{E\mathcal{A}}, \tag{4.19c}$$

$$\Delta_A + \Delta_B - \Delta_C - \Delta_D = 0. \tag{4.19d}$$

The fourth equation expresses the compatibility condition that defines the infinite stiffness of the plate, as discussed earlier.

Step 6 involves the solution of the system of linear equations, eqs. (4.19), to find the displacements of the attachment points of the four cables,

$$\begin{aligned}
\frac{\Delta_A}{h} &= \frac{1}{4}\left(1 - \frac{2x_{1p}}{\ell} - \frac{2x_{2p}}{\ell}\right)\frac{P}{E\mathcal{A}}, \\
\frac{\Delta_B}{h} &= \frac{1}{4}\left(1 + \frac{2x_{1p}}{\ell} + \frac{2x_{2p}}{\ell}\right)\frac{P}{E\mathcal{A}}, \\
\frac{\Delta_C}{h} &= \frac{1}{4}\left(1 - \frac{2x_{1p}}{\ell} + \frac{2x_{2p}}{\ell}\right)\frac{P}{E\mathcal{A}}, \\
\frac{\Delta_D}{h} &= \frac{1}{4}\left(1 + \frac{2x_{1p}}{\ell} - \frac{2x_{2p}}{\ell}\right)\frac{P}{E\mathcal{A}}.
\end{aligned} \tag{4.20}$$

In step 7, the deformations of the structure are recovered by introducing the displacement into the strain-displacement equations, eqs. (4.17), to find the elongations. The final step of the displacement method, step 8, recovers the forces in the cables by introducing the elongations into the constitutive laws, eqs. (4.16), to find

$$\begin{aligned}
\frac{F_A}{P} &= \frac{1}{4}\left(1 - \frac{2x_{1p}}{\ell} - \frac{2x_{2p}}{\ell}\right), \\
\frac{F_B}{P} &= \frac{1}{4}\left(1 + \frac{2x_{1p}}{\ell} + \frac{2x_{2p}}{\ell}\right), \\
\frac{F_C}{P} &= \frac{1}{4}\left(1 - \frac{2x_{1p}}{\ell} + \frac{2x_{2p}}{\ell}\right), \\
\frac{F_D}{P} &= \frac{1}{4}\left(1 + \frac{2x_{1p}}{\ell} - \frac{2x_{2p}}{\ell}\right).
\end{aligned} \tag{4.21}$$

Because the stiffness constants of all four cables are identical, the forces in the cables do not depend on the stiffnesses of the structure. Had the stiffnesses of the cables been different from each other, the final solution for the forces in the cables would depend on the relative stiffnesses of the cables.

4.3.3 The force or flexibility method

The force method focuses on the solution for the system internal forces. Compatibility equations are written in terms of a set of redundant forces. In contrast with the displacement method, the forces are determined first, and strains and displacements are then recovered. The procedure can be formalized in the following steps.

1. Write the *equilibrium equations of the system*. Equilibrium conditions express the vanishing of the sum of the forces and moments acting on the system. This step typically involves the construction of free body diagrams of the various sub-components of the system and then formulation of the equilibrium conditions.
2. Determine the *system degree of redundancy*, N_R, which equals the number of unknown internal forces minus the number of equilibrium equations.
3. *Cut the system* at N_R locations and define a single relative displacement for each of the cuts. With the N_R cuts, the *originally hyperstatic system is transformed into an isostatic system*.
4. Apply N_R *redundant forces* to the system, each acting along the relative displacement allowed by each of the N_R cuts. Express all internal forces of the system in terms of the applied loads and the N_R redundant forces by means of the equilibrium equations. Note that the choice of where to make the cuts is somewhat arbitrary, and some choices may lead to simpler solution processes. The key requirement in making the cuts is that the resulting system must be an isostatic system, not a mechanism.
5. Use the *constitutive laws* to express system deformations in terms of N_R redundant forces.
6. Use the *strain-displacement equations* to express the relative displacements at the N_R cuts in terms of the N_R redundant forces.
7. Impose the *vanishing of the relative displacements* at the N_R cuts, and use these N_R compatibility equations to solve for the N_R redundant forces.
8. Recover system deformations from the constitutive laws and system displacements from the strain-displacement equations.

The force method directly focuses on the determination of the redundant forces. All internal forces, system deformations and displacements are expressed in terms of redundant forces. The solution process involves the solution of a linear set of equations of size N_R, the degree of redundancy of the system. This contrasts with the displacement method that involves the solution of a system of linear equations of size equal to the number of unknown displacements, N_D. Depending on the relative values of N_R and N_D, the displacement or force methods can be more or less convenient to use.

As a final comment, note that while the force method can be applied quite effectively using good engineering judgement and experience, the displacement method is usually more amenable to automated solution processes using computers.

Example 4.6. Hyperstatic three-bar truss: force method solution

The three-bar truss problem treated in example 4.4 using the displacement method will now be solved using the force method. The truss is depicted in fig. 4.7, and here again, it is subjected to a vertical load P applied at point **O**, where the three bars are pinned together. The three bars will be identified by the points at which they are pinned to the ground at points **A**, **B**, and **C**. $\mathcal{A}_A$, $\mathcal{A}_B$ and $\mathcal{A}_C$ denote the cross-sectional areas of bars **A**, **B**, and **C**, respectively, and E_A, E_B and E_C denote their respective Young's moduli.

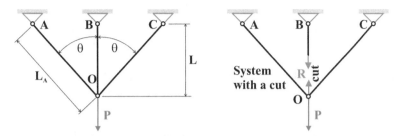

Fig. 4.7. Three bar truss.

The truss features geometric and material symmetry about vertical axis **OB**: the cross-sectional areas of bars **A** and **C** are equal, $\mathcal{A}_A = \mathcal{A}_C$, and so are their Young's moduli, $E_A = E_C$. Consequently, the forces acting in bars **A** and **C**, denoted F_A and F_C, respectively, are also equal, $F_A = F_C$.

Step 1 of the force method described in section 4.3.3 yields a single equation of vertical equilibrium for the problem based on the free body diagram shown in fig. 4.5

$$F_B + 2F_A \cos\theta = P. \tag{4.22}$$

Clearly, the two unknown forces, F_A and F_B, cannot be determined from equilibrium considerations alone. As required by step 2, the system degree of redundancy is determined as $N_R = 2 - 1 = 1$.

Step 3 calls for cutting the system at a single location because $N_R = 1$. As depicted in fig. 4.7, bar **B** is cut for this example, but cutting bars **A** or **C** would lead to a very similar procedure.

Next, in step 4, a single redundant force, R, is applied at the to sides of the cut. With R treated as a known load, it is now possible to solve the equilibrium eq. (4.22) for F_A and F_C as

$$F_A = F_C = \frac{(P-R)}{2\cos\theta}, \quad F_B = R. \tag{4.23}$$

In step 5, bar extensions are expressed in terms of the redundant force, R, using the constitutive laws, eq. (4.9), leading to

$$\frac{e_C}{L} = \frac{e_A}{L} = \frac{F_A}{(E\mathcal{A})_A \cos\theta} = \frac{(P-R)}{2(E\mathcal{A})_A \cos^2\theta}, \quad \frac{e_B}{L} = \frac{F_B}{(E\mathcal{A})_B} = \frac{R}{(E\mathcal{A})_B}. \tag{4.24}$$

Step 6 requires the determination of the relative displacement at the cut, and this is easily obtained from the strain-displacement equations and kinematics as

$$d_{\text{cut}} = \frac{e_A}{\cos\theta} - e_B = \frac{(P-R)L}{2(E\mathcal{A})_A \cos^3\theta} - \frac{RL}{(E\mathcal{A})_B}.$$

Step 7 enforces the vanishing of this relative displacement, $d_{\text{cut}} = 0$. This equation is then solved for the redundant force R to find

$$\frac{R}{P} = \frac{1}{1 + 2\bar{k}_A \cos^3 \theta},$$

where $\bar{k}_A = (E\mathcal{A})_A / (E\mathcal{A})_B$ is the non-dimensional stiffness of bar **A**. The internal forces in the bars then follow from eq. (4.23) as $F_B = R$ and $F_A = F_C = (P - R)/(2 \cos \theta)$.

In step 8, the bar elongations are recovered from eq. (4.24). As expected, the results obtained using the force method as presented here match those obtained in example 4.4 using the displacement method.

Example 4.7. Rigid plate supported by four cables: force method

The force method can also be used to find the forces and deformations in the rigid plate problem treated in example 4.5. The hyperstatic system is shown again in fig. 4.8. The square rigid square plate with sides of length ℓ is supported by four identical elastic vertical cables of length h, cross-sectional area $\mathcal{A}$, and Young's modulus E. A vertical load, P, is applied to the rigid plate at point **K** located at coordinates x_{1p} and x_{2p} as indicated in the figure.

The kinematics of the rigid plate require that all four corner points remain in a plane. Thus, only three of the vertical displacements of points **A**, **B**, **C**, and **D**, denoted Δ_A, Δ_B, Δ_C, and Δ_D, respectively, are independent, and the fourth can be computed from the other three. Again, it is easy to show that $\Delta_A + \Delta_B = \Delta_C + \Delta_D$ is the condition that expresses the infinite rigidity of the plate.

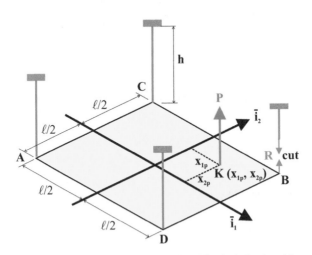

Fig. 4.8. A rigid plate supported by four identical elastic cables.

Step 1 of the force method is to derive the equations of equilibrium of the problem from the free body diagram shown in fig. 4.6. The vanishing of the sum of the forces and moments acting on the rigid plate leads to the equations of equilibrium given by eqs. (4.15). Clearly, the four unknown forces, F_A, F_B, F_C, and F_D, cannot be determined from those three equilibrium equations. As required by step 2, the system degree of redundancy is determined as $N_R = 4 - 3 = 1$.

Step 3 calls for cutting the system at a single location, since $N_R = 1$. As depicted in fig. 4.8, cable **B** is cut in this example, but cutting any one of the four cables would lead to a very similar procedure.

Next, in step 4, a single redundant force, R, is applied at the to sides of the cut. With the help of the equation of equilibrium, eq. (4.15), the internal forces in the cables are expressed in terms of the applied load, P, and the redundant force, R, to find

$$F_A = R - \left(\frac{x_{1p}}{\ell} + \frac{x_{2p}}{\ell}\right) P, \quad F_B = R,$$

$$F_C = \left(\frac{1}{2} + \frac{x_{2p}}{\ell}\right) P - R, \quad F_D = \left(\frac{1}{2} + \frac{x_{1p}}{\ell}\right) P - R. \tag{4.25}$$

In step 5, cable extensions are expressed in terms of the redundant force, R, by introducing the above forces into the constitutive laws, eqs. (4.16), to yield

$$\frac{EAe_A}{h} = R - \left(\frac{x_{1p}}{\ell} + \frac{x_{2p}}{\ell}\right) P, \quad \frac{EAe_B}{h} = R,$$

$$\frac{EAe_C}{h} = \left(\frac{1}{2} + \frac{x_{2p}}{\ell}\right) P - R, \quad \frac{EAe_D}{h} = \left(\frac{1}{2} + \frac{x_{1p}}{\ell}\right) P - R. \tag{4.26}$$

Step 6 requires determination of the relative displacement at the cut, and step 7 imposes the requirement that it vanish. The condition expressing the infinite rigidity of the plate is $\Delta_A + \Delta_B = \Delta_C + \Delta_D$. If this condition is satisfied, the relative displacement at the cut must vanish. Because the four cables are fixed at their bases, their tip displacements are equal to their elongations and hence, $e_A + e_B = e_C + e_D$. introducing eq. (4.26) into this compatibility equations leads to

$$R - \left(\frac{x_{1p}}{\ell} + \frac{x_{2p}}{\ell}\right) P + R = \left(\frac{1}{2} + \frac{x_{2p}}{\ell}\right) P - R + \left(\frac{1}{2} + \frac{x_{1p}}{\ell}\right) P - R.$$

This equation is now solved for the redundant force R to find

$$R = F_B = \frac{1}{4}\left(1 + \frac{2x_{1p}}{\ell} + \frac{2x_{2p}}{\ell}\right) P.$$

The other internal forces in the cables are then obtained by introducing the redundant force, R, into eqs. (4.25). In step 8, the cable elongations are recovered from eq. (4.26).

As expected, the results obtained using the force method presented here match those obtained in example 4.5 using the displacement method. It is interesting to note that the solution of this problem using the displacement method involves solving a linear system of four equations for the four unknown displacements of the cables, whereas the present force method requires the solution of a single compatibility equation for the unknown redundant force. In other words, the force method requires the solution of a linear system of size equal to the order of redundancy of the hyperstatic system, whereas the displacement method requires the solution of a larger linear system of size equal to the number of unknown displacements.

4.3.4 Problems

Problem 4.1. Simple hyperstatic bars - displacement method solution
Three axially loaded bars, each of length L and all constructed from a material of elasticity modulus E, are arranged as shown in fig. 4.9. Two bars are connected in parallel and one of these has a cross-sectional area that is twice that of the other. A third bar is connected in series at the common point. An axial load, P, is applied at the junction of the three bars. Using the displacement method, determine *(1)* the displacement, d, of the connecting point between the three bars and *(2)* the forces in each of the three bars.

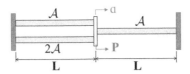

Fig. 4.9. Three bars in a parallel-series configuration.

Problem 4.2. Simple hyperstatic bars - force method solution
Solve problem 4.1 using using the force method.

Problem 4.3. Prestressed steel bar in an aluminum tube
A steel bar of cross-sectional area $A_s = 800$ mm^2 fits inside an aluminum tube of cross-sectional area $A_a = 1,500$ mm^2. The assembly is constructed in such a way that initially, the steel bar is prestressed with a compressive force, $-P$, while the aluminum tube is prestressed with a tensile load of equal magnitude, P. Next, the prestressed assembly is subjected to a tensile load F. *(1)* If no prestress is applied, *i.e.*, if $P = 0$, find the maximum external load, F, that can be applied to the assembly without exceeding allowable stress levels in either material. *(2)* Find the optimum prestress level to be applied. This optimum prestress is defined as that for which the allowable stress is reached simultaneously in both steel bar and aluminum tubes when subjected to the externally applied force, F. In other words, when optimally prestressed, both materials are used to their full capacity. *(3)* What improvement, in percent, is achieved by using the optimum prestress level as compared to not prestressing the assembly. Use the following data: $E_s = 210$ and $E_a = 73$ GPa; the yield stresses for steel and aluminum are $\sigma_y^s = 600$ and $\sigma_y^a = 400$ MPa, respectively.

Problem 4.4. Square plate supported by four cables
Consider the rigid square plate of side ℓ supported by four elastic cables each of length h, cross-sectional area A, and Young's modulus E, as depicted in fig. 4.10. A vertical load P is applied at point **K**, located at a distance d from the center of the plate along the line joining points **A** and **B**. *(1)* Determine the degree of redundancy of this system. *(2)* Determine the forces, F_A, F_B, F_C, and F_D, in bars **A**, **B**, **C**, and **D**, respectively. *(3)* On one graph, plot the four non-dimensional forces, F_A/P, F_B/P, F_C/P, and F_D/P, as functions of $\bar{d} = d/\ell$ for $\bar{d} \in [0, 1/\sqrt{2}]$. Hint: See example 4.5. Also note the symmetry of the problem with respect to line **AB**, which simplifies the moment equilibrium equations.

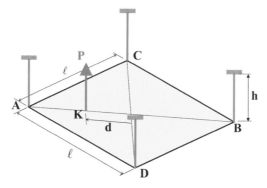

Fig. 4.10. Rigid square plate supported by four elastic cables.

Problem 4.5. Square plate supported by four cables

Consider the rigid square plate of side ℓ supported by four elastic cables each of length h and Young's modulus E, as depicted in fig. 4.10. The cross-sectional areas of the cables are $\mathcal{A}_A$, $\mathcal{A}_B$, $\mathcal{A}_C$, and $\mathcal{A}_D$, for cables **A**, **B**, **C**, and **D**, respectively. A vertical load P is applied at point **K**, located at a distance d from the center of the plate along the line joining points **A** and **B**. It is desired that the plate move straight down under the action of the load, *i.e.*, $\Delta_A = \Delta_B = \Delta_C = \Delta_D = \Delta$, where Δ_A, Δ_B, Δ_C, and Δ_D are the vertical displacements of points **A**, **B**, **C**, and **D**, respectively. *(1)* Determine the degree of redundancy of this system. *(2)* Determine the relationship(s) that must be satisfied by the cross-sectional areas of the cables for the plate to undergo the desired motion. Hints: The relationship between $\mathcal{A}_C$ and $\mathcal{A}_D$ should be obvious from inspection of the problem.

Problem 4.6. Rotor blade hub connection

Figure 4.11 shows a potential design for the attachment of a rotor blade to the rotorcraft hub. The yoke consists of two separate pieces each of which connects the rotor blade to the hub, and the spindle also connects the rotor blade to the hub through an elastomeric bearing. As the rotor blade spins, a large centrifugal force F is applied to the assembly, which can be idealized as three parallel bars of length L, which connect the blade to the hub. The two bars modeling the yoke each have an axial stiffness $(EA)_y$, while the spindle has an axial stiffness $(EA)_s$. The elastomeric bearing is idealized as a very short spring of stiffness k_b in series with the spindle. *(1)* Calculate and plot the non-dimensional forces in the yoke, F_y/F, and in the spindle, F_s/F, as a function of the non-dimensional bearing stiffness, $0 \le Lk_b/(EA)_s \le 25$. *(2)* For what value of the stiffness constant k_b is all the centrifugal load carried by the yoke? *(3)* Find the maximum load that can be carried by the spindle. What is the corresponding value of k_b? *(4)* For what value on $Lk_b/(EA)_s$ do the yoke and spindle carry equal loads? Use the following data: $(EA)_y/(EA)_s = 0.8$

4.3.5 Thermal effects in hyperstatic system

It is often the case that hyperstatic systems are more structurally efficient than isostatic systems. They potentially offer the additional advantage of redundant load paths. On the other hand, they present important drawbacks; one of them is sensitivity to *thermal effects*.

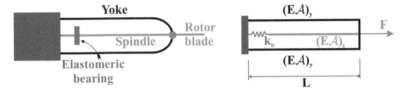

Fig. 4.11. Rotor blade connection to the hub by means of a yoke and spindle.

In isostatic structures, thermal strains simply cause additional deformations of the system, as implied by the modified constitutive laws that account for thermal strains, eqs. (2.22). For hyperstatic structures, however, the presence of thermal strains in the constitutive laws gives rise to additional stresses, called *thermal stresses*. This effect can be significant, even when the entire structure experiences a *uniform temperature change*, although the effect is usually more pronounced in the presence of *temperature gradients*, which result from non-uniform temperature fields, or when different portions of the structure are subjected to different temperatures.

Example 4.8. Series connected bars subjected to temperature change
Consider the system depicted in fig. 4.12 featuring two bars connected in series and constrained by rigid walls at points **A** and **C**. Load P is applied at point **B**, and in addition, both bars are subjected to a temperature change ΔT. Except for this thermal effect, the problem is identical to that treated in example 4.2.

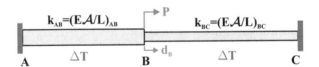

Fig. 4.12. Two bars connected in series with ends fixed.

The equilibrium equation of the system remains unchanged, $F_{AB} - F_{BC} = P$, and the displacement of point **B**, d_B, is still related to the elongations of the bars $d_B = e_{AB} = -e_{BC}$. The constitutive equations, however, must now be modified to account for the thermal strains. In view of eq. (2.22), the total strain in each bar is the sum of the mechanical and thermal strains, $\epsilon = \epsilon^m + \epsilon^t$, where the mechanical strain is related to the stress in the bar, $\epsilon^m = \sigma/E$, and the thermal strain depends on the temperature change, $\epsilon^t = \alpha \Delta T$. The extension in the bar now becomes

$$e_{AB} = \epsilon L_{AB} = \frac{\sigma_{AB}}{E_{AB}} L_{AB} + \alpha \Delta T L_{AB} = \frac{F_{AB}}{k_{AB}} + \alpha \Delta T L_{AB}.$$

A similar equation can also be developed for the elongation of the other bar, e_{BC}.

Following the steps of the displacement method, the internal forces are expressed in terms of deformations, then in terms of displacements, leading to the equilibrium equation expressed in terms of the displacement as

$$(k_{AB} + k_{BC})d_B = P + \alpha\Delta T \left[(E\mathcal{A})_{AB} - (E\mathcal{A})_{BC}\right].$$

The displacement of point **B** is then

$$d_B = \frac{P}{k_{AB} + k_{BC}} + \frac{\alpha\Delta T \left[(E\mathcal{A})_{AB} - (E\mathcal{A})_{BC}\right]}{k_{AB} + k_{BC}} = d_B^m + d_B^t.$$

This rather complex result shows that total displacement of point **B** is the superposition of the displacement d_B^m due to applied mechanical loads, and the displacement d_B^t due to thermal effects. This should not be unexpected, because mechanical and thermal effects are superposed in the constitutive law.

The internal forces are obtained by substituting the displacement back into the constitutive laws, to find

$$F_{AB} = \frac{k_{AB}}{k_{AB} + k_{BC}} \left[P - \alpha\Delta T \left(L_{AB} + L_{BC}\right) k_{BC}\right],$$

and

$$F_{BC} = \frac{k_{BC}}{k_{AB} + k_{BC}} \left[-P - \alpha\Delta T \left(L_{AB} + L_{BC}\right) k_{AB}\right].$$

It is interesting to consider the case when the two bars are identical, $k_{AB} = k_{BC} = k$. The displacement of point **B** simply becomes $d_B = P/(2k)$. In this case, the thermal displacement vanishes due to the symmetry of the problem, and the total displacement is due solely to the mechanical loads. The axial forces in the bars become $F_{AB} = P/2 - E\mathcal{A}\alpha\Delta T$ and $F_{BC} = -P/2 - E\mathcal{A}\alpha\Delta T$. Due to the symmetry of the problem, both bars share an equal burden in carrying the mechanical loads, $\pm P/2$, and are both subjected to the same compressive thermal stress, $E\mathcal{A}\alpha\Delta T$.

It is also interesting to consider the impact of thermal stresses on the load carrying capability of this system. The bars will yield when the yield stress is reached, that is when $F_{AB} = \pm\sigma_y\mathcal{A}_{AB}$ and when $F_{BC} = \pm\sigma_y\mathcal{A}_{BC}$. In the absence of thermal effects, the load carrying capacity of the system is then $P_{\max} = 2\mathcal{A}\sigma_y$, whereas in the presence of thermal effects, the load carrying capacity becomes $P_{\max} = 2\mathcal{A}(\sigma_y - E\alpha\Delta T) = 2\mathcal{A}\bar{\sigma}_y$. In other words, in the presence of thermal effects, the effective yield stress, $\bar{\sigma}_y$, is the yield stress of the material, σ_y, *reduced by the thermal stress*, $E\alpha\Delta T$.

Example 4.9. Hyperstatic three-bar truss subject to temperature change

The three-bar truss depicted in fig. 4.13 is assembled when all components are at common temperature T_0 and no initial stresses are present in the bars. The three bars will be identified by the points at which they are pinned to the ground at **A**, **B**, and **C**. $\mathcal{A}_A$, $\mathcal{A}_B$ and $\mathcal{A}_C$ denote the cross-sectional areas of bars **A**, **B**, and **C**, respectively, while E_A, E_B and E_C denote their respective Young's moduli. The truss features geometric and material symmetry about the vertical axis **OB**: the cross-sectional areas of bars **A** and **C** are equal, $\mathcal{A}_A = \mathcal{A}_C$, and so are their Young's moduli, $E_A = E_C$. Consequently, the forces acting in bars **A** and **C**, denoted F_A and

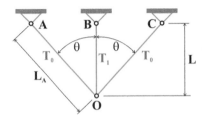

Fig. 4.13. Three bar truss subjected to temperature differentials.

F_C, respectively, are also equal, $F_A = F_C$. Assume that only bar **B** is now heated to a temperature $T_1 = T_0 + \Delta T$, thus preserving the symmetry of the problem.

Due to the heating, the center bar tries to expand by an amount $\alpha(T_1 - T_0)L = \alpha \Delta T\, L$ but is prevented from doing so by the other two bars. An equilibrium point will be reached where the truss expands, although less than the unconstrained bar would, and internal stresses will appear. Intuitively, bar **B** will be in compression, whereas bars **A** and **C** will be in tension.

This thermal problem will be treated using the force method. A a similar problem featuring the same three-bar truss subjected to external loading is treated using the same approach in example 4.6. The equation of equilibrium for this example is given by eq. (4.22) and remains valid for the present example: $F_B + 2F_A \cos\theta = 0$. Since the problem features a single degree of redundancy, a single cut is required. Here again, bar **B** is cut and an unknown redundant force, R, is assumed to act at the cut. The internal forces in the bars are expressed in terms of R, and eqs. (4.23) become $F_A = F_C = -R/(2\cos\theta)$ and $F_B = R$.

The constitutive laws are now used to express the non-dimensional bar elongation in terms of the unknown redundant force to find

$$\frac{e_C}{L} = \frac{e_A}{L} = -\frac{1}{2\bar{k}_A \cos^2\theta}\frac{R}{(E\mathcal{A})_B}, \qquad \frac{e_B}{L} = \frac{R}{(E\mathcal{A})_B} + \alpha\Delta T,$$

where $\bar{k}_A = (E\mathcal{A})_A/(E\mathcal{A})_B$ is the non-dimensional stiffness of bar **A**. These expressions are almost identical to those of eqs.(4.24), except for the thermal strain terms now contributing to the elongation of bar **B**. The relative displacement at the cut is now easily obtained

$$d_{\text{cut}} = \frac{e_A}{\cos\theta} - e_B = \frac{-L}{2\bar{k}_A \cos^3\theta}\frac{R}{(E\mathcal{A})_B} - L\frac{R}{(E\mathcal{A})_B} - L\alpha\Delta T.$$

The vanishing of this relative displacement implies $d_{\text{cut}} = 0$ and yields the unknown non-dimensional redundant force as

$$\frac{R}{(E\mathcal{A})_B} = \frac{F_B}{(E\mathcal{A})_B} = -\frac{2\bar{k}_A \cos^3\theta}{1 + 2\bar{k}_A \cos^3\theta}\alpha\Delta T.$$

The non-dimensional forces in bars **A** and **B** follow from the equilibrium equation as

$$\frac{F_A}{(E\mathcal{A})_B} = \frac{F_C}{(E\mathcal{A})_B} = \frac{\bar{k}_A \cos^2\theta}{1 + 2\bar{k}_A \cos^3\theta}\alpha\Delta T.$$

These internal forces, called *thermal forces*, are proportional to the thermal strain, $\alpha\Delta T$. As expected, the force in bar **B** is compressive, in contrast with the tensile forces present in bars **A** and **C**. Finally, the vertical displacement of the truss at point **O** is given by the elongation of bar **B**, and this is easily recovered as

$$\frac{d_B}{L} = \frac{e_B}{L} = \frac{\left[1 + 2(\bar{k}_A - 1)\cos^3\theta\right]}{1 + 2\bar{k}_A\cos^3\theta}\,\alpha\Delta T.$$

4.3.6 Manufacturing imperfection effects in hyperstatic system

An additional drawback of hyperstatic systems is their sensitivity to *dimensional* or *manufacturing imperfections*. Consider, here again, the three-bar truss depicted in fig. 4.13. Assume all bars to be at the same temperature, but due to manufacturing imperfections, bar **B** was made too long. It is impossible to assemble the system: if bars **A** an **C** are first connected together at point **O**, bar **B** is longer than the distance from point **B** to **O**. The only way to assemble the system is to compress bar **B** to the right length, pin the three bars together at point **O**, then release the compression in bar **B**. In the final assembly, *residual forces* will be present; intuitively, it follows that bar **B** is left under compression, whereas bars **A** and **C** have a residual tensile stress.

It is worth noting the close connection between thermal strain and manufacturing imperfections. In example 4.9, bar **B** is subjected to a temperature differential resulting in a thermal elongation $L\alpha\Delta T$. In other words, bar **B** is now too long by an amount $L\alpha\Delta T$. This is identical to a manufacturing imperfection where bar **B** is too long by an amount $\mu = L\alpha\Delta T$. This means that the *residual stress* due to thermal effects computed in example 4.9 are identical to the residual stress due to manufacturing imperfections in the same system, provided that $\alpha\Delta T$ is replaced by μ/L in all results of example 4.9.

Example 4.10. Rigid plate supported by four elastic bars

Consider the hyperstatic system depicted in fig. 4.14 in which a rigid square plate of side ℓ is supported by four identical elastic bars of length h, cross-sectional area $\mathcal{A}$, and Young's modulus E. This example is similar to the previous examples in which a rigid plate is suspended from four cables, but in this case, the support is provided by the four bars or legs. Assume that one of the bars is too short by an amount μ due, for example, to manufacturing imperfections.

Since the plate is assumed to be infinitely rigid, the vertical displacements of points **A**, **B**, **C**, and **D**, denoted Δ_A, Δ_B, Δ_C, and Δ_D, respectively, are not independent. Indeed, three points uniquely define a plane, hence the displacements of points **A**, **B**, and **C** uniquely define the configuration of the plate, and the displacement of the fourth point, **D**, follows. As in the previous examples, this constraint can be expressed for a square plate as, $\Delta_A + \Delta_B = \Delta_C + \Delta_D$.

In example 4.5, a similar configuration is considered, but a vertical load is applied at an arbitrary point on the plate as shown in fig. 4.6. The displacement method is used to solve the problem, and a similar procedure is used here. In the first step, the equations of equilibrium of the system are derived from the free body diagram

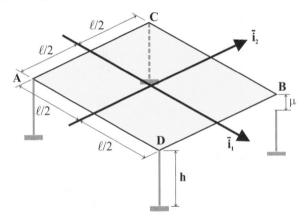

Fig. 4.14. A rigid plate supported by four identical elastic bars with a manufacturing imperfection.

shown in fig. 4.6, to find eqs. (4.15). Next, the constitutive laws relating the bar forces to the corresponding deformations are still given by eqs. (4.16). Finally, the strain displacement equations are still given by eqs. (4.17), except for bar **B** where, due to the manufacturing imperfection, $\Delta_B = e_B - \mu$.

The remaining steps of the displacement method closely follow the development presented in example 4.5 and lead to the following equations of equilibrium written in terms unknown displacements

$$\Delta_A + \Delta_B + \Delta_C + \Delta_D = -\mu, \tag{4.27a}$$

$$-\Delta_A + \Delta_B + \Delta_C - \Delta_D = -\mu, \tag{4.27b}$$

$$-\Delta_A + \Delta_B - \Delta_C + \Delta_D = -\mu, \tag{4.27c}$$

$$\Delta_A + \Delta_B - \Delta_C - \Delta_D = 0, \tag{4.27d}$$

where the last equation expresses the infinite stiffness of the plate as discussed earlier. The solution of this linear system yields the displacements of the corner points as

$$\Delta_A = \frac{\mu}{4}, \quad \Delta_B = -\frac{3\mu}{4}, \quad \Delta_C = \Delta_D = -\frac{\mu}{4}. \tag{4.28}$$

Finally, the bar forces are recovered as

$$F_A = F_B = \frac{1}{4}\frac{\mu}{h}E\mathcal{A}, \quad F_C = F_D = -\frac{1}{4}\frac{\mu}{h}E\mathcal{A}. \tag{4.29}$$

These are the *residual forces* due to manufacturing imperfections. The two opposite bars **A** and **B** are subjected to tension, whereas the two opposite bars **C** and **D** are under compression. The magnitudes of the forces in the four bars are equal and proportional to the manufacturing imperfection, μ.

Assume now that a vertical load, P, is applied at the center of the plate. The total forces in the bars are now the superpositions of the forces due to the applied loads, as given by eqs. (4.21), and the forces due to the manufacturing imperfection, as given by eqs. (4.29), to find

$$F_A = F_B = \frac{P}{4} + \frac{1}{4}\frac{\mu}{h}E\mathcal{A}, \quad F_C = F_D = \frac{P}{4} - \frac{1}{4}\frac{\mu}{h}E\mathcal{A}.$$

In view of the symmetry of the problem, the applied load is carried equally by the four bars, whereas the manufacturing imperfection put additional loads into bars **A** and **B**, but unloads bars **C** and **D**.

The maximum load the structure can carry is $P_{\max}/4 + E\mathcal{A}\mu/(4h) = \mathcal{A}\sigma_y$, where σ_y is the material yield stress. Hence, $P_{\max} = 4\mathcal{A}[\sigma_y - 1/4\,E\,\mu/h] = 4\mathcal{A}\bar{\sigma}_y$. Due to manufacturing imperfections, the effective yield stress, $\bar{\sigma}_y$, is the actual yield stress for the material, σ_y, reduced by $1/4\,E\,\mu/h$.

The residual forces are proportional to the magnitude of the manufacturing imperfections, as expected, but also to the Young's modulus of the material, see eqs. (4.29). Hence, the stiffer the system, the more sensitive it will be to manufacturing imperfections.

Example 4.11. Prestress in a bolt

Geometric incompatibility may also be created intentionally; indeed, it is sometimes desirable to introduce a prestress into a structural member. Consider, for instance, the prestress created in a bolt when tightened. Typically, a tensile force is created in the bolt to develop a compressive force acting on the bolted assembly. This situation is illustrated in fig. 4.15, which depicts a prestressed bolt-sleeve assembly.

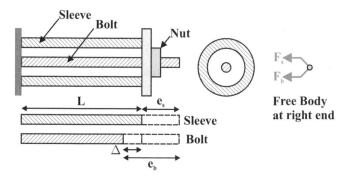

Fig. 4.15. Prestressed bolt-sleeve assembly.

The sleeve is assumed to be a hollow circular cylinder of cross-sectional area $\mathcal{A}_s$ and the bolt has a cross-sectional area $\mathcal{A}_b$; both are of initial length L. The Young's moduli of the sleeve and bolt are E_s and E_b, respectively. Assume that the nut on the bolt is turned until the entire assembly is snug, and then, the nut is rotated by an additional N turns. This will shorten the portion of the bolt between the end plates by an amount $\Delta = pN$, where p is the bolt's thread pitch or distance between successive threads.

The analysis follows a procedure similar to that used in example 4.10. The single equilibrium equation of the system is $F_s + F_b = 0$, where F_s and F_b are the forces in the sleeve and bolt, respectively, both assumed positive in tension. The constitutive equations for the sleeve and the bolt are simply, $F_s = k_s e_s$ and $F_b = k_b e_b$, where $k_s = (EA)_s/L$ and $k_b = (EA)_b/L$ are the equivalent sleeve and bolt stiffnesses, respectively.

Let the elongations of the sleeve and bolt be denoted e_s and e_b, respectively, as illustrated in the lower part of fig. 4.15. Due to the initial tightening of the nut, the bolt is shortened by an amount $\Delta = pN$, and hence, displacement compatibility requires $e_s = e_b - \Delta$.

Since only the prestress forces in the bolt and sleeve are to be determined, the force method provides the most direct solution procedure. Let the sleeve force be the redundant force in the system, and hence, $F_s = R$. The equilibrium equation then implies $F_b = -F_s = -R$, and substitution into the constitutive equations yields the sleeve and bolt extensions as $e_s = F_s/k_s = R/k_s$ and $e_b = F_b/k_b = -R/k_b$, respectively. Finally, introducing these results into the compatibility equation yields $R/k_s = -R/k_b - \Delta$. Solving this equation yields $R = -k_s k_b/(k_s + k_b)\Delta$. The forces in the bolt and sleeve are then

$$F_s = R = -\frac{k_s k_b}{k_s + k_b}\Delta, \quad \text{and} \quad F_b = -R = \frac{k_s k_b}{k_s + k_b}\Delta,$$

respectively. As expected, the bolt is in tension while the sleeve is in compression. From a practical point of view, the desired prestress level, F_s or F_b, would be specified first, and the required number of turns, N, would then be computed. For instance, for a prescribed compressive F_s, $N = (k_s + k_b)|F_s|/(pk_s k_b)$.

4.3.7 Problems

Problem 4.7. Constrained bar at uniform temperature
A uniform aluminum bar is constrained at its two end. If the bar is stress free for a temperature $T_0 = 20°$ C, find the compressive stress in the bar if the temperature is raised to value $T = 140°$ C. Note: $E_{al} = 73$ GPa, $\alpha_{al} = 16.5$ μ/C.

Problem 4.8. Steel bar inside a copper tube
A steel bar with a 750 mm^2 section is placed inside a copper tube with a section of 1250 mm^2. The bar and tube have a common length of 0.5 m and are connected at their ends. At the reference temperature, both elements are stress free. (1) If the assembly is heated up to 80° C, find the axial stresses in both elements. Note: $E_{steel} = 210$ GPa, $\alpha_{steel} = 12$ μ/C; $E_{copper} = 120$ GPa, $\alpha_{copper} = 17$ μ/C.

Problem 4.9. Bolt-sleeve assembly subjected to temperature rise
Consider the sleeve and bolt assembly shown in fig. 4.15, where the bolt is made of stainless steel, which presents a larger coefficient of thermal expansion than the titanium sleeve. Consequently, under a temperature rise $\Delta T = 100$ C, the bolt will extend more than the sleeve and will become loose, i.e., a gap will develop between the nut and washer plate. To prevent this, a pre-stress is applied to the assembly by turning the nut N turns before the temperature rise. Determine the number of turns N that must be used to create the required pre-stress for

the following conditions: $p = 0.5$ mm (bolt thread pitch), $L = 100$ mm, $A_b = 100$ mm^2, $A_s = 800$ mm^2, $E_b = 210$ GPa, $E_s = 120$ GPa, $\alpha_b = 18 \ \mu$/C, and $\alpha_s = 8 \ \mu$/C.

Problem 4.10. Three-bar truss

Consider the three-bar truss shown in fig. 4.7. The truss is not subjected to any external load, but due to a manufacturing imperfection, the middle bar is of length $L + \mu$ in its unstressed configuration. *(1)* Find the forces in bars **A**, **B**, and **C** as a function of the magnitude of the manufacturing imperfection, μ. *(2)* Find the displacement of point **O** as a function of μ.

4.4 Pressure vessels

This section briefly describes the behavior of structures operating under internal pressure such as rings, and cylindrical or spherical pressure vessels. Typically, these thin-walled structures are designed to contain fluids or gases under pressure. Two particular geometric shapes, the sphere and the cylinder with hemispherical end caps, are widely utilized, and for these shapes, a two-dimensional stress state develops in the thin walls.

4.4.1 Rings under internal pressure

Consider the thin-walled ring or tube of mean radius R and thickness t subjected to an internal pressure p_i, as depicted in fig. 4.16. Due of the internal pressure, a hoop stress, σ_h, will develop in the wall. This hoop stress is readily found by equilibrium consideration: fig. 4.16 shows a free body diagram for the half portion of the ring cut by a plane passing through the axis of the cylinder, revealing the hoop stress acting in the wall. The total vertical force per unit length of the ring due to the pressure acting on its upper half is $p2R$; this force is equilibrated by the hoop stress. Assuming that the hoop stress is uniformly distributed through the wall thickness, it follows that

$$\sigma_h = \frac{p2R}{2t} = \frac{pR}{t}. \tag{4.30}$$

The hoop stress is sometimes called the circumferential stress.

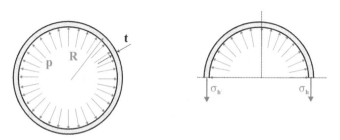

Fig. 4.16. Thin ring under internal pressure.

If the material is homogeneous and linearly elastic, the hoop strain, ϵ_h, is obtained from Hooke's law as $\epsilon_h = \sigma_h/E = pR/(tE)$ and the radius of the ring increases by an amount $\Delta R = R^2 p/(Et)$.

4.4.2 Cylindrical pressure vessels

Consider now a thin-walled pressure vessel consisting of a cylindrical tube of radius R, length L and thickness t closed by spherical end caps, as depicted in fig. 4.17. Pressure vessels operate under a multi-axial state of stress that includes a hoop, axial and radial stress components. The hoop stress is readily found from the same equilibrium arguments used for the ring; assuming the hoop stress to be uniformly distributed through the wall thickness, its magnitude then becomes

$$\sigma_h = \frac{pR}{t}. \tag{4.31}$$

The resultant axial force of the pressure loading on the end caps is independent of their shape and is equal to $p\pi R^2$. For a thin-walled pressure vessel, the stress along the axis of the vessel, σ_a, is assumed to be uniformly distributed through the wall thickness, and axial equilibrium reveals its magnitude to be

$$\sigma_a = \frac{p\pi R^2}{2\pi Rt} = \frac{pR}{2t} = \frac{\sigma_h}{2}. \tag{4.32}$$

This gives rise to a biaxial stress state where the hoop stress twice as large as the axial stress.

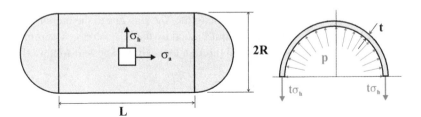

Fig. 4.17. Pressure vessel under internal pressure.

In addition, it should be noted that a radial stress component σ_r also exists. This stress acts along the radius of the cylindrical part of the vessel, and it varies from $\sigma_r = -p$ on the internal surface of the vessel to $\sigma_r = 0$ on the external surface. In most practical designs the ratio R/t is a large quantity and hence $\sigma_r \ll \sigma_h = 2\sigma_a$. Consequently, the radial stress is generally ignored.

If the material can be assumed to behave as a linearly elastic material, Hooke's law, eq. (2.4), implies

$$\epsilon_h = \frac{\sigma_h - \nu\sigma_a}{E} = \frac{\sigma_h}{E}\left(1 - \frac{\nu}{2}\right), \quad \epsilon_a = \frac{\sigma_a - \nu\sigma_c}{E} = \frac{\sigma_h}{E}\left(\frac{1}{2} - \nu\right).$$

Finally, the changes in vessel radial and longitudinal dimensions are

$$\Delta R = R\epsilon_h = \frac{R\sigma_h}{E}\left(1 - \frac{\nu}{2}\right), \quad \Delta L = L\epsilon_a = \frac{L\sigma_h}{E}\left(\frac{1}{2} - \nu\right),$$

respectively.

Since the hoop and axial stresses are the only stress components acting on the vessel, they are the principal stresses, $\sigma_{p1} = \sigma_h = pR/t$ and $\sigma_{p2} = \sigma_a = pR/2t$. According to Tresca's criterion, see eq. (2.29), the yield criterion reduces to $p_y R/t \leq \sigma_y$. This means that the internal pressure for which the yield stress is reached in the material is $p_y = t\sigma_y/R$. On the other hand, if von Mises' criterion is used, see eq. (2.32), the yield criterion becomes $\sigma_{eq} = \sqrt{3}/2\, p_y R/t \leq \sigma_y$. The internal pressure for which the yield stress is reached in the material is $p_y = 2/\sqrt{3}\, t\sigma_y/R$.

4.4.3 Spherical pressure vessels

Consider now a thin-walled sphere of radius R and thickness t subjected to an internal pressure p, as shown in fig. 4.18. This type of configuration is representative of spherical pressure vessels. To begin, the sphere is cut by a horizontal plane passing through its center, to reveal the free body diagram shown in the figure. Due to the symmetry of the problem, the pressure acting on the upper half of the sphere will be equilibrated by a hoop stress, σ_h, which is uniformly distributed around the circle at the intersection of the sphere with the plane of the cut. The total upward force generated by the pressure, $\pi R^2 p$, is equilibrated by the downward force generated by the distributed hoop stress, $2\pi R t \sigma_h$, where the hoop stress is assumed to be uniformly distributed through the wall thickness. This yields the following result

$$\sigma_h = \frac{pR}{2t}. \tag{4.33}$$

The hoop stress is half of that in a pressurized tube of equal radius and thickness, see eq. (4.31).

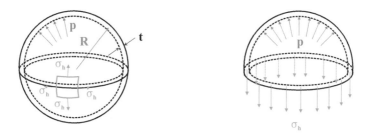

Fig. 4.18. Thin sphere under internal pressure.

Of course, in view of the spherical symmetry of the problem, the orientation of the plane of the cut is arbitrary. Hence, the hoop stress derived above is acting on a face with an arbitrary orientation. As shown in fig. 4.18, the stresses acting on an arbitrary differential element cut from the thin-walled sphere are σ_h in two orthogonal directions. Since the shear stress component vanish, these are the principal stresses, and hence, $\sigma_{p1} = \sigma_{p2} = \sigma_h$. Mohr's circle reduces to a single point at ordinate σ_h.

For a linearly elastic material, the hoop strain, ϵ_h, is obtained from Hooke's law, eq. (2.4), as

$$\epsilon_1 = \epsilon_2 = \epsilon_h = \frac{1 - \nu}{2} \frac{R}{t} \frac{p}{E}. \tag{4.34}$$

The deformation is identical in all directions, due to the spherical symmetry of the problem. Since the shear strain components vanish, the principal strains are $\epsilon_{p1} = \epsilon_{p2} = \epsilon_h$. The radius of the sphere increases by an amount $\Delta R = (1 - \nu)(pR^2)/(2Et)$.

4.4.4 Problems

Problem 4.11. Copper ring on a steel shaft
A copper ring is heated to a temperature of $150°$ C and then exactly fits onto a steel shaft at a uniform temperature of $25°$ C. *(1)* Find the hoop stress in the ring when the assembly has cooled down to a uniform temperature of $25°$ C. *(2)* Find the common temperature at which both ring and shaft must be brought to if the ring is to slip out of the shaft. Hint: since the steel cylinder is very stiff, it is reasonable to assume that it is remains rigid as the copper ring cools down. Of course, under heating, the steel cylinder will expand. Note: $\alpha_{\text{steel}} = 12.5\mu$/C; $E_{\text{copper}} = 110$ GPa, $\alpha_{\text{copper}} = 16.5\mu$/C.

Problem 4.12. Bi-material fly wheel
A fly wheel shown in fig. 4.19 is made of two concentric ring of metal: the inside ring, of thickness t_ℓ, is made of lead and the outside ring, of thickness t_s, is made of steel. The fly wheel has a radius R_m and $t_\ell \ll R_m$, $t_s \ll R_m$. It will be assumed that the lead ring provides little strength and stiffness to the assembly and hence, all stresses are carried in the steel ring. *(1)* Find the maximum angular velocity, Ω_{max}, the fly wheel can rotate at if the yield stress in the steel is σ_y. *(2)* Find the maximum kinetic energy that can be stored in the fly wheel. *(3)* Is this bi-material design a good concept for a high performance fly wheel?

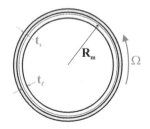

Fig. 4.19. Configuration of the bi-material fly wheel.

mance fly wheel? Use the following data: density of lead, $\rho_\ell = 11,300$ and of steel, $\rho_s = 7,700$ kg/m^3; thickness of lead, $t_\ell = 5$ and of steel $t_s = 3$ mm; radius of the fly wheel $R_m = 250$ mm, its width $b = 20$ mm; yield stress for steel $\sigma_y = 800$ MPa.

Problem 4.13. Cylindrical versus spherical pressure vessels
Spacecrafts often require pressure vessels to carry fuel under pressure. The question investigated here is the relative structural performance of cylindrical and spherical pressure vessels. Consider a cylindrical pressure vessel of radius R_c, length L_c and wall thickness t_c; $L_c = 2R_c$. On the other hand, consider a spherical pressure vessel of radius R_s and wall thickness t_s. The two vessels must carry the same amount of fluid, *i.e.*, must have the same volume; the two vessels are made of the same material with the yield stress σ_y, and must

be able to withstand the same internal pressure. *(1)* Find the ratio of the structural masses of the two vessels. *(2)* For weight sensitive applications such as spacecrafts, is it better to use cylindrical or spherical pressure vessels?

4.5 Saint-Venant's principle

An important concept in structural engineering concerns the effects of local loading and constraint conditions on the stresses and deformations that develop throughout a structure. An obvious example is a concentrated force, which is assumed to act at a point on the surface of a structure. Clearly, this will result in an infinite value for the stresses at the point of application, but yet the reactions and stresses at other parts of the structure are finite.

Consider a body subjected to a set of self-equilibrating loads, as depicted in fig. 4.20. In the vicinity of the applied loads, internal stresses will arise, as expected. However, since the net resultant of the applied load vanishes, it seems reasonable to expect their net effect to decrease away from their point of application. In other words, the effect of a set of self-equilibrating loads is expected to be localized. Typically,

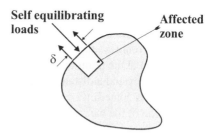

Fig. 4.20. Body subjected to a set of self-equilibrating loads

if the loads are applied over an area of characteristic dimension δ, the affected zone approximately extends a distance δ in all directions from the point of application.

This behavior has been observed experimentally, and is known as Saint-Venant's principle.

Principle 2 (Saint-Venant's principle) *If self-equilibrating loads are applied to a body over an area of characteristic dimension δ, the internal stresses resulting from these loads are only significant over a portion of the body of approximate characteristic dimension δ.*

Note that this principle is rather vague, as it deals with "approximate characteristic dimensions." It allows qualitative rather that quantitative conclusions to be drawn.

An important application of Saint-Venant's principle deals with end effects in bars and beams. In section 4.2, the stress distribution in bars subjected to end loads is studied. Clearly, the assumed uniform stress distribution over the cross-section of the bar is only valid far away from the end section of the bar. Consider fig. 4.21 where the end section of a bar of height h is subjected to a concentrated load P. This concentrated load is statically equivalent to a distributed load $p_0 = P/h$ plus a set of self-equilibrating loads, as depicted on the figure. Saint-Venant's principle implies that the self-equilibrating set of loads only affect a small zone of length h near the end of the bar.

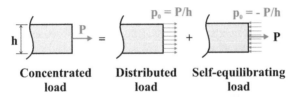

Concentrated   **Distributed**   **Self-equilibrating**
load   **load**   **load**

Fig. 4.21. Bar subjected to an end concentrated load.

According to Saint-Venant's principle, the stress distribution in the bar, namely the uniform axial stress distribution of eq. (4.1), is identical whether the bar is subjected to end distributed or concentrated loads, except in the two end zones of length h. If the end loads are applied as a uniform distribution, the axial stresses in the bar are uniformly distributed over the cross-section at all sections. On the other hand, if the end loads are concentrated loads, the axial stresses in the bar are uniformly distributed over the cross-section only in the central portion of the beam. Near the end points, a complex state of stress will arise; indeed, the axial stress should grow to infinity right at the point of application of the concentrated load. These end zones approximately extend a distance h at either end of the beam. The solution discussed in section 4.2 is sometimes called the *central solution*, *i.e.*. the solution valid in the central portion of the bar, away from the end zones.

Part II

Beams and thin-wall structures

5

Euler-Bernoulli beam theory

A beam is defined as a structure having one of its dimensions much larger than the other two. The axis of the beam is defined along that longer dimension, and a cross-section normal to this axis is assumed to smoothly vary along the span or length of the beam. Civil engineering structures often consist of an assembly or grid of beams with cross-sections having shapes such as T's or I's. A large number of machine parts also are beam-like structures: lever arms, shafts, etc. Finally, several aeronautical structures such as wings and fuselages can also be treated as thin-walled beams.

The solid mechanics theory of beams, more commonly referred to simply as "beam theory," plays an important role in structural analysis because it provides the designer with a simple tool to analyze numerous structures. Although more sophisticated tools, such as the finite element method, are now widely available for the stress analysis of complex structures, beam models are often used at a pre-design stage because they provide valuable insight into the behavior of structures. Such calculations are also quite useful when trying to validate purely computational solutions.

Several beam theories have been developed based on various assumptions, and lead to different levels of accuracy. One of the simplest and most useful of these theories was first described by Euler and Bernoulli and is commonly called Euler-Bernoulli beam theory. A fundamental assumption of this theory is that the cross-section of the beam is infinitely rigid in its own plane, *i.e.*, no deformations occur in the plane of the cross-section. Consequently, the in-plane displacement field can be represented simply by two rigid body translations and one rigid body rotation. This fundamental assumption deals only with in-plane displacements of the cross-section. Two additional assumptions deal with the out-of-plane displacements of the section: during deformation, the cross-section is assumed to remain plane and normal to the deformed axis of the beam. The implications of these assumptions are examined in the next section.

5.1 The Euler-Bernoulli assumptions

Figure 5.1 depicts the idealized problem of a long beam with constant properties along its span subjected only to two bending moments, both of magnitude M, applied at the ends. This type of loading is often referred to as "pure bending." The cross-section of the beam is assumed to be symmetric with respect to the plane of the figure, and bending takes place in that plane of symmetry.

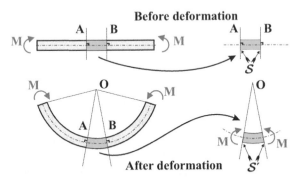

Fig. 5.1. Infinitely long beam under end bending moments.

The bending moment and physical properties are all constant along the beam's span. Hence, the deformation of the beam must be identical at all points along its axis resulting in a constant curvature. This means that the beam deforms into a curve of constant curvature, *i.e.*, a circle with center **O**. In the reference configuration, a cross-section of the beam consists of the ensemble of material particles at the intersection of the beam with a plane perpendicular to the axis of the beam. Figure 5.1 shows a small portion of the beam bounded by two cross-sections, denoted S, generated by two normal planes at points **A** and **B**.

Under the action of the bending moment, this segment deforms into a circular segment with ends defined by the cross-sections S' shown in fig. 5.1. After deformation, the beam is symmetric with respect to any plane perpendicular to its deformed axis. Because the deformed cross-section must satisfy this symmetry requirement, it must remain planar and perpendicular to the deformed axis of the beam.

For a more realistic problem, *e.g.* a finite length beam with specific boundary conditions and applied transverse loads, the bending moment distribution varies along the span and the symmetry arguments used for the above idealized problem no longer apply. By analogy, however, the following *kinematic assumptions* will now be made.

Assumption 1: The cross-section is infinitely rigid in its own plane.
Assumption 2: The cross-section of a beam remains plane after deformation.
Assumption 3: The cross-section remains normal to the deformed axis of the beam.

These assumptions are known as the Euler-Bernoulli assumptions for beams. Experimental measurements show that these assumptions are valid for long, slender beams made of isotropic materials with solid cross-sections. When one or more of theses conditions are not met, the predictions of Euler-Bernoulli beam theory can become inaccurate. The mathematical and physical implications of the Euler-Bernoulli assumptions will now be discussed in detail.

5.2 Implications of the Euler-Bernoulli assumptions

Consider a triad $\mathcal{I} = (\bar{\imath}_1, \bar{\imath}_2, \bar{\imath}_3)$ with coordinates x_1, x_2, and x_3. This set of axes is attached at a point of the beam cross-section; $\bar{\imath}_1$ is along the axis of the beam and $\bar{\imath}_2$ and $\bar{\imath}_3$ define the plane of the cross-section. Let $u_1(x_1, x_2, x_3)$, $u_2(x_1, x_2, x_3)$, and $u_3(x_1, x_2, x_3)$ be the displacement of an arbitrary point of the beam along directions $\bar{\imath}_1, \bar{\imath}_2$, and $\bar{\imath}_3$, respectively.

The first Euler-Bernoulli assumption states that the cross-section is undeformable in its own plane. Hence, the displacement field in the plane of the cross-section consists solely of two rigid body translations $\bar{u}_2(x_1)$ and $\bar{u}_3(x_1)$

$$u_2(x_1, x_2, x_3) = \bar{u}_2(x_1), \quad u_3(x_1, x_2, x_3) = \bar{u}_3(x_1). \tag{5.1}$$

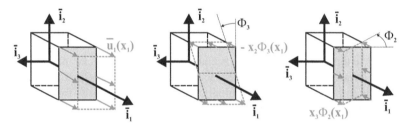

Fig. 5.2. Decomposition of the axial displacement field.

The second Euler-Bernoulli assumption states that the cross-section remains plane after deformation. This implies an axial displacement field consisting of a rigid body translation $\bar{u}_1(x_1)$, and two rigid body rotations $\Phi_2(x_1)$ and $\Phi_3(x_1)$, as depicted in fig. 5.2. The axial displacement is then

$$u_1(x_1, x_2, x_3) = \bar{u}_1(x_1) + x_3\Phi_2(x_1) - x_2\Phi_3(x_1), \tag{5.2}$$

where the location of the origin for the axis system on the cross-section is as yet undetermined. Note the sign convention: the rigid body translations of the cross-section $\bar{u}_1(x_1)$, $\bar{u}_2(x_1)$, and $\bar{u}_3(x_1)$ are positive in the direction of the axes $\bar{\imath}_1, \bar{\imath}_2$, and $\bar{\imath}_3$, respectively; the rigid body rotations of the cross-section, $\Phi_2(x_1)$ and $\Phi_3(x_1)$, are positive about axes $\bar{\imath}_2$ and $\bar{\imath}_3$, respectively. Figure 5.3 depicts these various sign conventions.

The third Euler-Bernoulli assumption states that the cross-section remains normal to the deformed axis of the beam. As depicted in fig. 5.4, this implies the equality of the slope of the beam and of the rotation of the section,

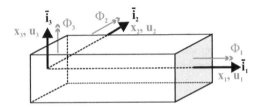

Fig. 5.3. Sign convention for the displacements and rotations of a beam.

$$\Phi_3 = \frac{\mathrm{d}\bar{u}_2}{\mathrm{d}x_1}, \quad \Phi_2 = -\frac{\mathrm{d}\bar{u}_3}{\mathrm{d}x_1}.$$ (5.3)

The minus sign in the second equation is a consequence of the sign convention for the sectional displacements and rotations.

Equations (5.3) can be used to eliminate the sectional rotation from the axial displacement field. The complete displacement field for Euler-Bernoulli beams is now

$$u_1(x_1, x_2, x_3) = \bar{u}_1(x_1) - x_3 \frac{\mathrm{d}\bar{u}_3(x_1)}{\mathrm{d}x_1} - x_2 \frac{\mathrm{d}\bar{u}_2(x_1)}{\mathrm{d}x_1},$$ (5.4a)

$$u_2(x_1, x_2, x_3) = \bar{u}_2(x_1),$$ (5.4b)

$$u_3(x_1, x_2, x_3) = \bar{u}_3(x_1).$$ (5.4c)

The complete three-dimensional displacement field of the beam can therefore be expressed in terms of three sectional displacements $\bar{u}_1(x_1)$, $\bar{u}_2(x_1)$, $\bar{u}_3(x_1)$ and their derivative with respect to x_1. This important simplification results from the Euler-Bernoulli assumptions and allows the development of a one-dimensional beam theory, *i.e.*, a theory in which the unknown displacements are functions of the span-wise coordinate, x_1, alone.

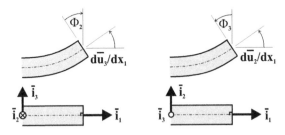

Fig. 5.4. Beam slope and cross-sectional rotation.

The strain field can be evaluated from the displacement field defined by eqs. (5.4) using eqs. (1.63) and (1.71) to find

$$\epsilon_2 = \frac{\partial u_2}{\partial x_2} = 0; \quad \epsilon_3 = \frac{\partial u_3}{\partial x_3} = 0, \quad \gamma_{23} = \frac{\partial u_2}{\partial x_3} + \frac{\partial u_3}{\partial x_2} = 0, \tag{5.5a}$$

$$\gamma_{12} = \frac{\partial u_1}{\partial x_2} + \frac{\partial u_2}{\partial x_1} = 0, \quad \gamma_{13} = \frac{\partial u_1}{\partial x_3} + \frac{\partial u_3}{\partial x_1} = 0, \tag{5.5b}$$

$$\epsilon_1 = \frac{\partial u_1}{\partial x_1} = \frac{d\bar{u}_1(x_1)}{dx_1} - x_3 \frac{d^2\bar{u}_3(x_1)}{dx_1^2} - x_2 \frac{d^2\bar{u}_2(x_1)}{dx_1^2}. \tag{5.5c}$$

At this point, it is convenient to introduce the following notation for the sectional deformations, which depend solely on the span-wise variable, x_1,

$$\bar{\epsilon}_1(x_1) = \frac{d\bar{u}_1(x_1)}{dx_1}, \quad \kappa_2(x_1) = -\frac{d^2\bar{u}_3(x_1)}{dx_1^2}, \quad \kappa_3(x_1) = \frac{d^2\bar{u}_2(x_1)}{dx_1^2}. \tag{5.6}$$

where $\bar{\epsilon}_1$ is the sectional axial strain, and κ_2 and κ_3 are the sectional curvature about the $\bar{\imath}_2$ and $\bar{\imath}_3$ axes, respectively. With the help of these sectional strains, the axial strain distribution over the cross-section, eq. (5.5c), becomes

$$\epsilon_1(x_1, x_2, x_3) = \bar{\epsilon}_1(x_1) + x_3\kappa_2(x_1) - x_2\kappa_3(x_1). \tag{5.7}$$

The vanishing of the in-plane strain field implied by eqs. (5.5a) is a direct consequence of assuming the cross-section to be infinitely rigid in its own plane. The vanishing of the transverse shearing strain field implied by eqs. (5.5b) is a direct consequence of assuming the cross-section to remain normal to the deformed axis of the beam. And finally, the linear distribution of axial strains over the cross-section expressed by eq. (5.7) is a direct consequence of assuming the cross-section to remain plane. Clearly, assuming a strain field of the form eqs. (5.5a), (5.5b), and (5.7) is the mathematical expression of the Euler-Bernoulli assumptions.

5.3 Stress resultants

The goal of beam theory is to develop a one-dimensional model of the three-dimensional beam structure involving only sectional quantities, *i.e.*, quantities solely dependent on the span-wise variable, x_1.

In the previous section, the Euler-Bernoulli assumptions are shown to allow description of the complete three-dimensional displacement field for the beam in terms of three sectional displacements $\bar{u}_1(x_1)$, $\bar{u}_2(x_1)$, and $\bar{u}_3(x_1)$ and their span-wise derivatives as expressed in eq. (5.4). Similarly, the complete three-dimensional strain field given by eqs. (5.5a), (5.5b), and (5.7) is expressed in terms of sectional strains and curvatures.

In this section, the three-dimensional stress field in the beam will be described in terms of sectional stresses called *stress resultants*. These stress resultants are equipollent to (not in equilibrium with) specific components of the stress field.

Three force resultants are defined: the axial force, $N_1(x_1)$, acting along axis $\bar{\imath}_1$ of the beam, and the transverse shearing forces, $V_2(x_1)$ and $V_3(x_1)$, acting along axes $\bar{\imath}_2$ and $\bar{\imath}_3$, respectively. They are defined as follows

$$N_1(x_1) = \int_{\mathcal{A}} \sigma_1(x_1, x_2, x_3) \, d\mathcal{A}. \tag{5.8}$$

$$V_2(x_1) = \int_{\mathcal{A}} \tau_{12}(x_1, x_2, x_3) \, d\mathcal{A}, \quad V_3(x_1) = \int_{\mathcal{A}} \tau_{13}(x_1, x_2, x_3) \, d\mathcal{A}, \tag{5.9}$$

where $\mathcal{A}$ is the cross-sectional area of the beam.

Next, two moment resultants are defined: the bending moments, $M_2(x_1)$ and $M_3(x_1)$, acting about axes $\bar{\imath}_2$ and $\bar{\imath}_3$, respectively, defined as

$$M_2(x_1) = \int_{\mathcal{A}} x_3 \, \sigma_1(x_1, x_2, x_3) \, d\mathcal{A}, \tag{5.10a}$$

$$M_3(x_1) = -\int_{\mathcal{A}} x_2 \, \sigma_1(x_1, x_2, x_3) \, d\mathcal{A}. \tag{5.10b}$$

Note the minus sign in the definition of $M_3(x_1)$, which is necessary to give a positive equipollent bending moment about axis $\bar{\imath}_3$. The sign convention for the forces and moments is depicted in fig. 5.5.

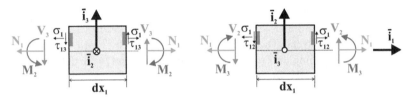

Fig. 5.5. Sign convention for the sectional stress resultants.

In the above definitions, the bending moments are computed with respect to the origin of the axes. In some cases, however, it will be advantageous to compute the bending moments about axes parallel to $\bar{\imath}_2$ and $\bar{\imath}_3$ passing through a specific point of the cross-section. The bending moments computed about point **P** of coordinates (x_{2p}, x_{3p}) on the cross-section are defined as

$$M_2^{\mathrm{P}}(x_1) = \int_{\mathcal{A}} (x_3 - x_{3p}) \, \sigma_1(x_1, x_2, x_3) \, d\mathcal{A}, \tag{5.11a}$$

$$M_3^{\mathrm{P}}(x_1) = -\int_{\mathcal{A}} (x_2 - x_{2p}) \, \sigma_1(x_1, x_2, x_3) \, d\mathcal{A}. \tag{5.11b}$$

5.4 Beams subjected to axial loads

Consider a beam subjected to distributed axial loads, $p_1(x_1)$, and a concentrated axial load, P_1, applied at the end of the beam, for instance, as depicted in fig. 5.6. The distributed axial loads have units of force per unit length (N/m in the SI system), whereas the concentrated axial loads have units of forces (N in the SI system). Under the effect of these loads, the beam will stretch, creating an axial displacement field, $\bar{u}_1(x_1)$. Furthermore, axial forces and axial stresses will be generated in the beam. This section focuses on the determination of these various quantities arising from the application of given axial loading to the beam. When only axial loads are applied to a beam, the structure is often called a "bar" rather than a "beam."

Fig. 5.6. Beam subjected to axial loads.

5.4.1 Kinematic description

The Euler-Bernoulli assumptions described above form the basis of the present analysis. Furthermore, it seems reasonable to assume that axial loads cause only axial displacement of the section. The general displacement field described by eq. (5.4) then reduces to

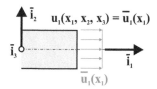

Fig. 5.7. Axial displacement distribution.

$$u_1(x_1, x_2, x_3) = \bar{u}_1(x_1), \qquad (5.12a)$$
$$u_2(x_1, x_2, x_3) = 0, \qquad (5.12b)$$
$$u_3(x_1, x_2, x_3) = 0, \qquad (5.12c)$$

and the corresponding axial strain field is now

$$\epsilon_1(x_1, x_2, x_3) = \bar{\epsilon}_1(x_1). \qquad (5.13)$$

The axial strain is uniform over the cross-section of the beam. These very simple results are illustrated in fig. 5.7.

5.4.2 Sectional constitutive law

At this point, the beam is assumed to be made of a linearly elastic, isotropic material that obeys Hooke's law, see eqs. (2.4). The stresses acting in the plane of the cross-section, σ_2 and σ_3, should remain much smaller than the axial stress component, σ_1: $\sigma_2 \ll \sigma_1$ and $\sigma_3 \ll \sigma_1$. Consequently, these transverse stress components are assumed to vanish, $\sigma_2 \approx 0$ and $\sigma_3 \approx 0$. For this stress state, the generalized Hooke's law, eqs. (2.4), reduce to

$$\sigma_1(x_1, x_2, x_3) = E\, \epsilon_1(x_1, x_2, x_3). \qquad (5.14)$$

Of course, the constitutive laws for shear stress and shear strain components, see eqs. (2.9), remain unchanged.

When describing the beam's kinematics, it is assumed that the cross-section does not deform in its own plane, and the strains in the plane of the cross-section vanish, see eqs. (5.5a). When dealing with the beam's constitutive laws, the transverse stress components are assumed to vanish. This is an inconsistency in Euler-Bernoulli beam theory that uses two contradictory assumptions, the vanishing of both the in-plane strain and transverse stress components. In view of Hooke's law, these two sets of quantities cannot vanish simultaneously. Indeed, if $\sigma_2 = \sigma_3 = 0$, eqs. (2.4b) and (2.4c) result in $\epsilon_2 = -\nu\sigma_1/E$ and $\epsilon_3 = -\nu\sigma_1/E$, which implies that the in-plane strains do not vanish due to Poisson's effect. Because this effect is very small, assuming the vanishing of these in-plane strain components when describing the beam's kinematics does not cause significant errors for most problems.

Introducing the axial strain distribution eq. (5.13) yields the axial stress distribution over the cross-section

$$\sigma_1(x_1, x_2, x_3) = E\,\bar{\epsilon}_1(x_1). \tag{5.15}$$

The axial force in the beam can be obtained by introducing this axial stress distribution into eq. (5.8) to find

$$N_1(x_1) = \int_{\mathcal{A}} \sigma_1(x_1, x_2, x_3)\,\mathrm{d}\mathcal{A} = \left[\int_{\mathcal{A}} E\,\mathrm{d}\mathcal{A}\right]\bar{\epsilon}_1(x_1) = S\,\bar{\epsilon}_1(x_1). \tag{5.16}$$

Since the sectional axial strain $\bar{\epsilon}_1(x_1)$ varies only along the span of the beam, it can be factored out of the integral over the section. The *axial stiffness*, S, of the beam is then defined as

$$S = \int_{\mathcal{A}} E\,\mathrm{d}\mathcal{A}. \tag{5.17}$$

If the section is made of a homogeneous material of Young's modulus E, the axial stiffness of the section becomes $S = E\int_{\mathcal{A}}\mathrm{d}\mathcal{A} = EA$.

Relationship (5.16) is the constitutive law for the axial behavior of the beam. It expresses the proportionality between the axial force and the sectional axial strain, with a constant of proportionality called the axial stiffness. This constitutive law is written at the sectional level, whereas Hooke's law, eq. (5.14), is written at the local, infinitesimal level.

5.4.3 Equilibrium equations

To complete the formulation, the equilibrium equations must be derived for this problem. An infinitesimal slice of the beam of length $\mathrm{d}x_1$ is depicted in fig. 5.8. In this figure, the axial force, $N_1(x_1)$, is shown acting on the face at location x_1. Using a Taylor series expansion, the axial force acting on the face at location $x_1 + \mathrm{d}x_1$ is found to be $N_1 + (\mathrm{d}N_1/\mathrm{d}x_1)\mathrm{d}x_1$; the remaining terms of the expansion are of higher differential order.

Summing up the force acting in the axial direction on the free body diagram depicted in fig. 5.8 yields the following equilibrium equation

$$\frac{dN_1}{dx_1} = -p_1. \tag{5.18}$$

This equation is a direct consequence of Newton's law. While the general equilibrium equations, eqs. (1.4), express the equilibrium conditions for a differential element of a three-dimensional solid, the present equation expresses the equilibrium of a slice of the beam of differential length, dx_1.

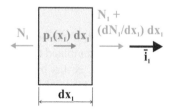

Fig. 5.8. Axial forces acting on an infinitesimal slice of the beam.

5.4.4 Governing equations

Finally, the governing equation of the problem is found by introducing the axial force, eq. (5.16), into the equilibrium, eq. (5.18), and recalling the definition of the sectional axial strain, eq. (5.6),

$$\frac{d}{dx_1}\left[S\frac{d\bar{u}_1}{dx_1}\right] = -p_1(x_1). \tag{5.19}$$

This second order differential equation can be solved for the axial displacement field, $\bar{u}_1(x_1)$, given the axial load distribution, $p_1(x_1)$.

Two boundary conditions are required for the solution of eq. (5.19), one at each end of the beam. Typical boundary conditions are:

1. A fixed (or clamped) end allows no axial displacement, *i.e.*,

$$\bar{u}_1 = 0;$$

2. A free (unloaded) end corresponds to $N_1 = 0$; using eq. (5.16), then leads to

$$\frac{d\bar{u}_1}{dx_1} = 0;$$

3. Finally, if the end of the beam is subjected to a concentrated load P_1, the boundary condition is $N_1 = P_1$, which implies

$$S\frac{d\bar{u}_1}{dx_1} = P_1.$$

5.4.5 The sectional axial stiffness

The axial stiffness, S, of the section characterizes the stiffness of the beam when subjected to axial loads. If the beam is made of a homogeneous material, Young's modulus is identical at all points of the section and can be factored out of integral, (5.17), to yield

$$S = E \, \mathcal{A}. \tag{5.20}$$

On the other hand, if the section is made of several different materials, the axial stiffness must be computed according to eq. (5.17).

An important case is that of a rectangular section of width b made of layered materials of different stiffness moduli, as depicted in fig. 5.9. It is assumed that the material is homogeneous within each of the n layers. In layer i, $E^{[i]}$ is Young's modulus, $\mathcal{A}^{[i]}$ the cross-sectional area, and $x_3^{[i]}$ and $x_3^{[i+1]}$ the coordinates of the bottom and top planes, defining the layer, respectively. Integration over the cross-section then yields the axial stiffness

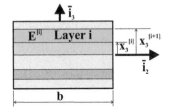

Fig. 5.9. Cross-section of a beam with various layered materials.

$$S = \int_{\mathcal{A}} E \, \mathrm{d}\mathcal{A} = \sum_{i=1}^{n} E^{[i]} \int_{\mathcal{A}^{[i]}} \mathrm{d}\mathcal{A}^{[i]} = \sum_{i=1}^{n} E^{[i]} b \left(x_3^{[i+1]} - x_3^{[i]} \right).$$

This expression clearly shows that the axial stiffness is a *weighted average* of the Young's modulus of the various layers. The weighting factor, $x_3^{[i+1]} - x_3^{[i]}$, is the thickness of the layer.

5.4.6 The axial stress distribution

The determination of the local axial stress, σ_1, for a given axial load, p_1, is of primary interest to designers. This can be readily obtained by eliminating the axial strain from eqs. (5.15) and (5.16) to find

$$\sigma_1(x_1, x_2, x_3) = \frac{E}{S} N_1(x_1) \tag{5.21}$$

If the beam is made of a homogeneous material, the axial stiffness is given by eq. (5.20), and eq. (5.21) then reduces to

$$\sigma_1(x_1, x_2, x_3) = \frac{N_1(x_1)}{\mathcal{A}}. \tag{5.22}$$

The axial stress is uniformly distributed over the section, and its value is independent of Young's modulus.

In contrast, the axial stress distribution for sections made of layers presenting different stiffness moduli will vary from layer to layer. Indeed, eq. (5.21) becomes

$$\sigma_1^{[i]}(x_1, x_2, x_3) = E^{[i]} \frac{N_1(x_1)}{S} \tag{5.23}$$

where $\sigma_1^{[i]}$ indicates the axial stress in layer i. This relationship implies that the axial stress in layer i is proportional to the modulus of that layer. Note that according to eq. (5.13) the axial strain distribution is uniform over the section, *i.e.*, each layer is equally strained. Layer with stiffer materials, however, will develop higher axial stresses. The axial stress distribution for homogeneous and layered sections are depicted in fig. 5.10.

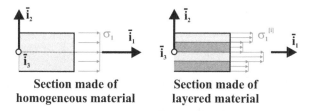

Section made of   **Section made of**
homogeneous material   **layered material**

Fig. 5.10. Axial stress distribution for sections made of homogeneous and layered materials.

Once the local axial stress is determined, a strength criterion can be applied to determine whether the structure can sustain the applied loads. Introducing eq. (5.21) into the strength criterion, eq. (2.28), yields $E/S\,|N_1(x_1)| \leq \sigma_{\text{allow}}^{\text{tens}}$ or $\sigma_{\text{allow}}^{\text{comp}}$. Because the axial force varies along the span of the beam, this condition must be checked at all points along the span. In practice, it is convenient to first determine the maximum tensile and compressive axial force, denoted $N_{1\,\text{max}}^{\text{tens}}$ and $N_{1\,\text{max}}^{\text{comp}}$, respectively, then apply the strength criterion

$$\frac{E}{S}|N_{1\,\text{max}}^{\text{tens}}| \leq \sigma_{\text{allow}}^{\text{tens}}, \quad \frac{E}{S}|N_{1\,\text{max}}^{\text{comp}}| \leq \sigma_{\text{allow}}^{\text{comp}}. \tag{5.24}$$

If the axial force is compressive, buckling of the beam becomes another possible failure mode. The maximum compressive load that a beam can sustain before lateral buckling occurs is discussed in chapter 14.

If the section consists of layers made of various materials, the strength of each layer will, in general, be different, and the strength criterion becomes

$$\frac{E^{[i]}}{S}|N_{1\,\text{max}}^{\text{tens}}| \leq \sigma_{\text{allow}}^{\text{tens}[i]}, \quad \frac{E^{[i]}}{S}|N_{1\,\text{max}}^{\text{comp}}| \leq \sigma_{\text{allow}}^{\text{comp}[i]}, \tag{5.25}$$

where $\sigma_{\text{allow}}^{\text{tens}[i]}$ and $\sigma_{\text{allow}}^{\text{comp}[i]}$ are the allowable stresses for layer i in tension and compression, respectively. The strength criterion must be checked for each material layer.

Example 5.1. Beam under a uniform axial load

Consider the uniform, clamped beam of length L subjected to a uniform axial loading $p_1(x_1) = p_0$, as depicted in fig. 5.6 . The governing differential equation is given by eq. (5.19), and for the particular case at hand, this becomes

$$S\frac{\mathrm{d}^2\bar{u}_1}{\mathrm{d}x_1^2} = -p_0.$$

The following boundary conditions apply: $\bar{u}_1 = 0$ at the root of the beam, whereas $S\mathrm{d}\bar{u}_1/\mathrm{d}x_1 = 0$ at its tip. The solution of this differential equation is then

$$\bar{u}_1 = \frac{p_0 L^2}{S}\left[\left(\frac{x_1}{L}\right) - \frac{1}{2}\left(\frac{x_1}{L}\right)^2\right]. \tag{5.26}$$

The axial force is obtained from eq. (5.16) as

$$N_1 = S\bar{\epsilon}_1 = S\frac{\mathrm{d}\bar{u}_1}{\mathrm{d}x_1} = p_0 L\left(1 - \frac{x_1}{L}\right).$$

This result can also be obtained by direct integration of the equilibrium eq. (5.18).

Example 5.2. Tapered beam under centrifugal load

A helicopter blade of length L is rotating at an angular velocity Ω about the $\bar{\imath}_2$ axis, as depicted in fig. 5.11. The blade is homogeneous and its cross-section linearly tapers from an area $\mathcal{A}_0$ at the root to $\mathcal{A}_1 = \mathcal{A}_0/2$ at the tip. The area can then be written as

$$\mathcal{A}(x_1) = \mathcal{A}_0 + (\mathcal{A}_1 - \mathcal{A}_0)\frac{x_1}{L} = \mathcal{A}_0\left(1 - \frac{x_1}{2L}\right).$$

Consequently, the axial stiffness varies along the beam span, $S(x_1) = E\mathcal{A}(x_1)$, where E is Young's modulus.

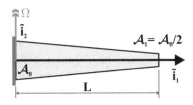

Fig. 5.11. A helicopter blade rotating at an angular speed Ω.

Due to the centrifugal loading associated with the rotation, the blade is subjected to a distributed load $p_1(x_1) = \rho\mathcal{A}(x_1)\Omega^2 x_1$, where ρ is the material density. The governing differential equation for this problem becomes

$$\frac{\mathrm{d}}{\mathrm{d}x_1}\left[E\mathcal{A}_0\left(1 - \frac{x_1}{2L}\right)\frac{\mathrm{d}\bar{u}_1}{\mathrm{d}x_1}\right] = -\rho\mathcal{A}_0\left(1 - \frac{x_1}{2L}\right)\Omega^2 x_1;$$

with the following boundary conditions: $\bar{u}_1 = 0$ at the root of the beam and $Sd\bar{u}_1/dx_1 = 0$ at its tip. It is convenient to use the non-dimensional span variable, $\eta = x_1/L$, to write these equations in a more compact form as $[(1 - \eta/2)\,\bar{u}_1']' = -\rho\Omega^2 L^3(\eta - \eta^2/2)/E$, where the notation $(\cdot)'$ denotes a derivative with respect to η. The boundary conditions are $\bar{u}_1 = 0$ and $\bar{u}_1' = 0$, at the root and tip of the beam, respectively. This differential equation can be integrated once, and with the help of the boundary condition at the tip of the blade becomes

$$\bar{u}_1' = \frac{\rho\Omega^2 L^3}{E} \frac{\left(\dfrac{1}{3} - \dfrac{\eta^2}{2} + \dfrac{\eta^3}{6}\right)}{1 - \eta/2} = \frac{\rho\Omega^2 L^3}{3E} \left[2 + \eta - \eta^2 - \frac{1}{1 - \eta/2}\right].$$

A second integration then yields

$$\bar{u}_1 = \frac{\rho\Omega^2 L^3}{3E} \left[2\eta + \frac{\eta^2}{2} - \frac{\eta^3}{3} + 2\ln\left(1 - \frac{\eta}{2}\right)\right], \tag{5.27}$$

where the boundary condition at the root of the blade is used to evaluate the integration constant.

Finally, the axial force in the blade is readily obtained from eq. (5.16)

$$N_1 = \rho A_0 \Omega^2 L^2 \left[\frac{1}{3} - \frac{\eta^2}{2} + \frac{\eta^3}{6}\right]. \tag{5.28}$$

Note the appearance of a transcendental function, the logarithm function, in the axial displacement expression. This is due to span-wise variation in axial stiffness. In practical applications, structures are subjected to complex loading conditions, and the structural properties vary dramatically along the span. Consequently, the integration of the governing differential equations becomes increasingly difficult, if not impossible.

5.4.7 Problems

Problem 5.1. Axial stress in a reinforced box beam
Figure 5.12 depicts an aluminum rectangular box beam of height $h = 0.30$ m, width $b = 0.15$ m, flange thickness $t_a = 12$ mm, and web thickness $t_w = 5$ mm. The beam is reinforced by two layers of unidirectional composite material of thickness $t_c = 4$ mm. The section is subjected to an axial load $N_1 = 600$ kN. The Young's moduli for the aluminum and unidirectional composite are $E_a = 73$ GPa and $E_c = 140$ GPa, respectively. (1) Find the distribution of axial stress over the cross-section and sketch the distribution around the perimeter of the section. (2) Find the magnitude and location of the maximum axial stress in the aluminum and composite layers. (3) Sketch the distribution of axial strain over the section. How does it vary over the full cross-section? (4) If the allowable stress for the aluminum and unidirectional composite are $\sigma_a^{\text{allow}} = 400$ MPa and $\sigma_c^{\text{allow}} = 1500$ MPa, respectively, find the maximum axial force the section can carry.

Problem 5.2. Axial stress in a reinforced I beam
Figure 5.13 depicts an aluminum I beam of height $h = 0.25$ m, width $b = 0.2$ m, flange thickness $t_a = 16$ mm, and web thickness $t_w = 12$ mm. The beam is reinforced by two layers

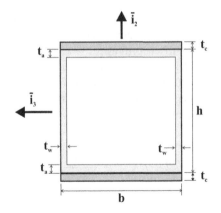

Fig. 5.12. Cross-section of a reinforced rectangular box beam.

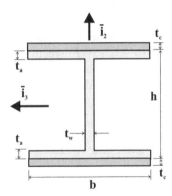

Fig. 5.13. Cross-section of a reinforced I beam.

of unidirectional composite material of thickness $t_c = 5$ mm. The section is subjected to an axial force $N_1 = 500$ kN. The Young's moduli for the aluminum and unidirectional composite are $E_a = 73$ GPa and $E_c = 140$ GPa, respectively. *(1)* Find the distribution of axial stress over the cross-section and sketch it along the $\bar{\imath}_2$ axis. *(2)* Find the magnitude and location of the maximum axial stress in the aluminum and composite layers. *(3)* Sketch the distribution of axial strain along the $\bar{\imath}_2$ axis, and describe how it varies over the entire cross-section. *(4)* If the allowable stress for the aluminum and unidirectional composite are $\sigma_a^{\text{allow}} = 400$ MPa and $\sigma_c^{\text{allow}} = 1500$ MPa, respectively, find the maximum axial force the section can carry.

5.5 Beams subjected to transverse loads

Figure 5.14 shows a beam subjected to a distributed load, $p_2(x_1)$, and a concentrated load, P_2, applied at the tip of the beam. Both load are applied in the *transverse direction*, *i.e.*, in the direction perpendicular to the beam's axis. The distributed loads have the units of force per unit length (N/m in the SI system), whereas the concentrated load have the units of force (N in the SI system). Under the action of these applied loads, bending moments, transverse shear forces, and axial and transverse shearing stresses will be generated in the beam. Moreover, the beam will bend, creating transverse displacement and curvature of the beam axis.

5.5.1 Kinematic description

To simplify the analysis, it is assumed that plane $(\bar{\imath}_1, \bar{\imath}_2)$ is a plane of symmetry of the structure. Since the loads are applied in this plane of symmetry, the response of the beam will be entirely contained in that plane. The three Euler-Bernoulli assumptions discussed in the previous sections are still applicable, and furthermore, it seems

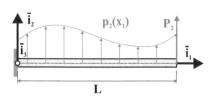

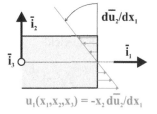

Fig. 5.14. Beam subjected to transverse loads.

Fig. 5.15. Axial displacement distribution on cross-section.

reasonable to assume that transverse loads only cause transverse displacement and curvature of the section. The general displacement field, eq. (5.4), then reduces to

$$u_1(x_1, x_2, x_3) = -x_2 \frac{d\bar{u}_2(x_1)}{dx_1}, \tag{5.29a}$$

$$u_2(x_1, x_2, x_3) = \bar{u}_2(x_1), \tag{5.29b}$$

$$u_3(x_1, x_2, x_3) = 0. \tag{5.29c}$$

This displacement field is depicted in fig. 5.15: it corresponds to a linear distribution of the axial displacement component over the cross-section. The only non-vanishing strain component from eq. (5.7) is

$$\epsilon_1(x_1, x_2, x_3) = -x_2 \kappa_3(x_1). \tag{5.30}$$

Here again, this describes a linear distribution of the axial strain over the cross-section.

5.5.2 Sectional constitutive law

It is assumed that the beam is made of a linearly elastic material. Hooke's law once again reduces to eq. (5.14), and the axial stress distribution becomes

$$\sigma_1(x_1, x_2, x_3) = -E x_2 \kappa_3(x_1). \tag{5.31}$$

The sectional axial force, given by eq. (5.8), is evaluated as

$$N_1(x_1) = \int_{\mathcal{A}} \sigma_1(x_1, x_2, x_3) \, d\mathcal{A} = -\left[\int_{\mathcal{A}} E \, x_2 \, d\mathcal{A}\right] \kappa_3(x_1). \tag{5.32}$$

Because the beam is subjected to transverse loads only, this axial force must vanish as can be proved by a simple equilibrium argument. On the other hand, the curvature, $\kappa_3(x_1)$, is not zero, and hence, the bracketed term must vanish, *i.e.*, $\int_{\mathcal{A}} E \, x_2 \, d\mathcal{A} = 0$. This requirement can be written as

$$x_{2c} = \frac{1}{S} \int_{\mathcal{A}} E \, x_2 \, d\mathcal{A} = \frac{S_2}{S} = 0, \tag{5.33}$$

where x_{2c} is the location of the *modulus-weighted centroid* of the cross-section. If the section is made of a homogeneous material, Young's modulus can be factored out of the integrals to yield

$$x_{2c} = \frac{E \int_A x_2 \, dA}{E \int_A dA} = \frac{1}{A} \int_A x_2 \, dA = 0, \tag{5.34}$$

where x_{2c} is now simply the area center of the section.

These results specify the location of the axis system on the cross-section. Equation (5.33) implies that the axis system is located at the modulus-weighted centroid of the section (or at the area center if the beam is constructed of a homogeneous material).

For a homogeneous material, the material density is also constant over the section, and hence, the location of the center of mass, $x_{2m} = (\rho \int_A x_2 \, dA)/(\rho \int_A dA) = (\int_A x_2 \, dA)/A = x_{2c}$. Clearly, when the section is made of a homogeneous material, the modulus-weighted centroid, the center of mass, and the area center all coincide. For simplicity, the terms *centroid* and *modulus-weighted centroid* will be used interchangeably.

The bending moment defined in eq. (5.10) can be evaluated by introducing the axial stress distribution, eq. (5.31), to find

$$M_3(x_1) = \left[\int_A E \, x_2^2 dA \right] \kappa_3(x_1) = H_{33}^c \, \kappa_3(x_1), \tag{5.35}$$

where the curvature, $\kappa_3(x_1)$, is factored out of the integral over the section. The *centroidal bending stiffness* about axis $\bar{\imath}_3$ is defined as

$$H_{33}^c = \int_A E \, x_2^2 \, dA. \tag{5.36}$$

The relationship given by eq. (5.35) is the constitutive law for the bending behavior of the beam. It expresses the proportionality between the bending moment and the curvature, with a constant of proportionality called the bending stiffness (also referred to as the flexural rigidity). It can be written as

$$M_3(x_1) = H_{33}^c \, \kappa_3(x_1). \tag{5.37}$$

Equation (5.37) is generally referred to as the *moment-curvature relationship* for a beam.

Finally, it should be noted that both the bending moment and bending stiffness are computed with respect to axis system $(\bar{\imath}_2, \bar{\imath}_3)$ with its origin at the centroid of the cross-section.

5.5.3 Equilibrium equations

Equilibrium equations are now derived to complete the formulation; an infinitesimal slice of the beam of length dx_1 is depicted in fig. 5.16. The bending moment,

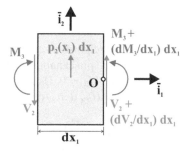

Fig. 5.16. Equilibrium of an infinitesimal slice of the beam.

$M_3(x_1)$, and transverse shear force, $V_2(x_1)$, are acting on the face at location x_1. The corresponding quantities acting on the face at location $x_1 + dx_1$ have been evaluated using a Taylor series expansion and higher differential order terms are ignored.

The free body diagram of this infinitesimal slice of the beam yields the following two equilibrium equations

$$\frac{dV_2}{dx_1} = -p_2(x_1), \tag{5.38a}$$

$$\frac{dM_3}{dx_1} + V_2 = 0, \tag{5.38b}$$

where the first equation expresses vertical force equilibrium and the second expresses moment equilibrium about point **O**.

The transverse shearing force, V_2, is readily eliminated from these two equilibrium equations to obtain a single equilibrium equation,

$$\frac{d^2 M_3}{dx_1^2} = p_2(x_1). \tag{5.39}$$

5.5.4 Governing equations

The governing equation for the transverse deflection of the beam are found by introducing the moment-curvature relation, eq. (5.37), into the equation of equilibrium, eq. (5.39), and recalling the expression for the curvature, eq. (5.6), to yield

$$\frac{d^2}{dx_1^2}\left[H_{33}^c \frac{d^2 \bar{u}_2}{dx_1^2} \right] = p_2(x_1). \tag{5.40}$$

This fourth order differential equation can be solved for the transverse displacement field, $\bar{u}_2(x_1)$, given the distribution of transverse loading, $p_2(x_1)$.

Four boundary conditions are required for the solution of eq. (5.40), two at each end of the beam. Typical boundary conditions are listed here.

1. A *clamped end* restricts both transverse displacement and rotation of the section. Since the rotation of the section and the slope of the beam are equal, see eq. (5.3), it follows that

$$\bar{u}_2 = 0, \quad \frac{\mathrm{d}\bar{u}_2}{\mathrm{d}x_1} = 0.$$

2. A *simply supported* (or pinned) end requires a zero transverse displacement, but the slope of the beam is arbitrary. The pin cannot support a bending moment implying a second boundary condition: $M_3 = 0$. Using eq. (5.37) and the definition of curvature, eq. (5.6), yields $M_3 = H_{33}^c \, \mathrm{d}^2\bar{u}_2/\mathrm{d}x_1^2 = 0$. Thus,

$$\bar{u}_2 = 0, \quad \frac{\mathrm{d}^2\bar{u}_2}{\mathrm{d}x_1^2} = 0.$$

3. At a *free (or unloaded) end*, both bending moment and shear force must vanish. In view of eq. (5.38), the vanishing of the shear force implies $V_2 = -\mathrm{d}M_3/\mathrm{d}x_1 = 0$, leading to

$$\frac{\mathrm{d}^2\bar{u}_2}{\mathrm{d}x_1^2} = 0, \quad -\frac{\mathrm{d}}{\mathrm{d}x_1}\left[H_{33}^c \frac{\mathrm{d}^2\bar{u}_2}{\mathrm{d}x_1^2}\right] = 0.$$

4. At an *end subjected to a concentrated transverse load*, P_2, the bending moment must still vanish, but the shear force must equal the applied load, *i.e.*, $P_2 = V_2 = -\mathrm{d}M_3/\mathrm{d}x_1$. This leads to the following conditions

$$\frac{\mathrm{d}^2\bar{u}_2}{\mathrm{d}x_1^2} = 0, \quad -\frac{\mathrm{d}}{\mathrm{d}x_1}\left[H_{33}^c \frac{\mathrm{d}^2\bar{u}_2}{\mathrm{d}x_1^2}\right] = P_2.$$

5. It is quite common for beams to feature end *rectilinear springs*, as depicted in fig. 5.17. The spring stiffness constant is denoted k and has units of force per unit length. Figure 5.17 shows a free body diagram of the spring. The shear force, V_2, acting on the beam's tip is positive up. Due to Newton's third law, a force of magnitude V_2 acts down on the spring. Vertical equilibrium of the forces acting on the spring then yields $-V_2(L) = k \, \bar{u}_2(L)$, where $\bar{u}_2(L)$ is the beam's tip transverse displacement, measured positive up. The minus sign in front of the shear force is a consequence of the sign conventions: the displacement is positive up, while the force is positive down. In view of eq. (5.38), the boundary conditions at the tip of the beam now become

$$\frac{\mathrm{d}}{\mathrm{d}x_1}\left[H_{33}^c \frac{\mathrm{d}^2\bar{u}_2}{\mathrm{d}x_1^2}\right]_{x_1=L} - k \, \bar{u}_2(L) = 0, \quad \frac{\mathrm{d}^2\bar{u}_2}{\mathrm{d}x_1^2} = 0,$$

where the second condition implies the vanishing of the tip bending moment. If the spring is located at the left end of the beam, the shear force on the spring will be positive upward and the sign of the second term of the first boundary condition will become positive. This can be readily verified by drawing a free body diagram of the system with the proper sign conventions.

6. In other cases, a *rotational spring* may be acting at the tip of the beam, as shown in fig. 5.18. The rotational spring stiffness constant is denoted k and has units of moment per radian. Figure 5.18 also shows a free body diagram of the spring.

Fig. 5.17. Free body diagram for the beam end linear spring of stiffness constant k.

The bending moment, M_3, acting at the beam's tip is positive counterclockwise. Due to Newton's third law, a clockwise bending moment of equal magnitude acts on the torsional spring. The moment equilibrium equation for the spring now becomes $-M_3(L) = k \, \Phi_3(L)$, where $\Phi_3(L)$ is the rotation of the tip spring. In view of eq. (5.3), the rotation of the spring equals the slope of the beam, and hence, $-M_3(L) = k \, \mathrm{d}\bar{u}_2(L)/\mathrm{d}x_1$. The minus sign in front of the bending moment is a consequence of the sign conventions: a positive rotation is counterclockwise, while a positive moment is clockwise. The boundary conditions at the tip of the beam now become

$$H_{33}^c \left.\frac{\mathrm{d}^2\bar{u}_2}{\mathrm{d}x_1^2}\right|_{x_1=L} + k \left.\frac{\mathrm{d}\bar{u}_2}{\mathrm{d}x_1}\right|_{x_1=L} = 0, \qquad -\frac{\mathrm{d}}{\mathrm{d}x_1}\left[H_{33}^c \frac{\mathrm{d}^2\bar{u}_2}{\mathrm{d}x_1^2}\right] = 0,$$

where the second condition implies the vanishing of the tip shear force. If the spring is located at the left end of the beam, the bending moment on the spring will be positive counterclockwise and the sign of the second term of the first boundary condition will become negative. This can be readily verified by drawing a free body diagram of the system with the proper sign conventions.

Fig. 5.18. Free body diagram for a beam with end rotational spring of stiffness constant k.

5.5.5 The sectional bending stiffness

The bending stiffness, H_{33}^c, of the section characterizes the stiffness of the beam when subjected to bending. If the beam is made of a homogeneous material Young's modulus can be factored out of the definition of the bending stiffness, eq. (5.36), to yield

$$H_{33}^c = E \, I_{33}^c, \tag{5.41}$$

where

$$I_{33}^c = \int_{\mathcal{A}} x_2^2 \, \mathrm{d}\mathcal{A}. \tag{5.42}$$

I_{33}^c is a purely geometric quantity known as the area second moment of the section computed about the area center.

On the other hand, if the section is made of several different materials, the bending stiffness must be computed according to eq. (5.36). An important case is that of a rectangular section of width b made of layered materials of different stiffnesses, as depicted in fig. 5.9. Assuming the material to be homogeneous within each layer with a Young modulus, $E^{[i]}$, in layer i, the bending stiffness becomes

$$H_{33}^c = \int_{\mathcal{A}} E x_2^2 \, \mathrm{d}\mathcal{A} = \sum_{i=1}^{n} E^{[i]} \int_{\mathcal{A}^{[i]}} x_2^2 \, \mathrm{d}\mathcal{A}^{[i]}.$$

When the integration is carried out for the rectangular areas, this expression reduces to

$$H_{33}^c = \frac{b}{3} \sum_{i=1}^{n} E^{[i]} \left[(x_2^{[i+1]})^3 - (x_2^{[i]})^3 \right]. \tag{5.43}$$

The bending stiffness is a *weighted average* of the Young's moduli of the various layers. The weighting factor, $\left[(x_2^{[i+1]})^3 - (x_2^{[i]})^3 \right]$, strongly biases the average in favor of the outermost layers, for which $x_2^{[i+1]}$ and $x_2^{[i]}$ are large, whereas the layers near the centroid, where $x_2^{[i+1]}$ and $x_2^{[i]}$ are nearly zero, contribute little to the overall bending stiffness.

Example 5.3. *The four-point bending test*
The bending stiffness of a beam can be computed from the geometry of the cross-section and the properties of the constituent materials, see eq. (5.41) for a beam made of a homogeneous, isotropic material, or eq. (5.43) for a beam made of composite materials. However, it is possible to directly measure the bending stiffness of a beam using a test setup that will subject a portion of the beam to pure bending.

The four-point bending test set-up depicted in fig. 5.19 accomplishes this over the test section between the inner supports. The load, P, applied by the testing machine is transmitted to the test sample through two rollers; the applied load is reacted underneath the test sample by two additional rollers. In view of the symmetry of the configuration, each of the four roller carries a load $P/2$. The test section of the beam is subjected to a bending moment $M_3 = Pd/2$, where d is the distance between the rollers.

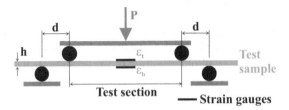

Fig. 5.19. Configuration of the four-point bending test.

The deformation of the test sample can be measured by two strain gauges, located one on top, the other on the bottom of the sample, as shown in fig. 5.19. Let ϵ_t and

ϵ_b be the strain measurements at the top and bottom locations, respectively. In view of eq. (5.30), these strains are related to the curvature of the beam: $\epsilon_t = -x_{2t}\kappa_3$ and $\epsilon_b = -x_{2b}\kappa_3$, where x_{2t} and x_{2b} are the x_2 coordinates of the locations of the top and bottom gauges, respectively. Subtracting these two relationships yields $\kappa_3 = (\epsilon_b - \epsilon_t)/h$, where $h = x_{2t} - x_{2b}$ is the depth of the beam.

The test procedure is as follows. The assembly is placed in the testing machine and a load P of increasing magnitude is applied. For each loading level, the corresponding deformation is measured by the strain gauges. The raw test data consists of loading levels, P_i, $i = 1, 2, \ldots, n$, where n is the number of data points, and the corresponding strains, ϵ_{ti} and ϵ_{bi}. From this raw data, the curvature of the beam is computed, $\kappa_{3i} = (\epsilon_{bi} - \epsilon_{ti})/h$, and the corresponding bending moment is evaluated, $M_{3i} = P_i d/2$. This computed data is then plotted with the curvature, κ_{3i}, along the abscissa and bending moment, M_{3i}, along the ordinate.

If the applied load remains moderate, the behavior of the beam is expected to be linear, $i.e.$, a linear relationship should be observed between bending moment and curvature, as expressed by eq. (5.37). The slope of the experimentally obtained moment versus curvature curve should yield the bending stiffness of the beam. This experimental technique can be used for beams made of homogeneous materials, or for complex constructions involving many layers of composite materials (although the relationship between ϵ_b and ϵ_t and κ_3 will depend on the location of the sectional centroid).

5.5.6 The axial stress distribution

The determination the local axial stress, σ_1, for a given transverse load, $p_2(x_1)$, is often of great interest to designers who must assure that this stress does not exceed an allowable value. This can be readily obtained by eliminating the curvature from eqs. (5.31) and (5.37) to find

$$\sigma_1(x_1, x_2, x_3) = -E\, x_2 \frac{M_3(x_1)}{H_{33}^c}. \tag{5.44}$$

If the beam is made of a homogeneous material, the bending stiffness is given by eqs. (5.41) and eq. (5.44) then reduces to

$$\sigma_1(x_1, x_2, x_3) = -x_2 \frac{M_3(x_1)}{I_{33}}. \tag{5.45}$$

This result shows that the axial stress is linearly distributed over the section, and is independent of Young's modulus. For a positive bending moment, the maximum tensile axial stress is found at the point of the section the farthest below the centroid, $i.e.$, at the point with the largest negative value of x_2, whereas the maximum compressive axial stress is found at the point on the section the farthest above the centroid, $i.e.$, at the point for which x_2 is maximum.

In contrast, the axial stress distribution for sections with various layers of materials will be linear only within each layer and will present a discontinuity at the interfaces. Indeed, eq. (5.44) becomes

$$\sigma_1^{[i]}(x_1, x_2, x_3) = -E^{[i]} x_2 \frac{M_3(x_1)}{H_{33}^c}. \tag{5.46}$$

According to eq. (5.30), the axial strain distribution is linear over the section, in contrast with the axial stress distribution, which is piece-wise linear. The axial stress distributions for homogeneous and layered sections are contrasted in fig. 5.20.

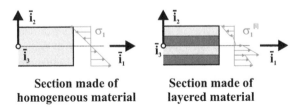

Section made of homogeneous material   **Section made of layered material**

Fig. 5.20. Axial stress distributions in homogeneous and layered sections.

Once the local axial stress is determined, a strength criterion can be applied to determine whether the structure can sustain the applied loads. If a positive positive bending moment is applied, combining the strength criterion, eq. (2.28), and eq. (5.44) leads to $|x_2^{\max}| EM_3/H_{33}^c \leq \sigma_{\text{allow}}^{\text{comp}}$ and $|x_2^{\min}| EM_3/H_{33}^c \leq \sigma_{\text{allow}}^{\text{tens}}$. The strength criterion becomes

$$\frac{|x_2^{\max}|}{H_{33}^c} E|M_3^{\max}| \leq \sigma_{\text{allow}}^{\text{comp}}, \qquad \frac{|x_2^{\min}|}{H_{33}^c} E|M_3^{\max}| \leq \sigma_{\text{allow}}^{\text{tens}},$$

where $|M_3^{\max}|$ is the maximum positive bending moment in the beam and

$$\frac{|x_2^{\max}|}{H_{33}^c} E|M_3^{\min}| \leq \sigma_{\text{allow}}^{\text{tens}}, \qquad \frac{|x_2^{\min}|}{H_{33}^c} E|M_3^{\min}| \leq \sigma_{\text{allow}}^{\text{comp}},$$

where $|M_3^{\min}|$ is absolute value of the minimum negative bending moment in the beam. If the section is such that $|x_2^{\min}| = |x_2^{\max}|$, and/or if the material presents equal tensile and compressive strengths, one or more of these four strength criteria might become redundant.

Of course, if the section consists of layers made of various materials, the strength of each layer will, in general, be different. Furthermore, the maximum stress does not necessarily occur at the points with the largest distance to the centroid, as illustrated in fig. 5.20. In such a case, the axial stress must be computed at the top and bottom locations of each ply, and then, the strength criterion is applied.

5.5.7 Rational design of beams under bending

The axial stress distribution of a beam under bending is given by eq. (5.31). The axial stress clearly vanishes anywhere along axis $\bar{\imath}_3$ of the beam, which passes through the section's centroid. This line on the cross-section is called the *neutral axis* of the beam. Consequently, the material located near the neutral axis carries almost no stresses and contributes little to the overall load carrying capability of the beam.

A similar conclusion can be drawn from examining the expression for the bending stiffness, eq. (5.36): the integrand vanishes along the neutral axis. This means that the material located near the neutral axis contributes little to the bending stiffness of the beam. Clearly, the rational design of a beam under bending calls for the removal of the material located at and near the neutral axis and its relocation away from that axis where it will contribute more significantly to the bending stiffness.

Consider first a beam made of homogeneous material. Two different cross-sections are depicted in fig. 5.21: the first section is a rectangle of width b and height h, and the second is composed of two flanges each of width b and height $h/2$ separated by a distance $2d$. Both sections have the same mass per unit span $m = bh\rho$ where ρ is the material density. The second section is an idealization since no material connects the two flanges. In practical designs, a thin web would be used to keep the two flanges in their respective positions.

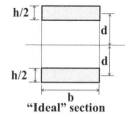

Rectangular section

"Ideal" section

Fig. 5.21. A rectangular section, and the ideal section.

The ratio of the bending stiffnesses of the two sections, denoted H_{ideal} and H_{rect}, for the ideal and rectangular sections, respectively, is

$$\frac{H_{\text{ideal}}}{H_{\text{rect}}} = \frac{E\,2\left[\dfrac{b(h/2)^3}{12} + \dfrac{bh}{2}d^2\right]}{E\,\dfrac{bh^3}{12}} = \frac{1}{4} + 12\left(\frac{d}{h}\right)^2.$$

When $d \gg h$ the bending stiffness of the ideal section is much larger than that of the rectangular section. Indeed, for $d/h = 10$, $H_{\text{ideal}}/H_{\text{rect}} \approx 12(d/h)^2 = 1200$.

The ratio of the maximum axial stresses in the two sections, denoted $\sigma_{\text{rect}}^{\max}$ and $\sigma_{\text{ideal}}^{\max}$ for the rectangular and ideal sections, respectively, is found as

$$\frac{\sigma_{\text{rect}}^{\max}}{\sigma_{\text{ideal}}^{\max}} = \frac{E\dfrac{h}{2}M_3\,I_{\text{ideal}}}{I_{\text{rect}}E\left(d + \dfrac{h}{4}\right)M_3} = \frac{\dfrac{1}{4} + 12\left(\dfrac{d}{h}\right)^2}{\dfrac{1}{2} + 2\left(\dfrac{d}{h}\right)}. \tag{5.47}$$

For $d/h = 10$, $\sigma_{\text{rect}}^{\max}/\sigma_{\text{ideal}}^{\max} \approx 6(d/h) = 60$. If the same material used for the two sections, the ideal section can carry a 60 times larger bending moment, although the two beams have the same amount of material (and therefore the same weight).

This example shows that the rational design of a beam in bending calls for a section with the largest possible height, and the concentration of all the material as far as possible from the neutral axis. In practical situations the ideal section cannot be used. A web is necessary to connect the two flanges resulting in what is called an "I beam" design, as shown in fig. 5.13. The maximum height of the section is often limited by other design considerations. Furthermore, as the height of the section increases, it becomes prone to instabilities such as web and flange buckling.

Example 5.4. Simply supported beam under a uniform load

Consider a simply supported, uniform beam of length L subjected to a uniform transverse loading $p_2(x_1) = p_0$, as depicted in fig. 5.22. Determine the deflected shape of the beam, and the bending moment and shear force distributions. This information will enable a designer to determine if the beam deflections are acceptable and, from the bending moment distribution, find the peak stresses value and compare it with the specified limit design value.

For this problem, since the bending stiffness is uniform along the span of the beam, the governing equation, eq. (5.40), reduces to

$$H_{33}^c \frac{d^4 \bar{u}_2}{dx_1^4} = p_0.$$

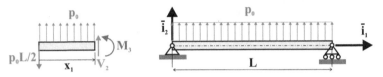

Fig. 5.22. Simply supported beam under a uniform transverse load.

The boundary conditions, $\bar{u}_2 = H_{33}^c d^2 \bar{u}_2 / dx_1^2 = 0$ at the beam's root and $\bar{u}_2 = H_{33}^c d^2 \bar{u}_2 / dx_1^2 = 0$ at its tip, express the vanishing of the transverse displacement and bending moment at the two end supports. The solution of this differential equation is

$$\bar{u}_2 = \frac{p_0 L^4}{24 H_{33}^c} \left[\left(\frac{x_1}{L} \right) - 2 \left(\frac{x_1}{L} \right)^3 + \left(\frac{x_1}{L} \right)^4 \right]. \tag{5.48}$$

The bending moment distribution is then computed from eq. (5.37)

$$M_3 = -\frac{p_0 L^2}{2} \frac{x_1}{L} \left(1 - \frac{x_1}{L} \right). \tag{5.49}$$

As expected, the bending moment is maximum at mid-span, $M_3^{\max} = p_0 L^2 / 8$. The same result can be obtained from simple statics considerations. The axial stress at any point in the beam can then be obtained from the formulæ developed in section 5.5.

A simpler solution of this problem can be obtained directly from equilibrium considerations. This problem is isostatic because all forces and moments can be determined solely from the equilibrium equations. A simple free body diagram of the beam reveals that the reaction forces at the ends of the beam are $p_0 L / 2$. The bending moment distribution then follows from the free body diagram shown in fig. 5.22 as $M_3 = -x_1 p_0 L / 2 + p_0 x_1^2 / 2$, which, as expected, is identical to eq. (5.49). The beam moment-curvature relation, eq. (5.37), and the relation between the curvature and second derivative of the transverse displacement, eq. (5.6), can be used to find

$$H_{33}^c \frac{d^2 \bar{u}_2}{dx_1^2} = M_3(x_1). \tag{5.50}$$

For the present case, this equation can be integrated directly to yield

$$\bar{u}_2(x_1) = \frac{p_0 L^4}{24 H_{33}^c} \left[-2 \left(\frac{x_1}{L} \right)^3 + \left(\frac{x_1}{L} \right)^4 \right] + C_1 x_1 + C_2,$$

where C_1 and C_2 are two integration constants which must be determined from two boundary conditions, one at each end of the beam. The transverse displacement must vanish at either end of the beam, $\bar{u}_2(0) = \bar{u}_2(L) = 0$, leading to

$$\bar{u}_2 = \frac{p_0 L^4}{24 H_{33}^c} \left[\left(\frac{x_1}{L} \right) - 2 \left(\frac{x_1}{L} \right)^3 + \left(\frac{x_1}{L} \right)^4 \right],$$

which is identical to eq. (5.48) above.

This alternative solution approach is easier to develop because it involves the solution of a second order differential equation, rather than a fourth order equation. This alternative solution, however, is only possible because this particular problem is isostatic, *i.e.*, the bending moment distribution can be determined from equilibrium considerations alone.

Example 5.5. Simply supported beam with concentrated load: approach 1
Consider now a simply supported, uniform beam of length L subjected to a concentrated load P acting at a distance αL from the left support, as depicted in fig. 5.23.

First, the solution of this problem might seem to be very similar to that presented in the previous example: the governing differential equation of the problem is $d^4 \bar{u}_2/dx_1^4 = 0$, and the boundary conditions are $\bar{u}_2 = d^2 \bar{u}_2/dx_1^2 = 0$ at both root and tip of the beam. But this approach cannot be possibly right, because the applied load P does not even appear in the governing equations!

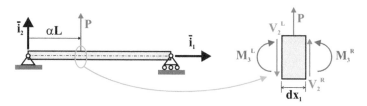

Fig. 5.23. Simply supported beam with one concentrated load.

The concentrated load, P, should normally appear in the statement of boundary conditions for the beam, but it is applied at an arbitrary location along the span of the beam, not at the ends. Hence, it is necessary to "create" a new set of boundary conditions at the point of application of the load. The beam is separated into two portions, one portion to the left of the applied load, the other to its right. For each of the two portions, the governing differential equation of the problem is $d^4\bar{u}_2/dx_1^4 = 0$, which integrates to

$$\bar{u}_2^L = A + Bx_1 + Cx_1^2 + Dx_1^3, \quad \text{and} \quad \bar{u}_2^R = E + Fx_1 + Gx_1^2 + Kx_1^3,$$

for the left and right portions of the beam, respectively.

The two solutions include 8 integration constants: A, B, C, and D, for the left portion of the beam and E, F, G, and K, for its right portion to be determined to complete the solution process. The boundary conditions at the two ends of the beam are still $\bar{u}_2 = d^2\bar{u}_2/dx_1^2 = 0$ at $x_1 = 0$ and L. Imposing these conditions leads to

$$\bar{u}_2^L = Bx_1 + Dx_1^3, \quad \text{and} \quad \bar{u}_2^R = F(x_1 - L) + K(x_1^3 - 3Lx_1^2 + 2L^3).$$

Four boundary conditions are imposed and four integration constant are determined. Clearly, the determination of the remaining four integration constants requires an additional four boundary conditions, which must be expressed at $x_1 = \alpha L$, the common end of the two beam portions. Because two different governing equations are written for the two portions of the beam, the left and right solutions are, as yet, unrelated: continuity conditions must be applied at $x_1 = \alpha L$. First, the displacement and slope of the beam must be continuous at this point: $\bar{u}_2^L(\alpha L) = \bar{u}_2^R(\alpha L)$ and $d\bar{u}_2^L(\alpha L)/dx_1 = d\bar{u}_2^R(\alpha L)/dx_1$. Furthermore, inspection of the free body diagram of the differential element located under the applied load depicted in fig. 5.23, reveals two equilibrium conditions: $M_3^L(\alpha L) = M_3^R(\alpha L)$ and $-V_2^L(\alpha L) + P + V_2^R(\alpha L) = 0$. These four continuity conditions will be used to evaluate the remaining integrations constants, B, D, F, and K.

The two equilibrium conditions yield the following two algebraic equations for the integration constants D and K: $6D\alpha L = 6K(\alpha L - L)$ and $6DH_{33}^c + P - 6KH_{33}^c = 0$. This leads to $D = -(1 - \alpha)P/(6H_{33}^c)$, and $K = \alpha P/(6H_{33}^c)$. Next, the continuity conditions for displacements and slope imply

$$B + 3D\alpha^2 L^2 = F + K(3\alpha^2 L^2 - 6\alpha L^2) \quad \text{and}$$
$$\alpha L + D\alpha^3 L^3 = F(\alpha L - L) + K(\alpha^3 L^3 - 3\alpha^2 L^3 + 2L^3).$$

Introducing the values for the constants found above, the last two continuity conditions become

$$\begin{bmatrix} 1 & -1 \\ \alpha & 1 - \alpha \end{bmatrix} \begin{Bmatrix} B/L^2 \\ F/L^2 \end{Bmatrix} = \frac{P}{6H_{33}^c} \begin{Bmatrix} -3\alpha^2 \\ 2\alpha(1 - \alpha^2) \end{Bmatrix}.$$

Finally, the solution of this linear system yields $B = \alpha(2 - \alpha)(1 - \alpha)PL^2/(6H_{33}^c)$ and $F = \alpha(2 + \alpha^2)PL^2/(6H_{33}^c)$. The deflected shape of the beam is now found as

$$\bar{u}_2(\eta) = \frac{PL^3}{6H_{33}^c} \begin{cases} -(1 - \alpha)\eta^3 + \alpha(2 - \alpha)(1 - \alpha)\eta, & 0 \leq \eta \leq \alpha, \\ \alpha(\eta^3 - 3\eta^2) + \alpha(2 + \alpha^2)\eta - \alpha^3, & \alpha < \eta \leq 1, \end{cases} \tag{5.51}$$

where the solutions for the left and right portions of the beam are indicated by their range of validity, $0 \leq \eta \leq \alpha$ and $\alpha < \eta \leq 1$, respectively, and $\eta = x_1/L$ is the non-dimensional span variable.

The bending moment distribution then follows

$$M_3(\eta) = PL \begin{cases} -(1-\alpha)\eta, & 0 \leq \eta \leq \alpha, \\ -\alpha(1-\eta), & \alpha < \eta \leq 1. \end{cases} \tag{5.52}$$

Finally, the shear force distribution is computed from the bending moment distribution to find

$$V_2(\eta) = P \begin{cases} (1-\alpha), & 0 \leq \eta \leq \alpha, \\ -\alpha, & \alpha < \eta \leq 1. \end{cases} \tag{5.53}$$

At $\eta = \alpha$, the shear force presents a discontinuity that is equal to the applied concentrated load at that point, as expected from the vertical equilibrium condition at $x_1 = \alpha L$.

Clearly, the presence of a concentrated load at an arbitrary point along the span of the beam considerably complicates the solution process: the problem must be split into two independent sub-problems, thereby creating a common "end point" for the two sub-problems where the applied load is introduced as a boundary condition. Continuity conditions must then be applied to enforce continuity conditions at the connection point.

Example 5.6. Simply supported beam with concentrated load: approach 2
Consider, once again, a simply supported, uniform beam of length L subjected to a concentrated load P acting at a distance αL from the left support, as depicted in fig. 5.24.

To avoid the complexity of the approach presented in the previous example, basic statics arguments are used determine the reaction forces at the two end points. A free body diagram of the entire beam reveals that these forces are $(1-\alpha)P$ and αP, at the left and right end supports, respectively. Next, fig. 5.24 shows the free body diagram of a portion of the beam extending from the left support to a location $0 \leq x_1 \leq \alpha L$ and yields the bending moment distribution, $M_3 = -(1-\alpha)Px_1 = -(1-\alpha)PL\eta$, where $\eta = x_1/L$ is the non-dimensional variable along the span of the beam. Similarly, fig. 5.24 also shows a free body diagram of a piece of the beam extending from location $\alpha L \leq x_1 \leq L$ to the right support; moment equilibrium of this free body diagram leads to $M_3 = -\alpha P(L - x_1) = -\alpha PL(1 - \eta)$. These results are identical to those found in eq. (5.52). The present process is much more expeditious: the bending moment distributions are readily obtained from equilibrium considerations alone. Here again, however, it is necessary to obtained distinct solutions for the left and right portions of the beam.

Next, the bending moment-curvature relationship, eq. (5.50), can be integrated twice to yield the displacement field as

$$\bar{u}_2(\eta) = \frac{PL^3}{H_{33}^c} \begin{cases} -(1-\alpha)\eta^3/6 + C_1\eta + C_2, & 0 \leq \eta \leq \alpha, \\ -\alpha\left(\eta^2/2 - \eta^3/6\right) + C_3\eta + C_4, & \alpha \leq \eta \leq 1. \end{cases}$$

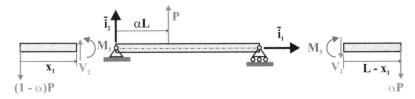

Fig. 5.24. Simply supported beam with one concentrated load.

where C_1, C_2, C_3 and C_4 are four integration constants to be evaluated from the boundary conditions.

Because the beam is simply supported at the two ends, $\bar{u}_2(0) = \bar{u}_2(1) = 0$, and furthermore, at $\eta = \alpha$, both displacement and slope of the beam must be continuous. These four conditions are sufficient to determine the four integration constants to yield the following solution for the deflected shape of the beam

$$\bar{u}_2(\eta) = \frac{PL^3}{6H_{33}^c} \begin{cases} -(1-\alpha)\eta^3 + \alpha(2-\alpha)(1-\alpha)\eta, & 0 \le \eta \le \alpha, \\ \alpha(\eta^3 - 3\eta^2) + \alpha(2+\alpha^2)\eta - \alpha^3, & \alpha < \eta \le 1. \end{cases}$$

As expected, the solution is identical to that found earlier, see eq. (5.51). Clearly, the present solution approach, based on the determination of the bending moment distribution from equilibrium considerations, is much more expeditious than the approach presented in the previous example.

Example 5.7. Cantilevered beam under uniform load

Consider now a cantilevered beam with a uniformly distributed transverse load, p_0, as shown in fig. 5.25. The first approach to this problem is to solve the governing differential equation of the problem, $d^4\bar{u}_2/dx_1^4 = p_0/H_{33}^c$, with the geometric boundary conditions $\bar{u}_2 = d\bar{u}_2/dx_1 = 0$ at the root of the beam and equilibrium boundary conditions, $M_3 = V_2 = 0$ at the tip of the beam.

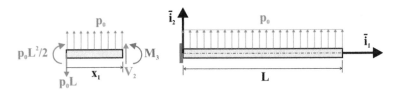

Fig. 5.25. Cantilevered beam under a uniform load.

The solution of the differential equation is $\bar{u}_2 = A + Bx_1 + Cx_1^2 + Dx_1^3 + p_0x_1^4/(24H_{33}^c)$, where the last term represent the particular solution for the non-zero right-hand side of the governing equation. The geometric boundary conditions at the root of the beam imply $A = B = 0$, and the solution reduces to $\bar{u}_2 = Cx_1^2 + Dx_1^3 + p_0x_1^4/(24H_{33}^c)$. Imposing the vanishing of the shear force at the tip of the beam leads to $D = -p_0 L/(6H_{33}^c)$, and the vanishing of the bending moment at the same location gives $C = p_0 L^2/(4H_{33}^c)$. The final solution is

$$\bar{u}_2(x_1) = \frac{p_0 L^4}{24 H_{33}^c} \left(6\eta^2 - 4\eta^3 + \eta^4\right), \qquad (5.54)$$

where $\eta = x_1/L$ is the non-dimensional variable along the span of the beam.

This problem is isostatic, therefore, the root reaction force and moment can be determined from equilibrium considerations alone. The free body diagram of a portion of the beam shown in fig. 5.25 then yields the bending moment distribution in the beam

$$M_3(x_1) = \frac{p_0 L^2}{2} - p_0 L x_1 + \frac{p_0 x_1^2}{2}.$$

Introducing this result into the bending moment-curvature relationship, eq. (5.50), and integrating twice yields the solution

$$\bar{u}_2(x_1) = \frac{1}{H_{33}^c} \left(\frac{p_0 L^2}{4} x_1^2 - \frac{p_0 L}{6} x_1^3 + \frac{p_0}{24} x_1^4\right) + C_1 x_1 + C_2,$$

where C_1 and C_2 are integration constants to be determined from the boundary conditions, $\bar{u}_2(0) = d\bar{u}_2(0)/dx_1 = 0$, to find the deflected shape of the beam as

$$\bar{u}_2(x_1) = \frac{p_0 L^4}{24 H_{33}^c} \left(6\eta^2 - 4\eta^3 + \eta^4\right).$$

Of course, this result matches that found earlier.

Example 5.8. Cantilevered beam under concentrated load
Consider a cantilevered, uniform beam of length L subjected to a concentrated load P acting at a distance αL from the left support, as depicted in fig. 5.26.

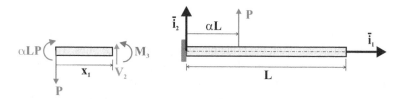

Fig. 5.26. Cantilevered beam under a concentrated load.

The bending moment distribution is readily obtained from simple equilibrium considerations. A free body diagram of the entire beam reveals that at the left support, the clamping force is P and the clamping moment $\alpha L P$. Next, the equilibrium condition for a free body diagram of a piece of the beam extending from the left support to a location $0 \le \eta \le \alpha$, as shown in fig. 5.26, yields $M_3 = PL(\alpha - \eta)$, where $\eta = x_1/L$ is the non-dimensional variable along the span of the beam. Similarly, a free body diagram of a piece of the beam extending from the left support to a location $\alpha \leq \eta \leq 1$ leads to $M_3 = 0$.

The bending moment-curvature relationship, eq. (5.50), for each segment of the beam can be integrated twice to yield the displacement field as

$$\bar{u}_2(\eta) = \frac{PL^3}{6H_{33}^c} \begin{cases} \eta^2(3\alpha - \eta), & 0 \leq \eta \leq \alpha, \\ \alpha^2(3\eta - \alpha), & \alpha < \eta \leq 1. \end{cases} \qquad (5.55)$$

The integration process involves a total of four integration constants that are evaluated from the boundary conditions: at the left support, the beam displacement and slope must vanish; at $\eta = \alpha$, both displacement and slope must be continuous.

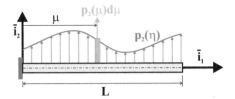

Fig. 5.27. Cantilevered beam under transverse loading.

This example suggests an approach to the computation of the deflection of a cantilevered beam subjected to a distributed loading, $p_2(\eta)$, depicted in fig. 5.27. First, a differential component of force, $p_2(\mu)d\mu$, acting at a non-dimensional distance μ from the left support is considered as a concentrated load applied at location μ. Equation (5.55) is used to evaluate the corresponding displacement field

$$\bar{u}_2(\eta) = \frac{p_2(\mu)d\mu L^4}{6H_{33}^c} \begin{cases} \eta^2(3\mu - \eta), & 0 \leq \eta \leq \mu, \\ \mu^2(3\eta - \mu), & \mu < \eta \leq 1. \end{cases}$$

Because the governing equation for beam transverse displacement, eq. (5.40), is a linear differential equation, the principle of superposition applies. This means that the displacement fields generated by two distinct loading conditions can be superposed to find the displacement field of the beam under the combined loading. In this case, the displacements generated by each differential loading component, $p_2(\mu)d\mu$, can be superposed to find the displacement of the beam under the distributed load. The following integral yields the displacement field

$$\bar{u}_2(\eta) = \frac{L^4}{6H_{33}^c} \left[\int_0^\eta p_2(\mu)\mu^2(3\eta - \mu)\,d\mu + \int_\eta^1 p_2(\mu)\eta^2(3\mu - \eta)\,d\mu \right],$$

where the integrals are used to sum up the contributions over $0 \leq \mu \leq \eta$ and $\eta \leq \mu \leq 1$.

The displacement field of the cantilevered beam under an arbitrary loading distribution, $p_2(\mu)$, can be obtained from the above expression by performing the indicated integrals. For instance, the displacement of a cantilevered beam under a uniform transverse loading p_0 becomes

$$\bar{u}_2(\eta) = \frac{p_0 L^4}{6H_{33}^c} \left[\int_0^\eta \mu^2(3\eta - \mu)\,d\mu + \int_\eta^1 \eta^2(3\mu - \eta)\,d\mu \right].$$

Integration then yields the desired displacement solution $\bar{u}_2(\eta) = p_0 L^4(\eta^4 - 4\eta^3 + 6\eta^2)/(24H_{33}^c)$. This solution is exactly the same as the solution developed for a uniformly loaded cantilevered beam in the previous example.

Example 5.9. The flexibility matrix: experimental determination

All the examples detailed thus far have focused on analytical solutions that predict a beam's transverse displacement field given the applied loads. In practice, however, these predictions must be validated through structural testing. For instance, to assess the structural behavior of an aircraft wing of length L, the following test, depicted in fig. 5.28, could be performed. The wing is cantilevered from a support structure and is subjected to various transverse loads, P_1, P_2, and P_3 applied at span-wise locations $\alpha_1 L$, $\alpha_2 L$, and $\alpha_3 L$, respectively. During the test, the displacements at those same points are monitored by means of displacement gauges, which measure the corresponding transverse displacements, denoted Δ_1, Δ_2, and Δ_3, respectively.

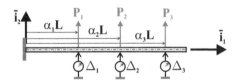

Fig. 5.28. Cantilevered wing under three concentrated loads.

Consider now the following test sequence. First, a single load is applied at location $\alpha_1 L$; this corresponds to a loading P_1, $P_2 = P_3 = 0$. The corresponding displacements are recorded and denoted Δ_{11}, Δ_{21}, Δ_{31}. Next, a single load is applied at location $\alpha_2 L$; this corresponds to a loading P_2, $P_1 = P_3 = 0$, and the corresponding displacements are Δ_{12}, Δ_{22}, Δ_{32}. Finally, a single load is applied at location $\alpha_3 L$; this corresponds to a loading P_3, $P_1 = P_2 = 0$, and the corresponding displacements are Δ_{13}, Δ_{23}, Δ_{33}.

Let q_1, q_2, and q_3, denote the transverse displacements of the wing at locations $\alpha_1 L$, $\alpha_2 L$, and $\alpha_3 L$, respectively. For each of the three loading cases, the experimental data can be presented in the following manner,

$$\begin{Bmatrix} q_1 \\ q_2 \\ q_3 \end{Bmatrix} = \begin{Bmatrix} \Delta_{11}/P_1 \\ \Delta_{21}/P_1 \\ \Delta_{31}/P_1 \end{Bmatrix} P_1, \quad \begin{Bmatrix} q_1 \\ q_2 \\ q_3 \end{Bmatrix} = \begin{Bmatrix} \Delta_{12}/P_2 \\ \Delta_{22}/P_2 \\ \Delta_{32}/P_2 \end{Bmatrix} P_2, \quad \begin{Bmatrix} q_1 \\ q_2 \\ q_3 \end{Bmatrix} = \begin{Bmatrix} \Delta_{13}/P_3 \\ \Delta_{23}/P_3 \\ \Delta_{33}/P_3 \end{Bmatrix} P_3.$$
(5.56)

At this point it is convenient to introduce the concept of *influence coefficient*, $\eta_{ij} = \Delta_{ij}/P_j$, which is the displacement at location $\alpha_i L$, when a single unit load, $P_j = 1$, is applied at location $\alpha_j L$. For the first loading case, $q_1 = (\Delta_{11}/P_1)P_1 = \eta_{11}P_1$. With the help of the influence coefficients, eqs. (5.56) can be restated as

$$\begin{Bmatrix} q_1 \\ q_2 \\ q_3 \end{Bmatrix} = \begin{Bmatrix} \eta_{11} \\ \eta_{21} \\ \eta_{31} \end{Bmatrix} P_1, \quad \begin{Bmatrix} q_1 \\ q_2 \\ q_3 \end{Bmatrix} = \begin{Bmatrix} \eta_{12} \\ \eta_{22} \\ \eta_{32} \end{Bmatrix} P_2, \quad \begin{Bmatrix} q_1 \\ q_2 \\ q_3 \end{Bmatrix} = \begin{Bmatrix} \eta_{13} \\ \eta_{23} \\ \eta_{33} \end{Bmatrix} P_3. \quad (5.57)$$

The influence coefficients are readily obtained from the experimental measurements: the measured displacements are divided by the magnitude of the known applied load. The first loading case provides three measurements, Δ_{11}, Δ_{21}, and Δ_{31}, which, after division by the magnitude of the applied load, P_1, give the influence coefficients appearing in the first equations. Each one of the three loading case gives one set of influence coefficients.

At this point, it is assumed that the wing behaves in a linearly elastic manner and therefore, the principle of superposition applies. If the three loading cases are combined, the resulting deflections can be obtained by adding eqs. (5.57). The result can be summarized in a single matrix relationship as

$$\begin{Bmatrix} q_1 \\ q_2 \\ q_3 \end{Bmatrix} = \begin{bmatrix} \eta_{11} & \eta_{12} & \eta_{13} \\ \eta_{21} & \eta_{22} & \eta_{23} \\ \eta_{31} & \eta_{32} & \eta_{33} \end{bmatrix} \begin{Bmatrix} P_1 \\ P_2 \\ P_3 \end{Bmatrix} = \underline{F} \begin{Bmatrix} P_1 \\ P_2 \\ P_3 \end{Bmatrix}, \quad \underline{F} = \begin{bmatrix} \eta_{11} & \eta_{12} & \eta_{13} \\ \eta_{21} & \eta_{22} & \eta_{23} \\ \eta_{31} & \eta_{32} & \eta_{33} \end{bmatrix}, \quad (5.58)$$

where $\underline{F}$ is the 3×3 *flexibility matrix* and the displacements, q_i, are those resulting from the superposition of the three loading cases. The flexibility matrix simply stores the influence coefficients in an orderly manner.

The process described above can be generalized to a situation where single loads are applied in sequence at N locations, $\alpha_i L$, $i = 1, 2, \ldots, N$, and the corresponding displacements, Δ_{ij}, are measured; Δ_{ij} corresponds to the displacement at location $\alpha_i L$, when a single load, P_j, is applied at location $\alpha_j L$. As before, the influence coefficient are $\eta_{ij} = \Delta_{ij}/P_j$ and eq. (5.58) becomes

$$\underline{q} = \underline{F}\,\underline{Q}, \quad (5.59)$$

where array $\underline{q} = \{q_1, q_2, \ldots, q_N\}^T$ stores the N displacements at locations, $\alpha_i L$, $i = 1, 2, \ldots, N$, resulting from the application of N loads at the same locations, and stored in array $\underline{Q} = \{P_1, P_2, \ldots, P_N\}^T$. The flexibility matrix, $\underline{F}$, now becomes an $N \times N$ matrix, and the N^2 measurements, Δ_{ij}, determine the N^2 entries of this matrix: indeed, the (i, j) entry of the flexibility matrix is $\underline{F}(i, j) = \eta_{ij} = \Delta_{ij}/P_j$. If a detailed study of structural behavior is necessary, displacements must be measured at a large number of points along the wing; of course, the cost of the experiment rapidly increases because the number of required measurements increases like N^2.

Example 5.10. The flexibility matrix: analytical determination

The developments presented in example 5.9 focus on the determination of the flexibility matrix based on experimental measurements. It is also possible to give closed form analytical expressions for each entry of the flexibility matrix.

Consider the cantilevered wing subjected to concentrated transverse loads at locations $\alpha_1 L$, $\alpha_2 L$, and $\alpha_3 L$, as depicted in fig. 5.28. The first loading case, P_1, $P_2 = P_3 = 0$, corresponds to the cantilevered beam problem treated in example 5.8. The displacements at location $\alpha_1 L$, $\alpha_2 L$, and $\alpha_3 L$, denoted Δ_{11}, Δ_{21}, Δ_{31}, respectively, can be expressed as $\Delta_{11} = \bar{u}_2(\eta = \alpha_1)$, $\Delta_{21} = \bar{u}_2(\eta = \alpha_2)$, and $\Delta_{31} = \bar{u}_2(\eta = \alpha_3)$. Equation (5.55) then gives the transverse displacement field at these points as

$$\Delta_{11} = \frac{P_1 L^3}{6H_{33}^c}2\alpha_1^3, \; \Delta_{21} = \frac{P_1 L^3}{6H_{33}^c}\alpha_1^2(3\alpha_2 - \alpha_1), \; \Delta_{31} = \frac{P_1 L^3}{6H_{33}^c}\alpha_1^2(3\alpha_3 - \alpha_1).$$

The influence coefficients are then found by dividing the corresponding displacements by P_1. Equation (5.55) can be used in a similar manner to find the displacements corresponding to the other two loading conditions, P_2, $P_1 = P_3 = 0$, and P_3, $P_1 = P_2 = 0$.

Collecting all the results then yields a closed form solution for the flexibility matrix, see eq. (5.58), as

$$\underline{F} = \frac{L^3}{6H_{33}^c}\begin{bmatrix} 2\alpha_1^3 & \alpha_1^2(3\alpha_2 - \alpha_1) & \alpha_1^2(3\alpha_3 - \alpha_1) \\ \alpha_1^2(3\alpha_2 - \alpha_1) & 2\alpha_2^3 & \alpha_2^2(3\alpha_3 - \alpha_2) \\ \alpha_1^2(3\alpha_3 - \alpha_1) & \alpha_2^2(3\alpha_3 - \alpha_2) & 2\alpha_3^3 \end{bmatrix}. \tag{5.60}$$

Flexibility matrices corresponding to a larger number of locations, say $\alpha_i L$, $i = 1, 2, \ldots, N$, are easily obtained by simple index manipulation of the above result. Although the above expression is complex, each entry of the flexibility matrix can be directly measured by a simple test, as discussed in example 5.9.

Example 5.11. *Clamped-simply supported beam under uniform load*
A beam subjected to a uniform loading p_0 is clamped at one end and is simply supported at the other, as depicted in fig. 5.29.

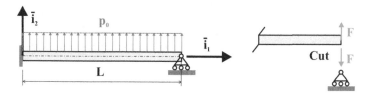

Fig. 5.29. Clamped - simply supported beam under uniform load.

First, this problem will be solved with the help of eq. (5.40). Since the bending stiffness is constant, the governing equation of the problem is $H_{33}^c \mathrm{d}^4\bar{u}_2/\mathrm{d}x_1^4 = p_0$. The necessary four boundary conditions for this problem are $\bar{u}_2 = \mathrm{d}\bar{u}_2/\mathrm{d}x_1 = 0$ at the beam's root and at its tip, $\bar{u}_2 = \mathrm{d}^2\bar{u}_2/\mathrm{d}x_1^2 = 0$. The solution of the differential equation is

$$\bar{u}_2(x_1) = \frac{p_0}{24H_{33}^c}x_1^4 + \frac{1}{6}C_1 x_1^3 + \frac{1}{2}C_2 x_1^2 + C_3 x_1 + C_4,$$

where the first term represents the particular solution associated with the non-vanishing right-hand side of the equation. The four integration constants, C_1, C_2, C_3, and C_4 are then determined with the help of the boundary conditions to yield

$$\bar{u}_2(\eta) = \frac{p_0 L^4}{48 H_{33}^c} \left(2\eta^4 - 5\eta^3 + 3\eta^2\right), \tag{5.61}$$

where $\eta = x_1/L$ is the non-dimensional variable along the span of the beam.

The bending moment is evaluated using the moment-curvature relation, eq. (5.50), to find $M_3(\eta) = p_0 L^2 \left[4\eta^2 - 5\eta + 1\right]/8$. Finally, the transverse shear force is found by using eq. (5.38) as $V_2(\eta) = -p_0 L \left[8\eta - 5\right]/8$.

This problem is hyperstatic: the reaction forces cannot be determined from equilibrium considerations alone. Indeed, for this two dimensional problem, statics provides two equations of equilibrium, but the problem involves three unknown reactions: a vertical force and a moment at the clamped end of the beam, and a vertical force at the simple support.

Using the nomenclature of section 4.3, the system is hyperstatic of order 1. Two solution methods are developed for hyperstatic problems: the displacement and force methods, see sections 4.3.2 and 4.3.3, respectively.

The force method provides a very expeditious approach to the solution of this problem. The system is cut at the simple support, as shown in fig. 5.29, and an unknown reaction force, F, is applied at the two sides of the cut. The problem is now transformed into a cantilevered beam subjected to a uniform loading, p_0, and a tip load, F. The solution is readily obtained by superposing the displacement field corresponding to these these two loadings to find

$$\bar{u}_2(\eta) = \frac{p_0 L^4}{24 H_{33}^c} \left(6\eta^2 - 4\eta^3 + \eta^4\right) + \frac{F L^3}{6 H_{33}^c} \left[\eta^2(3\alpha - \eta)\right],$$

where the first term represents the contribution of the uniform loading, see eq. (5.54), and the second term that of the tip concentrated load, see eq. (5.55).

The compatibility equation at the cut implies the vanishing of the beam's tip deflection, $\bar{u}_2(\eta = 1) = 0$, which leads to $p_0 L^4(6 - 4 + 1)/(24 H_{33}^c) + F L^3(3 - 1)/(6 H_{33}^c) = 0$. Solving this equation yields the reaction force at the support

$$F = -\frac{3 p_0 L}{8}, \tag{5.62}$$

and finally, the displacement field becomes

$$\begin{aligned}
\bar{u}_2(\eta) &= \frac{p_0 L^4}{24 H_{33}^c} \left(6\eta^2 - 4\eta^3 + \eta^4\right) - \frac{3}{8} p_0 L \frac{L^3}{6 H_{33}^c} \left[\eta^2(3\alpha - \eta)\right] \\
&= \frac{p_0 L^4}{48 H_{33}^c} \left(2\eta^4 - 5\eta^3 + 3\eta^2\right).
\end{aligned}$$

In this approach, the solution is obtained in an efficient manner using the force method together with known solutions to elementary problems. Solutions to complex problems can often be constructed in this way from solutions to simpler problems.

Example 5.12. Simply supported beam with concentrated loads

Consider a simply supported, uniform beam of length L subjected to two concentrated loads acting at a distance αL from the end supports, as depicted in fig. 5.30. This problem presents discontinuities in the transverse shear force distribution due the presence of the concentrated loads and for this reason, the solution will be derived from the moment-curvature relationship, eq. (5.37).

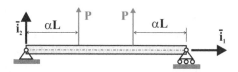

Fig. 5.30. Simply supported beam with two concentrated loads.

The bending moment distribution can be readily obtained from simple equilibrium considerations. A free body diagram of the entire beam reveals that the reaction forces at the end supports must each equal P. Next, the equilibrium condition for a free body diagram of a piece of the beam extending from the left support to a location $0 < x_1 < \alpha L$ yields $M_3 = -Px_1$. Similarly, a free body diagram of a segment of the beam extending from the left support to a location $\alpha L < x_1 < L/2$ leads to $M_3 = P(x_1 - \alpha L) - Px_1 = -P\alpha L$.

The bending moment-curvature relationship, eq. (5.37), is now integrated twice to yield

$$
H_{33}^c \bar{u}_2(x_1) = \begin{cases} -Px_1^3/6 + C_1 x_1 + C_2, & 0 \le x_1 \le \alpha L, \\ -\alpha L P x_1^2/2 + C_3 x_1 + C_4, & \alpha L < x_1 \le L/2, \end{cases}
$$

where C_1, C_2, C_3 and C_4 are four integration constants. Two of these constants can be determined from the boundary conditions: the simple support at the root implies $\bar{u}_2(0) = 0$, and the symmetry condition at mid-span requires $d\bar{u}_2/dx_1(L/2) = 0$. The other two integration constants are found by imposing the continuity of the displacement and slope at $x_1 = \alpha L$. The transverse displacement distribution becomes

$$
\bar{u}_2(\eta) = \frac{PL^3}{6H_{33}^c} \begin{cases} 3(\alpha - \alpha^2)\eta - \eta^3, & 0 \le \eta \le \alpha, \\ -\alpha^3 + 3\alpha\eta - 3\alpha\eta^2, & \alpha < \eta \le 1/2. \end{cases} \tag{5.63}
$$

The shear force distribution then follows from eqs. (5.38b) and (5.37)

$$
V_2(\eta) = P \begin{cases} -1, & 0 \le \eta \le \alpha, \\ 0, & \alpha < \eta \le 1/2. \end{cases}
$$

This result can be easily verified: the portion of the beam between the two concentrated loads is subjected to a constant bending moment αLP, and hence, the shear force vanishes. At $x_1 = \alpha L$ the shear force jumps from $V_2 = P$, immediately to the left, to $V_2 = 0$, immediately to the right of the applied concentrated force. The magnitude of this jump is of course equal to the force applied at that point.

The problem can also be solved using the governing differential equation, eq. (5.40). However, two separate problems must be solved: one for $0 \leq x_1 \leq \alpha L$, and another for $\alpha L < x_1 \leq L/2$. The integration of these two, fourth order equations will generate a total of eight integration constants to be determined from four boundary conditions, and four continuity conditions at $x_1 = \alpha L$. The boundary conditions at the root are $\bar{u}_2(0) = 0$ and $\mathrm{d}^2\bar{u}_2/\mathrm{d}x_1^2(0) = 0$, corresponding to the vanishing of the displacement and bending moment; at mid-span, $\mathrm{d}\bar{u}_2/\mathrm{d}x_1(L/2) = 0$ and $\mathrm{d}^3\bar{u}_2/\mathrm{d}x_1^3(L/2) = 0$, corresponding to the symmetry conditions of vanishing slope and shear force. At $x_1 = \alpha L$, continuity of displacement, slope and bending moment is required together with the enforcement of a jump in shear force. The approach used in this example is clearly less laborious.

Example 5.13. Simply supported beam with two elastic spring supports
A simply supported beam of span L is also supported by two spring of stiffness constant k located at stations $x_1 = \alpha L$ and $(1 - \alpha)L$, and is subjected to a uniform transverse loading p_0, as depicted in fig. 5.31.

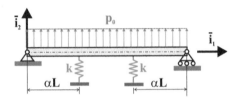

Fig. 5.31. Simply supported beam with two elastic spring supports.

First, the springs are replace by two unknown forces, F, acting downward at $x_1 = \alpha L$ and $(1 - \alpha)L$. The transverse displacement distribution is then readily obtained by superposing the displacement field for a beam under uniform loading and that of a beam under two concentrated forces, see eqs. (5.48) and (5.63), respectively, to find

$$\bar{u}_2(\eta) = \begin{cases} \dfrac{p_0 L^4}{24 H_{33}^c} \left(\eta - 2\eta^3 + \eta^4\right) - \dfrac{F L^3}{6 H_{33}^c} \left[3(\alpha - \alpha^2)\eta - \eta^3\right], & 0 \leq \eta \leq \alpha, \\[2ex] \dfrac{p_0 L^4}{24 H_{33}^c} \left(\eta - 2\eta^3 + \eta^4\right) - \dfrac{F L^3}{6 H_{33}^c} \left(-\alpha^3 + 3\alpha\eta - 3\alpha\eta^2\right), & \alpha < \eta \leq \tfrac{1}{2}, \end{cases}$$

where $\eta = x_1/L$ if the non-dimensional coordinate along the beam span. The unknown force, F, acts on the spring, and hence $F = k\bar{u}_2(\alpha)$.

Introducing the above expression for $\bar{u}_2(\alpha L)$ provides an additional equation to evaluate the non-dimensional force in the spring

$$\frac{F}{p_0 L} = \frac{1}{4} \frac{\bar{k}(\alpha^4 - 2\alpha^3 + \alpha)}{6 + \bar{k}(3\alpha^2 - 4\alpha^3)} = \frac{1}{4}\bar{p}, \tag{5.64}$$

where $\bar{k} = kL^3/H_{33}^c$ is the non-dimensional spring stiffness constant and $\bar{p} = \bar{k}(\alpha^4 - 2\alpha^3 + \alpha)/[6 + \bar{k}(3\alpha^2 - 4\alpha^3)]$ the non-dimensional load fraction. The transverse displacement distribution now becomes

$$\bar{u}_2(\eta) = \frac{p_0 L^4}{24 H_{33}^c} \begin{cases} \eta^4 - (2 - \bar{p})\,\eta^3 + [1 - 3\bar{p}(\alpha - \alpha^2)]\,\eta, & 0 \leq \eta \leq \alpha, \\ \eta^4 - 2\eta^3 + 3\alpha\bar{p}\,\eta^2 + (1 - 3\alpha\bar{p})\,\eta + \alpha^3\bar{p}, & \alpha < \eta \leq 1/2. \end{cases}$$
$$(5.65)$$

The bending moment distribution then follows from eq. (5.37)

$$M_3(\eta) = \frac{p_0 L^2}{2} \begin{cases} \eta^2 - (1 - \bar{p}/2)\,\eta, & 0 \leq \eta \leq \alpha, \\ \eta^2 - \eta + \alpha\bar{p}/2, & \alpha < \eta \leq 1/2. \end{cases} \qquad (5.66)$$

Finally, the shear force distribution is obtained from eq. (5.38)

$$V_2(\eta) = \frac{p_0 L}{2} \begin{cases} 2\eta - (1 - \bar{p}/2), & 0 \leq \eta \leq \alpha, \\ 2\eta - 1, & \alpha < \eta \leq 1/2. \end{cases} \qquad (5.67)$$

As anticipated, this distribution presents a discontinuity due to the concentrated forces the springs apply to the beam. Indeed, the shear forces immediately to the left and right of the spring located at $\eta = \alpha$, denoted F_{2l} and F_{2r}, respectively are such that $F_{2r} - F_{2l} = \bar{p}p_0 L/4 = F$.

Example 5.14. *Simply supported beam on an elastic foundation*
The last example deals with a simply supported beam of length L subjected to a uniform transverse load p_0, as depicted in fig. 5.32. The beam is supported by an elastic foundation of distributed stiffness constant k.

For this problem, the governing equation is $H_{33}^c \mathrm{d}^4\bar{u}_2/\mathrm{d}x_1^4 = p_2(x_1)$, and the boundary conditions are $\bar{u}_2 = \mathrm{d}^2\bar{u}_2/\mathrm{d}x_1^2 = 0$ at both root and tip of the beam. The total applied load $p_2(x_1) = p_0 - k\,\bar{u}_2(x_1)$, where the first term accounts for the applied load, and the second corresponds to the distributed restoring force of the elastic foundation.

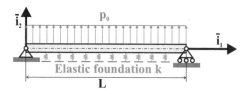

Fig. 5.32. Simply supported beam on an elastic foundation.

The governing equation can be recast as

$$\bar{u}_2'''' + \frac{kL^4}{H_{33}^c}\,\bar{u}_2 = \frac{p_0 L^4}{H_{33}^c},$$

where $\eta = x_1/L$ is the non-dimensional span-wise variable, and $(\cdot)'$ denotes a derivative with respect to η. The boundary conditions become $\bar{u}_2 = \bar{u}_2'' = 0$ at both $\eta = 0$ and $\eta = 1$.

This is a non-homogeneous differential equation and its solution consists of the sum of a homogeneous and particular solution. The solution of the homogeneous equation is of the form $\bar{u}_2(\eta) = \exp(z\eta)$, which yields the characteristic equation, $z^4 + kL^4/H_{33}^c = 0$, with roots

$$z = \pm\sqrt[4]{\frac{kL^4}{4H_{33}^c}}(1 \pm i) = \pm\beta\,(1 \pm i),$$

where $i = \sqrt{-1}$.

The general solution of the differential equation now becomes $\bar{u}_2(\eta) = A\,\exp[\beta(1+i)\eta] + B\,\exp[\beta(1-i)\eta] + C\,\exp[-\beta(1+i)\eta] + D\,\exp[-\beta(1-i)\eta] + p_0L^4/(4\beta^4 H_{33}^c)$, where A, B, C, and D are four integration constants. The particular solution is simply $\bar{u}_2 = p_0/k = p_0L^4/(4\beta^4 H_{33}^c)$. Using the relationships between the exponential function with imaginary and real exponents, and the trigonometric and hyperbolic functions, respectively, the complete solution can be recast as

$$\bar{u}_2 = C_1\,\cosh\beta\eta\cos\beta\eta + C_2\,\cosh\beta\eta\sin\beta\eta$$
$$+ C_3\,\sinh\beta\eta\cos\beta\eta + C_4\,\sinh\beta\eta\sin\beta\eta + \frac{p_0L^4}{\bar{k}H_{33}^c}, \tag{5.68}$$

where C_1, C_2, C_3, and C_4 form a different set of integration constants, and the non-dimensional elastic foundation stiffness is $\bar{k} = kL^4/H_{33}^c$.

A set of transcendental functions is now defined as

$$b_1(\beta\eta) = \cosh\beta\eta\cos\beta\eta, \quad b_2(\beta\eta) = \cosh\beta\eta\sin\beta\eta,$$
$$b_3(\beta\eta) = \sinh\beta\eta\cos\beta\eta, \quad b_4(\beta\eta) = \sinh\beta\eta\sin\beta\eta. \tag{5.69}$$

In terms of these functions, the solution becomes $\bar{u}_2 = C_1 b_1(\beta\eta) + C_2 b_2(\beta\eta) + C_3 b_3(\beta\eta) + C_4 b_4(\beta\eta) + p_0L^4/(\bar{k}H_{33}^c)$. A property of the newly defined transcendental functions is that their derivatives can be expressed as

$$
\begin{aligned}
b_1' &= \beta(b_3 - b_2), & b_1'' &= -2\beta^2 b_4, & b_1''' &= -2\beta^3(b_3 + b_2), & b_1'''' &= -4\beta^4 b_1 \\
b_2' &= \beta(b_1 + b_4), & b_2'' &= +2\beta^2 b_3, & b_2''' &= +2\beta^3(b_1 - b_4), & b_2'''' &= -4\beta^4 b_2, \\
b_3' &= \beta(b_1 - b_4), & b_3'' &= -2\beta^2 b_2, & b_3''' &= -2\beta^3(b_1 + b_4), & b_3'''' &= -4\beta^4 b_3, \\
b_4' &= \beta(b_3 + b_2), & b_4'' &= +2\beta^2 b_1, & b_4''' &= +2\beta^3(b_3 - b_2), & b_4'''' &= -4\beta^4 b_4.
\end{aligned}
\tag{5.70}
$$

The four integration constants appearing in eq. (5.68) are evaluated using the four boundary conditions to find

$$\bar{u}_2 = \frac{p_0L^4}{H_{33}^c}\frac{1}{\bar{k}}\left\{[1 - b_1(\beta\eta)] - \bar{B}\left[\sin\beta\,b_2(\beta\eta) - \sinh\beta\,b_3(\beta\eta)\right]\right\}, \tag{5.71}$$

where $\bar{B} = (\cosh\beta - \cos\beta)/(\sin^2\beta + \sinh^2\beta)$. The bending moment distribution now follows from eq. (5.37) as

$$M_3 = p_0 L^2 \frac{2\beta^2}{k} \left\{ b_4(\beta\eta) - \bar{B} \left[\sinh \beta \, b_2(\beta\eta) + \sin \beta \, b_3(\beta\eta) \right] \right\}. \tag{5.72}$$

This example clearly demonstrate that the solution of beam problems can rapidly become quite difficult. The simple addition of an elastic foundation to the beam significantly complicates the solution process, which often becomes unmanageable when realistic structural problems are considered.

5.5.8 Problems

Problem 5.3. Bending of reinforced box beam

Figure 5.12 depicts an aluminum rectangular box beam of height $h = 0.30$ m, width $b = 0.15$ m, flange thickness $t_a = 12$ mm, and web thickness $t_w = 5$ mm. The beam is reinforced by two layers of unidirectional composite material of thickness $t_c = 4$ mm. The section is subjected to an axial force $N_1 = 600$ kN and bending moment $M_3 = 120$ kN·m. The Young's moduli for the aluminum and unidirectional composite are $E_a = 73$ GPa and $E_c = 140$ GPa, respectively. *(1)* Compute the axial and bending stiffnesses of the cross-section. *(2)* Find the distribution of axial stress over the cross-section and sketch it along the perimeter of the section. *(3)* Find the magnitude and location of the maximum axial stress in the aluminum and composite layers. *(4)* Assume the applied loads grow in a proportional manner, *i.e.* the applied loads are λN_1 and λM_3^c. If the allowable stress for the aluminum and unidirectional composite are $\sigma_a^{\text{allow}} = 400$ MPa and $\sigma_c^{\text{allow}} = 1500$ MPa, respectively, find the maximum loading factor, λ_{Max}. *(5)* Sketch the distribution of axial strain along the perimeter of the cross-section, and describe its distribution over the entire cross-section.

Problem 5.4. Bending of reinforced I beam

Figure 5.13 depicts an aluminum I beam of height $h = 0.25$ m, width $b = 0.2$ m, flange thickness $t_a = 16$ mm, and web thickness $t_w = 12$ mm. The beam is reinforced by two layers of unidirectional composite material of thickness $t_c = 5$ mm. The section is subjected to an axial force $N_1 = 250$ kN and bending moment $M_3 = 200$ kN·m. The Young's moduli for the aluminum and unidirectional composite are $E_a = 73$ GPa and $E_c = 140$ GPa, respectively. *(1)* Compute the axial and bending stiffnesses of the cross-section. *(2)* Find the distribution of axial stress over the cross-section. Sketch it along the $\bar{\imath}_2$ axis. Sketch it across the tops of both flanges. *(3)* Find the magnitude and location of the maximum axial stress in the aluminum and composite layers. *(4)* Assume the applied loads grow in a proportional manner, *i.e.* the applied loads are λN_1 and λM_3. If the allowable stress for the aluminum and unidirectional composite are $\sigma_a^{\text{allow}} = 400$ MPa and $\sigma_c^{\text{allow}} = 1500$ MPa, respectively, find the maximum loading factor, λ_{Max}. *(5)* Sketch the distribution of axial strain along the $\bar{\imath}_2$ axis, and describe its distribution over the entire cross-section.

Problem 5.5. Various short questions

(1) Is it possible to use Euler-Bernoulli assumptions for a beam bent in such a manner that the material it is made out of goes into the plastic deformation range? Why? *(2)* Is it possible to use Euler-Bernoulli assumptions for a beam made of a laminated composite material? Why? *(3)* Consider a simply supported beam with a mid-span elastic spring, subjected to a uniform transverse loading p_0. Which one of the following quantities will present a discontinuity at mid-span: beam transverse deflection, beam slope, bending moment, and/or transverse shear force? Why? *(4)* Consider a cantilevered beam of length L, under a uniform transverse loading p_0. Does the root bending moment depend on the material Young's modulus? *(5)* Consider a

beam of length L, cantilevered at both ends and subjected to a uniform transverse loading p_0. Does the mid-span transverse deflection depend on the material Young's modulus? *(6)* Consider a beam of length L, cantilevered at both ends and subjected to a uniform transverse loading p_0. Does the mid-span bending moment depend on the material Young's modulus? Explain your answers to all the above questions; a YES/NO answer is not valid.

Problem 5.6. Bending of reinforced solid section beam

A rectangular cross-section made of a material of Young's modulus E_1 is reinforced by thin top and bottom plates made of a material of Young's modulus E_2, as depicted in fig. 5.33. M_3 is the bending moment applied to the section. $E_2/E_1 = 2$; $d/h = 0.96$. *(1)* Plot the non-dimensional axial strain distribution $H_{33}^c \epsilon_1/(M_3 h)$ versus $2x_2/h$. *(2)* Plot the non-dimensional axial stress distribution $H_{33}^c \sigma_1/(M_3 E_2 h)$ versus $2x_2/h$.

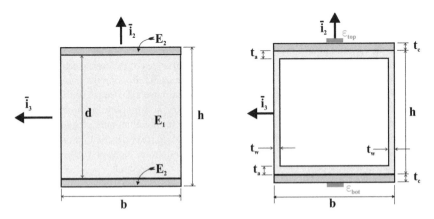

Fig. 5.33. Reinforced rectangular cross-section.

Fig. 5.34. Cross-section of a reinforced rectangular box beam.

Problem 5.7. Box beam with strain gauges

A cantilevered beam of length L is subjected to axial and transverse loads. Figure 5.34 depicts the cross-section of the beam: an aluminum rectangular box of height $h = 0.30$ m, width $b = 0.15$ m, flange thickness $t_a = 12$ mm, and web thickness $t_w = 5$ mm. The beam is reinforced by two layers of unidirectional composite material of thickness $t_c = 4$ mm. The Young's moduli for the aluminum and unidirectional composite are $E_a = 73$ GPa and $E_c = 140$ GPa, respectively. At a station along the span of the beam, an experimentalist has measured the axial strains on the top and bottom flanges of the beam as $\epsilon_{top} = -2560\mu$ and $\epsilon_{bot} = 3675\mu$, respectively. Find the bending moment and axial force acting at that station.

Problem 5.8. Cantilever with tip support and rotational spring

Consider a cantilevered beam of span L and bending stiffness H_{33}^c with a tip support and a rotational spring of stiffness constant k, as depicted in fig. 5.35. *(1)* Find and plot the transverse displacement distribution of this beam under a uniform transverse load p_0. *(2)* Find and plot the distribution of bending moment in the beam. *(3)* Find the location and magnitude of the maximum bending moment in the beam as a function of $\bar{k} = kL/H_{33}^c$. *(4)* Discuss your results when $\bar{k} \to 0$ and $\bar{k} \to \infty$.

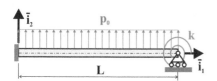

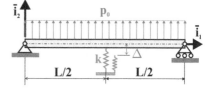

Fig. 5.35. Cantilevered beam with tip support and torsional spring.

Fig. 5.36. Simply supported beam with mid-span spring featuring a clearance Δ.

Problem 5.9. Simply supported beam with mid-span spring

Consider the uniform beam with simply supported ends as depicted in fig. 5.36. A spring of stiffness constant k is acting at mid-span, $\bar{k} = kL^3/H_{33}^c$. A uniform load p_0 is acting over the beam span of length L. The unstressed length of the spring is such that it fall a distance Δ short of reaching the beam, $\bar{\Delta} = \Delta H_{33}^c/(p_0 L^4)$. (1) Find and plot the transverse displacement distribution for this beam. (2) Find and plot the corresponding distribution of bending moment. (3) What value of $\bar{\Delta}$ that will minimize the maximum bending moment in the beam. Hint: replace the spring by an unknown force, F, acting at mid-span. This force can then be evaluated by equating the beam mid-span displacement with that of the spring. Use the following values for the plots: $\bar{k} = 600$, $\bar{\Delta} = 2.0 \; 10^{-3}$.

Problem 5.10. Cantilever beam with tip rotational spring

The uniform cantilevered beam of bending stiffness H_{33}^c and length L depicted in fig. 5.37 features a tip rotational spring of stiffness constant k. Due to manufacturing imperfections, the spring applies a restoring moment on the beam that is proportional to $(d\bar{u}_2/dx_1 - \theta_0)$, where θ_0 represents the imperfection magnitude. (1) Find the magnitude and location of the maximum bending moment in the beam due to the imperfection in the structure. (2) Plot the maximum bending moment $LM_3^{\max}/(\theta_0 H_{33}^c)$ as a function of the non-dimensional spring constant $\bar{k} = kL/H_{33}^c$. Explain your result in physical terms.

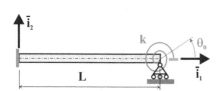

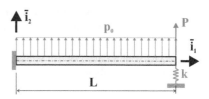

Fig. 5.37. Cantilevered beam with tip torsional spring.

Fig. 5.38. Cantilevered beam with tip spring.

Problem 5.11. Cantilever beam with tip spring

Consider the cantilevered beam of length L with a tip spring of stiffness k depicted in fig. 5.38. The beam is subjected to a uniform transverse load, p_0, and a tip concentrated load, P. (1) Write the governing differential equation and associated boundary conditions for this problem.

Problem 5.12. Bending of beam with nonuniform bending stiffness

A simply supported beam of span L is subjected to forces of magnitude P located at stations $x_1 = \alpha L$ and $(1 - \alpha)L$, as depicted in fig. 5.39. The beam has a bending stiffness H_0 and is reinforced in its central portion where its bending stiffness is H_1. *(1)* Find the transverse displacement field. *(2)* Plot the non-dimensional transverse displacement, $H_0\bar{u}_2(\eta)/L^3$. Use $H_1/H_0 = 20$ and $\alpha = 0.2$. *(3)* Plot the non-dimensional bending moment, $M_3(\eta)/(PL)$. *(4)* Plot the non-dimensional shear force, $V_3(\eta)/P$.

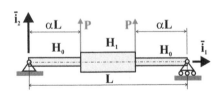

Fig. 5.39. Simply supported beam with varying bending stiffness.

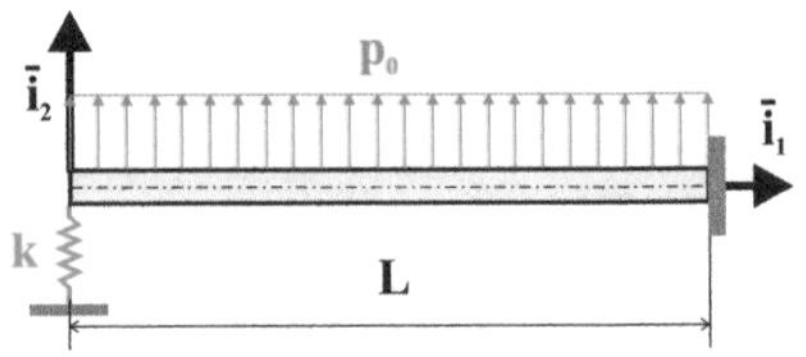

Fig. 5.40. Cantilevered beam under uniform load with a root spring.

Problem 5.13. Cantilever beam with uniform load and tip spring

The uniform cantilevered beam of bending stiffness H^c_{33} and length L depicted in fig. 5.40 features a root spring of stiffness constant k and is subjected to a uniform distributed load, p_0. *(1)* Find the transverse displacement distribution of the beam as a function of the non-dimensional spring constant, $\bar{k} = kL^3/H^c_{33}$. *(2)* Determine the location and magnitude of the maximum bending moment in the beam. *(3)* Plot the maximum bending moment, $M^{\max}_3/(p_0L^2)$ as a function of $\bar{k}$. Explain your result in physical terms.

Problem 5.14. Flexibility matrix of a simply supported beam

The concept of flexibility matrix is introduced in examples 5.9 and 5.10 for a cantilevered beam. *(1)* Determine the flexibility matrix for the simply supported beam depicted in fig. 5.24. The closed form expression of the flexibility matrix corresponding to concentrated loads applied at three locations, $\alpha_1 L$, $\alpha_2 L$, and $\alpha_3 L$ should be derived.

Problem 5.15. Experimental estimation of the bending stiffness

The procedure for the experimental determination of the flexibility matrix is described in example 5.9 for the cantilevered beam depicted in fig. 5.28. Table 5.1 lists the displacements measured on a cantilevered beam of uniform bending stiffness, H^c_{33}, under three loading cases. The first column of this table lists the displacements at locations $\alpha_1 L$, $\alpha_2 L$, and $\alpha_3 L$ for $P_1 = 1.5$ kN, $P_2 = P_3 = 0$. The next two columns list the corresponding data for $P_2 = 1.0$ kN, $P_1 = P_3 = 0$ and $P_3 = 0.5$ kN, $P_1 = P_2 = 0$, respectively. *(1)* Determine the experimental flexibility matrix of the system. *(2)* Determine the bending stiffness, H^c_{33}, of the beam. HINT: the analytical expression for the flexibility matrix is given by eq. (5.60). The bending stiffness of the beam can be determined by equating any entry of the analytical and experimentally determined flexibility matrices. The most accurate strategy is to use all nine entries to form a set of over-determined equations to be solved using a least-squares approach, see section A.2.10. Use the following data: $L = 15$ m, $\alpha_1 = 0.25$, $\alpha_2 = 0.50$, and $\alpha_3 = 0.75$.

Problem 5.16. Bending of two crossed simply supported beams

The lower beam depicted in fig. 5.41 is of length $2L$ and is simply supported at both ends. The upper beam of length $L + a$ is cantilevered at the root, supported by the lower beam at point **A**, and subjected to a uniform transverse loading, p_0. Both upper and lower beams have a uniform bending stiffness H_0. *(1)* Find the exact solution for the transverse deflection of the

Table 5.1. Measured displacements for the three loading cases. Load are measured in kN, displacements in mm.

	$P_1 = 1.5$	$P_2 = 1.0$	$P_3 = 0.5$
Δ_1	10.9	18.3	14.6
Δ_2	27.7	59.1	51.1
Δ_3	43.1	104.	98.5

lower beam under a mid-span concentrated load. Show that the lower beam can be replaced by a concentrated spring of stiffness constant $k_{\text{eq}} = 6H_0/L^3$. *(2)* Find the exact solution of the problem from the solutions of the governing differential equations and associated boundary conditions for both upper and lower beams. Replace the interaction between the beams by a force X, yet unknown. The magnitude of this force is found by equating the displacements of the upper and lower beams at point **A**. *(3)* Plot the distribution of transverse displacement for both beams. Use $L/a = 2$. *(4)* Plot the distribution of bending moment for both beams. *(5)* Plot the distribution of shear force for both beams.

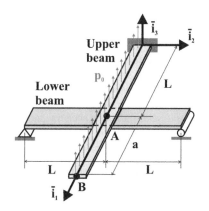

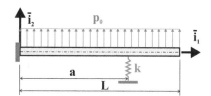

Fig. 5.41. Two beam assembly under transverse load.

Fig. 5.42. Cantilevered beam with concentrated spring.

Problem 5.17. Cantilever beam with uniform load and spring

The cantilevered beam depicted in fig. 5.42 is of length L, uniform bending stiffness H^c_{33}, and is subjected to a uniform distributed load p_0. A concentrated spring of stiffness constant k is connected to the beam at a distance a from its root. *(1)* Find the solution of the problem. It will be convenient to define the non-dimensional spring constant $\bar{k} = kL^3/H^c_{33}$. *(2)* Plot the distribution of transverse displacement for the beam. Use $L/a = 3$ and $\bar{k} = 100$. *(3)* Plot the distribution of bending moment for the beam. *(4)* Plot the distribution of shear force for the beam. *(5)* Find the value of $\bar{k}$ that will minimize the maximum bending moment in the beam.

Problem 5.18. Two simply supported beams interconnected by two springs

Figure 5.43 depicts a system consisting of two simply supported beams connected by two elastic springs of stiffness constant k. The upper and lower beam have the same bending stiffness, H^c_{33}, and the upper beam is subjected to a uniform load distribution, p_0. *(1)* Solve this problem: determine the deflection and bending moment distributions in the upper and lower beams, and the force in the connecting springs. *(2)* Plot the displacements for the upper and lower beams on the same graph. *(3)* Plot the bending moments for the upper and lower beams on the same graph. *(4)* Find the spring constant, $\bar{k}_{\rm opt}$, that will minimize the maximum bending moments in both upper and lower beams. Plot $\bar{k}_{\rm opt} = \bar{k}_{\rm opt}(\alpha)$. Hint: it will be convenient to replace the interconnecting springs by forces of unknown magnitude acting on the upper and lower beams, then enforcing a compatibility condition. Use the following data for the plots: $\alpha = 0.3$, $\bar{k} = kL^3/H_{33}^c = 10, 100, 1000$.

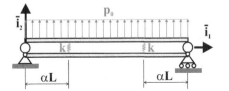

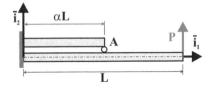

Fig. 5.43. Simply supported beam connected by spring.

Fig. 5.44. Cantilevered beam with intermediate support.

Problem 5.19. Two cantilever beams with intermediate support
The cantilevered beam depicted in fig. 5.44 is subjected to a tip load P. The tip of a second cantilevered beam contacts the first at point **A**. The lower and upper beams have a uniform bending stiffness H_{33}^c and are of length L and αL, respectively. *(1)* Find the displacement fields for the two beams. *(2)* Plot the distribution of transverse displacement, bending moment, and shear force for both beams. Use $\alpha = 1/2$. *(3)* Find the magnitude and location of the maximum bending moment in the beams. Plot these quantities as a function of α.

Problem 5.20. Two simply supported beams with intermediate support
The two simply supported beam shown in fig. 5.45 are connected by an intermediate roller located a distance αL from the left support. The upper beam is subjected to a uniform loading p_0. Both beams have a uniform bending stiffness, H_{33}^c. *(1)* Find the displacement fields for the two beams by solving of the governing differential equations and associated boundary conditions. *(2)* Plot the distribution of transverse displacement, bending moment, and shear force for both beams. Use $\alpha = 1/3$. *(3)* Find the magnitude and location of the maximum bending moment in the beams. Plot these quantities as a function of α.

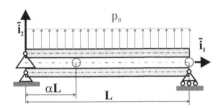

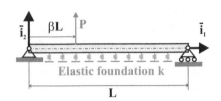

Fig. 5.45. Superposed simply supported beams.

Fig. 5.46. Beam with elastic foundation subjected to a concentrated load.

Problem 5.21. Simply supported beam on elastic foundation
Consider the simply supported beam of length L depicted in fig. 5.46. The beam rests on an elastic foundation of stiffness constant k and is subjected to a concentrated load, P, acting at a distance βL from the left support. *(1)* Find the displacement field for the beam by solving of the governing differential equation and associated boundary conditions. It is convenient to define $\bar{k} = kL^4/H_{33}^c$. *(2)* Plot the distribution of transverse displacement, bending moment, and shear force for the beams. Use $\beta = 1/2$ and $\bar{k} = 8 \times 10^3$.

5.6 Beams subjected to combined axial and transverse loads

In earlier sections, Euler-Bernoulli beam theory is developed separately for two distinct loading cases: beams under axial loads and beams under transverse loads, see sections 5.4 and 5.5, respectively. Based on equilibrium considerations, it is also shown in section 5.5.2 that it is convenient to locate the origin of the axes system at the centroid of the beam's cross-section. In fact, all the developments presented thus far assume that the origin of the axes is located at the centroid of the cross-section.

This section will generalize the theory developed thus far in two important ways: first, beams under combined axial and transverse loading will be considered, and second, the origin of the axes will not be located at the centroid.

5.6.1 Kinematic description

The starting point is a displacement field that combines the axial and transverse displacement fields found in beams subjected to axial and transverse loads, see eqs. (5.12) and (5.29), respectively,

$$u_1(x_1, x_2, x_3) = \bar{u}_1(x_1) - (x_2 - x_{2c})\frac{d\bar{u}_2(x_1)}{dx_1}, \qquad (5.73a)$$

$$u_2(x_1, x_2, x_3) = \bar{u}_2(x_1), \qquad (5.73b)$$

$$u_3(x_1, x_2, x_3) = 0, \qquad (5.73c)$$

where x_{2c} is the location of the centroid. The origin of the axis system is not located at the cross-section's centroid, however $\bar{u}_1$ is still the axial displacement of the centroid.

The corresponding strain field combines the characteristics of the fields associated with beams subjected to axial and transverse loads, see eqs. (5.13) and (5.30), respectively,

$$\epsilon_1(x_1, x_2, x_3) = \bar{\epsilon}_1(x_1) - (x_2 - x_{2c})\kappa_3(x_1). \qquad (5.74)$$

5.6.2 Sectional constitutive law

It is assumed that the beam is made of a linearly elastic material. As discussed in sections 5.4.2 and 5.5.2, Hooke's law then reduces to (5.14), and the axial stress distribution becomes

$$\sigma_1(x_1, x_2, x_3) = E \,\bar{\epsilon}_1(x_1) - E(x_2 - x_{2c})\kappa_3(x_1). \tag{5.75}$$

The axial force in the beam in now evaluated using eq. (5.8) to find

$$N_1 = \int_{\mathcal{A}} [E \,\bar{\epsilon}_1(x_1) - E(x_2 - x_{2c})\kappa_3(x_1)] \, d\mathcal{A}$$

$$= \left[\int_{\mathcal{A}} E \, d\mathcal{A}\right] \bar{\epsilon}_1(x_1) - \left[\int_{\mathcal{A}} E(x_2 - x_{2c}) \, d\mathcal{A}\right] \kappa_3(x_1).$$

The first bracketed term is the axial stiffness of the beam, S, as defined by eq. (5.17). The second bracketed term can be shown to vanish,

$$\int_{\mathcal{A}} E(x_2 - x_{2c}) \, d\mathcal{A} = \int_{\mathcal{A}} Ex_2 \, d\mathcal{A} - x_{2c} \int_{\mathcal{A}} E \, d\mathcal{A} = S_2 - S x_{2c} = 0, \tag{5.76}$$

where the last equality follows from the definition of the location of the centroid, see eq. (5.33). The axial force equation now reduces to $N_1 = S\bar{\epsilon}_1$.

Next, the beam's bending moment with respect to the centroid, M_3^c, is evaluated with the help of eq. (5.11) to find

$$M_3^c = -\int_{\mathcal{A}} (x_2 - x_{2c}) [E \,\bar{\epsilon}_1(x_1) - E(x_2 - x_{2c})\kappa_3(x_1)] \, d\mathcal{A}$$

$$= -\left[\int_{\mathcal{A}} E(x_2 - x_{2c}) \, d\mathcal{A}\right] \bar{\epsilon}_1(x_1) + \left[\int_{\mathcal{A}} E(x_2 - x_{2c})^2 \, d\mathcal{A}\right] \kappa_3(x_1),$$

The first bracketed term vanishes in view of eq. (5.76). The second bracketed terms is the beam's bending stiffness, H_{33}^c, computed with respect to the centroid. The previous equation now reduces to $M_3^c = H_{33}^c \kappa_3$.

In summary, the sectional constitutive laws reduce to $N_1 = S\bar{\epsilon}_1$ and $M_3^c = H_{33}^c \kappa_3$, which are identical to eqs. (5.16) and (5.37), respectively. Although the beam is subjected to combined axial and transverse loading, it is possible to derive *decoupled sectional constitutive laws*: one equation relates the axial force to the axial strain, the other the bending moment to the curvature. To achieve this decoupling, two crucial steps are required: first, the displacement field must be in the form of eq. (5.73), where x_{2c} is the location of the centroid, and second, the bending moment must be evaluated with respect to the centroid. The centroid thus plays a crucial role in decoupling the axial and bending responses of beams.

5.6.3 Equilibrium equations

To complete the formulation, the equilibrium equations must be derived for the combined problem. An infinitesimal slice of the beam of length dx_1 is depicted in fig. 5.47.

This figure shows the axial force, N_1, shear force, V_2, and bending moment, M_3, acting on the face at location x_1. The corresponding quantities acting on the face at location $x_1 + dx_1$ are obtained from a Taylor series expansion and terms of higher

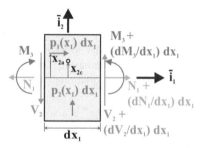

Fig. 5.47. Axial forces acting on an infinitesimal slice of the beam.

differential order are neglected in the expansion. Summation of the forces acting on the free body diagram in the horizontal direction yields the following equilibrium equation, $dN_1/dx_1 = -p_1$, which is identical to eq. (5.18). The vertical equilibrium equation is $dV_2/dx_1 = -p_2$, which is identical to eq. (5.38a). Finally, equilibrium of moments expressed about the centroid leads to

$$\frac{dM_3}{dx_1} + V_2 = (x_{2a} - x_{2c})p_1, \tag{5.77}$$

which should be compared with eq. (5.38b) obtained earlier. If the axial distributed load, $p_1(x_1)$, is not applied at the centroid, it generates a moment $(x_{2a} - x_{2c})p_1$, where x_{2a} is the coordinate of the point of application of the axial load and $(x_{2a} - x_{2c})$ its moment arm with respect to the centroid.

5.6.4 Governing equations

The governing equations for a beam subjected to combined axial and transverse loads are found by manipulating the equations developed in the previous sections to yield

$$\frac{d}{dx_1}\left[S\frac{d\bar{u}_1}{dx_1}\right] = -p_1(x_1), \tag{5.78a}$$

$$\frac{d^2}{dx_1^2}\left[H_{33}^c\frac{d^2\bar{u}_2}{dx_1^2}\right] = p_2(x_1) + \frac{d}{dx_1}\left[(x_{2a} - x_{2c})p_1(x_1)\right]. \tag{5.78b}$$

The first equation is identical to eq. (5.19), which describes the behavior of beams subjected to axial loads. The second equation is almost identical to eq. (5.40), which describes the behavior of beams subjected to transverse load. In the formulation developed in section 5.5, the beam is subjected to transverse loads only, and hence, the last term on the right-hand side of eq. (5.78b) does not appear in eq. (5.40).

The equations describing the behavior of beams under combined axial and transverse loads are *decoupled*, i.e., one equation, eq. (5.78a), can be solved to find the axial displacement field, $\bar{u}_1(x_1)$, and the other, eq. (5.78b), can be independently solved to find the transverse displacement field, $\bar{u}_2(x_1)$.

The term "decoupled" used in the previous paragraph can be misleading because it is often used to describe distinct concepts. In the previous paragraph, the term "decoupled" is used in a mathematical sense to indicate that eq. (5.78a) is a single equation in a single unknown, $\bar{u}_1(x_1)$, whereas eq. (5.78b) is also a single equation for a single unknown, $\bar{u}_2(x_1)$. The two equations can be solved independently of each other, they are "decoupled."

On the other hand, the term "decoupling" is also used in a more physical sense. As implied by the presence of the axial load, $p_1(x_1)$, on the right-hand side of both eqs. (5.78a) and (5.78b), axial loads generate both axial and transverse displacement of the beam. If the axial load is not applied at the centroid, *i.e.*, if $x_{2a} - x_{2c} \neq 0$, the beam bends. Hence the following statement: if axial loads are applied at the centroid, extension and bending are "decoupled," whereas when not applied at the centroid, extension and bending are "coupled."

Example 5.15. Bi-material cantilevered beam

Consider the bi-material cantilevered beam of length L with the rectangular cross-section shown in fig. 5.48. The beam is constructed by bonding together two strips of materials, each of width b and height $h/2$. The two materials, denoted material A and B, have Young's moduli E_a and E_b, respectively. The beam is subjected to a tip axial load, P, applied at the geometric center of the section and a tip bending moment, Q. It is convenient to select the origin of the axes at the geometric center of the section, rather than at the centroid.

The axial stiffness of the section is $S = (E_a + E_b)bh/2$, and the location of the centroid is given by $x_{2c}S = E_a(bh/2)(h/4) + E_b(bh/2)(-h/4)$, or

$$\frac{x_{2c}}{h} = \frac{E_a - E_b}{4(E_a + E_b)}.$$

Depending on the relative stiffnesses of the two materials, the centroid could be located above or below the geometric center of the section, as illustrated in fig. 5.48.

The beam's bending stiffness with respect to the centroid is then

$$H_{33}^c = E_a \left[\frac{b(h/2)^3}{12} + \frac{bh}{2} \left(\frac{h}{4} - x_{2c} \right)^2 \right] + E_b \left[\frac{b(h/2)^3}{12} + \frac{bh}{2} \left(\frac{h}{4} + x_{2c} \right)^2 \right]$$

$$= \frac{bh^3}{96} \frac{E_a^2 + E_b^2 + 14E_aE_b}{E_a + E_b}.$$

The axial problem is solved first. The governing equation is $d^2\bar{u}_1/dx_1^2 = 0$; the boundary conditions are $\bar{u}_1 = 0$ at the beam's root and $Sd\bar{u}_1/dx_1 = P$ at the beam's tip. The solution of this problem is

$$\bar{u}_1(\eta) = \frac{PL}{S}\eta,$$

where $\eta = x_1/L$ is the non-dimensional variable along the beam's span.

Next, the bending problem is solved. The governing equation is $d^4\bar{u}_2/dx_1^4 = 0$; the boundary conditions are $\bar{u}_1 = d\bar{u}_1/dx_1 = 0$ at the beam's root and $M_3^c =$

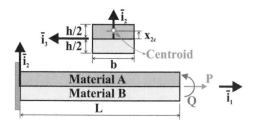

Fig. 5.48. Cantilevered bi-material beam under tip loads.

$Q + x_{2c}P$, $V_2 = 0$ at the beam's tip. Because the tip axial load is not applied at the centroid, it generates a tip bending moment, $x_{2c}P$, with respect to the centroid. The solution of this problem is

$$\bar{u}_2(\eta) = \frac{(Q + x_{2c}P)L^2}{2H_{33}^c}\eta^2.$$

Note that both tip moment and tip force generate a transverse displacement. In the presence of the tip axial force alone, the tip transverse displacement is

$$\bar{u}_2^{\text{tip}} = 12\frac{E_a - E_b}{E_a^2 + E_b^2 + 14E_aE_b}\frac{P}{b}\left(\frac{L}{h}\right)^2.$$

If materials A and B are identical, $E_a - E_b = 0$ and the tip transverse displacement vanishes. Indeed, if the two materials are identical, the centroid is at the geometric center of the cross-section, and the tip axial force generates no bending moment about the centroid. The tip axial force generates only axial displacements.

6

Three-dimensional beam theory

In the previous chapter, Euler-Bernoulli theory is developed for beams under axial and transverse loads. The analysis is limited, however, to deformations of the beam in plane $(\bar{\imath}_1, \bar{\imath}_2)$. This behavior can be observed, for instance, when the cross-section of the beam presents a plane of symmetry and the only applied loads are acting in this plane.

In numerous practical applications, the beam's cross-section presents no particular symmetries and is instead of arbitrary shape. In addition, the applied loads may act along several distinct directions and not just in plane $(\bar{\imath}_1, \bar{\imath}_2)$. Consider an aircraft wing: the cross-section is of a complex shape involving curved skins and two or more spars, and the wing is subjected lift and drag forces. In the case of a helicopter blade, large centrifugal forces generated by the rotation of the blade are also present. Similarly, machine components often operate in a complex, three-dimensional loading environment.

Figure 6.1 shows a beam of arbitrary cross-sectional shape subjected to a complex three-dimensional loading. This loading consists of distributed and concentrated axial and transverse loads, as well as distributed and concentrated moments. The axial and transverse distributed loads, $p_1(x_1)$, $p_2(x_1)$, and $p_3(x_1)$ act along directions, $\bar{\imath}_1$, $\bar{\imath}_2$, and $\bar{\imath}_3$, respectively. The same convention is used for the concentrated loads $P_1^{[k]}$, $P_2^{[k]}$, and $P_3^{[k]}$, but in this case it is necessary to add a second index to identify the kth concentrated load in the direction specified by the first index: $P_3^{[2]}$ is the second concentrated force acting along axis $\bar{\imath}_3$. Distributed moments, $q_2(x_1)$ and $q_3(x_1)$, acting about axes $\bar{\imath}_2$ and $\bar{\imath}_3$, respectively, can be introduced in a similar manner. Concentrated moments $Q_2^{[k]}$ and $Q_3^{[k]}$ act about the same axes.

Figure 6.1 depicts concentrated forces and moments acting at the tip of the beam, but in practical situations, such concentrated loads could be applied at any spanwise location. The notation used in this text for the various loads is summarized in table 6.1. The subscript indicates the direction of the loading component. If multiple concentrated loads are applied, a second subscript might be used to keep track of individual concentrated loads.

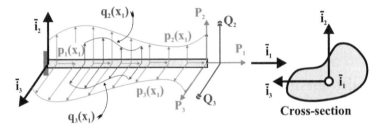

Fig. 6.1. Beam with arbitrary three-dimensional loading.

Table 6.1. Loading components acting on the beam.

Loading Type	Notation	Units
Distributed loads	$p_1(x_1), p_2(x_1), p_3(x_1)$	N/m
Concentrated loads	$P_1^{[k]}, P_2^{[k]}, P_3^{[k]}$	N
Distributed moments	$q_2(x_1), q_3(x_1)$	N.m/m
Concentrated moments	$Q_2^{[k]}, Q_3^{[k]}$	N.m

This three-dimensional loading is general, with an important exception: no torsional loads are applied, and the transverse loads are assumed to be applied in such a manner that the beam will bend without twisting. This important restriction will be removed in a later chapter after the study of the torsional behavior of beams. As mentioned earlier, the cross-section of the beam is of arbitrary shape. The origin of the axes has not yet been specified, and the orientation of axes $\bar{\imath}_2$ and $\bar{\imath}_3$ within the plane of the section is arbitrary, as depicted in fig. 6.1.

6.1 Kinematic description

The development of the three-dimensional beam theory starts with the three Euler-Bernoulli assumptions discussed in section 5.1. These assumptions are of a purely kinematic nature and are shown to imply the following displacements field

$$u_1(x_1, x_2, x_3) = \bar{u}_1(x_1) + x_3 \Phi_2(x_1) - x_2 \Phi_3(x_1), \tag{6.1a}$$

$$u_2(x_1, x_2, x_3) = \bar{u}_2(x_1), \tag{6.1b}$$

$$u_3(x_1, x_2, x_3) = \bar{u}_3(x_1). \tag{6.1c}$$

where the origin of the axis system on the cross-section is not yet specified. The corresponding strain field is shown to be

$$\varepsilon_2 = 0; \quad \varepsilon_3 = 0; \quad \gamma_{23} = 0, \tag{6.2a}$$

$$\gamma_{12} = 0; \quad \gamma_{13} = 0, \tag{6.2b}$$

$$\varepsilon_1(x_1, x_2, x_3) = \bar{\varepsilon}_1(x_1) + x_3 \kappa_2(x_1) - x_2 \kappa_3(x_1). \tag{6.2c}$$

6.2 Sectional constitutive law

Assume now that the beam is made of linearly elastic, isotropic material for which the stress-strain relationships are adequately described by Hooke's law, eq. (5.14). Because the cross-section does not deform in its own plane, the stress components, σ_2 and σ_3, acting in the plane of the section are far smaller than the axial stress component, σ_1, and Hooke's law is shown to reduce to eq. (5.14). The axial stress distribution is found by introducing eq. (6.2c) into eq. (5.14) to find

$$\sigma_1(x_1, x_2, x_3) = E\left[\bar{\epsilon}_1(x_1) + x_3\,\kappa_2(x_1) - x_2\,\kappa_3(x_1)\right] \tag{6.3}$$

The axial force, N_1, is now evaluated by introducing this axial stress distribution into eq. (5.8) to find

$$
\begin{aligned}
N_1(x_1) &= \int_A \sigma_1\,\mathrm{d}A = \int_A E\bar{\epsilon}_1\,\mathrm{d}A + \int_A Ex_3\kappa_2\,\mathrm{d}A - \int_A Ex_2\kappa_3\,\mathrm{d}A \\
&= \left[\int_A E\,\mathrm{d}A\right]\bar{\epsilon}_1 + \left[\int_A Ex_3\,\mathrm{d}A\right]\kappa_2 - \left[\int_A Ex_2\,\mathrm{d}A\right]\kappa_3 \\
&= S\,\bar{\epsilon}_1(x_1) + S_3\,\kappa_2(x_1) - S_2\,\kappa_3(x_1),
\end{aligned}
\tag{6.4}
$$

where the following sectional stiffness coefficients are defined

$$S = \int_A E\,\mathrm{d}A; \quad S_2 = \int_A Ex_2\,\mathrm{d}A; \quad S_3 = \int_A Ex_3\,\mathrm{d}A. \tag{6.5}$$

The bending moments, M_2 and M_3, acting about axes $\bar{\imath}_2$ and $\bar{\imath}_3$, respectively, are evaluated by introducing the axial stress distribution eq. (6.3) into eq. (5.11) to find

$$
\begin{aligned}
M_2 &= \int_A x_3\sigma_1\,\mathrm{d}A = \int_A x_3 E\bar{\epsilon}_1\,\mathrm{d}A + \int_A Ex_3^2\kappa_2\,\mathrm{d}A - \int_A Ex_2x_3\kappa_3\,\mathrm{d}A \\
&= \left[\int_A Ex_3\,\mathrm{d}A\right]\bar{\epsilon}_1 + \left[\int_A Ex_3^2\,\mathrm{d}A\right]\kappa_2 - \left[\int_A Ex_2x_3\,\mathrm{d}A\right]\kappa_3 \\
&= S_3\,\bar{\epsilon}_1(x_1) + H_{22}\,\kappa_2(x_1) - H_{23}\,\kappa_3(x_1),
\end{aligned}
\tag{6.6}
$$

and

$$
\begin{aligned}
M_3 &= -\int_A x_2\sigma_1\,\mathrm{d}A = -\int_A x_2 E\bar{\epsilon}_1\,\mathrm{d}A - \int_A x_2 Ex_3\kappa_2\,\mathrm{d}A + \int_A Ex_2^2\kappa_3\,\mathrm{d}A \\
&= -\left[\int_A Ex_2\,\mathrm{d}A\right]\bar{\epsilon}_1 - \left[\int_A Ex_2x_3\,\mathrm{d}A\right]\kappa_2 + \left[\int_A Ex_2^2\,\mathrm{d}A\right]\kappa_3 \\
&= -S_2\,\bar{\epsilon}_1(x_1) - H_{23}\,\kappa_2(x_1) + H_{33}\,\kappa_3(x_1),
\end{aligned}
\tag{6.7}
$$

where the following additional sectional stiffness coefficients are defined

$$H_{22} = \int_A E\,x_3^2\,\mathrm{d}A; \quad H_{33} = \int_A E\,x_2^2\,\mathrm{d}A; \tag{6.8}$$

$$H_{23} = \int_{\mathcal{A}} E\, x_2 x_3 \, \mathrm{d}\mathcal{A}. \tag{6.9}$$

The axial stiffness, S, is found in section 5.4 to characterize the axial stiffness of the beam and in section 5.5 the bending stiffness, H_{33}, is found to characterize the bending behavior of the beam about axis $\bar{\imath}_3$. The bending stiffness H_{22} plays the same role, but for bending about axis $\bar{\imath}_2$. A new bending stiffness coefficient, H_{23}, is called the *cross bending stiffness*.

Equations (6.4), (6.6) and (6.7) can be rewritten in a more compact matrix form as follows

$$\begin{Bmatrix} N_1(x_1) \\ M_2(x_1) \\ M_3(x_1) \end{Bmatrix} = \begin{bmatrix} S & S_3 & -S_2 \\ S_3 & H_{22} & -H_{23} \\ -S_2 & -H_{23} & H_{33} \end{bmatrix} \begin{Bmatrix} \bar{\epsilon}_1(x_1) \\ \kappa_2(x_1) \\ \kappa_3(x_1) \end{Bmatrix}. \tag{6.10}$$

These equations express a general linear relationship between the sectional stress resultants and the sectional strains. Thus, they are the constitutive laws for the cross-section of the beam, and the matrix on the right hand side of eq. (6.10) is called the *sectional stiffness matrix*. Clearly, these equations are fully coupled: all of the sectional strains affect the values of each of the sectional stress resultants. For example, the axial force $N_1(x_1)$ is not proportional only to the axial strain $\bar{\epsilon}_1(x_1)$, nor are the bending moments proportional only to curvatures. Instead, the behavior is fully coupled through the sectional coupling stiffness coefficients S_2, S_3 and H_{23} that appear in the off-diagonal entries of the sectional stiffness matrix. This means that an axial force, N_1, will appear as a result of an axial strain, $\bar{\epsilon}_1$, but also in the presence of curvatures, κ_2 or κ_3. Similarly, a bending moment appears as a result of either the κ_2 or κ_3 curvatures, but also in the presence of an axial strain, $\bar{\epsilon}_1$.

A general formulation of three dimensional Euler-Bernoulli beam theory can be developed based on the constitutive laws of eq. (6.10). Unfortunately, this leads to complex governing differential equations for the problem and this approach will not be pursued further. Rather, it will be shown that the sectional constitutive laws can be simplified by selecting the axis system appropriately. Indeed, in the formulation developed thus far, the origin of the axis system is arbitrary, and although the orientation of axis $\bar{\imath}_1$ is along the axis of the beam, the orientations of axes $\bar{\imath}_2$ and $\bar{\imath}_3$ within the plane of the cross-section are also arbitrary.

More specifically, the *origin of the axis system can be selected to coincide with the centroid of the section*, i.e.,

$$x_{2c} = \frac{1}{S} \int_{\mathcal{A}} E\, x_2 \, \mathrm{d}\mathcal{A} = \frac{S_2}{S} = 0; \quad x_{3c} = \frac{1}{S} \int_{\mathcal{A}} E\, x_3 \, \mathrm{d}\mathcal{A} = \frac{S_3}{S} = 0, \tag{6.11}$$

where the sectional coefficients, S_2 and S_3, are defined in eq. (6.5). The sectional constitutive laws, eq. (6.10), reduces to a partially uncoupled form

$$\begin{Bmatrix} N_1(x_1) \\ M_2(x_1) \\ M_3(x_1) \end{Bmatrix} = \begin{bmatrix} S & 0 & 0 \\ 0 & H_{22}^c & -H_{23}^c \\ 0 & -H_{23}^c & H_{33}^c \end{bmatrix} \begin{Bmatrix} \bar{\epsilon}_1(x_1) \\ \kappa_2(x_1) \\ \kappa_3(x_1) \end{Bmatrix}. \tag{6.12}$$

The bending stiffness coefficient, H_{22}, H_{33}, and H_{23}, are now replaced by their counterparts, H_{22}^c, H_{33}^c, and H_{23}^c, evaluated with respect to the centroid of the cross-section.

It is important to note that these partially uncoupled equations show that the axial force N_1 is now related to only the axial strain $\bar{\epsilon}_1$ and that the bending moments are related to the curvatures κ_2 and κ_3 only. This decoupling of the axial and bending behavior results from **locating the origin of the axis system the centroid of the cross-section**, rather than at an arbitrary point of the section. The two bending moments and corresponding curvatures, however, are still coupled due to the presence of the stiffness coefficient, H_{23}^c.

For most problems, the forces and moments are specified and it is required to find the resulting displacements and internal stresses. The sectional constitutive equations, eqs. (6.12), must therefore be inverted and solved for the sectional strain, $\bar{\epsilon}_1$, and curvatures κ_2 and κ_3, in terms of stress resultants, N_1, M_2 and M_3. This results in

$$
\begin{Bmatrix} \bar{\epsilon}_1(x_1) \\ \kappa_2(x_1) \\ \kappa_3(x_1) \end{Bmatrix} = \begin{bmatrix} 1/S & 0 & 0 \\ 0 & H_{33}^c/\Delta_H & H_{23}^c/\Delta_H \\ 0 & H_{23}^c/\Delta_H & H_{22}^c/\Delta_H \end{bmatrix} \begin{Bmatrix} N_1(x_1) \\ M_2(x_1) \\ M_3(x_1) \end{Bmatrix}, \tag{6.13}
$$

where $\Delta_H = H_{22}^c H_{33}^c - H_{23}^c H_{23}^c$.

The axial stress can now be found by substituting these results into eq. (6.3) to find

$$
\sigma_1 = E \left[\frac{N_1}{S} + x_3 \frac{H_{33}^c M_2 + H_{23}^c M_3}{\Delta_H} - x_2 \frac{H_{23}^c M_2 + H_{22}^c M_3}{\Delta_H} \right], \tag{6.14}
$$

or, with minor rearrangements,

$$
\sigma_1 = E \left[\frac{N_1}{S} - \frac{x_2 H_{23}^c - x_3 H_{33}^c}{\Delta_H} M_2 - \frac{x_2 H_{22}^c - x_3 H_{23}^c}{\Delta_H} M_3 \right]. \tag{6.15}
$$

This is a key result because it relates the axial stress distribution to the stress resultants which are, in turn, functions of the applied loads.

6.3 Sectional equilibrium equations

To complete the theory, equilibrium equations must also be derived. Consider an infinitesimal slice of the beam of length dx_1 as depicted in fig. 6.2. The axial force, N_1, acts on the face at span-wise location x_1. A Taylor's series expansion is then used to express this axial force at location $x_1 + dx_1$. Higher order differential terms are neglected, leading to the contribution shown in fig. 6.2. Summing all the forces in the axial direction yields the axial equilibrium equation

$$
\frac{dN_1}{dx_1} = -p_1(x_1). \tag{6.16}
$$

A similar approach can be applied to transverse force and moment equilibrium. The left portion of fig. 6.3 depicts the transverse loads and bending moments acting on an infinitesimal slice of the beam, focusing on plane $(\bar{\imath}_1, \bar{\imath}_2)$ in the left figure. Summation of the forces acting along axis $\bar{\imath}_2$ gives the transverse equilibrium equation

$$\frac{dV_2}{dx_1} = -p_2(x_1). \tag{6.17}$$

Summation of the moments taken about the centroidal axis $\bar{\imath}_3$ yields

$$\frac{dM_3}{dx_1} + V_2 = -q_3(x_1) + x_{2a}p_1(x_1), \tag{6.18}$$

where the last term arises because the line of action of the axial load, $p_1(x_1)$, passes through a point of coordinates (x_{2a}, x_{3a}). In general, there is no reason to believe that $x_{2a} = x_{3a} = 0$, i.e., that the line of action of the applied axial load passes through the origin of the axis system, which is selected to coincide with the centroid of the section. For instance, if the axial load is the centrifugal force acting on the cross-

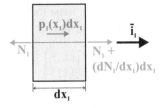

Fig. 6.2. Free body diagram for the axial forces.

section of a spinning beam, this axial load will be applied at the center of mass of the section, which might not coincide with its centroid.

Similarly, the right portion of fig. 6.3 depicts the transverse loads and bending moments acting on an infinitesimal slice of the beam, but now focusing on plane $(\bar{\imath}_1, \bar{\imath}_3)$. Summing the forces along axis $\bar{\imath}_3$ gives the second transverse equilibrium equation

$$\frac{dV_3}{dx_1} = -p_3(x_1), \tag{6.19}$$

and summing the moments about the centroidal axis $\bar{\imath}_2$ leads to

$$\frac{dM_2}{dx_1} - V_3 = -q_2(x_1) - x_{3a}p_1(x_1), \tag{6.20}$$

where x_{3a} defines the location at which the axial force p_1 acts on the cross-section.

The shear forces, V_2 and V_3, can be eliminated from the equilibrium equations by taking a derivative of eqs. (6.20) and (6.18), then introducing eqs. (6.19) and (6.17), respectively, to yield the equilibrium equations

$$\frac{d^2M_2}{dx_1^2} = -p_3(x_1) - \frac{d}{dx_1}[x_{3a}p_1(x_1) + q_2(x_1)], \tag{6.21a}$$

$$\frac{d^2M_3}{dx_1^2} = p_2(x_1) + \frac{d}{dx_1}[x_{2a}p_1(x_1) - q_3(x_1)]. \tag{6.21b}$$

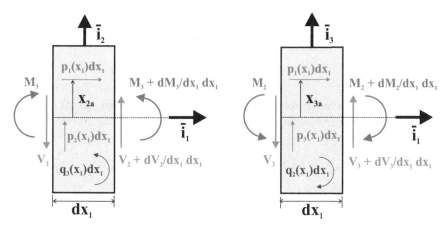

Fig. 6.3. Free body diagram for the transverse shear forces and bending moments. Left figure: view of the $(\bar{\imath}_1, \bar{\imath}_2)$ plane; right figure: view of the $(\bar{\imath}_1, \bar{\imath}_3)$ plane;

6.4 Governing equations

The governing equations for the beam transverse displacement field can be formulated as second order differential equations by introducing eqs. (5.6) into the sectional constitutive laws, eqs. (6.12) to find

$$
\begin{aligned}
H_{23}^c \frac{d^2 \bar{u}_2}{dx_1^2} + H_{22}^c \frac{d^2 \bar{u}_3}{dx_1^2} &= -M_2(x_1), \\
H_{33}^c \frac{d^2 \bar{u}_2}{dx_1^2} + H_{23}^c \frac{d^2 \bar{u}_3}{dx_1^2} &= M_3(x_1).
\end{aligned}
\tag{6.22}
$$

These differential equations can be used to solve for the beam transverse displacement field when the bending moments, $M_2(x_1)$ and $M_3(x_1)$, are known. For isostatic problems, the bending moment distribution can be expressed in terms of the externally applied loads based on equilibrium considerations alone.

For hyperstatic problems, another approach is necessary. Fourth order differential equations are obtained by introducing the sectional constitutive laws, eqs. (6.12), into the equilibrium equations, eqs. (6.16), (6.21a), and (6.21b), and then using the definition of the sectional strains, eq. (5.6), to find

$$
\frac{d}{dx_1}\left[S \frac{d\bar{u}_1}{dx_1} \right] = -p_1,
\tag{6.23a}
$$

$$
\frac{d^2}{dx_1^2}\left[H_{33}^c \frac{d^2 \bar{u}_2}{dx_1^2} + H_{23}^c \frac{d^2 \bar{u}_3}{dx_1^2} \right] = p_2 + \frac{d}{dx_1}[x_{2a}p_1 - q_3],
\tag{6.23b}
$$

$$
\frac{d^2}{dx_1^2}\left[H_{23}^c \frac{d^2 \bar{u}_2}{dx_1^2} + H_{22}^c \frac{d^2 \bar{u}_3}{dx_1^2} \right] = p_3 + \frac{d}{dx_1}[x_{3a}p_1 + q_2].
\tag{6.23c}
$$

These are second and fourth order, ordinary differential equations and their solution requires specification of a number of boundary conditions on $\bar{u}_1$, $\bar{u}_2$, and $\bar{u}_3$.

Using the beam example shown in fig. 6.1, the boundary conditions at the root of the beam are purely geometric,

$$\bar{u}_1 = \bar{u}_2 = \bar{u}_3 = 0, \quad \frac{d\bar{u}_2}{dx_1} = \frac{d\bar{u}_3}{dx_1} = 0, \tag{6.24}$$

which correspond to zero displacements and slopes at the clamped end. At the tip of the beam, the boundary conditions deal with the applied tip shear and axial loads, and bending moments

$$N_1 = P_1, \ M_3 = Q_3 - x_{2a}P_1, \ M_2 = Q_2 + x_{3a}P_1, \ V_2 = P_2, \ V_3 = P_3. \tag{6.25}$$

These boundary conditions must now be expressed in terms of the displacement components, $\bar{u}_1$, $\bar{u}_2$, and $\bar{u}_3$. Introducing the sectional constitutive laws, eq. (6.12), into eq. (6.25) and using the definition of the sectional strains, eq. (5.6), yields the boundary conditions expressed in terms of displacements as

$$S\frac{d\bar{u}_1}{dx_1} = P_1,$$

$$H_{33}^c \frac{d^2\bar{u}_2}{dx_1^2} + H_{23}^c \frac{d^2\bar{u}_3}{dx_1^2} = Q_3 - x_{2a}P_1,$$

$$H_{23}^c \frac{d^2\bar{u}_2}{dx_1^2} + H_{22}^c \frac{d^2\bar{u}_3}{dx_1^2} = -Q_2 - x_{3a}P_1, \tag{6.26}$$

$$-\frac{d}{dx_1}\left[H_{33}^c \frac{d^2\bar{u}_2}{dx_1^2} + H_{23}^c \frac{d^2\bar{u}_3}{dx_1^2}\right] = P_2 - [x_{2a}p_1 - q_3]_L,$$

$$-\frac{d}{dx_1}\left[H_{23}^c \frac{d^2\bar{u}_2}{dx_1^2} + H_{22}^c \frac{d^2\bar{u}_3}{dx_1^2}\right] = P_3 - [x_{3a}p_1 + q_2]_L.$$

In summary, the governing equations of the problem are in the form of the three coupled differential equations (6.23a), (6.23b), and (6.23c) for the three sectional displacements $\bar{u}_1$, $\bar{u}_2$, and $\bar{u}_3$. The equations are second order in the axial displacement $\bar{u}_1$, and fourth order in the transverse displacements $\bar{u}_2$, and $\bar{u}_3$. There are ten associated boundary conditions, five at each end of the beam, as specified in eqs (6.24) and (6.26). Boundary conditions corresponding to various end configurations can be easily derived, as described in section 5.5.4.

6.5 Decoupling the three-dimensional problem

The governing equations described in the previous section form a set of coupled differential equations, and as such, are more difficult to solve than the bending problems presented in chapter 5. The axial behavior, eq. (6.23a), is decoupled from the bending behavior governed by eqs. (6.23b) and (6.23c), hence, these two problems can be handled separately. The bending equations (6.23b) and (6.23c), however, are coupled and must be solved simultaneously. Stated in another way, the coupling between the two bending equations means that loads applied along axis $\bar{\imath}_2$ will not only cause deflection in that direction but can also produce deflection along axis $\bar{\imath}_3$.

The theory developed in the previous section requires the *axis system to be centroidal*, that is that axis $\bar{\imath}_1$ passes through the centroid of the cross-section. Although this choice decouples the axial behavior from bending, if an axial force is not applied at the centroid, it will contribute to the bending problem: see the terms $x_{2a}p_1$ and $x_{3a}p_1$, in eqs. (6.23b) and (6.23c), respectively. Similar contributions appear in the boundary conditions.

An important case of axial forces not applied at the centroid is found in air vehicles such as helicopters. The large centrifugal force generated by the rotation of the blade is an axial force *applied at the sectional center of mass*, which is, in general, distinct from its centroid. In such a case, $p_1(x_1)$ is the distributed centrifugal force applied on the blade, and (x_{2a}, x_{3a}) the coordinates of the sectional center of mass in a centroidal axis system. If a non-centroidal axis system is chosen, the resulting equations are considerably more complicated to solve, and in addition, the results are harder to understand and interpret.

6.5.1 Definition of the principal axes of bending

The question to be raised in this section is whether the governing equations can be further simplified by a judicious choice of the orientation of the centroidal axis system. The coupling between displacement components $\bar{u}_2$ and $\bar{u}_3$ in eqs. (6.23b) and (6.23c) arises from the presence of the cross bending stiffness coefficient, H_{23}^c, defined in eq. (6.9). This term can be made to vanish by an appropriate choice of the orientation or rotation of axes $\bar{\imath}_2$ and $\bar{\imath}_3$, within the plane of the cross-section. The *principal centroidal axes of bending* are defined as a set of axes with their origin at the centroid of the section and for which

$$H_{23}^c = \int_{\mathcal{A}} E x_2 x_3 \, \mathrm{d}\mathcal{A} = 0. \tag{6.27}$$

The actual procedure for determining the orientation of the principal centroidal axes of bending is described in section 6.6. The result is a new axis system, $\mathcal{I}^* = (\bar{\imath}_2^*, \bar{\imath}_3^*)$, that is rotated about the axis of the beam, $\bar{\imath}_1$, *i.e.*, leaving the axis of the beam unchanged, $\bar{\imath}_1 = \bar{\imath}_1^*$. The notation $(\cdot)^*$ will be used to indicate quantities resolved in the new reference frame.

In this frame of reference, the constitutive laws for the cross-section, eq. (6.13), take the following, fully decoupled form

$$\bar{\epsilon}_1^* = \frac{N_1^*}{S^*}, \quad \kappa_2^* = \frac{M_2^*}{H_{22}^{c*}}, \quad \kappa_3^* = \frac{M_3^*}{H_{33}^{c*}}. \tag{6.28}$$

The corresponding axial stress distribution, eq. (6.3), becomes

$$\sigma_1^* = E \left[\frac{N_1^*}{S^*} + x_3^* \frac{M_2^*}{H_{22}^{c*}} - x_2^* \frac{M_3^*}{H_{33}^{c*}} \right], \tag{6.29}$$

which is considerably simpler than eq. (6.14).

6.5.2 Decoupled governing equations

The use of the principal centroidal axis of bending also simplifies the governing equations of the problem, eqs. (6.23a) to 6.23c, which now decouple into *three independent equations* that describe the axial and bending behaviors of the beam. With reference to the particular beam configuration illustrated in fig. 6.1, three independent problems are now defined.

The axial problem

The axial problem is governed by eq. (6.23a), which now takes on the following form

$$\frac{d}{dx_1^*}\left[S^*\frac{d\bar{u}_1^*}{dx_1^*}\right] = -p_1^*. \tag{6.30}$$

For the problem shown in fig. 6.1, the boundary conditions are as follows: $\bar{u}_1^* = 0$ at the root of the beam, whereas at its tip, $S^*d\bar{u}_1^*/dx_1^* = P_1^*$. This extensional problem is identical to that discussed in section 5.4. Note that $S = S^*$ since the axial stiffness remains unaffected by a rotation of axes $\bar{\imath}_2$ and $\bar{\imath}_2$ about axis $\bar{\imath}_1$.

The first bending problem

The bending problem reduces to two independent equations. The first of these, eq. (6.23b), takes the following form

$$\frac{d^2}{dx_1^{*2}}\left[H_{33}^{c*}\frac{d^2\bar{u}_2^*}{dx_1^{*2}}\right] = p_2^* + \frac{d}{dx_1^*}[x_{2a}^*p_1^* - q_3^*], \tag{6.31}$$

which describes bending in plane $(\bar{\imath}_1^*, \bar{\imath}_2^*)$. This differential equation is subject to the following boundary conditions at the beam's root $\bar{u}_2^* = 0$ and $d\bar{u}_2^*/dx_1^* = 0$ and at its tip,

$$H_{33}^{c*}\frac{d^2\bar{u}_2^*}{dx_1^{*2}} = Q_3^* - x_{2a}^*P_1^*, \quad -\frac{d}{dx_1^*}\left[H_{33}^{c*}\frac{d^2\bar{u}_2^*}{dx_1^{*2}}\right] = P_2^* - [x_{2a}^*p_1^* - q_3^*].$$

The second bending problem

Finally, eq. (6.23c) takes the following form

$$\frac{d^2}{dx_1^{*2}}\left[H_{22}^{c*}\frac{d^2\bar{u}_3^*}{dx_1^{*2}}\right] = p_3^* + \frac{d}{dx_1^*}[x_{3a}^*p_1^* + q_2^*], \tag{6.32}$$

which describes bending in plane $(\bar{\imath}_1, \bar{\imath}_3^*)$. The differential equation is subject to the following boundary conditions at the beam's root $\bar{u}_3^* = 0$ and $d\bar{u}_3^*/dx_1^* = 0$ and at its tip,

$$H_{22}^{c*} \frac{d^2 \bar{u}_3^*}{dx_1^{*2}} = -Q_2^* - x_{3a}^* P_1^*, \quad -\frac{d}{dx_1^*} \left[H_{22}^{c*} \frac{d^2 \bar{u}_3^*}{dx_1^{*2}} \right] = P_3^* - [x_{3a}^* p_1^* + q_2^*].$$

Note that the two bending problems are identical problems written in the two orthogonal planes defined by the principal centroidal axes of bending. Each bending problem is identical to the bending problems discussed in section 5.5. It is clear that the rotation of the axes to the principal directions takes place about axis $\bar{\imath}_1$. Hence, $\bar{\imath}_1^* = \bar{\imath}_1$, $x_1^* = x_1$, and $\bar{u}_1^* = \bar{u}_1$. The notational difference is made to emphasize the fact that all quantities in the decoupled equations are resolved along the principal centroidal axes of bending.

6.6 The principal centroidal axes of bending

Consider an arbitrary set of axes, $\mathcal{I} = (\bar{\imath}_2, \bar{\imath}_3)$, with their origin at the centroid of the section, as depicted in fig. 6.4. Next, a new set of axes, $\mathcal{I}^* = (\bar{\imath}_2^*, \bar{\imath}_3^*)$, is defined by rotating the first set of axes by an angle α. Let (x_2, x_3) and (x_2^*, x_3^*) denote the coordinates of point **P** resolved in coordinate systems $\mathcal{I}$ and $\mathcal{I}^*$, respectively.

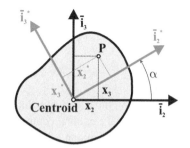

Coordinate transformations are discussed in appendix A.3.4, and the two sets of coordinates, (x_2, x_3) and (x_2^*, x_3^*), are related by eq. (A.43). The centroidal bending stiffnesses in system $\mathcal{I}^*$ can be computed using eq. (6.8) to find

Fig. 6.4. Rotation of the axes of the cross-section.

$$H_{22}^{c*} = \int_{\mathcal{A}} E \left(-x_2 \sin \alpha + x_3 \cos \alpha\right)^2 d\mathcal{A},$$

$$H_{33}^{c*} = \int_{\mathcal{A}} E \left(x_2 \cos \alpha + x_3 \sin \alpha\right)^2 d\mathcal{A},$$

$$H_{23}^{c*} = \int_{\mathcal{A}} E \left(x_2 \cos \alpha + x_3 \sin \alpha\right)(-x_2 \sin \alpha + x_3 \cos \alpha) d\mathcal{A}.$$

Expanding these expressions, and noting that centroidal axes are being used, gives

$$H_{22}^{c*} = H_{22}^c \cos^2 \alpha + H_{33}^c \sin^2 \alpha - 2H_{23}^c \sin \alpha \cos \alpha, \tag{6.33a}$$

$$H_{33}^{c*} = H_{22}^c \sin^2 \alpha + H_{33}^c \cos^2 \alpha + 2H_{23}^c \sin \alpha \cos \alpha, \tag{6.33b}$$

$$H_{23}^{c*} = (H_{22}^c - H_{33}^c) \sin \alpha \cos \alpha + H_{23}^c (\cos^2 \alpha - \sin^2 \alpha). \tag{6.33c}$$

With the help of basic double-angle trigonometric identities, these expressions can be rewritten as

$$H_{22}^{c*} = \frac{H_{22}^c + H_{33}^c}{2} + \frac{H_{22}^c - H_{33}^c}{2}\cos 2\alpha - H_{23}^c \sin 2\alpha; \tag{6.34a}$$

$$H_{33}^{c*} = \frac{H_{22}^c + H_{33}^c}{2} - \frac{H_{22}^c - H_{33}^c}{2}\cos 2\alpha + H_{23}^c \sin 2\alpha; \tag{6.34b}$$

$$H_{23}^{c*} = \qquad + \frac{H_{22}^c - H_{33}^c}{2}\sin 2\alpha + H_{23}^c \cos 2\alpha. \tag{6.34c}$$

Note the very close similarity between these equations, expressing the relationship between bending stiffnesses in two different coordinate system and eqs. (1.49) expressing the relationship between stress components in two different coordinate systems, or eqs. (1.94) expressing the relationship between strain components in two different coordinate systems. This is due to the fact that bending stiffnesses, stress components, and strain components, all form *second order tensors*. The components of second order tensors under an axis rotation all behave in the same manner, as expressed by eqs. (6.34), (1.49), or (1.94).

By definition (6.27), the principal centroidal axes of bending are such that $H_{23}^{*c} = 0$. Equation (6.34c) yields the following equation for the orientation of the principal axes

$$\tan 2\alpha^* = \frac{2H_{23}^c}{H_{33}^c - H_{22}^c}. \tag{6.35}$$

This equation presents two solutions, α^* and $\alpha^* + \pi/2$, corresponding to two mutually orthogonal principal centroidal axes directions. The maximum bending is found about one direction, and the minimum about the other. To define these orientations unequivocally, it is convenient to separately define the sine and cosines of angle $2\alpha^*$ as follows

$$\sin 2\alpha^* = \frac{H_{23}^c}{\Delta} \quad \text{and} \quad \cos 2\alpha^* = \frac{H_{33}^c - H_{22}^c}{2\Delta}, \tag{6.36}$$

where

$$\Delta = \sqrt{\left(\frac{H_{33}^c - H_{22}^c}{2}\right)^2 + (H_{23}^c)^2}. \tag{6.37}$$

This result is equivalent to eq. (6.35), but it gives a unique solution for α^* because both the sine and cosine of the angle are known. The minimum and maximum bending stiffnesses, denoted H_{22}^{c*} and H_{33}^{c*}, respectively, act about the directions α^* and $\alpha^* + \pi/2$, respectively. These minimum and maximum bending stiffnesses, called *principal centroidal bending stiffnesses*, are evaluated by introducing the orientation of the principal axes, eq. (6.36), into eqs. (6.34a) and (6.34b), to find

$$H_{22}^{c*} = \frac{H_{33}^c + H_{22}^c}{2} - \Delta; \quad H_{33}^{c*} = \frac{H_{33}^c + H_{22}^c}{2} + \Delta. \tag{6.38}$$

In summary, the orientation of the principal centroidal axes of bending is obtained according to the following procedure.

1. Compute the centroid of the section using the definition, eq. (6.11);
2. Compute the bending stiffnesses in this axis system using eqs. (6.8) and (6.9);
3. Compute the orientation of the principal axes of bending using eq. (6.36);

4. Compute the principal bending stiffnesses using eq. (6.38).

It is interesting to note that the principal axes of bending are axes about which the bending stiffnesses are extremal: minimum about $\bar{\imath}_2^*$, and maximum about $\bar{\imath}_3^*$. Indeed, the bending stiffness is expressed in terms of α^* in eq. (6.34a): the minimum value of $H_{22}^{c*}(\alpha^*)$ occurs when its derivative with respect to α^* vanishes

$$\frac{\mathrm{d}H_{22}^{c*}}{\mathrm{d}\alpha^*} = \frac{H_{33}^c - H_{22}^c}{2} 2\sin 2\alpha^* - H_{23}^c 2\cos 2\alpha^* = 0. \tag{6.39}$$

This condition is identical to eq. (6.35). Similarly, the value of α^* which maximizes $H_{33}^{c*}(\alpha^*)$ is the same value determined by eq. (6.35). In summary, the principal axes of bending are such that H_{23}^{*c} vanishes, and the corresponding bending stiffnesses are extremal.

6.6.1 The bending stiffness ellipse

It is noted in the previous section that bending stiffnesses, stress components, and strain components, all form second order tensors, and the components of second order tensors under an axis rotation all behave in the same manner, as expressed by eqs. (6.34), (1.49), or (1.94). Hence, it should not come as a surprise that Mohr's circle representation of stress components, as presented in section 1.3.6, or of strain components, as presented in section 1.6.4, can also be used to represent the bending stiffness components.

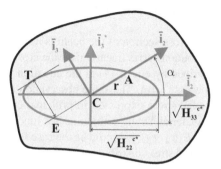

Fig. 6.5. The bending stiffness ellipse of a cross-section.

Bending stiffness components, however, afford another representation that is more informative than Mohr's circle. Figure 6.5 shows the arbitrarily shaped cross-section of a beam with its principal centroidal axes of bending, $\bar{\imath}_2^*$ and $\bar{\imath}_3^*$; point **C** is located at the centroid of the cross-section. An ellipse, called the *bending stiffness ellipse*, with semi-axes $\sqrt{H_{33}^{c*}}$ and $\sqrt{H_{22}^{c*}}$ is constructed with its center at the centroid of the section and its axes aligned with the principal centroidal axes of bending.

By construction, the equation of this ellipse is

$$\frac{x_2^{*2}}{H_{22}^{c*}} + \frac{x_3^{*2}}{H_{33}^{c*}} = 1. \tag{6.40}$$

Consider now an arbitrary axis system, $\mathcal{I} = (\bar{\imath}_2, \bar{\imath}_3)$, where $\bar{\imath}_2$ forms an angle α with respect to axis $\bar{\imath}_2^*$. Let point **A** be located at the intersection of axis $\bar{\imath}_2$ with the bending stiffness ellipse. The coordinates of point **A** are $x_2^* = r\cos\alpha$ and $x_3^* = r\sin\alpha$, where r is the length of segment **CA**. Since point **A** is on the bending stiffness ellipse, it follows that eq. (6.40) can be rewritten as $r^2(\cos^2\alpha/H_{33}^{c*} + \sin^2\alpha/H_{22}^{c*}) = 1$, and hence,

$$r^2 = \frac{H_{22}^{c*}H_{33}^{c*}}{H_{22}^{c*}\cos^2\alpha + H_{33}^{c*}\sin^2\alpha} = \frac{H_{22}^{c*}H_{33}^{c*}}{H_{22}^{c}},$$

where the last equality results from eq. (6.33a). A fundamental property of an ellipse is that the product of the lengths of segments $\mathbf{TE}$ and $\mathbf{CA}$ equals the product of the lengths of the semi-axes, *i.e.*, $r\,\mathbf{TE} = \sqrt{H_{22}^{c*}H_{33}^{c*}}$. Introducing the value of r computed above leads to

$$\mathbf{TE}^2 = H_{22}^{c}. \tag{6.41}$$

The interpretation of this result is as follows: the bending stiffness of the cross-section about an arbitrary axis $\bar{\imath}_2$ equals the square of the distance between this axis and the tangent to the bending stiffness ellipse that is parallel to $\bar{\imath}_2$. As axis $\bar{\imath}_2$ rotates around the centroid, the bending stiffness ellipse provides a convenient visualization of the variation of the bending stiffness about this axis.

6.7 The neutral axis

If the cross-section of the beam is made of a homogeneous material, the axial stress distribution varies linearly over the cross-section. Indeed, the axial stress distribution described by eq. (6.14) is the equation of a plane with terms in x_2, x_3, and an independent term. The same observation can be made by considering the distribution of axial stress expressed in principal centroidal axes of bending, see eq. (6.29). If the material Young's modulus is a function of position over the cross-section, *i.e.*, if $E = E(x_2, x_3)$, as would be the case for a beam made of layered composite material, the axial stress distribution over the cross-section is no longer linear.

For sections made of homogeneous material, three distinct types of the axial stress distribution are possible over the cross-section.

1. If the *axial force, N_1, has a sufficiently large tensile (positive) value*, the axial stress is tensile over the entire cross-section.
2. If the *axial force, N_1, has a sufficiently large compressive (negative) value*, the axial stress is compressive over the entire cross-section.
3. If the *axial force, N_1, assumes an intermediate value or vanishes*, the axial stress will vanish along a straight line intersecting the boundaries of the cross-section; the axial stress will be tensile on one side of this line and compressive on the other. The locus of zero axial stress is a straight line called the *neutral axis*.

Figure 6.6 illustrates the concept of the neutral axis, which divides the cross-section into two regions, one subjected to compressive stresses, the other to tensile stresses. Along the neutral axis, the axial stress vanishes, while along lines parallel to the neutral axis, the axial stress is constant. Consequently, axial stresses will increase or decrease most rapidly when moving along the direction perpendicular to the neutral axis: the maximum axial stress gradient direction is normal to the neutral axis. It then follows that the extremal values of the axial stress are found at the points of the cross-section that are at the largest perpendicular distance from the neutral axis, as

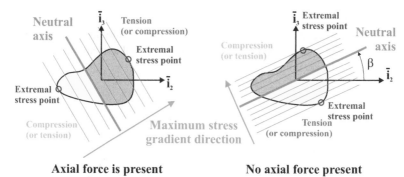

Fig. 6.6. Neutral axis on a cross-section: left portion, axial force is preset, right portion, the axial force vanishes. Along the red lines, the axial stresses remain constants.

illustrated in fig. 6.6. The neutral axis is an important concept that helps with visualizing the axial stress field over a cross-section subjected to axial forces and bending moments. It also facilitates the determination of the locations of the extremal axial stresses on the cross-section.

The neutral axis is a straight line, and its equation is readily found by imposing the vanishing of the axial stress in eq. (6.14) to find

$$\frac{N_1}{S} + \frac{H_{33}^c M_2 + H_{23}^c M_3}{\Delta_H} x_3 - \frac{H_{23}^c M_2 + H_{22}^c M_3}{\Delta_H} x_2 = 0. \qquad (6.42)$$

Clearly, this is the equation of a line in the plane of the cross-section for a given axial force, N_1, and bending moments, M_2 and M_3. The slope of this line is found as

$$\tan \beta = \frac{x_3}{x_2} = \frac{H_{23}^c M_2 + H_{22}^c M_3}{H_{33}^c M_2 + H_{23}^c M_3}. \qquad (6.43)$$

It is often convenient to work with the principal centroidal axes of bending. In that case, the equation of the neutral axis is found by imposing the vanishing of the axial stress in eq. (6.29) to find

$$\frac{N_1^*}{S^*} + x_3^* \frac{M_2^*}{H_{22}^{c*}} - x_2^* \frac{M_3^*}{H_{33}^{c*}} = 0. \qquad (6.44)$$

The slope of the neutral axis is simply $\tan \beta^* = x_3^*/x_2^* = (H_{22}^{c*} M_3^*)/(H_{33}^{c*} M_2^*)$.

As illustrated in fig. 6.6, when the axial force vanishes, the neutral axis passes through the origin of the axis system, which coincides with the centroid of the section.

Example 6.1. *Relationship between the bending stiffness ellipse and the neutral axis*

Consider a cross-section of arbitrary shape subjected to a bending moment of magnitude M, as depicted in fig. 6.7. Axes $\bar{\imath}_1^*$ and $\bar{\imath}_2^*$ are the principal centroidal axes of bending of the cross-section and the bending stiffness ellipse, as defined in section 6.6.1, has also been drawn on the figure. The bending moment vector is oriented at an angle γ^* with axis $\bar{i}_2^*$. Find the location of the neutral axis and of the maximum axial stresses in the section.

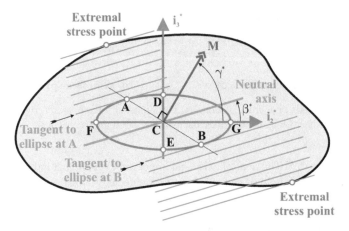

Fig. 6.7. Cross-section subjected to a bending moment, M.

Since axes $\bar{i}_1^*$ and $\bar{i}_2^*$ are the principal centroidal axes of bending of the cross-section, the distribution of axial stress is given by eq. (6.29) as $\sigma_1^*/E = x_3^* M_2^*/H_{22}^{c*} - x_2^* M_3^*/H_{33}^{c*}$. Clearly, $M_2^* = M\cos\gamma^*$ and $M_3^* = M\sin\gamma^*$, leading to the following distribution of axial stress, $\sigma_1^*/E = M(x_3^*\cos\gamma^*/H_{22}^{c*} - x_2^*\sin\gamma^*/H_{33}^{c*})$. The orientation of the neutral axis is

$$\tan\beta^* = \frac{x_3^*}{x_2^*} = \frac{H_{22}^{c*}}{H_{33}^{c*}}\tan\gamma^*.$$

This result implies that angles γ^* and β^* are, in general, not equal. Two notable exceptions exist: if $\gamma^* = 0$ or $\pi/2$, $\beta^* = 0$ or $\pi/2$, respectively. Because the selected axes are principal centroidal axes of bending, $\gamma^* = 0$ or $\pi/2$ implies that the bending moment is applied about one of the principal centroidal axes of bending directions, and its direction then coincides with that of the neutral axis. The other exception is when $H_{22}^{c*} = H_{33}^{c*}$, in which case any axis system with its origin at the centroid is a principal centroidal axis of bending system.

In summary, the direction of the neutral axis coincides with that of the applied bending moment if and only if the applied bending moment acts about a principal centroidal axes of bending direction.

The equation of the bending stiffness ellipse is given by eq. (6.40), and it is easy to show that the equation of the tangent to the ellipse at one of its points, **A**, with coordinates (x_{2A}, x_{3A}), is $x_2^* x_{2A}/H_{33}^{c*} + x_3^* x_{3A}/H_{22}^{c*} = 1$. Now, let points **A** and **B** be the intersections of the normal to the moment vector with the bending stiffness ellipse, as shown in fig. 6.7. The coordinates of point **A** become $x_{2A} = -\mathbf{AC}\sin\gamma^*$

and $x_{3A} = \mathbf{AC}\cos\gamma^*$, and the equation of the tangent to the ellipse at point $\mathbf{A}$ becomes $x_2^*\sin\gamma^*/H_{33}^{c*}+x_3^*\cos\gamma^*/H_{22}^{c*} = 1/\mathbf{AC}$. Clearly, the slope of this tangent is

$$\frac{x_3^*}{x_2^*} = \frac{H_{22}^{c*}}{H_{33}^{c*}}\tan\gamma^*.$$

The *slope of this tangent is identical to that of the neutral axis*. Hence, the neutral axis is parallel to the tangent to the bending stiffness ellipse at point $\mathbf{A}$.

The orientation of the neutral axis as the orientation of the applied bending moment vector changes is now easily visualized. First, let the direction of the applied bending moment coincide with the principal direction $\bar{\imath}_2^*$, *i.e.*, $\gamma^* = 0$. The neutral axis is parallel to the tangent to the ellipse at points $\mathbf{D}$ or $\mathbf{E}$, *i.e.*, it coincides with axis $\bar{\imath}_2^*$. Similarly, if the direction of the applied bending moment coincides with the principal direction $\bar{\imath}_3^*$, the neutral axis coincides again with $\bar{\imath}_3^*$. If the bending stiffness ellipse is very elongated, *i.e.*, if $H_{33}^{c*} \gg H_{22}^{c*}$, very rapid variations of the orientation of the neutral axis must be expected because of the very rapid variation of the tangent to the bending stiffness ellipse about points $\mathbf{F}$ or $\mathbf{G}$.

Finally, as explained in section 6.7, the orientation of the neutral axis gives the direction of the maximum axial stress gradient and the location of the maximum axial stress in the cross-section, as illustrated in fig. 6.7.

Example 6.2. Maximum bending moments for rectangular section
Consider a solid rectangular section of width b and height h subjected to an axial force, N_1, and bending moments, M_2 and M_3, as depicted in fig. 6.8. If the material has a yield strain ϵ_y, find the yield envelope for the section. In view of the symmetry of the cross-section, the axes shown in fig. 6.8 are the principal centroidal axes of bending.

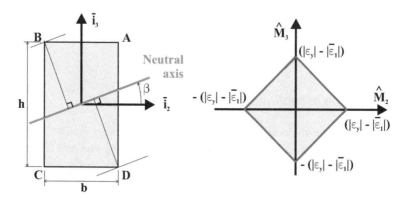

Fig. 6.8. Left figure: neutral axis for a rectangular section. Right figure: yield envelope for the rectangular section under combined bending moment and axial force.

First, assume that no axial force is applied to the section and that the applied bending moments give the neutral axis depicted in fig. 6.8. The extremal axial stresses will occur at the largest normal distance from the neutral axis, *i.e.*, at the corners of the section, points **B** and **D**. Note that for $0 < \beta^* < \pi/2$, points **B** and **D** remain the locations of the extremal stresses. On the other hand, for $-\pi/2 < \beta^* < 0$, the extremal axial stresses will occur at the other two corners of the section, points **A** and **C**. Note that $\beta^* = 0$ or $\beta^* = \pi/2$ are special cases: the extremal stresses are found along edge **AB** and **CD** or **BC** and **DA**, respectively. If an axial force is present, a *uniform axial stress* is added to the axial stress distribution due to the bending moments, but this does not affect the location of the extremal axial stresses. Clearly, yielding will initiate at one of the four corner points **A**, **B**, **C**, or **D**.

In view of eq. (6.29), the non-dimensional axial stress distribution can be written as

$$
\begin{aligned}
\frac{\sigma_1}{E} &= \frac{N_1}{S} + \frac{M_2}{H_{22}^c} x_3 - \frac{M_3}{H_{33}^c} x_2 \\
&= \frac{N_1}{S} + \frac{hM_2}{2H_{22}^c} \frac{2x_3}{h} - \frac{bM_3}{2H_{33}^c} \frac{2x_2}{b} = \bar{\epsilon}_1 + \bar{M}_2 \bar{x}_3 - \bar{M}_3 \bar{x}_2,
\end{aligned}
$$

where $\bar{M}_2 = hM_2/(2H_{22}^c)$ and $\bar{M}_3 = bM_3/(2H_{33}^c)$ are the non-dimensional bending moment, and $\bar{x}_2 = 2x_2/b$ and $\bar{x}_3 = 2x_3/h$ the non-dimensional coordinates.

The yield criterion is $\sigma_1/E = \epsilon_y$, which must be applied at points **A**, **B**, **C**, and **D**, because the extremal stresses occur at those locations. These four yield conditions are summarized as

$$
|\bar{M}_2 - \bar{M}_3| = |\epsilon_y| - |\bar{\epsilon}_1|, \quad |\bar{M}_2 + \bar{M}_3| = |\epsilon_y| - |\bar{\epsilon}_1|.
$$

These conditions correspond to four line segments in the bending moment plane $(\bar{M}_2, \bar{M}_3)$ that define the diamond-shaped zone shown in fig. 6.8. The inside of the diamond corresponds to safe loading conditions, and the material starts to yield for loading conditions falling on the edges of the diamond. As the axial force and hence, axial strain increases, the size of the diamond shrinks, which indicates that smaller bending moments can be applied. When $\bar{\epsilon}_1 = \epsilon_y$, the material yields under the axial force alone, and no bending moments can be applied.

6.8 Evaluation of sectional stiffnesses

The determination of the orientation of the principal centroidal axes of bending requires the computation of sectional stiffnesses. This section presents a number of tools that will ease this task.

6.8.1 The parallel axis theorem

The bending stiffness of a section are sometimes to be computed with respect to two axis systems that are parallel to each other, but have a different origin, as illustrated in fig. 6.9. The bending stiffnesses of the section with respect to axes $\bar{i}_2$ and $\bar{i}_2^c$ will be denoted H_{22} and H_{22}^c, respectively; H_{22}^c is called the centroidal bending stiffness.

The *parallel axis theorem* relates the distinct bending stiffnesses computed with respect to parallel axes, one of them centroidal. Let d_3 be the distance between the parallel axes, $\bar{\imath}_2$ and $\bar{\imath}_2^c$.

Bending stiffness H_{22} is given by eq. (6.8) as

$$H_{22} = \int_{\mathcal{A}} E \, (d_3 + x_3^c)^2 \, \mathrm{d}\mathcal{A},$$

where x_3^c is the coordinate of a point of the section measured in the centroidal system, $(\bar{\imath}_2^c, \bar{\imath}_3^c)$. Expanding this result then leads to

$$H_{22} = d_3^2 \left[\int_{\mathcal{A}} E \, \mathrm{d}\mathcal{A} \right] + 2d_3 \left[\int_{\mathcal{A}} E \, x_3^c \, \mathrm{d}\mathcal{A} \right] + \left[\int_{\mathcal{A}} E \, (x_3^c)^2 \, \mathrm{d}\mathcal{A} \right].$$

The first bracketed term is the axial stiffness of the section, S, see eq. (6.5); the second bracketed term vanishes because axis $\bar{\imath}_2^c$ is centroidal, see eq. (6.5); finally, the last bracketed term is the centroidal bending stiffness defined by eq. (6.8). Hence, the result simplifies to $H_{22} = Sd_3^2 + H_{22}^c$. A similar process can be applied to the bending stiffness H_{33} and cross bending stiffness H_{23} to find

$$H_{22} = H_{22}^c + Sd_3^2; \quad H_{33} = H_{33}^c + Sd_2^2; \tag{6.45}$$

and

$$H_{23} = H_{23}^c + Sd_2d_3. \tag{6.46}$$

Because the second term on the right hand side of eqs. (6.45), called the "transport term," is *strictly positive*, it follows that $H_{22} > H_{22}^c$ and $H_{33} > H_{33}^c$, that is, the bending stiffness always increases when moving away from the centroid. In other words, the minimum bending stiffness is obtained when computed with respect to the centroid. On the other hand, the second term on the hand side of eq. (6.46) can be *positive or negative*: cross bending stiffnesses can increase or decrease when moving away from the centroid.

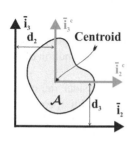

Fig. 6.9. Parallel axes.

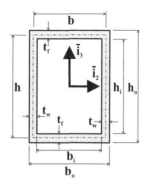

Fig. 6.10. Thin-walled rectangular section.

6.8.2 Thin-walled sections

Many beam sections involved aerospace structures are thin-walled sections, and this fact simplifies the evaluation of the bending stiffnesses. Consider the homogeneous, thin-walled rectangular box beam shown in fig. 6.10. Due to the symmetry of the section, the axes indicated on the figure are principal centroidal axes of bending. The inner and outer heights are h_i and h_o, respectively, whereas the inner and outer width are b_i and b_o, respectively. The thickness of the flange, t_f, and web, t_w, are written as $t_f = (h_o - h_i)/2$ and $t_w = (b_o - b_i)/2$, respectively. The height, h, and width, b, of the section, as measured from mid-wall lines are now $h = (h_o + h_i)/2$ and $b = (b_o + b_i)/2$.

These dimensions are the average height and width of the section. The bending stiffness of the section with respect to axis $\bar{\imath}_2$ can be computed by subtracting the bending stiffness of the inner rectangular area from that of the outer rectangular area to find

$$H_{22}^c = E\left(\frac{b_o h_o^3}{12} - \frac{b_i h_i^3}{12}\right), \tag{6.47}$$

where E is the material Young's modulus. This expression can be rewritten in terms of the average dimensions and wall thicknesses by noting that $b_o = b + t_w$, $b_i = b - t_w$, $h_o = h + t_f$, and $h_i = h - t_f$, to find

$$H_{22}^c = \frac{E}{12}\left[(b + t_w)(h + t_f)^3 - (b - t_w)(h - t_f)^3\right]. \tag{6.48}$$

Expanding the cubic power and regrouping terms then yields

$$H_{22}^c = \frac{E}{12}\left\{6bh^2 t_f\left[1 + \frac{1}{3}\left(\frac{t_f}{h}\right)^2\right] + 2h^3 t_w\left[1 + 3\left(\frac{t_f}{h}\right)\right]\right\}. \tag{6.49}$$

If the wall thickness is small, i.e., $t_f/h \ll 1$, this term is negligible compared to unity, and the bending stiffness reduces to

$$H_{22}^c \approx E\left[2\frac{t_w h^3}{12} + 2bt_f\left(\frac{h}{2}\right)^2\right]. \tag{6.50}$$

The first term represents the bending stiffnesses of the left and right webs, computed with the average height h, whereas the last term gives the contribution of the top and bottom flanges using their average width b.

To better understand the meanings of these terms, consider the calculation of H_{22}^c directly from the individual components of the section. First, compute bending stiffness of the left and right webs about their centroids: $t_w h^3/12$ for each web. Next the contributions of the flanges are added: $(bt_f^3/12 + bt_f h^2/4)$ for each flange; the first term represents the bending stiffness of the flange with respect to its own centroid, and the second term is the transport term according to the parallel axis theorem, eq. (6.45). Adding up the contributions of the various components yields the bending stiffness of the section as

$$H_{22}^c = E \left[2\frac{t_w h^3}{12} + 2 \left(\frac{bt_f^3}{12} + bt_f \left(\frac{h}{2} \right)^2 \right) \right].$$

If the wall thicknesses satisfy the thin-wall assumption, *i.e.*, $t_f/h \ll 1$, terms containing higher powers of the wall thickness can be ignored, and the result is identical to that shown in eq. (6.50) above. A similar reasoning can be used to evaluate the bending stiffness, H_{33}^c.

6.8.3 Triangular area equivalence method

Consider the homogeneous triangular area depicted in fig. 6.11. It can be shown that all area moment calculations can be performed based on lumping the area of the triangle, $\mathcal{A}$, at three points located at the midpoint of each side of the triangle. In other words, the triangular area is replaced by three concentrated areas, each of area $\mathcal{A}/3$, located at the midpoint of each side of the triangle, as illustrated on the figure. The area moment are evaluated based on these lumped areas to find

$$I_{22} = \frac{\mathcal{A}}{3} \sum_{i=1}^3 x_{3i}^2, \quad I_{33} = \frac{\mathcal{A}}{3} \sum_{i=1}^3 x_{2i}^2, \quad I_{23} = \frac{\mathcal{A}}{3} \sum_{i=1}^3 x_{2i} x_{3i}. \tag{6.51}$$

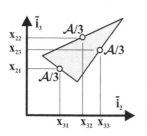

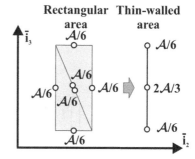

Fig. 6.11. Triangular area equivalent lumped areas.

Fig. 6.12. Rectangular area equivalent lumped areas.

An important special case is the rectangular area which, as shown in fig. 6.12, can be decomposed into two triangular areas, each with one-half of the area of the rectangle. This yields an equivalent lumped model for the rectangle with lumped areas at the midpoints of each of its sides and a fifth at its center. A very useful result is obtained by letting the width of the rectangle decrease to a vanishingly small value while retaining the height and area (in other words, the thin-wall assumption). This case is shown on the right side of fig. 6.12. Now, the representation collapses to a one-dimensional line with lumped areas of $\mathcal{A}/6$ at each end and $2\mathcal{A}/3$ at the midpoint. This is a particularly useful representation for computing the centroids, area moments and bending stiffnesses for thin-walled sections.

6.8.4 Useful results

Thin rectangular strip

The left portion of fig. 6.13 shows a thin rectangular strip of thickness t and height h where $t \ll h$. The centroid of this strip is located at distances d_2 and d_3 from axes $\bar{\imath}_3$ and $\bar{\imath}_2$, respectively. The bending stiffnesses of this strip are approximated as

$$
H_{22} = E \left(\frac{th^3}{12} + htd_3^2 \right), \quad H_{33} = E \, htd_2^2, \quad H_{23} = E \, htd_2 d_3. \tag{6.52}
$$

These results are obtained by first computing the bending stiffness in the principal centroidal axes of the thin strip, then using the parallel axis theorem to translate the bending stiffnesses to the required axis. Terms containing higher powers of the thickness are neglected.

Rotated thin rectangular strip

Similar results can be obtained for the same rectangular strip rotated of an angle, α, with respect to the $\bar{\imath}_2$ axis shown in the right portion fig. 6.13

$$
\begin{aligned}
H_{22} &= E \left(\frac{th^3}{12} \sin^2 \alpha + htd_3^2 \right), \quad H_{33} = E \left(\frac{th^3}{12} \cos^2 \alpha + htd_2^2 \right), \\
H_{23} &= E \left(\frac{th^3}{12} \sin \alpha \cos \alpha + htd_2 d_3 \right) = E \left(\frac{th^3}{24} \sin 2\alpha + htd_2 d_3 \right).
\end{aligned} \tag{6.53}
$$

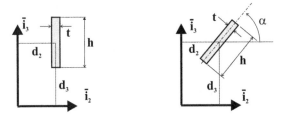

Fig. 6.13. A thin rectangular strip.

Example 6.3. Bending stiffness of a trapezoidal section - Approach 1
Consider the trapezoidal section shown on the left in fig. 6.14. Because axis $\bar{\imath}_2$ is an axis of symmetry of the section, the centroid of the section is located along axis, *i.e.*, $x_{3c} = 0$, and axis $\bar{\imath}_2$ is a principal centroidal axis of bending. Using eqs. (6.52) and (6.53), the centroidal bending stiffness about axis $\bar{\imath}_2$ becomes

$$
H_{22}^c = E \left[\frac{t(2h_1)^3}{12} + \frac{t(2h_2)^3}{12} + 2\frac{t\ell^3}{12} \sin^2 \alpha + 2t\ell(\frac{h_1 + h_2}{2})^2 \right], \tag{6.54}
$$

where $\ell = [b^2 + (h_2 - h_1)^2]^{1/2}$ is the length of the upper and lower flanges. The first two terms represent the contribution of the two webs evaluated with the help of eq. (6.52), whereas the last two terms give the contribution of the flanges obtained from eq. (6.53). It is clear that $\sin \alpha = (h_2 - h_1)/\ell$, and after simplification, the bending stiffness becomes

$$H_{22}^c = \frac{2Et}{3} \left[h_1^3 + h_2^3 + \ell(h_1^2 + h_2^2 + h_1 h_2) \right]. \tag{6.55}$$

The bending stiffness about axis $\bar{\imath}_3$ can be evaluated in a similar fashion. Note that the location of the centroid, x_{2c}, must be calculated first, because this quantity is required to evaluated H_{33}^c.

Example 6.4. Bending stiffness of a trapezoidal section - Approach 2
The problem treated in the previous examples can also be approached using the triangle area equivalence method depicted in the right part of fig. 6.12. Specifically, each straight segment of the cross-section is represented by three lumped areas located at the ends and midpoint of each segment. Figure 6.14 shows the thin-walled trapezoidal section and its lumped equivalent. Using the lumped areas, it follows that a calculation of H_{22}^c will require areas $\mathcal{A}_1, \mathcal{A}_2, \mathcal{A}_3, \mathcal{A}_5, \mathcal{A}_6, \mathcal{A}_7$; areas $\mathcal{A}_4$ and $\mathcal{A}_8$ are at a vanishing distance from axis $\bar{\imath}_2$ and hence, do not appear in the computation of the bending stiffness H_{22}^c. The other areas are $\mathcal{A}_1 = \mathcal{A}_7 = 1/6\ (2h_2t + \ell t)$, $\mathcal{A}_2 = \mathcal{A}_6 = 2/3\ \ell t$ and $\mathcal{A}_3 = \mathcal{A}_5 = 1/6\ (2h_1t + \ell t)$, leading to

$$H_{22}^c = 2E \left[\frac{1}{6}(2h_1t + \ell t)h_1^2 + \frac{2}{3}\ell t(\frac{h_1 + h_2}{2})^2 + \frac{1}{6}(2h_2t + \ell t)h_2^2 \right],$$

$$= \frac{Et}{3} \left[2h_1^3 + 2h_2^3 + \ell(h_1^2 + h_2^2) + \ell(h_1 + h_2)^2 \right]$$

$$= \frac{2Et}{3} \left[h_1^3 + h_2^3 + \ell(h_1^2 + h_2^2 + h_1 h_2) \right],$$

which is identical to eq. (6.55).

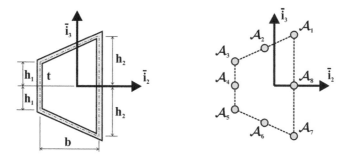

Fig. 6.14. Trapezoidal thin-walled section and lumped representation.

Example 6.5. *Principal centroidal axes of an "L" shaped section*

Consider the thin-walled, "L" shaped cross-section of a beam made of a homogeneous material of Young modulus E, as shown in fig. 6.15. Let $b = 0.25$ m , $h = 0.1$ m, and $t = 2.3$ mm. For convenience, a set of axes $(\bar{\imath}_2, \bar{\imath}_3)$ is defined, which is aligned with the flanges and has its origin at their intersection, point **A**; clearly, this axis system is not centroidal. The axial stiffness of the section is $S = Et(b + h)$ and the location of the centroid is then computed using eqs. (6.11), to find $x_{2c} = b^2/[2(b+h)]$ and $x_{3c} = h^2/[2(b + h)]$. A set of centroidal axes $(\bar{\imath}_2^c, \bar{\imath}_3^c)$ parallel to the flanges are shown in fig. 6.15.

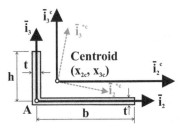

Fig. 6.15. Thin-walled, L shaped cross-section.

Next, the centroidal bending stiffnesses are computed with the help of the parallel axis theorem, eq. (6.45), as

$$H_{22}^c = E \left[\frac{th^3}{12} + ht \left(\frac{h}{2} - x_{3c} \right)^2 + bt\, x_{3c}^2 \right] = \frac{Eth^3}{3} \left[1 - \frac{3h}{4(b + h)} \right],$$

and

$$H_{33}^c = E \left[\frac{tb^3}{12} + bt \left(\frac{b}{2} - x_{2c} \right)^2 + ht\, x_{2c}^2 \right] = \frac{Etb^3}{3} \left[1 - \frac{3b}{4(b + h)} \right].$$

Although the centroidal axes $(\bar{\imath}_2^c, \bar{\imath}_3^c)$ are convenient to use because they are parallel to the flanges, *they are not principal axes*. Indeed, the cross bending stiffness, computed based on the parallel axis theorem, eq. (6.46), is

$$H_{23}^c = Eth\, x_{2c} \left[-\left(\frac{h}{2} - x_{3c} \right) \right] + Etb \left[-\left(\frac{b}{2} - x_{2c} \right) \right] x_{3c} = -\frac{Etb^2h^2}{4(b + h)}.$$

Using the numbers given above, the bending stiffnesses are evaluated as

$$\frac{H_{22}^c}{E} = 0.655 \times 10^{-6} \mathrm{m}^4, \quad \frac{H_{33}^c}{E} = 6.04 \times 10^{-6} \mathrm{m}^4, \quad \frac{H_{23}^c}{E} = -1.12 \times 10^{-6} \mathrm{m}^4.$$

The orientation of the principal centroidal axes then follows from eqs. (6.36)

$$\sin 2\alpha^* = \frac{H_{23}^c}{\Delta} = -0.3826; \quad \cos 2\alpha^* = \frac{H_{33}^c - H_{22}^c}{2\Delta} = 0.9239. \tag{6.56}$$

where $\Delta/E = \sqrt{(H_{33}^c - H_{22}^c)^2/4 + (H_{23}^c)^2} = 2.917 \times 10^{-6}$ m^4. Angle $2\alpha^*$ is in the fourth quadrant, and hence, $\alpha^* = -11.25$ deg. Finally, the principal centroidal bending stiffness are evaluated based on eq. (6.38) to find $H_{22}^{c*}/E = 432.8 \times 10^{-9}$ m^4 and $H_{33}^{c*}/E = 5.940 \times 10^{-6}$ m^4. The bending stiffness is minimum with respect to axis $\bar{\imath}_2^{*c}$ and maximum with respect to axis $\bar{\imath}_3^{*c}$.

6.8.5 Problems

Problem 6.1. Various questions on three-dimensional beam theory

(1) For a particular cross-section, the centroidal bending stiffnesses have been computed as H_{22}^c, H_{33}^c, and H_{23}^c. Next, the bending stiffnesses are computed about a set of parallel axes with their origin at an arbitrary point **D** and found to be H_{22}^d, H_{33}^d, and H_{23}^d. Is it possible to find a point **D** such that $H_{22}^d < H_{22}^c$? Why? *(2)* For a particular cross-section, the centroidal bending stiffnesses have been computed as H_{22}^c, H_{33}^c, and H_{23}^c. Next, the bending stiffnesses are computed about a set of parallel axes with their origin at an arbitrary point **D** and found to be H_{22}^d, H_{33}^d, and H_{23}^d. Is it possible to find a point **D** such that $H_{23}^d < H_{23}^c$? Why? *(3)* Consider a uniform cantilevered beam subjected to a uniform transverse loading distribution $p_0\bar{n}$, where $\bar{n}$ is a unit vector perpendicular to the axis of the beam, $\bar{\imath}_1$. Under what condition will the transverse deflection of the beam be oriented in the direction of $\bar{n}$? *(4)* For a particular cross-section, the principal centroidal bending stiffnesses have been computed as H_{22}^{c*} and H_{33}^{c*}, $H_{22}^{c*} \leq H_{33}^{c*}$. Next, the bending stiffnesses are computed about a set of non-principal axes, $\bar{\imath}_2$, $\bar{\imath}_3$ where axis $\bar{\imath}_2$ is at an arbitrary angle α with respect to $\bar{\imath}_2^*$, and found to be H_{22}^c, H_{33}^c, and H_{23}^c. Is it possible to find an angle α such that $H_{22}^c < H_{22}^{c*}$? Why? *(5)* A uniform cantilevered beam is subjected to a tip axial force. The beam is made of a homogeneous material. Under what condition will the strain distribution over the cross-section be uniform?

Problem 6.2. Axial stresses in a circular cross-section

Consider a solid circular section of radius R subjected to an axial force, N_1, and bending moments, M_2 and M_3. If the material has a yield strain ϵ_y, find the yield envelope for the section.

Problem 6.3. Three-dimensional bema theory

In section 5.6, the governing equations for a beam subjected to combined axial and transverse loads were developed. The origin of the axis system was located at an arbitrary point, *i.e.*, it was not coincident with the centroid of the cross-section. *(1)* Generalize the displacement field given by eq. (5.73) to accommodate general, three-dimensional deformations. *(2)* Find the corresponding strain field. *(3)* Develop the sectional constitutive laws. *(4)* Derive the equilibrium equations of the problem. *(5)* For the problem depicted in fig. 6.1, provide the governing equations and associated boundary conditions. *(6)* Clearly defined all the sectional stiffness coefficients appearing in your developments.

Problem 6.4. Principal axes of bending of a "Z" section

A beam made of a homogeneous material features the "Z" cross-section depicted in fig. 6.16. *(1)* Find the location of the centroid. *(2)* Find the bending stiffnesses in a coordinate system parallel to that shown on the figure, but with its origin at the centroid. *(3)* Find the orientation of the principal axes of bending. *(4)* Find the principal centroidal bending stiffnesses, H_{22}^{c*}, H_{33}^{c*}. Use $a/t = 10$.

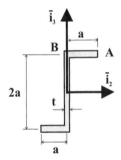

Fig. 6.16. "Z" shaped cross-section of a beam.

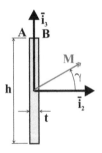

Fig. 6.17. Thin rectangular cross-section.

Problem 6.5. Neutral axis of a "Z" section

A beam made of a homogeneous material features the "Z" cross-section depicted in fig. 6.16. If a bending moment M_2 is applied to the section, find the equation of the neutral axis.

Problem 6.6. Stresses in a thin-walled rectangular cross-section

A beam made of a homogeneous material features the thin-walled rectangular cross-section depicted in fig. 6.17, with $h/t = 12$. A bending moment M is applied to the section and its axis is oriented at an angle γ with respect to axis $\bar{i}_2$. *(1)* Compute the axial stresses at points **A** and **B**. *(2)* On one graph, plot the non-dimensional stresses at points **A** and **B**, denoted $\sigma_1^{(A)}(\gamma)/\sigma_1^{(A)}(\gamma = 0)$ and $\sigma_1^{(B)}(\gamma)/\sigma_1^{(B)}(\gamma = 0)$, respectively, for $\gamma \in [0, \pi/2]$. *(3)* Let σ_y be the yield stress for the material. Plot the non-dimensional maximum bending moment the section can carry, $6M_{\max}/(th^2\sigma_y)$, as a function of $\gamma \in [-\pi/2, \pi/2]$. Comment on your results.

6.9 Summary of three-dimensional beam theory

Solving a three-dimensional beam problem involves determining the three components of displacement field of the beam, $\bar{u}_1(x_1), \bar{u}_2(x_1)$, and $\bar{u}_3(x_1)$ and the axial stress distribution, $\sigma_1(x_1, x_2, x_3)$, over the cross-section.

A solution for the displacement field can be developed by following either of two equivalent approaches described below.

☐ Deflection calculation: approach 1
   1. Compute the location of the centroid using eq. (6.11).
   2. Select a set of axes $\mathcal{I} = (\bar{i}_1, \bar{i}_2, \bar{i}_3)$, for which the $\bar{i}_1$ axis lies along the sectional centroids and project all applied loads along these axes.
   3. Compute the sectional stiffness coefficients eqs. (6.5), (6.8), and (6.9).
   4. Solve the axial problem (6.23a) and the two coupled bending differential equations (6.23b) to 6.23c, subjected to the boundary conditions (6.24) and (6.26).
☐ Deflection calculation: approach 2
   1. Compute the location of the centroid using eq. (6.11).

2. Compute the orientation of the principal centroidal axes of bending $\mathcal{I}^* = (\bar{i}_1^*, \bar{i}_2^*, \bar{i}_3^*)$, and the principal bending stiffnesses according to the procedure described in section 6.6.
3. Project all applied load along the principal centroidal axes of bending.
4. Solve the axial problem (6.30) and two uncoupled bending problems (6.31) and (6.32), subjected to the appropriate boundary conditions.

The two approaches will give identical results. The unknowns of the problem in the first approach are the displacement components $\bar{u}_1$, $\bar{u}_2$, and $\bar{u}_3$ along an arbitrary set of centroidal axes, whereas the displacement components $\bar{u}_1^*$, $\bar{u}_2^*$, and $\bar{u}_3^*$ along the principal centroidal axes of bending are the unknown of the second approach. The solution of the axial and two *coupled* differential equations of the first approach is, in general, quite difficult to obtain. In the second approach, additional work, namely the computation of the principal axes of bending orientation, is initially required. The solution phase then reduces to solving three *decoupled* differential equations.

Once the axial force and bending moment distributions are evaluated, the axial stress distribution is easily obtained. It is also possible to carry out the stress calculation using either the original centroidal axes or the principal centroidal axes of bending.

☐ Axial stress calculation: approach 1
  1. Compute the location of the centroid using eq. (6.11).
  2. Select an axis system set of axes $\mathcal{I} = (\bar{i}_1, \bar{i}_2, \bar{i}_3)$, for which the $\bar{i}_1$ axis lies along the section centroids and project all applied loads along these axes.
  3. Compute the sectional stiffness coefficients eqs. (6.5), (6.8), and (6.9).
  4. Determine the bending moments, $M_2(x_1)$ and $M_3(x_1)$, at a particular axial location, x_1, and use either eq. (6.15) or (6.14) to compute the axial stress, σ_1 at any location on the cross-section.

☐ Axial stress calculation: approach 2
  1. Compute the location of the centroid using eq. (6.11).
  2. Compute the orientation of the principal centroidal axes of bending $\mathcal{I}^* = (\bar{i}_1^*, \bar{i}_2^*, \bar{i}_3^*)$, and the principal bending stiffnesses according to the procedure described in section 6.6.
  3. Project all applied load along the principal centroidal axes of bending.
  4. Determine the bending moments, $M_2^*(x_1)$ and $M_3^*(x_1)$, at a particular axial location, x_1, and use eq. (6.29) to compute the axial stress, σ_1 at any location on the cross-section.

If the geometry of the cross-section is more easily expressed in axis system $\mathcal{I}$, approach 1will be more expeditious.

To demonstrate the use of these approaches for three-dimensional beams, a simple problem will be solved using both approaches, and the results will be shown to be identical.

Example 6.6. Bending of a Z section - Approach 1
Consider a thin-walled cantilevered beam subjected to a uniform transverse load, p_0, as depicted in fig. 6.18. The beam is clamped at the root and is unrestrained at the

tip. The beam is thin walled, *i.e.*, $t/a \ll 1$, and its "Z" shaped cross-section is made of a homogeneous material with a Young's modulus E. In approach 1, the solution will be developed in the axes aligned with the cross-section.

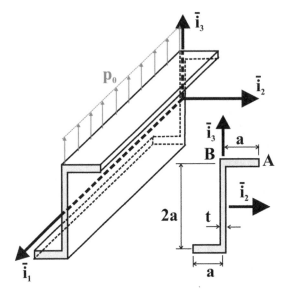

Fig. 6.18. Thin-walled cantilevered beam under a uniform transverse load.

Displacement calculations

A centroidal axis system is used with axis $\bar{\imath}_3$ aligned vertically as shown in fig. 6.18. This axis system makes it easy to locate different points on the cross-section. The first step is to compute the location of the centroid using eq. (6.11). Since the material is homogeneous, the location of the centroid is identical to that of the center of mass of the section and is located on the vertical web, midway between the upper and lower flanges. Axis system $\mathcal{I} = (\bar{\imath}_1, \bar{\imath}_2, \bar{\imath}_3)$ is located at the centroid, as shown in fig. 6.18. The next step is to compute the various sectional stiffnesses. The axial stiffness is computed first using eq. (6.5) to find $S = E\left[at + 2at + at\right] = 4atE$. The bending stiffnesses are computed from eq. (6.8)

$$H_{22}^c = E\left[\left(\frac{at^3}{12} + at\, a^2\right) + \frac{t(2a)^3}{12} + \left(\frac{at^3}{12} + at\, a^2\right)\right] \approx \frac{8a^3 t E}{3};$$

$$H_{33}^c = E\left[\left(\frac{ta^3}{12} + at\, \frac{a^2}{4}\right) + \frac{2a(t)^3}{12} + \left(\frac{ta^3}{12} + at\, \frac{a^2}{4}\right)\right] \approx \frac{2a^3 t E}{3},$$

where the thin wall approximation, $t/a \ll 1$, is used to simplify the results. Finally, the cross bending stiffness is obtained from eq. (6.9),

$$H_{23}^c = E \left[at \left(-\frac{a}{2} \right) (-a) + 2at(0)(0) + at \left(\frac{a}{2} \right) (a) \right] = a^3 tE.$$

Although the selected centroidal axis system conveniently describes the geometry of the problem, it does not coincide with the principal axes of bending, which are characterized by a vanishing cross bending stiffness.

The fourth step of this approach is the solution of the governing equations. The axial and two bending governing equations can be written as

$$S \frac{d^2 \bar{u}_1}{dx_1^2} = 0, \quad H_{33}^c \frac{d^4 \bar{u}_2}{dx_1^4} + H_{23}^c \frac{d^4 \bar{u}_3}{dx_1^4} = 0, \quad H_{23}^c \frac{d^4 \bar{u}_2}{dx_1^4} + H_{22}^c \frac{d^4 \bar{u}_3}{dx_1^4} = p_0.$$

The boundary conditions at the root are purely geometric and are given by eqs. (6.24), which specify that the axial displacement and the transverse displacements and slopes must all vanish.

The boundary conditions at the tip are a bit more complicated. Since no axial force is applied, the axial boundary condition at the tip requires $N_1 = 0$ or $N_1 = S d\bar{u}_1/dx_1 = 0$, which implies $d\bar{u}_1/dx_1 = 0$ at $x_1 = L$. Similarly, at the tip of the beam, the two bending moments must vanish, $M_2(L) = M_3(L) = 0$, and hence

$$\left[H_{23}^c \frac{d^2 \bar{u}_2}{dx_1^2} + H_{22}^c \frac{d^2 \bar{u}_3}{dx_1^2} \right]_{x_1=L} = \left[H_{33}^c \frac{d^2 \bar{u}_2}{dx_1^2} + H_{23}^c \frac{d^2 \bar{u}_3}{dx_1^2} \right]_{x_1=L} = 0.$$

Finally, the shear forces must also vanish at the beam's tip, $V_2(L) = V_3(L) = 0$, leading to

$$\left[-H_{33}^c \frac{d^3 \bar{u}_2}{dx_1^3} - H_{23}^c \frac{d^3 \bar{u}_3}{dx_1^3} \right]_{x_1=L} = \left[H_{23}^c \frac{d^3 \bar{u}_2}{dx_1^3} + H_{22}^c \frac{d^3 \bar{u}_3}{dx_1^3} \right]_{x_1=L} = 0.$$

The axial equation is decoupled from the two bending equations. Its solution for homogeneous boundary conditions is the trivial solution, $\bar{u}_1 = 0$, which means that there is no axial displacement of the beam's centroid.

The two bending equations are coupled, but a simple algebraic manipulation yields two uncoupled equations for this problem

$$\frac{d^4 \bar{u}_2}{dx_1^4} = -\frac{H_{23}^c p_0}{H_{22}^c H_{33}^c - H_{23}^c H_{23}^c} = -\frac{9p_0}{7a^3 tE},$$

$$\frac{d^4 \bar{u}_3}{dx_1^4} = \frac{H_{33}^c p_0}{H_{22}^c H_{33}^c - H_{23}^c H_{23}^c} = \frac{6p_0}{7a^3 tE}.$$

The boundary conditions can be decoupled in a similar manner to yield $\bar{u}_2 = d\bar{u}_2/dx_1 = 0$ and $\bar{u}_3 = d\bar{u}_3/dx_1 = 0$, at $x_1 = 0$ and $d\bar{u}_2^2/dx_1^2 = d\bar{u}_2^3/dx_1^3 = 0$ and $d\bar{u}_3^2/dx_1^2 = d\bar{u}_3^3/dx_1^3 = 0$ at $x_1 = L$. Solving these two decoupled, fourth order differential equations gives the solution of the problem

$$\bar{u}_2(x_1) = -\frac{3p_0 L^4}{56a^3 tE} \left(\eta^4 - 4\eta^3 + 6\eta^2 \right), \tag{6.57}$$

$$\bar{u}_3(x_1) = \frac{p_0 L^4}{28a^3 tE} \left(\eta^4 - 4\eta^3 + 6\eta^2 \right), \qquad (6.58)$$

where $\eta = x_1/L$ is the non-dimensional variable along the beam's span. The displacements at the tip of the beam are $\bar{u}_2^{\text{tip}} = -9/56 \, p_0 L^4/(a^3 tE)$ and $\bar{u}_3^{\text{tip}} = 6/56 \, p_0 L^4/(a^3 tE)$.

Bending stress calculation

The axial stress due to bending, σ_1, can be computed from eqs. (6.14) or (6.15), but eq. (6.14) is preferable because the coordinates of a point on the section, (x_2, x_3), explicitly appear in this equation. For this problem, the stress resultants are obtained from equilibrium considerations as $M_2 = p_0 L^2 (1 - \eta)^2/2$, $N_1 = 0$ and $M_3 = 0$, and hence,

$$\begin{aligned}
\sigma_1(\eta, x_2, x_3) &= \frac{E}{H_{22}^c H_{33}^c - H_{23}^c H_{23}^c} \left[-x_2 H_{23}^c M_2(\eta) + x_3 H_{33}^c M_2(\eta) \right] \\
&= \frac{9E}{7(a^3 tE)^2} \left[-x_2(a^3 tE) + x_3 \left(\frac{2}{3} a^3 tE \right) \right] M_2(\eta) \\
&= \frac{3}{7a^3 t} (-3x_2 + 2x_3) \frac{p_0 L^2}{2} (1 - \eta)^2.
\end{aligned}$$

A number of conclusions can be drawn from this result.

1. Axial stresses vary along the span of the beam because they depend on η. Stresses are maximum where $M_2(\eta)$ is a maximum, *i.e.*, at the root of the beam.
2. For this loading case, it is possible to define the neutral axis of the section. Setting $\sigma_1 = 0$, yields the equation for the neutral axis: $-3x_2 + 2x_3 = 0$. The neutral axis is a line in the plane of the cross-section that makes a $56°$ angle with axis $\bar{\imath}_2$. Axial stresses are positive on one side of this axis and negative on the other.
3. Axial stresses vary over the cross-section of the beam, *i.e.*, they depend on x_2 and x_3. At any given span-wise location, extremum axial stresses are found in non-dimensional form at points **A** $(x_2 = a, x_3 = a)$ or **B** $(x_2 = 0, x_3 = a)$, see fig. 6.18:

$$\frac{\sigma_1^{(A)} a^2 t}{M_2(x_1)} = -\frac{3}{7}, \quad \frac{\sigma_1^{(B)} a^2 t}{M_2(x_1)} = \frac{6}{7}.$$

The maximum magnitude is found at point **B**.

Example 6.7. Bending of a Z section - approach 2
In this example, the same problem treated in the previous example will be solved in the principal axes.

Displacement calculation

The first step of this second approach is once again the computation of the location of the centroid using eq. (6.11); it is located on the web, midway between the flanges. The next step is computation of the orientation of the principal centroidal axes of bending. Equation (6.36) yields

$$\sin 2\alpha^* = \frac{a^3 t E}{a^3 t E \sqrt{2}} = \frac{1}{\sqrt{2}}, \quad \cos 2\alpha^* = -\frac{a^3 t E}{a^3 t E \sqrt{2}} = -\frac{1}{\sqrt{2}}.$$

Thus, the principal axis of bending $\bar{\imath}_2^*$ is oriented at an angle $\alpha^* = 67.5°$ with respect to axis $\bar{\imath}_2$, as shown in fig. 6.19. The principal centroidal bending stiffnesses are found from eq. (6.38)

$$H_{22}^{c*} = \frac{5a^3 t E}{3} - a^3 t E \sqrt{2} = \left(\frac{5}{3} - \sqrt{2}\right) a^3 t E, \quad H_{33}^{c*} = \left(\frac{5}{3} + \sqrt{2}\right) a^3 t E.$$

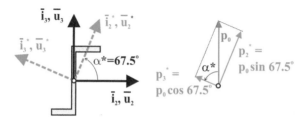

Fig. 6.19. The principal axes of bending for the thin walled section.

The applied load is now projected along the directions of the principal axes of bending to find

$$p_2^* = p_0 \sin 67.5°, \quad p_3^* = p_0 \cos 67.5°.$$

The last step consists of the solution of three independent problems. As in the first approach, the extensional problem yields $\bar{u}_1^* = 0$. The two decoupled bending problems are

$$H_{33}^{c*} \frac{d^4 \bar{u}_2^*}{dx_1^{*4}} = p_0 \sin 67.5°, \quad H_{22}^{c*} \frac{d^4 \bar{u}_3^*}{dx_1^{*4}} = p_0 \cos 67.5°.$$

subjected to the following boundary conditions at the root $\bar{u}_2^* = d\bar{u}_2^*/dx_1^* = 0$, $\bar{u}_3^* = d\bar{u}_3^*/dx_1^* = 0$ and at the tip $d^2\bar{u}_2^*/dx_1^{*2} = d^3\bar{u}_2^*/dx_1^{*3} = 0$, $d^2\bar{u}_3^*/dx_1^{*2} = d^3\bar{u}_3^*/dx_1^{*3} = 0$. The solution of these two decoupled equations is:

$$\bar{u}_2^*(\eta) = \frac{p_0 \sin 67.5°}{H_{33}^{c*}} \frac{L^4}{24} \left(\eta^4 - 4\eta^3 + 6\eta^2\right), \tag{6.59}$$

$$\bar{u}_3^*(\eta) = \frac{p_0 \cos 67.5°}{H_{22}^{c*}} \frac{L^4}{24} \left(\eta^4 - 4\eta^3 + 6\eta^2\right). \tag{6.60}$$

The corresponding non-dimensional deflections at the tip of the beam become

$$\frac{\bar{u}_{2tip}^* a^3 tE}{p_0 L^4} = \frac{\sin 67.5°}{8(5/3 + \sqrt{2})} = 0.0375, \quad \frac{\bar{u}_{3tip}^* a^3 tE}{p_0 L^4} = \frac{\cos 67.5°}{8(5/3 - \sqrt{2})} = 0.1895.$$

To compare the results obtained with approaches 1 and 2, their respective predictions must be expressed in the same axis system. Displacement components $\bar{u}_2$ and $\bar{u}_3$ obtained with approach 1 and given by eqs. (6.57) and (6.58), respectively, are the displacement components of along axes $\bar{\imath}_2$ and $\bar{\imath}_3$, respectively, whereas the displacements components $\bar{u}_2^*$ and $\bar{u}_3^*$ obtained with approach 2 and given by eqs. (6.59) and (6.60), respectively, are the displacement components along the principal centroidal axes of bending $\bar{\imath}_2^*$ and $\bar{\imath}_3^*$, respectively. These two results describe identical displacements of the beam. Indeed, fig. 6.19 shows that the two sets of displacement are related through the following transformations: $\bar{u}_2^* = \bar{u}_2 \cos 67.5° + \bar{u}_3 \sin 67.5°$ and $\bar{u}_3^* = -\bar{u}_2 \sin 67.5° + \bar{u}_3 \cos 67.5°$. Using these equations to compute the non-dimensional tip displacements yields

$$\frac{\bar{u}_{2tip}^* a^3 tE}{p_0 L^4} = \left(-\frac{9}{56}\cos 67.5° + \frac{3}{28}\sin 67.5°\right) = 0.0375,$$

$$\frac{\bar{u}_{3tip}^* a^3 tE}{p_0 L^4} = \left(\frac{9}{56}\sin 67.5° + \frac{3}{28}\cos 67.5°\right) = 0.1895,$$

which agree exactly with the above results.

Bending stress calculation

The axial stress due to bending, σ_1^*, is now computed using eq. (6.29) where $N_1^* = 0$. The bending moment components, M_2^* and M_3^*, about axes $\bar{\imath}_2^*$ an $\bar{\imath}_2^*$, respectively, are related to their counterparts, M_2 and M_3, about axes $\bar{\imath}_2$ and $\bar{\imath}_3$, respectively, as $M_2^*(\eta) = M_2(\eta)\cos 67.5°$ and $M_3^*(\eta) = -M_2(\eta)\sin 67.5°$. The axial stress distribution becomes

$$\frac{\sigma_1^* a t^3}{M_2} = x_3^* \frac{\cos 67.5°}{(5/3 - \sqrt{2})} + x_2^* \frac{\sin 67.5°}{(5/3 + \sqrt{2})}.$$

To reconcile these results with those obtained with approach 1, it is necessary to perform a coordinate transformation between the coordinate x_2^* and x_3^* of a point on the cross-section expressed in the principal centroidal axes of bending, $\mathcal{I}^* = (\bar{\imath}_2^*, \bar{\imath}_3^*)$, to the counterparts in coordinate system $\mathcal{I} = (\bar{\imath}_2, \bar{\imath}_3)$: $x_2^* = x_2 \cos 67.5° + x_3 \sin 67.5°$ and $x_3^* = -x_2 \sin 67.5° + x_3 \cos 67.5°$.

For instance, at point A, $x_2 = a$, $x_3 = a$, and the axial stress becomes

$$\frac{\sigma_1^{*A} a^2 t}{M_2(\eta)} = \left[\frac{(\cos 67.5° + \sin 67.5°)\sin 67.5°}{5/3 + \sqrt{2}}\right.$$
$$\left. +\frac{(-\sin 67.5° + \cos 67.5°)\cos 67.5°}{5/3 - \sqrt{2}}\right] = -0.43.$$

Similarly, at point **B**, $x_2 = 0, x_3 = a$, and the axial stress follows as

$$\frac{\sigma_1^{*B} a^2 t}{M_2} = \frac{\sin^2 67.5°}{5/3 + \sqrt{2}} + \frac{\cos^2 67.5°}{5/3 - \sqrt{2}} = 0.86.$$

As expected, these results are identical to those obtained using approach 1.

6.9.1 Discussion of the results

Although the applied load acts in the $\bar{\imath}_3$ direction only, the beam displaces along both axes $\bar{\imath}_3$ and $\bar{\imath}_2$. In fact, the tip displacement component along axis $\bar{\imath}_2$ is *larger* than that along axis $\bar{\imath}_3$. This is due to the fact that bending in planes $(\bar{\imath}_1, \bar{\imath}_2)$ and $(\bar{\imath}_1, \bar{\imath}_3)$ is coupled, as expressed by the coupled governing equations (6.23a), (6.23b), and (6.23c).

This behavior is more easily understood when considering the results of the second approach expressed in the principal centroidal axes of bending. Indeed, the bending behavior of the beam along the principal axes of bending is *decoupled*. This means that load p_2^*, applied along axis $\bar{\imath}_2^*$, produces a displacement *along axis $\bar{\imath}_2^*$ only*. Similarly, load p_3^*, applied along axis $\bar{\imath}_3^*$, produces a displacement *along axis $\bar{\imath}_3^*$ only*. The displacement along axis $\bar{\imath}_2^*$ is fairly small because the bending stiffness, H_{33}^{c*}, that characterizes bending about axis $\bar{\imath}_3^*$ is maximum. On the other hand, the displacement along axis $\bar{\imath}_3^*$ is large because the bending stiffness, H_{22}^{c*}, that characterizes bending about axis $\bar{\imath}_2^*$ is minimum. The resulting displacement, $\bar{u}_3^*$, when resolved along the axes $\bar{\imath}_2$ and $\bar{\imath}_3$, has the expected upward component, together with a *leftward* component. This explains the negative sign of the $\bar{u}_2$ in eq. (6.57).

6.10 Problems

Problem 6.7. Sectional bending stiffness
Consider the solid cross-section depicted in fig. 6.20. *(1)* Determine the location of the centroidal of the section. *(2)* Compute the sectional centroidal bending stiffnesses. *(3)* Determine the orientation of the principal centroidal axes of bending. *(4)* Compute the principal centroidal bending stiffnesses.

Problem 6.8. Beam with and "L" shaped cross-section
The "L" shaped cross-section beam shown in fig. 6.21 is subjected to a bending moment of magnitude M_b, which is acting in the direction indicated in the figure. Create and use a spreadsheet to accomplish the following tasks. Make your spreadsheet general so that different dimensions can be entered into the spreadsheet along with different values for the load and its orientation. The spreadsheet outputs should be in clearly labeled cells. *(1)* Determine the centroid location. *(2)* Determine the axial and bending stiffnesses in the centroidal axis system $\mathcal{I}^c = (\bar{\imath}_2^c, \bar{\imath}_3^c)$ indicated on the figure. *(3)* Using this axis system, compute the orientation of the neutral axis and compute magnitude and location of the maximum axial stress, $|\sigma_1^{\max}|$. *(4)* Find the orientation of the principal centroidal axes of bending, $\mathcal{I}^{c*} = (\bar{\imath}_2^{c*}, \bar{\imath}_3^{c*})$. *(5)* Using these axes, determine the magnitude and location of the maximum axial stress. Verify that this is the same result as in step *3*. Use the following data: $h = 150$ mm, $b = 100$ mm, $t_h = 10$ mm, $t_b = 14$ mm, $\theta = 30$ degrees, and $M_b = 10$ kN·m.

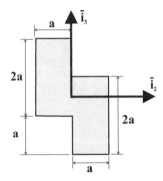

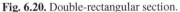

Fig. 6.20. Double-rectangular section.

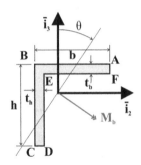

Fig. 6.21. "L" shaped cross-section.

Problem 6.9. Beam with "C" shaped cross-section

The "C" shaped cross-section beam shown in fig. 6.22 is subjected to a bending moment of magnitude M_b, which is acting in the direction indicated in the figure. Create and use a spreadsheet to accomplish the following tasks. Make your spreadsheet general so that different dimensions can be entered into the spreadsheet along with different values for the load and its orientation. The spreadsheet outputs should be in clearly labeled cells. *(1)* Determine the centroid location. *(2)* Determine the axial and bending stiffnesses in the centroidal axis system $\mathcal{I}^c = (\bar{i}_2^c, \bar{i}_3^c)$ indicated on the figure. *(3)* Using this axis system, compute the orientation of the neutral axis and compute magnitude and location of the maximum axial stress, $|\sigma_1^{\max}|$. *(4)* Find the orientation of the principal centroidal axes of bending, $\mathcal{I}^{c*} = (\bar{i}_2^{c*}, \bar{i}_3^{c*})$. *(5)* Using these axes, determine the magnitude and location of the maximum axial stress. Verify that this is the same result as in step 3. Use the following data: $t_a = 15$ mm, $t_b = 30$ mm, $b_a = 30$ mm, $b_b = 40$ mm, $t_w = 20$ mm, and $h = 100$ mm, $\theta = -45$ degrees, and $M_b = 20$ kN·m.

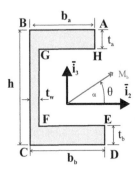

Fig. 6.22. "C" shaped cross-section.

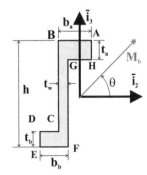

Fig. 6.23. "Z" shaped cross-section.

Problem 6.10. Beam with "Z" section

The "Z" shaped cross-section beam shown in fig. 6.23 is subjected to a bending moment of magnitude M_b, which is acting in the direction indicated in the figure. Create and use a spreadsheet to accomplish the following tasks. Make your spreadsheet general so that different dimensions can be entered into the spreadsheet along with different values for the load and its orientation. The spreadsheet outputs should be in clearly labeled cells. *(1)* Determine the centroid location. *(2)* Determine the axial and bending stiffnesses in the centroidal axis system $\mathcal{I}^c = (\bar{\imath}_2^c, \bar{\imath}_3^c)$ indicated on the figure. *(3)* Using this axis system, compute the orientation of the neutral axis and compute magnitude and location of the maximum axial stress, $|\sigma_1^{\max}|$. *(4)* Find the orientation of the principal centroidal axes of bending, $\mathcal{I}^{c*} = (\bar{\imath}_2^{c*}, \bar{\imath}_3^{c*})$. *(5)* Using these axes, determine the magnitude and location of the maximum axial stress. Verify that this is the same result as in step *3*. Use the following data: $h = 95$ mm, $b_a = 30$ mm, $b_b = 50$ mm, $t_w = 20$ mm, $t_a = 15$ mm, and $t_b = 30$ mm, $\theta = -45$ degrees, and $M_b = 10$ kN·m.

Problem 6.11. Thin-walled "L" section
Consider the thin-walled, "L" shaped cross-section of a beam as shown in fig. 6.15. Let $b = 0.25$ m , $h = 0.1$ m, and $t = 2.5$ mm. *(1)* Find the location of the centroid of the section. *(2)* Find the orientation of the principal centroidal axes of bending.

Problem 6.12. Beam with "T" shaped cross-section
The "T" shaped cross-section beam shown in fig. 6.24 is subjected to a bending moment of magnitude M_b, which is acting in the direction indicated in the figure. Create and use a spreadsheet to accomplish the following tasks. Make your spreadsheet general so that different dimensions can be entered into the spreadsheet along with different values for the load and its orientation. The spreadsheet outputs should be in clearly labeled cells. *(1)* Determine the centroid location. *(2)* Determine the axial and bending stiffnesses in the centroidal axis system $\mathcal{I}^c = (\bar{\imath}_2^c, \bar{\imath}_3^c)$ indicated on the figure. *(3)* Using this axis system, compute the orientation of the neutral axis and compute magnitude and location of the maximum axial stress, $|\sigma_1^{\max}|$. *(4)* Find the orientation of the principal centroidal axes of bending, $\mathcal{I}^{c*} = (\bar{\imath}_2^{c*}, \bar{\imath}_3^{c*})$. *(5)* Using these axes, determine the magnitude and location of the maximum axial stress. Verify that this is the same result as in step *3*. Use the following data: $h = 140$ mm, $b = 120$ mm, $t_h = 12$ mm, $t_b = 10$ mm, and $a = 15$ mm, $\theta = -45$ degrees, and $M_b = 10$ kN·m.

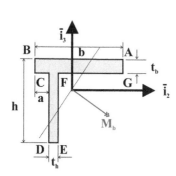

Fig. 6.24. "T" shaped cross-section.

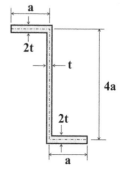

Fig. 6.25. Reversed "Z" shaped cross-section.

Problem 6.13. Thin-walled "Z" section
A beam is made of a homogeneous material of Young's modulus, E, and has the unsymmetric, thin-walled cross-section shown in fig 6.25. *(1)* Compute the centroidal stiffnesses in the coordinate system indicated on the figure. *(2)* Compute the orientation of the neutral axis for

the loading case where $M_2 \neq 0$, $M_3 = 0$. *(3)* Using the orientation of the neutral axis, determine the points on the section where the bending stress will have the maximum positive and negative values.

Problem 6.14. Thin-walled inverted "V" section

A thin-walled beam of length L and with cross-section shown in fig. 6.26 is simply supported at both ends and carries a distributed loading, p_0, acting upwards. *(1)* Find the maximum direct stress due to bending and where it acts. *(2)* Sketch the distribution of axial stress on the cross-section of the beam where the maximum bending stress acts.

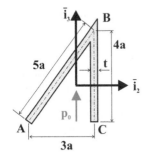

Fig. 6.26. Inverted "V" shaped cross-section.

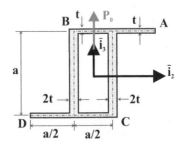

Fig. 6.27. Box Z shaped cross-section.

Problem 6.15. Thin-walled cantilever beam with Box-Z section

A thin-walled cantilevered beam of length L and elastic modulus E is constructed with a cross-section shown in fig. 6.27. A vertical load P is applied at the tip of the beam. *(1)* Determine the axial stress acting at the root of the cantilever at point **A** and **B**. *(2)* Determine the deflection of the tip using the given centroidal axes. *(3)* Determine the tip deflection using the principal axes of bending, and show that they are equivalent to the results obtained in *(2)*. Hint: this is a numerically tedious problem, and use of a spreadsheet or computer program can be very effective.

Problem 6.16. Cantilevered beam with a "T" shaped cross-section

Consider the cantilevered beam of length L with a thin-walled "T" cross-section as depicted in fig. 6.28. A tip axial load P acts at the left edge of the top flange. A transverse tip load R acts in the plane of the tip cross-section in the direction indicated on the figure. *(1)* Find the principal centroidal axes of bending, $\bar{\imath}_1^*$, $\bar{\imath}_2^*$ and $\bar{\imath}_3^*$, of the cross-section. *(2)* Write the three uncoupled equations governing this problem and the corresponding boundary conditions. *(3)* Compute all the stiffness constants appearing in the equations, but do not solve the problem.

Problem 6.17. Cantilevered beam with "Z" shaped cross-section

Figure 6.29 depicts a cantilevered beam with a thin-walled "Z" shaped cross-section subjected to an axial load P applied at point **A** located at the lower left corner of the cross-section. *(1)* Determine the location of centroid of the section and locate the axis system at this point (with axis $\bar{\imath}_3$ parallel to the web). *(2)* Determine the bending stiffnesses H_{22}^c, H_{33}^c, and H_{23}^c for the centroidal axis system. *(3)* Determine the orientation of the principal axes of bending, $\bar{\imath}_2^*$ and $\bar{\imath}_3^*$, and the principal centroidal bending stiffnesses H_{22}^{c*}, H_{33}^{c*}. *(4)* Solve this problem in the centroidal coordinate system to determine the lateral displacements of the cross-section,

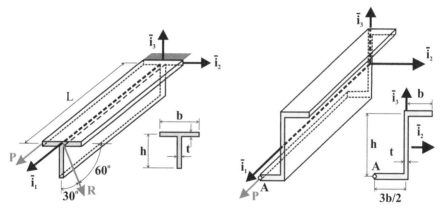

Fig. 6.28. Cantilevered beam with "T" shaped section under tip axial loads.

Fig. 6.29. Cantilevered beam with "Z" shaped section under tip axial load.

$\bar{u}_2(x_1)$ and $\bar{u}_3(x_1)$. *(5)* Solve this problem in the coordinate system defined by the principal axes of bending to determine the lateral displacements, $\bar{u}_2^*(x_1)$ and $\bar{u}_3^*(x_1)$. *(6)* Show that the two above solutions are identical. *(7)* Find the two components of displacement at the point of application of the load P which can be in non-dimensional terms as $Eb\bar{u}_2/P$ and $Eb\bar{u}_3/P$, respectively. *(8)* Find the axial stress distribution at the root of the beam. Plot this distribution along the web and flanges. Where does the maximum axial stress occur? Express this as a non-dimensional stress $b^2\sigma_1/P$. Use the following data: $L = 10b$, $h = 2b$ and $t = b/10$.

Problem 6.18. Cantilevered beam with a "U" shaped cross-section

Consider the cantilevered beam of length L with a thin-walled "U" shaped cross-section as depicted in fig. 6.30. A tip axial load, P, acts at the lower right corner of the section. Two transverse tip loads, both of magnitude R, act down in the plane of the tip cross-section. *(1)* Find the principal centroidal axes of bending, $\bar{\imath}_1^*$, $\bar{\imath}_2^*$ and $\bar{\imath}_3^*$, of the cross-section. *(2)* Write the three uncoupled equations governing this problem and the corresponding boundary conditions. *(3)* Compute all the stiffness constants appearing in the equations, but do not solve the problem.

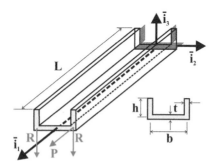

Fig. 6.30. Cantilevered beam with "U" shaped section under tip axial loads.

7

Torsion

In the previous chapters, the behavior of beams subjected to axial and transverse loads is studied in detail. In chapter 6, a fairly general, three dimensional loading is considered, with one important restriction: the beam is assumed to *bend without twisting*. Twisting, however, is often present in structures, and in fact, many important structural components are designed to carry torsional loads primarily.

Power transmission drive shafts are a prime example of structural components designed to carry a specific torque. Such components are designed with solid or thin-walled circular cross-sections. Numerous other structural components are designed to carry a combination of axial, bending, and torsional loads. For instance, an aircraft wing must carry the bending and torsional moments generated by the aerodynamic forces.

The behavior of structural components under torsional loads is the subject of this chapter. The focus is on long prismatic structures similar to the beams treated in the two previous chapters. When a long prismatic structure is subjected to torsion, it is often referred to as a "bar" rather than a "beam," but the two terms are often used interchangeably.

7.1 Torsion of circular cylinders

Consider an infinitely long, homogeneous, solid or hollow circular cylinder subjected to end torques, Q_1, of equal magnitude and opposite directions, as depicted in fig. 7.1. The cross-section of the cylinder can be a circle of radius R, or a circular annulus of inner and outer radii, R_i and R_o, respectively.

This problem is characterized by two types of symmetries. First, a cylindrical symmetry about axis $\bar{\imath}_1$: any rotation of the cylinder or tube about axis $\bar{\imath}_1$ leaves both the structure and the loading unchanged, and hence, the solution must remain unchanged. Second, as illustrated in fig. 7.2, the cylindrical structure is symmetric with respect to any plane, $\mathcal{P}$, passing through axis $\bar{\imath}_1$. Depicted on this figure are two points, **A** and **B**, both on a circle, $\mathcal{C}$, of radius $r < R$. The plane of symmetry, $\mathcal{P}$, is selected to be normal to the line segment joining these two points. Along circle $\mathcal{C}$, shear stresses will develop stemming from the application of torque Q_1. Because of the circular symmetry of the system, this shear stress must be of constant magnitude along circle $\mathcal{C}$, and tangent to it at all points. While the structure is *symmetric* with respect to plane $\mathcal{P}$, the loading is *antisymmetric* with respect to the same plane. Consequently, the solution must be antisymmetric with respect to plane $\mathcal{P}$.

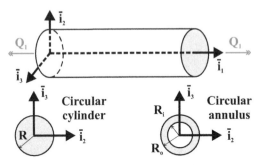

Fig. 7.1. Circular cylinder subjected to end torques.

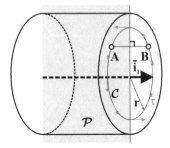

Fig. 7.2. A plane of symmetry, $\mathcal{P}$, of the circular cylinder.

First, consider the axial displacement components at points **A** and **B**, denoted u_1^A and u_1^B, respectively. The cylindrical symmetry of the problem implies that $u_1^A = u_1^B$. On the other hand, the antisymmetry of the problem with respect to plane $\mathcal{P}$ implies $u_1^A = -u_1^B$. The only solution consistent with these two requirements is $u_1^A = u_1^B = 0$. Because points **A** and **B** are arbitrary points on the cross-section, the axial displacement must vanish at all points of the cross-section: *the cross-section does not warp out-of-plane.*

Next, consider the in-plane displacements of the same two points. The only displacement field that is compatible with the cylindrical symmetry of the problem is a rigid body rotation of the cross-section about its own center. It is easy to show that this rigid body rotation also presents the required antisymmetry about any plane passing through axis $\bar{\imath}_1$.

In summary, for a circular cylinder or annulus, each cross-section *rotates about its own center like a rigid disk.* This is the only deformation compatible with the symmetries of the problem.

7.1.1 Kinematic description

Since the only deformation induced by torsion in a circular cylinder or annulus consists of rigid body rotation of each cross-section, its motion is fully described by a rotation angle, Φ_1, as shown in fig. 7.3. This rotation brings an arbitrary point **A** of the reference configuration to point **A′** in the deformed configuration. Figure 7.3 also shows polar coordinates r and α that define the position of point **A**. As usual, displacement, and rotations are assumed to remain small, and hence, the distance from

A to $\mathbf{A}'$ can be approximated as $r\,d\Phi_1$, as shown in the figure. The sectional in-plane displacement field can then be written as the projection of this displacement vector along directions $\bar{\imath}_2$ and $\bar{\imath}_3$, respectively, to find

$$u_2(x_1, r, \alpha) = -r\Phi_1(x_1)\sin\alpha, \quad u_3(x_1, r, \alpha) = r\Phi_1(x_1)\cos\alpha. \tag{7.1}$$

Because the cross-section does not deform out of its own plane, the axial displacement field must vanish, *i.e.*, $u_1(x_1, x_2, x_3) = 0$.

The out-of-plane displacement field describing the torsional deformation of the circular cylinder becomes

$$u_1(x_1, x_2, x_3) = 0, \tag{7.2}$$

whereas the in-plane displacement field given by eq. (7.1) becomes

$$
\begin{aligned}
u_2(x_1, x_2, x_3) &= -x_3\Phi_1(x_1), \\
u_3(x_1, x_2, x_3) &= x_2\Phi_1(x_1),
\end{aligned}
\tag{7.3}
$$

where the following transformation from polar to Cartesian coordinates is used: $x_2 = r\cos\alpha$ and $x_3 = r\sin\alpha$.

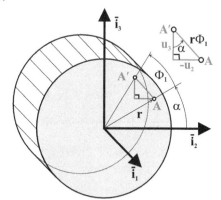

Fig. 7.3. In-plane displacements for a circular cylinder. The cross-section undergoes a rigid body rotation.

Using the strain-displacement relationships, the corresponding strain field is now obtained as

$$\epsilon_1 = \frac{\partial u_1}{\partial x_1} = 0, \tag{7.4}$$

$$\epsilon_2 = \frac{\partial u_2}{\partial x_2} = 0, \quad \epsilon_3 = \frac{\partial u_3}{\partial x_3} = 0, \quad \gamma_{23} = \frac{\partial u_2}{\partial x_3} + \frac{\partial u_3}{\partial x_2} = 0, \tag{7.5}$$

$$\gamma_{12} = \frac{\partial u_1}{\partial x_2} + \frac{\partial u_2}{\partial x_1} = -x_3\,\kappa_1(x_1), \quad \gamma_{13} = \frac{\partial u_1}{\partial x_3} + \frac{\partial u_3}{\partial x_1} = x_2\,\kappa_1(x_1), \tag{7.6}$$

where the *sectional twist rate* is defined as

$$\kappa_1(x_1) = \frac{d\Phi_1}{dx_1}. \tag{7.7}$$

The sectional twist rate, κ_1, measures the deformation of the circular cylinder. Note that a constant twist angle implies a rigid body rotation of the cylinder about its axis, but no deformation.

The axial strain field, eq. (7.4), vanishes because the section does not warp out-of-plane, and the in-plane strain field, eq. (7.5), vanishes because the in-plane motion of the section is a rigid body rotation. Under torsion, the only non-vanishing strain components are the out-of-plane shearing strains given by eq. (7.6).

This strain field is not easily visualized in rectangular coordinates because the Cartesian strain components, γ_{12} and γ_{13}, act in planes $(\bar{\imath}_1, \bar{\imath}_2)$ and $(\bar{\imath}_1, \bar{\imath}_3)$, respectively. In view of the cylindrical symmetry of the problem at hand, it is more natural to describe this strain field in the polar coordinate system, (r, α), shown in fig. 7.3. In this axis system, the corresponding strain components are γ_{r1} and $\gamma_{\alpha1}$, where the second index refers to axis $\bar{\imath}_1$. For simplicity, however, these strain components will be simply denoted γ_α and γ_r.

The relationship between the Cartesian and polar strain components can be expressed using eq. (1.81) for a rotation, α, about axis $\bar{\imath}_1$, so that $\bar{\imath}_1^* = \bar{\imath}_1$, $\bar{\imath}_2^* = \bar{\imath}_r$, and $\bar{\imath}_3^* = \bar{\imath}_\alpha$. In this case, $\ell_1 = 1$, $\ell_2 = \ell_3 = 0$ and $m_1 = 0$, $m_2 = \cos\alpha$, $m_3 = \sin\alpha$ and $n_1 = 0$, $n_2 = -\sin\alpha$, $n_3 = \cos\alpha$. Using these direction cosines, eq. (1.81) then yields

$$\gamma_r = \gamma_{12}\cos\alpha + \gamma_{13}\sin\alpha, \quad \gamma_\alpha = -\gamma_{12}\sin\alpha + \gamma_{13}\cos\alpha. \tag{7.8}$$

Introducing the Cartesian shear strain components, eqs. (7.6), leads to

$$\gamma_r(x_1, r, \alpha) = 0, \quad \gamma_\alpha(x_1, r, \alpha) = r\,\kappa_1(x_1). \tag{7.9}$$

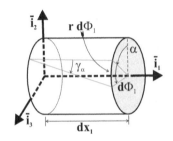

The only non-vanishing strain component is the circumferential shearing strain component, γ_α, which is proportional to the twist rate, κ_1, and varies linearly from zero at the center of the section to its maximum value, $R\kappa_1$, along the outer edge of the cylinder. It is of course independent of circumferential variable α, as required by the cylindrical symmetry of the problem.

Fig. 7.4. Visualization of out-of-plane shear strain in polar coordinates.

This strain component is depicted in fig. 7.4. Each circular cross-section retains its circular shape and experiences no in-plane or out-of-plane deformation: two adjacent sections experience a small differential rotation, $\mathrm{d}\Phi_1$, which gives rise to the circumferential shearing strain γ_α. As illustrated in fig. 7.4, the shearing strain is readily obtained as $\gamma_\alpha = r\,\mathrm{d}\Phi_1/\mathrm{d}x_1 = r\kappa_1$, in agreement with eq. (7.9).

7.1.2 The stress field

Let the cylinder be made of a linearly elastic material that obeys Hooke's law, eq. (2.9). In view of the strain field, eq. (7.6), the only non-vanishing stress components are

$$\tau_{12} = -Gx_3\,\kappa_1(x_1), \quad \tau_{13} = Gx_2\,\kappa_1(x_1), \tag{7.10}$$

where G is the shear modulus of the material. Once again, polar coordinates are more convenient to use in visualizing the stress field, which is obtained from eq. (7.9) and Hooke's law as

$$\tau_r(x_1, r, \alpha) = 0, \quad \tau_\alpha(x_1, r, \alpha) = Gr\, \kappa_1(x_1), \quad (7.11)$$

where τ_r and τ_α are the radial and circumferential shear stress components, respectively.

The distribution of the circumferential shear stress over the cross-section is shown in fig. 7.5. Two characteristics of this distribution should be noted. First, at all points, the shear stress acts in the circumferential direction, and the component in the radial direction vanishes. Second, the magnitude of the stress varies linearly along the radial direction: it is zero at the center and maximum at the largest radius. This implies that the central region of the bar does not experience very high stress values and is not very effective in resisting torsion. The peak stresses is reached at the outer radius of the bar.

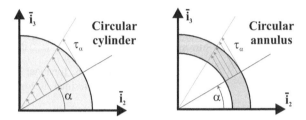

Fig. 7.5. Distribution of circumferential shearing stress over the cross-section.

7.1.3 Sectional constitutive law

The torque acting on the cross-section at a given span-wise location is readily obtained by integrating the circumferential shear stress, τ_α, multiplied by the moment arm, r, over the circular cross-section to find

$$M_1(x_1) = \int_{\mathcal{A}} \tau_\alpha r \, \mathrm{d}\mathcal{A}. \quad (7.12)$$

Introducing the circumferential shear stress, eq. (7.11) then yields

$$M_1(x_1) = \int_{\mathcal{A}} Gr^2 \kappa_1(x_1) \, \mathrm{d}\mathcal{A} = \left[\int_{\mathcal{A}} Gr^2 \, \mathrm{d}\mathcal{A}\right] \kappa_1(x_1) = H_{11}\, \kappa_1(x_1), \quad (7.13)$$

where the *torsional stiffness* of the section is defined as

$$H_{11} = \int_{\mathcal{A}} Gr^2 \, \mathrm{d}\mathcal{A}. \quad (7.14)$$

Relationship (7.13) is the constitutive law for the torsional behavior of the beam. It expresses the proportionality between the torque and the twist rate, with a constant of proportionality, H_{11}, called the torsional stiffness. Formula (7.14) is true *for circular cross-sections only*.

If the section is made of a homogeneous material of shear modulus G, the torsional stiffness then becomes $H_{11} = GJ$, where $J = \int_A r^2 \, \mathrm{d}\mathcal{A}$ is the purely geometric integral known as the area polar moment. The entire theory is developed for bars with circular cross-sections, and therefore this expression for the torsional stiffness is valid *for circular cross-sections only.*

7.1.4 Equilibrium equations

The equations of equilibrium associated with the torsional behavior can be obtained by considering the infinitesimal slice of the cylinder of length $\mathrm{d}x_1$ depicted in fig. 7.6. Using a Taylor series expansion, the moment acting on the right-hand face is $M_1(x_1 + \mathrm{d}x_1) = M_1(x_1) + (\mathrm{d}M_1/\mathrm{d}x_1)\mathrm{d}x_1$, where higher order differential terms have been neglected. Summing all the moments acting about axis $\bar{\imath}_1$ then yields the torsional equilibrium equation

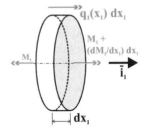

Fig. 7.6. Torsional loads acting on an infinitesimal slice of the bar.

$$\frac{\mathrm{d}M_1}{\mathrm{d}x_1} = -q_1. \tag{7.15}$$

7.1.5 Governing equations

Finally, the governing equation for the torsional behavior of circular cylinders is obtained by introducing the torque, eq. (7.13), into the equilibrium condition, eq. (7.15) and recalling the definition of the twist rate, eq. (7.7), to find

$$\frac{\mathrm{d}}{\mathrm{d}x_1}\left[H_{11}\frac{\mathrm{d}\Phi_1}{\mathrm{d}x_1}\right] = -q_1. \tag{7.16}$$

This second order differential equation can be solved for the twist distribution, $\Phi_1(x_1)$, given the applied torque distribution, $q_1(x_1)$.

Two boundary conditions involving the rotation, Φ_1, or the twist rate, κ_1, are required for the solution of eq. (7.16), one at each end of the cylinder. Typical boundary conditions are as follows.

1. A fixed (or clamped) end allows no rotation, *i.e.*, $\Phi_1 = 0$.
2. A free (unloaded) end corresponds to $M_1 = 0$, which, for eq. (7.13), can be expressed as $\kappa_1 = \mathrm{d}\Phi_1/\mathrm{d}x_1 = 0$.
3. Finally, if the end of the cylinder is subjected to a concentrated torque, Q_1, the boundary condition is $M_1 = Q_1$, which becomes $H_{11} \, \mathrm{d}\Phi_1/\mathrm{d}x_1 = Q_1$.

7.1.6 The torsional stiffness

The torsional stiffness of the section, H_{11}, characterizes the stiffness of the cylinder when subjected to torsion. If the cylinder is made of a homogeneous material, the shear modulus is identical at all points of the cross-section and can be factored out of eq. (7.14), which is then easily evaluated in polar coordinates

$$H_{11} = G \int_0^{2\pi} \int_0^R r^2 \, \mathrm{d}r \mathrm{d}\alpha = \frac{\pi}{2} G R^4. \tag{7.17}$$

For a circular tube the second integral extends from the inner radius, R_i, to the outer radius, R_o, to find

$$H_{11} = G \int_0^{2\pi} \int_{R_i}^{R_o} r^2 \, \mathrm{d}r \mathrm{d}\alpha = \frac{\pi}{2} G (R_o^4 - R_i^4). \tag{7.18}$$

A common situation of great practical importance is that of a thin-walled circular tube. Let the mean radius of the tube be $R_m = (R_o + R_i)/2$, and the wall thickness $t = R_o - R_i$. The thin wall assumption implies $t/R_m \ll 1$. The torsional stiffness of the thin-walled tube then becomes

$$H_{11} = \frac{\pi}{2} G (R_o^2 + R_i^2)(R_o + R_i)(R_o - R_i) \approx 2\pi G R_m^3 t. \tag{7.19}$$

Consider now a thin-walled circular tube consisting of N concentric layers of different materials through the thickness of the wall, as depicted in fig. 7.7. Assuming the material to be homogeneous within each layer, the torsional stiffness becomes

$$H_{11} = \frac{\pi}{2} \sum_{i=1}^N G^{[i]} \left[(R^{[i+1]})^4 - (R^{[i]})^4 \right],$$

where $G^{[i]}$ is the shear modulus in layer i. For a thin-walled tube, each layer will be thin, and the above approximation can be used once again to find

$$H_{11} = 2\pi \sum_{i=1}^N G^{[i]} t^{[i]} \left(\frac{R^{[i+1]} + R^{[i]}}{2} \right)^3. \tag{7.20}$$

The torsional stiffness is the *weighted average* of the shear moduli of the various layers. The weighting factor, $t^{[i]} \left[(R^{[i+1]} + R^{[i]})/2 \right]^3$, strongly biases the average in favor of the outermost layers.

7.1.7 Measuring the torsional stiffness

In the previous section, the torsional stiffness of a circular cylinder is computed from the geometry of the cross-section and the properties of the constituent materials. For example, eq. (7.17) gives the torsional stiffness for a cylinder made of a homogeneous, isotropic material, while eq. (7.20) gives the stiffness for a thin-walled tube

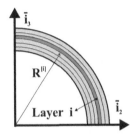

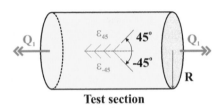

Fig. 7.7. Thin-walled tube made of layered materials.

Fig. 7.8. Configuration of the test to determine the torsional stiffness.

made of composite materials. It is possible to experimentally measure the torsional stiffness of a cylinder using the torsional test set-up depicted in fig. 7.8. The torque, Q_1, is applied to the test sample by a torsional testing machine.

The deformation of the test section can be measured by the chevron strain gauge shown in fig. 1.24. Two strain gauges oriented at ± 45 degree angles with respect to the axis of the cylinder, as shown in fig. 7.8, yield the shear strain at the outer surface of the cylinder. In view of eq. (1.102), $\gamma_{12} = e_{+45} - e_{-45}$, where e_{+45} and e_{-45} are the extensional strain measurements along these two directions. Using eq. (7.9), this shear strain can be related to the twist rate of the cylinder: $\gamma_{12} = \gamma_\alpha = R\kappa_1$, where R is the radius of the cylinder. It then follows that $\kappa_1 = (e_{45} - e_{-45})/R$.

The test procedure is as follows. The circular cylinder is placed in the torsional testing machine and a torque Q_1 of increasing magnitude is applied. For each loading level, the corresponding deformation is measured by the strain gauges. The raw test data consists of loading levels, Q_{1i}, $i = 1, 2, \ldots, n$, where n is the number of data points, and the corresponding strains, ϵ_{45i} and e_{-45i}. From this raw data, the deformation of the cylinder is computed, $\kappa_{1i} = (e_{45i} - e_{-45i})/R$. This computed data is then plotted in the following manner: deformation, κ_{1i}, along the abscissa and torque, Q_{1i}, along the ordinate.

If the applied load remains small, the behavior of the cylinder is expected to be linear as expressed by eq. (7.13), *i.e.*, a linear relationship should be observed between torque and twist rate. Hence, the slope of the experimentally obtained Q_{3i} versus κ_{1i} curve should yield the torsional stiffness of the cylinder. Note that this experimental technique is valid for cylinders made of homogeneous materials, or for complex constructions involving many layers of concentric composite materials, as long as the cylindrical symmetry of the sample is maintained.

7.1.8 The shear stress distribution

The local circumferential shear stress can be related to the sectional torque by eliminating the twist rate between eqs. (7.11) and (7.13) to find

$$\tau_\alpha = G \frac{M_1(x_1)}{H_{11}} r. \tag{7.21}$$

where G is the shear modulus at the location where the stress is computed.

The shear strain defined by eq. (7.9) increases linearly from zero at the center of the circular section to a maximum value at the outer radius. As discussed in section 7.1, this linear distribution of shear strain is a direct consequence of the symmetries of the problem, and is independent of the bar's constituent materials. If the bar is made of a homogeneous material, the linear distribution of shear strains results in a linear distribution of shear stresses, as implied by eq. (7.21) and depicted in fig. 7.5. On the other hand, if the section is made of concentric layers of distinct material as depicted in fig. 7.7, the shear stress in layer i, denoted $\tau_\alpha^{[i]}$, is still given by eq. (7.21) as $\tau_\alpha^{[i]} = G^{[i]}(M_1/H_{11})\, r$. Within each layer, the shear stress distribution is still linear, but discontinuities might appear at the interface between the various layers.

The maximum shear stress in a section of homogeneous material occurs at the largest value of r, *i.e.*, at the outer edge of the cylinder. For a circular cylinder, the torsional stiffness is given eq. (7.17) and the magnitude of maximum shear stress becomes

$$\tau_\alpha^{\max} = \frac{2M_1(x_1)}{\pi R^3}. \tag{7.22}$$

For a circular tube, the torsional stiffness is given eq. (7.18), and the magnitude of maximum shear stress is

$$\tau_\alpha^{\max} = \frac{2R_o M_1(x_1)}{\pi(R_o^4 - R_i^4)}. \tag{7.23}$$

Finally, for a thin-walled circular tube, the shear stress distribution becomes nearly uniform through-the-thickness of the wall,

$$\tau_\alpha^{\max} \approx \frac{M_1(x_1)}{2\pi R_m^2 t}. \tag{7.24}$$

Similarly, the shear stress distribution in a tube made of thin concentric layers of various materials will be nearly uniform within each layer

$$\tau_\alpha^{[i]} \approx G^{[i]} \frac{R^{[i+1]} + R^{[i]}}{2} \frac{M_1(x_1)}{H_{11}}, \tag{7.25}$$

where the torsional stiffness, H_{11}, is given by eq. (7.20).

Once the local shear stress is determined, a strength criterion is applied to determine whether the structure can sustain the applied loads. For a cylindrical bar, combining the strength criterion, eq. (2.28) and the shear stress distribution given by eq. (7.21) yields $GR|M_1(x_1)|/H_{11} \leq \tau_{\text{allow}}$, where τ_{allow} is the allowable shear stress for the material. Since the torque varies along the bar's span, this condition must be checked at all points along the span. In practice, it is convenient to first determine the maximum torque, denoted $M_1^{\max}$, then apply the strength criterion

$$\frac{GR}{H_{11}}|M_1^{\max}| \leq \tau_{\text{allow}}. \tag{7.26}$$

If the section consists of layers made of various materials, the strength of each layer will, in general, be different, and the strength criterion becomes

$$\frac{G^{[i]} R^{[i+1]}}{H_{11}} |M_1^{\max}| \leq \tau_{\text{allow}}^{[i]}, \tag{7.27}$$

where $\tau_{\text{allow}}^{[i]}$ is the allowable shear stresses for layer i. The strength criterion must be checked for each material layer.

7.1.9 Rational design of cylinders under torsion

The shear stress distribution in a cylinder subjected to torsion is shown in fig. 7.5. Clearly, the material near the center of the cylinder is not used efficiently because the shear stress becomes small in the central portion of the cylinder. A far more efficient design is the thin-walled tube. Indeed, the shear stress becomes nearly uniform through-the-thickness of the wall, and all the material is used at full capacity.

For a homogeneous, thin-walled tube, the mass of material per unit span is $\mu = 2\pi R_m t \rho$, where ρ is the material density, R_m the mean radius, and t the thickness. The torsional stiffness, eq. (7.19), now becomes

$$H_{11} = 2\pi G R_m^3 t = \frac{\mu}{\rho} G R_m^2.$$

Consider two thin-walled tubes made of identical materials, with identical masses per unit span, but with mean radii, R_m and R'_m, respectively, and thicknesses t and t', respectively. Because the mass per unit span are equal, the thicknesses of the two tubes will be in inverse proportion of their radii, $t/t' = R'_m/R_m$. The ratio of their torsional stiffnesses, denoted H_{11} and H'_{11}, respectively, is

$$\frac{H_{11}}{H'_{11}} = \frac{(\mu/\rho) G R_m^2}{(\mu/\rho) G R_m'^2} = \left(\frac{R_m}{R'_m}\right)^2. \tag{7.28}$$

For two tubes of equal mass, the torsional stiffness increases with the square of the mean radius.

When subjected to identical torques, the ratio of the shear stresses in the two tubes, denoted τ_α and τ'_α, respectively, becomes

$$\frac{\tau_\alpha}{\tau'_\alpha} = \frac{G M_1 R_m / H_{11}}{G M_1 R'_m / H'_{11}} = \frac{R_m H'_{11}}{R'_m H_{11}} = \frac{R'_m}{R_m}. \tag{7.29}$$

For two tubes of equal mass, the shear stress is inversely proportional to the mean radius.

The ideal structure to carry torsional loads is a thin-walled tube with a large mean radius, because this yields the highest torsional stiffness and lowest maximum shear stress for a given mass of material and applied torque. In specific applications, limits will be placed on how large the mean radius can be. Furthermore, very thin-walled tubes can become unstable through a phenomenon called *torsional buckling*. This type of instability puts a limit on how thin the wall can be.

7.1.10 Problems

Problem 7.1. Torsion of a bimetallic bar

A circular bar is constructed by bonding an aluminum shell around a solid steel cylinder. The radius of the steel cylinder is $R_S = 10$ mm, and the outer radius of the aluminum shell is $R_A = 20$ mm. The overall length of the bar is given by $L = 1$ m, and a torque $T = 1$ kN·m is applied at the ends. The shear moduli for the aluminum and steel are $G_A = 28$ GPa and $G_S = 76$ GPa, respectively. (1) Find the maximum shear stress in the steel and in the aluminum. (2) Determine the total twist angle of the bar. (3) Determine the torsional stiffness. (4) Find the allowable torque for a safety factor of 2 when the yield stresses for both materials is 300 MPa.

Problem 7.2. Torsion of a circular bar with hollow segment

The cylindrical bar shown in fig. 7.9 consists of two segments; the left segment is clamped at point **R**, $\Phi_1(0) = 0$. The left segment of length L is a solid circular bar of radius R_O, while the right segment of length L is a hollow circular bar of inner radius R_i. A moment Q_1 is applied at point **T**. (1) Determine the twist angle at point **T**. (2) Determine the equivalent torsional stiffness, H, for the complete bar, defined as $H = \Phi_1(2L)/Q_1$. (3) Determine ratio of maximum shear stress in the two sections.

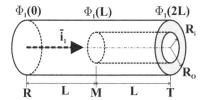

Fig. 7.9. Circular bar with hollow segment.

Problem 7.3. Torsion of a circular bar with hollow segment

The cylindrical bar shown in fig. 7.9 consists of two segments, clamped at point **R** and **T**, $\Phi_1(0) = \Phi_1(2L) = 0$. The left segment of length L is a solid circular bar of radius R_O, while the right segment of length L is a hollow circular bar of inner radius R_i. A moment Q_1 is applied at point **M**. (1) Determine the torque carried in each segment. (2) Determine the twist angle at point **M**. (3) Determine the equivalent torsional stiffness, H, at point **M**, defined as $H = \Phi_1(L)/Q_1$. (4) Determine the maximum shear stress in each segment.

Problem 7.4. Torsion of a hollow bar

A circular bar of radius $R = 200$ mm is replaced by a hollow bar of inner and outer radii R_i and R_o, respectively, with $R_O/R_i = 2$. If the two bars are made of the same material and can carry the same maximum torque, determine (1) the outer radius of the hollow bar, R_o, and (2) the mass ratio for the hollow and solid bars.

7.2 Torsion combined with axial force and bending moments

An aircraft propeller is connected to a homogeneous, circular shaft. The engine applies a torque to the shaft resulting in the shear stress distribution described in section 7.1.8. On the other hand, the propeller creates a thrust that generates a uniform axial stress distribution over the cross-section. If the torque acts alone, the yield criterion is $\tau < \tau_y$. If the axial force acts alone, the corresponding criterion is $\sigma < \sigma_y$. The question is now: what is the proper strength criterion to be used when both axial and shear stresses are acting simultaneously? The yield criteria developed in section 2.3 will be used to answer this question.

Propeller shaft under torsion and thrust

Consider an aircraft propeller connected to a homogeneous, circular shaft of radius R. The engine applies a torque M_1 to the shaft and the propeller exerts a thrust N_1; the corresponding stresses are

$$\tau = \frac{2M_1}{\pi R^3}, \quad \text{and} \quad \sigma = \frac{N_1}{\pi R^2}. \tag{7.30}$$

Clearly, the shaft is in a state of plane stress, and Tresca's criterion, eq. (2.31), requires the following inequalities to hold

$$\left| \frac{1}{2} \frac{N_1}{\pi R^2} \pm \sqrt{\frac{1}{4}\left(\frac{N_1}{\pi R^2}\right)^2 + 4\left(\frac{M_1}{\pi R^3}\right)^2} \right| \leq \sigma_y,$$

$$2\sqrt{\frac{1}{4}\left(\frac{N_1}{\pi R^2}\right)^2 + 4\left(\frac{M_1}{\pi R^3}\right)^2} \leq \sigma_y.$$

if the material is to be free of yielding. Of these three conditions, the last is the most stringent, and hence, Tresca's yield criterion corresponds to an ellipse,

$$\left(\frac{N_1}{\pi R^2 \sigma_y}\right)^2 + 16\left(\frac{M_1}{\pi R^3 \sigma_y}\right)^2 = 1.$$

Figure 7.10 shows the geometric interpretation of the criterion. The structure behaves in a linearly elastic manner under combined loadings represented by points inside an ellipse drawn in the non-dimensional load space. The non-dimensional torque is $M_1/(\pi R^3 \sigma_y)$ and the non-dimensional axial force is $N_1/(\pi R^2 \sigma_y)$.

If the von Mises criterion, eq. (2.36), is applied instead, the material will behave in a linearly elastic manner when the following condition is satisfied

$$\left[\left(\frac{N_1}{\pi R^2}\right)^2 + 3\left(\frac{2M_1}{\pi R^3}\right)^2\right]^{1/2} \leq \sigma_y.$$

Here again, the criterion is conveniently recast into a non-dimensional form as

$$\left(\frac{N_1}{\pi R^2 \sigma_y}\right)^2 + 12\left(\frac{M_1}{\pi R^3 \sigma_y}\right)^2 \leq 1.$$

where the terms in parentheses are non-dimensional loading components defined earlier. Figure 7.10 shows this ellipse in the non-dimensional loading space. As expected, the predictions of Tresca's and von Mises' criteria differ most when the loading primarily generates shear stresses, *i.e.*, along the applied torque axis, see section 2.3.3.

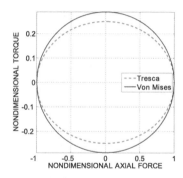

Fig. 7.10. Yield envelopes predicted by Tresca's and von Mises' criteria plotted in the non-dimensional loading space.

Shaft under torsion and bending

Consider now a circular shaft subjected to both bending and torsion, as would occur, for instance, in a cantilever shaft with a loaded tip pulley. Let M_3 and M_1 be the applied bending moment and torque, respectively. The corresponding axial and shear stress components are

$$\sigma = \frac{4M_3 r}{\pi R^4}, \text{ and } \tau = \frac{2M_1 r}{\pi R^4}, \tag{7.31}$$

respectively. The maximum values occur at the same location on the cross-section at the upper or lower edge where $\sigma = 4M_3/\pi R^3$ and $\tau = 2M_1/\pi R^3$. Clearly, the shaft is in a state of plane stress, and Tresca's criterion, eq. (2.31), requires the following inequalities to hold

$$\left| \frac{2M_3}{\pi R^3} \pm \sqrt{\left(\frac{2M_3}{\pi R^3} \right)^2 + 4 \left(\frac{M_1}{\pi R^3} \right)^2} \right| \leq \sigma_y,$$

$$2\sqrt{\left(\frac{2M_3}{\pi R^3} \right)^2 + 4 \left(\frac{M_1}{\pi R^3} \right)^2} \leq \sigma_y,$$

if the material is to be free of yielding. Of these three conditions, the last is the most stringent, and hence, Tresca's yield criterion corresponds to an ellipse

$$16 \left(\frac{M_3}{\pi R^3 \sigma_y} \right)^2 + 16 \left(\frac{M_1}{\pi R^3 \sigma_y} \right)^2 = 1.$$

Figure 7.11 shows the geometric interpretation of the criterion. The structure behaves in a linearly elastic manner for combined loadings represented by points inside an ellipse drawn in the non-dimensional loading space defined by non-dimensional torque, $M_1/(\pi R^3 \sigma_y)$, and non-dimensional bending moment, $M_3/(\pi R^3 \sigma_y)$.

If von Mises' criterion, eq. (2.36), is applied instead, the material will behave in a linearly elastic manner when the following condition is satisfied

$$\left[\left(\frac{4M_3}{\pi R^2} \right)^2 + 3 \left(\frac{2M_1}{\pi R^3} \right)^2 \right]^{1/2} \leq \sigma_y$$

Here again, the criterion is conveniently recast into a non-dimensional form as

$$16 \left(\frac{M_3}{\pi R^3 \sigma_y} \right)^2 + 12 \left(\frac{M_1}{\pi R^3 \sigma_y} \right)^2 \leq 1,$$

Figure 7.11 shows this ellipse in the non-dimensional loading space.

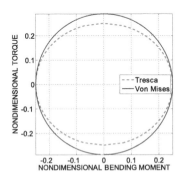

Fig. 7.11. Yield envelopes predicted by Tresca's and von Mises' criteria plotted in the non-dimensional loading space.

7.2.1 Problems

Problem 7.5. Pressure vessel subjected to combined loading
Consider the pressure vessel subjected to an internal pressure p_i and an external torque Q, as depicted in fig. 7.12. The pressure vessel is of radius R and wall thickness t. Use von Mises criterion to compute the failure envelope in the non-dimensional loading space defined by $Q/(tR^2\sigma_{\text{allow}})$ and $p_i R/(t\sigma_{\text{allow}})$.

Problem 7.6. Pressure vessel subjected to combined loading
The experimental set-up depicted in fig. 7.13 is aimed at studying the behavior of materials under complex stress states. A thin-walled pressure vessel of radius $R = 11$ mm and thickness $t = 2.0$ mm is subjected to an internal pressure p_i. At the same time, a normal force, N, and a torque, Q, are applied to the sample. In a specific experiment, the applied normal force is $N = 16$ kN and the internal pressure $p_i = 20$ MPa. The applied torque is slowly increased. The first permanent deformations are observed at the outer surface of the sample when $Q = 120$ N·m.
(1) Find the yield stress for the material if it is assumed to follow von Mises' yield criterion.
(2) Find the yield stress for the material if it is assumed to follow Tresca's yield criterion.
(3) Find and plot the yield surface in the space defined by the three loading components, the internal pressure, the applied axial force, and the applied torque.

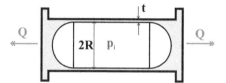

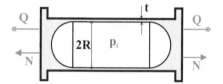

Fig. 7.12. Pressure vessel subjected to an external torque.

Fig. 7.13. Pressure vessel subjected to internal pressure, external torque and axial force.

Problem 7.7. Beam with circular section under bending and torsion
Consider a cantilevered beam of length $L = 1$ m with a circular cross-section of inner radius $R_i = 45$ mm and outer radius $R_o = 50$ mm. The beam is subjected to a tip torque $Q = 7$ kN·m and a tip transverse load P. Find the maximum allowable transverse load $P_{\max}$ if the allowable stress for the material is $\sigma_{\text{allow}} = 450$ MPa. Note: for a hollow circular section, $H_{22}^c = H_{33}^c = \pi E(R_o^4 - R_i^4)/4$.

7.3 Torsion of bars with arbitrary cross-sections

The theory of torsion presented in the two previous sections is valid for *bars with circular cross-sections only*. In this section, the theory of torsion will be generalized to bars presenting cross-sections of arbitrary shape.

7.3.1 Introduction

When analyzing the torsional behavior of circular cylinders, the circular symmetry of the problem leads to the conclusion that each cross-section rotates about its own center like a rigid disk. If this type of deformation is assumed to remain valid for a bar of arbitrary cross-section, the displacement field, eqs. (7.2) and (7.3), and the corresponding strain field, eqs. (7.4) to (7.6), will also describe the kinematics of bars with arbitrary sections. The only remaining stress component are the circumferential shear stress given by eq. (7.11).

Unfortunately, this assumption can lead to grossly erroneous results because the solution it implies violates the equilibrium equations of the problem along the edge of the section. Consider, for instance, torsion of the rectangular bar depicted in fig. 7.14. The circumferential shear stress, τ_α, given by eq. (7.11), is shown at an edge of the section, and it is resolved into its Cartesian components, τ_{12} and τ_{13}. In view of the principle of reciprocity of shear stresses, eq. (1.5), the existence of a stress component, τ_{13}, acting on the cross-section of the bar implies the existence of a shear stress component of equal magnitude acting on the orthogonal face, which happens to be the outer surface of the bar. Since the outer surface of the bar is stress free, the shear stress component, τ_{13}, must vanish on both faces. Consequently, the only shear stress component that can exist along the edge is component, τ_{12}, which is parallel to the edge. This reasoning can be applied to any point along the edge of the section, and consequently, *at any point along the edge of the bar's section, the shear stress*

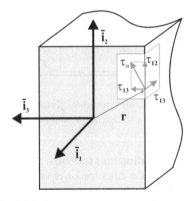

Fig. 7.14. Shearing stresses along the edge of a rectangular section.

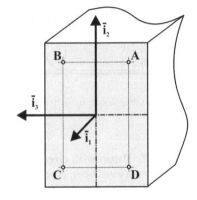

Fig. 7.15. Four points on a rectangular cross-section.

must be tangent to the edge. This condition is satisfied by the shear stress distribution acting on the circular section depicted in fig. 7.5, but the same circumferential shear stress distribution is not correct for the rectangular section shown in fig. 7.14.

As discussed in section 7.1, the symmetries associated with a circular cylinder imply that the bar's cross-section does not warp out-of-plane. No such conclusion can be reached for the rectangular section shown in fig. 7.15, because it presents fewer symmetries than the circular section. Indeed, the rectangular section is symmetric with respect to planes $(\bar{\imath}_1, \bar{\imath}_2)$ and $(\bar{\imath}_1, \bar{\imath}_3)$, but does not present the circular symmetry about axis $\bar{\imath}_1$ characteristic of a circular section. Since the section is symmetric with respect to plane $(\bar{\imath}_1, \bar{\imath}_2)$ but the torsional loading is antisymmetric with respect to the same plane, the solution must be antisymmetric with respect to this plane, *i.e.*, $u_1^A = -u_1^B$ and $u_1^C = -u_1^D$, where u_1^A, u_1^B, u_1^C and u_1^D, are the axial displacement components at points **A**, **B**, **C**, and **D**, respectively. Similarly, the antisymmetry of the solution with respect to plane $(\bar{\imath}_1, \bar{\imath}_3)$ implies $u_1^A = -u_1^D$ and $u_1^B = -u_1^C$. Combining the results then leads to $u_1^A = -u_1^B = u_1^C = -u_1^D$, which does not imply the vanishing of axial displacement at any of these points.

The same reasoning can be repeated for any set of four points symmetrically located with respect to the two planes of symmetry of the section. It follows that while the axial displacement field does present symmetries for the rectangular section, it does not vanish; in other words, *the section warps* out-of-plane. In general, *bars of arbitrary shaped cross-sections will warp*, in contrast with circular sections which do not.

7.3.2 Saint-Venant's solution

The solution to the problem of torsion of a bar with a cross-section of arbitrary shape was first given by Saint-Venant. The solution process provides a good application of basic elasticity theory and at the same time yields results of practical importance.

Kinematic description

Consider a solid bar with a cross-section of arbitrary shape. The area of the cross section is denoted $\mathcal{A}$, while its outer contour is defined by curve $\mathcal{C}$. The bar is of infinite length and is subjected to end torques. A closer look at the problem and experimental tests reveal that for a bar with an arbitrary section, each cross-section rotates like a rigid body, and warps out of its own plane. This type of deformation is described by the following assumed displacement field

$$u_1(x_1, x_2, x_3) = \Psi(x_2, x_3)\, \kappa_1(x_1), \tag{7.32a}$$

$$u_2(x_1, x_2, x_3) = -x_3 \Phi_1(x_1), \quad u_3(x_1, x_2, x_3) = x_2 \Phi_1(x_1). \tag{7.32b}$$

The in-plane displacement field, eq. (7.32b), describes a rigid body rotation of the cross-section, similar to the case for the circular cylinder, see eq. (7.3). The out-of-plane displacement field does not vanish, however. Instead, it is assumed to be proportional to the twist rate, κ_1, and has an arbitrary variation over the cross-section described by the unknown warping function, $\Psi(x_2, x_3)$. This warping function will be determined by enforcing equilibrium conditions for the resulting shear stress field. It will be further assumed that the twist rate is constant along the axis of the bar, $i.e.$, $\kappa_1(x_1) = \kappa_1$. This restriction results in what is known as the *uniform torsion* problem.

The strain field

Given the assumed displacement field defined by eqs. (7.32a) and (7.32b), the associated strain field can be evaluated based on the strain-displacement relationships, eqs. (1.63) and (1.71), to find

$$\varepsilon_1 = \Psi(x_2, x_3) \frac{\mathrm{d}\kappa_1}{\mathrm{d}x_1} = 0, \tag{7.33a}$$

$$\varepsilon_2 = 0, \quad \varepsilon_3 = 0, \quad \gamma_{23} = 0, \tag{7.33b}$$

$$\gamma_{12} = \left(\frac{\partial \Psi}{\partial x_2} - x_3 \right) \kappa_1, \quad \gamma_{13} = \left(\frac{\partial \Psi}{\partial x_3} + x_2 \right) \kappa_1. \tag{7.33c}$$

The vanishing of the axial strain, eq. (7.33a), is a direct consequence of the uniform torsion assumption, whereas the vanishing of the in-plane strains, eq. (7.33b), stems from the rigid body rotation assumption for the in-plane displacement field, eq. (7.32b). The only non-vanishing strain components, γ_{12} and γ_{13}, depend on the partial derivatives of the unknown warping function.

The stress field

For bars made of a linearly elastic, isotropic material, Hooke's law, eqs. (2.4) and (2.9), is assumed to apply. The stress field is then found from the strain field as

$$\sigma_1 = 0, \tag{7.34a}$$

$$\sigma_2 = 0, \quad \sigma_3 = 0, \quad \tau_{23} = 0, \tag{7.34b}$$

$$\tau_{12} = G\kappa_1 \left(\frac{\partial \Psi}{\partial x_2} - x_3 \right), \quad \tau_{13} = G\kappa_1 \left(\frac{\partial \Psi}{\partial x_3} + x_2 \right). \tag{7.34c}$$

Equilibrium equations

This stress field must satisfy the general equilibrium equations, eqs. (1.4), at all points of the section. Neglecting body forces, and in view of eq. (7.34b), two of the three equilibrium equations are identically satisfied and the remaining one reduces to

$$\frac{\partial \tau_{12}}{\partial x_2} + \frac{\partial \tau_{13}}{\partial x_3} = 0. \tag{7.35}$$

Introducing eqs. (7.34c), it follows that the warping function must satisfy the following partial differential equation

$$\frac{\partial^2 \Psi}{\partial x_2^2} + \frac{\partial^2 \Psi}{\partial x_3^2} = 0. \tag{7.36}$$

at all points of the cross-section.

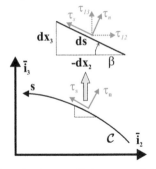

The relevant boundary conditions can be developed by requiring the satisfaction of the equilibrium conditions along the outer edge of the section that defines curve C. Figure 7.16 shows a portion of the outer contour, C, and a curvilinear variable, s, that measures length along this curve.

As illustrated in fig. 7.14, the normal component of shear stress must vanish at all points along C, i.e.,

$$\tau_n = 0, \tag{7.37}$$

Fig. 7.16. Equilibrium condition along the outer contour C.

whereas the component of shear stress, τ_s, tangent to the contour does not necessarily vanish. In terms of Cartesian components, the normal component of shear stress, see fig. 7.16, is

$$\tau_n = \tau_{12} \sin \beta + \tau_{13} \cos \beta = \tau_{12} \left(\frac{dx_3}{ds} \right) + \tau_{13} \left(-\frac{dx_2}{ds} \right) = 0. \tag{7.38}$$

Introducing eq. (7.34c) then yields the following boundary condition for the warping function

$$\left(\frac{\partial \Psi}{\partial x_2} - x_3 \right) \frac{dx_3}{ds} - \left(\frac{\partial \Psi}{\partial x_3} + x_2 \right) \frac{dx_2}{ds} = 0. \tag{7.39}$$

The warping function, $\Psi(x_2, x_3)$, is the solution of the following partial differential equation and associated boundary conditions,

$$\frac{\partial^2 \Psi}{\partial x_2^2} + \frac{\partial^2 \Psi}{\partial x_3^2} = 0, \text{ over } \mathcal{A}, \tag{7.40a}$$

$$\left(\frac{\partial \Psi}{\partial x_2} - x_3\right)\frac{dx_3}{ds} - \left(\frac{\partial \Psi}{\partial x_3} + x_2\right)\frac{dx_2}{ds} = 0, \text{ along } \mathcal{C}. \tag{7.40b}$$

This particular kind of partial differential equation is called *Laplace's equation*, and solution of this problem is rather complicated in view of the complex boundary condition that must hold along $\mathcal{C}$.

Prandtl's stress function

An alternative formulation of the problem that leads to simpler boundary conditions is found by introducing a *stress function*, ϕ, proposed by Prandtl. This function, $\phi(x_2, x_3)$, is defined as

$$\tau_{12} = \frac{\partial \phi}{\partial x_3}, \quad \tau_{13} = -\frac{\partial \phi}{\partial x_2}. \tag{7.41}$$

This shear stress field automatically satisfies the local equilibrium equation, as can be verified by introducing eq. (7.41) into eq. (7.35).

Next, the shear stresses, τ_{12} and τ_{13}, expressed in terms of the warping function by eq. (7.34c) must equal their counterparts expressed in terms of Prandtl's stress function by eq. (7.41) to find

$$\tau_{12} = G\kappa_1\left(\frac{\partial \Psi}{\partial x_2} - x_3\right) = \frac{\partial \phi}{\partial x_3}, \quad \tau_{13} = G\kappa_1\left(\frac{\partial \Psi}{\partial x_3} + x_2\right) = -\frac{\partial \phi}{\partial x_2}. \tag{7.42}$$

The warping function can be eliminated by taking a partial derivative of the first equation with respect to x_3 and a partial derivative of the second with respect to x_2. Subtracting these two equations then yields a single partial differential equation for Prandtl's stress function,

$$\frac{\partial^2 \phi}{\partial x_2^2} + \frac{\partial^2 \phi}{\partial x_3^2} = -2G\kappa_1. \tag{7.43}$$

The boundary conditions along $\mathcal{C}$ follow from eqs. (7.38) and (7.41)

$$\tau_n = \frac{\partial \phi}{\partial x_3}\frac{dx_3}{ds} + \frac{\partial \phi}{\partial x_2}\frac{dx_2}{ds} = \frac{d\phi}{ds} = 0. \tag{7.44}$$

which implies a constant value of ϕ along curve $\mathcal{C}$. If the section is bounded by several disconnected curves, the stress function must be a constant along each individual curve, although the value of the constant can be different for each curve. For solid cross-sections bounded by a single curve, the constant value of the stress function along that curve may be chosen to vanish because this choice has no effect on the resulting stress distribution.

The stress function is the solution of the following partial differential equation and associated boundary condition

$$\frac{\partial^2 \phi}{\partial x_2^2} + \frac{\partial^2 \phi}{\partial x_3^2} = -2G\kappa_1, \quad \text{on } \mathcal{A}, \tag{7.45a}$$

$$\frac{d\phi}{ds} = 0, \quad \text{along } \mathcal{C}. \tag{7.45b}$$

This partial differential equation is no longer homogeneous, a form referred to as *Poisson's equation*. The advantage of this formulation is that the boundary condition is much simpler than that obtained for the warping function, see eq. (7.40b).

Sectional equilibrium

The differential equations for the warping and stress functions are found from local equilibrium consideration. Global equilibrium of the section must also be verified. For a solid section bounded by a single contour, the resultant shear forces acting on the section are

$$V_2 = \int_{\mathcal{A}} \tau_{12} \, d\mathcal{A} = \int_{x_2} \int_{x_3} \frac{\partial \phi}{\partial x_3} \, dx_2 dx_3 = \int_{x_2} \left[\int_{x_3} \frac{\partial \phi}{\partial x_3} \, dx_3 \right] dx_2 = 0,$$

and

$$V_3 = \int_{\mathcal{A}} \tau_{13} \, d\mathcal{A} = \int_{x_2} \int_{x_3} -\frac{\partial \phi}{\partial x_2} \, dx_2 dx_3 = -\int_{x_3} \left[\int_{x_2} \frac{\partial \phi}{\partial x_2} \, dx_2 \right] dx_3 = 0,$$

where the last equalities follow from selecting a zero value for the stress function along the contour $\mathcal{C}$. This is the expected result because no shear forces are applied.

The total torque acting on the section is

$$M_1 = \int_{\mathcal{A}} (x_2 \tau_{13} - x_3 \tau_{12}) \, d\mathcal{A} = \int_{\mathcal{A}} \left(-x_2 \frac{\partial \phi}{\partial x_2} - x_3 \frac{\partial \phi}{\partial x_3} \right) d\mathcal{A}. \tag{7.46}$$

Integrating by parts then yields

$$M_1 = 2 \int_{\mathcal{A}} \phi \, d\mathcal{A} - \int_{x_3} [x_2 \phi]_{x_2} \, dx_3 - \int_{x_2} [x_3 \phi]_{x_3} \, dx_2. \tag{7.47}$$

For solid cross-sections bounded by a single curve, the constant value of the stress function along that curve may be chosen as zero, and the boundary terms disappear, leading to the simple result

$$M_1 = 2 \int_{\mathcal{A}} \phi \, d\mathcal{A}. \tag{7.48}$$

The applied torque equals twice the "volume" under the stress function. This formula applies only to solid cross-sections bounded by a single curve. Indeed, if the section is bounded by several disconnected curves, the stress function equals a different constant along each individual curve, and the boundary terms no longer vanish. For such sections, the applied torque should be evaluated with the help of eq. (7.46) rather than (7.48).

In summary, the stress distribution in a bar of arbitrary cross-section subjected to uniform torsion can be obtained by evaluating either the warping or stress function from eqs. (7.40) or (7.45), respectively. The stress field then follows from eqs. (7.34c) or (7.41), respectively. Since all governing equations are satisfied, this represents an exact solution of the problem.

Saint-Venant's solution procedure is an example of the semi-inverse solution technique. The displacement field is assumed to be of the form given by eqs. (7.32). It is shown, however, that based on this displacement field, all equations of elasticity are satisfied, and hence the assumed displacement field must be the exact solution of the problem.

Example 7.1. Torsion of an elliptical bar
Consider a bar with an elliptical cross-section as shown in fig. 7.17. The equation for curve $\mathcal{C}$ defining the section is $(x_2/a)^2 + (x_3/b)^2 = 1$. A stress function of the following form is assumed

$$
\phi = C_0 \left[\left(\frac{x_2}{a} \right)^2 + \left(\frac{x_3}{b} \right)^2 - 1 \right],
$$

where C_0 is an unknown constant. The boundary condition, eq. (7.45b), is clearly satisfied since $\phi = 0$ along $\mathcal{C}$. Substituting this in the governing differential equation, eq. (7.45), leads to the following equation for constant C_0: $C_0(2/a^2 + 2/b^2) = -2G\kappa_1$, or $C_0 = -a^2 b^2 G \kappa_1/(a^2 + b^2)$. The stress function then becomes

$$
\phi = -\frac{a^2 b^2}{a^2 + b^2} \left[\left(\frac{x_2}{a} \right)^2 + \left(\frac{x_3}{b} \right)^2 - 1 \right] G\kappa_1.
\tag{7.49}
$$

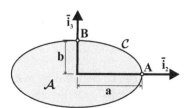

Fig. 7.17. A bar with an elliptical cross-section.

The torque can now be computed from eq. (7.48) to find

$$
M_1 = -\frac{2a^2 b^2}{a^2 + b^2} G\kappa_1 \int_{\mathcal{A}} \left[\left(\frac{x_2}{a} \right)^2 + \left(\frac{x_3}{b} \right)^2 - 1 \right] \mathrm{d}\mathcal{A} = G\frac{\pi a^3 b^3}{a^2 + b^2} \kappa_1 = H_{11}\kappa_1,
$$

where $\int_{\mathcal{A}} \mathrm{d}\mathcal{A} = \pi ab$, $\int_{\mathcal{A}} x_2^2 \, \mathrm{d}\mathcal{A} = \pi a^3 b/4$, and $\int_{\mathcal{A}} x_3^2 \, \mathrm{d}\mathcal{A} = \pi a b^3/4$ are the ellipse's area and second moments of area about axes $\bar{\imath}_2$ and $\bar{\imath}_3$, respectively. The torsional stiffness of the elliptical section is

$$H_{11} = G \frac{\pi a^3 b^3}{a^2 + b^2}.$$ (7.50)

Using these results, the stress function can be expressed in terms of the applied torque

$$\phi = -\frac{M_1}{\pi a b} \left[\left(\frac{x_2}{a} \right)^2 + \left(\frac{x_3}{b} \right)^2 - 1 \right].$$

The stress distribution then follows from eqs. (7.41),

$$\tau_{12} = -\frac{2x_3}{\pi a b^3} M_1, \quad \tau_{13} = \frac{2x_2}{\pi a^3 b} M_1.$$

The maximum shear stresses occur for the extreme values of x_2 and x_3, which are found along the section's boundary. The shear stress distributions along axes $\bar{\imath}_2$ and $\bar{\imath}_3$ are shown in fig. 7.18a: the maximum stresses are found points **B** and **A** as $\tau_{12}^B = -2M_1/(\pi a b^2)$ and $\tau_{13}^A = 2M_1/(\pi a^2 b)$, respectively. The maximum shear stress occurs at the end of the minor axis of the ellipse, *i.e.*, at point **B**, where

$$|\tau_{\max}| = \frac{2M_1}{\pi a b^2}.$$

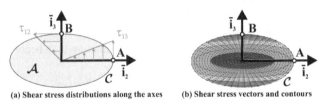

(a) Shear stress distributions along the axes   **(b) Shear stress vectors and contours**

Fig. 7.18. Shear stress distribution for an elliptical cross-section.

Figure 7.18b shows the shear stress vectors over the cross-section; as required by the principle of reciprocity of shear stresses, eq. (1.5), the shear stress vectors along curve C are tangent to this curve.

Finally, the warping function can be obtained by integrating eq. (7.42). Substituting the calculated stress function, eq. (7.49), into these equations yields

$$\frac{\partial \Psi}{\partial x_2} = -\frac{a^2 - b^2}{a^2 + b^2} x_3, \quad \frac{\partial \Psi}{\partial x_3} = -\frac{a^2 - b^2}{a^2 + b^2} x_2.$$

Integrating the first equation with respect to x_2 and the second with respect to x_3 yields $\Psi = -x_2 x_3 (a^2 - b^2)/(a^2 + b^2) + f(x_3)$, and $\Psi = -x_2 x_3 (a^2 - b^2)/(a^2 + b^2) + g(x_2)$, respectively. These two solutions are equal only if $f(x_3) = g(x_2) = 0$, which implies $\Psi = -(a^2 - b^2)/(a^2 + b^2) x_2 x_3$. Equation (7.32a) now yields the warping displacement as

$$u_1(x_2, x_3) = -\kappa_1 \frac{a^2 - b^2}{a^2 + b^2} x_2 x_3.$$ (7.51)

Note that the elliptic cross-section presents two planes of symmetry, planes $(\bar{\imath}_1, \bar{\imath}_2)$ and $(\bar{\imath}_1, \bar{\imath}_3)$. As discussed in section 7.3.1, this implies that the warping displacement must be antisymmetric with respect to these two planes. The left portion of fig. 7.19 depicts the warping displacement with a contour plot immediately below it. A separate contour plot is shown in the right portion of the same figure. As expected for an antisymmetric function, the warping displacement vanishes along axes $\bar{\imath}_2$ and $\bar{\imath}_3$, and is of equal magnitude but opposite signs at points symmetrically located with respect to axes $\bar{\imath}_2$ and $\bar{\imath}_3$.

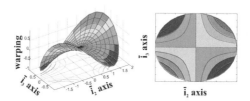

Fig. 7.19. Warping distribution for an elliptic cross-section.

For $a = b = R$, the bar with an elliptical section becomes a circular cylinder of radius R. The torsional stiffness for the elliptical section reduces to eq. (7.17), and the maximum shear stress to eq. (7.22). Finally, the warping function vanishes, and this is fully consistent with the symmetry arguments made for the circular cylinder proving that the warping displacement must vanish.

Example 7.2. Torsion of a thick cylinder

Consider a circular tube of inner radius R_i and outer radius R_o made of a homogeneous, isotropic material of shear modulus G, as shown in fig. 7.20. Note that this section is bounded by two curves, $\mathcal{C}_i$ and $\mathcal{C}_o$, as shown on the figure, that denote the inner and outer circles bounding the section.

The stress function for this problem is assumed to be in the following form: $\phi = Cr^2$, where $r^2 = x_2^2 + x_3^2$ and C is an unknown constant. The values of the stress function along curves $\mathcal{C}_i$ and $\mathcal{C}_o$ are $\phi_i = CR_i^2$ and $\phi_o = CR_o^2$, respectively. Since C, R_i and R_o are constants, this implies that the boundary conditions on the stress function, given by eq. (7.45b), are satisfied: $\mathrm{d}\phi_i/\mathrm{d}s_i = \mathrm{d}\phi_o/\mathrm{d}s_o = 0$, where s_i and s_o are curvilinear variables along $\mathcal{C}_i$ and $\mathcal{C}_o$, respectively. Note that the boundary condition requires ϕ to be *constant* along curves $\mathcal{C}_i$ and $\mathcal{C}_o$, but this does not imply that $\phi_i = 0$, or $\phi_o = 0$, or $\phi_i = \phi_o$.

Introducing the assumed stress function into the governing partial differential equation (7.45) yields $2C + 2C = -2G\kappa_1$. Hence, the stress function becomes $\phi = -G\kappa_1 r^2/2$. This represents the exact solution of the problem, because the stress function satisfies the governing partial differential equation and boundary conditions. The shear stress distribution then follows from eq. (7.41) as $\tau_{12} = -G\kappa_1 x_3$ and $\tau_{13} = G\kappa_1 x_2$. The torque generated by this shear stress distribution is evaluated with the help of eq. (7.46) to find

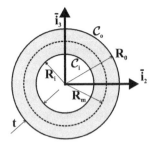

Fig. 7.20. Cross-section of a circular tube.

$$M_1 = \int_0^{2\pi} \int_{R_i}^{R_o} (x_2\tau_{13} - x_3\tau_{12})\, r dr d\alpha = \int_0^{2\pi} \int_{R_i}^{R_o} G\kappa_1(x_2^2 + x_3^2)\, r dr d\alpha$$
$$= \frac{\pi}{2} G\kappa_1(R_o^4 - R_i^4) = H_{11}\kappa_1,$$

where the Cartesian to polar coordinate transformation relationships, $x_2 = r\cos\alpha$ and $x_3 = r\sin\alpha$ are used. Using eq. (7.48) to evaluate the torque will yield incorrect results, as can be easily verified. This is because eq. (7.48) is derived assuming a *solid cross-section bounded by a single curve*; this is not the case for the present thick tube that is bounded by two curves, $\mathcal{C}_i$ and $\mathcal{C}_o$.

The torsional stiffness of the thick tube is $H_{11} = \pi G(R_o^4 - R_i^4)/2$, which matches the previously obtained result, eq. (7.18). It is left to the reader to show that the stress field obtained from the stress function matches that found in section 7.1.2.

7.3.3 Saint-Venant's solution for a rectangular cross-section

The formulation of the uniform torsion problem for bars of arbitrary cross-sectional shape is treated in section 7.3.2 and requires the solution of a partial differential equation for either the warping function, or the stress function, see eq. (7.40) or (7.45), respectively. Except for very simple geometries, such as the elliptical section treated in the previous example, the exact solution of the problem is arduous.

Two solutions of the uniform torsion problem for a rectangular cross-section are presented in this section. First, an approximate solution based on the co-location approach, then an exact solution based on Fourier series expansion.

Approximate solution

Consider a bar with a rectangular cross-section of width a and height b depicted in fig. 7.21. The following expression will be assumed for the stress function

$$\phi(\eta, \zeta) = C_0 \left(\eta^2 - \frac{1}{4}\right)\left(\zeta^2 - \frac{1}{4}\right),$$

where C_0 is an unknown constant, $\eta = x_2/a$ is the non-dimensional coordinate along axis $\bar{\imath}_2$, and $\zeta = x_3/b$ that along axis $\bar{\imath}_3$, as shown in fig. 7.21. This choice of the stress function implies that $\phi(\eta = \pm 1/2, \zeta) = 0$ and $\phi(\eta, \zeta = \pm 1/2) = 0$, i.e., ϕ vanishes along the edge, $\mathcal{C}$, of the section, as required by the boundary conditions of the problem, eq. (7.45b).

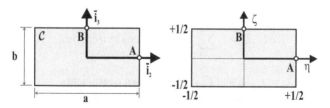

Fig. 7.21. Bar with a rectangular cross-section.

Using the chain rule for partial derivatives, $\partial/\partial x_2 = \partial/\partial \eta \; (\partial \eta/\partial x_2)$, where $\partial \eta/\partial x_2 = 1/a$; a similar expression holds for $\partial/\partial x_3$. Substituting the assumed stress function into the governing partial differential equation, eq. (7.43), then leads to $2C_0(\zeta^2 - 1/4)/a^2 + 2C_0(\eta^2 - 1/4)/b^2 = -2G\kappa_1$. This result shows that the assumed solution does not satisfy the partial differential equation.

A number of methods are available to construct approximate solutions, but one of the simplest is to satisfy this equation only a specific points of the cross-section, an approach called the *co-location method*. In this case, the governing partial differential equation will be satisfied at the center of the section, $(\eta, \zeta) = (0, 0)$, which implies $-C_0/(2a^2) - C_0/(2b^2) = -2G\kappa_1$. Solving for C_0 yields $C_0 = 4G\kappa_1 a^2 b^2/(a^2 + b^2)$. The stress function now becomes

$$\phi(\eta, \zeta) = \frac{4a^2 b^2 G\kappa_1}{a^2 + b^2} \left(\eta^2 - \frac{1}{4} \right) \left(\zeta^2 - \frac{1}{4} \right).$$

For this section bounded by a single curve, the externally applied torque is given by eq. (7.48) as

$$M_1 = 2 \int_{\mathcal{A}} \phi \, d\mathcal{A} = \frac{a^2 b^2 G\kappa_1}{2(a^2 + b^2)} \int_{\mathcal{A}} \left(\eta^2 - \frac{1}{4} \right) \left(\zeta^2 - \frac{1}{4} \right) d\mathcal{A} = \frac{2}{9} \frac{a^3 b^3 G\kappa_1}{a^2 + b^2}.$$

This result reveals the torsional stiffness, $H_{11} = M_1/\kappa_1$. The non-dimensional torsional stiffness, $\bar{H}_{11} = H_{11}/(ab^3 G)$, then becomes

$$\bar{H}_{11} = \frac{H_{11}}{ab^3 G} = \frac{2}{9} \frac{1}{1 + (b/a)^2}. \tag{7.52}$$

The stress function can be be expressed in terms of the applied torque as $\phi = 18M_1(\eta^2 - 1/4)(\zeta^2 - 1/4)/(ab)$. The shear stress field now follows from eqs. (7.41) as

$$\tau_{12} = \frac{1}{b} \frac{\partial \phi}{\partial \zeta} = \frac{36M_1}{ab^2} \left(\eta^2 - \frac{1}{4} \right) \zeta; \quad \tau_{13} = -\frac{1}{a} \frac{\partial \phi}{\partial \eta} = -\frac{36M_1}{a^2 b} \eta \left(\zeta^2 - \frac{1}{4} \right).$$

Open form exact solution using a Fourier series

Consider once again the bar with a rectangular cross-section of width a and height b, as depicted in fig. 7.21. A Fourier series expansion of the stress function will be assumed as the solution of the problem,

$$\phi(\eta, \zeta) = \sum_{i=\text{odd}}^{\infty} \sum_{j=\text{odd}}^{\infty} C_{ij} \cos i\pi\eta \cos j\pi\zeta,$$

where $\eta = x_2/a$, $\zeta = x_3/b$, and C_{ij} are unknown coefficients.

First, it is verified that this assumed solution satisfies the boundary conditions of the problem, eq. (7.45b). Indeed, at $\eta = \pm 1/2$, $\cos(i\pi\eta) = \cos(\pm i\pi/2) = 0$ for all odd values of i; similarly, ϕ vanishes at $\zeta = \pm 1/2$ for all odd values of j. The function ϕ does not vanish along the boundaries for even values of i or j, and this is why only odd values of i and j are included in the expression for the stress function.

Substituting the above expression into the governing partial differential equation, eq. (7.43), yields

$$\sum_{i=\text{odd}}^{\infty} \sum_{j=\text{odd}}^{\infty} C_{ij} \left[\left(\frac{i\pi}{a} \right)^2 + \left(\frac{j\pi}{b} \right)^2 \right] \cos i\pi\eta \cos j\pi\zeta = 2G\kappa_1.$$

This forms a set of equations for the unknown coefficients, C_{ij}. The evaluation of these coefficients relies on the orthogonality properties of cosine functions. The above equation is first multiplied by $\cos m\pi\eta \cos n\pi\zeta$, where m and n are arbitrary odd integers, then integrated over the cross-section to yield

$$\sum_{i=\text{odd}}^{\infty} \sum_{j=\text{odd}}^{\infty} C_{ij} \left[\left(\frac{i\pi}{a} \right)^2 + \left(\frac{j\pi}{b} \right)^2 \right] \left[\int_{-1/2}^{1/2} \cos m\pi\eta \cos i\pi\eta \, d\eta \right]$$
$$\left[\int_{-1/2}^{+1/2} \cos n\pi\zeta \cos j\pi\zeta \, d\zeta \right] = -2G\kappa_1 \left[\int_{-1/2}^{1/2} \cos m\pi\eta \, d\eta \right] \left[\int_{-1/2}^{1/2} \cos n\pi\zeta \, d\zeta \right].$$

The bracketed integrals can be evaluated in closed form with the help of eqs. (A.46b) and (A.47) and vanish when $m \neq i$ or $n \neq j$, thus eliminating the summations. The remaining terms are

$$C_{mn} \left[\left(\frac{m\pi}{a} \right)^2 + \left(\frac{n\pi}{b} \right)^2 \right] \frac{1}{4} = \frac{8}{mn\pi^2} (-1)^{(m-1)/2} (-1)^{(n-1)/2} G\kappa_1.$$

Solving for the unknown coefficients, C_{mn}, then yields the stress function as

$$\phi(\eta, \zeta) = \frac{32 G\kappa_1}{\pi^2} \sum_{i=\text{odd}}^{\infty} \sum_{j=\text{odd}}^{\infty} \frac{(-1)^{(i+j-2)/2}}{ij \left[(i\pi/a)^2 + (j\pi/b)^2 \right]} \cos i\pi\eta \cos j\pi\zeta. \quad (7.53)$$

Since the section is bounded by a single curve, the externally applied torque is given by eq. (7.48)

$$M_1 = \frac{2^8}{\pi^6} ab^3 G \kappa_1 \sum_{i=\text{odd}}^{\infty} \sum_{j=\text{odd}}^{\infty} \frac{1}{(ij)^2 \left[i^2 (b/a)^2 + j^2\right]} = H_{11} \kappa_1,$$

from which it follows that the non-dimensional torsional stiffness is

$$\bar{H}_{11} = \frac{H_{11}}{ab^3 G} = \frac{2^8}{\pi^6} \sum_{i=\text{odd}}^{\infty} \sum_{j=\text{odd}}^{\infty} \frac{1}{(ij)^2 \left[i^2 (b/a)^2 + j^2\right]}. \tag{7.54}$$

Although in the form of a doubly infinite series, this expression for the torsional stiffness converges rapidly. For a bar with a square cross-section, $a = b$, the torsional stiffness obtained using the double sine series is $\bar{H}_{11} = 0.140577$. Considering only a single term in the series, $i = j = 1$, results in $\bar{H}_{11} = 2^8 \left[1/2\right]/\pi^6 = 0.133$, a 5% error. The four term series generated by i and j taking values of 1 and 3 yields $\bar{H}_{11} = 2^8 \left[1/2 + 1/90 + 1/90 + 1/1458\right]/\pi^6 = 0.139$, a 1% error.

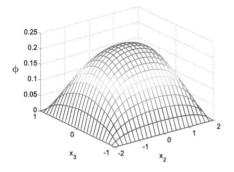

Fig. 7.22. Stress function, ϕ.

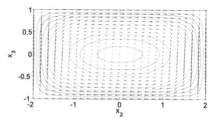

Fig. 7.23. Distribution of shear stress over cross-section. The arrows represent the shear stresses; the contours represent constant values of the stress function ϕ.

The shear stress field now follows from eqs. (7.41) as

$$\tau_{12} = -\frac{2^5}{\pi^3} \frac{bG}{H_{11}} M_1 \sum_{i=\text{odd}}^{\infty} \sum_{j=\text{odd}}^{\infty} \frac{(-1)^{(i+j-2)/2}}{i \left[i^2 (b/a)^2 + j^2\right]} \cos \frac{i\pi x_2}{a} \sin \frac{j\pi x_3}{b}, \tag{7.55a}$$

$$\tau_{13} = \frac{2^5}{\pi^3} \frac{b^2 G}{a H_{11}} M_1 \sum_{i=\text{odd}}^{\infty} \sum_{j=\text{odd}}^{\infty} \frac{(-1)^{(i+j-2)/2}}{j \left[i^2 (b/a)^2 + j^2\right]} \sin \frac{i\pi x_2}{a} \cos \frac{j\pi x_3}{b}. \tag{7.55b}$$

Here again, the results are in the form of a double sine series that is tedious to evaluate but converges rapidly. The stress function and shear stress distributions are shown in fig. 7.22 and 7.23, respectively, for $a = 4$ and $b = 2$. For the shear stress plot, the shear stress components, τ_{12} and τ_{13}, are converted into stress vectors and represented by arrows whose lengths are proportional to their magnitude.

Comparison of solutions

The Fourier series solution developed in the previous section converges to the exact solution to the problem as the number of terms used in the series increases. In practice, nearly exact solutions can be obtained by using a large but finite number of terms in all series; this will be referred to as the exact solution. The solution obtained from the co-location approach will be referred to as the approximate solution.

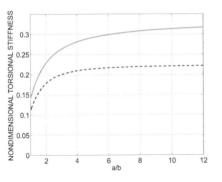

Fig. 7.24. Non-dimensional torsional stiffness, $\bar{H}_{11}$, versus aspect ratio, a/b. Exact solution: solid line; approximate solution: dashed line.

First, the non-dimensional torsional stiffness, $\bar{H}_{11}$, evaluated using the co-location method and Fourier series approaches, see eqs. (7.52) and (7.54), respectively, are compared in fig. 7.24. Both solutions are in fair agreement for aspect ratios near unity, but the approximate solution significantly under-predicts the stiffness for higher aspect ratios. For a very thin strip, $a/b \to \infty$, $\bar{H}_{11} = 1/3 = 0.333$ for the exact solution, but $\bar{H}_{11} = 2/9 = 0.222$ for the approximate solution, a 33% error.

The maximum values of the shear stress components, τ_{12} and τ_{13}, are found at points **B** and **A**, respectively, at the middle of the two sides, see fig. 7.21. The approximate solution gives $ab^2|\tau_{12}^B|/M_1 = 4.5$ and $ab^2|\tau_{13}^A|/M_1 = 4.5\, b/a$. The exact solution is obtained from the series in eqs. (7.55). Figures 7.25 and 7.26 show the shear stresses at points **B** and **A**, respectively, as a function of the aspect ratio, a/b. The maximum shear stress occurs at point **B**, the mid-point of the section's long side. For a thin strip, $ab^2|\tau_{12}^B|/M_1 = 3$.

Large discrepancies are observed between the two solutions. The approximate solution obtained with the co-location method is not good enough to accurately estimate the stress distribution in the section.

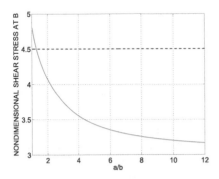

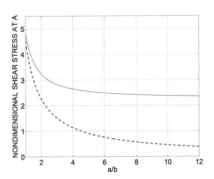

Fig. 7.25. Non-dimensional shear stress at point **B** versus aspect ratio a/b. Exact solution: solid line; approximate solution: dashed line.

Fig. 7.26. Non-dimensional shear stress at point **A** versus aspect ratio a/b. Exact solution: solid line; approximate solution: dashed line.

7.3.4 Problems

Problem 7.8. Bar with circular section and semi-circular keyway

Consider a circular shaft of radius a with a semi-circular keyway of radius b, as depicted in fig. 7.27. The shaft is subjected to torsion. A stress function of the following form will be used

$$\phi = A(x_2^2 + x_3^2 - 2ax_2)\left[1 - \frac{b^2}{(x_2^2 + x_3^2)}\right],$$

where A is an unknown constant. (1) Verify that the proposed stress function satisfies the required boundary conditions. (2) Determine the stress function for this problem, i.e., find the value of constant A. (3) Find the shear stress distribution $\tau_r = \tau_r(\alpha)$ and $\tau_\alpha = \tau_\alpha(\alpha)$ along the contour $\mathcal{C}_a$ of the shaft. (4) Find the shear stress distribution $\tau_r = \tau_r(\beta)$ and $\tau_\beta = \tau_\beta(\beta)$ along the contour $\mathcal{C}_b$ of the keyway. (5) Let $\tau_N = G\kappa_1 a$ be the shaft maximum shear stress *in the absence of keyway*. Find $\lim_{b \to 0} \tau_\alpha^A/\tau_N$ and $\lim_{b \to 0} \tau_\beta^B/\tau_N$. Comment on your results.

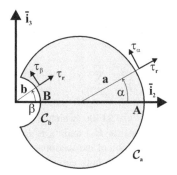

Fig. 7.27. Circular shaft with a circular keyway.

Problem 7.9. Torsion of bar with rectangular cross-section

A exact solution for the torsion of a bar with a rectangular cross-section depicted in fig. 7.21 is developed in section 7.3.3 using an open double trigonometric series. It is possible to develop a somewhat more efficient solution by assuming a trigonometric series solution in only one direction and an unknown function, $g_n(\eta)$, in the other. Consider the following single open series expansion for the stress function

$$\phi(\eta, \zeta) = \sum_{n=\text{odd}}^{\infty} g_n(\eta) \cos \alpha_n \zeta,$$

where $g_n(\eta)$ are unknown functions, $\alpha_n = n\pi/2$, $\eta = 2x_2/a$ is the non-dimensional coordinate along axis $\bar{\imath}_2$, and $\zeta = 2x_3/b$ is the non-dimensional coordinate along axis $\bar{\imath}_3$. Following the same approach used in section 7.3.3 and making use of the orthogonality of cosine functions, show that eq. (7.43) reduces to the following ordinary differential equations for $g_n(\eta)$

$$g_n'' - \beta_n^2 g_n = -\frac{Ga^2\kappa_1}{\alpha_n}(-1)^{(n-1)/2}, \quad \text{for n=odd}$$

where $\beta_n = \alpha_n a/b$, along with the boundary conditions $0 = g_n(\eta = \pm1)$. Next, solve these equations, and after substituting in the above expression for $\phi(\eta, \zeta)$, show that

$$\phi(\eta, \zeta) = b^2 G \kappa_1 \sum_{n=\text{odd}}^{\infty} \frac{(-1)^{(n-1)/2}}{\alpha_n^3} \left[1 - \frac{\cosh \beta_n \eta}{\cosh \beta_n} \right] \cos \alpha_n \zeta$$

From this result, show that the non-dimensional torsional stiffness can be written as

$$\bar{H}_{11} = \frac{H_{11}}{Gab^3} = 2 \sum_{n=\text{odd}}^{\infty} \left[\frac{1}{\alpha_n^4} - \frac{\tanh \beta_n}{\alpha_n^4 \beta_n} \right] = \frac{1}{3} - 2\frac{1}{a/b} \sum_{n=\text{odd}}^{\infty} \frac{\tanh \beta_n}{\alpha_n^5}.$$

Note that $\sum_{n=\text{odd}}^{\infty} 1/n^4 = \pi^4/96$, and hence, $2\sum_{n=\text{odd}}^{\infty} 1/\alpha_n^4 = 1/3$. For a thin rectangular strip, $a/b \to \infty$ and $\bar{H}_{11} \to 1/3$. Finally, show that the shear stress at point **B** is given by

$$\frac{ab^2 |\tau_B|}{M_1} = \frac{2}{\bar{H}_{11}} \sum_{n=\text{odd}}^{\infty} \left[\frac{1}{\alpha_n^2} - \frac{1}{\alpha_n^2 \cosh \beta_n} \right] = \frac{1}{\bar{H}_{11}} - \frac{2}{\bar{H}_{11}} \sum_{n=\text{odd}}^{\infty} \frac{1}{\alpha_n^2 \cosh \beta_n}.$$

Note that $\sum_{n=\text{odd}}^{\infty} 1/n^2 = \pi^2/8$, and hence, $2\sum_{n=\text{odd}}^{\infty} 1/\alpha_n^2 = 1$. The shear stress component at point **A** is given by

$$\frac{ab^2 |\tau_A|}{M_1} = \frac{2}{\bar{H}_{11}} \sum_{n=\text{odd}}^{\infty} \left[\frac{(-1)^{(n-1)/2}}{\alpha_n^2} - (-1)^{(n-1)/2} \frac{1 - \tanh \beta_n}{\alpha_n^2} \right]$$

$$= \frac{0.742454}{\bar{H}_{11}} - \frac{2}{\bar{H}_{11}} \sum_{n=\text{odd}}^{\infty} (-1)^{(n-1)/2} \frac{1 - \tanh \beta_n}{\alpha_n^2}.$$

Note that $\sum_{n=\text{odd}}^{\infty} (-1)^{(n-1)/2}/n^2 = 0.91596$, which is known as Catalan's constant. This particular arrangement of the equations for $\bar{H}_{11}$ and τ_A is done so that the series expressions can be more easily evaluated as a function of the sectional aspect ratio, a/b, as it approaches large values (at which $\tanh \beta_n \to 1$); in both cases the second term in the equations approaches zero. Note also the very fast convergence of all the series involved in this solution due to the powers of α_n appearing in the denominators.

7.4 Torsion of a thin rectangular cross-section

The torsion of a thin rectangular strip is an important problem that will form the basis for the analysis of beams with thin-walled cross-sections. An exact solution for the limiting case of a very thin rectangular strip can be easily developed. Consider the thin rectangular strip shown in fig. 7.28, where b is the long dimension of the cross-section, taken along axis $\bar{i}_3$, and t the thickness of the strip. If the thickness is much smaller than the length, i.e., if $t \ll b$, it is reasonable to assume that both stress function and associated shear stress distributions will be nearly constant along axis $\bar{i}_3$. This will imply that $\partial \phi / \partial x_3 \approx 0$.

The term $\partial^2 \phi / \partial x_3^2$ that appears in the governing equation for the stress function, eq. (7.43), now vanishes, and this governing equation reduces to the following ordinary differential equation,

$$\frac{d^2 \phi}{dx_2^2} = -2G\kappa_1. \tag{7.56}$$

This equation is easily integrated to find $\phi(x_2) = -G\kappa_1 x_2^2 + C_1 x_2 + C_2$, where C_1 and C_2 are two integration constants. The boundary condition, eq. (7.45b), requires that $\phi(x_2 = \pm t/2) = 0$, which implies $C_1 = 0$ and $C_2 = G\kappa_1 t^2/4$. The stress function then becomes

$$\phi(x_2) = -G\kappa_1 \left(x_2^2 - \frac{t^2}{4} \right). \tag{7.57}$$

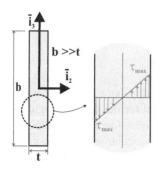

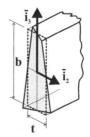

Fig. 7.28. Thin rectangular strip under torsion.

Fig. 7.29. Warping function for a thin rectangular strip.

The resulting torque is computed using eq. (7.48), to find

$$M_1 = 2 \int_{\mathcal{A}} \phi \, d\mathcal{A} = -2G\kappa_1 \int_{-t/2}^{t/2} \left(x_2^2 - \frac{t^2}{4} \right) b \, dx_2 = \frac{1}{3} G\kappa_1 bt^3.$$

This result reveals the torsional stiffness of the section, $H_{11} = M_1/\kappa_1$, as

$$H_{11} = \frac{1}{3} Gbt^3. \tag{7.58}$$

The shear stress distribution now follows from eq. (7.41) as

$$\tau_{12} = \frac{\partial \phi}{\partial x_3} = 0, \quad \tau_{13} = -\frac{\partial \phi}{\partial x_2} = 2G\kappa_1 x_2 = \frac{6M_1}{bt^3} x_2. \tag{7.59}$$

This distribution is depicted in the right portion of fig. 7.28. The maximum shear stress occurs all along the long edges of the section, where $x_2 = \pm t/2$, and is of magnitude $|\tau^{\max}| = 3M_1/(bt^2)$.

The warping function, Ψ, can be determined by substituting the stress function solution, eq. (7.57), into eq. (7.42) to find two partial differential equations

$$\frac{\partial \Psi}{\partial x_2} = \frac{1}{G\kappa_1} \frac{\partial \phi}{\partial x_3} + x_3 = x_3, \quad \frac{\partial \Psi}{\partial x_3} = -\frac{1}{G\kappa_1} \frac{\partial \phi}{\partial x_2} - x_2 = x_2,$$

the solutions of which are $\Psi = x_3 x_2 + f(x_3)$ and $\Psi = x_2 x_3 + g(x_2)$, respectively; $f(x_3)$ and $g(x_2)$ are two arbitrary functions. Because the problem must have a unique solution, the two expressions for Ψ must be equal. This is only possible if $f(x_3) = g(x_2) = 0$, leaving the warping function as $\Psi = x_2 x_3$. The axial displacement, $u_1(x_2, x_3)$, can be determined by substituting this result into eq. (7.32a) to find

$$u_1(x_2, x_3) = \Psi(x_2, x_3)\kappa_1 = \kappa_1 x_2 x_3. \tag{7.60}$$

As discussed in section 7.3.1, the warping function for a rectangular section must be antisymmetric with respect to both axes $\bar{\imath}_2$ and $\bar{\imath}_3$. The above solution does indeed satisfy this antisymmetry requirement, as illustrated in fig. 7.29.

7.5 Torsion of thin-walled open sections

The results presented in the previous section are readily extended to thin-walled open sections of arbitrary shape. The solution developed for the thin rectangular strip is based on the assumption that the gradient of the stress function vanishes in the direction tangential to the thin wall; for the thin rectangular strip shown in fig. 7.28, this means along axis $\bar{\imath}_3$. Of course, had the thin strip been rotated by 90 degrees, the gradient of the stress function would have been assumed to vanish along axis $\bar{\imath}_2$.

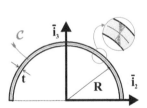

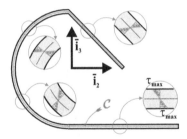

Fig. 7.30. Semi-circular thin-walled open section.

Fig. 7.31. Thin-walled open section composed of several curved.

More generally, the gradient of the stress function should vanish along the local tangent to the section's thin wall, and the corresponding shear stress distribution will then be linear through the wall thickness. For thin-walled open sections, the geometry of the cross-section can be represented by an open curve, $\mathcal{C}$, drawn along the wall's mid-thickness, as illustrated in fig. 7.30 for a semi-circular, thin-walled section.

The developments of the previous section still apply to a generally curved, thin-walled open section, and by extension of eq. (7.58), the torsional stiffness of such section becomes

$$H_{11} = G\,\frac{\ell t^3}{3}, \tag{7.61}$$

where ℓ is the length of curve $\mathcal{C}$ and t the wall thickness. For instance, the torsional stiffness of the semi-circular section shown in fig. 7.30 is $H_{11} = G\,\pi R t^3/3$.

For the thin rectangular section, the shear stress τ_{12} vanishes, leaving τ_{13} as the sole shear stress component, see eq. (7.59). For the present problem, the only non-vanishing stress component is the *tangential shear stress*, τ_s, acting in the direction tangent to curve $\mathcal{C}$. Here again, the shear stress is not uniform across the thickness, but instead, varies linearly from zero at the midline to maximum positive and negative values at the opposite edges of the wall, a distance $\pm t/2$ from the midline. At these points, the magnitude of the shear stress is

$$\tau_s^{\max} = Gt\,\kappa_1. \tag{7.62}$$

The maximum shear stress can also be expressed in terms of the applied torque as

$$\tau_s^{\max} = \frac{3M_1}{\ell t^2}. \tag{7.63}$$

A more general thin-walled open section could be composed of a number of straight and curved segments, such as the situation illustrated in fig. 7.31. In this case, the torsional stiffness of the cross-section is the sum of the torsional stiffnesses of the individual segments and can be expressed as,

$$H_{11} = \sum_i H_{11}^{(i)} = \frac{1}{3}\sum_i G_i \ell_i t_i^3, \tag{7.64}$$

where G_i, ℓ_i and t_i are the shear modulus, length and thickness of the ith segment, respectively. The shear stress along the edge of each segment is still given by eq. (7.62), where κ_1 is the twist rate of the cross-section. Hence, the maximum shear stress will be found in the segment featuring the largest thickness

$$\tau_s^{\max} = Gt_{\max}\frac{M_1}{H_{11}}, \tag{7.65}$$

where $t_{\max}$ is the thickness of the segment with the largest thickness.

Warping of a thin-walled open section is more complex and involves not only the warping behavior of a thin rectangular strip described in section 7.4 and defined by eq. (7.60), but it also includes a much larger warping of the overall cross-section. The warping of open thin-walled sections will be described in chapter 8 in section 8.7.

Example 7.3. Torsion of thin-walled section
Consider, as an example, the C-channel shown in fig. 7.32. The torsional stiffness of the section is given by eq. (7.64) as

$$H_{11} = \frac{G}{3}\left(bt_f^3 + ht_w^3 + bt_f^3\right) = \frac{G}{3}\left(ht_w^3 + 2bt_f^3\right). \tag{7.66}$$

The tangential shear stresses at the outer edges of the wall are given by eq. (7.62) as $\tau_w = Gt_w\kappa_1 = Gt_w M_1/H_{11}$ and $\tau_f = Gt_f\kappa_1 = Gt_f M_1/H_{11}$, for the stresses in the web and flanges, respectively. The maximum shear stress will be found in the segment featuring the maximum thickness.

7.5.1 Problems

Problem 7.10. Torsional stiffness of a section with variable thickness

Figure 7.32 depicts the cross-section of a thin-walled beam with different thicknesses. For this problem, assume that $t_w = t$ and $t_f = 2t$. *(1)* Find the torsional stiffness of the section. *(2)* Find the magnitude and location of the maximum shear stress if the section is subjected to a torque Q. *(3)* Sketch the distribution of shear stress through the thickness of the wall for the two regions with different thicknesses.

Problem 7.11. Torsional stiffness of a C-section

Consider the thin-walled, C-section of a beam depicted in fig. 7.32. The dimensions of the section are $b = 20$ mm, $h = 50$ mm, $t_w = 4$ mm and $t_f = 5$ mm. *(1)* Find the torsional stiffness of the section. *(2)* Compute the maximum shear stress in the section due to an applied torque Q. *(3)* Indicate the location of the maximum shear stress. *(4)* Sketch the distribution of shear stress through the thickness of the wall. The shear modulus for the material is $G = 30$ GPa and the applied torque is $Q = 120$ N·m.

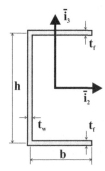

Fig. 7.32. A thin-walled C-channel section

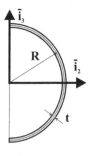

Fig. 7.33. Semi-circular open cross-section.

Problem 7.12. Torsional stiffness of a semi-circular section

Figure 7.33 depicts the thin-walled, semi-circular open cross-section of a beam. The wall thickness is t, and the material Young's and shear moduli are E and G, respectively. *(1)* Find the torsional stiffness of the section. *(2)* Find the distribution of shear stress due to an applied torque Q. *(3)* Indicate the location and magnitude of the maximum shear stress, $Rt^2\tau_{\max}/Q$.

Problem 7.13. Torsional stiffness of an "H" shaped cross-section

Figure 7.34 depicts the cross-section of a thin-walled beam with what is sometimes called an "H" shaped cross-section. For this problem, assume that $h_1 = b/2$ and $h_2 = b/4$. *(1)* Find the torsional stiffness of the section. *(2)* Find the magnitude and location of the maximum shear stress if the section is subjected to a torque Q. *(3)* Sketch the distribution of shear stress through the thickness of the wall.

Problem 7.14. Torsional stiffness of a "Y" shaped cross-section

Figure 7.35 depicts the "Y" shaped cross-section of a thin-walled beam. The horizontal leg of the cross-section has a thickness $2t$, whereas the other two legs are of thickness t. *(1)* Determine the torsional stiffness of the section. *(2)* Determine the magnitude and location of the maximum shear stress if a torque Q is applied to the beam. *(3)* Sketch the shear stress distribution through the wall thickness.

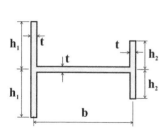

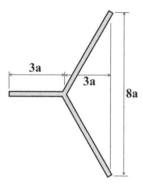

Fig. 7.34. "H" cross-section of a thin-walled beam.

Fig. 7.35. "Y" cross-section of a thin-walled beam.

Thin-walled beams

Typical aeronautical structures involve light-weight, thin-walled, beam-like structures that must operate in a complex loading environment where combined axial, bending, shearing, and torsional loads are present. These structures may consist of closed or open sections, or a combination of both. A closed cross-section is one for which the thin wall forms one or more closed paths; in the opposite case, it is an open section. This distinction has profound implications for the structural response of the beam, most importantly when it comes to shearing and torsion.

In the analysis of thin-walled beams, the specific geometric nature of the beam consisting of an assembly of thin sheets will be exploited to simplify the problem's formulation and solution process. Figures 8.1 to 8.4 show different types of thin-walled cross-sections. Figure 8.1 shows a beam with a closed section, as opposed to the open section of fig. 8.2. A combination of both types depicted in fig. 8.3 is also possible. Finally, multi-cellular sections such as shown in fig. 8.4 are very common in aeronautical constructions.

8.1 Basic equations for thin-walled beams.

The geometry of the section is described by a curve, $\mathcal{C}$, drawn along the mid-thickness of the wall, see figs. 8.1-8.4. A curvilinear variable, s, measuring length along this contour is defined with an arbitrary origin. This variable defines an orientation along $\mathcal{C}$ at all points. Of course, this orientation can be chosen arbitrarily. The wall thickness, $t(s)$, can vary from point to point along the contour. For multi-cellular sections, a number of different curves are used to completely describe the section, and a corresponding number of curvilinear variables define the length and orientation of these various curves.

8.1.1 The thin wall assumption

In thin-walled beams, the wall thickness is assumed to be much smaller than the other representative dimensions of the cross-section. Considering fig. 8.1, this means

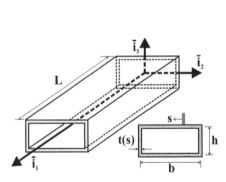

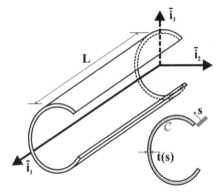

Fig. 8.1. Thin-walled beam with a closed, single cell section.

Fig. 8.2. Thin-walled beam with an open section.

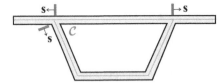

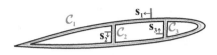

Fig. 8.3. Thin-walled beam with open and closed components.

Fig. 8.4. Thin-walled beam with a multi-cellular section.

$$\frac{t(s)}{b} \ll 1, \quad \frac{t(s)}{h} \ll 1, \quad \text{or} \quad \frac{t(s)}{\sqrt{b^2 + h^2}} \ll 1. \tag{8.1}$$

Of course, for beam theory to be a reasonable approximation to the structural behavior, the thin-walled beam must also be long, *i.e.*, $\sqrt{b^2 + h^2}/L \ll 1$.

8.1.2 Stress flows

As discussed in sections 5.4.2 and 5.5.2, the stress components acting in the plane of the cross-section are assumed to be negligible as compared to other stress components. This implies that $\sigma_2 \ll \sigma_1$ and $\sigma_3 \ll \sigma_1$ and furthermore, $\tau_{23} \ll \tau_{12}$ and $\tau_{23} \ll \tau_{13}$; it is then assumed that the only non-vanishing stress components are the axial stress, σ_1, and the transverse shear stresses, τ_{12} and τ_{13}.

Given the geometry of thin-walled beams described in the previous section, it is not convenient to work with the Cartesian components of transverse shear stress, rather, it is preferable to resolve the shear stress into its components parallel and normal to $\mathcal{C}$, denoted τ_s and τ_n, respectively, as illustrated in fig. 8.5. The relationship between these two sets of stress components is

$$\tau_n = \quad \cos\alpha\, \tau_{12} + \sin\alpha\, \tau_{13} = \tau_{12}\frac{\mathrm{d}x_3}{\mathrm{d}s} - \tau_{13}\frac{\mathrm{d}x_2}{\mathrm{d}s}, \tag{8.2a}$$

$$\tau_s = -\sin\alpha\, \tau_{12} + \cos\alpha\, \tau_{13} = \tau_{12}\frac{\mathrm{d}x_2}{\mathrm{d}s} + \tau_{13}\frac{\mathrm{d}x_3}{\mathrm{d}s}, \tag{8.2b}$$

where basic trigonometric relationships for triangle **PQR** reveal that $\cos \alpha = \mathrm{d}x_3/\mathrm{d}s$ and $\sin \alpha = -\mathrm{d}x_2/\mathrm{d}s$, and where the negative sign results from the sign convention for the curvilinear variable, s.

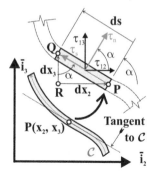

Fig. 8.5. Geometry of a differential element of the wall.

Using the principle of reciprocity of shear stresses, eq. (1.5), the normal shear stress component, τ_n, must vanish at the two edges of the wall because the outer surfaces of the beam are stress free. Furthermore, since the wall is very thin, no appreciable magnitude of this shear stress component can build up within the structure. As a result, it is assumed that τ_n vanishes through the wall thickness. The only non-vanishing shear stress component is the tangential shear stress component, τ_s, which is taken to be positive in the direction of s.

Inverting relations (8.2a) and (8.2b) yields $\tau_{12} = \tau_n \cos \alpha - \tau_s \sin \alpha$ and $\tau_{13} = \tau_n \sin \alpha + \tau_s \cos \alpha$. Because the normal shear stress component vanishes, $i.e.$, $\tau_n \approx 0$, the Cartesian components of stress can be expressed as

$$\tau_{12} \approx \tau_s \frac{\mathrm{d}x_2}{\mathrm{d}s}, \quad \tau_{13} \approx \tau_s \frac{\mathrm{d}x_3}{\mathrm{d}s}. \tag{8.3}$$

Finally, because the wall is very thin, it seems reasonable to assume that the non-vanishing stress component, τ_s, is *uniformly distributed across the wall thickness*. Figure 8.6 shows the axial and shear stress components through the thickness of the wall.

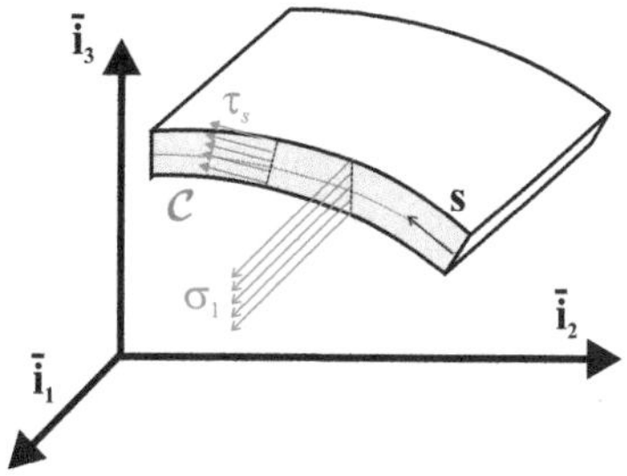

It is customary to introduce the concept of *stress flows* defined as

$$n(x_1, s) = \sigma_1(x_1, s)t(s), \tag{8.4a}$$
$$f(x_1, s) = \tau_s(x_1, s)t(s), \tag{8.4b}$$

Fig. 8.6. Uniform distributions of axial and shear stresses across the wall thickness.

where n is the *axial stress flow* or *axial flow*, and f is the *shearing stress flow* or *shear flow* taken *positive in the direction* of s. Using these definitions, instead of integrating a stress over an area to compute a force, it is only necessary to integrate a stress flow along curve $\mathcal{C}$. This will greatly simplify subsequent developments.

8.1.3 Stress resultants

The definitions of stress resultant in thin-walled beams are identical to those given in section 5.3 for beams with solid sections. Due to the thin wall assumption, integration over the beam's cross-sectional area reduces to an integration along curve $\mathcal{C}$. An infinitesimal area of the cross-section, $d\mathcal{A}$, can now be written as $d\mathcal{A} = t\,ds$, and the axial force, eq. (5.8), becomes

$$N_1(x_1) = \int_A \sigma_1 \, d\mathcal{A} = \int_C \sigma_1 t\,ds = \int_C n \, ds, \tag{8.5}$$

where the definition of the axial flow, eq. (8.4a), is used. The bending moments are found from eqs. (5.10) using a similar process

$$M_2(x_1) = \int_C n \, x_3 \, ds, \quad M_3(x_1) = -\int_C n \, x_2 \, ds. \tag{8.6}$$

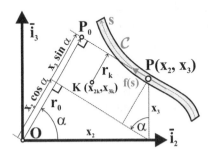

Fig. 8.7. Geometry of a differential element of the wall.

The shear forces acting along axes $\bar{\imath}_2$ and $\bar{\imath}_3$ can be calculated from the corresponding shear stress components, τ_{12} and τ_{13}, respectively, using eqs. (5.9) to find

$$V_2(x_1) = \int_C f \frac{dx_2}{ds} \, ds, \tag{8.7a}$$

$$V_3(x_1) = \int_C f \frac{dx_3}{ds} \, ds, \tag{8.7b}$$

where eq. (8.3) and the definition of the shear flow, eq. (8.4b), are used.

The torque computed about the origin, **O**, of the axis system can be expressed in the form of a vector cross product,

$$\underline{M}_O(x_1) = \int_C \underline{r}_P \times f \, d\underline{s},$$

where $\underline{r}_P = x_2\bar{\imath}_2 + x_3\bar{\imath}_3$ is the position vector of point **P**, see fig. 8.7. An increment in the curvilinear coordinate, s, can be written as $d\underline{s} = dx_2\bar{\imath}_2 + dx_3\bar{\imath}_3$, and the torque about the origin then becomes

$$\underline{M}_O(x_1) = \int_C (x_2 dx_3 - x_3 dx_2) f\bar{\imath}_1 = \int_C \left(x_2 \frac{dx_3}{ds} - x_3 \frac{dx_2}{ds} \right) f\bar{\imath}_1 \, ds.$$

Further inspection of fig. 8.7 reveals that the perpendicular distance from the origin, **O**, to point, **P₀**, on the tangent to curve C at point **P**, denoted r_o, is

$$r_o = x_2 \cos \alpha + x_3 \sin \alpha = x_2 \frac{dx_3}{ds} - x_3 \frac{dx_2}{ds}. \tag{8.8}$$

The magnitude of the torque evaluated with respect to the origin of the axes finally becomes

$$M_{1O}(x_1) = \int_C f r_o \, ds. \tag{8.9}$$

This result expresses the familiar formula for evaluating moments: the torque equals the magnitude of the force times the perpendicular distance from the point about which it is computed to the line of action of the force.

It will also be necessary to evaluate the torque about an arbitrary point of the cross-section, say point **K**, with coordinates (x_{2k}, x_{3k}), as shown in fig. 8.7, to find

$$M_{1k}(x_1) = \int_C f r_k \, ds, \tag{8.10}$$

where r_k is the perpendicular distance from point **K** to the line of action of the shear flow, f as shown in fig. 8.7. This distance is evaluated by replacing x_2 and x_3 in eq. (8.8) by $(x_2 - x_{2k})$ and $(x_3 - x_{3k})$, respectively, to find

$$r_k = (x_2 - x_{2k}) \cos \alpha + (x_3 - x_{3k}) \sin \alpha = r_o - x_{2k} \frac{dx_3}{ds} + x_{3k} \frac{dx_2}{ds}. \tag{8.11}$$

8.1.4 Sign conventions

Consider the thin wall segment depicted in fig. 8.8 extending from point **P** to point **Q**, located at distances a and b from origin **O** of the axis system, respectively. For this simple case, curve C is the straight line segment, **PQ**. Assume one analyst selects variable s describing curve C from point **P** to point **Q**, whereas another analyst selects variable s' describing the same curve from point **Q** to point **P**.

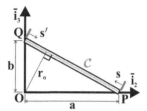

Fig. 8.8. Thin wall component.

An important sign convention is chosen for the shear flow: *the shear flow is positive in the direction of the curvilinear variable s.* This is an arbitrary *sign convention*, because the positive direction of the curvilinear coordinate is itself chosen arbitrarily. For the first analyst (using variable s), a positive shear flow is oriented from point **P** to point **Q**, whereas for the other (using variable s'), a positive shear flow is oriented from point **Q** to point **P**.

The geometry of curve C is described by its coordinates, $x_2(s)$ and $x_3(s)$, which are functions of the curvilinear variable s. In this case, curve C is simply a straight line, and its coordinates are

$$x_2(s) = a\left(1 - \frac{s}{\ell}\right), \quad x_3(s) = b\frac{s}{\ell},$$

where $\ell = \sqrt{a^2 + b^2}$ is the length of segment **PQ**. Using eq. (8.8), the perpendicular distance from the origin, **O**, to the tangent to curve C, denoted r_o, becomes

$$r_o = x_2 \frac{dx_3}{ds} - x_3 \frac{dx_2}{ds} = a\left(1 - \frac{s}{\ell}\right)\frac{b}{\ell} - b\frac{s}{\ell}\left(-\frac{a}{\ell}\right) = \frac{ab}{\ell}. \tag{8.12}$$

For the other analyst, the geometry of curve C is described in terms the curvilinear variable s', and its coordinates are

$$x_2(s') = a\frac{s'}{\ell}, \quad x_3(s') = b\left(1 - \frac{s'}{\ell}\right).$$

The perpendicular distance, from the origin, **O**, to the tangent to curve $\mathcal{C}$, denoted r'_o, now becomes

$$r'_o = x_2\frac{dx_3}{ds'} - x_3\frac{dx_2}{ds'} = a\frac{s'}{\ell}\left(-\frac{b}{\ell}\right) - b\left(1 - \frac{s'}{\ell}\right)\frac{a}{\ell} = -\frac{ab}{\ell}. \tag{8.13}$$

Because the two analysts describe the geometry of curve $\mathcal{C}$ in two different manners, the perpendicular distance from the origin, **O**, to the tangent to curve $\mathcal{C}$ is different. This distance becomes an algebraic quantity, *i.e.*, its sign depends on the selected direction of the curvilinear variable. Comparing eq. (8.12) to eq. (8.13), it is clear that $r'_o(s') = -r_o(s)$.

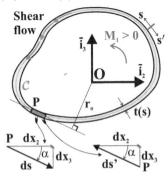

Fig. 8.9. Sign conventions for the shear flow, $f(s)$.

Figure 8.9 shows a thin-walled, closed section of arbitrary shape. Here again, one analyst selects variable s describing curve $\mathcal{C}$ in the counterclockwise direction, whereas another analyst selects variable s' describing curve $\mathcal{C}$ in the clockwise direction. The sign convention for the torque, M_1, is indicated on the figure and is defined as positive for a positive rotation about axis $\bar{\imath}_1$ according to the right-hand rule, see fig. 5.3. The sign convention for the torque is independent of the choice of the curvilinear variable.

Let $f(s)$ and $r_0(s)$ be the shear flow and normal distance associated with the choice of the curvilinear variable s, whereas $f'(s')$ and $r'_0(s')$ are the corresponding quantities associated with the choice of the curvilinear variable s'. Assuming that the shear flow arising from the application of a torque is physically oriented in the counterclockwise direction as indicated in fig. 8.9, it then follows that $f > 0$ whereas $f' < 0$, with $f' = -f$. Furthermore, The normal distance from point **O** to the tangent to curve $\mathcal{C}$ at point **P** is such that $r'_o(s') = -r_o(s)$.

The sign of both the shear flow and the normal distance are determined by the choice of direction for the curvilinear coordinate: $f'(s') = -f(s)$ and $r'_o(s') = -r_o(s)$. The resulting torque, however, is unaffected by this choice: $M_{1O} = \int_{\mathcal{C}} f r_o \, ds = \int_{\mathcal{C}} f' r'_o \, ds'$ because $f r_o = f' r'_o$. It is left to the reader to verify that the definition of the transverse shear forces, eq. (8.7), remains unaffected by the choice of the direction of the curvilinear variable.

8.1.5 Local equilibrium equation

Figure 8.10 shows a differential element of the thin-walled beam. The dimensions of the differential element are dx_1 along the axis of the beam, and ds along curve $\mathcal{C}$.

The axial and shear flows are acting on the faces at x_1 and s, and a Taylor's series expansion is used to evaluate the axial and shear flows on the opposite faces. The shear flow is positive in the increasing direction of the curvilinear variable, s. Body forces are neglected for this differential element of the thin-walled beam. Summing up all the forces acting on this free-body diagram along axis $\bar{\imath}_1$ yields

$$-n \, ds + \left(n + \frac{\partial n}{\partial x_1} \, dx_1 \right) \, ds - f \, dx_1 + \left(f + \frac{\partial f}{\partial s} \, ds \right) \, dx_1 = 0.$$

After simplification, the equilibrium condition becomes

$$\frac{\partial n}{\partial x_1} + \frac{\partial f}{\partial s} = 0. \tag{8.14}$$

This local equilibrium equation implies that any change in axial stress flow, n, along the beam axis must be equilibrated by a corresponding change in shear flow, f, along curve $\mathcal{C}$ that defines the cross-section.

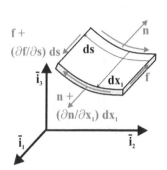
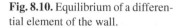
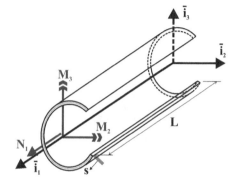

Fig. 8.10. Equilibrium of a differential element of the wall.

Fig. 8.11. Thin-walled beam subjected to axial forces and bending moments.

8.2 Bending of thin-walled beams

Consider a thin-walled beam subjected to axial forces and bending moments, as shown in fig. 8.11. Axes $\bar{\imath}_2$ and $\bar{\imath}_3$ are located at the centroid of the cross-section. The Euler-Bernoulli assumptions discussed in section 5.1 are equally applicable to the bending of thin-walled beams with either open or closed cross-sections. Hence, assuming a displacement field in the form of eq. (6.1) results in the strain field given by eqs. (6.2a) to eq. (6.2c), and the distribution of axial stresses given by eq. (6.15) follows

$$\sigma_1 = E \left[\frac{N_1}{S} - \frac{x_2 H_{23}^c - x_3 H_{33}^c}{\Delta_H} M_2 - \frac{x_2 H_{22}^c - x_3 H_{23}^c}{\Delta_H} M_3 \right], \tag{8.15}$$

where $\Delta_H = H_{22}^c H_{33}^c - (H_{23}^c)^2$. Using eq. (8.4a), the axial flow distribution over the cross-section now becomes

$$
n(x_1, s) = E(s)t(s) \left[\frac{N_1(x_1)}{S} - \frac{x_2(s)H_{23}^c - x_3(s)H_{33}^c}{\Delta_H} M_2(x_1) \right.
$$
$$
\left. - \frac{x_2(s)H_{22}^c - x_3(s)H_{23}^c}{\Delta_H} M_3(x_1) \right].
$$
(8.16)

8.2.1 Problems

Problem 8.1. Sign conventions
Verify that the definition of the transverse shear forces, eq. (8.7), remains unaffected by the choice of the direction of the curvilinear variable.

Problem 8.2. Thin-walled "Z" shaped cross-section beam
Figure 8.12 shows the cross-section of a thin-walled, "Z" shaped beam skewed at an angle α with respect to axis $\bar{\imath}_2$. (1) Find the centroidal bending stiffnesses. (2) For $M_2 = M_0$ and $M_3 = 0$, find the neutral axis orientation with respect to axis $\bar{\imath}_2$. (3) Determine the location and magnitude of the maximum axial stress. Use $b = h/2$ and $\sin \alpha = 4/5$.

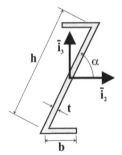

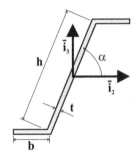

Fig. 8.12. "Z" shaped cross-section.   **Fig. 8.13.** "Z" shaped cross-section.

Problem 8.3. Thin-walled "Z" shaped cross-section beam
Figure 8.13 shows the cross-section of a thin-walled, "Z" shaped beam skewed at an angle α with respect to axis $\bar{\imath}_2$. (1) Find the centroidal bending stiffnesses. (2) For $M_2 = M_0$ and $M_3 = 0$, find the neutral axis orientation with respect to axis $\bar{\imath}_2$. (3) Determine the location and magnitude of the maximum axial stress. Use $b = h/2$ and $\sin \alpha = 4/5$.

Problem 8.4. Thin-walled "L" shaped cross-section beam
Figure 8.14 shows a thin-walled beam with an "L" shaped cross-section. The cantilevered beam is of length $L = 48$ in and carries a tip load, $P = 200$ lbs, applied along axis $\bar{\imath}_3$. (1) Determine the location of the centroid. (2) Find the centroidal bending stiffnesses. (3) Determine the orientation of the neutral axis. (4) Determine the axial stress distribution over the cross-section. Find the location and magnitude of the maximum axial stress. Use $b = h = 2.0$ in, $t_b = t_h = 0.100$ in and $E = 10.6 \times 10^6$ psi.

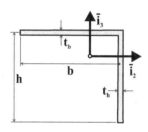

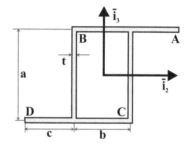

Fig. 8.14. "L" shaped cross-section.   **Fig. 8.15.** Box-Z shaped cross-section.

Problem 8.5. Thin-walled "Box-Z" shaped cross-section beam

A cantilevered beam of length L is constructed with the thin-walled cross-section shown in fig. 8.15. A concentrated load, P, is applied at the tip of the beam and acts along axis $\bar{\imath}_3$. *(1)* Determine the location of the centroid. *(2)* Find the centroidal bending stiffnesses. *(3)* Determine the axial stress acting in the root section at points **A** and **B**. *(4)* Determine the vertical and horizontal components of the deflection at the tip using the given centroidal axes shown in the figure. Use $b = c = a/2$.

Problem 8.6. Thin-walled angle section

A beam of length L with the thin-walled cross-section shown in fig. 8.16 is simply supported at both ends and carries a transverse distributed loading, p_0, acting along axis $\bar{\imath}_2$ at point **C**. *(1)* Determine the location of the centroid. *(2)* Find the centroidal bending stiffnesses. *(3)* Find the location and magnitude of the maximum axial stress. *(4)* Sketch the distribution of axial stress over the cross-section for the section where the maximum bending moment occurs.

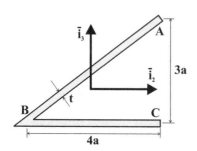

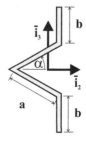

Fig. 8.16. Angled "L" shaped cross-section.

Fig. 8.17. Horizontal "V" section with vertical flanges.

Problem 8.7. Thin-walled "V" shaped cross-section beam

A beam of length L with the thin-walled cross-section shown in fig. 8.17 is simply supported at both ends and carries a transverse distributed loading, p_0, acting along axis $\bar{\imath}_2$ at point **C**. *(1)* Determine the location of the centroid. *(2)* Find the centroidal bending stiffnesses. *(3)* Find the location and magnitude of the maximum axial stress. *(4)* Sketch the distribution of axial stress over the cross-section for the section where the maximum bending moment occurs.

Problem 8.8. Skewed "I" shaped cross-section

A cantilevered beam of length L is constructed with the thin-walled, skewed "I" shaped cross-section shown in fig. 8.18. The wall thickness for both flanges and web is a constant, t. Axis $\bar{\imath}_2$ is an axis of symmetry of the section. A concentrated load, P, is applied at the tip of the beam and acts along axis $\bar{\imath}_3$. *(1)* Determine the location of the centroid. *(2)* Find the centroidal bending stiffnesses. *(3)* Determine the axial stress acting in the root section at points **A**, **B**, and **C**.

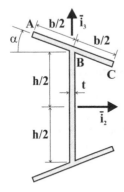
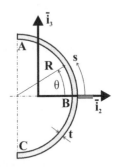

Fig. 8.18. Skewed "I" shaped cross-section.   **Fig. 8.19.** Semi-circular open cross-section.

Problem 8.9. Thin-walled semi-circular cross-section beam

A beam of length L with the thin-walled, semi-circular cross-section shown in fig. 8.19 is simply supported at both ends and carries a transverse distributed loading, p_0, acting along axis $\bar{\imath}_2$ at point **B**. *(1)* Determine the location of the centroid. *(2)* Find the centroidal bending stiffnesses. *(3)* Find the location and magnitude of the maximum axial stress. *(4)* Sketch the distribution of axial stress over the cross-section for the section where the maximum bending moment occurs. **Note:** It is more convenient to work with the angle θ as a variable describing the geometry of the section: $s = R\theta$, $\mathrm{d}s = R\mathrm{d}\theta$.

Problem 8.10. Thin-walled semi-circular cross-section with flanges

A beam of length L with the thin-walled, semi-circular cross-section with flanges shown in fig. 8.20 is simply supported at both ends and carries a transverse distributed loading, p_0, acting along axis $\bar{\imath}_2$ at point **C**. *(1)* Determine the location of the centroid. *(2)* Find the centroidal bending stiffness. *(3)* Find the location and magnitude of the maximum axial stress. *(4)* Sketch the distribution of axial stress over the cross-section for the section where the maximum bending moment occurs.

Problem 8.11. Thin-walled C-channel with variable thickness

A beam of length L with the thin-walled, C-channel cross-section shown in fig. 8.21 is simply supported at both ends and carries a transverse distributed loading, p_0, acting along axis $\bar{\imath}_2$ at point **C**. *(1)* Determine the location of the centroid. *(2)* Find the centroidal bending stiffnesses. *(3)* Find the location and magnitude of the maximum axial stress. *(4)* Sketch the distribution of axial stress over the cross-section for the section where the maximum bending moment occurs.

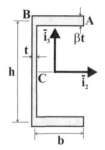

Fig. 8.20. Semi-circular section with vertical flanges.

Fig. 8.21. C-channel with variable flange thickness.

8.3 Shearing of thin-walled beams

In most practical cases, the bending moments considered in the previous section are accompanied by transverse shear forces, which give rise to shear flow distributions over the cross-section. This distribution is evaluated by introducing the axial flow given by eq. (8.16) into the local equilibrium equation, eq. (8.14), to find

$$\frac{\partial f}{\partial s} = -Et \left[\frac{1}{S} \frac{dN_1}{dx_1} - \frac{x_2 H_{23}^c - x_3 H_{33}^c}{\Delta_H} \frac{dM_2}{dx_1} - \frac{x_2 H_{22}^c - x_3 H_{23}^c}{\Delta_H} \frac{dM_3}{dx_1} \right]. \quad (8.17)$$

The sectional equilibrium equations, eqs. (6.16), (6.18), and (6.20), are introduced for the moment derivatives, and to simplify this expression, it is further assumed that the distributed axial loads, p_1, and moments, q_2 and q_3, are zero, to find

$$\frac{\partial f}{\partial s} = -E(s)t(s) \left[-\frac{x_2 H_{23}^c - x_3 H_{33}^c}{\Delta_H} V_3 + \frac{x_2 H_{22}^c - x_3 H_{23}^c}{\Delta_H} V_2 \right]. \quad (8.18)$$

Integration of this differential equation then yields the shear flow distribution arising from shear forces, V_2 and V_3, as

$$f(s) = c - \int_0^s Et \left[-\frac{x_2 H_{23}^c - x_3 H_{33}^c}{\Delta_H} V_3 + \frac{x_2 H_{22}^c - x_3 H_{23}^c}{\Delta_H} V_2 \right] ds, \quad (8.19)$$

where c is an integration constant corresponding to the value of the shear flow at $s = 0$. The procedure to determine this integration constant depends on whether the cross-section is open or closed. Because the bending stiffnesses and shear forces are functions of variable x_1 alone, they can be factored out of the integral, leading to

$$f(s) = c + \frac{Q_3(s)H_{23}^c - Q_2(s)H_{33}^c}{\Delta_H} V_3 - \frac{Q_3(s)H_{22}^c - Q_2(s)H_{23}^c}{\Delta_H} V_2, \quad (8.20)$$

where the *stiffness static moments*, also called *stiffness first moments*, are defined as

$$Q_2(s) = \int_0^s Ex_3(s) \, tds; \quad Q_3(s) = \int_0^s Ex_2(s) \, tds. \quad (8.21)$$

These integrals are the static moments for the portion of the cross-section from $s = 0$ to s, and thus, Q_2 and Q_3 are functions of s.

8.3.1 Shearing of open sections

For open sections, the principle of reciprocity of shear stresses, eq. (1.5), implies the vanishing of shear flow at the end points of curve $\mathcal{C}$. Indeed, if a shear flow does exist at those points, a non-vanishing shear stress must also act along the lateral edge of the beam, which is assumed to be stress free. For instance, the shear flow at points **A** and **D** of the cross-section of the C-channel depicted in fig. 8.24 must vanish, because edges **AE** and **DF** are stress free. If the origin of the curvilinear coordinate, s, is chosen to be located at such a stress free edge, the integration constant, c, in eq. (8.20) must vanish.

The shear flow distribution over the open cross-section of a thin-walled beam subjected to transverse shear forces can be determined using the following procedure.

1. Compute the location of the centroid of the cross-section and select a set of centroidal axes, $\bar{\imath}_1$ and $\bar{\imath}_2$, at this point; compute the sectional centroidal bending stiffnesses H_{22}^c, H_{33}^c and H_{23}^c. If desired, principal centroidal axes of bending may also be used, in which case $H_{23}^c = 0$.
2. Select suitable curvilinear coordinates, s, that describe the geometry of the cross-section. It will often be simpler to define several curvilinear coordinates to describe the entire contour, $\mathcal{C}$, of the cross-section.
3. Evaluate the first stiffness moments as functions of position, s, along contour, $\mathcal{C}$, of the cross-section, using eqs. (8.21).
4. The shear flow distribution, $f(s)$, then follows from eq. (8.20).

8.3.2 Evaluation of stiffness static moments

The stiffness static moments, Q_2 and Q_3, defined by eqs. (8.21), are key to the evaluation of the shear flow distribution over thin-walled cross-sections. Consider the homogeneous, thin-walled rectangular strip oriented at an angle, α, with respect to axis $\bar{\imath}_2$ as depicted in fig. 8.22. The stiffness static moment, $Q_2(s)$, is readily computed as

$$Q_2(s) = \int_0^s Ex_3 \, t ds = E \int_0^s (d_3 + s\sin\alpha) \, t ds = E \, st \left(d_3 + \frac{s}{2}\sin\alpha\right). \quad (8.22)$$

This result can be interpreted as follows: the stiffness static moment is the stiffness static moment of a portion of the strip from 0 to s and equals the product of Young's modulus times the area, st, times the value of the coordinate, $x_3 = d_3 + s/2 \sin\alpha$, which is the coordinate of the centroid of the local area, st (*i.e.*, at the area midpoint).

A similar result can be obtained for the other stiffness static moment,

$$Q_3(s) = E \, st \, (d_2 + s/2 \, \cos\alpha), \quad (8.23)$$

and can be interpreted in the same manner. Note that since the strip is made of a homogeneous material, Young's modulus factors out of the integral; hence, $Q_2(s) = E \int_0^s x_3 \, t ds$, where $\int_0^s x_3 \, t ds$ represents the *area static moment* of the strip.

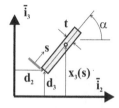

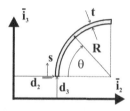

Fig. 8.22. Stiffness static moments for a thin-walled rectangular strip at an angle α.

Fig. 8.23. Stiffness static moments for a thin-walled circular arc.

Consider next the thin-walled homogeneous circular arc of radius R depicted in fig. 8.23. Length along the arc is measured by the curvilinear coordinate s, but in view of the circular geometry of the problem, it is preferable to work with angle θ. Noting that $ds = Rd\theta$, eq. (8.21) yields

$$Q_2(s) = \int_0^s Ex_3 \, t ds = Et \int_0^\theta (d_3 + R\sin\theta) \, R d\theta = EtR^2 \left(\frac{d_3}{R}\theta + 1 - \cos\theta \right).$$
(8.24)

A similar development yields $Q_3(s) = EtR^2[(1 + d_2/R)\theta - \sin\theta]$.

The key insight here is that the stiffness static moment of any arbitrary area with a given Young's modulus, E, is simply the product of the modulus, the area, and the distance to the area centroid. In other words, $Q_2 = EAx_{3c}$ or $Q_3 = EAx_{2c}$, where x_{2c} and x_{3c} are the distances to the centroid of the area, A. This result is essentially a statement of the parallel axis theorem, see section 6.8.1, applied to stiffness static moments, but in this case, the result is only the transport term because the static moment about the area centroid itself is zero by definition.

8.3.3 Shear flow distributions in open sections

The calculation of the shear flow in thin-walled open cross-sections will now be illustrated using several examples.

Example 8.1. Shear flow distribution in a C-channel
Evaluate the distribution of shear flow over the thin-walled C-channel[1] shown in fig. 8.24. The section has a uniform thickness, t, a vertical web height, h, a flange width, b, and is subjected to a vertical shear force, V_3, at the specific span-wise location where the shear flow is to be computed. The origin of the axes on the section is placed at the centroid, which is located at a distance $d = b/(2 + h/b)$ to the right of the web's mid-point. Because the section is symmetric about axis $\bar{\imath}_2$, these axes are principal centroidal axes of bending, *i.e.*, H_{23}^c vanishes.

Because the section is subjected to the shear force V_3 only, the shear flow distribution, given by eq. (8.20), reduces to

---

[1] This cross-section is referred to with various names: a "C" section, a channel section, or a C-channel.

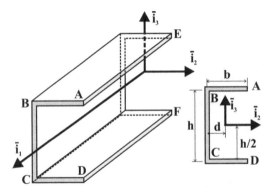

Fig. 8.24. Cantilevered beam with a C-channel cross-section.

$$f(s) = c - \frac{Q_2(s)}{H_{22}^c}V_3,$$
(8.25)

where the bending stiffness is easily evaluated as

$$H_{22}^c = E\left[\frac{th^3}{12} + 2bt\left(\frac{h}{2}\right)^2\right] = E\left(\frac{h^3}{12} + \frac{bh^2}{2}\right)t.$$

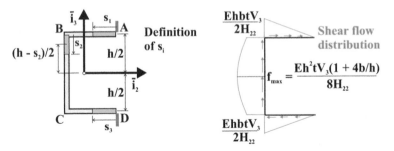

Fig. 8.25. Distribution of shear flow over the C-channel cross-section.

To simplify the algebra, the curvilinear coordinate describing the cross-section's geometry is broken into three parts: s_1 across the upper flange, s_2 down the vertical web, and s_3 across the lower flange, as depicted in fig. 8.25. For the section's upper flange, eq. (8.25) yields the shear flow distribution as

$$f(s_1) = c_1 - \frac{Q_2(s_1)}{H_{22}}V_3 = 0 - \frac{Ets_1h/2}{H_{22}^c}V_3 = -\frac{Ehts_1}{2}\frac{V_3}{H_{22}^c},$$
(8.26)

where the stiffness static moment, $Q_2(s)$, for the thin rectangular strip is evaluated with the help of eq. (8.22). The integration constant, c_1, vanishes because the shear flow must vanish at point **A**, where $s_1 = 0$.

Next, consider the section's vertical web. Equation (8.22) yields the stiffness static moment as $Q_2(s_2) = Ets_2(h - s_2)/2$, and the corresponding shear flow distribution then follows from eq. (8.25) as

$$f(s_2) = c_2 - \frac{h - s_2}{2} t s_2 \frac{EV_3}{H_{22}^c} = -\frac{1}{2} [bh + s_2(h - s_2)] \frac{tEV_3}{H_{22}^c}. \tag{8.27}$$

The integration constant, c_2, is evaluated by enforcing the continuity of the shear flow: $f(s_2 = 0) = f(s_1 = b)$, leading to $c_2 = -hb/2 \, EV_3/H_{22}^c$.

Finally, eq. (8.22) yields the stiffness static moment of the lower flange, where $x_3 = -h/2$, as $Q_2(s_3) = -E \, ts_3 \, h/2$. The corresponding shear flow then follows from eq. (8.25) as

$$f(s_3) = c_3 + \frac{E \, ts_3 \, h/2}{H_{22}^c} V_3 = \frac{hs_3}{2} \frac{tEV_3}{H_{22}^c}, \tag{8.28}$$

where the integration constant, c_3, vanishes because $f(s_3 = 0) = 0$.

The present solution also satisfies the shear flow continuity condition at point **C**, although this condition is not explicitly enforced. Indeed, the above results imply $f(s_2 = h) = -1/2 \, bhtEV_3/H_{22}^c$ and $f(s_3 = b) = 1/2 \, bhtEV_3/H_{22}^c$, i.e., $f(s_2 = h) + f(s_3 = b) = 0$. Note the algebraic nature of shear flows: the two shear flow add up to zero because curvilinear variables s_2 and s_3 both converge towards point **C**.

The shear flow distribution along the cross-section is plotted in the right portion of fig. 8.25. In this plot, the shear flow direction is indicated by arrows along the section contour and the magnitude of the shear flow is represented by the curve plotted along the contour; the shear flow's magnitude is proportional to the distance from the contour line to the curve, measured in the direction perpendicular to the cross-section contour. In the vertical web, the shear flow varies parabolically, and its algebraic value is negative, see eq. (8.27). Because the curvilinear coordinate, s_2, is positive down while the shear flow is negative, it implies that the shear flow and associated shear stresses are actually pointing up the vertical web; this physical direction of the shear flow is indicated by the arrows in fig. 8.25.

The shear flow has certain characteristics that deserve further discussion. First, the shear flows in the upper and lower flanges are linearly distributed along the flanges and vanish at the edges. Second, the shear flow in the vertical web varies in a quadratic manner. Finally, the maximum shear flow is found at the mid-point of the vertical web, and its magnitude is indicated in fig. 8.25.

Example 8.2. Shear flow continuity conditions

In the previous example, shear flow continuity conditions are imposed at point **B** and **C** of the C-channel depicted in fig. 8.24. The continuity condition can be obtained from simple equilibrium arguments: consider the free-body diagram of the two-wall joint, where two walls are connected at point **J**, as illustrated in fig. 8.26. This represents the configuration of the present problem at point **B**. Due to the principle of reciprocity of shear stresses, eq. (1.5), the shear flows acting in the plane of the cross-section must be equilibrated by shear flows acting on orthogonal faces. Equilibrium of forces acting along the beam's axis then yields $-f_1 + f_2 = 0$, or $f_1 = f_2$, *i.e.*, the shear flow must be continuous at the junction of the upper flange and vertical web, $f(s_1 = b) = f(s_2 = 0)$.

Figure 8.26 also illustrates the situation at the edge of the wall, labeled point **E**. Due to the principle of reciprocity of shear stresses, the shear flow at the edge of the wall must vanish, because the orthogonal face is a stress free edge of the beam.

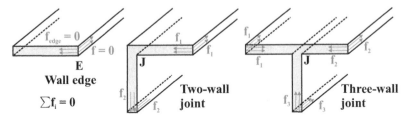

Fig. 8.26. Equilibrium condition at the junction of two or more thin walls.

Finally, fig. 8.26 depicts a more elaborate three-wall joint configuration, where multiple walls are connected together at point **J**. Equilibrium of forces acting along axis $\bar{\imath}_1$ of the beam yields $-f_1 - f_2 - f_3 = 0$, or more generally,

$$\sum f_i = 0. \tag{8.29}$$

This simple equilibrium argument implies that *the sum of the shear flows converging to a joint must vanish*. In practical applications of eq. (8.29), the shear flows must be interpreted as *algebraic quantities*: each shear flow is positive in the direction of the corresponding curvilinear coordinate. For instance, application of eq. (8.29) to the two-wall configuration illustrated in fig. 8.26 yields $(+f_1) + (-f_2) = 0$, or $f_1 = f_2$, as expected. Equation (8.29) applies at wall joints and edges; in the former case, the sum extends over all walls connected at the joint, whereas in the latter case, the sum reduces to a single term, enforcing the vanishing of the shear flow at the wall's edge.

Example 8.3. *Shear flow distribution in a C-channel*
The choice of the curvilinear coordinates is entirely arbitrary as long as each point of the cross-sectional contour is uniquely defined. For the C-channel treated in example 8.1, fig. 8.27 shows an alternative definition of the curvilinear variable: s_1 now runs across the lower flange, s_2 up the vertical web, starting from its mid-point, and s_3 across the upper flange, starting at point **B**. As in the previous example, eq. (8.25) defines the shear flow; in the lower flange, it leads to

$$f(s_1) = 0 + \frac{thb}{2} \left(\frac{s_1}{b} \right) \frac{EV_3}{H_{22}^c}, \tag{8.30}$$

where the term s_1/b is the non-dimensional length across the lower flange.

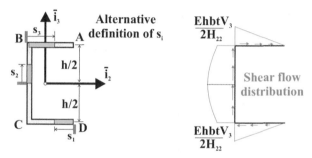

Fig. 8.27. Distribution of shear flow over the C-channel cross-section for an alternative definition of the curvilinear variable.

For the vertical web, eq. (8.22) yields the stiffness static moment as $Q_2 = E\, s_2 t\, s_2/2$, and eq. (8.25) then leads to the following shear flow distribution

$$
f(s_2) = c_2 - \frac{s_2^2 t}{2} \frac{EV_3}{H_{22}^c} = \frac{hbt}{2} \frac{EV_3}{H_{22}^c} + \frac{h^2 t}{8}\left[1 - \left(\frac{2s_2}{h}\right)^2\right]\frac{EV_3}{H_{22}^c}, \tag{8.31}
$$

where the integration constant is evaluated with the joint equilibrium condition, eq. (8.29). At point **C**, $f(s_1 = b) = f(s_2 = -h/2)$, leading to $hbt/2\; EV_3/H_{22}^c = c_2 - h^2 t/8\; EV_3/H_{22}^c$.

Finally, the shear flow distribution along the upper flange is found in a similar manner

$$
f(s_3) = c_3 - \frac{Q_2}{H_{22}^c} V_3 = c_3 - \frac{E\, s_3 t\, h/2}{H_{22}^c} V_3 = \frac{hbt}{2}\left(1 - \frac{s_3}{b}\right)\frac{EV_3}{H_{22}^c}. \tag{8.32}
$$

Here again, the integration constant is determined from the joint equilibrium condition, eq. (8.29). At point **B**, $f(s_2 = h/2) = f(s_3 = 0)$, which implies $hbt/2\; EV_3/H_{22}^c = c_3$. Equilibrium requires the shear flow to vanish at $s_3 = b$; this condition is satisfied by the present solution, although it is not explicitly enforced.

The overall distribution of shear flow for this alternative formulation is shown in fig. 8.27; it is, of course, identical to that found in example 8.1. For either approach, the maximum shear flow occurs at the web's mid-point and is

$$
f_{\max} = \frac{EV_3 h^2 t}{8 H_{22}^c}\left(1 + \frac{4b}{h}\right). \tag{8.33}
$$

While the results found here are physically identical to those found in example 8.1, sign differences will occur because of the different choices for the curvilinear variable, s. For instance, the shear flow distribution over the vertical web, see eqs. (8.27) and (8.31), are of opposite sign, reflecting the opposite choices for the direction of the curvilinear variable s_2, see fig. 8.25 and 8.27. Similar remarks can be made concerning the shear flow distributions in the upper and lower flanges.

Example 8.4. Shear flow distribution in an open triangular section - A

Figure 8.28 shows a thin-walled, homogeneous section in the shape of a triangle open at point **A** and subjected to a vertical shear force, V_3. The opening at point **A** is simply a small cut in the wall that does not affect the dimensions of the cross-section. The width and height of the section are specified in multiples of the wall thickness, and to simplify the computations, the upper and lower halves are right triangles, whose side lengths have ratios 5:12:13.

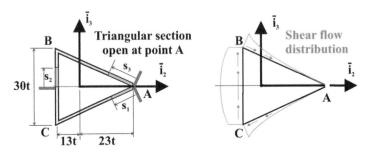

Fig. 8.28. Thin-walled, open triangular section

Since axis $\bar{\imath}_2$ is an axis of symmetry, the axes depicted on the figure are principal axes of bending, *i.e.*, $H_{23}^c = 0$. It is also readily verified that the origin of the centroidal axes is located as shown in fig. 8.28. The bending stiffness of the section can be computed using the triangle area equivalence method developed in section 6.8.3 as

$$H_{22}^c = 2E\left[\left(\frac{2}{3}39t^2\right)\left(\frac{15t}{2}\right)^2 + \left(\frac{30t^2}{6} + \frac{39t^2}{6}\right)(15t)^2\right] = 8100\ Et^4,\quad (8.34)$$

where it is noted that the section's upper flange is of length $39t$.

The curvilinear coordinate, s, along the cross-section is broken into three components starting at point **A**: s_1 runs down the lower flange, s_2 up the vertical web, starting from its mid-point, and s_3 along the upper flange, starting from point **A**, as shown in fig. 8.28.

From eq. (8.20), the shear flow distribution in the section's lower flange simplifies to $f(s_1) = c_1 - Q_2(s_1)V_3/H_{22}^c$. Equation (8.22) yields the stiffness static moment as $Q_2(s_1) = -E\ s_1 t\ s_1/2 \sin\alpha$, where α is the angle between the upper flange and axis $\bar{\imath}_2$, with $\sin\alpha = 15/39$. The shear flow distribution then becomes

$$f(s_1) = 0 + \frac{5}{26}s_1^2 t\frac{EV_3}{H_{22}^c} = \frac{13}{360}\left(\frac{s_1}{39t}\right)^2\frac{V_3}{t},\qquad (8.35)$$

where the integration constant is evaluated from the condition that $f(s_1 = 0) = 0$. The non-dimensional variable $s_1/39t$ is used to simplify the expression; it runs from 0 at point **A** to 1 at point **C**.

Next, the stiffness static moment for the section's vertical web is $Q_2(s_2) = E\, s_2 t\, s_2/2$, and the corresponding shear flow distribution becomes

$$f(s_2) = c_2 - \frac{1}{2}s_2^2 \frac{tEV_3}{H_{22}^c} = \frac{13}{360}\frac{V_3}{t} + \frac{1}{72}\left[1 - \left(\frac{s_2}{15t}\right)^2\right]\frac{V_3}{t}, \tag{8.36}$$

where the integration constant, c_2, is evaluated using the joint equilibrium condition, eq. (8.29), at point **C**: $f(s_1 = 39t) = f(s_2 = -15t)$.

Finally, due to symmetry, the shear flow distribution along the upper flange is identical to that along the lower flange, except for a change in sign due to the change in sign of x_3,

$$f(s_3) = -\frac{13}{360}\left(\frac{s_3}{39t}\right)^2 \frac{V_3}{t}. \tag{8.37}$$

Although not explicitly enforced, the joint equilibrium condition at point **B** is satisfied by the present solution; indeed, $f(s_2 = 15t) + f(s_3 = 39t) = 0$.

Figure 8.28 shows the computed shear flows distributions over the cross-section. The magnitude of the shear flow is represented by the curve plotted along the cross-sectional contour where the shear flow is proportional to the normal distance from the contour to the curve and where the direction of the shear flow is indicated by the arrows.

Example 8.5. Shear flow distribution in an open triangular section - B

The thin-walled triangular section of height h and width b treated in the previous example is considered again, but the section is now open at point **B**, the vertical web's mid-point, as illustrated in fig. 8.29.

First, the bending stiffness of the section is evaluated with the help of eq. (6.53) to find

$$H_{22}^c = E\frac{th^3}{12} + 2E\left[\frac{t\ell^3}{12}\sin^2\alpha + t\ell\left(\frac{h}{4}\right)^2\right] = \frac{Et\ell h^2}{6}(1 + \sin\alpha),$$

where ℓ is the length of the flange (note that $h/2 = \ell\sin\alpha$). This expression should be compared with eq. (8.34) found using the triangle area equivalence method; it is left to the reader to show that the two expressions are identical. In view of the symmetry of the problem, the axes shown on the figure are principal centroidal axes of bending. The centroid's location along axis $\bar{\imath}_2$ will not be required to evaluate the shear flow distribution, and hence, its computation is omitted.

The shear flow distribution generated by a vertical shear force, V_3, will be computed by directly integrating the governing differential equation, rather than using the stiffness static moment defined in eq. (8.22). For this problem, eq. (8.18) reduces to

$$\frac{df}{ds} = -\frac{Et}{H_{22}^c}x_3 V_3 = -\frac{6}{\ell h^2(1 + \sin\alpha)}x_3 V_3. \tag{8.38}$$

For convenience, the section is broken into four straight segments and a curvilinear variable is defined along each segment, as shown in fig. 8.29.

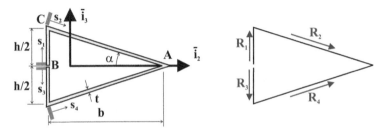

Fig. 8.29. Thin-walled, open triangular section.

For the upper part of the vertical web defined by s_1, $x_3 = s_1$, and integration of eq. (8.38) leads to

$$f(\bar{s}_1) = -\frac{6}{\ell h^2(1+\sin\alpha)}\frac{\bar{s}_1^2}{2}V_3 = -\frac{3\bar{s}_1^2}{4(1+\sin\alpha)}\frac{V_3}{\ell},$$

where $\bar{s}_1 = 2s_1/h$ is the non-dimensional variable along the vertical web.

For the upper flange defined by s_2, $x_3 = h/2 - s_2\sin\alpha$, and eq. (8.38) yields

$$f(\bar{s}_2) = -\frac{6}{\ell h^2(1+\sin\alpha)}\left(\frac{hs_2}{2} - s_2^2/2\sin\alpha\right)V_3 + c_2$$

$$= -\frac{3(\bar{s}_2 - \bar{s}_2^2/2)}{1+\sin\alpha}\frac{V_3}{\ell} - \frac{3}{4(1+\sin\alpha)}\frac{V_3}{\ell},$$

where $\bar{s}_2 = s_2/\ell$ is the non-dimensional curvilinear variable along the top flange. Integration constant, c_2, is evaluated by enforcing the joint equilibrium condition, eq. (8.29). At point **C**, $f(\bar{s}_1 = 1) - f(\bar{s}_2 = 0) = 0$. Due to the symmetry of the section, $f(\bar{s}_3) = -f(\bar{s}_1)$ and $f(\bar{s}_4) = -f(\bar{s}_2)$, where $f(\bar{s}_3)$ and $f(\bar{s}_4)$ are the shear flow distributions in the lower part of the vertical web and lower flange, respectively.

The net force resulting from the shear flow distribution in the upper part of the web is found by integration as

$$R_1 = \int_0^{h/2} f_1\,ds_1 = -\frac{3}{4(1+\sin\alpha)}\frac{V_3}{\ell}\frac{h}{2}\int_0^1 \bar{s}_1^2\,d\bar{s}_1 = -\frac{\sin\alpha}{4(1+\sin\alpha)}V_3.$$

Similarly, the net force resulting from the shear flow distribution in the upper flange is

$$R_2 = \int_0^\ell f_2\,ds_2 = -\frac{2+3\sin\alpha}{4\sin\alpha(1+\sin\alpha)}V_3.$$

The net force resultants in the lower part of the section are found by symmetry: $R_3 = -R_1$ and $R_4 = -R_2$. The shear flow resultants for the section's four segments are shown in fig. 8.29.

It is interesting to point out the counter-intuitive result that the shear flow distribution in the vertical web is pointing down, although the applied vertical shear force is pointing up. Indeed, $R_1 < 0$ and $R_3 > 0$, while s_1 is pointing up and s_3 is pointing down. To understand this result, it is necessary to consider the equilibrium of the entire section. First, summing all forces along axis $\bar{\imath}_2$ yields $R_2 \cos \alpha + R_4 \cos \alpha = 0$; this is expected because no external shear force is applied along this axis. Second, summation of all forces in the vertical direction gives $R_1 - R_3 - R_2 \sin \alpha + R_4 \sin \alpha = 2R_1 - 2R_2 \sin \alpha = V_3$, as can be verified by using the previously determined values of R_1 and R_2. This result is also expected: the resultant force in the vertical direction is equal to the externally applied shear force, V_3. The shear flow distributions in the upper and lower flanges have a net contribution in the upward vertical direction that overcomes the downward contribution of the shear flow in the vertical web to equilibrate the externally applied vertical shear force, V_3.

8.3.4 Problems

Problem 8.12. Thin-walled "Z" shaped cross-section beam
Consider the thin-walled "Z" shaped cross-section beam shown in fig. 8.12. *(1)* Determine the shear flow distribution in the section under a vertical shear force, V_3. *(2)* Verify that all joint and edge equilibrium conditions, eq. (8.29), are satisfied. *(3)* Find the magnitude and location of the maximum shear stress. Use $b = h/2$ and $\sin \alpha = 4/5$.

Problem 8.13. Thin-walled skewed "Z" shaped cross-section beam
Consider the thin-walled skewed "Z" shaped cross-section beam shown in fig. 8.13. *(1)* Determine the shear flow distribution in the section under a vertical shear force, V_3. *(2)* Verify that all joint and edge equilibrium conditions, eq. (8.29), are satisfied. *(3)* Find the magnitude and location of the maximum shear stress. Use $b = h/2$ and $\sin \alpha = 4/5$.

Problem 8.14. Thin-walled "L" shaped cross-section beam
Consider the thin-walled "L" section beam shown in fig. 8.14. *(1)* Determine the location of the centroidal axes and compute the centroidal bending stiffness. *(2)* Determine the shear flow distribution in the section under a vertical shear force, V_3. *(3)* Verify that all joint and edge equilibrium conditions, eq. (8.29), are satisfied. *(4)* Find the magnitude and location of the maximum shear stress. Use $t_h = t_b = t$ and $h = 2b$.

Problem 8.15. Thin-walled angle section beam
A cantilevered beam with the thin-walled angle section shown in fig. 8.16 carries a tip vertical load, P, applied along axis $\bar{\imath}_3$. *(1)* Determine the location of the centroidal axes and compute the centroidal bending stiffnesses. *(2)* Find the shear flow distribution in the root section. *(3)* Verify that all joint and edge equilibrium conditions, eq. (8.29), are satisfied.

Problem 8.16. Skewed "I" shaped cross-section
A beam has the thin-walled, skewed "I" shaped cross-section shown in fig. 8.18. The wall thickness for both flanges and web is a constant, t. Axis $\bar{\imath}_2$ is an axis of symmetry of the section. *(1)* Determine the location of the centroidal axes and compute the centroidal bending stiffnesses. *(2)* Determine the shear flow distribution in the section under a vertical shear force, V_3. *(3)* Verify that all joint and edge equilibrium conditions, eq. (8.29), are satisfied.

Problem 8.17. Skewed "I" shaped cross-section
Treat problem 8.16 when the section is subjected a horizontal shear force, V_2.

Problem 8.18.   Thin-walled semi-circular cross-section beam

Figure 8.19 depicts a thin-walled, semi-circular open cross-section. *(1)* Determine the location of the centroidal axes and compute the centroidal bending stiffnesses. *(2)* Determine the shear flow distribution in the section under a vertical shear force, V_3. *(3)* Verify that all joint and edge equilibrium conditions, eq. (8.29), are satisfied. *(4)* Indicate the location and magnitude of the maximum shear flow. **Note:** It is more convenient to work with the angle θ as a variable describing the geometry of the section: $s = R\theta, \mathrm{d}s = R\mathrm{d}\theta$.

Problem 8.19.   Thin-walled semi-circular cross-section beam

Work problem 8.18 when the section is subjected a horizontal shear force, V_2.

Problem 8.20.   Thin-walled semi-circular cross-section with flanges

Figure 8.20 depicts a thin-walled, semi-circular open cross-section with end flanges. *(1)* Determine the location of the centroidal axes and compute the centroidal bending stiffnesses. *(2)* Determine the shear flow distribution in the section under a vertical shear force, V_3. *(3)* Verify that all joint and edge equilibrium conditions, eq. (8.29), are satisfied. *(4)* Indicate the location and magnitude of the maximum shear flow.

Problem 8.21.   Thin-walled semi-circular cross-section with flanges

Work problem 8.20 when the section is subjected a horizontal shear force, V_2.

Problem 8.22.   Thin-walled C-channel with variable flange thickness

Figure 8.21 depicts a thin-walled C-channel with variable flange thickness. *(1)* Determine the location of the centroidal axes and compute the centroidal bending stiffnesses. *(2)* Determine the shear flow distribution in the section under a vertical shear force, V_3. *(3)* Verify that all joint and edge equilibrium conditions, eq. (8.29), are satisfied. *(4)* Indicate the location and magnitude of the maximum shear flow.

Problem 8.23.   Thin-walled C-channel with variable thickness

Treat problem 8.22 when the section is subjected a horizontal shear force, V_2.

8.3.5 Shear center for open sections

Section 8.3.1 focuses on the determination of the shear flow distribution in thin-walled open sections. The beam's cross-section is assumed to be subjected to transverse shear forces, V_2 and V_3, and it is shown that integration of the local equilibrium equation, eq. (8.18), over the cross-section yields the desired shear flow distribution. Intuitively, it is expected that integration over the cross-section of the shear flow components in the horizontal and vertical directions must yield the applied shear forces, V_2 and V_3, respectively.

In the statement of the problem, the beam is assumed to be subjected to the transverse shear forces, V_2 and V_3 alone, and no torque is applied, *i.e.*, $M_1 = 0$. Consequently, the net torque generated by the shear flow distribution is expected to vanish. It should be noted, however, that as stated, the problem is not precisely defined: whereas the magnitudes of the transverse shear forces are given, *their lines of action are not specified*. This precludes the computation of the torque generated by the applied shear forces, and hence, it is not possible verify the torque equilibrium of the cross-section.

Definition of the shear center

Consider a beam with the thin-walled, open cross-section depicted in fig. 8.30. At a particular span-wise location along the beam, the cross-section is subjected to horizontal and vertical shear forces of magnitudes V_2 and V_3, respectively, with lines of action passing through point **K**, with coordinates (x_{2k}, x_{3k}), which are, as yet, unknown. No external torque is applied with respect to point **K**, i.e., $M_{1k} = 0$.

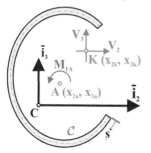

Fig. 8.30. Thin-walled open cross-section subjected to shear forces.

The shear flow distribution over the cross-section must satisfy the following three equipollence conditions. First, integration of the horizontal component of the shear flow over the cross-section must equal the applied horizontal shear force, i.e., $\int_C f\,(\mathrm{d}x_2/\mathrm{d}s)\,\mathrm{d}s = V_2$. This condition will be satisfied because it simply corresponds to the definition of the shear force, see eq. (8.7a). The second condition is similar to the first and requires the integration of the vertical component of the shear flow to equal the applied vertical shear force, i.e., $\int_C f\,(\mathrm{d}x_3/\mathrm{d}s)\,\mathrm{d}s = V_3$, which is identical to eq. (8.7b). The third condition requires the equivalence of the torque generated by the distributed shear flow with the externally applied torque, when computed about the same point. In summary, the shear flow distribution and the externally applied shear forces and torque must form *two equipollent systems* of forces.

Whereas the first two equipollence conditions do not require the knowledge of the line of action of the applied shear forces, the last condition does. The torque generated by the shear flow distribution about point **K** is given by eq. (8.10) as $M_{1k} = \int_C f r_k\,\mathrm{d}s$, where r_k is the perpendicular distance from point **K** to the line of action of the shear flow, as defined by eq. (8.11). The torque generated by the externally applied forces with respect to the same point vanishes: $M_{1k} = 0 + 0 \cdot V_2 + 0 \cdot V_3 = 0$. Indeed, no external torque is applied and the moment arms of the transverse shear forces with respect to point **K** both vanish because their lines of action both pass through point **K**, as illustrated in fig. 8.30.

The third equipollence condition now requires $M_{1k} = \int_C f r_k\,\mathrm{d}s = 0$. This means that point **K** cannot be an arbitrary point; rather, its coordinates must satisfy the torque equipollence condition,

$$M_{1k} = \int_C f r_k\,\mathrm{d}s = 0. \tag{8.39}$$

This torque equipollence condition provides the *definition of the shear center location*.

Alternative definition of the shear center

The perpendicular distance from point **K** to the line of action of the shear flow is given by eq. (8.11). Similarly, the perpendicular distance from an arbitrary point **A** to the line of action of the shear flow is $r_a = r_o - x_{2a}\, dx_3/ds + x_{3a}\, dx_2/ds$, where (x_{2a}, x_{3a}) are the coordinates of point **A**. Subtracting this equation from eq. (8.11) results in $r_k = r_a - (x_{2k} - x_{2a})\, dx_3/ds + (x_{3k} - x_{3a})\, dx_2/ds$. Introducing this result into the torque equipollence condition, eq. (8.39), then yields

$$\int_C fr_a\, ds - (x_{2k} - x_{2a})\left[\int_C f\frac{dx_3}{ds}\, ds\right] + (x_{3k} - x_{3a})\left[\int_C f\frac{dx_2}{ds}\, ds\right]$$
$$= \int_C fr_a\, ds - (x_{2k} - x_{2a})V_3 + (x_{3k} - x_{3a})V_2 = 0.$$

Since the torque generated about point **A** by the shear flow distribution is $M_{1a} = \int_C fr_a$, it follows that

$$M_{1a} = \int_C fr_a\, ds = (x_{2k} - x_{2a})V_3 - (x_{3k} - x_{3a})V_2. \qquad (8.40)$$

This result could also have been obtained from statics consideration by calculating the moment at point **A** due to force and moment resultants at point **K**. Indeed, as illustrated in fig. 8.30, $M_{1a} = M_{1k} + (x_{2k} - x_{2a})V_3 - (x_{3k} - x_{3a})V_2$, where $(x_{3k} - x_{3a})$ and $(x_{2k} - x_{2a})$ are the moment arms of shear forces V_2 and V_3 with respect to point **A**, respectively. Equation (8.40) then results from the fact that $M_{1k} = 0$, as required by eq. (8.39).

Equations (8.39) and (8.40) both express the same torque equipollence condition: *the torque generated by the shear flow distribution associated with transverse shear forces must vanish when computed with respect to the shear center.* Either equation can be used to evaluate the location of the shear center, as will be illustrated in the following examples. The choice of whether to use eq. (8.39) or (8.40) depends on familiarity or convenience. Equation (8.40) is often easier to apply when M_{1a} can be computed more easily, *i.e.*, by careful choice of point **A**, $M_{1a} == \int_C fr_a$ can be much easier to calculate.

Summary

The discussion presented in the previous paragraphs can be summarized as follows. Consider a beam subjected to transverse shear forces only, *i.e.*, no external torque is applied. Consequently, a shear flow distribution will arise in the cross-section, and its distribution is obtained by integrating eq. (8.18). This shear flow distribution must be equipollent to the externally applied shear forces. The equipollence of this shear flow distribution is only possible if the transverse shear forces have lines of action passing through point **K**, called the *shear center*, whose coordinates must satisfy eq. (8.39).

Because externally applied shear forces also generate bending of the beam, the following equivalent statement results: *a beam bends without twisting if and only if the transverse loads are applied at the shear center.* A corollary of this statement is that *if the transverse loads are not applied at the shear center, the beam will both bend and twist.* The analysis of coupled bending-twisting problems will be treated in section 8.6.

The computation of the location of the shear center is a four step procedure. First, evaluate the shear flow distribution associated with a transverse shear force, V_2. Second, determine the location, x_{3k}, of the line of action of this shear force with the help of eq. (8.39) or (8.40). Third, evaluate the shear flow distribution associated with a transverse shear force, V_3. Finally, determine the location, x_{2k}, of the line of action of this shear force with the help of eq. (8.39) or (8.40). Of course, if the cross-section features a plane of symmetry, the shear center must lie in that plane of symmetry.

Example 8.6. Shear center for a C-channel

The shear flow distribution generated by a vertical shear force applied to a C-channel is computed in example 8.3 and is shown in fig. 8.31. Axis $\bar{\imath}_2$ is an axis of symmetry for the C-channel, and hence, the shear center lies at a point along this axis. Thus, it is only necessary to evaluate the shear flow distribution generated by a vertical shear force, V_3, to determine the location of the shear center.

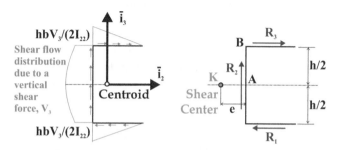

Fig. 8.31. Shear center in C-channel section.

The cross-section is made up of three straight line segments, the lower flange, the vertical web, and the upper flange. The resultant force in each segment is easily evaluated by integrating the associated shear flow distribution given in eqs. (8.30), (8.31) and (8.32), to find

$$
R_1 = \int_0^b f(s_1)\, ds_1 = \frac{hb^2 t}{4}\frac{EV_3}{H_{22}^c}, \quad R_2 = \int_{-h/2}^{h/2} f(s_2)\, ds_2 = V_3,
$$

$$
R_3 = \int_0^b f(s_3)\, ds_3 = \frac{hb^2 t}{4}\frac{EV_3}{H_{22}^c} = R_1.
$$

These three resultant forces are shown in fig. 8.31.

It is now possible to check the three equipollence conditions. The integration of the shear flow distribution component along axis $\bar{\imath}_2$ is simply $R_1 - R_1 = 0$; this resultant vanishes, as expected since no horizontal shear force is applied to the section. Next, integration of the shear flow distribution component along axis $\bar{\imath}_3$ is simply $R_2 = V_3$. Here again, this result is expected: the shear flow distribution is equipollent to the externally applied shear force, V_3.

The torque equipollence condition, eq. (8.39), yields $\int_C f r_k \, ds = -R_1 \, h/2 + R_2 \, e - R_1 \, h/2 = 0$, where e is the distance between the shear center and the vertical web, see fig. 8.31. Solving this equation leads to

$$e = \frac{hR_1}{R_2} = \frac{h^2 b^2 t}{4} \frac{E}{H_{22}^c} = \frac{3}{6 + h/b} b, \tag{8.41}$$

which gives the location of the shear center along axis $\bar{\imath}_2$.

The same result can be obtained by using the torque equipollence condition expressed by eq. (8.40). Selecting point **A** as shown in fig. 8.31, $\int_C f r_a \, ds = -hR_1$; note that the contribution of the vertical web vanishes because the line of action of resultant R_2 passes through point **A**. The torque generated by the externally applied forces is $(x_{2k} - x_{2a})V_3 - (x_{3k} - x_{3a})V_2 = (x_{2k} - x_{2a})V_3 = (-e)V_3$. Equation (8.40) then yields $-hR_1 = (-e)V_3$, or $e = hR_1/V_3 = Eh^2 b^2 t/(4H_{22}^c)$, the same result as that given by eq. (8.41).

Example 8.7. Shear center for an open triangular section

The shear flow distribution generated by a vertical shear force applied to an open triangular section is computed in example 8.4 and is shown in fig. 8.32. Axis $\bar{\imath}_2$ is an axis of symmetry for the section, and hence, the shear center lies at a point along this axis. Thus, it is only necessary to evaluate the shear flow distribution generated by a vertical shear force, V_3, to determine the location of the shear center.

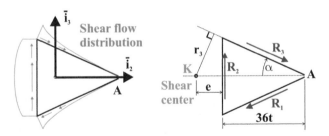

Fig. 8.32. Shear center in thin-walled open triangular section

First, the force resultants, R_1, R_2 and R_3, in the lower flange, vertical web, and upper flange, respectively, are evaluated by integrating the corresponding shear flow distributions given by eqs, (8.35), (8.36) and (8.37), respectively, to find

$$R_1 = \int_0^{39t} f(s_1) \, ds_1 = \frac{169}{360} V_3, \quad R_2 = \int_{-15t}^{15t} f(s_2) \, ds_2 = \frac{49}{36} V_3,$$

$$R_3 = -\int_0^{39t} f(s_3) \, ds_3 = \frac{169}{360} V_3 = R_1.$$

These three force resultants are shown in fig. 8.32, and to facilitate the computation, the positive directions of these resultants are shown in the figure.

The three equipollence conditions are now checked. Projection of the shear flow distribution along axis $\bar{\imath}_2$ implies $-R_1 \cos \alpha + R_3 \cos \alpha = -R_1 \cos \alpha + R_1 \cos \alpha = 0$, as expected since no horizontal shear force is applied to the section.

Next, projection of the shear flow distribution along axis $\bar{\imath}_3$ leads to $R_2 - R_1 \sin \alpha - R_3 \sin \alpha$ or

$$R_2 - 2R_1 \sin \alpha = +\frac{49}{36} V_3 - 2\frac{169}{360} V_3 \frac{15}{39} = V_3.$$

Here again, this result is expected: the shear flow distribution is equipollent to the externally applied shear force, V_3. The upward resultant acting in the vertical web, $R_2 = 49/36\, V_3$, is 36% larger than the applied shear force, V_3, to compensate for the downward components of the flange resultants, R_1 and R_3.

Finally, the torque equipollence condition, eq. (8.39), yields $\int_C f r_k\, ds = -r_3 R_1 + e R_2 - r_3 R_3 = 0$, where e is the distance between the shear center and the vertical web and $r_3 = (36t + e) \sin \alpha$ the normal distance from the shear center to the upper flange, see fig. 8.32. Solving this equation leads to

$$e = \frac{2 r_3 R_1}{R_2} = 13t, \tag{8.42}$$

which gives the location of the shear center along axis $\bar{\imath}_2$.

The same result can be obtained by using the torque equipollence condition expressed by eq. (8.40). Selecting point **A** in fig. 8.32, $\int_C f r_a\, ds = -36t R_2$, where the contributions of the upper and lower flanges vanish because the lines of action of resultants R_1 and R_3 pass through point **A**. The torque generated by the externally applied forces is $(x_{2k} - x_{2a})V_3 - (x_{3k} - x_{3a})V_2 = (x_{2k} - x_{2a})V_3 = (-e - 36t)V_3$. Equation (8.40) then yields $-36t R_2 = (-e - 36t)V_3$, or $e = 36t(R_2 - V_3)/V_3 = 13t$, the same result as that given by eq. (8.42).

This example demonstrates why it it sometimes expeditious to use the torque equipollence condition written as eq. (8.40) rather than eq. (8.39). When selecting point **A** indicated in fig. 8.32, the shear flow resultants in the flanges, R_1 and R_3, do not enter the computation because their moment arms with respect to point **A** vanish. Consequently, the evaluation of integral, $\int_C f r_a\, ds$, is simplified. In the present case, combining the torque equipollence condition, eq. (8.40), with the judicious selection point **A** at the intersection of the upper and lower flanges simplifies the computation of the shear center location.

Example 8.8. *Shear center for an angle section*

Consider the homogeneous angle section or "L" section depicted in fig. 8.33, which consists of two flanges connected at a 90 degree angle with respect to each other. Such "L" sections are commonly employed in building and aircraft structures, although in aircraft applications, the flanges might be connected at arbitrary angles with respect to each other.

A naive approach to the determination of the shear center location for this section is to follow the procedure described in section 8.3.5. This will first require the computation of the location of the section's centroid and centroidal bending stiffnesses.

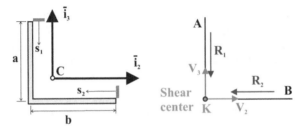

Fig. 8.33. Shear center in thin-walled right-angle section

Next, the shear flow distributions generated by applied shear forces, V_2 and V_3, must be evaluated, and finally, these distributions are used to locate the shear center using either eq. (8.39) or (8.40).

In this case, however, none of these developments are necessary, if the aim is only to determine the shear center location. Indeed, let the resultant of the shear flow distributions in the vertical and horizontal flanges be denoted R_1 and R_2, respectively, as indicated in fig. 8.33. The lines of actions of these two resultants will intersect at point **K**, and hence, the shear flow distribution produces no net torque about this point, which must then be the shear center.

8.3.6 Problems

Problem 8.24. Thin-walled angle section
Determine the location of the shear center for the thin-walled angle section shown in fig. 8.16.

Problem 8.25. Horizontal "V" shaped cross-section
Determine the location of the shear center of the thin-walled cross-section shown in fig. 8.17. Use $b = a/2$ and $\alpha = \arcsin(3/5)$.

Problem 8.26. Skewed "I" shaped cross-section
Determine the location of the shear center of the thin-walled, skewed "I" shaped cross-section shown in fig. 8.18. Axis $\bar{\imath}_2$ is an axis of symmetry of the section.

Problem 8.27. Thin-walled semi-circular cross-section beam
Determine the location of the shear center of the thin-walled, semi-circular open cross-section shown in fig. 8.19. **Note:** It is more convenient to work with the angle θ as a variable describing the geometry of the section: $s = R\theta$, $ds = Rd\theta$.

Problem 8.28. Semi-circular cross-section beam with vertical flanges
Determine the location of the shear center for the section shown in fig. 8.20. Use $a = R$.

Problem 8.29. Thin-walled C-channel with variable thickness
Determine the location of the shear center for the section shown in fig. 8.21.

Problem 8.30. Thin-walled "Y" shaped cross-section
Determine the location of the shear center for the section shown in fig. 7.35.

8.3.7 Shearing of closed sections

In the case of closed sections, the governing equation for the shear flow distribution, eq. (8.19), still applies, although no boundary condition is readily available to integrate this equation. One notable exception occurs when the section presents an axis of symmetry, such as the case shown in fig. 8.34. Plane $(\bar{\imath}_1, \bar{\imath}_3)$ is a plane of symmetry of the section, and if a shear force, V_3, acts in this plane, the solution must be symmetric with respect to this plane. Thus, the shear flow distribution for the left half of the section must be the mirror image of that for the right half.

Consider now the free-body diagram of a small portion of the thin wall in the neighborhood of point **A**, the intersection of the section with the plane of symmetry, as sketched in fig. 8.34. The joint equilibrium condition, eq. (8.29), implies $f_1 + f_2 = 0$, whereas the symmetry condition implies $f_1 = f_2$. The only possible solution is $f_1 = f_2 = 0$, *i.e.*, the shear flow must vanish at point **A**.

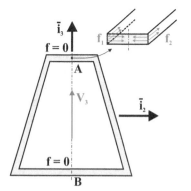

A similar reasoning will conclude that the shear flow also vanishes at point **B**, the other intersection of the section's wall with the plane of symmetry. Consequently, the section's left and right halves can be analyzed separately, as if they are two independent open sections.

If a horizontal shear force, V_2, is applied, the above symmetry argument is no longer applicable. While integration of eq. (8.19) would yield the shear flow distribution, no boundary condition is available to determine the integration constant.

Fig. 8.34. Trapezoidal section subjected to a shear force.

To overcome this problem, the following solution process is devised. Consider a closed section of arbitrary shape, as shown in part (A) of fig. 8.35. In the first step, the beam is cut along its axis at an arbitrary point of the cross-section as shown in part (B) of the figure, defining an "auxiliary problem." The points at the two edges of the cut are denoted E_1 and E_2.

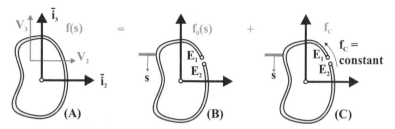

Fig. 8.35. (A): a general closed section. (B): the auxiliary problem created by cutting the section open. (C): the constant closing shear flow.

The shear flow distribution in this auxiliary problem is denoted $f_o(s)$, and is readily found using the procedure described in section 8.3.1 for computing shear flow distributions in open sections. As illustrated in fig. 8.36, this shear flow creates a shear strain, γ_s, which in turn, creates an infinitesimal axial displacement du_1. For small shearing angles, this axial displacement becomes

$$du_1 = \gamma_s \, ds = \frac{\tau_s}{G} \, ds = \frac{f_o(s)}{Gt} \, ds, \tag{8.43}$$

where Hooke's laws is used to characterize material behavior. The shear flow distribution over the entire section creates a finite relative axial displacement at the two edges of the cut, points E_1 and E_2, which can be evaluated by integrating around the section the infinitesimal axial displacement given by eq. (8.43). The total relative axial displacement at the cut, u_0, is thus

$$u_0 = \int_C \frac{f_o(s)}{Gt} \, ds.$$

In the last step of the solution process, a constant shear flow, denoted f_c, is applied to the section, as illustrated in part (C) of fig. 8.35. This constant shear flow must be adjusted to eliminate the relative axial displacement, u_0, between the edges of the cut, thereby returning the section to its original, closed state. The constant shear flow, f_c, is therefore called the *closing shear flow*.

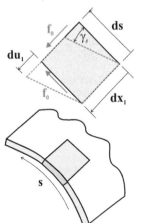

After addition of the closing shear flow, the total shear flow in the section becomes $f(s) = f_o(s) + f_c$, and the corresponding relative axial displacement must now vanish, which implies

$$u_t = \int_C \frac{f_o(s) + f_c}{Gt} \, ds = 0. \tag{8.44}$$

This condition is, in fact, the displacement compatibility equation for the closed section. Solving this equation then yields the closing shear flow as

$$f_c = -\frac{\displaystyle\int_C \frac{f_o(s)}{Gt} \, ds}{\displaystyle\int_C \frac{1}{Gt} \, ds}. \tag{8.45}$$

Fig. 8.36. Axial displacement arising from the shear flow f_o.

The procedure to compute the shear flow distribution in a closed section is summarized in the following steps.

1. Compute the shear flow distribution, $f_o(s)$, for an auxiliary problem obtained by cutting the beam along its axis at an arbitrary point of the section. The solution procedure described in section 8.3.1 can be used.
2. Compute the closing shear flow, f_c, using eq. (8.45).
3. The shear flow distribution in the closed section is then $f(s) = f_o(s) + f_c$.

Example 8.9. Shear flow distribution in a closed triangular section

Consider the triangular section depicted in fig. 8.28, but now assume that the section is *closed, i.e.,* there is no cut at point **A**. The location of the centroid and bending stiffnesses are identical in the open and closed sections.

The shear flow distributions given by eqs. (8.35), (8.36) and (8.37) now correspond to the shear flow distributions associated with the auxiliary problem, and hence,

$$f_o(s_1) = \frac{13}{360} \left(\frac{s_1}{39t}\right)^2 \frac{V_3}{t}, \quad f_o(s_3) = -\frac{13}{360} \left(\frac{s_3}{39t}\right)^2 \frac{V_3}{t},$$

$$f_o(s_2) = \frac{13}{360} \frac{V_3}{t} + \frac{1}{72} \left[1 - \left(\frac{s_2}{15t}\right)^2\right] \frac{V_3}{t}. \tag{8.46}$$

The constant closing shear flow f_c can now be calculated using eq. (8.45). Integrating in a clockwise direction around the section, the numerator is

$$\int_c \frac{f_o}{Gt} ds = \int_0^{39t} \frac{f_o(s_1)}{Gt} ds_1 + \int_{-15t}^{15t} \frac{f_o(s_2)}{Gt} ds_2 - \int_0^{39t} \frac{f_o(s_3)}{Gt} ds_3 = \frac{23V_3}{10Gt},$$

where is should be noted that the last integral has a negative sign because curvilinear variable s_3 is selected to be in the opposite direction of the clockwise orientation used to express the integral, as defined in fig. 8.28. The denominator in eq. (8.45) is

$$\int_c \frac{ds}{Gt} = \frac{1}{Gt}(39t + 30t + 39t) = \frac{108}{G}.$$

Thus, the closing shear flow becomes

$$f_c = -\frac{23V_3/(10Gt)}{108/G} = -\frac{23}{1080} \frac{V_3}{t}. \tag{8.47}$$

Finally, the resulting shear flow for the triangular section is $f(s) = f_o(s) + f_c$, where $f_o(s)$ is given by eq. (8.46). Note that both shear flow in the auxiliary section and the closing shear flow are positive when pointing along the local curvilinear variable. The non-dimensional shear flow distribution, $\bar{f}(s) = tf(s)/V_3$, is shown in fig. 8.37 where the arrows indicate the physical direction of the shear flow.

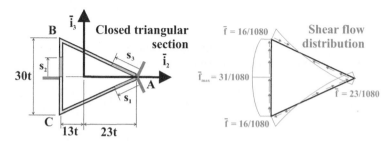

Fig. 8.37. Non-dimensional shear flow distribution in a closed triangular section.

Example 8.10. Shear flow distribution in a trapezoidal section

Consider the thin-walled trapezoidal section shown in fig. 8.38. The wall thickness, t, is constant, and the curvilinear variable is broken into four components: s_1 along the upper flange starting from point **A**, s_2 down the left web, with its origin at mid-point, s_3 across lower flange starting from point **C**, and s_4 up the right web with a mid-span origin. Axis $\bar{\imath}_2$ is an axis of symmetry for this section and hence, the cross bending stiffness vanishes, $H^c_{23} = 0$. This means that centroid lies on this axis somewhere between the two webs. For this example, a vertical shear force, V_3, is applied to the section and it is only necessary to determine the bending stiffness H^c_{22}, given by eq. (6.55), where ℓ is the distance between points **A** and **B**: $\ell^2 = b^2 + (h_2 - h_1)^2$.

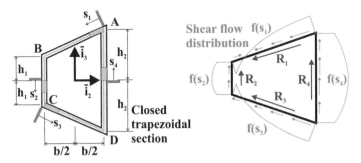

Fig. 8.38. Thin-walled trapezoidal section subjected to a vertical shear force, V_3.

The first step of the procedure is to cut the section at an arbitrary point, say **A**. The distribution of shear flow in the open section is found using eq. (8.20), and the boundary condition is $f(s_1 = 0) = 0$. The shear flow in the other segments of the section is obtained by integrating the same equation and enforcing the joint equilibrium condition, eq. (8.29), at points **B**, **C**, and **D**. The complete shear flow distribution is

$$
f_o(s_1) = \frac{EV_3}{H^c_{22}}\left[\frac{h_2 - h_1}{2\ell}s_1^2 - h_2 s_1\right], \quad f_o(s_2) = \frac{EV_3}{2H^c_{22}}\left[s_2^2 - h_1^2 - (h_1 + h_2)\ell\right],
$$

$$
f_o(s_3) = \frac{EV_3}{H^c_{22}}\left[\frac{h_2 - h_1}{2\ell}s_3^2 + h_1 s_3 - \frac{h_1 + h_2}{2}\ell\right], \quad f_o(s_4) = \frac{EV_3}{2H^c_{22}}\left[-s_4^2 + h_2^2\right].
$$
$$(8.48)$$

The next step of the procedure requires the evaluation of the closing shear flow according to eq. (8.45). The numerator can be evaluated as

$$
\int_C \frac{f_o}{Gt}\,ds = \int_0^l \frac{f_o}{Gt}\,ds_1 + \int_{-h_1}^{+h_1} \frac{f_o}{Gt}\,ds_2 + \int_0^\ell \frac{f_o}{Gt}\,ds_3 + \int_{-h_2}^{+h_2} \frac{f_o}{Gt}\,ds_4,
$$

$$
= -\frac{EV_3}{3GtH^c_{22}}\left[2(h_1^3 - h_2^3) + (h_1 + 2h_2)\ell^2 + 3(h_1 + h_2)\ell h_1\right],
$$

and the denominator as

$$\oint_c \frac{1}{Gt}\,ds = \int_0^\ell \frac{ds}{Gt} + \int_{-h_1}^{+h_1} \frac{ds}{Gt} + \int_0^\ell \frac{ds}{Gt} + \int_{-h_2}^{+h_2} \frac{ds}{Gt} = \frac{2(\ell + h_1 + h_2)}{Gt},$$

to yield the closing shear flow

$$f_c = \frac{EV_3}{H_{22}^c} \frac{2(h_1^3 - h_2^3) + (h_1 + 2h_2)\ell^2 + 3(h_1 + h_2)\ell h_1}{6(\ell + h_1 + h_2)}. \tag{8.49}$$

The final distribution of shear flow is found by adding this closing shear flow to the shearing flow distribution for the open section, eq. (8.48). The shear flow distribution in the closed section is depicted in fig. 8.38, which shows that the maximum shear flow is found at the mid-point of the right web. Also shown in this figure are the shear force resultants on each side of the section. Clearly, the horizontal forces must sum up to zero since no shear force is applied in that direction. The summation of the forces in the vertical direction yields, after a lengthy of algebraic manipulation,

$$\frac{EV_3}{H_{22}^c}\left\{\left[\left(\frac{h_1 + 2h_2}{6}\ell^2 - f_c\ell\right)\frac{h_2 - h_1}{\ell}\right] + \left[\frac{2h_1^3}{3} + (h_1 + h_2)\ell h_1 - 2f_c h_1\right]\right.$$
$$\left. + \left[\left(\frac{h_1 + 2h_2}{6}\ell^2 - f_c\ell\right)\frac{h_2 - h_1}{\ell}\right] + \left[\frac{2h_2^3}{3} + 2f_c h_2\right]\right\} = \frac{EV_3}{H_{22}^c}\frac{H_{22}^c}{E} = V_3.$$

Thus, the distributed shear flow exactly sums up to the applied shear force V_3.

8.3.8 Shearing of multi-cellular sections

Multi-cellular sections are common in aeronautical construction. Figure 8.39 shows a typical wing section with two closed cells. The shear flow distribution must satisfy eq. (8.18), but no boundary condition is available to evaluate the integration constant. To remedy this situation, a procedure similar to that used for a single closed section must be developed.

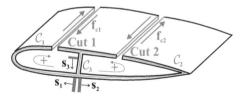

Fig. 8.39. A thin-walled, multi-cellular section.

The multi-cellular beam is cut along its axis at arbitrary points. One cut per cell is required to eliminate all the closed paths of the section, as illustrated in fig. 8.39 for the present example. The shear flow distribution in the resulting open section is evaluated using the procedure described in section 8.3.1. Let $f_o(s_1)$, $f_o(s_2)$, and $f_o(s_3)$ be the shear flow distributions along curves C_1, C_2, and C_3, respectively.

Next, closing shear flows are applied at each of the cuts; f_{c1} and f_{c1} for the front and aft cells, respectively. The shear flow distributions are now $f_o(s_1) + f_{c1}$, $f_o(s_2) + f_{c2}$, and $f_o(s_3) + (f_{c1} + f_{c2})$, along curves C_1, C_2, and C_3, respectively.

The two unknown closing shear flows will be evaluated by enforcing the displacement compatibility condition for each of the two cells. When enforcing these conditions, it is important to keep track of sign conventions. The front cell is described clockwise, leading to the following equation

$$u_{t1} = \int_{C_1} \frac{f_o(s_1) + f_{c1}}{Gt} \, ds_1 + \int_{C_3} \frac{f_o(s_3) + (f_{c1} + f_{c2})}{Gt} \, ds_3 = 0.$$

Since the curvilinear variables s_1 and s_3 are defined in the same clockwise direction, all integral have a positive sign. The aft cell is described counterclockwise, leading to

$$u_{t2} = \int_{C_2} \frac{f_o(s_2) + f_{c2}}{Gt} \, ds_2 + \int_{C_3} \frac{f_o(s_3) + (f_{c1} + f_{c2})}{Gt} \, ds_3 = 0.$$

These compatibility conditions can be recast as a set of two linear equations for the unknown closing shear flows,

$$\left[\int_{C_1+C_3} \frac{1}{Gt} \, ds \right] f_{c1} + \left[\int_{C_3} \frac{1}{Gt} \, ds \right] f_{c2} = - \int_{C_1+C_3} \frac{f_o(s)}{Gt} \, ds;$$

$$\left[\int_{C_3} \frac{1}{Gt} \, ds \right] f_{c1} + \left[\int_{C_2+C_3} \frac{1}{Gt} \, ds \right] f_{c2} = - \int_{C_2+C_3} \frac{f_o(s)}{Gt} \, ds.$$

Solving these two equations yields the two closing shear flows, f_{c1} and f_{c2}.

The total shear flow in the multi-cellular section is then found by adding the closing shear flows to the shear flows in the open section, *i.e.*, $f_o(s_1) + f_{c1}$, $f_o(s_2) + f_{c2}$, and $f_o(s_3) + (f_{c1} + f_{c2})$, along curves C_1, C_2, and C_3, respectively.

The procedure is readily extended to multi-cellular section possessing N closed cells. First, the multi-cellular section is transformed into an open section by creating N cuts, one per cell. The shear flow distribution in the resulting open section is then evaluated with the help of the procedure of section 8.3.1. Next, unknown closing shear flows are applied at each cut and displacement compatibility conditions are imposed for each of the N cells. These conditions yield a set of N simultaneous equations for the N closing shear flows. Finally, the total shear flow distribution in the multi-cellular section is found by adding the closing shear flows to the shear flows for the open section.

Example 8.11. Shear flow in thin-walled double-box section

Consider the closed multi-cellular, thin-walled, double-box section subjected to a vertical shear force, V_3, as shown in fig. 8.40. The section consists of two closed cells; the right cell has a wall thickness of $2t$, while the three remaining walls of the left cell have wall thicknesses of t. The vertical sides will be referred to as the vertical webs, whereas the horizontal sides will be called flanges.

Due to symmetry, the centered horizontal axis $\bar{\imath}_2$ is a principal axis of bending, and hence, $H_{23}^c = 0$. The centroid of the section will be located in the right cell, as indicated in the figure. Using thin-wall assumptions, the bending stiffness becomes

$$H_{22}^c = E \left[2 \left(\frac{2tb^3}{12} \right) + \frac{tb^3}{12} + 2(bt + b2t) \left(\frac{b}{2} \right)^2 \right] = \frac{23}{12} tb^3 E.$$

In the first step of the procedure, the closed multi-cellular section is transformed into an open section by cutting the two lower flanges at the location where they connect to the center web, as indicated in fig. 8.40. The locations of these cuts are

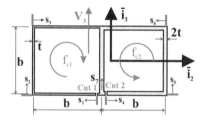

Fig. 8.40. A thin-walled double-box section.

arbitrary, but those selected here lead to simple definitions of the curvilinear coordinates, s_i, for each component of the section.

The stiffness static moments in each component of the section are found with the help of eq. (8.22). Equation (8.20) now reduces to $f(s) = Q_2(s)V_3/H_{22}^c$ and the shear flow distribution in the open section, $f_o(s_i)$, becomes

$$
f_o(s_1) = \frac{6V_3}{23b} \frac{s_1}{b}, \quad f_o(s_3) = \frac{6V_3}{23b}\left(1 - \frac{s_3}{b}\right), \quad f_o(s_4) = \frac{12V_3}{23b}\frac{s_4}{b},
$$

$$
f_o(s_2) = \frac{6V_3}{23b}\left[1 + \left(1 - \frac{s_2}{b}\right)\frac{s_2}{b}\right], \quad f_o(s_5) = \frac{12V_3}{23b}\left[1 + \left(1 - \frac{s_5}{b}\right)\frac{s_5}{b}\right],
$$

$$
f_o(s_6) = \frac{12V_3}{23b}\left(1 - \frac{s_6}{b}\right), \quad f_o(s_7) = -\frac{12V_3}{23b}\left(1 - \frac{s_7}{b}\right)\frac{s_7}{b}.
$$

Next, two closing shear flows, denoted f_{c1} and f_{c2}, are added to the left and right cells, respectively. The axial displacement compatibility condition for each of the two cells will be used to evaluate these two unknown closing shear flows. When enforcing these conditions, it is important to keep track of sign conventions. The left cell is described clockwise, leading to the following compatibility equation

$$
u_{t1} = \int_0^b \frac{f_o(s_1) + f_{c1}}{Gt}\, ds_1 + \int_0^b \frac{f_o(s_2) + f_{c1}}{Gt}\, ds_2 + \int_0^b \frac{f_o(s_3) + f_{c1}}{Gt}\, ds_3
$$

$$
- \int_0^b \frac{f_o(s_7) - f_{c1} - f_{c2}}{G2t}\, ds_7 = \frac{b}{Gt}\left(\frac{7f_{c1}}{2} + \frac{f_{c2}}{2} + \frac{12V_3}{23b}\right) = 0.
$$

Note the minus sign in front of the last integral because curvilinear variable s_7 is defined in the direction that opposes the clockwise description of the left cell. The right cell is described counterclockwise and the corresponding compatibility equation is

$$
u_{t2} = \int_0^b \frac{f_o(s_4) + f_{c2}}{G2t}\, ds_4 + \int_0^b \frac{f_o(s_5) + f_{c2}}{G2t}\, ds_5 + \int_0^b \frac{f_o(s_6) + f_{c2}}{G2t}\, ds_6
$$

$$
- \int_0^b \frac{f_o(s_7) - f_{c1} - f_{c2}}{G2t}\, ds_7 = \frac{b}{Gt}\left(\frac{f_{c1}}{2} + 2f_{c2} + \frac{12V_3}{23b}\right) = 0.
$$

Here again, note the minus sign in front of the last integral because curvilinear variable s_7 is defined in the direction that opposes the counterclockwise description of the right cell. Evaluation of the integrals and solution of these two simultaneous algebraic equations yields $f_{c1} = -8V_3/(69b)$ and $f_{c2} = -16V_3/(69b)$.

Finally, the total shear flow in each segment of the section can be computed by combining the open-section shear flows, $f_o(s_i)$, with the constant closing shear flows, f_{c1} and f_{c2}, to find

$$
f(s_1) = -\frac{2V_3}{69b}\left(4 - 9\frac{s_1}{b}\right), \quad f(s_2) = \frac{2V_3}{69b}\left[5 + 9\frac{s_2}{b} - 9\left(\frac{s_2}{b}\right)^2\right],
$$

$$
f(s_3) = \frac{2V_3}{69b}\left(5 - 9\frac{s_3}{b}\right), \quad f(s_4) = -\frac{4V_3}{69b}\left(4 - 9\frac{s_4}{b}\right),
$$

$$
f(s_5) = \frac{4V_3}{69b}\left[5 + 9\frac{s_5}{b} - 9\left(\frac{s_5}{b}\right)^2\right], \quad f(s_6) = \frac{4V_3}{69b}\left(5 - 9\frac{s_6}{b}\right),
$$

$$
f(s_7) = \frac{12V_3}{69b}\left[2 + 3\frac{s_7}{b} - 3\left(\frac{s_7}{b}\right)^2\right].
$$

(8.50)

The shear flows in the webs vary quadratically while those in the flanges vary linearly, as illustrated in fig. 8.41. This is consistent with previous examples for shear flow in open sections. The net resultant of the shear flows in the flanges must vanish because no shear force is externally applied in the horizontal direction. The resultant of the shear flows in the webs must equal the externally applied vertical shear force, V_3. This important check of the computations is left as an exercise.

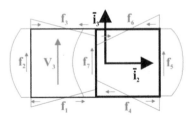

Fig. 8.41. Shear flow in the thin-walled double-box section.

8.3.9 Problems

Problem 8.31. Shear flow distribution in closed circular section
Consider a beam with a thin-walled, circular cross-section of radius R and thickness t. The section is subjected to a vertical shear force, V_3. (1) Determine the bending stiffnesses of the section. (2) Find the shear flow distribution in the section. (3) Find the location and magnitude of the maximum shear flow in the section.

Problem 8.32. Shear flow in a closed rectangular section
The thin-walled beam with a rectangular section depicted in fig. 8.42 is subjected to a vertical shear force V_3. (1) Determine the centroidal bending stiffnesses of the section. (2) Find the shear flow distribution in the section. (3) Verify that all joint and edge equilibrium conditions, eq. (8.29), are satisfied. (4) Find the location and magnitude of the maximum shear flow in the section. Use the following data: $\alpha = 1.0$

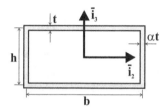

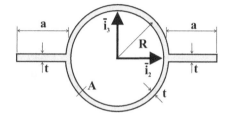

Fig. 8.42. Thin-walled beam with rectangular cross-section.

Fig. 8.43. Thin-walled, circular cross-section with flanges.

Problem 8.33. Shear flow in closed rectangular section

The thin-walled, rectangular beam section shown in fig. 8.42 is subjected to a horizontal shear force, V_2. The thickness of the right vertical web is $5t$, whereas that of the remaining walls is t. *(1)* Determine the centroidal bending stiffnesses of the section. *(2)* Find the shear flow distribution in the section. *(3)* Verify that all joint and edge equilibrium conditions, eq. (8.29), are satisfied. *(4)* Find the location and magnitude of the maximum shear flow in the section. Use the following data: $\alpha = 5.0$

Problem 8.34. Shear flow in closed circular tube with flanges

The thin-walled, circular tube with flanges in fig. 8.43 is subjected to a vertical shear force, V_3. *(1)* Determine the centroidal bending stiffnesses of the section. *(2)* Find the shear flow distribution in the section. *(3)* Verify that all joint and edge equilibrium conditions, eq. (8.29), are satisfied. *(4)* Find the location and magnitude of the maximum shear flow in the section.

Problem 8.35. Shear flow in closed circular tube with flanges

Treat problem 8.34 under a horizontal shear force, V_2.

Problem 8.36. Thin-walled "Box-Z" shaped cross-section beam

The thin-walled cross-section shown in fig. 8.15 is subjected to a vertical shear force, V_3. *(1)* Determine the centroidal bending stiffnesses of the section. *(2)* Find the shear flow distribution in the section. *(3)* Verify that all joint and edge equilibrium conditions, eq. (8.29), are satisfied. *(4)* Find the location and magnitude of the maximum shear flow in the section. Use with $b = c = a/2$.

Problem 8.37. Shear flow in high-lift device

The cross-section of a high lift device is shown in fig 8.44. The aerodynamic pressure acting on the lower panel of the device has a net resultant $V_3 = 100$ kN and its line of action is aligned with axis $\bar{\imath}_3$, as indicated on the figure. Material properties are: $E = 73$ GPa, $G = 30$ GPa. *(1)* Find the shear flow distribution in the section. *(2)* Sketch this shear flow distribution. *(3)* Verify that all joint and edge equilibrium conditions, eq. (8.29), are satisfied. *(4)* Determine the location and magnitude of the maximum shear stress.

Problem 8.38. Shear flow in a multi-cellular, thin-walled section

The cross-section of the multi-cellular thin-walled beam shown in fig. 8.45 is subjected to a vertical shear force V_3. *(1)* Find the shear flow distribution in the section. *(2)* Verify that all joint and edge equilibrium conditions, eq. (8.29), are satisfied. Use $a = b$ and $c = 2b$ and $t_1 = t_2 = t_w = t$.

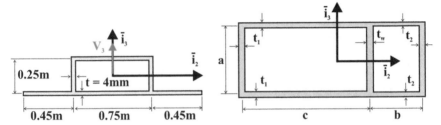

Fig. 8.44. High lift device subjected to a transverse shear force.

Fig. 8.45. Thin-walled beam with two cells.

8.4 The shear center

In the previous section, procedures for computing shear flow distributions in open and closed sections are developed. In addition, the concept of the shear center is introduced and defined as the point of the cross-section about which the torque equipollent to the shear flow distribution vanishes. The procedure for determining the shear center location is developed for open sections, but the concept of shear center also exists for closed sections.

In chapter 6, it is assumed that transverse loads are applied in "such a way that the beam will bend without twisting." This restriction can now be stated in a more precise manner: the lines of action of all transverse loads are assumed to pass through the shear center. Consequently, the results derived in chapter 6, and the shear flow distributions presented in section 8.3 are only valid if the transverse loads are applied at the section's shear center. Clearly, the determination of the shear center location is a crucial step in the analysis of beams. If the lines of action of the applied transverse shear forces pass through the shear center, the beam will bend without twisting. If the shear forces are not applied at the shear center, the beam will undergo both bending and twisting.

A general procedure for the determination of the location of the shear center for thin-walled cross-sections will now be described. It is based on the requirement first presented in section 8.3.5: when computed about the shear center, the torque equipollent to the shear flow distribution over the cross-section must vanish. Since the two coordinates of the shear center must be evaluated, this torque equipollence requirement must be applied, in general, to two linearly independent shear flow distributions.

8.4.1 Calculation of the shear center location

The general procedure for determining the shear center location involves two linearly independent loading cases and associated shear flow distributions. The first loading case, identified with a superscript $(\cdot)^{[2]}$, consists of a unit shear force, $V_2^{[2]} = 1$, acting along axis $\bar{\imath}_2$, while no shear force is acting along axis $\bar{\imath}_3$, i.e., $V_3^{[2]} = 0$. The shear flow associated with this loading case is denoted $f^{[2]}(s)$. The second loading case, identified with a superscript $(\cdot)^{[3]}$, consists of a unit shear force, $V_3^{[3]} = 1$, acting along axis $\bar{\imath}_3$, while no shear force is acting along axis $\bar{\imath}_2$, *i.e.*, $V_2^{[3]} = 0$. The associated shear flow is denoted $f^{[3]}(s)$.

From eq. (8.7), the shear forces equipollent to $f^{[2]}(s)$ are

$$
V_2^{[2]} = \int_C f^{[2]} \frac{\mathrm{d}x_2}{\mathrm{d}s} \, \mathrm{d}s = 1, \quad V_3^{[2]} = \int_C f^{[2]} \frac{\mathrm{d}x_3}{\mathrm{d}s} \, \mathrm{d}s = 0. \tag{8.51}
$$

This intuitive result is proved in a formal manner in example 8.16.

By definition, the shear center is located at point $\mathbf{K}$, whose coordinates, denoted (x_{2k}, x_{3k}), satisfy the torque equipollence condition expressed by eq. (8.39). Equation (8.10) then implies

$$
M_{1K} = \int_C f^{[2]} r_k \, \mathrm{d}s = \int_C f^{[2]} \left(r_o - x_{2k} \frac{\mathrm{d}x_3}{\mathrm{d}s} + x_{3k} \frac{\mathrm{d}x_2}{\mathrm{d}s} \right) \mathrm{d}s,
$$

where r_k is the distance from point $\mathbf{K}$ to the tangent to contour C, evaluated with the help of eq. (8.11). Rearranging this expression leads to

$$
-x_{2k} \left[\int_C f^{[2]} \frac{\mathrm{d}x_3}{\mathrm{d}s} \, \mathrm{d}s \right] + x_{3k} \left[\int_C f^{[2]} \frac{\mathrm{d}x_2}{\mathrm{d}s} \, \mathrm{d}s \right] = - \int_C f^{[2]} r_o \, \mathrm{d}s.
$$

In view of eq. (8.51), the two bracketed terms equal 0 and 1, respectively, and this equation reduces to

$$
x_{3k} = - \int_C f^{[2]} r_o \, \mathrm{d}s. \tag{8.52}
$$

A similar reasoning for the shear flow distribution $f^{[3]}$ yields the other coordinate of the shear center as

$$
x_{2k} = \int_C f^{[3]} r_o \, \mathrm{d}s. \tag{8.53}
$$

The torque equipollence condition given by eq. (8.39) is used in the above development. It is also possible to use the torque equipollence condition expressed by eq. (8.40). For the shear flow distribution, $f^{[2]}(s)$, associated with a unit shear force along axis $\bar{\imath}_2$, this condition yields

$$
x_{3k} = x_{3a} - \int_C f^{[2]} r_a \, \mathrm{d}s. \tag{8.54}
$$

A similar reasoning for shear flow distribution, $f^{[3]}(s)$, yields the other coordinate of the shear center as

$$
x_{2k} = x_{2a} + \int_C f^{[3]} r_a \, \mathrm{d}s, \tag{8.55}
$$

where (x_{2a}, x_{3a}) are the coordinates of an arbitrary point $\mathbf{A}$ of the cross-section, and r_a the normal distance from this point to the tangent to curve C. In some cases, the judicious choice of the location of point $\mathbf{A}$ can greatly simplify the evaluation of the integral, see example 8.7.

The general procedure for the determination of the location of the shear center is summarized in the following steps.

1. Compute the location of the section's centroid and select a set of centroidal axes. In some cases, it might be more convenient to work with the principal centroidal axes of bending, but this is not a requirement.
2. Compute the shear flow distribution, $f^{[2]}(s)$, corresponding to a unit shear force acting along axis $\bar{\imath}_2$, $V_2^{[2]} = 1$, $V_3^{[2]} = 0$;
3. Compute the shear flow distribution, $f^{[3]}(s)$, corresponding to a unit shear force acting along axis $\bar{\imath}_3$, $V_2^{[3]} = 0$, $V_3^{[3]} = 1$. The procedure used to determine these shear flow distributions is described in section 8.3.1 or 8.3.7 for open or closed sections, respectively.
4. Compute the coordinates of the shear center using either eqs. (8.52) and (8.53), or (8.54) and (8.55).

Note that if the cross-section exhibits a plane of symmetry, the procedure can be simplified. For instance, if plane $(\bar{\imath}_1, \bar{\imath}_2)$ is a plane of symmetry of the section, the shear center must be located in this plane of symmetry. Consequently, $x_{3k} = 0$, and the computation of the shear flow distribution, $f^{[2]}$, associated with the first loading case is not required since the use of eq. (8.52) can be bypassed. The computation of the shear flow distribution, $f^{[3]}$, will be required to evaluate the remaining coordinate, x_{2k}, of the shear center with the help of eq. (8.53). Of course, if the section presents two planes of symmetry, the shear center is at the intersection of those two planes.

Example 8.12. The shear center of a trapezoidal section
Consider the closed trapezoidal section depicted in fig. 8.38. The shear flow distribution generated by a vertical shear force, V_3, is evaluated in example 8.10 as the sum of the shear flow distribution in the auxiliary open section, denoted $f_o(s)$ and given by eqs. (8.48), and of the closing shear flow, denoted f_c and given by eq. (8.49).

The location of the shear center then follows from eq. (8.53)

$$
x_{2k} = \int_C \left(\bar{f}_o^{[2]}(s) + \bar{f}_c^{[2]} \right) r_o \, ds,
$$

where $\bar{f}_o^{[2]}(s) = f_o(s)/V_3$ and $\bar{f}_c^{[2]} = f_c/V_3$, *i.e.*, these quantities are the shear flow distributions associated with a unit shear force, $V_3 = 1$. Evaluation of the integral is quite tedious and yields

$$
x_{2k} = \frac{b}{4} \frac{h_2 - h_1}{\ell} \frac{1 - (h_1 + h_2)/\ell}{1 + (h_1 + h_2)/\ell} \frac{1 + \ell(h_2^2 - h_1^2)/(h_2^3 - h_1^3)}{1 + (h_2 - h_1)(h_2^3 + h_1^3)/(\ell(h_2^3 - h_1^3))}.
$$

Due to the symmetry of the problem, the other coordinate of the shear center is $x_{3k} = 0$. Of course, if $h_2 = h_1$, the trapezoidal section becomes rectangular, and $x_{2k} = 0$, as required by symmetry.

Example 8.13. Relationship between shear centers of open and closed sections
Consider two otherwise identical cross-sections, the first closed, and the second open. The open section is obtained by cutting the closed section along the axis of the beam at an arbitrary point of the section. Let (x_{2k}^o, x_{3k}^o) be the coordinates of

the shear center of the open section, and (x_{2k}^c, x_{3k}^c) those of the shear center of the closed section. The coordinates of the shear center of the open section are given by eqs. (8.53) and (8.52) as

$$x_{2k}^o = \int_C f_o^{[3]} r_o \, ds, \quad x_{3k}^o = -\int_C f_o^{[2]} r_o \, ds,$$

where $f_o^{[2]}$ and $f_o^{[3]}$ are the shear flow distributions in the open section, corresponding to unit shear forces acting along axes $\bar{\imath}_2$ and $\bar{\imath}_3$, respectively.

The coordinates of the shear center of the closed section are obtained in a similar manner as

$$x_{2k}^c = \int_C (f_o^{[3]} + f_c^{[3]}) r_o \, ds, \quad x_{3k}^c = -\int_C (f_o^{[2]} + f_c^{[2]}) r_o \, ds.$$

In these equations, $f_c^{[3]}$ and $f_c^{[2]}$ are the closing shear flows defined by eq. (8.45), and $f_o^{[2]}$ and $f_o^{[3]}$ are the shear flow distributions in the auxiliary open sections. The first of these two integrals can be expanded in the following manner,

$$x_{2k}^c = \int_C f_o^{[3]} r_o \, ds + \int_C f_c^{[3]} r_o \, ds = x_{2k}^o + f_c^{[3]} \int_C r_o \, ds,$$

where x_{2k}^o is the location of the shear center for the auxiliary open section. The remaining integral in this equation is a purely geometry quantity.

As illustrated in fig. 8.46, quantity $1/2\, r_o\, ds$ represents the area of a differential triangle of base ds and height r_o. As curvilinear variable s sweeps around the closed section, the integral represents the sum of these differential triangles and yields the total area, $\mathcal{A}$, enclosed by the closed curve C. Hence,

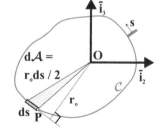

$$\int_C r_o \, ds = \int_C 2d\mathcal{A} = 2\mathcal{A}. \qquad (8.56)$$

It now follows that

Fig. 8.46. The area of a differential triangle, $d\mathcal{A} = 1/2\, r_o ds$.

$$x_{2k}^c - x_{2k}^o = 2\mathcal{A} f_c^{[3]}, \qquad (8.57a)$$

$$x_{3k}^c - x_{3k}^o = -2\mathcal{A} f_c^{[2]}. \qquad (8.57b)$$

This result shows that the coordinates of the shear center in the open and closed section are closely related to each other through the closing shear flows $f_c^{[3]}$ and $f_c^{[2]}$. As shown by eq. (8.8), r_o is an algebraic quantity: when the curvilinear variable describes C counterclockwise, r_o is a positive quantity, but it is negative if the curvilinear variable describes C clockwise. Consequently, the enclosed area, $\mathcal{A}$, appearing in eq. (8.57) must be understood as an algebraic quantity: positive is C is described counterclockwise, negative is C is described clockwise. Of course, the sign convention for the closing shear flow is also related to the direction of the curvilinear variable, and hence, the product, $\mathcal{A}f_c$, is independent of the arbitrary choice of the curvilinear coordinate direction, as should be expected.

Example 8.14. The shear center of a closed triangular section

In example 8.4, the shear flow distribution in an open triangular section is evaluated and the location of its shear center is determined in example 8.7. Finally, in example 8.9, the shear flow distribution in the corresponding thin-walled, closed triangular section is evaluated. In the present example, the location of the shear center of the closed section will be determined using different approaches.

The location of the centroid and the bending stiffnesses are identical for the open and closed sections. Since axis $\bar{\imath}_2$ is an axis of symmetry, the shear center must lie on this axis, *i.e.*, $x_{3k} = 0$ and therefore, according to the procedure described in section 8.4.1, it is only necessary to determine the shear flow distribution associated with a unit shear force acting along axis $\bar{\imath}_3$: $V_3 = 1$ and $V_2 = 0$. This shear flow distribution is computed in example 8.9 and consists of the superposition of the shear flow in the open section and a closing shear flow.

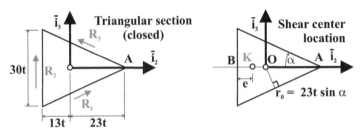

Fig. 8.47. Shear center for thin-walled triangular section.

First, the net force resultants in the flanges and vertical web will be evaluated. Starting with the lower flange, the net resultant of the shear flow distribution is

$$R_1 = \int_0^{39t} f_o(s_1)\, ds_1 + f_c \int_0^{39t} ds_1 = \frac{169}{360t} - \frac{23}{1080t} 39t = -\frac{13}{36},$$

where the shear flow in the open auxiliary section, $f_o(s_1)$, is given by eq. (8.46) and the closing shear flow, f_c, by eq. (8.47). In both equations, the applied shear force is set to $V_3 = 1$. Because the closing shear flow is constant, it can be factored out of the integral. The negative sign indicates that the net resultant is directed in the opposite direction of the curvilinear variable, s_1, see fig. 8.28. Figure 8.47 shows the physical direction of the resultant. Proceeding in a similar manner for the web and upper flange, the magnitudes of the resultants are found to be $R_1 = R_3 = 13/36$ and $R_2 = 26/36$, and their directions are indicated in fig. 8.47.

The shear force equipollence conditions can be verified. The net resultant along axis $\bar{\imath}_2$ is $V_2 = R_1 \cos\alpha - R_3 \cos\alpha = 0$, as expected since no shear force is applied along that axis. The net resultant along axis $\bar{\imath}_3$ is $V_3 = R_1 \sin\alpha + R_2 + R_3 \sin\alpha$.

This leads to $V_3 = 2(13/36)(15/39) + (26/36) = 5/18 + 13/18 = 1$, as expected, because a unit shear force is applied along that axis.

The shear center is now computed with the help of eq. (8.53) to find

$$x_{2k} = \int_C f^{[3]} r_o \, ds = 2R_1 23t \sin\alpha - R_2 13t = 2\frac{13}{36} 23t \frac{15}{39} - \frac{26}{36} 13t = -3t,$$

where $r_o = 23t \sin\alpha$ is the normal distance from the origin of the axes to the flanges. The shear center is located a distance $3t$ to the left of the origin of the axes, *i.e.*, a distance $e = 10t$ to the right of the vertical web as indicated in fig. 8.47.

The alternative manner of expressing the torque equipollence condition given by eq. (8.55) can also be used to find the location of the shear center. Using reference point **A** in fig. 8.47 leads to

$$x_{2k} = x_{2a} + \int_C f^{[3]} r_a \, ds = 23t - R_2 36t = 23t - \frac{26}{36} 36t = 23t - 26t = -3t,$$

which is the same result as that found above.

A final way to look at this problem is to consider the relationship between the shear centers of open and closed sections, as discussed in example 8.13. Equation (8.57a) yields

$$x_{2k}^c - x_{2k}^o = 2Af_c^{[3]} = 2\frac{30t}{2}\frac{36t}{1080}\frac{23}{t} = 23t,$$

where the closing shear flow, $f_c^{[3]}$, corresponding to a unit vertical shear force is given by eq. (8.47). Using eq. (8.57a) with $x_{2k}^o = -26t$ from example 8.4, results in $x_{2k}^c = x_{2k}^o + 23t = -26t + 23t = -3t$.

Example 8.15. Shear center location for a thin-walled, double-box section
Figure 8.40 depicts a closed multi-cellular, thin-walled, double-box section subjected to a vertical shear force, V_3. The shear flow distribution in this closed section is computed in example 8.11 and is given by eq. (8.50). Determine the location of the shear center.

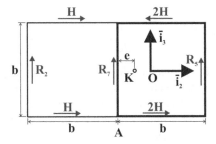

Fig. 8.48. Shear center for a thin-walled double-box section.

First, the resultant forces acting in each wall of the section under a unit vertical shear force, $V_3 = 1$, are computed by integrating eqs. (8.50), leading to the following results

$$H = \frac{1}{69}, \quad R_2 = \frac{13}{69}, \quad R_7 = \frac{30}{69}, \quad R_5 = \frac{26}{69}.$$

Figure 8.48 shows the magnitude and direction of these force resultants.

The shear force equipollence conditions can be verified as follows. The net resultant along axis $\bar{\imath}_2$ is $V_2 = H - H + 2H - 2H = 0$, as expected because no shear force is applied along that axis. The net resultant along axis $\bar{\imath}_3$ is $V_3 = R_2 + R_5 + R_7 = 13/69 + 26/69 + 30/69 = 1$, as expected, because a unit shear force is applied along that axis.

The shear center is calculated by expressing the torque equipollence condition at point **K**, leading to $2Hb - Hb - (b + e)R_2 - eR_7 + (b - e)R_5 = 0$. Solving this equation gives $e/b = H + R_5 - R_2 = 14/69$. The same result also can be obtained by expressing torque equipollence at point **A** at the bottom of the center web to find $bR_5 + bH - bR_2 = e$. It is left to the reader to explore alternative ways of determining the location of the shear center.

Example 8.16. Shear flow resultants

A key point of the development presented in section 8.4.1 is that the resultants of the shear flow distribution, $f^{[2]}$, along axes $\bar{\imath}_2$ and $\bar{\imath}_3$ are 1 and 0, respectively, as expressed by eq. (8.51). This result is intuitively correct because the shear flow distribution, $f^{[2]}$, is computed based on applied shear forces $V_2^{[2]} = 1$ and $V_3^{[2]} = 0$. In this example, these results are established in a formal manner.

Whether the section is open or closed, the shear flow distribution must satisfy the local equilibrium condition, eq. (8.18), which, when $V_2 = 1$ and $V_3 = 0$, becomes

$$\frac{\mathrm{d}f^{[2]}}{\mathrm{d}s} = -Et\frac{x_2 H_{22}^c - x_3 H_{23}^c}{\Delta_H}, \tag{8.58}$$

This equation can be integrated to find the shear flow distribution, $f^{[2]}$, using the procedures described in sections 8.3.1 or 8.3.7, for open or closed sections, respectively. The shear force, $V_2^{[2]}$, associated with this shear flow distribution is evaluated with the help of eq. (8.7) to find

$$V_2^{[2]} = \int_C f^{[2]} \frac{\mathrm{d}x_2}{\mathrm{d}s}\, \mathrm{d}s = -\int_C x_2 \frac{\mathrm{d}f^{[2]}}{\mathrm{d}s}\, \mathrm{d}s + \left[x_2 f^{[2]}\right]_{\text{boundary}},$$

where the second equality is obtained through an integration by parts. The boundary term always vanishes: if the section is open, $f^{[2]} = 0$ at the edges of the section, i.e., the boundaries of the integral, and if the section is closed or multi-cellular, no boundaries exist and this term vanishes. Introducing the governing eq. (8.58) for the shear flow distribution then leads to

$$V_2^{[2]} = \frac{H_{22}^c}{\Delta_H}\left[\int_C Ex_2^2\, t\mathrm{d}s\right] - \frac{H_{23}^c}{\Delta_H}\left[\int_C Ex_2 x_3\, t\mathrm{d}s\right] = \frac{H_{22}^c}{\Delta_H}H_{33}^c - \frac{H_{23}^c}{\Delta_H}H_{23}^c = 1,$$

where the bracketed terms are identified as the sectional bending stiffnesses computed with respect to the origin of the axes system, which is located at the section's centroid.

The shear force, $V_3^{[2]}$, associated with the shear flow distribution, $f^{[2]}$, is readily evaluated using a similar procedure. From eq. (8.7), this shear force becomes

$$V_3^{[2]} = \int_C f^{[2]} \frac{\mathrm{d}x_3}{\mathrm{d}s} \,\mathrm{d}s = -\int_C x_3 \frac{\mathrm{d}f^{[2]}}{\mathrm{d}s} \,\mathrm{d}s + \left[x_3 f^{[2]} \right]_{\text{boundary}},$$

where the boundary term vanishes for the reason given earlier. Introducing the governing equation (8.58) then yields

$$V_3^{[2]} = \frac{H_{22}^c}{\Delta_H} \left[\int_C E x_2 x_3 \, t \mathrm{d}s \right] - \frac{H_{23}^c}{\Delta_H} \left[\int_C E x_3^2 \, t \mathrm{d}s \right] = \frac{H_{22}^c}{\Delta_H} H_{23}^c - \frac{H_{23}^c}{\Delta_H} H_{22}^c = 0.$$

These results are expected. The shear force resultants associated with $f^{[2]}$ are $V_2^{[2]} = 1$ and $V_3^{[2]} = 0$ because $f^{[2]}$ is computed specifically for that applied loading. Similarly, it can be shown that the shear force resultants associated with $f^{[3]}$ are $V_2^{[3]} = 0$ and $V_3^{[3]} = 1$, as expected. The result presented in this example provide a formal proof of eq. (8.51).

8.4.2 Problems

Problem 8.39. Shear flow in closed rectangular section
The thin-walled rectangular beam section shown in fig. 8.42 is subjected to a vertical shear force, V_3. The thickness of the right-hand vertical web is $5t$, whereas that of the remaining walls is t. (1) Determine the bending stiffnesses of the section, H_{22}^c, H_{33}^c and H_{23}^c. (2) Find the shear flow distribution in the section. (3) Verify that all joint and edge equilibrium conditions, eq. (8.29), are satisfied. (3) Find the location of the shear center. Use the following data $\alpha = 5.0$

Problem 8.40. Shear center of thin-walled semi-circular section
Figure 8.19 depicts the thin-walled, semi-circular open cross-section of a beam. The wall thickness is t, and the material Young's and shearing moduli are E and G, respectively. Find the location of the shear center of the section. Note: It is more convenient to work with the angle θ as a variable describing the geometry of the section: $s = R\theta$, $\mathrm{d}s = R\mathrm{d}\theta$.

Problem 8.41. Shear center of a thin-walled "H" section
Figure 7.34 depicts the cross-section of a thin-walled "H" section beam. Compute the location of the shear center for this section.

Problem 8.42. Shear center of a thin-walled "L" section
Consider the thin-walled, "L" shaped cross-section of a beam as shown in fig. 8.14. Find the location of the shear center of this section.

Problem 8.43. Shear center of a multi-cellular cross-section
A thin-walled multi-cellular cross-section is shown in fig. 8.45. Determine the location of the shear center for this cross-section.

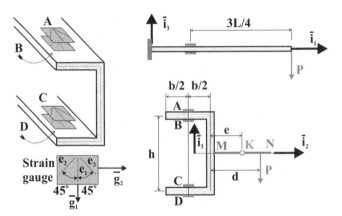

Fig. 8.49. Cantilever beam with C-section under offset tip load.

Problem 8.44. Cantilevered beam with a C-channel section

A thin-walled, C-section cantilevered beam of length L is subjected to a tip load, as depicted in fig. 8.49. The beam is loaded through a horizontal arm, **MN**, and a vertical load P is acting at a distance d from the vertical web; load P is allowed to slide along arm **MN**, $d/b \in [0, 1.5]$. The beam is instrumented with four strain gauge rosettes, denoted A, B, C, and D, and located at the beam's quarter span. Strain gauges A and B are located in the middle of the top flange, gauge A is located on top of the flange, whereas gauge B is located underneath the flange. Similarly, gauges C and D are located in the middle of the bottom flange, on top and underneath the flange, respectively. *(1)* Evaluate the bending moment, vertical shear force, and torque acting at the beam's quarter span. *(2)* On one graph, plot the three readings, e_1^A, e_2^A and e_3^A, of strain rosette A as a function of $d/b \in [0, 1.5]$. *(3)* On one graph, plot the three readings, e_1^B, e_2^B and e_3^B, of strain rosette B as a function of $d/b \in [0, 1.5]$. *(4)* On one graph, plot the three readings, e_1^C, e_2^C and e_3^C, of strain rosette C as a function of $d/b \in [0, 1.5]$. *(5)* On one graph, plot the three readings, e_1^D, e_2^D and e_3^D, of strain rosette D as a function of $d/b \in [0, 1.5]$. Use the following data: $P = 5.0$ kN; $E = 73.0$ GPa; $\nu = 0.3$; $L = 2.0$ m; $h = 0.4$ m; $b = 0.2$ m; $t = 4$ mm.

Problem 8.45. Strength of a cantilever beam with C-section under offset tip load

A thin-walled, C-section cantilevered beam of length L is subjected to a tip load, as depicted in fig. 8.49. The beam is loaded through a horizontal arm, **MN**, and a vertical load P is acting at a distance d from the vertical web; load P is allowed to slide along arm **MN**, $d/b \in [0, 1.5]$. *(1)* Find the location and magnitude of the maximum bending moment, vertical shear force, and torque acting in the beam. *(2)* Find the axial and shear stress distributions in the section where the maximum bending moment, vertical shear force, and torque occur. *(3)* Based on Von-Mises criterion, find the maximum load, $P_{\max}$, the beam can carry as a function of $d/b \in [0, 1.5]$. Use the following data: $P = 5.0$ kN; $E = 73.0$ GPa; $\nu = 0.3$; $\sigma_y = 600$ MPa; $L = 2.0$ m; $h = 0.4$ m; $b = 0.2$ m; $t = 4$ mm.

8.5 Torsion of thin-walled beams

In chapter 7, Saint-Venant's theory of torsion is developed for beams with cross-sections of arbitrary shape, see section 7.3.2. Unfortunately, Saint-Venant's approach requires the solution of a partial differential equation to evaluate the warping or stress function and the corresponding shear stress distribution over the cross-section. In the case of thin-walled beams, however, approximate solutions can be obtained without solving partial differential equations.

8.5.1 Torsion of open sections

The torsional behavior of beams with thin rectangular cross-sections is investigated in section 7.4 and in section 7.5, the results are extended to the more general case of thin-walled, open cross-sections of arbitrary shape. For thin-walled open sections, shear stresses are shown to be linearly distributed through the thickness of the wall, and the torsional stiffness is shown to be proportional to the cube of the wall thickness, see eq. (7.61). Hence, thin-walled open sections have very limited torque carrying capability.

8.5.2 Torsion of closed section

Consider now a thin-walled, closed cross-section of arbitrary shape subjected to an applied torque, as depicted in fig. 8.50. The cross-section consists of a single closed cell defined by a curve $\mathcal{C}$. As is the case for the Saint-Venant solution, eq. (7.3.2), the beam is assumed to be in a state of uniform torsion, *i.e.*, the twist rate is constant along the span. The axial strain and stress components vanish, and hence, the axial flow also vanishes, *i.e.*, $n(s) = 0$.

The local equilibrium equation for a differential element of the thin-walled beam, eq. (8.14), then implies

$$\frac{\partial f}{\partial s} = 0. \qquad (8.59)$$

Therefore the shear flow must remain constant along curve $\mathcal{C}$, *i.e.*,

$$f(s) = f = \text{constant}. \qquad (8.60)$$

This constant shear flow distribution generates a torque, M_1, about the origin of the axes given by eq. (8.9)

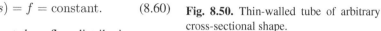

Fig. 8.50. Thin-walled tube of arbitrary cross-sectional shape.

$$M_1 = \int_{\mathcal{C}} f(s)\, r_o(s)\, ds = f \int_{\mathcal{C}} r_o(s)\, ds,$$

where the constant shear flow is factored out of the integral over curve $\mathcal{C}$. The last integral is a purely geometric quantity, which equals twice the area enclosed by curve $\mathcal{C}$, see eq. (8.56), and hence,

$$M_1 = 2\mathcal{A}f, \tag{8.61}$$

where $\mathcal{A}$ is the area enclosed by curve $\mathcal{C}$. This result is known as the *Bredt-Batho formula* and provides a simple relationship between the applied torque, M_1, to the resulting constant shear flow, f, for thin-walled, closed sections.

From the definition of the shear flow, eq. (8.4b), the shear stress, τ_s, resulting from the torque M_1 becomes

$$\tau_s(s) = \frac{M_1}{2\mathcal{A}t(s)}. \tag{8.62}$$

Next, the twist rate of the thin-walled closed section must be related to the applied torque. A simple energy argument will be used for this purpose. According to the first law of thermodynamics, the work done by the applied torque, M_1, must equal to strain energy, A, stored in the tube[2]. Under the effect of the applied shear stress, τ_s, each differential element of the wall undergoes a shear strain, γ_s. It will be shown in chapter 10that strain energy stored in a differential element of volume $tds\,dx_1$ is $1/2\,\gamma_s\tau_s\,tds\,dx_1$. The strain energy stored in a differential slice of the beam of length dx_1 is then found by integrating this expression over curve $\mathcal{C}$ to find

$$\mathrm{d}A = \left[\frac{1}{2}\int_{\mathcal{C}}\gamma_s\tau_s\,tds\right]\,\mathrm{d}x_1 = \left[\frac{1}{2}\int_{\mathcal{C}}\frac{\tau_s^2}{G}\,tds\right]\,\mathrm{d}x_1, \tag{8.63}$$

where Hooke's law, eq. (2.9), is used to relate the shear strain to the shear stress. Introducing the shear stress distribution, eq. (8.62), then yields

$$\mathrm{d}A = \left[\frac{1}{2}\frac{M_1^2}{4\mathcal{A}^2}\int_{\mathcal{C}}\frac{ds}{Gt(s)}\right]\,\mathrm{d}x_1. \tag{8.64}$$

On the other hand, the applied torque, M_1, produces an infinitesimal rotation, $\mathrm{d}\Phi_1$, of the same differential slice of the beam. The work done by the applied torque is then

$$\mathrm{d}W = \frac{1}{2}M_1\mathrm{d}\Phi_1 = \left[\frac{1}{2}M_1\frac{\mathrm{d}\Phi_1}{\mathrm{d}x_1}\right]\,\mathrm{d}x_1 = \left[\frac{1}{2}M_1\kappa_1\right]\,\mathrm{d}x_1, \tag{8.65}$$

where $\kappa_1 = \mathrm{d}\Phi_1/\mathrm{d}x_1$ is the section's twist rate.

The first law of thermodynamics now implies that the work done by the applied torque, eq. (8.65), must equal the strain energy stored in the structure, eq. (8.64). This leads to the following relationship between twist rate and applied torque,

$$\kappa_1 = \frac{M_1}{4\mathcal{A}^2}\int_{\mathcal{C}}\frac{ds}{Gt}. \tag{8.66}$$

This relationship expresses a proportionality between the applied torque, M_1, and the resulting twist rate, κ_1. The constant of proportionality is the *torsional stiffness*,

$$H_{11} = \frac{4\mathcal{A}^2}{\displaystyle\int_{\mathcal{C}}\frac{ds}{Gt}}. \tag{8.67}$$

---

[2] The concept of strain energy will be studied in a formal manner in chapter 10

For an arbitrary shaped closed section of constant wall thickness and made of a homogeneous material, the torsional stiffness reduces to

$$H_{11} = \frac{4Gt\mathcal{A}^2}{\ell},$$
(8.68)

where $\ell = \int_C \mathrm{d}s$ is the perimeter of curve C. This relationship proves that the cross-section of maximum torsional stiffness is the thin-walled circular tube. Indeed, a circle of radius R is the curve that encloses the largest area, $\mathcal{A} = \pi R^2$, for a given perimeter $\ell = 2\pi R$. This will maximize the numerator of eq. (8.68), and hence, will maximize the torsional stiffness.

Sign convention

The discussion of the torsional behavior of thin-walled, closed sections introduces an important geometric quantity, $\mathcal{A}$, the area enclosed by curve C that defines the section's configuration. This area is defined in eq. (8.56) as $2\mathcal{A} = \int_C r_o(s) \, \mathrm{d}s$. As discussed in section 8.1.4, the perpendicular distance from the origin, $\mathbf{O}$, of the axes to the tangent to C, denoted $r_o(s)$, is an algebraic quantity whose sign depends on the direction of the curvilinear variable, s. It follows that $\mathcal{A}$ is an *algebraic area*: $\mathcal{A}$ is positive when curvilinear variable describes curve C while leaving the area, $\mathcal{A}$, to the left; it is negative in the opposite case.

If the shear flow distribution has the direction indicated in fig. 8.50, $f > 0$, $\mathcal{A} > 0$ and eq. (8.61) yields $M_1 = 2\mathcal{A}f > 0$, as expected. If the curvilinear variable is selected in the clockwise direction, see variable s' in fig. 8.50, the corresponding shear flow and area is now negative, $f' = -f$, and $\mathcal{A}' = -\mathcal{A}$, but eq. (8.61) leaves the torque unchanged, $M_1 = 2\mathcal{A}'f' = 2\mathcal{A}f$, as expected.

8.5.3 Comparison of open and closed sections

The torsional behavior of closed sections contrasts sharply with that of open sections. For closed sections, the shear stress is *uniformly* distributed through the thickness of the wall, whereas a *linear* distribution through the wall thickness is found in open sections.

The torsional stiffness is proportional to the square of the enclosed area for a closed section, see eq. (8.67), in contrast with a thickness cubed proportionality for open sections, see eq. (7.64).

To illustrate these sharp differences, consider a thin strip of circular shape, and a thin-walled circular tube, both of identical mean radius R_m and thickness t, as de-

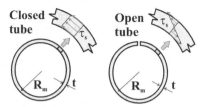

Fig. 8.51. A thin-walled open tube and a thin-walled closed tube.

picted in fig. 8.51. The torsional stiffness of the open and closed sections, denoted H_{11}^{open} and H_{11}^{closed}, respectively, are given by eqs. (7.64), and (7.19), respectively, as $H_{11}^{\mathrm{open}} = 2\pi G R_m t^3/3$ and $H_{11}^{\mathrm{closed}} = 2\pi G R_m^3 t$. Their ratio is

$$\frac{H_{11}^{closed}}{H_{11}^{open}} = 3\left(\frac{R_m}{t}\right)^2. \tag{8.69}$$

If the two section are subjected to the same torque, M_1, the maximum shear stresses in the open and closed sections, denoted τ_{max}^{open}, and τ_{max}^{closed}, respectively, are given by eqs. (7.65) and (7.24), respectively, as

$$\tau_{max}^{open} = G\kappa_1^{open}t = G\frac{M_1 t}{H_{11}^{open}} = \frac{3M_1}{2\pi R_m t^2},$$

$$\tau_{max}^{closed} = R_m G\kappa_1^{closed} = G\frac{M_1 R_m}{H_{11}^{closed}} = \frac{M_1}{2\pi R_m^2 t}.$$

Their ratio can then be expressed as

$$\frac{\tau_{max}^{open}}{\tau_{max}^{closed}} = 3\left(\frac{R_m}{t}\right). \tag{8.70}$$

Consider, for instance, a typical thin-walled construction for which $R_m = 20t$. The torsional stiffness of the closed section will be $1,200$ times larger than that of the open section. Under the same applied torque, the maximum shear stress in the open section will be 60 times larger than that of the closed section, or stated equivalently, the closed section can carry a 60 times larger torque for an equal shear stress level.

8.5.4 Torsion of combined open and closed sections

In the previous section, the behavior of open and closed sections is shown to contrast sharply. In practical situations, one is often confronted with cross-sections presenting a combination of open and closed curves. The section shown in fig. 8.52 combines a closed trapezoidal box and overhanging rectangular strips. All components have a constant thickness t, and the other dimensions are shown on the figure.

The twist rate is identical for the trapezoidal box and strips, whereas the torques they carry, denoted M_1^{box} and M_1^{strip}, respectively, are $M_1^{box} = H_{11}^{box}\kappa_1$ and $M_1^{strip} = H_{11}^{strip}\kappa_1$, respectively. The torsional stiffnesses of the trapezoidal box and strips are evaluated with the help of eqs. (8.68) and (7.64), respectively, to find $H_{11}^{box} = 4Gt\mathcal{A}^2/\ell$ and $H_{11}^{strip} = G\,wt^3/3$, respectively, where ℓ is the length of the perimeter of the trapezoidal box and $\mathcal{A} = h(b_1 + b_2)/2$ its enclosed area. The total torque, M_1, is the sum of the torques carried by the three components of the section, i.e., $M_1 = M_1^{box} + 2M_1^{strip}$, and hence,

$$M_1 = H_{11}^{box}\left(1 + 2\frac{H_{11}^{strip}}{H_{11}^{box}}\right)\kappa_1 = H_{11}^{box}\left[1 + \frac{2}{3}\frac{w\ell}{(b_1+b_2)^2}\left(\frac{t}{h}\right)^2\right]\kappa_1.$$

For thin-walled sections the last term in the bracket is clearly negligible because $t/h \ll 1$. It follows that $H_{11} \approx H_{11}^{box}$: *the torsional stiffness of the section is nearly equal to that of the closed trapezoidal box alone.* The contribution of the strips, i.e., of the open parts of the section, is negligible.

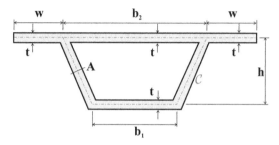

Fig. 8.52. Thin-walled trapezoidal beam with overhangs.

Using this simplification, the twist rate of the section is $\kappa_1 = M_1/H_{11}^{\text{box}}$, and the torques carried by the individual components of the section become

$$M_1^{\text{box}} = H_{11}^{\text{box}}\kappa_1 \approx H_{11}^{\text{box}}\frac{M_1}{H_{11}^{\text{box}}} = M_1, \quad M_1^{\text{strip}} = H_{11}^{\text{strip}}\kappa_1 \approx \frac{H_{11}^{\text{strip}}}{H_{11}^{\text{box}}}M_1.$$

Finally, the maximum shear stress in the trapezoidal box and strips are found from eqs. (8.62) and (7.65), respectively, as

$$\tau_{\text{max}}^{\text{box}} = \frac{M_1^{\text{box}}}{2\mathcal{A}t} \approx \frac{1}{2\mathcal{A}t}M_1, \quad \tau_{\text{max}}^{\text{strip}} = \frac{3M_1^{\text{strip}}}{wt^2} \approx \frac{3}{wt^2}\frac{H_{11}^{\text{strip}}}{H_{11}^{\text{box}}}M_1.$$

The ratio of these stresses is now

$$\frac{\tau_{\text{max}}^{\text{strip}}}{\tau_{\text{max}}^{\text{box}}} = \frac{\ell}{b_1 + b_2}\left(\frac{t}{h}\right).$$

Clearly, the maximum shear stress in the strips is far smaller than that in the trapezoidal box.

In summary, when dealing with a combination of open and closed sections under torsion, the contribution of the open portion of the section can be neglected when evaluating the torsional stiffness, and the shear stress acting in the open portion is far smaller than that acting in the closed portion of the section.

8.5.5 Torsion of multi-cellular sections

The analysis of thin-walled tubes under torsion developed in section 8.5.2 can be extended to the case of thin-walled cross-sections with multiple closed cells. Consider, for example, the four cell, thin-walled cross-section subjected to a torque, M_1, depicted in fig. 8.53, which shows an infinitesimal slice of the beam of span dx_1. The section is assumed to be under uniform torsion, and hence, the axial stress flow vanishes. The local equilibrium equation, eq. (8.14), then reduces to $\partial f/\partial s = 0$, which implies the constancy of the shear flow.

To study the shear flow distribution over the cross-section, free-body diagrams of different portions of the cross-section are shown in fig. 8.54. First, consider the free-body diagram of the portion of the section between points **A** and **B**, shown in fig. 8.54-(1). Since the axial stress flow acting on the element vanishes, equilibrium of forces acting along the beam's axis implies $f_A = f_B$, which is consistent with the constant shear flow requirement.

Next, consider the free-body diagram obtained by cutting the section at points **C** and **D**, as depicted in fig. 8.54-(2). Here again, axial equilibrium implies $f_C = f_D$. Finally, fig. 8.54-(3) shows the free-body diagram obtained by cutting the section around point **E**, where several walls connect to each other. This free-body diagram involves

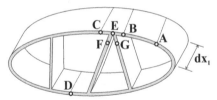

Fig. 8.53. A thin-walled, multi-cellular section under torsion.

shear flows f_B, f_C, f_F and f_G acting in the four walls that have been cut; axial equilibrium implies $f_C + f_F + f_G - f_B = 0$, or more generally,

$$\sum f_i = 0. \tag{8.71}$$

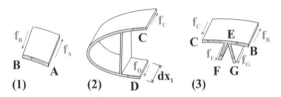

Fig. 8.54. Free-body diagrams of the thin-walled, multi-cellular section.

This simple equilibrium argument implies that *the sum of the shear flows going into a joint must vanish*. In practical applications of eq. (8.71), the shear flows must be interpreted as *algebraic quantities*: by convention, each shear flow is counted positive in the direction of the corresponding curvilinear coordinate. For instance, application of eq. (8.71) to the four-wall configuration illustrated in fig. 8.54-(3) yields $(-f_C) + (-f_F) + (-f_G) + (+f_B) = 0$, or $f_C + f_F + f_G - f_B = 0$, as derived earlier.

In summary, simple equilibrium arguments require the shear flow to remain constant along each wall of the multicellular section, and at each connection point, the sum of the flow going into the joint must vanish.

These continuity requirements are automatically satisfied if constant shear flows are assumed to act in each cell of the section, as shown in fig. 8.55. Shear flows circulating around each cell are denoted $f^{[1]}$, $f^{[2]}$, $f^{[3]}$ and $f^{[4]}$, and their assumed positive direction is indicated in the figure. As required, the shear flow remains constant in each wall, and furthermore, the sum of the flows going into each joint vanishes. Figure 8.55 illustrates the shear flows converging to joint **E**: the continuity condition, eq. (8.71), is satisfied because $(f^{[4]}) + (f^{[3]} - f^{[4]}) + (f^{[2]} - f^{[3]}) + (-f^{[2]}) = 0$.

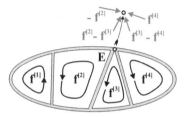

Fig. 8.55. Shear flows in each cell of a thin-walled, multi-cellular section.

The solution of the problem then requires the determination of the constant shear flows, one around each cell. The total torque, M_1, carried by the section equals the sum of the torques carried by each individual cell, $M_1^{[i]}$, where i indicates the cell number. The Bredt-Batho formula, eq. (8.61), leads to

$$M_1 = \sum_{i=1}^{N_{\text{cells}}} M_1^{[i]} = 2 \sum_{i=1}^{N_{\text{cells}}} \mathcal{A}^{[i]} f^{[i]}, \tag{8.72}$$

where N_{cells} is the number of cells and $\mathcal{A}^{[i]}$ the area enclosed by the ith cell. This single equation does not allow the determination of the shear flows in the N_{cells} cells.

Additional equations can be obtained by expressing the *compatibility conditions* requiring the twist rates of the various cells to be identical. In response to the shear flow, $f^{[i]}$, acting within the cell, a twist rate, $\kappa_1^{[i]}$, develops in the cell. Compatibility of the deformations of all cells provides $N_{\text{cells}} - 1$ additional equations,

$$\kappa_1^{[1]} = \kappa_1^{[2]} = \ldots = \kappa_1^{[i]} = \ldots = \kappa_1^{[N_{\text{cells}}]}. \tag{8.73}$$

Equation (8.66) expresses the relationship between this twist rate and the torque carried by the cell as

$$\kappa_1^{[i]} = \int_{\mathcal{C}^{[i]}} \frac{M_1^{[i]}}{4(\mathcal{A}^{[i]})^2} \frac{ds}{Gt} = \int_{\mathcal{C}^{[i]}} \frac{2\mathcal{A}^{[i]} f^{[i]}}{4(\mathcal{A}^{[i]})^2} \frac{ds}{Gt} = \frac{1}{2\mathcal{A}^{[i]}} \int_{\mathcal{C}^{[i]}} \frac{f^{[i]}}{Gt} \, ds. \tag{8.74}$$

Equations (8.72) and (8.73) provide the N_{cells} equations needed to solve for the N_{cells} shear flows in the cells of a multi-cellular section under torsion.

Example 8.17. Two-cell cross-section

The thin-walled cross-section shown in fig. 8.56 represents a highly idealized airfoil structure for which the curved portion is the leading edge, the thicker vertical web is the spar, and the trailing straight segments form the aft portion of the airfoil. Equation (8.72) implies that the total torque carried by the section is the sum of the torques carried in each cell

$$M_1 = 2 \sum_{i=1}^{N_{\text{cell}}} \mathcal{A}^{[i]} f^{[i]} = \pi R^2 f^{[1]} + 6R^2 f^{[2]}. \tag{8.75}$$

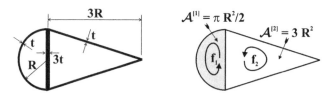

Fig. 8.56. A two-cell thin-walled section under torsion.

The compatibility condition, eq. (8.73), requires twist rates for the two cells to be identical. Equation (8.66) yields the twist rate for the front cell as

$$
\kappa_1^{[1]} = \frac{1}{2\mathcal{A}^{[1]}} \int_{\mathcal{C}_1} \frac{f}{Gt(s)}\,ds = \frac{1}{2\,G\,\pi\,R^2/2} \left[\frac{f^{[1]}}{t}\pi R + \frac{f^{[1]} - f^{[2]}}{3t}2R \right]
$$
$$
= \frac{1}{\pi GRt}\left[\pi f^{[1]} + \frac{2}{3}(f^{[1]} - f^{[2]})\right],
$$

and the twist rate for the aft cell is

$$
\kappa_1^{[2]} = \frac{1}{2\mathcal{A}^{[2]}} \int_{\mathcal{C}_2} \frac{f}{Gt(s)}\,ds = \frac{1}{2\,G\,3\,R^2}\left[\frac{f^{[2]} - f^{[1]}}{3t}2R + f^{[2]}\,2\sqrt{10}\frac{R}{t}\right]
$$
$$
= \frac{1}{6\,GRt}\left[\frac{2}{3}(f^{[2]} - f^{[1]}) + 2\sqrt{10}f^{[2]}\right].
$$

Equating the two twist rates yields the second equation for the shear flows

$$
\frac{1}{\pi}\left[\pi f^{[1]} + \frac{2}{3}(f^{[1]} - f^{[2]})\right] = \frac{1}{6}\left[\frac{2}{3}(f^{[2]} - f^{[1]}) + 2\sqrt{10}f^{[2]}\right],
$$

which simplifies to $f^{[1]} = 1.04\,f^{[2]}$.

This result, along with eq. (8.75), can be used to solve for $f^{[1]}$ and $f^{[2]}$. From eq. (8.75), it follows that $R^2 f^{[1]} = 1.04M_1/(6 + 1.04\pi)$ and $R^2 f^{[2]} = M_1/(6 + 1.04\pi)$. The shear flow in the front cell, $f^{[1]}$, is only about 4% greater than that in the aft cell, $f^{[2]}$, and hence, the shear flow in the spar, $R^2(f^{[1]} - f^{[2]}) = 0.04M_1/(6 + 1.04\pi)$, nearly vanishes.

Because the torsional stiffness of a closed section is proportional to the square of the enclosed area, the largest contribution to the torsional stiffness comes from the outermost closed section, which is the union of the front and aft cells. Consequently, the largest shear flow circulates in this outermost section, leaving the spar nearly unloaded.

The torsional stiffness is computed as the ratio of the torque, given by eq. (8.75), to the cell twist rate, given by eq. (8.74). Since the twist rates of the two cells are equal, either $\kappa_1^{[1]}$ or $\kappa_1^{[2]}$ can be used. For instance, using $\kappa_1^{[1]}$ yields

$$
H_{11} = \frac{M_1}{\kappa_1^{[1]}} = \frac{(\pi R^2\,1.04 + 6R^2)\,f^{[2]}}{1/(\pi GRt)\,[1.04\pi + 2/3(1.04 - 1)]\,f^{[2]}} = 2.81\pi GR^3 t.
$$

8.5.6 Problems

Problem 8.46. Torsion of a thin-walled trapezoidal section

Consider the beam with a trapezoidal cross-section shown in fig. 8.52. *(1)* If the depth of the cross-section, h, is doubled, how does its torsional stiffness vary? *(2)* If the width of the strips, w, are doubled, how does its torsional stiffness vary? *(3)* If the thickness of the cross-section, t, is doubled, how does its torsional stiffness vary? *(4)* If the section carries a torque Q, find the shear flow at point **A**. *(5)* Sketch the shear stress distribution through-the-thickness of the wall at point **A**, when the section is subjected to a torque Q.

Problem 8.47. Torsion of a closed, semi-circular thin-walled section

A beam has the closed, semi-circular thin-walled cross-section shown in fig. 8.57 and is sub-jected to a torque, Q_1. *(1)* Find the resulting shear flow distribution in the section. *(2)* Determine its torsional stiffness.

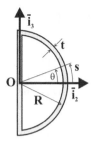

Fig. 8.57. Thin-walled closed semi-circular section.

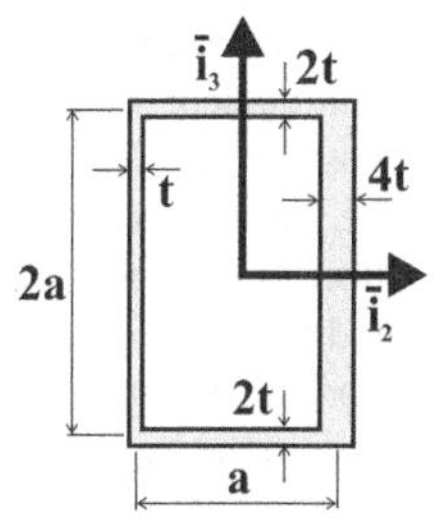

Fig. 8.58. Thin-walled closed rectangular section with variable wall thicknesses.

Problem 8.48. Torsion of rectangular box with variable thickness

A beam with the closed rectangular thin-walled cross-section shown in fig. 8.58 is subjected to a torque, Q_1. The walls have different thicknesses, as indicated in the figure. *(1)* Find the magnitude and location of the maximum shear stress in the section. *(2)* Determine its torsional stiffness.

Problem 8.49. Shearing and torsion of a high-lift device

The cross-section of a high lift device is shown in fig 8.59. The aerodynamic pressure acting on the lower panel of the device has a net resultant $V_3 = 100$ kN and its line of action is aligned with the left web, as indicated on the figure. Material properties are: $E = 73$ GPa, $G = 30$ GPa. *(1)* Find and sketch the shear stress distribution generated by the vertical shear force. *(2)* Find and sketch the shear stress distribution generated by the torque applied to the section. *(3)* Find and sketch the total shear stress distribution in the section. *(4)* Indicate the location and magnitude of the maximum shear stress in the section.

Problem 8.50. Torsion of a 2-cell rectangular cross-section

The cross-section of a thin-walled beam consists of two rectangular cells, as shown in fig. 8.60. The beam is subjected to a torque Q_1. *(1)* Determine the shear flow distribution in the cross-section. *(2)* Find the magnitude and location of the maximum shear stress in the section. *(3)* Determine its torsional stiffness. *(4)* Does the section's mid vertical web contribute signifi-cantly to the torsional stiffness? Explain.

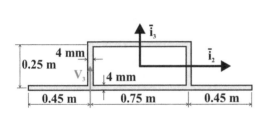

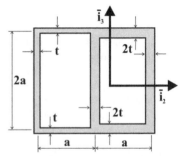

Fig. 8.59. High lift device subjected to a transverse shear force.

Fig. 8.60. Thin-walled 2-cell rectangular section with variable wall thicknesses.

Problem 8.51. Axial and shear flows in a thin-walled box beam

Consider the thin-walled, rectangular box beam of length $L = 2$ m shown in fig. 8.61. The cross-section has a width $b = 0.2$ m, a height $h = 0.1$ m, and a constant wall thickness $t = 5$ mm. The cantilevered beam is subjected to a tip load $P = 5$ kN acting in the plane of the tip section as indicated on the figure, a distributed transverse load $p_0 = 20$ kN/m, and a distributed torque $q_0 = 1.0$ kN. (1) Find the distribution of axial stress in the beam's root section. Find the magnitude and location of the maximum axial stress. (2) Find the shear stress distribution in the beam's root section generated by the transverse shear force. (3) Find the shear stress distribution in the beam's root section generated by torsion.

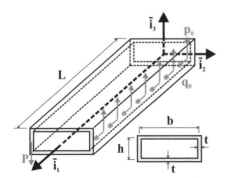

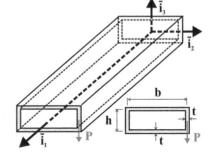

Fig. 8.61. Thin-walled, rectangular box beam.

Fig. 8.62. Rectangular box beam subjected to a tip load.

Problem 8.52. Thin-walled, cantilevered beam under tip load

Consider the thin-walled cantilevered beam of length L depicted in fig. 8.62. The rectangular cross-section is of width b, height h and thickness t. A tip transverse load P is applied at the section's right web, as indicated on the figure. Find the maximum load, $P_{\max}$, that the section can carry. Use the following data: $L = 2$ m, $b = 0.60$ m, $h = 0.15$ m, $t = 3$ mm, $\sigma_y = 620$ MPa.

Problem 8.53. Torsion of a 2-cell rectangular cross-section

The multi-cellular thin-walled cross-section depicted in fig. 8.45 is subjected to a torque, Q_1. *(1)* Determine the shear flow distribution in the section. *(2)* Determine the magnitude and location of the section's maximum shear flow. *(3)* Determine its torsional stiffness. *(4)* Sketch the distribution of shear stress through the thickness of the wall.

Problem 8.54. Shearing and torsion of a closed, semi-circular section

A beam with the closed semi-circular thin-walled cross-section shown in fig. 8.57 is subjected to a vertical shear force, V_3, with a line action passing through the section's vertical web. *(1)* Determine the location of the section's shear center. *(2)* Determine the shear flow distribution due to shearing. *(3)* Determine the shear flow distribution due to torsion. *(4)* Determine the total shear flow distribution.

Problem 8.55. Shearing and torsion of a rectangular section

A beam with the rectangular thin-walled cross-section shown in fig. 8.58 is subjected to a vertical shear force, V_3, with a line action passing through the section's right vertical web. *(1)* Determine the location of the section's shear center. *(2)* Determine the shear flow distribution due to shearing. *(3)* Determine the shear flow distribution due to torsion. *(4)* Determine the total shear flow distribution.

Problem 8.56. Shearing and torsion of a 2-cell rectangular cross-section

A beam with the 2-cell rectangular cross-section shown in fig. 8.45 is subjected to a vertical shear force, V_3, with a line action passing through the section's central vertical web. *(1)* Determine the location of the section's shear center. *(2)* Determine the shear flow distribution due to shearing. *(3)* Determine the shear flow distribution due to torsion. *(4)* Determine the total shear flow distribution.

Problem 8.57. Bending and shear in a 4-web flexure

Figure 8.63 depicts a flexure composed two rigid circular flanges and four flexible webs (homogeneous material of Young's modulus, E, and shearing modulus, G). The flexure is subjected to a tip axial load P_1, a tip torque, Q_1, and tip bending moments, Q_2 and Q_3. The resulting tip axial displacement is $\bar{u}_1(L)$, tip rotation $\Phi_1(L)$, $\Phi_2(L)$ and $\Phi_3(L)$. Find: *(1)* The shear center of the section. *(2)* The axial stiffness S_a of the flexure: $S_a = P_1/\bar{u}_1(L)$. Give the non-dimensional stiffness $LS_a/(EbR)$. *(3)* The torsional stiffness: $S_1 = Q_1/\Phi_1(L)$. Give the non-dimensional stiffness $LS_1/(EbR^3)$. *(4)* The bending stiffnesses: $S_2 = Q_2/\phi_2(L)$ and $S_3 = Q_3/\phi_3(L)$. Give the non-dimensional stiffnesses $LS_2/(EbR^3)$ and $LS_3/(EbR^3)$. Use an appropriate approximation of the torsional stiffness of individual webs.

Problem 8.58. Thin-walled, cantilevered beam under tip load

Consider the thin-walled cantilevered beam of length L depicted in fig. 8.64. The circular cross-section is of radius R and thickness t. A tip transverse load P is applied on the right edge of the section, as indicated on the figure. Find the maximum load, P_{max}, that the section can carry. Note: a circular section subjected to a vertical shear force V_3 will develop a shear flow distribution $f(\theta) = (V_3/\pi R)\cos\theta$. The bending stiffnesses of a thin-walled circular section are $H_{22}^c = H_{33}^c = E\pi t R^3$. Use the following data: $L = 2$ m, $R = 0.15$ m, $t = 5$ mm, $\sigma_y = 620$ MPa.

Problem 8.59. Thin-walled, circular cross-section with flanges

Consider the beam with a thin-walled, circular cross-section with flanges shown in fig. 8.43. *(1)* If the radius, R, is doubled, how does its torsional stiffness vary? *(2)* If the width of the

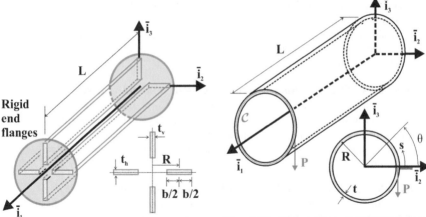

Fig. 8.63. Configuration of the four-web flexure.

Fig. 8.64. Thin-walled cantilevered beam
with circular cross-section under tip transverse load.

flanges, a, is doubled, how does its torsional stiffness vary? *(3)* If the thickness of the cross-
section, t, is doubled, how does its torsional stiffness vary? *(4)* If the section carries a torque
Q, find the shear flow at point **A**. *(5)* Sketch the shear stress distribution through-the-thickness
of the wall at point **A**, when the section is subjected to a torque Q.

8.6 Coupled bending-torsion problems

In chapter 6, the response of a beam with an arbitrary cross-section subjected to complex loading conditions is investigated. The loading involves distributed and concentrated axial and transverse loads, as well as distributed and concentrated moments.
Two important restrictions are made: no torques are applied, and the transverse shearing forces are assumed to be applied in such a way that the beam will bend without
twisting. The first restriction can now be removed. Indeed, if torques are applied,
the beam will twist and the analyses developed in chapter 7 for solid and open thin-
walled cross-sections and in section 8.5 for closed thin-walled sections can be applied to this problem.

The knowledge of the shear center location allows removal of the second restriction. Applied transverse forces will bend the beam without twisting it if and only if
their lines of action pass through the shear center. If all transverse loads are applied
at the shear center, the bending and shearing analyses developed in this chapter are
applicable. If a transverse load is not applied at the shear center, it can always be
replaced by an equipollent system consisting of an equal transverse load applied at
the shear center *plus a torque* equal to the moment of the transverse load about the
shear center.

Figure 8.65 shows a concentrated transverse load, P_2, acting at the beam's tip
and its point of application, point **A**, with coordinates (x_{2a}, x_{3a}). The points of application of all distributed loads, $p_1(x_1)$, $p_2(x_1)$, and $p_3(x_1)$, and of the concentrated

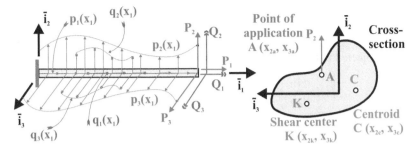

Fig. 8.65. Beam under a complex loading condition.

loads, P_1, P_2, and P_3, must be similarly defined. The transverse load, $p_2(x_1)$, applied at point **A** is equivalent to a transverse load of equal magnitude, $p_2(x_1)$, applied at the shear center, point **K**, plus a distributed torque $-(x_{3a} - x_{3k})p_2(x_1)$. A similar equivalence applies to all distributed and concentrated loads. Note the presence of distributed and concentrated torques, $q_1(x_1)$ and Q_1, respectively.

The above remarks lead to the following solution procedure.

1. Compute the location of the centroid, **C** (x_{2c}, x_{3c}), of the cross-section.
2. Compute the orientation of the principal axes of bending $\bar{\imath}_1^*$, $\bar{\imath}_2^*$, and $\bar{\imath}_3^*$, and the principal centroidal bending stiffnesses, see section 6.6.
3. Compute the location of the shear center, **K** (x_{2k}, x_{3k}), of the cross-section according to the procedure described in section 8.4.
4. Compute the torsional stiffness; see chapter 7, or section 8.5.2 for closed, thin-walled beams.
5. Solve the extensional problem, eq. (6.30), with appropriate boundary conditions.
6. Solve two decoupled bending problems eqs. (6.31) and (6.32), in principal centroidal axes of bending planes with appropriate boundary conditions.
7. Solve the torsional problem governed by the following differential equation

$$\frac{d}{dx_1^*}\left(H_{11}^* \frac{d\Phi_1^*}{dx_1^*}\right) = -\left[q_1^*(x_1^*) + (x_{2a}^* - x_{2k}^*)p_3^*(x_1^*) - (x_{3a}^* - x_{3k}^*)p_2^*(x_1^*)\right],$$
(8.76)

subjected to boundary conditions at the root, $\Phi_1^* = 0$, and at the tip

$$H_{11}^* \frac{d\Phi_1^*}{dx_1^*} = Q_1^* + (x_{2a}^* - x_{2k}^*)P_3^* - (x_{3a}^* - x_{3k}^*)P_2^*.$$

The two equations above are written in the axis system defined by the principal centroidal axes of bending. While this system simplifies the solution of the axial and bending problems, it is of little help to the solution of the torsional problem. Because $\Phi_1^* = \Phi_1$, $H_{11}^* = H_{11}$, and $x_1^* = x_1$, it is more convenient to recast the governing equation of the torsional problem in a coordinate system for which axis $\bar{\imath}_1$ is aligned with the axis of the beam,

$$\frac{\mathrm{d}}{\mathrm{d}x_1}\left(H_{11}\frac{\mathrm{d}\Phi_1}{\mathrm{d}x_1}\right) = -\left[q_1(x_1) + (x_{2a} - x_{2k})p_3(x_1) - (x_{3a} - x_{3k})p_2(x_1)\right],$$

$$(8.77)$$

and subjected to boundary conditions at the root, $\Phi_1 = 0$, and at the tip

$$H_{11}\frac{\mathrm{d}\Phi_1}{\mathrm{d}x_1} = Q_1 + (x_{2a} - x_{2k})P_3 - (x_{3a} - x_{3k})P_2. \qquad (8.78)$$

The knowledge of the centroid and shear center locations, as well as the orientation of the principal axes of bending allows a complete decoupling of the problem into four independent problems: an axial problem, two bending problems, and a torsional problem. Of course, as discussed in chapter 6, the two bending problems can also be treated as coupled problems; in that case, an arbitrary set of centroidal axes could be used instead of the principal centroidal axes of bending.

In the absence of externally applied torques, and if all transverse loads are applied at the shear center, the right hand side of eq. (8.77) vanishes, and its solution is then simply $\Phi_1(x_1) = 0$: *the beam does not twist.* When external torques are applied, or if any transverse load is not applied at the shear center, the right hand side of eq. (8.77) does not vanish and the beam twists, *i.e.*, each section of the beam undergoes a rigid body rotation of magnitude $\Phi_1(x_1)$ *about the shear center.*

Example 8.18. Wing subjected to aerodynamic lift and moment
An important example of this procedure is the wing coupled bending-torsion problem shown in fig. 8.66. The principal axes of bending, $\bar{\imath}_2$ and $\bar{\imath}_3$, are selected with their origin at the shear center. Axis $\bar{\imath}_1$ is along the locus of the shear centers of all the cross-sections assumed to form a straight line called the *elastic axis.* The aerodynamic loading consists of a lift per unit span, L_{AC}, applied at the aerodynamic center and an aerodynamic moment per unit span, M_{AC}.

According to the sign convention for torques, this nose-up aerodynamic moment is a negative quantity. The differential equation for bending in plane $(\bar{\imath}_1, \bar{\imath}_3)$ is

$$\frac{\mathrm{d}^2}{\mathrm{d}x_1^2}\left(H_{22}^c\frac{\mathrm{d}^2\bar{u}_3}{\mathrm{d}x_1^2}\right) = L_{AC}. \qquad (8.79)$$

The boundary conditions for a cantilevered wing of length L are $\bar{u}_3 = \mathrm{d}\bar{u}_3/\mathrm{d}x_1 = 0$ at the root and $\mathrm{d}^2\bar{u}_3/\mathrm{d}x_1^2 = \mathrm{d}^3\bar{u}_3/\mathrm{d}x_1^3 = 0$ at the unloaded tip.

The governing equation for torsion is

$$\frac{\mathrm{d}}{\mathrm{d}x_1}\left(H_{11}\frac{\mathrm{d}\Phi_1}{\mathrm{d}x_1}\right) = -(M_{AC} + eL_{AC}),$$

where e is the distance from the aerodynamic center to the shear center. The boundary condition at the wing's root is $\Phi_1 = 0$, and at its tip, $\mathrm{d}\Phi_1/\mathrm{d}x_1 = 0$. Note that for symmetric airfoils $M_{AC} = 0$, but the wing still twists because the lift is applied at the aerodynamic center, which does not coincide with the shear center for the case at hand. For typical transport aircraft, the aerodynamic and shear centers are located at 25% and 35% chord, respectively. Consequently, the lift generates a nose-up torque on the wing.

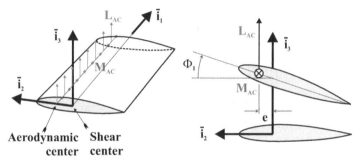

Fig. 8.66. The wing bending torsion coupled problem.

For aircraft wing analysis, it is convenient to select the origin of the axes at the shear center, rather than at the centroid as advised in chapter 6. The main advantage of selecting the centroid as the origin is that the bending problems decouple from the axial problem. If the beam is not subjected to any axial loads, the axial problem is of little interest, and hence it is more meaningful to select the shear center as the origin of the axis system. In that case, the beam rotates about the origin of the axis system. As illustrated in fig. 8.66, the rotation, $\Phi_1(x_1)$, of the section is, in fact, the geometric angle of attack of the airfoil. The lift, L_{AC}, is a function of this angle of attack; consequently, the aerodynamic problem, which involves the computation of the lift as a function of the angle of attack, and the elastic problem, which involves the computation of the wing deflection and twist as a function of the applied loads, become coupled. The study of this interaction is called *aeroelasticity*.

Example 8.19. Wing section subjected to transverse loading

Consider the highly idealized wing section depicted in fig. 8.67. The leading edge of the airfoil is semi-circular, of radius R and thickness t. Given the geometry of the section, $\tan \alpha = 2/3$. The lower portion of the trailing edge is horizontal, of length $a = 3R$ and thickness t. The vertical spar is of thickness ηt, where $\eta = 3$. The airfoil thickness is $2R$ and the chord length $4R$. Axis $\bar{i}_2$ is selected to be horizontal and passes through the section's centroid, located a distance d above the horizontal trailing edge and a distance c to the right of the vertical spar. Curvilinear coordinates s_1, s_2, s_3 and s_4 are defined as indicated on the figure.

The section is subjected to a vertical shear force, V_3, with its line of action passing through the vertical spar. In this case, this shear force is the aerodynamic lift, whose line of action passes through the section's quarter-chord point.

The solution to this problem involves a considerable amount of detailed algebraic computation, and for this reason, symbolic manipulation software is helpful. The length of the upper portion, **AT**, of the airfoil is found as $b = \sqrt{13}R$. The axial stiffness is $S = E(9Rt + \sqrt{13}Rt + \pi Rt)$ and the location of the centroid is

$$c = \frac{(5 + 3\sqrt{13})R}{2(9 + \sqrt{13} + \pi)} = 0.5022R \quad \text{and} \quad d = \frac{(6 + \sqrt{13} + \pi)R}{9 + \sqrt{13} + \pi} = 0.8094R.$$

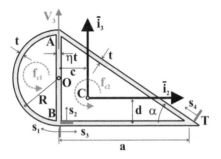

Fig. 8.67. Simplified 2-cell wing cross-section.

Next, the sectional centroidal bending stiffnesses are found through tedious algebra as

$$H_{22}^c = \frac{3\pi^2 + \pi(57 + 5\sqrt{13}) + 242 + 48\sqrt{13}}{6(\pi + 9 + \sqrt{13})} ER^3 t = 7.021 \ ER^3 t,$$

$$H_{33}^c = \frac{\pi^2 + \pi(27 + 7\sqrt{13}) + 169 + 57\sqrt{13}}{2(\pi + 9 + \sqrt{13})} ER^3 t = 17.42 \ ER^3 t,$$

$$H_{23}^c = \frac{\pi(9 + \sqrt{13}) + 79 + 9\sqrt{13}}{2(\pi + 9 + \sqrt{13})} ER^3 t = -4.796 \ ER^3 t.$$

The location of the shear center must be determined to study the torsional behavior of the structure; in particular, if the vertical shear force is not applied at the shear center, it will generate torsion of the section. To determine the shear flow distribution in the two-cell closed section, it will be cut at point **B**. Two cuts are needed, one to open the front cell and another to open the aft cell. Using eq. (8.21), the static stiffness moments in the various segments of the section are found as

$$Q_2(s_1) = \int_0^{s_1} Et(-d + R - R\cos s_1/R) \, ds_1, \quad Q_2(s_2) = E\eta ts_2(-d + s_2/2),$$

$$Q_3(s_1) = \int_0^{s_1} Et(-c - R\sin s_1/R) \, ds_1, \quad Q_3(s_2) = -E\eta tcs_2,$$

$$Q_2(s_3) = -Etds_3, \quad\quad Q_2(s_4) = \int_0^{s_4} Et(-d + s_4 \sin\alpha) \, ds_4,$$

$$Q_3(s_3) = Ets(-c + s_3/2), \quad Q_3(s_4) = \int_0^{s_4} Et(-c + a - s_4 \cos\alpha) \, ds_4.$$

The shear flow distribution in the open section can then be evaluated using eq. (8.20) to find

$$f_o(s_i) = f_o(0) + \frac{Q_{3i}(s_i)H_{23} - Q_{2i}(s_i)H_{33}}{\Delta_H} V_3,$$

where $\Delta_H = H_{22}^c H_{33}^c - (H_{23}^c)^2 = 102.41 \ E^2 R^6 t^2$. At point **A**, the shear flow must vanish in the open section: $f_o(s_1 = 0) = f_o(s_2 = 0) = f_o(s_3 = 0) = 0$. At points **T** and **B**, the joint equilibrium condition, eq. (8.29), must be satisfied.

To find the shear flow distribution in the closed section, closing shear flows are added: f_{c1} and f_{c2} in the front and aft cells, respectively, see fig. 8.67. The procedure described in section 8.3.8 is then used to evaluate these closing shear flows. In the present case, the compatibility conditions for the front and aft cells become

$$\int_0^{\pi R} \frac{f_o(s_1) - f_{c1}}{Gt} ds_1 - \int_0^{2R} \frac{f_o(s_2) + f_{c1} - f_{c2}}{G\eta t} ds_2 = 0,$$

$$\int_0^a \frac{f_o(s_3) + f_{c2}}{Gt} ds_3 + \int_0^b \frac{f_o(s_4) + f_{c2}}{Gt} ds_4 - \int_0^{2R} \frac{f_o(s_2) + f_{c1} - f_{c2}}{G\eta t} ds_2 = 0.$$

These conditions lead to two simultaneous equations for the closing shear flows, which are found as $f_{c1} = 0.06580\ V_3/R$ and $f_{c2} = -0.1458\ V_3/R$. The shear flow distribution in the closed section is then found by adding the shear flow in the open section and the closing shear flows to find $f(s_1) = f_o(s_1) - f_{c1}$, $f(s_2) = f_o(s_2) + f_{c1} - f_{c2}$, $f(s_3) = f_o(s_3) + f_{c2}$ and $f(s_4) = f_{o4}(s_4) + f_{c2}$. These distributions are plotted in fig. 8.68.

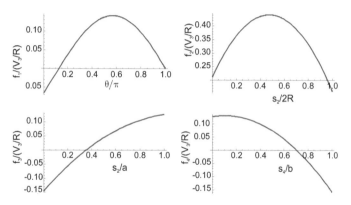

Fig. 8.68. Shear flow distribution in each segment of the 2-cell wing cross-section.

To determine the location of the shear center, the torque equipollence condition will be expressed about point **O**, the center of the leading edge semicircle, leading to

$$(x_{2k} + c)V_3 = -\int_0^{\pi R} f_1 R\, ds_1 + R\, R_3 + R\, R_4 \cos\alpha,$$

where $R_3 = \int_0^a f_3\, ds_3$ and $R_4 = \int_0^b f_4\, ds_4$ are the forces resulting from the shear flow distributions in the aft part of the section. Solving this equation yields the horizontal location of the shear center, $x_{2k} = -0.5290\ R$. The shear center location relative to the spar is given by $e = x_{2k} + c = -0.02682\ R$ which is to the left of the vertical spar. For real wing structures, the shear center is typically nearer the one-third chord location and aft of the aerodynamic center. For this example, the location is slightly forward of the spar, which is at the quarter-chord or aerodynamic center, so the lift will produce a decrease in angle of attack.

The torsional stiffness can be computed using the method developed in section 8.5.5. First, the areas enclosed by the two cells are $\mathcal{A}^{[1]} = \pi R^2/2$ and $\mathcal{A}^{[2]} = 3R^2$, for the front and aft cells, respectively. The counterclockwise shear flows due to torsion in each of the cells are defined as f_{t1} and f_{t2} for the leading and trailing edge cells, respectively. The sectional twisting moment, M_1, can be expressed in terms of the torsional moment developed in each cell using eq. (8.72) as

$$M_1 = 2(\mathcal{A}^{[1]} f^{[1]} + \mathcal{A}^{[2]} f^{[2]}) = 2\mathcal{A}^{[1]} f_{t1} + 2\mathcal{A}^{[2]} f_{t2}.$$

The twist rates for the two cells can be expressed using eq. (8.74) to find

$$\kappa_1^{[1]} = \frac{1}{2\mathcal{A}^{[1]}} \int_{\mathcal{C}^{[1]}} \frac{f^{[1]}}{Gt} \, ds = \frac{1}{2\mathcal{A}^{[1]}} \left[\frac{f_{t1}}{Gt} \pi R + \frac{(f_{t1} - f_{t2})}{G\eta t} 2R \right],$$

$$\kappa_1^{[2]} = \frac{1}{2\mathcal{A}^{[2]}} \int_{\mathcal{C}^{[2]}} \frac{f^{[2]}}{Gt} \, ds = \frac{1}{2\mathcal{A}^{[2]}} \left[\frac{f_{t2}}{Gt} (a + b) + \frac{(f_{t2} - f_{t1})}{G\eta t} 2R \right].$$

Compatibility of the twist rates for the two cells requires $\kappa_1^{[1]} = \kappa_1^{[2]} = \kappa_1$, where κ_1 is the section twist rate. Combining the above equations then yields the shear flow distributions due to torsion as

$$f_{t1} = \frac{\pi(11 + 3\sqrt{13}) + 12}{72 + 132\pi + \pi^2(11 + 3\sqrt{13})} \frac{M_1}{R^2} = 0.1147 \frac{M_1}{R^2},$$

$$f_{t2} = \frac{4(5\pi + 3)}{72 + 132\pi + \pi^2(11 + 3\sqrt{13})} \frac{M_1}{R^2} = 0.1066 \frac{M_1}{R^2}.$$

These resluts can now be used to calculate the section's torsional stiffness as

$$H_{11} = \frac{M_1}{\kappa_1} = \frac{72 + 132\pi + \pi^2(11 + 3\sqrt{13})}{6 + 2\sqrt{13} + \pi(11 + 3\sqrt{13})} GR^3t = 8.587 \ GR^3t.$$

To complete the solution, the decoupled bending and torsion problems will be solved using the procedure developed in example 8.18. First, the transverse deflection of the wing of length L under a uniformly distributed vertical load $p_3(x_1) = p_0$ will be determined. Since the axis system used in this problem is centroidal but not aligned with the principal axes of bending, bending in planes $(\bar{\imath}_1, \bar{\imath}_2)$ and $(\bar{\imath}_1, \bar{\imath}_3)$ is coupled, as expressed by eqs. (6.23b) and (6.23c), which become

$$\frac{d^2}{dx_1^2} \left[H_{23}^c \frac{d^2 \bar{u}_2}{dx_1^2} + H_{22}^c \frac{d^2 \bar{u}_3}{dx_1^2} \right] = p_0$$

$$\frac{d^2}{dx_1^2} \left[H_{33}^c \frac{d^2 \bar{u}_2}{dx_1^2} + H_{23}^c \frac{d^2 \bar{u}_3}{dx_1^2} \right] = 0.$$

The boundary conditions for the configuration shown in fig. 8.66 are $\bar{u}_2(0) = d\bar{u}_2(0)/dx_1 = \bar{u}_3(0) = d\bar{u}_3(0)/dx_1 = 0$ at the wing's root and $d^2\bar{u}_2(L)/dx_1^2 =$

$\mathrm{d}^3\bar{u}_2(L)/\mathrm{d}x_1^3 = \mathrm{d}^2\bar{u}_3(L)/\mathrm{d}x_1^2 = \mathrm{d}^3\bar{u}_3(L)/\mathrm{d}x_1^3 = 0$ at its tip. Using the bending stiffnesses computed earlier, the transverse displacement field is found as

$$\bar{u}_2(\eta) = 0.001951\,\eta^2(\eta^2 - 4\eta + 6)\frac{p_0 L^4}{ER^3 t},$$

$$\bar{u}_3(\eta) = 0.007086\,\eta^2(\eta^2 - 4\eta + 6)\frac{p_0 L^4}{ER^3 t},$$

where $\eta = x_1/L$ is the non-dimensional coordinate along the wing span.

Next, the torsion problem is governed by eq. (8.79), which in this case, becomes

$$H_{11}\frac{\mathrm{d}^2\Phi_1}{\mathrm{d}x_1^2} = -eV_3,$$

where e is the location of the shear center to the right of the vertical spar as computed earlier. The boundary conditions for the configuration shown in fig. 8.66 are $\Phi_1(0) = \mathrm{d}\Phi_1(L)/\mathrm{d}x_1 = 0$. The solution to the differential equation is

$$\Phi_1(x_1) = -0.001561\eta(2 - \eta)\frac{V_3 L^2}{GR^2 t}.$$

This result shows, as expected, that the section twists in a counterclockwise (nose down) direction.

8.6.1 Problems

Problem 8.60. Cantilevered beam under offset tip load
A thin-walled, C-section cantilevered beam of length L is subjected to a tip load, as depicted in fig. 8.69. The beam is loaded through a horizontal arm, **MN**, and a vertical load P is acting at point **Q**, located a distance d from the vertical web; load P is allowed to slide along arm **MN**, $d/b \in [0, 1.5]$. *(1)* Compute the tip vertical deflection of the beam due to bending. *(2)* Compute the tip rotation of the beam due to torsion. *(3)* Compute the transverse deflection of point **Q**, the point of application of the transverse load. *(4)* Plot this deflection as a function of $d/b \in [0, 1.5]$. *(3)* What is the value of d/b for which this transverse deflection is minimum. Use the following data: $P = 5.0$ kN; $E = 73.0$ GPa; $\nu = 0.3$; $L = 2.0$ m; $h = 0.4$ m; $b= 0.2$ m; $t = 4$ mm.

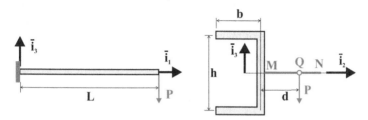

Fig. 8.69. Cantilever beam with C-section under offset tip load.

Problem 8.61. Simplified two-cell airfoil
Work example 8.19 for an airfoil of chord length $5R$, *i.e.*, for $a = 4R$. Compare your results with those presented in the example. This is a tedious calculation that is best done using a symbolic computation software tool.

8.7 Warping of thin-walled beams under torsion

When a thin-walled beam is subjected to an applied torque, shear stresses are generated, as discussed in the previous section. In turn, these shear stresses cause out-of-plane deformations of the cross-section called *warping*. Although the magnitude of these displacements is typically small, they have a dramatic effect on the torsional behavior of the structure.

Warping effects are particularly pronounced when dealing with non-uniform torsion of open sections. A beam is undergoing *non-uniform torsion* when its twist rate varies along the beam's span. This contrasts with Saint-Venant theory for torsion developed in section 7.3.2, which assumes the beam undergoes uniform torsion, *i.e.*, the twist rate is constant along the beam's span.

8.7.1 Kinematic description

Consider a thin-walled beam subjected to a tip concentrated torque, Q_1, as depicted in fig. 8.70. The formulation will be simplified if the axes are selected to be the principal centroidal axes of bending.

The analysis starts with an assumed displacement field similar to that for the Saint-Venant solution described in section 7.3.2. Under the action of the applied torque, each cross-section of the beam is assumed to rotate like a rigid body about point $\mathbf{R}$, called the *center of twist* whose coordinates, (x_{2r}, x_{3r}), are as yet unknown. The magnitude of the axial displacement component is assumed to be proportional to the twist rate, $\kappa_1(x_1)$, and is characterized by an unknown warping function, $\Psi(s)$, while the in-plane displacement field describes a rigid body rotation of magnitude $\Phi_1(x_1)$ about the center of twist

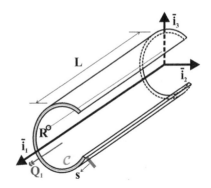

Fig. 8.70. Thin-walled beam subjected to an applied torque.

$$u_1(x_1, s) = \Psi(s)\, \kappa_1(x_1), \tag{8.80a}$$
$$u_2(x_1, s) = -(x_3 - x_{3r})\Phi_1(x_1), \tag{8.80b}$$
$$u_3(x_1, s) = \ \ (x_2 - x_{2r})\Phi_1(x_1). \tag{8.80c}$$

Because the section is thin-walled, the warping function is assumed to be a function only of the curvilinear variable, s, *i.e.*, the warping function does not vary through the thickness of the thin wall.

The strain field is now evaluated from this assumed displacement field

$$\varepsilon_1 = \frac{\partial u_1}{\partial x_1} = \Psi(s) \frac{\mathrm{d}\kappa_1}{\mathrm{d}x_1}, \tag{8.81a}$$

$$\varepsilon_2 = \frac{\partial u_2}{\partial x_2} = 0, \quad \varepsilon_3 = \frac{\partial u_3}{\partial x_3} = 0, \quad \gamma_{23} = \frac{\partial u_2}{\partial x_3} + \frac{\partial u_3}{\partial x_2} = 0, \tag{8.81b}$$

$$\gamma_{12} = \frac{\partial u_1}{\partial x_2} + \frac{\partial u_2}{\partial x_1} = \left[\frac{\mathrm{d}\Psi}{\mathrm{d}x_2} - (x_3 - x_{3r})\right]\kappa_1, \tag{8.81c}$$

$$\gamma_{13} = \frac{\partial u_1}{\partial x_3} + \frac{\partial u_3}{\partial x_1} = \left[\frac{\mathrm{d}\Psi}{\mathrm{d}x_3} + (x_2 - x_{2r})\right]\kappa_1. \tag{8.81d}$$

In the present analysis, the beam is assumed to undergo *non-uniform torsion*: the twist rate varies along the beam's span, and hence, $\mathrm{d}\kappa_1/\mathrm{d}x_1 \neq 0$. This contrasts with Saint-Venant's solution developed in section 7.3.2 and the analysis developed in section 8.5, where uniform torsion is assumed. Consequently, the axial strain, eq. (8.81a), does not vanish. The in-plane strain components, eq. (8.81b), vanish because of the assumed rigid body rotation of the section. The shear strain components, eqs. (8.81c) and (8.81d), depend on the partial derivatives of the warping function and are proportional to the twist rate.

8.7.2 Stress-strain relations

The non-vanishing components of the stress field are readily obtained from the constitutive laws as

$$\sigma_1 = E\varepsilon_1 = E\Psi(s) \frac{\mathrm{d}\kappa_1}{\mathrm{d}x_1}, \tag{8.82a}$$

$$\tau_{12} = G\gamma_{12} = \left[\frac{\mathrm{d}\Psi}{\mathrm{d}x_2} - (x_3 - x_{3r})\right]G\kappa_1, \tag{8.82b}$$

$$\tau_{13} = G\gamma_{13} = \left[\frac{\mathrm{d}\Psi}{\mathrm{d}x_3} + (x_2 - x_{2r})\right]G\kappa_1, \tag{8.82c}$$

where the linearly elastic, isotropic material is assumed to obey Hooke's law.

As discussed in section 8.1.2, the only non-vanishing shear stress component for thin-walled beams is component τ_s, tangent to curve $\mathcal{C}$. From eq. (8.2b), this shear stress component can be written as

$$\begin{aligned}
\tau_s &= \tau_{12} \frac{\mathrm{d}x_2}{\mathrm{d}s} + \tau_{13} \frac{\mathrm{d}x_3}{\mathrm{d}s} \\
&= \left[\frac{\partial \Psi}{\partial x_2}\frac{\mathrm{d}x_2}{\mathrm{d}s} + \frac{\partial \Psi}{\partial x_3}\frac{\mathrm{d}x_3}{\mathrm{d}s} + (x_2 - x_{2r})\frac{\mathrm{d}x_3}{\mathrm{d}s} - (x_3 - x_{3r})\frac{\mathrm{d}x_2}{\mathrm{d}s}\right]G\kappa_1.
\end{aligned}$$

In view of the chain rule for derivatives, the first two terms represent the total derivative of the warping function with respect to s. The last two terms evaluate the distance from the twist center to the tangent to curve $\mathcal{C}$, denoted r_r, see eq. (8.11). It then follows that

$$T_s = \left(\frac{\mathrm{d}\Psi}{\mathrm{d}s} + r_r\right) G\kappa_1. \tag{8.83}$$

To complete this analysis and determine the warping function, a distinction must now be made between open and closed sections.

8.7.3 Warping of open sections

The torsional behavior of thin-walled open section is investigated in section 8.5.1, and the shear stress distribution is found to be linearly distributed across the wall thickness and zero along the wall mid-line. Consequently, the shear stress, T_s, determined from eq. (8.83) vanishes along curve $\mathcal{C}$, leading to

$$T_s = \left(\frac{\mathrm{d}\Psi}{\mathrm{d}s} + r_r\right) G\kappa_1 = 0, \tag{8.84}$$

along curve $\mathcal{C}$. As a result, the warping function must satisfy the following differential equation

$$\frac{\mathrm{d}\Psi}{\mathrm{d}s} = -r_r = -\left(r_o - x_{2r}\frac{\mathrm{d}x_3}{\mathrm{d}s} + x_{3r}\frac{\mathrm{d}x_2}{\mathrm{d}s}\right). \tag{8.85}$$

To integrate this equation and determine the warping function, a purely geometric function, $\Gamma(s)$, is first defined as

$$\frac{\mathrm{d}\Gamma}{\mathrm{d}s} = -r_o. \tag{8.86}$$

Introducing this expression into eq. (8.85), it becomes possible to determine the warping function by integration to find

$$\Psi(s) = \Gamma(s) + x_{2r}x_3 - x_{3r}x_2 + c_1, \tag{8.87}$$

where c_1 is an integration constant. When integrating eq. (8.86), an arbitrary boundary condition must be used to evaluate $\Gamma(s)$, because integration constant c_1 is added in the expression for the warping function. To fully define the warping function, it is necessary to evaluate this integration constant, c_1, as well as the coordinates of the twist center, (x_{2r}, x_{3r}).

When dealing with uniform torsion, $\mathrm{d}\kappa_1/\mathrm{d}x_1 = 0$, and eq. (8.80a) implies that all span-wise sections undergo the same warping displacement; consequently, the axial strains, eq. (8.81a), and stresses, eq. (8.82a) both vanish. In this case, the integration constant and the location of the center of twist cannot be determined. The indeterminate part of the warping function, $x_{2r}x_3 - x_{3r}x_2 + c_1$, simply represents a rigid body displacement field that does not affect the state of strain or stress in the beam. In fact, any point can be selected as the center of twist.

Most practical problems, however, involve non-uniform torsion, either because the applied torque varies along the axis of the beam, or because warping displacement is constrained at a boundary or at some point along the beam's span. In such cases, two neighboring sections warp a different amount, giving rise to axial strains, which in turn, generate axial stresses *although the beam is acted upon by a torque alone*. The appearance of axial stresses under non-uniform torsion conditions is in sharp contrast with uniform torsion, which generates shear stresses only. Although axial stresses appear in the section, the axial force, N_1, and bending moments, M_2 and M_3, must vanish because no such loads are applied.

The vanishing of the axial force, N_1, defined by eq. (8.5), implies $\int_C \sigma_1 \, t ds = 0$. Introducing the axial stress, eq. (8.82a), and the warping function, eq. (8.87), then yields

$$\int_C E\Gamma \, t ds + x_{2r} \int_C E x_3 \, t ds - x_{3r} \int_C E x_2 \, t ds + c_1 \int_C E \, t ds = 0.$$

The second and third integrals in this expression vanish because the origin of the axes is selected to be at the section's centroid. The last integral is the axial stiffness, S, defined by eq. (5.17). The integration constant is then found as

$$c_1 = -\frac{1}{S} \int_C E\Gamma \, t ds. \tag{8.88}$$

Equation (8.6) defines the bending moment, M_2, as $\int_C \sigma_1 x_3 \, t ds = 0$. Imposing the vanishing of this quantity leads to

$$\int_C E\Gamma x_3 \, t ds + x_{2r} \int_C E x_3^2 \, t ds - x_{3r} \int_C E x_2 x_3 \, t ds + c_1 \int_C E x_3 \, t ds = 0.$$

The second integral in this expression is the bending stiffness, H_{22}^c. The third integral is the cross-bending stiffness, H_{23}^c, which vanishes because the axes are selected to be the principal centroidal axes of bending. The last integral also vanishes because the axes origin is at the centroid. The first coordinate of the twist center, x_{2r}, is thus found as

$$x_{2r} = -\frac{1}{H_{22}^c} \int_C E\Gamma x_3 \, t ds. \tag{8.89}$$

Finally, enforcing the vanishing of the other bending moment, M_3, yields the other coordinate of the twist center as

$$x_{3r} = \frac{1}{H_{33}^c} \int_C E\Gamma x_2 \, t ds. \tag{8.90}$$

The procedure to compute the warping function for thin-walled beams with open sections can be summarized by the following steps.

1. Compute the location of the section's centroid and select the axis system to be aligned with the principal centroidal axes of bending.
2. Compute the purely geometric function, $\Gamma(s)$, by integration of eq. (8.86). Use an arbitrary boundary condition.
3. Compute the integration constant, c_1, with the help of eq. (8.88);
4. Compute the coordinates of the twist center using eqs. (8.89) and (8.90).
5. The warping function is then fully defined by eq. (8.87).

The shape of the warping function describes the out-of-plane displacement field of the cross-section. Indeed, eq. (8.80a) gives the axial displacement distribution in terms of the warping function within a scaling factor, κ_1. For non-uniform torsion, the warping function also describes the axial strain distribution over the section, see eq. (8.81a). The warping function gives the axial strain distribution within a scaling factor, $d\kappa_1/dx_1$. Finally, for sections made of a homogeneous material, *i.e.*, if $E(s) = E$, the axial stress is also distributed according to the warping function, but this time with a scaling factor, $E \, d\kappa_1/dx_1$, see eq. (8.82a). These remarks help explain the importance of the warping function that describes the axial displacement, axial strain and axial stress distributions over the cross-section, although each with different scaling factors.

It is worthwhile noticing that the vanishing of the axial stress resultants, N_1, M_2, and M_3, implies the following properties of the warping function

$$\int_C E\Psi \, tds = \int_C E\Psi x_2 \, tds = \int_C E\Psi x_3 \, tds = 0. \tag{8.91}$$

Example 8.20. Warping of a C-channel
Let the beam with a C-channel cross-section depicted in fig. 8.24 be subjected to a tip torque. Determine the warping function. The axes shown in the figure are principal centroidal axes of bending because axis $\bar{\imath}_2$ is an axis of symmetry, and hence, is along a principal direction of bending. The first step of the procedure is to compute the purely geometric function, $\Gamma(s)$, defined in eq. (8.86), where r_o is the normal distance from the origin of the axes, point **O**, to the tangent to curve $\mathcal{C}$, and is given by eq. (8.8).

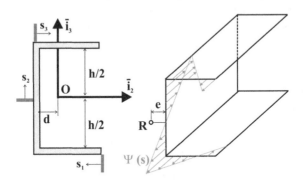

Fig. 8.71. The warping function for a C-channel.

As shown in fig. 8.71, curvilinear coordinate s_1 is used to describe the lower flange of the C-channel, where $r_o = -h/2$. According to the sign conventions, r_o is a negative quantity, see section 8.1.4. Function Γ then becomes $\Gamma(s_1) = hs_1/2 + c$, where c is an integration constant evaluated with the help of an arbitrary boundary condition, say $\Gamma(s_1) = 0$ at $s_1 = 0$, leading to $\Gamma(s_1) = hs_1/2$. A similar process is used for the vertical web and upper flange, where $r_o = -d$ and $r_o = -h/2$, respectively. Continuity of function Γ must be enforced at the corners and the process yields $\Gamma(s_2) = ds_2 + h(b + d)/2$ and $\Gamma(s_3) = hs_3/2 + h(b + 2d)/2$.

The next step is to evaluate the integration constant with the help of eq. (8.88) to find

$$c_1 = -\frac{Et}{S}\left[\int_0^b \Gamma(s_1)\,ds_1 + \int_{-h/2}^{+h/2} \Gamma(s_2)\,ds_2 + \int_0^b \Gamma(s_3)\,ds_3\right] = -\frac{h}{2}(b+d).$$

Finally, the coordinates of the twist center are determined by eqs. (8.89) and (8.90) to find

$$x_{2r} = -\frac{Et}{H_{22}^c}\left[\int_0^b \Gamma(s_1)\left(-\frac{h}{2}\right)ds_1 + \int_{-h/2}^{+h/2}\Gamma(s_2)s_2\,ds_2 + \int_0^b \Gamma(s_3)\frac{h}{2}\,ds_3\right]$$

$$= -d - \frac{h^2 b^2 t}{4}\frac{E}{H_{22}^c},$$

$$x_{3r} = \frac{Et}{H_{33}^c}\left[\int_0^b \Gamma(s_1)(b - d - s)\,ds_1 + \int_{-h/2}^{+h/2}\Gamma(s_2)(-d)\,ds_2\right.$$

$$\left. + \int_0^b \Gamma(s_3)(s - d)\,ds_3\right] = 0.$$

This last result could have been more easily obtained by invoking the symmetry of the problem.

The warping function then follows from eq. (8.87) as

$$\Psi(s_1) = \frac{h}{2}(s_1 + e - b); \quad \Psi(s_2) = -es_2; \quad \Psi(s_3) = \frac{h}{2}(s_3 - e), \qquad (8.92)$$

where $e = h^2 b^2 t E/(4H_{22}^c)$. Figure 8.71 shows the warping function and the location of the twist center, which is located at a distance e to the left of the vertical web. It is interesting to note that the location of shear center found in section 8.3.1 and that of the twist center coincide for this cross-section.

Example 8.21. Warping of a triangular section
Consider the thin-walled triangular section of height h and width b, open at point **B**, as depicted in fig. 8.72. Determine the warping function and the center of twist of the section. In view of the symmetry of the problem, the axes shown in the figure are principal centroidal axes provided that $d = b/[2(1 + \sin\alpha)]$. The bending stiffness of the section is evaluated with the help of eq. (6.53) to find

$$H_{22}^c = E\frac{th^3}{12} + 2E\left[\frac{t\ell^3}{12}\sin^2\alpha + t\ell\left(\frac{h}{4}\right)^2\right] = \frac{Et\ell h^2}{6}(1 + \sin\alpha),$$

where ℓ is the length of the flange; note that $h/2 = \ell\sin\alpha$.

Figure 8.72 shows the curvilinear coordinate, s_1, which runs along the lower flange, where $r_o = -(b-d)\sin\alpha$. According to the sign conventions, r_o is a negative

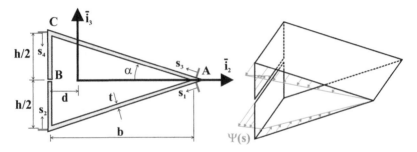

Fig. 8.72. Warping function for a open triangular section.

quantity, see section 8.1.4. Function Γ is then $\Gamma(s) = (b-d)s_1 \sin\alpha + c$, where c is an integration constant evaluated by means of an arbitrary boundary condition, say $\Gamma(s_1) = 0$ at $s_1 = 0$. Function Γ becomes

$$\Gamma(\bar{s}_1) = \frac{1 + 2\sin\alpha}{1 + \sin\alpha} \frac{bh}{4} \bar{s}_1,$$

where $\bar{s}_1 = s_1/\ell$ is the non-dimensional curvilinear variable across the lower flange.

A similar process is used for the vertical web where $r_o = -d$. Continuity of function Γ must be enforced at the corner and yields

$$\Gamma(\bar{s}_2) = \frac{1}{1 + \sin\alpha} \frac{bh}{4} (\bar{s}_2 + 1 + 2\sin\alpha),$$

where $\bar{s}_2 = 2s_2/h$ is the non-dimensional curvilinear variable for the vertical web.

The solution then proceeds with the last two segments of the section defined by curvilinear variables s_3 and s_4. Because of the symmetry of the problem, it is clear that $\Gamma(\bar{s}_3) = -\Gamma(\bar{s}_1)$ and $\Gamma(\bar{s}_4) = -\Gamma(\bar{s}_2)$.

The next step is to evaluate the integration constant with the help of eq. (8.88)

$$c_1 = -\frac{Et}{S}\left[\int_0^\ell \Gamma(s_1)ds_1 + \int_0^{h/2} \Gamma(s_2)ds_2 + \int_0^\ell \Gamma(s_3)ds_3 + \int_0^{h/2} \Gamma(s_4)ds_2\right].$$

In view of the symmetry of the problem, $c_1 = 0$.

Finally, the coordinate of the twist center is obtained from eq. (8.89) as

$$x_{2r} = -\frac{Et}{H_{22}^c}\left[\int_0^\ell \Gamma(s_1)(-s_1\sin\alpha)\,ds_1 + \int_0^{+h/2} \Gamma(s_2)\left(s_2 - \frac{h}{2}\right)\,ds_2\right.$$

$$\left. + \int_0^\ell \Gamma(s_3)(s_3\sin\alpha)\,ds_3 + \int_0^{+h/2} \Gamma(s_4)\left(\frac{h}{2} - s_4\right)\,ds_2\right]$$

$$= \frac{2Et}{H_{22}^c}\left[\int_0^\ell \Gamma(s_1)(s_1\sin\alpha)\,ds_1 + \int_0^{+h/2} \Gamma(s_2)\left(\frac{h}{2} - s_2\right)\,ds_2\right]$$

$$= (b-d) + \frac{\sin\alpha}{1 + \sin\alpha}\frac{b}{2}.$$

This means that the twist center is located at a distance $e = b \sin \alpha / [2(1 + \sin \alpha)]$ to the right of point **A**. The vertical coordinate of the twist center vanishes due to symmetry, as can be verified with the help of eq. (8.90).

The warping function then follows from eq. (8.87) as

$$\Psi_1 = -\Psi_3 = -\frac{\sin \alpha}{1 + \sin \alpha} \frac{bh}{4} \bar{s}_1, \ \Psi_2 = -\Psi_4 = \frac{(2 + 3\sin \alpha)\bar{s}_2 - \sin \alpha}{1 + \sin \alpha} \frac{bh}{4}. \quad (8.93)$$

Figure 8.72 shows the warping function for this open triangular section.

8.7.4 Problems

Problem 8.62. Alternative evaluation the warping function
In section 8.7, the procedure for the evaluation of the warping function is developed. The axes are selected to be the principal centroidal axes of bending. (1) Derive a procedure for evaluating the warping function when a set of centroidal axes is selected, *i.e.*, the axes are not the principal axes of bending. (2) Give explicit equations to compute the integration constant, a, and the coordinates of the twist center, *i.e.*, develop the equivalent of eqs. (8.88), (8.89), and (8.90), respectively.

Problem 8.63. Warping function of an "I" beam
Consider the thin-walled, "I" shaped cross-section depicted in fig. 8.73. (1) Compute and plot the warping function $\Psi(s)$ for this section. (2) Determine the location of the twist center. (3) Compute the torsional stiffness of the section.

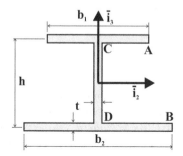

Fig. 8.73. Configuration of the I beam.

8.7.5 Warping of closed sections

The torsional behavior of thin-walled, closed sections is investigated in section 8.5.2. In open sections, the shear stress distribution is linear through the thickness of the wall, but the shear stress distribution is constant through the thin wall of closed sections. The shear stress is found to be $\tau_s = M_1/(2\mathcal{A}t) = H_{11}\kappa_1/(2\mathcal{A}t)$, where $\mathcal{A}$ is the area enclosed by curve $\mathcal{C}$, see eq. (8.62).

On the other hand, this shear stress is related to the warping function by eq. (8.83), leading to

$$\frac{d\Psi}{ds} = \frac{\tau_s}{G\kappa_1} - r_r = \frac{H_{11}}{2AGt} - r_r. \tag{8.94}$$

This is the governing equation for the warping function, and it should be compared to eq. (8.85), which applies to open sections.

The process of integration of eq. (8.94) closely follows that used for open sections. First, a purely geometric function, $\Gamma(s)$, is defined as

$$\frac{d\Gamma}{ds} = \frac{H_{11}}{2AGt} - r_o. \tag{8.95}$$

Here again, an arbitrary boundary condition is used to integrate this equation. Integrating eq. (8.94) then yields the warping function in the form of eq. (8.87). The integration constant, c_1, and the location of the twist center can then be found by enforcing the vanishing of the axial force, and bending moments, respectively.

In summary, the procedure for evaluating the warping function of closed sections is identical to that developed for open sections except that the governing equation for function Γ is now eq. (8.95), rather than eq. (8.86).

Example 8.22. Warping function for a thin-walled rectangular section
Consider the warping of the thin-walled rectangular beam shown in fig. 8.74. The width and height of the section are $2a$ and $2b$, respectively, and the thickness, t, is uniform. The torsional stiffness is found from eq. (8.67), and hence,

$$\frac{H_{11}}{2AGt} = \frac{2A}{t \displaystyle\int_c \frac{ds}{t}} = \frac{2ab}{a+b}.$$

Curvilinear coordinates s_1, s_2, s_3, and s_4 will be used along the four walls of the section as shown in fig. 8.74. Along the top flange, the governing equation for $\Gamma(s)$, eq. (8.95), becomes $d\Gamma/ds = 2ab/(a+b) - r_o$, where $r_o = b$. According to the sign conventions, r_o is a negative quantity, see section 8.1.4. Integration then yields $\Gamma(s_1) = 2abs_1/(a+b) - bs_1 + c$. Using an arbitrary boundary condition, say $\Gamma(s_1 = 0) = 0$, leads to $\Gamma(s_1) = bs_1(a-b)/(a+b)$.

The same process is repeated for the other three sides and continuity is enforced at the corners of the section to find $\Gamma(s_1) = \bar{d}bs_1$, $\Gamma(s_2) = -\bar{d}as_2$, $\Gamma(s_3) = \bar{d}bs_3$, and $\Gamma(s_4) = -\bar{d}as_4$, where $\bar{d} = (a-b)/(a+b)$. For obvious symmetry reasons, $c_1 = x_{2r} = x_{3r} = 0$, as can be verified with the help of eqs. (8.88), (8.89), and (8.90), respectively. The warping function, $\Psi = \Gamma$, now becomes

$$\Psi(s_1) = \bar{d}bs_1, \quad \Psi(s_2) = -\bar{d}as_2, \quad \Psi(s_3) = \bar{d}bs_3, \quad \Psi(s_4) = -\bar{d}as_4. \tag{8.96}$$

and is depicted in the right half of fig. 8.74.

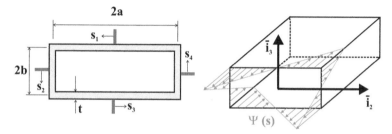

Fig. 8.74. Thin-walled beam with a rectangular cross-section.

8.7.6 Warping of multi-cellular sections

In the case of multi-cellular sections, the shear flow distribution, $f(s)$, due to an applied torque must be computed first with the help of the procedure described in section 8.5.5. This shear flow distribution is proportional to the twist rate, *i.e.*, $f(s) = \mathcal{G}(s)\,\kappa_1$, and the corresponding shear stress is $\tau_s = \mathcal{G}(s)\kappa_1/t$, where $\mathcal{G}(s)$ is a function of s determined by the analysis outlined in that section. Comparing this result with eq. (8.83) yields the governing equation for the warping function,

$$\frac{d\Psi}{ds} = \frac{\mathcal{G}(s)}{Gt} - r_r. \tag{8.97}$$

The procedure for the determination of the warping function of multi-cellular sections then exactly mirrors that for open and closed sections except that the governing differential equation for function Γ is now

$$\frac{d\Gamma}{ds} = \frac{\mathcal{G}(s)}{Gt} - r_o. \tag{8.98}$$

8.8 Equivalence of the shear and twist centers

The analysis of the shear flow distribution in thin-walled beams subjected to shear forces leads to the concept of shear center. The shear center is defined by the torque equipollence condition expressed by eq. (8.39). This important concept allows the decoupling of bending and twisting problems, as discussed in section 8.6. On the other hand, in the present section the center of twist is introduced for the analysis of thin-walled beams under torsion.

Consider eq. (8.53) for the coordinate, x_{2k}, of the shear center, and introduce function Γ defined by eq. (8.86) to find

$$x_{2k} = \int_{\mathcal{C}} f^{[3]} r_o \, ds = - \int_{\mathcal{C}} f^{[3]} \frac{d\Gamma}{ds} \, ds.$$

Integrating by parts then yields

$$x_{2k} = \int_C \Gamma \frac{\mathrm{d}f^{[3]}}{\mathrm{d}s} \, \mathrm{d}s - \left[f^{[3]} \Gamma \right]_{\text{boundary}}.$$

The boundary term vanishes because $f^{[3]} = 0$ at the boundaries.

Next, the governing equation for $f^{[3]}$, eq. (8.58), is introduced

$$x_{2k} = -\int_C \frac{Et}{H_{22}^c} x_3 \Gamma \, \mathrm{d}s = -\frac{1}{H_{22}^c} \int_C E\Gamma x_3 \, t \mathrm{d}s = x_{2r},$$

where the last equality follows from eq. (8.89). A similar reasoning leads to $x_{3k} = x_{3r}$, thus establishing the equivalence of the shear and twist centers for open sections. The equivalence also holds for closed section, as can be shown by a similar development. This equivalence is a direct consequence of Betti's reciprocity theorem, which will be developed in section 10.10.1, see eq. (10.117).

8.9 Non-uniform torsion

A thin-walled beam under non-uniform torsion develops a complex state of stress that involves both shear stresses and axial stresses generated by differential warping. The presence of the axial stress gives rise to a markedly different behavior from that observed in the case of uniform torsion.

The axial stresses generated by non-uniform torsion are uniformly distributed across the thickness of the wall, and the associated axial flow is denoted $n_w = t\sigma_1$. As discussed in section 8.7.3, although the axial stress does not vanish at all points of the section, the resulting axial force and bending moment do vanish. This condition implies the global equilibrium of the section, but the local equilibrium equation, eq. (8.14), is not necessarily satisfied. For this local equilibrium to hold, a shear flow, f_w, called the *warping shear flow* is generated to satisfy the local equilibrium condition

$$\frac{\partial n_w}{\partial x_1} + \frac{\partial f_w}{\partial s} = 0.$$

The implication of this new shear flow, f_w, is investigated for the case of open sections. Introducing eq. (8.82a) for the axial stress and solving for the warping shear flow yields

$$\frac{\partial f_w}{\partial s} = -Et\Psi \frac{\mathrm{d}^2\kappa_1}{x_1^2}. \tag{8.99}$$

This first order differential equation can be integrated to determine the warping shear flow, f_w, following the same procedure as that developed in section 8.3. The result is depicted in fig. 8.75 for the simple case of a C-channel.

In the presence of this new shear flow, the question of overall equilibrium arises once more: does the warping shear flow generate resultant transverse shear forces? The shear force resultant, V_{2w}, associated with the warping shear flow is evaluated with the help of eq. (8.7)

$$V_{2w} = \int_C f_w \frac{\mathrm{d}x_2}{\mathrm{d}s}\,\mathrm{d}s = -\int_C x_2 \frac{\partial f_w}{\partial s}\,\mathrm{d}s + [x_2 f_w]_{\text{boundary}},$$

where the second equality follows from integration by parts. The boundary term vanishes because $f_w = 0$ at the edges of the contour. Introducing eq. (8.99) then yields

$$V_{2w} = \frac{\mathrm{d}^2\kappa_1}{\mathrm{d}x_1^2}\int_C E\Psi x_2\,t\mathrm{d}s = 0,$$

where the last equality stems from property (8.91) of the warping function. It can be shown in a similar manner that $V_{3w} = 0$.

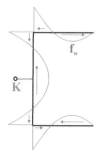

Fig. 8.75. Shearing flow, f_w, for a C-channel.

Next, the torque resultant about the shear center generated by the warping shear flow is evaluated with the help of eq. (8.10) to find

$$M_{1wK} = \int_C f_w r_k\,\mathrm{d}s = -\int_C f_w \frac{\mathrm{d}\Psi}{\mathrm{d}s}\,\mathrm{d}s, \qquad (8.100)$$

where the governing equation for the warping function, (8.85), is introduced. Integrating by parts then yields

$$M_{1wK} = \int_C \Psi \frac{\mathrm{d}f_w}{\mathrm{d}s}\,\mathrm{d}s - [f_w\Psi]_{\text{boundary}}, \qquad (8.101)$$

where the boundary term vanishes once more. Finally, introducing eq. (8.99) yields

$$M_{1wK} = -H_w \frac{\mathrm{d}^2\kappa_1}{\mathrm{d}x_1^2}, \qquad (8.102)$$

where

$$H_w = \int_C E\Psi^2\,t\mathrm{d}s, \qquad (8.103)$$

is called the *warping stiffness*.

The total torque is the sum of that generated by the twist rate and that due to warping,

$$M_{1K} = H_{11}\kappa_1 - H_w \frac{\mathrm{d}^2\kappa_1}{\mathrm{d}x_1^2}. \qquad (8.104)$$

The first torque component, $H_{11}\kappa_1$, is that generated by the shear stress distribution described in section 8.5.1, whereas the second torque component, $-H_w\,\mathrm{d}^2\kappa_1/\mathrm{d}x_1^2$, is the additional contribution arising from the warping shear flow. Note that the second contribution vanishes for the case of uniform torsion.

The equilibrium equation for a differential element of the beam under torsional loads is obtained in section 7.15. Introducing the torque expression, eq. (8.104), yields the governing equation for beams undergoing non-uniform torsion,

$$\frac{\mathrm{d}}{\mathrm{d}x_1}\left(H_{11}\frac{\mathrm{d}\Phi_1}{\mathrm{d}x_1} - H_w\frac{\mathrm{d}^3\Phi_1}{\mathrm{d}x_1^3}\right) = -q_1. \qquad (8.105)$$

This fourth order differential equation can be solved to find the beam twist given the applied distributed torque q_1.

Example 8.23. Torsion of a cantilevered beam with free root warping
Consider a uniform cantilevered beam of length L subjected to a tip torque, Q. At first, the root condition is such that no twisting is allowed, but warping is free to occur. This condition could be obtained by attaching the beam's root to a diaphragm that prevents any root rotation, but does not constrain axial displacements. The beam has uniform properties along its length. Hence, the governing equation (8.105) becomes

$$H_{11} \frac{d^2\Phi_1}{d^2x_1} - H_w \frac{d^4\Phi_1}{dx_1^4} = 0.$$

At the root of the beam, no twist occurs, *i.e.*, $\Phi_1 = 0$. Since warping is free at the root, the axial stress must vanish, and in view of eq. (8.82a), this implies $d^2\Phi_1/dx_1^2 = 0$. At the tip of the beam, the torque must equal the applied torque, Q, and the axial stress must vanish once again. From eq. (8.104), the first condition implies $Q = H_{11}d\Phi_1/dx_1 - H_w d^3\Phi_1/dx_1^3$, and the second condition again implies $d^2\Phi_1/dx_1^2 = 0$.

To ease the solution of this problem, a non-dimensional span-wise variable, $\eta = x_1/L$, is introduced, and the governing equation becomes

$$\Phi_1'''' - \bar{k}^2 \, \Phi_1'' = 0, \tag{8.106}$$

with the boundary conditions, $\Phi_1 = 0$ and $\Phi_1'' = 0$ at the beam's root, and $\Phi_1'' = 0$ and $\bar{k}^2\Phi_1' - \Phi_1''' = QL^3/H_w$ at its tip. The notation $(\cdot)'$ is used to denote a derivative with respect to η, and

$$\bar{k}^2 = \frac{H_{11}L^2}{H_w}, \tag{8.107}$$

is a non-dimensional parameter that characterizes the ratio of the torsional stiffness to the warping stiffness.

The general solution of the governing differential equation is

$$\Phi_1 = C_1 + C_2\eta + C_3 \cosh \bar{k}\eta + C_4 \sinh \bar{k}\eta, \tag{8.108}$$

where C_1, C_2, C_3, and C_4 are four integration constants. The boundary conditions at the root are used to evaluate C_1 and C_3 to find $\Phi_1 = C_2\eta + C_4 \sinh \bar{k}\eta$. The remaining two integration constants are found with the help of the boundary conditions at the tip of the beam. Hence,

$$\Phi_1 = \frac{QL}{H_{11}} \eta. \tag{8.109}$$

This solution is identical to the uniform torsion solution, and could have been obtained from the governing equation for uniform torsion developed in chapter 7. Indeed, the solution implies a twist rate $\kappa_1 = d\Phi_1/dx_1 = Q/L = $ constant. The torsional warping stiffness, H_w, disappears from the solution.

Example 8.24. Torsion of a cantilevered beam with constrained root warping
The cantilevered beam of the above example is considered here again, but the root
section is now solidly fixed to prevent any warping at the root. At this built-in end,
no twisting occurs, *i.e.*, $\Phi_1 = 0$, and no axial displacement is allowed. In view of
eq. (8.80), this last condition implies $\kappa_1 = \mathrm{d}\Phi_1/\mathrm{d}x_1 = 0$.

The governing equation of the problem is once again eq. (8.106), but the boundary conditions now become $\Phi_1 = 0$ and $\Phi_1' = 0$ at the beam's root and $\Phi_1'' = 0$ and
$\bar{k}^2\Phi_1' - \Phi_1''' = QL^3/H_w$ at its tip. The general solution of the governing equation
still has the form of eq. (8.108), and the boundary conditions at the root are used to
evaluate C_3 and C_2. Hence, $\Phi_1 = C_1(1 - \cosh \bar{k}\eta) + C_4(\sinh \bar{k}\eta - \bar{k}\eta)$. The tip
boundary conditions allow the evaluation of the remaining integration constants to
find

$$\Phi_1 = \frac{QL}{H_{11}}\left[\eta - \frac{\sinh \bar{k} - \sinh \bar{k}(1 - \eta)}{\bar{k}\cosh \bar{k}}\right].$$

The first term is the linear distribution of twist along the beam's span found in the
previous example and is characteristic of uniform torsion. The second term represents the influence of non-uniform torsion induced by the root warping constraint.
This constraint decreases the twist of the beam, and stiffens the beam.

Two types of sections will be considered here, the closed rectangular section
shown in fig. 8.74, and the open C-channel of fig. 8.71. The warping stiffness of
the rectangular section is found by introducing eq. (8.96) into eq. (8.103) and integrating to find $H_w = 4E/3\, a^2b^2t(a - b)^2/(a + b)$. The torsional stiffness follows
from eq. (8.67) as $H_{11} = G\, 16a^2b^2t/(a + b)$. Coefficient $\bar{k}$ defined in (8.107) then
becomes

$$\bar{k}^2 = \left(\frac{G}{E}\right)\left(\frac{2\sqrt{3}L}{a - b}\right)^2.$$

Note that for $a = b$, the thin-walled, rectangular section becomes square, the warping
function vanishes, as does the warping stiffness, and the uniform torsion solution is
recovered. Consider a rectangular section made of aluminum, and for which $a = 4b$.
Ratio $G/E = 1/2(1 + \nu)$ for isotropic, linearly elastic materials, see eq. (2.8).
Coefficient $\bar{k}$ then becomes proportional to the aspect ratio L/a. For a long beam,
$L/a = 10$, $\bar{k} = 16.54$, whereas for a shorter beam, $L/a = 5$, $\bar{k} = 8.27$. On the other
hand, if the beam is made of a unidirectional composite for which $G/E = 1/28$,
coefficient $\bar{k}$ becomes 5.04 and 2.52, for the long and short beams, respectively.
Figure 8.76 shows the twist distribution for the four cases considered.

For the C-channel, the warping stiffnesses is found by integrating the warping
function, given by eq. (8.92), to find $H_w = E/12\, h^2b^3t\, (3b + 2h)/(6b + h)$, and
the torsional stiffness is computed in eq. (7.66) as $H_{11} = Gt^3(2b + h)/3$. Finally,
coefficient $\bar{k}$ becomes

$$\bar{k}^2 = \left(\frac{G}{E}\right)\left(\frac{L}{h}\right)^2 \frac{2ht^2}{b^3}\frac{(1 + 2b/h)(1 + 6b/h)}{1 + 3b/2h}.$$

Let the C-channel be such that $h = 2b$ and $t = b/10$. If the C-channel is made of
aluminum, $\bar{k} = 2.65$ or 1.33 for the aspect ratios of $L/h = 10$ or 5, respectively. If

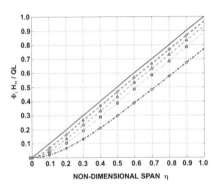

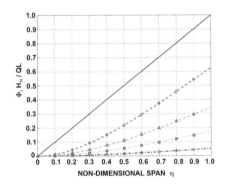

Fig. 8.76. Twist distribution for the closed rectangular section under non-uniform torsion. $\bar{k} = 16.54$ ($\diamond$), $\bar{k} = 8.27$ ($\triangle$), $\bar{k} = 5.04$ ($\square$), $\bar{k} = 2.52$ ($\circ$).

Fig. 8.77. Twist distribution for the C-channel section under non-uniform torsion. $\bar{k} = 2.65$ ($\diamond$), $\bar{k} = 1.33$ ($\triangle$), $\bar{k} = 0.808$ ($\square$), $\bar{k} = 0.404$ ($\circ$).

the beam is made of a unidirectional composite, the corresponding values are $\bar{k} = 0.808$ and 0.404, respectively. Figure 8.77 shows the twist distribution for the four cases.

The constrained warping effects become increasingly pronounced as $\bar{k}$ decreases. Coefficient $\bar{k}$ is proportional to $\sqrt{G/E}$. For most homogeneous, isotropic metal, Poisson's ratio is $\nu \approx 0.3$, implying $\sqrt{G/E} = \sqrt{(1+\nu)/2} \approx 0.62$. This ratio, however, can be far smaller for highly anisotropic materials. For instance, in the extreme case of a unidirectional graphite/epoxy material, $\sqrt{G/E} = 0.189$. Coefficient $\bar{k}$ is also proportional to the aspect ratio L/d, where d is a representative dimension of the cross-section; lower values of $\bar{k}$ will be associated with shorter beams. It is also important to note that *open sections* yield much smaller values of $\bar{k}$ than *closed section*.

In summary, the importance of non-uniform torsion is characterized by coefficient $\bar{k}$, which itself, depends on three factors. The first effect is a material effect characterized by the ratio of the shearing to Young's modulus. The second effect is a geometric effect characterized by the aspect ratio of the beam. The last factor is the geometric nature of the cross-section, which can be open or closed. Non-uniform torsion effects tend to become more pronounced for beams with the following characteristics: beams made of materials presenting lower shearing to Young's modulus ratios, beams of lower aspect ratio, and beams with open cross-sections.

8.9.1 Problems

Problem 8.64. Cantilever beam under uniform torque
A cantilevered beam is subjected to a uniformly distributed torque q_0. The root end of the beam is clamped, and a stiffening plate is welded at the tip section to prevent any warping. The thin-walled, I section of the beam is depicted in fig. 8.73. (1) Compute the torsional stiffness of the section. (2) Compute the twist distribution for this problem according to Saint-Venant solution for torsion. Denote this solution $\Phi_1^{SV}(x_1)$. (3) Compute and plot the warping function, $\Psi(s)$,

for this section. Determine the warping stiffness, H_w, for the section. *(4)* Show that

$$\bar{k}^2 = \frac{H_{11}L^2}{H_w} = 4\left(\frac{G}{E}\right)\left(\frac{L}{h}\right)^2 \frac{(b_1 + b_2 + h)(b_1^3 + b_2^3)t^2}{b_1^3 b_2^3}.$$

(5) Find the distribution of twist, $\Phi_1^{NU}(x_1)$, taking non-uniform torsion effects into account. Plot the twist distributions, $H_{11}\Phi_1^{SV}/(q_0L^2)$ and $H_{11}\Phi_1^{NU}/(q_0L^2)$, along the beam's span on the same graph. *(6)* On the same graph, plot the distribution of axial stress at points **A** and **B**, denoted $h^2\sigma_1^A/q_0$ and $h^2\sigma_1^B/q_0$, respectively, along the span of the beam. *(7)* Plot the distribution of shear stress due to torsion, $h^2\tau_s/q_0$, along the span of the beam. *(8)* On the same graph, plot the distribution of shear stress due to warping at points **C** and **D**, denoted $h^2\tau_w^C/q_0$ and $h^2\tau_w^D/q_0$, respectively, along the span of the beam. Use the following data: $b_1 = 0.2$, $b_2 = 0.3$, $h = 0.25$, $t = 0.02$ and $L = 1.5$ m. The Poisson's ratio for the material is $\nu = 0.3$.

Problem 8.65. C-channel with mid-span reinforcing plate
Consider the cantilevered beam of length L with a thin-walled C-channel cross-section as depicted in fig. 8.78. The beam is subjected to a tip torque, Q_1. At mid-span, a stiff plate connecting the top and bottom flanges is welded to the beam and prevents warping at that span-wise location. *(1)* Write the governing differential equation of the system and the associated boundary conditions. *(2)* Does the mid-span plate stiffen or soften the beam in torsion?

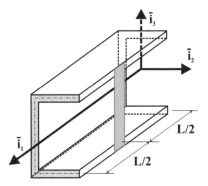

Fig. 8.78. Cantilevered beam with C-section with a mid-span reinforcing plate.

8.10 Structural idealization

The analysis of beams with thin-walled cross-sections is an extension of the basic Euler-Bernoulli theory developed in chapter 6. The details of the analysis, however, can quickly become quite cumbersome and tedious for all but the most basic cross-sectional configurations. This is particularly true when interactions among the bending, torsion and shear responses are considered, as discussed in section 8.9. Moreover, actual thin-walled beam structures rarely consist solely of of thin-walled components. Rather, these structures are usually reinforced with prismatic structural members running along the beam's axis. Such members are commonly called *stringers* and are added to increase the bending stiffness of the section and to provide attachment points on the beam for various hardware components.

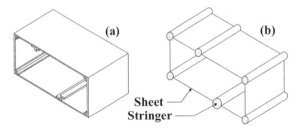

Fig. 8.79. (a) actual cross-section of a thin-walled beam. (b) Sheet-stringer idealization of the same section.

Thin-walled beam structures can be idealized by separating the axial stress and shear stress carrying components of the section into distinct entities called *stringers* and *sheets*, respectively. In this approach, illustrated in fig. 8.79, axial stresses are assumed to be carried only in the stringers which are idealized as *concentrated or lumped areas* with finite areas but vanishingly small cross-section dimensions. At the same time, shear stresses are carried entirely in the sheets which are assumed to be thin-walled components of vanishing thickness.

Figure 8.79 (a) depicts the cross-section of a thin-walled closed-section beam (sometimes called a "box beam") in which the four walls are connected at the corners by means of small "L" shaped longitudinal members. In addition, one or more additional longitudinal members, with "L," "T" or "Z" sectional shapes, are added to supplement the load carrying capability of the thin walls. These components usually constitute the largest fraction of the cross-sectional area and are located away from the centroid; hence, their contribution to the bending stiffness of the section is much larger than that of the sheets.

Figure 8.79 (b) shows a sheet-stringer idealization for this section. The stringers are assumed to carry all the axial stresses due to bending, while the sheets are assumed to carry only shear stresses due to both shearing and torsion. As explained in the following developments, this idealization leads to a considerably simplified analysis procedure for determining stress distributions over cross-sections subjected to complex combinations of bending, shear, and torsional loading.

8.10.1 Sheet-stringer approximation of a thin-walled section

If a thin-walled cross-section includes discrete stringers, such as the "L" shaped stringers shown in fig. 8.79, whose total cross-sectional area is large compared to that of the thin-walled portion, it can be assumed that these stringers alone carry bending stresses, while the thin-walls are sheets, which carry only shear stresses. This provides an immediate idealization of the cross-section.

If the cross-section does not include discrete stringers, or in the presence of a few stringers with a total cross-sectional area far smaller than that of the thin walls, it is still possible to construct a sheet-stringer model. In this case, an idealization process is used to create "virtual stringers" that are assumed to carry the axial stresses, whereas the thin-walled portions are assumed to carry only shear stresses. Such an idealization process is illustrated in fig. 8.80.

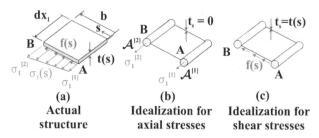

(a)   (b)   (c)
Actual   **Idealization for** **Idealization for**
structure   **axial stresses**   **shear stresses**

Fig. 8.80. (a): a portion of an actual cross-section; (b): idealization for axial stresses; (c): idealization for shear stresses. (b) and (c) are the sheet-stringer idealization of the actual structure represented in (a).

Consider a portion of a thin-walled beam of thickness t, width b along the contour C of the cross-section and span dx_1 along axis $\bar{\imath}_1$, as shown in fig. 8.80 (a). In the actual structure, the thin-walled section will carry both axial and shear stresses, whereas in the idealized structure, the various components of the section, skin and stringer, behave differently when it comes to carrying axial and shear stresses.

1. From the standpoint of the *axial stresses*, the structure is assumed to consist of stringers of finite area and sheets of vanishing thickness, as illustrated in fig. 8.80 (b). The axial stresses are carried solely by the stringers. The area of the stringers will be estimated using various approaches discussed below.
2. From the standpoint of the *shear stresses*, the structure is assumed to consist of sheets of thicknesses equal to those of the actual structure, as shown in fig. 8.80 (c). The shear stresses are carried solely by these sheets. The shear stresses will be evaluated based on equilibrium arguments applied to this idealized structure.

The first step of this idealization process is to estimate the areas of the stringers. Various procedures will lead to different stringer areas, and the idealization process does not lead to a unique solution. Problem characteristics or analysis accuracy requirements will dictate the particular approach to be taken.

One approach is to use the triangular equivalence method presented in section 6.8.3. This approach guarantees that the idealized structure will present the same bending stiffnesses and centroid location as the actual structure. Curved portions of the thin-walled cross-section can be approximated by a number of straight segments that, in turn, are represented by lumped areas, as illustrated in fig. 6.12.

In a second approach, the axial stresses are assumed to be linearly distributed across the width of the element, *i.e.*, $\sigma_1 = \sigma_1^{[1]} + (\sigma_1^{[2]} - \sigma_1^{[1]})s/b$, where $\sigma_1^{[1]}$ and $\sigma_1^{[2]}$ are the stresses at point **A** and **B**, respectively, and s is the local position along the contour of width b. To carry out the idealization, the areas, $\mathcal{A}^{[1]}$ and $\mathcal{A}^{[2]}$, of the stringers located at points **A** and **B**, respectively, must be determined. It is reasonable to enforce the following two constraints: *(1)* the axial stresses at points **A** and **B** are the same in the actual structure and in the sheet-stringer idealization, and *(2)* force and moment equivalences are maintained.

The axial force equivalence requires that

$$F_1 = \int_0^b \left[\sigma_1^{[1]} + (\sigma_1^{[2]} - \sigma_1^{[1]})s/b\right] t\,ds = \frac{1}{2}(\sigma_1^{[1]} + \sigma_1^{[2]})bt = \sigma_1^{[1]}\mathcal{A}^{[1]} + \sigma_1^{[2]}\mathcal{A}^{[2]},$$

where the last equality comes from the evaluation of the axial force in the idealized structure.

The bending moment equivalence with respect to an axis perpendicular to the plane of the sheet passing through point **A** implies

$$M_A = \int_0^b \left[\sigma_1^{[1]} + (\sigma_1^{[2]} - \sigma_1^{[1]})s/b\right] s\,t\,ds = \frac{b^2 t}{6}(\sigma_1^{[1]} + 2\sigma_1^{[2]}) = b\sigma_1^{[2]}\mathcal{A}^{[2]},$$

where the last equality comes from the evaluation of the bending moment in the idealized structure.

These two equations can be solved for the two unknown areas of the stringers to find

$$\mathcal{A}^{[1]} = \frac{bt}{6}\left(2 + \frac{\sigma_1^{[2]}}{\sigma_1^{[1]}}\right), \qquad \mathcal{A}^{[2]} = \frac{bt}{6}\left(2 + \frac{\sigma_1^{[1]}}{\sigma_1^{[2]}}\right). \tag{8.110}$$

In general, the areas of the idealized stringers will depend on the specific stress distribution in the real thin-walled section, but two cases deserve special attention.

1. **Uniform Axial Stress:** In this case it is assumed that $\sigma_1^{[1]} = \sigma_1^{[2]}$, and it follows that the stringers have equal areas given by

$$\mathcal{A}^{[1]} = \mathcal{A}^{[2]} = bt/2. \tag{8.111}$$

2. **Pure Bending:** For this case it is assumed that $\sigma_1^{[1]} = -\sigma_1^{[2]}$ (which would happen if the stringers are equidistant from and on opposite sides of the neutral axis for the section), and it follows that the stringers are of equal areas given by

$$\mathcal{A}^{[1]} = \mathcal{A}^{[2]} = bt/6. \tag{8.112}$$

If the real structure also possesses actual stringers, as shown in fig. 8.79, the actual stringer areas can be lumped at those locations. The areas computed by eq. (8.110) represent the axial stress carrying capability of the thin wall between the actual stringers, and these idealized areas would be added to the areas of the existing stringers, or possibly lumped into intermediate "virtual stringers." In many cases, the areas of the actual stringers will be considerably larger than the areas computed using eq. (8.110), and it might not even be necessary to carry out those computations.

Finally, it should be pointed out that the lumping of the axial stress-carrying portions of the thin-walled section into stringers using eq. (8.110) is based on an *a priori* assumption for the axial stress distribution in the section. If different distributions are considered that correspond to different loading conditions, equivalent idealized areas must be recomputed for each case. Thus, it is useful to keep in mind that the idealization process might be closely linked to the type of analysis that is being performed.

8.10.2 Axial stress in the stringers

The axial stress acting in the stringers can be computed using the same approach as that developed in chapter 6. More specifically, the axial stress, $\sigma_1^{[r]}$, acting in the rth stringer, is given by eq. (6.14), repeated here as

$$\sigma_1^{[r]} = E^{[r]} \left[\frac{N_1}{S} + x_3^{[r]} \frac{H_{33}^c M_2 + H_{23}^c M_3}{\Delta_H} - x_2^{[r]} \frac{H_{23}^c M_2 + H_{22}^c M_3}{\Delta_H} \right], \quad (8.113)$$

where $x_2^{[r]}$ and $x_3^{[r]}$ define the location of the rth stringer and $E^{[r]}$ its Young's modulus. The axial stiffness, S, bending stiffnesses, H_{22}^c and H_{33}^c, and cross bending stiffness, H_{23}^c, are computed based solely on the areas and locations of the stringers defined in the idealized cross-section. Because the stringers are assumed to be small "lumped areas," a uniform stress is assumed to act over this area and hence, the net axial force in the rth stringer is simply $\mathcal{A}^{[r]} \sigma_1^{[r]}$.

8.10.3 Shear flow in the sheet components

The sheet-stringer idealization greatly simplifies the computation of shear flow distributions. Indeed, since the sheet carry no axial stresses the local equilibrium condition, eq. (8.14), reduces to $\partial f / \partial s = 0$. This means that the shear flow must remain constant along the sheet

$$f = \text{constant.} \quad (8.114)$$

While the shear flow remains constant in a sheet, the situation is different at a point where two or more sheets connect to a stringer.

Stringer equilibrium

Figure 8.81 depicts a free-body diagram of an isolated portion of a stringer with two adjoining sheets. The axial stress in the stringer varies along the axis of the beam, due to nonzero values of the shear forces, V_2 and V_3. Let f_1 and f_2 denote the shear flows in the two neighboring sheets, respectively, as shown in fig. 8.81. The free-body diagram shows that the axial stress, σ_1, and the shear flows contribute force components along axis $\bar{\imath}_1$.

Axial equilibrium for the rth stringer yields $(\sigma_1 + \partial\sigma_1/\partial x_1 \, dx_1 - \sigma_1)\mathcal{A}^{[r]} + f_2 dx_1 - f_1 dx_1 = 0$, or

$$\Delta f^{[r]} = f_2 - f_1 = -\mathcal{A}^{[r]}\frac{\partial\sigma_1}{\partial x_1}, \qquad (8.115)$$

where $\Delta f^{[r]}$ is the change in shear flow across the rth stringer.

Introducing eq. (8.113) into eq. (8.115) and using the global equilibrium equations, eqs. (6.18) and (6.20), to express derivatives of the bending moments in terms of the shear forces yields the change in shear flow, $\Delta f^{[r]} = f_2 - f_1$, across the rth stringer as

$$\Delta f^{[r]} = -E^{[r]}\mathcal{A}^{[r]}\left[\frac{H_{22}^c V_2 - H_{23}^c V_3}{\Delta_H}x_2^{[r]} - \frac{H_{23}^c V_2 - H_{33}^c V_3}{\Delta_H}x_3^{[r]}\right], \qquad (8.116)$$

where $\Delta_H = H_{22}^c H_{33}^c - (H_{23}^c)^2$.

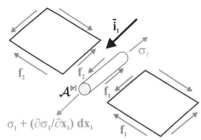

Fig. 8.81. Axial equilibrium of stringer with two joining sheets.

For general thin-walled cross-sections, the determination of the shear flow distribution associated with shear forces is governed by a differential equation, eq. (8.20). In the case of a sheet-stringer idealization, the shear flow distribution is governed instead by eq. (8.116), which is a *difference equation*: shear flows are constant within each sheet, but change across each stringer according to the difference equation.

For both thin-walled cross-sections or their sheet-stringer idealization, integration of the governing differential or difference equation, respectively, requires the determination of an integration constant. If the section is open, the shear flow must vanish at the stress-free edges of the section, as discussed in section 8.3.1. For closed sections, it is necessary to follow the procedure described in section 8.3.7; a similar procedure can be devised to determine shear flow distributions in sections modeled using the sheet-stringer idealization.

Finally, if N sheets are connected to a single stringer, the equilibrium argument used to establish eq. (8.116) can be generalized, leading to

$$\sum_{i=1}^{N} f_i = -E^{[r]}\mathcal{A}^{[r]}\left[\frac{H_{22}^c V_2 - H_{23}^c V_3}{\Delta_H}x_2^{[r]} - \frac{H_{23}^c V_2 - H_{33}^c V_3}{\Delta_H}x_3^{[r]}\right], \qquad (8.117)$$

In practical applications of this equation, the sheet shear flows must be interpreted as *algebraic quantities*: each shear flow is positive in the direction of the corresponding curvilinear coordinate used to describe the sheet.

Shear flow resultants

It is often necessary to compute force and moment resultants from shear flow distributions over the cross-section, for instance, to determine the location of the shear center for the section. For the sheet-stringer idealization, shear flows are constant in straight or curved sheets, but shear flow resultants are somewhat more difficult to evaluate for curved sheets. Consider the curved sheet carrying a constant shear flow, f_{12}, and connecting two stringers denoted 1 and 2, as shown in fig. 8.82. The shear stress resultant, V_3, along axis $\bar{\imath}_3$, is computed by integrating the shear flow distribution along the contour of the sheet to find

$$V_3 = \int_1^2 \bar{\imath}_3 \cdot f_{12} \, \mathrm{d}\underline{s} = f_{12} \int_1^2 \mathrm{d}x_3 = f_{12}(x_3^{[2]} - x_3^{[1]}).$$

where $\mathrm{d}\underline{s} = \bar{\imath}_2 \mathrm{d}x_2 + \bar{\imath}_3 \mathrm{d}x_3$ and $x_3^{[r]}$ is the coordinate of the rth stringer. Similarly, the shear stress resultant, V_2, along axis $\bar{\imath}_2$, becomes $V_2 = f_{12}(x_2^{[2]} - x_2^{[1]})$. Combining these results yields the shear force resultant, V, as

$$V = \sqrt{V_2^2 + V_3^2} = f_{12}\sqrt{(x_2^{[2]} - x_2^{[1]})^2 + (x_3^{[2]} - x_3^{[1]})^2} = f_{12}\, L_{12}, \qquad (8.118)$$

where the resultant acts in a direction parallel to the line connecting the two stringers, and L_{12} is the distance between the stringers, as indicated in fig. 8.82.

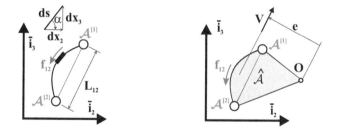

Fig. 8.82. Shear flow in curved sheet.

To find the line of action of this stress resultant, the moment resulting from the shear flow distribution is computed with respect to an arbitrary point **O** of the section, see fig. 8.82. Integrating the shear flow distribution times the moment arm and using eq. (8.56) yields

$$M_0 = \int_1^2 f_{12} r_o \, \mathrm{d}s = f_{12} \int_1^2 r_o \, \mathrm{d}s = f_{12} \int_{\hat{A}} 2 \, \mathrm{d}A = f_{12}\, 2\hat{A},$$

where $\hat{A}$ is the area of the sector defined by the two stringers, the curved sheet and point **O**. Thus, the shear resultant, V, given by eq. (8.118), must then have its line of action passing at a distance, e, from point **O** such that $V \cdot e = M_0 = 2\hat{A} f_{12}$ or

$$e = 2\hat{A}\frac{f_{12}}{V} = \frac{2\hat{A}}{L_{12}}. \qquad (8.119)$$

Equations (8.118) and (8.119) provide the shear flow resultant and its line of action for a curved sheet in sheet-stringer idealizations of a thin-walled cross-section.

If the sheet is a straight segment, eq. (8.118) provides the magnitude to the stress resultant and its line of action is the line that joins the two stringers.

8.10.4 Torsion of sheet-stringer sections

The calculation of stresses in sheet-stringer idealizations of cross-sections subjected to torsion is an extension of the approaches developed in sections 8.5.1 and 8.5.2, for open and closed sections, respectively. In both cases, only the sheet components are considered because they are the shear stresses carrying components. For uniform torsion, no axial stresses are present and hence, the stringers are ignored.

For an open section, the shear stresses are linearly distributed through the thickness of the sheets. The shear stresses assume extreme values at the edges of the sheet (*i.e.*, at $\pm t/2$ from the mid-line of the thin wall as illustrated in fig. 7.28) and vanish along the mid-line of the thin wall. Consequently, open sections are inefficient at carrying torsional loads. The torsional stiffness, H_{11}, of the section is given by eq. (7.64) as

$$H_{11} = G \frac{bt^3}{3},$$

where b is the curvilinear length of the cross-section and t its thickness. If individual sheets have different thicknesses, the overall torsional stiffness for the section is the sum of the stiffnesses for each sheet, or

$$H_{11} = \sum_{\text{sheets}} H_{11i} = \sum_{\text{sheets}} \frac{G_i b_i t_i^3}{3}. \tag{8.120}$$

For closed sections, the shear flow is constant, and the analysis is identical to that presented in section 8.5.2. The extension to multi-cellular sections follows the procedure detailed in section 8.5.5.

Example 8.25. *Shear flow in a sheet-stringer C-channel section*
Consider the C-channel section subjected to a shear load, V_3, and a bending moment, M_2, as shown in fig. 8.83. Axis $\bar{\imath}_2$ is an axis of symmetry, and hence, the axes shown on the figure are principal centroidal axes, provided their origin is located at the centroid of the section. The material is assumed to be uniform with a Young's Modulus, E.

A sheet-stringer idealization can be constructed using eq. (8.110). Under the effect of the bending moment, M_2, the axial stresses will be constant over the top and bottom flanges, but will vary linearly in the web. Hence, is seems logical to use eq. (8.111) to evaluate the areas of the two stringers that idealize the flanges, whereas eq. (8.112) is used to evaluate the areas of the stringers modeling the vertical web. With this approach, the areas of the various stringers become $A^{[1]} = 1/2\, bt$, $A^{[2]} = 1/2\, bt + 1/6\, ht$, $A^{[3]} = 1/2\, bt + 1/6\, ht$ and $A^{[4]} = 1/2\, bt$. It is easily verified that this sheet-stringer idealization yields the same bending stiffness as that computed for the thin-walled section,

$$H_{22}^c = \frac{1}{2}Ebth^2 + \frac{1}{12}Eth^3 = \frac{1}{12}Ebth^2\left(6 + \frac{h}{b}\right).$$

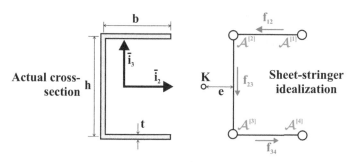

Fig. 8.83. Sheet-stringer model for C-channel section.

It will be convenient to denote the shear flow in each sheet as f_{ij}, where the subscripts indicate the stringers at the edge of the sheet; for instance, f_{12} is the shear flow in the sheet joining stringer $\mathcal{A}^{[1]}$ to stringer $\mathcal{A}^{[2]}$, counted positive in that direction. The equilibrium condition for stringer $\mathcal{A}^{[1]}$, given by eq. (8.116), yields $\Delta f^{[1]} = f_{12} - 0$ so that the shear flow in the upped flange is given as

$$f_{12} = \Delta f^{[1]} = -\frac{V_3}{H_{22}^c}E\mathcal{A}^{[1]}\frac{h}{2} = -\frac{3}{6 + h/b}\frac{V_3}{h}.$$

Next, the shear flow in the vertical web is computed from the equilibrium equation for stringer $\mathcal{A}^{[2]}$,

$$f_{23} = f_{12} - \frac{V_3}{H_{22}^c}E\mathcal{A}^{[2]}\frac{h}{2} = -\frac{3}{6 + h/b}\frac{V_3}{h} - \frac{3 + h/b}{6 + h/b}\frac{V_3}{h} = -\frac{V_3}{h}.$$

Finally, the equilibrium equation for stringer $\mathcal{A}^{[3]}$ gives the shear flow in the lower flange as $f_{34} = -3/(6 + h/b)\,V_3/h$, where it is clear that $f_{34} = f_{12}$, as expected. A last application of the equilibrium condition, eq. (8.116), this time for stringer $\mathcal{A}^{[4]}$, shows that the shear flow leaving this stringer must vanish as expected, because this is a free edge of the cross-section.

Several observations about this solution can be made. First, the shear flow is now constant in each sheet in contrast with the thin wall solution depicted in fig. 8.25 that features linear shear flow distributions in the top and bottom flanges and a parabolic distribution in the vertical web.

Second, the maximum shear flow in the sheet-stringer idealization is found in the web and its magnitude is $f_{\max} = V_3/h$; in the thin wall solution, the maximum shear flow is found at the mid-point of the web as $f_{\max} = 3/2\,(1 + 4b/h)/(1 + 6b/h)\,V_3/h$, (see eq. (8.33)). Note that the sheet-stringer idealization underestimates the true shear flow and therefore is not conservative.

Third, although the solution obtained for the sheet-stringer idealization is an approximation, it does exactly satisfy overall equilibrium requirements. The stress resultant along axis $\bar{\imath}_2$ vanishes, as expected, because no shear force is applied in that direction. On the other hand, the stress resultant along axis $\bar{\imath}_3$ exactly balances the applied shear, V_3.

Finally, torque equipollence about an arbitrary point of the section yields the location of the shear center, point **K**, located at $e = 3b/(6 + h/b)$ to the left of the web. This result exactly matches the location found using the thin wall solution, see eq. (8.41).

Example 8.26. Shear flow in a complex sheet-stringer section

The cross-section shown in fig. 8.84 is an idealization of a single-cell wing cross-section in which the areas carrying axial stress are lumped stringers. The sheets joining the stringers all are straight lines. A shear force $V_3 = 10$ kN is applied along the line connecting stringers 3 and 6, an axis that does not pass through the shear center of the section. The areas of the stringers are $A^{[1]} = 250$ mm^2, $A^{[2]} = A_7 = 300$ mm^2, $A^{[3]} = A^{[6]} = 500$ mm^2, $A^{[4]} = A^{[5]} = 150$ mm^2.

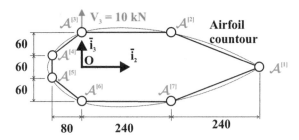

Fig. 8.84. Sheet-stringer model for wing cross-section (dimensions in mm).

First, in view of the symmetry of the section, the axes indicated on the section are principal axes of bending and $H_{23}^c = 0$. Since only a vertical shear force, V_3, is applied, the actual location of the centroid along the symmetry axis is not needed for this analysis. It is simply necessary to note that the centroid lies along axis $\bar{\imath}_2$. The only required bending stiffness of the section is then

$$H_{22}^c = \sum_{\text{stringers}} E_i A^{[i]} x_{3,i}^2 = 13.23 \times 10^{-6} \, E,$$

where it is assumed that the material is homogeneous and features a Young's Modulus of E.

As presented in section 8.3.7, the determination of the shear flow distribution due to a shear force in closed cross-sections proceeds in two steps: first, the section is cut at an arbitrary location and the shear flow in this open section is computed; next, a closing shear flow is evaluated. The shear flow distribution in the closed cross-section is the sum of these two shear flows. As the first step, the section is cut between stringers 2 and 3, and the shear flow in the open section is found to be

$f_{o,23} = 0,$

$f_{o,34} = f_{o,23} - (V_3/H_{22}^c) \, EA^{[3]} x_3^{[3]}$

$\qquad = 0 - (10^4/13.23 \times 10^{-6}) \cdot 400 \times 10^{-6} \cdot 90 \times 10^{-3} = -34.0 \text{ kN/m},$

$f_{o,45} = f_{o,34} - (V_3/H_{22}^c) \, EA^{[4]} x_3^{[4]} = -37.4 \text{ kN/m},$

$f_{o,56} = f_{o,45} - (V_3/H_{22}^c) \, EA^{[5]} x_3^{[5]} = -34.0 \text{ kN/m},$

$f_{o,67} = f_{o,56} - (V_3/H_{22}^c) \, EA^{[6]} x_3^{[6]} = 0.0 \text{ kN/m},$

$f_{o,71} = f_{o,67} - (V_3/H_{22}^c) \, EA^{[7]} x_3^{[7]} = 20.4 \text{ kN/m},$

$f_{o,12} = f_{o,71} - (V_3/H_{22}^c) \, EA^{[1]} x_3^{[1]} = 20.4 \text{ kN/m}.$

Next, the closing shear flow is evaluated using the compatibility condition for no axial displacements at the cut expressed by eq. (8.45). For this example, the sheet thicknesses are all the same so that Gt is a constant and can be factored out of both the numerator and denominator of eq. (8.45). In addition, since the shear flows are all constant between stringers the integrals in the numerator are simple and yield for the closing shear flow

$$f_c = - \left[\int_C \frac{f_o(s)}{Gt} \, ds \right] \Big/ \left[\int_C \frac{1}{Gt} \, ds \right]$$

$$\qquad = - \left(f_{o,12} \cdot 30\sqrt{73} + f_{o,23} \cdot 240 + f_{o,34} \cdot 100 + f_{o,45} \cdot 60 \right.$$

$$\qquad \left. + f_{o,56} \cdot 100 + f_{o,67} \cdot 240 + f_{o,71} \cdot 30\sqrt{73} \right) \Big/ 1252.6 = -1.129 \text{ kN/m}.$$

Finally, the shear flow in the cross-section due to the applied shear force is found by adding the shear flow in the open section to the closing shear flow: $f_{ij} = f_{o,ij} + f_c$.

For this case, the shear force is not applied at the shear center and therefore the section will also be subjected to a torsion. To evaluate the shear flow due to this torsion the shear center of the section must be determined. Symmetry arguments imply that $x_{3k} = 0$, and eq. (8.53) then yields

$$x_{2k} = \int_C f^{[3]}(s) r_o \, ds = \int_C (f_o(s) + f_c) \, r_o \, ds = 2\mathcal{A} f_c + \int_C f_o(s) r_o \, ds,$$

where $\mathcal{A}$ is the area enclosed by the cross-section, $f^{[3]}$ is the shear flow due to $V_3 = 1$ with $V_2 = 0$, and r_o is the perpendicular distance from point $\mathbf{O}$ to the sheet line. Because the shear flows computed above correspond to a shear force $V_3 = 10$ kN, these values must be reduced by a factor of $10,000$ to calculate the shear center. The enclosed area, $\mathcal{A}$, is readily computed as

$$\mathcal{A} = \frac{1}{2} 240 \cdot 180 + 240 \cdot 180 + \frac{1}{2}(60 + 180) \cdot 80 = 74,400 \text{ mm}^2 = 74.4 \times 10^{-3} \text{m}^2.$$

Using these values, the shear center is then given by

$$x_{2k} = 2 \cdot 74.4 \times 10^{-3} f_c + \left[2f_{o,12} 30\sqrt{73} \left(\frac{3}{5} 360 \right) + 2f_{o,34} 100 \left(\frac{4}{5} 90 \right) \right.$$
$$\left. + f_{o,45} 60 \cdot 80 \right] \times 10^{-6} = 0.093 \text{ m.}$$

Hence, the shear center is located a distance $x_{2k} = 93$ mm to the right of point **O**.

The torque about the shear center is now $M_{1K} = -V_3 x_{2k} = -930$ N·m. The shear flow due to torsion then follows from the Bredt-Batho formula, eq. (8.61)

$$f_t = \frac{M_1}{2A} = \frac{-930}{2 \cdot 74.4 \times 10^{-3}} = -6.222 \text{ kN/m.}$$

The total shear flow in the section is the superposition of the shear flows due to the shear force and the torque: $f_{s,ij} = f_{o,ij} + f_c + f_t$, and this leads to the final results

$$f_{s,12} = +20.41 - 1.13 - 6.22 = 13.06 \text{ kN/m,}$$
$$f_{s,23} = 0 - 1.13 - 6.22 = -7.35 \text{ kN/m,}$$
$$f_{s,34} = -34.01 - 1.13 - 6.22 = -41.36 \text{ kN/m,}$$
$$f_{s,45} = -37.41 - 1.13 - 6.22 = -44.76 \text{ kN/m,}$$
$$f_{s,56} = -34.01 - 1.13 - 6.22 = -41.36, \text{ kN/m,}$$
$$f_{s,67} = 0 - 1.13 - 6.22 = -7.35 \text{ kN/m,}$$
$$f_{s,71} = +20.41 - 1.13 - 6.22 = 13.06 \text{ kN/m.}$$

The above procedure for computing the total shear flow in the single closed cross-section follows the method described in section 8.3, but it is not necessarily the easiest procedure. For this example, the torque equipollence approach provides a more direct way to compute the shear flow due to the vertical shear force with a line of action passing through stringers 3 and 6. In this approach, a constant shear flow, f_{const}, is added to the shear flow distribution in the open section evaluated above, and the moment equipollence condition is then enforced.

Because the torque produced by the applied shear force, V_3, vanishes about point **O**, the torque produced by the total shear flow distribution about the same point must also vanish. This latter torque is the sum of the contribution from the constant shear flow, f_{const}, and those from the shear flow distribution in the open section, $f_{o,ij}$. The torque equipollence condition becomes

$$0 = 2A \, f_{const} + \left[2f_{o,12} \, 30\sqrt{73} \cdot \frac{3}{5} \cdot 360 + 2f_{o,23} \, 240 \cdot 90 \right.$$
$$\left. + 2f_{o,34} \, 100 \cdot \frac{4}{5} \cdot 90 + f_{o,45} \, 60 \cdot 80 \right] \times 10^{-6}.$$

Using the values for A and $f_{o,ij}$ from above yields $0 = 2 \cdot 74.4 \times 10^{-3} \cdot f_{const} + 1094.$, or $f_{const} = -7.35$ kN/m, which acts in the clockwise direction as a result of the negative sign. Using this, the total shear flow in each of the sheets is

$$f_{s,12} = 13.06, \quad f_{s,23} = -7.35, \quad f_{s,34} = -41.36, \quad f_{s,45} = -44.76,$$
$$f_{s,56} = -41.36, \quad f_{s,67} = -7.35, \quad f_{s,71} = 13.06 \text{ kN/m.}$$

These results are exactly the same as those developed using the cut-compatibility approach above.

As a final point, the bending stress in each of the stringers can be computed using eq. (8.113) if the bending moment is specified.

This example introduced a new way to compute the closing shear flow acting in the section. This approach is based on direct calculation of the torque equipollence condition between the applied shear force and the resulting shear flow distribution over the cross-section. The determination of the closing shear flow is expressed in a simpler manner. Unfortunately, this approach is suitable only when the cross-section consists of a single closed cell.

The equivalence of these two approaches for a single closed section can easily be demonstrated. Without loss of generality, assume a section similar to that shown in fig. (8.84) with one plane of symmetry. The section is subjected to a shear force, V_3, applied at point **O**. The first step of both approaches is the determination of the shear flow distribution in the open section, denoted $f_{o,ij}$.

In the torque equipollence method, the torque generated by the total shear flow, $f_{s,ij} = f_{o,ij} + f_{const}$, must equal that generated by the externally applied shear force. The torque will be computed with respect to point **O**; because the line of action of the applied shear forces passes through point **O**, the torque it generates vanishes, and hence, the torque generated by the total shear flow about the same point must also vanish. This latter condition implies $2\mathcal{A}f_{const} + M_{f_{o,ij}} = 0$, where the first term is the torque due to the constant shear flow expressed using the Bredt-Batho relationship, and $M_{f_{o,ij}}$ is the torque generated by $f_{o,ij}$ about point **O**. The total shear flow in the section can then be written as

$$f_{s,ij} = f_{o,ij} + f_{const} = f_{o,ij} - \frac{M_{f_{o,ij}}}{2\mathcal{A}}.$$

In the cut-compatibility method, the closing shear flow, f_c, is determined by enforcing warping compatibility at the cut, resulting in eq. (8.45). This closing shear flow develops when the section is subjected to the shear force, V_3, applied at the shear center. If the shear force is not applied at the shear center, the section is also subjected to torsion, and the constant shear flow, f_t, due to torsion, must be added to find the total shear flow distribution as $f_{s,ij} = f_{o,ij} + f_c + f_t$. Using Bredt-Batho relationship, $2\mathcal{A}f_t = -eV_3$, where e defines the location of the shear center with respect to point **O**.

The location of the shear center with respect to point **O** can easily be computed using moment equipollence, $V_3 e = 2\mathcal{A} f_c + M_{f_{o,ij}}$, and the shear flow due to torsion becomes

$$f_t = \frac{-eV_3}{2\mathcal{A}} = -\frac{2\mathcal{A} f_c + M_{f_{o,ij}}}{2\mathcal{A}}.$$

Finally, the full shear flow on the section can now be written as

$$f_{s,ij} = f_{o,ij} + f_c + f_t = f_{o,ij} + f_c - \frac{2\mathcal{A} f_c + M_{fo,ij}}{2\mathcal{A}} = f_{o,ij} - \frac{M_{fo,ij}}{2\mathcal{A}}.$$

This result is identical to that developed above using the moment equipollence condition. Note that the closing shear flow, f_c, cancels out. Hence, it is not required to evaluate this intermediate quantity, resulting in a more expeditious evaluation of the total shear flow.

8.10.5 Problems

Problem 8.66. Lumped sheet-stringer model development
Construct a lumped sheet-stringer model for the thin-walled cross-section shown in fig. 8.85. Assume that the stringers in the straight sections are spaced at 50 mm and at $45°$ in the curved portions. Assume the only loading is a moment, M_2. All dimensions are in millimeters.

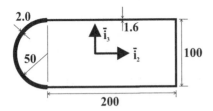
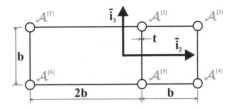

Fig. 8.85. Thin-walled beam cross-section (dimensions in millimeters).   **Fig. 8.86.** Sheet-stringer beam with two cells.

Problem 8.67. Torsional stiffness for a closed section
Determine the torsional stiffness, H_{11}, for the idealized section shown in fig. 8.85. Assume for this case that $E = 210$ GPa (steel) and that $G = 30$ GPa.

Problem 8.68. Shear flow in a multi-cellular sheet-stringer section
Consider the same thin-walled cross-section with two cells shown in fig. 8.45, but now construct an idealized sheet-stringer model as shown in fig. 8.86 where the stringer areas are computed under the assumption that only V_3 and M_2 are applied. Use this model to determine the shear flow in the sheets and to find the location of the shear center.

Problem 8.69. Shear flow in sheet-stringer section
Consider the same fuselage section shown in fig. 8.87 and calculate the shear flow due to the applied shear loading. First, decompose the shear load into a shear load applied at the shear center and a torsion about the $\bar{\imath}_1$ axis. Next, compute the shear flow due to the shear load in the closed section. Finally, compute the shear flow due to the torsion and combine your results to determine the final shear flow. Is the shear flow uniform across the thickness of the thin-walled section? Explain.

Problem 8.70. Bending stress in a sheet-stringer section
A fuselage cross-section is shown in fig. 8.87. The areas of the stringers 150 mm² and the applied loads are $P_0 = 50$ kN and $M_0 = 100$ kN·m. Determine the bending stress in the stringers. Use symmetry as much as possible to simplify the problem.

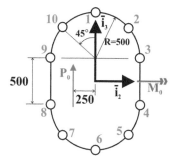

Fig. 8.87. Fuselage sheet-stringer cross-section (dimensions in millimeters).

Part III

Energy and variational methods

9

Virtual work principles

9.1 Introduction

The concept of mechanical work is fundamental to the study of mechanics. The mechanical work done by a force is defined as the scalar product of that force by the displacement through which it acts: work is a scalar quantity, in contrast with forces and displacements, which are vector quantities characterized by magnitudes and directions. Consequently, work quantities are simpler to manipulate than forces and displacements, and this simplification makes work based formulations of mechanics very attractive.

Newtonian mechanics is based on the concepts of forces and moments, which are vector quantities. The equilibrium conditions stated by Newton's law are expressed in their most general form as vector equations, and vector algebra is required for most practical applications. While it is customary to make a distinction between externally applied loads, internal forces and reaction forces, Newton's condition for equilibrium states that *the sum of all forces must vanish,* without making any distinction between them. It follows that all forces explicitly appear in the equilibrium equations of the problem and the solution process involves the determination of all forces, including internal and reaction forces. Newton's approach effectively determines forces and displacements, but it becomes increasingly difficult and tedious for problems of increasing complexity.

Formulations based on the concept of work are a part of what is generally referred to as *analytical mechanics* and provide powerful tools for dealing with complex problems. These methods are very attractive because they deal with quantities that are scalar rather than vector quantities, resulting in simpler analysis processes. Furthermore, specific types of forces, such as reaction forces, can often be eliminated from the solution process if the work they perform vanishes. Analytical mechanics formulations also enable the systematic development of procedures to obtain approximate solutions to very complex problems. In particular, the finite element method, a commonly used tool for structural analysis, has its roots in analytical mechanics.

If analytical mechanics is so powerful and versatile, why has it not completely eclipsed Newton's formulation? An important task in structural analysis is the determination of both magnitude and direction of all forces acting within a structure, which is required to estimate failure conditions. It is logical to use Newton's approach for this task because it is directly expressed in terms of the quantities that must be evaluated. Furthermore, forces are easily visualized as vectors of a specific magnitude and direction that can be directly measured in the laboratory. In contrast, because it can only be measured in an indirect manner, mechanical work is a more abstract concept, which is quite different from the concept of "human work," in the sense of "human labor."

The *Principle of Virtual Work* (PVW) is the most fundamental tool of analytical mechanics, and it will be shown to be entirely equivalent to Newton's law. Both the principle of virtual work and Newton's laws are statements of equilibrium, which must always be satisfied at any point in a structure. In this chapter, simple applications of the principle of virtual work will be presented, focusing on discrete, rather than continuous systems. In many case, both Newtonian and analytical mechanics approaches will be presented in parallel to highlight their respective features.

9.2 Equilibrium and work fundamentals

9.2.1 Static equilibrium conditions

Newton's first law of motion states that *every object in a state of uniform motion tends to remain in that state of motion unless an external force is applied to it.* The expression "state of uniform motion" means that the object moves at a constant velocity; for static problems, however, is is customary to focus on objects at rest. If several forces are applied to the object, the "external force" is, in fact, the resultant, *i.e.*, the vector sum, of all externally applied forces. Finally, the "object" mentioned in the law is to be understood as a "particle." With all these clarifications, Newton's law is then restated as *a particle at rest tends to remain at rest unless the sum of the externally applied force does not vanish.* This also implies that if the sum of the externally applied forces does not vanish, the particle is no longer at rest. A more mathematical statement of Newton's law is: *a particle is at rest if and only if the sum of the externally applied forces vanishes.* The expression "if and only if" is included in the statement because this is both a necessary and sufficient condition: a particle is at rest if the sum of the forces vanishes, and if the sum of the forces vanishes, then the particle is at rest.

In structural mechanics, a particle at rest is said to be in *static equilibrium.* Newton's first law then becomes

> *A particle is in static equilibrium if and only if the sum of the externally applied forces vanishes.*

Newton's first law gives the necessary and the sufficient condition for static equilibrium and can be stated in a mathematical form as: a particle is in static equilibrium if and only if

$$\sum \underline{F} = 0, \tag{9.1}$$

where $\underline{F}$ are the externally applied forces acting on the particle. From a vector algebra standpoint, this equation can be interpreted in various manners: *(1)* the vector sum of all forces acting on the particle must be zero, or *(2)* the vector force polygon must be closed, or *(3)* the components of the vector sum resolved in any coordinate system must vanish, *i.e.*, if $\sum \underline{F} = F_1 \bar{\imath}_1 + F_2 \bar{\imath}_2 + F_3 \bar{\imath}_3$, where $\mathcal{I} = (\bar{\imath}_1, \bar{\imath}_2, \bar{\imath}_3)$ is an arbitrary orthonormal basis, $F_1 = F_2 = F_3 = 0$.

Newton's third law is also of fundamental importance to statics; it states: *if particle A exerts a force on particle B, particle B simultaneously exerts on particle A a force of identical magnitude and opposite direction.* It is also postulated that *these two forces share a common line of action.* In a more compact manner, Newton's third law states that

> *Two interacting particles exert on each other forces of equal magnitude, opposite directions, and sharing a common line of action.*

Euler's first law

Newton's first and third laws only apply to particles, but they can be extended to a collection of interacting particles. Figure 9.1 shows a system consisting of N particles. Particle i is subjected to an external force, $\underline{F}_i$, and to $N - 1$ interaction forces, $\underline{f}_{ij}, j = 1, 2, \ldots, N, j \neq i$. Newton's first law, eq. (9.1), applied to particle i, then states that

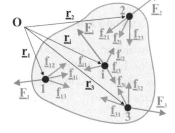

Fig. 9.1. A system of particles.

$$\underline{F}_i + \sum_{j=1, j \neq i}^{N} \underline{f}_{ij} = 0. \qquad (9.2)$$

Note that little has been said about the nature of the *system of particles,* or of the interaction forces. If the system of particles is a rigid body, the interaction forces are those that will ensure that the shape of the body remains unchanged by the externally applied loads. If the system of particles is an elastic body, the interaction forces are the stresses that will result from the deformation of the body. If the system of particles is planetary system, the interaction forces are the gravitational pull that each planet exerts on all others. Although of physically different natures, all interaction forces are assumed to obey Newton's third law, which implies that $\underline{f}_{ij} + \underline{f}_{ji} = 0$.

Since eq. (9.2) applies to each of the N particles of the system, summation of these N equations yields

$$\sum_{i=1}^{N} \underline{F}_i + \sum_{i=1}^{N} \sum_{j=1, j \neq i}^{N} \underline{f}_{ij} = 0.$$

In the double summation of the second term, interaction forces appear in pairs, $\underline{f}_{ij}$ and $\underline{f}_{ji}$. Newton's third law then implies the vanishing of each pair of interaction forces, $\underline{f}_{ij} + \underline{f}_{ji} = 0$, leading to

$$\sum_{i=1}^{N} \sum_{j=1,j\neq i}^{N} \underline{f}_{ij} = 0. \tag{9.3}$$

Since the second term vanishes, the above equation simplifies to

$$\sum_{i=1}^{N} \underline{F}_i = 0. \tag{9.4}$$

This statement is known as *Euler's first law* for a system of particles; it is very similar to Newton's first law, but now applies to a system of particle.

Note that eq. (9.4) is a necessary condition for the system of particles to be in static equilibrium, but is not a sufficient condition. As implied by Newton's first law, the necessary and sufficient conditions for the static equilibrium of the system are the satisfaction of eq. (9.2) for each of the N particles of the system; in all, this represents N vector equations that must be satisfied. Clearly, eq. (9.4) is a single vector equation, and therefore a much less stringent condition. The main advantage of eq. (9.4), however, is that all interaction forces are eliminated from this statement.

Euler's second law

It is possible to extract additional conditions for the static equilibrium of a system of particles. Let $\underline{r}_i$ be the position vector of particle i with respect to an arbitrary point **O**, see fig. 9.1. Taking a vector product of eq. (9.2) by $\underline{r}_i$, then summing over all particles leads to

$$\sum_{i=1}^{N} \underline{r}_i \times \underline{F}_i + \sum_{i=1}^{N} \sum_{j=1,j\neq i}^{N} \underline{r}_i \times \underline{f}_{ij} = 0.$$

In the double summation of the second term, interaction forces appear in pairs, $\underline{r}_i \times \underline{f}_{ij}$ and $\underline{r}_j \times \underline{f}_{ji}$. A property of the vector cross product is that $\underline{r}_i \times \underline{f}_{ij} = \underline{r}_\perp \times \underline{f}_{ij}$ and $\underline{r}_j \times \underline{f}_{ji} = \underline{r}_\perp \times \underline{f}_{ji}$, where $\underline{r}_\perp$ is the vector that joins point **O** to the point on the common line of action of the internal force pair that is at the shortest distance from point **O**. For Newton's third law, it then follows that $\underline{r}_i \times \underline{f}_{ij} + \underline{r}_j \times \underline{f}_{ji} = \underline{r}_\perp \times (\underline{f}_{ij} + \underline{f}_{ji}) = 0$, which yields

$$\sum_{i=1}^{N} \sum_{j=1,j\neq i}^{N} \underline{r}_i \times \underline{f}_{ij} = 0. \tag{9.5}$$

With this simplification, the above equation reduces to

$$\sum_{i=1}^{N} \underline{r}_i \times \underline{F}_i = \sum_{i=1}^{N} \underline{M}_i = 0, \tag{9.6}$$

where $\underline{M}_i$ is the moment of the external forces applied to particle i. The point about which moments of the externally applied forces is calculated is arbitrary. This statement is known as *Euler's second law* for a system of particles. Euler's first and second laws are both necessary conditions for the system of particles to be in static equilibrium, but are not a sufficient conditions. Taken together, they form two vector equations that clearly fall short of the N vector equations, eq. (9.2) for each of the N particles, required to guarantee static equilibrium of the system.

9.2.2 Concept of mechanical work

Mechanical work is defined as follows: *the work done by a force is the scalar product of the force by the displacement of its point of application.* At first, let the force and displacement vectors be collinear: the force vector is $\underline{F} = F\bar{u}$ and the displacement vector $\underline{d} = d\bar{u}$, where F is the magnitude of the force, d that of the displacement, and $\bar{u}$ is a unit vector along the common direction of the force and displacement. The work, W, done by the force becomes $W = F\bar{u} \cdot d\bar{u} = Fd$. Note that if the force and displacement are both in the same direction, the work is positive, whereas if the force and displacement are in opposite directions, the work is negative.

Next, consider the case where force and displacement vectors are not collinear: the force vector is $\underline{F} = F\bar{u}$ and the displacement vector $\underline{d} = d\bar{v}$, where $\bar{u}$ and $\bar{v}$ are unit vectors along the orientations of the force and displacement vectors, respectively. The work done by the force becomes $W = F\bar{u} \cdot d\bar{v} = Fd\,\bar{u} \cdot \bar{v} = Fd\cos\theta$, where θ is the angle between the unit vectors $\bar{u}$ and $\bar{v}$. If the force acts in the direction perpendicular to the displacement, $\cos\theta = \cos\pi/2 = 0$, and the work done by the force vanishes although both force and displacement vectors do not.

It is often the case that both force and displacement vectors change in time. To deal with this situation, the concept of *incremental* work is introduced as $dW = \underline{F} \cdot d\underline{r}$, where $d\underline{r}$ is the infinitesimal displacement vector. If the point of application of the force moves from $\underline{r}_i$ to $\underline{r}_f$, the total work is then found by integration of the incremental work

$$W = \int_{\underline{r}_i}^{\underline{r}_f} dW = \int_{\underline{r}_i}^{\underline{r}_f} \underline{F} \cdot d\underline{r}. \tag{9.7}$$

When the force and incremental displacement are three-dimensional vectors, their scalar product is easily computed by evaluating the components of the two vectors in a common orthonormal basis. The force and infinitesimal displacement vectors are written as $\underline{F} = F_1\bar{e}_1 + F_2\bar{e}_2 + F_3\bar{e}_3$ and $d\underline{r} = dr_1\bar{e}_1 + dr_2\bar{e}_2 + dr_3\bar{e}_3$, respectively, where $\mathcal{E} = (\bar{e}_1, \bar{e}_2, \bar{e}_3)$ forms an orthonormal basis, *i.e.*, a set of three mutually orthogonal unit vectors. The incremental work then becomes $dW = \underline{F} \cdot d\underline{r} = F_1 dr_1 + F_2 dr_2 + F_3 dr_3$. The force and displacement vectors must both be resolved in a common basis for this formula to apply, although this common basis can be chosen arbitrarily.

Only the component of the force vector acting along the differential displacement vector does work. Let the differential displacement be written as $d\underline{r} = dr\,\bar{u}$, where $\bar{u}$ is the unit vector in the direction of the differential displacement. Next, the force vector is written as $\underline{F} = F_\parallel \bar{u} + F_\perp \bar{v}$, where $\bar{v}$ is a unit vector perpendicular to $\bar{u}$, $F_\parallel$ the component of the force acting along the differential displacement vector, and $F_\perp$ that acting in the plane perpendicular to $\bar{u}$. The incremental work now becomes

$dW = (F_{\|}\bar{u} + F_{\perp}\bar{v}) \cdot dr\,\bar{u} = F_{\|}dr$. The component of the force acting along the displacement vector, $F_{\|}$, is the sole contributor to the work.

Work is a scalar product, and consequently, superposition holds. Let the applied force be written as $\underline{F} = \underline{F}_1 + \underline{F}_2$; the incremental work now becomes $dW = \underline{F} \cdot d\underline{r} = (\underline{F}_1 + \underline{F}_2) \cdot d\underline{r} = \underline{F}_1 \cdot d\underline{r} + \underline{F}_2 \cdot d\underline{r} = dW_1 + dW_2$. This implies that the sum of the work done by the two forces, $\underline{F}_1$ and $\underline{F}_2$, denoted dW_1 and dW_2, respectively, equals the work done by the resultant force, $\underline{F}$.

Structural analysis focuses on static problems, as opposed to structural dynamics, which broadens the scope of the investigation to include the dynamic response of structures to time dependent loads. The very definition of work involves a "force that displaces its point of application," which implies a dynamic problem. Why then is work a quantity of interest for the static analysis of structures? The answer to this question is found in the next section, which introduces the concept of *virtual work*, i.e., the work that would be done by a force *if it were to displace its point of application by a fictitious amount*.

9.3 Principle of virtual work

As discussed in the previous section, the static equilibrium condition for a particle, as stated by Newton's first law, is written as a vector equation that imposes the vanishing of the externally applied forces. In the present section, an alternative formulation will be developed, which results in the *Principle of Virtual Work* (PVW). Although expressed in terms of work rather than force vectors, the principle of virtual work will be shown to be entirely equivalent to Newton's first law. First, the principle will be developed for a single particle; next, it will be generalized to enable applications to systems of particles.

The principle of virtual work introduces the fundamental concept of "arbitrary virtual displacements" sometimes called "arbitrary test displacements," or also "arbitrary fictitious displacements," and all of these expressions will be used interchangeably. The word "arbitrary" is easily understood: it simply means that the displacements can be chosen in an arbitrary manner without any restriction imposed on their magnitudes or orientations. More difficult to understand are the words "virtual," "test," or "fictitious." All three imply that these are not real, actual displacements. More importantly, these fictitious displacements *do not affect the forces acting on the particle*. These important concepts will be explained in the following sections.

9.3.1 Principle of virtual work for a single particle

Consider a particle in static equilibrium under a set a externally applied loads, as depicted in fig. 9.2. According to Newton's first law, the sum of the externally applied load must vanish, as expressed by eq. (9.1). Next, consider a fictitious displacement of arbitrary magnitude and orientation, denoted $\underline{s}$ in fig. 9.2. Although the problem appears to be two-dimensional in the figure, both forces and fictitious displacements are three-dimensional quantities.

The virtual work done by the externally applied forces is now evaluated by computing the dot product of the externally applied loads by the fictitious displacement to find

$$W = \left[\sum \underline{F}\right] \cdot \underline{s} = 0. \qquad (9.8)$$

Because the particle is in static equilibrium, Newton's first law implies the vanishing of the bracketed term. It follows that the dot product vanishes *for any arbitrary fictitious displacement.*

This result sheds some light on the special nature of the fictitious, or virtual displacements. If the particle is in static equilibrium in a given configuration, the sum of the forces vanishes, *i.e.,* $\sum \underline{F} = 0$. Assume now that one of the externally applied forces, say $\underline{F}_1$, is the force acting in an elastic spring connected to the particle. If the particle undergoes a *real, but arbitrary displacement, $\underline{d}$,* the force in the spring will change to become $\underline{F}'_1$. All displacement dependent forces applied to the particle will change, and the sum of the externally applied loads

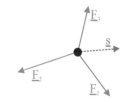

Fig. 9.2. A particle with applied forces subjected to a fictitious test displacement.

becomes $\sum \underline{F}'$. In the new configuration resulting from the application of the real displacement, $\underline{d}$, static equilibrium will not be satisfied, *i.e.,* $\sum \underline{F}' \neq 0$. Indeed, if the particle is in static equilibrium in the configuration resulting from the application of an *arbitrary displacement*, it will be in static equilibrium in *any* configuration, which makes little sense.

In contrast with real displacements, *virtual or fictitious displacements do not affect the loads applied to the particle.* This means that even in the presence of displacement dependent loads such as those arising within an elastic spring, if the particle is in static equilibrium, it remains in static equilibrium when virtual or fictitious displacements are applied. This is the reason why eq. (9.8) remains true for *all arbitrary virtual displacements*. The discussion thus far has thus established that if the particle is in static equilibrium, eq. (9.8) holds for all arbitrary fictitious displacements.

Next, the following question is asked: if eq. (9.8) holds, is the particle in static equilibrium? Consider fig. 9.2, and let the components of the applied forces be $\underline{F}_1 = F_{11}\bar{\imath}_1 + F_{12}\bar{\imath}_2 + F_{13}\bar{\imath}_3$, $\underline{F}_2 = F_{21}\bar{\imath}_1 + F_{22}\bar{\imath}_2 + F_{23}\bar{\imath}_3$, $\underline{F}_3 = F_{31}\bar{\imath}_1 + F_{32}\bar{\imath}_2 + F_{33}\bar{\imath}_3$, while the components of the virtual displacement are $\underline{s} = s_1\bar{\imath}_1 + s_2\bar{\imath}_2 + s_3\bar{\imath}_3$, where $\mathcal{I} = (\bar{\imath}_1, \bar{\imath}_2, \bar{\imath}_3)$ is an orthonormal basis. Equation (9.8) now states $(F_{11} + F_{21} + F_{31})s_1 + (F_{12} + F_{22} + F_{32})s_2 + (F_{13} + F_{23} + F_{33})s_3 = 0$.

At first, assume that the particle is not in static equilibrium, *i.e.,* $\sum \underline{F} \neq 0$. It is always possible to find *a particular virtual displacement* for which eq. (9.8) will be satisfied. Indeed, for a given set of forces, select s_1 and s_2 in an arbitrary manner, then solve eq. (9.8) for s_3 to find $s_3 = -[(F_{11} + F_{21} + F_{31})s_1 + (F_{12} + F_{22} + F_{32})s_2]/(F_{13} + F_{23} + F_{33})$. Consequently, the fact that eq. (9.8) is satisfied *for a particular virtual displacement* does not imply that it is in static equilibrium. In fact, even if it is satisfied *for many virtual displacements*, static equilibrium is still not guaranteed. Indeed, for each new arbitrary choice of s_1 and s_2, it is possible to compute an s_3 for which eq. (9.8) is satisfied.

Different conclusions are reached if eq. (9.8) is satisfied *for all arbitrary virtual displacements*. Indeed, if $(F_{11} + F_{21} + F_{31})s_1 + (F_{12} + F_{22} + F_{32})s_2 + (F_{13} + F_{23} + F_{33})s_3 = 0$ for all arbitrary values of independently chosen s_1, s_2 and s_3, it follows that $F_{11} + F_{21} + F_{31} = 0$, $F_{12} + F_{22} + F_{32} = 0$, and $F_{13} + F_{23} + F_{33} = 0$, is the only solution of eq. (9.8). In turn, this can be written as $(F_{11} + F_{21} + F_{31})\bar{\imath}_1 + (F_{12} + F_{22} + F_{32})\bar{\imath}_2 + (F_{13} + F_{23} + F_{33})\bar{\imath}_3 = 0$, and finally, $\sum \underline{F} = 0$. Thus, if eq. (9.8) is satisfied *for all arbitrary virtual displacements*, then $\sum \underline{F} = 0$, and the particle is in static equilibrium.

In conclusion, if a particle is in static equilibrium, the virtual work done by the externally applied forces vanishes for all arbitrary virtual displacements. Furthermore, it is also true that if the virtual work vanishes for all arbitrary fictitious test displacements, then the sum of the externally applied forces vanishes, and hence, the particle is in static equilibrium. These two facts can be combined into the statement of the principle of virtual work for a particle

Principle 3 (Principle of virtual work for a particle) *A particle is in static equilibrium if and only if the virtual work done by the externally applied forces vanishes for all arbitrary virtual displacements.*

Since the condition for static equilibrium is nothing but Newton's first law, it follows that the principle of virtual work, which states the condition for static equilibrium, is entirely equivalent to Newton's first law, and either statement provides a fundamental definition of static equilibrium. Simple examples will now be used to illustrate the principle of virtual work.

Example 9.1. Equilibrium of a particle
Consider the particle depicted in fig. 9.3, which is subjected to two vertical forces $\underline{F}_1 = 1\bar{\imath}_1$ and $\underline{F}_2 = -3\bar{\imath}_1$. The following question is asked: is the particle in static equilibrium? Rather than relying on Newton's first law, the principle of virtual work will used to answer the question. Consider the following arbitrary virtual displacement, $\underline{s} = s_1\bar{\imath}_1 + s_2\bar{\imath}_2$, and its associated virtual work

$$W = (1\bar{\imath}_1 - 3\bar{\imath}_1) \cdot (s_1\bar{\imath}_1 + s_2\bar{\imath}_2) = -2\bar{\imath}_1 \cdot (s_1\bar{\imath}_1 + s_2\bar{\imath}_2) = -2s_1 \neq 0.$$

The fact that $\underline{s}$ is an arbitrary virtual displacement implies that s_1 and s_2 are arbitrary scalars, and hence, $W = -2s_1 \neq 0$. Because the virtual work done by the externally applied forces does not vanish for all virtual displacements, the principle of virtual work, principle 3, implies that the particle is not in static equilibrium.

It is important to understand the implications of the last part of the principle of virtual work, "for all arbitrary virtual displacements." Consider the following arbitrary virtual displacement, $\underline{s} = s_2\bar{\imath}_2$, and its associated virtual work

$$W = (1\bar{\imath}_1 - 3\bar{\imath}_1) \cdot s_2\bar{\imath}_2 = -2\bar{\imath}_1 \cdot s_2\bar{\imath}_2 = 0.$$

This result is due to the fact that the sum of the externally applied loads, $-2\bar{\imath}_1$, is orthogonal to the virtual displacement, $s_2\bar{\imath}_2$, and hence, the virtual work vanishes.

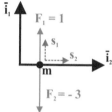

Fig. 9.3. A particle under the action of two forces.

Fig. 9.4. A particle suspended to an elastic spring.

One might be tempted to conclude from the above result that the particle is in static equilibrium because the virtual work vanishes. To satisfy the principle of virtual work, however, the virtual work must vanish *for all arbitrary virtual displacements*.

The above result shows that the virtual work may vanish for "a particular virtual displacement," but this is not a sufficient condition to guarantee static equilibrium. For the two-dimensional problem shown in fig. 9.3, an arbitrary fictitious displacement must span the plane of the problem, *i.e.*, must be of the form $\underline{s} = s_1 \bar{\imath}_1 + s_2 \bar{\imath}_2$. For three-dimensional problems, a three-dimensional virtual displacement must be selected, $\underline{s} = s_1 \bar{\imath}_1 + s_2 \bar{\imath}_2 + s_3 \bar{\imath}_3$, where s_1, s_2, and s_3 are three arbitrary scalars, and $\mathcal{I} = (\bar{\imath}_1, \bar{\imath}_2, \bar{\imath}_3)$ a basis that spans the three-dimensional space.

Example 9.2. Equilibrium of a particle connected to an elastic spring

Consider next a particle in static equilibrium under the effect of gravity and the restoring force of an elastic spring of stiffness constant k, as depicted in fig. 9.4. Find the displacement of the particle in its actual static equilibrium configuration.

For this two-dimensional problem, assume that the particle is at position u. An arbitrary fictitious displacement is selected as $\underline{s} = s_1 \bar{\imath}_1 + s_2 \bar{\imath}_2$, where s_2 and s_2 are two arbitrary scalars. The virtual work done by the externally applied loads becomes

$$W = (mg\bar{\imath}_1 - ku\bar{\imath}_1) \cdot (s_1 \bar{\imath}_1 + s_2 \bar{\imath}_2) = [mg - ku]s_1.$$

The principle of virtual work now implies that the particle is in static equilibrium at position u if and only if the virtual work done by the externally applied loads vanishes for all arbitrary virtual displacements, *i.e.*, if and only if $[mg - ku]s_1 = 0$ for all values of s_1. Equation $[mg - ku]s_1 = 0$ possesses two solutions, $[mg - ku] = 0$ or $s_1 = 0$; the second solution, however, is not valid because, as implied by the principle of virtual work, s_1 is arbitrary.

In conclusion, the vanishing of the virtual work for all arbitrary virtual displacements implies that $mg - ku = 0$, and the equilibrium configuration of the system is found as $u = mg/k$. Of course, the same conclusion can be drawn more expeditiously from a direct application of Newton's first law, which requires the sum of the externally applied forces to vanish, *i.e.*, $mg\bar{\imath}_1 - ku\bar{\imath}_1 = 0$, or $(mg - ku)\bar{\imath}_1 = 0$, and finally, $mg - ku = 0$.

This example involves the restoring force of an elastic spring, a displacement dependent force. Indeed, the elastic force in the spring is $-ku\bar{\imath}_1$, and if the particle undergoes a *real downward displacement* of magnitude d, the restoring force becomes $-k(u + d)\bar{\imath}_1$. In contrast, if the particle undergoes a *virtual downward displacement* of magnitude s_1, the restoring force *remains unchanged* as $-ku\bar{\imath}_1$. This difference has profound implications on the computation of work. First, consider the work done by the elastic force, $-ku\bar{\imath}_1 \cdot du\,\bar{\imath}_1$, under a *virtual displacement*, s_1,

$$W = \int_u^{u+s_1} -ku\,du = -ku \int_u^{u+s_1} du = -ku\,[u]_u^{u+s_1} = -kus_1. \tag{9.9}$$

It is possible to remove the elastic force, $-ku$, from the integral because this force remains unchanged by the virtual displacement, and hence, it can be treated as a constant.

In contrast, the work done by the same elastic force under a *real displacement*, d, is

$$W = \int_u^{u+d} -ku\,du = \left[-\frac{1}{2}ku^2\right]_u^{u+d} = -kud - \frac{1}{2}kd^2. \tag{9.10}$$

In this case, the real work includes an additional term that is quadratic in d and represents the work done by the change in force that develops due to the stretching of the spring. Even if the magnitude of the real displacement is equal to that of the virtual displacement, *i.e.*, even if $d = s_1$, the two expressions for the work done by the elastic restoring force are not identical.

These observations help explain the terminology used when dealing with the principle of virtual work. The concept of virtual displacement is key to the correct use of the principle of virtual work, which requires the virtual work done by displacement dependent forces to be evaluated according to eq. (9.9) rather than eq. (9.10). Of course, the *real* work done by the elastic force as it undergoes a real displacement is correctly evaluated by eq. (9.10).

Clearly, it is important to keep in mind the crucial difference between "real displacements" and "virtual" or "fictitious displacements." The words "virtual" or "fictitious" are used to emphasize the fact the forces remain unaffected by these displacements. In practice, the term "real displacement" is rarely used; real displacements are simply called displacements. The terms "virtual," "fictitious" or "test displacements" all imply that the forces acting on the system remain unaffected by the application of such displacements. The term "virtual displacement" is the most widely used.

Example 9.3. *Equilibrium of a particle sliding on a track*

Consider a particle of mass m that can slide on a track, as shown in fig. 9.5. The externally applied horizontal force is resited by friction between the particle and track. Newton's first law expresses the condition for static equilibrium as $mg\bar{\imath}_1 - R\bar{\imath}_1 + P\bar{\imath}_2 - F\bar{\imath}_2 = 0$, where $-R\bar{\imath}_1$ is the reaction force the track exerts on the particle, and $-F\bar{\imath}_2$ the friction force applies to the particle.

Note that the four forces applied to the particle are of different physical natures: $P\bar{\imath}_2$ is an externally applied force, $mg\bar{\imath}_1$ the force of gravity, $-R\bar{\imath}_1$ a reaction force, and $-F\bar{\imath}_2$ a friction force. Yet all forces play an equal role in Newton's law, which states that the sum of all forces must vanish. The law simply states "all forces" without making any distinction among them. Newton's first law is readily solved to find $(mg - R)\bar{\imath}_1 + (P - F)\bar{\imath}_2 = 0$, and finally $R = mg$ and $F = P$, as expected.

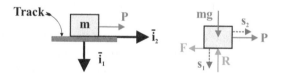

Fig. 9.5. A particle sliding on a track.

Next, the principle of virtual work will be used to solve the same problem. For this two dimensional problem, an arbitrary virtual displacement will be written as $\underline{s} = s_1\bar{\imath}_1 + s_2\bar{\imath}_2$, and the vanishing of the virtual work it performs implies

$$W = (mg\bar{\imath}_1 - R\bar{\imath}_1 + P\bar{\imath}_2 - F\bar{\imath}_2) \cdot (s_1\bar{\imath}_1 + s_2\bar{\imath}_2) = [mg - R]s_1 + [P - F]s_2 = 0. \quad (9.11)$$

Following a reasoning similar to that developed in the previous example, it is easy to show that the vanishing of the virtual work for all arbitrary scalars s_1 and s_2 implies the vanishing of the two bracketed terms in the above equation: $mg - R = 0$ and $P - F = 0$. This result is identical to that obtained from Newton's first law, as expected, since the principle of virtual work and Newton's first law are identical.

This example illustrates a crucial relationship between Newton's first law and the principle of virtual work. The projection of Newton's law along axes, $\bar{\imath}_1$ and $\bar{\imath}_2$, yields two scalar equilibrium equations, $mg - R = 0$ and $P - F = 0$, respectively. The same two equilibrium equations are obtained by imposing the vanishing of the factors multiplying the arbitrary virtual displacement components, s_1 and s_2, measured along the same axes, $\bar{\imath}_1$ and $\bar{\imath}_2$, respectively.

The principle of virtual work yields scalar equilibrium equations which are the projections of Newton's first law along the directions associated with the virtual displacement components. Because it is based on a scalar quantity, the virtual work, the principle of virtual work yields scalar equations of equilibrium, rather than their vector counterparts inherent to the application of Newton's first law.

9.3.2 Kinematically admissible virtual displacements

Example 9.3 illustrates an important feature of virtual displacements, which are selected to have components in the horizontal direction, $s_2\bar{\imath}_2$, and the vertical direction, $s_1\bar{\imath}_1$. This raises a basic question: how could the particle move in the vertical direction when it is constrained to remain on the track? The answer to the question lies in the nature of the virtual displacements which are not real, but rather are virtual or fictitious displacements. Of course, the particle cannot possibly undergo *real displacements* in the vertical direction because it must remain on the track, but *virtual* or *fictitious displacements* in that same direction are allowed.

In the derivation of the principle of virtual work, it is necessary to use completely arbitrary virtual displacements to prove that the vanishing of the virtual work implies Newton's first law. The completely arbitrary nature of the virtual displacements is key to the successful use of the principle of virtual work. The expression, "arbitrary virtual displacements" means *any virtual displacements, including those that violate the kinematic constraints of the problem.*

In fig. 9.5, the particle is confined to remain on the track; it can move along the track, but not in the direction perpendicular to it. The direction along the track is called the *kinematically admissible direction*, whereas the direction normal to it is called the *kinematically inadmissible direction*, or the *infeasible direction*.

It is sometimes convenient to introduce the concept of *kinematically admissible virtual displacements*. These are virtual displacements that satisfy the kinematic constraints of the problem.

For the problem depicted in fig. 9.5, the kinematic constraint enforces the particle to remain on the track. Arbitrary virtual displacements are written as $\underline{s} = s_1 \bar{\imath}_1 + s_2 \bar{\imath}_2$, but since these include a component in the vertical direction, *i.e.*, in a kinematically inadmissible direction, these are not kinematically admissible virtual displacements. On the other hand, virtual displacements of the form $\underline{s} = s_2 \bar{\imath}_2$, are kinematically admissible because these are oriented along the track.

At this point, the relationship between kinematic constraints and reaction forces should be clarified. Reaction forces are those forces arising from the enforcement of kinematic constraints. The particle depicted in fig. 9.5 is constrained to move along the track, and this kinematic constraint gives rise to a reaction force. Note that the reaction force acts along the kinematically inadmissible direction, *i.e.*, the direction normal to the track.

Consider now the virtual work done by the reaction force under arbitrary virtual displacements,

$$W = (-R\bar{\imath}_1) \cdot (s_1 \bar{\imath}_1 + s_2 \bar{\imath}_2) = -R s_1 \neq 0.$$

Next, consider the virtual work done by the same reaction force under arbitrary *kinematically admissible virtual displacements*,

$$W = (-R\bar{\imath}_1) \cdot (s_2 \bar{\imath}_2) = 0.$$

Because the reaction force acts along the infeasible direction, whereas the kinematically admissible virtual displacement is along the admissible direction, these two vectors are normal to each other, and hence, the virtual work done by the reaction force vanishes. In contrast, the work done by the same reaction force under arbitrary virtual displacements does not.

The vanishing of the virtual work done by reaction forces under kinematically admissible virtual displacements has profound implications for applications of the principle of virtual work. The principle is repeated here: "a particle is in static equilibrium if and only if the virtual work done by the externally applied forces vanishes *for all arbitrary virtual displacements*". Because this principle calls for the use of arbitrary virtual displacements, it is of crucial importance to treat reaction forces as externally applied forces. For instance, in example 9.3, the virtual work done by the reaction force must be included in the statement of the principle, as is done in eq. (9.11), because completely arbitrary virtual displacements are used.

Consider now a modified version of the principle of virtual work: "a particle is in static equilibrium if and only if the virtual work done by the externally applied forces vanishes *for all arbitrary kinematically admissible virtual displacements*". Rather than considering completely arbitrary virtual displacements, only kinematically admissible virtual displacements are considered now. Because the virtual work done by the constraint forces vanishes for kinematically admissible virtual displacements, constraint forces are automatically eliminated from this statement of the principle of virtual work. This often simplifies the statement of the principle because fewer terms are involved. On the other hand, because the constraint forces are eliminated from the formulation, this modified principle will not yield the equations required to evaluate the reaction forces, which are often quantities of great interest.

As pointed out earlier, Newton's first law requires the sum of all forces to vanish for static equilibrium to be achieved. The "sum of all forces" involves all forces without distinction. While the principle of virtual work is shown to be identical to Newton's first law, this principle creates an important distinction between reaction forces stemming from kinematic constraints, and all other forces. Indeed, reaction forces, also called forces of constraint, can be completely eliminated from the formulation by using kinematically admissible virtual displacements.

All other forces, such as those generated by springs, gravity, friction, temperature, electric or magnetic fields, are of a physical origin. It is easy to recognize such forces because their description involves physical constants that can only be determined by experiment. For instance, the stiffness constant of a spring, the universal constant of gravitation appearing in gravity forces, or the friction coefficient appearing in Coulomb's friction law. All these forces are referred to as *natural forces*, which can be further differentiated into *internal* and *external* forces. *Internal forces* are natural forces arising from and reacted within the structural system under consideration, whereas *external forces* are natural forces that act on the system but stem from outside it; these forces are also called *externally applied loads*.

Example 9.4. Equilibrium of a particle sliding on a track

Consider once again a particle of mass m resting on a track, as shown in fig. 9.5. For this simple problem, the kinematically admissible direction is along axis $\bar{\imath}_2$, while the infeasible direction is along axis $\bar{\imath}_1$. The free body diagram in the right part of fig. 9.5 shows the forces acting on the particle. The reaction force, $-R\bar{\imath}_1$, acts in the infeasible direction, as expected.

In contrast with example 9.3, which uses completely arbitrary virtual displacements, kinematically admissible virtual displacements will be used here so that $\underline{s} = s_2\bar{\imath}_2$. The vanishing of the virtual work then implies

$$W = (mg\bar{\imath}_1 - R\bar{\imath}_1 + P\bar{\imath}_2 - F\bar{\imath}_2) \cdot s_2\bar{\imath}_2 = [P - F]s_2 = 0.$$

Because s_2 is an arbitrary quantity, the bracketed term must vanish, leading to $F = P$.

First, note that the reaction force, R, is eliminated from the formulation: the statement of the principle of virtual work becomes simply $(P - F)s_2 = 0$ for all values of s_2. The reaction force does not appear in this statement. It is also possible to apply external loads along the infeasible direction: for instance, in this problem, gravity loads act in the infeasible direction and are also eliminated from the formulation. Of course, if gravity acts along the kinematically admissible direction, *i.e.*, along the track, this force will remain in the statement of the principle. In contrast, reaction forces always act along the infeasible direction and hence, are always eliminated from the formulation.

Second, note that less information about the system is obtained. In example 9.3 that uses virtual displacements, two equations are obtained: $F = P$ and $R = mg$. In contrast, the use of kinematically admissible virtual displacements yields a single equation, $F = P$. On the other hand, the solution process is simpler and involves one single equation; however, no information about the reaction force is available.

Finally, it is shown here that the modified version of the principle of virtual work stating "a particle is in static equilibrium if and only if the virtual work done by the externally applied forces vanishes *for all arbitrary kinematically admissible virtual displacements*," is not entirely correct. The vanishing of the virtual work for all kinematically admissible virtual displacements is a necessary condition, but it is not sufficient, because it does not guarantee equilibrium of the particle in the infeasible direction. Indeed, this latter condition, $R = mg$, is not recovered by the modified principle.

Example 9.5. *Equilibrium of a particle on a curved track*

Consider a particle of mass m constrained to move on a semi-circular track of radius R under the combined effects of gravity, friction, and a spring force, as depicted in fig. 9.6. Determine the equilibrium position of the particle and the forces acting on it in the equilibrium state.

The spring of stiffness constant k is pinned at point $\mathbf{C}$ located at coordinates $x_1 = c_1 R$ and $x_2 = c_2 R$ and its un-stretched length is zero. Force N is the reaction force acting on the particle due to its contact with the track and acts in direction $\bar{n}$, which is normal to the track. Force F is the force exerted by the track on the particle and acts in the tangential direction, $\bar{t}$; this force arises from friction between the particle and track.

The position of the particle on the track is conveniently given by angle θ. The unit vector tangent to the circular track is given by $\bar{t} = -\sin\theta\,\bar{\imath}_1 + \cos\theta\,\bar{\imath}_2$, whereas the normal to the track is $\bar{n} = -\cos\theta\,\bar{\imath}_1 - \sin\theta\,\bar{\imath}_2$. For this problem, the kinematically admissible direction is $\bar{t}$, and $\bar{n}$ the infeasible direction. In contrast with the previous example, the admissible direction is not a fixed direction in space, but instead, it depends on the position of the particle on the track, $\bar{t} = \bar{t}(\theta)$. The reaction force of magnitude N acts along the infeasible direction, as expected. The friction force of magnitude F acts in the admissible direction.

The force, $\underline{F}_s$, applied by the elastic spring to the particle is given by the spring stiffness constant times the distance between the particle and point $\mathbf{C}$ located at $(c_1 R, c_2 R)$ and is oriented in that same direction: $\underline{F}_s = kR[(c_1 - \cos\theta)\bar{\imath}_1 + (c_2 - $

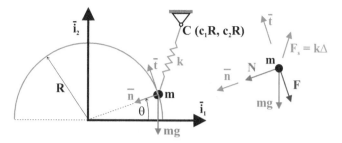

Fig. 9.6. Particle constrained to slide with friction on a circular track.

$\sin \theta)\bar{\imath}_2]$. This can be expressed in terms of admissible and infeasible directions, $\bar{t}$ and $\bar{n}$, respectively, as $\underline{F}_s = kR[(-c_1 \sin \theta + c_2 \cos \theta)\bar{t} + (1 - c_1 \cos \theta - c_2 \sin \theta)\bar{n}]$ where use is made of the following relationships: $\bar{\imath}_1 = -\sin \theta \, \bar{t} - \cos \theta \, \bar{n}$ and $\bar{\imath}_2 = \cos \theta \, \bar{t} - \sin \theta \, \bar{n}$.

An arbitrary virtual displacement of the form $\underline{s} = s_t \bar{t} + s_n \bar{n}$ is selected, where s_t and s_n are arbitrary numbers, and the virtual work done by the forces acting on the particle then becomes

$$
\begin{aligned}
W = \{ & kR\left[(-c_1 \sin \theta + c_2 \cos \theta)\bar{t} + (1 - c_1 \cos \theta - c_2 \sin \theta)\bar{n}\right] + N\bar{n} - F\bar{t} \\
& + mg(-\cos \theta \bar{t} + \sin \theta \bar{n}) \} \cdot (s_t \bar{t} + s_n \bar{n}) \\
= & \left[kR(-c_1 \sin \theta + c_2 \cos \theta) - F - mg \cos \theta\right] s_t \\
& + \left[kR(1 - c_1 \cos \theta - c_2 \sin \theta) + N + mg \sin \theta\right] s_n.
\end{aligned}
$$

Because the virtual work must vanish for arbitrary s_t and s_n, the two bracketed terms must vanish, leading to the two equilibrium equations of the problem,

$$
\begin{aligned}
F &= \quad kR(\ -c_1 \sin \theta + c_2 \cos \theta) - mg \cos \theta, & \text{(9.12a)} \\
N &= -kR(1 - c_1 \cos \theta - c_2 \sin \theta) - mg \sin \theta. & \text{(9.12b)}
\end{aligned}
$$

This forms a set of two equations for the three unknowns of the problem: the reaction force, N, the friction force, F, and the equilibrium position of the particle, θ.

One additional equation is required to solve the problem. Coulomb's law of static friction requires the friction force to be smaller than the normal contact force multiplied by the static friction coefficient, μ_s, i.e., $|F| \leq \mu_s |N|$. Substituting the friction and normal forces from eqs. (9.12a) and (9.12b), respectively, leads to

$$
\begin{aligned}
& kR(-c_1 \sin \theta + c_2 \cos \theta) - mg \cos \theta \\
& \leq \pm \mu_s \left[-kR(1 - c_1 \cos \theta - c_2 \sin \theta) - mg \sin \theta\right].
\end{aligned}
$$

This equation can be solved to find two solutions, θ_ℓ and θ_u: the particle is in equilibrium for all configurations, θ, such that $\theta_\ell \leq \theta \leq \theta_u$.

Next, kinematically admissible virtual displacements of the form $\underline{s} = s_t \bar{t}$ will be selected, where s_t is an arbitrary value. The virtual work done by the forces acting on the particle then becomes

$$W = \{kR\left[(-c_1 \sin\theta + c_2 \cos\theta)\bar{t} + (1 - c_1 \cos\theta - c_2 \sin\theta)\bar{n}\right] + N\bar{n} - F\bar{t}$$
$$+ mg(-\cos\theta\bar{t} + \sin\theta\bar{n})\} \cdot s_t\bar{t}$$
$$= \left[kR(-c_1 \sin\theta + c_2 \cos\theta) - F - mg\cos\theta\right]s_t.$$

Because the virtual work must vanish for all arbitrary s_t, the bracketed term must vanish, yielding a single equilibrium equation of the problem, which is the same as eq. (9.12a) above. As expected, the normal reaction force, N, is eliminated from the formulation. The problem still features three unknowns, N, F and θ, and the addition of the static friction law provides a second equation for the problem. Clearly, the principle of virtual work with kinematically admissible virtual displacements does not provide enough equations to solve this problem. This is because the static friction law establishes a relationship between friction and normal forces. By eliminating the normal contact force from the formulation, the use of kinematically admissible virtual displacements yields too little information to solve the problem.

Note that if friction is neglected, the friction force will vanish, $F = 0$, and the single equation stemming from the use of kinematically admissible virtual displacements yields the solution of the problem, $kR(-c_1 \sin\theta + c_2 \cos\theta) - mg\cos\theta = 0$, or $\tan\theta = (c_2 - mg/kR)/c_1$.

In summary, when using kinematically admissible virtual displacements, the principle of virtual work yields a reduced set of equilibrium equations from which the forces of constraints are eliminated. This often greatly simplifies and streamlines the solution process. In some cases, however, too few equations will be obtained, giving the impression that the problem cannot be solved. Arbitrary virtual displacements, *i.e.*, virtual displacements that violate the kinematic constraints must then be used to obtain the missing equations of equilibrium, which correspond to the projection of Newton's first law along the infeasible directions.

9.3.3 Use of infinitesimal displacements as virtual displacements

In the previous sections, three-dimensional virtual displacements are denoted $\underline{s} = s_1\bar{\imath}_1 + s_2\bar{\imath}_1 + s_3\bar{\imath}_3$, where s_1, s_2, and s_3 are arbitrary numbers. In view of the fundamental role they play in energy and variational principles, a special notation is commonly used to denote virtual displacements,

$$\underline{s} = \delta\underline{u}. \tag{9.13}$$

The symbol "δ" is placed in front of the displacement vector, $\underline{u}$, to indicate that it should be understood as a virtual displacement. Similarly, the virtual work done by a force undergoing a virtual displacement will be denoted δW to distinguish it from the real work done by the same force undergoing real displacements. The new notation changes nothing of the special nature of virtual displacements which are fictitious displacements that do not alter the applied forces.

In many applications of the principle of virtual work, it will also be convenient to use virtual displacements of infinitesimal magnitude. Because virtual displacements are of arbitrary magnitude, virtual displacements of infinitesimal magnitude qualify as valid virtual displacements. The infinitesimal magnitude of virtual displacements is a convenience that often simplifies algebraic developments, but is by no means a requirement.

Displacement dependent forces

A key simplification arising from the use of virtual displacements of infinitesimal magnitude is that displacement dependent forces automatically remain unaltered by their application, as illustrated in the following example.

Example 9.6. Equilibrium of a particle connected to an elastic spring
Consider a particle connected to an elastic spring, as illustrated in fig. 9.7. This is the same problem treated in example 9.2.

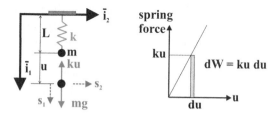

Fig. 9.7. Use of a differential displacement as a virtual displacement.

The principle of virtual work requires that

$$\delta W = (mg\bar{\imath}_1 - ku\bar{\imath}_1) \cdot (\delta u\bar{\imath}_1 + \delta v\bar{\imath}_2) = [mg - ku]\delta u = 0,$$

for all virtual displacements, δu, where the virtual displacements must leave the forces applied to the particle unchanged. Consider now a virtual displacement of infinitesimal magnitude, $\delta u = du$. The virtual work done by this virtual displacement of infinitesimal magnitude is still given by eq (9.10) as

$$\int_u^{u+du} -ku\,du = \left[-\frac{1}{2}ku^2\right]_u^{u+du} = -ku\,du - \frac{1}{2}k(du)^2 = -ku\,du,$$

where the last equality follows from neglecting the higher order differential quantity. The virtual work is now equal to the real work done by an infinitesimal displacement of magnitude $du = \delta u$. The right part of fig. 9.7 illustrates the differential work, dW, for a displacement of infinitesimal magnitude.

Rigid bodies

Next, the close relationship between infinitesimal displacements and virtual displacements of infinitesimal magnitude will be explored further in the context of rigid bodies. Consider two arbitrary points, **P** and **Q**, of a rigid body. When the rigid body undergoes arbitrary motions, the velocities of these two points are not independent and must satisfy the following well-known equation from rigid body dynamics, $\underline{v}_P = \underline{v}_Q + \underline{\omega} \times \underline{r}_{QP}$, where $\underline{v}_P$ and $\underline{v}_Q$ are the velocities of points **P** and **Q**, respectively, $\underline{\omega}$ is the angular velocity of the rigid body, and $\underline{r}_{QP}$ the position vector of point **P** with respect to **Q**. This relationship is now written as $d\underline{u}_P/dt = d\underline{u}_Q/dt + (d\underline{\psi}/dt) \times \underline{r}_{QP}$, where $d\underline{u}_P$ and $d\underline{u}_Q$ are the infinitesimal displacement vectors of points **P** and **Q**, respectively, and $d\underline{\psi}$ is the infinitesimal rotation vector for the rigid body. After multiplication by dt, the infinitesimal displacements are found to satisfy the following equation, $d\underline{u}_P = d\underline{u}_Q + d\underline{\psi} \times \underline{r}_{QP}$.

Because virtual displacements can be of infinitesimal magnitude, it is possible to write

$$\delta\underline{u}_P = \delta\underline{u}_Q + \delta\underline{\psi} \times \underline{r}_{QP}. \tag{9.14}$$

where $\delta\underline{u}_P$ and $\delta\underline{u}_Q$ are the virtual displacement vectors of arbitrary points **P** and **Q**, respectively, and $\delta\underline{\psi}$ is the virtual rotation vector for the rigid body. Equation (9.14) describes the field of kinematically admissible virtual displacements for a rigid body. Indeed, these virtual displacements satisfy the kinematic constraints for two points belonging to the same rigid body.

The discussion of the previous paragraph underlines the close relationship between infinitesimal quantities, denoted with symbol "d," and virtual quantities, denoted with symbol "δ." To obtain eq. (9.14) symbol "d" is replaced by "δ" in the last step of the reasoning. While this approach is correct, it must be emphasized that virtual displacements remain fictitious displacements, whereas infinitesimal displacements are real displacements. Furthermore, virtual displacements leave the forces unchanged, whereas no such requirement applies for real infinitesimal displacements. Finally, admissible virtual displacements are allowed to violate the kinematic constraints, whereas real displacement are not.

It is also important to note that eq. (9.14) shows that an infinitesimal rotation, $\delta\underline{\psi}$, is a vector quantity. This is not the case for finite rotations, as discussed in dynamics textbooks [3, 4].

Using virtual displacements of infinitesimal magnitude greatly simplifies the treatment of many problems. In the mathematical treatment of virtual quantities, a branch of mathematics called *calculus of variations,* virtual quantities are systematically assumed to be of infinitesimal magnitude [5, 6].

9.3.4 Principle of virtual work for a system of particles

Consider the system of N particles depicted in fig. 9.8; this problem is treated in section 9.2.1 using the classical Newtonian approach. Particle i is subjected to an external force, $\underline{F}_i$, and to $N-1$ interaction forces, $\underline{f}_{ij}$, $j = 1, 2, \ldots, N, j \neq i$. For particle i, the virtual work, denoted δW_i, done by all applied forces when subjected to a virtual displacement, $\delta\underline{u}_i$, is

$$\delta W_i = (\underline{F}_i + \sum_{j=1, j\neq i}^{N} \underline{f}_{ij}) \cdot \delta\underline{u}_i. \tag{9.15}$$

According to the principle of virtual work, this virtual work must vanish for all virtual displacements, $\delta\underline{u}_i$. The principle can be applied to each particle independently, leading to $\delta W_i = 0$, where δW_i is given by eq. (9.15), for $i = 1, 2, \ldots N$.

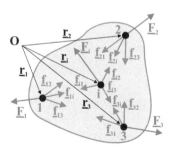

Fig. 9.8. A system of particles.

Because the virtual work must vanish for each particle independently, the sum of the virtual work for all particles must also vanish, leading to the following statement of the principle of virtual work for a system of N particles: a system of particle is in static equilibrium if and only if the virtual work,

$$\delta W = \sum_{i=1}^{N}\left\{\left[\underline{F}_i + \sum_{j=1, j\neq i}^{N} \underline{f}_{ij}\right]\cdot\delta\underline{u}_i\right\},$$
(9.16)

vanishes for all virtual displacements, $\delta\underline{u}_i$, $i = 1, 2, \ldots, N$. Because the N virtual displacements are all arbitrary and independent, the bracketed term in eq. (9.16) must vanish for $i = 1, 2, \ldots, N$, leading to equilibrium equations that are identical to those obtained from Newton's first law, eq. (9.2).

Because each of the N virtual displacement vectors involves three scalar components, the principle of virtual work yields $3N$ scalar equations for a system of N particles; all must be satisfied for the system to be in static equilibrium. The system is said to present $3N$ *degrees of freedom*. For a two-dimensional, or planar system, the number of scalar equations would reduce to $2N$, *i.e.*, $2N$ degrees of freedom.

The above developments have shown, once again, that the principle of virtual work is entirely equivalent to Newton's first law, and gives the necessary and sufficient conditions for the static equilibrium of the system. Equilibrium is the most fundamental requirement in structural analysis, and must always be satisfied. This means that Newton's first law, or the principle of virtual work since they are both equivalent, always applies. The system of particles considered above is very general; it could represent a rigid body, a flexible body deforming elastically or plastically, a fluid, or a planetary system. Yet, the same equilibrium requirements apply equally to all systems.

Internal and external virtual work

Eq. (9.16) also affords another important interpretation. The forces acting on the system are separated into two groups, the externally applied forces, $\underline{F}_i$, and the internal forces, $\underline{f}_{ij}$. The words "internal" and "external" should be understood with respect to the system of particles. *Internal forces* act and are reacted within the system, whereas *external forces* act on the system but are reacted outside the system. The virtual work done by the external and internal forces, denoted δW_E and δW_I, respectively, are defined as

$$\delta W_E = \sum_{i=1}^{N} \underline{F}_i \cdot \delta \underline{u}_i, \tag{9.17a}$$

$$\delta W_I = \sum_{i=1}^{N} \left[\sum_{j=1,j \neq i}^{N} \underline{f}_{ij} \right] \cdot \delta \underline{u}_i, \tag{9.17b}$$

respectively. With these definitions, eq. (9.16) is becomes

$$\delta W = \delta W_E + \delta W_I = 0, \tag{9.18}$$

for all arbitrary virtual displacements. This leads to the principle of virtual work for a system of particles.

Principle 4 (Principle of virtual work) *A system of particles is in static equilibrium if and only if the sum of the virtual work done by the internal and external forces vanishes for all arbitrary virtual displacements.*

Finally, it is interesting to note that because the virtual displacements are arbitrary, it is possible to choose them to be the actual displacements, and eq. (9.18) then implies

$$W = W_E + W_I = 0, \tag{9.19}$$

where W_E and W_I are the actual work done by the external and internal forces, respectively. Equation (9.19) states that *if a system of particles is in static equilibrium, the sum of the work done by the internal and external forces vanishes.*

Euler's laws

The $3N$ scalar equations implied by the vanishing of the virtual work expressed in eq. (9.16) are often cumbersome to use because they all involve the interaction forces between the particles of the system. To obtain equations that are more convenient to use, a special set of virtual displacements will be selected.

Inspired by eq. (9.14), the virtual displacement of particle i is written as

$$\delta \underline{u}_i = \delta \underline{u}_O + \delta \underline{\psi} \times \underline{r}_i, \tag{9.20}$$

where $\delta \underline{u}_O$ is the virtual displacement of an arbitrary point **O**, see fig. 9.8, $\delta \underline{\psi}$ the virtual rotation vector, and $\underline{r}_i$ the relative position vector of particle i with respect to point **O**. The virtual displacements of all particles are now expressed in terms of a virtual translation of the rigid body, $\delta \underline{u}_O$, and its virtual rotation, $\delta \underline{\psi}$, both chosen to be of infinitesimal magnitude. This corresponds to 6 independent virtual displacement components, far fewer than the original $3N$. The virtual work done by all forces acting on the system under these virtual displacement is

$$\delta W = \sum_{i=1}^{N} \left\{ \left[\underline{F}_i + \sum_{j=1,j \neq i}^{N} \underline{f}_{ij} \right] \cdot (\delta \underline{u}_O + \delta \underline{\psi} \times \underline{r}_i) \right\} = \left(\sum_{i=1}^{N} \underline{F}_i \right) \cdot \delta \underline{u}_O$$

$$+ \left(\sum_{i=1}^{N} \sum_{j=1,j \neq i}^{N} \underline{f}_{ij} \right) \cdot \delta \underline{u}_O + \sum_{i=1}^{N} \underline{F}_i \cdot (\delta \underline{\psi} \times \underline{r}_i) + \sum_{i=1}^{N} \sum_{j=1,j \neq i}^{N} \underline{f}_{ij} \cdot (\delta \underline{\psi} \times \underline{r}_i).$$

The last two terms of this expression can be simplified by using the triple scalar product identity: $\underline{a} \cdot (\underline{b} \times \underline{c}) = \underline{b} \cdot (\underline{c} \times \underline{a})$, which holds for any three vectors, $\underline{a}$, $\underline{b}$ and $\underline{c}$. The above equation now becomes

$$\delta W = \delta \underline{u}_O \cdot \left(\sum_{i=1}^{N} \underline{F}_i \right) + \delta \underline{u}_O \cdot \left(\sum_{i=1}^{N} \sum_{j=1, j \neq i}^{N} \underline{f}_{ij} \right)$$
$$+ \delta \underline{\psi} \cdot \left(\sum_{i=1}^{N} \underline{r}_i \times \underline{F}_i \right) + \delta \underline{\psi} \cdot \left(\sum_{i=1}^{N} \sum_{j=1, j \neq i}^{N} \underline{r}_i \times \underline{f}_{ij} \right).$$

In view of eqs. (9.3) and (9.5), the terms in the second and last sets of parenthesis now vanish, reducing the expression to

$$\delta W = \delta \underline{u}_O \cdot \left[\sum_{i=1}^{N} \underline{F}_i \right] + \delta \underline{\psi} \cdot \left[\sum_{i=1}^{N} \underline{r}_i \times \underline{F}_i \right].$$

Because the virtual work must vanish for all virtual displacements, $\delta \underline{u}_O$, and virtual rotations, $\delta \underline{\psi}$, the two bracketed terms must also vanish. Clearly, these two equations are identical to Euler's first and second laws obtained directly from Newtonian arguments, see eqs. (9.4) and (9.6).

These two vector equations are necessary but not sufficient conditions to guarantee static equilibrium. Indeed, static equilibrium requires a total of N vector equations to be satisfied; eqs. (9.4) and (9.6) are two linear combinations of those N equations. Only two vector equations are obtained from the principle of virtual work because the virtual displacement field, eq. (9.20), selected for the rigid body involves a single virtual displacement vector, $\delta \underline{u}_O$, and a single virtual rotation vector, $\delta \underline{\psi}$.

9.4 Principle of virtual work applied to mechanical systems

In the previous section, the principle of virtual work is discussed in a rather theoretical setting with applications to single particles and systems of particles. In the present section, the power and efficiency of the same principle will be demonstrated when applied to mechanical systems.

A rigid body is a particular case of the general system of N particles considered in the previous section. The configuration of a rigid body is determined by six parameters: the three components of the position vector of one of its points, and the three rotations that determine its orientation. Equivalently, a "rigid body motion," which is the only motion a body can undergo while remaining rigid, consists of a three-dimensional translation and a three-dimensional rotation. For a rigid body, the virtual displacement field given by eq. (9.20) is kinematically admissible, because it represents the superposition of a translation and a rotation, both in three dimensions.

This kinematically admissible virtual displacement field yields two vector equations, eqs. (9.4) and (9.6), or six scalar equations, which are just enough to determine the equilibrium configuration of the body. The internal forces, $\underline{f}_{ij}$, in the rigid body are the forces of constraint that maintain its shape unchanged and are entirely eliminated from the formulation, as expected, because a kinematically admissible displacement field is used in the application of the principle of virtual work.

When considering two-dimensional or planar mechanisms, the kinematically admissible displacement field defined by eq. (9.20) reduces to

$$\delta \underline{u}_i = \delta \underline{u}_O + \delta \phi \bar{\imath}_3 \times \underline{r}_i. \tag{9.21}$$

The three-dimensional virtual rotation vector, $\delta \underline{\psi}$, now simply becomes $\delta \phi \, \bar{\imath}_3$, where $\delta \phi$ is a virtual rotation of infinitesimal magnitude about axis $\bar{\imath}_3$. The planar mechanism is assumed to be entirely contained in plane $(\bar{\imath}_1, \bar{\imath}_2)$, and hence, rotations are only possible about axis $\bar{\imath}_3$.

Example 9.7. Equilibrium of a lever
Consider the simple lever subjected to two vertical end forces, F_a and F_b, acting at distances a and b, respectively, from the fulcrum, as shown in fig. 9.9.

First, this problem will be solved using the classical equations of statics, considering the free body diagram appearing in the right part of fig. 9.9. The equilibrium of forces in the horizontal and vertical directions yields $H = 0$ and $V = F_a + F_b$, respectively, whereas equilibrium of moments about point **A** leads to $aV \cos \phi = (a + b)F_b \cos \phi$. The Newtonian approach requires the explicit consideration of the horizontal and vertical reaction forces, H and V, respectively, at the lever's fulcrum. Solution of these equations leads to the familiar equilibrium conditions for a lever, $aF_a = bF_b$, $H = 0$, and $V = F_a + F_b$.

Fig. 9.9. Simple lever acted upon by two vertical end forces.

It is possible to eliminate the reaction forces from the formulation by writing a single moment equilibrium equation about the lever's fulcrum: $aF_a \cos \phi = bF_b \cos \phi$. Because the lines of action of the reaction forces pass through the fulcrum, they are automatically eliminated from the moment equilibrium equation.

Next, this single degree of freedom problem will be solved using the principle of virtual work. Because the lever is fixed at point **O**, a kinematically admissible virtual displacement field simply becomes $\delta \underline{u}_i = \delta \phi \bar{\imath}_3 \times \underline{r}_i$. The translation term, $\delta \underline{u}_O$, appearing in eq. (9.21), is omitted because a translation of point **O** violates the kinematic constraint at this point. The virtual displacement of point **A** now becomes $\delta \underline{u}_A = \delta \phi \bar{\imath}_3 \times \underline{r}_{OA}$, where $\underline{r}_{OA}$ is the position vector of point **A** relative to point **O**. Simple vector algebra then yields $\delta \underline{u}_A = a(\sin \phi \bar{\imath}_1 - \cos \phi \bar{\imath}_2)\delta \phi$. A similar reasoning reveals that $\delta \underline{u}_B = b(- \sin \phi \bar{\imath}_1 + \cos \phi \bar{\imath}_2)\delta \phi$. The virtual work is now

$$\delta W_E = (-F_a \bar{\imath}_2) \cdot \delta \underline{u}_A + (-F_b \bar{\imath}_2) \cdot \delta \underline{u}_B = \delta \phi \left[a F_a \cos \phi - b F_b \cos \phi \right].$$

The reaction forces are eliminated from the formulation because the virtual displacement is kinematically admissible, *i.e.*, it vanishes at point **O**, resulting in the vanishing of the virtual work done by the reaction forces at that point. Because the virtual displacement field is also compatible with the kinematic conditions required for the body to remain rigid (the virtual displacement field consists of a single rotation), all the internal forces that enforce the rigidity of the body are also eliminated.

Because the virtual work must vanish for all arbitrary virtual rotations, $\delta \phi$, the bracketed term in the above equation must vanish, leading to $(a F_a - b F_b) \cos \phi = 0$. The two solutions are $a F_a = b F_b$, the usual lever equilibrium equation, and $\cos \phi = 0$, which corresponds to the lever being in a vertical position. In this latter case, the lever is in equilibrium for any set of applied vertical forces.

As discussed earlier, it is also possible to use a virtual displacement field that violates the kinematic conditions, such as that given by eq. (9.21). The virtual displacement, $\delta \underline{u}_O = \delta u_1 \bar{\imath}_1 + \delta u_2 \bar{\imath}_2$, is a virtual displacement of point **O**, the lever's fulcrum. The virtual displacements of points **A** and **B** now become $\delta \underline{u}_A = \delta \underline{u}_O + a(\sin \phi \bar{\imath}_1 - \cos \phi \bar{\imath}_2)\delta \phi$, and $\delta \underline{u}_B = \delta \underline{u}_O + b(-\sin \phi \bar{\imath}_1 + \cos \phi \bar{\imath}_2)\delta \phi$, respectively. The virtual work done by the externally applied forces is now

$$\begin{aligned} \delta W_E &= (-F_a \bar{\imath}_2) \cdot \delta \underline{u}_A + (-F_b \bar{\imath}_2) \cdot \delta \underline{u}_B + (H \bar{\imath}_1 + V \bar{\imath}_2) \cdot \delta \underline{u}_O \\ &= \delta u_1 [H] + \delta u_2 [V - F_a - F_b] + \delta \phi [a F_a \cos \phi - b F_b \cos \phi]. \end{aligned}$$

Because the virtual work done by the reaction forces at the fulcrum does not vanish, is must be included in the formulation. Since the virtual work must vanish for all virtual displacements, δu_1 and δu_2, and rotation, $\delta \phi$, the three bracketed terms must vanish, leading to three equilibrium equations identical to those obtained using the Newtonian approach.

This underlines, once again, the complete equivalence of the principle of virtual work and Newton's first law. The use of a kinematically admissible virtual displacement field automatically eliminates the reactions forces when using the principle of virtual work. Although it is sometimes possible to achieve this elimination by a judicious choice of the point about which moment equilibrium equations are written in Newton's approach, the systematic approach stemming from the use of the principle of virtual work is more efficient and convenient.

Example 9.8. *Block and tackle system*

Consider the familiar two-pulley block and tackle shown in fig. 9.10. Determine the rope force, F, required to lift a weight, P. The system possesses a single degree of freedom defined by the rotation angle, ϕ, of the upper pulley.

Consider a virtual rotation, $\delta \phi$, of the pulley. The resulting virtual motion of the point of application of force F is $\delta b = R \delta \phi$, where R is the pulley's radius. The resulting motion of the lower block is $\delta a = -R \delta \phi / 2$. Because the virtual rotation is kinematically admissible, the only forces that perform work are the externally applied forces, F and P; the reaction forces need not be considered. The principle of virtual work is now simply

$$\delta W = F\delta b + P\delta a = FR\delta\phi - PR\delta\phi/2 = R[F - P/2]\,\delta\phi = 0.$$

Because the virtual work must vanish for all $\delta\phi$, the bracketed term must vanish, yielding the equilibrium equation of the system, $F = P/2$.

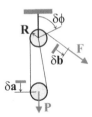

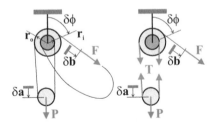

Fig. 9.10. Simple two-pulley block and tackle system.

Fig. 9.11. Differential pulley or "chain hoist".

Example 9.9. Differential pulley system

A differential pulley system is depicted in fig. 9.11 and is the basis for the common shop chain hoist. In this device, the two upper pulleys, of radii r_i and r_o, are constrained to rotate together about a common axis. The inextensible chain is not allowed to slip around the upper pulleys. The system features a single degree of freedom, defined by the rotation angle, ϕ, of the upper pulleys assembly.

A virtual rotation, $\delta\phi$, of the upper pulleys causes a virtual displacement of the point of application of force F of $\delta b = r_o\delta\phi$. The virtual motion of the lower pulley is $\delta a = -(r_o\delta\phi - r_i\delta\phi)/2$. This can be explained by noting that for a virtual rotation, $\delta\phi$, of the upper pulleys, the length of chain below these pulleys is decreased by $r_o\delta\phi$ (the length winding on the pulley of radius r_o), but at the same time, is increased by $r_i\delta\phi$ (the length unwinding from the pulley of radius r_i), and the lower pulley rotates to distribute the net shortening between the two segments of chain.

The principle of virtual work now requires

$$\delta W_E = F\delta b + P\delta a = Fr_o\delta\phi - P(r_o - r_i)\delta\phi/2 = [Fr_o - P(r_o - r_i)/2]\,\delta\phi = 0,$$

for all virtual rotations, $\delta\phi$, and hence, the bracketed term must vanish, revealing the equilibrium condition of the system $F = (1 - r_i/r_0)P/2$. The mechanical advantage of the differential pulley increases as radii r_i and r_0 approach equal values, at which point force F then vanishes. The load cannot be raised, however, because the amounts of chain winding around the pulley of radius r_0 and unwinding from the pulley of radius r_i become equal.

A kinematically admissible virtual rotation, $\delta\phi$, is used, and because the chain is inextensible, the virtual rotation of the upper pulleys determines the motion of the lower pulley. This is why the system presents a single degree of freedom.

It is also possible to use arbitrary virtual displacements that violate the kinematic constraint imposed by the inextensibility of the chain. Let the virtual displacement of the lower pulley, δa, be independent of the virtual rotation of the upper pulley, $\delta\phi$. In this case, the tension in the chain, the force of constraint that enforces its inextensibility, will perform virtual work. Referring to the right part of fig. 9.11, the virtual work done by the forces acting on the system is

$$\delta W_E = (Fr_o\delta\phi - Tr_o\delta\phi + Tr_i\delta\phi) + (P\delta a - 2T\delta a)$$
$$= [Fr_o - Tr_o + Tr_i]\,\delta\phi + [P - 2T]\,\delta a = 0.$$

The terms inside the two sets of parenthesis represent the virtual work done by the forces acting on the upper and lower pulleys, respectively. Because the virtual rotation, $\delta\phi$, and virtual displacement, δa, are both arbitrary and independent, the two bracketed term must vanish, yielding the equations of equilibrium of the system, $Fr_o = (r_o - r_i)T$, and $P = 2T$. Eliminating the tension in the chain leads to $F = (1 - r_i/r_0)P/2$, as before. Additionally, the tension in the chain, $T = P/2$, is also determined.

Example 9.10. The crank-slider mechanism

Consider the crank-slider mechanism depicted in fig. 9.12. The crank of length R is actuated by a torque, Q, and the link of length L transforms the rotary motion of the crank into a linear motion of the slider. Force F is applied to the slider. The crank angle is denoted ϕ and is measured positive in the counterclockwise direction. Determine the relationship between the torque, Q, and force, F.

For kinematically admissible virtual displacements, the principle of virtual work states that $\delta W_E = -Q\delta\phi - F\delta x = 0$, where x is the distance from the crank axis to the slider. The virtual displacement, δx, and virtual rotation, $\delta\phi$, are *arbitrary but not independent*; consequently, nothing can be concluded from the above statement.

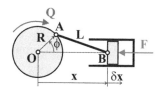

Fig. 9.12. The crank-slider mechanism.

The geometry of the problem links the two variables, x and ϕ, as shown in fig. 9.12. Projections of segments **OA** and **AB** onto the horizontal yield $x = R\cos\phi + \sqrt{L^2 - R^2\sin^2\phi}$. Taking a differential of this equation reveals the relationship between infinitesimal increments in variables x and ϕ as $dx = -R\sin\phi\,d\phi - R^2\sin\phi\cos\phi\,d\phi/\sqrt{L^2 - R^2\sin^2\phi}$. Since virtual displacements are of arbitrary magnitude, it is possible the select virtual displacements of infinitesimal magnitude, *i.e.*, $\delta x = dx$ and $\delta\phi = d\phi$. Kinematically admissible infinitesimal virtual displacements must then satisfy the following relationship

$$\delta x = -\left(1 + \frac{R\cos\phi}{\sqrt{L^2 - R^2\sin^2\phi}}\right) R\sin\phi\,\delta\phi.$$

The principle of virtual work now becomes

$$\delta W = -Q\delta\phi - F\delta x = -\left[Q - \left(1 + \frac{R\cos\phi}{\sqrt{L^2 - R^2\sin^2\phi}}\right)FR\sin\phi\right]\delta\phi = 0.$$

Because the virtual rotation, $\delta\phi$, is arbitrary, the bracketed term must vanish, yielding the desired relationship between the torque and force as

$$\frac{Q}{FR} = \left(1 + \frac{\cos\phi}{\sqrt{L^2/R^2 - \sin^2\phi}}\right)\sin\phi.$$

This expression yields the torque developed about the crank axis at a specific angular position, ϕ, of the crank in response to a force, F, applied to the slider.

In the Newtonian formulation, all reaction forces acting on the system must be considered. They include the horizontal and vertical components of the reaction forces at points **O**, **A**, and **B**, and the vertical reaction of the ground on the slider. The use of kinematically admissible virtual displacements with the principle of virtual work automatically eliminates all these forces from the formulation. In this example, only the torque applied to the crank and the force acting on the slider appear in the formulation.

9.4.1 Generalized coordinates and forces

The virtual work done by a force is defined as the scalar product of the force by a virtual displacement. When the principle of virtual work is introduced for a single particle in section 9.3.1, the virtual work is computed according to the definition: $\delta W = \underline{F} \cdot \delta\underline{u} = F_1\delta u_1 + F_2\delta u_2 + F_3\delta u_3$, where the force and virtual displacement vectors are represented by their components in a common orthonormal basis, $\mathcal{I} = (\bar{\imath}_1, \bar{\imath}_2, \bar{\imath}_3)$, as $\underline{F} = F_1\bar{\imath}_1 + F_2\bar{\imath}_2 + F_3\bar{\imath}_3$ and $\delta\underline{u} = \delta u_1\bar{\imath}_1 + \delta u_2\bar{\imath}_2 + \delta u_3\bar{\imath}_3$, respectively.

In many cases, however, it is not convenient to work with Cartesian coordinates. Consider, for instance, the crank-slider mechanism shown in fig. 9.12 and treated in example 9.10: the motion of the system is naturally expressed in terms of the crank angle, ϕ, and piston translation, x. Similarly, when dealing with pulleys in examples 9.8 and 9.9, the most natural way to describe the configuration of the system is in terms of the rotation angles of the pulleys.

In general, the configuration of a system will be represented in terms of N variables, called *generalized coordinates* and denoted $q_1, q_2, \ldots q_N$. These variables could be angles, relative motions, Cartesian coordinates, or a mixture thereof, hence the term "generalized coordinates." The displacement vector of any point of the system will be a function of these generalized coordinates, $\underline{u} = \underline{u}(q_1, q_2, \ldots q_N)$. Virtual displacements can now be evaluated using the chain rule for derivatives

$$\delta\underline{u} = \frac{\partial\underline{u}}{\partial q_1}\delta q_1 + \frac{\partial\underline{u}}{\partial q_2}\delta q_2 + \ldots + \frac{\partial\underline{u}}{\partial q_N}\delta q_N.$$

The virtual work done by a force, $\underline{F}$, undergoing this virtual displacement is found as

$$\delta W = \underline{F} \cdot \delta \underline{u} = \left(\underline{F} \cdot \frac{\partial \underline{u}}{\partial q_1} \right) \delta q_1 + \left(\underline{F} \cdot \frac{\partial \underline{u}}{\partial q_2} \right) \delta q_2 + \ldots + \left(\underline{F} \cdot \frac{\partial \underline{u}}{\partial q_N} \right) \delta q_N.$$

It is now convenient to define *generalized forces* as follows

$$Q_i = \underline{F} \cdot \frac{\partial \underline{u}}{\partial q_i}, \quad i = 1, 2, \ldots, N, \tag{9.22}$$

and the expression for the virtual work simply becomes

$$\delta W = Q_1 \delta q_1 + Q_2 \delta q_2 + \ldots + Q_N \delta q_N = \sum_{i=1}^{N} Q_i \delta q_i. \tag{9.23}$$

This result helps explain the term "generalized forces" used to denote the quantities defined by eq. (9.22); the virtual work is simply the product of the generalized forces and the corresponding generalized virtual displacements.

The above development is presented for a generic force, $\underline{F}$, which can be an externally applied load or an internal force. To distinguish between the two cases, the following notation is used

$$\delta W_I = \sum_{i=1}^{N} Q_i^I \delta q_i, \tag{9.24a}$$

$$\delta W_E = \sum_{i=1}^{N} Q_i^E \delta q_i, \tag{9.24b}$$

where Q_i^I and Q_i^E are the generalized forces associated with the internal forces and externally applied loads, respectively.

The principle of virtual work, expressed by eq. (9.18), can now be reformulated as

$$\delta W_I + \delta W_E = \sum_{i=1}^{N} Q_i^I \delta q_i + \sum_{i=1}^{N} Q_i^E \delta q_i = \sum_{i=1}^{N} \left[Q_i^I + Q_i^E \right] \delta q_i = 0,$$

for all virtual generalized displacements, δq_i. Clearly, because the virtual generalized displacements, δq_i, are arbitrary, each of the N bracketed terms under the summation sign must vanish, leading to

$$Q_i^I + Q_i^E = 0, \quad i = 1, 2, \ldots, N. \tag{9.25}$$

This equation represents yet another statement of the principle of virtual work.

As discussed in section 9.3.2, the principle of virtual work can be used with either arbitrary or kinematically admissible virtual displacements. Similarly, the present statement of the principle can be used with either arbitrary or kinematically admissible virtual generalized coordinates. When using arbitrary virtual generalized coordinates, the virtual work done by the reaction forces must be included in the evaluation of the virtual work done by the external forces; this implies that the generalized forces associated with the reaction forces must be included in Q_i^E. If the virtual generalized coordinates are kinematically admissible, the reaction forces are eliminated from the formulation.

To illustrate the concepts presented above, consider the pendulum with a torsional spring depicted in fig. 9.13.

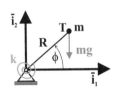

A rigid arm of length R connects the mass, m, to a pinned support point where a torsional spring of stiffness constant k acts between ground and the rod. The torsional spring is un-stretched when the arm is horizontal. The mass is subjected to gravity loading. The configuration of the system is conveniently represented by the angular position, ϕ, of the arm and is selected to be the single generalized coordinate for this one degree of freedom problem.

Fig. 9.13. Pendulum with torsional spring.

Consider first the virtual work done by the gravity load, $\delta W_E = -mg\bar{\imath}_2 \cdot \delta\underline{u}_T$, where $\delta\underline{u}_T$ is the virtual displacement at point **T**. Since $\underline{u}_T = R(\cos\phi\,\bar{\imath}_1 + \sin\phi\,\bar{\imath}_2)$, an infinitesimal virtual displacement of the same quantity is $\delta\underline{u}_T = R(-\sin\phi\,\bar{\imath}_1 + \cos\phi\,\bar{\imath}_2)\delta\phi$. It now follows that $\delta W_E = -mgR\cos\phi\,\delta\phi$, and by defining the generalized force as $Q_\phi^E = -mgR\cos\phi$, the virtual work becomes $\delta W_E = Q_\phi^E\delta\phi$. The same result can be obtained in a more expeditious manner by using eq. (9.22) to find $Q_\phi^E = -mg\bar{\imath}_2 \cdot \partial\underline{u}_T/\partial\phi = -mg\bar{\imath}_2 \cdot R(-\sin\phi\,\bar{\imath}_1 + \cos\phi\,\bar{\imath}_2) = -mgR\cos\phi$.

An even simpler interpretation is as follows. Because the virtual displacement is a rotation, $\delta\phi$, it must be multiplied by a moment to yield a virtual work; hence, the generalized force is simply the moment of the gravity load, $-mgR\cos\phi$.

For this problem, the virtual work done by the internal forces reduces to the virtual work done by the restoring moment of the elastic spring, $\delta W_I = -k\phi\,\delta\phi = Q_\phi^I\delta\phi$, where $Q_\phi^I = -k\phi$ is the generalized internal force of the system. The generalized force is, in this case, a moment, and hence, the expression "generalized force" must be interpreted carefully.

The principle of virtual work, eq. (9.25), yields the equilibrium equation for the system as $Q_\phi^I + Q_\phi^E = -mgR\cos\phi - k\phi = 0$. This is a transcendental equation, but if the angular displacement of the pendulum remains small, $\cos\phi \approx 1$, and the equilibrium configuration becomes $\phi = -mgR/k$.

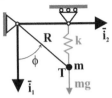

Consider next the modified system shown in fig. 9.14 where a rigid arm of length R connects mass m to a pinned support at ground. A linear spring of stiffness constant k supports the mass; this spring remains vertical because its support point is free to move horizontally on rollers. The spring is un-stretched when the arm is horizontal.

Fig. 9.14. Rotating mass with vertical spring.

As in the previous example, the virtual work done by the gravity load is easily found as $\delta W_E = mg\bar{\imath}_1 \cdot \delta\underline{u}_T$, where $\delta\underline{u}_T$ is the virtual displacement at point **T**. Since $\underline{u}_T = R(\cos\phi\,\bar{\imath}_1 + \sin\phi\,\bar{\imath}_2)$, an infinitesimal virtual displacement of the same quantity is $\delta\underline{u}_T = $

$R(-\sin\phi\ \bar{\imath}_1 + \cos\phi\ \bar{\imath}_2)\delta\phi$. The virtual work done by the gravity load now becomes $\delta W_E = -mgR\sin\phi\ \delta\phi$, and the corresponding generalized force is $Q_\phi^E = -mgR\sin\phi$. Next, the virtual work done by the restoring force in the spring is $\delta W_I = -kR\cos\phi\bar{\imath}_1 \cdot \delta\underline{u}_T$, which yields $Q_\phi^I = kR^2\cos\phi\sin\phi$.

The principle of virtual work as expressed in eq. (9.25) now implies

$$Q_\phi^I + Q_\phi^E = kR^2\cos\phi\sin\phi - mgR\sin\phi = R\sin\phi(kR\cos\phi - mg) = 0.$$

Two solutions are possible. First, $\sin\phi = 0$: this leads to $\phi = 0$ or π, *i.e.*, the arm is in the down or up vertical position, respectively. The second solution is $\cos\phi = mg/(kR)$. For $mg/(kR) > 1$, however, this solution no longer exists, leaving the first solution as the only valid solution of the problem.

Example 9.11. The crank-slider mechanism

Consider the crank-slider mechanism depicted in fig. 9.15. The crank of length R is actuated by a torque Q, and the link of length L transforms the rotary motion of the crank into a linear motion of the slider. A spring of stiffness constant k connects the slider to the ground and is un-stretched when $x = 0$. The configuration of the system is entirely determined by crank angle, ϕ, (measured positive in the counterclockwise direction) which is selected as the generalized coordinate for this problem.

The virtual work done by the externally applied torque, Q, is $\delta W_E = -Q\delta\phi$, and hence, the corresponding generalized force is $Q_\phi^E = -Q$. Similarly, the virtual work done by the internal force in the spring is $\delta W_I = -kx\ \delta x = -kx\ (\partial x/\partial\phi)\ \delta\phi = Q_\phi^I\ \delta\phi$. The position of the slider, x, can be expressed in terms of the generalized coordinate, ϕ. Indeed, projecting segments **OA** and **AB** onto the horizontal yields $x = R\cos\phi + \sqrt{L^2 - R^2\sin^2\phi}$. The generalized force associated with the force in the spring becomes

$$Q_\phi^I = -kx\frac{\partial x}{\partial\phi} = kx\left(1 + \frac{R\cos\phi}{\sqrt{L^2 - R^2\sin^2\phi}}\right)R\sin\phi.$$

The principle of virtual work expressed as eq. (9.25) implies $Q_\phi^I + Q_\phi^E = 0$ leading to the following expression for the applied torque

$$Q = \left(1 + \frac{R\cos\phi}{\sqrt{L^2 - R^2\sin^2\phi}}\right)kxR\sin\phi.$$

Example 9.12. Elastically supported aircraft

For the purpose of dynamic testing, an aircraft is suspended from a hangar's roof by means of three springs of stiffness constants k_L, k_R, and k_T, attached to the aircraft's left wing at point **L**, right wing at point **R**, and tail at point **T**, respectively, as depicted in fig. 9.16. The aircraft's total mass is M and the center of mass is located at point **C**. Determine the equilibrium position of the aircraft under gravity loads.

For simplicity, the aircraft is assumed to be rigid and all spring displacements under load are assumed to remain small (*i.e.*, the airplane remains nearly horizontal).

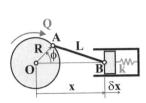

Fig. 9.15. Crank-slider mechanism with a spring.

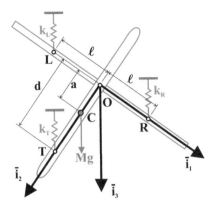

Fig. 9.16. Plan view of the elastically supported aircraft.

The generalized coordinates of the problem will be selected as follows: u is the displacement of point $\mathbf{O}$ along unit vector $\bar{\imath}_3$, and ϕ_1 and ϕ_2 are the rotations about axes $\bar{\imath}_1$ and $\bar{\imath}_2$, respectively. The displacements of the three suspension points, $\mathbf{L}$, $\mathbf{R}$, and $\mathbf{T}$ are now easily expressed as $x_L = u + \ell\phi_2$, $x_R = u - \ell\phi_2$, and $x_T = u + d\phi_1$, respectively. The displacement of the aircraft's center of mass is $x_C = u + a\phi_1$.

The virtual work done by the gravity load is $\delta W_E = Mg\delta x_C = Mg(\delta u + a\delta\phi_1)$, and hence, the corresponding generalized forces are $Q_u^E = Mg$, $Q_{\phi_1}^E = Mga$, and $Q_{\phi_2}^E = 0$. Similarly, the virtual work done by the internal forces in the three springs is $\delta W_I = -k_L x_L\, \delta x_L - k_R x_R\, \delta x_R - k_T x_T\, \delta x_T$, and the corresponding generalized forces become $Q_u^I = k_L x_L - k_R x_R - k_T x_T$, $Q_{\phi_1}^I = -dk_T x_T$, and $Q_{\phi_2}^I = -\ell k_L x_L + \ell k_R x_R$.

The principle of virtual work, eq. (9.25), then yields the three equilibrium equations of the problem as $Q_u^I + Q_u^E = 0$, $Q_{\phi_1}^I + Q_{\phi_1}^E = 0$, and $Q_{\phi_2}^I + Q_{\phi_2}^E = 0$, leading to

$$k_L x_L + k_R x_R + k_T x_T = Mg, \quad dk_T x_T = Mga, \quad \ell k_L x_L - \ell k_R x_R = 0.$$

The solution can be completed using the displacement method (see section 4.3.2). When the displacements of the suspension points are expressed in terms of the generalized coordinates, the three equilibrium equations can be recast as a system of three linear equations that can easily be solved for the generalized coordinates of the system

$$\begin{bmatrix} k_L + k_R + k_T & dk_T & \ell(k_L - k_R) \\ dk_T & d^2 k_T & 0 \\ \ell(k_L - k_R) & 0 & \ell^2(k_L + k_R) \end{bmatrix} \begin{Bmatrix} u \\ \phi_1 \\ \phi_2 \end{Bmatrix} = \begin{Bmatrix} Mg \\ Mga \\ 0 \end{Bmatrix}. \tag{9.26}$$

9.4.2 Problems

Problem 9.1. Rotating disk with spring restraint
A mechanism consists of the rotating circular disk pinned at its center as shown in fig. 9.17. A cable is wrapped around the outer edge and a force, P, is applied tangentially. The rotation is resisted by a spring of stiffness constant k attached to a pin on the disk's outer radius and fixed horizontally to a support that can move vertically, leaving the spring horizontal at all times. Use the principle of virtual work to determine the force, P, required to keep the disk in equilibrium as a function of disk angular position, θ.

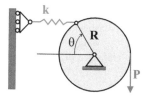
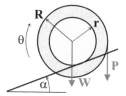

Fig. 9.17. Rotating disk with spring restraint.   **Fig. 9.18.** Double-radius wheel on incline.

Problem 9.2. Double-radius wheel on incline
The double radius wheel of weight W shown in fig. 9.18 is of inner radius r and outer radius R. A rope wrapped about the outer radius of the wheel applies a tangential force P. Determine the inclination angle, α, required to maintain the system in static equilibrium.

Problem 9.3. Lever with sliding pivots
Determine the magnitude of force P applied at point $\mathbf{N}$ required to equilibrate a downward force, F, applied at point $\mathbf{M}$, as shown in fig. 9.19. Rod $\mathbf{MN}$ is of length $a + b$ and is pinned to sleeves which slide along frictionless rods $\mathbf{AO}$ and $\mathbf{BO}$.

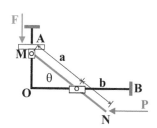

Fig. 9.19. Lever with sliding pivots.

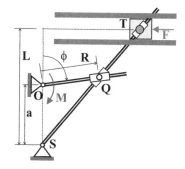

Fig. 9.20. Quick-release mechanism.

Problem 9.4. Quick-release mechanism
In the "quick release mechanism" shown in fig. 9.20, frictionless rod $\mathbf{ST}$ is a pivoted at point $\mathbf{S}$ and connected to a sliding hinge and piston at point $\mathbf{T}$. At point $\mathbf{Q}$, a sliding coupler connects

rods **ST** and **OQ**. Show that the moment, M, necessary to react the applied horizontal force, F, is $M/FL = (R/a)[(R/a) + \cos\phi]/[1 + (R/a)\cos\phi]^2$. Hint: use virtual displacements of infinitesimal magnitude.

Problem 9.5. Lever mechanism
A bar of length $3b$ is pinned at its lower end and supports a normal load, P, applied at its tip, as shown in fig. 9.21. A second bar, of length b, is pinned to the first bar as shown and to a slider that is constrained to move vertically on a frictionless rod. A weight, W, is supported at the slider. Use the principle of virtual work to determine the force, P, required to keep the weight, W, in equilibrium as a function of angle θ.

Problem 9.6. Spring-mass problem with nonlinear geometry
A spring of stiffness constant, k, and un-stretched length, L, is fastened to a support at point **A** and is connected to a weight, W, as shown in fig. 9.22. The weight slides on a friction-less vertical rod and the spring is un-stretched when horizontal. Determine the equilibrium configuration of the system, *i.e.*, position, u, of the weight.

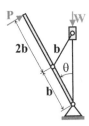

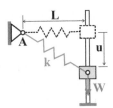

Fig. 9.21. Lever mechanism.

Fig. 9.22. Mechanism with nonlinear geom-etry.

Problem 9.7. Spring-mass system with nonlinear spring
Consider the spring-mass system depicted in fig. 9.4. Determine the equilibrium position of weight $W = mg$ supported by a vertical spring having a nonlinear stiffness $k = k_0[1 + a(x/L)^2]$, where k_0 is the initial stiffness, *i.e.*, the stiffness for small x, and a is a "hardening coefficient." Solve the problem assuming that $a = 0.5$ and $k_0 = W/8L$.

Problem 9.8. Lever with sliding pivots and spring
Bar **ABC** is of length $b + a$ and constrained to move vertically at point **A** and horizontally at point **B**, while a horizontal force, P, is applied at point **C**, as shown in fig. 9.23. A vertical spring is connected to bar **ABC** at point **A** and is un-stretched when angle $\theta = 0$. Use the principle of virtual work to determine the equilibrium relation(s) of the system.

Problem 9.9. Differential pulley with applied moment
A solid cylinder has two radii, a and b, and its axis is pinned but free to rotate as shown in fig. 9.24. A lever arm of length R is attached to the cylinder and a force, P, acts in the direction normal to the arm. A cable is attached at one end to the smaller radius, a, and at the other to the larger radius, b, and it supports a pulley which carries a vertical weight, W, as shown. Use the principle of virtual work to compute force, P, required to support (or lift) weight, W. Hint: this problem is different from the shop chain hoist discussed in example 9.9: here, as the upper cylinder rotates, cable is taken up on one radius and let out on the other.

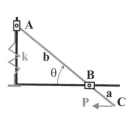

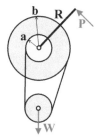

Fig. 9.23. Lever with sliding pivots and spring.

Fig. 9.24. Differential pulley with an applied moment.

Problem 9.10. Linked bars with lateral springs and forces

A mechanical system consists of two articulated bars pinned together at point **B** and to the ground at point **C**, as shown in fig. 9.25. Two springs of stiffness constants k_1 and k_2 support the bars at their mid-span and two forces, P and Q, are applied at points **B** and **A**, respectively. Let q_A and q_B, the downward deflection of points **A** and **B**, be the two generalized coordinates of the system. Use the principle of virtual work to determine the two equilibrium equations of the system. Assume small displacements: $|q_A| \ll L$ and $|q_B| \ll L$.

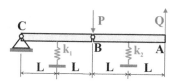

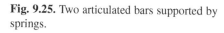

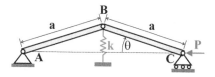

Fig. 9.25. Two articulated bars supported by springs.

Fig. 9.26. Axially loaded articulated bars with lateral spring restraint.

Problem 9.11. Axially loaded pinned bars with lateral spring restraint

Two rigid bars, **AB** and **BC**, are pinned together at point **B**, as shown in fig. 9.26. The end of the first bar is pinned to the ground at point **A**, whereas the end of the other bar is constrained to slide horizontally at point **C** under the action of load P. A lateral spring of stiffness constant k is attached at point **B**. Angle θ between bar **BC** and the horizontal is the generalized coordinate used to define the system's configuration. Use the principle of virtual work to develop an expression for $P = P(\theta)$. From your analysis, identify the buckling load of the problem.

Problem 9.12. Screw jack scissor lift

Consider the scissor lift (similar to an auto jack) shown in fig. 9.27. The configuration of the system is represented by a single generalized coordinate, θ, the angle between the jack legs and the horizontal. Determine the crank moment, M, required to lift a weight, W. The moment will depend on the configuration of the jack, *i.e.*, on angle θ. The threaded screw has a pitch of N threads per unit length. All bars of the jack are articulated.

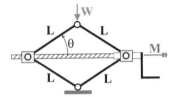

Fig. 9.27. Screw jack type of scissor lift.

9.5 Principle of virtual work applied to truss structures

In the previous section, the principle of virtual work is applied to mechanical systems consisting of rigid bodies and concentrated springs. Attention now turns to applications of the same principle to simple truss structures.

9.5.1 Truss structures

Trusses are a class of structures consisting of slender bars connected together at their ends by what are called *pinned joints*, which transmit forces but no moments. Consequently, the slender bars carry axial forces, in tension or compression, but no bending moments or transverse shear forces.[1] The simplest truss consists of only two members connected at a single joint to which a load may be applied. The resulting structure is isostatic; the addition of a third bar leads to a simple hyperstatic truss, called a "three-bar truss," which is analyzed in section 4.4.

Each bar of a truss acts like a simple rectilinear spring of stiffness constant $k = EA/L$, where L is the bar's length, A its cross-sectional area, and E the elastic modulus of the material making up its cross-section. In addition, if all bars lie in a plane, the truss is referred to as a *planar truss*; otherwise, it is called a three-dimensional truss or a *spatial truss*. For both configurations, analysis methods are identical, although the treatment of three-dimensional configurations is usually more cumbersome. This section focuses on planar trusses, but the methods developed here are equally applicable to spatial trusses.

Figure 9.28 illustrates a simple planar truss. A crude sketch of the truss shows the prismatic bars pinned together at their ends. In this illustration, the member widths have been exaggerated: in actual trusses, bars are quite slender. Actual pin joints were commonly employed in early planar truss designs, especially in trusses for railway bridges designed in the 19th century. With the development of higher strength alloys and use of thin-walled tubular sections, bar slenderness, the ratio of their length to diameter, can approach 100. For such truss members, it is practical to design rigid joints using welding or bolting, and although such joints introduce bending moments into the bars, the primary stresses are still almost entirely due to the axial forces, except in the immediate vicinity of the connections.

---

[1] If the bars are rigidly connected together at the joints so that no relative rotation is possible, bending moments will develop in the bars and bending deflections must be considered. Such structures are generally called *frames* and are more complicated to analyze.

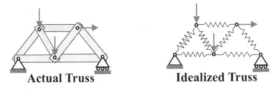

Fig. 9.28. Planar truss and its idealization as an assembly of rectilinear springs.

In simple truss design, each bar is fabricated from a homogeneous material and has a constant cross-section. Equation (4.3) then implies that each member can be represented as a rectilinear spring of stiffness constant $k = EA/L$. The complete truss can then be viewed as an assembly of springs connected to pinned joints, as illustrated in fig. 9.28.

The spring stiffnesses constants are quite large for a typical bars. For example, the axial stiffness of a bar of length $L = 0.75$ m, sectional area $A = 75$ mm^2 and modulus $E = 210$ GPa, is $k = 21$ MN/m. When subjected to a 5 kN force, the bar's elongation is $e = F/k = 0.24$ mm. In most practical designs, the maximum deflection of any joint of the truss is very small compared to the bar lengths. If self-weight is a significant component of the overall loading, as is the case for trusses used in civil engineering applications, the gravity force associated with each bar is lumped in two equal forces applied at the bar's two end joints.

Elongation-displacement equations

Consider the generic bar **AB** shown in fig. 9.29. Point **A** is assumed to remain pinned, while point **B** undergoes a displacement $\underline{\Delta} = \Delta_1 \bar{\imath}_1 + \Delta_2 \bar{\imath}_2$. The bar's original length is L, and its elongation e. Elementary geometry then yields $(L+e)^2 = (L_1+\Delta_1)^2+(L_2+\Delta_2)^2$, where L_1 and L_2 are the projections of the original length along axes $\bar{\imath}_1$ and $\bar{\imath}_2$, respectively.

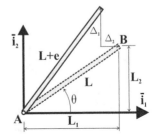

Fig. 9.29. Single bar of a planar truss.

The elongation-displacement relationship is nonlinear, but if the relatively joint displacements, Δ_1 and Δ_2, remain small compared to the bar's length, as is typically the case for engineered structures, this expression can be linearized as follows.

First, a division by the square of the bar's length yields a non-dimensional form of the equation,

$$\left(1 + \frac{e}{L}\right)^2 = \left(\frac{L_1}{L} + \frac{\Delta_1}{L}\right)^2 + \left(\frac{L_2}{L} + \frac{\Delta_2}{L}\right)^2.$$

Expanding all squares then leads to

$$1 + 2\frac{e}{L} + \frac{e^2}{L^2} = \frac{L_1^2}{L^2} + 2\frac{L_1}{L}\frac{\Delta_1}{L} + \frac{\Delta_1^2}{L^2} + \frac{L_2^2}{L^2} + 2\frac{L_2}{L}\frac{\Delta_2}{L} + \frac{\Delta_2^2}{L^2}.$$

It is now assumed that e, Δ_1, and Δ_2 are all small compared to L, and hence, terms e^2/L^2, Δ_1^2/L^2, and Δ_2^2/L^2 become negligible. Finally, noting that $L_1^2/L^2 + L_2^2/L^2 = 1$, the above equation reduces to

$$e \approx \Delta_1 \frac{L_1}{L} + \Delta_2 \frac{L_2}{L} = \Delta_1 \cos\theta + \Delta_2 \sin\theta. \tag{9.27}$$

This equation shows that the bar's elongation is the projection of the relative displacement of its end joint along its direction. Equation (9.27) is an approximate, linearized elongation-displacement relationship, which applies when joint displacements remain small compared to the bar's length.

Internal virtual work for a bar

Figure 9.30 shows a general planar truss member defined by its root and tip joints. Let $\bar{b}$ be the unit vector along the direction of the bar, and $\underline{u}^t$ and $\underline{u}^r$ the displacements of its tip and root joints, respectively. Let F be the magnitude of the force applied to the bar, and hence, forces $\underline{F}^r = -F\bar{b}$ and $\underline{F}^t = F\bar{b}$ are applied to the root and tip of the bar, respectively. The virtual work done by these two forces is $\delta W = \underline{F}^r \cdot \delta\underline{u}^r + \underline{F}^t \cdot \delta\underline{u}^t = F\bar{b} \cdot (\delta\underline{u}^t - \delta\underline{u}^r)$.

Because the internal and externally applied forces are of opposite sign, the virtual work done by the internal force becomes

$$\delta W_I = -\underline{F}^r \cdot \delta\underline{u}^r - \underline{F}^t \cdot \delta\underline{u}^t = -F\bar{b} \cdot (\delta\underline{u}^t - \delta\underline{u}^r) \tag{9.28}$$

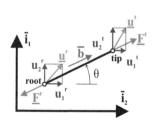

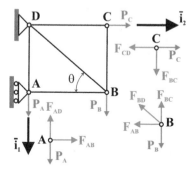

Fig. 9.30. Bar displacements and forces.

Fig. 9.31. Configuration of the 5-bar truss.

Using eq. (9.27), the elongation, e, of the bar is the projection of the relative displacements of its end points along its direction, and this is expressed by the following dot product, $e = \bar{b} \cdot (\underline{u}^t - \underline{u}^r)$. The virtual elongation then becomes $\delta e = \bar{b} \cdot (\delta\underline{u}^t - \delta\underline{u}^r)$, and the bar's internal virtual work, eq. (9.28), becomes

$$\delta W_I = -F\delta e. \tag{9.29}$$

It is possible to express the virtual elongation in terms of the end nodes virtual displacements as

$$
\begin{aligned}
\delta e &= (\sin\theta\,\bar{\imath}_1 + \cos\theta\,\bar{\imath}_2) \cdot (\delta u_1^t\,\bar{\imath}_1 + \delta u_2^t\,\bar{\imath}_2 - \delta u_1^r\,\bar{\imath}_1 - \delta u_2^r\,\bar{\imath}_2) \\
&= (\delta u_1^t - \delta u_1^r)\sin\theta + (\delta u_2^t - \delta u_2^r)\cos\theta.
\end{aligned}
\tag{9.30}
$$

9.5.2 Solution using Newton's law

Consider the five-bar planar truss depicted in fig. 9.31, subjected to two vertical loads, P_A and P_B, applied at joints **A** and **B**, respectively, and to a horizontal load, P_C, applied at joint **C**. The diagonal bar, **BD**, is inclined at an angle θ with respect to the horizontal. At joint **D**, the two components of displacements are constrained to be zero, whereas at joint **A**, the vertical component of displacement is allowed, but the horizontal is constrained to zero. In the development that follows, the vector notation employed in the previous examples will not be used because it is easier to write the work directly in scalar equations for these two-dimensional problems.

Each member in a truss transmits an axial force that is either tensile or compressive. By convention, tensile forces are defined as positive and compressive forces as negative. The truss can be treated as a system of particles where each particle is a joint of the truss to which two or more members are attached. Newton's law expresses the equilibrium condition at each of the 4 joints, **A**, **B**, **C** and **D** by eq. (9.2). For a planar truss, this yields two scalar equilibrium equations at each joint, a total of 8 equations for the present truss. These equilibrium equations involve the bar forces and reaction forces acting at each joint. The approach is commonly referred to as the *method of joints*.

Considering the free-body diagrams of each joint shown in fig. 9.31, the following 8 equations of equilibrium are obtained

$$P_A - F_{AD} = 0, \quad H_A + F_{AB} = 0 \tag{9.31a}$$
$$P_B - F_{BC} - F_{BD}\sin\theta = 0, \quad -F_{AB} - F_{BD}\cos\theta = 0, \tag{9.31b}$$
$$F_{BC} = 0, \quad P_C - F_{CD} = 0, \tag{9.31c}$$
$$V_D + F_{AD} + F_{BD}\sin\theta = 0, \quad H_D + F_{CD} + F_{BD}\cos\theta = 0. \tag{9.31d}$$

Equations (9.31a) are the vertical and horizontal equilibrium equations at joint **A**; eqs. (9.31b) are the vertical and horizontal equilibrium equations at joint **B**; eqs. (9.31c) are the corresponding equations at joint **C**; and finally, eqs. (9.31d) are the corresponding equations at joint **D**.

These 8 equilibrium equations can used to determine 8 independent forces. If the number of forces (bar and reaction forces) is larger than the number of equations, the problem is hyperstatic. If the number of forces equals the number of equations, the problem is isostatic and a complete solution can be obtained from the equilibrium equations alone. In this case, the problem is isostatic because the 8 equilibrium equations of the problem are sufficient to compute the 5 bar plus the 3 reaction forces.

9.5.3 Solution using kinematically admissible virtual displacements

In section 9.3.1, the principle of virtual work for a single particle is developed by multiplying the particle's equilibrium equation by an arbitrary virtual displacement, to obtain eq. (9.8), which is a statement of the principle of virtual work. A similar approach is followed here to develop the principle of virtual work for the five-bar truss depicted in fig. 9.31. Newton's law is used to obtain the joint equilibrium equations, eqs. (9.31); of these 8 equations, the 5 that correspond to equilibrium in an unconstrained direction are multiplied by virtual displacements to construct the following statement

$$
\begin{aligned}
&[P_A - F_{AD}] \, \delta u_1^A + [P_B - F_{BC} - F_{BD} \sin \theta] \, \delta u_1^B \\
&+ [-F_{AB} - F_{BD} \cos \theta] \, \delta u_2^B + [F_{BC}] \, \delta u_1^C + [P_C - F_{CD}] \, \delta u_2^C = 0,
\end{aligned}
\tag{9.32}
$$

where δu_1^A is a vertical virtual displacement at joint **A**, δu_1^B and δu_2^B vertical and horizontal virtual displacements at joint **B**, and δu_1^C and δu_2^C the corresponding quantities at joint **C**. These are kinematically admissible virtual displacements because they do not violate any of the geometric boundary conditions of the problem. A horizontal virtual displacement at joint **A**, or any virtual displacements at joint **D** would violate the geometric boundary conditions at those joints and are not considered here. If the truss is in equilibrium, eq. (9.32) vanishes for all kinematically admissible virtual displacements.

The various terms in eq. (9.32) are now regrouped in the following manner

$$
\begin{aligned}
&P_A \delta u_1^A + P_B \delta u_1^B + P_C \delta u_2^C - F_{AB} \delta u_2^B - F_{AD} \delta u_1^A \\
&- F_{BC} (\delta u_1^B - \delta u_1^C) - F_{BD} (\delta u_1^B \sin \theta + \delta u_2^B \cos \theta) - F_{CD} \delta u_2^C = 0.
\end{aligned}
\tag{9.33}
$$

The first 3 terms of this expression represent the virtual work done by the externally applied forces,

$$
\delta W_E = P_A \delta u_1^A + P_B \delta u_1^B + P_C \delta u_2^C,
\tag{9.34}
$$

where each loading component is multiplied by the virtual displacement in the direction of action of the load. Equation (9.33) now simplifies to

$$
\begin{aligned}
&\delta W_E - F_{AB} \delta u_2^B - F_{AD} \delta u_1^A - F_{BC} (\delta u_1^B - \delta u_1^C) \\
&- F_{BD} (\delta u_1^B \sin \theta + \delta u_2^B \cos \theta) - F_{CD} \delta u_2^C = 0.
\end{aligned}
$$

The last 5 terms of this equation represent the virtual work done by the internal forces in the 5 bars. From eq. (9.30), the virtual elongation of bar **AB** is $\delta e_{AB} = \delta u_2^B$, and eq. (9.29) then yields the bar's internal virtual work as $-F_{AB} \delta e_{AB} = -F_{AB} \delta u_2^B$. The virtual elongations in the other four bars are $\delta e_{AD} = \delta u_1^A$, $\delta e_{BC} = \delta u_1^B - \delta u_1^C$, $\delta e_{BD} = \delta u_1^B \sin \theta + \delta u_2^B \cos \theta$, and $\delta e_{CD} = \delta u_2^C$. The virtual work done by all internal forces now becomes

$$
\delta W_I = -F_{AB} \delta e_{AB} - F_{AD} \delta e_{AD} - F_{BC} \delta e_{BC} - F_{BD} \delta e_{BD} - F_{CD} \delta e_{CD}, \tag{9.35}
$$

and eq. (9.33) reduces to

$$\delta W = \delta W_E + \delta W_I = 0, \tag{9.36}$$

for all kinematically admissible virtual displacements.

The reasoning can be reversed: if eq. (9.36) holds for all kinematically admissible virtual displacements, eq. (9.32) then holds and the equilibrium equations of the problem follow. The results obtained here can be combined into another statement of the principle of virtual work.

Principle 5 (Principle of virtual work) *A structure is in static equilibrium if and only if the sum of the internal and external virtual work vanishes for all kinematically admissible virtual displacements.*

9.5.4 Solution using arbitrary virtual displacements

The development presented above is based on kinematically admissible virtual displacements, but this is not the only possible approach. Newton's law is used to obtain the 8 joint equilibrium equations, eqs. (9.31). Multiplying each of these equilibrium equations by a virtual displacement leads to the following statement

$$[P_A - F_{AD}] \, \delta u_1^A + [H_A + F_{AB}] \, \delta u_2^A + [P_B - F_{BC} - F_{BD} \sin \theta] \, \delta u_1^B$$
$$+ [-F_{AB} - F_{BD} \cos \theta] \, \delta u_2^B + [F_{BC}] \, \delta u_1^C + [P_C - F_{CD}] \, \delta u_2^C \tag{9.37}$$
$$+ [V_D + F_{AD} + F_{BD} \sin \theta] \, \delta u_1^D + [H_D + F_{CD} + F_{BD} \cos \theta] \, \delta u_2^D = 0.$$

If the truss is in equilibrium, eq. (9.37) vanishes for all virtual displacements.

In contrast to eqs. (9.32), these equations include the horizontal and vertical reaction forces acting at joint **D**, denoted V_D and H_D, respectively, and the horizontal reaction force at joint **A**, denoted H_A. In addition, δu_2^A is the horizontal virtual displacement component at point **A**, and δu_1^D and δu_2^D are the vertical and horizontal virtual displacement components at point **D**, respectively. These three virtual displacement components violate the geometric boundary conditions of the problem, *i.e.*, they are not kinematically admissible.

Regrouping terms in eq. (9.37) leads to

$$P_A \delta u_1^A + P_B \delta u_1^B + P_C \delta u_2^C + H_A \delta u_2^A + V_D \delta u_1^D + H_D \delta u_2^D$$
$$- F_{AB}(\delta u_2^B - \delta u_2^A) - F_{AD}(\delta u_1^A - \delta u_1^D) - F_{BC}(\delta u_1^B - \delta u_1^C)$$
$$- F_{BD} \left[(\delta u_1^B - \delta u_1^D) \sin \theta + (\delta u_2^B - \delta u_2^D) \cos \theta \right] - F_{CD}(\delta u_2^C - \delta u_2^D) = 0. \tag{9.38}$$

The first 6 terms of this expression represent the virtual work done by the externally applied forces,

$$\delta W_E = P_A \delta u_1^A + P_B \delta u_1^B + P_C \delta u_2^C + H_A \delta u_2^A + V_D \delta u_1^D + H_D \delta u_2^D, \tag{9.39}$$

where each loading component is multiplied by the virtual displacement in the direction of action of the load. Because virtual displacements that violate the geometric boundary conditions are used here, the virtual work done by the reaction forces does not vanish, and the reaction forces must be treated as externally applied forces.

Equation (9.38) now simplifies to

$$\delta W_E - F_{AB}(\delta u_2^B - \delta u_2^A) - F_{AD}(\delta u_1^A - \delta u_1^D) - F_{BC}(\delta u_1^B - \delta u_1^C)$$
$$- F_{BD}\left[(\delta u_1^B - \delta u_1^D)\sin\theta + (\delta u_2^B - \delta u_2^D)\cos\theta\right] - F_{CD}(\delta u_2^C - \delta u_2^D) = 0. \tag{9.40}$$

The last 5 terms of this equation represent the virtual work done by the internal forces in the 5 bars as can be shown by the following reasoning. The virtual elongations in the five bars are given with the help of eq. (9.30) as $\delta e_{AB} = \delta u_2^B - \delta u_2^A$, $\delta e_{AD} = \delta u_1^A - \delta u_1^D$, $\delta e_{BC} = \delta u_1^B - \delta u_1^C$, $\delta e_{BD} = (\delta u_1^B - \delta u_1^D)\sin\theta + (\delta u_2^B - \delta u_2^D)\cos\theta$, and $\delta e_{CD} = \delta_2^C - \delta u_2^D$, which are updated to reflect the presence of virtual displacements that violate the geometric constraints. Using these, the last 5 terms in eq. (9.40) become $-F_{AB}\delta e_{AB} - F_{AD}\delta e_{AD} - F_{BC}\delta e_{BC} - F_{BD}\delta e_{BD} - F_{CD}\delta e_{CD} = \delta W_I$ so that eq. (9.38) reduces to eq. (9.36).

The reasoning can be reversed: if eq. (9.36) holds for all virtual displacements, eq. (9.37) then holds and the equilibrium equations of the problem follow.

These results can be combined into the following statement of the principle of virtual work.

Principle 6 (Principle of virtual work) *A structure is in static equilibrium if and only if the sum of the internal and external virtual work vanishes for all virtual displacements.*

Although the above two principles are established for the simple truss structure depicted in fig. 9.31, it will be shown later that they are applicable to general structures. These two principles are nearly identical. When the principle of virtual work is used with kinematically admissible virtual displacements as principle 5, the virtual work done by the externally applied forces does not include the reaction forces; the virtual work they perform automatically vanishes because the corresponding virtual displacements are zero, see eq. (9.34). For the 5 bar truss, application of this principle yields the five equilibrium equations in eqs. (9.31) that do not involve the reaction forces.

When the principle of virtual work is used with arbitrary virtual displacements as principle 6, the virtual work done by the reactions forces must be included in the statement of the external virtual work, see eq. (9.39). For the 5 bar truss, application of this principle yields all 8 of the equilibrium equations, eqs. (9.31), two at each of the four joints in the truss.

These principles are derived from Newton's law to which they are entirely equivalent. When arbitrary virtual displacements are used, all equilibrium equations of the problem are recovered. If the virtual displacements are limited to those that are kinematically admissible, a subset of the equilibrium equations is recovered.

Example 9.13. Three-bar truss using principle of virtual work
The three-bar planar truss depicted in fig. 9.32 is a simple hyperstatic truss with a single free joint. It is subjected to a vertical load P at joint **O**, where the three bars are pinned together. The three bars are identified by the joints at which they are pinned

to the ground, denoted as **A**, **B**, and **C**. $\mathcal{A}_A$, $\mathcal{A}_B$ and $\mathcal{A}_C$ are the cross-sectional areas of bars **A**, **B**, and **C**, respectively, and E_A, E_B and E_C denote their respective Young's moduli. The axial stiffnesses of the three bars are $k_A = (E\mathcal{A})_A/L_A = (E\mathcal{A})_A \cos\theta/L$, $k_B = (E\mathcal{A})_B/L$, and $k_C = (E\mathcal{A})_C \cos\theta/L$, respectively.

This problem is hyperstatic of order 1, and it is solved using the displacement and force methods in examples 4.4 on page 147 and example 4.6 on page 152. In these two earlier examples, the stiffnesses of bars **A** and **C** are assumed to be identical to simplify the problem. This assumption is not made in the present example.

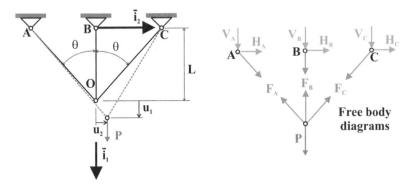

Fig. 9.32. Three-bar truss configuration with free-body diagram.

The virtual displacement vector for point **O** is $\delta\underline{u} = \delta u_1\bar{\imath}_1 + \delta u_2\bar{\imath}_2$. Equation (9.30) then gives the bar virtual elongations as $\delta e_A = \delta u_1 \cos\theta + \delta u_2 \sin\theta$, $\delta e_B = \delta u_1$, and $\delta e_C = \delta u_1 \cos\theta - \delta u_2 \sin\theta$, for bars **A**, **B** and **C**, respectively. The principle of virtual work now states that

$$
\begin{aligned}
\delta W &= \delta W_E + \delta W_I \\
&= P\delta u_1 - F_A(\delta u_1 \cos\theta + \delta u_2 \sin\theta) - F_B\delta u_1 - F_C(\delta u_1 \cos\theta - \delta u_2 \sin\theta) \\
&= -\left[F_A \cos\theta + F_B + F_C \cos\theta - P\right]\delta u_1 - \sin\theta\left[F_A - F_C\right]\delta u_2 = 0,
\end{aligned}
$$

for all arbitrary virtual displacement components, δu_1 and δu_2. Because these components are arbitrary, the two bracketed terms must vanish, leading to the two equilibrium equations of the problem, $F_A \cos\theta + F_B + F_C \cos\theta = P$ and $F_A = F_C$. Because these two equilibrium equations are not sufficient to evaluate the three bar forces, the problem is hyperstatic of order 1.

The solution of the problem can be completed using the displacement method, see section 4.3.2. The elongation-displacement equations are obtained by expressing the bar elongations in terms of the end-joint displacements using eq. (9.27). This yields $e_A = u_1 \cos\theta + u_2 \sin\theta$, $e_B = u_1$, and $e_C = u_1 \cos\theta - u_2 \sin\theta$, where e_A, e_B, and e_C are the elongations of bars **A**, **B** and **C**, respectively, and $\underline{u} = u_1\bar{\imath}_1 + u_2\bar{\imath}_2$ is the displacement vector of point **O**. The constitutive laws are $F_A = e_A(E\mathcal{A})_A \cos\theta/L$, $F_B = e_B(E\mathcal{A})_B/L$, and $F_C = e_C(E\mathcal{A})_C \cos\theta/L$, for bars **A**, **B** and **C**, respectively. Introducing the constitutive laws into the equilibrium equations, and the elongation-displacement equations into the resulting relationships yield two equations for the two displacement components. It is convenient to write these in matrix form as

$$
\cos^2 \theta \begin{bmatrix} 1 + (\bar{k}_A + \bar{k}_C) \cos \theta & (\bar{k}_A - \bar{k}_C) \sin \theta \\ (\bar{k}_A - \bar{k}_C) \sin \theta & (\bar{k}_A + \bar{k}_C) \sin \theta \end{bmatrix} \begin{Bmatrix} u_1 \\ u_2 \end{Bmatrix} = \begin{Bmatrix} \bar{P}L \\ 0 \end{Bmatrix},
$$

where $\bar{k}_A = (EA)_A/(EA)_B$ and $\bar{k}_C = (EA)_C/(EA)_B$ are non-dimensional bar stiffness ratios, and $\bar{P} = P/(EA)_B$ is the non-dimensional applied load. With simple matrix manipulations, this set of equations yields the non-dimensional displacement components of point **O** as

$$
\frac{u_1}{\bar{P}L} = \frac{\bar{k}_A + \bar{k}_C}{\bar{k}}, \qquad \frac{u_2}{\bar{P}L} = -\frac{\bar{k}_A - \bar{k}_C}{\bar{k}} \frac{\cos \theta}{\sin \theta}, \tag{9.41}
$$

where $\bar{k} = \bar{k}_A + \bar{k}_C + 4\bar{k}_A \bar{k}_C \cos^3 \theta$. With the help of the elongation-displacement equations, the bar non-dimensional elongations are found to be

$$
\frac{e_A}{\bar{P}L} = \frac{2\bar{k}_C \cos \theta}{\bar{k}}, \qquad \frac{e_B}{\bar{P}L} = \frac{\bar{k}_A + \bar{k}_C}{\bar{k}}, \qquad \frac{e_C}{\bar{P}L} = \frac{2\bar{k}_A \cos \theta}{\bar{k}}.
$$

Finally, the non-dimensional bar forces are obtained from the constitutive laws as

$$
\frac{F_A}{P} = \frac{F_C}{P} = \frac{2 \cos \theta^2 \bar{k}_A \bar{k}_C}{\bar{k}}, \qquad \frac{F_B}{P} = \frac{\bar{k}_A + \bar{k}_C}{\bar{k}}. \tag{9.42}
$$

Except for the fact that equilibrium equations are obtained from the principle of virtual work rather than from Newton's first law, the solution process presented here is identical to that of the displacement method presented in section 4.3.2.

In this example, the principle of virtual work is used in conjunction with kinematically admissible virtual displacements. Figure 9.32 also shows free body diagrams of the four nodes of the truss, and these involve the reaction forces at joints **A**, **B**, and **C**. These reaction forces do not appear in the above developments because the virtual displacements at joints **A**, **B**, and **C** are selected to be kinematically admissible, *i.e.*, all three are assumed to vanish. If arbitrary virtual displacements are selected, the virtual work done by the reaction forces no longer vanishes. The external virtual work is now

$$
\delta W_E = V_A \delta u_1^A + H_A \delta u_2^A + V_B \delta u_1^B + H_B \delta u_2^B + V_C \delta u_1^C + H_C \delta u_2^C + P \delta u_1^O,
$$

where δu_1^A and δu_2^A are the vertical and horizontal components of the virtual displacement at joint **A**, and similar notations are used for the corresponding virtual displacements at joints **B**, **C**, and **O**.

Next, the internal virtual work is evaluated as

$$
\begin{aligned}
\delta W_I = &-F_A(\cos \theta \bar{\imath}_1 + \sin \theta \bar{\imath}_2) \cdot \left[(\delta u_1^O - \delta u_1^A) \bar{\imath}_1 + (\delta u_2^O - \delta u_2^A) \bar{\imath}_2 \right] \\
&-F_B \bar{\imath}_1 \cdot \left[(\delta u_1^O - \delta u_1^B) \bar{\imath}_1 + (\delta u_2^O - \delta u_2^B) \bar{\imath}_2 \right] \\
&-F_C(\cos \theta \bar{\imath}_1 - \sin \theta \bar{\imath}_2) \cdot \left[(\delta u_1^O - \delta u_1^C) \bar{\imath}_1 + (\delta u_2^O - \delta u_2^C) \bar{\imath}_2 \right],
\end{aligned}
$$

where the expressions for the root and tip virtual displacements of the bars reflect the virtual displacements at joints **A**, **B**, and **C**, which violate the geometric boundary conditions.

Invoking the principle of virtual work, principle 6, then yields

$$[V_A + F_A \cos \theta] \, \delta u_1^A + [H_A + F_A \sin \theta] \, \delta u_2^A + [V_B + F_B] \, \delta u_1^B + [H_B] \, \delta u_2^B$$
$$+ [V_C + F_C \cos \theta] \, \delta u_1^C + [H_C - F_C \sin \theta] \, \delta u_2^C$$
$$+ [P - F_A \cos \theta - F_B - F_C \cos \theta] \, \delta u_1^O + [F_A \sin \theta - F_C \sin \theta] \, \delta u_2^O = 0.$$

In this expression, all virtual displacement components are arbitrary, and this implies that all bracketed terms must vanish. The last two terms yield the vertical and horizontal equilibrium equations at joint **O**, which are the only two equations obtained when kinematically admissible virtual displacements are used. The first six bracketed terms yield the six equilibrium equations involving the reaction forces at joints **A**, **B**, and **C**.

The complete solution process mirrors that used earlier. First, the displacements of joint **O** can be obtained from eq. (9.41). Next, the forces in bars **A**, **B**, and **C** follow from eq. (9.42). Finally, the three bar forces can be introduced into the six equilibrium equations at joints **A**, **B**, and **C** to obtain the six components of the reaction forces at the corresponding points.

9.6 Principle of complementary virtual work

The basic equations of linear elasticity are derived in chapter 1. As shown in fig. 9.33, these equations are divided into three groups: the equilibrium equations, the strain-displacement relationships, and the constitutive laws. Given the proper boundary conditions, these three groups of equations are sufficient to obtain solutions of elasticity problems.

In addition, the strain compatibility equations impose constraints on the body's strain field. Because they can be derived from the strain-displacement relationships, the compatibility equations do not form an independent set of equations and are not required to solve elasticity problems. Their importance, however, arises in situations where the displacement field is to be evaluated from the strain field. Because the six strain components are expressed in terms of three displacement components only, the problem is over-determined. The compatibility equations ensure that the strain field can be derived from a compatible displacement field, *i.e.*, a displacement field that creates no gaps or overlaps in the solid. Consequently, the compatibility equations can become a critical element of any solution procedure.

The principle of virtual work derived in the first part of this chapter is shown to be entirely equivalent to the equilibrium equations, which are themselves a mathematical restatement of Newton's laws. The equivalence of these three statements is indicated in fig. 9.33 by the double headed arrows joining the corresponding boxes.

Although expressed within markedly different formalisms, the principle of virtual work and Newton's laws are two entirely equivalent statements. Because New-

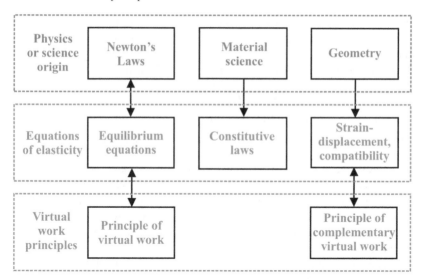

Fig. 9.33. Relationship between the equations of elasticity and virtual work principles.

ton's laws are the foundation of mechanics and elasticity, the principle of virtual work is an equivalent, alternative foundation of mechanics and elasticity.

As discussed in chapter 1, the solution of any elasticity problem requires the three groups of basic equations appearing in the middle row of fig. 9.33. Consequently, because it is equivalent to only the equilibrium equations, the principle of virtual work alone does not provide enough information to solve elasticity problems. To obtain complete solutions to elasticity problems, it must be complemented with strain-displacement relationships and constitutive laws.

In the next part of this chapter, a second virtual work principle will be developed: the *principle of complementary virtual work*. As indicated in fig. 9.33, this virtual work principle is entirely equivalent to the compatibility equations of the problem. The principle of complementary virtual work alone does not provide enough information to solve elasticity problems. It must be augmented with equilibrium equations and constitutive laws to derive complete solutions to elasticity problems.

Clearly, the strain-displacement relationships and compatibility equations are central to the understanding of the principle of complementary virtual work. Rather than consider the more abstract general case, these concepts will be examined in the next section for simple truss structures and in subsequent sections for more complex beam structures.

9.6.1 Compatibility equations for a planar truss

Before deriving the principle of complementary virtual work, the kinematics and compatibility equations for planar trusses will be investigated.

Compatibility conditions

Consider the two-bar planar truss depicted in fig. 9.34, consisting of two bars, denoted bars **A** and **C**, joined together at point **O**, connected to the ground at points
A and **C**, respectively, and of lengths L_A and L_C, respectively. Let the two bars
undergo arbitrary elongations of magnitudes e_A and e_C, respectively; the loads that
create these elongations are irrelevant to the present discussion and are not shown in
the figure.

The configuration of the truss that is compatible with these elongations is easily found with the help of a purely geometric reasoning. Draw two circles of radii
$L_A + e_A$ and $L_C + e_C$, centered at points **A** and **C**, respectively; the intersection of
these two circles, denoted point **O'**, is the connection point of the two bars in their deformed configuration. Given any two arbitrary elongations, the truss's configuration
is easily found, assuming, of course, that the two circles intersect.

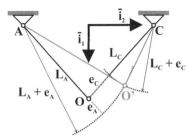

Fig. 9.34. Two-bar truss in the original and
deformed configurations.

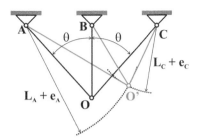

Fig. 9.35. Three-bar truss in the original and
deformed configurations.

Consider next the three-bar, planar truss shown in fig. 9.35 which is similar to the
two-bar truss considered in fig. 9.34 but with the addition of the middle bar, denoted
bar **B**, of length L_B. Let bars **A** and **C** undergo arbitrary elongations of magnitudes
e_A and e_C, respectively. Using the geometric construction described in the previous
paragraph, the configurations of bars **A** and **C** are readily found, and point **O'** is
obtained. The configuration of bar **B** is now uniquely defined, because it must join
points **B** and **O'**. If the deformed length of bar **B** is denoted L'_B, its elongation is then
$e_B = L'_B - L_B$. Clearly, the elongations of the three bars are no longer independent
because, given e_A and e_C, e_B can be obtained from simple geometric considerations.

Instead of using this purely a geometric reasoning, the same conclusions can be
reached by using the elongation-displacement equations of the problem. Consider
first the two-bar truss depicted in fig. 9.34. Let $\underline{u} = u_1 \bar{\imath}_1 + u_2 \bar{\imath}_2$ be the displacement
vector of point **O**. The elongations of bars **A** and **C** simply correspond to the projec-
tions of this displacement vector along the directions of the bars, see eq. (9.27), to
find

$$e_A = u_1 \cos\theta + u_2 \sin\theta, \quad e_C = u_1 \cos\theta - u_2 \sin\theta. \tag{9.43}$$

These equations relate the elongations, e_A and e_C, of the two bars to the two components of displacement of point **O**, u_1 and u_2, and can be inverted to find the dis-

placement components as a function of the elongations: $u_1 = (e_A + e_C)/(2\cos\theta)$ and $u_1 = (e_A - e_C)/(2\sin\theta)$. The final configuration of the system is uniquely defined if the two displacement components, u_1 and u_2, are given. Alternatively, if the two elongations are given, the two displacement components can be evaluated using eq. (9.43), and the final system configuration is obtained.

Consider now the three-bar truss shown in fig. 9.35. The elongations of bars **A**, **B** and **C**, correspond to the projections of the displacement vector of point **O** along the directions of the bars, see eq. (9.27), to find

$$e_A = u_1\cos\theta + u_2\sin\theta, \quad e_B = u_1, \quad e_C = u_1\cos\theta - u_2\sin\theta. \tag{9.44}$$

The present situation is quite different from that examined in the previous paragraph. Whereas the elongations of the three bars can readily be expressed in terms of the two displacement components, it is not possible to express the two displacement components in terms of the three elongations. This stems from the fact that given the three elongations, eqs. (9.44) form an over-determined set three equations for the two unknown displacement components. It is possible, however, to eliminate the two displacement components from eqs. (9.44) to find the compatibility equation of the problem

$$e_A - 2e_B\cos\theta + e_C = 0. \tag{9.45}$$

With this approach to the problem, it is easy to predict the number of compatibility equations: because the three bar elongations are expressed in terms of two displacement components, a single compatibility condition exists.

This simple example underlines a fundamental difference between the two- and three-bar trusses shown in figs. 9.34 and 9.35, respectively. For the two-bar truss, the deformed configuration of the system can be determined through purely geometric constructions, given the elongations of each bar. For the three-bar truss, the bar elongations are not independent of each other and must satisfy a compatibility condition. If this condition is satisfied, the deformed configuration of the system can be determined through purely geometric constructions.

Another fundamental difference between these two trusses exists: the two-bar truss is isostatic, whereas its three-bar counterpart is hyperstatic. In fact, the number of compatibility equations is equal to the order of redundancy of the hyperstatic problem, as defined in section 4.3. For isostatic problems, the order of redundancy is zero and therefore, the number of compatibility equations is zero.

To better understand the relationship between the number of compatibility equations and the order of a hyperstatic system, consider the following reasoning. The two-bar truss involves two force components (the forces in the two bars) that are linked by two equilibrium equations, the horizontal and vertical equilibrium equations for the forces acting at joint **O**. These two forces can be determined solely from the equilibrium equations, and hence, the system is isostatic. The same truss involves two elongations (the elongations in the two bars) that are related to the two displacement components of joint **O**. The two displacements are uniquely defined by the two elongations, leaving no compatibility conditions.

In contrast, the three-bar truss involves three force components (the forces in the three bars) that are linked by two equilibrium equations, the horizontal and vertical equilibrium equations for the forces acting at joint **O**. These three forces cannot be determined from the two equilibrium equations, and hence, the system is hyperstatic of degree 1. The same truss involves three elongations (the elongations in the three bars) that are related to the two displacement components of joint **O**. The three elongations must satisfy one condition to be compatible with the two displacement components.

In a general planar truss, the number of force components (one per bar) equals the number of elongations (one per bar). Each node of the truss introduces two independent displacement components and two independent equilibrium equations along two orthogonal directions. Starting from an isostatic configuration, the addition of one bar connecting existing joints results in a hyperstatic system of order 1 and creates one compatibility equation. Each additional bar connecting existing joints increase the order by one and creates a new compatibility equation. Therefore, the number of compatibility equations will always equal the order of the hyperstatic problem.

9.6.2 Principle of complementary virtual work for trusses

As illustrated in fig. 9.33, the principle of complementary virtual work focuses on a single group of equations, the strain-displacement relationships, instead of Newton's law, which the principle of virtual work focuses on. The strain-displacement equations simply provide a definition of the strains, and they are based on purely geometric arguments, as discussed in section 1.4.

Three-bar truss under applied load

Consider the three-bar truss depicted in fig. 9.36. It is assumed to undergo compatible deformations so that the three bar elongations satisfy the elongation-displacement relationships, eqs. (9.44). The following statement is now constructed

$$
\begin{aligned}
\delta W' = & - \left[e_A - u_1 \cos\theta - u_2 \sin\theta \right] \delta F_A - \left[e_B - u_1 \right] \delta F_B \\
& - \left[e_C - u_1 \cos\theta + u_2 \sin\theta \right] \delta F_C = 0,
\end{aligned}
\tag{9.46}
$$

where δF_A, δF_B, and δF_C are three arbitrary quantities, called *virtual forces*, and $\delta W'$ is the *complementary virtual work*.

The bracketed terms are the three elongation-displacement relationships of the system. Since the truss undergoes compatible deformations, the elongation-displacement equations are satisfied, and the bracketed terms in the statement vanish. Hence, the above statement is true for any arbitrary virtual forces, δF_A, δF_B, and δF_C. Simple algebraic manipulations then lead to

$$
\begin{aligned}
\delta W' = & - e_A \delta F_A - e_B \delta F_B - e_C \delta F_C \\
& + u_1 \left(\delta F_A \cos\theta + \delta F_B + \delta F_C \cos\theta \right) + u_2 \sin\theta \left(\delta F_A - \delta F_C \right) = 0,
\end{aligned}
\tag{9.47}
$$

for all virtual forces.

Figure 9.36 depicts a free body diagram of joint **O**, and the equilibrium equations are found as $F_A \cos \theta + F_B + F_C \cos \theta = P$ and $F_A - F_C = 0$. A set of forces that satisfies these equilibrium equations is said to be *statically admissible*.

Figure 9.36 also shows a set of virtual forces acting at joint **O**. These virtual forces are said to be *statically admissible virtual forces* if they satisfy the following equilibrium equations at the joint,

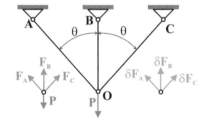

$$\delta F_A \cos \theta + \delta F_B + \delta F_C \cos \theta = 0,$$
$$\delta F_A - \delta F_C = 0. \qquad (9.48)$$

These equations do not include the externally applied loads because $\delta P = 0$ for a specified or given value of P. The geometry of the system is given, hence $\delta \theta = 0$.

Fig. 9.36. Three-bar truss with applied load.

Statement (9.47) is true for all arbitrary virtual forces. If the virtual forces, however, are required to be statically admissible, that is if eqs. (9.48) are satisfied, a much simpler statement results

$$\delta W' = -e_A \delta F_A - e_B \delta F_B - e_C \delta F_C = 0, \qquad (9.49)$$

for all statically admissible virtual forces. The internal virtual work done by a force, F, in a bar undergoing an elongation, e, is defined by eq. (9.29) as $\delta W_I = -F \delta e$. In statement (9.49), the three terms represent the *internal complementary virtual work* of the truss,

$$\delta W_I' = -e_A \delta F_A - e_B \delta F_B - e_C \delta F_C = -\sum_{i=1}^{N_b} e_i \delta F_i, \qquad (9.50)$$

where the last equality gives the general expression for internal complementary virtual work in a truss consisting of N_b bars. Statement (9.49) is now recast in a compact form as

$$\delta W' = \delta W_I' = 0, \qquad (9.51)$$

for all statically admissible virtual forces.

The reasoning developed in the previous paragraphs can be reversed. Equation (9.51) is equivalent to statement (9.49). If this statement holds for all statically admissible virtual forces, eq. (9.47) must also hold under the same conditions. Simple algebraic manipulations then lead to eq. (9.46) which implies the elongation-displacement relationships of the problem, because the virtual forces are arbitrary. Although the developments presented above apply to a three-bar truss, all the steps of the reasoning would still hold for trusses of arbitrary configuration.

Three-bar truss under prescribed displacement

The three-bar truss depicted in fig. 9.37 is similar to the truss discussed in the previous section, except that instead of being subjected to a concentrated load at joint **O**, the downward vertical displacement of joint **B** is now prescribed to be of magnitude Δ.

Prescribed displacements form an important class of problem parameters. Imagine that a displacement-controlled actuator, such as a screw-jack or a displacement-controlled servo-hydraulic actuator, is oriented downward at point **B**. The natural length of vertical bar **BO** is L, and displacement Δ is prescribed. The actuator will provide whatever force is required to obtain the specified displacement. The force required to obtain the specified displacement, often called the *"driving force,"* D, is as yet unknown.

A prescribed displacement of this type fundamentally affects the statement of the principle of complementary virtual work because it directly impacts the compatibility equations. Indeed, the elongation-displacement relationship for bar **B** now becomes $e_B = u_1 - \Delta$, instead of $e_B = u_1$.

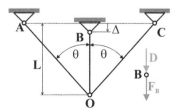

Fig. 9.37. Three-bar truss with prescribed displacement.

The following statement of complementary virtual work is constructed from the compatibility equations expressed in homogeneous form

$$\delta W' = - \left[e_A - u_1 \cos\theta - u_2 \sin\theta\right] \delta F_A - \left[e_B - u_1 + \Delta\right] \delta F_B \\ - \left[e_C - u_1 \cos\theta + u_2 \sin\theta\right] \delta F_C = 0, \tag{9.52}$$

where δF_A, δF_B, and δF_C are, here again, arbitrary virtual forces. This statement should be compared to eq. (9.46), written for the same truss in the absence of a prescribed displacement.

Because the truss undergoes compatible deformations, the elongation-displacement equations are satisfied, and the bracketed terms in the statement vanish. Simple algebraic manipulations then lead to

$$\delta W' = -e_A \delta F_A - e_B \delta F_B - e_C \delta F_C - \Delta\, \delta F_B \\ + u_1 \left(\cos\theta \delta F_A + \delta F_B + \cos\theta \delta F_C\right) + u_2 \sin\theta \left(\delta F_A - \delta F_C\right) = 0, \tag{9.53}$$

for all arbitrary virtual forces.

Consider now a set of statically admissible virtual forces that satisfy the following equilibrium equations,

$$\delta F_A \cos\theta + \delta F_B + \delta F_C \cos\theta = 0, \quad \delta F_A - \delta F_C = 0, \quad \delta F_B + \delta D = 0, \tag{9.54}$$

where the first two equations correspond to the equilibrium conditions of joint **O** and the third to that at joint **B**. The virtual driving force, δD, does not vanish because this force is unknown.

Statement (9.53) is true for all arbitrary virtual forces. If the virtual forces are required to be statically admissible, however, a simpler statement results,

$$\delta W' = \Delta\, \delta D - e_A \delta F_A - e_B \delta F_B - e_C \delta F_C = 0. \tag{9.55}$$

As observed earlier, the last three terms of this expression represent the internal complementary virtual work, $\delta W_I'$, as defined in eq. (9.50). The product of a force, D, by the displacement of its point of application in the direction of the force, Δ, is defined in section 9.2.2 as the work done by the force, $W = D\Delta$. The virtual work done by the same force is introduced in section 9.3.4 as the product of the real force by a virtual displacement, $\delta W = D \, \delta\Delta$. In statement 9.55, the first term is the product of the true displacement by a virtual force, and is called the *external complementary virtual work*,

$$\delta W_E' = \Delta \, \delta D. \tag{9.56}$$

Statement (9.55) is now recast in a compact form as

$$\delta W' = \delta W_E' + \delta W_I' = 0, \tag{9.57}$$

for all statically admissible virtual forces. In the absence of prescribed displacements, the external complementary virtual work vanishes and the simpler statement of eq. (9.51) remains. As before, the reasoning can be reversed: statement 9.52 can be recovered from statement 9.57. Consequently, the elongation-displacement are obtained and the truss undergoes compatible deformations.

In summary, the elongation-displacement relationships of the problem are satisfied if and only if the sum of the external and internal complementary virtual work vanishes for all statically admissible virtual forces. This leads to the principle of complementary virtual work.

Principle 7 (Principle of complementary virtual work) *A truss undergoes compatible deformations if and only if the sum of the internal and external complementary virtual work vanishes for all statically admissible virtual forces.*

In this principle, the virtual forces must be statically admissible. If the complementary virtual work is required to vanish for all arbitrary virtual forces, *i.e.*, for all independently chosen arbitrary δF_A, δF_B, δF_C, and δD, eq. (9.55) implies $e_A = e_B = e_C = \Delta = 0$, which in turn, implies that the truss cannot deform. This is clearly not correct.

Because the virtual forces are statically admissible, they must satisfy eq. (9.54), which forms a set of three equations for the four statically admissible virtual forces. This means that it is possible to express three of the virtual force components in terms of the fourth: $\delta F_B = -2\delta F_A \cos\theta$, $\delta F_C = \delta F_A$, and $\delta D = 2\delta F_A \cos\theta$, for example. The principle of complementary virtual work now becomes

$$\delta W' = \Delta(2\delta F_A \cos\theta) - e_A\delta F_A - e_B(-2\delta F_A \cos\theta) - e_C\delta F_A$$
$$= [2\Delta \cos\theta - e_A + 2e_B \cos\theta - e_C] \, \delta F_A = 0.$$

Since the remaining virtual force component, δF_A, is now entirely arbitrary, the bracketed term must vanish, yielding the compatibility equation of the problem: $e_A - 2(e_B + \Delta) \cos\theta + e_C = 0$.

Clearly, when the concept of statically admissible virtual forces is properly interpreted, the principle of complementary virtual work yields the correct compatibility equation of the problem.

9.6.3 Complementary virtual work

Complementary virtual work is defined as the work done by virtual forces acting through real displacements. In this sense, it is complementary to the virtual work done by real forces acting through virtual displacements. In both cases, the real quantities are assumed to remain fixed during the application of the virtual quantities. The meaning of the term "complementary" will be illustrated in the following discussion.

Consider a uniform bar fixed at one end and subjected to an axial load, F, at its tip and let the resulting tip deflection or elongation be denoted u. The material the bar is made of is not necessarily linearly elastic, and hence, the load-displacement curve is not a straight line, as illustrated in fig. 9.38.

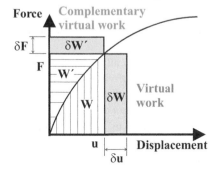

Fig. 9.38. Work and complementary work and their virtual counterparts.

As the displacement increases from 0 to u, the work done by the tip force is $W = \int_0^u F \, du$, see eq. (9.7). This integral corresponds to the area under the curve shown in fig. 9.38. If the material is linearly elastic, say $F = ku$, where k is the axial stiffness of the bar, the work simply becomes $W = \int_0^u ku \, du = ku^2/2 = Fu/2$.

The complementary work is defined as $W' = \int_0^F u \, dF$, and corresponds to the area to the left of the load-displacement curve, as indicated in fig. 9.38. Note that for a nonlinearly elastic material, the work and complementary work are not equal. For a linearly elastic material, however, $W' = \int_0^F F/k \, dF = F^2/(2k) = Fu/2 = W$. For either linearly or nonlinearly materials, $W + W' = Fu$, which explains why the complementary work is called "complementary."

The virtual work and complementary virtual work are also shown in fig. 9.38. The virtual work is the shaded area to the right of that representing the work itself. While computing the work done by the applied force, the force is a function of the displacement, $F = F(u)$, but when computing the virtual work, the force is held constant. As discussed in example 9.2, the term "virtual" used to qualify the virtual displacements indicates that the forces remain unchanged by these virtual displacements.

The complementary virtual work is the shaded area above that representing the complementary work itself. While computing the complementary work done by the applied force, the displacement is a function of the force, $u = u(F)$, but when computing the complementary virtual work, the displacement is held constant. Here again, the term "virtual forces" emphasizes the fact that the displacements remain unchanged by these virtual forces.

9.6.4 Applications to trusses

By now, the similarities and differences between the principle of virtual work and the principle of complementary virtual work are clear.

- The principle of virtual work focuses of the equilibrium equations of the system, whereas the principle of complementary virtual work focuses on the strain-displacement equations, see fig. 9.33.
- The principle of virtual work identifies the equilibrium state of the system from among all kinematically compatible configurations; the principle of complementary virtual work identifies the kinematically compatible configuration from among all statically admissible states.
- The concepts of virtual displacements and virtual forces are subjected to similar restrictions: the former leave the real system forces unchanged, whereas the latter leave real system displacements unchanged.
- The work and its complementary counterpart complement each other as illustrated in fig. 9.38.

To generalize these results, consider a planar truss consisting of a number of bars connected at N nodes. When using arbitrary virtual displacements, the principle of virtual work will provide $2N$ equilibrium equations corresponding to 2 equilibrium equations at each of the N nodes. On the other hand, the principle of complementary virtual work yields the compatibility equations of the problem: n equations will be produced for a hyperstatic truss of order n. If the truss is isostatic, however, no compatibility equations exist, and the principle of complementary virtual work yields no information about the system.

The principle of virtual work provides a statement of equilibrium. When supplemented with the strain-displacement relationships and constitutive laws, it enables the systematic development of the displacement method of solution first presented in section 4.3.2. This approach is easily implemented using computers that can solve the resulting large sets of linear equations.

The principle of complementary virtual work, on the other hand, provides a statement of compatibility. When supplemented with the constitutive laws and equilibrium equations, it enables the development of the force method first presented in section 4.3.3. More often than not, the hyperstatic order is much less than the number of nodes, $n \ll N$, and hence, the principle of complementary virtual work generates only a few ($n \ll N$) equations, which would seem to lead to a simpler solution process. The principle of complementary virtual work, however, suffers from a major drawback: the virtual forces must be statically admissible, *i.e.*, they must form a set of self-equilibrating virtual forces. This implies that the equilibrium equations must be derived at each joint of the truss before the principle can be applied. Thus, while the principle of complementary virtual work generates fewer equations, it requires much more extensive work for the generation of these equations. This problem hinders the systematic application of this principle, and helps explain why the principle of virtual work is used much more widely than its complementary counterpart.

Example 9.14. Three-bar truss under vertical load at joint
The three-bar truss problem depicted in fig. 9.36 is treated earlier in example 9.13 using the principle of virtual work. In this example, the principle of complementary virtual work will be used to solve the same problem, which involves the three forces, F_A, F_B, and F_C, acting in bars **A**, **B** and **C**, respectively. The equations of equilibrium of the problem are found earlier to be $F_A \cos\theta + F_B + F_C \cos\theta = P$ and $F_A = F_C$, and the statically admissible virtual forces satisfy eqs. (9.48).

The principle of complementary virtual work now can be written as

$$\delta W' = -e_A \delta F_A - e_B \delta F_B - e_C \delta F_C = 0, \tag{9.58}$$

for all statically admissible virtual forces. In this example, e_A, e_B, and e_C are the elongations of the three bars are the true displacements components of joint **O** along the directions of bars **A**, **B** and **C** respectively. It is easy to verify from kinematics that $e_A = u_1 \cos\theta + u_2 \sin\theta$, $e_B = u_1$, and $e_C = u_1 \cos\theta - u_2 \sin\theta$, where $\underline{u} = u_1 \bar{\imath}_1 + u_2 \bar{\imath}_2$ is the true displacement vector of joint **O**. The reaction forces at the attachment points **A**, **B**, and **C** do not appear in the above statements because the true displacements at those points are zero, and therefore the complementary virtual work they perform also vanishes.

Because the virtual forces must be statically admissible, the three virtual forces components are linked by the two equilibrium conditions, eqs. (9.48), and hence, it is possible to express two of them in terms of the third: $\delta F_A = \delta F_C$ and $\delta F_B = -2\cos\theta \delta F_C$. The complementary virtual work then becomes

$$\delta W' = -e_A \delta F_C + 2e_B \cos\theta \delta F_C - e_C \delta F_C = -\left[e_A - 2e_B \cos\theta + e_C\right]\delta F_C = 0.$$

The virtual force δF_C is arbitrary, and therefore the bracketed term must vanish, yielding the elongation compatibility equation for the problem

$$e_A - 2e_B \cos\theta + e_C = 0. \tag{9.59}$$

Using the force method, this equation, when combined with the two equilibrium equations, can be used to solve for the 3 unknown bar forces. The constitutive laws are $F_A = e_A \cos\theta (EA)_A/L$, $F_B = e_B (EA)_B/L$, and $F_C = e_C \cos\theta (EA)_C/L$, for bars **A**, **B** and **C**, respectively. Expressing the elongations in terms of forces in the compatibility equation, eq. (9.59), yields

$$\frac{F_A}{\bar{k}_A} - 2F_B \cos^2\theta + \frac{F_C}{\bar{k}_C} = 0,$$

where $\bar{k}_A = (EA)_A/(EA)_B$ and $\bar{k}_C = (EA)_C/(EA)_B$ are non-dimensional stiffness ratios.

The equilibrium equations of the problem are $F_A \cos\theta + F_B + F_C \cos\theta = P$ and $F_A = F_C$, and these two equilibrium equations, together with the compatibility equation expressed in terms of forces, form a set of three equations for the three force components, which are found to be

$$\frac{F_A}{P} = \frac{F_C}{P} = \frac{2\bar{k}_A \bar{k}_C \cos^2 \theta}{\bar{k}}, \quad \frac{F_B}{P} = \frac{\bar{k}_A + \bar{k}_C}{\bar{k}},$$

where $\bar{k} = \bar{k}_A + \bar{k}_C + 4\bar{k}_A \bar{k}_C \cos^3 \theta$ is the non-dimensional stiffness of the truss. This result is identical to that found with the principle of virtual work, see eq. (9.42).

Bar elongations are then evaluated with the help of the constitutive laws, and finally, the displacement can be obtained from the elongation-displacement relationships.

As expected, all results are identical to those found using the principle of virtual work. Except for the fact that the compatibility equations are obtained from the principle of complementary virtual work rather than geometric arguments, the solution process presented here mirrors that of the force method developed in section 4.3.3.

Example 9.15. Three-bar truss under prescribed displacement

Next, the same three-bar truss problem will be treated again, but the structure is now subjected to a prescribed displacement, Δ, at point **B**, as depicted in fig. 9.37. This problem involves involves the three forces, F_A, F_B, and F_C, acting in bars **A**, **B**, and **C**, respectively, and the driving force, D, applied at point **B** to achieve the specified displacement, Δ. The equilibrium equations of the problem are found earlier as $F_A \cos \theta + F_B + F_C \cos \theta = 0$, $F_A - F_C = 0$, and $F_B + D = 0$, and the statically admissible virtual forces must satisfy eqs. (9.54).

The principle of complementary virtual work is

$$\delta W' = \Delta \, \delta D - e_A \delta F_A - e_B \delta F_B - e_C \delta F_C = 0, \tag{9.60}$$

for all statically admissible virtual forces. The elongations of the three bars, e_A, e_B, and e_C, are associated with the true displacements of point **O** along the directions of bars **A**, **B** and **C** respectively.

Because the virtual forces must be statically admissible, the three virtual forces components are linked by the two equilibrium conditions, eqs. (9.54), and hence, it is possible to express three of them in terms of the fourth: $\delta F_A = \delta F_C$, $\delta F_B = -2 \cos \theta \, \delta F_C$, and $\delta D = 2 \cos \theta \, \delta F_C$. The complementary virtual work then becomes

$$\delta W' = [2\Delta \cos \theta - e_A + 2e_B \cos \theta - e_C] \, \delta F_C = 0.$$

The virtual force δF_C is arbitrary and therefore, the bracketed term must vanish. This yields the elongation compatibility equation for the problem

$$e_A - 2(e_B + \Delta) \cos \theta + e_C = 0. \tag{9.61}$$

Using the force method, this equation, when combined with the 3 equilibrium equations, can be used to solve for the 4 unknown bar forces. The constitutive laws are $F_A = e_A \cos \theta (EA)_A / L$, $F_B = e_B (EA)_B / L$, and $F_C = e_C \cos \theta (EA)_C / L$, for bars **A**, **B**, and **C**, respectively. Expressing the elongations in terms of forces in the compatibility equation, eq. (9.59), yields

$$\frac{F_A}{2\bar{k}_A \cos^2 \theta} - F_B + \frac{F_C}{2\bar{k}_C \cos^2 \theta} = \frac{\Delta}{L}(EA)_B,$$

where $\bar{k}_A = (EA)_A/(EA)_B$ and $\bar{k}_C = (EA)_C/(EA)_B$ are non-dimensional stiffness ratios for bars **A** and **C**, respectively.

The equilibrium equations of the problem are $F_A \cos\theta + F_B + F_C \cos\theta = 0$, $F_A - F_C = 0$, and $F_B + D = 0$. These three equilibrium equations, together with the above compatibility equation expressed in terms of forces, form a set of four equations for the four force components. The non-dimensional forces in bars **A** and **C** are found as

$$\frac{F_A}{(EA)_B} = \frac{F_C}{(EA)_B} = \frac{2\bar{k}_A\bar{k}_C\cos^2\theta}{\bar{k}}\frac{\Delta}{L},$$

whereas the non-dimensional driving force, D, and the force in bar **B** are

$$\frac{D}{(EA)_B} = -\frac{F_B}{(EA)_B} = \left(1 - \frac{\bar{k}_A + \bar{k}_C}{\bar{k}}\right)\frac{\Delta}{L},$$

where $\bar{k} = \bar{k}_A + \bar{k}_C + 4\bar{k}_A\bar{k}_C\cos^3\theta$ is the non-dimensional stiffness of the truss.

Bar elongations are evaluated with the help of the constitutive laws, and finally, the displacement can be obtained from the elongation-displacement relationships.

9.6.5 Problems

Problem 9.13. Three springs in series
(1) Use the principle of complementary virtual work to determine the forces in each of the three springs connected in series depicted in fig. 9.39. *(2)* Find the solution of the same problem using the force method developed in section 4.3.3. *(3)* Compare the two solution approaches.

Fig. 9.39. Three-spring hyperstatic system.

9.6.6 Unit load method for trusses

In section 9.6, the principle of complementary virtual work is shown to yield the compatibility equations of a system. It is also the basis for a general approach, called the *unit load method,* to determine deflections at specific points of structures. This simple and elegant method provides the displacement or rotation at any point of a structure. It will be presented here in the context of truss structures.

Consider the two-bar truss subjected to a load P, as depicted in fig. 9.40. Based on Newton's law or on the principle of virtual work, the equilibrium equations of the problem can be derived, and with the help of the free body diagram shown in the figure can be used to find the actual forces in the bars, F_A and F_B.

Next, the principle of complementary virtual work will be use to determine the displacement of point **O** in the direction of the applied load. To accomplish this, imagine that the displacement of point **O** is prescribed to be of magnitude Δ in the vertical direction, as indicated in the right part of fig. 9.40. Because the displacement at point **O** is prescribed to be Δ, the external complementary work is given by eq. (9.56) as $\delta W'_E = \Delta\delta D$, where δD is the virtual driving force that is applied to

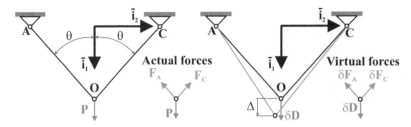

Fig. 9.40. The unit load method for a two-bar truss.

achieve the desired displacement, Δ. The principle of complementary virtual work, eq. (9.57), now implies $\delta W'_E + \delta W'_I = 0$ and can be recast as

$$\Delta \delta D = -\delta W'_I, \tag{9.62}$$

for all statically admissible virtual forces.

The internal complementary virtual work is given by eq. (9.50) as $\delta W'_I = -e_A \delta F_A - e_C \delta F_C$, where e_A and e_C are actual elongations of bars **A** and **C**, respectively. The principle of complementary virtual work, eq. (9.62), becomes $\Delta \delta D = e_A \delta F_A + e_C \delta F_C$. For a more general truss consisting of N_b bars, this statement can be written as

$$\Delta \delta D = \sum_{i=1}^{N_b} e_i \delta F_i, \tag{9.63}$$

for all statically admissible virtual forces.

Let δD, δF_A and δF_C be a set of statically admissible virtual forces; the free body diagram in the right part of fig. 9.40 leads to $\delta F_A - \delta F_C = 0$ and $\delta D - (\delta F_A + \delta F_C) \cos \theta = 0$. Because the two equilibrium equations of the system link the three virtual forces, δD, δF_A and δF_C, it is possible to select one arbitrarily, and the other two are then obtained from the equilibrium equations. In the unit load method, *the virtual driving force is select to be a unit load*, $\delta D = 1$, from which it follows that $\delta F_A = \delta F_C = \Delta \delta D / (2 \cos \theta) = 1 / (2 \cos \theta)$.

To simplify the notation, let $\delta D = 1$ be a unit virtual driving force and let $\delta F_A = \hat{F}_A$ and $\delta F_C = \hat{F}_C$ denote the corresponding statically admissible virtual forces. Equation (9.63) now becomes

$$\Delta = \sum_{i=1}^{N_b} \hat{F}_i \, e_i. \tag{9.64}$$

This equation yields the desired displacement at a point of the truss.

Two distinct sets of forces are involved in the unit load approach: F_i and $\hat{F}_i$. Forces F_i are the actual forces that develop in the bars under the externally applied loads. Because these forces are the actual forces acting in the system, they must satisfy all equilibrium conditions, and the associated elongations must be compatible. Forces $\hat{F}_i$ and the unit driving force form a set of statically admissible forces, as required by the principle of complementary virtual work. Because these forces are statically admissible, they must satisfy the equilibrium equations, but the associated elongations are not required to be compatible.

Equation (9.64) seems to be incorrect because the left-hand side has units of displacement, whereas the right-hand side has units of force times displacement. This is because the virtual driving force is selected to be of unit magnitude. Equation (9.64) should be written as $1 \cdot \Delta = \sum_{i=1}^{N_b} \hat{F}_i \, e_i$, where the term "1" has units of force, which reconciles the units on the two sides of the equation.

If the bars are made of a linearly elastic material, the actual, compatible elongations of the bars are obtained as $e_i = F_i L_i / (E_i A_i)$. Equation (9.64) now becomes

$$\Delta = \sum_{i=1}^{N_b} \frac{\hat{F}_i F_i L_i}{E_i A_i}. \tag{9.65}$$

The unit load method is based on the principle of complementary virtual work, which expresses compatibility conditions. Material constitutive laws are not considered in the derivation of this principle, which therefore remains true for all constitutive laws. Consequently, the unit load method for trusses as expressed by eq. (9.64) applies for all material behavior, whereas the use of eq. (9.65) is limited to linearly elastic materials.

The unit load method applied to truss structures can be summarized in the following three steps.

1. Determine the forces, F_i, $i = 1, 2, \ldots, N_b$, acting in the bars under the effect of the externally applied loads. From these determine e_i, $i = 1, 2, \ldots, N_b$ using the constitutive law for the material.
2. Determine the bar forces, $\hat{F}_i$, $i = 1, 2, \ldots, N_b$. These form a set of statically admissible forces in equilibrium with a unit load applied at the point and in the direction of the desired displacement component. This is called the *unit load system*.
3. Use eq. (9.64) to evaluate the displacement component at the point and in the direction of application of the unit load. If the truss is made of a linearly elastic material, use eq. (9.65).

The first two steps of the procedure require the evaluation of two sets of bar forces generated by two distinct loading conditions: first, the externally applied loads, and second, a unit load applied at a joint. If the truss is isostatic, bar forces can be evaluated based on the equilibrium equations at each joint of the truss. If the truss is hyperstatic, the unit load method is still applicable, although the evaluation of bars forces for these two loading conditions becomes more cumbersome; such cases will be treated in in section 9.8.1.

The unit load method can also be used to determine rotation at a point of the structure. A slight modification of the procedure presented above is then required: instead of applying a unit load, a unit moment is applied. For prescribed rotations, the complementary external virtual work is $\delta W'_E = \Phi \delta M$, where Φ is the prescribed rotation and δM the virtual driving moment. The principle of complementary virtual work, eq. (9.57), now implies $\delta W_E' + \delta W_I' = 0$, or

$$\Phi \delta M = -\delta W_I', \tag{9.66}$$

Instead of using a unit force, $\delta D = 1$, a unit moment is used, $\delta M = 1$, and the rest of the procedure is identical to that described above. For such cases, the method becomes the "unit moment method." Equation (9.64) now becomes $\Phi = \sum_{i=1}^{N_b} \hat{F}_i \, e_i$, where $\hat{F}_i$ are the forces acting in the bars under the action of this unit moment.

As pointed out earlier, the unit load method is not restricted to linearly elastic materials. If a nonlinear material is employed in the truss, eq. (9.64) still applies. For an isostatic truss, once the bar forces associated with the externally applied loads have been determined from the joint equilibrium equations, bar extensions can be evaluated from the nonlinear material constitutive law. In fact, it is also possible to consider bar extensions that are not due to mechanical loads. These include manufacturing imperfections and thermal deformations.

In the examples presented below, a tabular presentation will be used to keep track of the contributions of individual bars. This approach is convenient and minimizes computational errors.

Example 9.16. Joint deflection in a simple 2-bar truss
A simple two-bar truss shown in fig. 9.41 will be used to illustrate the unit load method. The truss member stiffnesses are $k_A = (E\mathcal{A}/L)_A$ and $k_C = (E\mathcal{A}/L)_C$.

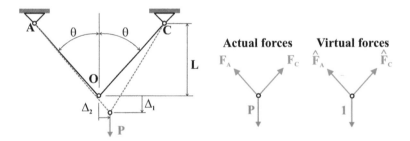

Fig. 9.41. Two-bar truss with unsymmetric properties and vertical load at joint.

Step 1 starts with the determination of the bar forces and extensions due to the externally applied loads. The equilibrium equations for the joint **O** yield $F_A = F_C = P/(2\cos\theta)$. The elongations associated with these bar forces are $e_A = F_A L_A/(E\mathcal{A})_A$ and $e_C = F_C L_C/(E\mathcal{A})_C$, and hence,

$$e_A = \frac{PL}{(E\mathcal{A})_A} \frac{1}{2\cos^2\theta}, \quad e_C = \frac{PL}{(E\mathcal{A})_C} \frac{1}{2\cos^2\theta}.$$

In step 2, a unit load is applied at the point and in the direction of the desired deflection component. Next, the bar forces arising from the application of this unit load are evaluated. This is illustrated in the second free body diagram in the right part of fig. 9.41. Because the unit virtual load acts at the same point and in the same direction as the applied load, the same equilibrium equation yields the desired forces as

$$\hat{F}_A = \frac{1}{2\cos\theta}, \quad \hat{F}_C = \frac{1}{2\cos\theta}.$$

The last step of the procedure uses eq. (9.65) to find the vertical displacement of joint **O** as

$$\Delta_1 = \sum_{i=1}^{N_b} \hat{F}_i e_i = \frac{1}{2\cos\theta} \frac{PL}{2\cos^2\theta(E\mathcal{A})_A} + \frac{1}{2\cos\theta} \frac{P}{2\cos^2\theta(E\mathcal{A})_C}$$

$$= \frac{PL}{4\cos^3\theta} \frac{(E\mathcal{A})_A + (E\mathcal{A})_C}{(E\mathcal{A})_A(E\mathcal{A})_C}. \tag{9.67}$$

Next, the horizontal deflection component of joint **O**, denoted Δ_2, will be evaluated. The first step of the procedure is identical to that developed above. In step 2, a unit load is applied in the horizontal direction at joint **O**, because the desired displacement component is in that direction. Based on equilibrium equations, the virtual forces associated with this horizontal unit load are $\hat{F}_A = 1/2\,\sin\theta$ and $\hat{F}_C = -1/2\,\sin\theta$. Equation (9.64) then yields the desired displacement components as

$$\Delta_2 = \sum_{i=1}^{N_b} \hat{F}_i e_i = \frac{1}{2\sin\theta} \frac{PL}{2\cos^2\theta(E\mathcal{A})_A} - \frac{1}{2\sin\theta} \frac{P}{2\cos^2\theta(E\mathcal{A})_C}$$

$$= \frac{PL}{4\sin\theta\cos^2\theta} \frac{(E\mathcal{A})_A - (E\mathcal{A})_C}{(E\mathcal{A})_A(E\mathcal{A})_C}. \tag{9.68}$$

Example 9.17. Joint deflection in a 2-bay planar truss

A more complicated planar truss will be analyzed next. Consider the two-bay cantilevered truss subjected to externally applied vertical loads depicted in fig. 9.42. Determine the vertical deflection of joint **F**. The following data are provided: $L = 30$ in., and for all bars, $E = 30 \times 10^6$ psi and $\mathcal{A} = 0.1$ in^2. The load applied at joints **B** and **C** are each 1,000 lbs.

The truss material is linearly elastic and therefore eq. (9.65) can be used. This is most conveniently carried out in a tabular form in table 9.1. The first column in the table lists the bars, and the second column lists the bar flexibility factors, $L_i/(E_i\mathcal{A}_i)$.

The first step of the unit load method calls for the determination of the bar forces under the externally applied loads; the results of this computation are listed in the third column of table 9.1. In the second step, a unit load is applied in the vertical direction at joint **F**. The bar forces generated by the unit load are listed in the fourth column of table 9.1. The last column of the table lists the products $\hat{F}_i F_i L_i/(E_i\mathcal{A}_i)$. In view of eq. (9.65), the sum of the numbers in that column yields the vertical displacement component at joint **F**, $\Delta = 0.185$ in.

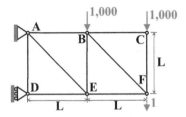

Fig. 9.42. Tip deflection of two-bay planar truss.

Table 9.1. Calculation of vertical deflection at joint **F** for the 2-bay planar truss.

Bar	$10^6 \times L_i/(E_i\mathcal{A}_i)$	F_i	$\hat{F}_i$	$\hat{F}_i F_i L_i/(E_i\mathcal{A}_i)$
AB	10	1,000	1	0.01
BC	10	0	0	0
DE	10	-3,000	-2	0.06
EF	10	-1,000	-1	0.01
AD	10	0	0	0
BE	10	-2,000	-1	0.02
CF	10	-1,000	0	0
AE	$10\sqrt{2}$	$2,000\sqrt{2}$	$\sqrt{2}$	0.056
BF	$10\sqrt{2}$	$1,000\sqrt{2}$	$\sqrt{2}$	0.028

Example 9.18. Rotation of a bar in a 2-bay planar truss

The unit load method also allows the determination of the rotation of an individual bar. To illustrate the process, the rotation of bar **CF** will be computed for the 2-bay planar truss depicted in fig. 9.43. The physical properties of the truss are identical to those used in example 9.17.

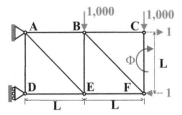

Fig. 9.43. Rotation of a bar in a two-bay planar truss.

As shown in fig. 9.43, a moment $\delta M = 1 \cdot L$ is applied to bar **CF**; this is provided by two unit load acting in opposite directions at joints **C** and **F**. Rather than a unit virtual moment, a virtual moment of magnitude $1 \cdot L$ is applied; this does not matter because the magnitude of the virtual moment is arbitrary.

The first step of the method calls for the evaluation of the actual forces in all bars when the truss is subjected to the externally applied loads. These results are listed in the third column of table 9.2 and are identical to the corresponding results listed in the third column of table 9.1, because the same external loads are applied to the truss. The fourth column of table 9.2 lists the bar virtual forces generated by the unit load system applied to bar **CF**. Equation (9.65) now yields

$$1 \cdot L \cdot \Phi = \sum_{i=1}^{N_b} \frac{\hat{F}_i F_i L_i}{E_i \mathcal{A}_i}, \quad \text{or} \quad \Phi = \frac{1}{L} \sum_{i=1}^{N_b} \frac{\hat{F}_i F_i L_i}{E_i \mathcal{A}_i}.$$

The last column of the table lists the individual contributions of each bar, and summing up all contributions yields $\Phi = 0.05/L = 0.00166 \text{ rad} = 0.96°$.

Table 9.2. Calculation of rotation of bar **CF** in the 2-bay planar truss.

Bar	$10^6 \times L_i/(E_i \mathcal{A}_i)$	F_i	$\hat{F}_i$	$\hat{F}_i F_i L_i/(E_i \mathcal{A}_i)$
AB	10	1,000	1	0.01
BC	10	0	1	0
DE	10	-3,000	-1	0.03
EF	10	-1,000	-1	0.01
AD	10	0	0	0
BE	10	-2,000	0	0
CF	10	-1,000	0	0
AE	$10\sqrt{2}$	$2,000\sqrt{2}$	0	0
BF	$10\sqrt{2}$	$1,000\sqrt{2}$	0	0

9.6.7 Problems

Problem 9.14. Deflection of a simple square truss

Consider the square planar truss shown in fig. 9.44 and assume that all bars are of cross-sectional area, $\mathcal{A}$, and modulus, E. Joints **D**, **E**, and **F** are pinned to the ground. This problem presents symmetries that may be helpful in simplifying the force calculations. *(1)* Find the vertical deflection at joint **A**. *(2)* Find the increase in horizontal distance between joints **B** and **C**.

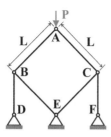

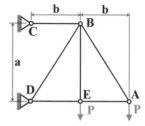

Fig. 9.44. Simple square planar truss with vertical load at joint **A**.

Fig. 9.45. Planar truss with nonlinear material properties.

Problem 9.15. Deflection of a truss with nonlinear material properties

The truss shown in fig. 9.45 is made of a material presenting a nonlinearly elastic behavior described by the following constitutive equation: $\sigma_1 = E_0 \epsilon_1^n$, where $E_0 = 500,000$ psi and $n = 1/3$. The truss dimensions are $a = 40$ in and $b = 30$ in, while the bar cross-sectional areas are $\mathcal{A}_{AB} = 2$, $\mathcal{A}_{BC} = 1$, $\mathcal{A}_{BD} = 2$, $\mathcal{A}_{BE} = 0.5$, $\mathcal{A}_{AE} = \mathcal{A}_{DE} = 1.5$ in^2, and the load $P = 10,000$ lbs. *(1)* Find the vertical and horizontal deflections of joint **A**.

Problem 9.16. Deflection of a planar truss

All bars of the planar truss shown in fig. 9.46 are of cross-sectional area, $\mathcal{A}$, and modulus, E. Joints **C** and **D** are pinned to the vertical wall. *(1)* Use the unit load method to find the vertical and the horizontal deflections of joint **A**, and *(2)* the rotation of bar **AB**.

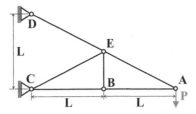

Fig. 9.46. Planar truss with tip load.

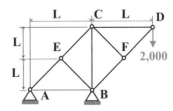

Fig. 9.47. Planar truss with tip load.

Problem 9.17. Deflection of a planar truss

Determine the vertical deflection of joint **D** in the planar truss shown in fig. 9.47. The truss is supported at point **A** and **B**, but no bar joins these two points. All bar are of cross-sectional area, $\mathcal{A}$, and modulus, E.

9.7 Internal virtual work in beams and solids

The previous sections of this chapter focus on simple mechanical problems such as particles, systems of particles, and trusses. In each case, expressions for work, virtual work, and complementary virtual work are developed. A formal proof of the principle of virtual work for three-dimensional solids will be presented in chapter 12, but by induction, it is assumed here that the principle developed for particles, systems of particles, and trusses, also holds for beams and three-dimensional solids. The key to this generalization is to develop an expression for the internal virtual work that is adapted to the structure under investigation.

Similarly, the principle of complementary virtual work is derived for trusses, but it will be shown in chapter 12 that it remains true for three-dimensional solids. By induction, the principle of complementary virtual work for trusses, stated as principle 7 and expressed by eq. (9.57), is generalized to state that "a structure undergoes compatible deformations if and only if the sum of the internal and external complementary virtual work vanishes for all statically admissible virtual stresses." Here again, the key to this generalization is to develop an expression for the complementary internal virtual work that is adapted to the structure under investigation.

Expressions for the virtual work and complementary virtual work in beams and three-dimensional solids will be developed in this section.

9.7.1 Beam bending

Euler-Bernoulli beam bending will be investigated first. The associated kinematic assumptions and their implications are presented in section 5.2. For simplicity, it is assumed that plane $(\bar{\imath}_1, \bar{\imath}_2)$ is a plane of symmetry of the problem. A bending moment, $M_3(x_1)$, acts, resulting in a rotation of the section, denoted $\Phi_3(x_1)$, and a transverse displacement, $\bar{u}_2(x_1)$.

Consider an infinitesimal slice of a beam depicted in fig. 9.48. Under the action of the bending moment, the two neighboring sections rotate by angles Φ_3 and $\Phi_3 + d\Phi_3$, at span-wise locations x_1 and $x_1 + dx_1$, respectively. The differential rotation of the two cross-sections generates the curvature of the differential element, $\kappa_3 = \Phi_3' = \bar{u}_2''$, see eq. (5.6).

The work done by a moment is the product of the moment by the rotation of its point of application. The work done by the moment acting on the left-hand side of the differential element of the beam is $-M_3\Phi_3$, where the minus sign is due to the fact that the moment and rotation are counted positive about opposite axes. The work done by the moment acting on the other side of the element is $M_3(\Phi_3 + d\Phi_3)$. Finally, the net work done by the two moments, dW, is found by summing up the two contributions to find $dW = M_3 d\Phi_3 = M_3(d\Phi_3/dx_1)dx_1$. The total internal work done by the moment distribution acting in the beam of length L is then found by integration,

$$W_I = -\int_0^L M_3 \frac{d\Phi_3}{dx_1}\, dx_1 = -\int_0^L M_3\kappa_3\, dx_1. \tag{9.69}$$

The minus sign is due to the fact that the internal work is that done by the internal moment, which is opposite in sign to the moment applied externally to the cross-section.

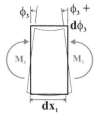

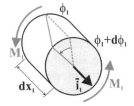

Fig. 9.48. Bending deformation of an infinitesimal segment of a beam.

Fig. 9.49. Torsional deformation of an infinitesimal segment of a beam.

The virtual internal work and its complementary counterpart are then found by considering the virtual work done by moments undergoing virtual curvatures, $\delta\kappa_3$, and the virtual work done by virtual moments, δM_3, undergoing actual curvatures, respectively. The internal virtual work and its complementary counterpart for Euler-Bernoulli beam bending can thus be written as

$$\delta W_I = -\int_0^L M_3 \delta \kappa_3 \, \mathrm{d}x_1, \tag{9.70a}$$

$$\delta W_I' = -\int_0^L \kappa_3 \delta M_3 \, \mathrm{d}x_1. \tag{9.70b}$$

Of course, similar expressions can be derived for the internal virtual work and its complementary counterpart for bending moments, M_2, and curvatures, κ_2, acting in the orthogonal plane.

9.7.2 Beam twisting

Next, the problem of torsion of a circular bar will be investigated. The associated kinematic assumptions and their implications are presented in section 7.1.1. A torque, $M_1(x_1)$, is acting, resulting in a rotation of the section, denoted $\Phi_1(x_1)$. Consider the infinitesimal slice of a beam depicted in fig. 9.49. Under the action of the torque, the two neighboring sections rotate by angles Φ_1 and $\Phi_1 + \mathrm{d}\Phi_1$, at span-wise locations x_1 and $x_1 + \mathrm{d}x_1$, respectively. The differential rotation of the two cross-sections generates the twist rate of the differential element, $\kappa_1 = \Phi_1'$, see eq. (7.7).

The work done by the torque acting on the left-hand side of the differential element of the beam is $-M_1 \Phi_1$, where the minus sign is due to the fact that the torque and rotation are counted positive about opposite axes. The work done by the torque acting on the other side of the element is $M_1(\Phi_1 + \mathrm{d}\Phi_1)$. Finally, the net work done by the two torques, $\mathrm{d}W$, is found by summing up the two contributions to find $\mathrm{d}W = M_1 \mathrm{d}\Phi_1 = M_1(\mathrm{d}\Phi_1/\mathrm{d}x_1)\mathrm{d}x_1$. The total internal work done by the torque distribution acting in the beam of length L is then found by integration,

$$W_I = -\int_0^L M_1 \frac{\mathrm{d}\Phi_1}{\mathrm{d}x_1} \, \mathrm{d}x_1 = -\int_0^L M_1 \kappa_1 \, \mathrm{d}x_1. \tag{9.71}$$

The minus sign is due to the fact that the internal work is that done by the internal torque, which is opposite in sign to the torque applied externally to the cross-section.

The virtual internal work and its complementary counterpart are then found by considering the virtual work done by torques undergoing virtual twist rates, $\delta \kappa_1$, and the virtual work done by virtual torques, δM_1, undergoing actual twist rates, respectively. The internal virtual work and its complementary counterpart for torsion of circular cylinders now become

$$\delta W_I = -\int_0^L M_1 \delta \kappa_1 \, \mathrm{d}x_1, \tag{9.72a}$$

$$\delta W_I' = -\int_0^L \kappa_1 \delta M_1 \, \mathrm{d}x_1. \tag{9.72b}$$

A similar development will reveal that identical expressions hold for the torsion of bars with cross-sections of arbitrary shape. This development is based on the kinematic assumptions of Saint-Venant's theory of uniform torsion, as presented in section 7.3.2.

9.7.3 Three-dimensional solid

The more general case of a three-dimensional solid will now be addressed. The strain-displacement equations for a three-dimensional solid are presented in section 1.4.1. At a specific point of the solid, three direct and three shear stress components are acting. To simplify the presentation, the work done by each of these six stress components will be computed separately, and because work is an additive quantity, the total work will be found by summing up the contributions of each stress component.

Axial stresses

Consider the infinitesimal differential element of a solid depicted in fig. 9.50. Under the effect of the axial stress component, σ_1, the displacement components of two neighboring edges become u_1 and $u_1 + du_1$, at locations x_1 and $x_1 + dx_1$, respectively. The differential displacement of the two edges generates the axial strain of the differential element, $\epsilon_1 = \partial u_1/\partial x_1$, see eq. (1.63).

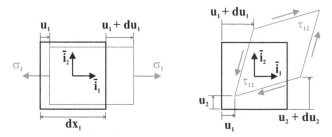

Fig. 9.50. Deformation of a differential element of a solid.

The work done by the force, $\sigma_1 dx_2 dx_3$, acting on the left-hand side of the differential element is $-(\sigma_1 dx_2 dx_3)u_1$ where the minus sign is due to the fact that the force and displacement are counted positive in opposite directions. The work done by the force acting on the other side of the element is $(\sigma_1 dx_2 dx_3)(u_1 + du_1)$. Finally, the net work done by the two forces, dW, is found by summing up the two contributions to find $dW = (\sigma_1 dx_2 dx_3)du_1 = (\sigma_1 dx_2 dx_3)(\partial u_1/\partial x_1)dx_1$. The total internal work done by the axial stress distribution acting in the solid of volume $\mathcal{V}$ is then found by integration,

$$W_I = -\int_{\mathcal{V}} \sigma_1 \frac{\partial u_1}{\partial x_1}\, dx_1 dx_2 dx_3 = -\int_{\mathcal{V}} \sigma_1 \epsilon_1 \, d\mathcal{V}. \tag{9.73}$$

Again, the minus sign is due to the fact that the internal work is that done by the internal axial stresses, which are opposite in sign to the axial stresses applied to the differential element. The differential element of volume is written as $dV = dx_1 dx_2 dx_3$. Following a similar reasoning, the internal work associated with the other two axial stress components, σ_2 and σ_3, can also be found.

Shear stresses

The work done by the three shear stress components requires further attention. Due to the principle of reciprocity of shear stresses, eq. (1.5), shear stress components will act on the right and left edges of the differential element but also on the top and bottom edges of the same element, as illustrated in the right part of fig. 9.50. The work done by the force, $\tau_{12} dx_1 dx_3$, acting on the bottom edge of the differential element of the solid is $-(\tau_{12} dx_1 dx_3) u_1$, where the minus sign is due to the fact that the force and displacement are counted positive in opposite directions. The work done by the force acting on the top edge of the differential element is $(\tau_{12} dx_1 dx_3)(u_1 + du_1)$. Finally, the net work done by these two forces, dW, is found by summing up the two contributions to find $dW = (\tau_{12} dx_1 dx_3) du_1 = (\tau_{12} dx_1 dx_3)(\partial u_1 / \partial x_2) dx_2$.

Next, the work done by the force, $\tau_{12} dx_2 dx_3$, acting on the left edge of the differential element of the solid is $-(\tau_{12} dx_2 dx_3) u_2$, where the minus sign is due to the fact that the force and displacement are counted positive in opposite directions. The work done by the force acting on the right edge of the differential element is $(\tau_{12} dx_2 dx_3)(u_2 + du_2)$. Finally, the net work done by these two forces, dW, is found by summing up the two contributions to find $dW = (\tau_{12} dx_2 dx_3) du_2 = (\tau_{12} dx_2 dx_3)(\partial u_2 / \partial x_1) dx_1$. The total internal work done by the shear stress distribution acting in the solid of volume V is then found by integration,

$$W_I = -\int_V \tau_{12} \left(\frac{\partial u_1}{\partial x_2} + \frac{\partial u_2}{\partial x_1} \right) dx_1 dx_2 dx_3 = -\int_V \tau_{12} \gamma_{12} \, dV. \qquad (9.74)$$

Again, the minus sign is due to the fact that the internal work is that done by the internal shear stresses, which are opposite in sign to the shear stresses applied to the differential element. Following a similar reasoning, the internal work associated with the other two shear stress components, τ_{23} and τ_{13}, can also be found.

The total work done by all six stress components is found by summing up all contributions to find

$$W_I = -\int_V (\sigma_1 \epsilon_1 + \sigma_2 \epsilon_2 + \sigma_3 \epsilon_3 + \tau_{23} \gamma_{23} + \tau_{13} \gamma_{13} + \tau_{12} \gamma_{12}) \, dV. \qquad (9.75)$$

To simplify the notation, the strain and stress arrays defined in eqs. (2.11a) and (2.11b), respectively, will be used here. The internal work then becomes

$$W_I = -\int_V \underline{\sigma}^T \underline{\epsilon} \, dV. \qquad (9.76)$$

The virtual internal work and its complementary counterpart are then found by considering the virtual work done by stress components undergoing a virtual strains,

$\delta\epsilon$, and the virtual work done by virtual stresses, $\delta\underline{\sigma}$, undergoing actual strains, respectively. The internal virtual work and its complementary counterpart for three-dimensional solids now become

$$\delta W_I = -\int_{\mathcal{V}} \underline{\sigma}^T \delta\underline{\epsilon}\, d\mathcal{V}, \tag{9.77a}$$

$$\delta W_I' = -\int_{\mathcal{V}} \underline{\epsilon}^T \delta\underline{\sigma}\, d\mathcal{V}. \tag{9.77b}$$

9.7.4 Euler-Bernoulli beam

The Euler-Bernoulli beam will be reexamined, but rather than using the procedure described in section 9.7.1, the beam will be viewed as a three-dimensional solid, and the results of section 9.7.3 will be used. Equations (5.5) give the complete strain field in Euler-Bernoulli beams: all strain components vanish, except for the axial strain, given by eq. (5.7). Using eq. (9.75), the total work done by the sole non-vanishing strain component becomes

$$W_I = -\int_V \sigma_1\epsilon_1\, d\mathcal{V} = -\int_0^L \int_{\mathcal{A}} \sigma_1(\bar{\epsilon}_1 + x_3\kappa_2 - x_2\kappa_3)\, d\mathcal{A}dx_1$$

$$= -\int_0^L \left\{ \left[\int_{\mathcal{A}} \sigma_1 d\mathcal{A}\right]\bar{\epsilon}_1 + \left[\int_{\mathcal{A}} \sigma_1 x_3 d\mathcal{A}\right]\kappa_2 + \left[-\int_{\mathcal{A}} \sigma_1 x_2 d\mathcal{A}\right]\kappa_3 \right\} dx_1.$$

The integration over the beam's volume is separated into an integration along the beam's length, L, followed by an integration over its cross-section, $\mathcal{A}$. The first bracketed term is the axial force, N_1, defined by eq. (5.8), whereas the next two bracketed terms are the two bending moments, M_2 and M_3, both defined by eqs. (5.10). The internal work done by the axial stress component in an Euler-Bernoulli beam now reduces to

$$W_I = -\int_0^L (N_1\bar{\epsilon}_1 + M_2\kappa_2 + M_3\kappa_3)\, dx_1. \tag{9.78}$$

Clearly, the internal work presented in equation (9.69) is a particular case of this more general result.

The virtual internal work and its complementary counterpart are then found by considering the virtual work done by stress resultants undergoing virtual deformations, $\delta\bar{\epsilon}_1$, $\delta\kappa_2$, and $\delta\kappa_3$, and the virtual work done by virtual stress resultants, δN_1, δM_2, and δM_3, undergoing actual deformations, respectively. The internal virtual work and its complementary counterpart for Euler-Bernoulli beam thus become

$$\delta W_I = -\int_0^L (N_1\delta\bar{\epsilon}_1 + M_2\delta\kappa_2 + M_3\delta\kappa_3)\, dx_1, \tag{9.79a}$$

$$\delta W_I' = -\int_0^L (\bar{\epsilon}_1\delta N_1 + \kappa_2\delta M_2 + \kappa_3\delta M_3)\, dx_1. \tag{9.79b}$$

9.7.5 Problems

Problem 9.18. Virtual work in circular tubes
(1) Starting from eq. (9.75) and using the kinematic assumption for the torsion of circular bars presented in section 7.1.1, develop a general expression for the internal work in circular bars. *(2)* Develop a general expression of the internal virtual work and complementary internal virtual work in circular bars.

Problem 9.19. Virtual work in torsion of bars
(1) Starting from eq. (9.75) and using the kinematic assumption for Saint-Venant's torsion theory of bars presented in section 7.3.2, develop a general expression for the internal work for the torsion of bars. *(2)* Develop a general expression of the internal virtual work and complementary internal virtual work for the torsion of bars.

9.7.6 Unit load method for beams

In section 9.6.6, the unit load method is presented for application to truss structures. Because the unit load method is a direct consequence of the principle of complementary virtual work, its application is easily generalized to Euler-Bernoulli beam structures. If the displacement at a point of the beam is prescribed to be of magnitude Δ, the principle of complementary virtual work, eq. (9.57), requires

$$\Delta \delta D + \delta W_I' = 0,$$

for all statically admissible virtual forces, where δD is the virtual driving force. The complementary internal virtual work in the beam, $\delta W_I'$, is given by eq. (9.79b), and the above equation can be written as

$$\Delta \delta D = \int_0^L \left(\bar{\epsilon}_1 \delta N_1 + \kappa_2 \delta M_2 + \kappa_3 \delta M_3 \right) \, dx_1, \tag{9.80}$$

where δD, δN_1, δM_2, and δM_3, are statically admissible virtual forces and moments, whereas $\bar{\epsilon}_1$, κ_2, and κ_3, are the actual deformations of the beam.

Following the procedure developed for truss structures, see section 9.6.6, the virtual driving force is selected to be a unit force, $\delta D = 1$, and $\delta N_1 = \hat{N}_1$, $\delta M_2 = \hat{M}_2$ and $\delta M_3 = \hat{M}_3$ are the resulting statically admissible axial forces and bending moments. Equation (9.80) now becomes

$$\Delta = \int_0^L \left(\hat{N}_1 \bar{\epsilon}_1 + \hat{M}_2 \kappa_2 + \hat{M}_3 \kappa_3 \right) \, dx_1, \tag{9.81}$$

Next, the beam is assumed to be made of a linearly elastic material. If the origin of the axis system is selected to be at the centroid of the cross-section, the sectional constitutive laws are given by eq. (6.13), and these can be used to eliminate the sectional strain and curvatures in eq. (9.81) to yield

$$\Delta = \int_0^L \left[\frac{\hat{N}_1 N_1}{S} + \frac{\hat{M}_2 (H_{33}^c M_2 + H_{23}^c M_3)}{\Delta_H} + \frac{\hat{M}_3 (H_{23}^c M_2 + H_{22}^c M_3)}{\Delta_H} \right] \, dx_1, \tag{9.82}$$

If the axes are selected to be the principal centroidal axes of bending, this result further simplifies to

$$\Delta = \int_0^L \left[\frac{\hat{N}_1 N_1}{S} + \frac{\hat{M}_2 M_2}{H_{22}^c} + \frac{\hat{M}_3 M_3}{H_{33}^c} \right] dx_1, \tag{9.83}$$

The unit load method applied to Euler-Bernoulli beams can be summarized in the following three steps.

1. Find the actual force and moment distributions acting in the beam under the action of the externally applied loads.
2. Apply a unit load at the point and in the direction of the desired displacement component. Evaluate the statically admissible force and moment distributions acting in the beam that are in equilibrium with this unit load; this is called the *unit load system*.
3. Use eq. (9.81) to evaluate the displacement component at the point and in the direction of application of the unit load. If the truss is made of a linearly elastic material and the origin of the axes is at the section's centroid, use eq. (9.82). If principal axes of bending are used, use eq. (9.83).

The first two steps of the procedure require the evaluation of two sets of force and moment distributions generated by two distinct loading conditions: first, the externally applied loads, and second, a unit load. If the beam is isostatic, these distributions can be evaluated based on the equilibrium equations. If the beam is hyperstatic, the unit load method is still applicable, although the evaluation of the force and moment distributions for these two loading conditions becomes more cumbersome.

The unit load method can also be used to determine rotation at a point of the beam. A slight modification of the procedure presented above is then required: instead of applying a unit load, a unit moment is applied. For prescribed rotations, the complementary virtual work is $\delta W'_E = \phi \delta M$, where ϕ is the prescribed rotation and δM the virtual driving moment. Instead of using a unit force, $\delta D = 1$, a unit moment is used, $\delta M = 1$. For such cases, the method becomes the "unit moment method."

The unit load method described above also applies to torsion problems. In this case, the relevant complementary internal virtual work expression is given by eq. (9.72b). If the beam is subjected to both bending moments and torques, the relevant complementary internal virtual work expression is the sum of those for bending and torsion, *i.e.*, the sum of eqs. (9.79b) and (9.72b).

Example 9.19. *Deflection of a tip-loaded cantilevered beam*
Consider the cantilevered beam of length L subjected to a concentrated load, P, as depicted in fig. 9.51. In this example, the load is applied at the beam's tip, $\alpha = 1$. Find the tip deflection of the beam. This problem is treated using the classical, differential equation approach in example 5.8 on page 201.

The first step of the unit load method calls for the evaluation of the bending moment distribution under the externally applied loads. Simple equilibrium arguments yield $M_3(x_1) = P(x_1 - L)$. Since the tip deflection is desired, a vertical unit load

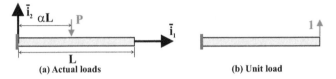

Fig. 9.51. Cantilevered beam under tip load.

is applied at the tip of the beam, as illustrated in the right part of fig. 9.51. Here again, equilibrium arguments yield the corresponding bending moment distribution as $\hat{M}_3(x_1) = -1(x_1 - L)$. The tip deflection, Δ, now follows from eq. (9.83) as

$$\Delta = \int_0^L \frac{\hat{M}_3 M_3}{H_{33}^c} \, dx_1 = \int_0^L \frac{[P(x_1 - L)][-(x_1 - L)]}{H_{33}^c} \, dx_1 = -\frac{PL^3}{3H_{33}^c}.$$

This result is in agreement with that found using the classical differential equation approach, see eq. (5.55). The minus sign in the present solution indicates that the tip displacement is *in the direction opposite to that of the unit load*, i.e., downward, as expected. Indeed, in the derivation of the unit load method, the external complementary virtual work is expressed as $\delta W_E' = \Delta \delta D$, where Δ is the desired displacement component and δD the virtual driving force. For his expression to be correct, both displacement and driving force must be counted as positive *along the same direction*. Hence, a positive displacement Δ is along the direction of the unit driving force.

The unit load method yields the desired result without requiring the solution of the governing differential equation, thereby considerably easing the solution process. Of course, if the transverse displacement distribution at all points along the beam is desired, the solution of the governing differential equation would be more expeditious, see example 5.8.

Example 9.20. Tip deflection of a cantilever beam with concentrated load
Consider now the cantilevered beam of length L subjected to a concentrated load, P, applied at a distance αL from the beam's root, as depicted in fig. 9.51. Find the tip deflection of the beam.

First, the bending moment distribution under the externally applied loads is obtained from equilibrium arguments as

$$M_3(\eta) = PL \begin{cases} -(\alpha - \eta), & 0 \leq \eta \leq \alpha, \\ 0, & \alpha < \eta \leq 1, \end{cases}$$

where $\eta = x_1/L$ is the non-dimensional variable along the beam's span.

Since the tip deflection is desired, a vertical unit load is applied at the tip of the beam, as illustrated in the right part of fig. 9.51. Here again, equilibrium arguments yield the corresponding bending moment distribution as $\hat{M}_3(\eta) = L(1 - \eta)$.

The tip deflection, Δ, now follows from eq. (9.83) as

$$\Delta = \int_0^\alpha \frac{[-PL(\alpha - \eta)][L(1 - \eta)]}{H_{33}^c} \, L \mathrm{d}\eta + \int_\alpha^1 \frac{[0][L(1 - \eta)]}{H_{33}^c} \, L \mathrm{d}\eta$$

$$= -\frac{PL^3}{H_{33}^c} \int_0^\alpha (\alpha - \eta)(1 - \eta) \, \mathrm{d}\eta = -\frac{PL^3}{6H_{33}^c} \alpha^2 (3 - \alpha).$$

Here again, this result is in agreement with that found using the classical differential approach, see eq. (5.55).

Example 9.21. Displacement field of a uniformly loaded cantilever beam
The unit load method can also be applied to situations involving distributed loading. Consider the case of a cantilever with a uniform load, p_0, as shown in fig. 9.52. Determine the tip displacement and the entire displacement field for the cantilevered beam.

Fig. 9.52. Tip deflection of a uniformly loaded cantilever.

The bending moment distribution caused by the uniform loading is found from equilibrium arguments as $M_3 = p_0 L^2 (1 - \eta)^2 / 2$, where $\eta = x_1/L$ is the non-dimensional variable along the beam's span.

First, the beam's tip deflection will be computed, and hence, a unit load is applied at its tip; the associated bending moment distribution is $\hat{M}_3 = L(1 - \eta)$. The tip deflection, Δ, now follows from eq. (9.83) as

$$\Delta = \int_0^1 \frac{[p_0 L^2 (1 - \eta)^2 / 2][L(1 - \eta)]}{H_{33}^c} \, L \mathrm{d}\eta = \frac{p_0 L^4}{2H_{33}^c} \int_0^1 (1 - \eta)^3 \, \mathrm{d}\eta = \frac{p_0 L^4}{8H_{33}^c},$$

The unit load method can also be used to compute the entire displacement field. For this purpose, a unit load is applied at location αL along the beam's span, as illustrated in fig. 9.52. The associated bending moment distribution is $\hat{M}_3 = L(\alpha - \eta)$ for $0 \le \eta \le \alpha$ and $\hat{M}_3 = 0$ for $\alpha \le \eta < 1$. The displacement field, $\Delta(\eta)$, now follows from eq. (9.83) as

$$\Delta(\alpha) = \int_0^\alpha \frac{[p_0 L^2 (1 - \eta)^2 / 2][L(\alpha - \eta)]}{H_{33}^c} \, L \mathrm{d}\eta + \int_\alpha^1 \frac{[p_0 L^2 (1 - \eta)^2 / 2][0]}{H_{33}^c} \, L \mathrm{d}\eta$$

$$= \frac{p_0 L^4}{2H_{33}^c} \int_0^\alpha (1 - \eta)^2 (\alpha - \eta) \, \mathrm{d}\eta = \frac{p_0 L^4}{24 H_{33}^c} \alpha^2 (6 - 4\alpha + \alpha^2).$$

As α varies along the beam's span, the entire displacement field is recovered. The present result matches that obtained with the classical differential equation approach, see eq. (5.54).

Example 9.22. Tip deflection of a simply supported beam

Consider a simply supported beam of length L with an overhang of length $L/2$. The first portion of the beam is subjected to a uniform loading, p_0. Determine the deflection and rotation at point **T**, as indicated in fig. 9.53.

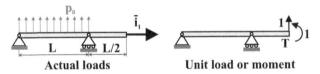

Actual loads   **Unit load or moment**

Fig. 9.53. Deflection at tip of a uniformly loaded simply-supported beam with overhang.

The bending moment distribution associated with the externally applied load found from statics as $M_3 = p_0 L^2(\eta^2 - \eta)/2$ for $0 \le \eta \le 1$ and $M_3 = 0$ for $1 \le \eta \le 3/2$, where $\eta = x_1/L$ is the non-dimensional variable along the beam's span.

To determine the deflection at point **T**, a vertical unit load is applied at that point, and the associated bending moment distribution is $\hat{M}_3 = L\eta/2$, for $0 \le \eta \le 1$; and $\hat{M}_3 = L(3/2 - \eta)$, for $1 \le \eta \le 3/2$. Equation (9.83) then yields the deflection, Δ, at point **T**, as

$$
\Delta = \int_0^1 \frac{[p_0 L^2(\eta^2 - \eta)/2][L\eta/2]}{H_{33}^c} \, L d\eta + \int_1^{3/2} \frac{[0][L(3/2 - \eta)]}{H_{33}^c} \, L d\eta
$$

$$
= \frac{p_0 L^4}{4H_{33}^c} \int_0^1 (\eta^3 - \eta^2) \, d\eta = -\frac{p_0 L^4}{48 H_{33}^c}.
$$

The negative sign means that the tip deflection is downward, *i.e.*, in the opposite direction of the unit load. Because the bending moment distribution associated with the externally applied loads vanishes in the overhang portion of the beam, the second integral in the above equation vanishes; consequently, it is not required to compute the bending moment distribution associated with the unit load over that portion of the beam, a further simplification of the procedure.

To determine the rotation at point **T**, a unit moment is applied at that point, and the associated bending moment distribution is $\hat{M}_3 = \eta$, for $0 \le \eta \le 1$; the rest of the bending moment distribution need not be computed. Equation (9.83) then yields the rotation, Φ, at point **T**, as

$$
\Phi = \int_0^1 \frac{[p_0 L^2(\eta^2 - \eta)/2][\eta]}{H_{33}^c} \, L d\eta = \frac{p_0 L^3}{2H_{33}^c} \int_0^1 (\eta^3 - \eta^2) \, d\eta = -\frac{p_0 L^3}{24 H_{33}^c}.
$$

This final result is non-dimensional, as should be expected for a rotation, which is measured in radians.

Example 9.23. Bent beam assembly under tip load

Consider the three-dimensional, bent beam assembly depicted in fig. 9.54. The beam's cross-section is assumed to be circular, and hence, $H_{11} = GI_{11}$ and

$H_{22}^c = H_{33}^c = EI_{11}/2$, where I_{11} is the section's second area moment. The beam consists of three segments **AB**, **BC**, and **CD** connected at right angles to each other. Load P is applied at point **A**, along the direction of segment **CD**. Find the deflection, Δ, of point **A**, in the direction of the applied load. To simplify the computation, a different coordinate system is assigned to each beam segment, as depicted in the right part of fig. 9.54.

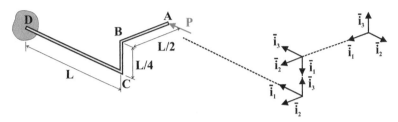

Fig. 9.54. Bent beam assembly under tip load.

Given the geometry of the system, segment **AB** is under bending in plane $(\bar{\imath}_1, \bar{\imath}_2)$, segment **BC** is under torsion and bending in plane $(\bar{\imath}_1, \bar{\imath}_3)$, and finally, segment **CD** is under bending in both planes $(\bar{\imath}_1, \bar{\imath}_2)$ and $(\bar{\imath}_1, \bar{\imath}_3)$. The bending and twisting moment distributions in each beam segment are readily found from equilibrium consideration for both the externally applied load and unit load.

For this problem, eq. (9.83) becomes

$$\Delta = \int_0^{L/2} \frac{M_3 \hat{M}_3}{H_{33}^c}\, dx_1$$
$$+ \int_0^{L/4} \frac{M_2 \hat{M}_2}{H_{22}^c}\, dx_1 + \int_0^{L/4} \frac{M_1 \hat{M}_1}{H_{11}}\, dx_1$$
$$+ \int_0^{L} \frac{M_2 \hat{M}_2}{H_{22}^c}\, dx_1 + \int_0^{L} \frac{M_3 \hat{M}_3}{H_{33}^c}\, dx_1.$$

The first line of this expression represents the contribution from the bending of segment **AB**, the second line provides the bending and torsion contributions for segment **BC**, and the third line gives the two bending contributions for segment **CD**. Introducing the bending and twisting moment distributions then leads to

$$\Delta = \int_0^{L/2} \frac{Px_1^2}{H_{33}^c}\, dx_1 + \int_0^{L/4} \frac{Px_1^2}{H_{22}^c}\, dx_1 + \int_0^{L/4} \frac{P(L/2)^2}{H_{11}}\, dx_1$$
$$+ \int_0^{L} \frac{P(L/4)^2}{H_{22}^c}\, dx_1 + \int_0^{L} \frac{P(L/2)^2}{H_{33}^c}\, dx_1.$$

Performing the integrals the yields the desired deflection as

$$\Delta = \frac{23}{64} \frac{PL^3}{H_{22}^c} + \frac{1}{16} \frac{PL^3}{H_{11}} = \frac{23}{64} \frac{PL^3}{H_{22}^c} \left[1 + \frac{2}{23} \frac{E}{G} \right] = \frac{23}{64} \frac{PL^3}{H_{22}^c} \left[1 + \frac{4(1+\nu)}{23} \right].$$

If the beam assembly is made of a material that obeys Hooke's law, eq. (2.8) implies $E = 2G(1 + \nu)$. In the last bracketed expression, the first term represents the contribution due to bending of the various beam segments, whereas the second term represents that of twisting of segment **BC**. The twisting of the middle segment of the assembly accounts for about 20% of the deflection at point **A**, assuming $\nu = 0.3$.

Example 9.24. *Bending of a cantilever with a "Z" cross-section*
Consider the cantilevered beam with a thin-walled "Z" shaped cross-section subjected to a uniform load, p_0, as shown in fig. 9.55. Find the beam's tip deflection along axes $\bar{\imath}_2$ and $\bar{\imath}_3$ using the unit load method. This problem is treated in example 6.6 on page 249 using the classical differential equation approach.

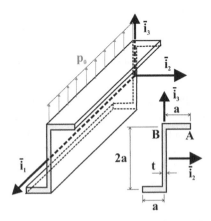

Fig. 9.55. Cantilevered Z-section beam under a uniform load.

The first step of the process is to compute the bending moment distribution due to the externally applied load, p_0: $M_2 = -p_0 L^2(1-\eta)^2/2$ and $M_3 = 0$. To compute the tip deflection along axis $\bar{\imath}_3$, the unit load must be applied at the tip along the same axis. The resulting bending moment distribution is found as $\hat{M}_2 = -L(1-\eta)$ and $\hat{M}_3 = 0$. Substituting these bending moment distributions into eq. (9.82) then yields

$$\Delta_3 = \int_0^L \frac{H_{33}^c}{\Delta_H} \hat{M}_2 M_2 \, dx_1 = \frac{6}{7Ea^3 t} \int_0^1 [-L(1-\eta)][-p_0 L^2(1-\eta)^2/2] \, L d\eta$$
$$= \frac{3p_0 L^4}{7Ea^3 t} \int_0^1 (1-\eta)^3 d\eta = \frac{3}{28} \frac{p_0 L^4}{Ea^3 t},$$

where $\eta = x_1/L$ is the non-dimensional variable along the beam's span, and the sectional bending stiffnesses, $H_{33}^c = 2Ea^3 t/3$ and $\Delta_H = 7(Ea^3 t)^2/9$, are evaluated in example 6.6. This result is in agreement with that found in example 6.6, see eq. (6.58).

To compute the tip deflection along axis $\bar{\imath}_2$, a tip unit load must be applied along that direction. The associated bending moment distribution is $\hat{M}_2 = 0$ and $\hat{M}_3 =$

$L(1 - \eta)$. The bending moment distribution due to the externally applied load is, of course, unchanged, and eq. (9.82) then leads to

$$\Delta_2 = \int_0^L \frac{H_{23}^c}{\Delta_H} \hat{M}_3 M_2 \, dx_1 = \frac{9}{7Ea^3t} \int_0^1 [L(1-\eta)][-p_0 L^2 (1-\eta)^2/2] \, L d\eta$$

$$= -\frac{9p_0 L^4}{14Ea^3t} \int_0^1 (1-\eta)^3 d\eta = -\frac{9}{56} \frac{p_0 L^4}{Ea^3t},$$

where the sectional bending stiffness, $H_{22}^c = 8Ea^3t/3$, is evaluated in example 6.6. Here again, this result matches that found in example 6.6, see eq. (6.57).

Evaluation of the beam's tip deflection is far easier when using the unit load method as compared to the classical approach that required the solution of coupled differential equations. If the complete displacement field of the beam is desired, unit loads should be applied at location αL along unit vectors $\bar{\imath}_2$ and $\bar{\imath}_3$.

Example 9.25. Torsion of a thin-walled tube with a closed section

The torsion of a thin-walled tube with a closed cross-section of arbitrary shape is investigated in section 8.5.2. The Bredt-Batho formula, $M_1 = 2\mathcal{A}f$, relates the applied torque, M_1, to the constant shear flow, f, in the thin wall, where $\mathcal{A}$ is the area enclosed by curve $\mathcal{C}$, which defines the shape of the cross-section, as depicted in fig. 9.56. To find the torsional stiffness of the structure, the first law of thermodynamics is invoked in section 8.5.2: the work done by the applied torque must equal the strain energy stored in the structure. In this example, the torsional stiffness of the structure is calculated using the unit load method. The structure is assumed to be fixed at one end, and a torque, M_1, is applied at the other, as shown in fig. 9.56.

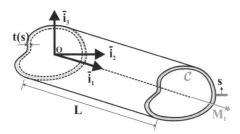

Fig. 9.56. Twisting of a thin walled tube of arbitrary cross-sectional shape.

The principle of complementary virtual work for a prescribed rotation, Φ, can be stated as

$$\Phi \delta M = -\delta W_I' = \int_{\mathcal{V}} \underline{\epsilon}^T \delta \underline{\sigma} \, d\mathcal{V} = \int_0^L \int_{\mathcal{C}} \gamma_s \delta \tau_s \, t ds dx_1.$$

In this expression, the complementary external virtual work, $\Phi \delta M$, is the product of the prescribed tip rotation, Φ, by a virtual tip torque, δM. The complementary internal virtual work is taken to be that of a general three-dimensional solid, given in eq. (9.77b). Finally, as discussed in section 8.5.2, the only vanishing stress component in a thin-walled tube undergoing uniform torsion is the tangential shear strain component, γ_s.

Next, the virtual driving moment is select to be of unit magnitude, $\delta M = 1$, and the corresponding statically admissible virtual stress field is denoted $\delta \tau_s = \hat{\tau}_s$. The Bredt-Batho formula then yields $\hat{\tau}_s = \hat{M}_1/(2\mathcal{A}t) = 1/(2\mathcal{A}t)$ and $\gamma_s = \tau_s/G = M_1/(2G\mathcal{A}t)$. The tip twist now becomes

$$\Phi = \int_0^L \int_{\mathcal{C}} \frac{M_1}{2G\mathcal{A}t} \frac{1}{2\mathcal{A}t} \, t \, ds \, dx_1 = \frac{M_1}{4\mathcal{A}^2} \int_0^L \left[\int_{\mathcal{C}} \frac{ds}{Gt} \right] dx_1 = \frac{M_1 L}{4\mathcal{A}^2} \int_{\mathcal{C}} \frac{ds}{Gt}.$$

Because this is a uniform torsion problem, the twist rate is simply $\kappa_1 = \Phi/L$ or

$$\kappa_1 = \frac{\Phi}{L} = \frac{M_1}{4\mathcal{A}^2} \int_{\mathcal{C}} \frac{ds}{Gt}.$$

The torsional stiffness, H_{11}, is the constant of proportionality between the torque and the twist rate, $H_{11} = M_1/\kappa_1$, which leads to $H_{11} = 4\mathcal{A}^2 / \left[\int_{\mathcal{C}} ds/(Gt) \right]$. This result is identical to that developed in section 8.5.2, see eq. (8.67). Here again, the principle of complementary virtual work provides an elegant solution of the problem.

9.7.7 Problems

Problem 9.20. Cantilevered beam subjected to two concentrated loads
Consider the cantilevered beam subjected to two concentrated loads of equal magnitude and opposite direction applied at points **M** and **T**, as shown in fig. 9.57. (1) Compute the beam's transverse deflection at point **M**. (2) Compute the beam's transverse deflection at point **T**.

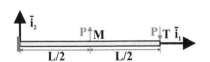

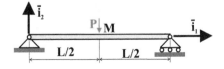

Fig. 9.57. Cantilevered beam subjected to two concentrated loads.

Fig. 9.58. Simply supported beam subjected to concentrated load.

Problem 9.21. Simply supported beam subjected to concentrated load
Consider the simply supported beam subjected to a mid-span concentrated load applied at point **M**, as shown in fig. 9.58. (1) Compute the beam's transverse deflection at point **M**. (2) Determine the beam's transverse displacement field, $\bar{u}_2(x_1)$.

Problem 9.22. Cantilevered beam subjected to triangular loading
Consider the cantilevered beam subjected to a distributed triangular loading of magnitude p_0 at the root and vanishing at the tip, as shown in fig. 9.59. (1) Compute the beam's transverse deflection at point **T**. (2) Determine the beam's transverse displacement field, $\bar{u}_2(x_1)$.

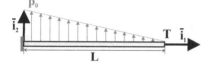

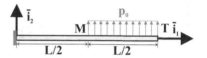

Fig. 9.59. Cantilevered beam subjected to tri-angular loading.

Fig. 9.60. Cantilevered beam under uniform loading.

Problem 9.23. Cantilevered beam subjected to uniform loading

Consider the cantilevered beam subjected to a uniform loading of magnitude p_0 extending from the beam's mid-span to its tip, as shown in fig. 9.60. *(1)* Compute the beam's transverse deflection at point **M**. *(2)* Compute the beam's transverse deflection at point **T**. *(2)* Determine the beam's transverse displacement field, $\bar{u}_2(x_1)$.

Problem 9.24. Pivoted beam supported by three-bar truss

A root pivoted beam carries a concentrated mid-span load, P, and is supported by a three-bar truss, as shown in fig. 9.61. *(1)* Combine the unit load method for beams and trusses to determine the midpoint deflection for the beam. *(2)* Determine the vertical deflection of point **A**.

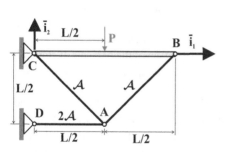

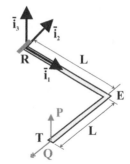

Fig. 9.61. Pivoted beam supported by three-bar truss.

Fig. 9.62. Beam under combined bending and torsion.

Problem 9.25. Beam under combined bending and torsion

The beam shown in fig. 9.62 has a circular cross-section with bending stiffnesses, $H_{22} = H_{33} = EI_{11}/2$, and torsional stiffness, $H_{11} = GI_{11}$. A torque, Q, and a vertical force, P, are applied at the beam's tip as indicated in the figure. Consider both bending and torsional deformation. *(1)* Determine the beam's tip deflection along axis $\bar{i}_3$. *(2)* Determine the beam's tip twist about axis $\bar{i}_2$.

Problem 9.26. Cantilevered beam under combined loads

A cantilevered beam of length $3L$ is subjected to a uniformly distributed loading, p_0, over its central portion and concentrated transverse loads of magnitude P acting in opposite directions at points **B** and **C**, as depicted in fig. 9.63. *(1)* Find the beam's deflection at point **A**. *(2)* Determine it rotation at point **C**.

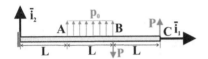

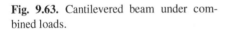

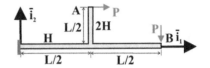

Fig. 9.63. Cantilevered beam under combined loads.

Fig. 9.64. Cantilevered beam under eccentric loading.

Problem 9.27. Cantilevered beam under eccentric loading

The cantilevered beam shown in fig. 9.64 is of length L and bending stiffness H, and carries a tip transverse load, P. A vertical beam of length $L/2$ and bending stiffness $2H$ is connected at its mid-span and carries an axial load, P. *(1)* Determine the rotation at point **A**. *(1)* Determine the transverse deflection at point **B**.

9.8 Application of the unit load method to hyperstatic problems

In the previous section, the unit load method is developed for trusses, beams, and solids. As explained in section 9.6.6, the approach calls for the determination of two sets of statically admissible forces corresponding to two distinct loading cases: the first associated with the externally applied loads, the second with the unit load. In all examples treated in the previous section, the structures are isostatic, and consequently, the two sets of statically admissible forces can always be determined solely from the equilibrium equations. The unit load method applies equally to iso- and hyperstatic system; in the latter case, however, the evaluation of the two sets of statically admissible forces is more arduous, because equilibrium equations are not sufficient for this task.

In chapter 4, two approaches are presented for the analysis of hyperstatic structures: the displacement or stiffness method and the force or flexibility method, see sections 4.3.2 and 4.3.3, respectively. The force method is particularly well-suited for dealing with hyperstatic problems because it focuses on the determination internal forces, moments and reactions. A key step of the procedure is the development of the compatibility equations that must complement the equilibrium equations to enable the solution of the problem. Because the principal of complementary virtual work is equivalent to the compatibility equations of the system, it seems logical to combine the force method with this principle.

The force method is intuitively described as the "method of cuts." For each cut made to the system, the order of the hyperstatic system is decreased by one because one internal force or moment then vanishes. For a hyperstatic system of nth order, n cuts are required to transform the original hyperstatic system into an isostatic system. Statically admissible forces in this isostatic system are then obtained solely from the equilibrium equations, and relative displacements at the cuts are evaluated. At each cut, sets of self-equilibrated forces are added, and their magnitudes are determined by enforcing the vanishing of the relative displacement at the cut.

The approach involves two crucial steps. First, determine the relative displacements at the cuts under the externally applied loads alone, and second, evaluate the internal forces applied at the cuts that are required to eliminate the relative displacements at the cuts. The principle of complementary virtual work is a powerful tool to solve both problems.

Fig. 9.65. Relative displacements and rotations.

The left part of fig. 9.65 depicts a single bar of a truss. The bar is cut to release the axial force, and a set of self-equilibrating forces of magnitude, R, is applied at the cut. Let d_1 and d_2 be the displacements at the two sides of the cut. If the displacements, d_1 and d_2, are prescribed, the associated complementary external virtual work is $\delta W'_E = d_1 \delta R - d_2 \delta R = (d_1 - d_2)\delta R$. The relative displacement at the cut is $\Delta = d_1 - d_2$, measured positive when the two segments of the cut bar overlap. The principle of complementary virtual work, eq. (9.57), stated as $\delta W'_E + \delta W'_I = 0$, now implies

$$\Delta \delta R = -\delta W'_I. \tag{9.84}$$

This result is very similar to eq. (9.62), but Δ now represents the relative displacement at a cut and δR the set of self-equilibrating virtual forces applied at the cut.

The right part of fig. 9.65 depicts a cantilevered beam with a cut to release the bending moment[2] and a set of self-equilibrating moments of magnitude, M, applied at the cut. Let θ_1 and θ_2 be the rotations at the two sides of the cut. If the rotations, θ_1 and θ_2, are prescribed, the associated complementary external virtual work is $\delta W'_E = \theta_1 \delta M - \theta_2 \delta M = (\theta_1 - \theta_2)\delta M$. The relative rotation at the cut is $\Phi = \theta_1 - \theta_2$. The principle of complementary virtual work, eq. (9.57), stated as $\delta W'_E + \delta W'_I = 0$, now implies

$$\Phi \delta M = -\delta W'_I. \tag{9.85}$$

This result is very similar to eq. (9.66), but Φ now represents the relative rotation at a cut and δM the set of self-equilibrating virtual moments applied at the cut.

9.8.1 Force method for trusses

In this section, the force method will be combined with the unit load method to find internal forces in hyperstatic trusses. The basic steps of the force method are presented in section 4.3.3, and the same procedure will be followed here. The approach will be described using the three-bar hyperstatic truss depicted in fig. 9.66 as an example. The truss carries a load, P, at joint **O**.

This hyperstatic system is of order 1, and hence, a single cut is required to transform it into an isostatic system. The middle bar is cut, and fig. 9.66 shows the resulting isostatic truss. The actual system is viewed as the superposition of two problems.

---

[2] The cut must release only the moment and not the shear. It can be imagined as a hinge.

First, the isostatic system obtained by cutting one member, subjected to the externally applied load, and second, the internal force system in which an internal force of unknown magnitude, R, is applied at the cut.

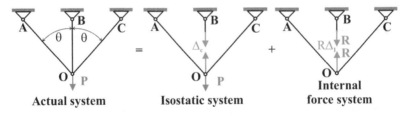

Actual system   **Isostatic system**   **Internal force system**

Fig. 9.66. Force method for the three-bar truss.

For the isostatic system, the unit load method developed in section 9.6.6 is directly applicable to compute the relative displacement, Δ_c, at the cut by using eq. (9.84), and results in

$$\Delta_c = \sum_{i=1}^{N_b} \frac{\hat{F}_i F_i L_i}{(EA)_i}, \tag{9.86}$$

where F_i are the bar forces in the isostatic truss subjected to the externally applied loads, and $\hat{F}_i$ the statically admissible virtual forces corresponding to the self-equilibrating unit load system applied at the cut. The bar forces corresponding to the externally applied load are $F_A = F_C = P/(2\cos\theta)$ and $F_B = 0$. The bar forces corresponding to a self-equilibrating unit load system applied at the cut are $\hat{F}_A = \hat{F}_C = -1/(2\cos\theta)$ and $\hat{F}_B = 1$. Equation (9.86) then yields the relative displacement at the cut as

$$\Delta_c = \frac{-1}{2\cos\theta} \frac{P}{2\cos\theta} \frac{L}{(EA)_A \cos\theta} + \frac{-1}{2\cos\theta} \frac{P}{2\cos\theta} \frac{L}{(EA)_C \cos\theta}$$

$$= -\left(\frac{1}{(EA)_A} + \frac{1}{(EA)_C}\right) \frac{PL}{4\cos^3\theta}.$$

The minus sign reflects the fact that the externally applied load opens the cut.

Next, the internal force system illustrated fig. 9.66 is investigated. The relative displacement at the cut, Δ_1, due to a unit internal force in bar **B** is computed using the unit load once again. Equation (9.84) now yields

$$\Delta_1 = \sum_{i=1}^{N_b} \frac{\hat{F}_i^2 L_i}{(EA)_i}, \tag{9.87}$$

In this case, the bar forces due to a self-equilibrating unit load system applied at the cut, $\hat{F}_i$, represent both the loads due to the external loading and the unit load system. This set of forces is the same as that computed in the previous step. For the three-bar truss,

$$\Delta_1 = \frac{1}{(2\cos\theta)^2}\frac{L}{(EA)_A\cos\theta} + \frac{L}{(EA_B)} + \frac{1}{(2\cos\theta)^2}\frac{L}{(EA)_C\cos\theta}$$

$$= \frac{L}{(EA)_B 4\cos^3\theta}\frac{\bar{k}_A + \bar{k}_C + 4\bar{k}_A\bar{k}_C\cos^3\theta}{\bar{k}_A\bar{k}_C}.$$

where $\bar{k}_A = (EA)_A/(EA)_B$ and $\bar{k}_C = (EA)_C/(EA)_B$ are the non-dimensional stiffnesses of bars **A** and **C**, respectively,

In the last step of this process, the results of the two loading cases are superposed. The sum of the relative displacements at the cut for the isostatic and internal force systems must vanish, because it is artificially introduced. This implies the following compatibility condition at the cut

$$\Delta_c + R\Delta_1 = 0, \tag{9.88}$$

where R is the internal force in bar **B**. Equation (9.88) is solved for the unknown force in the cut bar,

$$R = -\frac{\Delta_c}{\Delta_1}. \tag{9.89}$$

For the three-bar truss example, this yields

$$R = -\frac{\Delta_c}{\Delta_1} = \frac{\bar{k}_A + \bar{k}_C}{\bar{k}_A + \bar{k}_C + 4\bar{k}_A\bar{k}_C\cos^3\theta}P.$$

Bar forces are then found by superposition

$$F_i + R\hat{F}_i, \quad i = 1, 2, \ldots N_b. \tag{9.90}$$

Summary of the force method for hyperstatic trusses of order 1

The procedure described in the previous section, which combines the force and unit load methods, can be summarized by the following steps.

1. Transform the original, hyperstatic truss into an isostatic truss by cutting one bar or one support of the system. The cut must transform the original system into an isostatic system, not a mechanism. This can be achieved in different ways, although specific choices might be more or less cumbersome from an algebraic standpoint.
2. Determine the bar forces, F_i, in the isostatic system subjected to the externally applied loads.
3. Determine the bar forces, $\hat{F}_i$, in the isostatic system loaded by a pair of unit forces at the cut.
4. Determine the relative displacement at the cut, Δ_c, due to the externally applied loads using eq. (9.86). Determine the relative displacement at the cut, Δ_1, due to the pair of unit forces applied at the cut using eq. (9.87).
5. Impose the compatibility condition given by eq. (9.88), and find the internal force in the cut bar, eq. (9.89).

6. Find the forces in all bars by superposition using eq. (9.89).

Once the procedure is completed, displacements at selected points of the truss can be evaluated using the unit load method, and this will be illustrated in the examples below. Because this method uses the principle of superposition, it is only valid for structures made of linearly elastic materials undergoing small displacement.

Example 9.26. Six-bar hyperstatic truss
Consider the six-bar, hyperstatic truss shown in fig. 9.67. All bars have identical Young's modulus, E, and cross-sectional area, $\mathcal{A}$. Determine the forces in the bars of the truss. The combination of the force and unit load methods will be used to solve this problem. Because the two diagonal bars of the square bay are present, the truss is hyperstatic of order 1.

First, an isostatic truss is created by cutting one of the two diagonal members, as indicated in fig. 9.67; this truss is subjected to the externally applied loads. The internal force system consists of the isostatic truss loaded by unit forces at the cut.

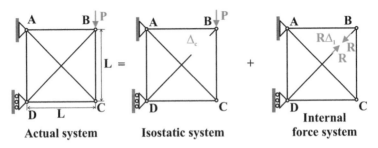

Fig. 9.67. Six-bar hyperstatic truss.

Table 9.3 presents the following information: the second column lists the bar flexibility factors, $L_i/(E\mathcal{A})$, the third lists the bar forces, F_i, in the isostatic system subjected to the externally applied loads, and the fourth lists the bar forces, $\hat{F}_i$, in the isostatic system subjected to a set of unit loads applied at the cut. The relative displacement at the cut, Δ_c, due to the externally applied loads is evaluated using eq. (9.86), and the intermediate results involved in the evaluation of this relative displacement are listed in the fifth column of the table. The relative displacement at the cut, Δ_1, due to a set of unit forces applied at the cut is computed using eq. (9.87), and intermediate results are presented in the sixth column. The set of unit forces applied at the cut is assumed to create a tensile force in the cut bar to comply with the customary sign convention of positive tensile bar forces.

The internal force in the cut bar is now evaluated using eq. (9.89). The two relative displacements at the cut, Δ_c and Δ_1, are found by adding the entries in columns 5 and 6 of table 9.3, respectively, to find

$$R = -\frac{\Delta_c}{\Delta_1} = -\frac{2(1+1/\sqrt{2})PL/(E\mathcal{A})}{(2+2\sqrt{2})L/(E\mathcal{A})} = -\frac{P}{\sqrt{2}}.$$

The bar forces are then found by superposition, see eq. (9.89), and are listed in the last column of the table.

Table 9.3. Calculation of Δ_c and Δ_1 for the six-bar hyperstatic truss.

Bar	$L_i/(E\mathcal{A})$	F_i	$\hat{F}_i$	$F_i\hat{F}_iL_i/(E\mathcal{A})$	$\hat{F}_i^2L_i/(E\mathcal{A})$	$F_i + R\hat{F}_i$
AB	1	0	$-1/\sqrt{2}$	0	$1/2$	$P/2$
BC	1	$-P$	$-1/\sqrt{2}$	$P/\sqrt{2}$	$1/2$	$-P/2$
CD	1	$-P$	$-1/\sqrt{2}$	$P/\sqrt{2}$	$1/2$	$-P/2$
DA	1	0	$-1/\sqrt{2}$	0	$1/2$	$P/2$
AC	$\sqrt{2}$	$P\sqrt{2}$	1	$2P$	$\sqrt{2}$	$P/\sqrt{2}$
BD	$\sqrt{2}$	0	1	0	$\sqrt{2}$	$-P/\sqrt{2}$

Example 9.27. Truss with redundant support

The two-bay truss depicted in fig. 9.68 is isostatic, but the presence of the tip support makes the complete system hyperstatic. All bars are of identical Young's modulus, E, and cross-sectional area, $\mathcal{A}$. Determine the forces in the bars of the truss.

Instead of cutting one of the bars, the tip support at point **F** will be removed to render the truss isostatic. The isostatic truss subjected to the externally applied loads is shown in fig. 9.68, and the same isostatic truss loaded by a set of internal forces of magnitude R at the support is also depicted.

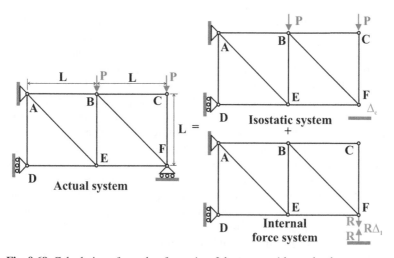

Fig. 9.68. Calculation of member forces in a 2-bay truss with a redundant support.

Table 9.4 presents the results of the analysis. The second column lists the bar flexibility factors, $L_i/(E\mathcal{A})$. The next two columns list the bar forces in the isostatic truss subjected to two loading cases: F_i for the isostatic truss subjected to the externally applied loads, and $\hat{F}_i$ for the same truss subjected to a set of unit loads at the cut. Columns 5 and 6 list the contributions of each bar to the relative displacements at the cut, Δ_c and Δ_1, according to eqs. (9.86) and (9.87), respectively. The reaction force at the support then follows from eq. (9.89) as $R = -\Delta_c/\Delta_1 = (10 + 6\sqrt{2})P/(7 + 4\sqrt{2})$. The bar forces are found by superposition and are listed in the last column of table 9.4.

Table 9.4. Calculation of Δ_c and Δ_1 for the two-bay truss. $\alpha = 7 + 4\sqrt{2}$.

Bar	$L_i/(EA)$	F_i	$\hat{F}_i$	$F_i\hat{F}_iL_i/(EA)$	$\hat{F}^2L_i/(EA)$	$F_i + R\hat{F}_i$
AB	1	P	-1	$-P$	1	$-(3+2\sqrt{2})P/\alpha$
BC	1	0	0	0	0	0
DE	1	$-3P$	2	$-6P$	4	$-P/\alpha$
EF	1	$-P$	1	$-P$	1	$(3+2\sqrt{2})P/\alpha$
AD	1	0	0	0	0	0
BE	1	$-2P$	1	$-2P$	1	$-(4+2\sqrt{2})P/\alpha$
CF	1	$-P$	0	0	0	$-P$
AE	$\sqrt{2}$	$2\sqrt{2}P$	$-\sqrt{2}$	$-4\sqrt{2}P$	$2\sqrt{2}$	$(4+4\sqrt{2})P/\alpha$
BF	$\sqrt{2}$	$\sqrt{2}P$	$-\sqrt{2}$	$-2\sqrt{2}P$	$2\sqrt{2}$	$-(4+3\sqrt{2})P/\alpha$

Example 9.28. *Deflection of a hyperstatic truss*

The previous examples focus on the determination of bar and reaction forces in hyperstatic trusses. In some cases, it is also necessary to compute displacements at specific points of hyperstatic trusses, as is done for isostatic trusses and beams in sections 9.6.6 and 9.7.6, respectively. Here again, the unit load method will be used for this task. In this case, the first step of the method will be to compute the bar forces in the hyperstatic system subjected to the externally applied loads.

Consider the two-bay truss depicted in fig. 9.69. The bar forces in this hyperstatic truss are computed in example 9.27. In this example, the vertical displacement at joint **E** will be evaluated using the unit load method. This approach requires the computation of two sets of bar forces: the bar forces, F_i, associated with the externally applied loads, and the bar forces, $\hat{F}_i$, generated by a unit load applied at joint **E**, as illustrated in fig. 9.69. The first set of forces, F_i, are listed in the last column of table 9.4 and repeated, for convenience, in the third column of table 9.5.

To complete the problem, it is necessary to evaluate a set of statically admissible bar forces that are in equilibrium with a unit load applied at joint **E**. At first it appears that a procedure similar to that developed in the previous example will yield the desired bar forces. Indeed, it is possible to compute the forces in all the bars of the hyperstatic truss when subjected to a unit load at joint **E**. But while this is feasible and will lead to the desired result, it is a cumbersome approach that can be easily bypassed using the following reasoning. The unit load method is a direct application of the principle of complementary virtual work for which the bar forces, $\hat{F}_i$, are required to be statically admissible, but are not necessarily those acting in the truss as it undergoes compatible deformations. In particular, the principle of complementary

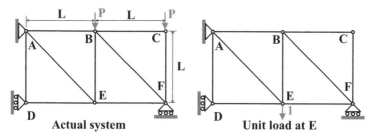

Fig. 9.69. Deflection of a joint in a two-bay truss with a redundant support.

virtual work holds "for all statically admissible virtual forces." Rather than selecting those statically admissible virtual forces acting in the hyperstatic truss as it undergoes compatible deformations, it is much simpler to select a set of statically admissible virtual forces corresponding to an arbitrary choice of the redundant force in the truss.

For the problem at hand, a particularly simple set of statically admissible bar forces, $\hat{F}_i$, is the set associated with a vanishing reaction at the support, point **F**, as listed in the fourth column of table 9.5. While this set of statically admissible forces is not equal to the set acting in the truss as it undergoes compatible deformations, it nonetheless is statically admissible, and hence is a valid set of forces for application of the unit load method.

Table 9.5. Calculation of vertical deflection at joint **E**. $\alpha = 7 + 4\sqrt{2}$

Bar	$L_i/(E\mathcal{A})$	F_i	$\hat{F}_i$	$F_i\hat{F}_iL_i/(E\mathcal{A})$
AB	1	$-(3+2\sqrt{2})P/\alpha$	0	0
BC	1	0	0	0
DE	1	$-P/\alpha$	-1	P/α
EF	1	$(3+2\sqrt{2})P/\alpha$	0	0
AD	1	0	0	0
BE	1	$-(4+2\sqrt{2})P/\alpha$	0	0
CF	1	$-P$	0	0
AE	$\sqrt{2}$	$(4+4\sqrt{2})P/\alpha$	$\sqrt{2}$	$8(1+\sqrt{2})P/\alpha$
BF	$\sqrt{2}$	$-(4+3\sqrt{2})P/\alpha$	0	0

The last column of table 9.5 lists the partial results necessary for the application of eq. (9.65): summing up the entries in the last column yields the vertical displacement at point **E** as

$$\Delta_E = \sum_{i=1}^{N_b} \frac{F_i\hat{F}_iL_i}{(E\mathcal{A})_i} = \frac{(9+8\sqrt{2})}{(7+4\sqrt{2})}\frac{PL}{E\mathcal{A}}.$$

It will be left to the reader to verify that an identical answer will be obtained by using other sets of statically admissible virtual forces that are in equilibrium with the applied unit load. Various sets of statically admissible forces are readily obtained by setting selected bar or reaction forces to zero and computing forces in the remaining bars based on equilibrium. This procedure automatically produces statically admissible virtual forces.

By simplifying the evaluation of the statically admissible bar forces in equilibrium with the unit load, the overall amount of effort required to compute the displacement at a point of the truss is dramatically reduced, as demonstrated in this example.

9.8.2 Force method for beams

In this section, the combined use of the force and unit load methods will be developed for the analysis of hyperstatic beams. The procedure closely follows that developed in section 9.8.1 for truss structures. As compared to trusses, relatively few beam structures involve internally redundant configurations. Examples of beam structures featuring internal redundancy include closed circular beams or rings and beam grillage. In most cases, however, beam structures become hyperstatic due to the presence of multiple supports.

Figure 9.70 depicts a cantilevered beam with an additional mid-span support. Without this additional support, the structure is isostatic and it is possible to compute the root reactions and the bending moment distribution from equilibrium considerations alone. When the mid-span support is added, an additional reaction, R, arises and equilibrium equations are no longer sufficient to determine the reaction forces and moment.

The essence of the force method described in section 4.3.3 is to transform the original, hyperstatic problem into an isostatic system. When the redundancy of the beam structure is due to multiple supports, this is achieved by eliminating, or cutting, the appropriate number of supports to render the beam isostatic. Reaction forces and moments, as well as shear force and bending moment diagrams are then readily obtained from statics.

For the simple example depicted in fig. 9.70, the mid-span support is eliminated, leaving an isostatic, cantilevered beam. The unit load method will be used to compute the deflection, Δ_c, at the location of the support that is eliminated. Equation (9.83) will be used for this purpose and yields

$$\Delta_c = \int_0^L \frac{M_3 \hat{M}_3}{H_{33}^c} \, dx_1, \tag{9.91}$$

where $M_3(x_1)$ is the bending moment distribution in the isostatic beam subjected to the externally applied loads, and $\hat{M}_3(x_1)$ the statically admissible bending moment distribution in the isostatic beam subjected to a set of self-equilibrating unit forces applied at the support.

Next, the unit load method is used to compute the relative deflection at the support due to a set of self-equilibrating, unit loads applied at that location, as illustrated in fig. 9.70. Equation (9.84) then yields the desired relative displacement, denoted Δ_1, as

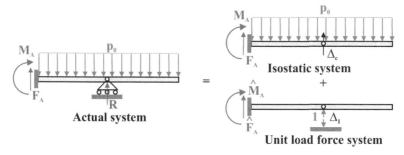

Fig. 9.70. Cantilever with a mid-span support. The isostatic system is obtained by eliminating the mid-span support.

$$\Delta_1 = \int_0^L \frac{\hat{M}_3^2}{H_{33}^c} \, dx_1. \qquad (9.92)$$

where $\hat{M}_3(x_1)$ is the statically admissible bending moment distribution in the isostatic beam subjected to a set of self-equilibrating unit forces applied at the support. This moment distribution is identical to that used in eq. (9.91).

The displacement compatibility equation at the support is now expressed as

$$\Delta_c + R \, \Delta_1 = 0, \qquad (9.93)$$

where $R\Delta_1$ is the deflection at the cut when the isostatic beam is subjected to a set of self-equilibrating loads of magnitude R applied at the cut. Equation (9.93) provides an additional relationship to evaluate the unknown reaction force at the support as

$$R = -\frac{\Delta_c}{\Delta_1}. \qquad (9.94)$$

Once the redundant reaction force, R, is computed, the other reaction forces and bending moments can be obtained from the principle of superposition as $F_A + R\hat{F}_A$ and $M_A + R\hat{M}_A$. Finally, superposition also yields the beam's bending moment distribution as $M_3(x_1) + R\hat{M}_3(x_1)$.

The essence of the force method is to transform the original, hyperstatic system into an isostatic system by cutting or eliminating one support. Usually, this can be done in several different ways, by cutting or eliminating any one of the beam's support. The only requirement is that after the cut, the structure must be isostatic and free of any mechanism.

For instance, fig. 9.71 depicts the cantilevered beam with a mid-span support treated in the previous paragraphs. To transform the system into an isostatic structure, a cut will be made at the root to allow rotation of the beam at this point. This is equivalent to transforming the root clamp into a simple support, as illustrated in fig. 9.71.

When dealing with trusses, this first step of the force method is adequately described as "cutting one of the bars." When dealing with supports, however, the expression "cutting one of the supports" is confusing. The expression "eliminating one support" more accurately describes the removal of the mid-span support illustrated in fig. 9.70. The expression "releasing one constraint" is more appropriate when describing the replacement of the root clamp by a simple support.

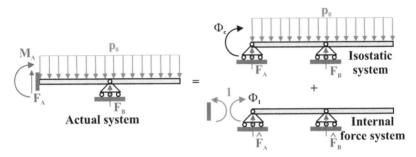

Fig. 9.71. Cantilever with a mid-span support. The isostatic system is obtained by eliminating the mid-span support.

After releasing the root rotation constraint, the unit load method is used to compute the relative root rotation, Φ_c, in the isostatic structure using eq. (9.85). Next, a set of self-equilibrating moments are applied at the root of the beam, as illustrated in fig. 9.71, and the unit load method is used once again to compute the associated root rotation, Φ_1. Finally, the root rotation compatibility condition is expressed as $\Phi_c + M_A\Phi_1 = 0$, where Φ_c is the root rotation when the isostatic beam is subjected to the externally applied loads, and $M_A\Phi_1$ the rotation of the isostatic beam at the same location where a set of self-equilibrating moments of magnitude M_A is applied at the beam's simply supported root. The compatibility condition implies the vanishing of the root rotation when the system is subjected to the combined loading, and provides an additional relationship to evaluate the unknown root reaction moment as $M_A = -\Phi_c/\Phi_1$. The other reactions forces and the beam's bending moment distribution are then obtained by superposition.

This discussion illustrates two ways of transforming the original hyperstatic beam problem into an isostatic system. Both approaches yield identical results. The choice between the two is entirely a matter of convenience: the approach that will minimize the burden of the solution process is the preferred course of action.

Example 9.29. Cantilevered beam with tip support
Consider a cantilevered beam of length L subjected to a uniform loading distribution, p_0, as illustrated in fig. 9.72. Find the bending moment distribution in the beam.

To transform this hyperstatic system into an isostatic problem, the tip support is eliminated, *i.e.*, the tip constraint is released. In the first step of the process, the beam's tip deflection is computed using the unit load method. The bending moment distribution in the isostatic beam generated by the externally applied load is $M_3(\eta) = -p_0L^2(1-\eta)2/2$, where $\eta = x_1/L$ is the non-dimensional variable along the beam's span. The statically admissible bending moment distribution associated with a unit load applied at the beam's tip is $\hat{M}_3(\eta) = L(1-\eta)$. The tip deflection of the isostatic beam is then

$$\Delta_c = \int_0^L \frac{M_3(x_1)\hat{M}_3(x_1)}{H_{33}^c}\, \mathrm{d}x_1 = -\frac{p_0 L^4}{2H_{33}^c} \int_0^1 (1-\eta)^3\, \mathrm{d}\eta = -\frac{p_0 L^4}{8H_{33}^c}.$$

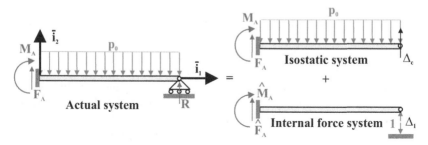

Fig. 9.72. Cantilever with a redundant support.

Next, the tip deflection of the isostatic beam subjected to a set of self-equilibrating tip unit loads is evaluated using the unit load method to find

$$\Delta_1 = \int_0^L \frac{\hat{M}_3^2(x_1)}{H_{33}^c}\, \mathrm{d}x_1 = \frac{L^3}{H_{33}^c} \int_0^1 (1-\eta)^2\, \mathrm{d}\eta = \frac{L^3}{3H_{33}^c}.$$

The compatibility condition, eq. (9.93), allows determination of the reaction force at the tip support as

$$R = -\frac{\Delta_c}{\Delta_1} = \frac{p_0 L^4}{8H_{33}^c} \frac{3H_{33}^c}{L^3} = \frac{3p_0 L}{8}.$$

The solution of the original, hyperstatic problem is now found by superposition. In particular, the bending moment distribution is

$$M_3 + R\hat{M}_3 = -\frac{p_0 L^2}{2}(1-\eta)^2 + \frac{3p_0 L^2}{8}(1-\eta) = \frac{p_0 L^2}{8}\left[3(1-\eta) - 4(1-\eta)^2\right].$$

Example 9.30. Tip rotation of a cantilevered beam with tip support
Consider a cantilevered beam of length L subjected to a uniform load, p_0, as illustrated in fig. 9.72. Determine the beam's tip rotation, Φ, using the unit load method.

The unit load method is equally applicable to iso- and hyperstatic problems; hence, the desired tip rotation is given as

$$\Phi = \int_0^L \frac{M_3(x_1)\hat{M}_3(x_1)}{H_{33}^c}\, \mathrm{d}x_1,$$

where $M_3(x_1)$ is the bending moment distribution in the hyperstatic beam subjected to the externally applied loads, and $\hat{M}_3(x_1)$ is any statically admissible bending moment distribution in equilibrium with a unit moment applied at the beam's tip.

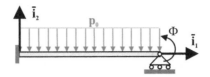

Fig. 9.73. Tip rotation for cantilever with a redundant tip support.

The bending moment distribution in the hyperstatic beam is evaluated in example 9.29 as $M_3 = p_0 L^2 \left[3(1 - \eta) - 4(1 - \eta)^2\right]/8$. Because the unit load method is a direct application of the principle of complementary virtual work, the bending moment distribution, $\hat{M}_3(x_1)$, can be selected as any statically admissible virtual bending moment distribution in equilibrium with the unit tip moment. Instead of trying the determine the statically admissible bending moment distribution acting in the beam undergoing compatible deformations, it is simpler to determine a statically admissible distribution acting in the beam undergoing incompatible deformations. This is acceptable because the principle of complementary virtual work calls for "any statically admissible bending moment distribution."

A simple, acceptable bending moment distribution is found by setting the tip reaction force to zero and evaluating the statically admissible bending moment distribution associated with a tip unit moment to find $\hat{M}_3(\eta) = 1$. By setting the tip reaction force to zero, the beam becomes isostatic, and the statically admissible bending moment distribution associated with a tip unit moment is then easily evaluated based on equilibrium considerations. Another suitable bending moment distribution is found by setting the root reaction moment to zero and evaluating the statically admissible bending moment distribution associated with a tip unit moment to find $\hat{M}_3(\eta) = \eta$. In this case, an isostatic problem with simple supports is obtained by releasing the beam's root rotation.

With the first bending moment distribution, the tip rotation becomes

$$\Phi = \int_0^L \frac{M_3(x_1)\hat{M}_3(x_1)}{H_{33}^c}\, dx_1 = \frac{p_0 L^3}{8 H_{33}^c} \int_0^1 \left[3(1 - \eta) - 4(1 - \eta)^2\right] 1\, d\eta = \frac{p_0 L^3}{48 H_{33}^c}.$$

Using the second bending moment distribution, the resulting tip rotation is

$$\Phi = \frac{p_0 L^3}{8 H_{33}^c} \int_0^1 \left[3(1 - \eta) - 4(1 - \eta)^2\right] \eta\, d\eta = \frac{p_0 L^3}{48 H_{33}^c}.$$

As expected, the results obtained with both bending moment distributions are identical. This surprising conclusion stems from the principle of complementary virtual work, which holds for "any statically admissible bending moment distribution." In fact, for hyperstatic problems of order 1, an infinite number of statically admissible bending moment distributions can be generated, each corresponding to an arbitrary choice of any one of the reaction forces of moments. All will generate the same tip rotation. The analyst should select the approach that simplifies computations as much as possible.

9.8.3 Combined truss and beam problems

The approach developed in the previous sections combines the force method with the unit load method, which itself is a direct application of the principle of complementary virtual work. In the previous examples, truss and beam structures are treated separately using expressions for the complementary internal virtual work given by eqs. (9.50) and (9.69), respectively. Because the complementary internal virtual work is an additive quantity, the complementary internal virtual work of a structure composed of a combination of beams and trusses is found by adding the contributions of each of the beams and bars. The rest of the unit load method remains unchanged.

Example 9.31. Beam with hyperstatic truss bracing
Consider the simply supported beam of bending stiffness H_{33}^c subjected to a uniform loading, as depicted in fig. 9.74. At mid-span, a vertical strut **CD** of infinite stiffness and height $h = L$ is pinned to the beam. Cable **ACB** braces the beam through the strut to provide additional support. Each of the two cable segments will be treated as bars of axial stiffness EA. Determine the bending moment distribution in the beam and the forces in all bars; also find the beam's mid-span vertical deflection.

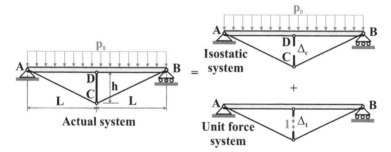

Fig. 9.74. Calculation of bending moment distribution in beam and forces in bars in a hyperstatic beam-truss problem.

First, the original, hyperstatic system is transformed into an isostatic system by cutting the vertical strut **CD**. This is not the only way to proceed: an isostatic system is also obtained if *(1)* cable **AC** is cut, *(2)* cable **CB** is cut, or *(3)* the beam's mid-span rotation is released by introducing a hinge at **D**.

For the isostatic system subjected to the externally applied loads, the beam's bending moment distribution is $M_3(\eta) = p_0 L^2(\eta - \eta^2/2)$, where $\eta = x_1/L$ is a non-dimensional variable along the beam's span, and the bar forces are $F_{AC} = F_{BC} = F_{CD} = 0$. The moment distribution is symmetric with respect to the beam's mid-span.

The deflection, Δ_c, at the cut in the strut is

$$\Delta_c = \frac{2}{H_{33}^c} \int_0^L M_3(x_1)\hat{M}_3(x_1)\, \mathrm{d}x_1 + \sum_{i=1}^{N_b} \frac{F_i \hat{F}_i L_i}{(EA)_i}.$$

The right-hand side represents the negative complementary internal virtual work in the structure and comprises two terms: the integral is the contribution of the flexible beam, while the sum represents the contributions from the individual bars. In view of the symmetry of the beam's bending moment distribution, the integral is computed from 0 to L and the result is then doubled. Introducing the bending moment distribution and bar forces then leads to

$$\Delta_c = \frac{2}{H_{33}^c} \int_0^L p_0 L^2 \left(\eta - \frac{\eta^2}{2} \right) \frac{L\eta}{2} \, L\mathrm{d}\eta = \frac{5}{24} \frac{p_0 L^4}{H_{33}^c}.$$

Note that in the isostatic system, the bar forces all vanish and hence, do not contribute to the displacement at the cut.

Analysis of the isostatic structure subjected to a system of self-equilibrating unit loads applied at the cut yields the following bending moment distribution in the beam $\hat{M}_3(\eta) = L\eta/2$, for $0 \leq \eta \leq 1$ and bar forces $\hat{F}_{AC} = \hat{F}_{BC} = -\sqrt{2}/2$, $\hat{F}_{CD} = 1$. Here again, the moment distribution is symmetric with respect to the beam's mid-span. The deflection at the cut, Δ_1, due to the unit load system is

$$\Delta_1 = \frac{2}{H_{33}^c} \int_0^L \hat{M}_3^2(x_1) \, \mathrm{d}x_1 + \sum_{i=1}^{N_b} \frac{\hat{F}_i^2 L_i}{(EA)_i}.$$

Introducing the beam's bending moment distribution and bar forces results in

$$\Delta_1 = \frac{2}{H_{33}^c} \int_0^1 \frac{L^2 \eta^2}{4} \, L\mathrm{d}\eta + \frac{L}{EA} \left[\left(-\frac{\sqrt{2}}{2} \right)^2 \sqrt{2} + 1 + \left(-\frac{\sqrt{2}}{2} \right)^2 \sqrt{2} \right]$$

$$= \frac{L^3}{6H_{33}^c} + (1 + \sqrt{2}) \frac{L}{EA}.$$

The compatibility condition at the cut, $\Delta_c + R\Delta_1 = 0$, then gives the force, R, in the strut

$$R = -\frac{\Delta_c}{\Delta_1} = -\frac{5p_0 L}{4} \frac{1}{1 + 6(1 + \sqrt{2})H_{33}^c/(EA\,L^2)}. \tag{9.95}$$

Finally, superposition yields the bar forces $F_{AC} = F_{BC} = -\sqrt{2}R/2$, $F_{CD} = R$, and beam bending moment distribution as $M_3(\eta) = p_0 L^2(\eta - \eta^2/2) + RL\eta/2$.

Next, the beam's mid-span deflection, Δ_D, is determined using the unit load method as follows

$$\Delta_D = \frac{2}{H_{33}^c} \int_0^L \hat{M}_3(x_1) M_3(x_1) \, \mathrm{d}x_1 + \sum_{i=1}^{N_b} \frac{\hat{F}_i F_i L_i}{(EA)_i}.$$

In this expression, the bending moment distribution, $M_3(x_1)$, and bar forces, F_i, are those acting in the hyperstatic structure subjected to the externally applied loads, which are computed in the first part of this example. The virtual bending moment distribution, $\hat{M}_3(x_1)$, and bar forces, $\hat{F}_i$, are any statically admissible internal forces in equilibrium with a unit load applied at point **D**. For this problem, it is convenient to compute a set of statically admissible forces acting in the isostatic system obtained by setting the strut force to zero. This leads the following statically admissible beam bending moment distribution $\hat{M}_3(\eta) = L\eta/2$ for $0 \leq \eta \leq 1$ and bar forces $F_{CD} = F_{AC} = F_{BC} = 0$. Using these results, the mid-span deflection becomes

$$\Delta_D = \frac{2}{H_{33}^c} \int_0^1 \left[p_0 L^2 \left(\eta - \frac{\eta^2}{2} \right) + \frac{RL\eta}{2} \right] \left[\frac{L\eta}{2} \right] L d\eta = \frac{5}{24} \frac{p_0 L^4}{H_{33}^c} + \frac{1}{6} \frac{RL^3}{H_{33}^c}$$

where R is given by eq. (9.95). Substituting for R and simplification then yields

$$\Delta_D = \frac{5}{24} \frac{p_0 L^4}{H_{33}^c} \left[1 - \frac{1}{1 + 6(1 + \sqrt{2})H_{33}^c/(EA\,L^2)} \right].$$

It is interesting to verify limiting cases for this mid-span deflection. First, if the stiffness of the cable becomes negligible compared to that of the beam, *i.e.*, if $EA\,L^2/H_{33}^c \to 0$, $\Delta_D \approx 5p_0L^4/(24H_{33}^c)$, as expected for a uniformly loaded, simply supported beam. Second, if the stiffness of the cable becomes very large compared to that of the beam, *i.e.*, if $EA\,L^2/H_{33}^c \to \infty$, then $\Delta_D \approx 0$, as expected because the truss essentially provides a pinned support at the beam's mid-span. For intermediate cable stiffness values, the bracketed term is always smaller than unity and $\Delta_D < 5p_0L^4/(24H_{33}^c)$. The cables stiffen the structure and reduce the beam's mid-span deflection.

9.8.4 Multiple redundancies

All hyperstatic problems treated in the previous sections are of order 1, *i.e.*, a single cut is sufficient to transform the hyperstatic system into an isostatic system. Many practical hyperstatic structures are of higher order. If a hyperstatic structure is of order N, then N cuts will be required to create an isostatic problem. The combination of the force and unit load methods still leads to an efficient but possible tedious solution process: N compatibility equations are generated that can be solved for the N unknown internal forces. The process will be illustrated in the following example.

Example 9.32. Cantilevered beam with redundant supports
Consider the cantilevered beam of length L subjected to a uniform load distribution, as depicted in fig. 9.75. The beam also features mid-span and tip supports, and this is therefore a hyperstatic system of order 2.

In the first step of the force method, the structure is transformed into an isostatic system by eliminating the two supports, as illustrated in fig. 9.75. The bending moment distribution in the isostatic structure subjected to the externally applied loads is found from statics as $M_3(\eta) = -p_0 L^2 (1 - \eta)^2/2$.

Next, the unit load method is used to compute the beam's deflections at the support locations, denoted Δ_{c1} and Δ_{c2}, for the mid-span and tip support locations,

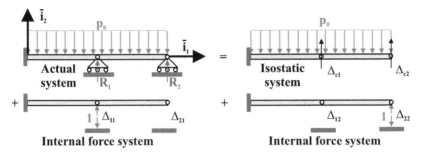

Fig. 9.75. Cantilevered beam with two support.

respectively. To accomplish this task, the statically admissible bending moment distribution in equilibrium with a set of self-equilibrating unit loads applied at the location of the mid-span support is found to be $\hat{M}_3^{[1]} = L(1/2 - \eta)$ for $0 \leq \eta \leq 1/2$ and $\hat{M}_3^{[1]} = 0$ for $1/2 \leq \eta \leq 1$. The corresponding bending moment distribution associated with a set of self-equilibrating unit loads applied at the location of the tip support is $\hat{M}_3^{[2]} = L(1 - \eta)$. The desired deflection at the mid-span support then follows as

$$\Delta_{c1} = \int_0^L \frac{M_3 \hat{M}_3^{[1]}}{H_{33}^c} \, \mathrm{d}x_1 = -\frac{p_0 L^4}{2 H_{33}^c} \int_0^{1/2} (1 - \eta)^2 (1/2 - \eta) \, \mathrm{d}\eta = -\frac{17 p_0 L^4}{384 H_{33}^c},$$

and the deflection at the tip support is found in a similar manner as

$$\Delta_{c2} = \int_0^L \frac{M_3 \hat{M}_3^{[2]}}{H_{33}^c} \, \mathrm{d}x_1 = -\frac{p_0 L^4}{2 H_{33}^c} \int_0^1 (1 - \eta)^2 (1 - \eta) \, \mathrm{d}\eta = -\frac{p_0 L^4}{8 H_{33}^c}.$$

Next, a set of self-equilibrating unit loads are applied at the location of the mid-span support, and the resulting deflections at the location of the mid-span and tip supports, denoted $\Delta_1^{[1]}$ and $\Delta_2^{[1]}$, respectively, are found

$$\Delta_1^{[1]} = \int_0^L \frac{\hat{M}_3^{[1]} \hat{M}_3^{[1]}}{H_{33}^c} \, \mathrm{d}x_1 = \frac{L^3}{H_{33}^c} \int_0^{1/2} (1/2 - \eta)(1/2 - \eta) \, \mathrm{d}\eta = \frac{L^3}{24 H_{33}^c},$$

$$\Delta_2^{[1]} = \int_0^L \frac{\hat{M}_3^{[1]} \hat{M}_3^{[2]}}{H_{33}^c} \, \mathrm{d}x_1 = \frac{L^3}{H_{33}^c} \int_0^{1/2} (1/2 - \eta)(1 - \eta) \, \mathrm{d}\eta = \frac{5 L^3}{48 H_{33}^c}.$$

Similarly, a set of self-equilibrating unit loads are applied at the location of the tip support, and the resulting deflections at the location of the mid-span and tip supports, denoted $\Delta_1^{[2]}$ and $\Delta_2^{[2]}$, respectively, are found

$$\Delta_1^{[2]} = \int_0^L \frac{\hat{M}_3^{[2]} \hat{M}_3^{[1]}}{H_{33}^c} \, \mathrm{d}x_1 = \frac{L^3}{H_{33}^c} \int_0^{1/2} (1 - \eta)(1/2 - \eta) \, \mathrm{d}\eta = \frac{5 L^3}{48 H_{33}^c}.$$

$$\Delta_2^{[2]} = \int_0^L \frac{\hat{M}_3^{[2]} \hat{M}_3^{[2]}}{H_{33}^c} \, dx_1 = \frac{L^3}{H_{33}^c} \int_0^1 (1-\eta)^2 \, d\eta = \frac{L^3}{3H_{33}^c}.$$

The two compatibility conditions impose the vanishing of the displacement at the mid-span support, $\Delta_{c1} + R_1 \Delta_1^{[1]} + R_2 \Delta_1^{[2]} = 0$, and at the tip support, $\Delta_{c2} + R_1 \Delta_2^{[1]} + R_2 \Delta_2^{[2]} = 0$, where R_1 and R_2 are the unknown reaction forces at the mid-span and tip supports, respectively. Introducing the various displacement components in these compatibility conditions yields a set of algebraic equations for the two unknown reaction forces

$$\begin{bmatrix} 1/24 & 5/48 \\ 5/48 & 1/3 \end{bmatrix} \begin{Bmatrix} R_1 \\ R_2 \end{Bmatrix} = p_0 L \begin{Bmatrix} 17/384 \\ 1/8 \end{Bmatrix}.$$

Solution of this set of two equations in two unknowns yields $R_1 = 4p_0 L/7$ and $R_2 = 11 p_0 L/56$. The bending moment distribution in the hyperstatic beam then follows from the principle of superposition as $p_0 L^2 (1-\eta)^2/2 + R_1 \hat{M}_3^{[1]} + R_2 \hat{M}_3^{[2]}$.

This example demonstrates the use of the force method for hyperstatic systems of higher order. As the order increases, the solution process becomes increasingly cumbersome. The solution of a hyperstatic system of order N will call for the solution of a system of N compatibility equations written in terms of N unknown reaction components. While this example presents the approach for a beam with multiple redundant supports, it can also be used for hyperstatic trusses, beams, combined beam and truss, or three-dimensional structures.

9.8.5 Problems

Problem 9.28. Redundant planar frame with tip load
Consider the cantilevered beam consisting of two segments of length L connected at a 90 degree angle, as shown in fig. 9.76. A simple support is located at point **B**, and a horizontal load, P, is applied at point **A**. *(1)* Find the magnitude and location of the maximum bending moment in the bent beam. *(2)* Find the horizontal deflection at point **A**.

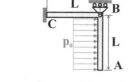

Fig. 9.76. Planar right angle frame with tip load.

Fig. 9.77. Planar right angle frame with distributed load.

Problem 9.29. Redundant planar frame with tip load
Consider the cantilevered beam consisting of two segments of length L connected at a 90 degree angle, as shown in fig. 9.76. A simple support is located at point **B**, and a horizontal load, P, is applied at point **A**. *(1)* Find the magnitude and location of the maximum bending moment in the bent beam. *(2)* Find the rotation at point **A**.

Problem 9.30. Redundant planar frame with distributed loading

Consider the cantilevered beam consisting of two segments of length L connected at a 90 degree angle, as shown in fig. 9.77. A simple support is located at point **B**, and a distributed horizontal load, p_0, is acting along segment **BA**. *(1)* Find the magnitude and location of the maximum bending moment in the bent beam. *(2)* Find the horizontal tip deflection at point **A**.

Problem 9.31. Cantilevered beam with truss bracing

A cantilevered beam of length, L, and bending stiffness, H, carries a tip load, P, as shown in fig. 9.78. At mid-span, the bean is braced by a bar, **BM**, of stiffness $S = E\mathcal{A}$, oriented at an angle $\phi = 60$ degrees. *(1)* Find the magnitude and location of the maximum bending moment in the beam. The effect of the axial load in portion **RM** of the beam is negligible because the beam's axial stiffness of very large. *(2)* Find the transverse deflection at point **T**.

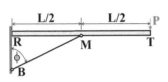

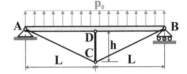

Fig. 9.78. Cantilevered beam with supporting truss.

Fig. 9.79. Simply supported beam with supporting truss.

Problem 9.32. Simply supported beam with truss bracing

The structure depicted in fig. 9.79 consists of a simply supported beam, **AB**, supported at its mid-point by cable **ACB** and a rigid vertical strut, **CD**, of length $h = L$ connecting points **D** and **C**. Cable **ACB** can be modeled as two bars with sectional stiffnesses, $E\mathcal{A}$, and the strut can be modeled as a bar of infinite stiffness. Ignore the axial force developed in the beam itself because the beam's axial stiffness is much larger than that of the cable. *(1)* Find the bending moment distribution in the beam and the forces in the cable segments and vertical strut. Hint: It will be convenient to cut the vertical strut. *(2)* Find the mid-span deflection of the beam.

Problem 9.33. Curved cantilevered beam with mid-support

The cantilevered beam shown in fig. 9.80 is straight from point **A** to point **B**. From point **B** to point **C**, the beam has the shape of a quarter circle of radius R. A horizontal load of magnitude P is applied at point **C**. *(1)* Determine the tip displacement of the beam at point **C**. Assume the the beam only undergoes bending deformations. For the beam's curved portion, express the bending moment as a function of $\theta \in [0, \pi/2]$ and use $dx_1 = R d\theta$.

Problem 9.34. Redundant truss

A vertical load, P_1, is applied to the six-bar hyperstatic planar truss depicted in fig. 9.81. Bars **AD**, **BD** and **CD** are of equal length and joint **D** is at the center of the triangle. *(1)* Determine all bar forces and the displacement at point **C** when load P_1 is acting alone. *(2)* If loads P_1 and P_2 are applied simultaneously, find the value of the horizontal load, P_2, for which the displacement at joint **B** vanishes. *(3)* Determine all corresponding bar forces. Use the following data: $S = E\mathcal{A} = 1 \times 10^6$ psi for all bars; $L = 48$ inches and $P_1 = 4,000$ lbs.

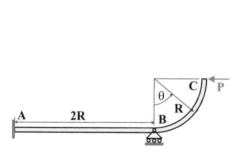

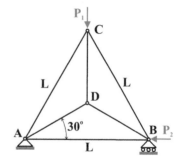

Fig. 9.80. Curved cantilevered beam with tip load.

Fig. 9.81. Triangular truss with internal redundancy.

10

Energy methods

The basic equations of linear elasticity are derived in chapter 1 and are conveniently divided into three groups: the equilibrium equations, the strain-displacement relationships, and the constitutive laws, as illustrated in fig. 9.33. In chapter 9, two virtual work principles are derived. First, the principle of virtual work is established and shown to be entirely equivalent to the equilibrium equations of the system; this principle, however, provides no information about the other two sets of equations, the strain-displacement relationships and constitutive laws, which must be obtained in the traditional manner. Second, the principle of complementary virtual work is established and shown to be entirely equivalent to the strain-displacement relationships of the system; this principle, however, provides no information about the other two sets of equations, the equilibrium equations and constitutive laws, which must obtained in the traditional manner. To remedy this situation, new principles will be developed in this chapter that are entirely equivalent to two of the three groups of equation of linear elasticity. The main tool used to achieve this generalization of the principles presented in chapter 9 is the concept of *conservative forces*.

Types of forces

Newton's first law states that for static equilibrium to be achieved, the "sum of all forces must vanish." The power of this law resides in its generality, and all forces, without any distinction, play an equal role in this equilibrium condition, as underlined in example 9.3 on page 404, for instance.

With virtual work principles, however, various categories of forces are defined. For instance, internal and external forces and the virtual work they perform are clearly separated in the statement of both the principle of virtual work and its complementary counterpart, see principles 6 and 7 on pages 434 and 444, respectively. Externally applied forces and reaction forces also warrant a different treatment in the principle of virtual work. Reaction forces can be eliminated from the formulation because the work they perform vanishes when using kinematically admissible virtual displacements; on the other hand, when arbitrary virtual displacements are used, the virtual work they perform does not vanish, and they become an integral part of the formulation.

Conservative forces

The developments presented in this chapter are rooted in the crucial distinction between conservative and non-conservative forces. This fundamental concept is introduced in elementary physics courses, and it will be examined in much more depth in this chapter. Conservative forces enjoy many remarkable properties. For instance, the work they perform always vanishes when the force undergoes displacements that form a closed path; in other words, forces return to their initial magnitudes when displacements are returned to their initial values. Furthermore, when a dynamical system is subjected only to conservative forces, the total mechanical energy of the system is preserved in time, and hence, the term "conservative forces." From a more mathematical view point, conservative forces are characterized by the existence of a scalar quantity called a *potential* from which they can be derived.

The principle of virtual work considerably simplifies the analysis procedure for elastic structures because it involves only the computation and manipulation of scalar work quantities. If the externally applied forces acting on the system are conservative, they can be derived from a potential, and this fact can be used to further simplify the calculation of the virtual work done by these externally applied forces. Similarly, if the *strain energy* of an elastic component exists, the corresponding elastic forces can be derived from this strain energy, thus further simplifying the evaluation of the virtual work done by the internal forces.

The combination of the principle of virtual work and the concepts of strain energy and potential of the externally applied loads leads to the principle of minimum total potential energy, which will further simplify the analysis of elastic structures. This principle, however, is not as general as the principle of virtual work because it assumes that both internal and externally applied forces are conservative. Clearly, not all externally applied forces are conservative; for instance, friction or aerodynamic forces are not conservative. Similarly, if a material is deformed beyond its elastic limit and into the plastic regime, no strain energy function exists.

Whereas the principle of virtual work is always valid because is is equivalent to Newton's law, the applicability of the principle of minimum total potential energy is limited to systems involving conservative forces.

10.1 Conservative forces

Let r denote the position vector of a particle, and let F be a force acting on this particle. Conservative forces are a class of forces that depend only upon the position of the particles on which they act, $F = F(r)$. Although these forces may vary with time if the system moves, they do not depend explicitly on time or velocity. Figure 10.1 shows two arbitrary paths, denoted **ACB** and **ADB**, along which the particle moves in space from point **A** to point **B**.

Definition

By definition, force $\underline{F}$ is conservative if and only if the work it performs along any path joining the same initial and final points is identical. This is expressed by the following equation

$$W = \int_{\text{Path ACB}} \underline{F} \cdot d\underline{r} = \int_{\text{Path ADB}} \underline{F} \cdot d\underline{r}. \qquad (10.1)$$

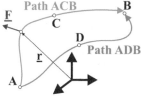

Fig. 10.1. Paths **ACB** and **ADB** join the same two points, **A** and **B**.

Fig. 10.2. Path enclosing a surface of area $\mathcal{A}$ with a normal $\bar{n}$.

Since reversing the limits of integration simply changes the sign of the integral, the work done by the force along path **ADB** is equal in magnitude and opposite in sign to that along path **BDA**. Equation (10.1) then implies the vanishing of the work done by the force over the closed path **ACBDA**. Because path **ACB** and **ADB** are arbitrary paths joining points **A** and **B**, it follows the a force is conservative if and only if the work it performs vanishes over any arbitrary closed path,

$$W = \oint_{\text{Any path}} \underline{F} \cdot d\underline{r} = \oint_{C} \underline{F} \cdot d\underline{r} = 0, \qquad (10.2)$$

where C is an arbitrary closed curve.

Potential of a conservative force

Based on the definition of conservative forces, eq. (10.2), Stokes' theorem [7] then implies that

$$\oint_{C} \underline{F} \cdot d\underline{r} = \int_{\mathcal{A}} \bar{n} \cdot \nabla \times \underline{F} \, d\mathcal{A} = 0, \qquad (10.3)$$

where $\mathcal{A}$ is an area enclosed by curve C and $\bar{n}$ the outward normal to area $\mathcal{A}$, as shown in fig. 10.2. If the force is conservative, the area integral must vanish for any area, $\mathcal{A}$, and this can only occur if the integrand vanishes, leading to $\nabla \times \underline{F} = 0$ for any curve, C, and area, $\mathcal{A}$. Textbooks on vector algebra [7], prove the following identity: $\nabla \times \nabla \Phi = 0$, where Φ is an arbitrary scalar function. It can then be shown that the solution of equation $\nabla \times \underline{F} = 0$ is simply

$$\underline{F} = -\nabla \Phi, \qquad (10.4)$$

where ∇ is the gradient operator.

If a vector field, $\underline{F}$, can be derived from a scalar function, Φ, this function is called a *potential*, and the vector function is said to "be derived from a potential." Because Φ is an arbitrary scalar function, the minus sign is redundant, but is, however, a convention that will be justified later.

It has now been established that if a force is conservative, it can be "derived from a potential." In more mathematical terms, a conservative force must be the gradient a scalar function, called the *potential of the force*. If $\mathcal{I} = (\bar{\imath}_1, \bar{\imath}_2, \bar{\imath}_3)$ is an orthonormal basis, conservative forces can be expressed as

$$\underline{F} = -\nabla \Phi = -\frac{\partial \Phi}{\partial x_1}\bar{\imath}_1 - \frac{\partial \Phi}{\partial x_2}\bar{\imath}_2 - \frac{\partial \Phi}{\partial x_3}\bar{\imath}_3. \tag{10.5}$$

The work done by a conservative force over an arbitrary path joining point **1** to point **2**, with position vectors $\underline{r}_1$ and $\underline{r}_2$, respectively, is then

$$W = \int_{\underline{r}_1}^{\underline{r}_2} \underline{F} \cdot \mathrm{d}\underline{r} = -\int_{\underline{r}_1}^{\underline{r}_2} \nabla \Phi \cdot \mathrm{d}\underline{r}$$

$$= -\int_{\underline{r}_1}^{\underline{r}_2} \left(\frac{\partial \Phi}{\partial x_1}\mathrm{d}x_1 + \frac{\partial \Phi}{\partial x_2}\mathrm{d}x_2 + \frac{\partial \Phi}{\partial x_3}\mathrm{d}x_3 \right) = -\int_{\underline{r}_1}^{\underline{r}_2} \mathrm{d}\Phi = \Phi(\underline{r}_1) - \Phi(\underline{r}_2).$$

Thus the work done by a conservative force *along any path* joining point **1** to point **2** depends only on the positions of these points and can be evaluated as the difference between the values of the potential function expressed at these two points,

$$W = \Phi(\underline{r}_1) - \Phi(\underline{r}_2) = -\Delta\Phi. \tag{10.6}$$

Summary

Conservative forces enjoy a number of remarkable properties. Initially, conservative forces are defined as forces that perform the same work along any path joining the same initial and final points, as expressed by eq. (10.1). Simple calculus reasoning is then used to prove that a force is conservative if and only if the work it performs vanishes over any arbitrary closed path, see eq. (10.2). Finally, conservative forces are shown to be derivable from a potential, as expressed by eq. (10.4). Consequently, the work done by a conservative force *along any path* joining two points can be evaluated as the difference between the potential function evaluated at these two points, see eq. (10.6).

Examples of conservative forces

To illustrate these concepts, consider the gravity force acting on a particle of mass m located in a gravity field characterized by an acceleration $-g\bar{\imath}_3$. It can easily be shown that an applied force that remains constant in magnitude and direction, such as a gravitational force, is conservative. Therefore, the scalar potential, Φ, of the gravity forces is $\Phi = mg\,\underline{r}\cdot\bar{\imath}_3 = mgx_3$, where $\underline{r} = x_1\bar{\imath}_1 + x_2\bar{\imath}_2 + x_3\bar{\imath}_3$ is the position vector of the particle. The gravity force, $\underline{F}_g$, acting on the particle can be obtained from this potential using eq. (10.5) to find $\underline{F}_g = -\nabla\Phi = -\partial\Phi/\partial x_3\,\bar{\imath}_3 = -mg\bar{\imath}_3$, and the gravity forces is said to be "derived from a potential." The work done by the gravity force as the particle moves from elevation x_{3a} to x_{3b} then becomes $W = \int_{x_{3a}}^{x_{3b}} \underline{F}_g \cdot \mathrm{d}\underline{r} = -\int_{x_{3a}}^{x_{3b}} \partial\Phi/\partial x_3\,\mathrm{d}x_3 = \Phi(x_{3a}) - \Phi(x_{3b})$. Clearly, this work depends only on the initial and final elevations but not on the particular path followed by the particle as it moved from the initial to the final elevation. If the particle moves along a closed path starting and ending at the same elevation, the work done by the gravity force vanishes.

As another example, consider the restoring force of an elastic spring of stiffness constant k. If the spring is stretched by an amount u, the restoring force is $-ku$, and can be derived from a potential of the form $A(u) = 1/2\,ku^2$. Indeed, using eq. (10.5), the elastic force in the spring becomes $F_s = -\partial A/\partial u = -ku$. This relationship is the constitutive law for the spring because it relates the force in the spring to its elongation. The quantity $A(u)$ is called the *strain energy* and it can be viewed as a "potential of the elastic forces" in the spring. Hence, the strain energy function implicitly defines the constitutive behavior of the component. Finally, the work done by the elastic restoring force as the spring stretches from u_a to u_b is $W = \int_{u_a}^{u_b} F_s\,\mathrm{d}u = -\int_{u_a}^{u_b} \partial A/\partial u\,\mathrm{d}u = A(u_a) - A(u_b)$. Here again, the work depends only on the initial and final positions.

At first glance, the potential, Φ, of a gravity force and the strain energy, A, of an elastic spring seem to be distinct, unrelated concepts. Both quantities, however, share a common property: forces can be derived from these scalar potentials. Consider a particle of mass m connected to an elastic spring of stiffness constant k and subjected to a gravity force acting in the direction of the spring. The downward displacement, u, of the mass measures both the spring stretch and the elevation of the particle. The externally applied gravity force can be derived from the potential, $\Phi = mgu$, as $F_g = -\partial\Phi/\partial u = -mg$; the restoring force in the spring can be derived from the strain energy, $A = 1/2\,ku^2$, which can also be viewed as the potential of the internal forces, as $F_s = -\partial A/\partial u = -ku$. The two forces acting on the particle can therefore be derived from a potential. Note that here again, a distinction is made between externally applied and internal forces, as is done for both the principle of virtual work and its complementary counterpart.

10.1.1 Potential for internal and external forces

In the development of the principle of virtual work, see section 9.4.1, a distinction is made between internal forces and externally applied loads. The same distinction will be made here: if external forces are conservative, they can be derived from a potential, called the "potential of the external loads," and if the internal forces are conservative, they can be derived from a potential, called the "potential of the internal forces."

When dealing with elastic systems, the internal forces are the stresses acting within the body, or the elastic forces acting in structural components such as springs or trusses. The potential of the internal forces is then more appropriately called *strain energy*, *deformation energy*, or *internal energy* and is denoted A. In view of eq. (10.6), it is possible to write

$$W_I = -\Delta A, \tag{10.7}$$

where W_I is the work done by the internal forces or stresses. Similarly, the potential of external forces is denoted Φ and eq. (10.6) then implies

$$W_E = -\Delta\Phi. \tag{10.8}$$

In future developments, it will be convenient to combine the strain energy and the potential of external forces into a single potential called the *total potential energy*, defined as

$$\Pi = A + \Phi. \tag{10.9}$$

The total work done by both internal and external forces then becomes

$$W = W_I + W_E = -\Delta A - \Delta\Phi = -\Delta\Pi. \tag{10.10}$$

In summary, if the internal forces in a body are conservative, a strain energy function exists, and the work done by these internal forces can be computed with the help of eq. (10.7). Similarly, if the loads externally applied to the body are conservative, a potential of the externally applied loads exists, and the work done by the externally applied loads can be computed with the help of eq. (10.8).

If both internal forces and externally applied loads are conservative, the system is called a *conservative system*. The result expressed by eq. (10.10) states: *for conservative systems, the work done by the internal and external forces equals the negative change in total potential energy of the system*. Note that since the work equals the change in the potential function, this potential function is defined only to within a constant, and so adding an arbitrary constant to the potential function will not alter the work done by the corresponding conservative force.

10.1.2 Calculation of the potential functions

To make use of potential functions, it is necessary to first evaluate them. To begin, consider the potential of the internal forces. The strain energy is a function of the deformation state in the body, $A = A(\underline{\epsilon})$, where the array of strain components is defined in eq. (2.11a). Because the strain energy is defined within a constant, it is convenient to select $A(\underline{\epsilon} = 0) = 0$, *i.e.*, the strain energy vanishes for the undeformed or unstrained state of the body. Equation (10.7) then becomes $W_I = -\Delta A = -[A(\underline{\epsilon}) - A(\underline{\epsilon} = 0)] = -A(\underline{\epsilon})$, and hence,

$$A(\underline{\epsilon}) = -W_I. \tag{10.11}$$

This formula provides a direct way to evaluate the strain energy by computing the work done by the internal forces. In many cases, however, it is cumbersome to compute the work done within a solid as the negative product of the internal stress component acting through strains or deformations. Consequently, an alternative approach is often used to determine the strain energy. In view of eq. (9.19), $W_I = -W_E$, and it follows that

$$A(\underline{\epsilon}) = W_E. \tag{10.12}$$

This result provides a convenient way of determining the strain energy stored in an elastic component by computing the work done by the externally applied loads as they deform the component.

Equation 10.12 can be interpreted as follows: if the internal forces in a solid are conservative, the work done by the externally applied forces is equal to the strain energy stored in the body. As the external forces are applied, the body deforms, the strain magnitudes increase, and so does the strain energy. Consequently, the work done by the externally applied loads is transformed into strain energy.

Of course, it is assumed that the forces are applied slowly, in a quasi-steady manner so that velocities remain very small and the associated kinetic energy is negligible. If the externally applied loads are slowly released, the body will return to its original, unstrained configuration. Because this corresponds to a motion of the internal forces along a closed path, the work done by the conservative internal forces will vanish for the entire cycle, and therefore, the work done during unloading will be the negative of that done during loading. This explains the term "conservative" used to characterize forces that can be derived from a potential.

The evaluation of the potential of the externally applied loads, Φ, is much more straightforward because it is simply the negative of the work done by the external forces acting through the displacements at their points of application. Consider a set of N_P forces, P_i, each of specified *constant magnitude* and each with a *line of action fixed in space*. Similarly, consider N_Q moments, Q_j, each of specified *constant magnitude* and each acting *about a fixed axis in space*. Such loads are sometimes called *dead loads*, because they remain unaffected by the motion of the body they act upon. The potential of these loads is then

$$\Phi = -W_E = -\sum_{i=1}^{N_P} P_i d_i - \sum_{j=1}^{N_Q} Q_j \phi_j, \tag{10.13}$$

where d_i and ϕ_j are the displacements and rotations, respectively, at the points of applications of the external forces and moments, respectively.

It is important to note that not all externally applied loads are conservative forces. For instance, aerodynamic loads vary with the motion of the structure they act upon. For thin airfoils at small angles of attack, the lift acting on the airfoil is proportional to this angle of attack, and therefore the lift depends on the rotation of the airfoil. Aerodynamic forces are non-conservative and cannot be derived from a potential. Another common class of non-conservative forces are follower forces. Such forces might be of constant magnitude, but the orientation of their line of action changes with the rotation of the structure upon which they act. Consider, for instance, the thrust of a rocket jet engine: if the rocket bends, the orientation of the engine thrust will change with the rotation of the structure at the point of attachment of the engine.

10.2 Principle of minimum total potential energy

Let a system be represented by N generalized coordinates, $q = \{q_1, q_2, \ldots, q_N\}^T$, as discussed in section 9.4.1. If the system is conservative, the strain energy of the system can now be viewed as a function of these generalized coordinates, $A = A(q)$, and similarly, $\Phi = \Phi(q)$. Using the chain rule for derivatives, infinitesimal increments in strain energy and potential of the externally applied loads can be written as

$$dA = \frac{\partial A}{\partial q_1} dq_1 + \frac{\partial A}{\partial q_2} dq_2 + \ldots + \frac{\partial A}{\partial q_N} dq_N = \sum_{i=1}^{N} \frac{\partial A}{\partial q_i} dq_i, \qquad (10.14a)$$

$$d\Phi = \frac{\partial \Phi}{\partial q_1} dq_1 + \frac{\partial \Phi}{\partial q_2} dq_2 + \ldots + \frac{\partial \Phi}{\partial q_N} dq_N = \sum_{i=1}^{N} \frac{\partial \Phi}{\partial q_i} dq_i. \qquad (10.14b)$$

If the internal forces are conservative, eq. (10.11) relates the work they perform to the strain energy as $W_I = -A(\epsilon) = -A(q)$ because the deformation field inside the body is a function of the generalized coordinates. Similarly, if the external forces are conservative, eq. (10.13) relates the work they perform to the potential of the externally applied loads as $W_E = -\Phi(q)$.

The virtual work done by the internal forces now becomes $\delta W_I = -\delta A(q)$, and for the external forces, $\delta W_E = -\delta \Phi(q)$. As discussed in section 9.3.1, operators "d" and "δ" are closely related, and by analogy with eqs. (10.14), it is possible to write

$$\delta W_I = -\delta A = -\sum_{i=1}^{N} \frac{\partial A}{\partial q_i} \delta q_i, \qquad (10.15a)$$

$$\delta W_E = -\delta \Phi = -\sum_{i=1}^{N} \frac{\partial \Phi}{\partial q_i} \delta q_i. \qquad (10.15b)$$

In section 9.4.1, the generalized forces associated with internal forces and externally applied loads, denoted Q_i^I and Q_i^E, respectively, are defined in eqs. (9.24a) and (9.24b), respectively. Identifying eq. (9.24a) with eq. (10.15a) and eq. (9.24b) with eq. (10.15b) then yields

$$Q_i^I = -\frac{\partial A}{\partial q_i}, \qquad (10.16a)$$

$$Q_i^E = -\frac{\partial \Phi}{\partial q_i}. \qquad (10.16b)$$

Since the internal forces in the body are assumed to be conservative, it follows that the internal generalized forces, Q_i^I, are themselves conservative because they can be derived from a potential, the strain energy of the structure. Similarly, the externally applied loads are assumed to be conservative, and their generalized counterparts are conservative as well because the can be derived from the potential of the externally applied loads.

The principle of virtual work, as expressed by eq. (9.25), implies $Q_i^I + Q_i^E = 0$, for all generalized coordinates. Introducing eqs. (10.16) then yields $-\partial A/\partial q_i - \partial\Phi/\partial q_i = \partial(A + \Phi)/\partial q_i = 0$. Finally, using the definition of the total potential energy, eq. (10.9), results in

$$\frac{\partial \Pi}{\partial q_i} = 0, \quad i = 1, 2, \ldots, N. \tag{10.17}$$

The same result can be obtained in a more expeditious manner by observing that when both internal forces and externally applied loads are conservative, the work done by these forces equals the negative change in total potential energy of the system, as expressed by eq. (10.10). Since the total potential energy, Π, is defined within a constant, it follows that the virtual work can be expressed as $\delta W = -\delta \Pi$. The principle of virtual work, principle 4, states that a system is in static equilibrium if and only if the sum of the virtual work done by the internal and external forces vanishes for all arbitrary virtual displacements, $i.e.$, $\delta W = -\delta \Pi = 0$, or

$$\delta \Pi = 0. \tag{10.18}$$

The total potential energy is a function of the generalized coordinates, $\Pi = \Pi(q)$, and hence, virtual changes in this quantity must vanish

$$\delta \Pi = \sum_{i=1}^{N} \left[\frac{\partial \Pi}{\partial q_i} \right] \delta q_i = 0. \tag{10.19}$$

Because the virtual changes in the generalized coordinates are arbitrary, the bracketed term must vanish, leading once again to eqs. (10.17). These observations lead to the following principle.

Principle 8 (Principle of stationary total potential energy) *A conservative system is in equilibrium if and only if virtual changes in the total potential energy vanish for all virtual displacements.*

Equation (10.17) expresses this principle in a somewhat more mathematical manner: *a conservative system is in equilibrium if and only if all partial derivatives of the total potential energy with respect to the generalized coordinates vanish.*

Because the principle of stationary total potential energy is derived directly from the principle of virtual work, it inherits many of its features. Sections 9.3.1 and 9.3.2 describe the use of the principle of virtual work with arbitrary virtual displacements and with kinematically admissible virtual displacements, respectively. When kinematically admissible virtual displacements are used, the virtual work done by the reaction forces vanishes, and these forces are eliminated from the formulation. On the other hand, when arbitrary virtual displacements are used, the virtual work done by the reaction forces does not vanish, and must be included in the virtual work done by the externally applied loads; the reaction forces must therefore be treated as externally applied loads.

The same distinction must be made when using the principle stationary total potential energy. If kinematically admissible virtual displacements are used, reaction forces are eliminated from the formulation, whereas if arbitrary virtual displacements are used, reaction forces must be treated as externally applied loads. In this latter case, reaction forces must be included in the potential of the externally applied loads when evaluating the total potential energy of the system.

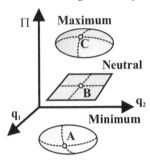

Fig. 10.3. Total potential energy.

For low dimensionality systems, it possible to give a graphical illustration of principle 8. Figure 10.3 shows the total potential energy as a function of two generalized coordinates, q_1 and q_2. Since it is always possible to select a virtual change in generalized coordinate as an actual, infinitesimal change in the same coordinate, eq. (10.18) implies $d\Pi = 0$. This means that at an equilibrium point, the total potential energy is stationary as shown by points **A**, **B**, and **C**.

As illustrated in fig. 10.3, however, this stationary point could correspond to a minimum (point **A**), a maximum (point **C**), or even a saddle point. To make a distinction between these various cases, changes in the total potential energy in the neighborhood of the stationary point must be studied. Increments in this energy are expanded using a Taylor series

$$d\Pi \approx \sum_{i=1}^{N} \frac{\partial \Pi}{\partial q_i} dq_i + \sum_{i=1}^{N} \sum_{j=1}^{N} \frac{\partial^2 \Pi}{\partial q_i \partial q_j} dq_i dq_j,$$

where the higher order terms are neglected. In the neighborhood of static equilibrium, the first term on the right-hand side vanishes in view of eq. (10.17), leaving

$$d\Pi \approx \sum_{i=1}^{N} \sum_{j=1}^{N} \frac{\partial^2 \Pi}{\partial q_i \partial q_j} dq_i dq_j, \tag{10.20}$$

Based on this result, three different cases are possible.

1. If $\partial^2 \Pi / (\partial q_i \partial q_j) dq_i dq_j > 0$ for all dq_i, the total potential energy is a minimum at equilibrium. The equilibrium is said to be *stable*, as illustrated by point **A** in fig. 10.3.
2. If $\partial^2 \Pi / (\partial q_i \partial q_j) dq_i dq_j = 0$ for all dq_i, the total potential energy remains constant around the equilibrium point. The equilibrium is said to be *neutrally stable*, as illustrated by point **B** in fig. 10.3.
3. If $\partial^2 \Pi / (\partial q_i \partial q_j) dq_i dq_j < 0$ for any dq_i, the total potential energy is a maximum at equilibrium. The equilibrium is said to be *unstable*, as illustrated by point **C** in fig. 10.3.

A minimum value of the total potential energy corresponds to a stable equilibrium configuration of the system because any perturbation from such an equilibrium configuration must increase the total potential energy. Since the work done by the externally applied loads is included in the total potential, the total potential cannot increase without an external source of energy, and the equilibrium configuration is stable. On the other hand, at a maximum point, any disturbance will decrease the total potential. Again, since the work done by the externally applied loads is included in the total potential, the released potential energy is converted into kinetic energy, leading to spontaneous motion of the system. This represents an unstable situation. The neutrally stable situation is the intermediate case for which a disturbance causes no change in the total potential.

Combining the principle of stationary total potential energy with the above discussion leads to the principle of minimum total potential energy.

Principle 9 (Principle of minimum total potential energy) *A conservative system is in a stable state of equilibrium if and only if the total potential energy is a minimum with respect to changes in the generalized coordinates.*

Practical applications of the principle of minimum total potential energy require the development of expressions for the strain energy stored in the structure and for the potential of the externally applied forces, the two quantities that make up the total potential energy. These will be described in sections 10.3 to 10.5. It is first useful, however, to consider the possibility that some of the external forces might be non-conservative, as discussed in the following section.

10.2.1 Non-conservative external forces

The principle of minimum total potential energy is based on two assumptions: first, the internal forces are conservative, and second, the externally applied loads are conservative. For important classes of problems, the first assumption is satisfied, but not the second. In this case, the principle of virtual work, principle 6 on page 434, implies

$$\delta W = \delta W_I + \delta W_E = -\delta A + \delta W_E^{nc} = 0,$$

where δW_E^{nc} represents the virtual work done by the *non-conservative forces*. This leads to the following principle.

Principle 10 *A system is in equilibrium if and only if virtual changes in the strain energy equal the virtual work done by the externally applied loads for all arbitrary virtual displacements.*

In other cases, externally applied forces are a mixture of conservative and non-conservative forces. It is then convenient to split the virtual work done by the externally applied forces, δW_E, into two parts: δW_E^c due to the conservative forces, and δW_E^{nc} due to the non-conservative forces, to find $\delta W_E = \delta W_E^c + \delta W_E^{nc}$. The principle of virtual work, principle 6, then implies $\delta W_I + \delta W_E = \delta W_I + \delta W_E^c + \delta W_E^{nc} = 0$. Introducing the strain energy and the potential of the conservative forces then yields

$$\delta(A + \Phi) = \delta W_E^{nc}.$$

It is important to note that the term δW_E^{nc} represents the virtual work done by the non-conservative forces, whereas $\delta(\Phi)$ represents the negative virtual work done by the conservative forces.

10.3 Strain energy in springs

The strain energy is a function of the deformation of the structure, $A = A(\epsilon)$; in turn, the deformation field is a function of the displacement field, or of the generalized coordinates, depending on the formulation of the problem. This section considers one of the simplest elastic structure: a spring. Two different types of springs will be considered. First, the *rectilinear spring* can be deformed in a rectilinear manner by a force that acts along the axis of the spring. Second, the *torsional* or *rotational spring* can be deformed in a rotation about its axis by a moment acting about this axis.

10.3.1 Rectilinear springs

Rectilinear springs are simple, elastic elements with two primary lumped properties: the stiffness constant and un-stretched length. For rectilinear springs, the applied force and the resulting deformation are along a common straight line, as depicted in fig. 10.4. The displacement of the spring is denoted u and its natural length, sometimes called the "un-stretched length," is denoted u_0. The force

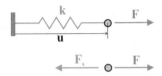

Fig. 10.4. Rectilinear spring subjected to a force F.

applied to the spring is denoted F and the force in the spring F_s. The constitutive behavior of the spring is typically given as $F = F(\Delta)$, where $\Delta = u - u_0$ is the extension of the spring, and $F(\Delta = 0) = F(u = u_0) = 0$.

Linearly elastic springs

If the relationship between an applied load and the resulting extension is linear, *i.e.*, if $F = k\Delta$, where k is the spring's stiffness constant, the spring is said to be linear. It is unfortunate that the term "linear" is often used to describe both the rectilinear motion of the spring as well as the linearity of the force-extension relationship.

If the spring exhibits a linear constitutive behavior, $F = k\Delta$, it implies that the spring resists both tensile and compressive external forces, and the spring stiffness constant, k, is the same constant value for all forces or extension magnitudes. The stiffness has units of force per length, or N/m in the SI system.

The strain energy in the spring is evaluated with the help of eq. (10.12) to find

$$A = W_E = \int_{u_0}^u F \, du = \int_{u_0}^u k\Delta \, du = \int_0^\Delta k\Delta \, d\Delta = \frac{1}{2}k\Delta^2 = \frac{1}{2}F\Delta. \quad (10.21)$$

The strain energy is a positive-definite function of the stretch, *i.e.*, $A > 0$ for any positive or negative value of the extension, Δ, and vanishes only when $\Delta = 0$. The internal force in the spring can be derived from the strain energy using eq. (10.16a) as $F_s = -\partial A/\partial u = -k\Delta$. The minus sign stems from the fact that the force in the spring opposes the externally applied force as shown in the free-body diagram in fig. 10.4.

The constitutive law for the spring is depicted as the straight line in the force versus extension plot shown in fig. 10.5. In view of eq. (10.21), the strain energy, A, is the shaded area under the curve.

The complementary strain energy, A', often called the stress energy, is defined as the shaded are to the left of the straight line and is computed as

$$A' = \int_0^F (u - u_0)\, \mathrm{d}F = \int_0^F \Delta\, \mathrm{d}F = \int_0^F \frac{F}{k}\, \mathrm{d}F = \frac{1}{2}\frac{F^2}{k} = \frac{1}{2}F\Delta. \quad (10.22)$$

The complementary strain energy is naturally expressed in terms of forces, and hence, its name, "stress energy," or less often used "force energy."

Using the spring's constitutive law, it follows that $A' = 1/2\, F^2/k = 1/2\, F\Delta = 1/2\, k\Delta^2 = A$. Thus, the strain energy and its complementary counterpart are equal for linearly elastic springs. In fig. 10.5, the rectangle of area $F\Delta$ is separated by its diagonal into two triangles of equal areas $A = A' = F\Delta/2$. It follows that A and A' are related through

$$A + A' = F\Delta. \quad (10.23)$$

This expression helps explain the term "complementary energy" used to denote this energy.

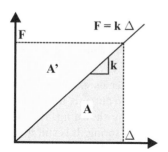

Fig. 10.5. Constitutive law for a linearly elastic spring.

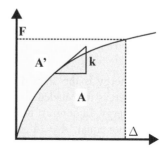

Fig. 10.6. Constitutive law for a nonlinearly elastic spring.

Nonlinearly elastic springs

The discussion has thus far focused on springs with a linear constitutive behavior. The concept of conservative forces, however, is not limited to elastic components presenting a linear behavior. Some metals, such as aluminum and copper, exhibit a slight amount of nonlinearly elastic behavior prior to their yield points. Many elastomers present quite obvious nonlinearly elastic behavior.

A number of analytical models have been developed to approximate the observed constitutive behavior, but perhaps the simplest is a law of the form

$$F = F_0 \tanh \left(\frac{\Delta}{u_0} \right),$$ (10.24)

where F_0 is a reference force and u_0 a reference displacement.

This type of law, which is shown in fig. 10.6, is representative of materials such as aluminum which do not exhibit a sharp transition from linear to nonlinear behavior. The stiffness of this spring is given by

$$k = \frac{\partial F}{\partial \Delta} = \frac{F_0}{u_0} \operatorname{sech}^2 \left(\frac{\Delta}{u_0} \right) = k_0 \operatorname{sech}^2 \left(\frac{\Delta}{u_0} \right),$$

where $k_0 = F_0/u_0$ is the stiffness of the spring at zero elongation.

The constitutive law, eq. (10.24), is now recast in a non-dimensional form by defining the non-dimensional force and extension as $\bar{F} = F/F_0$ and $\bar{\Delta} = \Delta/u_0$, respectively. The constitutive law then becomes $\bar{F} = \tanh(\bar{\Delta})$ and its inverse is $\bar{\Delta} = \operatorname{arctanh}(\bar{F})$.

The strain energy in the spring can be found by direct integration of the force over a differential displacement as

$$A = \int_0^{\Delta} F \, \mathrm{d}\Delta = F_0 u_0 \int_0^{\Delta} \tanh \bar{\Delta} \, \mathrm{d}\bar{\Delta} = F_0 u_0 \ln(\cosh \bar{\Delta}),$$

and the complementary strain energy, A', is given in a similar manner by

$$A' = \int_0^F \Delta \, \mathrm{d}F = F_0 u_0 \int_0^{\bar{F}} \operatorname{arctanh} \bar{F} \, \mathrm{d}\bar{F} = u_0 F_0 \left(\bar{F} \operatorname{arctanh} \bar{F} + \ln \sqrt{1 - \bar{F}^2} \right).$$

The strain energy is the shaded area under the force versus extension curve, see fig. 10.6, whereas the shaded area to the left of the same curve is the complementary strain energy. In contrast to the linearly elastic spring shown fig. 10.5, the two energies are not equal, $A \neq A'$, for a nonlinearly elastic spring. It is still true, however, that $A + A' = F\Delta$, as can be seen graphically as well as shown by using the non-dimensional forms for the strain and complementary strain energy as

$$\frac{A}{u_0 F_0} + \frac{A'}{u_0 F_0} = \ln \left(\cosh \bar{\Delta} \right) + \bar{F} \operatorname{arctanh} \bar{F} + \ln \sqrt{1 - \bar{F}^2}$$

$$= \ln \frac{1}{\sqrt{1 - \tanh^2 \bar{\Delta}}} + \bar{F} \operatorname{arctanh} \bar{F} + \ln \sqrt{1 - \bar{F}^2}$$

$$= -\ln \sqrt{1 - \bar{F}^2} + \bar{F} \operatorname{arctanh} \bar{F} + \ln \sqrt{1 - \bar{F}^2} = \bar{F} \bar{\Delta},$$

where the hyperbolic function identity, $\cosh^2 a = 1/(1 - \tanh^2 a)$, is used along with the non-dimensional constitutive law itself. This result shows that A and A' are truly complementary in the same way that they are for a linearly elastic spring.

The strain energy function incorporates the constitutive law for the material. Indeed, the elastic force in the spring can be derived from the strain energy,

$$F = \frac{\partial A}{\partial \Delta} = \frac{1}{u_0} \frac{\partial}{\partial \bar{\Delta}} \left[F_0 u_0 \ln(\cosh \bar{\Delta}) \right] = F_0 \tanh \left(\frac{\Delta}{u_0} \right). \qquad (10.25)$$

Figure 10.7 illustrates the difference between the nonlinearly and linearly elastic springs. The upper figure shows the strain energy or potential for both springs, the middle figure the force-extension relationship and the bottom figure the spring stiffness defined as the local tangent to the constitutive law curve; all three figures are plotted against the normalized spring extension. The spring studied here is a "softening spring," because it presents a decreasing stiffness at higher extensions and its strain energy is less than that of a linearly elastic spring at all extension magnitudes.

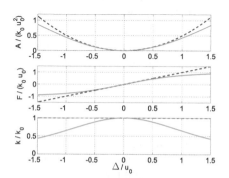

Fig. 10.7. Nonlinear spring with the constitutive law given by eq. (10.24). Top figure: strain energy; middle figure: force; bottom figure: stiffness. Solid line: nonlinear spring; dashed line: linear spring.

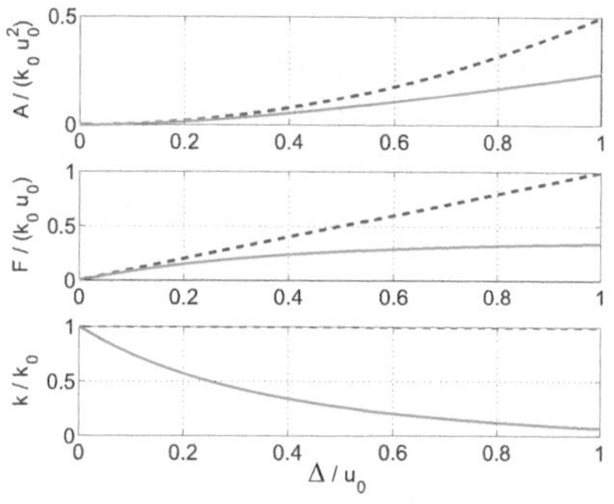

Fig. 10.8. Nonlinear "bungee" spring with the potential given by eq. (10.27). Top figure: strain energy; middle figure: force; bottom figure: stiffness. Solid line: nonlinear spring; dashed line: linear spring.

Example 10.1. Nonlinearly elastic "bungee cords"

The mathematical form of the force-extension relationship must reflect as accurately as possible the experimentally observed behavior of the spring. Typically, the force-extension curve is first obtained experimentally, then a curve fitting procedure is used to approximate the data using a carefully chosen analytical representation of the constitutive law.

An interesting example is a "bungee cord," which can undergo very large deformations without failing. The following equation uses the logarithmic function to approximate the experimentally measured behavior of bungee cords

$$F = \begin{cases} k_0 u_0 \dfrac{\ln(1 + \bar{\Delta})}{1 + \bar{\Delta}}, & \text{for } 0 \le \bar{\Delta} < 1, \\ 0, & \text{for } \bar{\Delta} < 0, \end{cases} \qquad (10.26)$$

where $\bar{\Delta} = (u - u_0)/u_0$ is the non-dimensional bungee stretch, u_0 its natural length, and k_0 is the initial elastic stiffness (*i.e.*, at $\Delta = 0$).

The relationship in eq. (10.26) closely approximates the experimental data for $\bar{\Delta} < 1$, *i.e.*, when the bungee cord extends to less than twice its natural length. When the bungee cord is in its un-stretched state, a force is required to increase its length, and this force is a nonlinear function of the extension. Another nonlinearity in the constitutive law is the unsymmetrical behavior of the bungee in tension and compression: the cord cannot support a compressive force, and therefore, the above constitutive law is not valid for negative extensions, $\bar{\Delta} < 0$.

The potential of the bungee cord for $\bar{\Delta} > 0$ is

$$A = \int_0^u F \, \mathrm{d}u = \int_0^u k_0 u_0 \frac{\ln(1 + \bar{\Delta})}{1 + \bar{\Delta}} \, \mathrm{d}u = \frac{k_0 u_0^2}{2} \ln^2(1 + \bar{\Delta}). \qquad (10.27)$$

The characteristics of the bungee cord are illustrated in fig. 10.8. The upper figure shows the spring's strain energy given by eq. (10.27), the middle figure the force-stretch relationship given by eq. (10.26) and the bottom figure the spring's apparent stiffness. For reference, the corresponding quantities for a linear spring with equal stiffness constant, k_0, are also depicted. The apparent stiffness, k, of the bungee cord is the tangent to the force-extension curve,

$$k = \frac{\mathrm{d}F}{\mathrm{d}\Delta} = k_0 \frac{1 - \ln(1 + \bar{\Delta})}{(1 + \bar{\Delta})^2}. \qquad (10.28)$$

As the stretch of the cord increases, its stiffness decreases and vanishes when $\ln(1 + \bar{\Delta}) = 1$, or $\bar{\Delta} \approx 1.718$. Clearly, this is not realistic, and therefore, the approximation to the force-stretch behavior given by eq. (10.26) is only valid for $\bar{\Delta} < 1$.

For the constitutive law expressed by eq. (10.26), the complementary strain energy cannot be easily computed. Indeed, it would be necessary to express the spring stretch in terms of the applied force, $\bar{\Delta} = \bar{\Delta}(F)$, but in view of the logarithmic function appearing in eq. (10.26), it is not easy to obtain this expression.

10.3.2 Torsional springs

Torsional springs are also simple elastic elements with lumped elastic properties. Instead of the rectilinear motion that characterizes the springs described in the previous section, torsional springs undergo an angular motion, θ, under the action of an externally applied torque, M, as depicted in fig. 10.9.

Fig. 10.9. Torsional spring subjected to a moment M.

For a linearly elastic torsional spring, the constitutive law is $M = k\theta$, where k is the stiffness constant of the spring. Note that although the same symbol, k, is often used to denote the stiffness constants of both rectilinear and torsional springs, their units are not the same: for a torsional spring the stiffness constant has units of moment per rotation, N·m/rad, or sometimes N·m/deg. The un-stretched rotation, θ_0, of the spring is not necessarily zero, and in such cases, the constitutive relationship should be written as $M = k(\theta - \theta_0)$.

Of course, the elastic behavior of the torsional spring could be nonlinear, as discussed in the previous section for rectilinear springs. In either case, if the force in the spring is conservative, it is possible to obtain an expression for its potential.

10.3.3 Bars

As discussed in section 9.5, each bar of a truss is assumed to behave like a rectilinear spring. It then follows from eq. (10.21) that the strain energy in a bar can be written as

$$A = \frac{1}{2}ke^2 = \frac{1}{2}\frac{E\mathcal{A}}{L}e^2, \tag{10.29}$$

where e is the bar elongation, and $k = E\mathcal{A}/L$ its stiffness.

Example 10.2. Spring-mass system
Consider the rectilinear spring with a weight, mg, attached as depicted in fig. 10.10. The spring is assumed to behave linearly, its natural length is denoted u_0, its final length is u, and its extension is $\Delta = u - u_0$.

The strain energy in the linear spring is given by eq. (10.21) as $A = 1/2\ k\Delta^2$, and the potential of the gravity force is $\Phi = -mg\Delta$. Note that since the potential is defined within a constant, it is also correct to write $\Phi = -mg(u_0 + \Delta) = -mgu$.

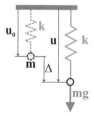

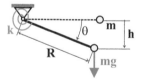

Fig. 10.10. Rectilinear spring supporting a weight.

Fig. 10.11. Torsional spring supporting a weight.

The total potential energy of the system is $\Pi = A + \Phi = 1/2\ k\Delta^2 - mg\Delta$. Because the problem presents a single generalized coordinate, Δ, the principle of minimum total potential energy, eq. (10.17), implies

$$\frac{\partial \Pi}{\partial \Delta} = k\Delta - mg = 0.$$

The solution of this linear equation gives the extension of the spring, $\Delta = mg/k$.

In the neighborhood of this equilibrium configuration, variation in the total potential energy is given by eq. (10.20) as $d\Pi \approx (\partial^2\Pi/\partial\Delta^2)\ d\Delta^2 = k\ d\Delta^2$. Because the spring stiffness constant is a positive number, it follows that $d\Pi > 0$ and the equilibrium configuration is stable.

Example 10.3. Rotational spring-mass system

Consider next a rigid bar of length R carrying a weight, mg, at its tip and pinned to the ground at the other end, as shown in fig. 10.11. At the pivot point, a linear torsional spring of stiffness constant k restrains the rotation of the bar. The spring is unstretched when $\theta = 0$, which corresponds to the horizontal position of the bar.

The strain energy of the spring is simply $A = 1/2 \, k\theta^2$. The potential of the externally applied load is computed with the help of eq. (10.13) as $\Phi = -W_E = -mgh$, where h is the motion of the point of application of the tip weight projected along the direction of the load. In this case, $h = R\sin\theta$, leading to $\Phi = -mgR\sin\theta$. The total potential energy of the system is now $\Pi = A + \Phi = 1/2 \, k\theta^2 - mgR\sin\theta$.

Because the problem presents a single generalized coordinate, θ, the principle of minimum total potential energy, eq. (10.17), implies

$$\frac{\partial \Pi}{\partial \theta} = k\theta - mgR\cos\theta = 0.$$

The equilibrium configuration is the solution of this transcendental equation for θ, which can be recast as $\bar{k}\theta = \cos\theta$, where $\bar{k} = k/(mgR)$ is the non-dimensional stiffness constant of the spring. The transcendental equation does not admit a closed form solution, and it must be solved graphically or iteratively using a procedure such as Newton's method. For $\bar{k} \to 0$, that is, for a spring of very low stiffness or for a very large tip mass, the equilibrium angle $\theta \approx \pi/[2(1 + \bar{k})]$. For $\bar{k} \to \infty$, that is, for a very stiff spring or very small tip mass, the equilibrium angle $\theta \approx 1/\bar{k}$.

Although the spring is assumed to be linear, the equilibrium equations of the problem are nonlinear because the motion of the bar is finite, *i.e.*, no limit is set on the magnitude of angle θ. When $\bar{k} \to \infty$, system deflections remain small, the equilibrium equation of the problem becomes linear, $\bar{k}\theta \approx 1$, and the solution is easily found as $\theta = 1/\bar{k} = mgR/k$. The nonlinearities introduced by large deflections are called *geometric nonlinearities*.

Example 10.4. Buckling of a rigid bar under compressive load

The study of the behavior of the total potential energy in the vicinity of an equilibrium state provides important information about the stability of the system. To illustrate this concept, consider the rigid bar of length L, connected to the ground through a pivot point and subjected to a tip compressive load, P, as shown in fig. 10.12. A rectilinear spring is attached at the tip of the bar and is assumed to remain horizontal at all times; the spring is unstretched when the bar is in the vertical position, *i.e.*, when $\theta = 0$.

The strain energy for the spring is $A = 1/2 \, k\Delta^2$, where the extension of the spring is $\Delta = L\sin\theta$. The potential of the externally applied load is computed with the help of eq. (10.13) as $\Phi = -W_E = -Ph$, where $h = L(1 - \cos\theta)$ is the motion of the point of application of the tip force projected along its line of action. The total potential energy of the system is $\Pi = A + \Phi = 1/2 \, kL^2\sin^2\theta - PL(1 - \cos\theta)$.

Since the problem presents a single generalized coordinate, θ, the principle of minimum total potential energy, eq. (10.17), implies

$$\frac{\partial \Pi}{\partial \theta} = kL^2\sin\theta\cos\theta - PL\sin\theta = L\sin\theta\,(kL\cos\theta - P) = 0.$$

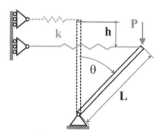

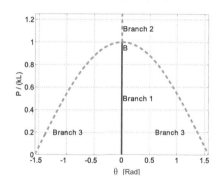

Fig. 10.12. Rigid bar with tip spring under compressive load. Original and deformed configurations.

Fig. 10.13. Response of the system: stable branch: solid line; unstable branches: dotted lines.

This equation possesses two distinct solutions: $\sin\theta = 0$ and $P = kL\cos\theta$. The first solution, $\sin\theta = 0$, yields $\theta = 0$ or $\theta = \pm\pi$, for any value of load, P. This corresponds to the configuration in which the bar remains vertical, straight up or down with respect to the pivot, for any applied load. The second solution, $P = kL\cos\theta$, can be solved for the deflection of the bar as a function of the applied load: $\theta = \arccos(P/kL)$. These two solutions are depicted in fig. 10.13.

To study the stability of these solutions, the variation of the total potential energy in the vicinity of an equilibrium configuration is evaluated with the help of eq. (10.20) as

$$\mathrm{d}\Pi \approx \frac{\partial^2\Pi}{\partial\theta^2}\mathrm{d}\theta^2 = \left[kL^2\left(\cos^2\theta - \sin^2\theta\right) - PL\cos\theta\right]\mathrm{d}\theta^2.$$

Consider the first solution, $\theta = 0$. Along this branch, labeled "branch 1" in fig. 10.13, the variation of the total potential energy is $\mathrm{d}\Pi \approx L(kL - P)\mathrm{d}\theta^2$. For $P < kL$, it follows that $\mathrm{d}\Pi > 0$, leading to a stable solution. A solid line is used in fig. 10.13 to indicate that this branch is stable. For $P > kL$, however, $\mathrm{d}\Pi < 0$, and the equilibrium solution becomes unstable; this unstable branch, labeled "branch 2," is shown as a dotted line.

Consider now the second solution, $P = kL\cos\theta$. In the vicinity of this equilibrium configuration, it is clear that $\mathrm{d}\Pi \approx -kL^2\sin^2\theta\,\mathrm{d}\theta^2 < 0$, and the solution, labeled "branch 3," is unstable. Finally, it is easily verified that the last equilibrium solution, $\theta = \pm\pi$, is stable for all $P > 0$.

It now becomes possible to describe the behavior of the system under an increasing load, P. For $P < kL$, the bar is in stable equilibrium in the vertical configuration. As load P increases, point **B** in fig. 10.13 is reached. At this point three distinct equilibrium solutions now become possible: $\theta = 0$, and $\theta = \arccos(P/kL)$, for either positive or negatives values of θ. These equilibrium solutions are labeled "branch 2" and "branch 3," and are all unstable, as indicated in fig. 10.13. Point **B** is called a *bifurcation point* because three equilibrium solutions emanate from it.

From a purely mathematical perspective, although all are unstable, each of the three solutions is a correct equilibrium solution of the problem. In practice, the branch to be followed by the bar depends on the imperfection present in the system. Indeed, a rigid bar is never "perfectly straight" nor "perfectly homogeneous," and load P can never be "exactly aligned" with the axis of the bar. If the bar is slightly canted to the right or load P leaning in that direction, the system will collapse to the right; conversely, a left leaning imperfection or loading will cause the bar to collapse to the left. Because imperfections are always present, "branch 2" is never observed in practice.

From this discussion, it is clear that the system can only sustain loads $P < kL$ with $\theta = 0$, at which point the system collapses. Point **B** is called the *buckling point*, and $P_{\text{cr}} = kL$ is the *critical load* or *buckling load*. More details about the buckling phenomenon can be found in chapter 14.

As a final point, it should be noted that the branch $\theta = 0$ is stable for any *negative* value of P; this correspond to the case when the load is pulling the vertical bar upward, an obviously stable configuration.

Example 10.5. Rigid aircraft suspended by "bungee cords"

For the purpose of dynamic testing, an aircraft is suspended from a hangar's roof by means of three bungee cords attached to the aircraft's left wing at point **L**, right wing at point **R**, and tail at point **T**, as depicted in fig. 9.16 on page 424. The aircraft's total mass is M and the center of mass is located at point **C**. For simplicity, the aircraft is assumed to be rigid and the displacements under the load are assumed to remain small. The generalized coordinates of the problem are selected as Δ_L, Δ_R, and Δ_T, the downward vertical distance of points **L**, **R**, and **T**, respectively, from the bungee cord attachment points. The strain energy for each of the bungee cords is given by eq. (10.27), and hence, the total strain energy of the system is

$$A = \frac{k_L u_L^2}{2} \ln^2(1 + \bar{\Delta}_L) + \frac{k_R u_R^2}{2} \ln^2(1 + \bar{\Delta}_R) + \frac{k_T u_T^2}{2} \ln^2(1 + \bar{\Delta}_T), \quad (10.30)$$

where $\bar{\Delta}_L = (\Delta_L - u_L)/u_L$, $\bar{\Delta}_R = (\Delta_R - u_R)/u_R$, and $\bar{\Delta}_T = (\Delta_T - u_T)/u_T$, are the non-dimensional extensions of the three bungee cords, u_L, u_R, and u_T are their natural lengths, and k_L, k_R, and k_T, are their apparent stiffness for small extensions.

Equation (10.30) illustrates one of the fundamental advantages of energy methods: because strain energy is an additive scalar quantity, *the strain energy of a system is simply the sum of the strain energies stored in each of its elastic components.* In this example, the total strain energy of the system is the sum of the strain energies in each of the three bungee cords. Because the aircraft is assumed to be rigid, it stores no strain energy.

The potential of the gravity load acting on the aircraft is $\Phi = -Mgh$, where h is the distance of the mass center below a reference plane and is given by, $h = (a/d)\Delta_T + (1 - a/d)(\Delta_L + \Delta_R)/2$. This expression is a direct consequence of the assumption of a rigid aircraft. The potential of the externally applied loads now becomes

$$\Phi = -Mg\left[\Delta_T \frac{a}{d} + \frac{\Delta_L + \Delta_R}{2}\left(1 - \frac{a}{d}\right)\right]. \quad (10.31)$$

The total potential energy of the system is $\Pi = A + \Phi$, and the equilibrium equations of the system, eqs. (10.17), become

$$k_L u_L \frac{\ln(1 + \bar{\Delta}_L)}{1 + \bar{\Delta}_L} = \frac{Mg}{2}\left(1 - \frac{a}{d}\right), \tag{10.32a}$$

$$k_R u_R \frac{\ln(1 + \bar{\Delta}_R)}{1 + \bar{\Delta}_R} = \frac{Mg}{2}\left(1 - \frac{a}{d}\right), \tag{10.32b}$$

$$k_T u_T \frac{\ln(1 + \bar{\Delta}_T)}{1 + \bar{\Delta}_T} = Mg\frac{a}{d}. \tag{10.32c}$$

These are three nonlinear equations to be solved for the non-dimensional extensions of the three bungee cords.

It is interesting to compare the present solution with the solution presented in example 9.12 on page 423. Nonlinear springs are used in the present example, but linear springs are used in example 9.12. The present example illustrates the concept of *material nonlinearities*, *i.e.*, nonlinear relationships for the material constitutive laws, as opposed to the *geometric nonlinearities* encountered in examples 10.3 and 10.4.

In example 9.12, the generalized coordinates are selected as u, ϕ_1, and ϕ_2, the vertical translation and two rotations of the rigid aircraft, whereas in the present case, the non-dimensional extensions of the three bungee cords are used. The use of the former generalized coordinates results in a system of coupled equations as shown in eq. (9.26), whereas the use of the latter yields the uncoupled equations (10.32). Both sets of generalized coordinates are equally valid because both uniquely define the configuration of the aircraft. The final form of the governing equations of the problem, however, does depend on the choice of a specific set of generalized coordinates. An interesting exercise would be to repeat the solution of the present problem by selecting u, ϕ_1, and ϕ_2 as the generalized coordinates.

Finally, it must be repeated that the solution presented here assumes the movement of the aircraft to remain small. If that assumption is violated, the kinematics of the problem will be more complicated. For instance, if the generalized coordinates are selected to be u, ϕ_1, and ϕ_2, the development of large displacements, and hence, of finite rotations, will introduce trigonometric functions of angles ϕ_1 and ϕ_2, which leads to geometric nonlinearities.

10.3.4 Problems

Problem 10.1. Rotating disk with spring restraint
Work problem 9.1 using the principle of minimum total potential energy.

Problem 10.2. Lever with sliding pivots
Bar **ABC** is of length $b + a$ and is constrained to move vertically at point **A** and horizontally at **B**, while a horizontal force, P, is applied at point **C**, as depicted in fig. 10.14. Point **A** is restrained by a vertical spring of stiffness constant k, which is relaxed when angle $\theta = 0$. Use the principle of minimum total potential energy to determine the equilibrium configurations of the system.

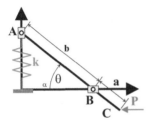

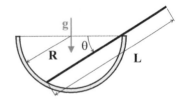

Fig. 10.14. Lever with spring-restrained sliding pivots.

Fig. 10.15. Rod in frictionless hemispherical bowl ($L > 2R$).

Problem 10.3. Rod in frictionless hemispherical bowl

A uniform rod of mass, m, and length, L, rests from inside to across the rim of a frictionless hemispherical bowl of radius, R, as shown in the cross-sectional view in fig. 10.15. Assume that $L > 2R$. Using the principle of minimum total potential energy, find the equilibrium angle of inclination, θ, of the rod to the horizontal plane.

Problem 10.4. Geometrically nonlinear spring-mass

Problem 9.6 describes a spring-mass system with a nonlinear geometry which arises from the large displacements that are developed, as shown in fig. 9.22. Use the principle of minimum total potential energy to compute the equilibrium configuration of the system.

Problem 10.5. Rigid aircraft suspended by bungee cords

Work example 10.5 with the generalized coordinates u, ϕ_1, and ϕ_2 defined in example 9.12.

10.4 Strain energy in beams

10.4.1 Beam under axial loads

Consider a beam subjected only to axial loads as discussed in section 5.4 and depicted in fig. 5.6 on page 179. Material constitutive laws are assumed to be linear and elastic. Focus now on an infinitesimal slice of the beam of span-wise length dx_1, acted upon by an axial force N_1. The left face of this differential element undergoes an axial displacement $\bar{u}_1$, whereas the displacement of its right face is $\bar{u}_1 + (d\bar{u}_1/dx_1)dx_1$. As the axial force acting on the left face increases from zero to its final value, N_1, the work it performs is $-1/2\, N_1\bar{u}_1$; the minus sign is due to the fact on the left face, displacement and force are counted positive in opposite directions. For linear constitutive laws, see fig. 10.5, the area under the force-displacement curve is the area of a triangle, $1/2\, N_1\bar{u}_1$.

The work done by the axial force acting on the right face as it increases from zero to N_1 is $1/2\, N_1\left[\bar{u}_1 + (d\bar{u}_1/dx_1)dx_1\right]$. The total work done by the axial force is found by adding the contributions from the two faces to find $1/2\, N_1(d\bar{u}_1/dx_1)dx_1 = 1/2\, N_1\bar{\epsilon}_1 dx_1$, where $\bar{\epsilon}_1$ is the sectional axial strain. The work done by the externally applied force, N_1, on a differential element of the beam is

$$dW_E = \frac{1}{2} N_1 \bar{\epsilon}_1 dx_1 = \frac{1}{2} S \bar{\epsilon}_1^2 \, dx_1, \tag{10.33}$$

where the linear sectional constitutive law, eq. (5.16), is used to obtain the last equality.

The quantity

$$a(\bar{\epsilon}_1) = \frac{1}{2} S \bar{\epsilon}_1^2, \tag{10.34}$$

is known as the *strain energy density function* and gives the strain energy per unit length of the beam. This strain energy density can be viewed as the potential of the axial force, which can be derived from this potential as $N_1 = -\partial a(\bar{\epsilon}_1)/\partial \bar{\epsilon}_1 = -S\bar{\epsilon}_1$. Again, the minus sign indicates that this is the internal force in the beam, not the axial force externally applied to the differential element.

The total strain energy developed by the axial force distribution over the beam's span is now obtained by integration of the strain energy density

$$A(\bar{\epsilon}_1) = \int_0^L a(\bar{\epsilon}_1) \, dx_1 = \frac{1}{2} \int_0^L S \bar{\epsilon}_1^2 \, dx_1. \tag{10.35}$$

Sometimes, it is preferable to express the strain energy stored in the beam in terms of the axial force by using eq. (5.6), to find

$$A(\bar{\epsilon}_1) = \int_0^L \frac{N_1^2}{2S} \, dx_1 = A'(N_1). \tag{10.36}$$

Here, $a'(N_1) = N_1^2/2S$ is known as the *stress energy density function*, or *complementary strain energy density*. $A'(N_1)$ is the total stress energy or complementary energy stored in the beam expressed in terms of the axial force distribution. As observed earlier, in the case of a linear constitutive law, the strain energy and its complementary counterpart are equal.

To illustrate these concepts, consider a bar fixed at its root end and subjected to only an axial tip force, P. Static equilibrium implies the $N_1 = P$ at all points along the beam's span, and hence, the axial strain is a constant, $\bar{\epsilon}_1 = \Delta/L$, where Δ is the bar's tip deflection and L its length. The strain energy, eq. (10.35), is then $A(\bar{\epsilon}_1) = 1/2 \int_0^L S \bar{\epsilon}_1^2 \, dx_1 = 1/2 \, S\Delta^2/L$. Clearly, a beam subjected to a tip axial load is equivalent to a rectilinear spring of stiffness constant $k = S/L$.

10.4.2 Beam under transverse loads

Beams subjected to transverse loads are discussed in section 5.5 and are depicted in fig. 5.14 on page 187. Material constitutive laws are assumed to be linear. Consider now an infinitesimal slice of the beam of span-wise length dx_1, acted upon by a bending moment M_3. The left face of this differential element rotates by an angle $d\bar{u}_2/dx_1$, whereas the rotation of its right face is $d\bar{u}_2/dx_1 + (d^2\bar{u}_2/dx_1^2)dx_1$. When the bending moment acting on the left face increases from zero to its final value, M_3, the work it performs is $-1/2 \, M_3 d\bar{u}_2/dx_1$; the minus sign is due to the fact on the left face, rotation and moment are counted positive in opposite directions. The work done by the bending moment acting on the right face is $1/2\ M_3\left[d\bar{u}_2/dx_1 + (d^2\bar{u}_2/dx_1^2)dx_1\right]$. The total work done by the bending moment is found by adding the contributions from the two faces to find $1/2\ M_3(d^2\bar{u}_2/dx_1^2)dx_1 = 1/2\ M_3\kappa_3 dx_1$, where κ_3 is the sectional curvature defined by eq. (5.6). The work done by the externally applied bending moment, M_3, on a differential element of the beam is

$$dW_E = \frac{1}{2}\ M_3\kappa_3\ dx_1 = \frac{1}{2}\ H_{33}^c\kappa_3^2\ dx_1. \tag{10.37}$$

where the linear sectional constitutive law, eq. (5.37), is used to obtain the last equality.

The quantity

$$a(\kappa_3) = \frac{1}{2}\ H_{33}^c\kappa_3^2, \tag{10.38}$$

is known as the strain energy density function and gives the strain energy per unit length of the beam. This strain energy density can be viewed as the potential of the bending moment, which can be derived from this potential as $M_3 = -\partial a(\kappa_3)/\partial\kappa_3 = -H_{33}^c\kappa_3$. Again, the minus sign indicates that this is the internal moment in the beam, not the bending moment externally applied to the differential element.

The total strain energy developed by the bending moment distribution in the beam is then obtained by integration of the strain energy density

$$A(\kappa_3) = \int_0^L a(\kappa_3)\ dx_1 = \frac{1}{2}\int_0^L H_{33}^c\kappa_3^2\ dx_1. \tag{10.39}$$

The curvature can also be expressed in terms of the transverse deflection using eq. (5.6) so that

$$A(u_2(x_1)) = \frac{1}{2}\int_0^L H_{33}^c\left(\frac{d^2\bar{u}_2}{dx_1^2}\right)^2\ dx_1. \tag{10.40}$$

The strain energy stored in the beam can also be expressed in terms of the bending moment by using eq. (5.37) in eq. (10.39) to find

$$A(M_3) = \int_0^L \frac{M_3^2}{2H_{33}^c}\ dx_1 = A'(M_3). \tag{10.41}$$

In this case, $a'(M_3) = M_3^2/2H_{33}^c$ is known as the stress energy density function. $A'(M_3)$ is the total complementary strain energy stored in the beam expressed in terms of the bending moment distribution.

10.4.3 Beam under torsional loads

Consider a circular cylindrical beam subjected to torsion as discussed in section 7.1. Material constitutive laws are assumed to be linear. An infinitesimal slice of the cylinder of span-wise length dx_1 is acted upon by a torque M_1. The left face of this differential element undergoes a rotation, ϕ_1, whereas the rotation of its right face is $\phi_1 + (d\phi_1/dx_1)dx_1$. As the torque acting on the left face increases from zero to its final value, M_1, the work it performs is $-1/2\ M_1\phi_1$; the minus sign is due to the fact on the left face, rotation and torque are counted positive in opposite directions. The work done by the torque acting on the right face as it increases from zero to M_1 is $1/2\ M_1\ [\phi_1 + (d\phi_1/dx_1)dx_1]$. The total work done by the torque is found by adding the contributions from the two faces to find $1/2\ M_1(d\phi_1/dx_1)dx_1 = 1/2\ M_1\kappa_1 dx_1$, where κ_1 is the sectional twist rate. The work done by the externally applied torque, M_1, on a differential element of the beam is

$$dW_E = \frac{1}{2} M_1\kappa_1 dx_1 = \frac{1}{2} H_{11}\kappa_1^2\ dx_1. \tag{10.42}$$

where the linear sectional constitutive law, eq. (7.13), is used to obtain the last equality.

The quantity

$$a(\kappa_1) = \frac{1}{2} H_{11}\kappa_1^2, \tag{10.43}$$

is known as the strain energy density function and gives the strain energy per unit length of the cylinder. This strain energy density can be viewed as the potential of the torque, which can be derived from this potential as $M_1 = -\partial a(\kappa_1)/\partial\kappa_1 = -H_{11}\kappa_1$. Again, the minus sign indicates that this is the internal torque in the beam, not the torque externally applied to the differential element.

The total strain energy developed in the cylindrical beam by the torque distribution over the cylinder's span is then obtained by integration

$$A(\kappa_1) = \int_0^L a(\kappa_1)\ dx_1 = \frac{1}{2} \int_0^L H_{11}\kappa_1^2\ dx_1. \tag{10.44}$$

Sometimes, it is preferable to express the strain energy stored in the cylindrical beam in terms of the torque by using eq. (7.13) to find

$$A(M_1) = \int_0^L \frac{M_1^2}{2H_{11}}\ dx_1 = A'(M_1). \tag{10.45}$$

$a'(M_1) = M_1^2/2H_{11}$ is known as the stress energy density function. $A'(M_1)$ is the total complementary strain energy stored in the cylinder expressed in terms of the torque distribution.

10.4.4 Relationship with virtual work

It is interesting to compare the results obtained in this section with those developed is section 9.7. The internal work done by a constant bending moment, M_3, undergoing a curvature, κ_3, is given by eq. (9.69) as $dW_I = -M_3\kappa_3\ dx_1$, for a slice of the beam of infinitesimal size, dx_1. Next, eq. (9.19) yields $dW_E = -dW_I = M_3\kappa_3\ dx_1$. This result seems to contradict eq. (10.37) in section 10.4.2, which states that $\mathrm{d}W_E = 1/2\ M_3\kappa_3\ \mathrm{d}x_1$. Fortunately, this is only an apparent contradiction. In section 9.7, the bending moment is assumed to remain constant in magnitude while undergoing a curvature; in section 10.4.1, however, the bending moment is assumed grow in proportion to the curvature. If the bending moment is kept constant, the work it performs as the curvature increases is

$$
\mathrm{d}W_E = \left[\int_0^{\kappa_3} M_3\ \mathrm{d}\kappa_3 \right] \mathrm{d}x_1 = \left[M_3 \int_0^{\kappa_3} \mathrm{d}\kappa_3 \right] \mathrm{d}x_1 = M_3\kappa_3\ \mathrm{d}x_1.
$$

In contrast, if the bending moment is proportional to the curvature, *i.e.*, if $M_3 = k\kappa_3$, where k is the constant of proportionality between the two quantities, the work becomes

$$
\mathrm{d}W_E = \left[\int_0^{\kappa_3} M_3\ \mathrm{d}\kappa_3 \right] \mathrm{d}x_1 = \left[\int_0^{\kappa_3} k\kappa_3\ \mathrm{d}\kappa_3 \right] \mathrm{d}x_1 = \frac{1}{2}k\kappa_3^2\mathrm{d}x_1 = \frac{1}{2}M_3\kappa_3\ \mathrm{d}x_1.
$$

Clearly, the difference between the two results can be directly attributed to the nature of the bending moment: if the bending moment *remains constant during the deformation*, the work it performs is $\mathrm{d}W_E = M_3\kappa_3\ \mathrm{d}x_1$, whereas if the bending moment *increases in proportion to the deformation*, the work becomes $\mathrm{d}W_E = 1/2\ M_3\kappa_3\ \mathrm{d}x_1$.

The same reasoning applies to bars under torsion. In section 9.7.2, the work done by a constant torque, M_1, undergoing a twist rate, κ_1, is found to be $\mathrm{d}W_E = -\mathrm{d}W_I = M_1\kappa_1\ \mathrm{d}x_1$, see eq. (9.71). In section 10.4.3, the work done by a torque that increases in proportion to the twist rate is found as $\mathrm{d}W_E = 1/2\ H_{11}\kappa_1^2\ \mathrm{d}x_1$, see eq. (10.42). Here again, the two results differ by a factor of one half, which is directly related to the nature of the torque.

All the results derived in section 9.7 for the work done by internal stresses or forces of constant magnitude in various types of structures can be readily used to obtain the work done by internal stresses or forces of magnitude proportional to the deformation in the same structures by simply multiplying the expression by a factor of one half.

The discussion of the previous paragraphs begs the following question: why is the moment assumed to be constant in the developments of section 9.7, whereas it is assumed to increase in proportion to the deformation in the present section? In section 9.7, the goal is to derive expressions for the virtual work and complementary virtual work. When computing the virtual work, virtual displacements do not affect the forces or stresses in the system, *i.e.*, the internal forces or stress remain constant, unaffected by virtual displacements. For instance, the work done by a constant moment in a beam is $W_I = -\int_0^L M_3\kappa_3\ \mathrm{d}x_1$, see eq. (9.69), where the bending moment, M_3, remains constant, unaffected by the curvature, κ_3. The virtual work is then $\delta W_I = -\int_0^L M_3\delta\kappa_3\ \mathrm{d}x_1$, see eq. (9.70a), where the bending moment remains constant, unaffected by the virtual curvature.

In contrast, the present study focuses on the determination of the strain energy stored in a structure. As the external loads are slowly applied to the solid, internal forces and moments increase in proportion to the deformation. Because the system is assumed to be conservative, the work done by the externally applied forces is now stored in the elastic body in the form of strain energy. Consequently, when the strain energy is computed from the evaluation of the work done by the externally applied loads, it must be assumed that the internal forces increase in proportion to the deformation, as is done in the present section.

10.5 Strain energy in solids

In this section, expressions for the strain energy in three-dimensional solids will be derived. The starting point of the development is the expression developed in section 9.7.3. Expressions for the strain energy in beams undergoing three-dimensional deformations will also be developed.

10.5.1 Three-dimensional solid

In section 9.7.3, the internal work done by constant stresses undergoing general, three-dimensional deformation is found to be $W_I = -\int_\mathcal{V} \underline{\sigma}^T \underline{\epsilon} \, d\mathcal{V}$, see eq. 9.76, where $\underline{\epsilon}$ and $\underline{\sigma}$ are the arrays of strain and stress components defined by eqs. (2.11a) and (2.11b), respectively. It follows that the work done by the constant, external stresses is $W_E = \int_\mathcal{V} \underline{\sigma}^T \underline{\epsilon} \, d\mathcal{V}$. Finally, if the stresses increase in proportion to the deformations, the work becomes

$$W_E = \frac{1}{2} \int_\mathcal{V} \underline{\sigma}^T \underline{\epsilon} \, d\mathcal{V}. \tag{10.46}$$

If the material behaves according to Hooke's law given by eqs. (2.4) and (2.9), the work can be expressed in terms of the strain components only as

$$W_E = \frac{1}{2} \int_\mathcal{V} \frac{E}{(1+\nu)(1-2\nu)} \left[(1-\nu)(\epsilon_1^2 + \epsilon_2^2 + \epsilon_3^2) + 2\nu(\epsilon_1\epsilon_2 + \epsilon_1\epsilon_3 + \epsilon_2\epsilon_3) \right.$$
$$\left. + \frac{1-2\nu}{2}(\gamma_{23}^2 + \gamma_{31}^2 + \gamma_{12}^2) \right] d\mathcal{V} = \int_\mathcal{V} a(\underline{\epsilon}) \, d\mathcal{V} = A(\underline{\epsilon}).$$

From this, the strain energy density function for a three-dimensional solid behaving according to Hooke's law becomes

$$a(\underline{\epsilon}) = \frac{1}{2} \frac{E}{(1+\nu)(1-2\nu)} \left[(1-\nu)(\epsilon_1^2 + \epsilon_2^2 + \epsilon_3^2) + 2\nu(\epsilon_1\epsilon_2 + \epsilon_1\epsilon_3 + \epsilon_2\epsilon_3) \right.$$
$$\left. + \frac{1-2\nu}{2}(\gamma_{23}^2 + \gamma_{31}^2 + \gamma_{12}^2) \right].$$
$$\tag{10.47}$$

This expression can be written in a more compact form as follows

$$a(\underline{\epsilon}) = \frac{1}{2} \frac{E}{(1+\nu)(1-2\nu)} [(1-\nu)I_1^2 - 2(1-2\nu)I_2], \tag{10.48}$$

where I_1 and I_2 are the first two invariants of the strain tensor defined by eqs. (1.86a) and (1.86b), respectively. It is also possible to write the strain energy density function in term of the strain array as

$$a(\underline{\epsilon}) = \frac{1}{2} \underline{\epsilon}^T \underline{\underline{C}} \underline{\epsilon}, \qquad (10.49)$$

where $\underline{\underline{C}}$ is the 6×6 stiffness matrix of the material defined by eq. (2.14).

Because Hooke's law is a linear stress-strain relationship, the strain energy and its complementary counterpart are equal, $a(\underline{\epsilon}) = a'(\underline{\sigma})$. The complementary strain energy density is expressed in terms of the stress components as

$$a'(\underline{\sigma}) = \frac{1}{2E} \left[\sigma_1^2 + \sigma_2^2 + \sigma_3^2 - 2\nu (\sigma_1 \sigma_2 + \sigma_1 \sigma_3 + \sigma_2 \sigma_3) \right. \\ \left. + 2(1+\nu) (\tau_{12}^2 + \tau_{23}^2 + \tau_{31}^2) \right]. \qquad (10.50)$$

A more compact expression can be obtained by making use of the invariants of the stress tensor, I_1 and I_2, given by eqs. (1.15a) and (1.15b), respectively, to find

$$a'(\underline{\sigma}) = \frac{1}{2E} \left[I_1^2 - 2(1+\nu)I_2 \right] = \frac{1}{2} \left[\frac{I_1^2}{E} - \frac{I_2}{G} \right]. \qquad (10.51)$$

Finally, it is also possible to write the complementary strain energy density function in term of the stress array as

$$a'(\underline{\sigma}) = \frac{1}{2} \underline{\sigma}^T \underline{\underline{S}} \underline{\sigma}, \qquad (10.52)$$

where $\underline{\underline{S}}$ is the 6×6 compliance matrix of the material defined by eq. (2.12).

10.5.2 Three-dimensional beams

The internal work done by constant stress resultants in three-dimensional beams undergoing deformation is derived in section 9.7.4, see eq. (9.78). From this result, the work done by the same stress resultants when they increase in proportion to the deformations becomes

$$W_E = \frac{1}{2} \int_0^L (N_1 \bar{\epsilon}_1 + M_2 \kappa_2 + M_3 \kappa_3) \, dx_1. \qquad (10.53)$$

If the beam is made of a linearly elastic material obeying Hooke's law, the sectional constitutive laws are given by eq. (6.12), assuming that the origin of the axis system is selected to be at the section's centroid. Eliminating the stress resultants from eq. (10.53) with the help of the sectional constitutive laws yields the strain energy in the beam as

$$A = \frac{1}{2} \int_0^L \left(S\bar{\epsilon}_1^2 + H_{22}^c \kappa_2^2 - 2H_{23}^c \kappa_2 \kappa_3 + H_{33}^c \kappa_3^2 \right) \, dx_1. \qquad (10.54)$$

Similarly, the complementary strain energy is obtained from eq. (10.53), where the sectional strains are expressed in terms of the stress resultants using the compliance form of the sectional constitutive laws, eqs. (6.13), to find

$$A' = \frac{1}{2} \int_0^L \left(\frac{N_1^2}{S} + \frac{H_{33}^c}{\Delta_H} M_2^2 + 2 \frac{H_{23}^c}{\Delta_H} M_2 M_3 + \frac{H_{22}^c}{\Delta_H} M_3^2 \right) \mathrm{d}x_1, \qquad (10.55)$$

where $\Delta_H = H_{22}^c H_{33}^c - H_{23}^c H_{23}^c$.

Equations (10.54) and (10.55) are general expression for the strain energy in three-dimensional beams and its complementary counterpart, respectively. They assume a linearly elastic material behavior characterized by Hooke's law, and the origin of the axis system must be located at the section's centroid.

These expressions can be simplified for specific applications. For instance, if the beam is undergoing axial deformations only, the first term only is kept and $A = 1/2 \int_0^L S \bar{\epsilon}_1^2 \, \mathrm{d}x_1$ whereas $A' = 1/2 \int_0^L N_1^2/S \, \mathrm{d}x_1$. If the axis system is selected to coincide with the principal centroidal axes of bending, $H_{23}^c = 0$, and

$$A = \frac{1}{2} \int_0^L \left(S \bar{\epsilon}_1^2 + H_{22}^c \kappa_2^2 + H_{33}^c \kappa_3^2 \right) \mathrm{d}x_1, \qquad (10.56a)$$

$$A' = \frac{1}{2} \int_0^L \left(\frac{N_1^2}{S} + \frac{M_2^2}{H_{22}^c} + \frac{M_3^2}{H_{33}^c} \right) \mathrm{d}x_1. \qquad (10.56b)$$

10.6 Applications to trusses and beams

The principle of minimum total potential energy leads to an elegant solution procedure for truss and beam problems, both of which will be addressed in the sections below.

10.6.1 Applications to trusses

To illustrate the application of the principle of minimum total potential energy to truss problems, a simple problem will be solved first, then, the general approach will be presented more formally, leading to a step-by-step procedure.

Consider the three-bar, hyperstatic truss depicted in fig. 10.16. All bars have the same cross-sectional area, $\mathcal{A}$, and modulus, E. Determine both the joint displacements and the member forces. In fig. 10.16, the three bars are labeled by a number indicated in a square box. The bar lengths are $L_1 = L_3 = L/\cos\theta$ and $L_2 = L$.

First, eq. (9.27) is used to find the bar elongations as $e_1 = u_1 \cos\theta + u_2 \sin\theta$, $e_2 = u_2$, and $e_3 = -u_1 \cos\theta + u_2 \sin\theta$. In view of eq. (10.29), the bar strain energy is written as $A = 1/2 \, ke^2$, where e is the bar elongation and $k = E\mathcal{A}/L$ its stiffness. The strain energy in the truss is then the sum of the bar strain energies

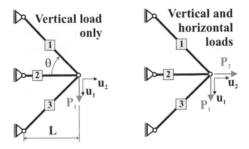

Fig. 10.16. Simple 3-bar truss.

$$A = \frac{1}{2}\left(\frac{EA\cos\theta}{L}e_1^2 + \frac{EA}{L}e_2^2 + \frac{EA\cos\theta}{L}e_3^2\right)$$

$$= \frac{1}{2}\frac{EA}{L}\left[(u_1\cos\theta + u_2\sin\theta)^2\cos\theta + u_2^2 + (-u_1\cos\theta + u_2\sin\theta)^2\cos\theta\right]$$

$$= \frac{1}{2}\frac{EA}{L}\left[2u_1^2\cos^3\theta + (1 + 2\sin^2\theta\cos\theta)u_2^2\right].$$

Based on eq. (10.13), the potential of the externally applied load, P_1, is given by $\Phi = -P_1u_1$. The total potential, Π, then becomes $\Pi = A + \Phi = A - P_1u_1$.

This problem has two degrees of freedom, u_1 and u_2, and the principle of minimum total potential energy, eq. (10.17), then requires

$$\frac{\partial\Pi}{\partial u_1} = \frac{EA}{L}2u_1\cos^3\theta - P_1 = 0,$$

$$\frac{\partial\Pi}{\partial u_2} = \frac{EA}{L}(1 + 2\sin^2\theta\cos\theta)u_2 = 0.$$

It is convenient to recast these equations in a matrix form to underline the fact that they form a set of two linear equations for the two generalized coordinates of the problem, u_1 and u_2,

$$\begin{bmatrix} 2\cos^3\theta & 0 \\ 0 & 1 + 2\sin^2\theta\cos\theta \end{bmatrix}\begin{Bmatrix} u_1 \\ u_2 \end{Bmatrix} = \frac{L}{EA}\begin{Bmatrix} P_1 \\ 0 \end{Bmatrix}.$$

Solving these equations then yields $u_1 = P_1L/(2EA\cos^3\theta)$ and $u_2 = 0$.

Once the displacements of the system have been evaluated, the elongation-displacement equations yield the non-dimensional elongations in each bar as

$$\frac{e_1}{L} = \frac{1}{2\cos^2\theta}\frac{P_1}{EA}, \quad e_2 = 0, \quad \frac{e_3}{L} = -\frac{1}{2\cos^2\theta}\frac{P_1}{EA}.$$

Next, the non-dimensional bar forces are obtained from the constitutive laws as

$$\frac{F_1}{P_1} = \frac{1}{2\cos\theta}, \quad F_2 = 0, \quad \frac{F_3}{P_1} = -\frac{1}{2\cos\theta}.$$

The approach presented here first finds the joint displacements, then evaluates bar elongations based on the elongation-displacement equations, and finally determines the bar forces with the help of the constitutive laws. The principle of minimum total potential energy enforces the equilibrium equations of the problem. The solution process does not make special provisions for the fact that the three-bar truss is a hyperstatic structure: it is equally applicable to both iso- and hyperstatic structures.

The response of a particular structure must often be evaluated under various loading conditions. The right portion of fig. 10.16 depicts the same three-bar truss subjected to two loads, P_1 and P_2, both applied at the common joint of the three bars. The only change in the above analysis is that the expression for the potential of the externally applied loads now becomes $\Phi = -P_1 u_1 - P_2 u_2$. Repeating the steps of the analysis leads to the following set of linear equations

$$
\begin{bmatrix} 2\cos^3\theta & 0 \\ 0 & 1 + 2\sin^2\theta\cos\theta \end{bmatrix} \begin{Bmatrix} u_1 \\ u_2 \end{Bmatrix} = \frac{L}{EA} \begin{Bmatrix} P_1 \\ P_2 \end{Bmatrix},
$$

and yields the joint displacements: $u_1 = P_1 L/(2EA\cos^3\theta)$ and $u_2 = P_2 L/[EA(1 + 2\sin^2\theta\cos\theta)]$.

General procedure

The general procedure for the solution of truss problems using the principle of minimum total potential energy can be summarized in the following steps.

1. Based on the geometry of the problem, find the length, L_i, of each of the N_b bars of the truss. The Young's modulus, E_i, and cross-section area, A_i, are given for each bar. Compute the stiffness, $k_i = (EA)_i/L_i$, of each bar.
2. Select the generalized coordinates of the problem to be the N joint displacements. Do not include the displacements at the supports because these are constrained to be zero.
3. Find the bar extensions, e_i, in terms of the joint displacements using eq. (9.27).
4. Determine the total strain energy of the system by adding up the contributions from the N_b bars,

$$
A = \frac{1}{2}\sum_{i=1}^{N_b} k_i e_i^2.
$$

5. Write the potential of the externally applied loads, Φ, using eq. (10.13). Because the externally applied loads, $P_j, j = 1, 2, \ldots N_P$, are assumed to act at the joints, the contribution of each load is $-P_j d_j$, where d_j is the displacement along the line of action of the force. The total potential of the externally applied loads is then

$$
\Phi = -\sum_{j=1}^{N_P} P_j d_j.
$$

6. The governing equations of the system are found by invoking the principle of minimum total potential energy expressed by eq. (10.17). Because the strain energy is a quadratic function of the joint displacements, and the potential of the externally applied loads a linear function of the same variables, the resulting equations form a linear set of N equations for the N generalized coordinates.

7. Solve the equations for the joint displacements.
8. Determine the bar elongations from the elongation-displacement equations.
9. Determine the bar forces from the constitutive laws, $F_i = k_i e_i$.

The procedure is unaffected by the nature of the truss: both iso- and hyperstatic problems can be solved in the same manner. For large trusses, the number of generalized coordinates increases, and the size of the set of linear equations for the generalized coordinates increases. Clearly, the approach is not suitable for hand calculations, but the solution of large systems of linear equations is easily obtained with the help of computers.

Example 10.6. Pentagonal truss
Consider the ten-bar pentagonal truss depicted in fig. 10.17. All bars have the same modulus, E, and cross-sectional area, $\mathcal{A}$, and a single vertical load, P, is applied at the top joint of the truss. Because both structure and loading are symmetric with respect to the vertical axis, the response of the truss must exhibit the same symmetry. The numbering of the bars reflects the symmetry of the problem: the behavior of the identically numbered bars must be identical. Only four independent joint displacement components are needed and are indicated in fig. 10.17; these will be selected as the generalized coordinates of the problem.

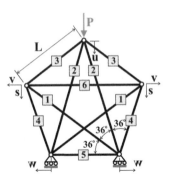

Fig. 10.17. A pentagonal truss.

The geometry of a regular pentagon implies that the angles between the diagonals of the pentagon and the sides are all $36°$. The bar lengths are found as $L_1 = L_2 = L_6 = L/(2\sin 18°) = 1.618\,L$, and $L_3 = L_4 = L_5 = L$. Equation (9.27) then yields the elongations of the bars as

$$
\begin{aligned}
e_1 &= (v+w)\cos(36°) - s\sin(36°) = 0.809(v+w) - 0.588s \\
e_2 &= -u\sin(72°) + w\cos(72°) = -0.951u + 0.309w \\
e_3 &= (s-u)\sin(36°) + v\cos(36°) = 0.588(s-u) + 0.809v \\
e_4 &= (v-w)\cos(72°) - s\sin(72°) = 0.309(v-w) - 0.951s \\
e_5 &= 2w \\
e_6 &= 2v.
\end{aligned}
\tag{10.57}
$$

The truss strain energy is found by summing up the individual bar contributions

$$A = \frac{1}{2} \frac{EA}{L} \left(\frac{2e_1^2}{1.618} + \frac{2e_2^2}{1.618} + 2e_3^2 + 2e_4^2 + e_5^2 + \frac{e_6^2}{1.618} \right).$$

Due to symmetry, the contributions of bars 1, 2, 3, and 4 are doubled. The total potential of the externally applied loads reduces to a single term, $\Phi = -Pu$.

The principle of minimum total potential energy expressed by eq. (10.17) then implies the vanishing of the derivatives of the total potential energy, $\Pi(u, v, w, s)$, with respect to each of the generalized coordinates,

$$\frac{\partial \Pi}{\partial u} = \frac{\partial \Pi}{\partial v} = \frac{\partial \Pi}{\partial w} = \frac{\partial \Pi}{\partial s} = 0.$$

Tedious algebraic manipulations yield a set of four simultaneous algebraic equations for the unknown joint displacements

$$\begin{bmatrix} 1.809 & -0.9511 & -0.3633 & -0.6910 \\ -0.9511 & 4.781 & 0.6180 & -0.2245 \\ -0.3633 & 0.6180 & 5.118 & 0 \\ -0.6910 & -0.2245 & 0 & 2.927 \end{bmatrix} \begin{Bmatrix} u \\ v \\ w \\ s \end{Bmatrix} = \frac{PL}{EA} \begin{Bmatrix} 1 \\ 0 \\ 0 \\ 0 \end{Bmatrix}.$$

These equations can solved numerically using a computer, and the result is

$$\frac{u}{L} = 0.703 \frac{P}{EA}, \quad \frac{v}{L} = 0.144 \frac{P}{EA}, \quad \frac{w}{L} = 0.0325 \frac{P}{EA}, \quad \frac{s}{L} = 0.177 \frac{P}{EA}.$$

Once the joint displacements are determined, the bar non-dimensional elongation are obtained from the elongation-displacement relationships, eqs. (10.57),

$$\frac{e_1}{L} = 0.0387 \frac{P}{EA}, \quad \frac{e_2}{L} = -0.6580 \frac{P}{EA}, \quad \frac{e_3}{L} = -0.1930 \frac{FL}{EA},$$
$$\frac{e_4}{L} = -0.1340 \frac{P}{EA}, \quad \frac{e_5}{L} = 0.0650 \frac{P}{EA}, \quad \frac{e_6}{L} = 0.2880 \frac{P}{EA}.$$

Finally, the non-dimensional bar forces are evaluated with the help of the constitutive laws to find

$$\frac{F_1}{P} = 0.0239, \quad \frac{F_2}{P} = -0.4070, \quad \frac{F_3}{P} = -0.1930,$$
$$\frac{F_4}{P} = -0.1340, \quad \frac{F_5}{P} = 0.0650, \quad \frac{F_6}{P} = 0.1780.$$

10.6.2 Problems

Problem 10.6. Planar 3-bar truss
The hyperstatic, three bar truss depicted in fig. 10.18 is subjected to a load, P, applied at joint **A**, with a line of action at an angle $\theta = 45$ degrees with respect to the horizontal. All bars have the same Young's modulus, E, and cross-sectional area, A. (1) Determine the displacement components, u_1 and u_2, of joint **A**. (2) Find the elongations in each bar. (3) Evaluate the forces in each bar.

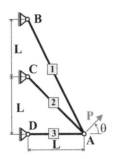

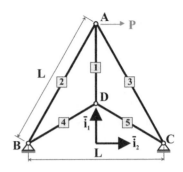

Fig. 10.18. Planar 3-bar truss with load applied at joint A.

Fig. 10.19. Hyperstatic planar triangular truss.

Problem 10.7. Hyperstatic planar triangular truss

The five-bar, hyperstatic truss shown in fig. 10.19 has the overall shape of an equilateral triangle, and is subjected to a horizontal load of magnitude P at joint **A**. All bars have the same Young's modulus, E, and cross-sectional area, $\mathcal{A}$. Note that this problem features four generalized coordinates: the horizontal and vertical displacement components at joints **A** and **D**. (1) Determine the generalized coordinates. (2) Find the elongations in each bar. (3) Evaluate the forces in each bar. Use the following data: $L = 2$ m, $\mathcal{A} = 100$ mm^2, $E = 70$ GPa, $P = 20$ kN. It will be necessary to use a computer to solve this problem.

Problem 10.8. Multi-cable truss structure

A vertical load, P, is supported by seven cables of equal cross-sectional area, $\mathcal{A}$, and elastic modulus, E as shown in fig 10.20. The angles between the cables and the vertical are 60°, 45°, 30°, 0°, −30°, −45°, and −60°. (1) Determine the generalized coordinates. (2) Find the elongations in each cable. (3) Evaluate the forces in each cable.

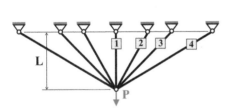

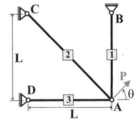

Fig. 10.20. Multiple cables supporting a single load point.

Fig. 10.21. Planar rectangular 3-bar truss with unequal axial stiffnesses.

Problem 10.9. Rectangular 3-bar planar truss

The three-bar, hyperstatic truss shown in fig. 10.21 is subjected to a load of magnitude P with a line of action at an angle $\theta = 45$ degrees with respect to the horizontal. All bars have the same elastic modulus, E; bars 1 and 3 have a cross-sectional area, $\mathcal{A}$, whereas that of bar 3 is $2\mathcal{A}$. (1) Determine the generalized coordinates. (2) Find the elongations in each bar. (3) Evaluate the forces in each bar.

10.6.3 Applications to beams

The principle of minimum total potential energy can also be applied to beam problems. Expressions for the strain energy stored in beams are developed in section 10.4 for beams under axial, transverse, and torsional loads.

Consider, for instance, a beam under a distributed transverse load, $p_2(x_1)$, as shown in fig. 5.14 on page 187. The transverse force applied on a differential element of the beam of length dx_1 is $p_2(x_1) \, dx_1$, and the work it performs is then $p_2(x_1) \bar{u}_2(x_1) \, dx_1$, where $\bar{u}_2(x_1)$ is the displacement of the force along its line of action. The work done by this distributed load applied along the beam's span is then found by integration, $W_E = \int_0^L p_2(x_1) \bar{u}_2(x_1) \, dx_1$. Equation (10.8) then yields the potential of this externally applied load as

$$\Phi = - \int_0^L p_2(x_1) \bar{u}_2(x_1) \, dx_1. \tag{10.58}$$

The total potential energy of the beam now follows from eq. (10.9) as

$$\Pi = A + \Phi = \frac{1}{2} \int_0^L H_{33}^c \left(\frac{d^2 \bar{u}_2}{dx_1^2} \right)^2 dx_1 - \int_0^L p_2 \bar{u}_2 \, dx_1,$$

where eq. (10.40) is used to express the beam's strain energy in bending in terms of the transverse displacement field, $u_2(x_1)$.

At first glance, this form for the total potential energy is similar to that developed earlier for mechanisms and trusses. A fundamental difference should be pointed out. Whereas in earlier developments the total potential energy is a function of the generalized coordinates, $\Pi = \Pi(\underline{q})$, the potential energy is now a function of another function, $\Pi = \Pi(\bar{u}_2(x_1))$, where $\bar{u}_2(x_1)$ is the beam's transverse displacement field. A "function of a function" is called a *functional*.

Beam problems are *infinite dimensional problems*, or *continuous problems*, because the solution to the problem requires the determination of the transverse displacements field, $\bar{u}_2(x_1)$, at all points $0 \leq x_1 \leq L$, and this is equivalent to an infinite number of unknowns. This contrasts with planar truss problems, for instance, that involve only $2N$ unknowns (*i.e.*, two displacement components at each of the truss' N joints) and are known as *finite dimensional problems* or *discrete problems*.

Minimization of the total potential energy is a standard calculus problem when it is a function of one or a finite number of variables such as the generalized coordinates in eqs. (10.17). When the total potential energy becomes a functional, new mathematical concepts are required to find the configuration of the system that minimizes this functional. The *calculus of variations* [6, 5] is the branch of mathematics that studies functionals, and elements of calculus of variations will be developed in chapter 12.

It is also possible to transform continuous problems into discrete problems by choosing specific functions for $u_2(x_1)$ whose amplitudes can then be determined using the principle of minimum total potential energy. This effectively reduces a problem with an infinite number of degrees of freedom to one with a finite number of degrees of freedom. As will be seen in chapter 11, the principle of minimum total potential energy is a powerful tool for constructing such approximate solutions.

Equation (10.58) gives the potential of the externally applied loads for a beam subjected to transverse distributed transverse loads, $p_2(x_1)$. As illustrated in fig. 6.1 on page 224, three-dimensional beam problems often involve complex loading conditions. In general, the beam can be subjected to distributed loading components, $p_1(x_1)$, $p_2(x_1)$, and $p_3(x_1)$, acting along axes $\bar{\imath}_1$, $\bar{\imath}_2$, and $\bar{\imath}_3$, respectively. Concentrated loads, P_1, P_2, and P_3, can also be applied along the same directions at any point along the span of the beam. Distributed moments, $q_1(x_1)$, $q_2(x_1)$, and $q_3(x_1)$, acting about axes $\bar{\imath}_1$, $\bar{\imath}_2$, and $\bar{\imath}_3$, respectively, can be applied. Finally, concentrated moments, Q_1, Q_2, and Q_3, can also be applied about the same directions at any point along the span of the beam. The potential of these externally applied loads becomes

$$
\begin{aligned}
\varPhi = &- \int_0^L p_1 \bar{u}_1 \, \mathrm{d}x_1 - P_1 \bar{u}_1(\alpha L) - \int_0^L q_1 \varPhi_1 \, \mathrm{d}x_1 - Q_1 \varPhi_1(\alpha L) \\
&- \int_0^L p_2 \bar{u}_2 \, \mathrm{d}x_1 - P_2 \bar{u}_2(\alpha L) + \int_0^L q_2 \frac{\mathrm{d}\bar{u}_3}{\mathrm{d}x_1} \, \mathrm{d}x_1 + Q_2 \frac{\mathrm{d}\bar{u}_3}{\mathrm{d}x_1}(\alpha L) \quad (10.59) \\
&- \int_0^L p_3 \bar{u}_3 \, \mathrm{d}x_1 - P_3 \bar{u}_3(\alpha L) - \int_0^L q_3 \frac{\mathrm{d}\bar{u}_2}{\mathrm{d}x_1} \, \mathrm{d}x_1 - Q_3 \frac{\mathrm{d}\bar{u}_2}{\mathrm{d}x_1}(\alpha L).
\end{aligned}
$$

The various terms appearing in this lengthy expression can be interpreted individually as follows.

For each concentrated load component, the potential is the negative product of the load by the displacement of its point of application projected along the line of action of the load. For instance, the potential of a concentrated load, P_1, applied at $x_1 = \alpha L$, is $-P_1 \bar{u}_1(\alpha L)$. For simplicity, all concentrated loads and moments are assumed to be applied at the same location, $x_1 = \alpha L$. In practical applications, however, each concentrated load must be multiplied by the displacement of its own point of application. For instance, if three concentrated loads, P_1, P_2, and P_3, acting along axes $\bar{\imath}_1$, $\bar{\imath}_2$, and $\bar{\imath}_3$, respectively, are applied at location $x_1 = \alpha L$, βL, and γL, respectively, the corresponding potential is $\varPhi = -P_1 \bar{u}_1(\alpha L) - P_2 \bar{u}_2(\beta L) - P_3 \bar{u}_3(\gamma L)$.

For each concentrated moment component, the potential is the negative product of the moment by the rotation of its point of application projected along the line of action of the moment. For instance, the potential of a concentrated torque, Q_1, applied at $x_1 = \alpha L$, is $-Q_1 \varPhi_1(\alpha L)$. Similarly, the potential of a concentrated moment, Q_3, is $-Q_3 \varPhi_3(\alpha L)$. According to the Euler-Bernoulli assumptions, the rotation of the section equals the slope of the beam, $\varPhi_3 = \mathrm{d}\bar{u}_2/\mathrm{d}x_1$, see eq. (5.3). The potential then becomes $-Q_3 \mathrm{d}\bar{u}_2(\alpha L)/\mathrm{d}x_1$. For rotation in the orthogonal plane, $\varPhi_2 = -\mathrm{d}\bar{u}_3/\mathrm{d}x_1$, see eq. (5.3); the corresponding potential then becomes $-Q_2 \varPhi_2(\alpha L) = Q_2 \mathrm{d}\bar{u}_3(\alpha L)/\mathrm{d}x_1$. Here again, if the concentrated moments are applied at different locations along the beam, the expression for the corresponding potential must be updated accordingly.

When dealing with distributed loads, a similar reasoning applies. The potential of a distributed axial force, $p_1(x_1)$, acting on an infinitesimal slice of the beam of length dx_1 is $-p_1(x_1)dx_1\,\bar{u}_1(x_1)$. The complete potential of the distributed load is then $-\int_0^L p_1(x_1)\bar{u}_1(x_1)\,dx_1$. Similar expression are readily derived for the other loading components, as indicated in eq. (10.59).

10.7 Development of a finite element formulation for trusses

The principle of minimum total potential energy provides a powerful tool for the analysis of trusses, as demonstrated in the previous section. While the approach is manageable for simple trusses consisting of only a few bars, it is clear that the algebraic manipulations become increasingly tedious as the number of bars increases. The method is, however, very systematic and reduces the problem to the solution of a set of simultaneous linear equations. While difficult to solve by hand, large sets of simultaneous linear equations are easily solved with the help of computers. In fact, powerful algorithms have been developed that routinely allow the accurate solution of very large systems of linear system, involving millions of degrees of freedom. Since computers take care of the solution phase, *i.e.*, the solution of large sets of linear equations, attention is directed in this section to the development of a systematic approach to generating the equilibrium equations of the problem.

A key to the approach presented here is to first focus on an individual truss member, *i.e.*, an axially loaded bar, rather than on the entire truss. The strain energy and potential of the externally applied loads are generated for each individual bar. Next, the total potential energy of the entire truss is obtained by summing up the contributions from each bar. Equilibrium equations of the problem are then generated by applying the principle of minimum total potential energy to the entire truss. This approach allows the development of *element oriented* methods, which focus on a single element of the structure at a time.

Another key to this approach is the additive property of strain energy: the total strain energy stored in a structure is the sum of the strain energy stored in all of its elastic components. More specifically, the total strain energy in a truss is equal to the sum of the strain energies in each of its bars. For trusses, each bar is an "element" of the system, and the heart of the approach is the evaluation of the strain energy in a generic bar or element of the truss. This simple computation is repeated for each element of the truss. The total strain energy in the system is then found by adding the contributions of the individual elements. This process, known as the *assembly process*, can be performed in an efficient manner through matrix operations that are readily implemented on computers.

From this cursory description of the approach, it is apparent that the process is exceedingly tedious if carried out by hand. It is, however, very systematic: the overall method is broken into a large number of step, each of which is rather simple to complete. The approach is ideally suited for computer implementation, and each step becomes a simple task to be efficiently performed by the computer. First, the strain energy in each bar is computed; next, the contribution of each bar is added to the total strain energy of the truss; finally, the large system of linear equations resulting from the application of the principle of minimum total potential energy is solved. The systematic use of linear algebra and matrix notation greatly simplifies the computer implementation of these procedures. Consequently, a brief summary of key concepts from linear algebra and the matrix and array notation used in this text is provided in appendix A.2. A quick review of the material presented in this appendix may prove useful in understanding the developments that follow.

The approach described in this section is basically an introduction to the *finite element method*, which has become the tool of choice for the solution of complex structural problems. While several key concepts of this method are present in this development, other distinctive features of the method are not required for truss problems. In particular, when applied to more complex structural components, the finite element method involves a discretization procedure that is not required for the problem at hand. This discretization procedure is needed for beams and will be described in section 11.5.

10.7.1 General description of the problem

Figure 10.22 depicts an 11-bar, 7-node planar truss that will be used to illustrate the development of the method. To avoid confusion, each *node number* is circled, and each *bar number* is indicated in a square box; the numbering sequence of both nodes and bars is otherwise arbitrary. The truss is in a plane defined by unit vectors $\bar{\imath}_1$ and $\bar{\imath}_2$ and is pinned to the ground at nodes 1 and 7. The geometry of the truss will be defined in a *global coordinate system* defined by orthonormal basis $\mathcal{I} = (\bar{\imath}_1, \bar{\imath}_2)$.

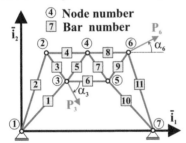

Fig. 10.22. Eleven-bar truss with node and element numbering.

Two concentrated loads are applied to the truss. Loads P_3 and P_6 are applied at nodes 3 and 6, respectively, and are acting at angles α_3 and α_6 with respect to the horizontal, respectively.

The stiffness properties of each bar are denoted $E_{(i)}\mathcal{A}_{(i)}/L_{(i)}$, where $E_{(i)}$, $\mathcal{A}_{(i)}$, and $L_{(i)}$ are the bar's Young's modulus, cross-sectional area, and length, respectively. Throughout this development, subscript $(\cdot)_{(i)}$ will be used to indicate quantities pertaining to the ith bar or element.

The geometry of the truss is defined by the coordinates of its 7 nodes. For instance, the components of the position vector of node 1 with respect to the origin of the coordinate system are denoted x_1 and y_1, along unit vectors $\bar{\imath}_1$ and $\bar{\imath}_2$, respectively, and stored in array $\underline{p}_1 = \{x_1, y_1\}^T$. Similar arrays[1] can be defined for all the nodes of the truss,

---

[1] This notation uses symbols x, y, and z, to denote position components, instead of x_1, x_2, and x_3, which are used throughout this book. Notations with multiple subscripts, such as x_{1i} to indicate the position component of node i along axis $\bar{\imath}_1$ are therefore avoided.

$$\underline{p}_1 = \begin{Bmatrix} x_1 \\ y_1 \end{Bmatrix}, \quad \underline{p}_2 = \begin{Bmatrix} x_2 \\ y_2 \end{Bmatrix}, \quad \ldots, \quad \underline{p}_7 = \begin{Bmatrix} x_7 \\ y_7 \end{Bmatrix}. \tag{10.60}$$

The subscript $(\cdot)_i$ will be used to indicate quantities pertaining to the ith node.

The generalized coordinates of the problem will be selected as the horizontal and vertical displacement components of each of the 7 nodes, denoted u_i and v_i, respectively. The following nodal displacement arrays will be used to contain these generalized coordinates,

$$\underline{q}_1 = \begin{Bmatrix} u_1 \\ v_1 \end{Bmatrix}, \quad \underline{q}_2 = \begin{Bmatrix} u_2 \\ v_2 \end{Bmatrix}, \quad \ldots, \quad \underline{q}_7 = \begin{Bmatrix} u_7 \\ v_7 \end{Bmatrix}. \tag{10.61}$$

Array $\underline{q}_1$ stores the two components of displacement at node 1, while array $\underline{q}_i$ stores those at node i. It will also be necessary to define a *global displacement array*, $\underline{q}$, that stores all the nodal displacement arrays in a single column as

$$\underline{q} = \{\underline{q}_1^T, \underline{q}_2^T, \underline{q}_3^T, \underline{q}_4^T, \underline{q}_5^T, \underline{q}_6^T, \underline{q}_7^T\}^T. \tag{10.62}$$

As mentioned earlier, the finite element method first focuses on a "generic element" of the system, in this case, a generic bar of the truss, to evaluate the strain energy stored in that specific element. Each bar is connected to two nodes: a root node, denoted *Node 1*, and a tip node, denoted *Node 2*. These nodes are referred to as "local nodes," and are used when focusing on a single bar of the system.

On the other hand, when the complete truss is considered, "global nodes" must be used. For instance, referring to fig. 10.22, bar 4 has two *local nodes*, denoted *Node 1* and *Node 2*, whereas its *global nodes* are nodes 2 and 4. Similarly, bar 9 has two *local nodes*, denoted *Node 1* and *Node 2*, whereas its *global nodes* are nodes 5 and 6. Since the local nodes are denoted *Node 1* and *Node 2* for each and every bar, they are not indicated on the figure as it would lead to confusion. This distinction between local and global nodes is important for the development of the method.

10.7.2 Kinematics of an element

The kinematics of a specific bar in the truss will be studied first. Figure 10.23 depicts a single bar with local nodes denoted *Node 1* and *Node 2*. To simplify the formulation of the problem, a *local coordinate system* is defined: unit vector $\bar{\jmath}_1$ is aligned with the axis of the bar, and $\bar{\jmath}_2$ is normal to the bar. The local coordinate system, $\mathcal{J} = (\bar{\jmath}_1, \bar{\jmath}_2)$, corresponds to a rotation of the global coordinate system, $\mathcal{I} = (\bar{\imath}_1, \bar{\imath}_2)$, by an angle $\hat{\theta}$, which is the angle between the bar and the horizontal axis, $\bar{\imath}_1$.

The position vectors of the two local nodes of the element are denoted as

$$\hat{\underline{p}}_1 = \begin{Bmatrix} \hat{x}_1 \\ \hat{y}_1 \end{Bmatrix}, \quad \text{and } \hat{\underline{p}}_2 = \begin{Bmatrix} \hat{x}_2 \\ \hat{y}_2 \end{Bmatrix}, \tag{10.63}$$

For clarity, the quantities pertaining to an element will be indicated with a caret $(\hat{\cdot})$, to distinguish them from their global counterparts. For example, it is important to distinguish the position vector of node 1, denoted $\underline{p}_1$ as defined by eq. (10.60), from $\hat{p}_1$, which indicates the position vector of *Node 1* of a generic bar element.

Similarly, the displacements of the two nodes of the elements, resolved in axis systems $\mathcal{I}$ and $\mathcal{J}$, are denoted

$$\hat{\underline{q}}_1 = \left\{ \begin{matrix} \hat{u}_1 \\ \hat{v}_1 \end{matrix} \right\}, \quad \hat{\underline{q}}_2 = \left\{ \begin{matrix} \hat{u}_2 \\ \hat{v}_2 \end{matrix} \right\}, \quad \text{and} \quad \hat{\underline{q}}_1^* = \left\{ \begin{matrix} \hat{u}_1^* \\ \hat{v}_1^* \end{matrix} \right\}, \quad \hat{\underline{q}}_2^* = \left\{ \begin{matrix} \hat{u}_2^* \\ \hat{v}_2^* \end{matrix} \right\}, \quad (10.64)$$

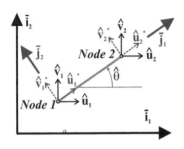

Fig. 10.23. General bar element.

respectively. For each of the two nodes, two sets of displacement components are thus defined. At *Node 1*, the components of the displacement vector resolved in the *global coordinate system* are denoted $\hat{u}_1$ and $\hat{v}_1$, whereas the corresponding components resolved in the *local coordinate system* are denoted $\hat{u}_1^*$ and $\hat{v}_1^*$, respectively. The superscript $(\cdot)^*$ will be used here to indicate the components of quantities resolved in the local coordinate system, $\mathcal{J}$.

The relationships between quantities resolved in two distinct orthonormal bases are discussed in appendix A.3. Since $\hat{\underline{q}}_1$ and $\hat{\underline{q}}_1^*$ are the components of the displacement vectors of *Node 1* resolved in two orthonormal bases, $\mathcal{I}$ and $\mathcal{J}$, eqs. (A.43) apply, and thus

$$\hat{\underline{q}}_1 = \hat{\underline{\underline{R}}}\, \hat{\underline{q}}_1^*, \quad (10.65)$$

where the element rotation matrix, $\hat{\underline{\underline{R}}}$, is similar to that defined in eq. (A.40),

$$\hat{\underline{\underline{R}}} = \begin{bmatrix} \cos\hat{\theta} & -\sin\hat{\theta} \\ \sin\hat{\theta} & \cos\hat{\theta} \end{bmatrix}. \quad (10.66)$$

A similar result can be developed for *Node 2*, $\hat{\underline{q}}_2 = \hat{\underline{\underline{R}}}\, \hat{\underline{q}}_2^*$, where the same rotation matrix is used.

The bar's length, $\hat{L}$ and orientation angle, $\hat{\theta}$, can be computed from the position vectors of its end nodes. The length is given by

$$\hat{L} = \|\hat{\underline{p}}_2 - \hat{\underline{p}}_1\| = \sqrt{(\hat{x}_2 - \hat{x}_1)^2 + (\hat{y}_2 - \hat{y}_1)^2}. \quad (10.67)$$

Angle $\hat{\theta}$ can be found from the nodal position vectors using the definition of the scalar product, $\bar{\imath}_1 \cdot (\hat{\underline{p}}_2 - \hat{\underline{p}}_1) = \|\bar{\imath}_1\|\,\|(\hat{\underline{p}}_2 - \hat{\underline{p}}_1)\| \cos\hat{\theta}$, and $\bar{\imath}_2 \cdot (\hat{\underline{p}}_2 - \hat{\underline{p}}_1) = \|\bar{\imath}_2\|\,\|(\hat{\underline{p}}_2 - \hat{\underline{p}}_1)\| \sin\hat{\theta}$. It then follows that

$$\cos\hat{\theta} = \frac{\bar{\imath}_1 \cdot (\hat{\underline{p}}_2 - \hat{\underline{p}}_1)}{\hat{L}}, \quad \sin\hat{\theta} = \frac{\bar{\imath}_2 \cdot (\hat{\underline{p}}_2 - \hat{\underline{p}}_1)}{\hat{L}}. \quad (10.68)$$

Finally, it will be convenient to combine the displacements of the element's two nodes into single array, called the *element displacement array*, which can be expressed in either global or local coordinates as

$$\underline{\hat{q}} = \left\{ \begin{matrix} \underline{\hat{q}}_1 \\ \underline{\hat{q}}_2 \end{matrix} \right\}, \quad \underline{\hat{q}}^* = \left\{ \begin{matrix} \underline{\hat{q}}_1^* \\ \underline{\hat{q}}_2^* \end{matrix} \right\}. \tag{10.69}$$

The relationship between these two arrays follows from eq. (10.65) as

$$\underline{\hat{q}} = \left\{ \begin{matrix} \underline{\hat{q}}_1 \\ \underline{\hat{q}}_2 \end{matrix} \right\} = \begin{bmatrix} \underline{\hat{R}} & \underline{0} \\ \underline{0} & \underline{\hat{R}} \end{bmatrix} \left\{ \begin{matrix} \underline{\hat{q}}_1^* \\ \underline{\hat{q}}_2^* \end{matrix} \right\} = \underline{\underline{\hat{T}}}\, \underline{\hat{q}}^*, \tag{10.70}$$

where $\underline{0}$ indicates a 2×2 null matrix, and the element coordinate transformation matrix, $\underline{\underline{\hat{T}}}$, is defined as

$$\underline{\underline{\hat{T}}} = \begin{bmatrix} \underline{\hat{R}} & \underline{0} \\ \underline{0} & \underline{\hat{R}} \end{bmatrix} \tag{10.71}$$

In view eq. (A.41), $\underline{\hat{R}}$ is an orthogonal matrix, and therefore, the element coordinate transformation matrix inherits the same property, $\underline{\underline{\hat{T}}}\,\underline{\underline{\hat{T}}}^T = I$. Consequently, it is possible to invert eq. (10.70) to find

$$\underline{\hat{q}}^* = \underline{\underline{\hat{T}}}^{-1}\, \underline{\hat{q}} = \underline{\underline{\hat{T}}}^T\, \underline{\hat{q}}. \tag{10.72}$$

10.7.3 Element elongation and force

Once the kinematics of an element are defined, it becomes possible to evaluate its elongation. From examination of fig. 10.23, the elongation, $\hat{e}$, of the bar is simply $\hat{e} = \hat{u}_2^* - \hat{u}_1^*$. It will be convenient to recast this expression as an array operation by writing $\hat{e} = \hat{u}_2^* - \hat{u}_1^* = \{-1, 0, 1, 0\}\, \underline{\hat{q}}^* = \underline{\hat{b}}^{*T}\, \underline{\hat{q}}^*$, where $\underline{\hat{b}}^* = \{-1, 0, 1, 0\}^T$ is an array that relates the element elongation to the nodal displacements, $\underline{\hat{q}}^*$.

Elongations are naturally expressed in terms of the displacement components expressed in the local coordinate system, but it is also possible to express them in terms of displacement components resolved in the global coordinate system as

$$\hat{e} = \underline{\hat{b}}^{*T}\, \underline{\hat{q}}^* = \underline{\hat{b}}^{*T}\, \underline{\underline{\hat{T}}}^T\, \underline{\hat{q}} = \underline{\hat{b}}^T\, \underline{\hat{q}}, \tag{10.73}$$

where eq. (10.72) is used to calculate the nodal displacement components expressed in the local coordinate system in terms of their global coordinate counterparts. Array $\underline{\hat{b}}$ is defined as $\underline{\hat{b}} = \underline{\underline{\hat{T}}}\, \underline{\hat{b}}^*$, where $\underline{\underline{\hat{T}}}$ is given by eq. (10.71).

The bar force, $\hat{F}$, is obtained by multiplying the elongation by the bar's axial stiffness to find

$$\hat{F} = \frac{\hat{E}\hat{A}}{\hat{L}}\, \hat{e} = \frac{\hat{E}\hat{A}}{\hat{L}}\, \underline{\hat{b}}^{*T}\, \underline{\hat{q}}^* = \frac{\hat{E}\hat{A}}{\hat{L}}\, \underline{\hat{b}}^T\, \underline{\hat{q}}. \tag{10.74}$$

10.7.4 Element strain energy and stiffness matrix

Next, the strain energy stored in a typical bar of the truss is evaluated. Because the stiffness of the bar is $\hat{E}\hat{A}/\hat{L}$, eq. (10.21) yields the element strain energy as

$$\hat{A} = \frac{1}{2}\frac{\hat{E}\hat{A}}{\hat{L}}\,\hat{e}^2 = \frac{1}{2}\frac{\hat{E}\hat{A}}{\hat{L}}\,\hat{e}\cdot\hat{e} = \frac{1}{2}\frac{\hat{E}\hat{A}}{\hat{L}}(\underline{b}^{*T}\underline{\hat{q}}^{*})^{T}(\underline{b}^{*T}\underline{\hat{q}}^{*}),$$

where the elongation is expressed in terms of the nodal displacement components in local coordinates using the first part of eq. (10.73). Regrouping the terms then leads to

$$\hat{A} = \frac{1}{2}\,\underline{\hat{q}}^{*T}\left[\frac{\hat{E}\hat{A}}{\hat{L}}(\underline{b}^{*}\underline{b}^{*T})\right]\underline{\hat{q}}^{*} = \frac{1}{2}\,\underline{\hat{q}}^{*T}\underline{\underline{\hat{k}}}^{*}\,\underline{\hat{q}}^{*},$$

where $\underline{\underline{\hat{k}}}^{*}$ is the *element stiffness matrix* expressed in the local coordinate system. Since $\underline{\hat{b}} = \{-1,0,1,0\}^{*T}$, the entries in this matrix become

$$\underline{\underline{\hat{k}}}^{*} = \frac{\hat{E}\hat{A}}{\hat{L}}(\underline{b}^{*}\underline{b}^{*T}) = \frac{\hat{E}\hat{A}}{\hat{L}}\begin{bmatrix} 1 & 0 & -1 & 0 \\ 0 & 0 & 0 & 0 \\ -1 & 0 & 1 & 0 \\ 0 & 0 & 0 & 0 \end{bmatrix}. \qquad (10.75)$$

It is also possible to evaluate the components of the same stiffness matrix expressed in the global coordinate system. Equation (10.70) expresses the nodal displacement components resolved in the local coordinate system in terms of their global coordinate counterparts as $\underline{\hat{q}}^{*} = \underline{\underline{\hat{T}}}^{T}\underline{\hat{q}}$. It then follows that

$$\hat{A} = \frac{1}{2}\,\underline{\hat{q}}^{*T}\underline{\underline{\hat{k}}}^{*}\,\underline{\hat{q}}^{*} = \frac{1}{2}\,(\underline{\hat{q}}^{T}\underline{\underline{\hat{T}}})\underline{\underline{\hat{k}}}^{*}(\underline{\underline{\hat{T}}}^{T}\underline{\hat{q}}) = \frac{1}{2}\,\underline{\hat{q}}^{T}(\underline{\underline{\hat{T}}}\,\underline{\underline{\hat{k}}}^{*}\,\underline{\underline{\hat{T}}}^{T})\underline{\hat{q}} = \frac{1}{2}\,\underline{\hat{q}}^{T}\underline{\underline{\hat{k}}}\,\underline{\hat{q}}, \qquad (10.76)$$

where $\underline{\underline{\hat{k}}} = \underline{\underline{\hat{T}}}\,\underline{\underline{\hat{k}}}^{*}\,\underline{\underline{\hat{T}}}^{T}$ stores the components of the element stiffness matrix expressed in the global coordinate system. Simple algebra reveals that

$$\underline{\underline{\hat{k}}} = \frac{\hat{E}\hat{A}}{\hat{L}}\left[\begin{array}{cc|cc} \cos^2\hat{\theta} & \sin\hat{\theta}\cos\hat{\theta} & -\cos^2\hat{\theta} & -\sin\hat{\theta}\cos\hat{\theta} \\ \sin\hat{\theta}\cos\hat{\theta} & \sin^2\hat{\theta} & -\sin\hat{\theta}\cos\hat{\theta} & -\sin^2\hat{\theta} \\ \hline -\cos^2\hat{\theta} & -\sin\hat{\theta}\cos\hat{\theta} & \cos^2\hat{\theta} & \sin\hat{\theta}\cos\hat{\theta} \\ -\sin\hat{\theta}\cos\hat{\theta} & -\sin^2\hat{\theta} & \sin\hat{\theta}\cos\hat{\theta} & \sin^2\hat{\theta} \end{array}\right]. \qquad (10.77)$$

In eqs. (10.75) and (10.77), the 4×4 element stiffness matrix is partitioned into four, 2×2 sub-matrices. The first two rows and columns of these matrices represent the stiffnesses associated with the two degrees of freedom, *i.e.*, the two displacement components, at *Node 1* of the element, whereas the last two rows and columns represent those associated with the two degrees of freedom at *Node 2* of the element. In eq. (10.75) the degrees of freedom are displacement components resolved in the *local* coordinate system, whereas in eq. (10.77) the degrees of freedom are resolved in the *global* coordinate system.

Not unexpectedly, the expression for element stiffness matrix expressed in the local system, eq. (10.75), is far simpler than its counterpart expressed in the global system, eq. (10.77). Why then is it desirable to derive element stiffness matrices in the global system? This question can be answered by considering fig. 10.22: bars

1, 3, 5, and 6 all connect to a single node, node 3. The local coordinate systems of these four bars are all different, and hence, the four corresponding element stiffness matrices expressed in their individual local systems are associated with local orthogonal displacement components resolved in four different systems. In contrast, when the four element stiffness matrices are expressed in the global system, they are associated with orthogonal displacement components all resolved in the same global system. This latter form of the stiffness matrix will considerable simplify the assembly procedure described below.

It is also important to note that $\hat{A}$ is a positive-definite quantity because the strain energy density for an axially loaded bar, see eq. (10.34), is positive-definite.

10.7.5 Element external potential and load array

When dealing with trusses, it is assumed that the externally applied loads act only at the nodes. Considering the single bar element depicted in fig. 10.23, let $\underline{\hat{F}}_1$ and $\underline{\hat{F}}_2$ be concentrated loads acting at local nodes, *Node 1* and *Node 2*, respectively. These two forces are resolved in the global coordinate system as $\underline{\hat{F}}_1 = \hat{f}_1 \bar{\imath}_1 + \hat{g}_1 \bar{\imath}_2$ and $\underline{\hat{F}}_2 = \hat{f}_2 \bar{\imath}_1 + \hat{g}_2 \bar{\imath}_2$, respectively, and their potential is easily evaluated using eq. (10.13) to find

$$\hat{\Phi} = -\left[\hat{f}_1 \hat{u}_1 + \hat{g}_1 \hat{v}_1\right] - \left[\hat{f}_2 \hat{u}_2 + \hat{g}_2 \hat{v}_2\right] = -\{\hat{f}_1, \hat{g}_1, \hat{f}_2, \hat{g}_2\} \, \underline{\hat{q}} = -\underline{\hat{f}}^T \underline{\hat{q}}, \quad (10.78)$$

where the *element load array* is defined as

$$\underline{\hat{f}} = \{\hat{f}_1, \hat{g}_1, \hat{f}_2, \hat{g}_2\}^T. \quad (10.79)$$

To illustrate this, consider a concentrated load of magnitude P and orientation α with respect to the horizontal, acting at *Node 1* of a bar element. The corresponding element load array is then $\hat{f}_1 = P \cos \alpha$, $\hat{g}_1 = P \sin \alpha$, and $\hat{f}_2 = \hat{g}_2 = 0$. If the same load is applied at *Node 2* instead, the element load array is $\hat{f}_1 = \hat{g}_1 = 0$, $\hat{f}_2 = P \cos \alpha$, and $\hat{g}_2 = P \sin \alpha$.

It is also possible to include the weight of the bar as an externally applied force. Let $\hat{m}$ be the bar's mass and $\hat{m}g$ the corresponding weight acting at its center of mass. For a homogeneous bar, the center of mass will be at its geometric center, and it makes sense to apply half of the weight at each of the element's two nodes. For example, if gravity acts along the negative axis $\bar{\imath}_2$ direction, the corresponding element load array is $\hat{f}_1 = 0$, $\hat{g}_1 = -\hat{m}g/2$, $\hat{f}_2 = 0$, and $\hat{g}_2 = -\hat{m}g/2$.

10.7.6 Assembly procedure

In the previous sections, attention is focused on a single, generic bar to determine its *element* stiffness matrix, see eq. (10.77), and *element* load array, see eq. (10.79). These two quantities are obtained from the element strain energy and external potential, respectively. In this section, attention shifts to the overall truss problem to determine the *global stiffness matrix* and *global load array*. These two quantities will be obtained from the system's total strain energy and total external potential, respectively. Since both strain energy and external potential are scalar quantities, their combined total will be evaluated simply by summing up the contributions from the individual elements.

The total strain energy, A, stored in the truss is the sum of the contributions of all bars. In eq. (10.76), the strain energy of a single, generic bar is $\hat{A}$, and this notation is not ambiguous because only a single element is considered. It now becomes necessary, however, to add the element identification using the subscript $(.)_{(i)}$ introduced earlier. Summing over all elements yields

$$A = \sum_{i=1}^{N_e} \hat{A}_{(i)} = \frac{1}{2} \sum_{i=1}^{N_e} \hat{\underline{q}}_{(i)}^T \underline{\underline{\hat{k}}}_{(i)} \hat{\underline{q}}_{(i)}, \qquad (10.80)$$

where N_e is the number of bars in the truss ($N_e = 11$ for the truss illustrated in fig. 10.23). In this case, it is also necessary to add the element identification subscript to both the element stiffness matrix, $\underline{\underline{\hat{k}}}_{(i)}$, and the nodal displacement array, $\hat{\underline{q}}_{(i)}$.

Equation (10.80) gives the total strain energy in the structure, but it is not easy to manipulate because each term in the sum is expressed in terms of a different set of degrees of freedom. For example, with reference to fig. 10.22, element 8 is connected to global nodes 4 and 6 which are local *Node 1* and *Node 2* for the element, respectively. The element stiffness, $\underline{\underline{\hat{k}}}_{(8)}$, is defined in terms of these global nodes, see eq. (10.77), and the corresponding element displacement array is $\hat{\underline{q}}_{(8)}^T = \left\{ \hat{\underline{q}}_1^T, \hat{\underline{q}}_2^T \right\} = \left\{ \underline{q}_4^T, \underline{q}_6^T \right\} = \left\{ u_4, v_4, u_6, v_6 \right\}^T$.

To remedy this situation, the *connectivity matrix*, $\underline{\underline{C}}_{(i)}$, for the ith element is introduced. This matrix is designed to extract the element displacement array, $\hat{\underline{q}}_{(i)}$, from the global displacement array, $\underline{q}$, defined by eq. (10.62). This operation can be written as

$$\hat{\underline{q}}_{(i)} = \underline{\underline{C}}_{(i)} \underline{q}. \qquad (10.81)$$

To best understand this abstract relationship, consider a specific element of the truss, say bar 6, as shown in fig. 10.22. Its local nodes, *Node 1* and *Node 2*, are associated with the global node numbers 3 and 5, respectively, so that $\hat{\underline{q}}_1 = \underline{q}_3$ and $\hat{\underline{q}}_2 = \underline{q}_5$. The element displacement array, $\hat{\underline{q}}_{(6)}$, can thus be written as

$$\hat{\underline{q}}_{(6)} = \left\{ \begin{matrix} \hat{\underline{q}}_1 \\ \hat{\underline{q}}_2 \end{matrix} \right\}_{(6)} = \left\{ \begin{matrix} \underline{q}_3 \\ \underline{q}_5 \end{matrix} \right\} = \begin{bmatrix} \underline{0} & \underline{0} & \underline{I} & \underline{0} & \underline{0} & \underline{0} & \underline{0} \\ \underline{0} & \underline{0} & \underline{0} & \underline{0} & \underline{I} & \underline{0} & \underline{0} \end{bmatrix} \left\{ \begin{matrix} \underline{q}_1 \\ \underline{q}_2 \\ \underline{q}_3 \\ \underline{q}_4 \\ \underline{q}_5 \\ \underline{q}_6 \\ \underline{q}_7 \end{matrix} \right\} = \underline{\underline{C}}_{(6)} \underline{q},$$

where $\underline{0}$ and $\underline{I}$ represent the 2×2 null and identity matrices, respectively. The connectivity matrix, $\underline{\underline{C}}_{(6)}$, is called a *Boolean matrix* because its entries consist solely of 0's and 1's. Matrix $\underline{\underline{C}}_{(6)}$ establishes the connections of bar 6 within the truss by indicating the nodes to which this bar is connected, and this explains its being named a "connectivity matrix."

Expressing the element nodal displacement arrays, $\hat{\underline{q}}_{(i)}$, in terms of the global displacement array, $\underline{q}$, with the help of eq. (10.81), the total strain energy of the truss given by eq. (10.80) now becomes

$$A = \frac{1}{2} \sum_{i=1}^{N_e} \left(\underline{q}^T \underline{\underline{C}}_{(i)}^T \right) \hat{\underline{\underline{k}}}_{(i)} \left(\underline{\underline{C}}_{(i)} \underline{q} \right) = \frac{1}{2} \underline{q}^T \left[\sum_{i=1}^{N_e} \underline{\underline{C}}_{(i)}^T \hat{\underline{\underline{k}}}_{(i)} \underline{\underline{C}}_{(i)} \right] \underline{q}.$$

This expression can be simplified to

$$A = \frac{1}{2} \underline{q}^T \underline{\underline{K}} \underline{q}, \tag{10.82}$$

by defining the *global stiffness matrix*, $\underline{\underline{K}}$, as

$$\underline{\underline{K}} = \sum_{i=1}^{N_e} \underline{\underline{C}}_{(i)}^T \hat{\underline{\underline{k}}}_{(i)} \underline{\underline{C}}_{(i)}. \tag{10.83}$$

The potential of the externally applied loads, Φ, is found by adding the contributions of all bars,

$$\Phi = \sum_{i=1}^{N_e} \hat{\Phi}_{(i)} = - \sum_{i=1}^{N_e} \hat{\underline{q}}_{(i)}^T \hat{\underline{f}}_{(i)}, \tag{10.84}$$

where $\hat{\underline{f}}_{(i)}$ is the load array for the ith element, as defined by eq. (10.79) for a generic bar element. Here again, it is convenient to use the connectivity matrix defined in eq. (10.81) to evaluate the potential,

$$\Phi = - \sum_{i=1}^{N_e} \left(\underline{\underline{C}}_{(i)} \underline{q} \right)^T \hat{\underline{f}}_{(i)} = -\underline{q}^T \left\{ \sum_{i=1}^{N_e} \underline{\underline{C}}_{(i)}^T \hat{\underline{f}}_{(i)} \right\}.$$

This expression can be simplified to

$$\Phi = -\underline{q}^T \underline{Q}, \tag{10.85}$$

by defining the *global load array*, $\underline{Q}$, as

$$\underline{Q} = \sum_{i=1}^{N_e} \underline{\underline{C}}_{(i)}^T \hat{\underline{f}}_{(i)}. \tag{10.86}$$

Finally, the total potential energy, Π, of the truss is obtained by adding the potential of the external loads, eq. (10.85), to the total strain energy, eq. (10.82), to find

$$\Pi = A + \Phi = \frac{1}{2} \underline{q}^T \underline{\underline{K}} \underline{q} - \underline{q}^T \underline{Q}. \tag{10.87}$$

This compact expression for the total potential energy of the complete system is only possible because the matrix notation encapsulates the nodal and element quantities in arrays and matrices.

The total strain energy is a quadratic form of the generalized coordinates, whereas the potential of the externally applied loads is a linear form of the same variables. It should also be noted that the strain energy of the truss is positive-definite because it is the sum of the positive-definite strain energies for each bar.

10.7.7 Alternative description of the assembly procedure

The assembly procedure described in terms of the connectivity matrix defined in eq. (10.81) is formally correct, but it is not easy to understand nor is it computationally efficient for realistic trusses with many members and nodes. The connectivity matrix, $\underline{\underline{C}}_{(i)}$, has four rows and $2N$ columns, where N is the total number of nodes. For large trusses consisting of many bars and nodes, this matrix becomes very large with a total of $8N$ entries, and yet, only four entries have a unit value while all $(8N - 4)$ others are zero. Furthermore, the evaluation of the global stiffness matrix involves a triple matrix product for each elements, see eq. (10.83). These become increasingly expensive to perform as the problem size increases, and they also are very wasteful because most operations actually are multiplications by zero.

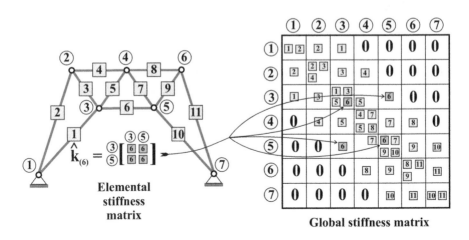

Fig. 10.24. Illustration of the assembly procedure.

It is possible to give a more graphical visualization of the assembly process. Figure 10.24 shows the 11-bar, 7-node truss under consideration. It also depicts the global stiffness matrix; the 7 rows and columns in the matrix are labeled with their corresponding node numbers. Each node has two degrees of freedom (the horizontal and vertical displacement components at that node), so each of the 49 entries is actually a 2×2 matrix and the size of the global stiffness matrix itself is 14×14.

Consider now a typical bar of the truss, say bar 6. Its local nodes, *Node 1* and *Node 2*, are associated with the global node numbers 3 and 5, respectively. The local stiffness matrix for this bar, $\underline{\hat{k}}_{(6)}$, can be partitioned into four 2×2 matrices, as shown in eq. (10.77). Bar 6 is connected to global nodes 3 and 5, and therefore, the four sub-matrices of the local stiffness matrix can simply be added to entries $\underline{K}(3,3)$, $\underline{K}(5,5)$, $\underline{K}(3,5)$, and $\underline{K}(5,3)$ in the global stiffness matrix, as indicated by the arrows in fig. 10.24. In this discussion, the notation $\underline{K}(i,j)$ refers to the 2×2 sub-matrix that appears as the (i,j) entry of the matrix depicted in fig. 10.24.

This procedure is repeated for each bar of the truss to give the final result shown in fig. 10.24. This figure requires careful interpretation. Each of the element numbers shown in square boxes defines a 2×2 sub-matrix extracted from the corresponding element stiffness matrix. These 2×2 sub-matrices are added together to produce the final result in the global stiffness matrix.

Another way to look at the same process is to consider the fully assembled global stiffness matrix in fig. 10.24. Diagonal entry $\underline{K}(2,2)$ collects contributions from bars 2, 3, and 4 because these three bars are all physically connected to node 2. Similarly, diagonal entry $\underline{K}(5,5)$ collects contributions from bars 6, 7, 9, and 10 because these four bars all connect to node 5.

The off-diagonal entries in the global stiffness matrix can be interpreted in a similar manner. For instance, entries $\underline{K}(1,3)$ and $\underline{K}(3,1)$ each collect the single contribution stemming from bar 1, because bar 1 connects nodes 1 and 3. Similarly, bar 8 connects nodes 4 and 6, and is the sole contributor to entries $\underline{K}(4,6)$ and $\underline{K}(6,4)$ in the global stiffness matrix. It is important to note that the symmetry of the local stiffness matrix, see eq. (10.77), and the symmetry of the assembly process, result in the global stiffness matrix also being a symmetric matrix.

At completion of the assembly process, many entries of the global stiffness matrix remain empty or null. For instance, entries $\underline{K}(2,6) = \underline{K}(6,2) = 0$, because no bar directly connects nodes 2 and 6. Similarly, $\underline{K}(1,4) = \underline{K}(4,1) = 0$ because nodes 1 and 4 are not directly connected by a bar. For the node numbering sequence selected in fig. 10.24, the non-zero entries in the global stiffness matrix concentrate near the diagonal, and the resulting matrix is called a "banded matrix." It should be obvious that other node and/or element numbering could lead to a more dispersed arrangement of the non-zero entries.

This alternative description of the element assembly process is more graphical than the initial description based on connectivity matrices. When implementing the approach in a computer program, this process of simply adding the entries of the element stiffness matrices to corresponding entries in the global stiffness matrix is the preferred approach, because it is easy to program and efficient to execute.

10.7.8 Derivation of the governing equations

The total potential energy of the truss is given by eq. (10.87), and application of the principle of minimum total potential energy, eq. (10.17), now implies

$$\frac{\partial \Pi}{\partial \underline{q}} = \frac{\partial}{\partial \underline{q}} \left(\frac{1}{2} \underline{q}^T \underline{K} \underline{q} - \underline{q}^T \underline{Q} \right) = \underline{K} \underline{q} - \underline{Q} = 0. \tag{10.88}$$

To compute the derivative of the total potential energy, eqs. (A.29) and (A.27) are used to evaluate the derivatives of the strain energy and potential of the externally applied loads, respectively. Appendix A.2.9 also proves that this solution is a minimum.

The governing equation of the system take the form of a linear system of equations,

$$\underline{\underline{K}}\,\underline{q} = \underline{Q}. \tag{10.89}$$

The global stiffness matrix, $\underline{\underline{K}}$, is computed from the given geometry and material properties of the truss, while the global load array, $\underline{Q}$, stems from the external loads applied at the nodes. The unknown quantities are the nodal displacements stored in array $\underline{q}$. The solution of eq. (10.89) yields the displacements at all joints of the truss.

The approach presented here is an element-oriented version of the displacement or stiffness method described in section 4.3.2. Each line of the matrix relationship, eq. (10.89), represents an equilibrium equation of the problem. For instance, the equation obtained by extracting the first line eq. (10.89) represents the equilibrium equation obtained by imposing the vanishing of the sum of the horizontal forces acting at node 1, whereas the second equation corresponds to the vertical equilibrium equation at the same node.

10.7.9 Solution procedure

Efficient algorithms are available for the solution of large sets of linear equations using computers. At this point, however, the linear system given in eq. (10.89) cannot be solved because the global stiffness matrix is singular.

This situation arises because the element stiffness matrices that make up the global stiffness matrix are each singular. Examination of eq. (10.75) reveals that the element stiffness matrix contains two rows of zeros and furthermore, the third row is simply -1 times the first row. Consequently, this 4×4 matrix is three times singular: it is of rank 1, presents a rank deficiency of 3, and has a zero determinant. The element stiffness matrix in global coordinates given by eq. (10.77) has the same rank deficiency because it is the same matrix, expressed in a different coordinate system. Finally, the global stiffness matrix, $\underline{\underline{K}}$, also presents a rank deficiency of 3, and because it is three times singular, the global stiffness matrix cannot be inverted.

Calculation of the eigenvectors and eigenvalues[2] of the element stiffness matrix, $\underline{\underline{\hat{k}}}$, given by eq. (10.77), reveals more information about this rank deficiency. The unit eigenvectors of this matrix are given by the arrays

$$\underline{n}_1 = \frac{1}{\sqrt{2}}\begin{Bmatrix} 1 \\ 0 \\ 1 \\ 0 \end{Bmatrix}, \quad \underline{n}_2 = \frac{1}{\sqrt{2}}\begin{Bmatrix} 0 \\ 1 \\ 0 \\ 1 \end{Bmatrix}, \quad \underline{n}_3 = \frac{1}{\sqrt{2}}\begin{Bmatrix} \sin\hat{\theta} \\ -\cos\hat{\theta} \\ -\sin\hat{\theta} \\ \cos\hat{\theta} \end{Bmatrix}, \quad \underline{n}_4 = \frac{1}{\sqrt{2}}\begin{Bmatrix} \cos\hat{\theta} \\ \sin\hat{\theta} \\ -\cos\hat{\theta} \\ -\sin\hat{\theta} \end{Bmatrix},$$

---

[2] See appendix A.2.4 for details on the calculation of eigenvalues and eigenvectors of symmetric, positive-definite matrices.

and the corresponding eigenvalues are $\lambda_1 = \lambda_2 = \lambda_3 = 0$, and $\lambda_4 = 2\hat{E}\hat{A}/\hat{L}$, respectively.

The first three eigenvectors, $\underline{n}_1$, $\underline{n}_2$, and $\underline{n}_3$, represent the horizontal rigid body translation of the bar, its vertical rigid body translation, and its rigid body rotation, respectively. The three associated eigenvalues vanish. By definition, rigid body motions create no deformation or straining of the element, and hence, no strain energy is associated with rigid body modes. Using eq. (10.76), the strain energy associated with the first eigenvector is $\hat{A} = 1/2\, \underline{n}_1^T \hat{\underline{k}}\, \underline{n}_1$, and as expected, $\hat{A} = 0$, because the definition of eigenvectors implies $\hat{\underline{k}}\, \underline{n}_1 = \lambda_1 \underline{n}_1 = 0$. Using a similar reasoning, it is easy to prove the vanishing of the strain energy associated with each of the three rigid body modes of the element. Clearly, the presence of three rigid body modes for the structure implies the rank deficiency of 3 for the element stiffness matrix. The entire truss also presents three rigid body modes, and hence, the global stiffness matrix also features a rank deficiency of 3.

The physical interpretation of this situation is that boundary conditions have not yet been applied to the truss, which is still free to translate and rotate in plane $(\bar{\imath}_1, \bar{\imath}_2)$. Figure 10.22 shows that nodes 1 and 7 are pinned to the ground, preventing any rigid body motion of the truss. These conditions, however, are not reflected in the global stiffness matrix, $\underline{K}$, given in eq. (10.83).

The boundary conditions require the vanishing of the displacements at nodes 1 and 7: $\underline{q}_1 = \underline{q}_7 = 0$. Consequently, the reaction forces arising at nodes 1 and 7, denoted $\underline{R}_1$ and $\underline{R}_7$, respectively, should be treated as externally applied forces. The equilibrium equations associated with those two nodes correspond to the first two and last two rows of the global stiffness matrix illustrated in fig. 10.24. These equations can be removed from eq. (10.89) and written separately as

$$\underline{\underline{K}}(1,1)\underline{q}_1 + \underline{\underline{K}}(1,2)\underline{q}_2 + \underline{\underline{K}}(1,3)\underline{q}_3 = \underline{R}_1, \tag{10.90a}$$

$$\underline{\underline{K}}(7,5)\underline{q}_5 + \underline{\underline{K}}(7,6)\underline{q}_6 + \underline{\underline{K}}(7,7)\underline{q}_7 = \underline{R}_7, \tag{10.90b}$$

where the indices on $\underline{\underline{K}}$ denote nodes and not degrees of freedom (therefore these are 2×2 sub-matrices from the global stiffness matrix). This leaves $14 - 4 = 10$ rows remaining in the set of equations. Since the displacements at nodes 1 and 7 vanish, the corresponding terms vanish in eq. (10.90), which can be solved for the unknown reaction forces

$$\underline{R}_1 = \underline{\underline{K}}(1,2)\underline{q}_2 + \underline{\underline{K}}(1,3)\underline{q}_3, \quad \underline{R}_7 = \underline{\underline{K}}(7,5)\underline{q}_5 + \underline{\underline{K}}(7,6)\underline{q}_6. \tag{10.91}$$

Because the displacements at nodes 1 and 7 are zero, the contributions from the terms appearing in the first two and last two columns of the global stiffness matrix vanish. Consequently, the first two and last two columns of $\underline{K}$, as well as the first two and last two entries in arrays $\underline{q}$ and $\underline{Q}$ can also be eliminated to create a reduced set of 10 equations that can now be solved for the remaining 10 unknown nodal displacements.

In summary, the boundary conditions can be imposed through the following generalized process. *(1)* Eliminate the rows and columns of the global stiffness matrix corresponding to constrained degrees of freedom to create its reduced counterpart, $\underline{\bar{K}}$. (2) Eliminate the row of the global displacement array corresponding to constrained degrees of freedom to create its reduced counterpart, $\underline{\bar{q}}$. (3) Finally, eliminate the row of the global load array corresponding to constrained degrees of freedom to create its reduced counterpart, $\underline{\bar{Q}}$. The system of equations for the truss then reduces to

$$\underline{\bar{K}}\,\underline{\bar{q}} = \underline{\bar{Q}}. \tag{10.92}$$

The reduced stiffness matrix will now be non-singular, and the solution of the problem is found by solving the linear system to find the remaining nodal displacements as $\underline{\bar{q}} = \underline{\bar{K}}^{-1}\underline{\bar{Q}}$. Finally, the reactions can be determined from the equations extracted in step 1.

10.7.10 Solution procedure using partitioning

The procedure developed in the previous paragraphs is very descriptive and consists of "eliminating rows and columns" in the global stiffness matrix, global displacement array, and global load array. A more mathematical description of the process is based on partitioning of the same quantities in a manner that allows separate treatment of the constrained and unconstrained nodes.

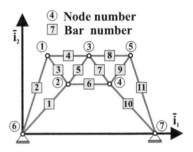

Fig. 10.25. Eleven-bar truss.

First, it should be observed that the node numbering sequence is arbitrary: in fig. 10.22, each node is assigned a number at random. Figure 10.25, shows the same truss, but with a different node numbering for which the two nodes where boundary conditions are to be applied are now numbered as nodes 6 and 7 and are the last in the series.

The global displacement array can now be partitioned into two sub-arrays (using a vertical bar $\{.|.\}$) as follows

$$\underline{q} = \left\{ \underline{q}_1^T, \underline{q}_2^T, \underline{q}_3^T, \underline{q}_4^T, \underline{q}_5^T, | \underline{q}_6^T, \underline{q}_7^T \right\}^T = \left\{ \underline{q}_u^T, \underline{q}_p^T \right\}^T. \tag{10.93}$$

Array $\underline{q}_u$ is of size N_u and stores the N_u unknown displacements at nodes 1 to 5, while array $\underline{q}_p$ is of size N_p and stores the N_p prescribed displacements at support nodes 6 and 7. For the truss depicted in fig. 10.25, $N_u = 10$ (two displacement components at each of the five nodes numbered 1 to 5), and $N_p = 4$ (two displacement components at both nodes 6 and 7). Figure 10.25 illustrates the case where nodes 6 and 7 have zero prescribed values, *i.e.*, they are pinned to the ground. In some case, the prescribed displacement at a node might be non-zero. For instance, if node 7 are prescribed to move by an amount Δ along axis $\bar{\imath}_1$, then $\underline{q}_7 = \{\Delta, 0\}^T$.

The node numbering sequence is arbitrary, and therefore, the formulation of the problem with the numbering sequence shown in fig. 10.25 is equivalent to that described earlier for the numbering sequence presented in fig. 10.22. It leads to the following partitioned governing equations

$$\begin{bmatrix} \underline{\underline{K}}_{uu} & \underline{\underline{K}}_{up} \\ \underline{\underline{K}}_{up}^T & \underline{\underline{K}}_{pp} \end{bmatrix} \begin{Bmatrix} \underline{q}_u \\ \underline{q}_p \end{Bmatrix} = \begin{Bmatrix} \underline{Q}_u \\ \underline{Q}_p \end{Bmatrix}, \tag{10.94}$$

where subscripts u and p refer to "unconstrained" nodes and nodes with "prescribed" displacements, respectively.

This system of equation corresponds to a partitioned version of the general governing equation for the truss given by eqs. (10.89). In these equations, the global load array is partitioned in the same manner as the global displacement array, *i.e.*,

$$\underline{Q} = \left\{ \underline{Q}_1^T, \underline{Q}_2^T, \underline{Q}_3^T, \underline{Q}_4^T, \underline{Q}_5^T, |\underline{R}_6^T, \underline{R}_7^T \right\}^T = \left\{ \underline{Q}_u^T, \underline{Q}_p^T \right\}^T. \tag{10.95}$$

Array $\underline{Q}_u$ is of size N_u and stores the known forces applied at nodes 1 to 5, while array $\underline{Q}_p$ is of size N_p and stores the reaction forces at nodes 6 and 7. Matrices $\underline{\underline{K}}_{uu}$, $\underline{\underline{K}}_{pp}$, and $\underline{\underline{K}}_{up}$ are of size $(N_u \times N_u)$, $(N_p \times N_p)$, and $(N_u \times N_p)$, respectively.

The first N_u equations of system (10.94) can be rewritten as

$$\underline{\underline{K}}_{uu} \underline{q}_u = \underline{Q}_u - \underline{\underline{K}}_{up} \underline{q}_p. \tag{10.96}$$

Because the prescribed displacements, $\underline{q}_p$, are known, their contribution is moved to the right-hand side of the equations. The unknown nodal displacements are evaluated as $\underline{q}_u = \underline{\underline{K}}_{uu}^{-1}(\underline{Q}_u - \underline{\underline{K}}_{up} \underline{q}_p)$. If the boundary conditions consist solely of nodes rigidly connected to the ground, all prescribed displacement vanish, $\underline{q}_p = 0$, and the system reduces to $\underline{q}_u = \underline{\underline{K}}_{uu}^{-1} \underline{Q}_u$, which is equivalent to eqs. (10.92).

Once the unknown displacements have been evaluated, the last N_p equations of system (10.94) can be rewritten to evaluate the reactions as

$$\underline{Q}_p = \underline{\underline{K}}_{up}^T \underline{q}_u + \underline{\underline{K}}_{pp} \underline{q}_p. \tag{10.97}$$

Here again, all known quantities have been moved to the right-hand side of the equations. If the boundary conditions consist solely of nodes rigidly connected to the ground, all prescribed displacement vanish, $\underline{q}_p = 0$, and $\underline{Q}_p = \underline{\underline{K}}_{up}^T \underline{q}_u$ are the reaction forces at the nodes pinned to the ground.

On the other hand, if some nodal displacements are prescribed to non-vanishing values, eq. (10.96) can still be used to find the unconstrained nodal displacements, $\underline{q}_u$, and eq. (10.97) then yields the reaction forces, $\underline{Q}_p$. These are still reaction forces because they arise from either zero or non-zero prescribed nodal displacements. Those acting at nodes where the displacements are prescribed are sometimes called the "driving forces," *i.e.*, the forces that must be applied at a node to achieve the prescribed displacement. Nodes with prescribed displacements can also be used to represent misalignments in the supports due to non-ideal truss geometry.

In the partitioned system of eq. (10.94), the reduced stiffness matrix, $\underline{\underline{K}}_{uu}$, is obtained by eliminating the last N_p rows and columns of the global stiffness matrix and is equivalent to the reduced stiffness matrix, $\underline{\underline{K}}$, in eq. (10.92). The present approach, which is based on partitioning of the reordered system of equations, gives a rigorous justification of the procedure introduced in the previous section.

It should also be noted that both approaches to enforcing the boundary conditions may require re-numbering of the rows and columns in the original set of equations. While tedious to do by hand, such manipulations are easily handled using computer programs.

10.7.11 Post-processing

The last step in the solution process is to determine the bar elongations and forces. The elongation of a bar is given by eq. (10.73) as $\hat{e} = \hat{\underline{b}}^T \hat{\underline{q}}$. For the ith element, this can be written in a formal manner as

$$\hat{e}_{(i)} = \hat{\underline{b}}_{(i)}^T \hat{\underline{q}}_{(i)} = \hat{\underline{b}}_{(i)}^T \underline{\underline{C}}_{(i)} \underline{q}, \qquad (10.98)$$

where eq. (10.81) is used to express the element nodal displacement array, $\hat{\underline{q}}_{(i)}$, in terms of the global displacement array, $\underline{q}$. Once the bar's elongation is obtained, the constitutive law is used to evaluate the bar force as

$$\hat{F}_{(i)} = \frac{\hat{E}_{(i)} \hat{A}_{(i)}}{\hat{L}_{(i)}} \hat{e}_{(i)}. \qquad (10.99)$$

To illustrate the process, consider bar 6 of the truss shown in fig. 10.22; for this bar, local *Node 1* and *Node 2* correspond global nodes 3 and 5, respectively. Because this bar orientation is parallel to axis $\bar{\imath}_1$, $\hat{\theta}_{(6)} = 0$, the element coordinate transformation matrix, $\underline{\underline{\hat{T}}}_{(6)}$ becomes an identity matrix, and $\hat{\underline{b}} = \underline{\underline{\hat{T}}}\hat{\underline{b}}^* = \hat{\underline{b}}^*$. The bar elongation then becomes

$$\hat{e}_{(6)} = \hat{\underline{b}}_{(6)}^T \underline{\underline{C}}_{(6)} \underline{q} = \hat{\underline{b}}_{(6)}^T \left\{ \begin{matrix} \hat{\underline{q}}_3 \\ \hat{\underline{q}}_5 \end{matrix} \right\} = \hat{\underline{b}}^{*T} \left\{ \begin{matrix} \hat{\underline{q}}_3 \\ \hat{\underline{q}}_5 \end{matrix} \right\} = \{-1, 0, 1, 0\} \left\{ \begin{matrix} \hat{\underline{q}}_3 \\ \hat{\underline{q}}_5 \end{matrix} \right\} = -u_3 + u_5.$$

The corresponding bar force is now

$$\hat{F}_{(6)} = \frac{\hat{E}_{(6)} \hat{A}_{(6)}}{\hat{L}_{(6)}} \hat{e}_{(6)} = \frac{\hat{E}_{(6)} \hat{A}_{(6)}}{\hat{L}_{(6)}} (-u_3 + u_5).$$

Example 10.7. Pentagonal truss revisited

The finite element formulation will be applied to the pentagonal truss depicted in fig. 10.26 and analyzed previously in example 10.6. All bars have the same elastic modulus, E, and cross-sectional area, A. This five node, ten-bar, hyperstatic truss involves a total of 10 generalized coordinates, 2 displacement components at each of the five nodes. Three displacement components are prescribed to zero: the vertical displacement component at node 4 and the two displacement components at node 5. These three constraints will eliminate the three rigid body motions of this planar truss. To facilitate partitioning in the solution procedure, the nodes are numbered as indicated in fig. 10.26: the nodes with constraints, *i.e.*, nodes 4 and 5, appear last. The global coordinate system, $\mathcal{I} = (\bar{\imath}_1, \bar{\imath}_2)$, is also shown in the figure.

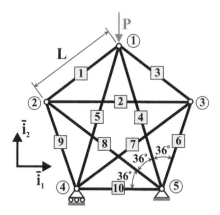

Fig. 10.26. Pentagonal truss with nodes and members defined for analysis using finite element approach.

The local coordinates for each element must be defined with respect to *Node 1* and *Node 2* for the element. Table 10.1 lists these nodes for each element, provides their orientation angle, $\hat{\theta}$, and their length, $\hat{L}$.

Table 10.1. Definition of local nodes and element geometry for pentagonal truss.

Element	Node 1	Node 2	$\hat{\theta}$	$\hat{L}$
1	2	1	$36°$	L
2	2	3	$0°$	$2L \cos 36°$
3	1	3	$-36°$	L
4	1	5	$-72°$	$2L \cos 36°$
5	1	4	$-108°$	$2L \cos 36°$
6	5	3	$72°$	L
7	4	3	$36°$	$2L \cos 36°$
8	2	5	$-36°$	$2L \cos 36°$
9	4	2	$108°$	L
10	4	5	$0°$	L

Based on the data listed in table 10.1, the stiffness matrices for each element is computed using eq. (10.77). The non-dimensional element stiffness matrices, defined as $\underline{\bar{k}}_{(i)} = \underline{\hat{k}}_{(i)} L_{(i)}/E\mathcal{A}$, are given here for the four bars connected to node 2,

$$\underline{\bar{k}}_{(1)} = \begin{bmatrix} 0.65 & 0.48 & -0.65 & -0.48 \\ 0.48 & 0.35 & -0.48 & -0.35 \\ -0.65 & -0.48 & 0.65 & 0.48 \\ -0.48 & -0.35 & 0.48 & 0.35 \end{bmatrix}, \quad \underline{\bar{k}}_{(2)} = \begin{bmatrix} 1.0 & 0.0 & -1.0 & 0.0 \\ 0.0 & 0.0 & 0.0 & 0.0 \\ -1.0 & 0.0 & 1.0 & 0.0 \\ 0.0 & 0.0 & 0.0 & 0.0 \end{bmatrix},$$

$$\underline{\bar{k}}_{(8)} = \begin{bmatrix} 0.65 & -0.48 & -0.65 & 0.48 \\ -0.48 & 0.35 & 0.48 & -0.35 \\ -0.65 & 0.48 & 0.65 & -0.48 \\ 0.48 & -0.35 & -0.48 & 0.35 \end{bmatrix}, \quad \underline{\bar{k}}_{(9)} = \begin{bmatrix} 0.10 & -0.29 & -0.10 & 0.29 \\ -0.29 & 0.90 & 0.29 & -0.90 \\ -0.10 & 0.29 & 0.10 & -0.29 \\ 0.29 & -0.90 & -0.29 & 0.90 \end{bmatrix}.$$

During the assembly process, these four element matrices will all contribute to the entries in the global stiffness matrix corresponding to the two degrees of freedom at node 2. Given the node numbering shown in fig. 10.26, the top left 2×2 sub-matrix from $\hat{\underline{k}}_{(1)}$, $\hat{\underline{k}}_{(2)}$, and $\hat{\underline{k}}_{(8)}$, and the lower right 2×2 sub-matrix in $\hat{\underline{k}}_{(9)}$ will be added together to form the 2×2 sub-matrix in $\underline{K}$ for node 2. Adding the various contributions, this sub-matrix becomes

$$\frac{EA}{L} \begin{bmatrix} 1.7725 & -0.1123 \\ -0.1123 & 1.4635 \end{bmatrix}.$$

The other entries in the global stiffness matrix are constructed in the same manner.

The boundary conditions impose constraints on the degrees of freedom at nodes 4 and 5. At node 5, both horizontal and vertical displacement components must vanish, $u_5 = v_5 = 0$, whereas at node 4, the sole vertical component vanishes, $v_4 = 0$. In view of the node numbering depicted fig. 10.26, those constrained degrees of freedom correspond to the last 3 entries of the global displacement array. The global equations are partitioned into the form given in eq. (10.94) to give

$$\begin{bmatrix} \underline{K}_{uu} & \underline{K}_{up} \\ \underline{K}_{up}^T & \underline{K}_{pp} \end{bmatrix} \begin{Bmatrix} \underline{q}_u \\ \underline{q}_p \end{Bmatrix} = \begin{Bmatrix} \underline{Q}_u \\ \underline{Q}_p \end{Bmatrix}, \tag{10.100}$$

where $\underline{K}_{uu}$, $\underline{K}_{up}$ and $\underline{K}_{pp}$, of size (7×7), (7×3), and (3×3), respectively, define a partition of the global stiffness matrix. Array $\underline{q}_u$, of size (7×1), stores the 7 unconstrained degrees of freedom, while array $\underline{q}_p$, of size (3×1), stores degrees of freedom u_5, v_5, and v_4. It follows that the boundary conditions of the problem imply that $\underline{q}_p = 0$. Array $\underline{Q}_u$, of size (7×1), stores the externally loads applied at the unconstrained nodes, and array $\underline{Q}_p$, of size (3×1), stores the reaction forces at the constrained nodes.

Since $\underline{q}_p = 0$, eq. (10.96) can be solved for the unconstrained nodal displacements as

$$\underline{q}_u = \underline{K}_{uu}^{-1} \underline{Q}_u = \underline{K}_{uu}^{-1} \begin{Bmatrix} 0 \\ -P \\ 0 \\ 0 \\ 0 \\ 0 \\ 0 \end{Bmatrix} = \frac{PL}{EA} \begin{Bmatrix} -0.0325 \\ -0.7025 \\ -0.1763 \\ -0.1769 \\ 0.1114 \\ -0.1769 \\ -0.0650 \end{Bmatrix},$$

where the reduced global stiffness matrix, $\underline{\underline{K}}_{uu}$, is

$$\underline{\underline{K}}_{uu} = \frac{EA}{L} \begin{bmatrix} 1.4271 & 0 & -0.6545 & -0.4755 & -0.6545 & 0.4755 & -0.0590 \\ 0 & 1.8090 & -0.4755 & -0.3455 & 0.4755 & -0.3455 & -0.1816 \\ -0.6545 & -0.4755 & 1.7725 & -0.1123 & -0.6180 & 0 & -0.0955 \\ -0.475 & -0.3455 & -0.1123 & 1.4635 & 0 & 0 & 0.2939 \\ -0.6545 & 0.4755 & -0.6180 & 0 & 1.7725 & 0.1123 & -0.4045 \\ 0.4755 & -0.3455 & 0 & 0 & 0.1123 & 1.4635 & -0.2939 \\ -0.0590 & -0.1816 & -0.0955 & 0.2939 & -0.4045 & -0.2939 & 1.5590 \end{bmatrix}.$$

Now that displacements at all nodes have been computed, the bar elongation can be computed, and finally, eq. (10.99) yields the bar forces as $F_{(1)} = -0.1926\, P$, $F_{(2)} = 0.1778\, P$, $F_{(3)} = -0.1926\, P$, $F_{(4)} = -0.4067\, P$, $F_{(5)} = -0.4067\, P$, $F_{(6)} = -0.1338\, P$, $F_{(7)} = 0.0239\, P$, $F_{(8)} = 0.0239\, P$, $F_{(9)} = -0.1338\, P$, $F_{(10)} = 0.0650\, P$.

Even for this relatively simple problem, the formulation of the individual element stiffness matrices in global coordinates is a tedious numerical exercise. The procedure, however, is systematic and well suited for implementation on computers. In particular, the linear algebra formalism and matrix notation ease the transfer of the different mathematical entities into computer data structures. The various stiffness matrices, displacement and load arrays, are all easily implemented as data arrays in high level computing languages.

The finite element approach described here is particularly well suited for computer implementation because many crucial operations are performed at the element level. When dealing with trusses, this means that many operations, such as the generation of the element stiffness matrix, require only the data associated with a single element. And although developed for planar trusses to simplify the presentation, the approach is readily generalized to three-dimensional truss problems.

In chapter 11, the finite element method introduced here will be extended to deal with beam structures. An additional discretization step will be required to deal with beams, but the assembly process and solution method remain identical to those presented here.

10.7.12 Problems

Problem 10.10. Three-dimensional element stiffness matrix
Section 10.7.4 presents the derivation of the element stiffness matrix for a bar, leading to eq. (10.77). The presentation is limited to planar trusses; the goal of this problem is to generalize the formulation to three-dimensional (3D) problems. *(1)* Generalize the kinematic description of the element give in section 10.7.2. Generalize the element position vector, displacement vectors, and rotation matrix given eqs. (10.63), (10.64), and (10.66) respectively, to 3D. Select the rotation matrix as

$$\underline{\underline{\hat{R}}} = \begin{bmatrix} \ell_1 & -(\ell_2 + \ell_3)/\Delta & \ell_1(\ell_2 - \ell_3)/\Delta \\ \ell_2 & \ell_1/\Delta & [\ell_2(\ell_2 - \ell_3) - 1]/\Delta \\ \ell_3 & \ell_1/\Delta & [\ell_3(\ell_2 - \ell_3) + 1]/\Delta \end{bmatrix},$$

where $\Delta = \sqrt{2\ell_1^2 + (\ell_2 + \ell_3)^2}$. Prove that matrix $\underset{=}{\hat{R}}$ is orthogonal. Express the bar's length and direction cosines in terms of the element nodal coordinates. Generalize the element coordinate transformation matrix, $\underset{=}{\hat{T}}$, defined by eq. (10.71). *(2)* Generalize the expressions derived in section 10.7.3 for the element elongation and axial force, see eqs. (10.73) and (10.74), respectively. *(3)* Generalize the expressions for the element strain energy and stiffness matrix given in section 10.7.4. Give the expression for the stiffness matrix in the local and global coordinate systems, see eqs. (10.75) and (10.77), respectively. *(4)* The stiffness matrix expressed in the global coordinate system is of size 6×6. Of the six eigenvalues of this matrix, how many are zero? Discuss the nature of the corresponding eigenvectors.

Problem 10.11. Global stiffness matrix assembly process
Figure 10.24 gives a pictorial representation of the assembly process for the truss and node numbering sequence shown in fig. 10.22. Give the corresponding representation of the assembly process for the same truss using the node numbering shown in fig. 10.25.

10.8 Principle of minimum complementary energy

In section 10.2, the principle of minimum total potential energy is derived from the principle of virtual work. Two assumptions are used in this derivation: *(1)* the internal forces are assumed to be conservative and *(2)* the external forces are also assumed to be conservative. This means that the internal forces can be derived from a potential, called the strain energy, and the external forces can also be derived from a potential, called the potential of the externally applied loads.

Figure 10.27 shows the relationship between the principle of minimum total potential energy and the principle of virtual work. The arrow linking the constitutive relationship to the strain energy is unidirectional to indicate that an assumption is made: the internal forces in the solid must be conservative for the strain energy to exist. The figure does not indicate the second assumption that is required to obtain the principle of minimum total potential energy: the externally applied loads must be conservative.

Figure 10.27 also shows how the principle of minimum complementary energy is related to the principle of complementary virtual work developed in section 9.6. Here again, two assumptions are required in the derivation. First, the internal forces are assumed to be conservative. This implies the existence of a strain energy function, and hence, of a complementary strain energy function, see section 10.3. Second, it is assumed that the prescribed displacements can be derived from a potential; this new concept is introduced in the next section. The initial development will focus on a simple three-bar hyperstatic truss considered earlier in chapter 9, but the methodology is general.

10.8.1 The potential of the prescribed displacements

The principle of complementary virtual work is derived in section 9.6, using the three-bar truss depicted in fig. 10.28 to present the concepts associated with this principle. In that development, the vertical displacement of point **B** is prescribed to

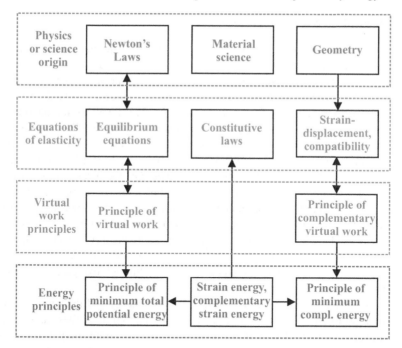

Fig. 10.27. Filiation from elasticity equations to virtual work and energy principles.

be of magnitude Δ. The force required to obtain this desired displacement, denoted D and often called the driving force, is an unknown quantity.

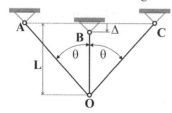

Fig. 10.28. Three-bar truss with prescribed displacement.

The statement of the principle of complementary virtual work, expressed by eq. (9.57), involves the external complementary virtual work, $\delta W'_E = \Delta \, \delta D$, which is directly related to the prescribed displacement. This term is the product of the true prescribed displacement, Δ, and a virtual force, δD. It is now assumed that the prescribed displacement can be derived from a potential, Φ', as

$$\Delta = -\frac{\partial \Phi'(D)}{\partial D}, \tag{10.101}$$

where $\Phi'(D)$ is called the *potential of the prescribed displacement*, or sometimes the *dislocation potential*. With this assumption, the external complementary virtual work becomes

$$\delta W'_E = \Delta \, \delta D = -\frac{\partial \Phi'}{\partial D} \delta D = -\delta \Phi'(D). \tag{10.102}$$

10.8.2 Constitutive laws for elastic materials

Consider a truss consisting of bars made of a linearly elastic material. If a uniform bar of length L, elastic modulus, E, and cross-sectional area, $\mathcal{A}$, is subjected to a force F, an elongation, e, results. Equation (10.29) gives the strain energy in the bar as $A = 1/2\ ke^2$, where $k = E\mathcal{A}/L$. The force applied to the bar can be derived from this strain energy function as $F = \partial A(e)/\partial e = ke$. This result shows that once the strain energy is known, the material's constitutive law follows. Thus, defining the strain energy function for a material is equivalent to defining the constitutive law.

The complementary strain energy for the same bar is given by eq. (10.22) as $A' = 1/2\ F^2/k$, where $1/k$ is the bar's compliance, *i.e.*, the inverse of its stiffness. The bar's elongation can be derived from the complementary strain energy as $e = \partial A'(F)/\partial F = F/k$. Once the complementary strain energy is known, the material's constitutive law follows. The strain energy yields the constitutive law in stiffness form, $F = ke$, whereas the complementary strain energy yields the same relationship in compliance form, $e = F/k$. For a linearly elastic material, the strain energy and its complementary counterpart are equal, $A = A'$, although expressed in terms of different variables: $A(e) = 1/2\ ke^2$, and $A'(F) = 1/2\ F^2/k$.

If the material is elastic but not linear, the strain energy and its complementary counterpart still yield the constitutive laws for the material in stiffness and compliance forms, respectively. The relationship between the two strain energies is given by eq. (10.23) as $A(e) + A'(F) = e\ F$. Taking the differential of this equation leads to $(\partial A/\partial e)de + (\partial A'/\partial F)dF = Fde + edF$. Regrouping the terms in this differential form leads to

$$\left[F - \frac{\partial A}{\partial e}\right]de + \left[e - \frac{\partial A'}{\partial F}\right]dF = 0.$$

Since the differential in elongation and force are arbitrary and independent, the two bracketed terms must vanish, revealing the following relationships

$$F = \frac{\partial A(e)}{\partial e}, \tag{10.103a}$$

$$e = \frac{\partial A'(F)}{\partial F}, \tag{10.103b}$$

which both express the same constitutive law for the material, one in stiffness, the other in compliance form. The existence of the strain energy function guarantees that of its complementary counterpart. Hence, both stiffness and compliance forms of the constitutive law are entirely equivalent.

To illustrate the role of the strain energy and of its complementary counterpart for elastic materials, consider the following strain energy expression for a bar, $A(e) = kL^2(1 - \cos\bar{e})$, where $\bar{e} = e/L$ is the bar's axial strain. The material's constitutive law in stiffness form is readily obtained as $F = \partial A(e)/\partial e = kL\sin\bar{e}$. Due to the periodic nature of the cosine function, this particular strain energy is only meaningful for bar strains of moderate value, *i.e.*, $|\bar{e}| < 1$.

The compliance form of the same constitutive law is obtained by inversion as $\bar{e} = \arcsin\bar{F}$, where $\bar{F} = F/(kL)$ is a non-dimensional force. The complementary strain energy is then obtained from its definition as $A'(F) = \int_0^F e \, dF = \int_0^{\bar{F}} \bar{e} \, L \, kL \, d\bar{F} = kL^2(\bar{F} \arcsin \bar{F} + \sqrt{1 - \bar{F}^2} - 1)$. The same result can also be obtained more directly from the relationship the strain energy and its complementary counterpart, eq. (10.23), as $A'(F) = eF - A(e) = kL^2(\bar{F} \arcsin \bar{F} + \sqrt{1 - \bar{F}^2} - 1)$. The material's constitutive law can also be obtained from this expression of the complementary strain energy as

$$e = \frac{\partial A'(F)}{\partial F} = L \arcsin \bar{F},$$

which expresses the same constitutive law as that derived from the strain energy function.

Finally, it must be noted that the existence of the strain energy function or of its complementary counterpart is an *assumption* equivalent to the assumption of a constitutive law. Indeed, as discussed in section 10.1.1, if the material's internal forces are assumed to be conservative, they can be derived from a potential, called the strain energy function. In other words, the existence of a strain energy function, the existence of a complementary energy function, or the fact that the material's internal forces are conservative are three entirely equivalent assumptions.

10.8.3 The principle of minimum complementary energy

The principle of complementary virtual work is introduced for truss structures in section 9.6.2, on page 441. The principle is summarized by eq. (9.55) as $\delta W' = \delta W'_E + \delta W'_I = 0$, and states that a truss undergoes compatible deformations if and only if the sum of the internal and external complementary virtual work vanishes for all statically admissible virtual forces. The internal complementary virtual work for the three-bar truss depicted in fig. 10.28 is given by eq. (9.49) as $\delta W'_I = -e_A \delta F_A - e_B \delta F_B - e_C \delta F_C$.

At this point, it is assumed that the material the bars are made of is elastic, *i.e.*, the existence of a complementary strain energy function is assumed. The material's constitutive law is now expressed in compliance form by eq. (10.103b) and the complementary virtual work becomes

$$\delta W'_I = -\frac{\partial A'_A(F_A)}{\partial F_A} \delta F_A - \frac{\partial A'_B(F_B)}{\partial F_B} \delta F_B - \frac{\partial A'_C(F_C)}{\partial F_C} \delta F_C,$$

where A'_A, A'_B, and A'_C are the complementary strain energies of bars **A**, **B**, and **C**, respectively. Treating δ as a differential, this expression readily simplifies to

$$\delta W'_I = -\delta A'_A - \delta A'_B - \delta A'_C = -\delta A',$$

where $A' = A'_A + A'_B + A'_C$ is the total complementary strain energy for the three-bar truss.

Next, it is assumed that the prescribed displacement at point **B**, see fig. 10.28, can be derived from a potential. As discussed in section 10.8.1, the external complementary virtual work can then be written as $\delta W'_E = -\delta \Phi'(D)$, where Φ' is the potential of the prescribed displacement. The *potential of the prescribed displacements*, Φ', is different from the *potential of the externally applied loads*, Φ, defined in eq. (10.13).

Given these two assumptions, the existence of the material's complementary strain energy function and of the prescribed displacement potential, the principle of complementary virtual work, eq. (9.55), becomes

$$\delta W' = \delta W'_E + \delta W'_I = -\delta A' - \delta \Phi' = -\delta(A' + \Phi') = 0.$$

It is convenient to define the *total complementary energy*, Π' as

$$\Pi' = A' + \Phi', \tag{10.104}$$

and the above statement then further simplifies to

$$\delta \Pi' = 0. \tag{10.105}$$

These developments lead to the principle of stationary complementary energy.

Principle 11 (Principle of stationary complementary energy) *A   conservative system undergoes compatible deformations if and only if the total complementary energy vanishes for all statically admissible virtual forces.*

It can be shown in a manner similar to that for the principle of minimum total potential energy that this stationary value is also a minimum value for stable equilibrium. With this, it is now possible to state the principle of minimum complementary energy.

Principle 12 (Principle of minimum complementary energy) *A conservative system undergoes compatible deformations if and only if the total complementary energy is a minimum with respect to arbitrary changes in statically admissible forces.*

Example 10.8. Three-bar truss with prescribed displacement
Consider the hyperstatic three-bar truss treated previously in example 9.15 on page 448 using the principle of complementary virtual work. The configuration is shown again in fig. 10.29. Assume that support joint **B** is given a prescribed displacement, Δ (perhaps due to an initial assembly imperfection). Determine the resulting force in each of the bars.

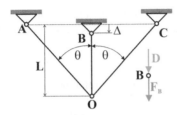

Fig. 10.29. Three-bar truss with prescribed displacement (see fig. 9.37).

For this hyperstatic problem, the three unknown bar forces cannot be evaluated based on the two equilibrium equations for joint **O**. Horizontal equilibrium yields $F_A = F_C$, and vertical equilibrium results in $F_A \cos\theta + F_B + F_C \cos\theta = 0$. The complementary strain energy written in terms of the bar forces, F_A, F_B, and F_C, is

$$A' = \frac{1}{2}\left(\frac{F_A^2}{k_A \cos\theta} + \frac{F_B^2}{k_B} + \frac{F_C^2}{k_C \cos\theta}\right),$$

where $k_A = (EA)_A/L$, $k_B = (EA)_B/L$, and $k_C = (EA)_C/L$ are the bar stiffnesses. With the help of the two equilibrium equations, the three bar forces can be expressed in terms one, say F_C, to find

$$A' = \frac{1}{2}\left(\frac{F_C^2}{k_A \cos\theta} + \frac{(2F_C \cos\theta)^2}{k_B} + \frac{F_C^2}{k_C \cos\theta}\right) = \frac{\bar{k}F_C^2}{2\bar{k}_A k_B \bar{k}_C \cos\theta},$$

where $\bar{k}_A = k_A/k_B$, $\bar{k}_C = k_C/k_B$, and $\bar{k} = \bar{k}_A + \bar{k}_C + 4\bar{k}_A\bar{k}_C \cos^3\theta$ are non-dimensional stiffness coefficients.

The potential of the prescribed displacement, Δ, at joint **B** is $\Phi' = -D\Delta$. The equilibrium equation at joint **B** states that $D + F_B = 0$ and using the equilibrium equation $F_B = -2F_C \cos\theta$, the potential can now be expressed in terms of bar force F_C as $\Phi' = -2\Delta F_C \cos\theta$. The total complementary potential energy thus takes the following form

$$\Pi' = A' + \Phi' = \frac{\bar{k}F_C^2}{2\bar{k}_A k_B \bar{k}_C \cos\theta} - 2\Delta F_C \cos\theta,$$

and the principle of minimum complementary energy, principle 12, requires that

$$\frac{\partial \Pi'}{\partial F_C} = \frac{\bar{k}F_C}{\bar{k}_A k_B \bar{k}_C \cos\theta} - 2\Delta \cos\theta = 0,$$

which expresses the requirement that the truss must undergo a compatible deformation. This equation yields the bar force, F_C, and the other two bar forces, F_A and F_B, are then obtained from the two equilibrium equations as

$$F_A = F_C = \frac{2\bar{k}_A\bar{k}_C \cos^2\theta}{\bar{k}}k_B\Delta \quad \text{and} \quad F_B = D = \left(1 - \frac{\bar{k}_A + \bar{k}_C}{\bar{k}}\right)k_B\Delta.$$

Finally, the displacement of joint **O** can be found from the extension of bar **B** as
$$u_1^{(B)} = e_B + \Delta = (\bar{k}_A + \bar{k}_C)\Delta/\bar{k}.$$

10.8.4 The principle of least work

The principle of minimum complementary energy developed in the previous section involves the total complementary energy, which is the sum of two scalar quantities: the system's complementary strain energy, and the potential of the prescribed displacements, see eq. (10.104). In the absence of prescribed displacements, the total complementary energy reduces to the complementary strain energy alone. The *principle of least work*, a corollary of the principle of minimum complementary energy, states the following.

Principle 13 (Principle of least work) *In the absence of prescribed displacements, a conservative system undergoes compatible deformations if and only if the complementary strain energy is a minimum with respect to arbitrary changes in statically admissible forces.*

If the system is made of a linearly elastic material, the complementary strain energy is equal the strain energy, see section 10.3. The principle of least work then takes on the following form.

Principle 14 (Principle of least work) *In the absence of prescribed displacements, a linearly elastic system undergoes compatible deformations if and only if the strain energy is a minimum with respect to arbitrary changes in statically admissible forces.*

When using the principle of least work, the system's complementary strain energy or its strain energy must be expressed in terms of the statically admissible forces rather than deformations.

Example 10.9. Three-bar truss with tip load

To illustrate the use of the least work principle, the hyperstatic, three-bar truss treated in example 9.14 on page 447 using the principle of complementary virtual work will be re-examined. As shown in fig. 10.30, a vertical downward load P is applied at joint **O**. The objective is to determine the resulting bar forces.

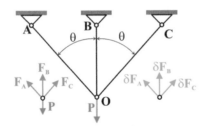

Fig. 10.30. Three-bar truss with applied load (fig. 9.36).

Since it is not necessary to determine the reaction forces at the support joints, the only relevant equilibrium equations are those at joint **O**; the horizontal and vertical equilibrium equations are $F_A = F_C$ and $F_A \cos\theta + F_B + F_C \cos\theta = P$, respectively. The strain energy is first written in terms of the bar forces F_A, F_B and F_C, as

$$A = \frac{1}{2}\left(\frac{F_A^2}{k_A \cos\theta} + \frac{F_B^2}{k_B} + \frac{F_C^2}{k_C \cos\theta}\right),$$

where $k_A = (EA)_A/L$, $k_B = (EA)_B/L$, and $k_C = (EA)_C/L$ are the bar stiffnesses. Next, using the two equilibrium equations, the three bar forces are expressed in terms of one, say F_C, to find

$$A = \frac{1}{2}\left[\frac{F_C^2}{k_A \cos\theta} + \frac{(P - 2F_C \cos\theta)^2}{k_B} + \frac{F_C^2}{k_C \cos\theta}\right].$$

To impose the condition that the truss must undergo compatible deformations, the least work principle, principle 14, is applied,

$$\frac{\partial A}{\partial F_C} = \left[\frac{F_C}{k_A \cos \theta} - \frac{(P - 2F_C \cos \theta) 2 \cos \theta}{k_B} + \frac{F_C}{k_C \cos \theta} \right] = 0.$$

This equation can be solved for the bar force, F_C, and the equilibrium equations then yield the forces in the other two bars as

$$\frac{F_A}{P} = \frac{F_C}{P} = \frac{2 \bar{k}_A \bar{k}_C \cos^2 \theta}{\bar{k}}, \quad \frac{F_B}{P} = \frac{\bar{k}_A + \bar{k}_C}{\bar{k}},$$

where $\bar{k}_A = k_A/k_B$, $\bar{k}_C = k_C/k_B$, and $\bar{k} = \bar{k}_A + \bar{k}_C + 4 \bar{k}_A \bar{k}_C \cos^3 \theta$ are the bar non-dimensional stiffness factors.

Application of the principle of minimum complementary energy, or of the principle of least work in the absence of prescribed displacement, leads to a solution process that is very similar to that of the force or flexibility method first developed in section 4.3.3. In this earlier presentation, the compatibility conditions are developed from simple geometric arguments, whereas the principle of minimum complementary energy is used here to derive the same conditions in a more abstract but also more systematic manner.

Example 10.10. Beam on 3 supports
The simply supported beam with an additional mid-span support depicted in fig. 10.31 is subjected to two concentrated loads of equal magnitude, P. Determine the location and magnitude of the maximum bending moment in the beam.

Due to the additional mid-span support, the system is hyperstatic of order 1. Any one of the three support reaction forces, denoted R_1, R_2, and R_3, respectively, can be selected as the redundant force; in this example, R_1 is selected to be the redundant quantity, and therefore will be treated as the unknown.

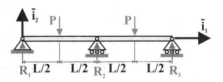

Fig. 10.31. A beam on 3 simple supports.

The overall moment and vertical force equilibrium equations result in $R_3 = R_1$ and $R_2 = 2P - 2R_1$, respectively. The bending moment distribution in the left segment of the beam is found from equilibrium considerations as

$$M_3(\eta) = \begin{cases} R_1 L \eta, & 0 \leq \eta \leq 1/2, \\ R_1 L \eta - PL(\eta - 1/2), & 1/2 \leq \eta \leq 1, \end{cases} \qquad (10.106)$$

where $\eta = x_1/L$ is the non-dimensional variable along the beam's span. Due to the symmetry of the problem it is only necessary to compute the strain energy in the beam by integrating over the range 0 to L and then doubling the result. Thus

$$A = \int_0^{2L} \frac{M_3^2(x_1)}{2H_{33}^c} \, dx_1 = 2 \int_0^L \frac{M_3^2(x_1)}{2H_{33}^c} \, dx_1 = 2L \int_0^1 \frac{M_3^2(\eta)}{2H_{33}^c} \, d\eta$$

$$= \frac{R_1^2 L^3}{H_{33}^c} \int_0^{1/2} \eta^2 \, d\eta + \frac{L}{H_{33}^c} \int_{1/2}^1 [R_1 L\eta - PL(\eta - 1/2)]^2 \, d\eta.$$

After integration, $A = \left(8R_1^2 - 5PR_1 + P^2\right) L^3/(24H_{33}^c)$, and using the least work principle then yields

$$\frac{\partial A}{\partial R_1} = \frac{L^3}{24H_{33}^c}(16R_1 - 5P) = 0.$$

This equation yields $R_1 = 5P/16$, and the overall equilibrium equations then reveal the remaining reaction forces as $R_1 = R_3 = 5P/16$ and $R_2 = 11P/8$. The beam's bending moment distribution is obtained by substituting these forces in eq. (10.106). The bending moment at the left support vanishes, as expected. At the point of application of the concentrated load, i.e., at $\eta = 1/2$, the bending moment is $M_3 = 5PL/32$; at the mid-span support, i.e., at $\eta = 1$, $M_3 = -6PL/32$. The maximum bending moment is found at the mid-span support, and its magnitude is $|M_3| = 6PL/32$.

Example 10.11. Simply supported beam with a mid-span elastic spring
Consider the simply supported bean of length L with a mid-span spring of stiffness constant k, as depicted in fig. 10.32. The beam carries a uniformly distributed vertical load, p_0. Determine the load in the spring and the reaction forces at the two supports.

Let R_1 and R_2 be the two support reaction forces, and F_s the force that the spring applies on the beam, counted positive downward. The overall equilibrium equations are $R_1 + R_2 + F_s = p_0 L$ and $R_1 L + F_s L/2 - p_0 L^2/2 = 0$. The problem is hyperstatic of order 1, because the two equilibrium equations involve three unknown reaction forces. Two of the reactions can be expressed in terms of the third; for instance, $F_s = p_0 L - 2R_1$ and $R_2 = R_1$, where R_1 is treated as the redundant quantity.

Based on equilibrium considerations, the bending moment distribution in the beam is now expressed in terms of the unknown reaction force, R_1, as

$$M_3(\eta) = \begin{cases} p_0 L^2 \eta^2/2 - R_1 L\eta, & 0 \leq \eta \leq 1/2, \\ p_0 L^2 (1 - \eta)^2/2 - R_1 L(1 - \eta), & 1/2 \leq \eta \leq 1, \end{cases}$$

where $\eta = x_1/L$ is the non-dimensional variable along the beam.

The strain energy in the system is

$$A = \frac{1}{2} \int_0^L \frac{M_3^2(x_1)}{H_{33}^c} \, dx_1 + \frac{1}{2} \frac{F_s^2}{k},$$

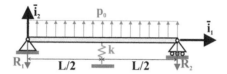

Fig. 10.32. Simply supported beam with a mid-span elastic spring.

where the first term represents the strain energy stored in the beam, and the second represents that stored in the elastic spring.

Rather than substitute for M_3 and F_s in terms of R_1 and then integrate, the principle of least work is applied first,

$$\frac{\partial A}{\partial R_1} = \int_0^L \frac{M_3}{H_{33}^c} \frac{\partial M_3}{\partial R_1} \, dx_1 + \frac{F_s}{k} \frac{\partial F_s}{\partial R_1} = 0.$$

Introducing the bending moment distribution and the force in the spring, both expressed in terms of the unknown reaction force, R_1, then leads to

$$2 \int_0^{1/2} \left[\frac{p_0 L^2 \eta^2 / 2 - R_1 L \eta}{H_{33}^c} \right] (-L\eta) \; L d\eta + \left[\frac{p_0 L - 2 R_1}{k} \right] (-2) = 0.$$

In this expression, the strain energy in the left half of the beam is computed and then doubled, based on symmetry. After integration, this becomes

$$\frac{2L^3}{H_{33}^c} \left(-\frac{p_0 L}{128} + \frac{R_1}{24} \right) - 2\frac{p_0 L - 2 R_1}{k} = 0.$$

The solution of this equation yields the non-dimensional reaction forces at the supports and the spring force as

$$\frac{R_1}{p_0 L} = \frac{R_2}{p_0 L} = \frac{1\,384 + 3\bar{k}}{2\,384 + 8\bar{k}}, \quad \frac{F_s}{p_0 L} = \frac{5\bar{k}}{384 + 8\bar{k}},$$

where $\bar{k} = kL^3 / H_{33}^c$ is the non-dimensional spring stiffness constant expressing the spring stiffness relative to the beam bending stiffness.

The force in the spring is obtained from the overall equilibrium equation, $F_s = p_0 L - 2 R_1$. When $\bar{k} \to 0$, i.e., in the absence of mid-span spring, the reaction forces become $R_1 = R_2 = p_0 L/2$, as expected from symmetry, and $F_s = 0$. For $\bar{k} \to \infty$, i.e., for a mid-span support, $R_1 = R_2 = 3 p_0 L/16$ and the mid-span reaction force is $F_s = 5 p_0 L/8$.

Example 10.12. Simply supported beam with a misaligned mid-span support
Figure 10.33 shows a simply supported beam with a misaligned mid-span support. In the unloaded configuration, the mid-span support is at a distance d below the beam. As the loading increases, the beam will touch the mid-span support. For the analysis, it is assumed that the beam is touching the support because the applied loads are high

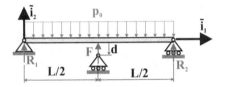

Fig. 10.33. Simply supported beam with a misaligned mid-span support.

enough. Of course, for small loads, the beam does not touch the support, which can then be ignored.

Let R_1 and R_2 be the reaction forces at the two end support and F the force that the mid-span support applies to the beam. The overall equilibrium equations are $R_1 + R_2 + F = p_0 L$ and $R_1 L + FL/2 - p_0 L^2/2 = 0$. The problem is hyperstatic of order 1, because the two equilibrium equations involve three unknown reaction forces. Two of the reactions can be expressed in terms of the third; for instance, $F = p_0 L - 2R_1$ and $R_2 = R_1$. Based on equilibrium considerations, the bending moment distribution in the beam is now expressed in terms of the unknown reaction force, R_1, as

$$M_3(\eta) = \begin{cases} p_0 L^2 \eta^2/2 - R_1 L\eta, & 0 \le \eta \le 1/2, \\ p_0 L^2 (1-\eta)^2/2 - R_1 L(1-\eta), & 1/2 \le \eta \le 1, \end{cases}$$

where $\eta = x_1/L$ is the non-dimensional variable along the beam.

For this linearly elastic structure, the total complementary energy is $\Pi' = A' + \Phi' = A + \Phi'$. The potential of the prescribed displacements is $\Phi' = -F(-d) = Fd = d(p_0 L - 2R_1)$, where the minus sign stems from the fact that the force and displacements are counted positive in opposite direction.

The principle of minimum complementary energy now states

$$\frac{\partial \Pi'}{\partial R_1} = \int_0^L \frac{M_3}{H_{33}^c} \frac{\partial M_3}{\partial R_1} \, dx_1 + \frac{\partial \Phi'}{\partial R_1} = 0.$$

Introducing the above bending moment distribution and potential of the prescribed displacements yields

$$2 \int_0^{1/2} \left[\frac{p_0 L^2 \eta^2/2 - R_1 L\eta}{H_{33}^c} \right] [-L\eta] \, L d\eta + \frac{\partial}{\partial R_1} d(p_0 L - 2R_1) = 0.$$

In this expression, the strain energy in the left half of the beam is computed and then doubled, based on symmetry. After integration, this condition becomes

$$\frac{2L^3}{H_{33}^c} \left(-\frac{p_0 L}{128} + \frac{R_1}{24} \right) - 2d = 0.$$

The solution of this equation yields the reaction forces at the supports and the mid-span reaction force as

$$R_1 = R_2 = \frac{3p_0 L}{16} + 24\frac{H_{33}^c d}{L^3}, \quad F = \frac{5p_0 L}{8} - 48\frac{H_{33}^c d}{L^3}.$$

The result can be interpreted in different manner. First, if the three supports are on the same level, $d = 0$, and the end point reaction forces are $R_1 = R_2 = 3p_0 L/16$ and the mid-span reaction force is $F = 5p_0 L/8$. If the mid-span support is misaligned and below the beam, *i.e.*, $d > 0$, the loading level for which the beam will just touch the mid-span support is found by setting $F = 0$, *i.e.*, a reaction force is just about to vanish at the support. This leads to $5p_{cr}L/8 - 48H_{33}^c d/L^3 = 0$ and $p_{cr} = 384H_{33}^c d/(5L^4)$. Thus for $p_0 \leq p_{cr}$, the beam does not reach the support and for $p_0 > p_{cr}$, the reaction force give above develops in the misaligned, mid-span support.

The analysis is also valid if $d < 0$. This means that the mid-span support is protruding and pushing the beam upwards. Even in the absence of applied loading, end point reaction forces, $R_1 = R_2 = 24H_{33}^c d/L^3$, and a mid-span reaction force, $F = -48H_{33}^c d/L^3$, will develop. If $d < 0$, it follows that $R_1 < 0$, $R_2 < 0$, and $F > 0$, as expected.

10.8.5 Problems

Problem 10.12. Cantilevered beam with intermediate spring support
The cantilevered beam shown in fig. 10.34 carries a uniform loading distribution, p_0, and is supported by a spring of stiffness constant k, located at a distance a from the root of the beam. *(1)* Use the least work principle to determine the force in the spring. *(2)* Find the bending moment distribution in the beam.

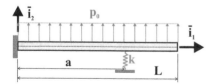

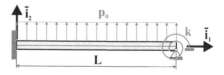

Fig. 10.34. Cantilevered beam with intermediate spring support.

Fig. 10.35. Cantilevered beam with tip rotational spring.

Problem 10.13. Cantilevered beam with tip rotational spring
A cantilevered beam of span L is subjected to a uniform loading distribution, p_0, as depicted in fig. 10.35. An additional support is located at the beam's tip, and a rotational spring of stiffness constant k acts at the same point. *(1)* Use the least work principle to determine the tip support reaction force. *(2)* Find the bending moment distribution in the beam.

Problem 10.14. 3-bar truss with unequal bar stiffness properties
The three-bar, hyperstatic truss shown in fig. 10.36 is subjected to a tip vertical load P. The three bars have a Young's modulus E, bar 1 is of cross-sectional area $\mathcal{A}$, while that bars 2 and 3 is $2\mathcal{A}$. *(1)* Use the principle of complementary energy to find the bar forces.

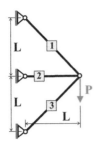

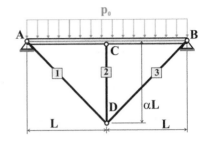

Fig. 10.36. 3-bar truss with unequal bar stiffness properties.

Fig. 10.37. Simply supported beam with uniform load and truss bracing.

Problem 10.15. Combined beam and truss problem

The structure shown in fig. 10.37 consists of a simply supported, continuous beam, **AB**, subjected to a uniformly distributed load, p_0. Additional support is provided by a truss consisting
of bars **AD**, **CD**, and **BD**, which are pinned together to provide a mid-span support for the
beam. Bar **CD** is of length αL. *(1)* Using the principle of least work, find the forces in the
three bars. *(2)* Determine the bending moment distribution in the beam. Use the following
data: $\alpha = 1$. Ignore the axial forces in the beam.

Problem 10.16. Cantilevered beam with simple support and concentrated load

A cantilevered beam with a mid-span support carries a tip concentrated load, P, as depicted
in fig. 10.38. *(1)* Using the principle of least work, determine the reaction forces. *(2)* Find the
bending moment distribution in the beam.

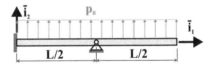

Fig. 10.38. Cantilevered beam with simple
support and concentrated load.

Fig. 10.39. Cantilevered beam with simple
support under uniform loading.

Problem 10.17. Cantilevered beam with simple support and uniform load

A cantilevered beam with a mid-span support carries a uniformly distributed load, p_0, as depicted in fig. 10.39. *(1)* Using the principle of least work, determine the reaction forces. *(2)*
Find the bending moment distribution in the beam.

10.9 Energy theorems

A number of important energy theorems will be developed in this section. These theorems are corollaries of the fundamental energy principles developed earlier. Consequently, all theorems are valid for elastic structures only. The application of two

of these theorems, Clapeyron's theorem and Castigliano's second theorem, is further limited to linearly elastic materials.

A properly constrained[3] elastic body subjected to various concentrated loads and couples is shown in fig. 10.40. The first loading type consists of N concentrated loads, $P_i, i = 1, 2, \ldots, N$, and the displacements of their points of application projected in the direction of the loads are denoted $\Delta_i, i = 1, 2, \ldots, N$, respectively. The second type of loading consists of M couples, $Q_j, j = 1, 2, \ldots, M$, and the rotations at their points of application about the axis of the couple are denoted Φ_j, $i = 1, 2, \ldots, M$, respectively.

A special case of these loading conditions consists of two forces sharing the same line action, such as forces P_3 and P_4 in fig. 10.40. In some cases, the two forces are of equal magnitude and opposite direction, $P_3 = P_4 = P$, and the relative displacement of their points of application is then denoted $\Delta_0 = \Delta_3 + \Delta_4$. A similar situation could occur with two couples of equal magnitude and opposite direction sharing a common axis, say $Q_3 = Q_4 = Q$, and the relative rotation at their points of application is $\Phi_0 = \Phi_3 + \Phi_4$.

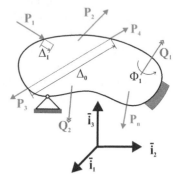

Fig. 10.40. Elastic body subjected to various loads.

10.9.1 Clapeyron's theorem

For an elastic system, eq. (10.12) implies that the strain energy stored in the body equals the work done by the external forces as they are increased quasi-statically from zero to their final values. Consider a suitably restrained body that is subjected to N external loads, P_i, and M external couples, Q_j. Equation (10.12) now implies

$$A = W_E = \sum_{i=1}^{N} \int_0^{\Delta_i} P_i \, du_i + \sum_{j=1}^{N} \int_0^{\Phi_j} Q_j \, d\theta_j,$$

where u_i are the displacements of the external load projected along their line of action and Δ_i are the final displacements, θ_j are the rotations of the external moments about their axes, and Φ_j are the final rotations.

Next, the body is assumed to be linearly elastic and hence, the applied loads are proportional to the displacements of their point of application, $P_i \propto u_i$, and the applied couples are proportional to the rotation of their point of application, $Q_j \propto \theta_j$. It now becomes possible to evaluate the two integrals to find

$$A = W_E = \sum_{i=1}^{N} \frac{P_i \Delta_i}{2} + \sum_{j=1}^{N} \frac{Q_j \Phi_j}{2}. \tag{10.107}$$

This result is known as Clapeyron's theorem.

---

[3] Properly constrained means that the body cannot rotate or translate freely but may be subjected to a hyperstatic set of reactions.

Theorem 10.1 (Clapeyron's theorem). *The strain energy stored in a linearly elastic structure equals the sum of the half product of the applied loads by the displacements of their respective points of applications projected along their lines of action.*

While Clapeyron's theorem is useful for evaluating the strain energy, it can also be used to compute the deflection, Δ, at the point of application of a load, P, when this single load is the only load applied. In such a case, eq. (10.107) becomes $\Delta = 2A/P$. A similar result is obtain for the rotation at the point of application of a single moment.

It is interesting to compare eqs. (10.107) and (10.13), which differ by a factor of two. In the derivation of eq. (10.107), load P is assumed to grow in proportion to the displacement, whereas for eq. (10.13), load P is assumed to remain constant. As discussed in section 10.4.4, this difference in the nature of the applied loading explains the factor of two in the work they perform.

Clapeyron's theorem can also be proved by the following alternative reasoning. First, the total potential energy of the system is written as

$$\Pi = A - \sum_{i=1}^{N} P_i \Delta_i,$$

where $A = 1/2 \int_V \underline{\epsilon}^T \underline{\underline{C}} \, \underline{\epsilon} \, dV$ is the strain energy for the general three-dimensional linearly elastic body, see eq. (10.49). The principle of minimum total potential energy implies the stationarity of the total potential energy, which can be expressed as[4] $\delta\Pi = \int_V \underline{\epsilon}^T \underline{\underline{C}} \, \delta\underline{\epsilon} \, dV - \sum_{i=1}^{N} P_i \delta\Delta_i = 0$. Since this relationship must hold for all arbitrary virtual displacement and associated compatible strain fields, a valid choice is $\delta\Delta_i = \Delta_i$ and $\delta\underline{\epsilon} = \underline{\epsilon}$, where Δ_i and $\underline{\epsilon}$ correspond to the displacement and strain fields for the equilibrium configuration of the structure, respectively. It follows that $\int_V \underline{\epsilon}^T \underline{\underline{C}} \, \underline{\epsilon} \, dV = \sum_{i=1}^{N} P_i \Delta_i$ and finally, $A = \sum_{i=1}^{N} P_i \Delta_i/2$, which is the statement of Clapeyron's theorem, see eq. (10.107). Note that the order in which the forces are applied is immaterial, and the strain energy stored in the structure depends only on the magnitude of the forces and the resulting projected displacements.

Consider now the case of two forces, P_3 and P_4, of equal magnitude and opposite sign sharing a common line of action, as shown in fig. 10.40. Clapeyron's theorem yields $A = (P_3\Delta_3 + P_4\Delta_4)/2$; let P denote the intensity of the forces, $P_3 = P_4 = P$, and hence, $A = P(\Delta_3 + \Delta_4)/2 = P\Delta_0/2$, where $\Delta_0 = \Delta_3 + \Delta_4$ is the relative distance between the points of application of the two forces. This relative distance is a positive quantity if the forces pull away from each other, and is negative in the opposite case.

A similar reasoning yields $A = Q\Phi_0/2$, where Q is the common magnitude of two couples of equal magnitude and opposite direction sharing a common axis and Φ_0 is the relative rotation at their points of application. In summary, each of the loading terms appearing in Clapeyron's theorem, eq. (10.107), could be of either of the following four types: $P_i\Delta_i/2$, $Q_j\Phi_j/2$, $P\Delta_0/2$, or $Q\Phi_0/2$.

---

[4] See appendix A.2.7 for details on taking the differential of a quadratic form.

*Example 10.13. **Simply supported beam with concentrated load***
Consider a simply supported, uniform beam of length L subjected to a concentrated
load P acting at a distance αL from the left support, as depicted in fig. 10.41. This
problem is treated using classical approaches in examples 5.5 and 5.6 on pages 197
and 199, respectively. The bending moment distribution is readily obtained from
equilibrium considerations and is given by eq. (5.52).

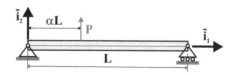

Fig. 10.41. Simply supported beam with concentrated load.

The strain energy stored in the beam is now obtained from eq. (10.41) as

$$A = \frac{1}{2} \int_0^L \frac{M_3^2}{H_{33}^c} \, dx_1 = \frac{P^2 L^2}{2H_{33}^c} \left[\int_0^\alpha (1-\alpha)^2 \eta^2 \, L d\eta + \int_\alpha^1 \alpha^2 (1-\eta)^2 \, L d\eta \right],$$

where $\eta = x_1/L$ is a non-dimensional variable along the beam's span. Performing
the integrations leads to

$$A = \frac{P^2 L^3}{2H_{33}^c} \left[(1-\alpha)^2 \frac{\alpha^3}{3} + \alpha^2 \frac{(1-\alpha)^3}{3} \right] = \frac{P^2 L^3}{6H_{33}^c} \alpha^2 (1-\alpha)^2.$$

Clapeyron's theorem, eq. (10.1), yields $A = P\Delta/2 = P^2 L^3 \alpha^2 (1-\alpha)^2/(6H_{33}^c)$,
and solving for the displacement, Δ, at the point of application of the load leads to
$\Delta = PL^3 \alpha^2 (1 - \alpha)^2/(3H_{33}^c)$. This result is identical to that obtained earlier with
the classical approach. Indeed, $\Delta = \bar{u}_2(\alpha)$, where $\bar{u}_2(\eta)$ is the beam's transverse
displacement field given by eq. (5.51) and $\bar{u}_2(\alpha)$ its value at $\eta = \alpha$.

Clapeyron's theorem yields the deflection under the load in a very expeditious
manner. A disadvantage of this theorem, however, is that it is useful only when a
single concentrated load is applied. In contrast, the principle of virtual work developed in section 9.7.6 is nearly as simple to formulate and can be used for any type or
combination of loading. Finally, the classical approaches presented in section 5.5 and
demonstrated in examples 5.6 and 5.6 require the solution of the governing differential equation of the problem but yield the distributions of transverse displacements,
bending moments, and shear forces over the beam's entire span. More detailed information is obtained, but at a higher cost.

10.9.2 Castigliano's first theorem

Consider, again, a properly constrained elastic body subjected to various concentrated loads and couples as shown in fig. 10.40. The total potential energy, see

eq. (10.10), can be written as $\Pi = A + \Phi = A - \sum_{i=1}^{N} P_i \Delta_i$, where P_i is an externally applied load and Δ_i the displacements of its point of application projected along its line of action.

The principle of minimum total potential energy now implies the stationarity of the total energy, eq. (10.17), and hence,

$$\frac{\partial \Pi}{\partial \Delta_j} = \frac{\partial A}{\partial \Delta_j} - \frac{\partial}{\partial \Delta_j} \sum_{i=1}^{N} P_i \Delta_i = \frac{\partial A}{\partial \Delta_j} - P_j = 0.$$

This equation leads to *Castigliano's first theorem*,

$$P_i = \frac{\partial A}{\partial \Delta_i}. \tag{10.108}$$

Theorem 10.2 (Castigliano's first theorem). *For an elastic system, the magnitude of the load applied at a point is equal to the partial derivative of the strain energy with respect to the projected load's displacement.*

To make use of this theorem the strain energy in the structure must be expressed in term of the projected displacements, Δ_i. Because this theorem is derived directly from the principle of minimum total potential energy, eq. (10.108) is simply an equilibrium statement for the problem.

Castigliano's first theorem is easily extended to other loading conditions such as applied couples, loads of equal magnitude and opposite directions sharing a common line of action, or couples of equal magnitude and opposite directions sharing a common axis.

10.9.3 Crotti-Engesser theorem

Clapeyron's and Castigliano's first theorems are corollaries of the principle of minimum total potential energy. Not unexpectedly, parallel developments based on the principle of minimum complementary energy will lead to similar results.

The total complementary energy, Π', is defined in eq. (10.104) as the sum of the complementary strain energy, A', and potential of the prescribed displacements, Φ'. If the system is subjected to N prescribed displacements, Δ_i, $i = 1, 2, \ldots, N$, the potential of the prescribed displacements is $\Phi' = -\sum_{i=1}^{N} P_i \Delta_i$, where P_i, $i = 1, 2, \ldots, N$, are the driving forces required to obtain the prescribed displacements. The total complementary energy now becomes

$$\Pi' = A' + \Phi' = A' - \sum_{i=1}^{N} P_i \Delta_i.$$

Next, the statically admissible stress field in the elastic body is expressed in terms of the driving forces, *i.e.*, $A' = A'(P_i)$. The principle of minimum complementary energy, principle 12, then implies

$$\frac{\partial \Pi'}{\partial P_j} = \frac{\partial A'}{\partial P_j} - \frac{\partial}{\partial P_j} \sum_{i=1}^{N} P_i \Delta_i = \frac{\partial A'}{\partial P_j} - \Delta_j = 0.$$

It now follows that

$$\Delta_i = \frac{\partial A'}{\partial P_i}, \tag{10.109}$$

where, clearly, the complementary energy in the structure must be expressed in terms of the driving forces, P_i. This result is known as the *Crotti-Engesser theorem* which can be stated as follows.

Theorem 10.3 (Crotti-Engesser theorem). *For an elastic structure, the prescribed deflection at a point is given by the partial derivative of the complementary energy with respect to the driving force.*

Unlike Clapeyron's theorem, 10.1, the Crotti-Engesser theorem can be applied to problems with multiple applied loads.

10.9.4 Castigliano's second theorem

In the derivation of the Crotti-Engesser theorem, the existence of the complementary energy is assumed for the elastic material. If the material is assumed to be linearly elastic, the strain energy and its complementary counterpart become equal, $A = A'$, and the Crotti-Engesser theorem, eq. (10.109), then leads to *Castigliano's second theorem*,

$$\Delta_i = \frac{\partial A}{\partial P_i}. \tag{10.110}$$

Theorem 10.4 (Castigliano's second theorem). *For a linearly elastic structure, the prescribed deflection at a point is given by the partial derivative of the strain energy with respect to the driving force.*

Note the obvious symmetry between eq. (10.110) and eq. (10.108). It should be noted, however, that Castigliano's first theorem applies to any elastic system, whereas Castigliano's second theorem only applies to linearly elastic structures.

10.9.5 Applications of energy theorems

The energy theorems are useful for determining deflections at specific points of a structure. In particular, Castigliano's second theorem yields structural deflections under applied loads.

Castigliano's second theorem is also useful when dealing with hyperstatic systems. Imagine a cantilevered beam with a tip support. One way to look at this problem is to consider a cantilevered beam with a prescribed tip displacement, which is required to vanish. The driving force, in this case, is the reaction force at the support. If P_i denotes this reaction force and $\Delta_i = 0$ the prescribed tip displacement, Castigliano's second theorem, eq. (10.110), requires $\partial A / \partial P_i = 0$. This equation is the compatibility equation at the tip support. It is interesting to note that in this case, Castigliano's second theorem reduces to the principle of least work, principle 13 on page 554.

Example 10.14. Deflection of a cantilever under transverse load
Consider a cantilevered beam of length L subjected to a tip transverse load P acting at a distance αL from the left support, as depicted in fig. 5.26 on page 201. This problem is treated using the classical approach in example 5.8 on page 201.

Simple equilibrium arguments yield the following bending moment distribution: $M_3 = PL(\alpha - \eta)$, for $\eta \leq \alpha$, and $M_3 = 0$, for $\alpha \leq \eta \leq 1$, where $\eta = x_1/L$ is the non-dimensional variable along the beam's span. The strain energy can be expressed in terms of the bending moment distribution as

$$A = \frac{1}{2H_{33}^c} \int_0^L M_3^2 \, dx_1 = \frac{1}{2H_{33}^c} \int_0^\alpha (PL)^2 (\alpha - \eta)^2 \, Ld\eta = \frac{P^2(\alpha L)^3}{6H_{33}^c}.$$

Castigliano's second theorem then yields the deflection under the load as $\Delta = \partial A/\partial P = P(\alpha L)^3/(3H_{33}^c)$. This result is identical to that found with the classical approach, see eq. (5.55).

Example 10.15. Rotation of a cantilever under couple
Consider a cantilevered beam of length L subjected to a concentrated couple Q acting at a distance αL from the left support. Find the rotation at the point where Q acts.

Simple equilibrium arguments yield the following bending moment distribution: $M_3 = Q$, for $\eta \leq \alpha$, and $M_3 = 0$, for $\alpha \leq \eta \leq 1$, where $\eta = x_1/L$ is the non-dimensional variable along the beam's span. The strain energy can be expressed in terms of the bending moment distribution as

$$A = \frac{1}{2H_{33}^c} \int_0^L M_3^2 \, dx_1 = \frac{1}{2H_{33}^c} \int_0^\alpha Q^2 \, Ld\eta = \frac{Q^2(\alpha L)}{2H_{33}^c}.$$

Castigliano's second theorem then yields the rotation at the point of application of the couple as $\Phi = \partial A/\partial Q = Q\alpha L/H_{33}^c$.

Example 10.16. Simply supported beam with concentrated load
Consider a simply supported, uniform beam of length L subjected to a concentrated load P acting at a distance αL from the end supports, as depicted in fig. 10.41. This problem is treated using classical approaches in examples 5.6 on page 199, and in example 10.13 using Clapeyron's theorem. The use of Clapeyron's theorem will be contrasted here with Castigliano's second theorem.

The algebra associated with the use of Castigliano's second theorem is somewhat simplified if the following manipulation is performed first

$$\Delta = \frac{\partial A}{\partial P} = \frac{\partial}{\partial P} \int_0^L \frac{M_3^2}{2H_{33}^c} \, dx_1 = \int_0^L \frac{M_3}{H_{33}^c} \frac{\partial M_3}{\partial P} \, dx_1.$$

The bending moment distribution is given by eq. (5.52), or it can be obtained directly from equilibrium considerations as

$$M_3(\eta) = PL \begin{cases} -(1-\alpha)\eta, & 0 \le \eta \le \alpha, \\ -\alpha(1-\eta), & \alpha < \eta \le 1, \end{cases}$$

where $\eta = x_1/L$ is the non-dimensional variable along the beam's span. Substituting this in the previous equation results in

$$\Delta = \frac{PL^2}{H_{33}^c} \left[\int_0^\alpha (1-\alpha)^2 \eta^2 \, Ld\eta + \int_\alpha^1 \alpha^2 (1-\eta)^2 \, Ld\eta \right],$$

and performing the integrations then yields

$$\Delta = \frac{PL^3}{H_{33}^c} \left[(1-\alpha)^2 \frac{\alpha^3}{3} + \alpha^2 \frac{(1-\alpha)^3}{3} \right] = \frac{PL^3}{3H_{33}^c} \alpha^2 (1-\alpha)^2.$$

This result is identical to that obtained in example 10.13 using Clapeyron's theorem.

Example 10.17. Ring under internal forces

Consider the open circular ring of radius R shown in fig. 10.42. The ring is cut at one location and two opposite tangential forces of equal magnitude P are applied along the same tangential line of action in the plane of the ring. Evaluate the relative displacement, Δ, of the points of application of the two forces using Castigliano's second theorem.

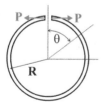

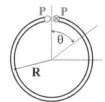

Fig. 10.42. Ring subjected to internal forces acting *in the plane of the ring*.

Fig. 10.43. Ring subjected to internal forces acting *out of the plane of the ring*.

For this configuration, the ring is subjected to both bending and axial loading, and hence, the total strain energy in the system is the sum of the strain energies due to bending and extension, given eqs. (10.41) and (10.36), respectively. Castigliano's second theorem now becomes

$$\Delta = \frac{\partial A}{\partial P} = \int_0^{2\pi} \frac{M_3}{H_{33}^c} \frac{\partial M_3}{\partial P} \, Rd\theta + \int_0^{2\pi} \frac{N_1}{S} \frac{\partial N_1}{\partial P} \, Rd\theta,$$

where $Rd\theta$ is the infinitesimal axial distance around the ring.

The ring's bending moment and axial force distributions are evaluated by considering the equilibrium conditions of a segment of the beam of length $R\theta$. They are found to be $M_3(\theta) = -PR(1 - \cos\theta)$ and $N_1 = -P\cos\theta$, respectively. Substituting these expression into the statement of Castigliano's theorem then yields

$$\Delta = \frac{PR^3}{H_{33}^c} \int_0^{2\pi} (1 - \cos\theta)^2 \, d\theta + \frac{PR}{S} \int_0^{2\pi} \cos^2\theta \, d\theta = \frac{3\pi PR^3}{H_{33}^c} + \frac{\pi PR}{S}.$$

If the ring has a rectangular cross-section of radial thickness h and width b, the bending stiffness becomes $H_{33}^c = Ebh^3/12$, and it then follows that

$$\Delta = \frac{36\pi PR^3}{Ebh^3} + \frac{\pi PR}{Ebh} = \frac{\pi PR}{Ebh}\left[36\left(\frac{R}{h}\right)^2 + 1\right].$$

For a thin ring, $R/h \gg 1$, and the first term becomes dominant, implying that bending rather than extension of the ring is the principal contributor to the relative displacement of the points of application of the two forces.

The problem can be modified to introduce torsional deformation into the ring by changing the orientation of the applied forces, P, to now act along the same line of action but *normal to the plane of the ring*. This configuration is shown in fig. 10.43, where the two forces of magnitude P are acting normal to the plane of the figure. The total strain energy in the system is now the sum of the strain energies due to bending and torsion, given by eqs. (10.41) and (10.45), respectively. Castigliano's second theorem can now be written as

$$\Delta = \frac{\partial A}{\partial P} = \int_0^{2\pi} \frac{M_3}{H_{33}^c} \frac{\partial M_3}{\partial P} R d\theta + \int_0^{2\pi} \frac{M_1}{H_{11}} \frac{\partial M_1}{\partial P} R d\theta,$$

where $M_3 = PR\sin\theta$ is the bending moment distribution in the ring and $M_1 = PR(1 - \cos\theta)$ the torsion moment distribution. Note that Δ is the relative displacement of the points of application of the forces projected along their line of action, *i.e.*, measured in the direction perpendicular to the plane of the ring. This relative displacement is then given by

$$\Delta = \frac{PR^3}{H_{33}^c} \int_0^{2\pi} \sin^2\theta \, d\theta + \frac{PR^3}{H_{11}} \int_0^{2\pi} (1 - \cos\theta)^2 \, R d\theta = \frac{\pi PR^3}{H_{33}} + \frac{3\pi PR^3}{H_{11}}.$$

If the ring has a circular cross-section of radius a, and is made of a linearly elastic, homogeneous material so that $G = E/[2(1+\nu)]$, the relative displacement becomes $\Delta = 4PR^3[1+3(1+\nu)]/(Ea^4)$. For this configuration, torsional deformation in the ring contributes $[3(1+\nu)]/[1+3(1+\nu)] \approx 80\%$ of the total relative displacement, for $\nu = 0.3$.

Example 10.18. Cantilevered beam with intermediate support

The cantilevered beam subjected to a uniform loading, p_0 and with an intermediate support located a distance $x_1 = \alpha L$ from the left clamp is depicted in fig. 10.44. This is a hyperstatic problem of order 1, and the reaction force at the support will be determined using Castigliano's second theorem.

Within the framework of Castigliano's second theorem, the reaction force, R, at the support is the driving force that prescribes a vanishing displacement at the intermediate support. This theorem, eq. (10.110), now implies $\Delta = \partial A/\partial R = 0$

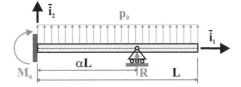

Fig. 10.44. Cantilevered beam with an intermediate support at location αL.

and this equation will be used to compute the unknown driving force, which is the desired reaction force.

The bending moment distribution in the beam can be found from equilibrium considerations as

$$M_3 = \begin{cases} p_0 L^2 (1-\eta)^2/2 - RL(\alpha - \eta), & 0 \le \eta \le \alpha, \\ p_0 L^2 (1-\eta)^2/2, & \alpha < \eta \le 1, \end{cases}$$

where $\eta = x_1/L$ is the non-dimensional span-wise variable.

The compatibility condition at the support now follows from the application of Castigliano's second theorem, to find

$$\Delta = \frac{\partial A}{\partial R} = \int_0^\alpha \left[\frac{p_0 L^2}{2} (1-\eta)^2 - RL(\alpha - \eta) \right] \left[-L(\alpha - \eta) \right] L d\eta = 0.$$

The second part of this integral, extending from α to 1, vanishes because $\partial M_3/\partial R$ vanishes over that portion of the beam. This equation can be integrated and solved to determine the unknown reaction force

$$R = \frac{p_0 L}{8} \frac{6 - 4\alpha + \alpha^2}{\alpha}.$$

A free body diagram of the entire beam reveals that the reaction moment at the clamped end of the beam is $M_0 = p_0 L^2/2 - \alpha RL$, and substituting for R, it follows that $M_0 = -(p_0 L^2/8)(\alpha^2 - 4\alpha + 2)$.

The same hyperstatic problem can be handled in a different manner within the framework of Castigliano's second theorem by choosing another reaction as the driving force. In this case, the reaction moment, M_0, at the root clamp is the driving moment that prescribes a vanishing rotation at the clamp. Castigliano's second theorem now implies $\Phi = \partial A/\partial M_0 = 0$ and this equation will be used to compute the unknown driving moment, which is the desired reaction moment.

The bending moment distribution in what is now a simply supported beam is found from equilibrium considerations as

$$M_3 = \begin{cases} p_0 L^2 \left[\eta^2 + (1/\alpha - 2)\eta \right]/2 + M_0(1 - \eta/\alpha), & 0 \le \eta \le \alpha, \\ p_0 L^2 (1-\eta)^2/2, & \alpha < \eta \le 1. \end{cases}$$

The compatibility condition at the clamp now follows from the application of Castigliano's second theorem, to find

$$\Phi = \int_0^\alpha \frac{1}{H_{33}^c} \left\{ \frac{p_0 L^2}{2} \left[\eta^2 + (\frac{1}{\alpha} - 2)\eta \right] + M_0(1 - \frac{\eta}{\alpha}) \right\} \left(1 - \frac{\eta}{\alpha} \right) L d\eta = 0.$$

This equation can be solved to determine the reaction moment $M_0 = -p_0 L^2(\alpha^2 - 4\alpha + 2)/8$, which is identical to that found earlier.

Clearly, the two approaches are identical, and the choice between the two is dictated by simplicity of the required algebra. Again, it should be noted that this example can also be solved in almost the same way using the principle of least work, principle 13 on page 554

10.9.6 The dummy load method

Castigliano's second theorem as expressed in eq. (10.110) gives the deflection at the point of application of a concentrated load. This prompts the following question: is it possible to use Castigliano's second theorem to compute the deflection of a structure at a point where no load is applied? The *dummy load method* is a procedure that enables the use of Castigliano's second theorem to compute the deflection at any point of a structure whether or not a concentrated load is applied at that point.

In the first step of the procedure, a fictitious or "dummy load," $\mathcal{P}$, is applied to the structure at the point where the displacement is to be computed. Furthermore, the line of action of this dummy load is aligned with the direction of the desired displacement component. In the second step of the procedure, the displacement component, $\hat{\Delta}$, is computed using Castigliano's second theorem as $\hat{\Delta} = \partial A/\partial \mathcal{P}$. In the last step, the dummy load is removed by setting it equal to zero to find the desired displacement, $\Delta = \lim_{\mathcal{P} \to 0} \hat{\Delta}$. Load $\mathcal{P}$ is just an artifact that enables the use of Castigliano's second theorem, and this observation explains why load $\mathcal{P}$ is called a dummy load, and why the method is called the dummy load method.

The dummy load method can be summarized by the following equation

$$\Delta = \lim_{\mathcal{P} \to 0} \frac{\partial A}{\partial \mathcal{P}}. \tag{10.111}$$

The strain energy, A, must be determined as a function of the applied loads, including the dummy load $\mathcal{P}$, and any redundant quantities. If the material the structure is made of is elastic, but nonlinear, the complementary strain energy, A', must be used instead of the strain energy.

Example 10.19. Tip deflection of a cantilevered beam
Consider a cantilevered beam of length L subjected to a uniform loading p_0, as shown in fig. 10.45. Determine the beam's tip deflection, Δ, using the dummy load method.

Since no concentrated load is applied at the beam's tip, the dummy load method described in section 10.9.6 will be used. In the first step of the procedure, a dummy load, $\mathcal{P}$, is applied at the beam's tip, as illustrated in fig. 10.45. The bending moment distribution in the beam is readily obtained from equilibrium considerations as $M_3(x_1) = p_0(L - x_1)^2/2 + \mathcal{P}(L - x_1)$.

The strain energy in the structure can then be written as

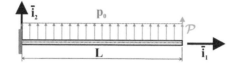

Fig. 10.45. Cantilevered beam under uniform loading.

$$
A = \frac{1}{2H_{33}^c} \int_0^1 \left[\frac{p_0 L^2}{2}(1-\eta)^2 + \mathcal{P}L(1-\eta) \right]^2 L d\eta
$$

$$
= \frac{1}{2H_{33}^c} \left(\frac{p_0^2 L^5}{20} + \frac{\mathcal{P}p_0 L^4}{4} + \frac{\mathcal{P}^2 L^3}{3} \right).
$$

As expected, the strain energy explicitly depends on the dummy load. The second step of the procedure uses Castigliano's second theorem to compute the tip deflection due to all loads as

$$
\hat{\Delta} = \frac{\partial A}{\partial \mathcal{P}} = \frac{1}{2H_{33}^c} \left(\frac{p_0 L^4}{4} + \frac{2\mathcal{P}L^3}{3} \right).
$$

Finally, the last step of the procedure reveals the desired tip displacement as

$$
\Delta = \lim_{\mathcal{P} \to 0} \hat{\Delta} = \frac{p_0 L^4}{8H_{33}^c}.
$$

This result is identical to that obtained using the classical approach; indeed, $\Delta = \bar{u}_2(\eta = 1)$, where the transverse displacement field is given by eq. (5.54). The same result is also obtained using the unit load method described in section 9.7.6.

The solution just presented has scrupulously followed the dummy load procedure described in section 10.9.6. While easily understood, this procedure involves unnecessarily complicated integrations. The desired displacement can be obtained more directly by carrying out the derivative with respect to the dummy load and taking the limit before carrying out the integrations. This is illustrated as follows,

$$
\Delta = \left[\frac{\partial}{\partial \mathcal{P}} \int_0^L \frac{M_3^2}{2H_{33}^c} dx_1 \right]_{\mathcal{P}=0} = \int_0^L \frac{M_3}{H_{33}^c} \left[\frac{\partial M_3}{\partial \mathcal{P}} \right]_{\mathcal{P}=0} dx_1. \tag{10.112}
$$

For the present problem, this yields

$$
\Delta = \int_0^1 \left[\frac{p_0 L^2}{2H_{33}^c}(1-\eta)^2 \right] L(1-\eta) L d\eta = \frac{p_0 L^4}{2H_{33}^c} \int_0^1 (1-\eta)^3 d\eta = \frac{p_0 L^4}{8H_{33}^c},
$$

which is the same result as before.

Example 10.20. Deflection of a simply supported beam

Consider the simply supported beam of length L subjected to a uniform transverse loading, p_0, as depicted in fig. 10.46. Determine the transverse deflection of the beam at location αL.

Using the dummy load method, a dummy load, $\mathcal{P}$, is added at location αL. To evaluate the strain energy in the structure, the bending moment distribution is computed first as

$$
M_3(\eta) = \begin{cases} -p_0 L^2 \eta (1 - \eta)/2 - \mathcal{P} L(1 - \alpha)\eta, & 0 \leq \eta \leq \alpha, \\ -p_0 L^2 \eta (1 - \eta)/2 - \mathcal{P} L \alpha (1 - \eta), & \alpha < \eta \leq 1, \end{cases}
$$

where $\eta = x_1/L$ is a non-dimensional span-wise variable. The first term represents the contribution of the distributed load, p_0, and the second term represents that due to the dummy load, $\mathcal{P}$.

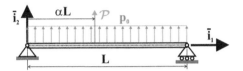

Fig. 10.46. Simply supported beam under uniform loading.

Using the simplified approach given by eq. (10.112), the deflection under the dummy load at $x_1 = \alpha L$ becomes

$$
\Delta(\alpha) = \frac{p_0 L^4}{2 H_{33}^c} \left[\int_0^\alpha \eta (1 - \eta)(1 - \alpha)\eta \, d\eta + \int_\alpha^1 \eta (1 - \eta)\alpha(1 - \eta) \, d\eta \right].
$$

Performing the integrations and simplifying leads to

$$
\begin{aligned}
\Delta(\alpha) &= \frac{p_0 L^4}{2 H_{33}^c} \left[(1 - \alpha)\left(\frac{\alpha^3}{3} - \frac{\alpha^4}{4}\right) + \alpha\left[\frac{(1 - \alpha)^3}{3} - \frac{(1 - \alpha)^4}{4}\right] \right] \\
&= \frac{p_0 L^4}{24 H_{33}^c} \left(\alpha^4 - 2\alpha^3 + \alpha \right).
\end{aligned}
$$

This result gives the transverse displacement at an arbitrary point along the beam and is, in fact, the transverse displacement field for the beam, $\bar{u}_2(\alpha)$, $0 \leq \alpha \leq 1$. This result is identical to that obtained with the classical approach in eq. (5.48), but note that with the present procedure, the transverse displacement field is obtained without having to integrate the governing differential equation of the problem.

10.9.7 Unit load method revisited

The unit load method is developed in section 9.7.6 based on the principle of complementary virtual work. In this section, the same method will be derived from the dummy load method presented in section 10.9.6. The close relationship between the two methods should not come as a surprise: the unit load method is a direct consequence of the principle of complementary virtual work, whereas the dummy load method is derived from Castigliano's second theorem, which itself, stems from the principle of minimum complementary energy. Clearly, both unit and dummy load methods have their roots in the principle of complementary virtual work, and hence, are statements of compatibility conditions.

When using the dummy load method, the expression for the strain energy in an isostatic beam is

$$A = \int_0^L \frac{\mathcal{M}_3^2}{2H_{33}^c}\, dx_1,$$

where $\mathcal{M}_3(x_1)$ is the bending moment distribution generated by the externally applied loads and the dummy load. Castigliano's second theorem now implies that the deflection at the point of application of the dummy load is

$$\Delta = \lim_{\mathcal{P} \to 0} \frac{\partial A}{\partial \mathcal{P}} = \lim_{\mathcal{P} \to 0} \int_0^L \frac{\mathcal{M}_3}{H_{33}^c} \frac{\partial \mathcal{M}_3}{\partial \mathcal{P}}\, dx_1. \tag{10.113}$$

As the dummy load tends to zero, the quantities appearing in this equation can be interpreted as follows.

$$\lim_{\mathcal{P} \to 0} \mathcal{M}_3 = M_3 = \text{bending moment due to externally applied loads only,}$$

$$\lim_{\mathcal{P} \to 0} \frac{\partial \mathcal{M}_3}{\partial \mathcal{P}} = \hat{M}_3 = \text{bending moment due to a unit load only.}$$

With this notation, eq. (10.113) becomes the familiar statement of the unit load method, see eq. (9.83),

$$\Delta = \int_0^L \frac{\hat{M}_3 M_3}{H_{33}^c}\, dx_1. \tag{10.114}$$

Although this expression seems to be identical to that derived for the unit load method, important differences exist. The bending moment distribution due to externally applied loads, denoted M_3, is identical for both unit and dummy load methods, and is the bending moment distribution acting in the actual structure under the action of the externally applied loads.

A subtle difference exists, however, between the bending moment distribution, $\hat{M}_3$, defined in the two methods. For the dummy load method, $\hat{M}_3$ is the bending moment acting in the structure subjected to a unit dummy load. For the unit load method, $\hat{M}_3$ is *any statically admissible* bending moment distribution in equilibrium with the unit load. In this latter case, $\hat{M}_3$ is not necessarily the actual bending moment distribution acting in the structure subjected to the unit load, but rather, any statically admissible bending moment distribution in equilibrium with the unit load. Consequently, the unit load method is more versatile, and the fact that any statically admissible bending moment distribution can result in a significant simplification of the procedure.

Example 10.21. Deflection of a hyperstatic beam (Dummy load method)
Consider the cantilevered beam with a mid-span support subjected to a uniformly distributed loading, p_0, as depicted in fig. 10.47. Determine the beam's tip deflection using the dummy load method.

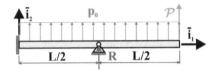

Fig. 10.47. Application of Castigliano's second theorem to calculate deflections in a hyperstatic beam.

Because of the presence of the mid-span support, the system is hyperstatic of order 1. The mid-span reaction force, R, is selected as a redundant quantity, and the bending moment distribution in the bean is then obtained from equilibrium considerations as

$$M_3(\eta) = \begin{cases} \mathcal{P}L(1-\eta) + p_0 L^2 (1-\eta)^2/2 + RL(1/2-\eta), & 0 \le \eta \le 1/2, \\ \mathcal{P}L(1-\eta) + p_0 L^2 (1-\eta)^2/2, & 1/2 \le \eta \le 1. \end{cases}$$

First, Castigliano's second theorem (or the principle of least work) is used to find the unknown reaction force, R, by imposing the vanishing of the displacement at the mid-span support, leading to

$$\frac{\partial A}{\partial R} = \frac{\partial}{\partial R} \int_0^L \frac{1}{2} \frac{M_3^2}{H_{33}^c} \mathrm{d}x_1 = \int_0^L \frac{M_3}{H_{33}^c} \frac{\partial M_3}{\partial R} \, \mathrm{d}x_1 = 0.$$

Introducing the above bending moment distribution, this compatibility condition leads to

$$\int_0^{1/2} \left[\frac{\mathcal{P}L(1-\eta) + p_0 L^2 (1-\eta)^2/2 + RL(1/2-\eta)}{H_{33}^c} \right] \left[L\left(\frac{1}{2}-\eta\right) \right] L \mathrm{d}\eta = 0.$$

Evaluating the integrals then yields the reaction force at the mid-span support as

$$R = -\frac{5\mathcal{P}}{2} - \frac{17 p_0 L}{16}.$$

Next, the beam's tip deflection, Δ, is computed using the dummy load method as

$$\Delta = \frac{\partial A}{\partial \mathcal{P}} = \int_0^L \frac{M_3}{H_{33}^c} \frac{\partial M_3}{\partial \mathcal{P}} \, \mathrm{d}x_1.$$

Introducing the above bending moment distribution then yields

$$\Delta = \int_0^{1/2} \left[\frac{p_0 L^2 (1-\eta)^2/2 + RL(1/2-\eta)}{H_{33}^c} \right] [L(1-\eta)] \, L \mathrm{d}\eta$$
$$+ \int_{1/2}^1 \left[\frac{p_0 L^2 (1-\eta)^2/2}{H_{33}^c} \right] [L(1-\eta)] \, L \mathrm{d}\eta.$$

Regrouping and evaluating these integrals gives the beam's tip deflection as $\Delta = p_0 L^4/(8 H_{33}^c) + 5 R L^3/(48 H_{33}^c)$, and introducing the value of the mid-span reaction force leads to

$$\Delta = \frac{11p_0 L^4}{768 H_{33}^c}.$$

Example 10.22. Deflection of a hyperstatic beam (Unit load method)
Consider the same cantilevered beam with a mid-span support subjected to a uniformly distributed loading, p_0, treated in example 10.21 and shown in fig. 10.47. Determine the beam's tip deflection, but now using the unit load method.

A cut is made at the mid-span support and the mid-span reaction force, R, is selected as the redundant quantity. The bending moment distribution in the beam due to the externally applied loads is obtained from equilibrium considerations as $M_3(\eta) = p_0 L^2 (1-\eta)^2/2$. Next, a statically admissible bending moment distribution is evaluated that is in equilibrium with a unit load applied upwards to the beam at the cut support point: $\hat{M}_3 = L(1/2 - \eta)$ for $0 \le \eta \le 1/2$ and $\hat{M}_3 = 0$ for $1/2 \le \eta \le 1$. The deflection at the support under the externally applied loads is

$$\Delta_c = \int_0^L \frac{M_3 \hat{M}_3}{H_{33}^c} \, \mathrm{d}x_1 = \int_0^{1/2} \left[\frac{p_0 L^2 (1-\eta)^2/2}{H_{33}^c} \right] [L(1/2 - \eta)] \, L \mathrm{d}\eta = \frac{17 p_0 L^4}{384 H_{33}^c}.$$

The deflection at mid-span support due to the unit load alone is

$$\Delta_1 = \int_0^L \frac{\hat{M}_3^2}{H_{33}^c} \, \mathrm{d}x_1 = \int_0^{1/2} \frac{L^2 (1/2 - \eta)^2}{H_{33}^c} \, L \mathrm{d}\eta = \frac{L^3}{24 H_{33}^c}.$$

The reaction force at the support is now $R = -\Delta_c/\Delta_1 = -17 p_0 L/16$, from which

$$M_3(\eta) = \begin{cases} p_0 L^2 (1-\eta)^2/2 + RL(1/2 - \eta), & 0 \le \eta \le 1/2, \\ p_0 L^2 (1-\eta)^2/2, & 1/2 \le \eta \le 1, \end{cases}$$

which is identical to the result found with the dummy load method, provided that the dummy load is set to zero.

To determine the beam's tip deflection, a unit load is applied at its tip. A statically admissible bending moment distribution that is in equilibrium with this tip unit load is found from equilibrium consideration as $\hat{M}_3 = L(1 - \eta)$. When evaluating this bending moment distribution, the mid-span reaction force is set to zero, because all that is required of this distribution is that it be statically admissible for the tip unit load. The beam's tip deflection, Δ, now becomes

$$\Delta = \int_0^L \frac{M_3 \hat{M}_3}{H_{33}^c} \mathrm{d}x_1 = \int_0^{1/2} \left[\frac{p_0 L^2 (1-\eta)^2/2 + RL(1/2 - \eta)}{H_{33}^c} \right] [L(1 - \eta)] \, L \mathrm{d}\eta$$
$$+ \int_{1/2}^1 \left[\frac{p_0 L^2 (1-\eta)^2/2}{H_{33}^c} \right] [L(1 - \eta)] \, L \mathrm{d}\eta = \frac{11 p_0 L^4}{768 H_{33}^c}.$$

The solution is identical to that found in the previous example.

The choice between the unit and the dummy load methods is largely a matter of convenience, although a hybrid approach using Castigliano's second theorem (or the principle of least work) to find R and the unit load method to find Δ will often lead to simpler integrals.

10.9.8 Problems

Problem 10.18. Cantilevered beam subjected to distributed load

Consider the cantilevered beam subjected to a uniformly distributed load over half its span, as depicted in fig. 10.48. *(1)* Use the dummy load method to compute the deflection of the beam at point **M**.

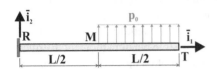

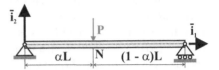

Fig. 10.48. Cantilevered beam subjected to half-span loading.

Fig. 10.49. Simply-supported beam under concentrated load

Problem 10.19. Simply supported beam with concentrated load

Consider the simply supported beam subjected to a concentrated load applied at a distance αL from the left support, as depicted in fig. 10.49. *(1)* Use the dummy load method to compute the deflection of the beam at point **N**.

Problem 10.20. Cantilevered beam subjected to triangular loading

Consider the cantilevered beam subjected to a triangular loading, as depicted in fig. 10.50. *(1)* Use the dummy load method to compute the deflection of the beam at point **T**.

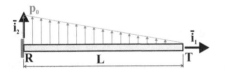

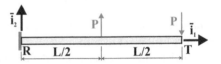

Fig. 10.50. Cantilevered beam subjected to half-span loading.

Fig. 10.51. Cantilevered beam subjected to two concentrated loads.

Problem 10.21. Simply supported beam with concentrated load

Consider the cantilevered beam subjected to two concentrated loads, each of magnitude P, as depicted in fig. 10.51. *(1)* Use the dummy load method to compute the deflection of the beam at point **T**.

Problem 10.22. Semi-circular beam with rigid arm

Consider the uniform, semi-circular beam with a rigid arm attached at its tip, as shown in fig. 10.52. The beam is made of a linearly elastic material and the radius of its centerline is R. A load of magnitude P acts at the tip of the rigid arm in the plane of the beam, but its orientation in this plane is otherwise arbitrary. Prove that: *(1)* The displacement, Δ, of point **O** is in the direction of the applied load *for any arbitrary orientation* of P, and *(2)* the spring constant $k = P/\Delta$ is independent of the orientation of the load P. *Hint*: At first, study the behavior of the beam under a horizontal force, H. Next, turn to a vertical force, V. The behavior of the system under a general loading is then obtained by invoking the principle of superposition for a linear system. You should assume that only bending deformations will contribute to the strain energy in the beam, *i.e.*, ignore axial deformation.

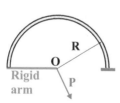

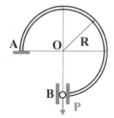

Fig. 10.52. Semi-circular beam with a rigid arm.

Fig. 10.53. Uniform spring under a vertical load P.

Problem 10.23. Circular beam with vertical tip load

The uniform circular beam with centerline radius R shown in fig. 10.53 is clamped at point **A** and constrained to move in the only in the vertical direction at point **B**, where it is also subjected to an applied vertical load, P. *(1)* Find the displacement, Δ, in the direction of the applied load. *(2)* Find the horizontal reaction Q at point **B**. *(3)* Find the equivalent spring constant $k = P/\Delta$. Assume that only bending deformations are significant, *i.e.*, ignore axial deformation.

Problem 10.24. Deflection of 3-bar truss with different member properties

The three-bar, hyperstatic truss shown in fig. 10.36 is subjected to a tip vertical load P. The three bars have a Young's modulus E, bar 1 is of cross-sectional area $\mathcal{A}$, while that bars 2 and 3 is $2\mathcal{A}$. *(1)* Determine the vertical deflection of the loaded joint. *(1)* Determine the horizontal deflection of the loaded joint.

Problem 10.25. Deflection of cantilevered beam with simple support and concentrated load

A cantilevered beam with a mid-span support carries a tip concentrated load, P, as depicted in fig. 10.38. *(1)* Compute the deflection at the beam's tip.

10.10 Reciprocity theorems

For linearly elastic structures, a useful reciprocity exists between loads applied at one set of locations and deflections produced at another set of locations. This reciprocity can be stated in the form of two theorems that are developed in the following sections.

10.10.1 Betti's theorem

Consider a properly constrained elastic body subjected to various concentrated loads and couples, as shown in fig. 10.40. If the displacements of the points of application of these loads projected along their lines of action are denoted $\Delta_i^{[1]}$, Clapeyron's theorem, eq. 10.107, implies $A^{[1]} = \sum_{i=1}^N P_i^{[1]} \Delta_i^{[1]}/2$, where $A^{[1]}$ is the total strain energy of the system under this loading condition, denoted *state 1*. Let the magnitude of the applied loads be changed to $P_i^{[2]}$ and the corresponding projected displacements will then be $\Delta_i^{[2]}$. In this new state, denoted *state 2*, the points of application of the loads and their lines of action are identical to those in *state 1*. The strain energy in *state 2* is $A^{[2]} = \sum_{i=1}^N P_i^{[2]} \Delta_i^{[2]}/2$. When going from *state 1* to *state 2*, the added work done by the applied loads equals the change in total strain energy

$$A^{[2]} - A^{[1]} = \sum_{i=1}^N \frac{P_i^{[2]} \Delta_i^{[2]}}{2} - \sum_{i=1}^N \frac{P_i^{[1]} \Delta_i^{[1]}}{2}. \tag{10.115}$$

This difference can be computed in an alternative manner. It is possible to go from *state 1* to *state 2* by gradually adding to the forces $P_i^{[1]}$ of *state 1* the forces $P_i^{[2]} - P_i^{[1]}$ with unchanged lines of action. During this transition, additional work will be done by the forces $P_i^{[1]}$ which remain constant, and the forces, $P_i^{[2]} - P_i^{[1]}$, which are allowed to increase gradually to *state 2*. The work done by $P_i^{[1]}$ is $\sum_{i=1}^N P_i^{[1]}(\Delta_i^{[2]} - \Delta_i^{[1]})$, and the work done by the gradually increasing forces, $P_i^{[2]} - P_i^{[1]}$, is $\sum_{i=1}^N (P_i^{[2]} - P_i^{[1]})(\Delta_i^{[2]} - \Delta_i^{[1]})/2$. Thus, the change in strain energy between the two states is equal to the work done by these forces, and this can be written as

$$A^{[2]} - A^{[1]} = \sum_{i=1}^N P_i^{[1]}(\Delta_i^{[2]} - \Delta_i^{[1]}) + \sum_{i=1}^N \frac{(P_i^{[2]} - P_i^{[1]})(\Delta_i^{[2]} - \Delta_i^{[1]})}{2}$$
$$= \frac{1}{2} \sum_{i=1}^N (P_i^{[2]} + P_i^{[1]})(\Delta_i^{[2]} - \Delta_i^{[1]}). \tag{10.116}$$

Comparing eqs. (10.115) and (10.116) then yields

$$\sum_{i=1}^N P_i^{[1]} \Delta_i^{[2]} = \sum_{i=1}^N P_i^{[2]} \Delta_i^{[1]}. \tag{10.117}$$

This result be interpreted as follows

Theorem 10.5 (Reciprocity theorem or Betti's theorem). *A linearly elastic body is subjected to two loading states characterized by loads of different magnitudes but identical points of applications and lines of action. The sum of the product of the loads in one state by the projected displacements of the other is identical to that obtained when the two states are interchanged.*

Because Betti's theorem is a direct consequence of Clapeyron's theorem, theorem 10.1, both theorems are valid for the same loading cases. The loadings defining *states 1* and *2* can involve one or more of the following *(1)* a concentrated load and the projected displacement of its point of application, *(2)* a couple and the projected rotation at its point of application, *(3)* two opposite forces of identical magnitude sharing a common line of action and the projected relative displacement of their points of application, and *(4)* two opposite couples of identical magnitude sharing a common line of action and the projected relative rotation at their points of application.

10.10.2 Maxwell's theorem

Let the simply supported beam depicted in fig. 10.54 be subjected to two loading states. *State 1* consists of load $P^{[1]}$ applied at point **1**, whereas *state 2* consists of load $P^{[2]}$ applied at point **2**. For *state 1*, the displacements at points **1** and **2** will be denoted $\Delta_1^{[1]}$ and $\Delta_2^{[1]}$, respectively. Similarly, the corresponding displacements for *state 2* are $\Delta_1^{[2]}$ and $\Delta_2^{[2]}$, respectively.

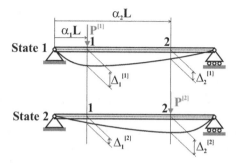

Fig. 10.54. Simply supported beam under two loading states.

If the structure is made of a linearly elastic material, Betti's theorem, theorem 10.5, is applicable and implies

$$P^{[1]}\,\Delta_1^{[2]} = P^{[2]}\,\Delta_2^{[1]}, \quad \text{or} \quad \frac{\Delta_1^{[2]}}{P^{[2]}} = \frac{\Delta_2^{[1]}}{P^{[1]}}. \tag{10.118}$$

The *influence coefficient* is defined as the displacement at a point due to the application of a unit load at another point. For instance, influence coefficient η_{12} gives the displacement at point **1** due to the application of a unit load at point **2**. Clearly, $\eta_{12} = \Delta_1^{[2]}/P^{[2]}$ and η_{21}, the displacement at point **2** due to a unit load applied at point **1**, is $\eta_{21} = \Delta_2^{[1]}/P^{[1]}$. Eq. (10.118) becomes

$$\eta_{12} = \eta_{21}. \tag{10.119}$$

This result be interpreted as follows.

Theorem 10.6 (Maxwell's theorem). *For a linearly elastic structure, the influence coefficient of point **1** on point **2** equals that of point **2** on point **1**, for any choice of points **1** and **2**.*

Maxwell's theorem is a simple corollary of Betti's theorem, and applies to all the loading conditions for which Betti's theorem is valid. Hence, the concept of influence coefficient should be understood as the projected displacement or rotation at a point, relative displacement of two points, or relative rotation at two points due to any of the following four loading types applied at another location: a concentrated load or couple, or two opposite forces of identical magnitude sharing a common line, or two opposite couples of identical magnitude sharing a common line of action.

Example 10.23. Simply supported beam

Consider a simply supported, uniform beam of length L as shown in fig. 10.54. Let points **1** and **2** be located at distances $\alpha_1 L$ and $\alpha_2 L$ from the left end support.

The influence coefficient η_{12} can be evaluated from the exact solution of the problem given in example 5.5. In this case, η_{12} is the displacement at point **1** when the beam is subjected to a unit load at point **2**. Equation (5.51) gives this displacement as

$$\eta_{12} = \frac{L^3}{6H_{33}^c} \left[-(1 - \alpha_2)\alpha_1^3 + \alpha_2(2 - \alpha_2)(1 - \alpha_2)\alpha_1 \right]$$

$$= \frac{L^3}{6H_{33}^c} \left[-\alpha_1^3 + \alpha_1^3\alpha_2 + 2\alpha_1\alpha_2 - 3\alpha_1\alpha_2^2 + \alpha_1\alpha_2^3 \right].$$

The following values are used in eq. (5.51): $P = 1$ because a unit load is applied, $\alpha = \alpha_2$ because this load is applied at $\eta = \alpha_2$, and $\eta = \alpha_1$ to find the displacement $\bar{u}_2(\alpha_1)$.

The influence coefficient η_{21}, corresponding to the displacement at point **2** when the beam is subjected to a unit load at point **1** is evaluated in a similar manner, to find

$$\eta_{21} = \frac{L^3}{6H_{33}^c} \left[\alpha_1(\alpha_2^3 - 3\alpha_2^2) + \alpha_1(2 + \alpha_1^2)\alpha_2 - \alpha_1^3 \right]$$

$$= \frac{L^3}{6H_{33}^c} \left[\alpha_1\alpha_2^3 - 3\alpha_1\alpha_2^2 + 2\alpha_1\alpha_2 + \alpha_1^3\alpha_2 - \alpha_1^3 \right].$$

The following values are used in eq. (5.51): $P = 1$ because a unit load is applied, $\alpha = \alpha_1$ because this load is applied at $\eta = \alpha_1$, and $\eta = \alpha_2$ to find the displacement $\bar{u}_2(\alpha_2)$. As expected, $\eta_{12} = \eta_{21}$, in accordance with Maxwell's theorem.

Example 10.24. Symmetry of the flexibility matrix

The concept of flexibility matrix is introduced in example 5.9 on page 203. Since the flexibility matrix simply stores the influence coefficients, see eq. (5.58), Maxwell's theorem implies the *symmetry of the flexibility matrix*. In example 5.10 on page 204, the flexibility matrix of a cantilevered beam is determined analytically, see eq. (5.60), and is found to be symmetric, as expected.

Of course, if the flexibility matrix is determined experimentally according to the procedure described in example 5.9, the symmetry condition will only be satisfied within the bounds of experimental errors. Consider the case of the cantilevered beam depicted in fig. 5.28 on page 203. Table 10.2 lists the displacements measured on a cantilevered beam subjected to three loading cases. The first column of

Table 10.2. Measured displacements for the three loading cases. Load are measured in kN, displacements in mm.

	$P_1 = 1.5$	$P_2 = 1.0$	$P_3 = 0.5$
Δ_1	10.9	18.3	14.6
Δ_2	27.7	59.1	51.1
Δ_3	43.1	104.	98.5

this table lists the displacements at locations $\alpha_1 L$, $\alpha_2 L$, and $\alpha_3 L$ for $P_1 = 1.5$ kN, $P_2 = P_3 = 0$. The next two columns list the corresponding data for $P_2 = 1.0$ kN, $P_1 = P_3 = 0$ and $P_3 = 0.5$ kN, $P_1 = P_2 = 0$, respectively. The influence coefficients are easily obtained from the experimental data: for instance, $\eta_{11} = \Delta_{11}/P_1 = 10.9 \ 10^{-3}/1.5 \ 10^3 = 7.27 \ 10^{-6}$ m/N. Proceeding similarly with all other influence coefficients, the flexibility matrix is found as

$$\underline{\underline{F}} = \begin{bmatrix} 0.0073 & 0.0183 & 0.0292 \\ 0.0185 & 0.0591 & 0.1022 \\ 0.0287 & 0.1040 & 0.1970 \end{bmatrix} 10^{-3} \text{ m/N},$$

which is not exactly symmetric. From the measurements, the influence coefficient, η_{23}, can be estimated by averaging the corresponding entries of the measured flexibility matrix as $\bar{\eta}_{23} = \bar{\eta}_{32} \approx (0.1022 \ 10^{-3} + 0.1040 \ 10^{-3})/2 = 0.1031 \ 10^{-3}$ m/N. The estimated displacements now becomes $\bar{\Delta}_{23} = \bar{\eta}_{23} P_3 = 51.6$ mm and $\bar{\Delta}_{32} = \bar{\eta}_{23} P_2 = 103.1$ mm. The relative experimental error is now $e = |\bar{\Delta}_{23} - \Delta_{23}|/\bar{\Delta}_{23} \approx 0.9\%$ or $e = |\bar{\Delta}_{32} - \Delta_{32}|/\bar{\Delta}_{32} \approx 0.9\%$. An error estimation using the other off-diagonal terms of the flexibility matrix reveals a relative error of approximatively the same magnitude. Based on Maxwell's theorem, the experimental error is of the order of one percent.

10.10.3 Problems

Problem 10.26. Direct proof of Maxwell's theorem
Prove Maxwell's theorem directly from the unit load method. Use the simply supported beam depicted in fig. 10.54to support your reasoning.

Problem 10.27. Direct proof of Maxwell's theorem
Prove Maxwell's theorem directly from the dummy load method. Use the simply supported beam depicted in fig. 10.54to support your reasoning.

11

Variational and approximate solutions

11.1 Approach

In chapter 1, the fundamental equations of linear elasticity are developed, and fifteen equations fully describe the mechanics of deformable bodies. Unfortunately, these governing equations are partial differential equations in three dimensions, and although of first order only, their solution cannot be completed in closed form for most practical problems. Open form or series solutions have been developed for a limited number of applications, but no general approach exists for solving these equations in closed form. A barrier to the development of closed-form solutions to partial differential equations is the fact the arbitrary integration constants involved in the solution of ordinary differential equations are now replaced by arbitrary integration functions. Consequently, boundary conditions often play a greater role in the solution process for partial differential equations.

A very successful approach for dealing with complex problems is to reduce their geometric dimensionality from three to one, thereby replacing the governing partial differential equations by ordinary differential equations for which general solution procedures are available. A important example of this dimensional reduction procedure is beam theory, which leads to the ordinary differential equations presented in chapters 4 through 8. The reduction is based on the assumption that long, slender beams have one dimension, their span, which is much larger than the cross-sectional dimensions. Another important example of dimensional reduction is plate theory, presented in chapter 16, which transforms the three-dimensional elasticity equations into two-dimensional partial differential equations. In this case, the basic assumption is that one of the plate's geometric dimensions, its thickness, is much smaller than the other two.

The geometric dimension of a problem refers to the number of variables used to represent the displacement field: beam problems are *one-dimensional* because the displacement field is expressed in terms of a single variable along the beam's span. On the other hand, beam problems are sometimes referred to as *infinite-dimensional* because the solution of the problem requires the knowledge of the displacement field at all points, *i.e.* at an *infinite number points*, along the beam's span. To minimize confusion, this latter situation will also be referred to as a problem with an *infinite number of degrees of freedom* for the displacement field.

Even within the framework of beam theory, closed-form solutions cannot readily be developed for many practical problems. For instance, it is arduous to find closed-form solutions for beams presenting sectional properties with arbitrary variations along their span, a situation commonly encountered for many practical aircraft structures. Consequently, considerable effort has been devoted to the development of approximate solution procedures.

In most approximate solution procedures, infinite degree of freedom problems are reduced to finite degree of freedom problems. Three main approaches are used to achieve this type of dimensional reduction. In the first approach, the solution is sought at a finite number of discrete points of the structure; this approach is essentially a *discretization procedure,* because it transforms the original problem, expressed in terms of continuous, infinite degree of freedom functions, into a discrete problem involving the values of these functions at a finite number of points. The derivatives appearing in the governing equations are then approximated using finite difference techniques. The original equations are transformed into a set of algebraic equations that is easily solved.

In the second approach, the solution of the problem is approximated by a finite sum of continuous functions, each weighted by an unknown coefficient. The solution of the problem then reduces to the determination of the unknown coefficients. It will be shown in the present chapter that the combination of this approximation technique with the energy methods developed in the previous two chapters yields powerful tools for the systematic derivation of approximate solutions.

Finally, the last approach, called the *finite element method,* combines aspects of the previous two. In this widely used approach, the solution domain is first divided into a finite number of sub-domains called *finite elements*. Within each element, the solution is then approximated by a finite number of continuous functions, based on the value of these functions at discrete points, often called *nodes*, associated with the element. The main advantage of this two-step approximation process is that many aspects of the solution procedure can be carried out at the element level, *i.e.*, by considering one single element at a time, independently of all others. The continuity of the solution across elements can be guaranteed by the fact that neighboring elements share common nodes. Here again, energy methods provide a systematic way of obtaining algebraic equations for the unknown values of the solution at the nodes. For complex problems, very large sets of linear algebraic are obtained. To a large extent, the success of the finite element method is due to the fact that computers can easily solve these large sets of equations.

This chapter focuses on the development of approximate solutions for the types of problems that are formulated in earlier chapters. The treatment begins with a re-examination of the energy methods introduced in chapter 10 with emphasis on the principle of minimum total potential energy. Specifically, several methods of deriving approximate solutions from this principle are investigated. Fundamental concepts of the calculus of variations [5, 6] are useful to streamline these formulations, which are also called variational formulations. While a general formulation of the finite element method is beyond the scope of this book, the rudiments of the approach will be presented for beam structures as an extension of the element-oriented approach for trusses developed in section 10.7.

Finite element analysis is now a well established method, which has been the subject of intense study over the past several decades, largely because of its many successful applications. Although approximate, finite element procedures often yield very accurate and reliable solutions. Powerful commercial software tools are now widely available, and complex analysis can readily be handled on personal computers. In fact, contemporary structural analysis heavily relies on finite element analysis.

11.2 Rayleigh-Ritz method for beam bending

Approximate solutions of structural problems can be obtained through various approaches. The Rayleigh-Ritz approach is one of the simplest and will be introduced first. In this approach, the displacement field is represented by a linear combination of preselected displacement shapes, or shape functions, defined over the entire structure. This transforms the original, infinite dimensional problem into a finite dimensional problem. The principle of minimum total potential energy is then used to obtain an approximate solution of the problem.

11.2.1 Statement of the problem

Consider the simply supported beam of length L subjected to a distributed transverse loading, $p_2(x_1)$, and concentrated transverse loads, P_a and P_b, applied at locations $x_1 = a$ and $x_1 = b$, respectively, along the beam's span, as shown in fig. 11.1. The Euler-Bernoulli formulation developed in chapter 5 is used to model transverse bending deformations of the beam. Based on eq. (10.40), the strain energy stored in the beam is expressed in terms of the transverse displacement field, $\bar{u}_2(x_1)$, as

$$A = \frac{1}{2} \int_0^L H_{33}^c \left(\frac{d^2 \bar{u}_2}{dx_1^2} \right)^2 dx_1. \tag{11.1}$$

The potential of the externally applied loads is obtained from eq. (10.59), which in this case, reduces to

$$\Phi = - \int_0^L p_2 \, \bar{u}_2(x_1) \, dx_1 - P_a \, \bar{u}_2|_{x_1=a} - P_b \, \bar{u}_2|_{x_1=b} \, .$$

The total potential energy of the system, $\Pi = A + \Phi$, becomes

$$\Pi = \frac{1}{2} \int_0^L H_{33}^c \left(\frac{d^2 \bar{u}_2}{dx_1^2} \right)^2 dx_1 - \int_0^L p_2 \, \bar{u}_2(x_1) \, dx_1 - P_a \bar{u}_2(a) - P_b \bar{u}_2(b), \tag{11.2}$$

where $\bar{u}_2(x_1)$ is the unknown displacement field of the beam.

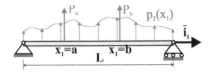

Fig. 11.1. Simply-supported beam subjected to distributed lateral load and.

The total potential energy is now a function of a function, $\bar{u}_2(x_1)$, *i.e.*, it is a *functional*. As mentioned earlier, this is an infinite degree of freedom problem, or a problem of infinite dimensionality because the definition of the transverse deflection field, $\bar{u}_2(x_1)$, requires the knowledge of this function for all points $0 \leq x_1 \leq L$, *i.e.*, for an infinite number of points. Rather than attempting to determine the exact solution of the problem that minimizes the total potential energy, as will be done in chapter 12, the present chapter seeks to construct approximate solutions of the problem.

11.2.2 Description of the Rayleigh-Ritz method

The main steps of what is known as the *Rayleigh-Ritz method* will be described here in a cursory manner. The first step of the solution procedure is to reduce the infinite degree of freedom problem into a finite degree of freedom problem. This can be done here by writing the solution for the transverse displacement field as a linear combination of N suitably chosen shape functions

$$\bar{u}_2(x_1) = \sum_{i=1}^{N} h_i(x_1)q_i. \tag{11.3}$$

In this expression, the functions $h_i(x_1)$, $i = 1, 2, \ldots, N$, are **known** functions called *shape functions*. Because the solution of the problem must satisfy the geometric boundary conditions of the problem, it is convenient to impose this condition on each of the shape functions which are otherwise arbitrarily chosen functions. The **unknown** coefficients, q_i, $i = 1, 2, \ldots, N$, are called *degrees of freedom*. Expression (11.3) now only involves N unknown coefficients: it is a finite degree of freedom approximation to the exact, infinite degree of freedom solution of the problem.

The second step of the solution process is to introduce the approximate solution, eq. (11.3), into the expression for the total potential energy, eq. (11.2), and perform all indicated integrations over the span of the beam. This is now possible because the solution is expressed in terms of shape function of known analytical form. Once these integrations are performed, the total potential energy becomes a function of the degrees of freedom, q_i, $i = 1, 2, \ldots, N$, *i.e.*, $\Pi(\bar{u}_2(x_1)) = \Pi(q_1, q_2, \ldots, q_N)$. Because the expression for the total potential energy is a quadratic function of the displacement field, it now becomes a quadratic function of the degrees of freedom, q_i, $i = 1, 2, \ldots, N$.

The last step of the process is to invoke the principle of minimum total potential energy, requiring the total potential energy to be minimum. The total potential energy is now a function of N independent, unconstrained variables, and therefore calculus requires its derivatives to vanish, see eq. (10.17), leading to

$$\frac{\partial \Pi}{\partial q_i} = 0, \quad i = 1, 2, 3, \ldots, N. \tag{11.4}$$

Because the total potential energy is a quadratic expression of the degrees of freedom, its first derivatives are linear functions of the same variables, and hence, the above equations form a set of linear equations that can be solved for the values of the degrees of freedom that minimize the total potential energy.

11.2.3 Discussion of the Rayleigh-Ritz method

The solution obtained from eqs. (11.4) is not an exact solution. Indeed, by selecting the solution to be of the form given by eq. (11.3), the ability of the structure to deform is restricted; it can only deform in a finite number of allowable deformation shapes, $h_i(x_1)$, $i = 1, 2, \ldots, N$. In effect, the structure is made artificially stiffer than the real structure by limiting its deformation to be the linear combination of a finite number of arbitrarily preselected deformation mode shapes. Of course, the real structure is able to deform in an infinite number of deformation shapes.

In the above procedure, the shape functions are required only to satisfy the geometric boundary conditions and are otherwise arbitrary. If a different set of shape functions is selected in eq. (11.3), a different solution will be found. These remarks prompt the following question: how good is the approximate solution obtained from this process?

Let Π be the total potential energy for the exact solution of the problem. Next, let $\widetilde{\Pi}$ be the total potential energy corresponding to an approximate solution, i.e., $\widetilde{\Pi} = \Pi(q_1, q_2, \ldots, q_N)$, where q_i, $i = 1, 2, \ldots, N$ are the solution of the linear system defined by eqs. (11.4). Because Π is the minimum of the exact, infinite dimensional problem, whereas $\widetilde{\Pi}$ is the minimum of the approximate, finite dimensional problem, it is clear that $\Pi \leq \widetilde{\Pi}$. As the number of degrees of freedom of the approximate solution increases, better and better solutions should be obtained and $\widetilde{\Pi} \to \Pi$, but $\widetilde{\Pi}$ always remains larger or equal to Π.

Unfortunately there is no way, short of knowing the exact solution, of ascertaining how close $\widetilde{\Pi}$ is to Π, and hence, how good the approximate solution will be. Furthermore, from a structural designer's viewpoint, a good approximation is probably a safe approximation, i.e., a conservative solution, which over-estimates deflections and stresses. Unfortunately, the procedure described above guarantees only that $\Pi \leq \widetilde{\Pi}$, but little can be said about other characteristics of the solution.

In practice, this shortcoming is overcome by performing a convergence study: a series of solutions is generated that involves an increasing number of degrees of freedom. As N increases, $\widetilde{\Pi} \to \Pi$ and the solution converges to the exact solution. Typically, the displacement and stress fields are monitored, and when further increase in N has little effect on the solution, it is said to be converged. These statements can be made more precise using advanced mathematical concepts that are beyond the scope of this book. More details can be found in references [8, 9].

From a practical viewpoint, it is desirable to select shape functions that closely approximate the actual displacement field. This is, of course, not easily done, because the solution is, in general, unknown. From a mathematical viewpoint, the shape functions should form a *complete set of functions*. The precise mathematical definition of this concept is beyond the scope of this book, but loosely speaking, it implies that as $N \to \infty$ a series of complete functions must be able to exactly reconstruct an arbitrary continuous function. For instance, it is well known from Fourier expansion theory that an arbitrary continuous function can be represented by an infinite series of trigonometric functions; this implies that trigonometric functions form a complete set. Selecting shape function that present orthogonality properties will also simplify the solution process.

The discussion presented in the previous paragraphs underlines the importance of a judicious choice of the shape function. In the procedure described here, the principle of minimum total potential energy is used with kinematically admissible virtual displacements. This means that all virtual displacements must satisfy the constraints of the problem, and in particular, the geometric boundary conditions. Each shape function is, in fact, a virtual displacement field. In eq. (11.3), consider the case where all degrees of freedom vanish except q_1. Shape function $h_1(x_1)$ then becomes the only virtual displacement field and, hence, must satisfy the geometric boundary conditions. By induction, it is easy to conclude that all shape functions must then individually satisfy these conditions. It is not required, however, that the shape functions satisfy the natural boundary conditions.

The following examples will focus on approximate solutions based on polynomial and trigonometric expansions. Polynomials are, in fact, solutions of certain beam problems. Trigonometric series are often convenient to use because they enable the term-by-term satisfaction of many types of geometric boundary, and in addition, the orthogonality properties of these series simplify the calculations of the degrees of freedom.

Example 11.1. *Polynomial solution for a cantilever beam with uniform load*
Consider a cantilever beam of length L and bending stiffness H_{33}^c, subjected to a uniform transverse loading distribution, p_0, as shown in fig. 11.2. This problem is treated using the classical differential equation approach in example 5.7 on page 200.

For this cantilevered beam, the geometric boundary conditions require both deflection and slope to vanish at the root of the beam: $\bar{u}_2(0) = \mathrm{d}\bar{u}_2(0)/\mathrm{d}x_1 = 0$. A monomial approximation will be selected,

$$\bar{u}_2(x_1) = q_2 x_1^2,$$

where the first two terms of the series, q_0 and $q_1 x_1$, cannot be used because the corresponding shape functions, 1 and x_1, do not satisfy the geometric boundary conditions.

Using this approximation, the total potential energy of the system, eq. (11.2), becomes

$$\Pi = \frac{1}{2} \int_0^L H_{33}^c (2q_2)^2 \, \mathrm{d}x_1 - \int_0^L p_0 q_2 x_1^2 \, \mathrm{d}x_1 = \frac{1}{2} H_{33}^c (2q_2)^2 L - p_0 q_2 \frac{L^3}{3}.$$

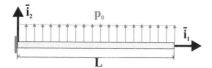

Fig. 11.2. Cantilevered beam subjected to a uniform lateral load.

The principle of minimum total potential energy now requires $\Pi(q_2)$ to be a minimum for the system to be in equilibrium. The necessary condition is: $\partial\Pi/\partial q_2 = 4H_{33}^c q_2 L - p_0 L^3/3 = 0$, and solving for q_2 yields $q_2 = (p_0 L^2)/(12 H_{33}^c)$. The resulting solution for the transverse displacement field is then

$$\bar{u}_2(x_1) = \frac{p_0 L^4}{12 H_{33}^c} \eta^2,$$

where $\eta = x_1/L$ is the non-dimensional variable along the beam's span.

A more accurate approximate solution can be developed by adding an additional term to the approximation. To simplify the computation, it will be convenient to write the approximation as

$$\bar{u}_2(\eta) = q_2 \eta^2 + q_3 \eta^3,$$

where the shape functions are written in terms of the non-dimensional variable η. Performing the change of variable $x_1 = \eta L$, the strain energy in the beam, given by eq. (11.1), becomes

$$A = \frac{1}{2} \int_0^1 \frac{H_{33}^c}{L^3} (\bar{u}_2'')^2 \, d\eta, \tag{11.5}$$

where the notation $(\cdot)'$ is used to indicate a derivative with respect to η. Introducing the assumed displacement field then yields the total potential energy as

$$\Pi = \frac{1}{2} \frac{H_{33}^c}{L^3} \int_0^1 (4q_2^2 + 36\eta^2 q_3^2 + 24\eta q_2 q_3) \, d\eta - p_0 L \int_0^1 (q_2 \eta^2 + q_3 \eta^3) \, d\eta.$$

After integration, this expression becomes

$$\Pi = \frac{1}{2} \frac{H_{33}^c}{L^3} \left(4q_2^2 + \frac{36}{3} q_3^2 + \frac{24}{2} q_2 q_3 \right) - p_0 L \left(\frac{q_2}{3} + \frac{q_3}{4} \right).$$

The total potential energy is now a function of the two degrees of freedom, q_2 and q_3, i.e., $\Pi = \Pi(q_2, q_3)$. Minimization of the total potential then requires $\partial\Pi/\partial q_2 = 0$ and $\partial\Pi/\partial q_3 = 0$, yielding $(H_{33}^c/L^3)[4q_2 + 6q_3] - p_0 L/3 = 0$ and $(H_{33}^c/L^3)[12q_3 + 6q_2] - p_0 L/4 = 0$, respectively. These two algebraic equations form a set of linear equations; this becomes more obvious when the two equations are recast in matrix form as

$$\begin{bmatrix} 4 & 6 \\ 6 & 12 \end{bmatrix} \begin{Bmatrix} q_2 \\ q_3 \end{Bmatrix} = \frac{p_0 l^4}{H_{33}^c} \begin{Bmatrix} 1/3 \\ 1/4 \end{Bmatrix}.$$

Solving this system of linear equations yields the two unknown coefficients, q_2 and q_3, and hence, the following approximate solution for the non-dimensional transverse displacement field is obtained

$$\frac{H_{33}^c \bar{u}_2(\eta)}{p_0 L^4} = \frac{1}{24}(5\eta^2 - 2\eta^3).$$

A still more accurate approximate solution can be developed by including another monomial in the approximation,

$$\bar{u}_2(\eta) = q_2\eta^2 + q_3\eta^3 + q_4\eta^4.$$

Note once again that each monomial individually satisfies the geometric boundary conditions of the problem. Introducing this assumed displacement field in the expression for the total potential energy leads to

$$\Pi = \frac{1}{2}\frac{H_{33}^c}{L^3}\int_0^1 (4q_2 + 6\eta q_3 + 12\eta^2 q_4)^2\, d\eta - p_0 L\int_0^1 (q_2\eta^2 + q_3\eta^3 + q_4\eta^4)\, d\eta.$$

The first task is to expand the square under the first integral, and the second to evaluate all integrals. The total potential energy then becomes a quadratic expression of the degrees of freedom, i.e., $\Pi = \Pi(q_2, q_3, q_4)$. Minimization of the total potential requires conditions (11.4), leading to a set of three simultaneous linear equations of the three degrees of freedom of the problem, q_2, q_3 and q_4. Here again, it is convenient to recast the resulting equations in a matrix form as

$$\begin{bmatrix} 4 & 6 & 8 \\ 6 & 12 & 18 \\ 8 & 18 & 144/5 \end{bmatrix} \begin{Bmatrix} q_2 \\ q_3 \\ q_4 \end{Bmatrix} = \frac{p_0 L^4}{H_{33}^c} \begin{Bmatrix} 1/3 \\ 1/4 \\ 1/5 \end{Bmatrix}.$$

Solving this system of linear equations yields the following approximate solution for the transverse displacement field

$$\bar{u}_2(\eta) = \frac{1}{24}\frac{p_0 L^4}{H_{33}^c}(6\eta^2 - 4\eta^3 + \eta^4).$$

Comparing this expression with that obtained in example 5.7 on page 200 using the classical differential equation approach, it appears that the exact solution of the problem has now been obtained. The principle of minimum total potential energy does not actually indicate that an exact solution has been obtained; it is only possible to ascertain this fact here because the exact solution for this problem was previously obtained by another method. If additional terms are taken in the series solution, it will be found that their coefficients are all zero. For instance, assuming a solution such as $\bar{u}_2(\eta) = q_2\eta^2 + q_3\eta^3 + q_4\eta^4 + q_5\eta^5$ will yield identical results to those found above but with $q_5 = 0$. The fact that an exact solution is found here is fortuitous; the exact solution happens to be of a polynomial form, and that same form is used for the assumed solution.

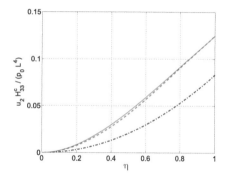

Fig. 11.3. Non-dimensional transverse displacement field for the cantilevered beam. Solid line: three-term polynomial solution (exact solution); dashed line: two-term approximation; dashed-dotted line: single-term approximation.

It is interesting to compare the solutions obtained from the three approximations investigated here. Figure 11.3 shows the non-dimensional transverse displacement fields, $\bar{u}_2 H_{33}^c/(p_0 L^4)$, obtained with the three approximations. As expected, the single term approximation is very inaccurate; the stiffness of the beam is overestimated because it is limited to deform into the parabolic shape implied by the shape function x_1^2. In fact, the tip deflection is 33% smaller than that of the exact solution. The two term approximation is noticeably better; at the tip of the beam, the transverse deflection is exact, although at all other points, the solution is not exact. The fact that the exact solution is obtained at the tip of the beam is purely fortuitous.

Although the two-term approximation produces a transverse displacement field that is in close agreement with the exact solution, as shown in fig. 11.3, the associated predictions for the internal bending moment and shear force distributions are not nearly as good. Figure 11.4 shows the distributions of non-dimensional bending moments, $M_3/(p_0 L^2)$, obtained with the three approaches. Large errors are observed over the entire span of the beam for the two approximate solutions. Furthermore, the natural boundary condition at the tip of the beam, $M_3 = 0$, is not satisfied by the approximate solutions. The same observations can be made concerning the distribution of non-dimensional shear force, $T_2/(p_0 L)$, depicted in fig. 11.5.

Example 11.2. *Simply supported beam under uniform loading*

A simply supported beam of length L and bending stiffness H_{33}^c is subjected to a uniformly distributed load p_0, as depicted in fig. 11.6. This problem is treated using the classical differential equation approach in example 5.7 on page 200, which gives the exact solution for the transverse displacement field in a simple polynomial form.

The geometric boundary conditions for this problem are $\bar{u}_2(0) = \bar{u}_2(L) = 0$, and hence, the shape functions to be selected must satisfy the conditions $h_i(0) = h_i(L) = 0$. The monomial shape functions selected in the previous example, $h_i(x_1) = x_1^i$, are not suitable for this problem because while $h_i(0) = 0^i = 0$, it is clear that $h_i(L) = L^i \neq 0$. A convenient way to satisfy this requirement is to select as the shape functions a set of periodic functions such as sine functions with

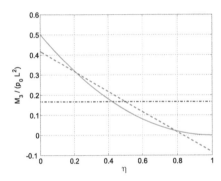

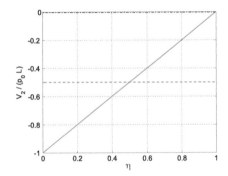

Fig. 11.4. Non-dimensional bending moment field for the cantilevered beam. Solid line: three-term polynomial solution (exact solution); dashed line: two-term approximation; dash-dotted line: single-term approximation.

Fig. 11.5. Non-dimensional shear force field for the cantilevered beam. Solid line: three-term polynomial solution (exact solution); dashed line: two-term approximation; dash-dotted line: single-term approximation.

increasing wave numbers

$$\bar{u}_2(x_1) = \sum_{n=1}^{N} q_n \sin \frac{n\pi x_1}{L} = \sum_{n=1}^{N} q_n \sin n\pi\eta, \tag{11.6}$$

where $\eta = x_1/L$ is the non-dimensional variable along the beam's span.

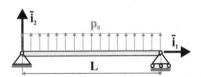

Fig. 11.6. Simply-supported beam subjected to a uniform lateral load.

With the help of eq. (11.5) and using the notation $(\cdot)'$ to indicate a derivative with respect to η, the strain energy in the beam becomes

$$
\begin{aligned}
A &= \frac{H_{33}^c}{2L^3} \int_0^1 (\bar{u}_2'')^2 \, d\eta = \frac{H_{33}^c}{2L^3} \int_0^1 \left[-\sum_{n=1}^{N} q_n (n\pi)^2 \sin n\pi\eta \right]^2 d\eta \\
&= \frac{H_{33}^c}{2L^3} \int_0^1 \left[-\sum_{m=1}^{N} q_m (m\pi)^2 \sin m\pi\eta \right] \left[-\sum_{n=1}^{N} q_n (n\pi)^2 \sin n\pi\eta \right] d\eta \\
&= \frac{H_{33}^c}{2L^3} \sum_{m=1}^{N} \sum_{n=1}^{N} q_m q_n (m\pi)^2 (n\pi)^2 \int_0^1 \sin m\pi\eta \sin n\pi\eta \, d\eta.
\end{aligned}
$$

At first glance, this expression appears to be very complicated because of the presence of the double summation. The sine functions, however, enjoy the orthogonality property stated in eq. (A.45a), and hence, the strain energy in the beam reduces to a single summation,

$$A = \frac{H_{33}^c}{4L^3} \sum_{n=1}^{N} (n\pi)^4 q_n^2.$$

The potential of the externally applied loads, $p_2(x_1) = p_0$, is the negative of the work done and can be written as

$$\Phi = -\int_0^L p_0 \,\bar{u}_2 \, \mathrm{d}x_1 = -p_0 L \int_0^1 \sum_{n=1}^{N} q_n \sin n\pi\eta \; \mathrm{d}\eta = -2p_0 L \sum_{n=\mathrm{odd}}^{N} \frac{q_n}{n\pi}.$$

Note that for even wave numbers, *i.e.*, for even values of n, the integrals of the sine function from 0 to 1 vanish. Hence, only the odd values of n remain in the expression for the potential.

The total potential energy, Π, now simply becomes a function of the degrees of freedom, q_i, $i = 1, 2, \ldots, N$, and the principle of minimum total potential energy requires that

$$\frac{\partial \Pi}{\partial q_i} = \frac{\partial A}{\partial q_i} + \frac{\partial \Phi}{\partial q_i} = \begin{cases} H_{33}^c/(2L^3)(i\pi)^4 q_i = 0, & i \text{ even}, \\ H_{33}^c/(2L^3)(i\pi)^4 q_i - (2p_0 L)/(i\pi) = 0, & i \text{ odd}. \end{cases}$$

This means that $q_i = 0$ for all even values of i, whereas $q_i = 4(p_0 L^4/H_{33}^c)/(i\pi)^5$ for all odd values of i. The transverse displacement field is then

$$\bar{u}_2(\eta) = \frac{4}{\pi^5} \frac{p_0 L^4}{H_{33}^c} \sum_{n=\mathrm{odd}}^{N} \frac{1}{n^5} \sin n\pi\eta.$$

Since $|\sin(n\pi\eta)| \leq 1$, the convergence of the series is very rapid, due to the presence of the factor n^5 in the denominator. It is interesting to note that the exact solution of the problem given by eq. 5.48 is a simple polynomial, whereas the present approximate solution is in the form of an infinite series.

Intuitively, the maximum transverse deflection is found at the beam's mid-span, *i.e.*, at $\eta = 0.5$, where the exact solution gives $H_{33}^c \bar{u}_2(0.5)/(p_0 L^4) = 5/384 = 0.01302$. Considering a single term in the above series yields $H_{33}^c \bar{u}_2(0.5)/(p_0 L^4) = 4/\pi^5 = 0.01307$, which is only 0.39% from the exact solution. As additional terms of the series are added, the solution rapidly converges to the exact answer, as shown in fig. 11.7, which plots the relative error associated with the approximate solution as N increases. Clearly, very accurate approximations are obtained with just a few terms of the series. It should be noted that the shape functions selected here each satisfy the natural boundary conditions of the problem, $M_3(0) = M_3(L) = 0$. While this is not a requirement for the solution process, it generally leads to closer agreement with the exact solution.

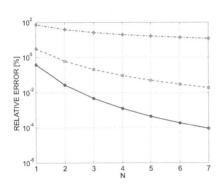

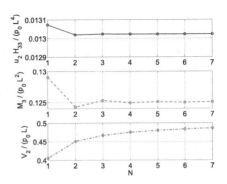

Fig. 11.7. Relative error of the approximate solution compared to the exact solution. Solid line: mid-span transverse displacement (○); dashed line: mid-span bending moment (□); dash-dotted line: root shear force (◇).

Fig. 11.8. Convergence of approximate solution as a function of the number of degrees of freedom. Solid line: mid-span transverse displacement (○); dashed line: mid-span bending moment (□); dash-dotted line: root shear force (◇).

Once the transverse displacement field is found, the moment distribution is readily obtained as

$$M_3(\eta) = \frac{H_{33}^c}{L^2} \bar{u}_2''(\eta) = -\frac{4}{\pi^3} p_0 L^2 \sum_{n=\text{odd}}^{N} \frac{1}{n^3} \sin n\pi\eta.$$

The maximum bending moment is found at the beam's mid-span, where the exact solution gives $M_3(0.5)/(p_0 L^2) = -1/8 = -0.125$. Considering a single term in the series yields $M_3(0.5)/(p_0 L^2) = -4/\pi^3 = -0.129$, which is 3.2% from the exact solution. As additional terms of the series are used, the error is reduced as shown in fig. 11.7, although the convergence is not nearly as rapid as that observed for the transverse deflection. Finally, the shear force distribution in the beam is obtained from eq. (5.38) as

$$V_2(\eta) = -\frac{1}{L} M_3'(\eta) = \frac{4}{\pi^2} p_0 L \sum_{n=\text{odd}}^{N} \frac{1}{n^2} \cos n\pi\eta.$$

The maximum shear force is found at the end supports; for instance, at $\eta = 0$, the exact shear force is $V_2/(p_0 L) = 0.5$, whereas the single term in the above series yields $V_2/(p_0 L) = 4/\pi^2 = 0.405$, which is 18.9% from the exact solution. Figure 11.7 shows the reduction in error of the root shear force predictions as an increasing number of terms is used in the series.

Clearly, the convergence rates of the bending moment and shear force predictions are far slower than those observed for the transverse displacement. This is a general feature of the approximate solutions obtained with the Rayleigh-Ritz method. This is easily understood by noting that the internal forces and moments are obtained by taking derivatives of the approximated displacement field. The bending moment is a second derivative of the displacement field, the shear force a third derivative. The accuracy of the predictions decreases as the order of the derivative increases. The results shown in fig. 11.7 clearly demonstrate this effect.

Figure 11.7 is referred to as a *convergence plot* because it demonstrates the convergence of the approximate solution to the exact solution of the problem. Such plot, however, assumes that the exact solution is known, because the relative error, the difference between the approximate and exact solutions, normalized by the exact solution, must be evaluated. In practical situations, the exact solution is not known, and hence, it is not possible to compute a relative error. In that case, approximate solutions are evaluated based on an increasing number of shape functions and are plotted against the number of degree of freedom. Figure 11.8 shows such a plot for the problem at hand. As the number of degrees of freedom increases, the solutions stabilize to a horizontal asymptote, which is presumed to be the exact solution.

If $\bar{u}_2^{(N)}$ is the approximate mid-span transverse displacement obtained with N degrees of freedom, the accuracy of the approximate solution can be assessed in an ad-hoc manner by considering the following convergence criterion $(\bar{u}_2^{(N)} - \bar{u}_2^{(N-1)})/\bar{u}_2^{(N)} < \epsilon$, which compares the solutions obtained with $N - 1$ and N degrees of freedom; if ϵ is a small number, the satisfaction of the criterion implies that increasing the number of degrees of freedom has little effect on the solution. It is important to understand that such a criterion provides a good indication that the approximate solution is close to the exact solution, but is by no means a proof.

Example 11.3. Discussion of the requirements for shape functions
The Rayleigh-Ritz method presented here relies on the principle of minimum total potential energy using kinematically admissible virtual displacements. Consequently, the shape functions must satisfy the geometric boundary conditions of the problem. If these conditions are not satisfied by the shape functions, erroneous solutions will result.

Consider once again the simply supported beam of length L and bending stiffness H_{33}^c subjected to a uniformly distributed load p_0, as depicted in fig. 11.6. In view of the exact solution of this problem given by eq. (5.48), it is tempting to explore a solution in the following polynomial form with $\eta = x_1/L$

$$\bar{u}_2(\eta) = q_1\eta + q_3\eta^3 + q_4\eta^4.$$

Introducing this assumed displacement field in the expression for the total potential energy leads to

$$\Pi = \frac{1}{2}\frac{H_{33}^c}{L^3}\int_0^1 36(q_3\eta + 2q_4\eta^2)^2 \, d\eta - p_0 L \int_0^1 (q_1\eta + q_3\eta^3 + q_4\eta^4) \, d\eta.$$

Performing all integrations and imposing the conditions (11.4) for the minimization of the total potential energy leads to the following set of linear equations

$$\begin{bmatrix} 0 & 0 & 0 \\ 0 & 12 & 18 \\ 0 & 18 & 144/5 \end{bmatrix} \begin{Bmatrix} q_1 \\ q_3 \\ q_4 \end{Bmatrix} = \frac{p_0 L^4}{H_{33}^c} \begin{Bmatrix} 1/2 \\ 1/4 \\ 1/5 \end{Bmatrix}.$$

Obviously, it is not possible to solve this linear system because the system matrix is singular: indeed, first row and column vanish. This arises because the first shape function, η, represents a rigid body motion for the beam corresponding to a rotation of the beam about the left support.

By definition, rigid body motions generate no strains, and hence, no strain energy. Indeed, the associated curvature of the beam is $\kappa_3 = \bar{u}_2''/L^2 = (\eta)''/L^2 = 0$, and hence, the first degree of freedom, q_1, does not appear in the expression for the strain energy, resulting in the vanishing of the first row and column of the system matrix. Of course, selecting this rigid body motion as a shape function is not correct, because this rigid body motion does not satisfy the geometric boundary condition $h(\eta = 1) = 0$.

Consider next a solution of the following form: $\bar{u}_2(\eta) = q_3\eta^3 + q_4\eta^4$, in which the rigid body mode is ignored. Proceeding as above will lead to a system of two linear equations for degrees of freedom q_3 and q_4. The corresponding system of equations is obtained by eliminating the first row and column of the above system, which is then readily solved for the unknowns and leads to the following approximate solution: $H_{33}^c\bar{u}_2(\eta)/(p_0L^4) = (12\eta^3 - 7\eta^4)/72$. This solution is obviously incorrect because it violates the geometric boundary condition at the beam's tip: $H_{33}^c\bar{u}_2(\eta = 1)/(p_0L^4) = 5/72 \neq 0$. Clearly, when deriving approximate solutions using the principle of minimum total potential energy, the solution process is unaware of the geometric boundary conditions applied to the problem unless they are specifically imposed on each of the shape functions.

The exact solution of the problem, see eq. (5.48), can be written as $H_{33}^c\bar{u}_2/(p_0L^4) = (\eta - 2\eta^3 + \eta^4)/24$. Note that the individual terms, η, η^3, and η^4, do not satisfy the geometric boundary conditions, while the complete solution does: $\bar{u}_2(0) = \bar{u}_2(1) = 0$. It is interesting to factor the exact solution as $H_{33}^c\bar{u}_2/(p_0L^4) = \eta(1-\eta)(1+\eta-\eta^2)/24$, which now shows the satisfaction of the geometric boundary conditions, $\bar{u}_2(0) = \bar{u}_2(1) = 0$.

This suggests that the following polynomial approximation is suitable for the application of the principle of minimum total potential energy to simply supported beam problems: $\bar{u}_2(\eta) = \eta(1-\eta)q_1 + \eta^2(1-\eta)q_2 + \eta^3(1-\eta)q_3 + \ldots$, where the shape functions are selected as

$$h_i = \eta(1-\eta)\,\eta^{i-1}, \quad i = 1, 2, \ldots, N. \tag{11.7}$$

Clearly, each shape function now individually satisfies the requirements $h_i(0) = h_i(1) = 0$ for all values of i. It will be left to the reader to verify that good solutions are obtained with the above polynomial shape functions, and for $N = 3$, the exact solution is recovered.

Consider next a beam clamped at both end. The associated geometric boundary conditions are $\bar{u}_2(0) = \bar{u}_2'(0) = 0$, and $\bar{u}_2(1) = \bar{u}_2'(1) = 0$. Note that the sine functions, $h_i(\eta) = \sin i\pi\eta$, are not admissible because although $h_i(0) = h_i(1) = 0$, the slope is not zero: $h_i'(0) \neq 0$ and $h_i'(1) \neq 0$. Using cosine functions, $h_i(\eta) = \cos i\pi\eta$, is not valid either, because $h_i(0) \neq 0$ and $h_i(1) \neq 0$, although $h_i'(0) = h_i'(1) = 0$. By analogy to the polynomial shape functions proposed in eq. (11.7) for simply supported beams, a suitable set of shape functions for a beam clamped at both ends is

$$h_i = \eta^2(1 - \eta)^2 \, \eta^{i-1}, \quad i = 1, 2, \ldots, N. \tag{11.8}$$

If trigonometric shape functions are desired, it is easy to verify that the following expressions satisfy the geometric boundary conditions

$$h_i = \cos(i - 1)\pi\eta - \cos(i + 1)\pi\eta, \quad i = 1, 2, \ldots, N. \tag{11.9}$$

It will be left to the reader to combine the shape function given in the previous equations to find suitable shape functions to analyze problems involving a combination of the boundary conditions discussed above, for example, a cantilevered beam with a tip support.

Example 11.4. Simply supported beam with concentrated load

Consider now a simply supported beam of length L subjected to a concentrated load P at a distance αL from its root, as shown in fig. 11.9. This problem is treated using the classical differential equation approach in example 5.5 on page 197, and then again in example 5.6 on page 199 by first evaluating the bending moment distribution. The concentrated load introduces a discontinuity in the shear force diagram which complicates the classical differential equation approach by requiring the solution to be separately computed over two regions of the beam.

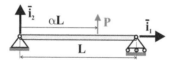

Fig. 11.9. Simply-supported beam subjected to a concentrated transverse load.

For this problem, the sine series,

$$\bar{u}_2(\eta) = \sum_{n=1}^{N} q_n \sin n\pi\eta,$$

used in example 11.2 is suitable because each sine function satisfies the geometric boundary conditions. The total potential energy of the system now becomes

$$\Pi = \frac{H_{33}^c}{2L^3} \int_0^1 \left[-\sum_{n=1}^{N} q_n (n\pi)^2 \sin n\pi\eta \right]^2 \mathrm{d}\eta - P \sum_{n=1}^{N} q_n \sin n\pi\alpha.$$

Rather than performing the integrals indicated in this expression, it is also possible to first write the conditions for minimization of the total potential energy, eqs. (11.4). This then leads to

$$\frac{\partial \Pi}{\partial q_i} = \frac{H_{33}^c}{2L^3} \int_0^1 2 \left[\sum_{n=1}^{N} q_n (n\pi)^2 \sin n\pi\eta \right] \left[(i\pi)^2 \sin i\pi\eta \right] \mathrm{d}\eta - P \sin i\pi\alpha = 0,$$

where the order of the partial derivative and integral operators have been interchanged according to Leibniz' integral rule. Rearranging the individual terms then yields

$$\frac{H_{33}^c}{L^3}(i\pi)^2 \left[\sum_{n=1}^{N} q_n (n\pi)^2 \int_0^1 \sin n\pi\eta \sin i\pi\eta \; d\eta \right] - P \sin i\pi\alpha = 0.$$

Again, because the sine functions enjoy the orthogonality properties expressed by eq. (A.45a), the integral reduces to $\delta_{ni}/2$, where δ_{ni} is the Kronecker delta defined in eq. (A.44). Because the Kronecker delta vanishes for all values of $n \neq i$, only a single term of the sum remains and the above expression reduces to

$$\frac{H_{33}^c}{2L^3}(i\pi)^4 q_i - P \sin i\pi\alpha = 0, \quad i = 1, 2, \dots N.$$

These equations are readily solved to find q_i, and the transverse deflection field is now

$$\bar{u}_2(x_1) = \frac{2}{\pi^4} \frac{PL^3}{H_{33}^c} \sum_{n=1}^{N} \frac{1}{n^4} \sin n\pi\alpha \sin n\pi\eta.$$

Figure 11.10 shows the transverse displacement field calculated using the above series for $N = 1, 3$ and 24, when $\alpha = 0.25$. Note the very rapid convergence of the results to the exact solution provided by eq. (5.51).

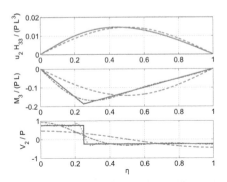

Fig. 11.10. Simply-supported beam subjected to a concentrated transverse load. Top figure: transverse displacement, $\bar{u}_2 H_{33}^c/(PL^3)$; middle figure: bending moment, $M_3/(PL)$; lower figure: shear force, V_2/P. Exact solution: solid line; single term solution: dashed line; three-term solution: dash-dotted line; 24-term solution: dotted line.

The bending moment and shear force distributions can be obtained from derivatives of the displacement field and are also shown in fig. 11.10. As expected, the convergence rates for the predictions of the internal bending moment and shear force distributions are slower than those observed for the displacement field.

The shear force distribution presents a discontinuity at the location of application of the concentrated force. The approximate solution is, of course, unable to capture this discontinuity, because the displacement field is assumed to be the superposition of continuous, trigonometric functions; the bending moment and shear force distributions, obtained by successive derivatives of the displacement field, are continuous functions as well. Even when $N = 24$, large errors are still observed near the discontinuity of the shear force. The slow convergence of Fourier series near a discontinuity is known as the Gibbs phenomenon.

Example 11.5. Cantilever beam with tip support

Consider a cantilevered beam of length L with a tip support, subjected to a uniformly distributed load, p_0, as shown in fig. 11.11. This problem is treated using the classical differential equation approach in example 5.11 on page 205. Unlike the previous examples, this problem is hyperstatic, but this does not affect the Rayleigh-Ritz approximation procedure.

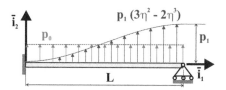

Fig. 11.11. Cantilever beam with tip support subjected to two different loading conditions.

In view of the discussion presented in example 11.3, an appropriate set of shape functions that satisfy the geometric boundary conditions for this problem is given by $h_i = \eta^2(1 - \eta)\eta^{i-1}$, $i = 1, 2, \ldots, N$. Two terms of this series will be used for this problem

$$\bar{u}_2(\eta) = \eta^2(1 - \eta)\, q_1 + \eta^2(1 - \eta)\eta\, q_2.$$

The total potential energy of the system is now evaluated using the approximate solution to find

$$\Pi = \frac{1}{2}\frac{H^c_{33}}{L^3} \int_0^1 \left[(2 - 6\eta)q_1 + (6\eta - 12\eta^2)q_2\right]^2 d\eta$$

$$-p_0 L \int_0^1 \left[(\eta^2 - \eta^3)q_1 + (\eta^3 - \eta^4)q_2\right] d\eta.$$

After expanding the square under the first integral, and evaluating all integrals, the total potential energy becomes a quadratic expression of the degrees of freedom.

Minimization of the total potential leads to the conditions given in eq. (11.4), and the following set of linear equations results

$$\begin{bmatrix} 4 & 4 \\ 4 & 24/5 \end{bmatrix} \begin{Bmatrix} q_1 \\ q_2 \end{Bmatrix} = \frac{p_0 L^4}{H^c_{33}} \begin{Bmatrix} 1/12 \\ 1/20 \end{Bmatrix}.$$

Solving this system of linear equations yields the following approximate solution for the transverse displacement field

$$\bar{u}_2(\eta) = \frac{1}{48}\frac{p_0 L^4}{H_{33}^c}\eta^2(1-\eta)(3-2\eta).$$

In fact, this is the exact solution of the problem, as can be ascertained by comparing this result to the exact solution given in eq. (5.61).

It is often the case that a particular structural problem must be solved for a variety of loading conditions. For instance, the stresses in an aircraft wing must be computed for level flight loading, but also for a variety of maneuver cases. The procedure developed here provides an elegant solution to this problem because if only the loading is changed, only the external potential energy must be recalculated. To illustrate, let the same beam be subjected to a new transverse distributed load, $p_2(\eta) = p_1(3\eta^2 - 2\eta^3)$, depicted in fig. 11.11. The total potential energy of the system is now evaluated as

$$\Pi = \frac{1}{2}\frac{H_{33}^c}{L^3}\int_0^1 \left[(2-6\eta)q_1 + (6\eta - 12\eta^2)q_2\right]^2 d\eta$$
$$-p_0 L \int_0^1 (3\eta^2 - 2\eta^3)\left[(\eta^2 - \eta^3)q_1 + (\eta^3 - \eta^4)q_2\right] d\eta.$$

Note that the expression for the strain energy in the structure is unchanged, and the only difference is in the potential of the externally applied load. Proceeding as before, the following set of linear equations for the degrees of freedom, q_1 and q_2, is found

$$\begin{bmatrix} 4 & 4 \\ 4 & 24/5 \end{bmatrix}\begin{Bmatrix} q_1 \\ q_2 \end{Bmatrix} = \frac{p_0 L^4}{H_{33}^c}\begin{Bmatrix} 11/210 \\ 1/28 \end{Bmatrix}.$$

Comparing this system of equations to that obtained above, it is clear that changing the externally applied load only modifies the right-hand side of the equations, which, for obvious reasons, is often called the *load array*. Within the framework of this approach, different loading cases are associated with different load arrays. The left-hand side system matrix needs to be inverted only once, and multiplication of the inverse by the various different load arrays then yields the desired approximate solutions for each of the loading cases. For the loading case at hand, the approximate solution is

$$\bar{u}_2(\eta) = \frac{1}{1680}\frac{p_1 L^4}{H_{33}^c}\eta^2(1-\eta)(57-35\eta).$$

The exact solution of the problem can be obtained using the classical differential equation approach as $\bar{u}_2 = p_1 L^4/H_{33}^c\, \eta^2(1-\eta)(2\eta^4 - 5\eta^3 - 5\eta^2 - 5\eta + 24)/840$. It is left to the reader to compare the exact and approximate solutions of this problem. If necessary, the approximation could be improved by increasing the number of degrees of freedom.

Example 11.6. *Simply supported beam with two elastic springs*
A simply supported beam of span L is supported by two springs of stiffness constant k located at stations equidistant from the two ends, and it is subjected to a uniform transverse loading, p_0, as depicted in fig. 11.12. To simplify the formulation of the shape functions, it will be convenient to locate the origin of the axes at the beam's

mid-span. The springs are located at stations $x_1 = \pm(1 - 2\alpha)L/2 = \pm\beta L/2$, where $\beta = 1 - 2\alpha$ is the non-dimensional location of the spring. The two springs model intermediate supports for the beam that are not infinitely rigid, but rather, present a flexibility that is modeled by the spring stiffness constant k; as k approaches infinity, these intermediate supports become rigid supports.

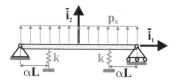

Fig. 11.12. Simply-supported beam with two elastic springs.

This hyperstatic problem is treated using the classical differential equation approach in example 5.13 on page 208. When using the Rayleigh-Ritz method, springs are treated as elastic components much as the beam itself. The strain energy in the structure is now the sum of the strain energy due to bending of the beam, A_b, and that due to deformation of the springs, A_s. The total strain energy of the structure becomes

$$A = A_b + A_s = \frac{1}{2}\int_0^L H_{33}^c \left(\frac{d^2\bar{u}_2}{dx_1^2}\right)^2 dx_1 + \frac{1}{2}k\bar{u}_2^2\bigg|_{-\beta L/2} + \frac{1}{2}k\bar{u}_2^2\bigg|_{\beta L/2},$$

where the last two terms represent the strain energy in the springs computed using eq. (10.21).

For this problem, an approximate solution of the following form will be used

$$\bar{u}_2(\eta) = (1 - \eta^2)q_1 + (1 - \eta^2)\eta^2 q_2 + (1 - \eta^2)\eta^4 q_3,$$

where $\eta = 2x_1/L$ is the non-dimensional variable along the beam's span. In view of the symmetry of the problem, only the even powers of η are included. The strain energy of the structure now becomes

$$A = \frac{1}{2}\frac{8H_{33}^c}{L^3}\int_{-1}^{+1} \left[-2q_1 + (2 - 12\eta^2)q_2 + (12\eta^2 - 30\eta^4)q_3\right]^2 d\eta$$

$$+ \frac{1}{2}k(1 - \beta^2)^2 \left(q_1 + \beta^2 q_2 + \beta^4 q_3\right)^2 + \frac{1}{2}k(1 - \beta^2)^2 \left(q_1 + \beta^2 q_2 + \beta^4 q_3\right)^2.$$

Note that the strain energies for the left and right springs are identical because of the symmetry of the problem. The potential energy of the externally applied load can be evaluated as is done in the previous examples.

After expanding the square under the first integral and evaluating all integrals, the total potential energy becomes a quadratic expression of the three degrees of freedom, i.e., $\Pi = \Pi(q_1, q_2, q_3)$. Minimization of the total potential energy leads to eqs. (11.4), and finally, to the following set of linear equations

$$
\left[16 \begin{bmatrix} 4 & 4 & 4 \\ 4 & \frac{84}{5} & \frac{1956}{105} \\ 4 & \frac{1956}{105} & \frac{8172}{315} \end{bmatrix} + 2\bar{k}(1-\beta^2)^2 \begin{bmatrix} 1 & \beta^2 & \beta^4 \\ \beta^2 & \beta^4 & \beta^6 \\ \beta^4 & \beta^6 & \beta^8 \end{bmatrix} \right] \begin{Bmatrix} q_1 \\ q_2 \\ q_3 \end{Bmatrix} = \frac{2p_0 L^4}{H_{33}^c} \begin{Bmatrix} \frac{1}{3} \\ \frac{1}{15} \\ \frac{1}{35} \end{Bmatrix},
$$

where $\bar{k} = kL^3/H_{33}^c$ is the non-dimensional spring stiffness constant.

Figure 11.13 shows the transverse displacement field for $\alpha = 0.3$ and $\bar{k} = 10^2$; a good correlation with the exact solution, see eq. (5.65), is obtained with the three-term approximate solution derived here. As the stiffness constant increases, the approximation becomes increasingly poorer, as shown in fig. 11.14, which gives the transverse displacement field for $\alpha = 0.3$ and $\bar{k} = 10^4$. To remedy the situation, a larger number of degrees of freedom would be needed in the approximate solution.

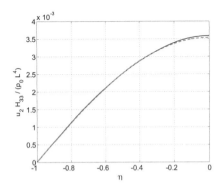

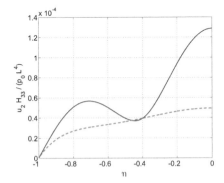

Fig. 11.13. Transverse displacement over left half-span of beam under uniform load for $\bar{k} = 10^2$. Solid line: exact solution; dashed line: three-term approximate solution.

Fig. 11.14. Transverse displacement over left half-span of beam under uniform load for $\bar{k} = 10^4$. Solid line: exact solution; dashed line: three-term approximate solution.

It is left to the reader to compute the bending moment and shear force distributions in the beam. Intuitively, the shear force distribution presents a discontinuity at the location of the spring, and the discrete change in the shear force from the right to the left of the spring equals the force in the spring. The approximate solution developed here is based on continuous shape functions, and hence, very slow convergence should be expected for the approximate shear force distribution.

11.2.4 Problems

Problem 11.1. Beam with various end conditions
Consider a beam of length L extending from $\eta = 0$ to $\eta = 1$, where $\eta = x_1/L$. At each end of the beam, the boundary conditions could be cantilevered, simply-supported, or free, for a total of nine possible combinations. *(1)* Write a set of polynomial shape functions that is suitable for each of the nine cases. *(2)* If the beam of length L extends from $\eta = -1$ to $\eta = 1$, where $\eta = 2x_1/L$, write the corresponding shape functions for each case.

Problem 11.2. Cantilever with tip load: polynomial solution

Consider a cantilever beam of length L and bending stiffness H_{33}^c subjected to a transverse concentrated load, P, acting at the beam's tip. *(1)* Construct a one-term monomial solution, $\bar{u}_2(\eta) = \eta^2 q_1$, and compare the computed tip displacement to the exact value of $PL^3/(3H_{33}^c)$. *(2)* Construct a two-term monomial solution, $\bar{u}_2(\eta) = \eta^2 q_1 + \eta^3 q_2$, and compare this with the exact value. *(3)* Compute the bending moment M_3 and the shear V_2 at the root using the two-term solution and compare these values to the exact values, which can readily be determined from statics.

Problem 11.3. Cantilever with tip load: trigonometric solution

Consider a cantilever beam of length L and bending stiffness H_{33}^c subjected to a transverse concentrated load, P, acting at the beam's tip. *(1)* Explain why the series, $\bar{u}_2(\eta) = \sum_{i=1}^N q_i(1 - \cos i\pi\eta/2)$, provides a good approximate solution. *(2)* Compute an approximate solution to the problem based on this series for arbitrary N. *(3)* Compare your approximate solution with $N = 1$ and 2 with the exact solution at the beam's tip. *(4)* Compare the root bending moments computed from the approximate solutions with the exact solution computed directly from statics.

Problem 11.4. Cantilever beam with elliptical pressure load

Consider a cantilever beam of length L and bending stiffness H_{33}^c subjected to a transverse distributed load $p_2(\eta) = p_0\sqrt{1 - \eta^2}$ that simulates the aerodynamic load acting on an aircraft wing of semi-span L. *(1)* Develop a one-term approximate solution and compare the tip deflection with the exact result determined using the unit load method. *(2)* Repeat the development for a two-term solution. Hint: follow the approach and shape functions used in example 11.1

Problem 11.5. Cantilever with tip support and rotational tip spring

Consider a cantilever beam of length L and bending stiffness H_{33}^c featuring a pinned tip support and rotational spring of stiffness constant k, as shown in fig. 11.15. *(1)* Develop a three-term approximate solution. Use the non-dimensional tip rotational spring stiffness constant $\bar{k} = kL/H_{33}^c$. Hint: follow the approach and shape functions used in example 11.5.

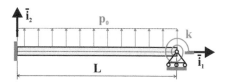

Fig. 11.15. Cantilevered beam with tip support and rotational spring under uniform load.

11.3 The strong and weak statements of equilibrium

When dealing with elastic structures, Newtonian equilibrium conditions typically are in the form of differential equations which impose the vanishing of the sums of all forces and moments on a differential element of the structure. This direct application of Newton's law leads to equilibrium equations that are referred to the *strong statement of equilibrium*. Equations (1.4) represent the strong statement of equilibrium for a three-dimensional solid, and the strong statement of equilibrium for a beam under axial loads is given by eq. (5.18). In the next section, the *weak statement of the equilibrium* will be developed for a beam under axial loading. The term "weak" does not imply a less rigorous formulation, but rather refers to the fact that solutions may be found with less demanding, or *weaker continuity requirements*.

11.3.1 The weak form for beams under axial loads

Consider a beam fixed at one end and subjected to distributed axial loads, $p_1(x_1)$, and a concentrated load, P_1, at the free end as depicted in fig 11.16. In section 5.4, the differential equation of equilibrium of a beam under axial loads is derived as eq. (5.18) which holds for all points over the span of the beam, *i.e.*, for $0 \leq x_1 \leq L$. At the loaded end of the beam, equilibrium requires the internal force to equal the externally applied load, $N_1(L) = P_1$. These two equilibrium requirements, one applicable over the entire span of the beam the other at its tip, are known as the *strong statement of equilibrium*,

$$\frac{dN_1}{dx_1} = -p_1, \quad \text{for} \quad 0 \leq x_1 \leq L, \tag{11.10a}$$

$$N_1 = P_1, \quad \text{for} \quad x_1 = L. \tag{11.10b}$$

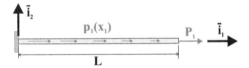

Fig. 11.16. Beam subjected to distributed axial load and tip load.

The following statement is now constructed

$$\int_0^L w(x_1) \left[p_1 + \frac{dN_1}{dx_1} \right] dx_1 + w(L) \left[P_1 - N_1 \right]_{x_1=L} = 0, \tag{11.11}$$

where $w(x_1)$ is an arbitrary function referred to as a *weighting function* or *test function*. If the beam is in equilibrium, eqs. (11.10) must hold, and therefore, eq. (11.11) is satisfied *for all arbitrary functions* $w(x_1)$. Indeed, the two bracketed terms, when set equal to zero, express the equilibrium equations of the problem.

Next, an integration by parts is performed on the first term appearing in the integral

$$\int_0^L w(x_1) \frac{dN_1}{dx_1} dx_1 = - \int_0^L \frac{dw}{dx_1} N_1 dx_1 + [wN_1]_0^L, \tag{11.12}$$

and introducing this result into eq. (11.11) leads to

$$-\int_0^L \frac{dw}{dx_1} N_1\ dx_1 + \int_0^L w p_1\ dx_1 + w(L)P_1 - w(0)N_1(0) = 0. \qquad (11.13)$$

The last term in this statement is the product of the root reaction force, $N_1(0)$, by the value of the test function at the same location, $w(0)$. To eliminate the reaction force from the formulation, the test function is required to vanish at the beam's root, $w(0) = 0$. The statement now reduces to

$$-\int_0^L \frac{dw}{dx_1} N_1\ dx_1 + \int_0^L w p_1\ dx_1 + w(L)P_1 = 0,$$

for all arbitrary $w(x_1)$ such that $w(0) = 0$. $\qquad (11.14)$

This integral is known as the *weak statement of equilibrium*. The strong statement, eq. (11.10), implies the weak statement, eq. (11.14). On the other hand, it is easily shown that the weak statement implies the strong statement. Indeed, the weak statement implies eq. (11.13), which in turn, implies eq. (11.11) by reversing the integration by parts process. Finally, the strong statement of equilibrium is implied by eq. (11.11) because if the test function, $w(x_1)$, is entirely arbitrary, the bracketed terms must vanish.

In summary, the strong statement, eq. (11.10), and the weak statement, eq. (11.14), are two entirely equivalent statements that both express the equilibrium conditions for the beam under axial loads. The weak statement of equilibrium is often referred to as a *variational* statement.

Comparison of the strong and weak statements

At this point, it is not clear what the advantages the weak statement might present over the strong statement. If the goal is to determine the *exact* solution for the internal forces and transverse displacement fields, little difference exists between these two entirely equivalent statements. If the goal, however, is to develop approximate solutions, the weak statement provides significant advantages.

When using the strong statement, it is necessary for the derivative of the axial force to exists because it appears in this statement. Consequently, the axial force must be continuous to use the strong statement. When using the weak statement, however, the only requirement is for the product of axial force by the derivative of the weighting function be integrable over the beam's span. Hence, the weak statement can be used with an axial force field that satisfies weaker continuity requirements.

The integration by parts is an essential part of the derivation of the weak statement of equilibrium. It is responsible for the decreased (or weakened) continuity requirement for the axial forces, but this is achieved at the expense of increasing the continuity requirements on the test function. The boundary terms generated by the integration by parts also affect the formulation of the boundary conditions of the problem.

Sign conventions

The particular sign convention employed in this development deserves further comment. When formulating the weak statement in eq. (11.11), the bracketed expressions must both vanish, and therefore, an arbitrary sign can be assigned to each expression. Either positive or negative sign is possible, but to simplify the development, the following sign convention will be adopted. Each bracketed equilibrium equations appearing in the weak statement, eq. (11.11), will be constructed using a positive sign for the products of the test functions by externally applied loads acting in a positive axis direction. For instance, the term under the integral is written as $+w(x_1) [p_1 + \ldots]$ and the boundary terms as $+w(L) [P_1 + \ldots]$.

Geometric boundary conditions

In the preceding development, the test function is chosen to satisfy the condition $w(0) = 0$. This choice eliminates the reaction force at the root of the beam from the weak statement of equilibrium. Reaction forces are the forces that appear at the locations where geometric boundary conditions are enforced. More generally, *geometric boundary conditions*, sometimes referred to as *essential boundary conditions*, are defined as those boundary conditions that restrict allowable displacements, such as the clamping of the beam at a point. For the problem depicted in fig. 11.16, the geometric boundary condition is $\bar{u}_1(0) = 0$, and the associated reaction force is the root reaction force, $N_1(0)$. If the test function is chosen to satisfy the geometric boundary condition, $w(0) = 0$, the corresponding reaction force is eliminated from the weak statement of equilibrium.

Beam fixed at both ends

To further illustrate the concept of geometric boundary conditions, consider a beam fixed at both ends and subjected to a distributed axial load $p_1(x_1)$. The differential equation of equilibrium is still given by eq. (5.18), however, because the beam is fixed at both ends, no additional equilibrium conditions can be stated at these ends. The following statement is now constructed

$$\int_0^L w(x_1) \left[\frac{dN_1}{dx_1} + p_1 \right] \, dx_1 = 0,$$

that must be satisfied *for all arbitrary functions $w(x_1)$*. Next, an integration by parts is performed on the first term appearing in the integral, and the above statement becomes

$$-\int_0^L \frac{dw}{dx_1} N_1 \, dx_1 + \int_0^L w p_1 \, dx_1 + [w N_1]_0^L = 0.$$

Because $w(x_1)$ is an entirely arbitrary function, it is chosen to vanish at the points where *geometric boundary conditions* are imposed, *i.e.*, $w(0) = w(L) = 0$. The above statement then reduces to

$$-\int_0^L \frac{\mathrm{d}w}{\mathrm{d}x_1} N_1 \,\mathrm{d}x_1 + \int_0^L wp_1 \,\mathrm{d}x_1 = 0,$$

for all arbitrary $w(x_1)$ such that $w(0) = w(L) = 0$.

This integral is the weak statement of equilibrium for this problem.

Concentrated loads at an interior point

In the previous cases, a concentrated load, P_1, is applied at the beam's tip, and is introduced in the strong statement as an equilibrium condition at that location. Concentrated loads, however, may also be applied at any point along the span of the beam. To illustrate this situation, consider a beam fixed at both ends, subjected to a distributed axial load, $p_1(x_1)$, and a concentrated axial load, P_1, applied at location $x_1 = \alpha L$, as shown in fig. 11.17.

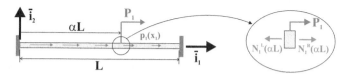

Fig. 11.17. Beam fixed at both ends and subjected to distributed axial load and concentrated load.

For this configuration, the presence of the concentrated load, P_1, creates a discontinuity in the axial loading applied along the beam, and consequently, the strong statement of equilibrium can no longer be formulated as a single differential equation over the entire span of the beam. Instead, the strong statement, eq. (11.10a), must be split into two separate differential equations over the left and right portions of the beam, leading to

$$\frac{\mathrm{d}N_1^L}{\mathrm{d}x_1} = -p_1, \quad 0 \le x_1 \le \alpha L, \tag{11.15a}$$

$$\frac{\mathrm{d}N_1^R}{\mathrm{d}x_1} = -p_1, \quad \alpha L \le x_1 \le L, \tag{11.15b}$$

$$N_1^R(\alpha L) + P_1 - N_1^L(\alpha L) = 0, \tag{11.15c}$$

where the symbols N_1^L and N_1^R denote the axial forces over the left and right portions of the beam, respectively. Equation (11.15c) expresses axial force equilibrium at the point of application of the concentrated load, as illustrated in the expanded detail shown in fig. 11.17. The boundary conditions at the ends are both geometric constraints, $\bar{u}_1(0) = \bar{u}_1(L) = 0$.

A weak statement of equilibrium is now constructed

$$\int_0^{\alpha L} w(x_1) \left[p_1 + \frac{\mathrm{d}N_1^L}{\mathrm{d}x_1} \right] \mathrm{d}x_1 + \int_{\alpha L}^L w(x_1) \left[p_1 + \frac{\mathrm{d}N_1^R}{\mathrm{d}x_1} \right] \mathrm{d}x_1$$
$$+ w(\alpha L) \left[N_1^R(\alpha L) + P_1 - N_1^L(\alpha L) \right] = 0,$$

that must be satisfied for all arbitrary $w(x_1)$. As before, the bracketed expressions are simply the equilibrium equations of the problems, eqs. (11.15).

Next, an integration by parts is performed for the terms involving the derivatives of the axial force to find

$$-\int_0^{\alpha L} \frac{dw}{dx_1} N_1^L dx_1 + \left[w N_1^L\right]_0^{\alpha L} - \int_{\alpha L}^L \frac{dw}{dx_1} N_1^R dx_1 + \left[w N_1^R\right]_{\alpha L}^L + \int_0^L p_1 dx_1$$
$$+ w(\alpha L)\left[N_1^R(\alpha L) + P_1 - N_1^L(\alpha L)\right] = 0.$$

Because the test function is arbitrary, it can be chosen to satisfy the geometric boundary conditions, $w(0) = w(L) = 0$, and then all boundary terms vanish except for $w(\alpha L) P_1$. The two integral can be recombined to yield the following statement

$$-\int_0^L \frac{dw}{dx_1} N_1 dx_1 + \int_0^L p_1 dx_1 + w(\alpha L) P_1 = 0, \tag{11.16}$$

for all arbitrary $w(x_1)$ such that $w(0) = w(L) = 0$.

Because the derivatives of the axial force have been eliminated through the integration by parts, it is no longer necessary to distinguish between the left and right portions of the beam, as is the case for the strong statement, see eqs. (11.15).

The weak statement of equilibrium for the present configuration is very similar to that obtained for a beam with an axial tip load, see in eq. (11.14). For the present configuration, the applied concentrated load is multiplied by the test function evaluated at the point of application of the concentrated load. For the strong formulation, the discontinuity of the axial force requires splitting the problem into two separate portions, see eqs. (11.15). In the weak formulation, the continuity requirement on the axial force is relaxed and the presence of concentrated loads has little effect on the weak statement of equilibrium, see eq. (11.16).

11.3.2 Approximate solutions for beams under axial loads

The weak statement of equilibrium takes the form of a weighted integral that is particularly well suited for obtaining approximate solutions. In the case of the axially loaded beam examined above, the product of the axial force distribution by the test function needs to be integrable over of the beam's span and the test functions must satisfy the geometric boundary conditions. These are weaker requirements than those imposed by the differential equation in the strong statement of equilibrium.

At this point, a subtle but important distinction must be made. While the strong and weak statements of equilibrium are equivalent (meaning that one can be derived from the other), not all solutions to the weak statement are solutions to the strong statement. In particular, those solutions that do not meet the continuity requirements imposed by the strong statement may, in fact, be acceptable solutions to the weak form. In this sense, these solutions, while exactly satisfying the weak form, are nonetheless approximate solutions to the strong form. This versatility makes the weak form attractive for developing approximate solutions.

As discussed in chapter 3, the solution of elasticity problems requires the simultaneous solution of three groups of equations: the strain-displacement equations, the constitutive laws, and the equilibrium equations. The weak and strong statements of equilibrium are shown in section 11.3.1 to be entirely equivalent to Newton's law. Both strain-displacement equations and constitutive laws must be added to the weak statement to solve general elasticity problems. For a beam under axial loading, the constitutive law is given by eq. (5.16) as $N_1 = S \bar{\epsilon}_1$, and the strain-displacement relationship by eq. (5.6) as $\bar{\epsilon}_1 = \mathrm{d}\bar{u}_1/\mathrm{d}x_1$. Substituting these equations into the weak statement of equilibrium, eq. (11.14), yields

$$-\int_0^L \frac{\mathrm{d}w}{\mathrm{d}x_1} S \frac{\mathrm{d}\bar{u}_1}{\mathrm{d}x_1}\, \mathrm{d}x_1 + \int_0^L w p_1\, \mathrm{d}x_1 + w(L)P_1 = 0. \tag{11.17}$$

Note that the use of the strong statement of equilibrium leads to a second order differential equation for the axial displacement field, see eq. (5.19), whereas only first order derivatives of the same displacement field appear in the above statement, implying weaker continuity requirements.

An approximate solution of the problem is now selected in the following form

$$\bar{u}_1(x_1) = \sum_{i=1}^N h_i(x_1)q_i, \tag{11.18}$$

where the $h_i(x_1)$, $i = 1, 2, \ldots, N$, are known shape functions, and q_i, $i = 1, 2, \ldots, N$, unknown degrees of freedom. This approximation is identical to that used earlier in the Rayleigh-Ritz method, see eq. (11.3). Here again, the shape functions must individually satisfy the geometric boundary conditions.

The weak statement of equilibrium, see eq. (11.14), also involves a set of arbitrary functions, $w(x_1)$, which must satisfy the geometric boundary conditions. In a similar manner, these test functions can be approximated as

$$w(x_1) = \sum_{i=1}^N g_i(x_1)w_i, \tag{11.19}$$

where the $g_i(x_1)$, $i = 1, 2, \ldots, N$, are known shape functions, and w_i, $i = 1, 2, \ldots, N$, a set of *arbitrary coefficients*.

The test functions, $w(x_1)$, must satisfy the geometric boundary conditions. If each of the shape functions, $g_i(x_1)$, individually satisfies the same conditions, the coefficients, w_i, become *entirely arbitrary coefficients*, i.e., are not subjected to any restriction.

The shape functions, $h_i(x_1)$, used to approximate the axial displacement field in eq. (11.18), and $g_i(x_1)$, used to approximate the test functions must all satisfy the geometric boundary conditions, but are otherwise arbitrary and unrelated. It is possible, and often convenient, to select $h_i(x_1) = g_i(x_1)$ but this is not a requirement. When selecting $h_i(x_1) = g_i(x_1)$, the procedure is called *Galerkin's method*.

Example 11.7. Beam under a uniform axial load

Consider the uniform, cantilevered beam of length L subjected to a uniform axial loading, $p_1(x_1) = p_0$, as depicted in fig. 11.16. For simplicity, no concentrated load is applied to the beam, *i.e.*, $P_1 = 0$. The solution of this problem presented in section 5.4 is based on the strong statement of equilibrium.

The solution of the same problem will now be derived with the help of the weak statement of equilibrium, eq. (11.17), which is now written as

$$- \int_0^L \frac{dw}{dx_1} S \frac{d\bar{u}_1}{dx_1} \, dx_1 + \int_0^L w p_0 \, dx_1 = 0. \tag{11.20}$$

This statement must vanish for all arbitrary functions, $w(x_1)$, that satisfy the geometric boundary condition, $w(0) = 0$. A suitable approximation to the solution, see eq. (11.18), is selected as

$$\bar{u}_1(x_1) = x_1 q_1 + x_1^2 q_2, \tag{11.21}$$

where $h_1(x_1) = x_1$ and $h_2(x_1) = x_1^2$ are the shape functions, and q_1 and q_2 the two degrees of freedom. The two shape functions individually satisfy the geometric boundary condition. Next, the test functions are approximated in the form of eq. (11.19), using shape functions $g_1(x_1) = x_1$ and $g_2(x_1) = x_1^2$ as

$$w(x_1) = x_1 w_1 + x_1^2 w_2. \tag{11.22}$$

Note that $h_1 = g_1$ and $h_2 = g_2$, *i.e.*, Galerkin's method is used here.

Given these approximations, separate expressions of the weak statement can be written for test functions h_1 and h_2 as

$$- \int_0^L 1 \, S(q_1 + 2q_2 x_1) \, dx_1 + \int_0^L p_0 x_1 \, dx_1 = 0,$$

$$- \int_0^L 2x_1 S(q_1 + 2q_2 x_1) \, dx_1 + \int_0^L p_0 x_1^2 \, dx_1 = 0,$$

respectively. The weak statement must be satisfied for each of the two test functions, h_1 and h_2, because because coefficients w_1 and w_2 are entirely arbitrary. Also, for this reason, the number of test functions must be equal to the number of shape functions.

After carrying out the integrations, these two equations are cast in the following matrix form

$$S \begin{bmatrix} 1 & 1 \\ 1 & 4/3 \end{bmatrix} \begin{Bmatrix} q_1 \\ Lq_2 \end{Bmatrix} = p_0 L \begin{Bmatrix} 1/2 \\ 1/3 \end{Bmatrix}.$$

Solving this set of linear equations yields $q_1 = p_0 L/S$ and $Lq_2 = -p_0 L/(2S)$, and the axial displacement field now becomes

$$\bar{u}_1(x_1) = \frac{p_0 L^2}{S} \left(\eta - \frac{1}{2} \eta^2 \right), \tag{11.23}$$

where $\eta = x_1/L$ is a non-dimensional variable along the beam's span.

The solution is identical to that found with the strong equilibrium statement, eq. (5.26). Nevertheless, the solution processes for the two approaches are strikingly different. The strong equilibrium statement leads to a second order differential equation that must be solved to obtain the axial displacement. On the other hand, solution of the problem based on the weak statement involves the evaluation of integrals over the beam's span and the solution of a set of linear, algebraic equations. In this case, the two solutions are identical because the approximate solution selected for this example is, in fact, the exact solution.

Example 11.8. Bar with a concentrated axial load

Consider the bar fixed at both ends and carrying a concentrated axial load, P_1, at location $x_1 = \alpha L$, as shown in fig. 11.17. An exact solution of this problem using the classical differential equation approach developed in section 5.4 will require separate axial displacement fields to be evaluated over the left and right portions of the beam, denoted $\bar{u}_1^L(x_1)$ and $\bar{u}_1^R(x_1)$, respectively. Boundary conditions are used at the two fixed ends, and at the point of application of the concentrated load, two compatibility conditions must be imposed between the two separate solutions. The first compatibility condition is the continuity of displacement, $\bar{u}_1^L(\alpha L) = \bar{u}_1^R(\alpha L)$ and the second is the equilibrium condition, $N_1^L(\alpha L) = P_1 + N_1^R(\alpha L)$.

The weak statement given by eq. (11.16) reduces the solution process to the evaluation of a much simpler integral form over the beam's span. A single degree of freedom approximation is selected for this problem, $\bar{u}_1(\eta) = q_1 h_1(\eta)$, with the following shape function,

$$
h_1(\eta) = \begin{cases} \eta/\alpha, & 0 \le \eta \le \alpha, \\ (1-\eta)/(1-\alpha), & \alpha \le \eta \le 1, \end{cases}
$$

where $\eta = x_1/L$ is the non-dimensional variable along the beam's span. As required, this shape function satisfies the geometric boundary conditions of the problem. Using Galerkin's method, the test function is selected to be identical to the shape function, $g_1(\eta) = h_1(\eta)$, and $w(\eta) = w_1 g_1(\eta)$. The weak statement, eq. (11.16), becomes

$$
-\int_0^L \frac{dw}{dx_1} S \frac{d\bar{u}_1}{dx_1}\, dx_1 + w(\alpha L)P_1 =
$$
$$
-\int_0^1 w' \frac{S}{L} \bar{u}_1'\, d\eta + w(\alpha L)P_1 = w_1\left[-\frac{S}{L}q_1\frac{1}{\alpha(1-\alpha)} + P_1\right] = 0,
$$

where the notation $(\cdot)'$ is used to indicate a differentiation with respect to η.

Since coefficient w_1 is entirely arbitrary, the bracketed expression must vanish, leading to $q_1 = \alpha(1-\alpha)P_1L/S$, and finally

$$
\bar{u}_1(\eta) = \frac{P_1 L}{S}\begin{cases} (1-\alpha)\eta, & 0 \le \eta \le \alpha, \\ \alpha(1-\eta), & \alpha \le \eta \le 1. \end{cases}
$$

The axial force is given by $N_1 = S d\bar{u}_1/dx_1 = S\bar{u}_1'/L$ and this leads to

$$N_1(\eta) = P_1 \begin{cases} (1 - \alpha), & 0 \le \eta \le \alpha, \\ -\alpha, & \alpha \le \eta \le 1. \end{cases}$$

This solution consists of a linearly varying displacement in each portion of the beam. The axial strain, $\bar{\epsilon}_1$, and the axial force, N_1 are constant with each portion of the beam. At $\eta = \alpha$, the axial force is discontinuous, $N_1^L(\alpha L) = P_1 + N_1^R(\alpha L)$, as required by equilibrium. A complete solution of the problem using the classical differential equation approach will reveal that the present solution is, in fact, the exact solution of the problem.

Example 11.9. Beam under a uniform axial load: a more formal presentation

In the previous examples, Galerkin's method is used to solve simple problems that only required limited algebraic manipulations. As the number of degrees of freedom increases, a more systematic approach will be needed to obtain a streamlined procedure. A compact matrix notation will be introduced; in particular, the important concepts of stiffness matrix and load array will enable a more formal presentation of Galerkin's method. Furthermore, this matrix algebra presentation of the method is readily implemented on computers.

The polynomial forms used for approximate displacements in the previous examples will be used here again, see eq. (11.21) for the assumed solution and eq. (11.22) for the assumed test functions. The degrees of freedom of the problem are q_1 and q_2, whereas w_1 and w_2 are arbitrary coefficients. With these approximations, the weak statement of equilibrium, eq. (11.20), becomes

$$-\int_0^L S(w_1 + 2w_2 x_1)(q_1 + 2q_2 x_1) \, dx_1 + \int_0^L p_0(w_1 x_1 + w_2 x_1^2) \, dx_1 = 0.$$

Expansion and integration then leads to

$$-S\left(w_1 q_1 L + 2w_1 q_2 \frac{L^2}{2} + 2w_2 q_1 \frac{L^2}{2} + 4w_2 q_2 \frac{L^3}{3}\right) + p_0\left(w_1 \frac{L^2}{2} + w_2 \frac{L^3}{3}\right) = 0.$$

It is customary to write this equation in a matrix form as

$$-\{w_1, w_2\} \, SL \begin{bmatrix} 1 & L \\ L & 4L^2/3 \end{bmatrix} \begin{Bmatrix} q_1 \\ q_2 \end{Bmatrix} + \{w_1, w_2\} \, p_0 L^2 \begin{Bmatrix} 1/2 \\ L/3 \end{Bmatrix} = 0. \quad (11.24)$$

The 2×2 matrix of stiffness coefficients is called the *stiffness matrix*, $\underline{\underline{K}}$, and the array of loading coefficients the *load array*, $\underline{Q}$, which are defined as

$$\underline{\underline{K}} = SL \begin{bmatrix} 1 & L \\ L & 4L^2/3 \end{bmatrix}, \quad \underline{Q} = p_0 L^2 \begin{Bmatrix} 1/2 \\ L/3 \end{Bmatrix}.$$

The array of degrees of freedom is called the *solution array*, $\underline{q}$, and the array of arbitrary coefficients the *test array*, $\underline{w}$,

$$\underline{q} = \begin{Bmatrix} q_1 \\ q_2 \end{Bmatrix}, \quad \underline{w} = \begin{Bmatrix} w_1 \\ w_2 \end{Bmatrix}.$$

With these definitions, the weak statement, eq. (11.24), takes the following compact form

$$\underline{w}^T \left[\underline{\underline{K}} \, \underline{q} - \underline{Q} \right] = 0. \tag{11.25}$$

Coefficients w_1 and w_2 are entirely arbitrary, *i.e.*, can be assigned any value. In particular, the following choices will be used here: $\underline{w}^T = \{1,0\}$ and $\underline{w}^T = \{0,1\}$, leading to

$$\{1,0\} \left[\underline{\underline{K}} \, \underline{q} - \underline{Q} \right] = 0, \quad \{0,1\} \left[\underline{\underline{K}} \, \underline{q} - \underline{Q} \right] = 0.$$

Combining these two equations yields

$$\underline{\underline{I}} \left[\underline{\underline{K}} \, \underline{q} - \underline{Q} \right] = \underline{0},$$

where $\underline{\underline{I}}$ is the 2×2 identity matrix. Of course, because the identity matrix is never singular, it can be dropped to yield a set of algebraic equations for the solution array

$$\underline{\underline{K}} \, \underline{q} = \underline{Q}. \tag{11.26}$$

The solution of this equation is simply $\underline{q} = \underline{\underline{K}}^{-1} \underline{Q}$, or

$$\begin{Bmatrix} q_1 \\ q_2 \end{Bmatrix} = \frac{1}{SL} \begin{bmatrix} 1 & L \\ L & 4L^2/3 \end{bmatrix}^{-1} p_0 L^2 \begin{Bmatrix} 1/2 \\ L/3 \end{Bmatrix},$$

which yields the solution array as $q_1 = p_0 L/S$, and $q_2 = -p_0/(2S)$. Substituting these coefficients into the assumed solution, eq. (11.21), leads to

$$\bar{u}_1 = \frac{p_0 L}{S} x_1 - \frac{p_0}{2S} x_1^2 = \frac{p_0 L^2}{S} \left(\eta - \frac{1}{2} \eta^2 \right),$$

where $\eta = x_1/L$ is a non-dimensional variable along the beam's span. As expected, this solution is identical to that obtained in the previous example 11.8.

Example 11.10. Tapered beam under centrifugal load
In the previous examples, the solutions developed based on the weak statement of equilibrium are exact solutions. This occurs because the assumed form of the solution could represent the exact solution. In general, the exact solution is not known, and the assumed form of the solution cannot represent the exact solution.

Consider a helicopter blade of length L rotating at an angular velocity Ω about axis $\bar{\imath}_2$, as depicted in fig. 11.18. This problem is treated using the classical differential equation approach in example 5.2 on page 184. The rotor blade is homogeneous and its cross-section tapers linearly from an area $\mathcal{A}_0$ at the root to $\mathcal{A}_1 = \mathcal{A}_0/2$ at the tip, and hence, its axial stiffness is $S = E\mathcal{A}(x_1) = \mathcal{A}_0 \left(1 - x_1/2L\right)$.

For this problem, the weak statement of equilibrium becomes

$$-\int_0^L \frac{dw}{dx_1} S(x_1) \frac{d\bar{u}_1}{dx_1} \, dx_1 + \int_0^L w\rho\mathcal{A}(x_1)\Omega^2 x_1 \, dx_1 = 0.$$

Introducing the non-dimensional span variable, $\eta = x_1/L$, then leads to

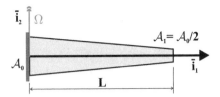

Fig. 11.18. A helicopter blade rotating at an angular speed Ω.

$$-\frac{EA_0}{L} \int_0^1 w' \left(1 - \frac{\eta}{2}\right) \bar{u}_1' \, d\eta + \rho A_0 \Omega^2 L^2 \int_0^1 w \left(\eta - \frac{\eta^2}{2}\right) d\eta = 0,$$

where the notation $(\cdot)'$ indicates a derivative with respect to η.

An approximate solution for the axial displacement field is assumed in the following simple polynomial form, which satisfies the only geometric boundary condition of the problem, $\bar{u}_1(0) = 0$,

$$\bar{u}_1(\eta) = q_1 \eta + q_2 \eta^2. \tag{11.27}$$

Using Galerkin's method, an identical form is selected for the weighting function, $w(\eta) = w_1 \eta + w_2 \eta^2$, and the weak statement becomes

$$-\frac{EA_0}{L} \int_0^1 (w_1 + 2w_2\eta)(1 - \frac{\eta}{2})(q_1 + 2q_2\eta) \, d\eta$$

$$+\rho A_0 \Omega^2 L^2 \int_0^1 (w_1\eta + w_2\eta^2) \left(\eta - \frac{\eta^2}{2}\right) d\eta = 0.$$

After expansion and integration, this expression can be recast into a matrix form following a procedure similar to that used in example 11.9, leading to

$$- \{w_1, w_2\} \frac{EA_0}{L} \begin{bmatrix} 3/4 & 2/3 \\ 2/3 & 5/6 \end{bmatrix} \begin{Bmatrix} q_1 \\ q_2 \end{Bmatrix} + \{w_1, w_2\} \rho A_0 \Omega^2 L^2 \begin{Bmatrix} 5/24 \\ 3/20 \end{Bmatrix} = 0.$$

As before, it is convenient to define a stiffness matrix, $\underline{\underline{K}}$, and a loading array, $\underline{Q}$, as

$$\underline{\underline{K}} = \frac{EA_0}{L} \begin{bmatrix} 3/4 & 2/3 \\ 2/3 & 5/6 \end{bmatrix}, \quad \text{and} \quad \underline{Q} = \rho A_0 \Omega^2 L^2 \begin{Bmatrix} 5/24 \\ 3/20 \end{Bmatrix}.$$

With these definitions, the weak statement is again in the form of eq. (11.25).

Following the same steps as those detailed in the previous example, the weak statement then leads to a set of linear equations, eq. (11.26). These can be solved to yield the solution array

$$\begin{Bmatrix} q_1 \\ q_2 \end{Bmatrix} = \underline{\underline{K}}^{-1}\underline{Q} = \frac{L}{EA_0} \begin{bmatrix} 3/4 & 2/3 \\ 2/3 & 5/6 \end{bmatrix}^{-1} \rho A_0 \Omega^2 L^2 \begin{Bmatrix} 5/24 \\ 3/20 \end{Bmatrix}.$$

Inverting the 2×2 stiffness matrix yields the degrees of freedom as $q_1 = 53\rho\Omega^2 L^3/(130E)$ and $q_2 = -19\rho\Omega^2 L^3/(130E)$. Introducing these coefficients into the assumed solution, eq. (11.27), leads to

$$\bar{u}_1 = \frac{\rho \Omega^2 L^3}{E} \left(\frac{53}{130} \eta - \frac{19}{130} \eta^2 \right). \tag{11.28}$$

This solution is clearly different from that obtained based on the strong statement of equilibrium, eq. (5.27). Indeed, eq. (5.27) is an *exact solution* of the problem, whereas eq. (11.28) is *an approximate solution*. The approximation is introduced by assuming the form of the solution and weighting function in eq. (11.27).

Clearly, the assumed polynomial form of the solution cannot possibly represent the exact solution of the problem, eq. (5.27), which involves transcendental functions. Furthermore, the weak statement, eq. (11.14), requires the vanishing of an integral for all arbitrary choices of the test function, but the two polynomials selected here to represent the test function cannot possibly represent all arbitrary choices of this function.

Within the framework of this approximation, the solution process determines the degrees of freedom, q_1 and q_2, which provide the "best overall match" with the exact solution. Detailed mathematical analysis of the solution procedure [8, 9] can be used to prove that under certain restrictions, the approximate solution does converge to the exact solution as the number of degrees of freedom increases.

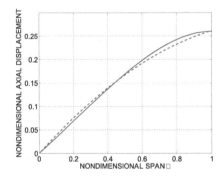

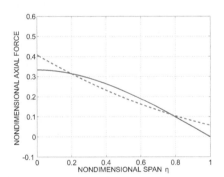

Fig. 11.19. Non-dimensional axial displacement, $\bar{u}_1 E/(\rho A_0^2 L^3)$, versus η. Strong statement of equilibrium: solid line; weak statement: dashed line.

Fig. 11.20. Non-dimensional axial force, $N_1/(\rho A_0 \Omega^2 L^2)$, versus η. Strong statement of equilibrium: solid line; weak statement: dashed line.

Although only two unknown coefficients are used here, it is interesting to note that the approximate solution is in close agreement with the exact solution, as shown in fig. 11.19, which depicts the distribution of non-dimensional axial displacement over the beam's span. Table 11.1 compares the predictions of the two approaches at two locations along the span of the blade.

Finally, the axial force in the blade is obtained by introducing eq. (11.28) into eq. (5.16) to find

$$N_1 = \frac{\rho A_0 \Omega^2 L^2}{260} \left(106 - 129 \eta + 38 \eta^2 \right). \tag{11.29}$$

Table 11.1. Comparison of the exact and approximate solutions.

| | $\bar{u}_1 E/(\rho\Omega^2 L^3)$ | $\bar{u}_1 E/(\rho\Omega^2 L^3)$ | $N_1/(\rho\mathcal{A}_0\Omega^2 L^2)$ | $N_1/(\rho\mathcal{A}_0\Omega^2 L^2)$ |
	$\eta = 0.8$	$\eta = 1.0$	$\eta = 0.5$	$\eta = 0.8$
Strong statement	0.2426	0.2601	0.2292	0.0987
Weak statement	0.2326	0.2615	0.1962	0.1043
Error (%)	-4.1%	0.54%	-14%	5.6%

Figure 11.20 compares the axial forces predicted by the two approaches. This result is, of course, different from that obtained from the strong statement, eq. (5.28). The two solutions, however, are in reasonable agreement, as shown in fig. 11.20, which depicts the distribution of non-dimensional axial force along the span of the blade. Table 11.1 lists the predictions of the two approaches at two locations along the span of the blade.

11.3.3 Problems

Problem 11.6. Rotating helicopter blade with tip mass

A helicopter blade of length L and with a tip mass M_0 is rotating at an angular velocity Ω about axis $\bar{\imath}_2$, see fig 11.21. The blade is homogeneous and its cross-section linearly tapers from an area $\mathcal{A}_0$ at the root to $\mathcal{A}_1$ at the tip so that $\mathcal{A}(x_1) = \mathcal{A}_0 + (\mathcal{A}_1 - \mathcal{A}_0)x_1/L$. Select $\mathcal{A}_0 = 2\mathcal{A}_1$. The tip mass $M_0 = \zeta\rho\mathcal{A}_0 L$, where $\zeta = 0.2$ and ρ is the material mass density. (1) Solve the governing differential equations of this problem to find the axial displacement $\bar{u}(x_1)$ and the axial load $N_1(x_1)$. (2) Find an approximate solution for the axial displacement $\bar{u}_1(x_1)$ using a weak formulation. Select the following forms for the displacement field, $\bar{u}_1(x_1) = q_1 x_1 + q_2 x_1^2$, and weighting function, $w(x_1) = w_1 x_1 + w_2 x_1^2$. (3) Determine the axial force $N_1(x_1)$. (4) On the same graph, plot the non-dimensional displacement fields for the exact and approximate solutions. (5) On the same graph, plot the non-dimensional axial force for the exact and approximate solutions. (6) How would you improve the approximate solution? **Hint**: A mass M rotating about axis $\bar{\imath}_2$ at an angular velocity Ω is subjected to a centrifugal force $F_c = M\Omega^2 r$, where r is the distance between the mass and the axis of rotation. Hence, the helicopter blade is subjected to an axial load per unit span $p_1(x_1) = \rho\mathcal{A}(x_1)\Omega^2 x_1$, where ρ is the material density. In a similar way, the tip mass M_0 creates a concentrated tip force $M_0\Omega^2 L$.

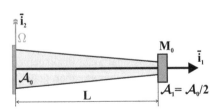

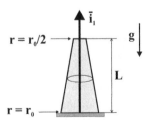

Fig. 11.21. A helicopter blade with a tip mass rotating at an angular speed Ω.

Fig. 11.22. Tapered beam subjected to gravity loads

Problem 11.7. Tapered beam under gravity load

Consider the tapered beam of circular cross-section subjected to gravity loads, as shown in fig. 11.22. The root section is of radius r_0, whereas the tip section has a radius $r_0/2$. The radius linearly tapers along the length of the beam, $r(x_1) = r_0[1-x_1/(2L)]$. The acceleration of gravity is g, the material Young's modulus E, and the material density ρ. (1) Solve the governing differential equations of this problem to find the axial displacement $\bar{u}(x_1)$ and the axial load $N_1(x_1)$. (2) Find an approximate solution of the problem using a weak formulation. Select the following forms for the displacement field $\bar{u}_1(x_1) = q_1 x_1 + q_2 x_1^2$ and test function $w(x_1) = w_1 x_1 + w_2 x_1^2$. (3) On the same graph, plot the non-dimensional displacement fields for the exact and approximate solutions. (4) On the same graph, plot the non-dimensional axial force for the exact and approximate solutions.

Problem 11.8. Tapered beam with mid-span axial load

Consider the axially loaded beam shown in fig. 11.17, but assume now that the beam is tapered uniformly from a radius, r_0, at $x_1 = 0$ to a radius, $r_0/2$, at the right end, $x_1 = L$. The concentrated axial load, P_1, is applied at $x_1 = \alpha L$. (1) Using the same assumed linear form for the axial displacement field used in example 11.8, develop an approximate solution for the axial displacement, $\bar{u}_1(x_1)$ in each portion of the beam. (2) Calculate the axial force, N_1, in each portion of the beam. (3) Determine if equilibrium is satisfied at the load point by this approximate solution and comment on your results.

11.3.4 The weak form for beams under transverse loads

The differential equation of equilibrium of a cantilever beam subjected to distributed transverse loading and a concentrated transverse tip load, P_2, as shown in fig. 11.23, is derived in section 5.5.3, and the equilibrium equation is given by eq. (5.39). This equation holds at all points over the span of the beam, i.e., for $0 \le x_1 \le L$.

At the loaded tip of the beam, equilibrium requires the shear force to equal the applied transverse load, $V_2(L) = P_2$, and the bending moment must vanish, $M_3 = 0$. These three equilibrium requirements are summarized as

$$\frac{d^2 M_3}{dx_1^2} = p_2, \quad \text{for} \quad 0 \le x_1 \le L, \tag{11.30a}$$

$$V_2 = P_2, \quad M_3 = 0, \quad \text{for} \quad x_1 = L, \tag{11.30b}$$

and are known as the *strong statement of equilibrium* for this problem.

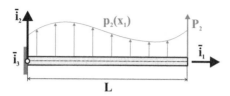

Fig. 11.23. Beam subjected to transverse loads.

These three equilibrium equations are used to construct the following integral statement using the sign convention described in section 11.3.1

$$-\int_0^L w(x_1) \left[\frac{\mathrm{d}^2 M_3}{\mathrm{d}x_1^2} - p_2 \right] \mathrm{d}x_1 + w(L) \left[P_2 - V_2 \right]_L + \frac{\mathrm{d}w(L)}{\mathrm{d}x_1} \left[0 - M_3 \right]_L = 0.$$
(11.31)

The second boundary equilibrium equation is multiplied by the derivative of the test function to preserve the dimensionality of the overall equation. If the beam is in equilibrium, eqs. (11.30) hold, and eq. (11.31) must be satisfied *for all arbitrary functions* $w(x_1)$.

Next, to reduce or weaken the continuity requirements on the bending moment distribution and to reveal the boundary conditions, an integration by parts is performed on the first term appearing in the integral

$$\int_0^L w(x_1) \frac{\mathrm{d}^2 M_3}{\mathrm{d}x_1^2} \, \mathrm{d}x_1 = -\int_0^L \frac{\mathrm{d}w}{\mathrm{d}x_1} \frac{\mathrm{d}M_3}{\mathrm{d}x_1} \, \mathrm{d}x_1 + \left[w \frac{\mathrm{d}M_3}{\mathrm{d}x_1} \right]_0^L.$$

The integration by parts is repeated for the first term on the right-hand side of the equation and eq. (5.38) is introduced in the boundary term to find

$$\int_0^L w(x_1) \frac{\mathrm{d}^2 M_3}{\mathrm{d}x_1^2} \, \mathrm{d}x_1 = \int_0^L \frac{\mathrm{d}^2 w}{\mathrm{d}x_1^2} M_3 \, \mathrm{d}x_1 - \left[\frac{\mathrm{d}w}{\mathrm{d}x_1} M_3 \right]_0^L - [wV_2]_0^L.$$
(11.32)

Introducing this result into eq. (11.31) leads to

$$-\int_0^L \frac{\mathrm{d}^2 w}{\mathrm{d}x_1^2} M_3 \, \mathrm{d}x_1 + \int_0^L w p_2 \, \mathrm{d}x_1 + w(L) P_2 + \frac{\mathrm{d}w}{\mathrm{d}x_1}(0) M_3(0) + w(0) V_2(0) = 0.$$
(11.33)

At the root, the beam is clamped, giving rise to two geometric boundary conditions, $\bar{u}_2(0) = 0$ and $\mathrm{d}\bar{u}_2(0)/\mathrm{d}x_1 = 0$. Since $w(x_1)$ is an entirely arbitrary function, it is possible to choose $w(0) = \mathrm{d}w/\mathrm{d}x_1(0) = 0$. This choice causes $w(x_1)$ to satisfy the geometric boundary conditions and eliminates the root reaction force and moment from the formulation. This leads to

$$-\int_0^L \frac{\mathrm{d}^2 w}{\mathrm{d}x_1^2} M_3 \, \mathrm{d}x_1 + \int_0^L w p_2 \, \mathrm{d}x_1 + w(L) P_2 = 0,$$
(11.34)

for all arbitrary $w(x_1)$ *such that* $w(0) = \mathrm{d}w/\mathrm{d}x_1(0) = 0$.

This integral is known as the *weak statement of equilibrium*. The strong statement, eq. (11.30), implies the weak statement, eq. (11.34). On the other hand, it is easily shown that the weak statement implies the strong statement. Indeed, the weak statement implies eq. (11.33), which in turn, implies eq. (11.31) by reversing the integration by parts processes. Finally, the strong statement of equilibrium is implied by eq. (11.31) because this equation must hold for all arbitrary choices of the test function, $w(x_1)$.

The above reasoning still holds if different boundary conditions are imposed to the problem. The test function, or its derivative, must vanish at those locations where the transverse displacement, or rotation, of the beam are prescribed, respectively. More generally, the integral appearing in the weak statement must vanish for all test functions that satisfy the *geometric boundary conditions*, *i.e.*, for all arbitrary test functions that vanish at the points where geometric boundary conditions are applied. As before, these are also referred to as the *essential boundary conditions*.

In summary, the strong statement, eq. (11.30), and the weak statement, eq. (11.34), are two entirely equivalent statements that both express the equilibrium conditions for the beam under transverse loads.

Simply supported beam under a uniform load

To illustrate the effect of geometric boundary conditions on the weak form statement, consider a simply supported beam subjected to a distributed transverse load $p_2(x_1) = p_0$, as depicted in fig 11.24 and treated earlier in example 11.2. The differential equilibrium equation is still given by eq. (5.39), and the equilibrium conditions at the ends of the beam are $M_3(0) = M_3(L) = 0$.

The following statement is now constructed using the sign convention defined in section 11.3.1

$$-\int_0^L w(x_1)\left[\frac{d^2M_3}{dx_1^2} - p_0\right] dx_1 - \left[\frac{dw}{dx_1}M_3\right]_0^L = 0. \qquad (11.35)$$

that is satisfied *for all arbitrary functions*, $w(x_1)$. The last term is a statement of moment equilibrium at the ends, *i.e.*, $M_3(0) = M_3(L) = 0$, while $\bar{u}_2(0) = \bar{u}_2(L) = 0$ are the geometric boundary conditions.

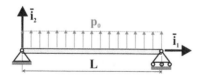

Fig. 11.24. Simply supported beam under a uniform transverse load.

Next, two integrations by parts are performed on the first term appearing in the integral, and the above statement becomes

$$-\int_0^L \frac{d^2w}{dx_1^2}M_3\,dx_1 + \int_0^L wp_0\,dx_1 + \left[\frac{dw}{dx_1}M_3\right]_0^L + [wV_2]_0^L - \left[\frac{dw}{dx_1}M_3\right]_0^L = 0.$$

Since $w(x_1)$ is an entirely arbitrary function, it is selected to vanish at the points where *geometric boundary conditions* are applied, *i.e.*, $w(0) = w(L) = 0$, to yield

$$-\int_0^L \frac{d^2w}{dx_1^2}M_3\,dx_1 + \int_0^L wp_0\,dx_1 = 0,$$

for all arbitrary $w(x_1)$ *such that* $w(0) = w(L) = 0$.

This integral is the weak statement of equilibrium for this problem.

Simply supported beam under a concentrated load

Consider the simply supported beam subjected to a concentrated load, P, applied at location $x_1 = \alpha L$, as shown in fig. 11.25. When using the classical differential equation approach, the beam must be split into two portions, one to the left, the other to the right of the point of application of the concentrated load, as is done in example 5.5 on page 197. Superscripts $(\cdot)^L$ and $(\cdot)^R$ will be used to indicate quantities associated with the left and right portions of the beam, respectively.

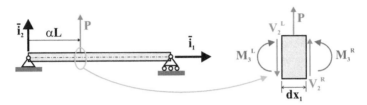

Fig. 11.25. Simply supported beam with one concentrated load.

At the point of application of the concentrated load, two conditions arise, $M_3^L = M_3^R$ and $V_2^R + P - V_2^L = 0$, corresponding to moment and vertical force equilibrium equations of the differential element shown in fig 11.25, respectively. The geometric boundary conditions of the problem are $\bar{u}_2(0) = \bar{u}_2(L) = 0$.

The strong statement of equilibrium for this problem is

$$\frac{\mathrm{d}^2 M_3^L}{\mathrm{d}x_1^2} = 0, \quad \text{for} \quad 0 \le x_1 \le \alpha L,$$

$$\frac{\mathrm{d}^2 M_3^R}{\mathrm{d}x_1^2} = 0, \quad \text{for} \quad \alpha L \le x_1 \le L.$$

Equilibrium conditions at the beam's root and tip are $M_3^L = 0$ and $M_3^R = 0$, respectively, and at $x_1 = \alpha L$, $M_3^L = M_3^R$ and $V_2^R + P - V_2^L = 0$.

The following statement is now constructed,

$$-\int_0^{\alpha L} w(x_1) \left[\frac{\mathrm{d}^2 M_3^L}{\mathrm{d}x_1^2}\right] \mathrm{d}x_1 - \int_{\alpha L}^L w(x_1) \left[\frac{\mathrm{d}^2 M_3^R}{\mathrm{d}x_1^2}\right] \mathrm{d}x_1$$

$$+ \frac{\mathrm{d}w(\alpha L)}{\mathrm{d}x_1} \left[M_3^R - M_3^L\right] + w(\alpha L) \left[V_2^R + P - V_2^L\right] - \left[\frac{\mathrm{d}w}{\mathrm{d}x_1} M_3\right]_0^L = 0.$$

Each of the bracketed terms represents one the equilibrium equations of the problem. If the beam is in equilibrium, the above statement is satisfied *for all arbitrary functions, $w(x_1)$.*

Next, two integrations by parts are performed on the first two integrals to weaken the continuity requirements on the bending moment distribution and reveal the remaining boundary conditions, leading to

$$-\int_0^{\alpha L} \frac{\mathrm{d}^2 w}{\mathrm{d}x_1^2} M_3^L \, \mathrm{d}x_1 + \left[\frac{\mathrm{d}w}{\mathrm{d}x_1} M_3\right]_0^{\alpha L} + \left[w V_2^L\right]_0^{\alpha L}$$

$$-\int_{\alpha L}^{L} \frac{\mathrm{d}^2 w}{\mathrm{d}x_1^2} M_3^R \, \mathrm{d}x_1 + \left[\frac{\mathrm{d}w}{\mathrm{d}x_1} M_3\right]_{\alpha L}^{L} + \left[w V_2^R\right]_{\alpha L}^{L}$$

$$+\frac{\mathrm{d}w(\alpha L)}{\mathrm{d}x_1} \left[M_3^R - M_3^L\right] + w(\alpha L) \left[V_2^R + P - V_2^L\right] - \left[\frac{\mathrm{d}w}{\mathrm{d}x_1} M_3\right]_0^{L} = 0.$$

The first two integrals can be combined and many of the boundary terms cancel out, leaving the following statement

$$-\int_0^{\alpha L} \frac{\mathrm{d}^2 w}{\mathrm{d}x_1^2} M_3 \, \mathrm{d}x_1 + w(\alpha L)P - w(0)V_2^L(0) + w(L)V_2^R(L) = 0.$$

Because $w(x_1)$ is an entirely arbitrary function, it can be chosen to satisfy the geometric boundary conditions, $w(0) = w(L) = 0$, thereby eliminating the reaction forces at the two end supports from the formulation. The only remaining terms are

$$-\int_0^{L} \frac{\mathrm{d}^2 w}{\mathrm{d}x_1^2} M_3 \, \mathrm{d}x_1 + w(\alpha L)P = 0.$$

This is the weak statement of equilibrium for the problem. In contrast with the strong statement of equilibrium, it is not necessary to write two distinct statements over the left and right portions of the beam, because the continuity requirements for the bending moment distribution have been weakened.

11.3.5 Approximate solutions for beams under transverse loads

Approximate solutions for the transverse displacement field of a beam under transverse loads can be developed from the weak statement of equilibrium in a manner similar to that used for axially loaded beams in section 11.3.2. Because the objective is to determine the displacement field, the equilibrium equations must be expressed first in terms of sectional strains using the sectional constitutive laws, eq. (5.37), then in terms of transverse displacement using the strain-displacement relationship, eq. (5.6).

The weak statement of equilibrium, eq. (11.34), applied to the cantilever beam shown in fig. 11.23 can now be written as

$$-\int_0^{L} \frac{\mathrm{d}^2 w}{\mathrm{d}x_1^2} H_{33} \frac{\mathrm{d}^2 \bar{u}_2}{\mathrm{d}x_1^2} \, \mathrm{d}x_1 + \int_0^{L} w p_2 \, \mathrm{d}x_1 + w(L)P_2 = 0. \tag{11.36}$$

Because the beam is cantilevered, this integral must vanish for all arbitrary test functions that satisfy the geometric boundary conditions, $w(0) = \mathrm{d}w(0)/\mathrm{d}x_1 = 0$.

At this point, the continuity requirements on the test function and displacement field are identical. Whereas the strong statement of equilibrium leads to a fourth order differential equation for the displacement field, see eq. (5.40), the weak statement involves only second order derivatives for both the transverse displacement field and test functions.

Approximate solutions for the transverse displacement field are constructed as before,

$$\bar{u}_2(x_1) = \sum_{i=0}^{N} q_i h_i(x_1),\tag{11.37}$$

where the assumed shape functions, $h_i(x_1)$, must individually satisfy the geometric boundary conditions. Similarly, the test functions will be approximated as

$$w(x_1) = \sum_{i=0}^{N} w_i g_i(x_1),\tag{11.38}$$

where the assumed shape functions, $g_i(x_1)$, must also individually satisfy the geometric boundary conditions. Otherwise, the two sets of shape functions are unrelated. In Galerkin's method, both sets of shape functions are selected to be identical, *i.e.*, $h_i(x_1) = g_i(x_1)$.

Example 11.11. *Simply supported beam under a uniform load*

Consider a simply supported, uniform beam of length L subjected to a uniform transverse loading $p_2(x_1) = p_0$, as depicted in fig. 11.24. The exact solution of this problem, based on the classical differential equation approach, is presented in example 5.4 on page 196.

An approximate solution of the same problem will now be derived using the weak statement of equilibrium, eq. (11.36),

$$-\int_0^L \frac{\mathrm{d}^2 w}{\mathrm{d}x_1^2} H_{33} \frac{\mathrm{d}^2 \bar{u}_2}{\mathrm{d}x_1^2} \,\mathrm{d}x_1 + \int_0^L w p_0 \,\mathrm{d}x_1 = 0.$$

This integral must vanish for all test functions that satisfy the geometric boundary conditions, $w(0) = w(L) = 0$.

Using Galerkin's method, the transverse displacement field and test function are assumed to be of the following respective forms

$$\bar{u}_2(x_1) = q_1 \sin\frac{\pi x_1}{L} + q_3 \sin\frac{3\pi x_1}{L}, \quad w(x_1) = w_1 \sin\frac{\pi x_1}{L} + w_3 \sin\frac{3\pi x_1}{L},$$

where q_1 and q_3 are two degrees of freedom, and w_1 and w_3 two arbitrary coefficients. The selection of sine functions for the shape functions guarantees the satisfaction of the geometric boundary conditions of the problem. With these approximations, the weak statement becomes

$$-\int_0^L \left[-w_1 \left(\frac{\pi}{L}\right)^2 \sin\frac{\pi x_1}{L} - w_3 \left(\frac{3\pi}{L}\right)^2 \sin\frac{3\pi x_1}{L} \right]$$

$$H_{33} \left[-q_1 \left(\frac{\pi}{L}\right)^2 \sin\frac{\pi x_1}{L} - q_3 \left(\frac{3\pi}{L}\right)^2 \sin\frac{3\pi x_1}{L} \right] \mathrm{d}x_1$$

$$+ \int_0^L p_0 \left(w_1 \sin\frac{\pi x_1}{L} + w_3 \sin\frac{3\pi x_1}{L} \right) \mathrm{d}x_1 = 0.$$

After expanding, integrating, and using the orthogonality properties of the sine functions, see eqs. (A.45a), this equation can be recast into a matrix form as

$$-\{w_1, w_2\} \frac{\pi^4 H_{33}}{2L^3} \begin{bmatrix} 1 & 0 \\ 0 & 81 \end{bmatrix} \begin{Bmatrix} q_1 \\ q_3 \end{Bmatrix} + \{w_1, w_2\} \frac{2p_0 L}{\pi} \begin{Bmatrix} 1 \\ 1/3 \end{Bmatrix} = 0.$$

If the stiffness matrix, $\underline{\underline{K}}$, and the load array, $\underline{Q}$, are defined as

$$\underline{\underline{K}} = \frac{\pi^4 H_{33}}{2L^3} \begin{bmatrix} 1 & 0 \\ 0 & 81 \end{bmatrix}, \quad \text{and} \quad \underline{Q} = \frac{2p_0 L}{\pi} \begin{Bmatrix} 1 \\ 1/3 \end{Bmatrix},$$

respectively, the weak statement can again be written in the form of eq. (11.25), leading to a set of linear equations, eq. (11.26). These can be solved for the solution array to find

$$\begin{Bmatrix} q_1 \\ q_3 \end{Bmatrix} = \frac{2L^3}{\pi^4 H_{33}} \begin{bmatrix} 1 & 0 \\ 0 & 81 \end{bmatrix}^{-1} \frac{2p_0 L}{\pi} \begin{Bmatrix} 1 \\ 1/3 \end{Bmatrix}.$$

The degrees of freedom are then $q_1 = 4p_0 L^4/(\pi^5 H_{33})$ and $q_3 = 4p_0 L^4/(243\pi^5 H_{33})$. Introducing these coefficients into the assumed solution leads to

$$\bar{u}_2 = \frac{4p_0 L^4}{\pi^5 H_{33}} \left(\sin\frac{\pi x_1}{L} + \frac{1}{243}\sin\frac{3\pi x_1}{L} \right). \tag{11.39}$$

This solution is clearly different from that obtained from the classical differential equation approach, eq. (5.48), which is the exact solution of the problem, whereas eq. (11.39) is an approximate solution.

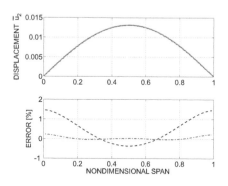

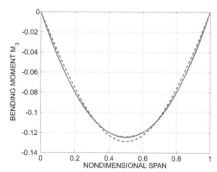

Fig. 11.26. Top figure: transverse displacement $\bar{u}_2$; bottom figure: error. Exact solution: solid line; Approximate solution: dashed line (*case 1*), dash-dotted line (*case 2*).

Fig. 11.27. Bending moment M_3 versus nondimensional span. Exact solution: solid line; Approximate solution: dashed line (*case 1*), dash-dotted line (*case 2*).

Figure 11.26 depicts the distribution of non-dimensional transverse displacement $\bar{u}_2 H_{33}/(p_0 L^4)$ over the span of the beam for both exact and approximate solutions. Two approximate solutions are shown: *Case 1* includes only the term $\sin\pi x_1/L$ in eq. (11.39), whereas *Case 2* includes both terms. At the scale of the figure, the exact and approximate solutions are indistinguishable. The lower portion of the figure shows the discrepancy between the approximate and exact solutions. The excellent agreement between the various solutions is also apparent in table 11.2.

Table 11.2. Comparison of the exact and approximate solutions at the quarter- and mid-points.

	$\bar{u}_2 H_{33}/(p_0 L^4)$ $\eta = 0.25$	$\bar{u}_2 H_{33}/(p_0 L^4)$ $\eta = 0.5$	$M_3/(p_0 L^2)$ $\eta = 0.25$	$M_3/(p_0 L^2)$ $\eta = 0.5$
Exact solution	0.009277	0.01302	0.09375	0.125
Approximation (*Case 1*)	0.009243	0.01307	0.09122	0.129
Relative error	-0.4%	0.4%	-2.7%	3.2%
Approximation (*Case 2*)	0.009281	0.01302	0.0946	0.124
Relative error	0.04%	-0.03%	0.9%	-0.8%

Finally, the bending moment distribution is obtained by introducing eq. (11.39) into eq. (5.37) to find

$$M_3 = -\frac{4p_0 L^2}{\pi^3}\left(\sin\frac{\pi x_1}{L} + \frac{1}{27}\sin\frac{3\pi x_1}{L}\right). \tag{11.40}$$

Figure 11.27 compares the bending moments predicted by the two approaches. Here again, excellent agreement is observed between the various solutions, as confirmed by table 11.2.

11.3.6 Problems

Problem 11.9. Uniformly loaded simply supported beam
Consider a simply supported, uniform beam of length L subjected to a uniform transverse loading $p_2(x_1) = p_0$, as depicted previously in fig. 11.24. (1) Solve the governing differential equations of this problem to find the transverse displacement $\bar{u}_2(x_1)$, the bending moment $M_3(x_1)$, and the shear force $V_2(x_1)$. (2) Find an approximate solution of the problem using a weak formulation. Select the following forms for the displacement field $\bar{u}_2(x_1) = \sum_{i=1}^{N} q_i \sin(2i-1)\pi x_1/L$ and test function $w(x_1) = \sum_{i=1}^{N} w_i \sin(2i-1)\pi x_1/L$. (3) Plot the exact and approximate transverse displacement fields $\bar{u}_2(x_1)$ on the same plot. For the approximate solutions use $N = 1, 2, 3, 4$, and 5. (4) Plot the exact and approximate bending moments $M_3(x_1)$ on the same plot. (5) Plot the exact and approximate shear forces $V_2(x_1)$ on the same plot.

Problem 11.10. Simply supported beam with concentrated load
Consider a simply supported beam with a concentrated load, P, applied at a point $x_1 = \alpha L$ from the left support as illustrated in fig. 5.23 on page 197. This configuration is solved using the classical differential equation approach in example 5.5, and the transverse displacement is found to be given by eq. (5.51). The solution presents a discontinuity in the transverse shear force, and solutions are developed separately for the portions of the beam to the left and right of the concentrated load. Using the weak statement, it is possible to develop a single expression that approximates the deflection over the entire span of the beam, because the continuity

requirements associated with this approach are lower than those required for the differential equation approach. *(1)* Find an approximate solution of the problem using a weak formulation. Select the following forms for the displacement field $\bar{u}_2(x_1) = \sum_{i=1}^{N} q_i \sin i\pi x_1/L$ and test function $w(x_1) = \sum_{i=1}^{N} w_i \sin i\pi x_1/L$. *(2)* On one graph, plot the exact and approximate transverse displacement fields, $H_{33}^c \bar{u}_2(x_1)/(PL^3)$. For the approximate solutions, use $N = $ 1, 2, 3, 4, and 5. *(3)* On one graph, plot the exact and approximate bending moment distributions, $M_3(x_1)/(PL)$. *(4)* On one graph, plot the exact and approximate shear force diagrams, $V_2(x_1)/P$.

11.3.7 Equivalence with energy principles

The weak statement of equilibrium developed in the previous sections is equivalent to the statement of equilibrium cast in the form of the principle of virtual work developed in chapter 9 and recast in terms of work and energy in chapter 10. This equivalence will be demonstrated in the following sections. Consequently, it is clear that all approaches are equally valid, and all lead to effective ways of developing approximate solutions to structural problems.

The principle of virtual work

The weak statement of equilibrium is introduced in section 11.3.1 through a purely mathematical process, and consequently, its physical interpretation is not clear at this point. In particular, this approach is based on the introduction of a test or weighting function, $w(x_1)$, that is entirely arbitrary except that it must satisfy the geometric boundary conditions for the problem.

To help cast the weak statement of equilibrium in a more physical framework, the test functions are now interpreted as follows

$$w(x_1) \equiv \delta u(x_1), \tag{11.41}$$

where the right-hand side can be read as: *arbitrary variation in displacement* or, using the concepts introduced in chapter 9, as *virtual displacement*. The equivalence stated in eq. (11.41) is purely a matter of notation: the test functions are arbitrary functions that must satisfy only the geometric boundary conditions, and equivalently, the virtual displacements defined in section 9.3.4, are arbitrary displacements that must satisfy only the geometric boundary conditions. The two notations are entirely equivalent, although virtual displacements afford a more direct physical interpretation.

In chapter 9, the principle of virtual work for a single particle is introduced by using an arbitrary fictitious displacement, $\underline{s}$, multiplying the strong statement of equilibrium, $\sum \underline{F} = 0$, to find $(\sum \underline{F}) \cdot \underline{s} = 0$, see eq. (9.8). Later in that chapter, the concept of virtual displacement is introduced by setting $\underline{s} = \delta \underline{u}$, see eq. (9.13), leading to a more physical interpretation of the principle of virtual work based on the concepts of internal and external virtual work. A similar thought process is followed here. The weak statement of equilibrium is introduced by using an arbitrary test function (equivalent to the arbitrary fictitious displacement, $\underline{s}$), multiplying the strong statement of equilibrium of the problem. The weak statement of equilibrium is difficult to interpret physically, but by introducing the virtual displacement defined in eq. (11.41) it will become possible establish the equivalence of the weak statement of equilibrium with the principle of virtual work.

Beam under axial load

To illustrate this equivalence, the equilibrium of an axially loaded beam examined in section 11.3.1 will now be recast using virtual displacements rather than weighting functions. Introducing this notation into the weak statement of equilibrium, eq. (11.14), yields

$$-\int_0^L \frac{\mathrm{d}\delta\bar{u}_1}{\mathrm{d}x_1} N_1 \, \mathrm{d}x_1 + \int_0^L \delta\bar{u}_1 p_1 \, \mathrm{d}x_1 + \delta\bar{u}_1(L)P_1 = 0, \tag{11.42}$$

for all arbitrary $\delta\bar{u}_1$ satisfying the geometric boundary conditions.

The first integral in this statement can be written as $-\int_0^L N_1 \delta\bar{\epsilon}_1 \, \mathrm{d}x_1$ and represents the internal virtual work, δW_I, in a beam subjected to axial loads only, see eq. (9.79a).

The last two terms in this equation are the product of forces by virtual displacements and hence, are naturally interpreted as virtual work quantities. More specifically, the last term is the work done by the concentrated tip force, P_1, acting through a virtual displacement, $\delta\bar{u}_1(L)$. Similarly, the middle term is the work done by the distributed axial load, $p_1(x_1)\mathrm{d}x_1$, acting through the virtual axial displacement, $\delta\bar{u}_1(x_1)$; the integral then sums up the virtual work done by the force distributed all along the beam's span to find the total virtual work done by the distributed force. The sum of these two terms defines the virtual work done by the externally applied loads acting on the beam, and following the notation in chapter 9, it is called the external virtual work, δW_E.

With these interpretations, eq. (11.42) is recast in a more physically meaningful manner as

$$\delta W_I + \delta W_E = \delta W = 0, \tag{11.43}$$

which is immediately recognized as a restatement of the principal of virtual work. The principle of work is therefore entirely equivalent to the weak statement of equilibrium, which in turn, is entirely equivalent to the equilibrium conditions for the beam.

Beam under transverse load

In a similar manner, a more physical interpretation of the weak statement of equilibrium for beams under transverse loads given by eq. (11.34) can be obtained by introducing virtual transverse displacements, $\delta\bar{u}_2(x_1)$, in place of test functions, $w(x_1)$. This leads to

$$-\int_0^L \frac{d^2\delta\bar{u}_2}{dx_1^2}M_3\,dx_1 + \int_0^L \delta\bar{u}_2 p_2\,dx_1 + \delta\bar{u}_2(L)P_2 = 0, \tag{11.44}$$

for all arbitrary $\delta\bar{u}_2$ satisfying the geometric boundary conditions.

The first integral in this statement can be written as $-\int_0^L M_3\delta\kappa_3\,dx_1$ and represents the internal virtual work, δW_I, in a beam subjected to transverse loads only, see eq. (9.79a).

Using the same reasoning as before, the second and third terms in eq. (11.44) represent the virtual work done by the applied distributed transverse load, p_2, and concentrated tip load, P_2. This is the total work done by the externally applied loads, δW_E.

In summary, eq. (11.44) can now be restated as, $\delta W_I + \delta W_E = \delta W = 0$, which is simply a statement of the principal of virtual work. Although the expressions of the principle of virtual work for beams under axial and transverse loads, eqs. (11.42) and (11.44), respectively, are different, their physical interpretation is identical.

Approximate solutions

Approximate solutions for beam problems can be obtained from the principle of virtual work by following a procedure identical to that used with the weak statement of equilibrium in section 11.3.2. Consider, for instance, the beam under a uniform axial load treated in example 11.7. The principal of virtual work can be written for this problem as

$$\delta W = -\int_0^L \frac{d\delta\bar{u}_1}{dx_1}S\frac{d\bar{u}_1}{dx_1}\,dx_1 + \int_0^L \delta\bar{u}_1(x_1)p_0\,dx_1 = 0.$$

The following forms are then assumed for the solution, $\bar{u}_1(x_1)$, and for the virtual displacements, $\delta\bar{u}_1(x_1)$,

$$\bar{u}_1(x_1) = q_1 x_1 + q_2 x_1^2; \quad \delta\bar{u}_1(x_1) = \delta q_1 x_1 + \delta q_2 x_1^2, \tag{11.45}$$

where q_1 and q_2 are two unknown coefficients, and δq_1 and δq_2 are arbitrary virtual coefficients since the virtual displacement is itself arbitrary.

These expressions should be compared with eq. (11.21). The only difference is one of notation: $w_1 = \delta q_1$ and $w_2 = \delta q_2$. The rest of the procedure is identical to that outlined in section 11.3.4, and identical results are obtained.

A similar approach can also be taken for bending of a beam subjected to transverse loads as treated in section 11.3.5.

11.3.8 The principle of minimum total potential energy

The principle of virtual work is solely a statement of equilibrium. Equations (11.42) and (11.44) are statements of this principle for beams under axial and transverse loads, respectively, and do not involve the beam's sectional strains or stiffness characteristics. Clearly, the principle is unaware of the strain-displacement relationships and constitutive laws. These two sets of equations will now be combined with the principle of virtual work.

Consider an axially loaded beam. Introducing the axial constitutive law, eq. (5.16), and the axial strain-displacement relationship, eq. (5.6), into eq. (11.42) yields

$$-\int_0^L \frac{\mathrm{d}\,\delta\bar{u}_1}{\mathrm{d}x_1} S \frac{\mathrm{d}\bar{u}_1}{\mathrm{d}x_1}\,\mathrm{d}x_1 + \int_0^L \delta\bar{u}_1(x_1)p_1\,\mathrm{d}x_1 + \delta\bar{u}_1(L)P_1 = 0. \tag{11.46}$$

The beam is now assumed to be made of a linearly elastic material, resulting in the sectional constitutive law give by eq. (5.16). Hence, the developments that follow apply only to linearly elastic materials, whereas the principle of virtual work, which is equivalent to Newton's law, is not limited by any such restrictions.

The last two terms in eq. (11.46) can still be interpreted as the virtual work done by the externally applied loads, whereas the first term is the virtual work done by the internal axial forces. As observed before, the variational operator, δ, and the derivative operator, d, commute, and hence, the internal virtual work can be manipulated as follows

$$\int_0^L \left(\frac{\mathrm{d}\delta\bar{u}_1}{\mathrm{d}x_1}\right) S \frac{\mathrm{d}\bar{u}_1}{\mathrm{d}x_1}\,\mathrm{d}x_1 = \int_0^L \delta\left(\frac{\mathrm{d}\bar{u}_1}{\mathrm{d}x_1}\right) S \frac{\mathrm{d}\bar{u}_1}{\mathrm{d}x_1}\,\mathrm{d}x_1$$
$$= \int_0^L \delta\left[\frac{1}{2} S \left(\frac{\mathrm{d}\bar{u}_1}{\mathrm{d}x_1}\right)^2\right]\,\mathrm{d}x_1 = \delta\int_0^L \frac{1}{2} S\bar{\epsilon}_1^2\,\mathrm{d}x_1.$$

The second equality is the direct result of treating the variational operator, δ, as a differential operator, and the third equality follows from the definition of the sectional axial strain, $\bar{\epsilon}_1$, see eq. (5.6). Note that in the third equality, the integral sign and the variational operator are assumed to commute, by analogy with the differential operator, which enjoys this property according to Leibnitz' integral rule.

The quantity $1/2\,S\bar{\epsilon}_1^2$ is the strain energy density function, $a(\bar{\epsilon}_1)$, defined in eq. (10.34), which represents the elastic energy stored in a deformed differential slice of the beam, as discussed in section 10.4.1. The total elastic energy stored in the deformed beam, $A(\bar{\epsilon}_1)$, is then found by integrating the strain energy density function over the beam's span. The previous equation can now be written as

$$\int_0^L \delta\left(\frac{\mathrm{d}\bar{u}_1}{\mathrm{d}x_1}\right) S \frac{\mathrm{d}\bar{u}_1}{\mathrm{d}x_1}\,\mathrm{d}x_1 = \delta\int_0^L a(\bar{\epsilon}_1)\,\mathrm{d}x_1 = \delta A(\bar{\epsilon}_1).$$

Using this result, eq. (11.46) now becomes

$$\delta A(\bar{\epsilon}_1) - \delta W_E = 0, \tag{11.47}$$

and is interpreted as follows. *A beam is in equilibrium if and only if virtual changes in the total strain energy equal the virtual work done by the externally applied loads, for all arbitrary virtual displacements that satisfy the geometric boundary conditions.*

An even more compact statement can be obtained if the externally applied loads are assumed to be conservative and can therefore be derived from a potential,

$$p_1(x_1) = -\frac{\partial \phi}{\partial \bar{u}_1} \quad \text{and} \quad P_1 = -\frac{\partial \Psi}{\partial \bar{u}_1},\qquad(11.48)$$

where ϕ is the *potential of the distributed loads*, and Ψ the *potential of the concentrated load*. The virtual work done by the externally applied loads now becomes

$$\int_0^L p_1\,\delta\bar{u}_1(x_1)\,\mathrm{d}x_1 + P_1\delta\bar{u}_1(L) = -\int_0^L \frac{\partial\phi}{\partial\bar{u}_1}\,\delta\bar{u}_1(x_1)\,\mathrm{d}x_1 - \frac{\partial\Psi}{\partial\bar{u}_1}\delta\bar{u}_1(L)$$
$$= -\delta\left[\int_0^L \phi\,\mathrm{d}x_1 + \Psi\right] = -\delta\Phi,$$

where Φ is the *total potential of the externally applied loads*. Introducing this result in eq. (11.47) yields $\delta A(\epsilon_1) + \delta\Phi = 0$.

Finally, if the *total potential energy*, Π, of the system is defined as

$$\Pi = A + \Phi,\qquad(11.49)$$

it then follows that

$$\delta\Pi = 0.\qquad(11.50)$$

This result is the statement of the principle of stationary total potential energy developed in section 10.2 as principle 8. It states: *a beam is in equilibrium if and only virtual changes in the total potential energy vanish for all virtual displacements.*

For beams subjected transverse loads, the total potential energy is still given by eq. (11.49). The strain energy is now due to curvature, and the total strain energy in the beam is given by eq. (10.39). The total potential of the applied load combines the potentials of the distributed and concentrated transverse loads as follows

$$p_2(x_1) = -\frac{\partial\phi}{\partial\bar{u}_2}; \quad P_2 = -\frac{\partial\Psi}{\partial\bar{u}_2},\qquad(11.51)$$

respectively. The statement of the principle of stationary total potential energy, principle 8, remains unchanged.

11.3.9 Treatment of the boundary conditions

In the weak statement of equilibrium and in the principle of virtual work, the geometric and natural boundary conditions are treated in a different manner. Test functions or virtual displacements are introduced, which are required to satisfy the geometric boundary conditions. On the other hand, the natural boundary conditions are not mentioned in the statement of these principles.

Although this distinction is generally made in presentations of energy and variational principles, it is an unnecessary distinction as will be demonstrated in the following sections through a number of examples. Beams under both axial and transverse loads will be considered.

Beams under axial loads

Consider a cantilever beam subjected to distributed axial loads, $p_1(x_1)$, and a concentrated axial load, P_1, at the free end as depicted in fig 11.28. In section 11.3.1, the weak statement of equilibrium is formulated for this problem. A similar reasoning will be presented here; the concept of virtual displacements rather than that of test functions will be used. Furthermore, the treatment of the boundary conditions will be fundamentally altered.

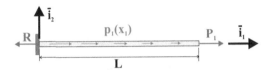

Fig. 11.28. Simple cantilevered beam.

The equations of equilibrium of the problem are in two parts: the differential equilibrium condition that holds for all points along the beam's span, eq. (11.10a), and the equilibrium conditions that apply at the beam's boundaries. At the beam's tip, eq. (11.10b) states that $N_1(L) = P_1$. As indicated in fig. 11.28, a reaction force, R, acts at the root end of the beam, and equilibrium implies $N_1(0) = R$.

The forces acting at the two ends of the beam are often regarded as being of a different nature, and distinct names are used for the two types of forces: the root force, R, is called a *reaction force*, whereas the tip load, P_1, is called an *externally applied load*. When it comes to Newton's law, however, this distinction is irrelevant: the "sum of the forces must vanish" applies to all forces including reaction forces and externally applied loads.

Equilibrium is the most fundamental condition in structural analysis, and it must always be satisfied. The two equilibrium equations, $N_1(0) = R$ and $N_1(L) = P_1$, are both correct, and equally important. In view of this discussion, the strong statement of equilibrium is recast as follows,

$$\frac{\mathrm{d}N_1}{\mathrm{d}x_1} + p_1 = 0, \quad \text{for} \quad 0 \leq x_1 \leq L, \tag{11.52a}$$

$$N_1 = R, \quad \text{for} \quad x_1 = 0, \tag{11.52b}$$

$$N_1 = P_1, \quad \text{for} \quad x_1 = L. \tag{11.52c}$$

The following integral statement can now be constructed, again using the sign convention defined in section 11.3.1

$$\int_0^L \delta\bar{u}_1 \left[\frac{\mathrm{d}N_1}{\mathrm{d}x_1} + p_1 \right] \mathrm{d}x_1 + \delta\bar{u}_1(0) \left[N_1 - R \right]_0 - \delta\bar{u}_1(L) \left[N_1 - P_1 \right]_L = 0,$$

where $\delta\bar{u}_1(x_1)$ is an arbitrary virtual displacement field.

If the beam is in equilibrium, eqs. (11.52) must hold, and therefore the above equation is satisfied *for all arbitrary virtual displacements*. Indeed, the three bracketed terms are simply the equilibrium equations of the problem.

Next, an integration by parts is performed on the first term appearing in the above integral, see eq. (11.12), leading to the following statement

$$
-\int_0^L \delta\frac{\mathrm{d}\bar{u}_1}{\mathrm{d}x_1}N_1 \,\mathrm{d}x_1 + \int_0^L \delta\bar{u}_1 p_1 \,\mathrm{d}x_1 - \delta\bar{u}_1(0)R + \delta\bar{u}_1(L)P_1 = 0,
\tag{11.53}
$$

for all arbitrary virtual displacements.

As discussed in section 11.3.7, the first integral can be interpreted as the internal virtual work, δW_I, done by the axial force acting within the beam. The last three terms form the virtual work, δW_E, done by the externally applied loads. The integral is the virtual work done by the distributed load, p_1, and the last two terms are the virtual work done by the root reaction and tip load, respectively.

Clearly, the above statement is, once again, a statement of the principle of virtual work. This new statement should be compared to that given in eq. (11.42). Two crucial differences can be observed. In eq. (11.53), the virtual work done by the externally applied forces includes the work done by the root reaction, and the virtual displacements are entirely arbitrary. In contrast, in eq. (11.42), the root reaction does not appear in the expression for the virtual work done by the externally applied forces, and the virtual displacements must satisfy the geometric boundary conditions but are otherwise arbitrary.

The principle of virtual work as stated in eq. (11.53) is more general than that expressed by eq. (11.42). In eq. (11.53), virtual displacements are entirely arbitrary, and hence it is always possible to select $\delta\bar{u}_1(0) = 0$, *i.e.*, to restrict the virtual displacements to those that satisfy the geometric boundary conditions. The second virtual work term now vanishes, $\delta\bar{u}_1(0)R = 0$, and the principle of virtual work stated by eq. (11.42) is recovered.

Clearly, the vanishing of the virtual work done by the root reaction force does by no means imply the vanishing of the reaction force itself. Restricting virtual displacements to those that satisfy the geometric boundary conditions, *i.e.*, setting $\delta\bar{u}_1(0) = 0$, implies $\delta\bar{u}_1(0)R = 0$ although $R \neq 0$. Restricting the virtual displacements to those that satisfy the geometric boundary conditions does, however, eliminate the reaction force from the statement of the principle of virtual work because the virtual work it performs vanishes.

This discussion underlines an important feature of the principle of virtual work as stated by eq. (11.42): because virtual displacements are restricted to those satisfying the geometric boundary conditions, the reaction forces are eliminated from the statement of the principle. This simplifies the problem because it is no longer necessary to even identify the reaction forces which therefore, do not enter the formulation of the problem. On the other hand, the principle provides no information about the reaction forces which are often quantities of primary interest to structural analysts.

At this point in the discussion, the relationship between geometric boundary conditions and reaction forces can be clarified: reaction forces are those arising from the enforcement of geometric boundary conditions. At any point on a structure's outer surface, it is possible to prescribe either a displacement or an externally applied force. It is, however, impossible to prescribe both at the same time. If the displacement is prescribed at a point *i.e.*, a geometric boundary condition is imposed, an unknown force (the reaction force) will arise at that point. If a force is applied at a point (an externally applied force), an unknown displacement will arise at that point. The principle of virtual work as stated by eq. (11.42) will provide equations to evaluate the displacements at all points where external forces are applied, but it will yield no information concerning the reaction forces.

In contrast, the principle of virtual work stated by eq. (11.53) allows the use of any arbitrary virtual displacements, *including those that violate the geometric boundary conditions*. The reactions forces, however, must be included the formulation of the virtual work done by the externally applied loads and the *reaction forces should be treated as externally applied loads*.

This view is consistent with Newtonian mechanics: when formulating equilibrium equations, no distinction is made between externally applied loads and reaction forces. Since the principle of virtual work is entirely equivalent to Newton's law, it should also treat externally applied loads and reaction forces in identical ways. This slightly complicates problem formulations because all reaction forces must be properly identified, and the virtual work they perform must be accurately accounted for in the statement of the principle. On the other hand, the principle will then provide the necessary equations to compute these reaction forces.

This discussion mirrors the developments presented in chapter 9, where the use of kinematically admissible virtual displacements and arbitrary virtual displacements is contrasted in sections 9.5.3 and 9.5.4, respectively. Kinematically admissible virtual displacements satisfy the geometric boundary conditions, whereas arbitrary virtual displacements do not.

Beams under transverse loads

The reasoning presented in the previous section for beams under axial loads will be repeated here for beam under transverse loads. Consider a cantilevered beam with a tip support subjected to a uniform transverse load, p_0, and a tip moment, Q_T, as depicted in fig. 11.29. In section 11.3.4, the weak statement of equilibrium is developed for this problem. A very similar reasoning will be presented here, with special attention devoted to the treatment of the boundary conditions.

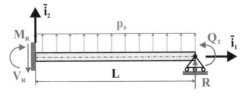

Fig. 11.29. Cantilevered beam with a tip support.

The differential equation of equilibrium of a beam under bending is developed in section 5.5.3, and the equilibrium equation is given by eq. (5.39). This equation holds at all points over the span of the beam, *i.e.*, for $0 \le x_1 \le L$.

As indicated in fig. 11.29, several reaction forces will also develop. At the beam's root, a shear reaction, V_R, and a bending moment, M_R, will appear, while at the beam's tip, a vertical force, R, arises. As expected, each of these forces is associated with a specific geometric boundary condition. At the beam's root, the vanishing of the displacement and rotation generates the reaction shear force, V_R, and moment, M_R, respectively, while at the beam's tip, the vanishing of the vertical displacement causes the vertical reaction, R. The complete set of equilibrium equations is now

$$\frac{d^2 M_3}{dx_1^2} = p_2, \quad \text{for} \quad 0 \le x_1 \le L, \tag{11.54a}$$

$$V_2 = V_R, \quad M_3 = M_R, \quad \text{for} \quad x_1 = 0, \tag{11.54b}$$

$$V_2 = R, \quad M_3 = Q_T, \quad \text{for} \quad x_1 = L. \tag{11.54c}$$

Next, the following integral statement is constructed using the sign convention defined in section 11.3.1

$$\int_0^L -\delta \bar{u}_2 \left[\frac{d^2 M_3}{dx_1^2} - p_2 \right] dx_1 + \delta \bar{u}_2(0) \left[V_2 - V_R \right]_0 + \delta \left(\frac{d\bar{u}_2}{dx_1} \right)_0 \left[M_3 - M_R \right]_0$$
$$- \delta \bar{u}_2(L) \left[V_2 - R \right]_L - \delta \left(\frac{d\bar{u}_2}{dx_1} \right)_L \left[M_3 - Q_T \right]_L = 0.$$

where $\delta \bar{u}_2(x_1)$ is an arbitrary virtual displacement field.

If the beam is in equilibrium, eqs. (11.54) must hold, and therefore, the above equation is satisfied *for all arbitrary virtual displacements*. Indeed, the five bracketed terms set equal to zero are simply the equilibrium equations of the problem.

Next, two integrations by parts are performed on the first term appearing in the above integral, and the δ and d operators are interchanged, see eq. (11.32), leading to the following statement

$$-\int_0^L \delta \left(\frac{d^2 \bar{u}_2}{dx_1^2} \right) M_3 \, dx_1 + \int_0^L \delta \bar{u}_2 p_2 \, dx_1$$
$$-\delta \bar{u}_2(0) V_R - \delta \left(\frac{d\bar{u}_2}{dx_1} \right)_0 M_R + \delta \bar{u}_2(L) R + \delta \left(\frac{d\bar{u}_2}{dx_1} \right)_L Q_T = 0, \tag{11.55}$$

for all arbitrary virtual displacements.

As discussed in section 11.3.7, the first integral can be interpreted as the virtual work, δW_I, done by the bending moment acting within in the beam. The remaining five terms form the virtual work, δW_E, done by the externally applied loads. The integral is the virtual work done by the distributed load, p_2, and the last four terms are the virtual work done by the root reaction shear force, root reaction bending moment, tip reaction and applied tip moment, respectively.

The above statement once again expresses the principle of virtual work and should be compared to that given by eq. (11.44). Here again, two crucial differences are observed. In eq. (11.55), the virtual work done by the externally applied forces includes the work done by all reaction forces and moments, and the virtual displacements are entirely arbitrary. In contrast, in eq. (11.44), the reaction forces and moments do not appear in the expression for the virtual work done by the externally applied forces, because the virtual displacements must satisfy the geometric boundary conditions and are therefore zero at the reaction points.

It is easy to derive statement (11.44) from eq. (11.55). Indeed, in eq. (11.55), virtual displacements are entirely arbitrary, and hence, it is always possible to select $\delta \bar{u}_2(0) = \delta \left(d\bar{u}_2(0)/dx_1 \right) = \delta \bar{u}_2(L) = 0$, i.e., to restrict the virtual displacements to those that satisfy the geometric boundary conditions. Introducing these three conditions into eq. (11.55) then leads to statement (11.44).

Restricting the virtual displacements to only those that satisfy the geometric boundary conditions implies that the reaction forces and moments no longer appear in the statement of the principle of virtual work, because the virtual work they performs does vanish.

In summary, the principle of virtual work stated by eq. (11.53) allows the use of any arbitrary virtual displacements, *including those that violate the geometric boundary conditions*. The reactions forces and moments, however, must be included in the statement of the virtual work done by the externally applied loads, *i.e., reaction forces and moments should be treated as externally applied loads*.

Example 11.12. *Cantilevered beam with tip support. Case 1*

To illustrate the treatment of the boundary conditions discussed in the previous sections, consider a cantilevered beam with a tip support, subjected to a uniform transverse load, p_0, as depicted in fig. 11.29. The tip moment, Q_T, will be set to zero in this example. The geometric boundary conditions for this problem are $\bar{u}_2(0) = d\bar{u}_2/dx_1(0) = \bar{u}_2(L) = 0$. This problem is treated in example 5.11 on page 205 using the classical differential equation approach.

The principle of virtual work will be used to find an approximate solution of this problem, starting with the following assumed displacement field

$$\bar{u}_2(\eta) = h_1(\eta)q_1 + h_2(\eta)q_3 + h_3(\eta)q_3 = \eta^2(1-\eta)q_1 + \eta^2(1-\eta)\eta q_2 + \eta^2 q_3, \quad (11.56)$$

where $\eta = x_1/L$ is the non-dimensional variable along the beam's span. Note that the first two shape functions, $h_1 = \eta^2(1 - \eta)$ and $h_2 = \eta^2(1 - \eta)\eta$, satisfy all three boundary conditions, whereas the last shape function, $h_3 = \eta^2$, satisfies the first two, but not the last, because $h_3(1) = 1 \neq 0$. Consequently, the appropriate statement of the principle of virtual work is

$$-\int_0^L \delta\left(\frac{d^2\bar{u}_2}{dx_1^2}\right) M_3 \, dx_1 + \int_0^L \delta\bar{u}_2 p_2 \, dx_1 + \delta\bar{u}_2(L)R = 0.$$

Because all shape functions satisfy the two geometric boundary conditions at the beam's root, the virtual work done by the root reaction force and bending moment vanish, and hence, do not appear in the above statement. On the other hand, because one of the shape functions violates the geometric boundary condition at the beam's tip, the virtual work done by the tip reaction force does not vanish and must be included in the principle.

Introducing the assumed displacement field along with the moment-curvature expression, $M_3 = H_{33}^c \bar{u}_2'' / L^2$, and using Galerkin's approach then leads to

$$- \frac{H_{33}^c}{L^3} \int_0^1 [h_1'' \delta q_1 + h_2'' \delta q_2 + h_3'' \delta q_3] [h_1'' q_1 + h_2'' q_2 + h_3'' q_3] \, d\eta$$

$$+ p_0 L \int_0^1 [(\eta^2 - \eta^3)\delta q_1 + (\eta^3 - \eta^4)\delta q_2 + \eta^2 \delta q_3] \, d\eta + \delta q_3 R = 0.$$

Introducing the derivatives of the shape functions, $h_1'' = 2 - 6\eta$, $h_2'' = 6\eta - 12\eta^2$, and $h_3'' = 2$, and evaluating all integrals leads to

$$- \{\delta q_1, \delta q_2, \delta q_3\} \frac{H_{33}^c}{L^3} \begin{bmatrix} 4 & 4 & -2 \\ 4 & \frac{24}{5} & -2 \\ -2 & -2 & 4 \end{bmatrix} \begin{Bmatrix} q_1 \\ q_2 \\ q_3 \end{Bmatrix} + \{\delta q_1, \delta q_2, \delta q_3\} \begin{Bmatrix} p_0 L/12 \\ p_0 L/20 \\ p_0 L/3 + R \end{Bmatrix} = 0.$$

Because the virtual displacements are arbitrary coefficients, they can be selected as $\{\delta q_1, \delta q_2, \delta q_3\} = \{1, 0, 0\}$, $\{\delta q_1, \delta q_2, \delta q_3\} = \{0, 1, 0\}$, and $\{\delta q_1, \delta q_2, \delta q_3\} = \{0, 0, 1\}$. The resulting three equations then form a set of linear equations

$$\frac{H_{33}^c}{L^3} \begin{bmatrix} 4 & 4 & -2 \\ 4 & \frac{24}{5} & -2 \\ -2 & -2 & 4 \end{bmatrix} \begin{Bmatrix} q_1 \\ q_2 \\ q_3 \end{Bmatrix} = \begin{Bmatrix} p_0 L/12 \\ p_0 L/20 \\ p_0 L/3 + R \end{Bmatrix}. \tag{11.57}$$

This system is a set of three equations for four unknowns, q_1, q_2, q_3, and the tip reaction force, R.

At this point it is important to remember that the assumed solution, eq. (11.56), does not satisfy the geometric boundary condition at the beam's tip, $\bar{u}_2(1) = q_3 \neq 0$. In other words, the solution process is not "aware" of the fact that the beam's tip deflection must vanish.

To proceed further, the tip boundary condition, *i.e.*, $q_3 = 0$, must now be enforced. It then becomes possible to solve the first two equations of system (11.57) to find $q_1 = 3p_0 L^4 / (48 H_{33}^c)$ and $q_2 = -2p_0 L^4 / (48 H_{33}^c)$. The transverse displacement field, eq. (11.56), is found as

$$\bar{u}_2 = \frac{1}{48} \frac{p_0 L^4}{H_{33}^c} \eta^2 (1 - \eta)(3 - 2\eta).$$

This result matches the exact solution given by eq. (5.61).

Next, the last equation of system (11.57) is solved for the tip reaction force to yield

$$R = -\frac{p_0 L}{3} - 2\frac{H_{33}^c}{L^3}(q_1 + q_2) = -\frac{3p_0 L}{8},$$

which matches the exact solution, eq. (5.62). The use of a virtual displacement that does not satisfy the geometric boundary condition at the beam's tip provides an additional equation that can be solved for the reaction force, R.

As mentioned earlier, at any point along the beam's span, either displacement or external force can be prescribed, but not both at the same time. The structure of system (11.57) reflect this fact: it features three equations linking four variables. If the tip deflection is prescribed, $q_3 = 0$, and the remaining three variables can be found, as is done in the previous paragraph. If the tip force is prescribed, R is a known quantity, and the solution of system (11.57) for q_1, q_2, and q_3, yields the transverse displacement field as

$$\bar{u}_2 = \frac{p_0 L^4}{48 H_{33}^c} \eta^2 (1 - \eta)(3 - 2\eta) + \frac{R L^3}{6 H_{33}^c} \eta^2 (3 - \eta).$$

The first term corresponds to the deflection of the cantilevered beam under the uniform load, and the second is that due to the externally applied tip load, R. In this case, the tip support is not present, and the tip boundary condition is not imposed.

Finally, it should be noted that system (11.57) can also be used to find the response of the structure under a prescribed displacement, Δ, at the tip. In this case, $q_3 = \Delta$, and the system is solved for the remaining variables to find the transverse displacement field

$$\bar{u}_2 = \frac{p_0 L^4}{48 H_{33}^c} \eta^2 (1 - \eta)(3 - 2\eta) + \frac{\Delta}{2} \eta^2 (3 - \eta).$$

The last equation of system (11.57) is then solved for the reaction force $R = 3 p_0 L / 8 + 3 H_{33}^c \Delta / L^3$.

Example 11.13. Cantilevered beam with tip support. Case 2

Consider a cantilevered beam with a tip support subjected to a uniform transverse load, p_0, as depicted in fig. 11.29. The tip moment, Q_T, shown in the figure will again be set to zero in this example. The geometric boundary conditions for this problem are $\bar{u}_2(0) = d\bar{u}_2/dx_1(0) = \bar{u}_2(L) = 0$.

The principle of virtual work will be used to find an approximate solution of this problem, starting with the following assumed displacement field that is slightly different from the one used in the previous example

$$\bar{u}_2(\eta) = h_1 q_1 + h_2 q_2 + h_3 q_3 = \eta^2 (1 - \eta) q_1 + \eta^2 (1 - \eta)\eta q_2 + \eta(1 - \eta) q_3, \quad (11.58)$$

where $\eta = x_1/L$ is the non-dimensional variable along the beam's span. Note that the first two shape functions, $h_1 = \eta^2(1 - \eta)$ and $h_2 = \eta^2(1 - \eta)\eta$, satisfy all three boundary conditions, whereas the last shape function, $h_3 = \eta(1 - \eta)$, satisfies the zero deflection conditions at the beam's root and tip, but the vanishing of the root rotation is not satisfied because $h_3'(0) = 1 \neq 0$.

An appropriate statement of the principle of virtual work is, therefore,

$$-\int_0^L \delta \left(\frac{d^2 \bar{u}_2}{dx_1^2} \right) M_3 \, dx_1 + \int_0^L \delta \bar{u}_2 p_2 \, dx_1 - \delta \left(\frac{d\bar{u}_2}{dx_1} \right)_0 M_R = 0.$$

In this case, the virtual work done by the root reaction moment must be included in the statement of the principle because the third shape function does not satisfy the zero slope condition at the beam's root.

Introducing the assumed displacement field and using Galerkin's approach then leads to

$$
-\frac{H_{33}^c}{L^3} \int_0^1 [h_1'' \delta q_1 + h_2'' \delta q_2 + h_3'' \delta q_3] [h_1'' q_1 + h_2'' q_2 + h_3'' q_3] \, d\eta
$$

$$
+ p_0 L \int_0^1 [(\eta^2 - \eta^3)\delta q_1 + (\eta^3 - \eta^4)\delta q_2 + \eta(1 - \eta)\delta q_3] \, d\eta - \delta q_3 M_R/L = 0.
$$

After introducing the derivatives of the shape functions, $h_1'' = 2 - 6\eta$, $h_2'' = 6\eta - 12\eta^2$, and $h_3'' = -2$, and evaluating of all integrals, the following system of equations is found

$$
\frac{H_{33}^c}{L^3}
\begin{bmatrix} 4 & 4 & 2 \\ 4 & \frac{24}{5} & 2 \\ 2 & 2 & 4 \end{bmatrix}
\begin{Bmatrix} q_1 \\ q_2 \\ q_3 \end{Bmatrix}
=
\begin{Bmatrix} p_0 L/12 \\ p_0 L/20 \\ p_0 L/6 - M_R/L \end{Bmatrix}.
\tag{11.59}
$$

This system is a set of three equations for four unknowns, q_1, q_2, q_3, and the root reaction moment, M_R.

The assumed solution, eq. (11.58), does not satisfy one of the geometric boundary conditions at the beam's root, $d\bar{u}_2(0)/dx_1 = q_3/L \neq 0$. To proceed, this root boundary condition, i.e., $q_3 = 0$, must now be enforced. It then becomes possible to solve the first two equations of system (11.59) to find $q_1 = 3p_0 L^4/(48 H_{33}^c)$ and $q_2 = -2p_0 L^4/(48 H_{33}^c)$. Once again, this leads to the exact solution for the transverse displacement field, which is given by eq. (5.61). Finally, the last equation of system (11.59) is solved for the root reaction moment, to yield

$$
M_R = \frac{p_0 L^2}{6} - 2\frac{H_{33}^c}{L^3} L(q_1 + q_2) = \frac{p_0 L^2}{8}.
$$

Elementary statics arguments reveal that this result is exact. Clearly, the use of a virtual displacement that does not satisfy the rotation geometric boundary condition at the beam's root yields an additional equilibrium equation that can be solved for the root reaction moment, M_R.

If the root moment, M_R, is assumed to be known, system 11.59 can then be solved directly for q_1, q_2, and q_3. In this case, the vanishing of the root rotation is not enforced, and the problem now consists of a beam simply supported at both ends and subjected to a uniform transverse loading and the root bending moment, M_R. The transverse displacement field is found as

$$
\bar{u}_2 = \frac{p_0 L^4}{24 H_{33}^c} \eta(1 - \eta)(1 + \eta - \eta^2) - \frac{M_R L^2}{6 H_{33}^c} \eta(1 - \eta)(2 - \eta).
$$

The first term corresponds to the deflection of the simply supported beam under the uniform load, see eq. (5.48), and the second is that due to the externally applied root moment, M_R.

11.3.10 Summary

Equilibrium formulations

The three equilibrium formulations derived thus far all express the equilibrium conditions for a beam because all are derived from the weak statement of equilibrium. However, these formulations are not all equivalent.

The weak statement of equilibrium and the principle of virtual work are entirely equivalent to Newton's law and express the equilibrium conditions for the beam; they provide no information about either constitutive laws or strain-displacement relationships. Newton's law expresses the vanishing of all forces and moments acting on each differential element of the structure, and so do the weak statement of equilibrium and the principle of virtual work. Because Newton's law always applies to any structure, so do the weak statement of equilibrium and the principle of virtual work.

The principle of minimum total potential energy is the third statement of equilibrium, but both constitutive laws and strain-displacement relationships are incorporated into this principle. The principle can be expressed in a single, concise statement that encapsulates the three groups of equations required for the solution of structural problems. It is, however, important to note that *two fundamental assumptions* are made in the derivation of this principle and restrict its applicability. First, the existence of a strain energy density function is assumed and second, the externally applied loads are assumed to be conservative. These two assumptions are discussed in section 10.2. A hybrid form, principle 10, is useful when a strain energy function exists but applied loads are nonconservative and therefore cannot be derived from a potential.

Variational operator

In making the connection between the weak statement of equilibrium and the principle of virtual work, the variational operator, δ, is introduced, and several properties of the this operator are used in this section. Clearly, this operator can be used in much the same way as the differential operator, d, although its physical interpretation is quite different. Using the variational operator, virtual displacements are introduced in eq. (11.41) as being simply equivalent to arbitrary functions. A formal mathematical treatment of the variational operator can be found in several textbooks [6, 5] and is beyond the scope of this book.

11.4 Formal procedures for the derivation of approximate solutions

The examples presented in the previous sections demonstrate that approximate solutions of structural problem can be derived from either the weak statement of equilibrium, the principle of virtual work, or the principle of minimum total potential energy. To extend this approach to more complex structures, a formal procedure for constructing approximate solutions based on these concepts is now introduced.

Approximate solutions are generally implemented in computer programs, and for this reason, it will be convenient to recast all quantities in the form of arrays or matrices to enable the systematic use of linear algebra methods.

After presentation of the basic approximation approach, procedures based on the principle of virtual work and principle of minimum total potential energy will then be described separately.

11.4.1 Basic approximations

The problem of a beam under axial loads will be used once again to illustrate the process. The first step of the solution procedure is to assume the displacements and virtual displacements fields to be of the following forms

$$\bar{u}_1(x_1) = \sum_{i=1}^{N} h_i(x_1)q_i, \tag{11.60a}$$

$$\delta\bar{u}_1(x_1) = \sum_{i=1}^{N} g_i(x_1)w_i, \tag{11.60b}$$

respectively, where the coefficients q_i, $i = 1, 2, \ldots, N$, are unknown coefficients, often called *degrees of freedom*, which determine the solution of the problem, and the coefficients w_i, $i = 1, 2, \ldots, N$, a set of arbitrary coefficients reflecting the arbitrary nature of the virtual displacements.

Functions $h_i(x_1)$ and $g_i(x_1)$ are sets of arbitrary functions called *shape functions*, each of which must satisfy the geometric boundary conditions of the problem. Polynomials or trigonometric functions can be selected as shape functions; transcendental functions can also be used as long as they form a set of linearly independent functions that each satisfy the geometric boundary conditions.

In Galerkin's approach, the same shape functions are selected for both displacements and virtual displacements, *i.e.*, $h_i(x_1) = g_i(x_1)$. Although this is a common and convenient choice, it is not required by any of the approaches.

Equation (11.60a) represents an approximate solution of the problem because it combines a finite number, N, of preselected shape functions. Each coefficient, q_i, indicates how much the corresponding shape function contributes to the final solution; hence, these coefficients are also called *participation factors*. If a complete series of shape functions is selected, the approximate solution should converge to the exact solution as an increasing number of shape functions is used. The finite series limit in eq. (11.60a) reduces the number of degrees of freedom from infinity to N and results in an approximate solution. The same remarks can be made about the assumed form of the virtual or test displacements, eq. (11.60b).

All three principles considered here require an integral statement to hold "for all arbitrary choices of the virtual displacements that satisfy the geometric boundary conditions." This calls for virtual displacements or test functions involving an infinite number of degrees of freedom: "for all arbitrary choices" clearly means "for an infinite number of arbitrary choices." Here again, the assumption implied by eq. (11.60b) reduces the number of choices for the virtual displacements from infinity to N; hence, "for all arbitrary choices" is replaced by "for N arbitrary choices."

It is convenient to recast the expressions for the assumed displacements and virtual displacements, eqs. (11.60), into a matrix form as

$$\bar{u}_1(x_1) = \underline{H}^T(x_1)\underline{q}; \quad \text{and} \quad \delta\bar{u}_1(x_1) = \underline{H}^T(x_1)\underline{w}, \tag{11.61}$$

where $\underline{q} = \{q_1, q_2, \cdots, q_N\}^T$ is an array of size N that stores the participation factors, and $\underline{w} = \{w_1, w_2, \cdots, w_N\}^T$ an array of arbitrary coefficients. The assumed displacements and virtual or test displacements are scalar quantities expressed as the scalar product of a row and a column array. The *displacement interpolation array*, $\underline{H}(x_1)$, also of size N, stores the selected shape functions,

$$\underline{H} = \{h_1(x_1), h_2(x_1), h_3(x_1), \cdots, h_N(x_1)\}^T. \tag{11.62}$$

When using Galerkin's approach, identical shape functions are chosen for both displacements and virtual displacements, leading to identical interpolation arrays for both quantities.

Next, the assumed displacements are introduced in the strain-displacement relationship for a beam under axial load, eq. (5.6), to find the corresponding axial strain distribution

$$\bar{\epsilon}_1(x_1) = \frac{\partial \bar{u}_1}{\partial x_1} = \frac{\partial}{\partial x_1} \underline{H}^T(x_1)\,\underline{q} = \underline{B}^T(x_1)\,\underline{q}. \tag{11.63}$$

The *strain interpolation array* of size N is defined as

$$\underline{B}(x_1) = \left\{ \frac{dh_1}{dx_1}, \frac{dh_2}{dx_1}, \frac{dh_3}{dx_1}, \cdots, \frac{dh_N}{dx_1} \right\}^T, \tag{11.64}$$

The virtual strains can be written in a similar manner as

$$\delta\bar{\epsilon}_1(x_1) = \delta\left(\frac{d\bar{u}_1}{dx_1}\right) = \frac{d}{dx_1}\delta\bar{u}_1 = \underline{B}^T(x_1)\,\underline{w}. \tag{11.65}$$

11.4.2 Principle of virtual work

All terms appearing in the principle of virtual work are virtual work quantities expressed as the product of forces by virtual displacements or stresses by virtual strains. In the present formalism, the virtual work is expressed as the scalar product of a row, representing forces, and a column, representing virtual displacements, *i.e.*, $\underline{F}^T\underline{d}$, where $\underline{F}$ and $\underline{d}$ represent the force and virtual displacement array, respectively. The factors of a scalar product commute, and the virtual work can be expressed as either $\underline{F}^T\underline{d}$ or as $(\underline{F}^T\underline{d})^T = \underline{d}^T\underline{F}$.

Beam under axial load

Consider a beam under axial loads and the corresponding statement of the principle of virtual work given by eq. (11.42), repeated here for reference

$$-\int_0^L N_1 \delta\bar{e} \, \mathrm{d}x_1 + \int_0^L \delta\bar{u}_1(x_1)p_1 \, \mathrm{d}x_1 + \delta\bar{u}_1(L)P_1 = 0.$$

Introducing the approximations presented in the previous section, and using the appropriate constitutive law, $N_1 = S\bar{e}_1$, leads to

$$-\int_0^L \left(\underline{B}^T\underline{w}\right)^T S\left(\underline{B}^T\underline{q}\right) \, \mathrm{d}x_1 + \int_0^L \left(\underline{H}^T\underline{w}\right)^T p_1 \, \mathrm{d}x_1 + \left(\underline{H}^T(L)\underline{w}\right)^T P_1$$

$$= -\int_0^L \underline{w}^T\underline{B}S\underline{B}^T\underline{q} \, \mathrm{d}x_1 + \int_0^L \underline{w}^T\underline{H}p_1 \, \mathrm{d}x_1 + \underline{w}^T\underline{H}(L)P_1 = 0,$$

where each term in this expression evaluates to a scalar, as expected.

The arrays of coefficients, $\underline{q}$ and $\underline{w}$, do not depend on variable x_1 and can be extracted from the integrals, to yield

$$-\underline{w}^T\left[\int_0^L \underline{B}S\underline{B}^T \, \mathrm{d}x_1\right]\underline{q} + \underline{w}^T\left[\int_0^L \underline{H}p_1 \, \mathrm{d}x_1\right] + \underline{w}^T\underline{H}(L)P_1 = 0. \quad (11.66)$$

To simplify the above equation, two important quantities are defined. First, the *stiffness matrix*, which is the bracketed matrix in the first term of the equation, is defined as

$$\underline{\underline{K}} = \int_0^L \underline{B}(x_1)S(x_1)\underline{B}^T(x_1) \, \mathrm{d}x_1, \quad (11.67)$$

and is of size $N \times N$. Each entry of the stiffness matrix is an average of the axial stiffness distribution weighted by a product of the strain interpolation array. For a given choice of the shape functions, each entry of the stiffness matrix can be obtained by integration along the beam's span. Moreover, the stiffness matrix is symmetric, $\underline{\underline{K}}^T = \underline{\underline{K}}$.

Second, the *load array* of size N is defined as

$$\underline{Q} = \int_0^L \underline{H}(x_1)p_1(x_1) \, \mathrm{d}x_1 + \underline{H}(L)P_1. \quad (11.68)$$

Here again, for a given choice of the shape functions, the entries of the load array can be obtained by integration along the beam's span. Each entry of the load array corresponds to an average of the applied load distribution weighted by the displacement interpolation array.

With these two definitions, eq. (11.66) can be recast in a compact form as

$$\underline{w}^T\left[\underline{\underline{K}}\underline{q} - \underline{Q}\right] = \underline{0}. \quad (11.69)$$

Because the N entries of the array of arbitrary coefficients, $\underline{w}$, can be selected at will, N convenient choices will be used: $\underline{w} = \{1, 0, 0, \cdots, 0\}^T$, $\underline{w} = \{0, 1, 0, \cdots, 0\}^T$, $\underline{w} = \{0, 0, 1, \cdots, 0\}^T$, $\cdots$, and finally $\underline{w} = \{0, 0, 0, \cdots, 1\}^T$. Each new choice gives rise to a new equation, and this collection of N equations can be written as

$$\underline{\underline{I}} \left[\underline{\underline{K}} \, \underline{q} - \underline{Q} \right] = \underline{0},$$

where $\underline{\underline{I}}$ is the $N \times N$ identity matrix. Because the identity matrix is never singular, N algebraic equations for the solution array are obtained

$$\underline{\underline{K}} \, \underline{q} = \underline{Q}. \tag{11.70}$$

This equation expresses the relationship between the externally applied loads represented by the load array, $\underline{Q}$, and the resulting structural displacements represented by the solution array, $\underline{q}$. Because the structure is assumed to behave in a linear manner, a linear relationship exists between the applied loads and the resulting displacements. In the present linear algebra formalism, this proportionality takes the form of set of linear equations.

The solution of the above set of equations is $\underline{q} = \underline{\underline{K}}^{-1}\underline{Q}$. Once the solution array is found, the displacement field follows from eq. (11.60a). Next, the strain field is evaluated with the help of eq. (11.63), and finally, the constitutive law, $N_1 = S\bar{\epsilon}_1$, yields the internal force.

The general solution procedure using the principle of virtual work can be summarized by the following steps,

1. Select a set of N shapes functions that satisfy the geometric boundary conditions.
2. Construct the displacement interpolation array, eq. (11.62), and strain interpolation array, eq. (11.64).
3. Compute the stiffness matrix according to eq. (11.67) and load array from eq. (11.68).
4. Solve the set of N simultaneous linear equations, $\underline{\underline{K}} \, \underline{q} = \underline{Q}$, for the solution array, $\underline{q}$.
5. Determine the strain distribution from eq. (11.63), and the internal forces from the constitutive law, eq. (5.16).

From a mathematical standpoint, this procedure involves the following types of operation: integrations over the beam's span for evaluating the stiffness matrix and load array, the solution of a set of linear algebraic equations to obtain the solution array, and linear algebra operations for the determination of the strain and internal force distributions. Of course, the process becomes increasingly tedious as the number of degrees of freedom increases. All the required operations, however, are easily performed on computers using numerical integration procedures for the evaluation of the stiffness matrix and load array, and standard linear algebra software packages for all remaining operations. This makes such methods very attractive for implementation on a computer.

Beam under transverse load

A process nearly identical to that presented above can be developed for beams subjected to transverse loads. To start, specific forms for the transverse displacements and virtual displacements fields are assumed to be in the following form

$$\bar{u}_2(x_1) = \sum_{i=1}^{N} h_i(x_1)q_i; \quad \text{and} \quad \delta\bar{u}_2(x_1) = \sum_{i=1}^{N} h_i(x_1)w_i. \qquad (11.71)$$

The displacement field can then be expressed as $\bar{u}_2(x_1) = \underline{H}^T(x_1)\underline{q}$, where the displacement interpolation array is given by eq. (11.62). Next, the curvatures are computed from eq. (5.6) to find $\kappa_3 = \underline{B}^T(x_1)\underline{q}$ where the curvature interpolation array is

$$\underline{B}(x_1) = \left\{ \frac{d^2h_1}{dx_1^2}, \frac{d^2h_2}{dx_1^2}, \frac{d^2h_3}{dx_1^2}, \cdots \frac{d^2h_N}{dx_1^2} \right\}^T. \qquad (11.72)$$

The rest of the procedure mirrors that presented above for the beam under axial load with the stiffness matrix and load array defined as

$$\underline{\underline{K}} = \int_0^L \underline{B}(x_1)H_{33}(x_1)\underline{B}^T(x_1)\,dx_1, \qquad (11.73)$$

and

$$\underline{Q} = \int_0^L \underline{H}(x_1)p_2(x_1)\,dx_1 + \underline{H}(L)P_2, \qquad (11.74)$$

respectively.

It is interesting to note that the entire solution process can be automated once the particular shape functions have been selected. For instance, example 11.7 is an axial load problem and is characterized by a displacement interpolation array $\underline{H}(x_1) = \{x_1, x_1^2\}^T$ and corresponding strain interpolation array $\underline{B}(x_1) = \{1, 2x_1\}^T$. On the other hand, example 11.11 is a transverse load (bending) problem with corresponding quantities $\underline{H}(x_1) = \{\sin \pi x_1/L, \sin 3\pi x_1/L\}^T$ and $\underline{B}(x_1) = -(\pi/L)^2 \{\sin \pi x_1/L, 9\sin 3\pi x_1/L\}^T$. The remaining solution steps are identical for both problems.

11.4.3 The principle of minimum total potential energy

A formal solution procedure can also be developed based on the principle of minimum total potential energy to obtain approximate solutions of structural problem. The methodology closely mirrors that used in conjunction with the principle of virtual work for axially loaded beams and for beams under transverse loads.

Axially loaded beams will be considered first. The first step of the solution procedure is to assume a specific form for the displacement field

$$\bar{u}_1(x_1) = \sum_{i=1}^{N} h_i(x_1)\, q_i = \underline{H}^T \underline{q}.$$

Next, the strains are expressed in terms of the assumed displacements, resulting in eq. (11.63), with the strain interpolation array defined in eq. (11.64). Using these, the total strain energy in the beam can now be written as

$$A(\bar{\epsilon}_1) = \frac{1}{2}\int_0^L S\bar{\epsilon}_1^2 \, dx_1 = \frac{1}{2}\int_0^L S(\underline{B}^T\underline{q})^T (\underline{B}^T\underline{q}) dx_1$$
$$= \frac{1}{2}\underline{q}^T \left[\int_0^L S\,\underline{B}\,\underline{B}^T \, dx_1\right]\underline{q} = \frac{1}{2}\underline{q}^T\underline{\underline{K}}\,\underline{q}, \tag{11.75}$$

where the stiffness matrix is identical to that defined in eq. (11.67).

Next, the total potential of the externally applied loads becomes

$$\varPhi = \int_0^L \phi \, dx_1 + \varPsi = -\left[\int_0^L \underline{H}^T(x_1)p_1(x_1)\, dx_1\right]\underline{q} - \underline{H}^T(L)P_1\underline{q} = -\underline{Q}^T\underline{q}. \tag{11.76}$$

Here again, the load array, $\underline{Q}$, is the same as that found earlier, eq. (11.68). The total potential energy of the system, $\varPi = A + \varPhi$, can now be written as

$$\varPi(\underline{q}) = \frac{1}{2}\underline{q}^T\underline{\underline{K}}\,\underline{q} - \underline{Q}^T\underline{q}. \tag{11.77}$$

According to the principle of stationary total potential energy, principle 8, expressed by eq. (10.17), the derivatives of the total potential energy with respect to the degrees of freedom, $\underline{q}$, must vanish, leading to

$$\frac{\partial \varPi}{\partial \underline{q}} = \frac{\partial}{\partial \underline{q}}\left(\frac{1}{2}\underline{q}^T\underline{\underline{K}}\,\underline{q} - \underline{Q}^T\underline{q}\right) = \underline{\underline{K}}\,\underline{q} - \underline{Q} = \underline{0}. \tag{11.78}$$

where eqs. (A.29) and (A.27) are used to compute the derivatives of the strain energy and potential of the externally applied loads, respectively. The final equilibrium equations of the problem are

$$\underline{\underline{K}}\,\underline{q} - \underline{Q} = \underline{0}. \tag{11.79}$$

The general solution procedure using the principle of minimum total potential energy can be summarized by the following steps.

1. Select N shapes functions that satisfy the geometric boundary conditions.
2. Construct the displacement interpolation array, eq. (11.62), and strain interpolation array, eq. (11.64).
3. Compute the stiffness matrix according to eq. (11.67) and load array from eq. (11.68).
4. Compute the total potential of the externally applied loads, $\varPhi(\underline{q}) = -\underline{q}^T\underline{Q}$, where the load array, $\underline{Q}$, is given by eq. (11.68).

5. Solve the set of N simultaneous linear equations, $\underline{\underline{K}}\,\underline{q} = \underline{Q}$, for the solution array, $\underline{q}$.
6. Determine the strain distribution from eq. (11.63), and the internal forces from the constitutive law, eq. (5.16).

Clearly, the solution procedures based on the principle of virtual work and principle of minimum total potential energy are closely related. Although the physical interpretation of the intermediate quantities is different, the major elements of the two procedures, the stiffness matrix and load array, are identical, and so are the final solutions.

The general method presented here for beams under axial loads can be readily applied to beams subjected to transverse loads by using the appropriate expressions for the stiffness matrix, eq. (11.73), and load array, eq. (11.74).

When the structural system to be analyzed comprises several elastic components, such as beams and springs, the strain energies of the various elastic elements are simply added together to find the total strain energy. This additive property of strain energy is one of the key simplifications inherent to variational and energy methods.

Consider, for instance, a cantilevered beam with a spring of stiffness constant k connected to the ground at location $x_1 = \alpha L$. The elastic components of the system are the beam and spring, and the total strain energy can be written as

$$A = \frac{1}{2} \int_0^L H_{33} \left(\frac{d^2 \bar{u}_2}{dx_1^2} \right)^2 dx_1 + \frac{1}{2} k \bar{u}_2^2(\alpha L),$$

where the first term corresponds to the strain energy stored in the beam, and the second represents that stored in the spring. The stiffness matrix is now $\underline{\underline{K}} = \underline{\underline{K}}_b + \underline{\underline{K}}_s$, where $\underline{\underline{K}}_b$ is associated with the strain energy of the beam and K_s with that stored in the spring. Matrix $\underline{\underline{K}}_b$ is given by eq. (11.73), and the strain energy in the spring gives rise to $\underline{\underline{K}}_s$

$$\frac{1}{2} k \bar{u}_2^2(\alpha L) = \frac{1}{2} k (\underline{H}^T(\alpha L)\underline{q})^T (\underline{H}^T(\alpha L)\underline{q})$$

$$= \frac{1}{2} \underline{q}^T \left[\underline{H}(\alpha L) k \underline{H}^T(\alpha L) \right] \underline{q} = \frac{1}{2} \underline{q}^T \underline{\underline{K}}_s \, \underline{q}.$$

Several examples will now be used to illustrate the formal solution procedure for beams under transverse loading.

Example 11.14. Simply supported beam with two elastic spring supports
Figure 5.31 on page 208 depicts a simply supported beam of span L supported by two spring of stiffness constant k located at stations $x_1 = \alpha L$ and $(1 - \alpha)L$, and subjected to a uniform transverse loading p_0. The exact solution of this problem is obtained using the classical differential equation approach in example 5.13 on page 208. This problem will now be analyzed with the principle of minimum total potential energy.

The system under consideration consists of an elastic beam and two elastic spring. The strain energy for beam in bending is given by eq. (10.39), and the strain energy for the springs is $A_s = 1/2\, k\bar{u}_2^2(\alpha L) + 1/2\, k\bar{u}_2^2[(1-\alpha)L]$. The strain energy for the entire system is then the sum to the strain energies for the various components of the system

$$A = \frac{1}{2}\int_0^L H_{33}^c \left(\frac{\mathrm{d}^2\bar{u}_2}{\mathrm{d}x_1^2}\right)^2 \mathrm{d}x_1 + \frac{1}{2}k\bar{u}_2^2(\alpha L) + \frac{1}{2}k\bar{u}_2^2((1-\alpha)L).$$

Next, the displacement interpolation array is selected as

$$\underline{H} = \left\{\sin\pi\eta,\ \sin 3\pi\eta,\ \sin 5\pi\eta\right\}^T,$$

where $\eta = x_1/L$ is a non-dimensional span variable. The shape functions each satisfy the geometric boundary conditions $h_i(0) = h_i(1) = 0$.

The corresponding strain interpolation array becomes

$$\underline{B} = -\pi^2\left\{\sin\pi\eta,\ 3^2\sin 3\pi\eta,\ 5^2\sin 5\pi\eta\right\}^T.$$

and the stiffness matrix for the beam is then

$$\underline{\underline{K}}_b = \frac{H_{33}^c}{L^3}\int_0^1 \underline{B}(\eta)\underline{B}^T(\eta)\,\mathrm{d}\eta = \frac{H_{33}^c\pi^4}{2L^3}\begin{bmatrix} 1 & 0 & 0 \\ 0 & 3^4 & 0 \\ 0 & 0 & 5^4 \end{bmatrix} \qquad (11.80)$$

The stiffness matrix associated with the springs is

$$\underline{\underline{K}}_s = k\,\underline{H}(\alpha)\,\underline{H}^T(\alpha) + k\,\underline{H}(1-\alpha)\,\underline{H}^T(1-\alpha)$$

$$= 2k\begin{bmatrix} \sin^2\pi\alpha & \sin\pi\alpha\sin 3\pi\alpha & \sin\pi\alpha\sin 5\pi\alpha \\ \sin 3\pi\alpha\sin\pi\alpha & \sin^2 3\pi\alpha & \sin 3\pi\alpha\sin 5\pi\alpha \\ \sin 5\pi\alpha\sin\pi\alpha & \sin 5\pi\alpha\sin 3\pi\alpha & \sin^2 5\pi\alpha \end{bmatrix}.$$

Finally, the stiffness matrix for the entire structure, $\underline{\underline{K}} = \underline{\underline{K}}_b + \underline{\underline{K}}_s$, becomes

$$\underline{\underline{K}} = \frac{H_{33}^c}{L^3}\begin{bmatrix} \pi^4/2 + 2\bar{k}\sin^2\pi\alpha & 2\bar{k}\sin\pi\alpha\sin 3\pi\alpha & 2\bar{k}\sin\pi\alpha\sin 5\pi\alpha \\ 2\bar{k}\sin 3\pi\alpha\sin\pi\alpha & 3^4\pi^4/2 + 2\bar{k}\sin^2 3\pi\alpha & 2\bar{k}\sin 3\pi\alpha\sin 5\pi\alpha \\ 2\bar{k}\sin 5\pi\alpha\sin\pi\alpha & 2\bar{k}\sin 5\pi\alpha\sin 3\pi\alpha & 5^4\pi^4/2 + 2\bar{k}\sin^2 5\pi\alpha \end{bmatrix},$$

where $\bar{k} = kL^3/H_{33}^c$ is the non-dimensional spring stiffness constant.

Next, the load array associated with the uniform transverse load is computed as

$$\underline{Q} = p_0 L\int_0^1 \underline{H}(\eta)\,\mathrm{d}\eta = \frac{2p_0 L}{\pi}\left\{1, 1/3, 1/5\right\}^T.$$

The solution of this problem is obtained by solving the linear set of equations $\underline{\underline{K}}\,\underline{q} = \underline{Q}$ to find the solution array. This step is most easily performed numerically because the inversion of the 3×3 stiffness matrix is a rather arduous task to do by hand.

If a single shape function is selected, $i.e.$, if $\underline{H} = \left\{\sin\pi\eta\right\}^T$, then $\underline{q}$ and $\underline{Q}$ contain only a single term and $\underline{\underline{K}}$ is a 1×1 matrix leading to

$$q_1 = \frac{p_0 L^4}{H_{33}^c} \frac{2}{\pi(\pi^4/2 + 2\bar{k}\sin^2 \pi\alpha)}.$$

The exact solution of the problem, given by eq. (5.65), will now be compared with this approximate solution. Three cases, denoted *cases 1, 2,* and *3* correspond to the following displacement interpolation arrays: $\underline{H} = \{\sin \pi\eta\}^T$, $\underline{H} = \{\sin \pi\eta, \sin 3\pi\eta\}^T$, and $\underline{H} = \{\sin \pi\eta, \sin 3\pi\eta, \sin 5\pi\eta\}^T$, respectively.

Figure 11.30 shows the non-dimensional transverse displacement $\bar{u}_2 H_{33}^c/(p_0 L^4)$ for the exact and approximate solutions with the following choice of the parameters: $\alpha = 0.3$ and $\bar{k} = 10^4$. Due to the symmetry of the problem, the solution is presented over a half-span only. Excellent correlation is observed between the exact and approximate solutions. At the scale of the figure, the predictions for *cases 2* and *3* are in close agreement with the exact solution. Table 11.3 quantifies the observed errors for the various approximations.

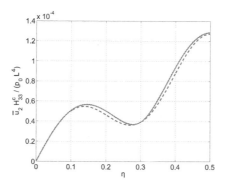

Fig. 11.30. Transverse displacement $\bar{u}_2 H_{33}^c/(p_0 L^4)$ for the exact and approximate solutions versus non-dimensional span. Exact solution: solid line; approximate solution *case 1*: dashed line, *case 2*: dash-dotted line, *case 3*: dotted line.

The exact distribution of bending moment is given by eq. (5.66), and fig. 11.31 depicts the non-dimensional bending moment $M_3/(p_0 L^2)$ for the various solutions. Note that the exact solution presents a cusp at $\eta = 0.3$. This feature is not reproduced by the approximate solutions that consist of a sum of smooth functions. As the number of degrees of freedom increases, the quality of the approximation improves, as detailed in table 11.3. The errors in bending moment predictions are much larger than those observed for the transverse displacements.

Finally, fig. 11.32 shows the non-dimensional shear force $V_2/(p_0 L)$ for the exact solution, eq. (5.67), and the approximate solutions. The exact shear force distribution presents a discontinuity at $\eta = 0.3$ corresponding to the concentrated force the spring applies to the beam. Here again, this feature cannot be reproduced by the approximate solutions that consist of a sum of smooth functions. As the number of degrees of freedom increases, the approximate solution converges to shear force value corresponding to the average of the exact shear forces to the left and right of

Table 11.3. Comparison of the exact and approximate solutions.

	Exact solution	*Case 1* error	*Case 2* error	*Case 3* error
Transverse displacement [10^{-05}]				
$\eta = 0.15$	5.6351	3.9%	-0.50%	-0.050%
$\eta = 0.30$	3.9068	0.03%	0.003%	0.0001%
$\eta = 0.45$	11.882	-2.6%	-0.3%	-0.08%
Bending moment [10^{-03}]				
$\eta = 0.15$	-5.1476	-1.9%	-7.5%	3.6%
$\eta = 0.30$	12.205	-42.%	-19.%	-14.%
$\eta = 0.45$	-6.5452	6.1%	-3.6%	2.4%
Shearing force [10^{-02}]				
$\eta = 0.15$	-4.0683	64.%	-4.9%	5.7%
$\eta = 0.30$	0.9320	-160.%	14.%	35.%
$\eta = 0.45$	5.000	95.%	6.9%	9.75%

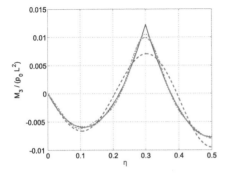

Fig. 11.31. Bending moment $M_3/(p_0 L^2)$ for the exact and approximate solutions versus non-dimensional span. Exact solution: solid line; approximate solution *case 1*: dashed line, *case 2*: dash-dotted line, *case 3*: dotted line.

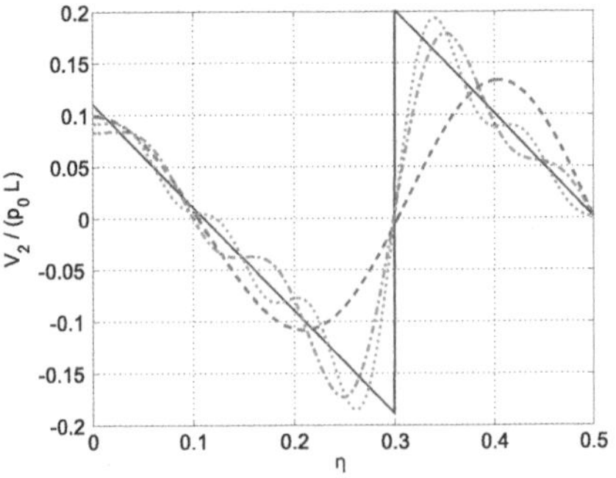

Fig. 11.32. Shear force $V_2/(p_0 L)$ for the exact and approximate solutions versus non-dimensional span. Exact solution: solid line; approximate solution *case 1*: dashed line, *case 2*: dash-dotted line, *case 3*: dotted line.

the discontinuity. This convergence, however, is quite slow, as detailed in table 11.3. The larger errors observed in the predictions of bending moments and shear forces are expected since these quantities are obtained from higher order derivatives of the approximate displacement field.

Example 11.15. Simply supported beam on an elastic foundation

Consider a simply supported beam of length L subjected to a uniform transverse load p_0 and supported by an elastic foundation of distributed stiffness constant k, as depicted in fig. 5.32 on page 209. The exact solution of the problem, obtained from the classical differential approach, is presented in example 5.14 on page 209.

The strain energy for the complete system is

$$A = A_b + A_{ef} = \frac{1}{2} \int_0^L H_{33}^c \left(\frac{d^2 \bar{u}_2}{dx_1^2} \right)^2 dx_1 + \frac{1}{2} \int_0^L k \bar{u}_2^2(x_1) \, dx_1.$$

where the first term is the strain energy in the beam due to bending given by eq. (10.39), and the second term is the strain energy in the elastic foundation.

Here again, the displacement interpolation array is selected as $\underline{H} = \{\sin \pi \eta, \sin 3\pi \eta, \sin 5\pi \eta\}^T$, where η is the non-dimensional variable along the beam's span. Note that each of the shape functions satisfies the geometric boundary conditions, $h_i(0) = h_i(1) = 0$. The corresponding strain interpolation array becomes $\underline{B} = -\pi^2 \left[\sin \pi \eta, \; 3^2 \sin 3\pi \eta, \; 5^2 \sin 5\pi \eta \right]^T$.

The beam's stiffness matrix is identical to that found in the previous example, see eq. (11.80), and the stiffness matrix associated with the elastic foundation is

$$\underline{\underline{K}}_{ef} = \int_0^L k(\underline{H}^T \underline{q})^T (\underline{H}^T \underline{q}) dx_1 = kL \int_0^1 \underline{H}(\eta) \underline{H}^T(\eta) \, d\eta = \frac{kL}{2} \begin{bmatrix} 1 & 0 & 0 \\ 0 & 1 & 0 \\ 0 & 0 & 1 \end{bmatrix}.$$

The stiffness matrix for the entire structure is now

$$\underline{\underline{K}} = \underline{\underline{K}}_b + \underline{\underline{K}}_{ef} = \frac{H_{33}^c}{L^3} \frac{1}{2} \begin{bmatrix} \pi^4 + \bar{k} & 0 & 0 \\ 0 & 3^4 \pi^4 + \bar{k} & 0 \\ 0 & 0 & 5^4 \pi^4 + \bar{k} \end{bmatrix},$$

where $\bar{k} = kL^4/H_{33}^c$ is the non-dimensional stiffness constant of the elastic foundation and expresses this relative to the bending stiffness.

Next, the load array associated with the uniform transverse load is found as

$$\underline{Q} = p_0 L \int_0^1 \underline{H}(\eta) \, d\eta = \frac{2p_0 L}{\pi} \{1, 1/3, 1/5\}^T.$$

The solution of the problem is then obtained by solving the linear set of equations $\underline{\underline{K}} \underline{q} = \underline{Q}$. The solution for the transverse displacement is

$$\bar{u}_2(\eta) = \frac{4}{\pi} \frac{p_0 L^4}{H_{33}^c} \left[\frac{\sin \pi \eta}{\pi^4 + \bar{k}} + \frac{\sin 3\pi \eta}{3(3^4 \pi^4 + \bar{k})} + \frac{\sin 5\pi \eta}{5(5^4 \pi^4 + \bar{k})} \right].$$

The exact solution of the problem, given by eq. (5.71), will now be compared with the above approximate solution. Case 1 and 2, corresponding to the displacement interpolation arrays $\underline{H} = \{\sin \pi \eta, \sin 3\pi \eta\}^T$, and $\underline{H} = \{\sin \pi \eta, \sin 3\pi \eta, \sin 5\pi \eta\}^T$, respectively, will be examined. Figure 11.33 shows the non-dimensional transverse displacement, $\bar{u}_2 H_{33}^c/(p_0 L^4)$, for the exact and approximate solutions when $\bar{k} = 8 \times 10^3$. Good correlation is observed and quantitative results are listed in table 11.4.

The exact distribution of bending moment is given by eq. (5.72), and fig. 11.34 depicts the non-dimensional bending moment, $M_3/(p_0 L^2)$, for the various solutions. Here again, the errors in bending moment predictions are much larger than those observed for those of the transverse displacement. As the number of degrees of freedom increases, the quality of the approximation improves, as shown in table 11.4.

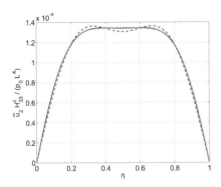

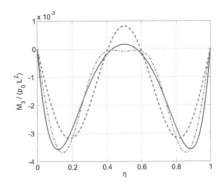

Fig.   11.33.   Transverse   displacement $\bar{u}_2 H_{33}^c/(p_0 L^4)$ for the exact and approximate solutions versus non-dimensional span. Exact solution: solid line; approximate solution *case 1*: dashed line, *case 2*: dash-dotted line.

Fig. 11.34. Bending moment $M_3/(p_0 L^2)$ for the exact and approximate solutions versus non-dimensional span. Exact solution: solid line; approximate solution *case 1*: dashed line, *case 2*: dash-dotted line.

Table 11.4. Comparison of the exact and approximate solutions.

	Exact solution	Case 1 error	Case 2 error
Transverse displacement $\eta = 0.50$	1.3364×10^{-4}	-2.3%	0.4%
Bending moment $\eta = 0.10$	-3.5381×10^{-3}	-32.%	-6.4%

11.4.4 Problems

Problem 11.11. Cantilever with nonuniform bending stiffness

Consider the cantilevered beam subjected to a tip load P as shown in fig. 11.35. The bending stiffness of the beam's left half is $3H_0$, while that of the right half is H_0, as shown in the figure. Develop an approximate solution for the transverse deflection of the entire beam using the principle of minimum total potential energy with a two-term polynomial. Compare your solution at the tip with the exact solution computed using the unit load method.

Problem 11.12. Simply-supported beam with nonuniform bending stiffness

Consider the cantilever beam shown in fig. 11.35 but now assume that both ends are simply supported instead. The bending stiffness of the beam's right half is H_0 while that of the left half is βH_0 where $\beta = 3$. Develop an approximate solution for the transverse deflection of the entire beam using the principle of minimum total potential energy with a two-term trigonometric approximate solution. Compare your solution at the mid-span with the exact solution computed using the unit load method.

Problem 11.13. Simply-supported beam with two mid-span springs

A simply supported beam of span L is supported by two spring of stiffness k located at stations $x_1 = \alpha L$ and $(1 - \alpha)L$, and is subjected to a uniform transverse loading p_0, as depicted in fig. 11.36. (1) Find the exact solution of the problem from the solution of the governing

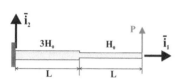

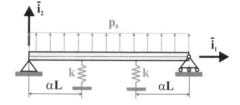

Fig. 11.35. Cantilevered beam with nonuniform bending stiffness.

Fig. 11.36. Simply supported beam with two mid-span springs.

differential equation and associated boundary conditions. *(2)* Find approximate displacement solutions for the problem using the principle of minimum total potential energy with the following assumed shape functions: $h_i(x_1) = \sin(2i-1)\pi x_1/L$. Construct 5 separate solutions for $n = 1, 2, 3$ term approximations. *(3)* On a single graph, plot the exact solution and the 3 approximate solutions. Also construct a single plot of the error in maximum displacement for each of the 3 cases. Use $\bar{k} = kL^3/H_{33}^c = 10^4$ and $\alpha = 0.3$. *(4)* Find the bending moment distribution for each of the approximate solutions. On a single graph, plot the exact solution and the approximate solutions. Also construct a single plot of the error in maximum bending moment for each of the 3 cases. *(5)* Check the overall equilibrium of the problem for both the exact and approximate solutions. In view of the symmetry of the problem, overall equilibrium implies $p_0L = 2R + 2k\bar{u}_2(L/3)$, where R is the reaction at either end of the beam, and $k\bar{u}_2(L/3)$ the force either elastic spring. Comment on your results. *(6)* Based on a simple free body diagram, show that for the exact solution the shear force presents a discontinuity at the elastic springs. What happens in your approximate solution? Comment and explain your results. On a single graph, plot the exact solution and the approximate shear force distribution. Quantify the error in maximum bending moment as the number of terms in the approximate solution increases.

Problem 11.14. Simply-supported beam with two mid-span springs

Consider a simply supported, uniform beam of length L with two end point torsional springs of stiffness k_1 and a mid-span spring of stiffness k_2. The beam, shown in fig. 11.37, is subjected to a uniform transverse loading $p_2(x_1) = p_0$. *(1)* Solve the governing differential equations of this problem to find the transverse displacement $\bar{u}_2(x_1)$, the bending moment $M_3(x_1)$, and the shear force $V_2(x_1)$. *(2)* Find an approximate solution of the problem using the principle of minimum total potential energy. Select the following form for the displacement field: $\bar{u}_2(x_1) = q_1 \sin \pi x_1/L + q_3 \sin \pi 3x_1/L$. *(3)* On the same graph, plot the exact and approximate transverse displacement fields, $H_{33}^c \bar{u}_2/(p_0 L^4)$. *(4)* On the same graph, plot the exact and approximate bending moment distributions, $M_3/(p_0 L^2)$. *(5)* On the same graph, plot the exact and approximate shear force distributions, $V_2/(p_0 L)$. *(6)* Explain why the approximation is so poor. Hint: look at the bending moment plots. It will be convenient to work with non-dimensional spring stiffnesses $\bar{k}_1 = k_1 L/H_{33}^c$ and $\bar{k}_2 = k_2 L^3/H_{33}^c$. For your plots, select $\bar{k}_1 = 10.0$ and $\bar{k}_2 = 100.0$.

Problem 11.15. Two parallel simply supported beams with interconnecting springs

Figure 11.38 depicts a system consisting of two simply supported beams connected by two elastic springs of stiffness k. The upper and lower beam have the same bending stiffness H_{33} and the upper beam is subjected to a uniform load distribution p_0. *(1)* Find the exact solution of the problem. Determine the deflection and bending moment distributions in the upper

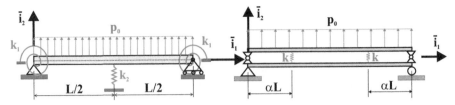

Fig. 11.37. Simply supported beam with mid-span and end point springs.

Fig. 11.38. Two simply supported beams with interconnecting springs.

and lower beams and the force in the connecting springs. *(2)* Find an approximate solution of the problem based on the principle of minimum total potential energy. Used the following interpolation array $\underline{H} = \{\sin \pi \eta, \sin 3\pi \eta, \sin 5\pi \eta\}^T$, where $\eta = x_1/L$, for the upper and lower beams. This gives a total of six degrees of freedom. *(3)* Plot the exact and approximate displacements for the upper and lower beams on the same graph. *(4)* Plot the exact and approximate bending moments for the upper and lower beams on the same graph.

Use the following data for the plots: $\alpha = 0.3$, $\bar{k} = kL^3/H_{33}^c = 10, 100, 1000$.

Problem 11.16. Simply supported beam with variable bending stiffness

A simply supported beam of span L is subjected to forces of magnitude P located at stations $x_1 = \alpha L$ and $(1 - \alpha)L$, as depicted in fig. 11.39. The beam has a bending stiffness H_0 and is reinforced in its central portion where its bending stiffness is H_1. *(1)* Find the exact solution of the problem from the solution of the governing differential equation and associated boundary conditions. *(2)* Use the principle of minimum total potential energy to find approximate solutions for this problem using the following shape functions: $h_i(x_1) = \sin(2i - 1)\pi x_1/L$ using the first 1, 2 and 3 terms. On a single graph, plot the exact solution and the 3 approximate solutions. Also, construct a single plot of the error in maximum displacement for the 3 approximate solutions. Use $H_1/H_0 = 2$ and $\alpha = 0.3$. *(3)* Find the bending moment distribution for the problem. On a single graph, plot the exact solution and the 3 approximate solutions using. Also, construct a single plot of the error in maximum bending moment for the approximate solutions. *(4)* Based on a simple free body diagram, show that for the exact solution the shear force presents a discontinuity at the point of application of the transverse loads P. What happens in your approximate solution? Comment and explain your results. On a single graph, plot the exact solution and the approximate shear force distribution for the approximate solutions.

Problem 11.17. Cross-supported beams

The lower beam depicted in fig. 11.40 is of length $2L$ and is simply supported at both ends. The upper beam of length $L + a$ is cantilevered from **C**, supported by the lower beam at point **A**, and subjected to a uniform transverse loading p_0. Both upper and lower beams have a uniform bending stiffness H_0. *(1)* Find the exact solution for the transverse deflection of the lower beam under a mid-span concentrated load. Show that the lower beam can be replaced by a spring of stiffness $k_{eq} = 6H_0/L^3$. *(2)* Find the exact solution for the transverse deflection of the upper beam from the governing differential equation and associated boundary conditions. Replace the lower beam by the equivalent spring of stiffness constant given above. Model the interaction between the upper beam and equivalent spring by a force X, yet unknown. The magnitude of this force is found by equating the displacements of the upper beam and spring at point **A**. *(3)* Find an approximate displacement solution for the upper beam. Use the principle of minimum total potential energy with the following assumed shape functions: $h_i(x_1) = x_1^{2+i}$. Use $L/a = 2$. On a single graph, plot the exact and approximate solutions

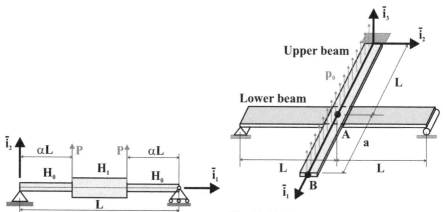

Fig. 11.39. Simply supported beam with variable bending stiffness.

Fig. 11.40. Simply supported beam with cantilever beam crossed and pinned at midspans.

for 3 cases with the first 1, 2, 3 and 4 terms. Quantify the error in maximum displacement for each case. *(4)* Find the approximate bending moment distribution for the problem. On a single graph, plot the exact and approximate solutions for each case. Quantify the error in maximum bending moment for each case. *(5)* Based on a simple free body diagram, show that for the exact solution the shear force presents a discontinuity at point A. What happens in your approximate solution? Comment and explain your results. On a single graph, plot the exact and approximate shear force distributions for each case. Quantify the error in maximum shear force for each case.

Problem 11.18. Cantilever beam with uniform load and spring

The cantilever beam depicted in fig. 11.41 is of length L, uniform bending stiffness, $H_{33}^c = H_0$, and is subjected to a uniform distributed load p_0. A spring of stiffness k is connected to the beam at a distance a from its root. *(1)* Find the exact solution of the problem from the solution of the governing differential equations and associated boundary conditions for the cantilever beam. It will be necessary to solve the problem in two parts for the segments to the right and left of the spring using the continuity conditions at this point. It will be convenient to define the non-dimensional spring constant $\bar{k} = ka^3/H_0$ *(2)* Find an approximate displacement solution for the beam. Use the principle of minimum total potential energy with the following shape functions: $h_i(x_1) = x_1^{2+i}$. Use $L/a = 3$ and $\bar{k} = 100$. On a single graph, plot the exact and approximate solutions using the first 1, 2, 3 and 4 terms. Quantify the error in maximum displacement for each case. *(3)* Find the approximate bending moment distribution for the problem. On a single graph, plot the exact and approximate solutions for each case. Quantify the error in maximum bending moment for each case. *(4)* Based on a simple free body diagram, show that for the exact solution the shear force presents a discontinuity at the connection point for the spring. What happens in your approximate solution? Comment and explain your results. On a single graph, plot the exact and approximate shear force distributions for each case. Quantify the error in maximum shear force for each case.

Problem 11.19. Simply supported beam on elastic foundation

Consider the simply supported beam of length L depicted in fig. 11.42. The beam rests on an elastic foundation of stiffness k and is subjected to a concentrated load P acting

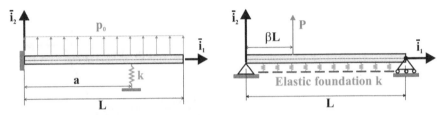

Fig. 11.41. Cantilever beam with uniform load and spring at intermediate point.

Fig. 11.42. Beam with elastic foundation subjected to a concentrated load.

at a distance βL from the left support. Use the principle of minimum total potential energy to find an approximate solution for this problem using the following shape functions: $h_i(x_1) = \sin(2i - 1)\pi x_1/L$. (1) Find the exact solution of the problem from the solution of the governing differential equation and associated boundary conditions. (2) Find the approximate displacement solution for the problem. On a single graph, plot the exact solution and the approximate solutions using 1, 2 and 3 terms in the shape function series. Quantify the error in maximum displacement for each case. Use $\bar{k} = kL^4/H_{33}^c = 8\ 10^3$ and $\beta = 0.5$. (3) Find the approximate bending moment distribution for the problem. On a single graph, plot the exact solution and the approximate solutions for each case. Quantify the error in maximum bending moment for each case. (4) Based on a simple free body diagram, show that for the exact solution the shear force presents a discontinuity at point of application of the load. What happens in your approximate solution? Comment and explain your results. On a single graph, plot the exact solution and the approximate shear force distribution for each case. Quantify the error in maximum bending moment for each case.

11.5 A finite element formulation for beams

In the previous sections, approximate solutions for the axial and lateral transverse displacement fields of beams are developed using both the weak statement of equilibrium, and work and energy methods. In each case, the starting point of the approach is the selection of shape functions used to approximate the beam's displacement field; typically, these shape functions are required to satisfy the geometric boundary conditions. For the simple problems considered thus far, the selection of appropriate shape functions is a relatively simple task because geometric boundary conditions are typically imposed at one or both ends of the beam.

In practice, however, a variety of more complex problem must be solved. Typical beam structures involve multiple supports or sectional properties with span-wise variations. Furthermore, loading conditions often involve complex distributed loads or multiple concentrated forces. In such cases, the classical differential equation approach becomes very tedious, if not impossible to manage, and more often than not, no closed-form solutions exist to practical problems of interest. Even the approximate solution procedures presented earlier in this chapter become more difficult to apply because the selection of a set of suitable shape function becomes very arduous. Furthermore, discrete supports or applied concentrated loads generate discontinuities in the solution. If continuous shape functions are used, the resulting approximate solutions are expected to yield poor accuracy in the neighborhood of the discontinuities.

The work and energy methods developed earlier lack the generality required to solve the various types of problems listed above. The main reason for this weakness is that the selection of the shape functions is problem dependent. Indeed, the presence of specific boundary conditions, multiple supports, or applied concentrated loads impacts the choice of suitable shape functions. Many of these problems stem from the fact that the shape functions used thus far are continuous functions defined over the entire span of the beam.

To overcome these deficiencies, the complete beam structure is first broken into a finite number of short segments, and simple polynomial shape functions are used to approximate the beam's displacement over that segment only. Of course, appropriate conditions must be imposed to preserve the continuity of the displacement and rotation fields between neighboring segments. These segments are commonly called *finite elements*; the complete structure is said to be divided in a number of finite elements. The approach outlined here is called the *finite element method*.

The approach to be presented here is very similar to that developed in section 10.7 for planar trusses. When dealing with trusses, each bar is a finite element, and within this element, the strain field is assumed to be constant. A stiffness matrix and a load array are then computed for each bar, and the contributions of all bars are then assembled into a global problem. The finite element method for beams follows the same pattern, although a preliminary discretization step is required. Because the displacement and strain fields vary within each element, a procedure must be devised to approximate theses fields within each element.

To underline the close connection between the finite element method for trusses developed in section 10.7 and that for beams, the notation used here echoes that used for truss problems.

The finite element formulation for beam bending is based on the Euler-Bernoulli assumptions presented in chapter 5 and for simplicity, axial deformations are ignored. Figure 11.43 depicts the problem under investigation. The axis of the bean coincides with axis $\bar{\imath}_1$, and plane $(\bar{\imath}_1, \bar{\imath}_2)$ is a plane of symmetry of the problem. Consequently, the beam deflects in the transverse direction along axis $\bar{\imath}_2$, and the cross-section rotates about axis $\bar{\imath}_3$. The transverse displacement field is denoted $\bar{u}_2(x_1)$, and the cross-sectional rotation field is denoted $\Phi_3(x_1)$. The beam is subjected to distributed loads and concentrated forces. Of course, the method can be generalized to deal with three-dimensional, anisotropic, curved beams with unsymmetrical cross-sections under arbitrary loading, but this is beyond the scope of this introductory treatment.

11.5.1 General description of the problem

In the first step of the approach, the beam is subdivided into a number of segments or finite elements, as depicted in fig. 11.43. Two neighboring elements are connected together at common point called a *node*. The illustration shown in fig. 11.43 depicts seven finite elements and the eight nodes that connect them. While nodes can be located at any span-wise location along the axis of the beam, it will be convenient to locate nodes at the locations of the supports, and at the points of application of the concentrated loads. As illustrated in fig. 11.43, nodes 1, 6 and 8 are located at the supports, while the locations of nodes 3 and 5 coincide with the point of applications of the two concentrated forces. Additional node locations are selected to create a nearly uniform distribution of nodes along the beam's span.

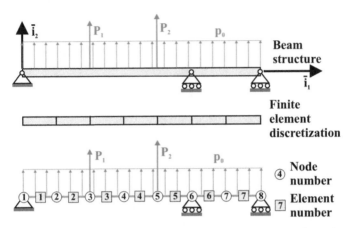

Fig. 11.43. Discretization of a beam into finite elements connected together at nodes.

The geometry of the beam is defined by the coordinates of its eight nodes. For instance, the components of the position vector of node 1 with respect to the origin of the coordinate system are denoted x_1 and y_1, along unit vectors $\bar{\imath}_1$ and $\bar{\imath}_2$, respectively, and stored in array $\underline{p}_1 = \{x_1, \; y_1\}^T$. Similar arrays[1] can be defined for all the nodes of the beam,

$$\underline{p}_1 = \begin{Bmatrix} x_1 \\ y_1 \end{Bmatrix}, \quad \underline{p}_2 = \begin{Bmatrix} x_2 \\ y_2 \end{Bmatrix}, \quad \dots, \quad \underline{p}_8 = \begin{Bmatrix} x_8 \\ y_8 \end{Bmatrix}. \tag{11.81}$$

The subscript $(\cdot)_i$ will be used to indicate quantities pertaining to the ith node.

The generalized coordinates of the problem will be selected as the vertical displacement and rotation components of each of the 8 nodes, denoted v_i and ϕ_i, respectively. The following nodal displacement arrays will be used

$$\underline{q}_1 = \begin{Bmatrix} v_1 \\ \phi_1 \end{Bmatrix}, \quad \underline{q}_2 = \begin{Bmatrix} v_2 \\ \phi_2 \end{Bmatrix}, \quad \dots, \quad \underline{q}_8 = \begin{Bmatrix} v_8 \\ \phi_8 \end{Bmatrix}. \tag{11.82}$$

Array $\underline{q}_1$ stores the two degrees of freedom at node 1, while array $\underline{q}_i$ stores those at node i. It will also be necessary to define a *global displacement array*, $\underline{q}$, that stores

---

[1] This notation uses symbols x, y, and z, to denote position components, instead of x_1, x_2, and x_3, which are used throughout this book. Notations with multiple subscripts, such as x_{1i} to indicate the position component of node i along axis $\bar{\imath}_1$ are thus avoided.

all the nodal displacement arrays in a single column as

$$q = \{q_1^T, q_2^T, q_3^T, q_4^T, q_5^T, q_6^T, q_7^T, q_8^T\}^T .$$ (11.83)

As mentioned earlier, the finite element method first focuses on a generic finite element of the system, in this case, a generic element of the beam, to evaluate the strain energy stored in that specific element. Each element is connected to two nodes: a root node, denoted *Node 1*, and a tip node, denoted *Node 2*. These nodes are referred to as local nodes, and are used when focusing on a single element of the system.

On the other hand, when the complete beam is considered, global nodes must be used. For instance, referring to fig. 11.43, element 3 has two *local nodes*, denoted *Node 1* and *Node 2*, whereas its *global nodes* are nodes 3 and 4. Similarly, element 7 has two *local nodes*, denoted *Node 1* and *Node 2*, whereas its *global nodes* are nodes 7 and 8. Since the local nodes are denoted *Node 1* and *Node 2* for each and every element, they are not indicated on the figure as it would lead to confusion. This distinction between local and global nodes is important for the development of the method.

11.5.2 Kinematics of an element

The kinematics of a specific element of the beam will be studied first. Figure 11.44 depicts a single beam element of length $\hat{\ell}$ with local nodes at each end denoted as *Node 1* and *Node 2*. A *local coordinate system* is centered at the mid-point of the element and defined by unit vector $\bar{\jmath}_1$ aligned with the axis of the beam and $\bar{\jmath}_2$ normal to the beam. Only horizontal beams in two dimensions will be considered in this development, and therefore the local coordinate system, $\mathcal{J} = (\bar{\jmath}_1, \bar{\jmath}_2)$, is aligned with the global coordinate system, $\mathcal{I} = (\bar{\imath}_1, \bar{\imath}_2)$. If the beam is not aligned with the global axis system, the two systems will differ, requiring a coordinate transformation similar to that used for bar elements, see section 10.7.2.

The position vectors of the two local nodes of an element are denoted as

$$\hat{p}_1 = \begin{Bmatrix} \hat{x}_1 \\ \hat{y}_1 \end{Bmatrix}, \text{ and } \hat{p}_2 = \begin{Bmatrix} \hat{x}_2 \\ \hat{y}_2 \end{Bmatrix},$$ (11.84)

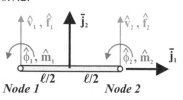

Fig. 11.44. Kinematic of a beam element.

where in this case $\hat{y}_1 = \hat{y}_2 = 0$. For clarity, the quantities pertaining to an element will be indicated with a caret $(\hat{\cdot})$ to distinguish them from their global counterparts. For example, it is important to distinguish the position vector of node 1, denoted p_1 as defined by eq. (11.81), from $\hat{p}_1$, which indicates the position vector of *Node 1* of a generic bar element.

Similarly, the displacements and rotations of the two nodes of an element can be expressed in global axis system, $(\bar{\imath}_1, \bar{\imath}_2)$ as

$$\hat{q}_1 = \begin{Bmatrix} \hat{v}_1 \\ \hat{\phi}_1 \end{Bmatrix}, \quad \hat{q}_2 = \begin{Bmatrix} \hat{v}_2 \\ \hat{\phi}_2 \end{Bmatrix}.$$ (11.85)

Because the global and local coordinate coincide, the displacements and rotations resolved in the two systems are identical. It will be convenient to combine the displacements and rotations of the element's two nodes into single array, called the *element displacement array*,

$$\hat{\underline{q}} = \left\{ \begin{matrix} \hat{\underline{q}}_1 \\ \hat{\underline{q}}_2 \end{matrix} \right\}. \tag{11.86}$$

Finally, the length of the beam element, $\hat{\ell}$, can be computed from the position vectors of its end nodes as follows,

$$\hat{\ell} = \|\hat{\underline{p}}_2 - \hat{\underline{p}}_1\| = \sqrt{(\hat{x}_2 - \hat{x}_1)^2 + (\hat{y}_2 - \hat{y}_1)^2} = \sqrt{(\hat{x}_2 - \hat{x}_1)^2} = |\hat{x}_2 - \hat{x}_1|. \tag{11.87}$$

11.5.3 Element displacement field

The displacement and slope fields in the beam element must be continuous over the entire span of the element. At this point, however, displacements and rotations have been specified only at the nodes: eq. (11.85) defines the displacements and rotations at the element's two end points, but the displacement and rotation fields within the element remain unknown. If $-\hat{\ell}/2 \le \hat{x} \le \hat{\ell}/2$ is the variable that describes position along the axis of the beam element, the displacement field within the element, $\hat{v}(\hat{x})$, is as yet unknown. The displacement and rotation fields will be continuous over the entire span of the beam if the nodal displacement and rotation values for elements that share a common node match. Consequently, the displacement and rotation fields within the element must interpolate the values specified at its two nodes, as expressed by the following relationship

$$\hat{v}(\hat{\eta}) = \hat{v}_1 h_1(\hat{\eta}) + \frac{\hat{\ell}\hat{\phi}_1}{2} h_2(\hat{\eta}) + \hat{v}_2 h_3(\hat{\eta}) + \frac{\hat{\ell}\hat{\phi}_2}{2} h_4(\hat{\eta}), \tag{11.88}$$

where $\hat{\eta} = 2\hat{x}/\hat{\ell}$ is a non-dimensional variable along the span of the beam element, and $h_i(\hat{\eta})$ are *shape functions*. The four degrees of freedom are the two nodal displacements, $\hat{v}_1$ and $\hat{v}_2$, and the two nodal rotation, $\hat{\phi}_1$ and $\hat{\phi}_2$; the factor $\hat{\ell}/2$ is used to keep the shape functions non-dimensional. Variable $\hat{x}$ is dimensional: $-\hat{\ell}/2 \le \hat{x} \le \hat{\ell}/2$ between *Node 1* and *Node 2* of a typical beam element. Variable $\hat{\eta}$ is non-dimensional: $-1 \le \hat{\eta} \le +1$ between the same nodes. Note that $d\hat{x}/d\hat{\eta} = \hat{\ell}/2$.

Equation (11.88) defines the displacement field within the element based solely on the four degrees of freedom. For eq. (11.88) to be correct, the interpolated displacement field must yield the nodal displacements, $\hat{v}_1$ and $\hat{v}_2$, and rotations, $\hat{\phi}_1$ and $\hat{\phi}_2$, when evaluated at the nodes, $\hat{\eta} = \pm 1$. These requirements define 4 boundary conditions for the displacement field of the element

$$\hat{v}(-1) = \hat{v}_1, \text{ and } \left. \frac{d\hat{v}}{d\hat{x}} \right|_{-\hat{\ell}/2} = \frac{2}{\hat{\ell}} \hat{v}'(-1) = \hat{\phi}_1 \text{ at } Node\ 1, \text{ and}$$

$$\hat{v}(+1) = \hat{v}_2, \text{ and } \left. \frac{d\hat{v}}{d\hat{x}} \right|_{+\hat{\ell}/2} = \frac{2}{\hat{\ell}} \hat{v}'(+1) = \hat{\phi}_2 \text{ at } Node\ 2,$$

$$\tag{11.89}$$

where the notation $(\cdot)'$ indicates differentiation with respect to η.

At this point, the shape functions are not yet defined. All that is known, is that the displacement field, $\hat{v}(\hat{x})$, and rotation field, $\hat{\phi}(\hat{x}) = d\hat{v}/d\hat{x}$, must satisfy the four boundary conditions expressed by eqs. (11.89). These conditions alone do not uniquely define the shape functions. It is, however, convenient to select the displacement field, $\hat{v}(\hat{x})$, in the form of a cubic polynomials because a cubic polynomial presents four unknown coefficients which can be uniquely determined by the four boundary conditions expressed by eqs. (11.89).

The element displacement field is assumed to be a cubic polynomial

$$\hat{v}(\hat{x}) = c_1\hat{\eta}^3 + c_2\hat{\eta}^2 + c_3\hat{\eta} + c_4, \tag{11.90}$$

where c_1, c_2, c_3, and c_4 are four unknown coefficients. The boundary conditions expressed by eqs. (11.89) imply the following equations

$$\hat{v}(-1) = -c_1 + c_2 - c_3 + c_4 = \hat{v}_1, \ \hat{v}'(-1) = 3c_1 - 2c_2 + c_3 = \frac{\hat{\ell}}{2}\hat{\phi}_1,$$

$$\hat{v}(+1) = c_1 + c_2 + c_3 + c_4 = \hat{v}_2, \ \hat{v}'(+1) = 3c_1 + 2c_2 + c_3 = \frac{\hat{\ell}}{2}\hat{\phi}_2, \tag{11.91}$$

which form a set of four linear equations for the four unknown coefficients. The solution of this system yields the coefficients c_1, c_2, c_3, and c_4, and eq. (11.90) then gives the displacement field in the element as

$$\hat{v}(\hat{x}) = \frac{1}{4}(-1+\hat{\eta})^2(2+\hat{\eta})\hat{v}_1 + \frac{1}{4}(-1+\hat{\eta})^2(1+\hat{\eta})\frac{\hat{\ell}}{2}\hat{\phi}_1$$

$$+ \frac{1}{4}(1+\hat{\eta})^2(2-\hat{\eta})\hat{v}_2 + \frac{1}{4}(1+\hat{\eta})^2(-1+\hat{\eta})\frac{\hat{\ell}}{2}\hat{\phi}_2 \tag{11.92}$$

$$= h_1(\hat{\eta})\hat{v}_1 + h_2(\hat{\eta})\frac{\hat{\ell}}{2}\hat{\phi}_1 + h_3(\hat{\eta})\hat{v}_2 + h_4(\hat{\eta})\frac{\hat{\ell}}{2}\hat{\phi}_2.$$

The shape functions of an element, $h_i(\hat{\eta})$, are given as

$$h_1(\hat{\eta}) = \frac{1}{4}(1-\hat{\eta})^2(2+\hat{\eta}), \quad h_2(\hat{\eta}) = \frac{1}{4}(1-\hat{\eta})^2(1+\hat{\eta}),$$

$$h_3(\hat{\eta}) = \frac{1}{4}(1+\hat{\eta})^2(2-\hat{\eta}), \quad h_4(\hat{\eta}) = -\frac{1}{4}(1+\hat{\eta})^2(1-\hat{\eta}). \tag{11.93}$$

The derivatives of the *shape functions* are

$$h_1'(\hat{\eta}) = -\frac{3}{4}(1-\hat{\eta}^2), \quad h_2'(\hat{\eta}) = -\frac{1}{4}(1-\hat{\eta})(1+3\hat{\eta}),$$

$$h_3'(\hat{\eta}) = \frac{3}{4}(1-\hat{\eta}^2), \quad h_4'(\hat{\eta}) = -\frac{1}{4}(1+\hat{\eta})(1-3\hat{\eta}). \tag{11.94}$$

The polynomial shape functions and their derivatives are shown in figs. 11.45 and 11.46, respectively.

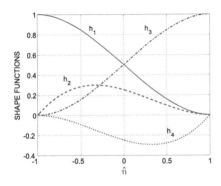

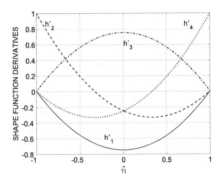

Fig. 11.45. Beam element shape functions. h_1: solid line; h_2: dashed line; h_3: dash-dotted line; h_4: dotted line.

Fig. 11.46. Derivatives of the beam element shape functions. h_1': solid line; h_2': dashed line; h_3': dash-dotted line; h_4': dotted line.

11.5.4 Element curvature field

The distribution of curvature over the element is obtained from its definition, eq. (5.6), as

$$
\hat{\kappa}_3(\hat{\eta}) = \frac{\mathrm{d}^2 \hat{v}(\hat{\eta})}{\mathrm{d}\hat{x}^2} = \left(\frac{2}{\hat{\ell}}\right)^2 \hat{v}''(\hat{\eta})
$$

$$
= \left(\frac{2}{\hat{\ell}}\right)^2 \left[\hat{v}_1 h_1''(\hat{\eta}) + \frac{\hat{\ell}\hat{\phi}_1}{2} h_2''(\hat{\eta}) + \hat{v}_2 h_3''(\hat{\eta}) + \frac{\hat{\ell}\hat{\phi}_2}{2} h_4''(\hat{\eta}) \right].
$$

This expression can be recast in a compact format as

$$
\hat{\kappa}_3(\hat{\eta}) = \hat{\underline{b}}^T(\hat{\eta})\hat{\underline{q}}, \tag{11.95}
$$

where the curvature interpolation array, $\hat{\underline{b}}(\hat{\eta})$, is defined as

$$
\hat{\underline{b}}(\hat{\eta}) = \left(\frac{2}{\hat{\ell}}\right)^2 \left\{ h_1''(\hat{\eta}), \frac{\hat{\ell}}{2} h_2''(\hat{\eta}), h_3''(\hat{\eta}), \frac{\hat{\ell}}{2} h_4''(\hat{\eta}) \right\}^T. \tag{11.96}
$$

11.5.5 Element strain energy and stiffness matrix

The strain energy stored in a beam subjected to bending is given by eq. (10.39). For the beam element under consideration, the strain energy is obtained by integration over the span of the element to find

$$
\hat{A} = \frac{1}{2} \int_0^{\hat{\ell}} H_{33}^c \hat{\kappa}_3^2(\hat{\eta}) \, \mathrm{d}\hat{x} = \frac{1}{2} \int_{-1}^{+1} H_{33}^c \left[\hat{\underline{q}}^T \hat{\underline{b}}(\hat{\eta}) \right] \left[\hat{\underline{b}}^T(\hat{\eta})\hat{\underline{q}} \right] \frac{\hat{\ell}}{2} \mathrm{d}\hat{\eta},
$$

where the curvature field is expressed using eq. (11.95) to find $\hat{\kappa}_3^2 = [\hat{\underline{b}}^T \hat{\underline{q}}]^T [\hat{\underline{b}}^T \hat{\underline{q}}] = [\hat{\underline{q}}^T \hat{\underline{b}}][\hat{\underline{b}}^T \hat{\underline{q}}]$. The nodal degrees of freedom stored in array $\hat{\underline{q}}$ are independent of $\hat{\eta}$, and therefore, they can be placed outside the integral to find

$$\hat{A} = \frac{1}{2} \hat{\underline{q}}^T \left[\frac{\hat{\ell}}{2} \int_{-1}^{+1} \underline{\hat{b}}(\hat{\eta}) H_{33}^c(\hat{\eta}) \underline{\hat{b}}^T(\hat{\eta}) \, d\hat{\eta} \right] \hat{\underline{q}} = \frac{1}{2} \hat{\underline{q}}^T \underline{\hat{k}} \, \hat{\underline{q}}, \tag{11.97}$$

where $\underline{\hat{k}}$ is the *element stiffness matrix*.

If the bending stiffness is constant over the span of the element, the stiffness matrix can be evaluated to find

$$\underline{\hat{k}} = \frac{\hat{\ell}}{2} \int_{-1}^{+1} \underline{\hat{b}}(\hat{\eta}) H_{33}^c \underline{\hat{b}}^T(\hat{\eta}) \, d\hat{\eta} = \frac{H_{33}^c}{\hat{\ell}^3} \begin{bmatrix} 12 & 6\hat{\ell} & -12 & 6\hat{\ell} \\ & 4\hat{\ell}^2 & -6\hat{\ell} & 2\hat{\ell}^2 \\ & & 12 & -6\hat{\ell} \\ sym & & & 4\hat{\ell}^2 \end{bmatrix}. \tag{11.98}$$

This 4×4 element stiffness matrix describes the stiffness of the beam element between *Node 1* and *Node 2*. In particular, the first and third rows and columns represent the stiffness associated with the displacement degrees of freedom, $\hat{v}_1$ and $\hat{v}_2$, respectively. The second and fourth rows and columns represent the stiffness associated with the rotational degrees of freedom, $\hat{\phi}_1$ and $\hat{\phi}_2$, respectively. Because, the displacement and rotation degrees of freedom are of different units, the entries of the stiffness matrix are also of different units.

11.5.6 Element external potential and load array

As illustrated in fig. 11.43, the externally applied loading consists of concentrated and distributed loads. For a typical element, concentrated loads, $\hat{f}_1$ and $\hat{f}_2$, are applied at *Node 1* and *Node 2*, respectively. Concentrated moments, $\hat{m}_1$ and $\hat{m}_2$, are applied at the same nodes, respectively. Finally, a distributed transverse load, $p_2(\hat{x})$, is applied over the span of the element. The potential of these externally applied loads then follows from eq. (10.59) as

$$\hat{\Phi} = -\hat{f}_1 \hat{v}_1 - \hat{m}_1 \hat{\phi}_1 - \hat{f}_2 \hat{v}_2 - \hat{m}_2 \hat{\phi}_2 - \int_0^{\hat{\ell}} \hat{p}_2(\hat{x}) \hat{v}(\hat{x}) \, d\hat{x}.$$

Introducing the interpolated displacement field, eq. (11.88), in the last term yields the following expression

$$\hat{\Phi} = -\hat{\underline{q}}^T \underline{\hat{f}}, \tag{11.99}$$

where the element load array is

$$\underline{\hat{f}} = \begin{Bmatrix} \hat{f}_1 + \dfrac{\hat{\ell}}{2} \displaystyle\int_{-1}^{+1} p_2(\hat{\eta}) h_1(\hat{\eta}) \, d\hat{\eta} \\ \hat{m}_1 + \dfrac{\hat{\ell}^2}{4} \displaystyle\int_{-1}^{+1} p_2(\hat{\eta}) h_2(\hat{\eta}) \, d\hat{\eta} \\ \hat{f}_2 + \dfrac{\hat{\ell}}{2} \displaystyle\int_{-1}^{+1} p_2(\hat{\eta}) h_3(\hat{\eta}) \, d\hat{\eta} \\ \hat{m}_2 + \dfrac{\hat{\ell}^2}{4} \displaystyle\int_{-1}^{+1} p_2(\hat{\eta}) h_4(\hat{\eta}) \, d\hat{\eta} \end{Bmatrix}. \tag{11.100}$$

For a given applied distributed load, the integral can be evaluated to find the element load array, $\hat{f}$. For instance, if the element is subjected to a uniform distributed load of magnitude $\hat{p}_0$, the element load array becomes

$$\hat{f} = \left\{ \frac{\hat{p}_0 \hat{\ell}}{2}, \frac{\hat{p}_0 \hat{\ell}^2}{12}, \frac{\hat{p}_0 \hat{\ell}}{2}, -\frac{\hat{p}_0 \hat{\ell}^2}{12} \right\}^T.$$

The first and third terms represent the nodal loads equivalent to the applied distributed load. The total load applied to the element is $p_0 \hat{\ell}$, and half of this load is applied to each of the two end nodes. The second and fourth terms represent the nodal moments equivalent to the applied distributed load. Equal and opposite moments of magnitude $\hat{p}_0 \hat{\ell}^2 / 12$ are applied to each of the two end nodes.

The work done by these nodal forces and moments is identical to that done by the distributed loading, within the approximation of the interpolated displacement field given by eq. (11.88). For this reason, these nodal forces and moments are sometimes referred to as "work-equivalent" nodal forces.

11.5.7 Assembly procedure

In the previous sections, attention is focused on a single, generic beam element to determine its *element* stiffness matrix, eq. (11.98), and *element* load array, eq. (11.100). These two quantities are obtained from the element strain energy and external potential, respectively. In this section, attention shifts to the overall beam problem to determine the *global stiffness matrix* and *global load array*. These two quantities will be obtained from the system's total strain energy and total external potential, respectively. Because both strain energy and external potential are scalar quantities, their combined total will be evaluated simply by summing up the contributions from the individual elements.

The total strain energy stored in the beam is the sum of the contributions of all elements. In eq. (11.97), the strain energy of a single, generic beam element is denoted $\hat{A}$, and this notation is not ambiguous because only a single element is considered. It now becomes necessary, however, to add the element identification using the subscript $(.)_{(i)}$ introduced earlier. Summing over all elements yields

$$A = \sum_{i=1}^{N_e} \hat{A}_{(i)} = \frac{1}{2} \sum_{i=1}^{N_e} \hat{\underline{q}}_{(i)}^T \hat{\underline{k}}_{(i)} \hat{\underline{q}}_{(i)}, \tag{11.101}$$

where N_e is the number of elements in the beam ($N_e = 7$ for the beam illustrated in fig. 11.43). In this case, it is also necessary to add the element identification subscript to both the element stiffness matrix, $\hat{\underline{k}}_{(i)}$, and the nodal displacement array, $\hat{\underline{q}}_{(i)}$.

Equation (11.101) gives the total strain energy in the structure, but it is not easy to manipulate because each term in the sum is expressed in terms of a different set of degrees of freedom. For example, with reference to fig. 11.43, element 3 is connected to global nodes 3 and 4 which are local *Node 1* and *Node 2* for the element, respectively. The element stiffness, $\hat{\underline{k}}_{(3)}$, is defined in terms of these global nodes, see eq. (11.98), and the corresponding element displacement array is $\hat{\underline{q}}_{(3)}^T = \left\{\hat{\underline{q}}_1^T, \hat{\underline{q}}_2^T\right\} = \left\{\underline{q}_3^T, \underline{q}_4^T\right\}^T = \left\{v_3, \phi_4, v_4, \phi_4\right\}^T$.

To remedy this situation, a *connectivity matrix*, $\underline{\underline{C}}_{(i)}$, for the ith element is introduced following the same approach used for a truss in section 10.7.6. This matrix is designed to extract the specific terms of the element displacement array from the global displacement array defined by eq. (11.83). This operation can be written as

$$\hat{\underline{q}}_{(i)} = \underline{\underline{C}}_{(i)}\underline{q}. \tag{11.102}$$

To best understand this abstract relationship, consider a specific element of the beam, say element 3, as shown in fig. 11.43. Its local nodes, *Node 1* and *Node 2*, are associated with the global node numbers 3 and 4, respectively, so that $\hat{\underline{q}}_1 = \underline{q}_3$ and $\hat{\underline{q}}_2 = \underline{q}_4$. The element displacement array, $\hat{\underline{q}}_{(3)}$, can thus be written as

$$\hat{\underline{q}}_{(3)} = \left\{\begin{matrix} \hat{\underline{q}}_1 \\ \hat{\underline{q}}_2 \end{matrix}\right\}_{(3)} = \left\{\begin{matrix} \underline{q}_3 \\ \underline{q}_4 \end{matrix}\right\} = \begin{bmatrix} \underline{0} & \underline{0} & \underline{I} & \underline{0} & \underline{0} & \underline{0} & \underline{0} & \underline{0} \\ \underline{0} & \underline{0} & \underline{0} & \underline{I} & \underline{0} & \underline{0} & \underline{0} & \underline{0} \end{bmatrix} \left\{\begin{matrix} q_1 \\ q_2 \\ q_3 \\ q_4 \\ q_5 \\ q_6 \\ q_7 \\ q_8 \end{matrix}\right\} = \underline{\underline{C}}_{(3)}\underline{q},$$

where $\underline{0}$ and $\underline{I}$ represent the 2×2 null and identity matrices, respectively. The connectivity matrix, $\underline{\underline{C}}_{(3)}$, is called a *Boolean matrix* because its entries consist solely of 0's and 1's. Matrix $\underline{\underline{C}}_{(3)}$ establishes the connections of beam element 3 within the entire beam by indicating the nodes to which this beam is connected, and this explains its name of "connectivity matrix."

Expressing the element nodal displacement arrays, $\hat{\underline{q}}_{(i)}$, in terms of the global displacement array, $\underline{q}$, with the help of eq. (11.102), the total strain energy of the truss given by eq. (11.101) now becomes

$$A = \frac{1}{2}\sum_{i=1}^{N_e}\left(\underline{q}^T\underline{\underline{C}}_{(i)}^T\right)\hat{\underline{\underline{k}}}_{(i)}\left(\underline{\underline{C}}_{(i)}\underline{q}\right) = \frac{1}{2}\underline{q}^T\left[\sum_{i=1}^{N_e}\underline{\underline{C}}_{(i)}^T\hat{\underline{\underline{k}}}_{(i)}\underline{\underline{C}}_{(i)}\right]\underline{q}.$$

This expression can be simplified to

$$A = \frac{1}{2}\underline{q}^T\underline{\underline{K}}\underline{q}, \tag{11.103}$$

where the *global stiffness matrix*, $\underline{\underline{K}}$, is defined as

$$\underline{\underline{K}} = \sum_{i=1}^{N_e}\underline{\underline{C}}_{(i)}^T\hat{\underline{\underline{k}}}_{(i)}\underline{\underline{C}}_{(i)}. \tag{11.104}$$

The potential of the externally applied loads, Φ, is found by adding the contributions of all beam elements

$$\Phi = \sum_{i=1}^{N_e} \hat{\Phi}_{(i)} = -\sum_{i=1}^{N_e} \hat{\underline{q}}_{(i)}^T \hat{\underline{f}}_{(i)}, \tag{11.105}$$

where $\hat{\underline{f}}_{(i)}$ is the load array for the ith element, as defined by eq. (10.79) for a generic beam element. Here again, it is convenient to use the connectivity matrix defined in eq. (10.81) to evaluate the potential,

$$\Phi = -\sum_{i=1}^{N_e} \left(\underline{\underline{C}}_{(i)} \underline{q} \right)^T \hat{\underline{f}}_{(i)} = -\underline{q}^T \left\{ \sum_{i=1}^{N_e} \underline{\underline{C}}_{(i)}^T \hat{\underline{f}}_{(i)} \right\}.$$

This expression can be simplified to

$$\Phi = -\underline{q}^T \underline{Q}, \tag{11.106}$$

by defining the *global load array*, $\underline{Q}$, as

$$\underline{Q} = \sum_{i=1}^{N_e} \underline{\underline{C}}_{(i)}^T \hat{\underline{f}}_{(i)}. \tag{11.107}$$

Finally, the total potential energy, Π, of the complete beam is obtained by adding the potential of the external loads, eq. (11.106), to the total strain energy, eq. (11.103), to find

$$\Pi = A + \Phi = \frac{1}{2} \underline{q}^T \underline{\underline{K}} \, \underline{q} - \underline{q}^T \underline{Q}. \tag{11.108}$$

This compact expression for the total potential energy of the complete system is only possible because the matrix notation encapsulates the nodal and element quantities in arrays and matrices. The total strain energy is a quadratic form of the generalized coordinates, whereas the potential of the externally applied loads is a linear form of the same variables. It should also be noted that the total strain energy is a positive-definite quantity because it is the sum of positive-definite strain energies for each beam element.

11.5.8 Alternative description of the assembly procedure

The assembly procedure described in terms of the connectivity matrix defined in eq. (11.102) is formally correct, but it is not easy to understand nor is it computationally efficient for realistic beams with many nodes. The connectivity matrix, $\underline{\underline{C}}_{(i)}$, has four lines and $2N$ columns, where N is the total number of nodes. For beam modeled with many nodes, this matrix becomes very large with a total of $8N$ entries, and yet, only four entries have a unit value while all $(8N - 4)$ others are zero. Furthermore, the evaluation of the global stiffness matrix involves a triple matrix

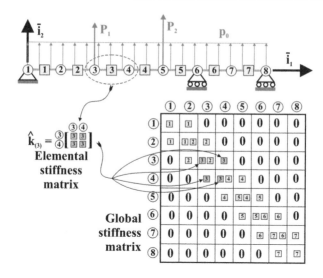

Fig. 11.47. Illustration of the assembly procedure.

product for each element, see eq. (11.104). These become increasingly expensive to perform as the problem size increases, and they also are very wasteful because most operations actually are multiplications by zero.

It is possible to give a more graphical visualization of the assembly process. Figure 11.47 depicts the seven element, eight node beam problem under consideration, together with a pictorial representation of the global stiffness matrix. The 8 rows and columns in the matrix are labeled with their corresponding node numbers. Each node has two degrees of freedom (the vertical displacement and rotation component at that node), so each of the entries is actually a 2×2 matrix and the size of the global stiffness matrix itself is 16×16.

Consider now a typical element of the beam, say element 3. Its local nodes, *Node 1* and *Node 2*, are associated with the global node numbers 3 and 4, respectively. The stiffness matrix for this beam element, $\hat{\underline{\underline{k}}}_{(3)}$, can be partitioned into four 2×2 matrices, as shown in eq. (11.98). Beam element 3 is connected to global nodes 3 and 4, and therefore, the four sub-matrices of the local stiffness matrix can simply be added to entries $\underline{\underline{K}}(3,3)$, $\underline{\underline{K}}(4,4)$, $\underline{\underline{K}}(3,4)$, and $\underline{\underline{K}}(4,3)$ in the global stiffness matrix, as indicated by the arrows in fig. 11.47. Note that the indices used with $\underline{\underline{K}}$ in this discussion refer to the nodes shown in fig. 11.47 and not to the individual degrees of freedon actually used to index $\underline{\underline{K}}$.

This procedure is repeated for each beam element to give the final result shown in fig. 11.47. The final figure requires careful interpretation. Each of the 64 squares represents a 2×2 matrix and may contain 0 or more element numbers. Each of the element numbers shown in square boxes defines a 2×2 matrix extracted from the corresponding element stiffness matrix. These 2×2 matrices are added together to produce the final result in the global stiffness matrix.

Another way to look at the same process is to consider the fully assembled global stiffness matrix in fig. 11.47. Diagonal entry $\underline{\underline{K}}(2,2)$ collects contributions from elements 1 and 2, because these two beam elements are all physically connected to node 2. Similarly, diagonal entry $\underline{\underline{K}}(5,5)$ collects contributions from elements 4, and 5, because these two beam elements connect to node 5.

At completion of the assembly process, many entries of the global stiffness matrix remain empty or null. For instance, entries $\underline{\underline{K}}(2,6) = \underline{\underline{K}}(6,2) = 0$, because no beam element directly connects nodes 2 and 6. Similarly, $\underline{\underline{K}}(1,4) = \underline{\underline{K}}(4,1) = 0$ because nodes 1 and 4 are not directly connected by a beam element.

The procedure described here is identical to that presented in sections 10.7.7 and 10.7.6 for truss structures. Although the stiffness matrices for bar and beam elements are different, their assembly process is identical.

11.5.9 Derivation of the governing equations

The total potential energy of the beam is given by eq. (11.108), and application of the principle of minimum total potential energy, eq. (10.17), now implies

$$\frac{\partial \Pi}{\partial \underline{q}} = \frac{\partial}{\partial \underline{q}} \left(\frac{1}{2} \underline{q}^T \underline{\underline{K}} \, \underline{q} - \underline{q}^T \underline{Q} \right) = \underline{\underline{K}} \, \underline{q} - \underline{Q} = 0. \tag{11.109}$$

To compute the derivative of the total potential energy, eqs. (A.29) and (A.27) are used to evaluate the derivatives of the strain energy and potential of the externally applied loads, respectively. The particular form of this result is due to the fact that the stiffness matrix is symmetric, as described in section A.2.9.

The governing equation of the system take the form of a linear system of equations,

$$\underline{\underline{K}} \, \underline{q} = \underline{Q}. \tag{11.110}$$

The process used to establish the governing equations for beam problems is identical to that used for trusses. The linear algebra formalism hides the fact that the entries of the stiffness matrix and load arrays are different for beam and truss problems. Once the total potential energy is evaluated in terms of a finite number of degrees of freedom, the derivation of the governing equations is formally identical for both types of structures.

11.5.10 Solution procedure

The linear system given in eq. (11.110) cannot be solved because the global stiffness matrix is singular.

This situation arises because the element stiffness matrices that make up the global stiffness matrix are each singular. Calculation of the eigenvectors and eigenvalues of the element stiffness matrix, $\underline{\underline{\hat{k}}}$, given by eq. (11.98), reveals more information about this rank deficiency. Two of the four unit eigenvectors of this matrix are

$$\underline{n}_1 = \frac{1}{\sqrt{2}} \begin{Bmatrix} 1 \\ 0 \\ 1 \\ 0 \end{Bmatrix}, \quad \underline{n}_2 = \frac{\ell}{\sqrt{8 + 2\ell^2}} \begin{Bmatrix} -1 \\ 2/\ell \\ 1 \\ 2/\ell \end{Bmatrix},$$

and the corresponding eigenvalues are $\lambda_1 = \lambda_2 = 0$. The last two eigenvalues of the stiffness matrix do not vanish. Consequently, each element stiffness matrix is two times singular. These two eigenvectors represent the two rigid body motion of the beam element: its vertical translation and rotation, corresponding to $\underline{n}_1$ and $\underline{n}_2$, respectively. By definition, rigid body motions create no deformation or straining of the element, and hence, no strain energy is associated with rigid body modes. Clearly, the presence of two rigid body modes for the structure implies the rank deficiency of 2 for the element stiffness matrix. The entire beam also presents two rigid body modes, and hence, the global stiffness matrix also features a rank deficiency of 2.

The physical interpretation of this situation is that boundary conditions have not yet been applied to the beam, which is still free to translate vertically and rotate in plane $(\bar{\imath}_1, \bar{\imath}_2)$. Figure 11.47 shows that nodes 1, 6 and 8 are pinned to the ground, preventing any rigid body motion of the beam. These conditions, however, are not reflected in the global stiffness matrix given by eq. (11.104).

The boundary conditions can be imposed through the following process: *(1)* eliminate the rows and columns of the stiffness matrix corresponding to constrained degrees of freedom to create its reduced counterpart, $\underline{\bar{K}}$; *(2)* eliminate the corresponding entries of the global displacement array, $\underline{q}$, to create its reduced counterpart, $\underline{\bar{q}}$; and finally, *(3)* eliminate the corresponding entries of the global load array, $\underline{Q}$, to create its reduced counterpart, $\underline{\bar{Q}}$. The system of equations for the truss then reduces to

$$\underline{\bar{K}}\,\underline{\bar{q}} = \underline{\bar{Q}}. \tag{11.111}$$

The reduced stiffness matrix will now be non-singular, and the solution of the problem is found by solving the linear system to find the remaining nodal displacements as $\underline{\bar{q}} = \underline{\bar{K}}^{-1}\underline{\bar{Q}}$. A more detailed justification of the procedure is described in section 10.7.9.

Although the stiffness matrices for truss and beam elements are quite different, see sections 10.7.4 and 11.5.5, respectively, many aspects of the formulation of the finite element method for the two types of structures are very similar, and often identical. In fact, once cast within the formalism of linear algebra, the governing equations for both problems are identical, see eqs. (10.89) and (11.110), for trusses and beams, respectively. The treatment of the boundary conditions, discussed in sections 10.7.9 and 11.5.10 for truss and beam structures, respectively, is also identical. A formal treatment of the boundary conditions based on partitioning is described in details for truss structures in section 10.7.10 and applies to beam structures as well.

The above discussion underlines one of the important advantages of the finite element method. Different types of structural components generate stiffness matrices and load arrays that reflect the specific nature of each structural component. For instance, the strain energy of a bar is based on the extensional strain of the component, whereas that of a beam is based on its curvature. Once the stiffness matrices and load arrays of all components have been generated, the remainder of the process does not distinguish between the various types of structural components. Consequently, the assembly procedure, the generation of the governing equations, and the various details of the solution procedure are identical for all types of structural elements because they correspond to generic, linear algebra operations. This very systematic approach to the solution of general structural problems is one of the key reasons for the immense success of the finite element method.

Example 11.16. Cantilevered beam with a mid-span support

Consider the uniform cantilevered beam with a mid-span support shown in fig. 11.48. For this simple problem, two finite elements will be used: the first extends from the left clamp to the mid-span support, the second from the mid-span support to the beam's tip. The two element are delimited by three nodes, for a total of 6 degrees of freedom.

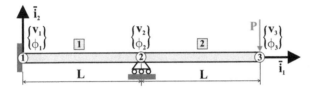

Fig. 11.48. Two-element model of cantilever with mid-span support.

Because the two elements are of equal length, $\hat{\ell} = L$, and bending stiffness, the stiffness matrix for each element is given by eq. (11.98). The 6×6 global stiffness matrix then consists of the assembly of the two 4×4 element stiffness matrices. Proceeding as described in section 11.5.8, the global stiffness matrix is found as

$$
\underline{\underline{K}} = \frac{H_{33}^c}{L^3}
\begin{bmatrix}
12 & 6L & -12 & 6L & 0 & 0 \\
 & 4L^2 & -6L & 2L^2 & 0 & 0 \\
 & & (12+12) & (-6L+6L) & -12 & 6L \\
 & & & (4L^2+4L^2) & -6L & 2L^2 \\
 & & & & 12 & -6L \\
 & sym & & & & 4L^2
\end{bmatrix} .
$$

The partitioning indicated in the above equation corresponds to the contributions of the degrees of freedom associated with the three nodes: the first two rows and columns correspond to the degrees of freedom of node 1, the next two rows and columns to those of node 2, and the last two rows and columns to those of node 3. The 4×4 stiffness matrix of the first element is assembled in the first four rows and columns of the global stiffness matrix, while the 4×4 stiffness matrix of the second element is assembled in the last four rows and columns of the global stiffness matrix. The two elements are connected at a common node, node 2. It follows that the middle two rows and columns of the global stiffness matrix store the sum of contributions from both elements.

The governing equations of the problem are in the form of eq. (11.110). In this case, the nodal load array is given as $Q = \{R_1, M_1, R_2, 0, -P, 0\}^T$ where R_1 and R_2 are the reaction forces at nodes 1 and 2, respectively, and M_1 is the root clamping moment.

The solution phase of the problem will follow the partitioning approach developed in section 10.7.10. The first three degrees of freedom, v_1, ϕ_1, and v_2, are the prescribed degrees of freedom, i.e., $q_p = \{v_1, \phi_1, v_2\}^T = 0$. The corresponding load array stores the reaction forces, $Q_p = \{R_1, M_1, R_2\}^T$. The last three degrees of freedom, ϕ_2, v_3, and ϕ_3, are the unconstrained degrees of freedom, $q_u = \{\phi_2, v_3, \phi_3\}^T$. The corresponding load array stores the externally applied loads, $Q_u = \{0, -P, 0\}^T$.

The global stiffness matrix is partitioned accordingly to find

$$\underline{\underline{K}}_{uu} = \frac{H_{33}^c}{L^3}\begin{bmatrix} 8L^2 & -6L & 2L^2 \\ -6L & 12 & -6L \\ 2L^2 & -6L & 4L^2 \end{bmatrix}, \quad \underline{\underline{K}}_{pu} = \frac{H_{33}^c}{L^3}\begin{bmatrix} 6L & 0 & 0 \\ 2L^2 & 0 & 0 \\ 0 & -12 & 6L \end{bmatrix}.$$

Matrix $\underline{\underline{K}}_{uu}$ corresponds to the lower right 3×3 partition of the global stiffness matrix, whereas matrix $\underline{\underline{K}}_{pu}$ corresponds to the upper right 3×3 partition of the same matrix.

Since $q_p = 0$, eq. (10.96) reduces to $\underline{\underline{K}}_{uu} q_u = Q_u$ and the unknown degrees of freedom are $q_u = \underline{\underline{K}}_{uu}^{-1} Q_u$, whose components are

$$\phi_2 = -\frac{PL^2}{4H_{33}^c}, \quad v_3 = -\frac{7PL^3}{12H_{33}^c}, \quad \text{and} \quad \phi_3 = -\frac{3PL^2}{4H_{33}^c}.$$

The reaction forces are now computed with the help of eq. (10.97), which reduces to $Q_p = \underline{\underline{K}}_{up}^T q_u$, with components given by

$$R_1 = -\frac{3P}{2}, \quad M_1 = -\frac{PL}{2}, \quad \text{and} \quad R_2 = \frac{5P}{2}.$$

Next, the displacement field within each element can be computed with the help of eq. (11.88). For the first element, $v_1 = \phi_1 = v_2 = 0$, and the displacement field reduces to $\hat{v}(\hat{\eta}) = \ell\hat{\phi}_2 h_4(\hat{\eta})/2$. Introducing the shape functions defined in eq. (11.93) then yields

$$\hat{v}(\hat{\eta}) = \frac{PL^3}{32H_{33}^c}(1+\hat{\eta})^2(1-\hat{\eta}).$$

Note that $\hat{\eta} = -1$ at node 1 and $\hat{\eta} = +1$ at node 2. For the second element, the displacement field reduces to $\hat{v}(\hat{\eta}) = L\phi_2 h_2/2 + v_3 h_3 + L\phi_3 h_4/2$, and introducing the shape function defined in eq. (11.93) then yields

$$\hat{v}(\hat{\eta}) = -\frac{PL^3}{96H_{33}^c}(1+\hat{\eta})\left[3(1-\hat{\eta})^2 + 14(1+\hat{\eta})(2-\hat{\eta}) - 9(1+\hat{\eta})(1-\hat{\eta})\right].$$

Note that $\hat{\eta} = -1$ at node 2 and $\hat{\eta} = +1$ at node 3, because $\hat{\eta}$ is a local variable defined within each element. The non-dimensional displacement field over the entire span of the beam is shown in fig. 11.49 as a function of a global non-dimensional variable, $\eta = x_1/(2L)$.

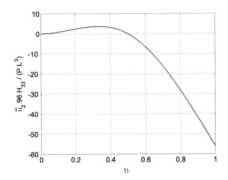

Fig. 11.49. Deflection of beam elements 1 and 2.

This deflected shape looks quite reasonable with zero displacement and slope at the root end and zero deflection at the mid-span support. It is, in fact, the exact solution to this problem. This is because in the element formulation, a cubic polynomial is assumed for the deflected shape and this is the exact form of the solution for a beam segment with only concentrated forces and/or moments applied at the ends.

11.5.11 Summary

The finite element approach presented in this section addresses the challenging problem of selecting good shape functions for complex beam problems. Instead of considering the entire beam, an approximate solution is created for a finite number of beam elements. Within each beam element, it is a easy to choose shape functions that easily satisfy the constraints imposed at the end nodes.

The solution to the full problem is then constructed by assembling the governing equations for each of the small elements into a formulation for the entire beam. This assembly process is systematic and lends itself to computer implementation. A set of linear algebraic equations results that can be solved easily. For this reason, the finite element method for developing approximate solutions is preferred over approaches that attempt to select approximations over the entire span of the beam.

The development of the finite element analysis method is a rich area to explore, and a considerable amount of research has been performed over the past decades. The finite element method has been incorporated into a number of large commercial software packages, which can be applied to solve a wide range of structural engineering and "multi-physics" problems. This chapter provide only the most basic introduction to this fascinating field.

11.5.12 Problems

Problem 11.20. Cantilever with mid-span load and tip support

Consider the cantilever beam of length $2L$ shown in fig. 11.50. A concentrated load is applied at mid-span and the tip is pinned. Construct a finite element solution to this problem using two elements of length L. *(1)* Determine the nodal displacements and rotations and compare to the exact results obtained using the unit load method. *(2)* Determine the nodal reactions for the constrained degrees of freedom. *(3)* Construct a plot of the deflected shape for the beam.

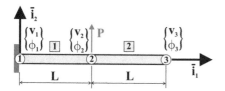 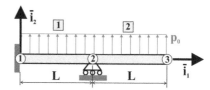

Fig. 11.50. Cantilever with tip support and concentrated load applied at mid-span.

Fig. 11.51. Cantilever with mid-span support and a uniform load.

Problem 11.21. Cantilever with mid-span support and uniform load

A cantilever beam is supported at mid-span and carries a uniform load p_0 as shown in fig. 11.51. (Note that this is very similar to example above.) Construct a finite element solution to this problem using two elements of length L. *(1)* determine the nodal displacements and rotations and compare to the exact results obtained using the unit load method. *(2)* Determine the nodal reactions for the constrained degrees of freedom. *(3)* Construct a plot of the deflected shape for the beam.

Problem 11.22. Simply supported beam with nonuniform bending stiffness

In this problem you are to reconsider the simply supported beam with nonuniform bending stiffness treated in problem 11.12. The present solution is to be developed using the finite element approach. Construct a 2-element solution using elements for the left and right halves of the beam. *(1)* Compute the nodal displacements and rotations at each node. *(2)* Compare the solution for the mid-span deflection with the exact solution computed using the unit load method. *(3)* Compare the solution for the mid-span deflection with the solution from problem 11.12.

Variational and energy principles

Chapter 9 presents the principle of virtual work and its complementary counterpart for particles, systems of particles, and trusses. These principles are introduced by means of simple examples and no attempt is made to formally derive them for three-dimensional solids.

Chapter 10 follows a similar pattern for the derivation of the principle of minimum total potential energy and of its complementary counterpart. Simple applications are presented focusing on mechanical systems, and trusses. Basic concepts of the finite element method applied to truss structures are presented.

Chapter 11 is devoted entirely to the development of approximate solutions for beam problems. The key to this approach is the ability to recast the differential equations of equilibrium into integral forms. The equivalence between the weak statement of equilibrium and the principle of virtual work is demonstrated for simple beam problems. Basic concepts of the finite element method applied to beam structures are presented.

In this chapter, the problem of determining stationary values of functionals (*i.e.,* functions of functions) will be addressed. The basic concepts from the *calculus of variations* [5, 6] that are required for this task will be reviewed first. Next, the principles of virtual and complementary virtual work, the principles of minimum total potential energy and total complementary energy, the Hu-Washizu principle, and the Hellinger-Reissner principle each will be formally presented for three-dimensional solids. Selected structural mechanics problems will then be examined to illustrate the use of these different principles.

12.1 Mathematical preliminaries

The basic equations of elasticity developed in chapter 1 use the formalisms of differential calculus and partial differential equations. Elements of the calculus of variations will be presented in this section.

12.1.1 Stationary point of a function

Consider a function of n variables, $F = F(u_1, u_2, \ldots, u_n)$. The stationary points [7] of this function are defined as those for which

$$\frac{\partial F}{\partial u_i} = 0, \quad i = 1, 2, \ldots, n. \tag{12.1}$$

For a function of a single variable, this condition corresponds to a horizontal tangent to the curve, as illustrated in fig. 12.1. At a stationary point, the function can present a minimum, a maximum, or a saddle point.

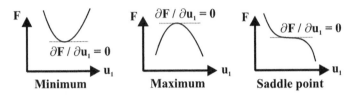

Fig. 12.1. Stationary points of a function.

If a function is stationary at a point, conditions (12.1) hold, and the following statement is then true

$$\frac{\partial F}{\partial u_1} w_1 + \frac{\partial F}{\partial u_2} w_2 + \ldots + \frac{\partial F}{\partial u_n} w_n = 0,$$

where $w_1, w_2, \ldots, w_n$ are arbitrary quantities. It is convenient to use a special notation for these arbitrary quantities, $w_i = \delta u_i$, where δu_i is called a *virtual change* in u_i. The above statement now becomes

$$\frac{\partial F}{\partial u_1} \delta u_1 + \frac{\partial F}{\partial u_2} \delta u_2 + \ldots + \frac{\partial F}{\partial u_n} \delta u_n = 0.$$

Comparison of this result with a similar expression for the differential, dF, of the same function implies that differentials can be used as virtual changes. Consequently, the virtual change operator, denoted "δ," behaves in a manner similar to the differential operator, denoted "d".

The *variation in* F, denoted δF, is defined as

$$\delta F = \frac{\partial F}{\partial u_1} \delta u_1 + \frac{\partial F}{\partial u_2} \delta u_2 + \ldots + \frac{\partial F}{\partial u_n} \delta u_n, \tag{12.2}$$

and it then follows that

$$\delta F = 0 \tag{12.3}$$

at a stationary point.

The differential condition, eq. (12.1), and the variational condition, eq. (12.3) must both hold at a stationary point. From the above developments, it is clear that eq. (12.1) implies eq. (12.3) and since the above reasoning can be reversed, it is simple to prove that eq. (12.3) implies eq. (12.1). Hence, the two conditions are entirely equivalent.

To determine whether a stationary point is a minimum, a maximum, or a saddle point it is necessary to consider the second derivatives [7] of the functions. If

$$\sum_{i,j=1,n} \frac{\partial^2 F}{\partial u_i \partial u_j} du_i du_j > 0 \tag{12.4}$$

at a stationary point for all increments du_i and du_j, the function presents a minimum. If, on the other hand, the same quantity is negative for all du_i and du_j, the function presents a maximum. Finally, if the same quantity can be positive or negative depending on the choice of the increments, the function presents a saddle point.

From the definition of δF, eq. (12.2), the second variation of function F is defined as

$$\delta^2 F = \sum_{i,j=1,n} \frac{\partial^2 F}{\partial u_i \partial u_j} \delta u_i \delta u_j.$$

It is now clear that a stationary point is a minimum if

$$\delta^2 F > 0, \tag{12.5}$$

for all arbitrary variations δu_i and δu_j. It is a maximum if $\delta^2 F < 0$ for all variations, and a saddle point occurs if the sign of the second variation depends on the choice of the variations of the independent variables.

12.1.2 Lagrange multiplier method

Consider once more the problem of determining a stationary point of a function of several variables, $F = F(u_1, u_2, \ldots, u_n)$, where the variables are not independent. Rather, they are subjected to a constraint of the form

$$f(u_1, u_2, \ldots, u_n) = 0. \tag{12.6}$$

Conceptually the constraint can be used to express one variable, say u_n, in terms of the others. Then, u_n can be eliminated from F to obtain a function of $n - 1$ independent variables, $F = F(u_1, u_2, \ldots, u_{n-1})$, which is a problem identical to that treated in the previous section. In many practical situations, however, it might be cumbersome, or even impossible, to completely eliminate one variable of the problem. For example, the constraint equation could be a transcendental equation with no closed-form solution for u_n, or an implicit equation with no simple solution.

This elimination-of-variable process can be avoided by using an alternative approach. At a stationary point, the variation of F vanishes

$$\delta F = \frac{\partial F}{\partial u_1} \delta u_1 + \frac{\partial F}{\partial u_2} \delta u_2 + \ldots + \frac{\partial F}{\partial u_n} \delta u_n = 0. \tag{12.7}$$

This statement, however, does not imply $\partial F/\partial u_i = 0$, for $i = 1, 2, \ldots, n$, because the variations, δu_i, cannot be chosen arbitrarily since they must satisfy the constraint, eq. (12.6).

The relation among the variations can be written explicitly by taking a variation of the constraint to find

$$\delta f = \frac{\partial f}{\partial u_1} \delta u_1 + \frac{\partial f}{\partial u_2} \delta u_2 + \ldots + \frac{\partial f}{\partial u_n} \delta u_n = 0. \tag{12.8}$$

A linear combination of eqs. (12.7) and (12.8) can be constructed to find

$$\frac{\partial F}{\partial u_1} \delta u_1 + \ldots + \frac{\partial F}{\partial u_n} \delta u_n + \lambda \left[\frac{\partial f}{\partial u_1} \delta u_1 + \ldots + \frac{\partial f}{\partial u_n} \delta u_n \right] = 0,$$

where λ is an arbitrary function of variables $u_1, u_2, \ldots, u_n$, called the *Lagrange multiplier*. Regrouping the various terms then leads to

$$\sum_{i=1}^{n} \left[\frac{\partial F}{\partial u_i} + \lambda \frac{\partial f}{\partial u_i} \right] \delta u_i = 0. \tag{12.9}$$

Conceptually, variation δu_n could now be express in term of the $n - 1$ other variations, δu_i, leaving the $n-1$ remaining variations to be independent and arbitrary. To avoid this cumbersome algebraic step, the arbitrary Lagrange multiplier is chosen such that

$$\frac{\partial F}{\partial u_n} + \lambda \frac{\partial f}{\partial u_n} = 0.$$

With this choice, the last term of the sum in eq. (12.9) vanishes for all variations δu_n. Hence, it is not necessary to express this variation in terms of the $n - 1$ others, which can now be treated as independent, arbitrary quantities. Equation (12.9) then implies

$$\frac{\partial F}{\partial u_i} + \lambda \frac{\partial f}{\partial u_i} = 0, \quad i = 1, 2, \ldots, n - 1. \tag{12.10}$$

Combining the last two equations then leads to the condition that

$$\delta F + \lambda \delta f = 0,$$

where all variations, δu_i, $i = 1, 2, \ldots, n$, are *independent*.

Because of the constraint, eq. (12.6), it is clear that $f \delta \lambda = 0$ for *any arbitrary* $\delta \lambda$. Hence, the stationarity condition can be written as

$$\delta F + \lambda \delta f = \delta F + \lambda \delta f + f \delta \lambda = \delta (F + \lambda f) = 0.$$

A modified function, $F^+ = F + \lambda f$ is introduced and the above statement now implies the vanishing of the variation in F^+ for *all arbitrary variations* δu_i, $i = 1, 2, \ldots, n$, *and* $\delta \lambda$.

In summary, the initial constrained problem can be replaced by an *unconstrained problem*

$$\delta F^+ = 0, \quad \text{where} \quad F^+ = F + \lambda f. \tag{12.11}$$

The modified function, F^+, involves $n + 1$ variables, u_i, $i = 1, 2, \ldots, n$ and λ. The vanishing of the variation in F^+ then implies

$$\sum_{i=1}^{n} \left[\frac{\partial F}{\partial u_i} + \lambda \frac{\partial f}{\partial u_i} \right] \delta u_i + f \, \delta \lambda = 0.$$

Because δu_i, $i = 1, 2, \ldots, n$, and $\delta \lambda$ are all independent, arbitrary variations, it follows that

$$\frac{\partial F}{\partial u_i} + \lambda \frac{\partial f}{\partial u_i} = 0, \quad i = 1, 2, \ldots, n; \quad \text{and} \quad f = 0.$$

These form $n + 1$ equations to be solved for the $n + 1$ unknowns. Note that the Lagrange multiplier method results in an unconstrained problem, but *increases* the number of unknowns from n to $n + 1$; the additional unknown is the Lagrange multiplier. On the other hand, if the constraint is used to eliminate one of the unknowns, the resulting problem will be an unconstrained problem for $n - 1$ unknowns.

The Lagrange multiplier methods can be readily generalized to problems involving multiple constraints, $f_i = 0$, $i = 1, 2, \ldots, m$. In the presence of m constraints, m Lagrange multipliers, λ_i, $i = 1, 2, \ldots, m$, are introduced. The modified function then becomes

$$F^+ = F + \sum_{i=1}^{m} \lambda_i f_i. \tag{12.12}$$

12.1.3 Stationary point of a definite integral

Next, the determination of the stationary point of the following definite integral

$$I = \int_a^b F(y, y', x) \, dx \tag{12.13}$$

is considered, where the notation $(\cdot)'$ is used to indicate a derivatives with respect to x, and $y(x)$ is an unknown function of x subject to boundary conditions, $y(a) = \alpha$ and $y(b) = \beta$.

This problem seems to be of a completely different nature from those treated in the previous sections. Indeed, I is a *functional* or a "function of a function", *i.e.*, the value of the definite integral I depends on the choice of the unknown function $y(x)$. Since there are an infinite number of values of y between a and b, functional I is equivalent to a function of an infinite number of variables.

This problem will be treated using the variational formalism introduced in section 12.1.1. First, the concept of *variation of a function*, denoted δf, is introduced. Figure 12.2 shows two functions $f(x)$ and $\hat{f}(x)$ such that

$$\delta f = \hat{f}(x) - f(x) = \psi(x),$$

where $\psi(x)$ is a continuous and differentiable, but otherwise arbitrary function such that $\psi(a) = \psi(b) = 0$. In other words, δf is a virtual change that brings the function $f(x)$ to a new, arbitrary function $\hat{f}(x)$. Note that $\delta f(a) = \delta f(b) = 0$ which means that δf does not violate the boundary conditions of the problem.

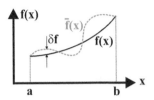

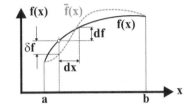

Fig. 12.2. The concept of variation of a function.

Fig. 12.3. The difference between an increment df and a variation δf.

The stationarity of functional I requires

$$\delta I = \delta \int_a^b F(y,\, y',\, x)\, dx = \int_a^b \delta F(y,\, y',\, x)\, dx = 0.$$

With the help of eq. (12.2) and treating δ as a differential, this results in

$$\delta I = \int_a^b \left[\frac{\partial F}{\partial y}\, \delta y + \frac{\partial F}{\partial y'}\, \delta y' \right] dx = 0.$$

Integration by parts is now applied to the second term in the square bracket

$$\int_a^b \frac{\partial F}{\partial y'}\, \delta \left(\frac{dy}{dx} \right) dx = \int_a^b \frac{\partial F}{\partial y'}\, \frac{d}{dx}\, (\delta y)\, dx = -\int_a^b \frac{d}{dx} \left(\frac{\partial F}{\partial y'} \right) \delta y\, dx + \left[\frac{\partial F}{\partial y'} \delta y \right]_a^b.$$

The boundary term vanishes because $\delta y(a) = \delta y(b) = 0$, and the stationarity condition then becomes

$$\delta I = \int_a^b \left[\frac{\partial F}{\partial y} - \frac{d}{dx} \left(\frac{\partial F}{\partial y'} \right) \right] \delta y\, dx = 0.$$

The bracketed term must vanish because the integral must go to zero for *all arbitrary variations* δy. This yields

$$\frac{\partial F}{\partial y} - \frac{d}{dx} \left(\frac{\partial F}{\partial y'} \right) = 0 \qquad (12.14)$$

which is known as the *Euler-Lagrange equation* for the problem.

Here again, the above reasoning can be reversed. Starting from eq. (12.14), and performing the integration by parts in the reversed order implies $\delta I = 0$. In summary, *the necessary and sufficient condition for the definite integral to be at a stationary point is that eq. (12.14) be satisfied.*

The variational formalism introduced in this section will be systematically applied to elasticity problems in the rest of this chapter. It will be shown that the equations of elasticity can be viewed as the Euler-Lagrange equations associated with the stationarity condition of definite integrals. Various forms of the equations of elasticity can be easily obtained by direct manipulations of these definite integrals. It is therefore important to understand the variational formalism and its implications.

A crucial difference exists between an increment, df, of a function $f(x)$ and a variation, δf, of the same function, as depicted in fig. 12.3. The differential, df, is an infinitesimal change in $f(x)$ resulting from an infinitesimal change, dx, in the independent variable, and df/dx represents the rate of change or tangent at the point. On the other hand, δf is an arbitrary virtual change that brings $f(x)$ to $\hat{f}(x)$. The two quantities, df and δf, are clearly unrelated, the former is positive in fig. 12.3, and the latter is negative.

Although the concepts associated with the notation df and δf are clearly distinct, manipulations of the two symbols are quite similar. For instance, the order of application of the two operations can be interchanged, indeed,

$$\frac{d}{dx}(\delta f) = \frac{d}{dx}(\hat{f} - f) = \frac{d\hat{f}}{dx} - \frac{df}{dx} = \delta\left(\frac{df}{dx}\right). \qquad (12.15)$$

Similarly, the order of the integration and variation operations commute

$$\delta \int_a^b F \, dx = \int_a^b \hat{F} \, dx - \int_a^b F \, dx = \int_a^b (\hat{F} - F) \, dx = \int_a^b \delta F \, dx. \qquad (12.16)$$

12.2 Variational and energy principles

Consider a general elasticity problem consisting of an elastic body of arbitrary shape subjected to surface tractions and body forces as well as geometric boundary conditions such as prescribed displacements at a point or over a portion of its outer surface, as depicted in fig. 12.4.

The volume of the body is de-

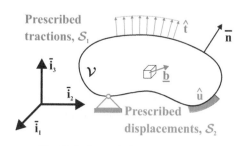

Fig. 12.4. General elasticity problem.

noted $\mathcal{V}$ and its outer surface $\mathcal{S}$. The outer normal to $\mathcal{S}$ is the unit vector $\bar{n}$. $\mathcal{S}_1$ and $\mathcal{S}_2$ denote the portions of the outer surface where prescribed tractions $\hat{\underline{t}}$ and prescribed displacements $\hat{\underline{u}}$ are applied, respectively. At a point of the outer surface, either tractions or displacements can be prescribed, but it is impossible to prescribe both. Consequently, $\mathcal{S}_1$ and $\mathcal{S}_2$ share no common points and $\mathcal{S} = \mathcal{S}_1 + \mathcal{S}_2$. Note that a point of the outer surface that is traction free belongs to $\mathcal{S}_1$ because vanishing traction conditions, $\hat{\underline{t}} = 0$, are prescribed at that point.

Body forces $\underline{b}$ might also be applied over the entire volume of the body. Gravity forces are a typical example of body forces, but such forces can also arise as a result of electric or magnetic fields. In dynamic problems, inertial forces are also applied as body forces in accordance with D'Alembert's principle.

The basic equations of elasticity developed in chapter 1 form a set of partial differential equations that can be solved to find the displacement, strain, and stress fields at all points in $\mathcal{V}$. These equations will be reviewed in section 12.2.1 where several important definitions are also introduced. In the subsequent sections, a number of variational and energy principles are presented that provide an alternative formalism for the solution of elasticity problems.

12.2.1 Review of the equations of linear elasticity

As depicted in fig. 3.1 on page 101, the equations of elasticity can be broken into three groups. The solution of an elasticity problem involves *(1)* a statically admissible stress field, *(2)* a kinematically admissible displacement field and the corresponding compatible strain field, and *(3)* a constitutive law satisfied at all points in volume $\mathcal{V}$. These concepts are explained below.

Equilibrium equations

The *equations of equilibrium* are the most fundamental equations. They are derived in sections 1.1.2 and 1.1.3 from Newton's law stating that the sum of all the forces acting on a differential element of the structure should vanish.

For reference, the equilibrium equations for a differential element of the body, eqs. (1.4), are rewritten here

$$\frac{\partial \sigma_1}{\partial x_1} + \frac{\partial \tau_{21}}{\partial x_2} + \frac{\partial \tau_{31}}{\partial x_3} + b_1 = 0,$$

$$\frac{\partial \tau_{12}}{\partial x_1} + \frac{\partial \sigma_2}{\partial x_2} + \frac{\partial \tau_{32}}{\partial x_3} + b_2 = 0, \qquad (12.17)$$

$$\frac{\partial \tau_{13}}{\partial x_1} + \frac{\partial \tau_{23}}{\partial x_2} + \frac{\partial \sigma_3}{\partial x_3} + b_3 = 0,$$

and must be satisfied at all points of volume $\mathcal{V}$.

The traction equilibrium equations are

$$t_1 = \hat{t}_1, \quad t_2 = \hat{t}_2, \quad t_3 = \hat{t}_3, \qquad (12.18)$$

where the surface tractions are defined in eq. (1.9). The surface equilibrium equations are also called the *force*, or *natural boundary conditions*. The compact stress array, $\underline{\sigma}$, defined be eq. (2.11b), will be used simplify the notation.

Definition 12.1. *A stress field, $\underline{\sigma}$, is said to be statically admissible if it satisfies the equilibrium equations, eqs. (12.17), at all points of volume $\mathcal{V}$ and the surface equilibrium equations, eqs. (12.18), at all points of surface $\mathcal{S}_1$.*

Strain-displacement relationships

The *strain-displacement equations* merely define the strain components that are used for the characterization of the deformation at a point of the body. The strain-displacement relationships are derived in section 1.4.1 from purely geometric considerations.

When the displacements are small, it is convenient to use the engineering strain components to measure the deformation at a point. From section 1.4, axial and shearing strain components are related to the displacements as

$$
\epsilon_1 = \frac{\partial u_1}{\partial x_1}, \quad \epsilon_2 = \frac{\partial u_2}{\partial x_2}, \quad \epsilon_3 = \frac{\partial u_3}{\partial x_3},
$$
$$
\gamma_{23} = \frac{\partial u_2}{\partial x_3} + \frac{\partial u_3}{\partial x_2}, \quad \gamma_{13} = \frac{\partial u_1}{\partial x_3} + \frac{\partial u_3}{\partial x_1}, \quad \gamma_{12} = \frac{\partial u_1}{\partial x_2} + \frac{\partial u_2}{\partial x_1}.
$$
(12.19)

To compute strain components, the displacements field must be continuous and differentiable. Furthermore, the displacements must be equal to the prescribed displacements over surface S_2

$$
u_1 = \hat{u}_1, \quad u_2 = \hat{u}_2, \quad u_3 = \hat{u}_3; \tag{12.20}
$$

these are called the *geometric boundary conditions.*The compact strain array, ϵ, defined be eq. (2.11a), will be used simplify the notation.

Definition 12.2. *A displacement field, u, is said to be kinematically admissible if it is continuous and differentiable at all points in volume V and satisfies the geometric boundary conditions, eqs. (12.20), at all points on surface S_2.*

Definition 12.3. *A strain field, ϵ, is said to be compatible if it is derived from a kinematically admissible displacement field through the strain-displacement relationships, eqs. (12.19).*

Constitutive laws

The *constitutive laws* relate the stress and strain components. They consist of a mathematical idealization of the experimentally observed behavior of materials. The homogeneous, isotropic, linearly elastic material behavior described in section 2.1.1 is a frequently used highly idealized constitutive law. Many materials can present one or more of the following features: anisotropy, plasticity, visco-elasticity, or creep, to name just a few commonly observed material behaviors.

The stress and strain fields are related by the constitutive laws at all points in volume V. For linearly elastic materials, Hooke's law, eq. (2.10), provides a simple linear relationship between the two fields. The positive-definite, symmetric stiffness matrix, $\underline{C}$, and the positive-definite, symmetric compliance matrix,$\underline{S}$ are given by eqs. (2.12) and (2.14), respectively.

12.2.2 The principle of virtual work

Consider an elastic body that is in equilibrium under applied body forces and surface tractions. This implies that the stress field is statically admissible, *i.e.*, the equilibrium equations, eqs. (12.17), are satisfied at all points in $\mathcal{V}$ and the surface equilibrium equations, eqs. (12.18), at all points on $\mathcal{S}_1$. The following statement is now constructed

$$\int_{\mathcal{V}} \left\{ \left[\frac{\partial \sigma_1}{\partial x_1} + \frac{\partial \tau_{21}}{\partial x_2} + \frac{\partial \tau_{31}}{\partial x_3} + b_1 \right] \delta u_1 + \left[\frac{\partial \tau_{12}}{\partial x_1} + \frac{\partial \sigma_2}{\partial x_2} + \frac{\partial \tau_{32}}{\partial x_3} + b_2 \right] \delta u_2 \right.$$
$$\left. + \left[\frac{\partial \tau_{13}}{\partial x_1} + \frac{\partial \tau_{23}}{\partial x_2} + \frac{\partial \sigma_3}{\partial x_3} + b_3 \right] \delta u_3 \right\} \, \mathrm{d}\mathcal{V} - \int_{\mathcal{S}_1} \left[\underline{t} - \hat{\underline{t}} \right]^T \delta \underline{u} \, \mathrm{d}\mathcal{S} = 0. \quad (12.21)$$

In this statement each of the three equilibrium equations is multiplied by an arbitrary, virtual change in displacement, then integrated over the range of validity of the equation, volume $\mathcal{V}$. Similarly, each of the three surface equilibrium equations is multiplied by an arbitrary, virtual change in displacement, then integrated over the range of validity of the equation, surface $\mathcal{S}_1$.

Because the stress field is statically admissible, each bracketed term vanishes, and multiplication by an arbitrary quantity results in a vanishing product. Each of the two integral then vanishes, as does their sum.

Next, integration by parts is performed. Using Green's theorem [7], the first term of the volume integral becomes

$$\int_{\mathcal{V}} \frac{\partial \sigma_1}{\partial x_1} \delta u_1 \, \mathrm{d}\mathcal{V} = - \int_{\mathcal{V}} \sigma_1 \frac{\partial \delta u_1}{\partial x_1} \, \mathrm{d}\mathcal{V} + \int_{\mathcal{S}} n_1 \sigma_1 \delta u_1 \, \mathrm{d}\mathcal{S}, \quad (12.22)$$

where n_1 is the component of the outward unit normal along $\bar{\imath}_1$, see fig. 12.4. A similar operation is performed on each stress derivative terms appearing in eq. (12.21).

Finally, the stress and strain arrays are introduced to obtain a compact result,

$$- \int_{\mathcal{V}} \underline{\sigma}^T \delta \underline{\epsilon} \, \mathrm{d}\mathcal{V} + \int_{\mathcal{V}} \underline{b}^T \delta \underline{u} \, \mathrm{d}\mathcal{V} + \int_{\mathcal{S}} \underline{t}^T \delta \underline{u} \, \mathrm{d}\mathcal{S} - \int_{\mathcal{S}_1} (\underline{t} - \hat{\underline{t}})^T \delta \underline{u} \, \mathrm{d}\mathcal{S} = 0, \quad (12.23)$$

where $\delta \underline{\epsilon}$ denotes a *virtual, compatible strain field* defined as

$$\delta \epsilon_1 = \frac{\partial \delta u_1}{\partial x_1}, \quad \delta \epsilon_2 = \frac{\partial \delta u_2}{\partial x_2}, \quad \delta \epsilon_3 = \frac{\partial \delta u_3}{\partial x_3},$$
$$\delta \gamma_{23} = \frac{\partial \delta u_2}{\partial x_3} + \frac{\partial \delta u_3}{\partial x_2}, \quad \delta \gamma_{13} = \frac{\partial \delta u_1}{\partial x_3} + \frac{\partial \delta u_3}{\partial x_1}, \quad \delta \gamma_{12} = \frac{\partial \delta u_1}{\partial x_2} + \frac{\partial \delta u_2}{\partial x_1}. \quad (12.24)$$

The virtual displacements are now chosen to be kinematically admissible, which implies $\delta \underline{u} = 0$ on $\mathcal{S}_2$, and expression (12.23) reduces to

$$- \int_{\mathcal{V}} \underline{\sigma}^T \delta \underline{\epsilon} \, \mathrm{d}\mathcal{V} + \int_{\mathcal{V}} \underline{b}^T \delta \underline{u} \, \mathrm{d}\mathcal{V} + \int_{\mathcal{S}_1} \hat{\underline{t}}^T \delta \underline{u} \, \mathrm{d}\mathcal{S} = 0. \quad (12.25)$$

The first term on the left hand side of this expression can be interpreted as the virtual work done by the internal stresses, δW_I, see eq. (9.77a). The remaining two terms correspond to the virtual work done by the externally applied body forces and surface tractions, δW_E. Equation (12.25) therefore becomes $\delta W_I + \delta W_E = 0$.

It can also be shown that if $\delta W_I + \delta W_E = 0$ holds, the stress field must be statically admissible. Indeed, this principle implies eq. (12.23), which in turn implies eq. (12.21) by reversing the integration by parts process. Finally, the volume and surface equilibrium equations are recovered because eq. (12.21) must hold for all arbitrary, kinematically admissible virtual displacements fields. These results imply the principle of virtual work.

Principle 15 (Principle of virtual work) *A body is in equilibrium if and only if the sum of the internal and external virtual work vanishes for all arbitrary kinematically admissible virtual displacements fields and corresponding compatible strain fields.*

In summary, the equations of equilibrium, eqs. (12.17) and (12.18), and the principle of virtual work are two entirely equivalent statements. Because the principle of virtual work is solely a statement of equilibrium, it is always true. For the solution of specific elasticity problems, however, it must be complemented with stress-strain relationships and constitutive laws.

It is interesting to compare the present statement of the principle of virtual work with that derived in chapter 11 for beams under axial and transverse loads and given by eqs.(11.42) and (11.44), respectively. The statements are different because the present formulation deals with general, three-dimensional stress states, whereas the formulation in chapter 11 deals with the stress resultants associated with beam theory. The physical interpretation of these statements, however, is identical in all cases.

12.2.3 The principle of complementary virtual work

Consider an elastic body undergoing kinematically admissible displacements and compatible strains. This implies that the strain-displacement relationships, eqs. (12.19), are satisfied at all points in volume $\mathcal{V}$ and the geometric boundary conditions, eqs. (12.20), at all points on surface $\mathcal{S}_2$. The following statement is now constructed

$$
\begin{aligned}
- \int_{\mathcal{V}} & \left\{ \left[\epsilon_1 - \frac{\partial u_1}{\partial x_1} \right] \delta\sigma_1 + \left[\epsilon_2 - \frac{\partial u_2}{\partial x_2} \right] \delta\sigma_2 + \left[\epsilon_3 - \frac{\partial u_3}{\partial x_3} \right] \delta\sigma_3 \right. \\
& + \left[\gamma_{23} - \frac{\partial u_2}{\partial x_3} - \frac{\partial u_3}{\partial x_2} \right] \delta\tau_{23} + \left[\gamma_{13} - \frac{\partial u_1}{\partial x_3} - \frac{\partial u_3}{\partial x_1} \right] \delta\tau_{13} \\
& \left. + \left[\gamma_{12} - \frac{\partial u_1}{\partial x_2} - \frac{\partial u_2}{\partial x_1} \right] \delta\tau_{12} \right\} \, \mathrm{d}\mathcal{V} - \int_{\mathcal{S}_2} [\underline{u} - \hat{\underline{u}}]^T \, \delta\underline{t} \, \mathrm{d}\mathcal{S} = 0. \quad (12.26)
\end{aligned}
$$

This statement is constructed in the following manner. Each of the six strain-displacement relationships is multiplied by an arbitrary, virtual change in stress, then integrated over the range of validity of the equations, volume $\mathcal{V}$. Similarly, each of the three geometric boundary conditions is multiplied by an arbitrary, virtual change in surface traction, then integrated over the range of validity of the equation, surface $\mathcal{S}_2$.

Because the strain field is compatible and the displacement field kinematically admissible, each bracketed term vanishes, and multiplication by an arbitrary quantity results in a vanishing product. Each of the two integral then vanishes, as does their sum.

Next, integration by parts is performed. Using Green's theorem [7], the first term of the volume integral becomes

$$\int_{\mathcal{V}} \frac{\partial u_1}{\partial x_1} \delta\sigma_1 \, d\mathcal{V} = - \int_{\mathcal{V}} u_1 \frac{\partial \delta\sigma_1}{\partial x_1} \, d\mathcal{V} + \int_{\mathcal{S}} u_1 n_1 \delta\sigma_1 \, d\mathcal{S}, \qquad (12.27)$$

where n_1 is the component of the outward unit normal along $\bar{\imath}_1$, see fig. 12.4. A similar operation is performed on each displacement derivative terms appearing in eq. (12.26) to yield

$$- \int_{\mathcal{V}} \underline{\epsilon}^T \delta\underline{\sigma} \, d\mathcal{V} - \int_{\mathcal{V}} \left[\left(\frac{\partial \delta\sigma_1}{\partial x_1} + \frac{\partial \delta\tau_{21}}{\partial x_2} + \frac{\partial \delta\tau_{31}}{\partial x_3} \right) u_1 \right.$$
$$+ \left(\frac{\partial \delta\tau_{12}}{\partial x_1} + \frac{\partial \delta\sigma_2}{\partial x_2} + \frac{\partial \delta\tau_{32}}{\partial x_3} \right) u_2 + \left. \left(\frac{\partial \delta\tau_{13}}{\partial x_1} + \frac{\partial \delta\tau_{23}}{\partial x_2} + \frac{\partial \delta\sigma_3}{\partial x_3} \right) u_3 \right] d\mathcal{V}$$
$$+ \int_{\mathcal{S}} \underline{u}^T \delta\underline{t} \, d\mathcal{S} - \int_{S_2} (\underline{u} - \hat{\underline{u}})^T \delta\underline{t} \, d\mathcal{S} = 0. \quad (12.28)$$

Next, a *statically admissible virtual stress field* is defined as a virtual stress field that satisfies equilibrium equations in volume $\mathcal{V}$,

$$\frac{\partial \delta\sigma_1}{\partial x_1} + \frac{\partial \delta\tau_{21}}{\partial x_2} + \frac{\partial \delta\tau_{31}}{\partial x_3} = 0,$$
$$\frac{\partial \delta\tau_{12}}{\partial x_1} + \frac{\partial \delta\sigma_2}{\partial x_2} + \frac{\partial \delta\tau_{32}}{\partial x_3} = 0, \qquad (12.29)$$
$$\frac{\partial \delta\tau_{13}}{\partial x_1} + \frac{\partial \delta\tau_{23}}{\partial x_2} + \frac{\partial \delta\sigma_3}{\partial x_3} = 0,$$

and the surface traction equilibrium equation, $\delta\underline{t} = 0$ on surface S_1.

Because the virtual stresses are arbitrary, they can be chosen to be statically admissible and eq. (12.28) reduces to

$$- \int_{\mathcal{V}} \underline{\epsilon}^T \delta\underline{\sigma} \, d\mathcal{V} + \int_{S_2} \hat{\underline{u}}^T \delta\underline{t} \, d\mathcal{S} = 0. \qquad (12.30)$$

The first term on the left hand side of this expression can be interpreted as the complementary virtual work done by the internal stresses, $\delta W_I'$, see eq. (9.77b). The remaining term corresponds to the complementary virtual work done by the prescribed displacements, $\delta W_E'$. Equation (12.30) therefore becomes $\delta W_I' + \delta W_E' = 0$.

It can also be shown that if $\delta W_I' + \delta W_E' = 0$ holds, the displacement field must be kinematically admissible and the strain field compatible. Indeed, this principle implies eq. (12.28), which in turn implies eq. (12.26) by reversing the integration by parts process. Finally, the strain-displacement relationships and geometric boundary conditions are recovered because eq. (12.26) must hold for all arbitrary stress virtual stress fields. These results imply the principle of complementary virtual work

Principle 16 (Principle of complementary virtual work) *A body is undergoing kinematically admissible displacements and compatible strains if and only if the sum of the internal and external complementary virtual work vanishes for all statically admissible virtual stress fields.*

In summary, the strain-displacement relationships and the geometric boundary conditions, eqs. (12.19) and (12.20), respectively, and the principle of complementary virtual work are two entirely equivalent statements. In addition, comparison of eq. (12.30) with the principle of complementary virtual work, principle 7, developed in chapter 9, shows that principle 16 above is simply a more general statement of principle 7.

12.2.4 Strain and complementary strain energy density functions

In section 10.5, on page 519, the strain and complementary strain energy density functions are developed for a linearly elastic, isotropic material governed by Hooke's law. The strain and complementary strain energy density functions are given by eqs. (10.47) and (10.50), respectively.

If the internal forces in the solid are assumed to be conservative, they can be derived from a potential, as discussed in section 10.1. In this case, the internal forces are the components of stress, and the potential is the strain energy density function. If the stresses in a solid can be derived from a strain energy density function, $a(\underline{\epsilon})$,

$$\underline{\sigma} = \frac{\partial a(\underline{\epsilon})}{\partial \underline{\epsilon}}, \tag{12.31}$$

the material is said to be an *elastic material*. Assuming the material to be elastic or assuming the existence of a strain energy density function are two equivalent assumptions. Linearly elastic materials are elastic materials for which the stress-strain relationship is linear.

If the material is elastic, the work done by the internal stresses when the system is brought from one state of deformation to another depends only on the two states of deformations, but not on the specific path that the system followed from one deformation state to the other. This restricts the types of material constitutive laws that can be expressed in terms of a strain energy density function. For instance, if a material is deformed in the plastic range, the work of deformation will depend on the specific deformation history; hence, there exists no strain energy density function that describes material behavior when plastic deformations are involved.

The concept of complementary strain energy is first introduced for springs in section 10.3.1. For nonlinearly elastic materials, the complementary strain energy density function is defined by the following identity

$$a(\underline{\epsilon}) + a'(\underline{\sigma}) = \underline{\epsilon}^T \underline{\sigma}, \tag{12.32}$$

which explains the term "complementary strain energy." Taking a differential of this identity yields

$$\left(\frac{\partial a(\underline{\epsilon})}{\partial \underline{\epsilon}} - \underline{\sigma}\right)^T d\underline{\epsilon} + \left(\frac{a'(\underline{\sigma})}{\partial \underline{\sigma}} - \underline{\epsilon}\right)^T d\underline{\sigma} = 0.$$

The term in the first parenthesis vanishes because of eq. (12.31). Because the differentials are arbitrary, the second parenthesis must vanish, leading to

$$\underline{\epsilon} = \frac{a'(\underline{\sigma})}{\partial \underline{\sigma}}. \tag{12.33}$$

In view of eq. (12.32), the existence of the strain energy density function implies the existence of the complementary strain energy density function. The strain energy density function allows the definition of the stresses by eqs. (12.31), which can be viewed as the constitutive laws for the elastic material because they define stresses as a function of strains. Similarly, the complementary strain energy density function allows the definition of the strains by eqs. (12.33), which can be viewed as the constitutive laws for the elastic material because they define strains as a function of stresses. Clearly, the strain and complementary strain energy density functions define the constitutive laws for elastic materials. The stiffness form of the constitutive laws, eqs. (12.31), stems from the strain energy density function, whereas the complementary strain energy density function yields the compliance form of the same constitutive laws, eqs. (12.33).

12.2.5 The principle of minimum total potential energy

Consider a general elastic body that is in equilibrium under applied body forces and surface tractions, and therefore, the principle of virtual work, eq. (12.25), must apply. It is now assumed that the constitutive law for the material can be expressed in terms of a strain energy density function, eq. (12.31). The virtual work done by the internal stresses appears in the first term of eq. (12.25), and it is readily evaluated as

$$-\int_{\mathcal{V}} \delta\underline{\epsilon}^T \underline{\sigma} \, d\mathcal{V} = \int_{\mathcal{V}} \delta\underline{\epsilon}^T \frac{\partial a(\underline{\epsilon})}{\partial \underline{\epsilon}} \, d\mathcal{V} = \int_{\mathcal{V}} \delta a(\underline{u}) \, d\mathcal{V} = \delta \int_{\mathcal{V}} a(\underline{u}) \, d\mathcal{V} = \delta A(\underline{u}),$$

where the chain rule for derivatives is used at the second equality.

The strain energy density and the *total strain energy* of the body, $A = \int_{\mathcal{V}} a \, d\mathcal{V}$, must be expressed in terms of the displacement field $\underline{u}$ using the strain displacement relationships because the principle of virtual work requires a compatible strain field. The principle of virtual work, eq. (12.25), now becomes

$$-\delta A(\underline{u}) + \int_{\mathcal{V}} \underline{b}^T \delta\underline{u} \, d\mathcal{V} + \int_{\mathcal{S}_1} \hat{\underline{t}}^T \delta\underline{u} \, d\mathcal{S} = 0. \tag{12.34}$$

Next, the body forces and surface tractions are assumed to be derivable from potential functions

$$\underline{b} = -\frac{\partial \phi}{\partial \underline{u}}; \quad \hat{\underline{t}} = -\frac{\partial \psi}{\partial \underline{u}},$$

where ϕ is the *potential of the body forces*, and ψ the *potential of the surface tractions*.

With these definitions, second and third terms (*i.e.*, the external work terms) in eq. (12.34) become

$$
\int_{\mathcal{V}} \underline{b}^T \delta\underline{u} \, d\mathcal{V} + \int_{\mathcal{S}_1} \hat{\underline{t}}^T \delta\underline{u} \, d\mathcal{S} = -\int_{\mathcal{V}} \frac{\partial\phi}{\partial\underline{u}}^T \delta\underline{u} \, d\mathcal{V} - \int_{\mathcal{S}_1} \frac{\partial\psi}{\partial\underline{u}}^T \delta\underline{u} \, d\mathcal{S}
$$

$$
= -\int_{\mathcal{V}} \delta\phi(\underline{u}) \, d\mathcal{V} - \int_{\mathcal{S}_1} \delta\psi(\underline{u}) \, d\mathcal{S} = -\delta \int_{\mathcal{V}} \phi(\underline{u}) \, d\mathcal{V} - \delta \int_{\mathcal{S}_1} \psi(\underline{u}) \, d\mathcal{S}
$$

$$
= -\delta\Phi(\underline{u}),
$$

where $\Phi(\underline{u}) = \int_{\mathcal{V}} \phi(\underline{u}) \, d\mathcal{V} + \int_{\mathcal{S}_1} \psi(\underline{u}) \, d\mathcal{S}$ is the *total potential the externally applied loads*.

Introducing this result into the principle of virtual work expressed in eq. (12.34) leads to

$$
-\delta A(\underline{u}) - \delta\Phi(\underline{u}) = 0, \quad \text{or} \quad \delta\left(A(\underline{u}) + \Phi(\underline{u})\right) = 0. \tag{12.35}
$$

The *total potential energy* of the body is now defined as

$$
\Pi(\underline{u}) = A(\underline{u}) + \Phi(\underline{u}), \tag{12.36}
$$

and it follows that

$$
\delta\Pi(\underline{u}) = 0. \tag{12.37}
$$

This statement expresses the requirement that the total potential energy must assume a stationary value with respect to compatible deformations when the body is in equilibrium. As discussed in section 12.1.1, the sign of the second variation, $\delta^2\Pi$, will determine whether the stationary point is actually a minimum. The first variation in Π is

$$
\delta\Pi(\underline{u}) = \int_{\mathcal{V}} \left(\frac{\partial a}{\partial\underline{\epsilon}}\right)^T \delta\underline{\epsilon} \, d\mathcal{V} - \int_{\mathcal{V}} \underline{b}^T \delta\underline{u} \, d\mathcal{V} - \int_{\mathcal{S}_1} \hat{\underline{t}}^T \delta\underline{u} \, d\mathcal{S},
$$

and its second variation is then

$$
\delta^2\Pi(\underline{u}) = \int_{\mathcal{V}} \delta\underline{\epsilon}^T \frac{\partial^2 a}{\partial\underline{\epsilon}\partial\underline{\epsilon}} \delta\underline{\epsilon} \, d\mathcal{V}.
$$

Based on physical reasoning, the strain energy density function must be a positive-definite function of the strain components, which implies $\delta\underline{\epsilon}^T \partial^2 a/(\partial\underline{\epsilon}\partial\underline{\epsilon}) \, \delta\underline{\epsilon} \geq 0$ for all $\delta\underline{\epsilon}$. Indeed, if the strain energy function is not positive-definite, strain states will exist that generate a negative strain energy, *i.e.* the elastic body will generate energy under deformation, a situation that is physically impossible. It follows that $\delta^2\Pi \geq 0$, and hence, Π presents an absolute minimum at its stationary point. These results can be interpreted as follows.

Principle 17 (Principle of minimum total potential energy) *Among all kinematically admissible displacements fields, the actual displacement field that corresponds to the equilibrium configuration of the body makes the total potential energy an absolute minimum.*

The reverse is also true: if the principle of minimum total potential energy holds, the total potential energy must present a stationary point, implying eq. (12.35). In turn, this equation implies the principle of virtual work in which the stresses are expressed in terms of the strains using constitutive laws of the form of eq. (12.31), and strains are themselves expressed in terms of displacements using the strain-displacement relationships. The principle of minimum total potential energy implies the equations of equilibrium of the problem expressed in terms of the displacement field. From section 12.1.3, it is clear that these equations are the Euler-Lagrange equations arising from the stationarity condition for the total potential energy.

The principle of minimum total potential energy implies the principle of virtual work, but the principle of virtual work only implies the principle of minimum total potential energy under restrictive assumptions on existence of a strain energy density function and of potentials of the body forces and surface tractions. In other words, the principle of virtual work is a more general but possibly less useful statement.

12.2.6 The principle of minimum complementary energy

Consider an elastic body undergoing kinematically admissible displacements and compatible strains. In this case, the principle of complementary virtual work, eq. (12.30), must apply. It is now assumed that the constitutive law for the material can be expressed in terms of a stress energy density function, eq.(12.33). The virtual work done by the internal strains appears in the first term of eq. (12.30) and is now readily evaluated as

$$\int_V \delta\underline{\sigma}^T \underline{\epsilon} \, d\mathcal{V} = \int_V \delta\underline{\sigma}^T \frac{\partial b(\underline{\sigma})}{\partial \underline{\sigma}} \, d\mathcal{V} = \int_V \delta b(\underline{\sigma}) \, d\mathcal{V} = \delta \int_V b(\underline{\sigma}) \, d\mathcal{V} = \delta A'(\underline{\sigma}),$$

where the chain rule for derivatives is used at the second equality. The quantity $A'(\underline{\sigma})$ is the *total stress energy* in the body.

The principle of complementary virtual work, eq. (12.30), can now be written as

$$-\delta A'(\underline{\sigma}) + \int_{S_2} \hat{\underline{u}}^T \delta\underline{t} \, d\mathcal{S} = 0 \tag{12.38}$$

Next, the prescribed displacements are *assumed* to be derivable from a potential function

$$\hat{\underline{u}} = -\frac{\partial \chi(\underline{t})}{\partial \underline{t}},$$

where $\chi(\underline{t})$ is the *potential of the prescribed displacements*. For instance, the potential of prescribed displacements is simply $\chi = -\hat{\underline{u}}^T \underline{t}$. It is important to note that potential functions do not exist for all types of prescribed displacements. For example, potential functions do not always exist for displacements that depend on the surface tractions, although such cases are not common in practice.

The second term in eq. (12.38) now becomes

$$\int_{S_2} \hat{\underline{u}}^T \delta\underline{t} \, d\mathcal{S} = -\int_{S_2} \frac{\partial \chi}{\partial \underline{t}}^T \delta\underline{t} \, d\mathcal{S} = -\int_{S_2} \delta\chi(\underline{t}) \, d\mathcal{S} = -\delta \int_{S_2} \chi(\underline{t}) \, d\mathcal{S} = -\delta\Phi'.$$

where $\Phi'(\underline{t}) = \int_{S_2} \chi(\underline{t}) \, dS$ is the total potential the prescribed displacements. Introducing this result into eq. (12.38) leads to

$$-\delta A'(\underline{\sigma}) - \delta \Phi'(\underline{t}), \quad \text{or} \quad \delta \left[A'(\underline{\sigma}) + \Phi'(\underline{t}) \right] = 0. \tag{12.39}$$

The *total complementary energy* of the body is now defined as

$$\Pi'(\underline{\sigma}) = A'(\underline{\sigma}) + \Phi'(\underline{t}), \tag{12.40}$$

and it follows that

$$\delta \Pi'(\underline{\sigma}) = 0. \tag{12.41}$$

This statement can be interpreted as follows

Principle 18 (Principle of minimum complementary energy) *Among all statically admissible stress fields, the actual stress field that corresponds to the compatible deformations of the body makes the total complementary energy an absolute minimum.*

Equation (12.41) only proves that for compatible deformations, the total complementary energy presents a stationary point. As discussed in section 12.1.1, the sign of the second variation $\delta^2 \Pi'$ will determine whether the stationary point actually is a minimum. The first variation in Π' is

$$\delta \Pi'(\underline{\sigma}) = \int_{\mathcal{V}} \sum_{i=1}^{6} \frac{\partial a'}{\partial \sigma_i} \delta \sigma_i \, d\mathcal{V} - \int_{S_2} \hat{\underline{u}}^T \delta \underline{t} \, dS, \tag{12.42}$$

and its second variation is then

$$\delta^2 \Pi'(\underline{\sigma}) = \int_{\mathcal{V}} \sum_{i,j=1}^{6} \frac{\partial^2 a'}{\partial \sigma_i \partial \sigma_j} \delta \sigma_i \delta \sigma_j \, d\mathcal{V}. \tag{12.43}$$

Just as for the strain energy density function, the stress energy density function must be a positive-definite function of the stress components, which implies $\sum_{i,j=1}^{6} \partial^2 a'/(\partial \sigma_i \partial \sigma_j) \, \delta \sigma_i \, \delta \sigma_j \geq 0$ for all $\delta \sigma_i$. Indeed, if the stress energy function is not positive-definite, stress states will exist that generate a negative stress energy, *i.e.* the elastic body will generate energy under stress, a situation that is physically impossible. It follows that $\delta^2 \Pi' \geq 0$, and hence, Π' presents an absolute minimum at its stationary point.

If the principle of minimum complementary energy holds, the complementary energy must present a stationary point, implying eq. (12.39). In turn, this equation implies the principle of complementary virtual work, in which the strains are expressed in terms of the stresses using constitutive laws of the form of eq. (12.33). As a result, the principle of minimum complementary energy implies the strain-displacement relationships of the problem expressed in terms of the stress field, which must satisfy equilibrium equations. From section 12.1.3, it follows that these equations are the Euler-Lagrange equations arising from the stationarity condition for the complementary energy.

The principle of minimum complementary energy implies the principle of complementary virtual work, but the principle of complementary virtual work only implies the principle of minimum complementary energy under restrictive assumptions on existence of a stress energy density function and of a potential for the prescribed displacements.

12.2.7 Energy theorems

In section 10.9, a number of energy theorems are presented that all are corollaries of the fundamental energy principles developed above. Clapeyron's theorem, theorem 10.1, and Castigliano's first theorem, theorem 10.2, are corollaries of the principle of minimum total potential energy. The principle of least work, principle 14, Crotti-Engesser theorem, theorem 10.3, and Castigliano's second theorem, theorem 10.4, are corollaries of the principle of complementary total potential energy. Finally, the reciprocity theorems of Betti and Maxwell, theorems 10.5 and 10.6, respectively, are direct consequences of these theorem. Because the principle of minimum total potential energy and its complementary counterpart have now been established for general, three-dimensional structures, the theorems listed above are also valid for the same three-dimensional structures.

12.2.8 Hu-Washizu's principle

The principle of virtual work developed in section 9.3 is shown to be entirely equivalent to the equations of equilibrium of a three-dimensional solid, eqs. (12.17) and (12.18). Because this principle is solely a statement of equilibrium, it must be complemented with stress-strain relationships and constitutive laws in order to obtain a complete set of equations for the solution of specific elasticity problems.

On the other hand, the principle of complementary virtual work developed in section 12.30 is equivalent to the strain-displacement relationships and the geometric boundary conditions, eqs. (12.19) and (12.20), respectively. This principle must be complemented with equilibrium equations and constitutive laws in order to obtain a complete set of equations for the solution of specific elasticity problems.

In summary, the principle of virtual work is a statement of equilibrium whereas the principle of complementary virtual work is a statement of compatibility. Clearly, these principles are equivalent to a subset of all the equations required for the solution of elasticity problems. Hu-Washizu's principle remedies this shortcoming, and it is equivalent to the complete set of equations required to solve elasticity problems.

Consider an elastic body that is in equilibrium under the applied body forces and surface tractions, that is undergoing compatible strains whose displacement field is kinematically admissible, and for which the stress and strain fields satisfy the material constitutive laws. This implies that the stress field is statically admissible, $i.e.$, the equilibrium equations, eqs. (12.17), are satisfied at all points in V and the surface equilibrium equations, eqs. (12.18), at all points on S_1. This further implies that the strain-displacement relationships, eqs. (12.19), are satisfied at all points in V and the geometric boundary conditions, eqs. (12.20), at all points on S_2. Finally, the constitutive equations, assumed to be expressed in terms of a strain energy density function, eq. (12.31), must hold at all points in $\mathcal{V}$.

The following statement is now constructed by combining eqs. (12.21), (12.26) and (12.31) into a single integral equation

$$
\begin{aligned}
\int_{\mathcal{V}} & \left\{ \left[\frac{\partial \sigma_1}{\partial x_1} + \frac{\partial \tau_{21}}{\partial x_2} + \frac{\partial \tau_{31}}{\partial x_3} + b_1 \right] \delta u_1 + \left[\frac{\partial \tau_{12}}{\partial x_1} + \frac{\partial \sigma_2}{\partial x_2} + \frac{\partial \tau_{32}}{\partial x_3} + b_2 \right] \delta u_2 \right. \\
& \left. + \left[\frac{\partial \tau_{13}}{\partial x_1} + \frac{\partial \tau_{23}}{\partial x_2} + \frac{\partial \sigma_3}{\partial x_3} + b_3 \right] \delta u_3 \right\} \, \mathrm{d}\mathcal{V} - \int_{S_1} \left[\underline{t} - \hat{\underline{t}} \right]^T \delta \underline{u} \, \mathrm{d}S \\
& - \int_{\mathcal{V}} \left\{ \left[\epsilon_1 - \frac{\partial u_1}{\partial x_1} \right] \delta\sigma_1 + \left[\epsilon_2 - \frac{\partial u_2}{\partial x_2} \right] \delta\sigma_2 + \left[\epsilon_3 - \frac{\partial u_3}{\partial x_3} \right] \delta\sigma_3 \right. \\
& + \left[\gamma_{23} - \frac{\partial u_2}{\partial x_3} - \frac{\partial u_3}{\partial x_2} \right] \delta\tau_{23} + \left[\gamma_{13} - \frac{\partial u_1}{\partial x_3} - \frac{\partial u_3}{\partial x_1} \right] \delta\tau_{13} \\
& \left. + \left[\gamma_{12} - \frac{\partial u_1}{\partial x_2} - \frac{\partial u_2}{\partial x_1} \right] \delta\tau_{12} \right\} \, \mathrm{d}\mathcal{V} - \int_{S_2} \left[\underline{u} - \hat{\underline{u}} \right]^T \delta\underline{t} \, \mathrm{d}S \\
& + \int_{\mathcal{V}} \left\{ \left[\frac{\partial a}{\partial \epsilon_1} - \sigma_1 \right] \delta\epsilon_1 + \left[\frac{\partial a}{\partial \epsilon_2} - \sigma_2 \right] \delta\epsilon_2 + \left[\frac{\partial a}{\partial \epsilon_3} - \sigma_3 \right] \delta\epsilon_3 \right. \\
& \left. + \left[\frac{\partial a}{\partial \gamma_{23}} - \tau_{23} \right] \delta\gamma_{23} + \left[\frac{\partial a}{\partial \gamma_{13}} - \tau_{13} \right] \delta\gamma_{13} + \left[\frac{\partial a}{\partial \gamma_{12}} - \tau_{12} \right] \delta\gamma_{12} \right\} \, \mathrm{d}\mathcal{V} = 0.
\end{aligned}
$$

$$(12.44)$$

This lengthy statement can be manipulated in several different ways. *(1)* The terms appearing in the equilibrium equations could be integrated by parts (as is done for the derivation of the principle of virtual work, section 12.2.2), *(2)* the terms appearing in the strain-displacement relationships could be integrated by parts (as is done for the derivation of the principle of complementary virtual work, section 12.2.3), or *(3)* both integrations by parts could be carried out. These three approaches will give rise to three different statements of Hu-Washizu's principle.

First statement of Hu-Washizu's principle

In the first approach, the terms appearing in the equations of equilibrium are integrated by parts using eq. (12.22). After regrouping all terms, this yields the *first statement of Hu-Washizu's principle*

$$
\begin{aligned}
\delta \int_{\mathcal{V}} & \left[a(\underline{\epsilon}) - \left(\epsilon_1 - \frac{\partial u_1}{\partial x_1} \right) \sigma_1 - \left(\epsilon_2 - \frac{\partial u_2}{\partial x_2} \right) \sigma_2 - \left(\epsilon_3 - \frac{\partial u_3}{\partial x_3} \right) \sigma_3 \right. \\
& - \left(\gamma_{23} - \frac{\partial u_2}{\partial x_3} - \frac{\partial u_3}{\partial x_2} \right) \tau_{23} - \left(\gamma_{13} - \frac{\partial u_1}{\partial x_3} - \frac{\partial u_3}{\partial x_1} \right) \tau_{13} \\
& \left. - \left(\gamma_{12} - \frac{\partial u_1}{\partial x_2} - \frac{\partial u_2}{\partial x_1} \right) \tau_{12} \right] \, \mathrm{d}\mathcal{V} \\
& - \int_{\mathcal{V}} \underline{b}^T \delta\underline{u} \, \mathrm{d}\mathcal{V} - \int_{S_1} \hat{\underline{t}}^T \delta\underline{u} \, \mathrm{d}S - \int_{S_2} (\underline{u} - \hat{\underline{u}})^T \delta\underline{t} \, \mathrm{d}S = 0.
\end{aligned}
$$

$$(12.45)$$

This principle involves three independent fields: the strain, stress, and displacement fields. Hence, Hu-Washizu's principle is a *three field principle*, whereas the principle of minimum total potential energy and the principle of minimum complementary energy both are single field principles, involving only the displacement and stress fields, respectively.

This principle is closely related to the principle of minimum total potential energy, eq. (12.34). Indeed, starting from eq. (12.45), the strain field is assumed to be compatible and the displacement field kinematically admissible. Hence, the strain displacement relationships are satisfied and the last six terms in the left hand side integrand vanish; furthermore, the displacement field satisfies the geometric boundary conditions on S_2 and the last integral on the right hand side vanishes as well. The remaining terms then express the principle of minimum total potential energy, eq. (12.34).

The first statement of Hu-Washizu's principle can also be obtained by starting from the principle of minimum total potential energy, eq. (12.34). This principle is a statement of equilibrium, because it is derived from the principle of virtual work, and the constitutive laws of the material are also included in the principle by means of the strain energy density function. However, this principle provides no information about the strain displacement relationships of the problem. Consequently, the principle of minimum total potential energy can be viewed as a constrained minimization problem that yields all the equations of elasticity: minimization of the total potential energy yields the equations of equilibrium and the constitutive laws, while the external constraints, the strain displacement equations, then yield the last set of equations.

This *constrained* minimization problem is then transformed into an *unconstrained* minimization problem using the Lagrange multiplier technique described in section 12.1.2. The modified function to be minimized is now in the form of eq. 12.12, where the f_i, $i = 1, 2, \ldots 6$, are the strain displacement relationships and the λ_i are six Lagrange multipliers. This is exactly the form of Hu-Washizu's principle, eq. (12.45), where the Lagrange multipliers are identified to be the stress components. This leads to an interesting interpretation of the stress field: the six stress components are the Lagrange multipliers used to enforce the corresponding compatibility equations.

Second statement of Hu-Washizu's principle

In the second approach, the terms appearing in the strain-displacement relationships of eq. 12.44 are integrated by parts using eq. 12.27. After regrouping all terms, this yields the *second statement of Hu-Washizu's principle*

$$\delta \int_{\mathcal{V}} \left[\left(a(\underline{\epsilon}) - \underline{\epsilon}^T \underline{\sigma} \right) + \left(\frac{\partial \sigma_1}{\partial x_1} + \frac{\partial \tau_{21}}{\partial x_2} + \frac{\partial \tau_{31}}{\partial x_3} + b_1 \right) u_1 \right.$$
$$+ \left(\frac{\partial \tau_{12}}{\partial x_1} + \frac{\partial \sigma_2}{\partial x_2} + \frac{\partial \tau_{32}}{\partial x_3} + b_2 \right) u_2 + \left. \left(\frac{\partial \tau_{13}}{\partial x_1} + \frac{\partial \tau_{23}}{\partial x_2} + \frac{\partial \sigma_3}{\partial x_3} + b_3 \right) u_3 \right] d\mathcal{V}$$
$$- \int_{\mathcal{S}_1} (\underline{t} - \underline{\hat{t}})^T \delta \underline{u} \, d\mathcal{S} - \int_{\mathcal{S}_2} \underline{\hat{u}}^T \delta \underline{t} \, d\mathcal{S} = 0.$$

$$(12.46)$$

This principle is closely related to the principle of minimum complementary energy, eq. (12.38). Indeed, starting from the above statement, the stress field is assumed to be statically admissible. Hence, the equations of equilibrium relationships are satisfied and the last three terms in the volume integral vanish; furthermore, equilibrium of the surface tractions is satisfied on $\mathcal{S}_1$ and the corresponding surface integral vanishes as well. The remaining terms then express the principle of minimum complementary energy, eq. (12.38).

The second statement of Hu-Washizu's principle can also be obtained by starting from the principle of minimum complementary energy, eq. 12.38. This principle is a statement of compatibility because it is derived from the principle of complementary virtual work, and the constitutive laws of the material are also included in the principle by means of the stress energy density function. This principle provides no information about the equilibrium equations of the problem. Consequently, the principle of minimum complementary energy can be viewed as a constrained minimization problem that yields all the equations of elasticity: minimization of the complementary energy yields the compatibility equations and the constitutive laws while the external constraints, the equilibrium equations, then yield the last set of equations.

This *constrained* minimization problem is then transformed into an *unconstrained* minimization problem using the Lagrange multiplier technique described in section 12.1.2. The modified function to be minimized is now in the form of eq. 12.12, where the f_i, $i = 1, 2, 3$, are the equilibrium equations and the λ_i are three Lagrange multipliers. This is exactly the form of Hu-Washizu's principle, eq. (12.46), where the Lagrange multipliers are identified to be the displacement components. This leads to an interesting interpretation of the displacement field: the three displacement components are the Lagrange multipliers used to enforce the corresponding equilibrium equations.

Third statement of Hu-Washizu's principle

Finally, in the last approach the terms appearing in both the equations of equilibrium and strain-displacement relationships of eq. (12.44) are integrated by parts using eqs. (12.22 and eq. (12.27), respectively. After regrouping all terms, this yields the *third statement of Hu-Washizu's principle*

$$\int_{\mathcal{V}} \left\{ \delta \left[a(\underline{\epsilon}) - \underline{\epsilon}^T \underline{\sigma} \right] + \sigma_1 \frac{\partial \delta u_1}{\partial x_1} + \sigma_2 \frac{\partial \delta u_2}{\partial x_2} + \sigma_3 \frac{\partial \delta u_3}{\partial x_3} + \tau_{23} \left[\frac{\partial \delta u_2}{\partial x_3} + \frac{\partial \delta u_3}{\partial x_2} \right] \right.$$

$$+ \tau_{13} \left[\frac{\partial \delta u_1}{\partial x_3} + \frac{\partial \delta u_3}{\partial x_1} \right] + \tau_{12} \left[\frac{\partial \delta u_1}{\partial x_2} + \frac{\partial \delta u_3}{\partial x_2} \right] - u_1 \left[\frac{\partial \delta \sigma_1}{\partial x_1} + \frac{\partial \delta \tau_{12}}{\partial x_2} + \frac{\partial \delta \tau_{13}}{\partial x_3} \right]$$

$$\left. - u_2 \left[\frac{\partial \delta \tau_{12}}{\partial x_1} + \frac{\partial \delta \sigma_2}{\partial x_2} + \frac{\partial \delta \tau_{23}}{\partial x_3} \right] - u_3 \left[\frac{\partial \delta \tau_{13}}{\partial x_1} + \frac{\partial \delta \tau_{23}}{\partial x_2} + \frac{\partial \delta \sigma_3}{\partial x_3} \right] \right\} \, \mathrm{d}\mathcal{V}$$

$$- \int_{\mathcal{V}} \underline{b}^T \delta \underline{u} \, \mathrm{d}\mathcal{V} - \int_{\mathcal{S}_1} \hat{\underline{t}}^T \delta \underline{u} \, \mathrm{d}\mathcal{S} + \int_{\mathcal{S}_2} \hat{\underline{u}}^T \delta \underline{t} \, \mathrm{d}\mathcal{S} = 0.$$

$$(12.47)$$

The main advantage of this third statement of Hu-Washizu's principle is that no derivatives of the three unknown fields are present; derivatives only show up in the variations. In numerical applications, this observation has important implications on the way in which the unknown fields can be approximated, because minimal continuity requirements are imposed.

12.2.9 Hellinger-Reissner's principle

Due to the complexity of the three-field Hu-Washizu's principle, a simpler, two-field principle is preferred for some applications. Hellinger-Reissner's principle is such a principle, and it is easily derived from Hu-Washizu's principle by eliminating the strain field.

Starting from the first statement of Hu-Washizu's principle, eq. (12.45), eq. (12.32) is used to eliminate the strain field: $\delta[a(\underline{\epsilon}) - \underline{\epsilon}^T \underline{\sigma}] = -\delta a'(\underline{\sigma})$. This simple operation yields the *first statement of Hellinger-Reissner's principle*

$$\delta \int_{\mathcal{V}} \left[\frac{\partial u_1}{\partial x_1} \sigma_1 + \frac{\partial u_2}{\partial x_2} \sigma_2 + \frac{\partial u_3}{\partial x_3} \sigma_3 + \left(\frac{\partial u_2}{\partial x_3} + \frac{\partial u_3}{\partial x_2} \right) \tau_{23} \right.$$

$$\left. + \left(\frac{\partial u_1}{\partial x_3} + \frac{\partial u_3}{\partial x_1} \right) \tau_{13} + \left(\frac{\partial u_1}{\partial x_2} + \frac{\partial u_2}{\partial x_1} \right) \tau_{12} - a'(\underline{\sigma}) \right] \mathrm{d}\mathcal{V} \qquad (12.48)$$

$$- \int_{\mathcal{V}} \underline{b}^T \delta \underline{u} \, \mathrm{d}\mathcal{V} + \int_{\mathcal{S}_1} \hat{\underline{t}}^T \delta \underline{u} \, \mathrm{d}\mathcal{S} - \int_{\mathcal{S}_2} (\underline{u} - \hat{\underline{u}})^T \delta \underline{t} \, \mathrm{d}\mathcal{S} = 0.$$

Next, starting from the second statement of Hu-Washizu's principle, eq. (12.46), the strain field is eliminated in a similar manner to obtain the *second statement of Hellinger-Reissner's principle*

$$\delta \int_{\mathcal{V}} \left[\left(\frac{\partial \sigma_1}{\partial x_1} + \frac{\partial \tau_{21}}{\partial x_2} + \frac{\partial \tau_{31}}{\partial x_3} + b_1 \right) u_1 + \left(\frac{\partial \tau_{12}}{\partial x_1} + \frac{\partial \sigma_2}{\partial x_2} + \frac{\partial \tau_{32}}{\partial x_3} + b_2 \right) u_2 \right.$$

$$\left. + \left(\frac{\partial \tau_{13}}{\partial x_1} + \frac{\partial \tau_{23}}{\partial x_2} + \frac{\partial \sigma_3}{\partial x_3} + b_3 \right) u_3 - a'(\underline{\sigma}) \right] \mathrm{d}\mathcal{V}$$

$$- \int_{\mathcal{S}_1} (\underline{t} - \hat{\underline{t}})^T \delta \underline{u} \, \mathrm{d}\mathcal{S} - \int_{\mathcal{S}_2} \hat{\underline{u}}^T \delta \underline{t} \, \mathrm{d}\mathcal{S} = 0.$$

$$(12.49)$$

Finally, a similar procedure starting from the third statement of Hu-Washizu's principle, eq. 12.47, leads to the the *third statement of Hellinger-Reissner's principle*

$$
\int_{\mathcal{V}} \left[\delta a'(\underline{\sigma}) + u_1 \left(\frac{\partial \delta \sigma_1}{\partial x_1} + \frac{\partial \delta \tau_{12}}{\partial x_2} + \frac{\partial \delta \tau_{13}}{\partial x_3} \right) + u_2 \left(\frac{\partial \delta \tau_{12}}{\partial x_1} + \frac{\partial \delta \sigma_2}{\partial x_2} + \frac{\partial \delta \tau_{23}}{\partial x_3} \right) \right.
$$

$$
+ u_3 \left(\frac{\partial \delta \tau_{13}}{\partial x_1} + \frac{\partial \delta \tau_{23}}{\partial x_2} + \frac{\partial \delta \sigma_3}{\partial x_3} \right) - \sigma_1 \frac{\partial \delta u_1}{\partial x_1} - \sigma_2 \frac{\partial \delta u_2}{\partial x_2} - \sigma_3 \frac{\partial \delta u_3}{\partial x_3}
$$

$$
\left. - \tau_{23} \left(\frac{\partial \delta u_2}{\partial x_3} + \frac{\partial \delta u_3}{\partial x_2} \right) - \tau_{13} \left(\frac{\partial \delta u_1}{\partial x_3} + \frac{\partial \delta u_3}{\partial x_1} \right) - \tau_{12} \left(\frac{\partial \delta u_1}{\partial x_2} + \frac{\partial \delta u_3}{\partial x_2} \right) \right] \, \mathrm{d}\mathcal{V}
$$

$$
+ \int_{\mathcal{V}} \underline{b}^T \delta \underline{u} \, \mathrm{d}\mathcal{V} + \int_{\mathcal{S}_1} \hat{\underline{t}}^T \delta \underline{u} \, \mathrm{d}\mathcal{S} - \int_{\mathcal{S}_2} \hat{\underline{u}}^T \delta \underline{t} \, \mathrm{d}\mathcal{S} = 0. \tag{12.50}
$$

12.3 Applications of variational and energy principles

The general formulation of the equations of linear elasticity as a series of variational problem provides powerful capabilities to analyze more complex structural problems than those considered in the previous chapters. As illustrated in chapter 11, this is particularly important when approximate solutions are sought because the integral forms appearing the variational equations involve lower order derivatives than those appearing in the corresponding differential equations, thereby decreasing the continuity requirements.

The equivalence of the differential equation and variational approaches is proven at several points in the above developments. This equivalence can also be demonstrated by deriving the governing differential equations from the variational formulation. This approach provides the additional benefit of yielding the associated boundary conditions.

The beam bending problem

Beam bending under transverse loading is discussed extensively in chapters 5, 6 and 8. In this section, the governing differential equations for beam bending will be derived from the principle of minimum total potential energy.

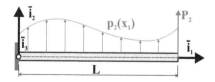

Fig. 12.5. Cantilevered beam with distributed transverse load.

Consider a general beam bending problem consisting of a uniform cantilevered beam with a symmetric cross-section subjected to a distributed transverse load, $p_2(x_1)$, and a concentrated tip load, P_2, as illustrated in fig. 12.5. The total potential energy of the system is given by eq. (10.40) as

$$
\Pi = \frac{1}{2} \int_0^L H_{33}^c \left(\frac{\mathrm{d}^2 \bar{u}_2}{\mathrm{d}x_1^2} \right)^2 \mathrm{d}x_1 - \int_0^L p_2(x_1) \bar{u}_2(x_1) \, \mathrm{d}x_1 - P_2 \bar{u}_2(L), \tag{12.51}
$$

where it is clear that $\Pi = \Pi(\bar{u}_2(x_1))$ is a functional.

The principle of minimum total potential energy, principle 17, implies that the total potential energy must be stationary, $\delta\Pi = 0$. Section 12.1.3 outlines the procedure to determine the stationary point of a functional in the form given by eq. (12.13), and the stationarity condition leads the the Euler-Lagrange equation, eq. (12.14). The same procedure will be followed here to determine the stationary point of the total potential energy, as required by the principle of minimum total potential energy.

Variation of of the total potential energy, eq. (12.51), can be written as

$$\delta\Pi = \frac{1}{2}\int_0^L H_{33}^c \, 2\frac{\mathrm{d}^2\bar{u}_2}{\mathrm{d}x_1^2}\, \delta\left(\frac{\mathrm{d}^2\bar{u}_2}{\mathrm{d}x_1^2}\right)\mathrm{d}x_1 - \int_0^L p_2\,\delta\bar{u}_2\,\mathrm{d}x_1 - P_2\delta\bar{u}_2(L) = 0.$$

Using eq. (12.15), it is possible to interchange the order of the variational and partial differential operators in the third term in the first integral to find

$$\delta\Pi = \frac{1}{2}\int_0^L H_{33}^c \, 2\frac{\mathrm{d}^2\bar{u}_2}{\mathrm{d}x_1^2}\frac{\mathrm{d}^2}{\mathrm{d}x_1^2}(\delta\bar{u}_2)\mathrm{d}x_1 - \int_0^L p_2\,\delta\bar{u}_2\,\mathrm{d}x_1 - P_2\delta\bar{u}_2(L) = 0.$$

To eliminate the higher differential order of the virtual displacement field appearing in the first integral, two integration by parts are carried out, leading to

$$\delta\Pi = \int_0^L \left[\frac{\mathrm{d}^2}{\mathrm{d}x_1^2}\left(H_{33}^c\frac{\mathrm{d}^2\bar{u}_2}{\mathrm{d}x_1^2}\right) - p_2\right]\delta\bar{u}_2\mathrm{d}x_1 + \left[H_{33}^c\frac{\mathrm{d}\bar{u}_2^2}{\mathrm{d}x_1^2}\,\delta\left(\frac{\mathrm{d}\bar{u}_2}{\mathrm{d}x_1}\right)\right]_0^L$$
$$- \left[\frac{\mathrm{d}}{\mathrm{d}x_1}\left(H_{33}^c\frac{\mathrm{d}\bar{u}_2^2}{\mathrm{d}x_1^2}\right)\delta\bar{u}_2\right]_0^L - P_2\delta\bar{u}_2(L) = 0. \tag{12.52}$$

This equation must be satisfied for all arbitrary displacements, $\delta\bar{u}_2(x_1)$ for $0 \leq x_1 \leq L$. This can only happen if the first bracketed term vanishes, leading to the following differential equation

$$\frac{\mathrm{d}^2}{\mathrm{d}x_1^2}\left(H_{33}^c\frac{\mathrm{d}\bar{u}_2^2}{\mathrm{d}x_1^2}\right) - p_2 = 0, \tag{12.53}$$

which is the governing equation for the lateral deflection of a beam under load, first developed using the classical differential equation approach, see eq. (5.40). Within the present formalism, the governing differential equation is the *Euler-Lagrange equation* associated with the stationary point of the total potential energy.

The stationarity condition of the total potential energy also yields the boundary conditions of the problem. First, the stationary condition, eq. (12.52) is rewritten as

$$\delta\Pi = \int_0^L \left[\frac{\mathrm{d}^2}{\mathrm{d}x_1^2}\left(H_{33}^c\frac{\mathrm{d}^2\bar{u}_2}{\mathrm{d}x_1^2}\right) - p_2\right]\delta\bar{u}_2\mathrm{d}x_1 + [M_3\delta\Phi_3]_0^L$$
$$+ [V_2\delta\bar{u}_2]_0^L - P_2\delta\bar{u}_2(L) = 0, \tag{12.54}$$

where the definitions of the bending moment, shear force, and sectional rotation are used.

The last two terms of eq. (12.54) can be written as $[V_2(L) - P_2]\delta\bar{u}_2(L) - V_2(0)\delta\bar{u}_2(0)$. The virtual displacement field must satisfy the geometric boundary condition. Because the beam is clamped at the root, $\delta\bar{u}_2(0) = 0$, and the last two terms of eq. (12.54) reduce to $[V_2(L) - P_2]\delta\bar{u}_2(L)$. This term must vanish for all arbitrary virtual displacements at the tip of the beam, and therefore the bracketed term must vanish, leading to $V_2(L) = P_2$. This equation is the first natural boundary condition at the tip of the beam.

The second bracketed term of eq. (12.54), written as $M_3(L)\delta\Phi_3(L) - M_3(0)\delta\Phi_3(0)$ must also vanish. Because the beam is clamped at the root, $\delta\Phi_3(0) = 0$, and this reduces to $[M_3(L)]\delta\Phi_3(L)$. This term must vanish for all arbitrary virtual rotations at the tip of the beam, and therefore the bracketed term must vanish, leading to $M_3(L) = 0$. This equation is the second natural boundary condition at the tip of the beam.

In summary, the classical Euler-Bernoulli governing differential equation for beam bending problems is derived as the Euler-Lagrange equation associated with the stationarity condition of the total potential energy. Note that both differential equation and natural boundary conditions are recovered. Once again, the equivalence of the various formalisms is demonstrated.

12.3.1 The shear lag problem

The axial and shear flows in thin-walled beams are computed in chapter 8 for both bending and torsional loads. In addition, the cross-sectional warping of both open and closed-section thin-walled beams is examined in section 8.7. In the treatment of warping, it is tacitly assumed that the warping is free of any constraints that might be present at the beam ends, for instance. If the beam is cantilevered at its root, warping must vanish at the root but will develop along the beam's span, causing a redistribution of the stresses in the beam that can become significant under certain conditions.

A good example of this effect is *shear lag*, which is present in thin-walled beams that are fixed in a manner that constrains warping of a cross-section. The result can be a significant redistribution of the axial and shear flows in that cross-section and along the beam's span.

Consider a thin-walled, rectangular box-beam clamped at its root and subjected to a downward tip load, P_0, as shown in fig. 12.6. According to Euler-Bernoulli beam theory, the stress distribution in the upper flange (top panel) of the beam consists of a uniform distribution of tensile axial stress across the width, $2b$, and equal magnitude compressive stresses arise in the lower flange. If concentrated stiffeners are present at the four corners of the section, the top panel will be loaded by shear flows at its edges, as illustrated in fig. 12.7.

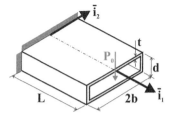

Fig. 12.6. Box beam subjected to tip load.

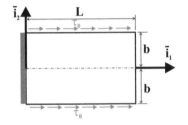

Fig. 12.7. Upper panel of a box beam subjected to edge shear forces.

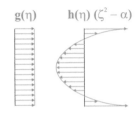

Fig. 12.8. The distribution of axial displacement.

An investigation of the shear lag phenomenon will be performed by considering only the rectangular panel loaded by constant shear stresses, τ_0, at its two edges, as shown in lower portion of fig. 12.7. This configuration is a crude approximation of the top flange of a thin-walled beam with four corner stiffeners. The width of the panel is $2b$, its thickness t, and its length L. At the root of the panel, the axial displacement must vanish, the tip section of the panel is stress free.

The solution procedure begins with kinematic assumptions: the displacement field in the panel is assumed to take the following form

$$u_1(\eta, \zeta) = g(\eta) + h(\eta)(\zeta^2 - \alpha); \quad u_2(\eta, \zeta) = 0, \tag{12.55}$$

where $\eta = x_1/L$ is a non-dimensional variable along the span of the panel and $\zeta = x_2/b$ that across its width. The axial displacement distribution is illustrated in fig. 12.8 and consists of two components. The first term corresponds to a uniform distribution across the width of the panel and gives rise to uniform displacements, strains, and stresses across the width, as predicted by beam theory. The second term describes the variation of the displacement field across the width of the panel. Due to the symmetry of the problem, a symmetric distribution across the width of the panel is selected: a symmetric parabolic distribution defined by parameter α.

This parabolic displacement distribution will result in uneven stress and strain distribution across the width of the panel and will characterize the importance of the shear lag effect. Because a parabolic distribution is arbitrarily selected at the onset of the analysis, an approximate solution is expected.

The strain field is obtained from the assumed displacement field using the strain-displacement relations, see eqs. (1.63) and (1.71), to find

$$\epsilon_1(\eta, \zeta) = \frac{g'}{L} + \frac{h'}{L}(\zeta^2 - \alpha), \quad \epsilon_2(\eta, \zeta) = 0; \quad \gamma_{12}(\eta, \zeta) = \frac{h}{b}2\zeta, \tag{12.56}$$

where the notation $(\cdot)'$ indicates a derivative with respect to η.

The panel is assumed to be made of a linearly elastic material in a plane stress state, and the material stiffness matrix is given by eq. (2.16). Furthermore, the stress in the transverse direction is assumed to be much smaller than the axial stress, $\sigma_2 \ll \sigma_1$, and the constitutive laws reduce to

$$\sigma_1 = E\epsilon_1 = E\left[\frac{g'}{L} + \frac{h'}{L}(\zeta^2 - \alpha)\right], \quad \sigma_2 \approx 0, \quad \tau_{12} = G\gamma_{12} = G\frac{h}{b}2\zeta. \tag{12.57}$$

The strain energy in the panel can now be written in terms of the axial and shear strains as

$$A = \int_0^L \int_{-b}^{+b} \int_{-t/2}^{t/2} \frac{1}{2} (E\epsilon_1^2 + G\gamma_{12}^2) \, dx_1 dx_2 dx_3 = \frac{Lbt}{2} \int_0^1 \int_{-1}^{+1} (E\epsilon_1^2 + G\gamma_{12}^2) \, d\eta d\zeta.$$

Introducing the strain distributions, eq. (12.56), and reordering the term leads to

$$A = \frac{Lbt}{2} \int_0^1 \left\{ \frac{Eg'^2}{L^2} \left[\int_{-1}^{+1} d\zeta \right] + \frac{Eh'^2}{L^2} \left[\int_{-1}^{+1} (\zeta^2 - \alpha)^2 \, d\zeta \right] \right.$$
$$\left. + 2\frac{Eg'h'}{L^2} \left[\int_{-1}^{+1} (\zeta^2 - \alpha) \, d\zeta \right] + \frac{4Gh^2}{b^2} \left[\int_{-1}^{+1} \zeta^2 \, d\zeta \right] \right\} d\eta.$$

The next step is to evaluate the integrals appearing in the brackets. To simplify the analysis, the free parameter α will be selected to make the third bracketed term vanish: $\int_{-1}^{+1} (\zeta^2 - \alpha) \, d\zeta = 0$, which leads to $\alpha = 1/3$. With this choice, the third term of the strain energy vanishes, eliminating the coupling between the two deformation modes characterized by functions $g(\eta)$ and $h(\eta)$. Using $\alpha = 1/3$, the strain energy now becomes

$$A = \frac{Ebt}{L} \int_0^1 \left[g'^2 + \frac{4}{45} h'^2 + \frac{4}{3} \left(\frac{L}{b} \right)^2 \left(\frac{G}{E} \right) h^2 \right] d\eta.$$

The potential of the applied loads consists of the potential of the shear loads applied along the left and right side edges of the panel,

$$\Phi = - \int_{t/2}^{t/2} \int_0^L \left[\tau_0 u_1 (x_1, x_2 = -b) + \tau_0 u_1 (x_1, x_2 = +b) \right] \, dx_1 dx_2$$
$$= 2tL \int_0^1 \tau_0 (g + \frac{2h}{3}) \, d\eta,$$

and the total potential energy therefore becomes

$$\Pi = \frac{Ebt}{L} \int_0^1 \left[g'^2 + \frac{4}{45} h'^2 + \frac{4}{3} \left(\frac{L}{b} \right)^2 \left(\frac{G}{E} \right) h^2 - \frac{2\tau_0 L^2}{Eb} (g + \frac{2h}{3}) \right] d\eta.$$

The principle of minimum total potential energy requires Π to be stationary, $\delta \Pi = 0$. Taking the first variation of Π leads to

$$\int_0^1 \left[g' \delta g' + \frac{4}{45} h' \delta h' + \frac{4}{3} \left(\frac{L}{b} \right)^2 \left(\frac{G}{E} \right) h \delta h - \frac{\tau_0 L^2}{Eb} (\delta g + \frac{2}{3} \delta h) \right] d\eta = 0.$$

Integration by parts is performed on the first two terms and regrouping yields

$$\int_0^1 \left\{ \delta g \left[-g'' - \frac{\tau_0 L^2}{Eb} \right] + \delta h \left[-\frac{4}{45} h'' + \frac{4}{3} \left(\frac{L}{b} \right)^2 \left(\frac{G}{E} \right) h - \frac{2\tau_0 L^2}{3Eb} \right] \right\} d\eta$$

$$+ [g' \delta g]_0^1 + \left[\frac{4}{45} h' \delta h \right]_0^1 = 0.$$

Because the expression must vanish for all arbitrary variations δg and δh, each of the bracketed terms must vanish individually. The first bracketed term leads to the following differential equation for $g(\eta)$: $g'' = -\tau_0 L^2/(Eb)$. The boundary conditions are $g(0) = 0$ at the root of the panel and $g'(1) = 0$ at its tip. The second bracketed term leads to the differential equation for $h(\eta)$: $h'' - \mu^2 h = -15\tau_0 L^2/(2Eb)$. The boundary conditions are $h(0) = 0$ at the root of the panel and $h'(1) = 0$ at its tip. The non-dimensional parameter μ is defined as

$$\mu^2 = 15 \left(\frac{L}{b} \right)^2 \left(\frac{G}{E} \right). \tag{12.58}$$

These second order differential equations can be solved to obtain the axial displacement field as

$$u_1(\eta, \zeta) = \frac{\tau_0}{E} \frac{L^2}{b} \left\{ \left(\eta - \frac{1}{2} \eta^2 \right) + \frac{15}{2\mu^2} \left[1 - \frac{\cosh \mu(1-\eta)}{\cosh \mu} \right] \left(\zeta^2 - \frac{1}{3} \right) \right\}.$$

The first term, $(\eta - \eta^2/2)$, represents the axial displacement, which is constant across the width of the panel as if it were a beam of axial stiffness $S = E2bt$ subjected to a uniform axial load $p_0 = 2\tau_0 t$, see eq. (5.26). The second term represents the displacement variation across the width of the panel and characterizes the shear lag effect that is controlled by the non-dimensional parameter μ.

The stress distribution in the panel can be obtained from the constitutive relationships, eq. (12.57)

$$\sigma_1 = \tau_0 \frac{L}{b} \left[(1 - \eta) + \frac{15}{2\mu} \frac{\sinh \mu(1-\eta)}{\cosh \mu} \left(\zeta^2 - \frac{1}{3} \right) \right],$$

$$\tau_{12} = \tau_0 \left[1 - \frac{\cosh \mu(1-\eta)}{\cosh \mu} \right] \zeta. \tag{12.59}$$

The first term describes the constant axial stress distribution across the panel width which is predicted by bean theory. The second term describes the axial stress redistribution due to shear lag.

Consider an aluminum panel with an aspect ratio $L = 8b$. The parameter $\mu = 19.2$ when $E/G = 2(1 + \nu) = 2.6$ for a homogeneous, isotropic material with a Poisson's ratio $\nu = 0.3$. On the other hand, $\mu = 6.41$ for a panel with the same aspect ratio but made of a medium modulus graphite fiber reinforced epoxy matrix composite material for which $E = 140 \, GPa$, $G = 6 \, GPa$ and $\nu = 0.3$. Figure 12.9 shows the non-dimensional axial stress $h\sigma_1/(L\tau_0)$ distribution for both materials at two locations across the panel width: $\zeta = 0$ (corresponding to the panel mid-line) and $\zeta = 1$ (corresponding to its right hand edge).

Figure 12.10 shows the distribution of axial stress across the width at the root of the panel for both materials. In contrast with the uniform distribution predicted by classical beam theory, the present results show a significant over-stress occurring at the root corners of the panel and significant under-stress along its mid-line. The magnitudes of the over- and under-stresses are $5\tanh\mu/\mu$ and $5\tanh\mu/(2\mu)$, respectively, as obtained from eqs.(12.59). For the isotropic panel this translates to 26% and 13% for the over- and under-stress, respectively, and the corresponding numbers are 78% and 39% for the composite panel.

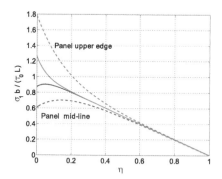

Fig. 12.9. Distribution of axial stress along the panel span. Isotropic material: solid line; anisotropic material: dashed line.

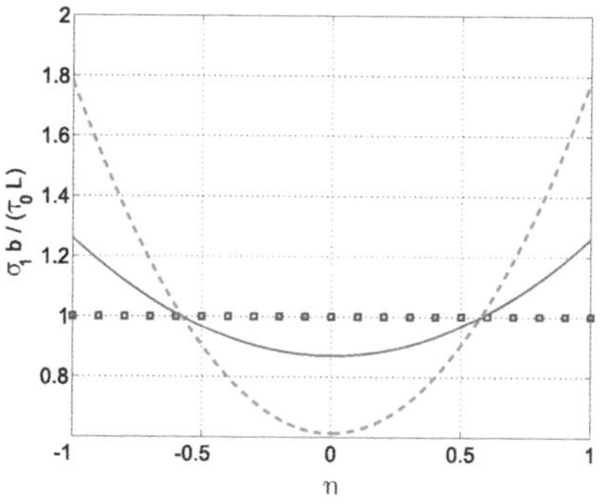

Fig. 12.10. Distribution of axial stress across the panel width at the root of the panel. The uniform distribution predicted by beam theory is indicated by the square boxes. Isotropic material: solid line; anisotropic material: dashed line.

The uniform stress distribution results in optimum structural efficiency because the material is equally stressed at all points across the width of the panel. The shear lag phenomenon considerably redistributes the stresses, creating undesirable over-stressed areas and decreasing the structural efficiency. The magnitude of the shear lag effect is controlled by the parameter μ defined in eq. (12.58). The smaller this parameter, the larger the shear lag effect. Clearly, shorter panel (smaller L/b values), made of shear deformable materials (smaller G/E values) will experience the most significant shear lag effects.

A more complicated configuration that includes both panels (sheets) and stringers under similar loading with warping restrained at one end is treated in problem 12.12.

12.3.2 The Saint-Venant torsion problem

Torsion of a beam with an arbitrary cross-section is examined in chapter 7 in section 7.3 using an analysis based on the Newtonian statement of equilibrium. The same problem will formulated here as a variational problem.

For a beam with arbitrary cross-section, Saint-Venant theory assumes that under torsion, each cross-section rotates like a rigid body, and warps out of its own plane.

The resulting assumed displacement field is described by eqs. (7.32a) and (7.32b), which are repeated

$$u_1(x_1, x_2, x_3) = \Psi(x_2, x_3)\,\kappa_1(x_1),$$ (12.60a)

$$u_2(x_1, x_2, x_3) = -x_3\Phi_1(x_1), \quad u_3(x_1, x_2, x_3) = x_2\Phi_1(x_1).$$ (12.60b)

where $\Phi(x_1)$ is the angle of twist about axis $\bar{\imath}_1$ and $\Psi(x_2, x_3)$ the warping function.

Under the assumption of uniform torsion, The strain field is then computed using the strain-displacement relations, and is given by eqs. (7.33a) to (7.33c). The only nonzero strains are the two shear strains components, γ_{12} and γ_{13}, given by eqs. (7.33c). The only non-vanishing stress components are τ_{12} and τ_{13}, given by eqs. (7.34c).

For uniform torsion, the total potential energy can be evaluated for a slice of the beam of unit length only. Indeed, for uniform torsion, the deformation of all slices are identical. The strain energy per unit length of the beam is

$$A = \frac{1}{2}\int_A (\tau_{12}\gamma_{12} + \tau_{13}\gamma_{13})\ \mathrm{d}A.$$

The potential of the externally applied load is the negative of the work done by the torque, M_1, acting along the unit length segment, $\Phi = -M_1\kappa_1 \cdot 1$. Using eqs. (7.34c) and (7.33c) for the stress and strains fields, respectively, the total potential energy becomes

$$\Pi = \frac{1}{2}\int_A \left[\left(\frac{\partial\Psi}{\partial x_2} - x_3\right)^2 + \left(\frac{\partial\Psi}{\partial x_3} + x_2\right)^2\right] G\kappa_1^2\ \mathrm{d}A - M_1\kappa_1.$$

The principle of minimum total potential energy requires the total potential energy to be stationary with respect to admissible displacement fields, leading to

$$\delta\Pi = G\kappa_1^2 \int_A \left[\left(\frac{\partial\Psi}{\partial x_2} - x_3\right)\frac{\partial\delta\Psi}{\partial x_2} + \left(\frac{\partial\Psi}{\partial x_3} + x_2\right)\frac{\partial\delta\Psi}{\partial x_3}\right]\ \mathrm{d}A - 0 = 0.$$ (12.61)

To eliminate the derivatives of variation in the warping function, Green's theorem [7] will now be used, leading to

$$\delta\Pi = -G\kappa_1^2 \iint_A \left[\frac{\partial^2\Psi}{\partial x_2^2} + \frac{\partial^2\Psi}{\partial x_3^2}\right]\delta\Psi\ \mathrm{d}A$$
$$+ G\kappa_1^2 \oint_C \left[\left(\frac{\partial\Psi}{\partial x_2} - x_3\right)n_2 + \left(\frac{\partial\Psi}{\partial x_3} + x_2\right)n_3\right]\delta\Psi\mathrm{d}s = 0,$$

where $n_2 = \mathrm{d}x_3/\mathrm{d}s$ and $n_3 = -\mathrm{d}x_2/\mathrm{d}s$ are the direction cosines of the unit normal to curve $\mathcal{C}$. Because the variation in the warping function are arbitrary, the bracketed terms in the expression above must vanish. The first bracketed term yields the governing partial differential equation of the problem, and the second, the boundary conditions along curve $\mathcal{C}$,

$$\frac{\partial^2 \Psi}{\partial x_2^2} + \frac{\partial^2 \Psi}{\partial x_3^2} = 0 \quad \text{over } \mathcal{A} \qquad (12.62a)$$

$$\left(\frac{\partial \Psi}{\partial x_2} - x_3\right) \frac{\mathrm{d}x_3}{\mathrm{d}s} - \left(\frac{\partial \Psi}{\partial x_3} + x_2\right) \frac{\mathrm{d}x_2}{\mathrm{d}s} = 0 \quad \text{along } \mathcal{C}. \qquad (12.62b)$$

Equations (12.62) are identical to eqs. (7.40) obtained using the classical Newtonian approach. The boundary condition of the problem, eq. (12.62b), corresponds to the vanishing of the shear stress components acting in the direction normal to curve $\mathcal{C}$, $\tau_n = 0$. In the variational approach, this natural boundary condition results from the application of Green's theorem.

12.3.3 The Saint-Venant torsion problem using the Prandtl stress function

The Saint-Venant uniform torsion problem is formulated in the previous section as a variational problem in terms of the warping function, Ψ, which defines the deformation of the beam. To avoid the complicated boundary condition expressed by eq. (12.62b), the problem is reformulated using Prandtl's stress function as developed in section 7.3.2. In this case, however, the principle of minimum total potential energy is no longer appropriate, and the principle of minimum complementary energy must be employed instead.

In Prandtl's stress function formulation, the non-vanishing shear stress components are expressed in terms of the stress function by eqs. (7.41), $\tau_{12} = \partial\phi/\partial x_3$ and $\tau_{13} = \partial\phi/\partial x_2$, where $\phi(x_2, x_3)$ is Prandtl's stress function. The complementary strain energy stored in a unit slice of the beam now becomes

$$A' = \frac{1}{2} \iint_{\mathcal{A}} \frac{1}{G} \left(\tau_{12}^2 + \tau_{13}^2\right) \mathrm{d}\mathcal{A} = \frac{1}{2} \iint_{\mathcal{A}} \frac{1}{G} \left[\left(\frac{\partial\phi}{\partial x_2}\right)^2 + \left(\frac{\partial\phi}{\partial x_3}\right)^2\right] \mathrm{d}\mathcal{A}.$$

The potential of the prescribed displacements, Φ', see section (12.2.6), requires more careful consideration. In this case, the beam of unit length is fixed at one end, and a rotation, Φ_1, is prescribed at the other. The associated potential is $\Phi' = -M_1\kappa_1 \cdot 1 = -\iint_{\mathcal{A}} 2\phi\kappa_1 \mathrm{d}\mathcal{A}$ where use is made of eq. (7.48) for a cross-section bounded by a single curve, $\mathcal{C}$. The total complementary potential energy is then[1]

$$\Pi' = A' + \Phi' = \frac{1}{2} \iint_{\mathcal{A}} \frac{1}{G} \left[\left(\frac{\partial\phi}{\partial x_2}\right)^2 + \left(\frac{\partial\phi}{\partial x_3}\right)^2\right] \mathrm{d}\mathcal{A} - \iint_{\mathcal{A}} 2\phi\kappa_1 \mathrm{d}\mathcal{A}. \quad (12.63)$$

The principle of stationary complementary energy, principle 18, requires the total complementary energy to be stationary value with respect to arbitrary choices of statically admissible stress fields, leading to

$$\delta\Pi' = \iint_{\mathcal{A}} \frac{1}{G} \left[\frac{\partial\phi}{\partial x_2}\frac{\partial\delta\phi}{\partial x_2} + \frac{\partial\phi}{\partial x_3}\frac{\partial\delta\phi}{\partial x_3} - 2G\kappa_1\delta\phi\right] \mathrm{d}\mathcal{A} = 0,$$

---

[1] The notation is treacherous: Φ_1 is a rotation about axis $\bar{\imath}_1$; Φ' is the potential of prescribed displacements; and ϕ is Prandtl's stress function.

To eliminate the derivatives of variations of the stress function, Green's theorem is applied to the first two terms in the integrand to find

$$\iint_A \frac{1}{G} \left[-\frac{\partial^2 \phi}{\partial x_2^2} - \frac{\partial^2 \phi}{\partial x_3^2} - 2G\kappa_1 \right] \delta\phi \mathrm{d}A + \frac{1}{G} \oint_C \left[\frac{\partial \phi}{\partial x_2} n_2 + \frac{\partial \phi}{\partial x_3} n_3 \right] \delta\phi \, \mathrm{d}s = 0.$$

Equation (7.44) implies that the stress function, ϕ, must remain constant along curve C. Hence, $\delta\phi = 0$ along the same curve and the second integral vanishes. The first integral must vanish for all arbitrary variation, $\delta\phi$. Consequently, the first bracketed term must vanish, and the Euler-Lagrange equation for this variational problem is

$$\frac{\partial^2 \phi}{\partial x_2^2} + \frac{\partial^2 \phi}{\partial x_3^2} = -2G\kappa_1, \tag{12.64}$$

This result is identical to that obtained with the classical approach, see eq. (7.45). Here again, the governing differential of the problem is recovered as the Euler-Lagrange equation of a variational problem

Example 12.1. Torsion of rectangular section - a crude solution
Consider a bar with a rectangular cross-section of width $2a$ and depth $2b$ as depicted in fig. 12.11. Determine the stress function and the sectional torsional stiffness using the principle of complementary strain energy.

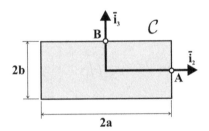

Fig. 12.11. Bar with a rectangular cross-section.

The following expression will be assumed for the stress function

$$\phi(\eta, \zeta) = c_0(\eta^2 - 1)(\zeta^2 - 1), \tag{12.65}$$

where c_0 is an unknown constant, $\eta = x_2/a$ is the non-dimensional coordinate along axis $\bar{\imath}_2$, and $\zeta = x_3/b$ is that along axis $\bar{\imath}_3$. This choice of the stress function satisfies the boundary conditions of the problem along C: $\phi(\eta = \pm 1, \zeta) = 0$ and $\phi(\eta, \zeta = \pm 1) = 0$.

The principle of minimum complementary energy will be used to determine the constant, c_0, that minimize the total complementary energy. The total complementary energy is given by eq. (12.63), and introducing the approximation of the stress function given by eq. (12.65) yields

$$\Pi' = \int_A \frac{c_0^2}{2G} \left[\frac{4\eta^2}{a^2}(\zeta^2 - 1)^2 + \frac{4\zeta^2}{b^2}(\eta^2 - 1)^2 \right] \mathrm{d}A$$
$$- 2c_0\kappa_1 \int_A (\eta^2 - 1)(\zeta^2 - 1) \,\mathrm{d}A. \tag{12.66}$$

After integration over the cross-section, this becomes

$$\Pi' = \frac{c_0^2}{2G} \left[\frac{8}{3a}\frac{16b}{15} + \frac{16a}{15}\frac{8}{3b} \right] - 2c_0\kappa_1 \frac{4a}{3}\frac{4b}{3}.$$

The total complementary energy must assumes a stationary value with respect to arbitrary statically admissible stress fields, which implies

$$\delta\Pi' = \left[\frac{2c_0}{2G}\left(\frac{8}{3a}\frac{16b}{15} + \frac{16a}{15}\frac{8}{3b} \right) - 2\kappa_1 \frac{16ab}{9} \right] \delta c_0 = 0.$$

This expression must vanish for arbitrary δc_0, and hence, the bracketed term must vanish. Solving the resulting equation for constant c_0 yields the stress function as

$$\phi(\eta, \zeta) = \frac{5}{4}\frac{a^2b^2}{a^2 + b^2} G\kappa_1 (\eta^2 - 1)(\zeta^2 - 1). \tag{12.67}$$

For this section bounded by a single curve, the externally applied torque is given by eq. (7.48),

$$M_1 = 2\int_A \phi \,\mathrm{d}A = \frac{5}{2}\frac{a^2b^2}{a^2 + b^2} G\kappa_1 \int_A (\eta^2 - 1)(\zeta^2 - 1) \,\mathrm{d}A = \frac{40}{9}\frac{a^3b^3}{a^2 + b^2}G\kappa_1.$$

The sectional torsional stiffness, H_{11}, then follows as

$$H_{11} = \frac{40}{9}\frac{a^3b^3}{a^2 + b^2}G. \tag{12.68}$$

The stress field is easily found from the derivatives of the stress function. This solution should be compared to the development in section 7.3.3 based on solutions to the governing partial differential equations of linear elasticity.

Example 12.2. Torsion of rectangular section, a refined solution
Consider once again a bar with a rectangular cross-section of width $2a$ and depth $2b$ analyzed in example 12.1 and shown in fig. 12.11. Study the behavior of this section in uniform torsion using the following approximation for the stress function,

$$\phi(\eta, \zeta) = (\zeta^2 - 1)g(\eta), \tag{12.69}$$

where $g(\eta)$ is an unknown function, $\eta = x_2/a$ the non-dimensional coordinate along axis $\bar{\imath}_2$, and $\zeta = x_3/b$ that along axis $\bar{\imath}_3$. This choice for the stress function explicitly satisfies the boundary conditions along two edges of the section, $\phi(\eta, \zeta = \pm 1) = 0$, but to satisfy the boundary conditions along the other two edges, the following conditions are required, $g(\eta = \pm 1) = 0$.

The total complementary energy for the problem is given by eq. (12.63), and introducing the approximation of the stress function given by eq. (12.69) yields

$$\Pi' = \int_{\mathcal{A}} \left[(\zeta^2 - 1)^2 \frac{g'^2}{a^2} + \frac{4\zeta^2}{b^2} g^2 \right] d\mathcal{A} - 2\kappa_1 \int_{\mathcal{A}} (\zeta^2 - 1) g(\eta) \, d\mathcal{A},$$

where the notation $(.)'$ indicates a derivative with respect to η. The integration over variable ζ can be performed to yield

$$\Pi' = \frac{ab}{2G} \int_{-1}^{+1} \left[\frac{16}{15a^2} g'^2 + \frac{8}{3b^2} g^2 \right] d\eta - 2\kappa_1 ab \int_{-1}^{+1} \left(-\frac{4}{3} \right) g \, d\eta.$$

When the total complementary energy is a minimum, Π' assumes a stationary value with respect to statically admissible stresses which implies

$$\delta \Pi' = \frac{ab}{2G} \int_{-1}^{+1} \left[\frac{16}{15a^2} 2g' \delta g' + \frac{8}{3b^2} 2g \delta g + 4G\kappa_1 \frac{4}{3} \delta g \right] d\eta = 0.$$

Performing an integration by parts of the first term then leads to

$$\int_{-1}^{+1} \left[-\frac{16}{15a^2} g'' + \frac{8}{3b^2} g + G\kappa_1 \frac{8}{3} \right] \delta g \, d\eta + \left[\frac{16}{15a^2} g' \delta g \right]_{-1}^{+1} = 0.$$

The variation, δg, is arbitrary, and hence, the first bracketed term must vanish. The Euler-Lagrange equation of this stationarity problem now becomes $g'' - \mu^2 g = \mu^2 b^2 G\kappa_1$, where $\mu = \sqrt{5/2} \, a/b$. Since the boundary conditions of the problem require $g(\eta = \pm 1) = 0$, the variations at the end points vanish, $\delta g(\eta = \pm 1) = 0$, and the second bracketed terms vanishes.

The general solution of the differential equation is $g(\eta) = C_1 \sinh \mu \eta + C_2 \cosh \mu \eta - b^2 G\kappa_1$, where C_1 and C_2 are two integration constants that must be evaluated with the help of the boundary conditions, $g(\eta = \pm 1) = 0$. The stress function then becomes

$$\phi(\eta, \zeta) = \left(\frac{\cosh \mu \eta}{\cosh \mu} - 1 \right) (\zeta^2 - 1) \, b^2 G\kappa_1. \tag{12.70}$$

For this section bounded by a single curve, the externally applied torque is given by eq. (7.48), leading to the following expression for the torsional stiffness

$$H_{11} = \frac{16ab^3}{3} \left(1 - \frac{\tanh \mu}{\mu} \right) G. \tag{12.71}$$

The stress field can be found from the derivatives of the stress function. This solution should be compared to the development in section 7.3.3 based on solutions to the governing partial differential equations on linear elasticity.

12.3.4 The non-uniform torsion problem

The equations governing the non-uniform torsion problem for a beam with a thin-walled cross-section are derived in section 8.9 using classical arguments. The derivation is complex, relying on a number of equilibrium arguments. It is possible to obtain these governing equations based solely on variational and energy arguments. The development starts with the following expression for the assumed displacement field in a general thin-walled beam,

$$u_1(x_1, s) = \Psi(s)\, \kappa_1(x_1),$$
$$u_2(x_1, s) = -(x_3 - x_{3k})\, \Phi_1(x_1), \quad u_3(x_1, s) = (x_2 - x_{2k})\, \Phi_1(x_1). \tag{12.72}$$

This displacement field is similar to that of the Saint-Venant solution described in section 7.3.2. The axial displacement component is proportional to the twist rate and distributed over the cross-section according to the warping function, $\Psi(s)$, assumed to be that found in section 8.7. The in-plane displacement components describe a rigid body rotation of the section about the center of twist (or shear center) which is computed using the procedure described in section 8.7.

The strain field is now computed from this assumed displacement field to find

$$\epsilon_1 = \Psi(s)\, \frac{\mathrm{d}\kappa_1}{\mathrm{d}x_1}, \quad \gamma_s = \left(\frac{\mathrm{d}\Psi}{\mathrm{d}s} + r_k\right) \kappa_1,$$

where r_k is defined by eq. (8.11). The remaining strain components all vanish, $\epsilon_2 = \epsilon_3 = \gamma_{23} = 0$, as should be expected since the in-plane displacement components describe a rigid body rotation.

The strain energy of the beam under non-uniform torsion reduces to the following expression

$$A = \frac{1}{2} \int_0^L \int_{\mathcal{A}} (\tau_s \gamma_s + \sigma_1 \epsilon_1)\, \mathrm{d}\mathcal{A}\, \mathrm{d}x_1.$$

Assuming the beam to be made of homogeneous, linearly elastic material, the strain energy can be written in terms of deformation as

$$A = \frac{1}{2} \int_0^L \int_{\mathcal{A}} \left[G\left(\frac{\mathrm{d}\Psi}{\mathrm{d}s} + r_k\right)^2 \kappa_1^2 + E\Psi^2 \left(\frac{\mathrm{d}\kappa_1}{\mathrm{d}x_1}\right)^2 \right] \mathrm{d}\mathcal{A}\, \mathrm{d}x_1.$$

Recognizing that the twist rate, κ_1, and its spatial derivative are functions only of the span-wise variable, x_1, this expression can be recast as

$$A = \frac{1}{2} \int_0^L \left\{ \left[\int_{\mathcal{C}} G\left(\frac{\mathrm{d}\Psi}{\mathrm{d}s} + r_k\right)^2 t\, \mathrm{d}s \right] \kappa_1^2 + \left[\int_{\mathcal{C}} E\Psi^2\, t\, \mathrm{d}s \right] \left(\frac{\mathrm{d}\kappa_1}{\mathrm{d}x_1}\right)^2 \right\} \mathrm{d}x_1.$$

For uniform torsion problems, the twist rate is a constant along the span of the beam, and the second term in the integral vanishes. The first term then represents the strain energy associated with this uniform torsion problem, $A = \frac{1}{2} \int_0^L H_{11} \kappa_1^2\, \mathrm{d}x_1$, where H_{11} is the torsional stiffness of the section. For closed section, the torsional stiffness is given by eq. (8.67), whereas for open sections, expressions are given in section 7.5.

The second term in the integral represents the strain energy associated the axial strain component that arises in non-uniform torsion problems. The bracketed term is the *warping stiffness* that is identified in the classical approach, see eq. (8.103). Hence, the strain energy becomes

$$A = \frac{1}{2} \int_0^L \left[H_{11} \left(\frac{\mathrm{d}\Phi_1}{\mathrm{d}x_1} \right)^2 + H_w \left(\frac{\mathrm{d}^2\Phi_1}{\mathrm{d}x_1^2} \right)^2 \right] \mathrm{d}x_1.$$

Consider a beam clamped at the root and subjected to a distributed torque, $q_1(x_1)$, and a concentrated torque, Q_1, at the tip. The total potential energy of the system is then

$$\Pi = \frac{1}{2} \int_0^L \left[H_{11} \left(\frac{\mathrm{d}\Phi_1}{\mathrm{d}x_1} \right)^2 + H_w \left(\frac{\mathrm{d}^2\Phi_1}{\mathrm{d}x_1^2} \right)^2 \right] \mathrm{d}x_1 - \int_0^L q_1 \Phi_1 \, \mathrm{d}x_1 - Q_1 \Phi_1(L).$$

The principle of minimum total potential energy requires this expression to be a minimum with respect to all the possible choices of the twist distribution, $\Phi_1(x_1)$, that are compatible with the geometric boundary conditions. This occurs when $\delta\Pi = 0$ and can be expressed as

$$\int_0^L \left[H_{11} \frac{\mathrm{d}\Phi_1}{\mathrm{d}x_1} \left(\frac{\mathrm{d}\delta\Phi_1}{\mathrm{d}x_1} \right) + H_w \frac{\mathrm{d}^2\Phi_1}{\mathrm{d}x_1^2} \left(\frac{\mathrm{d}^2\delta\Phi_1}{\mathrm{d}x_1^2} \right) - q_1 \delta\Phi_1 \right] \mathrm{d}x_1 - Q_1 \delta\Phi_1(L) = 0$$

where the order of the variational and differential operators are interchanged. The first term is integrated by parts once and the second term twice to yield

$$\int_0^L \left[-\frac{\mathrm{d}}{\mathrm{d}x_1} \left(H_{11} \frac{\mathrm{d}\Phi_1}{\mathrm{d}x_1} \right) + \frac{\mathrm{d}^2}{\mathrm{d}x_1^2} \left(H_w \frac{\mathrm{d}^2\Phi_1}{\mathrm{d}x_1^2} \right) - q_1 \right] \delta\Phi_1 \, \mathrm{d}x_1 + \left[H_{11} \frac{\mathrm{d}\Phi_1}{\mathrm{d}x_1} \delta\Phi_1 \right]_0^L$$
$$+ \left[H_w \frac{\mathrm{d}^2\Phi_1}{\mathrm{d}x_1^2} \left(\frac{\mathrm{d}\delta\Phi_1}{\mathrm{d}x_1} \right) \right]_0^L - \left[\frac{\mathrm{d}}{\mathrm{d}x_1} \left(H_w \frac{\mathrm{d}^2\Phi_1}{\mathrm{d}x_1^2} \right) \delta\Phi_1 \right]_0^L - Q_1 \delta\Phi_1(L) = 0.$$

Because the integral must vanish for all arbitrary variations $\delta\Phi_1$, the first bracketed term must vanish, leading to the governing differential equation of the problem

$$\frac{\mathrm{d}}{\mathrm{d}x_1} \left(H_{11} \frac{\mathrm{d}\Phi_1}{\mathrm{d}x_1} \right) - \frac{\mathrm{d}^2}{\mathrm{d}x_1^2} \left(H_w \frac{\mathrm{d}^2\Phi_1}{\mathrm{d}x_1^2} \right) = -q_1. \tag{12.73}$$

This equation is identical to eq. (8.105) obtained using the classical approach. At the root of the beam, the geometric boundary conditions imply $\Phi_1(0) = 0$ and $\mathrm{d}\Phi_1(0)/\mathrm{d}x_1 = 0$, where the second condition stems from the required vanishing of the axial displacement component, see eq. (12.72). Because the variations $\delta\Phi_1(L)$ and $\delta(\mathrm{d}\Phi_1(L)/\mathrm{d}x_1)$ are arbitrary, the boundary conditions at the tip of the beam are the natural conditions obtained from the boundary terms as

$$\left[H_{11} \frac{d\Phi_1}{dx_1} - \frac{d}{dx_1} \left(H_w \frac{d^2\Phi_1}{dx_1^2} \right) \right]_{x_1=L} = Q_1, \quad \text{and} \quad H_w \frac{d^2\Phi_1}{dx_1^2} \bigg|_{x_1=L} = 0.$$

(12.74)

The first condition requires the torque in the beam to be equal to the externally applied tip torque, Q_1. The second condition implies the vanishing of the tip axial stresses.

Clearly, the classical and energy approaches lead to the same governing equation and boundary conditions for non-uniform torsion. It should be noted, however, that the energy approach is much more convenient because the governing equation and boundary conditions are both obtained from the expression of the total potential energy by means of a purely algebraic procedure.

12.3.5 The non-uniform torsion problem (closed sections)

The variational and energy approach to the non-uniform torsion problem developed in the previous section is based on the displacement field given by eq. (12.72). When using an energy approach, different approximations to a problem can be easily obtained by starting from different displacement fields.

In this section, the following displacement field will be investigated

$$u_1(x_1, s) = \Psi(s)\, \alpha(x_1),$$
$$u_2(x_1, s) = -(x_3 - x_{3k})\, \Phi_1(x_1), \quad u_3(x_1, s) = (x_2 - x_{2k})\, \Phi_1(x_1),$$

(12.75)

where $\alpha(x_1)$ is an unknown function that characterizes the amplitude of the axial displacement, and Φ_1 is the rigid body rotation of the cross-section. This contrasts with the displacement field of the previous approach, eq. (12.72), where the amplitude of the axial displacement is taken to be proportional to the twist rate. Here again, the warping function, $\Psi(s)$, is assumed to be that found in section 8.7, and the in-plane displacement components describe a rigid body rotation of the section about the shear center.

The strain field is now computed from this assumed displacement field as

$$\epsilon_1 = \Psi(s) \frac{d\alpha}{dx_1}, \quad \gamma_s = \frac{d\Psi}{ds}\alpha + r_k \frac{d\Phi_1}{dx_1},$$

(12.76)

where r_k is defined by eq. (8.11). The remaining strain components all vanish, $\epsilon_2 = \epsilon_3 = \gamma_{23} = 0$, as should be expected because the in-plane displacement components describe a rigid body rotation.

Assuming the beam has a closed, single-cell, thin-walled cross-section made of homogeneous, linearly elastic material, the strain energy becomes

$$A = \frac{1}{2} \int_0^L \int_C \left[E\Psi^2 \left(\frac{d\alpha}{dx_1} \right)^2 + G \left(\frac{d\Psi}{ds}\alpha + r_k \frac{d\Phi_1}{dx_1} \right)^2 \right] t\, ds\, dx_1.$$

Expanding this expression leads to a number of sectional integrals. The first two are

$$H_p = \int_C G r_k^2 \, tds; \quad \text{and} \quad H_w = \int_C E \Psi^2 \, tds,$$

where the second integral defines the torsional warping of the section. Two other integrals are encountered,

$$\int_C G \frac{d\Psi}{ds} r_k \, tds = \int_C G \left[\frac{H_{11}}{2AGt} - r_k \right] r_k \, tds = H_{11} - H_p,$$

and

$$\int_C G \left(\frac{d\Psi}{ds} \right)^2 tds = \int_C G \left[\frac{H_{11}}{2AGt} - r_k \right]^2 tds = H_{11} + H_p - 2H_{11} = H_p - H_{11}.$$

The derivative of the warping function is expressed in terms of eq. (8.94), and the torsional stiffness of the closed section is given by eq. (8.67). Using these, the strain energy of the beam now becomes

$$A = \frac{1}{2} \int_0^L \left[H_w \left(\frac{d\alpha}{dx_1} \right)^2 + (H_p - H_{11})\alpha^2 + H_p \left(\frac{d\Phi_1}{dx_1} \right)^2 \right. $$
$$\left. - 2 (H_p - H_{11})\alpha \frac{d\Phi_1}{dx_1} \right] dx_1.$$

Consider a cantilevered beam subjected to a distributed torque, $q_1(x_1)$, and a tip torque, Q. The total potential energy of the beam is then

$$\Pi = \frac{1}{2} \int_0^L \left[H_w \left(\frac{d\alpha}{dx_1} \right)^2 + (H_p - H_{11})\alpha^2 + H_p \left(\frac{d\Phi_1}{dx_1} \right)^2 \right. $$
$$\left. - 2 (H_p - H_{11})\alpha \frac{d\Phi_1}{dx_1} \right] dx_1 - \int_0^L q_1 \Phi_1 \, dx_1 - Q\Phi_1(L).$$

Invoking the principle of stationary total potential energy, the total potential energy must be a stationary quantity,

$$\delta\Pi = \int_0^L \delta\alpha \left[-\frac{d}{dx_1} \left(H_w \frac{d\alpha}{dx_1} \right) + (H_p - H_{11})\alpha - (H_p - H_{11}) \frac{d\Phi_1}{dx_1} \right] dx_1$$
$$- Q\Phi_1(L) + \left[H_w \frac{d\alpha}{dx_1} \delta\alpha \right]_0^L$$
$$+ \int_0^L \delta\Phi_1 \left[-\frac{d}{dx_1} \left(H_p \frac{d\Phi_1}{dx_1} \right) + \frac{d}{dx_1}(H_p - H_{11})\alpha - q_1 \right] dx_1$$
$$+ \left[H_p \frac{d\Phi_1}{dx_1} \delta\Phi_1 \right]_0^L - [(H_p - H_{11})\alpha\delta\Phi_1]_0^L = 0.$$

Because this expression must vanish for all arbitrary variations, $\delta\alpha$ and $\delta\Phi_1$, the bracketed terms must be zero, leading to the two differential equations of the problem

$$\frac{d}{dx_1}\left(H_w \frac{d\alpha}{dx_1}\right) + (H_p - H_{11})\left(\frac{d\Phi_1}{dx_1} - \alpha\right) = 0, \tag{12.77a}$$

$$\frac{d}{dx_1}\left[H_p \frac{d\Phi_1}{dx_1} - (H_p - H_{11})\alpha\right] = -q_1. \tag{12.77b}$$

At the root of the beam, the geometric boundary conditions, $\Phi_1 = 0$, and $\alpha = 0$, must be imposed; the second condition stems from the required vanishing of the axial displacement component, see eq. (12.75). At the tip of the beam, variations $\delta\alpha(L)$ and $\delta\Phi_1(L)$ are arbitrary, leading to the natural boundary conditions, $H_w \, d\alpha/dx_1 = 0$, and $H_p \, d\Phi_1/dx_1 - (H_p - H_{11})\alpha = Q$. The first condition implies the vanishing of the axial stresses at the tip of the beam, the second the equilibrium of the moment in the beam with the externally applied torque.

Consider next the case of a cantilevered beam under a tip torque alone, *i.e.*, $q_1 = 0$. Integration of the second equation of the problem, eq. (12.77b), yields

$$H_p \frac{d\Phi_1}{dx_1} - (H_p - H_{11})\alpha = Q, \tag{12.78}$$

where the second boundary condition at the beam's tip is used to evaluate the integration constant. This result is used to substitute for $d\Phi_1/dx_1$ in terms of α in the first governing equation to find

$$H_w \frac{d^2\alpha}{dx_1^2} - H_{11}\left(1 - \frac{H_{11}}{H_p}\right)\alpha = -\left(1 - \frac{H_{11}}{H_p}\right)Q.$$

This second order, ordinary differential equation is readily solved to find

$$\alpha = \frac{Q}{H_{11}}\left[1 - \frac{\cosh \bar{k}(1 - \eta)}{\cosh \bar{k}}\right], \tag{12.79}$$

where $\eta = x_1/L$ is a non-dimensional variable along the span of the beam, and coefficient $\bar{k}$ is defined as

$$\bar{k}^2 = \frac{H_{11}L^2}{H_w}\left(1 - \frac{H_{11}}{H_p}\right). \tag{12.80}$$

Finally, eq. (12.79) is introduced into eq. (12.78) to determine the beam's twist,

$$\Phi_1 = \frac{QL}{H_{11}}\left[\eta - \left(1 - \frac{H_{11}}{H_p}\right)\frac{\sinh \bar{k} - \sinh \bar{k}(1 - \eta)}{\bar{k}\cosh \bar{k}}\right].$$

Of course, the stresses in the beam can now be obtained from the strain field, eq. (12.76), as

$$\sigma_1(\eta, s) = E\frac{QL}{H_{11}}\frac{\Psi(s)}{L^2}\frac{\bar{k}\sinh \bar{k}(1 - \eta)}{\cosh \bar{k}}.$$

for the axial stress and

$$\tau_s(\eta, s) = \frac{Q}{2\mathcal{A}t}\left[1 - \left(1 - \frac{2\mathcal{A}tG}{I_p}r_k\right)\frac{\cosh \bar{k}(1 - \eta)}{\cosh \bar{k}}\right],$$

for the shear flow.

12.3.6 The non-uniform torsion problem (open sections)

The previous section is focused on beams with closed, single-cell, thin-walled cross-sections made of homogeneous, linearly elastic material. If the section is an open section, the same developments will apply except for the fact that the warping function now satisfies eq. 8.85, rather than eq. 8.94. Consequently, the strain field now simplifies to

$$\epsilon_1 = \Psi(s)\frac{d\alpha}{dx_1}, \quad \gamma_s = \left(\frac{d\Phi_1}{dx_1} - \alpha\right) r_k, \tag{12.81}$$

and the strain energy in the beam becomes

$$A = \frac{1}{2}\int_0^L \left[H_w\left(\frac{d\alpha}{dx_1}\right)^2 + H_p\left(\frac{d\Phi_1}{dx_1} - \alpha\right)^2\right] dx_1.$$

This approximation does not take into account the linear through-the-thickness shear strain distribution that develops under torsion, see section 7.5. Indeed, the shear strain defined by eq. (12.81) is *uniform through-the-thickness*. To account for this effect, the strain energy expression is corrected as follows

$$A = \frac{1}{2}\int_0^L \left[H_w\left(\frac{d\alpha}{dx_1}\right)^2 + H_p\left(\frac{d\Phi_1}{dx_1} - \alpha\right)^2 + H_{11}\left(\frac{d\Phi_1}{dx_1}\right)^2\right] dx_1,$$

where H_{11} is the classical torsional stiffness for the thin-walled, open sections, see eq. (7.64).

Consider a cantilevered beam subjected to a distributed torque $q_1(x_1)$ and a tip torque Q. The total potential energy of the beam can be written

$$\Pi = \frac{1}{2}\int_0^L \left[H_w\left(\frac{d\alpha}{dx_1}\right)^2 + H_p\left(\frac{d\Phi_1}{dx_1} - \alpha\right)^2 + H_{11}\left(\frac{d\Phi_1}{dx_1}\right)^2\right] dx_1$$
$$- \int_0^L q_1\Phi_1 \, dx_1 - Q\Phi_1(L).$$

Invoking the principle of minimum total potential energy and following a procedure similar to that presented in the previous section yields the solution of the problem as

$$\alpha = \frac{Q}{H_{11}}\left[1 - \frac{\cosh k(1-\eta)}{\cosh k}\right], \tag{12.82}$$

$$\Phi_1 = \frac{QL}{H_{11}}\left[\eta - \left(1 + \frac{H_{11}}{H_p}\right)\frac{\sinh \bar{k} - \sinh \bar{k}(1-\eta)}{\bar{k}\cosh \bar{k}}\right].$$

where $\eta = x_1/L$ is the non-dimensional variable along the beam's span, and coefficient $\bar{k}$ is defined as

$$\bar{k}^2 = \frac{H_{11}H_pL^2}{H_w(H_p + H_{11})} = \frac{H_{11}L^2}{H_w}\frac{1}{1 + H_{11}/H_p}. \tag{12.83}$$

For open section, $H_{11} \ll H_p$, and coefficient $\bar{k}$ becomes nearly equal to its counterpart in the classical formulation of the non-uniform torsion problem, see eq. (8.107). Of course, the stress field can be recovered using the strain field given by eq. (12.81).

12.3.7 Problems

Problem 12.1. Cantilevered beam with elastic foundation

A cantilevered beam of length L is subjected to a tip load P_2, a tip bending moment Q_3, a transverse distributed load $p_2(x_1)$, and a distributed bending moment $q_3(x_1)$, as shown in fig. 12.12. The cantilevered beam is supported by an elastic foundation of stiffness k, not shown on the figure, for clarity. The total potential energy of the system is

$$\Pi = \int_0^L \left[\frac{1}{2} H_{33} \left(\frac{d^2 u_2}{dx_1^2} \right)^2 + \frac{1}{2} k\, u_2^2 \right] dx_1 - \int_0^L \left(p_2\, u_2 + q_3\, \frac{du_2}{dx_1} \right) dx_1$$
$$- P_2\, u_2(L) - Q_3\, \frac{du_2}{dx_1}\bigg|_L .$$

(1) Find the governing differential equations and boundary conditions for this problem using the principle of minimum total potential energy. *(2)* Derive the same equations and boundary conditions based on simple free body diagrams for a differential element of the beam.

Problem 12.2. Comparing solutions for torsion of a rectangular section

In examples 12.1 and 12.2, two solutions are developed for the uniform torsion of the rectangular section depicted in fig 12.11, leading to the stress functions given by eqs. (12.67) and (12.70), respectively. *(1)* On one graph, plot the non-dimensional torsional stiffnesses, $H_{11}/(4abG)$, predicted by the two solutions as a function of $a/b \in [1, 12]$. *(2)* On one graph, plot the non-dimensional shear stress at point **B**, $8ab^2\tau^B/M_1$, predicted by the two solutions as a function of $a/b \in [1, 12]$. *(3)* On one graph, plot the non-dimensional shear stress at point **A**, $8ab^2\tau^A/M_1$, predicted by the two solutions as a function of $a/b \in [1, 12]$.

Problem 12.3. Cantilevered beam with various loading

The uniform cantilevered beam of span L depicted in fig. 12.13 has a bending stiffness H_{33} and is supported by an elastic foundation of stiffness k over its first half. A concentrated spring of stiffness k_1 supports the beam at its free end. A mid-span concentrated load P is applied together with a uniform distributed load p_0 that acts over the second half of the beam span. Write the principle of minimum total potential energy for this system.

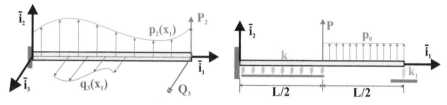

Fig. 12.12. Cantilevered beam with concentrated an distributed moments.

Fig. 12.13. Simply supported beam with partial elastic foundation.

Problem 12.4. Simply supported beam with concentrated load

Consider the uniform simply supported beam of span L subjected to a concentrated load acting at a distance αL from the left support, as shown in fig. 5.23. Write the principle of minimum total potential energy for the system. From this principle, derive the governing differential equations of the problem and the associated boundary conditions.

Problem 12.5. Cantilevered beam with tip spring

The uniform cantilevered beam of span L shown in fig. 12.14 features a tip spring of stiffness k and a tip concentrated load P. Write the principle of minimum total potential energy for the system. From this principle, derive the governing differential equations of the problem and the associated boundary conditions. Explain the physical meaning of the boundary conditions at $x_1 = L$ using a free body diagram.

Problem 12.6. Simply supported beam with end torsional springs

Consider a simply supported, uniform beam of length L with two end point torsional springs of stiffness k_1 and a mid-span spring of stiffness k_2. The beam, shown in fig. 12.15, is subjected to a uniform transverse loading $p_2(x_1) = p_0$. Write the principle of minimum total potential energy for the system. From this principle, derive the governing differential equations of the problem and the associated boundary conditions. Explain the physical meaning of the boundary conditions at $x_1 = L/2$ using a free body diagram.

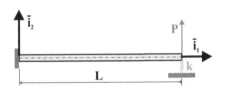

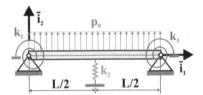

Fig. 12.14. Cantilevered beam with tip concentrated load and elastic spring.

Fig. 12.15. Simply supported beam with mid-span and end point springs.

Problem 12.7. Torsion of a beam with circular cross-section

Consider a uniform beam of length, L, with a circular cross-section of area, $\mathcal{A}$, and modulus, G. It is fixed at one end and loaded with both a distributed twisting moment, $q_1(x_1)$, and a concentrated moment, Q_1, at the other end. (1) Use the principle of minimum total potential energy to develop the governing differential equation. (2) Determine the possible boundary conditions that could be applied at each end of a general beam. (3) Indicate which of these boundary conditions apply for the present problem.

Problem 12.8. Torsion of a beam with concentrated torque applied at a mid-point

A uniform beam of length, L, with a circular cross-section of area, $\mathcal{A}$, and modulus, G, is clamped at both ends and subjected to a concentrated torque, Q_0, applied at $x_1 = a$. (1) Use the principle of minimum total potential energy to develop the governing differential equation. (2) Determine the possible boundary conditions that could be applied at each end of a general beam and at the point of application of the torque, Q_0. (3) Indicate which of these boundary conditions apply for the present problem.

Problem 12.9. Torsion stress in a thin-walled box beam

Consider the thin-walled beam with a rectangular cross-section depicted in fig. 12.16. The beam is clamped at the root and subjected to a tip torque Q_1. *(1)* Solve this problem using the following assumed displacement field $u_1(x_1, x_2, x_3) = 0$; $u_2(x_1, x_2, x_3) = -x_3\,\Phi_1(x_1)$; $u_3(x_1, x_2, x_3) = x_2\,\Phi_1(x_1)$. This leads to the classical theory for torsion: the solution is denoted Φ_1^{SV}. *(2)* Develop a beam theory for torsion based on the following assumed displacement field $u_1(x_1, x_2, x_3) = \Psi(s)\,\alpha(x_1)$; $u_2(x_1, x_2, x_3) = -x_3\,\Phi_1(x_1)$; $u_3(x_1, x_2, x_3) = x_2\,\Phi_1(x_1)$, where $\Psi(s)$ is the warping function obtained from thin-walled beam theory, and $\alpha(x_1)$ an unknown function. The warping function is given as

$$\Psi_1(s) = \frac{a-b}{a+b}\,bs; \quad \Psi_2(s) = -\frac{a-b}{a+b}\,as; \quad \Psi_3(s) = \frac{a-b}{a+b}\,bs; \quad \Psi_4(s) = -\frac{a-b}{a+b}\,as.$$

This leads to a theory for torsion that takes into account the effects of nonuniform torsion: the solution is denoted Φ_1^{NU}. *(3)* Plot the twist distributions Φ_1^{SV} and Φ_1^{NU} along the span of the beam on the same graph. *(4)* Sketch the shear stress flow distribution f^{SV} and f^{NU} over the cross-section of the beam at $x_1 = 0$ and L. *(5)* Plot the shear flow distribution along the span of the beam at points **A** and **B**, denoted f_A^{SV} and f_B^{SV}, respectively, for the classical theory, and f_A^{NU}, and f_B^{NU}, respectively, for the non-uniform torsion theory. *(6)* Plot the axial stress flow distribution n^{SV} and n^{NU} over the cross-section of the beam at $x_1 = 0$ and $L/2$. *(7)* Predict the failure loads $Q_{\text{fail}}^{\text{SV}}$ and $Q_{\text{fail}}^{\text{NU}}$ according to Von Mises strength criterion. Compute $Q_{\text{fail}}^{\text{SV}}/Q_{\text{fail}}^{\text{NU}}$.

Use the following parameters: $a = 4b$; $L = 6a$; $E/G = 20$; $n = Et\epsilon_{11}$; $f = Gt\gamma$.

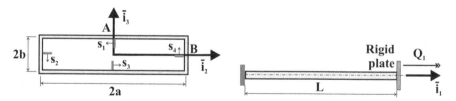

Fig. 12.16. Rectangular cross-section of a thin-walled beam.

Fig. 12.17. Clamped beam with a rigid tip plate.

Problem 12.10. Axial stress in a thin-walled box beam

Repeat the previous problem. However, the beam is now clamped at the root and a rigid plate prevents any warping deformation at the tip, see fig. 12.17

Problem 12.11. Beam with bending, axial and shear deformations

NOTE: this problem requires material on shear deformation in beams from chapter 15. Consider a uniform cantilevered beam of span L subjected to distributed axial and transverse loads $p_1(x_1)$ and $p_2(x_1)$, respectively, to tip concentrated axial and transverse loads P_1 and P_2, respectively, and to distributed and concentrated bending moments $q_3(x_1)$ and Q_3, respectively. The beam possesses a strain energy density function $a = 1/2\,(S\epsilon_1^2 + K_{22}\gamma_{12}^2 + H_{33}\kappa_3^2)$, where the axial strain $\epsilon_1 = du_1/dx_1$, the transverse shearing strain $\gamma_{12} = du_2/dx_1 - \Phi_3$, and the curvature $\kappa_3 = d\Phi_3/dx_1$. respectively. The equilibrium equations of the problem are $dF_1/dx_1 + p_1 = 0$; $dF_2/dx_1 + p_2 = 0$; and $dM_3/dx_1 + F_2 + q_3 = 0$, where F_1 is the axial force, F_2 the transverse shearing force, and M_3 the bending moment. *(1)* Derive the principle of virtual work for this problem. *(2)* Derive the principle of minimum total potential energy. *(3)* Derive Hu-Washizu's principle. *(4)* Derive Hellinger-Reissner's principle.

Consider now a cantilever beam subjected to a single transverse tip load $P_2 = P$. *(5)* Solve the problem using the principle of minimum total potential energy with the following assumed modes $u_2 = \eta u_T$ and $\Phi_3 = \eta \Phi_T$, where $\eta = x_1/L$, and u_T and Φ_T are unknown coefficients. *(6)* Solve the problem using Hellinger-Reissner's principle with the following assumed modes $u_2 = \eta u_T$, $\Phi_3 = \eta \Phi_T$, $F_2 = F_0$, and $M_3 = M_R + \eta M_T$, where u_T, Φ_T, F_0, M_R, and M_T are unknown coefficients. *(7)* On a single graph, plot the non-dimensional displacement fields $(H_{33}u_2)/(PL^3)$ and $(H_{33}\Phi_3)/(PL^2)$ versus η for the exact solution and each approximate solution. Choose $s^2 = 2.0 \ 10^{-3}$, where $s^2 = H_{33}/K_{22}l^2$. *(8)* On a single graph, plot the non-dimensional internal force fields M_3/PL and F_2/P versus η for the exact solution and each approximate solution. *(9)* On a single graph, plot the non-dimensional strain fields $(H_{33}\kappa_3)/(PL)$ and $(K_{22}\gamma_{12})/(P)$ versus η for the exact solution and each approximate solution. *(10)* For each approximate solution check how well, or how poorly, the equilibrium equations, the strain displacement equations, and the constitutive laws are satisfied. *(11)* Plot the approximate non-dimensional tip displacements normalized by the exact non-dimensional tip displacement versus $1/s^2$. Use a log scale to plot $1/s^2$. Comment on your results.

Problem 12.12. Axial stress in a reinforced panel

The thin panel depicted in fig. 12.18 is reinforced by stiffeners of axial stiffness S. The panel has a length L and the distance between two stiffeners is $2b$. At one end of the panel, loads P are applied to each stiffener, and the panel is clamped along its other end. The concentrated loads applied to the stiffener diffuses in to the panel and the purpose of the analysis is to determine how fast this diffusion process takes place. It can be assumed that at a large distance from the point of application of the concentrated loads, the axial strain and axial stress distributions become uniform across the width of the panel, *i.e.*, $\epsilon_1(x_1/L \rightarrow \infty, x_2) = \epsilon_f$ and $\sigma_1(x_1/L \rightarrow \infty, x_2) = \sigma_f$.

The number of stiffeners is assumed to be large so that a typical cell can be studied. In that typical cell, a panel of width $2b$ and length $L/b = \infty$ is attached to stiffeners of stiffness $S/2$ along each edge, and the stiffeners are subjected to concentrated loads $P/2$. At $x_1 = \infty$ the stress in the panel is σ_f and the loads in the stiffeners are $P_f/2$.

To analyze this problem, use an energy approach with the following assumed displacement field: $u_1(\eta, \zeta) = g(\eta) + h(\eta)(\zeta^2 - \alpha)$, $u_2(\eta, \zeta) = 0$, where $\eta = x_1/L$ and $\zeta = x_2/b$. The following geometric boundary conditions can be selected: $g(x_1 = 0) = 0$, implying the vanishing of the average axial displacement at $x_1 = 0$, and $h(x_1 = \infty) = 0$, implying a uniform axial displacement at $x_1 = \infty$.

(1) Express σ_f and $P_f/2$ as a function of the applied load P. *(2)* Write the total strain energy in the structure. Calculations will greatly simplify if you select the coefficient α so that the coupling term between g and h vanishes. Note that the strain energy of a stiffener is $1/2 \int_0^\infty S \, (du_1/dx_1)^2_{x_2=h} \, dx_1$. *(3)* Write the total potential of the externally applied loads. Do not forget to include the work done by the loads applied at $x_1 = \infty$. *(4)* Solve the problem for the unknown displacement functions g and h. *(5)* Determine the non-dimensional axial stress distribution in the panel, σ_1/σ_f, and the non-dimensional load in the stiffener, $F_s/(P_f/2)$. *(6)* Plot $\sigma_1/\sigma_f(\eta, \zeta = 1)$ and $\sigma_1/\sigma_f(\eta, \zeta = 0)$ versus η, for $E/G = 2.6$ and $E/G = 28$, on the same graph. *(7)* $F_s/(P_f/2)$ versus η, for $E/G = 2.6$ and $E/G = 28$, on the same graph. *(8)* Determine the non-dimensional shear stress distribution in the panel, $h\tau_{12}/P$. *(9)* Plot $h\tau_{12}/P(\eta, \zeta = 1)$ versus η, for $E/G = 2.6$ and $E/G = 28$, on the same graph. *(10)* On the same graph, plot the distribution of non-dimensional axial stress across the

width of the panel for $\eta = 0$, 2, 4 and 6, for $E/G = 2.6$; same question for $E/G = 28$. *(11)*
On the same graph, plot the distribution of non-dimensional shear stress across the width of
the panel for $\eta = 0$, 2, 4 and 6, for $E/G = 2.6$; same question for $E/G = 28$. *(12)* Find the
diffusion length d for the panel, *i.e.*, the distance it takes for the maximum stress to decrease
to within 1% of σ_f. *(13)* Plot the diffusion length as a function of $k \in [0, 1]$. Interpret your
results.

Use the following data: The panel and stiffener are made of a homogeneous, linearly
elastic material; $\sigma_1 = E\,\epsilon_1$, $\tau_{12} = G\,\gamma_{12}$; $F_s(x_1) = S\,\epsilon_1(x_1, x_2 = \pm b)$ for the right and left
stiffeners, respectively. The stiffener axial stiffness is $S = E\mathcal{A}$, where $\mathcal{A}$ is its cross-sectional
area; $k = \mathcal{A}/(2bt) = 0.15$ is the ratio of the stiffener area to the panel area.

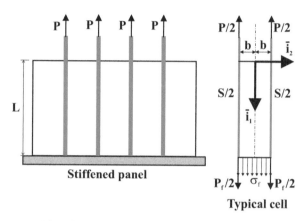

Fig. 12.18. Configuration of the stiffened panel.

Part IV

Advanced topics

13

Introduction to plasticity and thermal stresses

This chapter will address two important problems often encountered in the design of structure. In section 2.1, constitutive laws for linearly elastic, isotropic and anisotropic materials are presented. Although linearly elastic material behavior is often assumed in preliminary design work, it is often necessary to study the behavior of structures when the material they are made of yields and begins to deform inelastically.

Two issues will be addressed in this chapter. First, when a material is subjected to a complex state of stress, criteria for predicting the onset of yielding will be presented in section 13.1. Next, the behavior of simple structures operating in the plastic range will be discussed in section 13.2.

The second problem to be addressed in this chapter is the behavior of structures under thermal loading, see section 13.3. Two approaches to this problem will be presented: the direct method, see section 13.3.1, and the constraint method, see section 13.3.3. Applications to various structural configurations are presented in section 13.4.

13.1 Yielding under combined loading

The concept of allowable stress discussed in section 2.2, focuses on the highly idealized case of a structural component subjected to a *single stress component*. The yield criterion is then simply expressed in terms of the single stress component as eq. (2.28).

As depicted in fig. 1.3 on page 6, a differential element of material can be subjected to a number of stress components simultaneously. The question is now: what is the proper yield criterion to be used when *multiple stress components are acting simultaneously?* Consider an aircraft propeller connected to a homogeneous, circular shaft. The engine applies a torque that creates a distribution of shear stress throughout the shaft. On the other hand, the propeller creates a thrust that generates a uniform axial stress distribution over the cross-section. If the torque acts alone, the yield criterion is $\tau_{\max} < \tau_y$, where $\tau_{\max}$ is the maximum shear stress acting in the shaft;

if the axial force acts alone, the corresponding criterion is $\sigma_{max} < \sigma_y$, where σ_{max} is the maximum axial stress acting in the shaft. In the actual structure, both stress components are acting simultaneously, and it is natural to ask: what is the proper criterion to apply?

13.1.1 Introduction to yield criteria

The yield criteria to be presented in this section are applicable to isotropic, homogeneous materials subjected to general three-dimensional states of stress. Since the material is isotropic, the direction of application of the stress is irrelevant. If the material is subjected to a single stress component, it should yield under the same stress level regardless of the direction in which this stress component is applied. In contrast, if the material is anisotropic, the direction of application of stress is now relevant. For instance, consider a composite material consisting of long fibres, all aligned in a single direction and embedded in a matrix material. Intuitively, if a single stress component is applied along the fiber direction, the material response will be dramatically different from that observed when the stress is applied in the direction transverse to the fiber direction.

For isotropic materials, there is no directional dependency of the yield criterion, even when subjected to a combined state of stress. An arbitrary state of stress can be represented by the six stress components defining the stress tensor at that point, for example, see eq. (1.3). Alternatively, the state of stress can be represented by the three principal stresses, σ_{p1}, σ_{p2}, and σ_{p3} and the orientations of the faces on which they act, see section 1.2.2. If the yield criterion must be independent of directional information because of material isotropy, it is clear that *only the values of the principal stress* should appear in its expression. Alternatively, the yield criterion can be expressed in terms of the three stress invariants defined in eq. (1.21).

It is now convenient to represent a state of stress in the geometric space shown in fig. 13.1 where the magnitudes of the principal stresses are plotted as coordinates in a Cartesian system. For instance, point **S** defined by vector $\underline{S}$ represents the state of stress defined by the principal stresses σ_{p1}, σ_{p2}, and σ_{p3}.

An important experimental finding is that *the application of a hydrostatic state of stress has little effect on the yield condition of a material*. The hydrostatic state of stress is the state of stress that a solid experiences when it is immersed in a pressurized liquid. Clearly, the principal stresses associated with the hydrostatic state of stress are $\sigma_{p1} = \sigma_{p2} = \sigma_{p3} = p$, where p is the *hydrostatic pressure*.

If the material is subjected to an arbitrary state of stress, the stress tensor can be decomposed in the following manner

$$
\begin{bmatrix} \sigma_{p1} & 0 & 0 \\ 0 & \sigma_{p2} & 0 \\ 0 & 0 & \sigma_{p3} \end{bmatrix} = p \begin{bmatrix} 1 & 0 & 0 \\ 0 & 1 & 0 \\ 0 & 0 & 1 \end{bmatrix} + \begin{bmatrix} \sigma_{p1} - p & 0 & 0 \\ 0 & \sigma_{p2} - p & 0 \\ 0 & 0 & \sigma_{p3} - p \end{bmatrix}, \tag{13.1}
$$

where the first term on the right-hand side of the equation represents a state of hydrostatic stress associated with pressure $p = (\sigma_{p1} + \sigma_{p2} + \sigma_{p3})/3$, and the second term is the *deviatoric stress tensor*. The deviatoric stress tensor is denoted with an over-bar and defined as

$$\begin{bmatrix} \bar{\sigma}_{p1} & 0 & 0 \\ 0 & \bar{\sigma}_{p2} & 0 \\ 0 & 0 & \bar{\sigma}_{p3} \end{bmatrix} = \begin{bmatrix} \sigma_{p1}-p & 0 & 0 \\ 0 & \sigma_{p2}-p & 0 \\ 0 & 0 & \sigma_{p3}-p \end{bmatrix}. \tag{13.2}$$

By construction, the hydrostatic pressure associated with the deviatoric stress tensor vanishes, $(\bar{\sigma}_{p1} + \bar{\sigma}_{p2} + \bar{\sigma}_{p3})/3 = 0$.

In example 1.3 on page 18 it is shown that the direct stress acting on the octahedral face is given by eq. (1.23) as $\sigma_{oc} = (\sigma_{p1} + \sigma_{p2} + \sigma_{p3})/3$; hence, the hydrostatic stress or pressure is also the direct stress acting on the octahedral face.

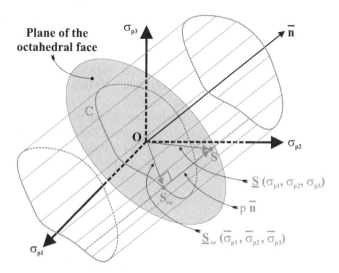

Fig. 13.1. The geometric representation of stress states defined in terms of principal stresses.

In the geometric representation of stress states depicted in fig. 13.1, the plane of the octahedral face is a plane equally inclined with respect to the Cartesian axis system. The equation of this plane is given by $\sigma_{p1} + \sigma_{p2} + \sigma_{p3} = 0$, and unit vector $\bar{n} = \{1,1,1\}^T/\sqrt{3}$ is normal to this plane. Let the projection of vector $\underline{S}$ onto the octahedral plane be denoted $\underline{S}_{oc}$. It then follows that the vector equation, $\underline{S} = \underline{S}_{oc} + p\bar{n}$, corresponds to the decomposition of the stress tensor into its hydrostatic and deviatoric parts, as expressed by eq. (13.2). Indeed, vector $\underline{S}_{oc}$ is associated with the stress state $(\sigma_{p1} - p, \sigma_{p2} - p, \sigma_{p3} - p)$.

A important conclusion can now be drawn from this geometric representation. Since the yield condition for the material is unaffected by the addition of a hydrostatic state of stress, the fact that stress state $\underline{S}$ is a yield point implies that $\underline{S}_{oc}$ is also a yield point. In fact, if $\underline{S}$ is a yield point, all the stress points on the line passing through point $\underline{S}$ and parallel to $\bar{n}$ are also yield points. It follows that all yield points form a cylinder with an axis parallel to $\bar{n}$, and the intersection of this cylinder with the octahedral plane is a curve, denoted $\mathcal{C}$ in fig. 13.1. The locus of the yield points is called the *yield envelope*. This yield envelope is entirely defined by the shape of curve $\mathcal{C}$ that lies in the plane of the octahedral face.

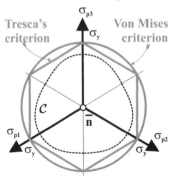

Fig. 13.2. View of the geometric stress space from along unit vector $\bar{n}$; the octahedral face is in the plane of the figure.

The shape of curve $\mathcal{C}$ can be further defined by considering a view of the geometric stress space from along unit vector $\bar{n}$, as depicted in fig. 13.2. In this view, the octahedral face is in the plane of the figure, and the Cartesian axes now appear 120 degrees apart. Because the material is isotropic, if $\underline{S}(\sigma_{p1}, \sigma_{p2}, \sigma_{p3})$ is a yield point, then $\underline{S}'(\sigma_{p2}, \sigma_{p1}, \sigma_{p3})$ is also a yield point. This implies the symmetry of curve $\mathcal{C}$ with respect to axis σ_{p3}. A similar reasoning with the other stress components implies to the symmetry of curve $\mathcal{C}$ with respect to all three axes, σ_{p1}, σ_{p2}, and σ_{p3}. A curve that satisfies these requirement is sketched with a dashed line in fig. 13.2.

In the next two sections, two yield criteria for homogeneous, isotropic materials will be presented. The only difference between the two criteria is the specific shape of curve $\mathcal{C}$, which should be selected to match as closely as possible experimental observation of yield points of such materials under combined stress states.

13.1.2 Tresca's criterion

For Tresca's yield criterion, curve $\mathcal{C}$ is selected to be the regular hexagon shown in fig. 13.2. The complete yield surface is now a regular hexagonal prism with its axis along unit vector $\bar{n}$. In the space of the principal stresses, σ_{p1}, σ_{p2}, and σ_{p3}, the stress states for which the material operates in the linearly elastic range are those stress points falling within this hexagonal prism. This condition can be stated in the following three inequalities derived from the 3 pairs of parallel lines defining the hexagon

$$|\sigma_{p1} - \sigma_{p2}| \le \sigma_y, \quad |\sigma_{p2} - \sigma_{p3}| \le \sigma_y, \quad |\sigma_{p3} - \sigma_{p1}| \le \sigma_y, \qquad (13.3)$$

where σ_y is the yield stress observed in a uniaxial test such as described in fig. 2.5. This result is identical to that expressed by eq. (2.29). Applications of Tresca's criterion to simple stress states are discussed in section 2.3.1.

13.1.3 Von Mises' criterion

For von Mises' yield criterion, curve $\mathcal{C}$ is selected to be the circle shown in fig. 13.2. The complete yield surface is now a circular cylinder with its axis along unit vector

$\bar{n}$. In the space of the principal stresses, σ_{p1}, σ_{p2}, and σ_{p3}, the stress states for which the material operates in the linearly elastic range are the stress points falling within this circular cylinder, and therefore the stress states satisfy the following inequality

$$\sigma_{\text{eq}} = \frac{1}{\sqrt{2}}\sqrt{[(\sigma_{p1} - \sigma_{p2})^2 + (\sigma_{p2} - \sigma_{p3})^2 + (\sigma_{p3} - \sigma_{p1})^2]} \leq \sigma_y, \qquad (13.4)$$

where the first equality defines the *equivalent stress*, σ_{eq}.

Von Mises' criterion now states that *the yield condition is reached under the combined loading, when the equivalent stress, σ_{eq}, reaches the yield stress for a uniaxial stress state, σ_y*. This result is identical to that expressed by eq. (2.32). Applications of Von Mises' criterion to simple stress states are discussed in section 2.3.2.

13.1.4 Problems

Problem 13.1. Alternative formulation of Von Mises' criterion
Consider yield criteria for homogeneous isotropic materials. Justify the following statements: *(1)* material isotropy implies that the yield envelope should be a function only of the invariants of the stress state, and *(2)* the independence of the yield envelope on the addition of a hydrostatic stress state implies that it should be a function only of the invariants of the deviatoric stress tensor. *(3)* Show that the invariant of the deviatoric stress tensor are $\bar{I}_1 = 0$, $\bar{I}_2 = \bar{\sigma}_{p1}\bar{\sigma}_{p2} + \bar{\sigma}_{p2}\bar{\sigma}_{p3} + \bar{\sigma}_{p3}\bar{\sigma}_{p1} = -(\bar{\sigma}_{p1}^2 + \bar{\sigma}_{p2}^2 + \bar{\sigma}_{p3}^2)/2$, and $\bar{I}_3 = \bar{\sigma}_{p1}\bar{\sigma}_{p2}\bar{\sigma}_{p3} = (\bar{\sigma}_{p1}^3 + \bar{\sigma}_{p2}^3 + \bar{\sigma}_{p3}^3)/3$. *(4)* Show that the yield envelope has the form $y(\bar{I}_2, \bar{I}_3) = 0$. *(5)* Show that Von Mises' criterion can be recast in that form.

Problem 13.2. Material sample in cylindrical tube
A cylindrical sample of material is put under a multi-axial stress state by confining it in a thin-walled, elastic tube, then applying a pressure, p, to the sample, as depicted in fig. 13.3. The sample is made of a linearly elastic, isotropic material of Young's modulus E_s, Poisson's ratio ν_s and radius R, whereas the tube has a modulus E_t, Poisson's ratio ν_t and thickness t. It is assumed that the friction forces between the sample and the cylinder are negligible. *(1)* Determine the pressure, q, that arises between the sample and the cylinder. *(2)* Determine the hoop stress in the cylinder. *(3)* If the material sample and confining tube are made of the same material, where will the system yield first? Use von Mises' criterion as a yield criterion. Assume $t/R \ll 1$.

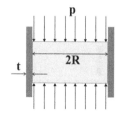

Fig. 13.3. Material sample in cylindrical tube.

13.2 Applications of yield criteria to structural problems

In this section, the yield criteria developed in the previous section will be applied to a number of structural problems. A simple two bar system will be used first to illustrate the basic concepts. Next, the spread of plasticity in a thick tube will be discussed, followed by beam plastic bending and torsion.

Example 13.1. Hyperstatic two-bar system with inelastic behavior

Hyperstatic systems are discussed in section 4.3 and are particularly useful when a structure is expected to be loaded beyond the elastic limit of some of its components. In such cases, the presence of multiple load paths allows the loads to be redistributed among the members that remain elastic, thereby delaying the collapse of the structure.

Consider the two-bar, hyperstatic system depicted in fig. 13.4. Bar **AB** is of length $2L$ and cross-sectional area $\mathcal{A}$, whereas bar **BC** is of length L and cross-sectional area $3\mathcal{A}/2$. Load P is applied at the common point, **B**, of the two bars. Both bars feature the same material with Young's modulus E and yield stress σ_y. The material is assumed to be elastic-perfectly plastic, *i.e.*, the stress-strain diagram is given in fig. 2.7 on page 64.

Intuitively, the system will first operate in the elastic regime with the stress in both bar smaller than σ_y. The solution of this problem is given in example 4.2 on page 141. As the applied load increases, the yield stress will be reached in one of the bar; this load, denoted P^y, is the *elastic limit* of the system. If the load is increased above P^y, one bar operates in the plastic range, *i.e.*, it deforms under constant load, while the other bar still operates in the elastic range. This elastic bar is able to carry additional load because it operates in the elastic regime. If the applied load is increased further, the yield stress is finally reached in the second bar, which now also operates in the plastic range; this load, denoted P^p, is the *plastic limit* of the system. Once this load level is reached, the entire structure now deforms under constant load, and it cannot carry any additional load. Failure occurs when the strain in one of the bar reaches the failure strain, ϵ^f, see fig. 2.7.

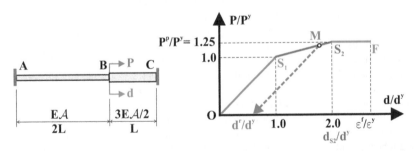

Fig. 13.4. Load-displacement diagram for a two-bar hyperstatic structure with yield stress σ_y.

The displacement method will be used here and follows the steps enumerated in example 4.2. When both bars behave elastically, the equilibrium of the system implies $F_{AB} - F_{BC} = P$, the constitutive laws can be written for each bar as $F_{AB} = e_{AB}\, E\mathcal{A}/(2L)$ and $F_{BC} = e_{BC}\, 3E\mathcal{A}/(2L)$, and finally, the strain-displacement equations imply that $d = e_{AB} = -e_{BC}$, where d is the displacement of point **B**.

The equilibrium equation, written in terms of the unknown displacement, d, becomes $P = [E\mathcal{A}/(2L) + 3E\mathcal{A}/(2L)]d$. This leads to $d = PL/(2E\mathcal{A})$, and the forces in the bars are then

$$F_{AB} = \frac{P}{4}, \quad F_{BC} = -\frac{3P}{4}, \tag{13.5}$$

where the negative sign indicates compression.

Finally, the stresses in the two bar are determined: $\sigma_{AB} = F_{AB}/A = P/(4A)$ and $\sigma_{BC} = F_{BC}/(3A/2) = -P/(2A)$. Clearly, bar **BC** is the first to reach the yield level, and hence, the elastic limit of the system is $\sigma_{BC} = -P^y/(2A) = -\sigma_y$, or $P^y = 2A\sigma_y$; the corresponding displacement is $d^y = L\sigma_y/E$. Figure 13.4 shows the load-displacement diagram of the system, and portion OS_1 of the curve represents the elastic regime.

If the applied load is increased further, the system enters the elastic-plastic regime: bar **AB** is still elastic, whereas bar **BC** operates in the plastic regime, and $\sigma_{BC} = -\sigma_y$ and $F_{BC} = -3A\sigma_y/2$. The equilibrium of the system still implies $F_{AB} - F_{BC} = P$, the constitutive laws are $F_{AB} = e_{AB} EA/(2L)$ and $F_{BC} = -3A\sigma_y/2$, and finally, the strain-displacement equations imply that $d = e_{AB} = -e_{BC}$. The equilibrium equation, written in terms of the unknown displacement becomes $P = 3A\sigma_y/2 + EAd/(2L)$, which can be written as

$$\frac{P}{P^y} - 1 = \frac{1}{4}\left(\frac{d}{d^y} - 1\right). \tag{13.6}$$

This is the load-displacement relationship in the elastic-plastic regime represented by portion $S_1 S_2$ of the curve in figure 13.4. The loads in the bar are readily found as

$$\frac{F_{AB}}{P^y} = \frac{P}{P^y} - \frac{3}{4}, \quad \frac{F_{BC}}{P^y} = -\frac{3}{4}. \tag{13.7}$$

If the applied load is increased further, the stress level in bar **AB** now reaches the yield stress, and this happens when the applied load equals P^p, the plastic limit of the structure. At this point, $F_{AB} = A\sigma_y = P^y/2$, and using eq. (13.7), $1/2 = P^p/P^y - 3/4$, hence

$$\frac{P^p}{P^y} = \frac{5}{4}. \tag{13.8}$$

The corresponding displacement, d^p, is found by introducing the plastic limit into eq. (13.6) to find $d^p/d^y = 2$. The plastic limit point corresponds to point S_2 in fig. 13.4.

If the structure is allowed to operate into the elastic-plastic range up to plastic limit, it can carry a load up to $P^p = 1.25P^y$ which is a 25% increase over the elastic limit load. On the other hand, the apparent stiffness of the structure in the elastic range is $k^y = 2EA/L$, whereas in the elastic-plastic range, it is reduced to $k^{ep} = EA/(2L)$, a fourfold decrease.

Above the plastic limit, both bars operate in the plastic regime, the structure deforms continuously under a constant load P^p, and no further increase in applied load is therefore possible. Segment $S_2 F$ of the curve in figure 13.4 represents this fully plastic regime. The structure finally fails when the strain in bar **BC** reaches the compressive failure strain, $-\epsilon^f$, and this happens when the displacement of point **B** equals d^f such that $d^f/d^y = \epsilon^f/\epsilon^y$.

If the structure is loaded into the elastic-plastic range and then unloaded, it behaves elastically because *the stress-strain diagram is linear upon unloading*, see segment **DG** in fig. 2.7. Consider the following scenario: first, the system in loaded to a maximum load, P_m in the elastic-plastic range at point **M** in fig. 13.4, so that $P^y < P_m < P^p$, and then load P_m is released, leaving the structure unloaded. The unloading is equivalent to the application of a reversed load $P_u = -P_m$ at point **M**. Clearly, the system is then unloaded, since the applied load has vanished, $P_m + P_u = 0$. The forces in the bars do not vanish because the initial plastic flow creates permanent deformations.

The remaining forces in the bars after unloading are called *residual forces* and are found by superposing the forces created by the loading in the elastic-plastic range, given in eq. (13.7), with those created by the unloading, given by the negative of eq. (13.5). Note that the elastic solution of eq. (13.5) applies to the unloading because *the material behaves linearly in this regime*. The non-dimensional residual forces in bars **AB** and **BC**, denoted F^r_{AB} and F^r_{BC}, respectively, become

$$
\begin{aligned}
\frac{F^r_{AB}}{P^y} &= \left(\frac{P_m}{P^y} - \frac{3}{4} \right) - \left(\frac{1}{4} \frac{P_m}{P^y} \right) = \frac{3}{4} \left(\frac{P_m}{P^y} - 1 \right), \\
\frac{F^r_{BC}}{P^y} &= \left(-\frac{3}{4} \right) - \left(-\frac{3}{4} \frac{P_m}{P^y} \right) = \frac{3}{4} \left(\frac{P_m}{P^y} - 1 \right).
\end{aligned}
\tag{13.9}
$$

The residual forces in bars **AB** and **BC** are tensile and equal. This can be explained as follows. When deformed in the plastic range, bar **BC** undergoes compressive plastic flow. Upon unloading, a permanent residual strain remains in the bar: the bar is now "too short." This is equivalent to a manufacturing imperfection that is created by the permanent plastic deformation of bar **BC**. Imagine that the system is loaded to its plastic limit before unloading, *i.e.*, $P_m = P^p = 1.25P^y$; the resulting residual forces are $F^r_{AB}/P^y = F^r_{BC}/P^y = 3/16$; the residual forces thus represent 18.75% of the elastic limit forces for the system.

It is also interesting to evaluate the residual displacement of point **M** by superposing, once again, the displacements associated with the loading and unloading phases to find

$$
\frac{d^r}{d^y} = \left[1 + 4 \left(\frac{P_m}{P^y} - 1 \right) \right] - \left(\frac{P_m}{P^y} \right) = 3 \left(\frac{P_m}{P^y} - 1 \right).
$$

This is a positive displacement, *i.e.*, in the direction of the applied load P_m, and it is consistent with the permanent shortening of bar **BC** due to plastic deformations. If the system is loaded to its plastic limit before unloading, the residual displacement is $d^r/d^y = 0.75$; the residual displacement therefore represents 75% of the displacement of the structure under the elastic limit load.

Example 13.2. Behavior of a thick-walled tube in the plastic regime
Consider a thick-walled tube with inner and outer radii denoted R_i and R_e, respectively, subjected to an inner pressure p_i. This problem is discussed in example 3.4 on page 122 for a cylinder in plane stress state. Figure 3.8 shows the stress distribution through the thickness of the wall and indicates that the highest stress component is the hoop stress at the inner bore of the cylinder, *i.e.*, at $r = R_i$.

As the inner pressure increases, a critical value is reached for which the material reaches the yield condition at that location. If the pressure is further increased, plasticity will spread in the tube as illustrated in fig. 13.5 where the material is in the plastic regime for $R_i \leq r \leq R_p$ and in the elastic regime for $R_p \leq r \leq R_e$. The radial location, R_p, of the interface between the plastic and elastic regions is an unknown function of the applied pressure.

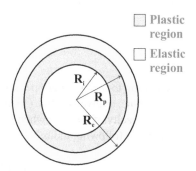

Fig. 13.5. Thick tube under internal pressurization showing the development of concentric plastic and elastic regions.

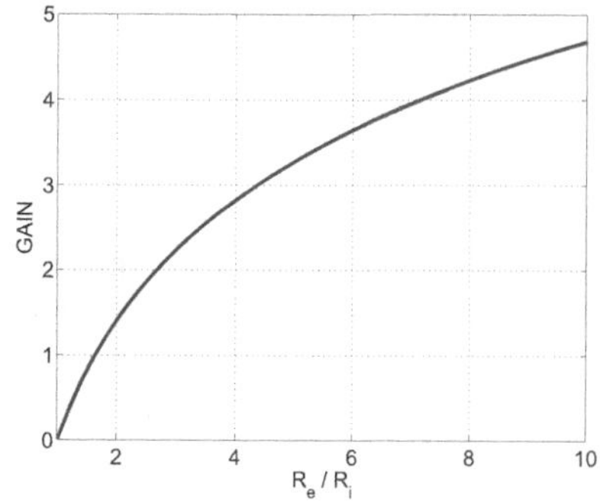

Fig. 13.6. Gain in maximum pressure for a fully plastic tube as compared to the tube operating entirely in the elastic regime.

It will now be assumed that the yield condition is given by Tresca's criterion, see eq. (13.3). Since the stress distribution inside the tube consists of only the radial and hoop stress components without any shear stress, these must be the principal stresses, and hence, $\sigma_{p1} = \sigma_\theta$ and $\sigma_{p2} = \sigma_r$, leading to the following statement for the yield point: $\sigma_\theta - \sigma_r = \sigma_y$, where σ_y is the yield stress for the material.

First, consider the case where the tube is operating entirely in the elastic regime, but it has just reached the yield point for $r = R_i$, i.e., $\sigma_\theta(r = R_i) - \sigma_r(r = R_i) = \sigma_y$. Because the tube is in the elastic regime, the stress field given by eqs. (3.56) is still correct, and introducing these stress components into the yield criterion yields

$$\frac{p_i^y}{\sigma_y} = \frac{1}{2}\left(1 - \frac{R_i^2}{R_e^2}\right). \tag{13.10}$$

This is the maximum internal pressure for which the entire tube remains in the elastic regime.

Next, if pressure is increased further, plasticity spreads into the tube wall. The material is assumed to be elastic-perfectly plastic, i.e., it features the stress-strain diagram illustrated in fig. 2.7 on page 64. The equilibrium condition for the stress components is given by eq. (3.39a), and implies

$$\frac{d\sigma_r}{dr} = \frac{\sigma_\theta - \sigma_r}{r} = \frac{\sigma_y}{r}, \tag{13.11}$$

because Tresca's criterion must be exactly satisfied at each point where the elastic-perfectly plastic material is yielding. This equation can be integrated to $\sigma_r/\sigma_y =$

$C_1 + \ln r$, and the boundary condition, $\sigma_r(r = R_i) = -p_i$, is used to evaluate the integration constant, C_1, to find the non-dimensional stress distribution in the plastic zone as

$$\frac{\sigma_r^p}{\sigma_y} = \ln \rho - \frac{p_i}{\sigma_y}, \quad \frac{\sigma_\theta^p}{\sigma_y} = 1 + \ln \rho - \frac{p_i}{\sigma_y}. \tag{13.12}$$

where $\rho = r/R_i$.

Finally, assume that the entire tube operates in the plastic regime, $i.e.$, $R_p = R_e$. An additional boundary condition must then be imposed, $\sigma_r^p(r = R_e) = 0$, to find the corresponding pressure as

$$\frac{p_i^p}{\sigma_y} = \ln \bar{R}_e, \tag{13.13}$$

where $\bar{R}_e = R_e/R_i$.

It is interesting to compare the pressures corresponding to the cylinder operating entirely in the elastic regime, or entirely in the plastic regime, as given by eqs. (13.10) or (13.13), respectively. The increase in load carrying capacity of the tube when it is allowed to operate in the plastic regime is measured by the following index

$$\frac{p_i^p - p_i^y}{p_i^y} = \frac{\ln \bar{R}_e^2}{1 - 1/\bar{R}_e^2}. \tag{13.14}$$

Figure 13.6 shows the gain in maximum pressure as a function of the ratio of external to internal radius; note that for $\bar{R}_e = 2$, the tube can carry an 85% higher internal pressure when operating in the fully plastic range as compared to the elastic range.

The last case to examine is when the tube operates in the mixed elastic-plastic regime. In the elastic zone, the stress field is given by eqs. (3.54)

$$\sigma_r^y = E\left[\frac{C_1}{1-\nu} - \frac{C_2}{(1-\nu)r^2}\right], \quad \sigma_\theta^y = E\left[\frac{C_1}{1-\nu} + \frac{C_2}{(1+\nu)r^2}\right],$$

while the stress field in the plastic zone is given by eqs. (13.12).

At the elastic-plastic interface radius $r = R_p$, two conditions must be imposed. First, to satisfy equilibrium, the radial stress components must match across interface: $\sigma_r^p(\rho = \bar{R}_p) = \sigma_r^y(\rho = \bar{R}_p)$, where $\bar{R}_p = R_p/R_i$. Next, from the definition of this interface, the stress fields on either side must satisfy Tresca's yield condition: $\sigma_\theta^p(\rho = \bar{R}_p) - \sigma_\theta^p(\rho = \bar{R}_p) = \sigma_y$ and $\sigma_\theta^y(\rho = \bar{R}_p) - \sigma_\theta^y(\rho = \bar{R}_p) = \sigma_y$, which, in view of the previous condition, implies $\sigma_\theta^p(\rho = \bar{R}_p) = \sigma_\theta^y(\rho = \bar{R}_p)$, Finally, the boundary condition at the outer edge of the cylinder must also be satisfied: $\sigma_r^y(\rho = \bar{R}_e) = 0$. Expressing these three conditions yields the following three equations

$$\ln \bar{R}_p - \frac{p_i}{\sigma_y} = \frac{EC_1}{(1-\nu)\sigma_y} - \frac{EC_2}{(1+\nu)\sigma_y}\frac{1}{\bar{R}_p^2},$$

$$1 + \ln \bar{R}_p - \frac{p_i}{\sigma_y} = \frac{EC_1}{(1-\nu)\sigma_y} + \frac{EC_2}{(1+\nu)\sigma_y}\frac{1}{\bar{R}_p^2},$$

$$\frac{EC_1}{(1-\nu)} = \frac{EC_2}{(1+\nu)\bar{R}_e^2},$$

which can be solved for the two integration constants, C_1 and C_2, and the unknown interface radius, $\bar{R}_p$. The non-dimensional stress field in the elastic region now becomes

$$\frac{\sigma_r^y}{\sigma_y} = \frac{\bar{R}_p^2}{2}\left(\frac{1}{\bar{R}_e^2} - \frac{1}{\rho^2}\right), \quad \frac{\sigma_\theta^y}{\sigma_y} = \frac{\bar{R}_p^2}{2}\left(\frac{1}{\bar{R}_e^2} + \frac{1}{\rho^2}\right). \tag{13.15}$$

Furthermore, the relationship between the applied pressure and the interface radius, R_p, is

$$\frac{p_i}{\sigma_y} = \frac{1}{2}\left[\ln \bar{R}_p^2 + 1 - \frac{\bar{R}_p^2}{\bar{R}_e^2}\right]. \tag{13.16}$$

In summary, the procedure for computing the response of a thick tube to an increasing internal pressure is as follows. For pressures below the elastic limit given by eq. (13.10), the stress field is that derived in example 3.3, see eq. (3.56). If the applied pressure is above the elastic limit, the tube features both plastic and elastic zones. For a give interface radius, $R_i \leq R_p \leq R_e$, the internal pressure the tube can carry is evaluated with the help of eq. (13.16). The plastic stress field given by eq. (13.12) then applies for $R_i \leq r \leq R_p$, whereas the elastic stress field given by eqs. (13.15) applies for $R_p \leq r \leq R_e$. Finally, the maximum load carrying capability of the tube is the pressure given by eq. (13.13); when that pressure is reached, the tube undergoes continuous plastic flow under constant pressure.

13.2.1 Problems

Problem 13.3. Three bar truss in plastic range
Consider the three bar truss depicted in fig. 4.5. The cross-sectional areas of the homogeneous bars are $\mathcal{A}_A$, $\mathcal{A}_B$, and $\mathcal{A}_C$, for the bars attached at points **A**, **B**, and **C**, respectively, with $\mathcal{A}_A = \mathcal{A}_C$. The three bars are made of an elastic, perfectly plastic material, with material behavior as depicted in fig. 2.7. Let P^y and Δ^y be the maximum load and deflection, respectively, for which the system remains in the linearly elastic range; Let P^p and Δ^p be the load and deflection, respectively, for which the system becomes fully plastic. *(1)* Determine P^y and the corresponding Δ^y. *(2)* Determine P^p and the corresponding elongation on a non-dimensional scale P/P^y vs. Δ/Δ^y. *(4)* Determine the the loading, P^f/P^y, and elongation, Δ^f/Δ^y, of the system at failure. *(5)* Plot the load deformation curve, P/P^y versus Δ/Δ^y, up to failure. *(6)* Assume the load $(P^y + P^p)/2$ is applied to the system, then released. Find the residual stresses in the bar, σ_B^r/σ_y, and σ_C^r/σ_y. Find the residual elongation of the system, Δ^r/Δ^y.

Problem 13.4. Thick-walled cylinder pressurized in the plastic regime
Consider a thick-walled cylinder of internal and external radii R_i and R_e, respectively, in a state of plane stress subjected to an internal pressure p_i. The tube is allowed to operate in the plastic regime; the material is elastic-perfectly plastic and the yield condition is given Tresca's criterion; $R_e/R_i = 2$. *(1)* Plot the non-dimensional interface radius, R_p/R_i, as a function of the non-dimensional pressure, p_i/p_i^E. *(2)* On one graph, plot the radial stress distribution through the thickness of the tube for different values of the interface radius. *(3)* On one graph, plot the hoop stress distribution through the thickness of the tube for different values of the interface radius. For the last two questions, let $R_p = R_i + \alpha(R_e - R_i)$; present your results for $\alpha = 0, 0.1, 0.2, 0.4, 0.6$ and 1.

13.2.2 Plastic bending

The beam theory developed in chapters 5, 6 and 8 assumes that the material a beam is made of behaves in a *linearly elastic* manner following Hooke's law, eq. (5.14). Assuming the beam is made of a ductile material presenting a stress-strain diagram similar to that shown in fig. 2.5 on page 63, once the bending moment applied to the beam generates axial stresses larger than the limit of proportionality, Hooke's law is no longer an appropriate approximation to the constitutive behavior of the material. In this section, beam bending theory is generalized to deal with materials that do not behave in a linearly elastic fashion.

It is important to understand that the Euler-Bernoulli assumptions presented in section 5.1 are *purely kinematic assumptions*. This means that these assumptions imply a specific displacement field for the beam, see eq. (5.4), or equivalently, its strain field, see eq. (5.7). Clearly, these assumptions are independent of the constitutive laws selected to represent the physical behavior of the material.

To simplify the development of the theory, the beam is assumed to present a rectangular section of width b and height h, as depicted in fig. 13.7. The strain distribution is assumed to be linear over the cross-section, as implied by the the Euler-Bernoulli assumptions, see eq. (5.30). The strains at the bottom and top locations of the section are denoted ϵ_b and ϵ_t, respectively, and the corresponding stresses are denoted σ_b and σ_t, respectively. In view of the linear distribution of axial strain, eq. (5.30), it follows that

$$\Delta\epsilon = \epsilon_t - \epsilon_b = -x_{2t}\kappa_3 + x_{2b}\kappa_3 = -(x_{2t} - x_{2b})\kappa_3 = -h\kappa_3. \tag{13.17}$$

where x_{2b} and x_{2t} are the coordinates of the bottom and top locations of the section, respectively.

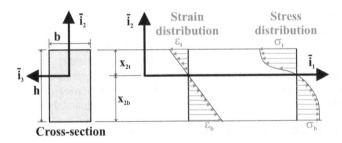

Fig. 13.7. Beam with a rectangular cross-section undergoing plastic bending.

Since the material is not linearly elastic, the stress distribution associated with the assumed linear strain distribution is no longer linear through the cross-section. Figure 13.7 shows the stress distribution through the cross-section. Because the strain is a linear function of vertical position on the cross-section, the vertical stress distribution is, in fact, identical in shape to the stress-strain diagram of the material, rotated by 90 degrees (*i.e.*, the strain axis lies along the vertical axis).

The stress distribution must be in equilibrium with the externally applied loads, and hence, the axial force must vanish, *i.e.*, $N_1 = \int_{\mathcal{A}} \sigma_1 \, d\mathcal{A} = 0$. Because the section is rectangular, $d\mathcal{A} = b \, dx_2$, and furthermore, the linearity of the axial strain distribution, eq. (5.30), implies $d\epsilon_1 = -\kappa_3 dx_2$, so that $d\mathcal{A} = -b/\kappa_3 \, d\epsilon_1$. The vanishing of the axial force now requires $N_1 = \int_{\mathcal{A}} \sigma_1 \, d\mathcal{A} = -(b/\kappa_3) \int_{\epsilon_b}^{\epsilon_t} \sigma_1 \, d\epsilon_1 = 0$. Because the stress is a function of the strain, as implied by the stress-strain diagram for the material, this condition becomes

$$\int_{\epsilon_b}^{\epsilon_t} \sigma_1(\epsilon_1) \, d\epsilon_1 = 0. \tag{13.18}$$

The physical interpretation of this result is as follows: the area under the stress-strain diagram from ϵ_b to ϵ_t must vanish. If the stress-strain diagram is symmetric for tension and compression, this requirement will be automatically met if $\epsilon_b = -\epsilon_t$, and both strain and stress components will vanish at the geometric center of the section. On the other hand, if the material does not behave in the same manner in tension and compression, the stress-strain diagram is no longer symmetric. Consequently, the strain and stress will then vanish at a point away from the geometric center of the section, and the location of the modulus-weighted centroid defined in eq. 5.33 will now be a function of the applied bending moment.

Next, the bending moment is computed from the axial stress distribution using eq. (5.10) to find

$$M_3 = \int_{\mathcal{A}} \sigma_1 x_2 \, d\mathcal{A} = \frac{b}{\kappa_3^2} \int_{\epsilon_b}^{\epsilon_t} \sigma_1 \epsilon_1 \, d\epsilon_1. \tag{13.19}$$

The second integral represents the static moment of the stress-strain diagram computed with respect to the origin. Equation (13.19) gives the relationship between the bending moment and the curvature of the beam in the plastic bending regime.

In practice, the relationship between the bending moment and the curvature is constructed as follows.

1. Select an arbitrary value of the strain ϵ_b. Then, determine the strain ϵ_t such that eq. (13.18) is satisfied. It is now possible to compute the curvature of the beam from $\Delta_\epsilon = \epsilon_t - \epsilon_b = -h\kappa_3$ and hence,

$$\kappa_3 = -\frac{\Delta_\epsilon}{h}. \tag{13.20}$$

2. Determine the location of the centroid. The linearity if the strain distribution implies

$$\frac{|x_{2t}|}{|x_{2b}|} = \frac{|\epsilon_t|}{|\epsilon_b|}. \tag{13.21}$$

3. Compute the bending moment using eq. (13.19).
4. Repeat the above procedure for a number of strain levels, ϵ_b. For each new strain level, a new point of the moment-curvature diagram is obtained. It is then possible to plot $M_3 = M_3(\kappa_3)$; this relationship is nonlinear as a result of the nonlinearity inherent to the stress-strain diagram.

Usually, the stress-strain diagram is obtained empirically, and the above procedure must be carried out numerically for each point defining the diagram. If an analytical expression is available for the stress-strain diagram, the above procedure will yield an analytical expression for the sectional moment-curvature diagram as will be illustrated in the following example.

Example 13.3. Plastic bending for an elastic-perfectly plastic material

In this example, the plastic bending of a rectangular cross-section of width b and height h, made of an elastic-perfectly plastic material is investigated.

The stress-strain diagram for an elastic-perfectly plastic material is given in fig. 2.7 on page 64. In view of the symmetry of this stress-strain diagram, the centroid of the section remains at the geometric center of the rectangular section, as implied by eq. (13.18).

Intuitively, the section will operate in three distinct regimes that are illustrated in fig. 13.8. As the applied bending moment increases, the stress levels will increase until the axial stresses at the top and bottom locations of the section reach the yield stress, σ_y. For stress levels below σ_y, the material behaves in a linearly elastic manner, and the axial stress distribution is linear through the section, see fig. 13.8a. The bending moment that will generate axial stress levels of $|\sigma_y|$ at the top and bottom locations of the section is easily found to be $M_3^y = bh^2\sigma_y/6$, and the corresponding curvature of the beam is evaluated with the help of eq. (13.20) as $\kappa_3^y = -2\epsilon_1^y/h$ where ϵ_1^y is the maximum elastic strain which occurs at the bottom edge of the section.

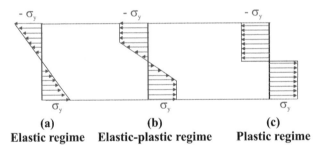

$$
\begin{array}{ccc}
\text{(a)} & \text{(b)} & \text{(c)} \\
\textbf{Elastic regime} & \textbf{Elastic-plastic regime} & \textbf{Plastic regime}
\end{array}
$$

Fig. 13.8. The three regimes of the cross-section: the purely elastic, elastic-plastic, and fully plastic regimes.

If larger bending moments, $M_3 > M_3^y$, are applied, a portion of the section will operate in the plastic regime as shown in fig. 13.8b. In this elastic-perfectly plastic regime, the bending moment is computed with the help of eq. (13.19); for a strain level ϵ_1, the static moment of the stress-strain diagram is

$$\int_{-\epsilon_1}^{\epsilon_1} \sigma_1 \epsilon_1 \, d\epsilon_1 = \int_{-\epsilon_1}^{-\epsilon_1^y} (-\sigma_y) \, \epsilon_1 \, d\epsilon_1 + \int_{-\epsilon_1^y}^{\epsilon_1^y} \left(\sigma_y \frac{\epsilon_1}{\epsilon_1^y} \right) \epsilon_1 \, d\epsilon_1 + \int_{\epsilon_1^y}^{\epsilon_1} (\sigma_y) \, \epsilon_1 \, d\epsilon_1$$

$$= \sigma_y \left(\epsilon_1^2 - \frac{(\epsilon_1^y)^2}{3} \right) = \sigma_y \epsilon_1^2 \left[1 - \frac{1}{3} \left(\frac{\epsilon_1^y}{\epsilon_1} \right)^2 \right].$$

The moment-curvature relationship then follows from eq. (13.19) as

$$M_3 = b\sigma_y \frac{\epsilon_1^2}{\kappa_3^2} \left[1 - \frac{1}{3} \left(\frac{\epsilon_1^y}{\epsilon_1} \right)^2 \right] = \frac{bh^2}{4} \sigma_y \left[1 - \frac{1}{3} \left(\frac{\kappa_3^y}{\kappa_3} \right)^2 \right]$$

$$= \frac{3 M_3^y}{2} \left[1 - \frac{1}{3} \left(\frac{\kappa_3^y}{\kappa_3} \right)^2 \right],$$

(13.22)

where the relationship between strain and curvature, eq. (13.20), is used to eliminate the strains, $h/2 = \epsilon_1^y/\kappa_3^y = \epsilon_1/\kappa_3$.

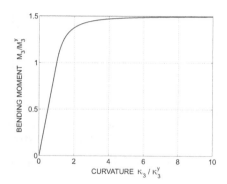

Fig. 13.9. The moment-curvature diagram for a section made of elastic-perfectly plastic material.

As the applied bending moment increases, plasticity spreads in the section until a limit bending moment, M_3^p, is reached for which the associated axial stress distribution is depicted in fig. 13.8c. It is easily verified from this figure that $M_3^p = bh^2/4 \, \sigma_y = 3/2 \, M_3^y$. At this point, no further increase in applied bending moment can be supported, and any attempt to increase the bending moment will result in an unbounded increase in the curvature. This is illustrated in the complete moment-curvature diagram depicted in fig. 13.9.

In section 2.1.4, it is pointed out that upon unloading, ductile materials tend to behave elastically. For the elastic-perfectly plastic material considered here, if deformed under a stress $|\sigma_1| > \sigma_y$, residual stresses and strains will remain after unloading. The portion of the section for which $|\epsilon_1| > \epsilon_1^y$ will have residual strains after unloading, whereas no residual strains will remain in the portion of the section for which $|\epsilon_1| \le \epsilon_1^y$. The total strain, which is the sum of the elastic and residual strains, must remain linearly distributed over the section in accordance with the Euler-Bernoulli kinematics and will not vanish upon unloading.

The unloading can be represented by the application of an unloading moment, M_u, that is equal and opposite to the applied moment, and the residual stresses are evaluated using the following process.

1. First, the beam is loaded into the elastic-plastic range with a moment $M_3^y \leq M_m \leq M_3^p$, and a corresponding curvature, κ_m, develops. The associated axial stress distribution is depicted in fig. 13.10a.
2. Next, an unloading bending moment, denoted $M_u = -M_m$, is applied to the beam. Because the material behaves elastically upon unloading, the stress distribution associated with this unloading moment is the linear distribution depicted in fig. 13.10b.
3. Finally, the residual stresses are obtained by combining these two stress distributions as shown in fig. 13.10c.

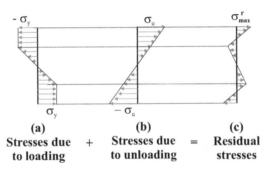

(a)   (b)   (c)
Stresses due **+** **Stresses due** **=** **Residual**
to loading   **to unloading**   **stresses**

Fig. 13.10. Residual stresses after plastic bending in a section made of elastic-perfectly plastic material.

At the end of this process the beam is unloaded, and the bending moment associated with the residual stress distribution must vanish. From eq. (13.22) the bending moment-curvature relationship is $M_m = 3/2\,(bh^2\sigma_y/6)[1 - (\kappa_3^y/\kappa_m)^2/3]$. The unloading bending moment is related to the maximum unloading axial stress as $M_u = -bh^2/6\,\sigma_u$. Since the total applied load must vanish, $M_u = -M_m$, and the maximum unloading stress becomes

$$\sigma_u = \frac{3\sigma_y}{2}\left[1 - \frac{1}{3}\left(\frac{\kappa_3^y}{\kappa_m}\right)^2\right].$$

The maximum residual stress is found in the outermost point of the section as $\sigma_u - \sigma_y$, see fig. 13.10c; hence

$$\sigma_{max}^r = \frac{\sigma_y}{2}\left[1 - \left(\frac{\kappa_3^y}{\kappa_m}\right)^2\right] = \sigma_y\left(\frac{M_m}{M_3^y} - 1\right).$$

If the section is unloaded after applying the maximum bending moment $M_m = M_3^p = 3/2\, M_3^y$, the maximum residual stress is $\sigma_{max}^r = \sigma_y/2$.

13.2.3 Problems

Problem 13.5. Plastic bending of beam with diamond cross-section

Consider a beam with a diamond shaped cross-section of size a by a as shown in fig. 13.11 and made of an elastic-perfectly plastic material. *(1)* Determine the maximum bending moment, M_3^y, for which the section remains in the elastic range and the corresponding curvature κ_3^y. *(2)* Plot the bending moment-curvature diagram for this beam; use non-dimensional abscissa κ_3/κ_3^y and ordinate M_3/M_3^y. *(3)* What is the maximum bending moment, M_3^p/M_3^y, the section can carry when all the material enters the plastic range? *(4)* Find the maximum residual stress in the cross-section if the beam is unloaded after application of a bending moment M_3^p.

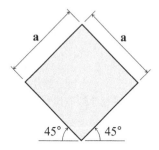

Fig. 13.11. Beam diamond cross-section.

Problem 13.6. Plastic bending of beam with strain hardening

The stress-strain relationship for strain hardening materials can be approximatively represented by Ludwik's power law, $\sigma/C = \epsilon^n$, where C and n are material parameters. Consider a beam with a rectangular cross-section of width b and height h, made of a material that follows Ludwik's power law. *(1)* On one graph, plot the non-dimensional stress, σ/C, versus strain diagrams for $n = 0.1, 0.2, 0.3$, and 0.5. *(2)* Compute the bending moment versus curvature relationship for the beam. *(3)* On one graph, plot the non-dimensional bending moment, $M_3/(Cbh^2)$, versus non-dimensional curvature, $h\kappa_3$, for $n = 0.1, 0.2, 0.3$, and 0.5. The material is assumed to have identical behavior in tension and compression, and hence, $\int_{-\epsilon_1}^{\epsilon_1} \sigma_1 \epsilon_1 \, d\epsilon_1 = 2 \int_0^{\epsilon_1} \sigma_1 \epsilon_1 \, d\epsilon_1$.

13.2.4 Plastic torsion

The theory developed for torsion of beams with circular cross-sections in the section 7.1 is based on the assumption that the material behaves in a *linearly elastic* manner, and Hooke's law, eq. (2.9), is assumed to apply. Assuming the circular bar is made of a ductile material presenting a shear stress-shear strain diagram similar to that presented in fig. 2.6 on page 64, once the torque applied to the bar generates shear stresses larger than the limit of proportionality, Hooke's law is no longer an appropriate approximation to the constitutive behavior of the material.

The theory of torsion of cylindrical bars is based on the kinematic description developed in section 7.1.1. Since this kinematic description is obtained from symmetry arguments and does not involve any consideration of constitutive laws, it remains valid for the present case involving materials deformed past their limit of proportionality. The only non-vanishing strain component is the circumferential shear strain,

γ_α, given by eq. (7.9), which is linearly distributed over the circular cross-section as depicted in fig. 7.4. Let γ_M be the maximum circumferential shear strain at the outer edge of the section, and in view of eq. (7.9), $\gamma_M = R\kappa_1$.

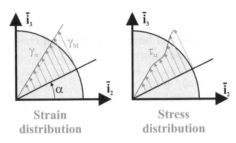

Strain
distribution

Stress
distribution

Fig. 13.12. Bar with a circular cross-section undergoing plastic torsion.

Because the material is not linearly elastic, the stress distribution associated with the linear shear strain distribution is no longer the linear stress distribution shown in fig. 7.5. Instead, fig. 13.12 shows the linear shear strain distribution and the associated nonlinear shear stress distribution. This shear stress distribution must be in equilibrium with the externally applied torque

$$M_1 = \int_\mathcal{A} \tau_\alpha r \, d\mathcal{A} = \int_0^{2\pi} \int_0^R \tau_\alpha r \, r \, dr \, d\theta$$

$$= 2\pi \int_0^R \tau_\alpha r^2 \, dr = \frac{2\pi}{\kappa_1^3} \int_0^{\gamma_M} \tau_\alpha(\gamma_\alpha) \gamma_\alpha^2 \, d\gamma_\alpha.$$

(13.23)

In this equation, the linear distribution of shear strain, $\gamma_\alpha = r\kappa_1$, is used to eliminate the radial variable r and to express $dr = d\gamma_\alpha/\kappa_1$, to facilitate the evaluation of the last integral.

Finally, the shear stress is explicitly a function of the shear strain, $\tau_\alpha = \tau_\alpha(\gamma_\alpha)$, through the material stress-strain diagram. The last integral in eq. (13.23) can be evaluated directly from the stress-strain diagram.

In practice, the relationship between the torque and the twist rate is constructed as follows.

1. Select an arbitrary value of the maximum shear strain γ_M and compute the associated twist rate as $\kappa_1 = \gamma_M/R$.
2. Compute the torque using eq. (13.23) and the material stress-strain diagram.
3. Repeat the above procedure for a number of maximum shear strain levels, γ_M. For each new maximum strain level, a new point of the torque-twist rate diagram is obtained. This relationship will be nonlinear in view of the nonlinearity inherent in the stress-strain diagram.

If an analytical expression is available for the shear stress-shear strain diagram, the above procedure will yield an analytical expression for the sectional torque-twist rate diagram. This situation is illustrated in the example below.

Example 13.4. Plastic torsion for an elastic-perfectly plastic material
Section 13.2.4 describes the general procedure for evaluating the torque-twist rate
diagram of a circular bar subjected plastic torsion. This procedure will be applied
in this example to a bar made of an elastic-perfectly plastic material presenting the
stress-strain diagram given in fig. 2.7 on page 64. The yield shear stress and strain
will be denoted τ_y and γ_y, respectively, while γ_f denotes the shear strain at failure.

Intuitively, the section will develop three distinct regimes that are illustrated in
fig. 13.13. As the applied torque is increased, the shear stress levels will increase
until the stress around the outer edge of the section reaches the yield stress, τ_y. For
stress levels below τ_y, the material behaves in a linearly elastic manner, and the shear
stress distribution is linear as shown in fig. 13.13a.

The torque that will generate a shear stress level of τ_y around the outer edge of
the section is easily found from eq. (7.21) to be

$$M_1^y = \frac{H_{11}\tau_y}{GR} = \pi/2\, R^3\tau_y$$

and the corresponding twist rate for the bar is $\kappa_1^y = \gamma_y/R$.

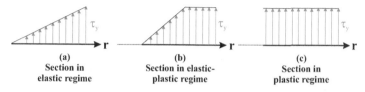

(a)   (b)   (c)
Section in   Section in elastic-   Section in
elastic regime   plastic regime   plastic regime

Fig. 13.13. The three stress regimes of the cross-section: (a) purely elastic, (b) elastic-plastic,
and (c) fully plastic.

If larger torques, $M_1 > M_1^y$, are applied, a portion of the section will operate
in the plastic regime as shown in fig. 13.13b. In this regime, the torque is computed
using eq. (13.23). For a shear strain level $\gamma > \gamma_y$, the torque becomes

$$
\begin{aligned}
M_1 &= \frac{2\pi}{\kappa_1^3}\left[\int_0^{\gamma_y} \tau_y\frac{\gamma}{\gamma_y}\gamma^2\,\mathrm{d}\gamma + \int_{\gamma_y}^{\gamma} \tau_y\gamma^2\,\mathrm{d}\gamma\right] \\
&= \frac{2\pi}{3}\tau_y\left(\frac{\gamma}{\kappa_1}\right)^3\left[1 - \frac{1}{4}\left(\frac{\gamma_y}{\gamma}\right)^3\right] = \frac{4M_1^y}{3}\left[1 - \frac{1}{4}\left(\frac{\kappa_1^y}{\kappa_1}\right)^3\right].
\end{aligned}
$$
(13.24)

As the twist rate increases, the torque increases more slowly than before, and
eventually as $\kappa_1 \to \infty$, it reaches an upper limit, M_1^p. At this point, the entire section
operates in the plastic regime as depicted in fig. 13.13c, and the shear stress at all
points in the cross section is now at the yield level, τ_y. It is easily verified from this
figure that $M_1^p = 2\pi R^3\tau_y/3 = 4/3\, M_1^y$.

A plot of the torque-twist rate relationship (*i.e.*, the force-deformation relationship) is shown in fig. 13.14. Initially, this is a linear relationship because Hooke's

law applies, but for $\kappa_1 > \kappa_1^y$ more and more material of the section reaches the plastic range, and the twist rate increases without bound under a nearly constant torque, M_1^p.

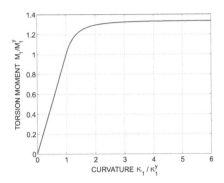

Fig. 13.14. The moment-curvature diagram for a section made of elastic-perfectly plastic material.

Example 13.5. Plastic torsion unloading for an elastic-perfectly plastic material
Consider the unloading of a circular bar twisted beyond its yield stress by the application of a torque $M_1^y \leq M_m \leq M_1^p$. Determine the residual stresses in the section.

In section 2.1.4, it is pointed out that upon unloading, ductile materials tend to behave elastically. For the elastic-perfectly plastic material considered here, if deformed under a stress $|\tau| > \tau_y$, residual stresses and strains will remain after unloading. The portion of the section for which $|\gamma| > \gamma^y$ will present residual strains after unloading, whereas no residual strains will remain in the portion of the section for which $|\gamma| \leq \gamma^y$. The total strain, which is the sum of the elastic and residual strains, must remain linearly distributed over the section in accordance with the kinematic assumptions and will not vanish upon unloading.

The unloading can be represented by the application of an unloading torque, M_u, that is equal and opposite to the applied torque, and the residual stresses are evaluated using the following process.

1. It is first assumed that the bar is loaded into the elastic-plastic range with a torque M_m, and a corresponding twist rate, κ_m, results. The associated shear stress distribution is depicted in fig. 13.13b.
2. Next, an opposing torque (*i.e.*, an unloading torque), denoted $M_u = -M_m$, is applied to the bar. Since the material behaves elastically upon unloading, the shear stress distribution associated with this unloading torque is linear.
3. The residual stresses are obtained by combining these two shear stress distributions. Since at the end of this process the bar is unloaded, the torque associated with the residual stress distribution must vanish.

From eq. (13.24), the relationship between applied torque and curvature is $M_m = 4M_1^y [1 - 1/4 (\kappa_3^y/\kappa_m)^3]/3$. The unloading torque is related to the maximum unloading shear stress as, $M_u = -\pi/2R^3 \tau_u$, where τ_u is the maximum shear stress at the outer edge of the circular section. Since the net applied torque must vanish, $M_u = -M_m$ and the maximum unloading stress becomes

$$\tau_u = \frac{4\tau_y}{3} \left[1 - \frac{1}{4} \left(\frac{\kappa_3^y}{\kappa_m} \right)^3 \right].$$

The maximum residual shear stress is found at the outer edge of the section as $\tau_{max}^r = \tau_u - \tau_y$,

$$\tau_{max}^r = \frac{\tau_y}{3} \left[1 - \left(\frac{\kappa_3^y}{\kappa_m} \right)^3 \right] = \tau_y \left(\frac{M_m}{M_1^y} - 1 \right).$$

If the section is unloaded after applying the maximum torque $M_{max} = M_1^p = 4M_1^y/3$, the maximum residual stress is then $\tau_{max}^r = \tau_y/3$.

13.3 Thermal stresses in structures

The evaluation of thermal stresses in structures subjected to thermal loading is an important part of structural analysis. Structures such as heat exchangers, jet engine turbine blades, supersonic aircraft and missiles, or space structures must all be designed to withstand significant thermal loading. The thermal loading can have a multitude of effects on structures ranging from induced thermal strain to accelerated viscoelasticity and plasticity.

Simple concepts about thermal stresses are introduced in section 2.1.2, and examples are presented for simple bars as well as isostatic and hyperstatic trusses. This chapter focuses on the computation of thermal stresses in structures arising from thermally induced strains using the theory of linear elasticity. Such problems are generally referred to as *thermoelastic problems.*

Thermal stresses are induced by three main sources: *(1)* nonuniform temperature distributions that create nonuniform strains within a structure, *(2)* external constraints that prevent the free deformation of a structure, and *(3)* differences in coefficients of thermal expansion that appear in heterogeneous structures.

In most practical applications, thermoelastic problems can be treated as *quasistatic, uncoupled problems.* The term "quasi-static" refers to the fact that the temperature variations are slow, and hence, inertia effects associated with the acceleration of the structure under time-dependent temperature fields can be neglected. Thermal stresses are evaluated for different steady temperature distributions, which represent the thermal loading at different instants in time. The term "uncoupled" refers to the fact that generation of heat in the structure is not taken into account. If a structure is subjected to repeated loading, heat is generated through hysteresis, resulting in a nonuniform increase in temperature. Here again, these changes in temperature are slow. The time constants of the resulting heat conduction problem are far slower than the time constants associated with the dynamic response of the structure, and decoupling the two problems has little effect on the accuracy of the solution.

In a free, unconstrained structure, any thermal stresses must form a system of *self-equilibrating stresses*; indeed, a free body diagram of any portion of the structure reveals that all stress resultants must vanish. If the structure is constrained so that its thermal deformation is prevented by boundary conditions, the thermal stresses will be in equilibrium with the reaction forces at those boundaries.

Two main approaches to the evaluation of thermal stresses are possible. First, the *direct method*, discussed in section 13.3.1, treats thermal problems as basic elasticity problems. While this approach is always possible, it can be cumbersome to apply when dealing with structural components such as beams and plates. In section 13.3.3, the *constraint method* is presented. This method is often much easier to apply to beam and plate problems.

13.3.1 The direct method

When a sample of material is heated, its dimensions will change. Under heat, homogeneous isotropic materials will expand equally in all directions, generating *thermal strains*, $\epsilon^t = f(\Delta T)$, where $f(\Delta T)$ is a function of the change in temperature ΔT. The volume of most materials increases when the material is subjected to increased temperatures, whereas temperature decreases generally cause the material volume to shrink. There are, however, notable exceptions: the transition from water to ice under decreasing temperature is accompanied by a volume increase. For moderate temperature changes, it is often adequate to assume that $f(\Delta T)$ is a linear function of the temperature change so that $f(\Delta T) = \alpha \Delta T$, where α is the *coefficient of thermal expansion*, a positive number if the material expands under increased temperature. The thermal strain now becomes

$$\epsilon^t = \alpha \Delta T. \tag{13.25}$$

Coefficients of thermal expansion for commonly used structural materials are listed in table 13.1.

Table 13.1. Coefficients of thermal expansion and Young's moduli for commonly used structural materials.

Material	Coefficient of thermal expansion [μ/C]	Young's modulus [GPa]
Aluminum	23	73
Copper	17	120
Steel	11	210
Titanium	8.6	110

Two important aspects of thermal deformations must be emphasized. First, thermal strains are purely extensional: temperature changes induce no shear strains. Second, thermal strains do not generate any internal stresses, in contrast with mechanical strains that are related to internal stresses through the material constitutive law. Consequently, an unconfined material sample subjected to a uniform temperature change simply expands, but no internal stresses are developed.

Thermal strains are the consequence of temperature changes, whereas mechanical strains result from the application of stresses. Hence, it is simpler to state the constitutive law with thermal effects by superposing these strains. The total strains are the sum of the mechanical strains given by eq. (2.4) and their thermal counterparts given by eq. (13.25)

$$\epsilon_1 = \frac{1}{E} \left[\sigma_1 - \nu(\sigma_2 + \sigma_3) \right] + \alpha \Delta T, \tag{13.26a}$$

$$\epsilon_2 = \frac{1}{E} \left[\sigma_2 - \nu(\sigma_1 + \sigma_3) \right] + \alpha \Delta T, \tag{13.26b}$$

$$\epsilon_3 = \frac{1}{E} \left[\sigma_3 - \nu(\sigma_1 + \sigma_2) \right] + \alpha \Delta T. \tag{13.26c}$$

Because temperature changes induce no shear strains, the constitutive laws for shear strain, given by eq. (2.9), remain unchanged and are repeated here

$$\gamma_{23} = \frac{1}{G} \tau_{23}, \quad \gamma_{13} = \frac{1}{G} \tau_{13}, \quad \gamma_{12} = \frac{1}{G} \tau_{12}. \tag{13.27}$$

When dealing with constrained material samples, temperature changes will indirectly generate stresses in the material. For example, consider a bar constrained at its two ends by rigid walls that prevent any extension of the bar. When subjected to a temperature change, ΔT, the bar will expand in all directions, but the rigid walls prevent expansion of the bar along its axis, $\bar{\imath}_1$. The stress components in the transverse direction, σ_2 and σ_3, must vanish because the bar is free to expand in those directions, whereas the axial strain, ϵ_1, must vanish due to the presence of the rigid walls. Equation (13.26a) then implies

$$\epsilon_1 = \frac{1}{E} \sigma_1 + \alpha \Delta T = 0,$$

and hence, $\sigma_1 = -E\alpha \Delta T$; the temperature change thus induces a compressive stress in the bar. Such stresses are called *thermal stresses*. If the same bar is allowed to expand freely, *i.e.*, if the end walls are removed, axial equilibrium of the bar implies $\sigma_1 = 0$ and eq. (13.26a) then yields $\epsilon_1 = \alpha \Delta T$. In this case, the temperature change induces thermal strains but no thermal stresses.

The equations of the theory of elasticity consist of three groups: equilibrium equations, strain-displacement equations, and constitutive laws. As expected, equilibrium equations are unaffected by the presence thermal strains because equilibrium conditions involve forces, not deformations. Strain-displacement equations also remain unchanged, although the strains are now the sum of the mechanical and thermal strains. Only the constitutive laws are directly affected by the presence of thermal strains, see eqs. (13.26).

In summary, the evaluation of thermal stresses based on the theory of three-dimensional elasticity relies on the following three sets of equations: equilibrium equations, see eqs. (1.4); strain-displacement, see eqs. (1.63) and (1.71); and finally, the constitutive laws, eqs. (13.26) and (13.27), for linearly elastic, isotropic materials.

The direct method will be illustrated by means of several problems that are solved in the following examples.

Example 13.6. Thick circular tube in plane strain state

Consider a thick tube of internal and external radii denoted as R_i and R_e, respectively, subjected to a radially varying temperature field, as shown in fig. 13.15. The temperature of the inner and outer walls of the tube are denoted T_i and T_e, respectively, and heat flow through the tube is assumed to maintain the temperature differential. At thermal equilibrium, the temperature distribution through the thickness of the tube can be shown to be of the form[1]

$$T(r) - T_e = -\frac{T_e - T_i}{\ln \bar{R}_i} \ln \rho, \tag{13.28}$$

where $\rho = r/R_e$ is the non-dimensional radial variable, and $\bar{R}_i = R_i/R_e$.

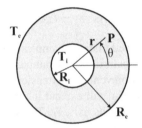

Fig. 13.15. Thick tube in plane strain state subjected to a temperature field.

Both the geometry and thermal loading are axisymmetric and therefore partial derivatives with respect to the circumferential coordinate vanish. The tube is assumed to be in a state of plane strain, *i.e.*, $\epsilon_3 = 0$, which would occur in a long tube constrained at both ends. Imposing the vanishing of the axial strain in eq. (13.26c) then yields $\sigma_3 = \nu(\sigma_r + \sigma_\theta) - E\alpha T$. Introducing this result into eqs. (13.26a) and (13.26b) gives the radial and circumferential stress components as

$$\sigma_r = \frac{E\left[(1-\nu)\epsilon_r + \nu\epsilon_\theta - (1+\nu)\alpha T\right]}{(1-2\nu)(1+\nu)},$$

$$\sigma_\theta = \frac{E\left[\nu\epsilon_r + (1-\nu)\epsilon_\theta - (1+\nu)\alpha T\right]}{(1-2\nu)(1+\nu)}.$$

The strain-displacement equations of the problem are given by eqs. (3.37a) and (3.37b) as $\epsilon_r = du_r/dr$ and $\epsilon_\theta = u_r/r$. The single radial equilibrium equation is $d\sigma_r/dr + (\sigma_r - \sigma_\theta)/r = 0$, see eq. (3.39a).

---

[1] The radial temperature distribution is determined by solution to a heat flow problem that is beyond the scope of this text.

Introducing the above expressions for the stress components into the equilibrium equation and expressing the strains in terms of the radial component of displacement yields the governing equation of the problem as

$$\frac{d^2 u_r}{dr^2} + \frac{1}{r}\frac{du_r}{dr} + \frac{u_r}{r^2} - \frac{1+\nu}{1-\nu}\alpha\frac{dT}{dr} = 0,$$

which can be put into more compact form as

$$\frac{d}{dr}\left[\frac{1}{r}\frac{d}{dr}(ru_r)\right] = \frac{1+\nu}{1-\nu}\alpha\frac{dT}{dr}.$$

Integrating this equation twice gives the radial displacement distribution as

$$u_r = C_1 r + \frac{C_2}{r} + \frac{1+\nu}{1-\nu}\frac{\alpha}{r}\int_{R_i}^{r} rT\, dr,$$

where C_1 and C_2 are integration constants.

Because the boundary conditions are expressed in terms of stress components at the inner and outer surfaces, the above solution must now be expressed in terms of stresses. The strains are readily found using the strain-displacement equations, and using the constitutive laws, the stress components become

$$\sigma_r = \frac{E}{1+\nu}\left(\frac{C_1}{1-2\nu} - \frac{C_2}{r^2}\right) - \frac{\alpha E}{1-\nu}\frac{1}{r^2}\int_{R_i}^{r} rT\, dr,$$

$$\sigma_\theta = \frac{E}{1+\nu}\left(\frac{C_1}{1-2\nu} + \frac{C_2}{r^2}\right) + \frac{\alpha E}{1-\nu}\frac{1}{r^2}\int_{R_i}^{r} rT\, dr - \frac{E\alpha T}{1-\nu}.$$

The two integration constants are determined from the two boundary conditions: $\sigma_r = 0$ at $r = R_i$ and R_e, as

$$\frac{C_1}{1-2\nu} = \frac{C_2}{R_i^2} = \frac{\alpha}{1-\bar{R}_i^2}\frac{1+\nu}{1-\nu}A,$$

where $A = \int_{\bar{R}_i}^{1}\rho T(\rho)d\rho$.

The thermal stresses in the thick tube now become

$$\sigma_r = \frac{\alpha E}{1-\nu}\left[\frac{A}{1-\bar{R}_i^2}\left(1 - \frac{\bar{R}_i^2}{\rho^2}\right) - \frac{B(\rho)}{\rho^2}\right],$$

$$\sigma_\theta = \frac{\alpha E}{1-\nu}\left[\frac{A}{1-\bar{R}_i^2}\left(1 + \frac{\bar{R}_i^2}{\rho^2}\right) + \frac{B(\rho)}{\rho^2} - T\right],$$

where $B(\rho) = \int_{\bar{R}_i}^{\rho}\rho T(\rho)d\rho$.

The axial displacement field is then readily found by substituting C_1 and C_2 into the above expression for u_r as

$$u_r = \alpha R_e\frac{1+\nu}{1-\nu}\left[\frac{A}{1-\bar{R}_i^2}\left(1 - 2\nu + \frac{\bar{R}_i^2}{\rho^2}\right) + \frac{B(\rho)}{\rho^2}\right].$$

To complete the solution process, the integrals A and $B(\rho)$ must be evaluated for the temperature distribution given by eq. (13.28) to find

$$\frac{A}{1-\bar{R}_i^2} = \frac{T_e - T_i}{2\ln \bar{R}_i^2}\left[1 + \frac{\bar{R}_i^2 \ln \bar{R}_i^2}{1 - \bar{R}_i^2}\right],$$

$$\frac{B(\rho)}{\rho^2} = \frac{T_e - T_i}{2\ln \bar{R}_i^2}\left[1 - \ln \rho^2 - \frac{\bar{R}_i^2(1 - \ln \bar{R}_i^2)}{\rho^2}\right].$$

The results of this analysis are shown in fig. 13.16 for $T_e - T_i < 0$, meaning that the interior surface is hotter than the exterior surface. This occurs when the tube contains a hot fluid. Note the large compressive hoop stress at the inner radius of the cylinder. Since this is a plane strain problem, the axial stress, σ_3, is compressive, except near the outer radius of the cylinder where it is relieved by Poisson's effects.

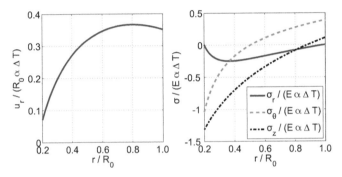

Fig. 13.16. Thermal stresses in a thick tube. Left figure: non-dimensional axial displacement field. Right figure: non-dimensional stress components. $\Delta T = T_e - T_i < 0$.

13.3.2 Problems

Problem 13.7. Thick circular tube in plane stress state
Example 13.6 focused on the determination of the thermal stress field in a thick cylinder in a state of plane strain subjected to the temperature map give by eq. (13.28). Repeat the development presented in example 13.6 for a thick cylinder in a state of plane stress. *(1)* find the radial displacement field. *(2)* Find the radial and hoop stress fields. *(3)* Plot the non-dimensional displacement and stress fields in the format of fig. 13.16.

13.3.3 The constraint method

The constraint method will be developed for a body made of a homogeneous, isotropic material. If the body is free to deform, the thermal strain at any point of the body will be $\epsilon_1^t = \epsilon_2^t = \epsilon_3^t = \alpha T$.

In the first step of the method, the structure is assumed to be *fully constrained*, so that *the thermal deformations are not allowed to take place*. This implies that a set of mechanical strains must appear that exactly compensate for the above thermal strains: $\epsilon_1^m = \epsilon_2^m = \epsilon_3^m = -\alpha T$.

In contrast with the thermal strains that generate no stresses, these mechanical strains are associated with a state of stress given by Hooke's law, eqs. (2.4), such that

$$\sigma_1 = \sigma_2 = \sigma_3 = -\frac{E\alpha T}{1 - 2\nu}. \tag{13.29}$$

This is a hydrostatic stress state, and introducing these stresses into eq. (2.4a), the first mechanical strain component is found as $\epsilon_1^m = [\sigma_1 - \nu(\sigma_2 + \sigma_3)]/E = -\alpha T$, as expected. Thus, if all thermal deformations are inhibited, the hydrostatic state of stress given by eq. (13.29) must develop at all points of the structure.

To maintain this hydrostatic state of stress in the structure, a set of surface tractions must be applied at the external surface, S, of the body, and body forces must be applied at each point in its volume, $\mathcal{V}$. The constrained surface tractions, $\underline{t}^c$, are readily found from the stress state as

$$\underline{t}^c = -\frac{E\alpha T}{1 - 2\nu}\bar{n}, \tag{13.30}$$

where $\bar{n}$ is the outer normal to S.

The body forces are easily found by considering the equilibrium equation at a point of the body, eq. (1.4a). Because the shear stresses vanish, this equation reduces to $\partial\sigma_1/\partial x_1 + b_1 = 0$, and the body force component becomes $b_1 = \partial\left[(E\alpha T)/(1 - 2\nu)\right]/\partial x_1$. Eqs. (1.4b) and (1.4c) yield the other two body force components, and combining these, the constrained body force vector becomes

$$\underline{b}^c = \frac{E\alpha}{1 - 2\nu}\nabla T, \tag{13.31}$$

where ∇ is the gradient operator, $\nabla = \bar{\imath}_1\partial/\partial x_1 + \bar{\imath}_2\partial/\partial x_2 + \bar{\imath}_3\partial/\partial x_3$.

The second step of the procedure calls for the solution of an elasticity problem where the structure is subjected, not to the thermal load but instead, to the set of surface tractions and body forces *opposite* to those required to inhibit thermal deformations. Thus, the structure must be subjected to the following equivalent body forces and surface tractions

$$\underline{b}^e = -\frac{E\alpha}{1 - 2\nu}\nabla T, \tag{13.32a}$$

$$\underline{t}^e = \frac{E\alpha T}{1 - 2\nu}\bar{n}, \tag{13.32b}$$

respectively. In the sequel, these body forces and surface tractions will be called the *equivalent thermal body forces* and *equivalent thermal surface tractions* and denoted with a superscript $(\cdot)^e$.

In third step of the procedure, the solution of the thermal stress problem is found by superposition of the results found in the previous steps. The displacements are identical to those of the structure subjected to the equivalent thermal loading defined in eq. (13.32), and the thermal stresses are the sum of those generated by the equivalent thermal loading and the constrained hydrostatic stresses given by eq. (13.29).

In summary, the constraint method consists of the following steps.

1. Assume that all thermal deformations are inhibited by a suitable set of body forces and surface tractions given by eqs. (13.31) and (13.30), respectively.
2. Solve an elasticity problem where the structure is subjected to the equivalent thermal body forces and surface tractions given by eqs. (13.32). In this step, the structure is not subjected to thermal effects.
3. Superpose the states of the structures in steps 1 and 2. The displacement of the structure are those found in step 2. The thermal stresses are the sum of those found in step 2 plus the hydrostatic state of stress given by eq. (13.29).

The advantages of the constraint method should be clear. Step 1 is easy, and step 2 is a standard elasticity problem for which all solution procedures developed for elasticity problems can be applied. In fact, the constraint method reduces the evaluation of thermal stresses to a standard isothermal elasticity problem. The equivalent thermal loading given by eqs. (13.32) provides a more intuitive understanding of the response of the structure to a temperature field. Because the constraint method reduces the evaluation of thermal stresses to a standard elasticity problem, it is not very different from the direct method presented in section 13.3.1. Its real advantage becomes evident when applied to bar, beam and plate problems, as discussed in the sections below.

13.4 Application to bars, trusses and beams

The constraint method will be used to solve a number of examples involving axially loaded bars, simple planar trusses, and beams in the following sections and a comparison will be made to the direct method, when appropriate.

13.4.1 Applications to bars and trusses

In section 4.2, the analysis of homogeneous bars with constant properties along their span and subjected to end loads is developed. It is shown that after deformation, the cross-sections remain plane and normal to the axis of the bar. The axial strain also remains uniform over the cross-section. Finally, since the bar is homogeneous, the axial stress is uniformly distributed over the cross-section, and all other stress components vanish.

For such simple structures, the constraint method develop in section 13.3.3 becomes particularly simple. It will be assumed here that each bar is heated to a uniform temperature, T. In step 1 of the approach, thermal deformations are assumed to be inhibited by a suitable set of body forces and surface tractions. Because the stress components σ_2 and σ_2 vanish, Hooke's law implies that the constraint stresses, eq. (13.29), reduce to $\sigma_1 = -E\alpha T$, and $\sigma_2 = \sigma_2 = 0$. In view of eq. (13.31), the constraint body forces, $\underline{b}^c$, vanish because the temperature field in the bar is uniform, and hence, $\nabla T = 0$. Finally, from eq. (13.30), the constraint surface tractions become forces applied to the bar's ends, $P^c = -E\mathcal{A}\alpha T$, where $\mathcal{A}$ is the cross-sectional area of the bar.

In step 2 of the approach, a bar problem is solved with the thermal equivalent loading consisting of only the thermal equivalent forces, $P^e = E\mathcal{A}\alpha T$, applied to the bar's ends. For trusses constructed from identical members, equivalent forces of magnitude $P^e = E\mathcal{A}\alpha T$ are applied at the ends of each bar. If additional mechanical loads are applied, they are included as well, and the solution proceeds using the procedures developed in previous chapters.

In step 3 of the approach, the solution of the thermal problem is obtained by superposition. The displacements of the bar or truss are those obtained in step 2, and the thermal forces in the bars are obtained superposing the forces obtained in step 2 and the constraint forces, P^c.

Example 13.7. *Bar subjected to a uniform temperature*
Consider a uniform bar clamped at one end and free at the other, as depicted in the left part of fig. 13.17. The bar is uniformly heated to a temperature T. Find the thermally induced stresses and the deformation in the bar.

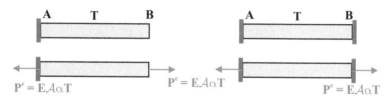

Fig. 13.17. Bar subjected to a uniform temperature. Left figure, the bar is clamped at one end. Right figure, the bar is clamped at both ends.

The constraint method will be used to solve this very simple problem. In step 1 of the approach, the thermal deformations are inhibited by an axial force, $P^c = -E\mathcal{A}\alpha T$. In step 2, equivalent thermal end forces, $P^e = E\mathcal{A}\alpha T$, are applied at points **A** and **B**. The force applied at point **A** is equal to the reaction force at this point. The equivalent problem is a clamped bar subjected to a tip force P^e, and the solution is trivial. The elongation of the bar is given by eq. (4.2) as $e = P^e L/(E\mathcal{A}) = \alpha T L$, and the load in the bar is simply P^e. Finally, in step 3, the deformation of the bar is that found in step 2 for which the displacement at point **B** is $d_B = e = \alpha T L$. The equivalent thermal forces are the superposition of the forces found in steps 1 and 2, $P = P^c + P^e = 0$. As expected, the thermal loads vanish because the bar is free to expand.

If the bar is clamped at both ends, as depicted in the right part of fig. 13.17, the first step of the procedure remains unchanged. In the second step, the equivalent thermal forces P^e are applied at points **A** and **B**, but because these are fixed the bar undergoes no deformation and no stress, and P^e appears as reactions. Finally, in the last step, the solution of the thermal problem is obtained by superposition of the results from steps 1 and 2. No displacements develop in the system and the thermal loads are the end reactions, $P = P^c + 0 = -E\mathcal{A}\alpha T$. As expected, the thermal forces are compressive.

Example 13.8. Three-bar truss subjected to temperature changes
Consider the three-bar truss depicted in fig. 13.18; the center bar is raised to a temperature T_B, and the two side bars are each raised to a temperature T_A. The side bars have identical Young's modulus E_A, sectional area, $\mathcal{A}_A$, and coefficient of thermal expansion, α_A. The corresponding quantities for the middle bar are E_B, $\mathcal{A}_B$, and α_A, respectively. Find the thermal forces in the system.

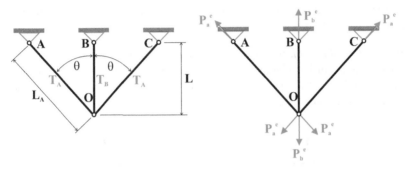

Fig. 13.18. Three-bar truss subjected to temperature changes.

The three-step constraint method is used here again. In the first step, all thermal deformations are inhibited, giving rise to constraint forces in the three bars of $P_A^c = P_C^c = -(E\mathcal{A})_A(\alpha T)_A$ and $P_B^c = -(E\mathcal{A})_B(\alpha T)_B$, respectively.

In the second step, end loads are applied to each bar, and as shown on the right half of fig. 13.18, equivalent thermal end forces $P_A^e = P_C^e = (E\mathcal{A})_A(\alpha T)_A$ and $P_B^e = (E\mathcal{A})_B(\alpha T)_B$ are applied to the side and middle bars, respectively. Clearly, the equivalent thermal forces at the upper ends of the bars are the reaction forces carried at the supports. At point **O**, the total equivalent thermal force is the vector sum of the forces in the three bar, with a net vertical resultant $P_O^e = P_B^e + 2P_A^e \cos\theta$, applied downwards. Step two of the procedure consists of finding the displacements and internal forces in a three-bar truss subjected to the load P_O^e at point **O**. This problem is treated in example 4.4 on page 147, and the vertical deflection, Δ, of point **O** is given by eq. (4.12) as

$$\frac{\Delta}{L} = \frac{(\alpha T)_B + 2\bar{k}_A(\alpha T)_A \cos\theta}{1 + 2\bar{k}_A \cos^3\theta}.$$

while the forces in the bars are given by eq. (4.14) as

$$\frac{F_A^e}{(E\mathcal{A})_B} = \frac{\bar{k}_A \cos^2\theta \left[(\alpha T)_B + 2\bar{k}_A(\alpha T)_A \cos\theta\right]}{1 + 2\bar{k}_A \cos^3\theta},$$

$$\frac{F_B^e}{(E\mathcal{A})_B} = \frac{(\alpha T)_B + 2\bar{k}_A(\alpha T)_A \cos\theta}{1 + 2\bar{k}_A \cos^3\theta}.$$

In step 3 the solution of the thermal problem is found by superposition. The displacement of point **O** is given by the expression above, and the thermal forces are $F_A^t = P_A^c + F_A^e$ and $F_B^t = P_B^c + F_B^e$ for the side and middle bars, respectively,

$$\frac{F_A^t}{(E\mathcal{A})_B} = \frac{\bar{k}_A \left[(\alpha T)_B \cos^2\theta - (\alpha T)_A\right]}{1 + 2\bar{k}_A \cos^3\theta},$$

$$\frac{F_B^t}{(E\mathcal{A})_B} = \frac{-2\bar{k}_A \cos\theta \left[(\alpha T)_B \cos^2\theta - (\alpha T)_A\right]}{1 + 2\bar{k}_A \cos^3\theta}.$$

As expected, these forces form a set of self-equilibrating forces; it is easily verified that $F_B^t + 2F_A^t \cos\theta = 0$.

First, note that even if the entire structure is heated to a uniform temperature, i.e., if $T_A = T_B$, non-vanishing forces will develop in all three bars. This is a direct consequence of the hyperstatic nature of this problem. If one bar is removed, the problem becomes isostatic and thermal forces will vanish under uniform temperature changes. Second, a three-bar truss problem under simpler thermal loading is treated in example 4.9. Checking that the results obtained in this example reduce to those obtained earlier in example 4.9 on page 159 is left to the reader.

Example 13.9. *Deflection of planar truss under thermal loading*

The unit load method is developed in section 9.6.6 to evaluate the deflections of truss joints. Equation (9.64) gives the deflection, Δ, at one node of the truss,

$$\Delta = \sum_{i=1}^{N_b} \hat{F}_i \, e_i, \tag{13.33}$$

where $\hat{F}_i$ a set of statically admissible bar forces in equilibrium with a unit load applied at the joint where Δ is to be determined and acting in the direction of Δ. The bar extensions, e_i, are those due to the externally applied loads. Section 9.6.6 presents many examples of application of the unit load method when trusses are subjected to externally applied loads. If thermal deformations are created by changing the temperature of one or more bars in the truss, the unit load method can still be used to determine deflections, but the bar elongations must reflect the induced thermal strains.

If a bar of length L is subjected to end loads, F, and a uniform temperature change, ΔT, its elongation is $e = FL/(E\mathcal{A}) + \alpha\Delta TL$, where the first term is due to the applied load, and the second to the thermal strains. In the presence of thermal deformation, eq. (13.33) now becomes

$$\Delta = \sum_{i=1}^{N_b} \hat{F}_i \, e_i = \sum_{i=1}^{N_b} \hat{F}_i \left(\frac{F_i L_i}{E_i \mathcal{A}_i} + L_i \alpha_i \Delta T_i\right). \tag{13.34}$$

Consider the five-bar, isostatic planar truss subjected to a single external load applied at joint **B**, as depicted in fig. 13.19. All bar have the same physical properties, $E\mathcal{A}$. The tabular format developed in section 9.6.6 is here again. The second column of table 13.2 list the bars flexibility factors and the third the statically admissible forces that are in equilibrium with a unit load acting downward at joint **E**. Column 4 gives the bar forces resulting from the application of the externally applied load, P. Column 5 lists the thermal elongations in each bar, and the last column gives the partial results for the application of eq. (13.34).

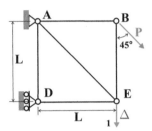

Fig. 13.19. Planar 1-bay truss with thermal loading.

Table 13.2. Calculation of member forces for truss with thermal loading.

Bar	$L_i/(E\mathcal{A})_i$	$\hat{F}_i$	F_i	$L_i\alpha_i\Delta T_i$	$\hat{F}_i[L_i/(E\mathcal{A})_i + L_i\alpha_i\Delta T_i]$
AB	1	0	$P/\sqrt{2}$	1	0
BE	1	0	$-P/\sqrt{2}$	1	0
DE	1	-1	$-P/\sqrt{2}$	1	$-1[-PL/(\sqrt{2}E\mathcal{A}) + L\alpha\Delta T]$
AD	1	0	0	1	0
AE	$\sqrt{2}$	$\sqrt{2}$	P	$\sqrt{2}$	$\sqrt{2}[PL/(E\mathcal{A}) + \sqrt{2}L\alpha\Delta T]$

The vertical deflection of joint **E** is obtained by summing up the entries in the last column of table 13.2 to find

$$\Delta = \frac{3}{\sqrt{2}}\frac{PL}{S} + L\alpha\Delta T.$$

The total deflection is the sum of the joint deflection due to the applied load P (the first term) and of that due to the change in temperature, ΔT, of all bars of the truss (the second term). The thermal deformations of bars **DE** and **AE** are the only contributors to the vertical deflection at joint **E** and are of opposite sign. If the temperature changes for bars **DE** and **AE** are ΔT_{DE} and ΔT_{AE}, respectively, the deflection at joint **E** becomes

$$\Delta = \frac{3}{\sqrt{2}}\frac{PL}{S} + L\alpha(2\Delta T_{AE} - \Delta T_{DE}).$$

It is possible to eliminate the thermal deflection at joint **E** by selecting $2\Delta T_{AE} = \Delta T_{DE}$.

Because this truss is isostatic, bar forces can be evaluated from the equilibrium equations alone; temperature changes do not affect bar forces, only the joint deflections.

13.4.2 Problems

Problem 13.8. Steel bar inside a copper tube
A steel bar with a 750 mm^2 section is placed inside a copper tube with a section of 1250 mm^2. The bar and tube have a common length of 0.5 m and are connected at their ends. At the reference temperature, both elements are stress free. *(1)* If the assembly is heated up of 80° C, find the thermal stresses in both elements using the constraint method. Use the data listed in table 13.1.

Problem 13.9. Rigid plate supported by four elastic bars under thermal loading
Consider the hyperstatic system depicted in fig. 13.20 and consisting of a rigid square plate of side ℓ supported by four identical elastic bars of height h, cross-sectional area $\mathcal{A}$, Young's modulus E, and coefficient of thermal expansion α. The four bars are raised to temperatures T_A, T_B, T_C, and T_D, respectively. Find the thermal forces in the bars using the constraint method. Hint: first study example 4.5.

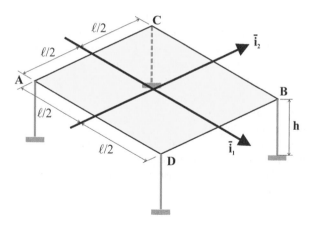

Fig. 13.20. A rigid plate supported by four identical elastic bars.

13.4.3 Application to beams

The constraint method developed in section 13.3.3 will now be applied to beam problems. Rather than starting from the basic equations of elasticity, the Euler-Bernoulli assumptions presented in section 5.1 will form the basis for this development. To simplify the problem, it will be assumed that plane $(\bar{\imath}_1, \bar{\imath}_2)$ is a plane of symmetry for the beam configuration and of the temperature distribution. This implies that the deformation of the beam will take place in this plane of symmetry. The theory presented in this section could be easily generalized to a beam presenting a cross-section of arbitrary shape and subjected to an arbitrary temperature field. Such developments would be based on the three-dimensional beam theory presented in chapter 6.

As discussed in section 5.4.2, the stresses, σ_2 and σ_3, acting in the plane of the cross-section should remain much smaller than the axial stress component, σ_1. In fact, these stress components can be assumed to vanish, $\sigma_2 \approx 0$ and $\sigma_3 \approx 0$. This leads to the reduced Hooke's law given by eq. (5.14), $\sigma_1 = E \, \epsilon_1$.

In step 1 of the constraint method, thermal deformations are assumed to be fully inhibited. This requires the presence of the following stress system

$$\sigma_1 = -E\alpha T, \quad \sigma_2 = \sigma_3 = 0, \tag{13.35}$$

instead of the hydrostatic state of stress given by eq. (13.29).

Beam theory deals with the stress resultants presented in section 5.3, rather than local stresses. The stress resultants that will fully inhibit thermal deformations consist of an axial force and a bending moment given by eqs. (5.8) and (5.10) as

$$N_1^c = -\int_A E\alpha T \, \mathrm{d}A, \quad \text{and} \quad M_3^c = \int_A x_2 E\alpha T \, \mathrm{d}A, \tag{13.36}$$

respectively. Of course, near the end sections of the beam, the actual distribution of axial stress will not be exactly equivalent to this axial force and bending moment. According to Saint-Venant's principle, (principle 2 on page 169), this mismatch is expected to affect only a small portion of the beam, near its end sections.

To equilibrate this distribution of axial forces and bending moments, a set of support loads and bending moments at the boundaries together with distributed axial and transverse loads must be applied to the beam. The end axial loads are simply $P_1^c = N_1^c$ and the end bending moments are $Q_3^c = M_3^c$.

The required distributed axial and transverse loads are found from equilibrium conditions. The axial equilibrium equation for the beam, eq. (5.18), is $p_1 = -\mathrm{d}N_1/\mathrm{d}x_1$, and this implies that the distributed axial load, p_1^c, that equilibrates the constraint axial force is

$$p_1^c = -\frac{\mathrm{d}N_1^c}{\mathrm{d}x_1} = \frac{\mathrm{d}}{\mathrm{d}x_1} \left[\int_A E\alpha T \, \mathrm{d}A \right]. \tag{13.37}$$

Similarly, the bending equilibrium equation for the beam, eq. (5.39), is $p_2 = \mathrm{d}^2 M_3/\mathrm{d}x_1^2$, leading to the following distribution of transverse loads

$$p_2^c = \frac{\mathrm{d}^2 M_3^c}{\mathrm{d}x_1^2} = \frac{\mathrm{d}^2}{\mathrm{d}x_1^2} \left[\int_A x_2 E\alpha T \, \mathrm{d}A \right]. \tag{13.38}$$

Step 2 of the constraint method calls for the solution of a beam problem subjected to a set of equivalent end forces and distributed loads opposite to those required to inhibit thermal deformations. Consequently, the beam must then be subjected to the following equivalent thermal end axial forces

$$P_1^e = \int_{\mathcal{A}} E\alpha T \, \mathrm{d}\mathcal{A}. \tag{13.39}$$

and bending moments

$$Q_3^e = -\int_{\mathcal{A}} x_2 E\alpha T \, \mathrm{d}\mathcal{A}. \tag{13.40}$$

Furthermore, the following equivalent thermal distributed axial load

$$p_1^e = -\frac{\mathrm{d}}{\mathrm{d}x_1} \left[\int_{\mathcal{A}} E\alpha T \, \mathrm{d}\mathcal{A} \right], \tag{13.41}$$

and transverse load

$$p_2^e = -\frac{\mathrm{d}^2}{\mathrm{d}x_1^2} \left[\int_{\mathcal{A}} x_2 E\alpha T \, \mathrm{d}\mathcal{A} \right]. \tag{13.42}$$

must also be applied to the beam.

In step 3, the solution of the thermal problem is found by superposition. The displacements of the beam are those found by solving the problem subjected to the equivalent thermal loads. The thermal stresses then are the sum of those found for the same loading, and the stresses required to inhibit thermal deformations. The constraint procedure will be illustrated in the following examples.

Example 13.10. *Relationship between the direct and constraint methods*

Show that the governing equations for a beam under thermal loading obtained from the constraint method in section 13.4.3 are identical to those obtained by the direct method applied to the classical Euler-Bernoulli beam theory developed in sections 5.4 and 5.5 for beams under axial and transverse loads, respectively.

If the direct method is applied to Euler-Bernoulli beam theory, the kinematic assumptions underpinning the theory remain unchanged, see section 5.1. The equilibrium conditions also remain unchanged, see sections 5.4.3 and 5.5.3 for beams under axial and transverse loads, respectively. The constitutive laws, however, must now reflect the thermal deformation of the beam.

The constitutive laws for beams under axial and transverse loads are discussed in sections 5.4.2 and 5.5.2, respectively. With the addition of thermal effects, eq. (5.14) becomes

$$\sigma_1(x_1, x_2, x_3) = E\,\epsilon_1(x_1, x_2, x_3) - E\alpha\Delta T(x_1, x_2, x_3).$$

Note that in Euler-Bernoulli beam theory, the transverse stress components, σ_2 and σ_3, are assumed to remain much smaller than the axial stress component: $\sigma_2 \ll \sigma_1$ and $\sigma_3 \ll \sigma_1$. In fact, these stress components are assumed to be vanishingly small, leading the reduced version of Hooke's law used here.

In this example, plane $(\bar{\imath}_1, \bar{\imath}_2)$ is assumed to be a plane of symmetry for both the structure and the thermal loading; hence, the deformation of the beam will be entirely contained in this plane. The purely kinematic Euler-Bernoulli assumptions discussed in section 5.1 lead to the same displacement field given by eq. (5.4), which in this case, reduces to $u_1(x_1, x_2, x_3) = \bar{u}_1(x_1) - x_2 \mathrm{d}\bar{u}_2/\mathrm{d}x_1$, $u_2(x_1, x_2, x_3) =$

$\bar{u}_2(x_1)$, and $u_3(x_1, x_2, x_3) = 0$. The corresponding strain field, eq. (5.5), features a single non-vanishing component, the axial strain, which reduces to $\epsilon_1(x_1, x_2, x_3) = \bar{\epsilon}_1(x_1) - x_2\kappa_3(x_1)$. The axial stress now becomes

$$\sigma_1(x_1, x_2, x_3) = E\left[\bar{\epsilon}_1(x_1) - x_2\kappa_3(x_1)\right] - E\alpha\Delta T(x_1, x_2, x_3). \tag{13.43}$$

The axial force in the beam, eq. (5.8), is now readily found as

$$N_1(x_1) = \int_{\mathcal{A}} \left[E\bar{\epsilon}_1 - Ex_2\kappa_3 - E\alpha\Delta T\right]\, d\mathcal{A} = S\bar{\epsilon}_1(x_1) - P_1^e(x_1), \tag{13.44}$$

where S is the axial stiffness of the beam, eq. (5.17), and P_1^e the equivalent thermal axial force defined by eq. (13.39).

As discussed in section 5.5.2, the origin of the axes system is selected to be at the centroid of the cross-section, implying eq. (5.33). The bending moment in the beam, see eq. (5.10), is found to be

$$M_3(x_1) = -\int_{\mathcal{A}} x_2 \left[E\bar{\epsilon}_1 - Ex_2\kappa_3 - E\alpha\Delta T\right]\, d\mathcal{A} = H_{33}^c\kappa_3(x_1) - Q_3^e(x_1), \tag{13.45}$$

where the bending stiffness of the cross-section, H_{33}^c, is given by eq. (5.36), and the equivalent thermal bending moment, Q_3^e, by eq. (13.40).

To complete the theory, the sectional constitutive laws given above are introduced into the equilibrium equations of the beam to find the governing differential equations of the problem. Introducing the axial force into the axial equilibrium equation, eq. (5.18), leads to

$$\frac{d}{dx_1}\left[S\frac{d\bar{u}_1}{dx_1}\right] = -\left[p_1(x_1) + p_1^e(x_1)\right], \tag{13.46}$$

where p_1^e is defined by eq. (13.41). When comparing the above equation with eq. (5.19), it is clear that thermal effects introduce an "equivalent thermal axial load," p_1^e, that is simply added to the externally applied axial load, p_1.

Similarly, introducing the bending moment into the transverse equilibrium equation, eq. (5.39), leads to

$$\frac{d^2}{dx_1^2}\left[H_{33}^c\frac{d^2\bar{u}_2}{dx_1^2}\right] = \left[p_2(x_1) + p_2^e(x_1)\right], \tag{13.47}$$

where p_2^e is defined by eq. (13.42). When comparing the above equation with eq. (5.40), it is clear that thermal effects introduce an equivalent thermal transverse load, p_2^e, that is simply added to the externally applied transverse load, p_2.

Finally, the boundary conditions of the problem must be investigated. Consider a cantilevered beam subjected to a tip axial force, P_1, and a tip bending moment, Q_3. The boundary conditions at the root of the beam are purely kinematic conditions, $\bar{u}_1 = 0$ and $\bar{u}_2 = d\bar{u}_2/dx_1 = 0$, that remain unaffected by thermal effects. On the other hand, the natural boundary conditions at the beam's tip are $N_1 = P_1$,

$V_2 = 0$, and $M_3 = Q_3$, that also remain unaffected by thermal effects. When these tip boundary conditions are expressed in terms of the sectional deformation using the sectional constitutive laws, they become $N_1 = S\bar{\epsilon}_1 - P_1^e = P_1$, and $M_3 = H_{33}^c \kappa_3 - Q_3^e = Q_3$. Expressing the sectional deformations in terms of the displacement field then yields the tip boundary conditions as

$$S\frac{d\bar{u}_1}{dx_1} = [P_1 + P_1^e], \quad \text{and} \quad H_{33}^c \frac{d^2\bar{u}_2}{dx_1^2} = [Q_3 + Q_3^e]. \tag{13.48}$$

Here again, it is clear that thermal effects introduce equivalent thermal axial loads and bending moments, P_1^e and Q_3^e, respectively, that are simply added to the externally applied tip axial force and bending moment.

In conclusion, when the direct method is applied in within the framework of the Euler-Bernoulli assumptions, the governing equations are found to be identical to those of the classical Euler-Bernoulli beam theory, except for the fact that a set of equivalent loads are added to the externally applied loads: the equivalent distributed axial load in eq. (13.46), the equivalent distributed transverse load in eq. (13.47), and the axial force and bending moment at the tip of the beam, see eqs. (13.48). Once this classical beam problem is solved, the axial stresses in the beam are recovered using eq. (13.43). Clearly, this approach is fully consistent with that developed based on the constraint method in section 13.4.3.

Example 13.11. Cantilevered beam under thermal gradient

Consider the cantilevered beam subjected to a parabolic distribution of temperature through the depth, h, of its rectangular cross-section, as shown in fig. 13.21. The temperature of the lower surface of the beam is the reference temperature, $T = 0$, and the temperature of the top surface is $T = T_0$; the temperature profile is $T(x_2) = (x_2/h + 1/2)^2 T_0$.

The equivalent thermal loading shown in fig. 13.21 consists of the end axial load given by eq. (13.39),

$$P_1^e = \int_{\mathcal{A}} E\alpha T \, d\mathcal{A} = \int_{\mathcal{A}} E\alpha \left(\frac{x_2}{h} + \frac{1}{2}\right)^2 T_0 \, d\mathcal{A}$$
$$= bhE\alpha T_0 \int_{-1/2}^{+1/2} \left(\zeta + \frac{1}{2}\right)^2 d\zeta = \frac{bh}{3} E\alpha T_0.$$

and an end bending moment given by eq. (13.40),

$$Q_3^e = -\int_{\mathcal{A}} x_2 E\alpha T \, d\mathcal{A} = -\int_{\mathcal{A}} x_2 E\alpha \left(\frac{x_2}{h} + \frac{1}{2}\right)^2 T_0 \, d\mathcal{A}$$
$$= -bh^2 E\alpha T_0 \int_{-1/2}^{+1/2} \zeta \left(\zeta + \frac{1}{2}\right)^2 d\zeta = -\frac{bh^2}{12} E\alpha T_0,$$

where $\zeta = x_2/h$ is the non-dimensional coordinate through the depth of the beam.

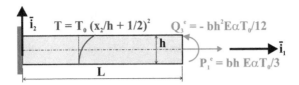

Fig. 13.21. Cantilevered beam under parabolic thermal gradient.

Because the temperature profile is independent of the axial coordinate, the equivalent thermal distributed axial and transverse loads given by eqs. (13.41) and (13.42), respectively, both vanish.

The axial displacement can be computed with the help of eq. (5.16) as $d\bar{u}_1/dx_1 = P_1^e/S = E\alpha T_0/3$, leading to the following displacement field

$$\bar{u}_1 = \frac{\alpha T_0 L}{3}\frac{x_1}{L}.$$

The tip axial deflection of the beam under thermal loading is $\bar{u}_{1(\text{tip})} = \alpha T_0 L/3$.

Similarly, the transverse deflection of the beam is found based on eq. (5.37) as $d^2\bar{u}_2/dx_1^2 = Q_3^e/H_{33}^c = -\alpha T_0/h$, leading to the following transverse displacement field

$$\bar{u}_2 = -\frac{\alpha T_0 L^3}{2h}\left(\frac{x_1}{L}\right)^2.$$

The tip transverse deflection of the beam under the thermal loading is $\bar{u}_{2(\text{tip})} = -\alpha T_0 L^2/(2h)$.

Finally, the stress state is found by superposition of the stresses required to inhibit thermal deformations, $\sigma_1^c = -E\alpha T$, and the stresses, σ_1^e, generated by the equivalent thermal loading, leading to

$$\sigma_1^t = -E\alpha\left(\frac{x_2}{h}+\frac{1}{2}\right)^2 T_0 + \frac{1}{3}E\alpha T_0 + E\alpha T_0\frac{x_2}{h} = \left(\frac{1}{12}-\frac{x_2^2}{h^2}\right)E\alpha T_0.$$

The first term of this expression represents the stresses that inhibit the thermal deformations, the second term is the axial stress distribution associated with the equivalent axial thermal load, P_1^e, and the last term is the axial stress distribution associated with the equivalent thermal moment, Q_3^e. The maximum axial stress is found at the top and bottom edges of the cross-section, $\sigma_{1(\text{top})}^t = \sigma_{1(\text{bot})}^t = -E\alpha T_0/6$, a compressive stress, whereas the axial stress in the middle of the section is $\sigma_{1(\text{mid})}^t = E\alpha T_0/12$, a tensile stress.

Because only thermal loads are applied to the beam, the thermal stress field is a self-equilibrating stress field. This means that although the axial stress does not vanish, the axial force and bending moment at any cross-section do vanish; indeed, it is easy to check that

$$E\alpha T_0\int_{-h/2}^{+h/2}\left(\frac{1}{12}-\frac{x_2^2}{h^2}\right)dx_2 = 0, \quad E\alpha T_0\int_{-h/2}^{+h/2}x_2\left(\frac{1}{12}-\frac{x_2^2}{h^2}\right)dx_2 = 0.$$

*Example 13.12. **Bi-material beam under uniform temperature field***

Consider the cantilevered beam of length L made of two bars, each of width b and height $h/2$, bonded together along the beam's mid-plane, as depicted in fig. 13.22. The top bar is made of a material with Young's modulus E_a and coefficient of thermal expansion α_a, and the corresponding quantities for the lower bar material are E_b and α_b, respectively. After fabrication, the temperature of the entire beam is raised by an amount T. Find the resulting axial and transverse tip deflections.

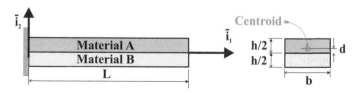

Fig. 13.22. Cantilevered bi-material beam under uniform temperature field.

The axial stiffness of the beam is $S = (E_a + E_b)bh/2$, and the centroid of the cross-section is located at a distance $d/h = (\bar{E}_a - \bar{E}_b)/4$, where $\bar{E}_a = E_a/(E_a + E_b)$ and $\bar{E}_b = E_b/(E_a + E_b)$.

The equivalent thermal loading consists of the end axial load given by eq. (13.39),

$$P_1^e = \int_A E\alpha T \, \mathrm{d}\mathcal{A} = \int_{A_a} E_a\alpha_a T \, \mathrm{d}\mathcal{A}_a + \int_{A_b} E_b\alpha_b T \, \mathrm{d}\mathcal{A}_b = (E_a\alpha_a + E_b\alpha_b)\frac{bhT}{2},$$

and the end bending moment computed with respect to the geometric center of the section, given by eq. (13.40),

$$Q_3^e = -\int_A x_2 E\alpha T \, \mathrm{d}\mathcal{A} = -\int_{A_a} x_2 E_a\alpha_a T \, \mathrm{d}\mathcal{A}_a - \int_{A_b} x_2 E_b\alpha_b T \, \mathrm{d}\mathcal{A}_b$$
$$= -(E_a\alpha_a - E_b\alpha_b)\frac{bh^2T}{8}.$$

Because the temperature profile is independent of variable x_1, the equivalent thermal distributed axial and transverse loads given by eqs. (13.41) and (13.42), respectively, both vanish.

Under the effect of the equivalent thermal axial tip load, the axial displacement can be computed from eq. (5.16), leading to the following displacement field

$$\bar{u}_1 = (\bar{E}_a\alpha_a + \bar{E}_b\alpha_b)TL\eta,$$

where $\eta = x_1/L$ is the non-dimensional variable along the beam's span. The tip axial deflection of the beam's centroid under thermal loading is $\bar{u}_{1(\text{tip})} = (\bar{E}_a\alpha_a + \bar{E}_b\alpha_b)TL$.

Similarly, the transverse deflection of the beam is found using eq. (5.37), leading to the following transverse displacement field

$$\bar{u}_2 = \frac{Q_3^e + dP_1^E}{2H_{33}^c} L^2 \eta^2.$$

Note that the applied equivalent thermal bending moment is computed with respect to the centroid of the section as $Q_3^e + dP_1^e$. The bending stiffness of the cross section, computed with respect to its centroid, is $H_{33}^c = bh^3(E_a^2 + E_b^2 + 14E_a E_b)/96(E_a + E_b)$. With these results, the transverse displacement field of the beam becomes

$$\bar{u}_2 = -12\frac{E_a E_b(\alpha_a - \alpha_b)TL}{E_a^2 + E_b^2 + 14E_a E_b}\frac{L}{h}\eta^2.$$

Finally, the thermal stresses can be computed using the process described in the previous example.

13.4.4 Problems

Problem 13.10. Bi-material beam under uniform temperature field
Consider the cantilevered beam of length L made of two half-beam, each of width b and height $h/2$, welded together along the beam's mid-plane, as depicted in fig. 13.22. The top beam is made of a material with Young's modulus E_a and coefficient of thermal expansion α_a, whereas the corresponding quantities for the lower beam material are E_b and α_b, respectively. After assembly, the uniform temperature of the beam is raised by an amount T. *(1)* Find the axial force, N_1, transverse shear force, V_2, and bending moment, M_3, at the beam's mid-span. *(2)* Consider now the same bi-material beam, but fully restrained at both ends. Find the axial force, N_1, transverse shear force, V_2, and bending moment, M_3, at the beam's mid-span.

Problem 13.11. Fully restrained beam under parabolic temperature field
Consider a beam fully restrained at both ends and subject to a parabolic temperature distribution. This problem is similar to that depicted in fig. 13.21, except that the beam is now fully restrained at both ends. *(1)* Find the axial and transverse deflection of the beam under the thermal loading. *(2)* Find the thermal stress distribution in the beam. *(3)* Is this state of stress self-equilibrating?

Problem 13.12. Non-uniform fully restrained beam under parabolic temperature field
Consider the beam with a sudden change in cress-section geometry at an intermediate point and fully restrained at both ends as depicted in fig 13.23. The left portion of the beam is of length L_1 and the rectangular cross-section has a width b and height h_1, whereas the corresponding dimensions for the right portion of the beam are L_2, b and h_2, respectively. Both portions of the beam are subjected to a parabolic thermal gradient, as indicated on the figure. *(1)* Find the axial and transverse deflection fields for the beam under the thermal loading. *(2)* Plot the non-dimensional transverse displacement field, $\bar{u}_2/(L\alpha T_0)$, over the beam. *(3)* Plot the non-dimensional bending moment, $M_3/(E\alpha T_0 bL^2)$, over the beam. *(4)* Plot the non-dimensional shear force, $V_2/(E\alpha T_0 bL)$, over the beam. *(5)* Find the thermal stress distributions in the beam. *(6)* Plot the non-dimensional axial stress distribution, $\sigma_1/(E\alpha T_0)$, at $x_1 = L_1/2$ and $x_1 = L - L_2/2$. Use the following data: $\hat{h}_1 = h_1/L = 0.05$; and $\hat{h}_2 = h_2/L = 0.03$; $L = L_1 + L_2$. Consider two cases. *Case A*: $\eta_1 = L_1/L = 0.3$; $\eta_2 = L_2/L = 0.7$ and *Case B*: $\eta_1 = 0.7$; $\eta_2 = 0.3$.

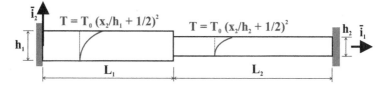

Fig. 13.23. Clamped-clamped beam subjected to parabolic thermal gradients.

Problem 13.13. Non-uniform simply supported beam under parabolic temperature field

Consider the simply supported beam with a sudden change in cress-section geometry, as depicted in fig 13.24. The left portion of the beam is of length L_1 and the rectangular cross-section has a width b and height h_1, whereas the corresponding dimensions for the right portion of the beam are L_2, b and h_2, respectively. Both portions of the beam are subjected to a parabolic thermal gradient, as indicated on the figure. *(1)* Find the axial and transverse deflection fields for the beam under the thermal loading. *(2)* Plot the non-dimensional transverse displacement field, $\bar{u}_2/(L\alpha T_0)$, over the beam. *(3)* Plot the non-dimensional bending moment, $M_3/(E\alpha T_0 b L^2)$, over the beam. *(4)* Plot the non-dimensional shear force, $V_2/(E\alpha T_0 b L)$, over the beam. *(5)* Find the thermal stress distributions in the beam. *(6)* Plot the non-dimensional axial stress distribution, $\sigma_1/(E\alpha T_0)$, at $x_1 = L_1/2$ and $x_1 = L - L_2/2$. Use the following data: $\hat{h}_1 = h_1/L = 0.05$; and $\hat{h}_2 = h_2/L = 0.03$; $L = L_1 + L_2$. Consider two cases. *Case A:* $\eta_1 = L_1/L = 0.3$; $\eta_2 = L_2/L = 0.7$ and *Case B:* $\eta_1 = 0.7$; $\eta_2 = 0.3$.

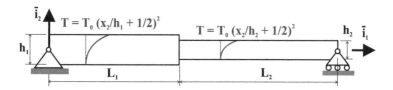

Fig. 13.24. Simply supported beam subjected to parabolic thermal gradients.

Problem 13.14. Bi-material cantilevered beam

Consider the cantilevered beam of length L made of two half-beam, each of width b and height $h/2$, welded together along the beam's mid-plane, as depicted in fig. 13.22. The top beam is made of a material of Young's modulus E_a and coefficient of thermal expansion α_a, whereas the corresponding quantities for the lower beam material are E_b and α_b, respectively. The entire assembly is heated to a uniform temperature T. *(1)* Find the transverse displacement field for the bi-material beam. *(2)* On a single graph, plot the transverse displacement field, $\bar{u}_2/(TL)$, for the six combinations of materials chosen from the materials listed in table 13.1. *(3)* What is the best combination of materials if the beam's tip deflection per degree of heating is to be maximized? *(4)* Find the thermal stress distribution in the beam. *(5)* On a single graph, plot the axial stress distribution, σ_1/T, over the cross-section the six combinations of materials chosen from the materials listed in table 13.1. *(6)* Find the location of the maximum axial stress. For what combination of materials are the thermal stresses maximized? *(7)* Check that the thermal stress field is a self-equilibrating stress field. Use $L/h = 10$.

Problem 13.15. Thermal effects in beams with unsymmetric cross-section

In example 13.10, the governing equations for beams with symmetric cross-sections are derived based on the direct method and Euler-Bernoulli kinematic assumptions. Generalize the governing equations for beam with unsymmetric cross-sections developed in chapter 6 to the case where such beams are subjected to arbitrary thermal gradients. Note that for such problems the kinematic description given in section 6.1 remains unchanged, and the equilibrium equations of the problem, see section 6.3, are still valid as well. However, the sectional constitutive laws derived in section 6.2 must be updated to accommodate thermal effects. *(1)* Derive the governing equations when principal centroidal axes of bending are used. *(2)* Derive the governing equations when centroidal axes are used that are not aligned with the principal axes of bending.

Consider now the problem of a cantilevered beam with a "Z" section as treated in example 6.6. Figure 13.25 shows the cantilevered beam subjected to the following temperature field: the top and bottom flanges are at temperatures T_t and T_b, respectively, whereas the temperature profile in the vertical web is $T = (T_t - T_b)\zeta/2 + (T_t + T_b)/2$, where $\zeta = x_3/a$. *(3)* Find the tip deflections of the beam. *(4)* Determine and plot the axial stress distribution over the cross-section of the beam. Use the following data: $T_t = \lambda T_b$, $\lambda = 3$.

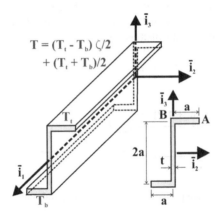

Fig. 13.25. Cantilevered beam subjected to thermal gradient.

14

Buckling of beams

When a structure is subjected to loading, it can fail because local stresses exceed the maximum allowable stress for the material. There exist, however, another type of failure mode where the entire structure suddenly collapses. The critical value of the applied load that triggers this failure mode primarily depends on the geometry of the structure and the stiffness of the material, not its strength. The study of this catastrophic failure mode is known as the theory of *elastic stability*.

The best known problem of elastic stability undoubtedly is the transverse buckling of a beam. Consider a straight cantilevered beam subjected to an end axial compressive load. If this load is applied at the centroid of the cross-section of the beam, it creates only an axial straining of the beam. As the axial compressive load is increased, a critical value is reached when the beam buckles sideways and collapses.

The basic concepts involved in the study of elastic stability will be introduced with the simple problem of a rigid bar with a root torsional spring subjected to a compressive load. This will serve as an introduction to the problem of buckling of beams. Beams, when subjected to compressive axial loads, are often called *columns*, although this designation will not be used in this text.

14.1 Rigid bar with root torsional spring

14.1.1 Analysis of a perfect system

Consider a rigid bar of length L articulated at the root as depicted in fig. 14.1. A torsional spring of constant k is acting at the root, and is un-stretched when the bar is vertical. A compressive axial load, P, constantly acting in the vertical direction, is applied at the bar's tip. Let the lateral deflection of the bar be defined by angle θ measured from the vertical. The equilibrium of moments about point **O** implies

$$M - k\theta = 0, \tag{14.1}$$

where M is the moment of the applied load, and $k\theta$ the elastic restoring force in the spring. The moment of the applied load is $M = PL\sin\theta$ resulting in the following

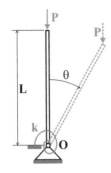

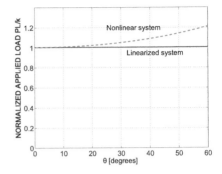

Fig. 14.1. Bar with a root torsional spring.   **Fig. 14.2.** Behavior of a perfect system.

equilibrium equation

$$PL \sin \theta - k\theta = 0 \tag{14.2}$$

In the theory of elastic stability, the determination of the critical load for the onset of instability plays a central role. Since the elastic spring is un-stretched when $\theta = 0$, the rigid bar will remain vertical as the applied load P is increased. At the onset of buckling, the bar will start to move sideways and θ will increase. If the onset of buckling is to be determined, angle θ can be assumed to be a small quantity,

$$\sin \theta \approx \theta. \tag{14.3}$$

Hence, eq. (14.2) becomes

$$(PL - k)\theta = 0 \tag{14.4}$$

This equation of equilibrium is satisfied if $\theta = 0$. This represents, however, the trivial solution where the bar remains vertical. Indeed, when the bar is vertical, the line of action of the applied force passes through point **O**, the moment in the spring vanishes, and equation (14.1) is then identically satisfied for any value of the applied load, P.

An important point is to determine whether equation (14.4) admits a non-trivial solution. In fact, the *buckling load*, sometimes called the *critical load* of the system and denoted P_{cr}, is defined as that load for which a non-trivial solution of eq. (14.4) exists. Clearly, if $(PL - k) = 0$,

$$P_{cr} = \frac{k}{L}. \tag{14.5}$$

When $P = P_{cr}$, equation of equilibrium, eq. (14.4), is satisfied for an arbitrary value of angle θ. The behavior of the system is depicted in fig. 14.2. For $P < P_{cr}$, the only solution of eq. (14.4) is $\theta = 0$. When the applied load reaches the buckling load, *i.e.*, when $P = P_{cr}$, another solution of eq. (14.4) exists for which angle θ is arbitrary, and this is shown by the horizontal line labeled "linearized system."

Strictly speaking, the solution described in fig. 14.2 is only valid for small θ, because assumption (14.3) is made. When θ becomes large, the post-buckling range starts, and eq. (14.2) must be used. This equation is recast as

$$\frac{P}{P_{cr}} = \frac{\theta}{\sin\theta} \tag{14.6}$$

This nonlinear relationship is also depicted in fig. 14.2. Both linearized and nonlinear solutions are in close agreement for small angles. The buckling load characterizes the onset of buckling, *i.e.*, the loading for which lateral displacement begins.

From a design standpoint, it is often imperative to keep the applied load well below the buckling load, because the collapse of the structure at the buckling load is a sudden and catastrophic phenomenon. The buckling load depends on k, the spring stiffness constant, and L, the length of the bar. The strength of the system components are irrelevant in this analysis.

14.1.2 Analysis of an imperfect system

The system considered above is a *perfect system* in the sense that the rigid bar is perfectly straight, the line of action of the applied load exactly passes through the pivot point, and the un-stretched position of the spring corresponds to $\theta = 0$. In practical situations, however, a certain level of *imperfection* is always present. The actual imperfection of the system is often unknown, as it comes from manufacturing inaccuracies or load misalignment.

A convenient way of introducing imperfection in the system is to assume that the un-stretched position of the spring corresponds to $\theta = \theta_0$, where θ_0 is now a measure of the initial imperfection of the system. The equilibrium eq. (14.2) then becomes

$$PL\sin\theta - k(\theta - \theta_0) = 0. \tag{14.7}$$

The onset of buckling can be determined assuming θ to be a small quantity, implying (14.3), to find

$$(PL - k)\theta = -k\theta_0. \tag{14.8}$$

This equation possesses a non-vanishing right-hand side, in contrast with the homogeneous equation, eq. (14.2), for the perfect system. The solution of eq. (14.8) is

$$\theta = \frac{\theta_0}{1 - P/P_{cr}} \tag{14.9}$$

The response of the system is very different from that of the perfect system and is depicted in fig. 14.3 for various levels of the initial imperfection, θ_0. Unlike the perfect system, rotation of the bar begins from the onset of loading and grows as the load is increased. When the initial imperfection is very small, the response of the system is very small except when the applied load approaches P_{cr}. When the applied load reaches P_{cr} the response rapidly grows to infinity. For larger initial imperfections, a large response of the system is observed even for applied loads well below the critical load.

In all cases, the response of the system is unbounded when the applied load approaches the buckling load, P_{cr}. In fact, the buckling load can be defined as the load for which the response of an initially imperfect system grows without bounds.

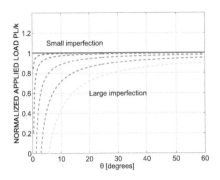

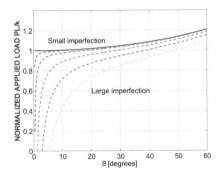

Fig. 14.3. System linearized response with various levels of imperfection.

Fig. 14.4. System nonlinear response with various levels of imperfection.

For the simple system discussed here, the analysis of both *perfect* and *imperfect* systems give the same buckling load, eq. (14.5).

The analysis developed here is valid for small values of θ. For larger values of θ equation (14.7) must be used. It can be recast as

$$\frac{P}{P_{cr}} = \frac{\theta - \theta_0}{\sin \theta} \tag{14.10}$$

This relationship is depicted in fig. 14.4 for various levels of the initial imperfection. For angles less than 10 degrees, the curves are nearly identical to those in fig. 14.3, but as the angle increases, the curves bend upwards and asymptotically approach the behavior of the perfect system at large angles as shown in fig. 14.2

Although the buckling loads obtained for the *perfect* and *imperfect* systems are the same, their respective behaviors are quite different for applied loads below the buckling load. Indeed, comparing figs. 14.2 and 14.3 shows that the *perfect* system presents no lateral deflection for applied loads below P_{cr}, whereas the *imperfect* system presents lateral deflections even for small applied load. Failure can occur for applied loads far smaller that the buckling load. Assuming that the spring fails when $\theta = \theta_{fail}$, the failure load is readily computed from eq. (14.9) as

$$\frac{P_{fail}}{P_{cr}} = 1 - \frac{\theta_0}{\theta_{fail}}. \tag{14.11}$$

Note the fundamental difference between the failure load associated with a displacement or stress reaching an allowable limit for the material ($\theta = \theta_{fail}$ in this simple example), and the buckling load associated with the instability of a structure, which depends solely on its elastic and geometric characteristics.

In this simple example, two conceptually different definitions of the buckling load are given. First, the buckling load is defined as the critical load for which a *perfect* system admits a non-trivial solution. Second, the buckling load is defined as the load for which the response of an *imperfect* system grows without bounds. Although conceptually different, these two definitions give the same value of the buckling load for the simple rigid bar problem considered here.

14.2 Buckling of beams

Consider now the simply-supported, uniform beam acted upon by a compressive load, P, depicted in fig. 14.5. This loading is assumed to be applied exactly at the centroid of the section. According to three dimensional beam theory, all the conditions for decoupling the problem are met. This means that three simpler, independent problems can be solved: first, an extensional problem giving rise to extension of the beam (compression in this case), and second, two uncoupled bending problems along the principal axes of bending. Because the axial load is applied exactly at the centroid and no transverse loads are applied, the beam does not bend. Consequently, the transverse deflections will remain zero, independently of the level of the applied axial load P.

Fig. 14.5. Simply-supported beam with end compressive load.

This simple reasoning shows that the beam theory developed in the previous chapters is unable to predict the buckling phenomenon. The basic equations of Euler-Bernoulli beam theory, eqs. (5.19), and (5.40) must be modified to account for the effect of the large compressive load that causes the instability. The key to this modification is a restatement of the equilibrium equations for a deformed configuration of the beam.

14.2.1 Equilibrium equations

Consider an infinitesimal slice of the beam subjected to axial and transverse forces, as well as bending moments. In contrast with earlier developments, see sections 5.4.3 or 5.5.3, a free body diagram of a *deformed slice of the beam* will be analyzed. Figure 14.6 shows the deformed slice of the beam subjected to the same force and moment components considered in chapter 5. At the onset of buckling, a large axial force is present in the beam, but the transverse shear force and deflection are still very small. Consequently, the following assumption will be made

$$N_1 \gg V_2; \quad \frac{d\bar{u}_2}{dx_1} \ll 1. \tag{14.12}$$

The first equilibrium equation is obtained by summing up forces along axis $\bar{\imath}_1$ to find

$$-N_1 \cos \frac{d\bar{u}_2}{dx_1} + \left(N_1 + \frac{dN_1}{dx_1} dx_1 \right) \cos \left(\frac{d\bar{u}_2}{dx_1} + \frac{d^2\bar{u}_2}{dx_1^2} dx_1 \right)$$

$$+V_2 \sin \frac{d\bar{u}_2}{dx_1} - \left(V_2 + \frac{dV_2}{dx_1} dx_1 \right) \sin \left(\frac{d\bar{u}_2}{dx_1} + \frac{d^2\bar{u}_2}{dx_1^2} dx_1 \right) + p_1 dx_1 = 0.$$

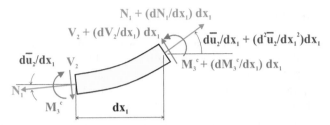

Fig. 14.6. Free body diagram of a deformed slice of the beam.

In view of assumption (14.12), $\cos(\mathrm{d}\bar{u}_2/\mathrm{d}x_1) \approx 1$, and $\sin \mathrm{d}\bar{u}_2/\mathrm{d}x_1 \approx \mathrm{d}\bar{u}_2/\mathrm{d}x_1$. Neglecting higher order differential terms in this equations, it reduces to

$$\frac{\mathrm{d}}{\mathrm{d}x_1}\left(N_1 - V_2\frac{\mathrm{d}\bar{u}_2}{\mathrm{d}x_1}\right) = -p_1.$$

Finally, assumption (14.12) implies that the second term in the parentheses is negligible as compared to the first, resulting in the following axial equilibrium equation $\mathrm{d}N_1/\mathrm{d}x_1 = -p_1$. This equation of axial force equilibrium is identical to that obtained when considering a differential element of the beam in its original, reference configuration, eq. (5.19).

The second equation of equilibrium is obtained by summing up the forces along axis $\bar{\imath}_2$ to find

$$-V_2 \cos\frac{\mathrm{d}\bar{u}_2}{\mathrm{d}x_1} + \left(V_2 + \frac{\mathrm{d}V_2}{\mathrm{d}x_1}\mathrm{d}x_1\right)\cos\left(\frac{\mathrm{d}\bar{u}_2}{\mathrm{d}x_1} + \frac{\mathrm{d}^2\bar{u}_2}{\mathrm{d}x_1^2}\mathrm{d}x_1\right)$$
$$-N_1 \sin\frac{\mathrm{d}\bar{u}_2}{\mathrm{d}x_1} + \left(N_1 + \frac{\mathrm{d}N_1}{\mathrm{d}x_1}\mathrm{d}x_1\right)\sin\left(\frac{\mathrm{d}\bar{u}_2}{\mathrm{d}x_1} + \frac{\mathrm{d}^2\bar{u}_2}{\mathrm{d}x_1^2}\mathrm{d}x_1\right) + p_2\mathrm{d}x_1 = 0.$$

Here again, this equation simplifies considerably when assumptions (14.12) are taken into account, leading to

$$\frac{\mathrm{d}}{\mathrm{d}x_1}\left(V_2 + N_1\frac{\mathrm{d}\bar{u}_2}{\mathrm{d}x_1}\right) = -p_2.$$

The two terms in the parentheses are now of the same order of magnitude. The equation cannot be further simplified and is recast as

$$\frac{\mathrm{d}V_2}{\mathrm{d}x_1} + \frac{\mathrm{d}}{\mathrm{d}x_1}\left(N_1\frac{\mathrm{d}\bar{u}_2}{\mathrm{d}x_1}\right) = -p_2. \tag{14.13}$$

This equilibrium equation differs from that derived earlier, eq. (5.38). The second term represents the contribution of the large axial force N_1 to the transverse equilibrium of the beam. The presence of this term stems from the fact that equilibrium conditions are expressed for a differential element of the beam in its deformed configuration.

The last equilibrium equation is obtained by summing the moments about axis $\bar{\imath}_3$ to find

$$-M_3^c + \left(M_3^c + \frac{\mathrm{d}M_3^c}{\mathrm{d}x_1}\mathrm{d}x_1\right) + V_2\mathrm{d}x_1 = 0. \qquad (14.14)$$

After simplification this equation becomes $\mathrm{d}M_3^c/\mathrm{d}x_1 + V_2 = 0$, and is identical to that obtained earlier as eq. (5.38).

The governing equation of the problem is obtained by eliminating the shear force, V_2, from eqs. (14.13) and (5.38), then introducing the constitutive law for the bending moment, eq. (5.37), to find

$$\frac{\mathrm{d}^2}{\mathrm{d}x_1^2}\left[H_{33}^{*c}\frac{\mathrm{d}^2\bar{u}_2}{\mathrm{d}x_1^2}\right] - \frac{\mathrm{d}}{\mathrm{d}x_1}\left[N_1\frac{\mathrm{d}\bar{u}_2}{\mathrm{d}x_1}\right] = p_2. \qquad (14.15)$$

The second term in this equation is a new term which is absent in previous developments, see eq. (5.40). The governing equation is now a fourth order differential equation for the transverse displacement field.

Four boundary conditions are required, two at each end of the beam. The boundary conditions are identical to those discussed in section 5.5.4, except when a large axial force is applied at an unsupported end of the beam. In this case, the natural boundary conditions must be derived from equilibrium of the beam in its deformed configuration. Consider, for instance, a cantilevered beam with an end tip load P_1 acting in a fixed, horizontal direction, as depicted in fig. 14.7. The boundary conditions at the tip end are

$$M_3^c = 0; \quad V_2 = -P_1\frac{\mathrm{d}\bar{u}_2}{\mathrm{d}x_1}. \qquad (14.16)$$

Governing equation (14.15) can be applied when large axial forces are present in the beam, whether in compression (N_1 is negative) or in tension (N_1 is positive). If the axial force is compressive, the beam will buckle when this compressive force reaches a critical level; if the axial force is tensile, the beam will not buckle, although its behavior will be affected by the presence of the large axial force which will tend to reduce the transverse deflection.

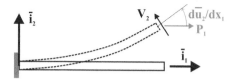

Fig. 14.7. Cantilevered beam with a tip load.

14.2.2 Buckling of a simply-supported beam (equilibrium approach)

Consider a uniform, simply-supported beam subjected to an end compressive load of magnitude P, as shown in fig. 14.5. The axial force in the beam, N_1, is constant along its span, and $N_1 = -P$. Assuming axis $\bar{\imath}_2$ to be a principal axis of bending, the uncoupled governing equation that accounts for the presence of the large compressive load is

$$H_{33}^{*c}\frac{\mathrm{d}^4\bar{u}_2}{\mathrm{d}x_1^4} + P\frac{\mathrm{d}^2\bar{u}_2}{\mathrm{d}x_1^2} = 0. \qquad (14.17)$$

with the following boundary conditions, $\bar{u}_2 = \mathrm{d}^2\bar{u}_2/\mathrm{d}x_1^2 = 0$, at the beam's root, and $\bar{u}_2 = \mathrm{d}^2\bar{u}_2/\mathrm{d}x_1^2 = 0$ at its tip. The governing equations and associated boundary conditions are homogeneous. Hence, the trivial solution, $\bar{u}_2 \equiv 0$, is a solution of the problem. The buckling load is defined as the lowest load for which a non-trivial solution of the governing equations exists.

For simplicity, the governing equation is recast in a non-dimensional form as

$$\bar{u}_2'''' + \lambda^2\bar{u}_2'' = 0, \qquad (14.18)$$

where

$$\lambda^2 = \frac{PL^2}{H_{33}^{*c}}, \qquad (14.19)$$

is a non-dimensional loading parameter, and $(\cdot)'$ denotes a derivative with respect to the non-dimensional span-wise variable $\eta = x_1/L$. The boundary conditions at the beam's root are $\bar{u}_2 = \bar{u}_2'' = 0$ and at its tip, $\bar{u}_2 = \bar{u}_2'' = 0$.

The solution of eq. (14.18) is

$$\bar{u}_2 = A + B\eta + C\cos\lambda\eta + D\sin\lambda\eta, \qquad (14.20)$$

where A, B, C, and D, are four integration constants to be determined from the boundary conditions. The two root boundary conditions imply $A+C = 0$ and $\lambda^2 C = 0$, which in turn, implies $A = C = 0$. The solution now reduces to $\bar{u}_2 = B\eta + D\sin\lambda\eta$. The two tip boundary conditions yield

$$\begin{bmatrix} 1 & \sin\lambda \\ 0 & \sin\lambda \end{bmatrix}\begin{Bmatrix} B \\ D \end{Bmatrix} = 0. \qquad (14.21)$$

This is a set of algebraic equations for the last two integration constants, B and D. Since the system is homogeneous, the solution is $B = D = 0$, which corresponds to the trivial solution of the problem. A linear system of homogeneous algebraic equations admits a non-trivial solution if and only if the determinant of the system vanishes, *i.e.*, when

$$\det\begin{bmatrix} 1 & \sin\lambda \\ 0 & \sin\lambda \end{bmatrix} = 0. \qquad (14.22)$$

Expanding this determinant yield the *buckling equation*

$$\sin\lambda = 0. \qquad (14.23)$$

The governing equation of the problem admits non-trivial solutions for the discrete values, λ_n, which satisfy eq. (14.23), and the lowest solution is the buckling load. The roots of eq. (14.23) are $\lambda_n = n\pi$, $n = 1, 2, ...\infty$, which, in view of eq. (14.19), can be written as

$$(P_{\text{cr}})_n = \frac{n^2\pi^2 H_{33}^{*c}}{L^2}; \quad n = 1, 2, ...\infty. \tag{14.24}$$

The lowest root corresponds to $n = 1$, resulting in a buckling load

$$P_{\text{cr}} = \frac{\pi^2 H_{33}^{*c}}{L^2}. \tag{14.25}$$

The deflected shape of the beam can also be determined from this analysis. The first equation in (14.21) is $BL + D\sin\lambda = 0$, which, in view of eq. (14.23), yields $B = 0$. The deflected shape is then $\bar{u}_2 = D\sin\lambda_n\eta$. The *buckling mode shape* corresponding to the lowest buckling load is now

$$\bar{u}_2 = D\sin\pi\eta. \tag{14.26}$$

When the beam is loaded, its transverse displacement remains zero until $P = P_{\text{cr}}$, at which point it buckles with the transverse deflection assuming the sinusoidal shape given by eq. (14.26). The integration constant, D, remains undetermined, *i.e.*, the transverse displacement is of arbitrary amplitude, indicating a lateral collapse of the beam. This is the same fundamental behavior presented by the rigid bar examined in section 14.1, and the unknown deflection amplitude is a direct result of the inherent linearization in the governing differential equation, eq. (14.15). A more precise description of the behavior of the structure at and beyond buckling cannot be determined within the framework of this linearized theory.

The above analysis is based on the uncoupled governing equation (14.17) pertaining to bending in plane $(\bar{\imath}_1, \bar{\imath}_2)$. A buckling analysis using the uncoupled governing equation in plane $(\bar{\imath}_1, \bar{\imath}_3)$ could be performed in the same manner, leading to the following buckling load

$$P_{\text{cr}} = \frac{\pi^2 H_{22}^{*c}}{L^2}, \tag{14.27}$$

assuming, of course that the boundary conditions in planes $(\bar{\imath}_1, \bar{\imath}_2)$ and $(\bar{\imath}_1, \bar{\imath}_3)$ are identical. For the case at hand, this would mean that the beam is pinned in both directions by a ball-and-socket joint. In such cases, the buckling load is based on the lowest principal bending stiffness, and the buckling load, called the *Euler buckling load* is

$$P_{\text{Euler}} = \frac{\pi^2 H^{*c}}{L^2}, \tag{14.28}$$

where H^{*c} is the lowest principal centroidal bending stiffness. This buckling load is proportional to this lowest principal centroidal bending stiffness, and inversely proportional to the square of the beam length.

14.2.3 Buckling of a simply-supported beam (imperfection approach)

In the previous section, a *perfect system* is analyzed. The beam is perfectly uniform and straight, and the line of action of the load passes exactly through the centroid. In practical situations, no system is ever perfect: manufacturing imperfections always result in non-uniform beams, and no experimental set-up can apply a compressive load exactly at the centroid. Imperfections are always present, although their exact nature and magnitude are generally unknown.

To investigate the effect of these imperfections, a pinned-pinned beam with eccentrically applied end compressive loads will be analyzed. Figure 14.8 depicts the geometry of the problem, and it shows the compressive loads are applied at a distance e from the centroid. The governing equation of the problem is eq. (14.17) once more. The boundary conditions at the beam's root now become $\bar{u}_2 = 0$, $H_{33}^{*c}\mathrm{d}^2\bar{u}_2/\mathrm{d}x_1^2 = -Pe$ and at its tip, $\bar{u}_2 = 0$, $H_{33}^{*c}\mathrm{d}^2\bar{u}_2/\mathrm{d}x_1^2 = -Pe$.

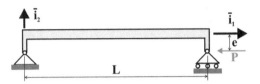

Fig. 14.8. Pinned-pinned beam with eccentric end loads.

The present problem is no longer homogeneous, because non-zero right-hand side terms appear in the boundary conditions. The trivial solution, $\bar{u}_2(x_1) \equiv 0$, is not a solution of the problem. Introducing once more the non-dimensional span and loading variables, η and λ, respectively, the differential equation can be written as eq. (14.18). The beam's root boundary conditions now become $\bar{u}_2 = 0$, $\bar{u}_2'' = -\lambda^2 e$ and at its tip, $\bar{u}_2 = 0$, $\bar{u}_2'' = -\lambda^2 e$.

The solution of eq. (14.18) is once more given by eq. (14.20). The root boundary conditions imply $A + C = 0$ and $-\lambda^2 C = -\lambda^2 e$, which yields $-A = C = e$. Proceeding with the tip boundary conditions yields $B + D\sin\lambda = e(1 - \cos\lambda)$ and $-\lambda^2 D\sin\lambda = -e\lambda^2(1 - \cos\lambda)$. Solving for the last two integration constants gives $D = e(1 - \cos\lambda)/\sin\lambda$ and $B = 0$. The final solution for the transverse displacement then becomes

$$\bar{u}_2 = e\left[\frac{1 - \cos\lambda}{\sin\lambda}\sin\lambda\eta + \cos\lambda\eta - 1\right] = e\left[\frac{\cos\lambda(\eta - 1/2)}{\cos\lambda/2} - 1\right]. \quad (14.29)$$

The buckling load for the structure can be defined as the load for which the response of an initially imperfect structure becomes unbounded. Clearly, $\bar{u}_2 \to \infty$ when $\cos\lambda/2 \to 0$, *i.e.*, when $\lambda = (2n - 1)\pi$, $n = 1, 2, ...\infty$. Using eq. (14.19), the dimensional buckling load now becomes

$$(P_{\mathrm{cr}})_n = \frac{(2n - 1)^2\pi^2 H_{33}^{*c}}{L^2}; \quad n = 1, 2, ... \infty. \quad (14.30)$$

The lowest buckling load corresponds to $n = 1$, leading to the same result given for the perfect beam in eq. (14.25). The solutions for the perfect and imperfect simply supported beams give the same buckling load.

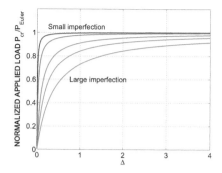

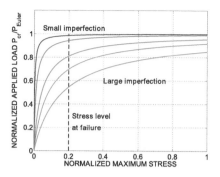

Fig. 14.9. Mid-span transverse deflection of an imperfect pinned-pinned beam for various levels of imperfection.

Fig. 14.10. Maximum stress in an imperfect pinned-pinned beam versus applied compressive load.

Insight into the behavior of imperfect structures can be gained by computing the mid-span deflection, Δ, of the beam. With $\eta = 1/2$, eq. (14.29) becomes

$$\Delta = \bar{u}_2\left(\eta = \frac{1}{2}\right) = e\left(\frac{1}{\cos \lambda/2} - 1\right)$$

The non-dimensional loading parameter is expressed in terms of the applied compressive load with the help of eqs. (14.19) and (14.28)

$$\lambda = \left(\frac{PL^2}{H^{*c}}\right)^{1/2} = \left(\frac{\pi^2 P}{P_{\text{Euler}}}\right)^{1/2} = \pi \left(\frac{P}{P_{\text{Euler}}}\right)^{1/2}. \tag{14.31}$$

The non-dimensional mid-span transverse deflection now becomes

$$\frac{\Delta}{e} = \frac{1}{\cos\left(\frac{\pi}{2}\sqrt{P/P_{\text{Euler}}}\right)} - 1 = \sec\left[\cos\left(\frac{\pi}{2}\sqrt{P/P_{\text{Euler}}}\right)\right] - 1. \tag{14.32}$$

Figure 14.9 depicts the mid-span deflection as a function of the applied compressive load, P, for various levels of the initial imperfection e. For all levels of imperfection, the mid-span deflection grows to infinity as the applied load approaches the buckling load. Although large mid-span deflections can occur for applied load levels far smaller than the buckling load when large imperfections are present, the buckling load itself is unaffected by the presence of imperfections.

The bending moment distribution in the beam is obtained from the transverse displacement field, eq. (14.29), as

$$M_3^c = H_{33}^{*c}\frac{\mathrm{d}^2\bar{u}_2}{\mathrm{d}x_1^2} = Pe\frac{\cos \lambda(\eta - \frac{1}{2})}{\cos \lambda/2}. \tag{14.33}$$

The maximum bending moment occurs in the middle of the beam and is $M_{\max} = Pe/\cos(\lambda/2)$. Considering a beam with a rectangular cross-section of height, h, made of a homogeneous material with a modulus of elasticity, E, the corresponding maximum axial (compressive) stress in the beam is the sum of the uniform stress due to the axial force and of that arising from the bending moment,

$$
|\sigma_{\max}| = \frac{P}{A} + E\frac{M_{\max}h/2}{H_{33}^{*c}} = \frac{P}{A}\left[1 + \frac{1}{2}\frac{e}{h}\frac{Sh^2}{H_{33}^{*c}}\sec\left(\frac{\pi}{2}\sqrt{\frac{P}{P_{\text{Euler}}}}\right)\right], \quad (14.34)
$$

where Sh^2/H_{33}^{*c} is a non-dimensional ratio of the axial to bending stiffnesses, and e/h a measure of the imperfection's magnitude.

The beam deflects until the maximum axial stress reaches the allowable stress level, σ_{allow},

$$
\frac{\sigma_{\text{allow}}A}{P_{\text{Euler}}} = \frac{P_{\text{allow}}}{P_{\text{Euler}}}\left[1 + \frac{1}{2}\frac{e}{h}\frac{Sh^2}{H_{33}^{*c}}\sec\left(\frac{\pi}{2}\sqrt{\frac{P_{\text{allow}}}{P_{\text{Euler}}}}\right)\right]. \quad (14.35)
$$

This is a transcendental equation for $P_{\text{allow}}/P_{\text{Euler}}$ and must be solved numerically. For structures with large imperfections, the allowable load, P_{allow}, can be substantially lower than the buckling load, P_{Euler}, as depicted in fig. 14.10.

14.2.4 Work done by the axial force

The analysis of the buckling behavior of simply-supported beams is developed in the previous section as the solution of a differential equation. The buckling load is a solution of the buckling equation, eq. (14.23), which is, in general, a transcendental equation. In view of the difficulties associated with this solution process, an alternative approach, such as an energy approach, is desirable. General procedures for obtaining approximate predictions of the static deflection of beams under transverse loading are developed in chapter 11, based on the principle of virtual work and the principle of minimum total potential energy. The latter approach requires the evaluation of the strain energy stored in the elastic system and of the work done by the externally applied forces.

Consider the problem of a cantilevered beam subjected to an axial force. When dealing with buckling problems, the strain energy stored in the deformed beam is a function of the sole curvature, as given by eq. (10.39). The work done by the axial force that causes buckling is a key aspect of the problem.

Figure 14.11 shows a differential element of the beam subjected to an axial force, N_1. In the deformed, buckled configuration of the beam, the point of application of the axial force displaces an amount δ along the line of action of the force. Consequently, the work, dW, done by the axial force is $dW = -N_1\delta$; the minus sign reflects the fact that displacement δ occurs in the direction opposite to that of the force. Figure 14.11 indicates that $a + \delta = dx_1$. Because the bending and axial loading problems are assumed to be decoupled, it follows that when undergoing bending deformation, the length of the beam remains unchanged. If the differential beam element is inextensible, $a = dx_1 \cos(d\bar{u}_2/dx_1)$, and hence,

$$\delta = \mathrm{d}x_1 - \mathrm{d}x_1 \cos\frac{\mathrm{d}\bar{u}_2}{\mathrm{d}x_1} = \mathrm{d}x_1 - \mathrm{d}x_1\left[1 - \frac{1}{2}\left(\frac{\mathrm{d}\bar{u}_2}{\mathrm{d}x_1}\right)^2 + \ldots\right], \qquad (14.36)$$

where the second equality results from a Taylor series expansion of the cosine function. If the slope of the beam is assumed to remain much smaller than unity, see eq. (14.12), the higher order terms in the expansion can be neglected and $\delta = 1/2\,(\mathrm{d}\bar{u}_2\mathrm{d}x_1)^2\,\mathrm{d}x_1$. The total work done by the *internal axial forces* is then found by integration along the span of the beam

$$\Phi = \frac{1}{2}\int_0^L N_1(x_1)\left(\frac{\mathrm{d}\bar{u}_2}{\mathrm{d}x_1}\right)^2\mathrm{d}x_1. \qquad (14.37)$$

The total potential energy of the structure now becomes

$$\Pi = \frac{1}{2}\int_0^L H_{33}^{*c}\left(\frac{\mathrm{d}^2\bar{u}_2}{\mathrm{d}x_1^2}\right)^2\mathrm{d}x_1 + \frac{1}{2}\int_0^L N_1\left(\frac{\mathrm{d}\bar{u}_2}{\mathrm{d}x_1}\right)^2\mathrm{d}x_1 - \int_0^L p_2\,\bar{u}_2\,\mathrm{d}x_1$$
(14.38)

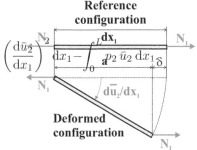

Reference configuration

Deformed configuration

Fig. 14.11. A differential element of the beam in the deformed configuration.

The first term represents the strain energy associated with the bending deformation of the beam, the second the work done by the externally applied axial load, N_1, and the third the work done by the externally applied transverse load, $p_2(x_1)$.

The principle of minimum total potential energy requires that the total potential energy of the system is a minimum with respect to all arbitrary variations of the displacement field, and hence the total potential energy must be stationary

$$\delta\Pi = \int_0^L H_{33}^{*c}\frac{\mathrm{d}^2\bar{u}_2}{\mathrm{d}x_1^2}\,\delta\frac{\mathrm{d}^2\bar{u}_2}{\mathrm{d}x_1^2}\,\mathrm{d}x_1 + \int_0^L N_1\frac{\mathrm{d}\bar{u}_2}{\mathrm{d}x_1}\,\delta\frac{\mathrm{d}\bar{u}_2}{\mathrm{d}x_1}\,\mathrm{d}x_1 - \int_0^L p_2\,\delta\bar{u}_2\,\mathrm{d}x_1 = 0.$$

The first term is integrated by parts twice and the second once to yield

$$\int_0^L \delta\bar{u}_2\left[\frac{\mathrm{d}^2}{\mathrm{d}x_1^2}\left(H_{33}^{*c}\frac{\mathrm{d}^2\bar{u}_2}{\mathrm{d}x_1^2}\right) - \frac{\mathrm{d}}{\mathrm{d}x_1}\left(N_1\frac{\mathrm{d}\bar{u}_2}{\mathrm{d}x_1}\right) - p_2\right]\mathrm{d}x_1$$
$$+ \left[H_{33}^{*c}\frac{\mathrm{d}^2\bar{u}_2}{\mathrm{d}x_1^2}\,\delta\left(\frac{\mathrm{d}\bar{u}_2}{\mathrm{d}x_1}\right)\right]_0^L + \left\{\left[N_1\frac{\mathrm{d}\bar{u}_2}{\mathrm{d}x_1} - \frac{\mathrm{d}}{\mathrm{d}x_1}\left(H_{33}^{*c}\frac{\mathrm{d}^2\bar{u}_2}{\mathrm{d}x_1^2}\right)\right]\delta\bar{u}_2\right\}_0^L = 0.$$

This equation is satisfied for arbitrary variations, $\delta\bar{u}_2(x_1)$, only if the integral and each of the boundary terms vanishes independently. The integral term vanishes if the integrand in brackets vanishes leading to the Euler-Lagrange equation of the problem,

$$\frac{\mathrm{d}^2}{\mathrm{d}x_1^2}\left(H_{33}^{*c}\frac{\mathrm{d}^2\bar{u}_2}{\mathrm{d}x_1^2}\right) - \frac{\mathrm{d}}{\mathrm{d}x_1}\left(N_1\frac{\mathrm{d}\bar{u}_2}{\mathrm{d}x_1}\right) = p_2.$$

This is the governing differential equation of the problem. It is identical to eq. (14.15) that is obtained from the equilibrium conditions for a differential element of the beam in the deformed configuration, see section 14.2.2.

For a cantilevered beam, the geometric boundary conditions at the beam's root imply $\bar{u}_2 = 0$ and $\mathrm{d}\bar{u}_2/\mathrm{d}x_1 = 0$. Because the variations at the beam's tip, $\delta(\mathrm{d}\bar{u}_2(L)/\mathrm{d}x_1)$ and $\delta\bar{u}_2(L)$, are arbitrary, $H_{33}^{*c}\mathrm{d}^2\bar{u}_2(L)/\mathrm{d}x_1^2 = 0$ and $N_1\mathrm{d}\bar{u}_2(L)/\mathrm{d}x_1 - \mathrm{d}/\mathrm{d}x_1(H_{33}^{*c}\mathrm{d}^2\bar{u}_2/\mathrm{d}x_1^2) = 0$, respectively. The first condition corresponds to the vanishing of the bending moment at the tip of the beam, as expected; the second condition corresponds to $N_1 \, \mathrm{d}\bar{u}_2(L)/\mathrm{d}x_1 + V_2(L) = 0$. These boundary conditions are identical to those derived in section 14.2.2, see eq. (14.16), and correspond to the tip equilibrium equations of the beam in its deformed configuration.

The energy approach developed in this section leads to the same governing differential equation and boundary conditions obtained from the equilibrium approach in section 14.2.1. This result should be expected because the principle of minimum total potential energy is derived from the equations of equilibrium.

14.2.5 Buckling of a simply-supported beam (energy approach)

Consider a uniform, simply-supported beam of length L subjected to an end compressive load of magnitude P, as depicted in fig. 14.5. The beam is assumed to be pinned about axis $\bar{\imath}_3$ which is perpendicular to the plane of symmetry of the beam. The axial force in the beam, N_1, is constant along the span, and $N_1 = -P$. The total potential energy of the system, eq. (14.38), becomes

$$\Pi = \frac{1}{2}\int_0^L H_{33}^{*c}\left(\frac{\mathrm{d}^2\bar{u}_2}{\mathrm{d}x_1^2}\right)^2 \, \mathrm{d}x_1 - \frac{1}{2}\int_0^L P\left(\frac{\mathrm{d}\bar{u}_2}{\mathrm{d}x_1}\right)^2 \, \mathrm{d}x_1.$$

An approximate solution can be developed by expressing $\bar{u}_2(x_1)$ as a sum of shape functions, each of which satisfies the geometric boundary conditions, see section 11.4.

Solution using a single shape function

The geometric boundary conditions for a simply supported beam require the displacement to vanish at both end. An assumed transverse displacement field that satisfies these conditions is

$$\bar{u}_2(x_1) = q_j \sin\frac{j\pi x_1}{L} \tag{14.39}$$

where q_j is the single degree of freedom of the problem and j an unspecified integer.

Substituting this assumed displacement field into the expression for the total potential energy, eq. (14.38), and carrying out the integrations yields the total potential energy as a function of the degree of freedom

$$\Pi = \left[H_{33}^c\left(\frac{j\pi}{L}\right)^2 - P\right]\left(\frac{j\pi}{L}\right)^2\frac{L}{2}q_j^2.$$

The principle of minimum total potential energy requires the total potential energy to be a minimum with respect to the choice of the degrees of freedom, leading to

$$
\frac{\partial \Pi}{\partial q_j} = \left[H_{33}^c \left(\frac{j\pi}{L} \right)^2 - P \right] \left(\frac{j\pi}{L} \right)^2 L q_j = 0.
$$

This homogeneous equation admits a trivial solution, $q_j = 0$, which implies $\bar{u}_2(x_1) \equiv 0$. A nontrivial solution exists if $(j^2\pi^2 H_{33}^c/L^2 - P) = 0$, and this condition yields the critical load,

$$
P_{\mathrm{cr}} = j^2 \pi^2 \frac{H_{33}^c}{L^2}.
$$

The lowest critical load is obtained when $j = 1$ and defines the buckling load, $P_{\mathrm{cr}} = \pi^2 H_{33}^c/L^2$. The buckling shape is given by eq. (14.39) as $\bar{u}_2(x_1) = q_1 \sin(\pi x_1)/L^2$.

This result is identical to that obtained in section 14.2.2. This should not be surprising because the shape function given in eq. (14.39) is, in fact, the exact solution. An infinite number of critical loads are obtained for all values of integer j, but the lowest critical load is the buckling load.

Solution using multiple shape functions

The following transverse displacement field containing two shape functions will be assumed next

$$
\bar{u}_2(x_1) = q_1 h_1(x_1) + q_2 h_2(x_1) = q_1 \sin \frac{\pi x_1}{L} + q_2 \sin \frac{2\pi x_1}{L} = \underline{H}^T(x_1)\underline{q}, \quad (14.40)
$$

where h_1 and h_2 are the assumed shape functions, $\underline{H}$ the displacement interpolation array defined by eq. (11.62), and $\underline{q} = \{q_1, q_2\}^T$ the solution array. In this case, the shape functions are selected as sine functions. In general, N shape functions could be used, but the assumed shape functions must all satisfy the geometric boundary conditions of the problem.

Calculation of the integrals in eq. (14.38) requires the first and second derivatives of the transverse deflection which can be expressed in matrix form as follows

$$
\frac{d\bar{u}_2}{dx_1} = q_1 \frac{\pi}{L} \cos \frac{\pi x_1}{L} + q_2 \frac{2\pi}{L} \cos \frac{2\pi x_1}{L} = \underline{G}^T(x_1)\underline{q}, \quad (14.41a)
$$

$$
\frac{d^2\bar{u}_2}{dx_1^2} = -q_1 \left(\frac{\pi}{L} \right)^2 \sin \frac{\pi x_1}{L} - q_2 \left(\frac{2\pi}{L} \right)^2 \sin \frac{2\pi x_1}{L} = \underline{B}^T(x_1)\underline{q}. \quad (14.41b)
$$

The *displacement gradient interpolation array*, $\underline{G}(x_1)$, is defined as

$$
\underline{G}(x_1) = \left\{ \frac{dh_1}{dx_1}, \frac{dh_2}{dx_1}, \dots \frac{dh_N}{dx_1} \right\}^T, \quad (14.42)
$$

and the *curvature interpolation array*, $\underline{B}(x_1)$, is given by eq. (11.72).

The strain energy in the structure is readily evaluated using the general procedure described in section 11.4.3. The strain energy is expressed as $A = 1/2\,\underline{q}^T\underline{\underline{K}}\,\underline{q}$, see eq. (11.75). The stiffness matrix is given by eq. (11.67) as

$$\underline{\underline{K}} = \int_0^L \underline{B}(x_1)\,H_{33}^c(x_1)\underline{B}^T(x_1)\,\mathrm{d}x_1. \tag{14.43}$$

For the shape functions defined in eq. (14.40), the stiffness matrix becomes

$$\underline{\underline{K}} = \frac{\pi^4 H_{33}^c}{2L^3}\begin{bmatrix}1 & 0 \\ 0 & 16\end{bmatrix}. \tag{14.44}$$

The second integral in eq. (14.38) represent the work done by the applied axial load. Using eq. (14.41a), it can be evaluated as follows

$$\Phi = -\frac{1}{2}\int_0^L P\left(\frac{\mathrm{d}\bar{u}_2}{\mathrm{d}x_1}\right)^2\,\mathrm{d}x_1 = -\frac{P}{2}\int_0^L \left(\underline{G}^T\underline{q}\right)^T\left(\underline{G}^T\underline{q}\right)\,\mathrm{d}x_1$$

$$= \frac{1}{2}\underline{q}^T P\left[\int_0^L \underline{G}\,\underline{G}^T\,\mathrm{d}x_1\right]\underline{q} = \frac{1}{2}\underline{q}^T P\underline{\underline{K}}_G\,\underline{q},$$

where the *geometric stiffness matrix*, $\underline{\underline{K}}_G$, is defined as

$$\underline{\underline{K}}_G = \int_0^L \underline{G}(x_1)\underline{G}^T(x_1)\,\mathrm{d}x_1. \tag{14.45}$$

For the problem at hand, this matrix is

$$\underline{\underline{K}}_G = \frac{\pi^2}{2L}\begin{bmatrix}1 & 0 \\ 0 & 4\end{bmatrix}. \tag{14.46}$$

The total potential energy of the structure is now found by combining the strain energy and the work done by the axial force as

$$\Pi = \frac{1}{2}\underline{q}^T\underline{\underline{K}}\,\underline{q} - \frac{1}{2}\underline{q}^T P\underline{\underline{K}}_G\,\underline{q} = \frac{1}{2}\underline{q}^T\left[\underline{\underline{K}} + P\underline{\underline{K}}_G\right]\underline{q}.$$

This is a positive-definite quadratic form in $\underline{q}$, and from the principle of minimum total potential energy, it must assume a minimum when the system is in equilibrium. It is shown in appendix A.2.9 that this quadratic form is minimum when the solution array, $\underline{q}$, satisfies the following condition

$$\left[\underline{\underline{K}} + P\underline{\underline{K}}_G\right]\underline{q} = 0. \tag{14.47}$$

This is a set of homogeneous, algebraic equations for the solution array, $\underline{q}$; the solution of this homogeneous system is the trivial solution, $\underline{q} = 0$, which corresponds to the vanishing of the transverse displacement field.

In the equilibrium approach developed in section 14.2.5, a trivial solution is also found. The buckling load of the system is defined as the lowest load for which a non-trivial solution of the governing equations exists. A nontrivial solution of the homogeneous, algebraic system given by eq. (14.47) exists if and only if the determinant of the system vanishes,

$$\det \left(\underline{\underline{K}} - P \underline{\underline{K}}_G \right) = 0. \tag{14.48}$$

This is the *buckling equation* for the system, and the values of the axial compressive load, P, for which this determinant vanishes are the critical loads. Introducing the stiffness matrix, eq. (14.44), and the geometric stiffness matrix, eq. (14.46), leads to an explicit expression for the buckling equation

$$\det \left\{ \frac{\pi^2}{2L} \left[\begin{array}{cc} \dfrac{\pi^2 H_{33}^{*c}}{L^2} - P & 0 \\ 0 & 16 \dfrac{\pi^2 H_{33}^{*c}}{L^2} - 4P \end{array} \right] \right\} = 0. \tag{14.49}$$

The solutions of this buckling equation are $P = \pi^2 H_{33}^{*c}/L^2$ and $4\pi^2 H_{33}^{*c}/L^2$. The lowest solution is the buckling load $P_{\mathrm{cr}} = \pi^2 H_{33}^{*c}/L^2$. This buckling load is identical to that found from the equilibrium approach, see section 14.2.5. Again, this should be expected because the exact buckling mode shape, $\sin \pi x_1/L$, see eq. (14.26), is one of the assumed mode shapes in eq. (14.40).

The solutions of the buckling equation give rise to non-trivial solutions of the problem. Since the stiffness and geometric stiffness matrices are, in general, matrices of size $N \times N$, the expansion of the determinant equation leads to an Nth order algebraic equation in P, and hence, solutions P_i, $i = 1, 2, \ldots, N$. For large values of N, the coefficients of this algebraic equation become very large, and the solution process becomes increasingly difficult, leading to inaccurate solutions. In this case, it is preferable to recast eq. (14.48) as $\underline{\underline{K}}\, \underline{q} = P \underline{\underline{K}}_G\, \underline{q}$. Multiplying through by $\underline{\underline{K}}^{-1}$ then yields

$$\underline{\underline{D}}\, \underline{q} = \lambda\, \underline{q}, \tag{14.50}$$

where $\underline{\underline{D}} = \underline{\underline{K}}^{-1} \underline{\underline{K}}_G$ and $\lambda = 1/P$. This is an eigenvalue problem of the type described in appendix A.2.4, and the eigenvalues, λ, are the reciprocals of the critical loads.

Matrix $\underline{\underline{D}}$ is of size $N \times N$ and it possesses N eigenvalues λ_i, $i = 1, 2, \cdots, N$, see appendix A.2.4. Let λ_1 be the highest eigenvalue of $\underline{\underline{D}}$; $P_{\mathrm{cr}} = 1/\lambda_1$ is then an approximation to the lowest critical load of the system, which is the buckling load. If $\underline{q}_1$ is the corresponding eigenvector, the buckling mode shape is $\bar{u}_2(x_1) = \underline{H}^T(x_1)\, \underline{q}_1$.

The procedure described in this section can be summarized by the following steps.

1. Select N shape functions that satisfy the geometric boundary conditions. Construct the corresponding displacement, displacement gradient, and strain interpolation arrays.

2. Compute the strain energy of the structure, leading to the $N \times N$ stiffness matrix, $\underline{K}$, see eq. (14.43).
3. Compute the work done by the axial forces, leading to the $N \times N$ geometric stiffness matrix, $\underline{\underline{K}}_G$, see eq. (14.45).
4. Solve the eigenproblem $\underline{\underline{D}}\, q = \lambda q$, where $\underline{\underline{D}} = \underline{\underline{K}}^{-1}\underline{\underline{K}}_G$, to find the highest eigenvalue λ_1, and corresponding eigenvector q_1.
5. The approximate buckling load of the structure is $P_{\mathrm{cr}} = 1/\lambda_1$ and the buckling mode shapes is $\bar{u}_2(x_1) = \underline{H}^T(x_1)q_1$.

If externally applied transverse loads are also applied, the above procedure must be modified as follows.

1. Select N shape functions that satisfy the geometric boundary conditions. Construct the corresponding displacement, displacement gradient, and strain interpolation arrays.
2. Compute the strain energy of the structure, leading to the $N \times N$ stiffness matrix, $\underline{K}$.
3. Compute the work done by the externally applied transverse loads, leading to the load array, $\underline{Q}$.
4. Compute the work done by the axial forces, leading to the $N \times N$ geometric stiffness matrix, $\underline{\underline{K}}_G$.
5. Solve the linear system $(\underline{\underline{K}} - P\underline{\underline{K}}_G)q = \underline{Q}$ to find the solution array, q.

In this case, the solution procedure assumes that matrix $(\underline{\underline{K}} - P\underline{\underline{K}}_G)$ is non-singular, i.e., $\det(\underline{\underline{K}} - P\underline{\underline{K}}_G) \neq 0$. If the axial force is such that $\det(\underline{\underline{K}} - P\underline{\underline{K}}_G) = 0$, the linear system cannot be solved, and P has reached the buckling load.

14.2.6 Applications to beam buckling

A number of examples will be worked to illustrate the use of both the equilibrium and energy approaches.

Example 14.1. Cantilevered beam with tip support (equilibrium approach)
Consider a uniform, cantilevered beam with a tip support as depicted in fig. 14.12. The beam is subjected to an axial compressive load P at the tip. The resulting axial force in the beam, N_1, is constant along the span, and $N_1 = -P$. The governing differential equation of the problem is given by eq. (14.17), with the following boundary conditions at the beam's root $\bar{u}_2 = d\bar{u}_2/dx_1 = 0$ and at its tip $\bar{u}_2 = d^2\bar{u}_2/dx_1^2 = 0$. The non-dimensional span-wise variable $\eta = x_1/L$ is defined and the governing equation then reduces to eq. (14.18), where $(\cdot)'$ denotes a derivative with respect to η and the non-dimensional loading parameter, λ^2, is defined by eq. (14.19). The boundary conditions at the beam's root become $\bar{u}_2 = \bar{u}_2' = 0$ and at its tip $\bar{u}_2 = \bar{u}_2'' = 0$.

The solution of eq. (14.18) is

$$\bar{u}_2 = A + B\eta + C\cos\lambda\eta + D\sin\lambda\eta$$

where A, B, C, and D, are four integration constants to be determined from the boundary conditions. The first two boundary conditions imply $A + C = 0$ and $B +$

$\lambda D = 0$. Eliminating constants B and C yields $\bar{u}_2 = A(1-\cos\lambda\eta)+D(\sin\lambda\eta-\lambda\eta)$.
The last two boundary conditions then yield

$$\begin{bmatrix} 1 - \cos\lambda & \sin\lambda - \lambda \\ \cos\lambda & -\sin\lambda \end{bmatrix} \begin{Bmatrix} A \\ D \end{Bmatrix} = 0.$$

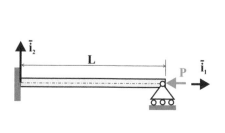

Fig. 14.12. Cantilevered beam with tip support under axial load.

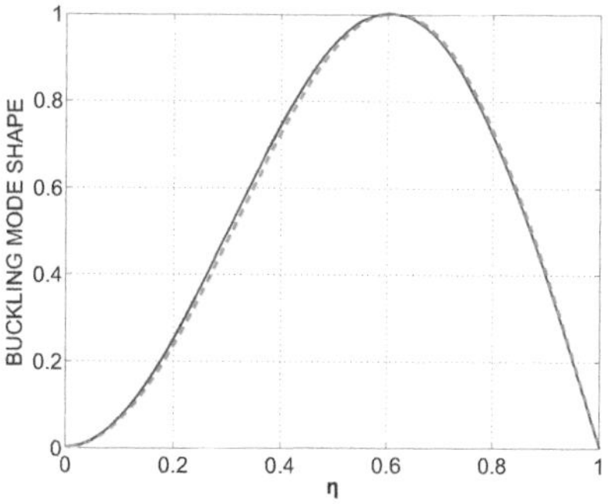

Fig. 14.13. Buckling mode shape for the cantilevered beam with tip support. Exact solution: solid line; approximate solution ($n = 1$): dashed line.

This is a set of algebraic equations for the last two integration constants, A and D. Because the system is homogeneous, the solution is $A = D = 0$, which corresponds to the trivial, non-buckled, solution of the problem. A linear system of homogeneous algebraic equations admits a non-trivial solution if and only if the determinant of the system vanishes,

$$\det \begin{bmatrix} 1 - \cos\lambda & \sin\lambda - \lambda \\ \cos\lambda & -\sin\lambda \end{bmatrix} = 0.$$

Expanding this determinant yield the *buckling equation*, $\tan\lambda - \lambda = 0$. The buckling equation is a transcendental equation, and the lowest solution is $\lambda = 4.4934$. The buckling load of the system is $P_{cr} = (4.4934)^2 H^{*c}_{33}/L^2$. For comparison, this can be recast as $P_{cr} = 2.0457\pi^2 H^{*c}_{33}/L^2 = 2.0457 P_{Euler}$, or about twice the buckling load for a simply supported beam of the same length.

The buckling mode shape can be determined from this analysis. The buckling equation implies $A\cos\lambda = D\sin\lambda$ or $A = D\tan\lambda = \lambda D$. the buckling mode shape is then $\bar{u}_2(\eta) = D[\lambda(1 - \eta) + \sin\lambda\eta - \lambda\cos\lambda\eta]$. The integration constant, D, remains undetermined, which means that the transverse displacement is of arbitrary amplitude, indicating a lateral collapse of the beam.

Example 14.2. Cantilevered beam with tip support (energy approach)
Consider the same cantilevered beam with a tip support depicted in fig. 14.12. An energy approach will be used to find an approximate solution to the problem. The following transverse displacement field will be assumed

$$\bar{u}_2(x_1) = \sum_{n=1}^{N} q_n \, h_n(\eta) = \sum_{n=1}^{N} q_n \frac{1}{2} \left[\cos a_n \eta - \cos b_n \eta\right], \tag{14.51}$$

where $\eta = x_1/L$ is the non-dimensional span-wise variable, $a_n = (n - 1/2)\pi$ and $b_n = (n + 1/2)\pi$. Note that each assumed shape function satisfies the geometric boundary conditions of the problem: $h_n(0) = 0$, $h'_n(0) = 0$, and $h_n(1) = 0$, where $(\cdot)'$ denotes a derivative with respect to η.

The displacement field is written as $\bar{u}_2(x_1) = \underline{H}^T(\eta) \, \underline{q}$, where $\underline{H}$ is the displacement interpolation array, and $\underline{q}^T$ the solution array. The displacement gradient and the curvature interpolation arrays are given by eqs. (14.41) as

$$\frac{d\bar{u}_2}{dx_1} = \sum_{n=1}^{N} \frac{q_n}{2L} \left[-a_n \sin a_n \eta + b_n \sin b_n \eta\right] = \underline{G}^T(\eta) \, \underline{q},$$

$$\frac{d^2 \bar{u}_2}{dx_1^2} = \sum_{n=1}^{N} \frac{q_n}{2L^2} \left[-a_n^2 \cos a_n \eta + b_n^2 \cos b_n \eta\right] = \underline{B}^T(\eta) \, \underline{q}.$$

The bending and geometric stiffness matrices are computed using eqs. (14.43) and (14.45), and using the first four terms in eq. (14.51) results in

$$\underline{K} = \frac{H_{33}^{*c}}{4L^3} \left(\frac{\pi}{2}\right)^4 \begin{bmatrix} 1^4 + 3^4 & -3^4 & 0 & 0 \\ -3^4 & 3^4 + 5^4 & -5^4 & 0 \\ 0 & -5^4 & 5^4 + 7^4 & -7^4 \\ 0 & 0 & -7^4 & 7^4 + 9^4 \end{bmatrix},$$

and

$$\underline{K}_G = \frac{1}{4L} \left(\frac{\pi}{2}\right)^2 \begin{bmatrix} 1^2 + 3^2 & -3^2 & 0 & 0 \\ -3^2 & 3^2 + 5^2 & -5^2 & 0 \\ 0 & -5^2 & 5^2 + 7^2 & -7^2 \\ 0 & 0 & -7^2 & 7^2 + 9^2 \end{bmatrix}.$$

The buckling equation is given by eq. (14.48). If a single term is taken in the assumed displacement field, *i.e.*, if $N = 1$, the buckling load is found to be $P_{cr} = 82/10(\pi/2)^2 H_{33}^{*c}/L^2 = 2.0500\pi^2 H_{33}^{*c}/L^2$. This compares very favorably with the exact solution found in the previous example as $P_{cr} = 2.0457\pi^2 H_{33}^{*c}/L^2$. A two term approximation yields $P_{cr} = 2.0467\pi^2 H_{33}^{*c}/L^2$ that overestimates the exact solution by 0.05% only. For larger values of N more accurate solutions are obtained, but it is the preferable to rely on the solution of the eigenproblem defined by eq. (14.50).

Figure 14.13 depicts the buckling mode shape for the exact solution developed in the previous example and the present approximate solution using a single assumed mode, *i.e.*, with $N = 1$. For $N = 2$, the exact and approximate buckling mode shapes are almost coincident.

Example 14.3. Cantilevered beam with linearly tapered depth

During buckling of a cantilevered beam, the region near the root undergoes the largest bending moment. To increase the buckling load, it is reasonable to increase the depth of the section in the root region to provide a greater bending stiffness.

To investigate this effect, consider a cantilevered beam with a depth in plane $(\bar{\imath}_2, \bar{\imath}_1)$ that is tapered linearly from h_r at the root to h_t at the tip, as shown in fig. 14.14. From a practical point of view, assume also that either the beam is restrained from deflection in the $\bar{\imath}_3$ direction or else $H_{22}^c \gg H_{33}^c$ so that buckling in plane $(\bar{\imath}_1, \bar{\imath}_3)$ will not occur.

For a beam with a rectangular cross-section having a linearly tapered depth, the bending stiffness can be written as

$$H_{33}^c = H_0 \left[\bar{h} - (\bar{h} - 1)\eta \right]^3, \tag{14.52}$$

where $\eta = x_1/L$ is a non-dimensional span-wise variable, H_0 the tip bending stiffness, and $\bar{h} = h_r/h_t > 1$ the root depth factor.

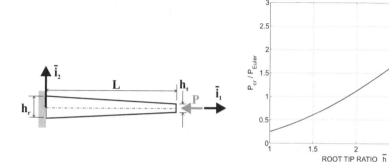

Fig. 14.14. Tapered cantilevered beam under axial load.

Fig. 14.15. Buckling load for tapered cantilevered beam as a function of the root depth to the tip depth, $\bar{h}$.

The following transverse displacement field satisfies the geometric boundary conditions

$$\bar{u}_2(x_1) = \sum_{n=1}^{N} q_n \, \eta^{n+1} = \sum_{n=1}^{N} q_n \, h_n(\eta), \tag{14.53}$$

where the shape functions are $h_n(\eta) = \eta^{n+1}$. For $N = 2$, the displacement gradient and curvature interpolation arrays are $\underline{G} = [2\eta, \ 3\eta^2]/L$ and $\underline{B} = [2, \ 6\eta]/L^2$, respectively.

The bending stiffness matrix, $\underline{\underline{K}}$, and geometric stiffness matrix, $\underline{\underline{K}}_G$, are found from eqs. (14.43) and (14.45), respectively, as

$$\underline{\underline{K}} = \frac{H_0}{5L^2} \begin{bmatrix} 5(1 + \bar{h} + \bar{h}^2 + \bar{h}^3) & 3(4 + 3\bar{h} + 2\bar{h}^2 + \bar{h}^3) \\ 3(4 + 3\bar{h} + 2\bar{h}^2 + \bar{h}^3) & 3(10 + 6\bar{h} + 3\bar{h}^2 + \bar{h}^3) \end{bmatrix},$$

and

$$\underline{\underline{K}}_G = \frac{1}{30L} \begin{bmatrix} 40 & 45 \\ 45 & 54 \end{bmatrix},$$

From eq. (14.50), the critical loads are the reciprocals of the two eigenvalues of the matrix $\underline{\underline{D}} = \underline{\underline{K}}^{-1}\underline{\underline{K}}_G$. The buckling load is the lowest of these. Figure 14.15 shows the increase in buckling load as the root depth factor, $\bar{h}$, increases. For $\bar{h} = 1$, the beam is a cantilever with uniform depth, and the buckling load is $P_{cr} = 0.2519 P_{\text{Euler}}$ where $P_{\text{Euler}} = \pi^2 H^{*c}/L^2$. This compares very favorably with the exact buckling load, $P_{cr} = 0.2500 P_{\text{Euler}}$. For $\bar{h} = 2$, the buckling load is approximately the same as that obtained for cantilevered beam with tip support, see example 14.1. For $\bar{h} = 3.7$, the buckling load is equal to that of a uniform, simply supported beam of the identical length and tip bending stiffness.

14.2.7 Buckling of beams with various end conditions

The analysis method developed in the previous section can be repeated for beams with various end conditions. The buckling load for various configurations are summarized in table 14.1 which lists the non-dimensional buckling parameter k, such that

$$P_{cr} = \bar{k}\pi^2 \frac{H^{*c}_{33}}{L^2} = \bar{k}\, P_{\text{Euler}}.$$

Table 14.1. Buckling loads for beams with various end conditions.

Boundary Conditions	Buckling parameter k
Clamped - Free	1/4
Clamped - Clamped	4
Clamped - Pinned	2.0457
Pinned - Pinned	1

Finally, it should be noted that in all the previous examples, it is assumed that the beam under consideration is supported in such a way that it is free to buckle in plane $(\bar{\imath}_1, \bar{\imath}_2)$. This can be assured if this is a plane of symmetry or contains one of the principal axes of bending and if the bending stiffness is the minimum for the given cross-section.

14.2.8 Problems

Problem 14.1. Uniform cantilevered beam
Consider a uniform cantilevered beam of length L subjected to an axial compressive load P. Find the lowest buckling load of the system and the associated buckling mode shape. Check your predictions with the results listed in table 14.1.

Problem 14.2. Simply-supported beam with end torsional springs
Consider a simply-supported beam with end torsional springs of stiffness k subjected to an axial compressive load P, as depicted in fig. 14.16. It will be convenient to use the non-dimensional load P/P_{Euler} where $P_{\text{Euler}} = \pi^2 H^{*c}_{33}/L^2$ and non-dimensional spring constant $\bar{k} = kL/H^{*c}_{33}$. (1) Find the lowest buckling load P_{cr} of the system. Plot P_{cr} as a function of

$\bar{k}$. Discuss the limiting cases $\bar{k} = 0$ and $\bar{k} = \infty$. *(2)* Use an energy method to estimate the lowest buckling load of the system. Select the following assumed displacements $\bar{u}_2(x_1) = q_1 \sin \pi x_1/L + q_2 \sin 2\pi x_1/L$. *(3)* Plot the exact and approximate buckling loads on the same graph. Comment on the accuracy of the approximate solution. How would you improve its accuracy?

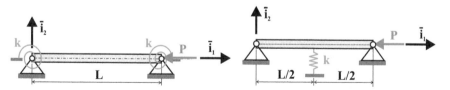

Fig. 14.16. Simply-supported beam with end torsional springs.

Fig. 14.17. Simply-supported beam with a mid-span spring.

Problem 14.3. Simply-supported beam with a mid-span spring

Study the effect of a mid-span spring of stiffness constant k on the buckling load of the simply-supported beam of span L depicted in fig. 14.17. To investigate this problem, use an energy approach with the following assumed mode shapes $\bar{u}_2(\eta) = q_1 \sin \pi\eta + q_2 \sin 2\pi\eta + a_3 \sin 3\pi\eta$. It is convenient to use the following notation: $P_{\text{Euler}} = \pi^2 H_{33}^{*c}/L^2$ is the Euler buckling load for the beam in the absence of the mid-span spring, $\bar{k} = kL^3/H_{33}^{*c}$ the non-dimensional stiffness of the spring, and $\eta = x_1/L$ the non-dimensional span variable. *(1)* Find the buckling loads of the system. *(2)* How does the lowest buckling load vary when $\bar{k}$ increases? Plot the non-dimensional buckling loads P/P_e as a function of $\bar{k}$. *(3)* How much improvement in buckling load can be expected from the mid-span spring?

Problem 14.4. Cantilevered beam with axial and transverse loads

Consider the cantilevered beam of length L subjected to tip axial load in tension, T, and tip transverse load, P, as depicted in fig. 14.18. *(1)* Write the governing differential equation of the problem and the associated boundary conditions. *(2)* Find the transverse displacement field of the beam.

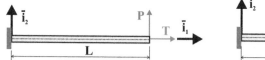

Fig. 14.18. Cantilevered beam subjected to tip axial and transverse loads.

Fig. 14.19. Cantilevered beam subjected to tip axial load with an off-set.

Problem 14.5. Various short questions

Consider a cantilevered beam subjected to a tip compressive load applied with an off-set e, as depicted in fig. 14.19. Let P_{Euler} denote the Euler buckling load on the system. *(1)* For the perfect system, *i.e.*, for $e = 0$, is the buckling load of the system affected by the strength of the material the beam is made of? *(2)* For the imperfect system, *i.e.*, for $e \neq 0$, is the buckling load of the system affected by the strength of the material the beam is made of? *(3)* For the perfect system, *i.e.*, for $e = 0$, does the beam experience transverse deflections when $P < P_{\text{Euler}}$? *(4)* For the imperfect system, *i.e.*, for $e \neq 0$, does the beam experience transverse deflections when $P < P_{\text{Euler}}$? *(5)* Consider an imperfect system, *i.e.*, $e \neq 0$, under increasing load P. Does the material allowable stress play a role in predicting the allowable applied load P? Explain your answers to all the above questions; a YES/NO answer is not valid.

Problem 14.6. Cantilevered beam with tip support

A uniform cantilevered beam of span L has a tip support, as depicted in fig. 14.20. The beam is subjected to a tip compressive axial force P and a uniform transverse loading p_0. *(1)* Find the exact solution of the problem when the beam is *subjected to the sole transverse load p_0*. *(2)* Find an approximate solution for the problem when the beam is *subjected to the combined loading, i.e.*, both transverse loading p_0 and tip compressive load P are applied. Use an energy method with a single assumed mode selected to be the solution to part 1. *(3)* Determine the buckling load P_{cr} for this problem. *(4)* How is the buckling load affected by the transverse loading p_0. *(5)* Under the combined loading condition, find the failure envelope in the two-dimensional space $P/P_{\text{cr}}, p_0/(b\sigma_{\text{ult}})$. Failure is reached when $|\sigma_{\text{ben}}^{\text{max}}| + |\sigma_{\text{axi}}^{\text{max}}| = \sigma_{\text{ult}}$, where $\sigma_{\text{ben}}^{\text{max}}$ and $\sigma_{\text{axi}}^{\text{max}}$ are the the maximum stresses due to bending and axial force, respectively, and σ_{ult} the ultimate allowable stress for the material. Use the following data $E = 73.0$ GPa, $\sigma_{\text{ult}} = 620$ MPa. Plot two failure envelopes for $L/h = 20$ and 10 on the same graph. Assume a rectangular section of width b and hight h.

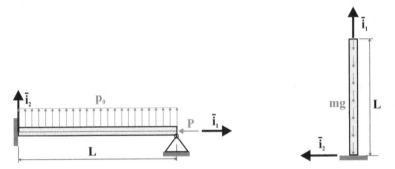

Fig. 14.20. Cantilevered beam with tip support subjected to a compressive load.

Fig. 14.21. Flag pole standing under its own weight.

Problem 14.7. Flag pole standing under its own weight

A flagpole of uniform mass per unit span m is standing up against gravity, as depicted in fig. 14.21. What is the critical weight mg_{cr} such that this flagpole will buckle under its own weight? *(1)* Use an energy method to solve this problem. Use the following assumed mode $\bar{u}_2(\eta) = a\,(1 - \cos \pi\eta/2)$. *(2)* If the flagpole is made of steel and has a square cross-section (1cm× 1cm), what is the critical length at which it will buckle under its own weight. (For steel: material density $\rho = 7700$kg/m^3, Young's modulus $E = 210$ GPa; $g = 9.81$m/sec^2).

Problem 14.8. Cantilevered beam with tip spring

An axial compressive load P is applied at the centroid of a uniform, cantilevered beam of span L with a tip spring of stiffness k, as depicted in fig. 14.22. *(1)* Derive the exact buckling equation for this problem in terms of the following non-dimensional parameters $\lambda^2 = PL^2/H_{33}^{*c}$, $\bar{k} = kL^3/H_{33}^{*c}$. *(2)* Plot the buckling load of the system for $\bar{k} \in [0, 75]$. *(3)* Find an approximate solution of the problem using an energy approach with the following assumed mode $\bar{u}_2 = q_1(\eta^4 - 4\eta^3 + 6\eta^2) + q_2(2\eta^4 - 5\eta^3 + 3\eta^2)$, where $\eta = x_1/L$. *(4)* Compare your results by plotting the exact and approximate buckling loads of the system for $\bar{k} \in [0, 75]$.

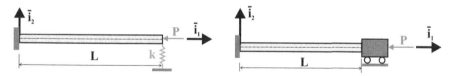

Fig. 14.22. Cantilevered beam with tip spring.

Fig. 14.23. Clamped-clamped beam under tip load.

Problem 14.9. Clamped-clamped beam under tip load

Consider the clamped-clamped beam of length L depicted in fig. 14.23. The beam is subjected to a compressive load P. *(1)* Find the lowest buckling load of this system using the differential equation approach. *(2)* Find the lowest buckling load of the system using an energy approach with the following assumed mode $\bar{u}_2(\eta) = a(\eta^2 - 2\eta^3 + \eta^4)$, where $\eta = x_1/L$ is the non-dimensional variable along the beam span.

Problem 14.10. Portal frame subjected to compressive loads

Consider the portal frame shown in fig. 14.24. Note that the problem is symmetric, and hence it is sufficient to consider one of the vertical beams with a tip torsional spring of stiffness constant k and subjected to a compressive load P. *(1)* Verify that $k = 6J_{33}/w$. *(2)* Find the lowest buckling load of the system. Show that the non-dimensional critical load depends on the non-dimensional parameter $\beta = (H_{33}w)/(J_{33}h)$. *(3)* Plot the lowest buckling load as a function of β. Discuss the meaning of your results when $\beta = 0$ and ∞.

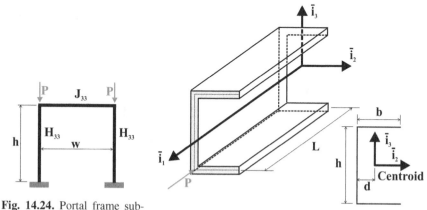

Fig. 14.24. Portal frame subjected to compressive loads.

Fig. 14.25. Cantilevered beam with a C-channel cross-section.

Problem 14.11. Cantilevered beam with a C-channel cross-section

A uniform cantilevered beam with a thin-walled, C-channel cross-section is subjected to an end compressive load P in the lower corner, as depicted in fig. 14.25. Let $L = 10b$, $h = 2b$, and $b = 10t$, where t is the constant wall thickness. *(1)* Find the distributions of transverse displacements $\bar{u}_2(x_1)$ and $\bar{u}_3(x_1)$. *(2)* Determine the buckling load P_{cr} of the system. *(3)* Plot P/P_{cr} versus $\bar{u}_2^{tip}/b$ and P/P_{cr} versus $\bar{u}_3^{tip}/b$, both on the same graph. $\bar{u}_2^{tip}$ and $\bar{u}_3^{tip}$ are the tip transverse displacements of the beam along the $\bar{i}_2$ and $\bar{i}_3$, respectively.

Problem 14.12. Cantilevered beam under tip tensile force

Consider a cantilevered beam of length L and bending stiffness H_{33}^c subjected to a tip transverse load, P, and a tip *tensile* axial force, F. The tip deflection of the beam will be denoted $\delta = \bar{u}_2(x_1 = L)$. The beam features a rectangular cross-section of width b and height h. *(1)* Compute the non-dimensional transverse displacement field $H_{33}^c \bar{u}_2(\eta)/PL^3$, where $\eta = x_1/L$ is the non-dimensional span-wise variable. Express your result in terms if the non-dimensional axial force factor $\lambda = \pi/2 \sqrt{F/P_{Euler}}$, where $P_{Euler} = \pi^2 H_{33}^c/4L^2$ is the buckling load for the cantilevered beam. *(2)* Compute the non-dimensional tip transverse displacement $H_{33}^c \delta/PL^3$. Show that for vanishing axial force, $F = 0$, the tip deflection becomes $\delta_0 = PL^3/(3H_{33}^c)$, as expected. *(3)* Plot the non-dimensional tip deflection δ/δ_0 as a function of $F/P_{Euler} \in [0, 5.0]$. *(4)* On one graph, plot the non-dimensional transverse displacement field, $H_{33}^c \bar{u}_2(\eta)/(PL^3)$, for $F/P_{Euler} = 0.0, 1.0, 2.0, 3.0, 4.0$ and 5.0. *(5)* Plot the non-dimensional root bending moment $M_{root}/(PL)$ as a function of $F/P_{Euler} \in [0, 5.0]$. *(6)* On one graph, plot the non-dimensional bending moment distribution, $M_3(\eta)/(PL)$, for $F/P_{Euler} = 0.0, 1.0, 2.0, 3.0, 4.0$ and 5.0. *(7)* Based on simple statics arguments, prove that $M_{root} = PL - F\delta$. Does your solution satisfy this relationship? *(8)* Plot the maximum root axial stress, σ/σ_0, as a function of $F/P_{Euler} \in [0, 5.0]$. $\sigma_0 = 6PL/(bh^2)$ is the maximum root axial stress for a vanishing axial force, $F = 0$. Your result should depend on δ_0/h; select values of $\delta_0/h = 0.5, 1.0, 2.0, 3.0, 4.0$ and 5.0. Is it possible to reduce the maximum axial stress in the cantilevered beam by applying the axial force F? Discuss your results.

Problem 14.13. Simply-supported beam under thermal loading

Consider a simply-supported beam of length L subjected to the parabolic thermal field depicted in fig. 14.26. The beam features a rectangular cross-section of height h and width b, where $h < b$. *(1)* Find the transverse deflection of the beam. *(2)* Find the critical thermal strain, αT_{cr}, at which the beam buckles. *(3)* Plot T_0/T_{cr} as a function of the beam's mid-span deflection, $\bar{u}_2^{mid}/h$. *(4)* Plot T_0/T_{cr} as a function of the mid-span bending moment, hM_3^{mid}/H_{33}^c. *(5)* How does the situation change if $h > b$?

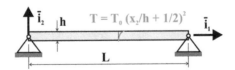

Fig. 14.26. Simply-supported beam under thermal loading.

14.3 Buckling of sandwich beams

Beams with sandwich cross-section are very common in aerospace constructions due to their high stiffness and strength to weight ratios. A typical sandwich cross-section is depicted in fig. 14.27. Two thin faces of thickness t_f and Young's modulus E_f sandwich a rather thick core of thickness t_c and Young's modulus E_c. The stiffness of the core is, in general, much smaller than that of the faces, $E_c \ll E_f$.

The bending stiffness of the complete sandwich section is readily found as

$$H_{33}^s = 2\frac{bt_f^3}{12}E_f + 2bt_f\frac{h^2}{4}E_f + \frac{bt_c^3}{12}E_c = bt_f\frac{h^2}{2}E_f\left[1 + \frac{1}{3}\frac{t_f^2}{h^2} + \frac{1}{6}\frac{t_c}{t_f}\frac{t_c^2}{h^2}\frac{E_c}{E_f}\right].$$

For typical constructions, the faces are much thinner than the core, $t_f/h \ll 1$, and the core material is very much softer than that of the faces, $E_c/E_f \ll 1$. Consequently, the bending stiffness reduces to

$$H_{33}^s \approx bt_f\frac{h^2}{2}E_f. \tag{14.54}$$

Very stiff constructions can be obtained by selecting deep cores (large values of h) and a stiff material for the faces (such as composite materials). Note that for a given amount of material, *i.e.*, for a specific face thickness t_f, the bending stiffness of the sandwich increases quadratically with the depth h.

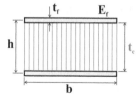

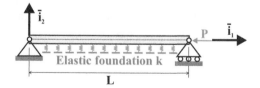

Fig. 14.27. Cross-section of a sandwich beam.

Fig. 14.28. Upper face of a sandwich structure under end compressive loads. The core of the sandwich is modeled by an elastic foundation.

If sandwich beams are subjected to axial compressive loads, buckling will occur when the critical load level is reached, *i.e.*, when $P_{cr} = \pi^2 H_{33}^s/L^2$ for a simply-supported boundary conditions. Because sandwich structures are designed to achieve large bending stiffnesses, they are efficient at sustaining large in-plane compressive loads.

This general instability of the structure is not the only failure mode of sandwich constructions. Because the core material is typically very soft, it carries a negligible portion of the axial force, and hence, the axial compressive load is primarily carried by the thin faces. Figure 14.28 shows an idealized model of the upper face of the sandwich subjected to an axial compressive load P. If a sandwich beam is subjected to an axial force N_1, the load in each of the upper and lower faces will be $P \approx N_1/2$, assuming identical faces and a negligible contribution of the core to supporting the axial force.

Although the core material is of little help in carrying the axial compressive load, it does provide an elastic support for the faces. This elastic support can be approximated by an elastic foundation of stiffness constant k, as depicted in fig. 14.28. The problem is now to determine the buckling load of the system shown in fig. 14.28, which consists a beam acted upon by axial compressive forces and resting on an elastic foundation of stiffness k. The bending stiffness of the face, $H_{33}^f = E_f bt_f^3/12$, is much smaller than that of the sandwich beam given by eq. (14.54), and hence its buckling load can be considerably lower than that of the sandwich beam.

The buckling problem for the thin face resting on an elastic foundation will be treated using an energy approach with the following assumed displacement field for the face

$$\bar{u}_2(x_1) = \sum_{n=1}^{\infty} q_n \sin \frac{n\pi x_1}{L},$$

where q_n are unknown displacement parameters. Note that each assumed displacement mode, $\sin n\pi x_1/L$ satisfies the geometric boundary conditions of the problem.

The total potential energy of the system is

$$\Pi = \frac{1}{2} \int_0^L H_{33}^f \left(\frac{d^2\bar{u}_2}{dx_1^2} \right)^2 dx_1 + \frac{1}{2} \int_0^L k\,\bar{u}_2^2\, dx_1 - \frac{1}{2} \int_0^L P \left(\frac{d\bar{u}_2}{dx_1} \right)^2 dx_1,$$

where the first term represents the strain energy associated with the bending of the face, the second term the strain energy in the elastic foundation, and the last term the work done by the axial compressive load P. Introducing the assumed displacement field in this expression leads to

$$\Pi = \frac{H_{33}^f}{2} \sum_{m=1}^{\infty} \sum_{n=1}^{\infty} q_m q_n \left(\frac{m\pi}{L} \right)^2 \left(\frac{n\pi}{L} \right)^2 \int_0^L \sin \frac{m\pi x_1}{L} \sin \frac{n\pi x_1}{L} \, dx_1$$

$$+ \frac{k}{2} \sum_{m=1}^{\infty} \sum_{n=1}^{\infty} q_m q_n \int_0^L \sin \frac{m\pi x_1}{L} \sin \frac{n\pi x_1}{L} \, dx_1$$

$$- \frac{P}{2} \sum_{m=1}^{\infty} \sum_{n=1}^{\infty} q_m q_n \left(\frac{m\pi}{L} \right) \left(\frac{n\pi}{L} \right) \int_0^L \cos \frac{m\pi x_1}{L} \cos \frac{n\pi x_1}{L} \, dx_1.$$

In view of the orthogonality of trigonometric functions, see appendix A.4, the total potential energy reduces to

$$\Pi = \frac{1}{2} \sum_{m=1}^{\infty} \left[H_{33}^f \left(\frac{m\pi}{L} \right)^4 + k - P \left(\frac{m\pi}{L} \right)^2 \right] \frac{L}{2} q_m^2.$$

The principle of minimum total potential energy implies that Π is a minimum with respect to the choice of the displacement parameters q_m. Hence $\partial \Pi/\partial q_n = 0$, for $n = 1, 2, \ldots \infty$ or

$$\left[H_{33}^f \left(\frac{n\pi}{L} \right)^4 + k - P \left(\frac{n\pi}{L} \right)^2 \right] q_n = 0, \qquad n = 1, 2, \ldots \infty.$$

This represents a set of homogeneous, uncoupled algebraic equations for the unknown displacement parameters q_n. The solution of this system is $q_n = 0$, $n = 1, 2, \ldots \infty$, corresponding to the trivial solution of the problem. A non-trivial solution exists if and only if the determinant of the system of equations vanishes, leading to the following condition

$$H_{33}^f \left(\frac{n\pi}{L} \right)^4 + k - P \left(\frac{n\pi}{L} \right)^2 = 0, \quad n = 1, 2, \ldots \infty.$$

Solving for the critical load yields

$$P_{\text{cr}\,n} = H_{33}^f \left(\frac{n\pi}{L} \right)^2 + k \left(\frac{L}{n\pi} \right)^2 = \frac{\pi^2 H_{33}^f}{L^2} \left[n^2 + \frac{kL^4}{\pi^4 H_{33}^f} \frac{1}{n^2} \right], \quad n = 1, 2, \ldots \infty.$$

In this expression, $P_{\text{Euler}} = \pi^2 H_{33}^f / L^2$ is the Euler buckling load for the simply-supported face in the absence of elastic foundation. If $\bar{k} = kL^4 / \pi^4 H_{33}^f$ is defined as the non-dimensional stiffness of the elastic foundation, the critical loads of the system become

$$\frac{P_{\text{cr}\,n}}{P_{\text{Euler}}} = n^2 + \frac{\bar{k}}{n^2}, \qquad n = 1, 2, \ldots \infty.$$

Each of the above critical loads gives rise to a non-trivial solution of the problem, and the lowest critical load is the buckling load. The lowest critical load, however, is not always obtained for $n = 1$; indeed, as n increases, the first term, n^2, increases, but the second term, $\bar{k}/n^2$, decreases. This implies that the wave number, n, that yields the lowest critical load is itself a function of $\bar{k}$.

To illustrate this point, fig 14.29 shows the non-dimensional critical loads $P_{\text{cr}\,n}/P_{\text{Euler}}$ for $n = 1$, 2, and 3. For a wave number $n = 1$, the critical load is $P_{\text{cr}\,1}/P_{\text{Euler}} = 1 + \bar{k}$, a linear relationship. For $n = 2$, the critical load is $P_{\text{cr}\,2}/P_{\text{Euler}} = 4 + \bar{k}/4$, which is also a linear relationship, and defines a dashed line with intermediate slope. A third dashed line is shown for $n = 3$. It follows that when $0 \leq \bar{k} \leq 4$, the lowest critical load is obtained for a wave number $n = 1$ but when $4 \leq \bar{k} \leq 36$, the lowest critical load is obtained with $n = 2$. The buckling load is therefore a function of $\bar{k}$, and it consists of an envelope of straight line segments, each with a different value of the wave number.

Let $\bar{k}_{\text{co}}$ be the cross-over value of $\bar{k}$ at which the wave number switches from n to $n+1$. This cross-over point is the intersection of two straight lines, $P_{\text{cr}\,(n)}/P_{\text{Euler}} = n^2 + \bar{k}/n^2$ and $P_{\text{cr}\,(n+1)}/P_{\text{Euler}} = (n+1)^2 + \bar{k}/(n+1)^2$, which is readily found as $\bar{k}_{\text{co}} = n^2(n+1)^2$. It follows that the range of values of $\bar{k}$ for which buckling occurs with wave number n is

$$(n-1)^2 n^2 \leq \bar{k} \leq n^2 (n+1)^2.$$

When looking at a large range of values of $\bar{k}$, the various straight line segments blend together into a smooth curve, as illustrated in fig. 14.30. The buckling load can then

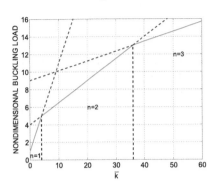

Fig. 14.29. Non-dimensional critical loads $P_{\mathrm{cr}\,n}/P_{\mathrm{Euler}}$ for $n = 1, 2,$ and 3.

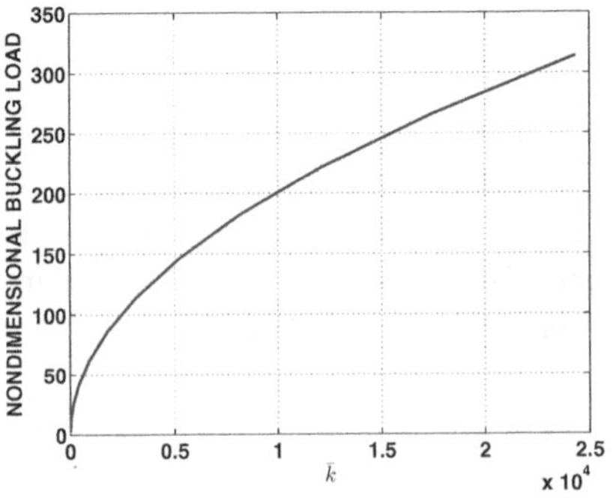

Fig. 14.30. Non-dimensional buckling load $P_{\mathrm{cr}}/P_{\mathrm{Euler}}$ versus $\bar{k}$.

be approximated as $P_{\mathrm{cr}}/P_{\mathrm{Euler}} \approx 2\sqrt{\bar{k}}$. This result, however, hides the constantly changing wave number as the elastic foundation stiffness $\bar{k}$ changes.

15

Shearing deformations in beams

15.1 Introduction

Euler-Bernoulli beam theory is developed in chapter 5 based on the purely kinematic assumptions discussed in section 5.1. In particular, the cross-section of the beam is assumed to remain plane after deformation, and furthermore, this plane is assumed to remain normal to the deformed axis of the beam. This second assumption implies the vanishing of the transverse shear strains, $\gamma_{12} = 0$, and leads to the following result for a beam made from a linearly elastic, homogeneous and isotropic material

$$V_2 = \int_{\mathcal{A}} \tau_{12} \, \mathrm{d}\mathcal{A} = \int_{\mathcal{A}} G\gamma_{12} \, \mathrm{d}\mathcal{A} = 0, \qquad (15.1)$$

where the second integral is the result of using the constitutive law relating shear stresses to shear strains, $\tau_{12} = G\gamma_{12}$.

On the other hand, equilibrium conditions require a non-vanishing transverse shear force, V_2, to equilibrate the distributed transverse load, $p_2(x_1)$, applied to the beam, see eq. (5.38). This apparent contradiction with eq. (15.1) can be resolved through the following reasoning: as required by equilibrium, the shear stress, τ_{12}, does not vanish, but the corresponding shear strain is vanishingly small. This implies a very large shearing modulus, $G \to \infty$, so that a vanishing shear strain $\gamma_{12} \to 0$, results in a product, $G\gamma_{12} = \tau_{12}$, that becomes a finite, non-vanishing quantity.

In view of this reasoning, the assumption "plane sections remain normal to the deformed axis of the beam," which implies the vanishing of the transverse shear strains, could be replaced by "the beam is made of a material with an infinite shear modulus." Because such a constitutive law is awkward, the transverse shear force (the stress resultant associated with the shear stress), is not evaluated from this constitutive law but from equilibrium considerations instead. In fact, the shear force is altogether eliminated from Euler-Bernoulli beam theory using equilibrium considerations, see eq. (5.39), and can be recovered from the bending moment as $V_2 = -\mathrm{d}M_3^c/\mathrm{d}x_1$, eq. (5.38).

In reality, the shear modulus is of the order of Young's modulus and for isotropic materials, $G = E/(2(1 + \nu))$. To investigate the effects of the additional flexibility of the beam introduced by shear deformation, a formulation must be developed that allows for non-vanishing transverse shear strains. In this case, the assumption "plane sections remain normal to the deformed axis of the beam" can no longer be made.

Two important questions must be addressed by this new formulation: (1) how do shearing deformations affect the transverse displacement of the beam, and (2) what is the distribution of shear stresses over the cross-section of the beam? The resulting theoretical description for shear deformable beams is generally referred to as *Timoshenko beam theory*.

15.1.1 A simplified approach

Consider a beam with a rectangular cross-section of width b and depth h subjected to a shear force, V_2. In this very simplified approach, the shear stresses are assumed to be uniformly distributed over the cross-section of the beam as depicted in fig. 15.1 which shows a differential segment along the length of a beam. The shear resultant can be calculated as

$$V_2 = \int_A \tau_{12} \, d\mathcal{A} = \tau_{12} \int_A d\mathcal{A} = \mathcal{A}\tau_{12}, \qquad (15.2)$$

and hence, the shear stress is $\tau_{12} = V_2/\mathcal{A}$.

If the beam is made of a homogeneous, linearly elastic material, the transverse shear strain is also uniformly distributed over the cross-section, $\gamma_{12} = \tau_{12}/G = V_2/(G\mathcal{A})$; hence, the local shear strain, γ_{12}, can also be interpreted as a sectional shear strain. The sectional constitutive law is then

$$V_2 = G\mathcal{A}\,\gamma_{12}, \qquad (15.3)$$

where $G\mathcal{A}$ is the sectional *shear stiffness*.

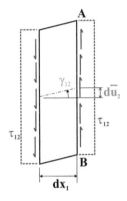

Fig. 15.1. Simplified deformation under shear.

Figure 15.1 shows that $d\bar{u}_1/dx_2 = 0$, and the transverse shear strain therefore becomes $\gamma_{12} = d\bar{u}_1/dx_2 + d\bar{u}_2/dx_1 = d\bar{u}_2/dx_1$. Equation (15.3) now implies

$$\frac{d\bar{u}_2}{dx_1} = \frac{V_2}{G\mathcal{A}}. \qquad (15.4)$$

This equation can be integrated to yield the transverse displacement field, $\bar{u}_2(x_1)$, associated with shear deformation.

Finally, the strain energy due to the shear deformation of a differential segment of the beam is

$$dA = \frac{1}{2}\int_A \tau_{12}\gamma_{12} \, d\mathcal{A}dx_1 = \frac{1}{2}\int_A \frac{\tau_{12}^2}{G} \, d\mathcal{A}dx_1 = \frac{1}{2}\frac{\tau_{12}^2}{G}\mathcal{A} \, dx_1 = \frac{1}{2}\frac{V_2^2}{G\mathcal{A}} \, dx_1. \qquad (15.5)$$

These developments provide a crude description of shear deformation effects in beams. The shear stress distribution is uniform over the cross-section, the sectional shear stiffness is GA, and the transverse displacements, $\bar{u}_2$, stemming from shear deformation can be obtained by integrating eq. (15.4).

Unfortunately, this simplified description is wrong because it violates a basic equilibrium condition. The principle of reciprocity of shear stresses, eq. (1.5), requires the shear stress, τ_{12}, to vanish at points **A** and **B**, see fig. 15.1, because the upper and lower faces of the beam are stress free. It follows that the shear stress distribution cannot possibly be uniform over the cross-section of the beam.

15.1.2 An equilibrium approach

To remedy the shortcomings of the simplified representation developed in the previous section, equilibrium conditions for the problem must be established. Figure 15.2 depicts a differential element of the beam with a rectangular cross-section of width b and depth h. Loading is applied in plane $(\bar{\imath}_2, \bar{\imath}_1)$, and consequently, stresses are uniform across the width of the section. Consider now a differential element in depth, as highlighted in fig. 15.2. Summing forces along axis $\bar{\imath}_1$ yields the following equilibrium equation

$$\frac{d\sigma_1}{dx_1} + \frac{d\tau_{12}}{dx_2} = 0. \tag{15.6}$$

This equation indicates that the distribution of shear stress through the depth of the section is related to the distribution of axial stress along the span of the beam. Consequently, simply assuming a certain shear stress distribution through the depth as is done in the previous section is unlikely to satisfy the basic equilibrium condition expressed by eq. (15.6).

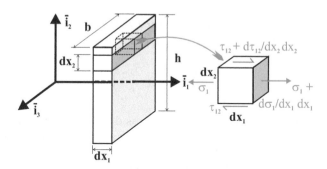

Fig. 15.2. Configuration of a differential element of the beam.

The purely kinematic Euler-Bernoulli assumptions state that the cross-section of the beam remains plane after deformation, and furthermore, this plane remains normal to the deformed axis of the beam. In the present development, the second assumption cannot be made, as it implies the vanishing of the transverse shear strains. The first assumption implies an axial displacement field in the form of eq. (5.2): $u_1(x_1, x_2, x_3) = -x_2 \Phi_3(x_1)$, where $\Phi_3(x_1)$ is the rotation of the section about axis $\bar{\imath}_3$.

For simplicity, the origin of the axis system is assumed to be located at the centroid of the section, *i.e.*, $x_{2c} = 0$. The axial strain distribution is then obtained as $\epsilon_1 = -x_2 d\Phi_3/dx_1$. The axial stress distribution follows from Hooke's law as $\sigma_1 = -Ex_2 d\Phi_3/dx_1$.

The basic equation of equilibrium, eq. (15.6), can now be used to solve for the shear stress distribution

$$\frac{d\tau_{12}}{dx_2} = -\frac{d\sigma_1}{dx_1} = Ex_2 \frac{d^2\Phi_3}{dx_1^2}. \tag{15.7}$$

Integration yields $\tau_{12} = Ex_2^2/2 \, d^2\Phi_3/dx_1^2 + c$, where c is an integration constant that can be evaluated by imposing the boundary condition, $\tau_{12}(x_2 = \pm h/2) = 0$, to find

$$\tau_{12} = \frac{1}{2}E\left(\frac{h}{2}\right)^2 \left[\left(\frac{2x_2}{h}\right)^2 - 1\right] \frac{d^2\Phi_3}{dx_1^2}. \tag{15.8}$$

This result is a parabolic distribution of the shear stress through the depth of the cross-section.

Next, the resultant shear force is obtained by integration of shear stress distribution over the cross-section, eq. (5.9), to find

$$V_2 = \frac{1}{2}E\left(\frac{h}{2}\right)^2 \frac{d^2\Phi_3}{dx_1^2} \int_{-h/2}^{h/2}\int_{-b/2}^{b/2} \left[\left(\frac{2x_2}{h}\right)^2 - 1\right] dx_2 dx_3$$

$$= -\frac{1}{3}EA\left(\frac{h}{2}\right)^2 \frac{d^2\Phi_3}{dx_1^2},$$

Finally, the sectional rotation, Φ_3, is eliminated with the help of eq. (15.8) to obtain the shear stress distribution in terms of the applied shear force,

$$\tau_{12} = \frac{3V_2}{2A}\left[1 - \left(\frac{2x_2}{h}\right)^2\right]. \tag{15.9}$$

Figure 15.3 shows this parabolic shear stress distribution through the depth of the section. For reference, the uniform distribution postulated in the simplified representation is also shown in the figure. The maximum shear stress, $\tau_{max} = 3V_2/2A$, occurs at $x_2 = 0$ and is 50% higher than that obtained for the uniform distribution, $\tau_{12} = V_2/A$. Clearly, the simplified representation is erroneous and grossly underestimates the maximum shear stress.

The strain energy associated with the shear deformation in a differential segment of the beam is

$$dA = \frac{1}{2}\int_A \frac{\tau_{12}^2}{G} \, dA dx_1 = \frac{1}{2}\int_{-h/2}^{h/2} \frac{b}{G}\left(\frac{3V_2}{2A}\right)^2 \left[1 - \left(\frac{2x_2}{h}\right)^2\right]^2 dx_2 dx_1.$$

After integration, the strain energy becomes

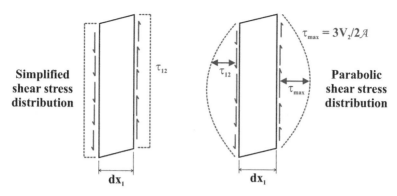

Fig. 15.3. Shear stress distribution for a rectangular section.

$$dA = \frac{1}{2} \frac{V_2^2}{GA} \frac{6}{5} = \frac{1}{2} \frac{V_2^2}{K_{22}} \, dx_1, \qquad (15.10)$$

where the sectional *shear stiffness*, K_{22}, is defined as the denominator in the expression for the strain energy in eq. (15.10). For for the rectangular cross-section,

$$K_{22} = \frac{5}{6} GA. \qquad (15.11)$$

This result should be contrasted with the sectional shear stiffness $K_{22} = GA$ found for the simplified solution that overestimates the shear stiffness by about 20%.

The deformation of the cross-section associated with the present representation is more complex than that obtained with the simplified model. Indeed, the parabolic shear stress distribution is associated with a parabolic distribution of shear strain. The following reasoning then implies that the axial displacement field must present the "S" shape shown in the right hand side of fig. 15.4. First, the vanishing of the shear strain at points **A** and **B** implies that the initial right angle between the beam's upper or lower surfaces and the cross-section plane must remain a right angle. Second, the change in the initially right angle between the beam axis and the cross-section plane, which is a direct measure of γ_{12}, will be maximum at the midpoint ($x_2 = 0$) of the cross-section. This "S" shaped axial displacement field directly contradicts the kinematic assumption underlying the present analysis: plane sections remain plane. Hence, the present solution is inconsistent.

Equivalent shear deformation model

Although the simplified deformation shown in fig. 15.4 is incorrect, it has the advantage of conveying a very simple picture of the deformation pattern. This raises the question as to whether it is possible to make the simplified and actual representations equivalent in some sense. One approach is to find the average shear strain, γ_{ave}, in the simplified deformation that will be equivalent to the more complex shear strain distribution obtained from equilibrium analysis.

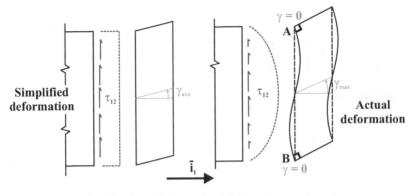

Fig. 15.4. The simplified and actual deformation configurations.

This equivalence can be obtained in an energy sense: the shear strain, γ_{ave}, can be selected so that the strain energies associated with the two deformation patterns are equal. From eqs. (15.5) and (15.10), this requires $1/2\ V_2^2/K_{22} = 1/2\ V_2\gamma_{ave}$ which can be solved for the equivalent average shear strain

$$\gamma_{ave} = \frac{V_2}{K_{22}}. \tag{15.12}$$

For the simplified model the strain-displacement relations reduces to $\gamma_{ave} = d\bar{u}_2/dx_1$, and hence,

$$\frac{d\bar{u}_2}{dx_1} = \frac{V_2}{K_{22}}. \tag{15.13}$$

This equation can be integrated to yield the transverse displacement field associated with shear deformation.

In summary, the following facts about shear deformation are established. The axial and shear stress distributions are related through an equilibrium condition, eq. (15.6). Assuming plane sections to remain plane, a linear distribution of axial stress is obtained for beam made of a linearly elastic, homogeneous material. The shear stress profile that is in equilibrium with this axial stress distribution can then be obtained from the equilibrium equation. A parabolic shear stress distribution, eq. (15.9), is found for a rectangular cross-section. The deformation of the section due to the parabolic variation of shear strain results in warping of the cross-section and violates the basic kinematic assumption of plane sections remaining plane. Consequently, the development presented here leads to an inconsistent and therefore approximate solution of the problem. Finally, the simplified deformation presented in the previous section, and the more realistic deformation presented here can be made equivalent in an energy sense by selecting the average sectional shear strain to be that given by eq. (15.12). The shear stiffness is found to be $K_{22} = 5/6\ GA$ for a rectangular cross-section.

Example 15.1. Shear distribution in a sandwich section

Consider the cross-section of a sandwich beam depicted in fig. 15.5. Such structures are typically made of two thin faces of thickness t_f and Young's modulus E_f that are designed to carry most of the bending stresses. The faces sandwich a core of much lighter material that is often highly anisotropic. The core material usually has a very low Young's modulus in the directions parallel to the faces, a moderate Young's modulus in the direction perpendicular to the faces, and a moderate shear modulus, G_c. A reasonable assumption is that Young's modulus for the core material, E_c, is orders of magnitude smaller than that of the faces; hence, setting $E_c \approx 0$ is a very reasonable approximation.

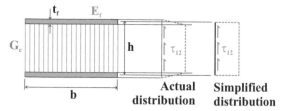

Fig. 15.5. Shear stress distribution on the cross-section of a sandwich section.

Under these assumptions, the bending stiffness of the sandwich section is due to the faces alone and is given by

$$H_{33}^c = 2 \left[\frac{bt_f^3 E_f}{12} + bt_t E_f \left(\frac{h}{2} \right)^2 \right] = \frac{1}{2} bt_t h^2 E_f \left[1 + \frac{1}{3} \left(\frac{t_f}{h} \right)^2 \right] \qquad (15.14)$$

where the first term represents the bending stiffness of the thin face with respect to its own centroid, and the second term is the transport term. The core does not significantly contribute to the bending stiffness because its intrinsic stiffness is negligible. In typical constructions, the face thickness is much smaller than the sandwich depth, i.e., $t_f/h \ll 1$, and the bending stiffness reduces to

$$H_{33}^c = \frac{1}{2} bt_t h^2 E_f. \qquad (15.15)$$

This results explains why sandwich structures are efficient lightweight structures in bending: the bending stiffness is proportional to the square of the sandwich depth, rather than the cube of the face thickness, while the low density core material contributes to a low overall density. In effect, the core material contributes little to the bending stiffness but keeps the two faces a distance h apart, dramatically increasing their contribution to the overall bending stiffness. These concepts are discussed in more details in section 5.5.7.

If the sandwich beam is subjected to shear forces, it is necessary to determine the shear stress distribution over the cross-section. Proceeding along the same path as in the previous section, the basic equation of equilibrium, eq. (15.6), is used to solve for the shear stress distribution in the lower face, to find

$$\tau_f = \frac{1}{2} E_f \left(\frac{h + t_f}{2} \right)^2 \left[\left(\frac{2x_2}{h + t_f} \right)^2 - 1 \right] \frac{d^2 \Phi_3}{dx_1^2}.$$

Because the axial stiffness of the core is negligible, the axial stress vanishes, and the basic equation of equilibrium, eq. (15.6), yields $d\tau_{12}/dx_2 = 0$. Hence, the shear stress is constant through the thickness of the core.

The shear stress must be continuous at the face/core interface implying that

$$\tau_c = \frac{1}{2} E_f \left(\frac{h + t_f}{2} \right)^2 \left[\left(\frac{h - t_f}{h + t_f} \right)^2 - 1 \right] \frac{d^2 \Phi_3}{dx_1^2} = -\frac{1}{2} E_f h t_f \frac{d^2 \Phi_3}{dx_1^2}.$$

In view of the symmetry of the problem, the shear stress in the top face is a mirror image of that in the lower face, as depicted in fig. 15.5. The shear force is then obtained by integrating this shear stress distribution over the cross-section to find

$$V_2 = -\frac{1}{2} E_f b h^2 t_f \left[1 + \frac{1}{3} \left(\frac{t_f}{h} \right)^2 \right] \frac{d^2 \Phi_3}{dx_1^2} \approx -\frac{1}{2} E_f b h^2 t_f \frac{d^2 \Phi_3}{dx_1^2}.$$

Eliminating $d^2\Phi_3/dx_1^2$ between the previous two equations yields the shear stress in the core in term of the applied shear force

$$\tau_c = \frac{V_2}{bh}.$$

This result indicates that the shear force is carried entirely by a uniform shear stress in the core. The shear stress in the faces contribute little to the load carrying capability of the sandwich, although the faces carry most of the bending stress, σ_1.

Finally, the strain energy associated with the shear deformation in the sandwich beam is

$$dA = \frac{1}{2} \int_A \frac{\tau_{12}^2}{G} \, dA dx_1 = \frac{b}{2} \int_{-h/2}^{h/2} \frac{\tau_c^2}{G_c} \, dx_2 dx_1 = \frac{1}{2} \frac{\tau_c^2}{G_c} bh \, dx_1 = \frac{1}{2} \frac{V_2^2}{bhG_c} \, dx_1.$$

It follows that the shear stiffness of the sandwich is

$$K_{22} = bhG_c. \tag{15.16}$$

This analysis reveals an important aspect of the structural behavior of sandwich structures. Whereas the bending stiffness and strength of sandwich beams are inherited solely from the stiffness and strength characteristics of the faces, the shear stiffness and strength of these structures are inherited from the stiffness and strength characteristics of the core. Because the mechanical properties of the core are generally much lower than those of the faces, shear failure in the core can possibly occur for applied loads far below those that would result in failure of the faces. Consequently, the evaluation of shear stress distributions plays an important role in the analysis and design of sandwich structures.

15.1.3 Problems

Problem 15.1. Simply supported beam under uniform load

Consider a beam with a solid rectangular cross-section under shear loads. In the process of developing a theory to model the deformation of the beam based on the assumption that plane sections remain plane, the two different shear stress distributions and associated shear strain distributions have been obtained and are depicted in fig. 15.4. *(1)* Why is the simplified deformation mode an incorrect solution? *(2)* Why is the improved deformation mode an incorrect solution?

Problem 15.2. Simply supported beam under uniform load

Consider a simply supported beam of length L subjected to a uniform transverse load p_0. The beam has a rectangular section of width b and depth h and is made of a homogeneous material. *(1)* Find the magnitude and location of the maximum axial stress in the structure. Sketch the axial stress distribution for this section. *(2)* Find the magnitude and location of the maximum shear stress in the structure. Sketch the shear stress distribution for this section. *(3)* Is the mid-span deflection of the beam affected by shearing deformations? How can you assess the importance of this effect? *(4)* Is the mid-span bending moment in the beam affected by shearing deformations? How can you assess the importance of this effect? *(5)* If the beam is cantilevered at both ends, how would your answers to the last two questions change?

Problem 15.3. Cross-section made of two materials

The cross-section of a sandwich beam is depicted in fig. 15.6. The subscript $(\bullet)_c$ refers to core quantities, whereas the subscript $(\bullet)_f$ refers to facing quantities. *(1)* Determine the shear stress distribution over the cross-section. *(2)* Evaluate the corresponding shear force V_2. *(3)* Plot the shear stress distribution $\tau_{12}bh/V_2$ and compute the shearing stiffness of the section $K_{22}/(G_cbh)$. Use $h_c/h = 2/3.6$, $E_c/E_f = 70/140$, $G_c/G_f = 70/140$. *(4)* Plot the shear stress distribution $\tau_{12}bh/V_2$ and compute the shearing stiffness of the section $K_{22}/(G_cbh)$. Con-

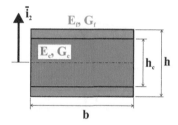

Fig. 15.6. Cross-section made of two materials.

sider a sandwich with a soft core $E_c \approx 0$ and very thin facings $t/h \ll 1$, $t = (h - h_c)/2$.

15.2 Shear deformable beams: an energy approach

Consider the cantilevered beam subjected to distributed transverse loads, $p_2(x_1)$, distributed bending moments, $q_3(x_1)$, a concentrated tip load, P_2, and a bending moment, Q_3, as depicted in fig. 15.7. The transverse displacement is assumed to take place in plane $(\bar{\imath}_2, \bar{\imath}_1)$. The strain energy stored in the beam can be evaluated using the general expression for the strain energy in a three-dimensional solid, eq. (10.46),

$$A = \frac{1}{2} \int_0^L \int_{\mathcal{A}} \underline{\sigma}^T \underline{\epsilon} \, d\mathcal{A} dx_1,$$

where L is the length of the beam and $\mathcal{A}$ its cross-sectional area.

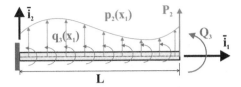

Fig. 15.7. Cantilevered beam under distributed and concentrated transverse loads and bending moments.

The kinematics are shown in fig. 15.8 where a cross-section containing point **P** is shown before and after deformation. Plane sections of the beam are assumed to remain plane, leading to an axial displacement field in the form of eq. (5.2): $u_1(x_1, x_2, x_3) = -x_2 \Phi_3(x_1)$, where $\Phi_3(x_1)$ is the rotation of the section about axis $\bar{\imath}_3$ as shown in fig. 15.8. For simplicity, the centroid of the section is assumed to be located of the origin of the axis system, *i.e.*, $x_{2c} = 0$. The cross-section of the beam is also assumed to remain rigid in its own plane, leading to a transverse displacement field in the form of eq. (5.1), $u_2(x_1, x_2, x_3) = \bar{u}_2(x_1)$.

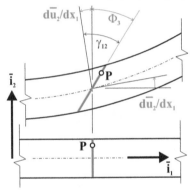

Fig. 15.8. Kinematic assumptions for shear deformable beams.

The axial strain distribution is then obtained as $\epsilon_1 = \partial \bar{u}_1 / \partial x_1 = -x_2 \, \mathrm{d}\Phi_3 / \mathrm{d}x_1$, and the transverse shear strain is

$$\gamma_{12} = \frac{\partial u_1}{\partial x_2} + \frac{\partial u_2}{\partial x_1} = -\Phi_3 + \frac{\mathrm{d}\bar{u}_2}{\mathrm{d}x_1}. \quad (15.17)$$

In Euler-Bernoulli theory, the transverse shear strain is assumed to vanish, $\gamma_{12} = 0$, and hence the rotation of the section, Φ_3, is equal to the slope of the beam, $\mathrm{d}\bar{u}_2 / \mathrm{d}x_1$. In the present development, illustrated in fig. 15.8, the shear strain given by eq. (15.17) is the difference between the slope of the beam and the rotation of its cross-section. The beam slope and sectional rotation are the independent variables of the theory.

The strain energy expression now reduces to

$$A = \frac{1}{2} \int_0^L \int_{\mathcal{A}} (\sigma_1 \varepsilon_1 + \tau_{12} \gamma_{12}) \, \mathrm{d}\mathcal{A} \mathrm{d}x_1.$$

The stress components, σ_2, σ_3, and τ_{23} acting in the plane of the cross-section are assumed to be much smaller than the axial stress, σ_1, and hence, Hooke's law implies $\sigma_1 = E\epsilon_1$ and $\tau_{12} = G\gamma_{12}$. Introducing these constitutive relationships and strain fields into the strain energy yields

$$A = \frac{1}{2} \int_0^L \left\{ \left[\int_{\mathcal{A}} E x_2^2 \, \mathrm{d}\mathcal{A} \right] \left(\frac{\mathrm{d}\Phi_3}{\mathrm{d}x_1} \right)^2 + \left[\int_{\mathcal{A}} G \, \mathrm{d}\mathcal{A} \right] \left(\frac{\mathrm{d}\bar{u}_2}{\mathrm{d}x_1} - \Phi_3 \right)^2 \right\} \, \mathrm{d}x_1.$$

The term in the first square bracket is the bending stiffness of the beam, H_{33}^c, see eq. (5.36). The curvature of the beam is still $d\Phi_3/dx_1$, but in view of eq. (15.17) it does not equal $d^2\bar{u}_2/dx_1^2$ as is the case for Euler-Bernoulli beam theory. The two expressions are different now because the slope of the beam no longer equals the rotation of the section. The term in the second square bracket is the shear stiffness, and for beams made of homogeneous, linearly elastic material, this stiffness is GA, as developed in section 15.1.1. With these results, the strain energy becomes

$$A = \frac{1}{2} \int_0^L \left[H_{33}^c \left(\frac{d\Phi_3}{dx_1} \right)^2 + GA \left(\frac{d\bar{u}_2}{dx_1} - \Phi_3 \right)^2 \right] dx_1.$$

At this point, the transverse shear strain is assumed to be uniformly distributed over the cross-section, see eq. (15.17). This is a direct consequence of assuming plane sections to remain plane after deformation, and therefore, the strain distribution is identical to that of the simplified representation discussed in section 15.1.1. This explains why the shear stiffness, GA, in the present formulation is identical to that for the simplified representation.

As discussed in section 15.1.2, the simplified and equilibrium based representations can be made equivalent from an energy perspective by selecting the average shear strain to be $\gamma_{ave} = V_2/K_{22}$, see eq. (15.12). With the help of eqs. (15.10) and (15.17), the shear strain energy for the equilibrium based representation is then $dA = 1/2\, K_{22}\gamma_{ave}^2\, dx_1 = 1/2\, K_{22}(d\bar{u}_2/dx_1 - \Phi_3)^2$. Thus, it is only necessary to replace the shear stiffness, GA, in the above equation by its more accurate counterpart, K_{22}, to obtain the shearing strain energy corresponding to the equilibrium based representation. Consequently, the following expression gives the strain energy stored in a beam subjected to bending and shearing

$$A = \frac{1}{2} \int_0^L \left[H_{33}^c \left(\frac{d\Phi_3}{dx_1} \right)^2 + K_{22} \left(\frac{d\bar{u}_2}{dx_1} - \Phi_3 \right)^2 \right] dx_1, \tag{15.18}$$

where the first term represents the strain energy associated with bending of the beam and the second is that associated with shearing. The total potential energy of the complete beam system can now be expressed as

$$\Pi = \frac{1}{2} \int_0^L \left[H_{33}^c \left(\frac{d\Phi_3}{dx_1} \right)^2 + K_{22} \left(\frac{d\bar{u}_2}{dx_1} - \Phi_3 \right)^2 \right] dx_1$$
$$- \int_0^L (p_2\bar{u}_2 + q_3\Phi_3)\, dx_1 - P_2\bar{u}_2(L) - Q_3\Phi_3(L). \tag{15.19}$$

Given the total potential energy, the principle of minimum total potential energy can be used to derive the governing differential equations of the problem and associated boundary conditions. This principle requires the total potential energy to be stationary for all arbitrary choices of the displacement fields $\bar{u}_2(x_1)$ and $\Phi_3(x_1)$, leading to

$$\delta\Pi = \int_0^L \left[H_{33}^c \frac{\mathrm{d}\Phi_3}{\mathrm{d}x_1} \delta\left(\frac{\mathrm{d}\Phi_3}{\mathrm{d}x_1}\right) + K_{22}\left(\frac{\mathrm{d}\bar{u}_2}{\mathrm{d}x_1} - \Phi_3\right)\delta\left(\frac{\mathrm{d}\bar{u}_2}{\mathrm{d}x_1} - \Phi_3\right)\right]\mathrm{d}x_1$$

$$- \int_0^L (p_2\,\delta\bar{u}_2 + q_3\delta\Phi_3)\,\mathrm{d}x_1 - P_2\delta\bar{u}_2(L) - Q_3\delta\Phi_3(L) = 0.$$

Integration by parts and interchanging the variational and differential operators then yields the following expression

$$\delta\Pi = \int_0^L \delta\Phi_3 \left\{ -\frac{\mathrm{d}}{\mathrm{d}x_1}\left(H_{33}^c \frac{\mathrm{d}\Phi_3}{\mathrm{d}x_1}\right) - K_{22}\left(\frac{\mathrm{d}\bar{u}_2}{\mathrm{d}x_1} - \Phi_3\right) - q_3 \right\}\mathrm{d}x_1$$

$$+ \int_0^L \delta\bar{u}_2 \left\{ -\frac{\mathrm{d}}{\mathrm{d}x_1}\left[K_{22}\left(\frac{\mathrm{d}\bar{u}_2}{\mathrm{d}x_1} - \Phi_3\right)\right] - p_2 \right\}\mathrm{d}x_1 + \left[H_{33}^c \frac{\mathrm{d}\Phi_3}{\mathrm{d}x_1}\delta\Phi_3\right]_0^L$$

$$+ \left[K_{22}\left(\frac{\mathrm{d}\bar{u}_2}{\mathrm{d}x_1} - \Phi_3\right)\delta\bar{u}_2\right]_0^L - P_2\delta\bar{u}_2(L) - Q_3\delta\Phi_3(L) = 0.$$

$$(15.20)$$

This expression must vanish for all arbitrary variations $\delta\Phi_3$ and $\delta\bar{u}_2$, and therefore the two integrand terms in braces must vanish, leading to the Euler-Lagrange equations for the problem

$$\frac{\mathrm{d}}{\mathrm{d}x_1}\left(H_{33}^c \frac{\mathrm{d}\Phi_3}{\mathrm{d}x_1}\right) + K_{22}\left(\frac{\mathrm{d}\bar{u}_2}{\mathrm{d}x_1} - \Phi_3\right) = -q_3, \qquad (15.21\mathrm{a})$$

$$\frac{\mathrm{d}}{\mathrm{d}x_1}\left[K_{22}\left(\frac{\mathrm{d}\bar{u}_2}{\mathrm{d}x_1} - \Phi_3\right)\right] = -p_2. \qquad (15.21\mathrm{b})$$

For the cantilevered beam depicted in fig. 15.7, the geometric boundary conditions at the root of the beam are $\bar{u}_2 = 0$ and $\Phi_3 = 0$. At the tip of the beam, the variations $\delta\Phi_3(L)$ and $\delta\bar{u}_2(L)$ are arbitrary, and the boundary terms of eq. (15.20) yield the following boundary conditions

$$H_{33}^c \frac{\mathrm{d}\Phi_3}{\mathrm{d}x_1} = Q_3, \text{ and } K_{22}\left(\frac{\mathrm{d}\bar{u}_2}{\mathrm{d}x_1} - \Phi_3\right) = P_2. \qquad (15.22)$$

These two natural boundary conditions are readily interpreted as $M_3^c = Q_3$ and $V_2 = P_2$, which are the equilibrium equations at the loaded end of the beam.

In summary, the problem of a shear deformable beam is governed by two second order, coupled ordinary differential equations, eqs. (15.21), for the two unknown displacement fields, $\bar{u}_2(x_1)$ and $\Phi_3(x_1)$. Four boundary conditions are required to solve these equations. Once these differential equations are solved, the bending moment and shear force distributions can be recovered as $M_3^c = H_{33}^c\,\mathrm{d}\Phi_3/\mathrm{d}x_1$ and $V_2 = K_{22}\,(\mathrm{d}\bar{u}_2/\mathrm{d}x_1 - \Phi_3)$, respectively.

The governing equations, eqs. (15.21), can be rewritten in terms of stress resultants as $\mathrm{d}M_3^c/\mathrm{d}x_1 + V_2 = -q_3$, and $\mathrm{d}V_2/\mathrm{d}x_1 = -p_2$. They represent the equilibrium conditions for a differential element of the beam and are identical to those obtained for Euler-Bernoulli beam theory, see eq. (5.38). This should be expected because equilibrium equations always apply, no matter what kinematic assumptions are made.

While the equilibrium equations are identical for Euler-Bernoulli and shear deformable beam theories, the relationships that involve kinematic quantities are different. For instance, in shear deformable beams, the bending moment-curvature relationship is $M_3^c = H_{33}^c d\Phi_3/dx_1$. Therefore, it would be erroneous to use the corresponding relationship from Euler-Bernoulli beam theory, *i.e.*, $M_3^c \neq H_{33}^c d^2\bar{u}_2/dx_1^2$.

Finally, eq. (15.19) for the total potential energy can also be used to obtain approximate solutions for shear deformable beam problems in a manner similar to that described in section 11.4.2 for Euler-Bernoulli beams. Both transverse displacement *and* sectional rotation fields must be approximated with the help of suitable sets of shape functions. This technique will be demonstrated in later sections.

15.2.1 Shearing effects on beam deflections

Consider the uniform, cantilevered beam of length L subjected to a concentrated transverse load P acting at a distance αL from its root, as depicted in fig. 15.9. The differential equilibrium equations for the beam are given by eqs. (15.21). For the first segment of the beam with $0 \leq x_1 \leq \alpha L$, the differential equations are

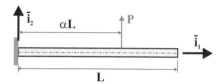

Fig. 15.9. Cantilevered beam with concentrated transverse load.

$$H_{33}^c \frac{d^2\Phi_3}{dx_1^2} + K_{22}\left(\frac{d\bar{u}_2}{dx_1} - \Phi_3\right) = 0, \quad \frac{d}{dx_1}\left[K_{22}\left(\frac{d\bar{u}_2}{dx_1} - \Phi_3\right)\right] = 0.$$

The boundary conditions at the beam's root are $\bar{u}_2 = 0$ and $\Phi_3 = 0$, and at $x_1 = \alpha L$, they are $K_{22}(d\bar{u}_2/dx_1 - \Phi_3) = P$ and $H_{33}^c d\Phi_3/dx_1 = 0$.

Integration of the second governing equation yields $K_{22}(d\bar{u}_2/dx_1 - \Phi_3) = c_1$, where c_1 is an integration constant which is determined from the boundary condition at $x_1 = \alpha L$ to be $c_1 = P$. Hence $K_{22}(d\bar{u}_2/dx_1 - \Phi_3) = P$. Because $K_{22}(d\bar{u}_2/dx_1 - \Phi_3) = K_{22}\gamma_{\text{ave}} = V_2$, this means that $V_2 = P$, and therefore the transverse shear force is constant along the span of the beam, a result that could have been obtained from simple equilibrium considerations.

Introducing this result into the first governing equation yields $H_{33}^c d^2\Phi_3/dx_1^2 + P = 0$. With the help of the boundary conditions, this integrates to

$$\Phi_3 = \frac{PL^2}{H_{33}^c}\left(-\frac{1}{2}\eta^2 + \alpha\eta\right),$$

where $\eta = x_1/L$ is the non-dimensional variable along the span of the beam.

Finally, the transverse displacement field is found by integrating $d\bar{u}_2/dx_1 = \Phi_3 + P/K_{22}$ and using the boundary condition, $\Phi_3(0) = 0$, to find

$$\bar{u}_2 = \frac{PL^3}{6H_{33}^c} \left(-\eta^3 + 3\alpha\eta^2\right) + \frac{PL}{K_{22}}\eta.$$

A similar process can be followed to find the displacement field over the second segment of the beam, $\alpha L \leq x_1 \leq L$. The boundary conditions at the beam's tip are $V_2 = K_{22} \left(\mathrm{d}\bar{u}_2/\mathrm{d}x_1 - \Phi_3\right) = 0$ and $M_3^c = H_{33}^c \, \mathrm{d}\Phi_3/\mathrm{d}x_1 = 0$, whereas continuity conditions on $\bar{u}_2$ and Φ_3 are applied at $x_1 = \alpha L$.

The transverse displacement field for the complete beam becomes

$$\bar{u}_2(\eta) = \frac{PL^3}{6H_{33}^c} \begin{cases} \eta^2(3\alpha - \eta) + 6\bar{s}^2\eta, & 0 \leq \eta \leq \alpha, \\ \alpha^2(3\eta - \alpha) + 6\bar{s}^2\alpha, & \alpha \leq \eta \leq 1, \end{cases} \tag{15.23}$$

where

$$\bar{s}^2 = \frac{H_{33}^c}{K_{22}L^2} \tag{15.24}$$

is a non-dimensional *shear flexibility parameter* defining the shear flexibility relative to the bending flexibility.

The sectional rotation is

$$\Phi_3(\eta) = \frac{PL^2}{2H_{33}^c} \begin{cases} \eta(-\eta + 2\alpha), & 0 \leq \eta \leq \alpha, \\ \alpha^2, & \alpha \leq \eta \leq 1. \end{cases} \tag{15.25}$$

The non-dimensional shear flexibility parameter, $\bar{s}^2$, measures the importance of shear deformations relative to bending deformations. For an Euler-Bernoulli beam, the shear modulus is assumed to be infinite, the shear stiffness is then also infinite, and the shear flexibility parameter vanishes. Using $\bar{s}^2 = 0$ in eq. (15.23), the transverse displacement for an Euler-Bernoulli beam is recovered, see eq. (5.55).

Example 15.2. Shear deformation in a tip loaded cantilevered beam
Consider the case of a cantilevered beam with a tip load, P. The tip deflection is obtained by introducing $\alpha = 1$ and $\eta = 1$ into eq. (15.23) to find

$$\bar{u}_2(1) = \frac{PL^3}{3H_{33}^c} \left(1 + 3\bar{s}^2\right). \tag{15.26}$$

The first term represents the tip deflection due to bending, denoted δ_b, and the second is the additional contribution due to shear deformation, denoted δ_s. The ratio of these two contributions is

$$\frac{\delta_s}{\delta_b} = 3\bar{s}^2. \tag{15.27}$$

If the shear stiffness decreases, the shear flexibility parameter increases and the tip deflection due to shear deformations becomes more and more pronounced compared to that due to bending.

To relate the shear flexibility parameter to the physical characteristics of beams, the two cross-sections depicted in fig. 15.10 will be investigated. Consider first a solid rectangular section of width b and depth h made of a linearly elastic, homogeneous material. The shear flexibility parameter is

$$\bar{s}^2 = \frac{bh^3 E}{12} \frac{6}{5bhG} \frac{1}{L^2} = \frac{1}{10} \left(\frac{E}{G} \right) \left(\frac{h}{L} \right)^2. \tag{15.28}$$

The shear flexibility parameter is the product of two ratios: a material property ratio, E/G, and a geometric aspect ratio, h/L. The material property ratio is the ratio of Young's modulus to the shearing modulus. For a linearly elastic, homogeneous and isotropic material, this ratio is $E/G = 2(1 + \nu)$. The geometric aspect ratio is the ratio of the depth of the beam to its length, and it is a powerful contributor to the shear flexibility parameter because the ratio is squared.

For many engineering materials such as steel, aluminum, or titanium, Poisson's ratio is about 0.3, and the shear flexibility parameter becomes solely a function of the geometric aspect ratio, $\bar{s}^2 = 0.26(h/L)^2$. For long beams, the shear flexibility parameter quickly tends to zero, and shear deformation effects quickly become negligible. It is perhaps more useful to consider the reverse reasoning. As the beam's length decreases, bending deformations decrease fast, and shear deformations to become relatively more pronounced.

Example 15.3. Shear deformation of a sandwich beam
Consider next a sandwich section such as that depicted in the left part of fig. 15.10 as "cross-section A." The bending and shearing stiffnesses are given by eqs. (15.15) and (15.16), respectively, and hence, the shear flexibility parameter becomes

$$\bar{s}^2 = \frac{tbh^2 E_f}{2} \frac{1}{bhG_c} \frac{1}{L^2} = \frac{1}{2} \left(\frac{E_f}{G_c} \right) \left(\frac{t_f}{h} \right) \left(\frac{h}{L} \right)^2. \tag{15.29}$$

The shear flexibility parameter is again the product of two ratios: a material property ratio, E_f/G_c, and a geometric aspect ratio, h/L. The ratio t_f/h characterizes the relative thickness of the faces. The material property ratio is the ratio of Young's modulus of the faces, E_f, to the shearing modulus of the core, G_c. There is, of course, no intrinsic relationship between these two moduli because they are the properties of two different materials.

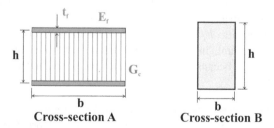

Fig. 15.10. Two different cross-sections for a beam.

Consider, for instance, a sandwich with aluminum faces ($E_f = 73$ GPa), and an aluminum honeycomb core ($G_c = 1$ GPa). The material ratio is then $E_f/G_c = 73$, as compared to $E/G = 2.6$ for a homogeneous aluminum beam. This implies much larger shear flexibility parameters for sandwich constructions as compared solid sections made of homogeneous material. The geometric aspect ratio is the ratio of the depth of the beam to its length and its effect is similar to that discussed earlier.

To quantitatively assess the importance of shear deformations, the following two cases will be considered. The first is an aluminum beam ($E/G = 2.6$) with a rectangular cross-section shown as "cross-section B" in fig. 15.10. Two beam aspect ratios will be investigated, a long beam, $h/L = 1/10$, and a shorter beam, $h/L = 1/5$.

The second case is a sandwich beam with thin composite faces, $t_f/h = 1/10$, and an aluminum honeycomb core, $E_f/G_c = 70 \times 10^9/1 \times 10^9 = 70$, shown as "cross-section A" in fig. 15.10. Here again, both long and short beams will be considered.

Table 15.1 summarizes the results for the two cases, listing the values of the shear flexibility parameters given by eqs. (15.28) and (15.29), respectively, and the corresponding relative magnitude of the transverse displacement due to shear deformations at the tip of the beam, see eq. (15.27).

Table 15.1. Effect of shear deformation on the tip deflection of a tip-loaded cantilevered beam.

	Rectangular section		Sandwich section	
h/L	1/10	1/5	1/10	1/5
$\bar{s}^2$	2.6×10^{-3}	10.4×10^{-3}	35.0×10^{-3}	140.0×10^{-3}
$\delta_s/\delta_b(\%)$	0.78%	3.1%	10.5%	42%

For the beam with a solid rectangular section, the results indicate a rather small influence of shearing deformations. Only for very short beams do shearing effects become significant; beam theory itself, however, is valid only for structures having one dimension larger than the other two, i.e., $L/h \gg 1$. Hence, shearing effects are unlikely to be very significant for such structures.

This contrasts with the case of sandwich beams. As indicated in table 15.1, much larger values of the shear flexibility parameter are found for these structures, resulting in much larger contributions of shear deformations to the total transverse displacement. With a 42% contribution to the total transverse displacement for a sandwich beam with an aspect ratio $h/L = 1/5$, shear deformations cannot be ignored. Of course, the stress resultants in the beam are identical to those obtained from Euler-Bernoulli beam theory, because stress resultants can be evaluated from equilibrium considerations alone.

Example 15.4. Cantilevered beam with intermediate support

A cantilevered beam of span L features an intermediate support at location $x_1 = \alpha L$ and is subjected to a tip load, P, as depicted in fig. 15.11. This is a hyperstatic configuration, and hence the reactions cannot be determined without considering the deformation of the beam. The force method from section 4.3.3 will be employed and both bending and shearing deflections will be included.

To begin, the intermediate support is replaced by a reaction force, R, of unspecified magnitude. The transverse displacement field under this combined loading then follows from eq. (15.23)

$$\bar{u}_2(\eta) = \frac{L^3}{6H_{33}^c} \begin{cases} R\left[\eta^2(3\alpha - \eta) + 6\bar{s}^2\eta\right] + P\left[\eta^2(3 - \eta) + 6\bar{s}^2\eta\right], & 0 \le \eta \le \alpha, \\ R\left[\alpha^2(3\eta - \alpha) + 6\bar{s}^2\alpha\right] + P\left[\eta^2(3 - \eta) + 6\bar{s}^2\eta\right], & \alpha \le \eta \le 1, \end{cases}$$

where $\eta = x_1/L$ is the non-dimensional variable along the beam's span.

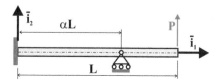

Fig. 15.11. Cantilevered beam with intermediate support.

The unknown reaction force at the support point is found by imposing the vanishing of the transverse displacement at $\eta = \alpha$. From $\bar{u}_2(\alpha) = 0$, it follows that

$$R = -\frac{\alpha(3 - \alpha) + 6\bar{s}^2}{2\alpha^2 + 6\bar{s}^2} P = -\mu P. \tag{15.30}$$

Because this problem is hyperstatic, the reaction force depends on the deformations of the system, and in this case, it depends on the magnitude of shearing deformations as is evident from the presence of the shear flexibility parameter in the above expression. Substituting for the support reaction force, R, from eq. (15.30) yields the transverse displacement in terms of the applied load

$$\bar{u}_2(\eta) = \frac{PL^3}{6H_{33}^c} \begin{cases} \left[-(1 - \mu)\eta^3 + 3(1 - \alpha\mu)\eta^2 + 6(1 - \mu)\bar{s}^2\eta\right], & 0 \le \eta \le \alpha, \\ \left[-\eta^3 + 3\eta^2 - 3\mu\alpha^2\eta + \mu\alpha^3 + 6(\eta - \alpha\mu)\bar{s}^2\right], & \alpha \le \eta \le 1, \end{cases}$$

Next, the rotation of the section is found with the help of eq. (15.25)

$$\Phi_3(\eta) = \frac{L^2}{2H_{33}^c} \begin{cases} R\left(-\eta^2 + 2\alpha\eta\right) + P\left(-\eta^2 + 2\eta\right), & 0 \le \eta \le \alpha, \\ R\left(\alpha^2\right) + P\left(-\eta^2 + 2\eta\right), & \alpha \le \eta \le 1, \end{cases}$$

and introducing the support reaction force yields the sectional rotation in terms of the applied load

$$\Phi_3(\eta) = \frac{PL^2}{2H_{33}^c} \begin{cases} \left[-(1 - \mu)\eta^2 + 2(1 - \alpha\mu)\eta\right], & 0 \le \eta \le \alpha, \\ \left(-\eta^2 + 2\eta - \mu\alpha^2\right), & \alpha \le \eta \le 1. \end{cases}$$

The bending moment distribution in the beam then follows from the sectional constitutive law, $M_3 = d\Phi_3/dx_1$, as

$$M_3(\eta) = PL \begin{cases} [-(1-\mu)\eta + (1-\alpha\mu)], & 0 \leq \eta \leq \alpha, \\ (1-\eta), & \alpha \leq \eta \leq 1. \end{cases}$$

Finally, the shear force distribution is found from the sectional constitutive law, $V_2 = K_{22}(d\bar{u}_2/dx_1 - \Phi_3)$, as

$$V_2(\eta) = P \begin{cases} (1-\mu), & 0 \leq \eta \leq \alpha, \\ 1, & \alpha \leq \eta \leq 1. \end{cases}$$

Figure 15.12 shows the transverse displacement distribution and sectional rotation of the beam for $\alpha = 0.6$. The first case shown on the figure is an Euler-Bernoulli solution obtained by setting $\bar{s}^2 = 0$ in the above expressions. For the second case, a beam with an aspect ratio, $h/L = 1/10$, is assumed with a solid rectangular cross-section made of aluminum. The shear flexibility parameter, $\bar{s}^2$, given by eq. (15.28), is $\bar{s}^2 = 0.0026$. The third case features a sandwich beam with $t_f/h = 1/10$ and $E_f/G_c = 70$. The shear flexibility parameter, $\bar{s}^2$, given by eq. (15.29), is $\bar{s}^2 = 0.035$, assuming $h/L = 1/10$. Little difference is observed between the Euler-Bernoulli solution ($\bar{s}^2 = 0$) and the shear deformable solution with $\bar{s}^2 = 0.0026$. For the sandwich structures, however, the beam's tip displacement is markedly larger, 67%, as compared to Euler-Bernoulli predictions.

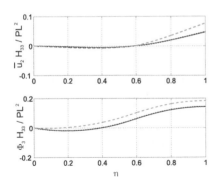

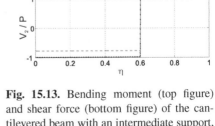

Fig. 15.12. Transverse displacement (top figure) and sectional rotation (bottom figure) of the cantilevered beam with an intermediate support, $\alpha = 0.6$. Shear flexibility parameter $\bar{s}^2 = 0$, solid line; $\bar{s}^2 = 0.0026$, dotted line; $\bar{s}^2 = 0.035$, dashed line.

Fig. 15.13. Bending moment (top figure) and shear force (bottom figure) of the cantilevered beam with an intermediate support, $\alpha = 0.6$. Shear flexibility parameter $\bar{s}^2 = 0$, solid line; $\bar{s}^2 = 0.0026$, dotted line; $\bar{s}^2 = 0.035$, dashed line.

Figure 15.13 shows the bending moment and shear force distributions along the beam for the same three cases. Here again, little difference is observed between the Euler-Bernoulli solution ($\bar{s}^2 = 0$) and the shear deformable solution with $\bar{s}^2 = 0.0026$. For the overhanging portion of the beam, $\alpha \leq \eta \leq 1$, the stress resultants are readily obtained from equilibrium considerations and hence, are identical for all values of the shear flexibility parameter.

For the remaining portion of the beam, $0 \le \eta \le \alpha$, the stress resultants are dependent on the reaction force at the support. This reaction force is obtained from a kinematic conditions, $\bar{u}_2(\eta = \alpha) = 0$, and is a function of the deformation of the beam. Consequently, in this portion of the beam, both displacements *and* stress resultants are affected by shearing deformations.

Two features of the shear deformable solution are worth noting. First, the root slope of the beam does not vanish; indeed, at $\eta = 0$, $d\bar{u}_2/dx_1 = PL^2/H_{33}^c (1 - \mu)\bar{s}^2 = (1 - \mu)P/K_{22} = V_2/K_{22} = \gamma_{12}$. Clearly, the root shear strain does not vanish, and hence, the slope of the beam is not zero, although the sectional rotation does indeed vanish. Second, the slope of the beam is discontinuous at the intermediate support. To see this, let $\alpha^- = \alpha - \varepsilon$ and $\alpha^+ + \varepsilon$ where $\varepsilon \to 0$ denote stations of the beam just before and after the intermediate support, respectively. It is then readily found that $d\bar{u}_2(\alpha^+)/dx_1 - d\bar{u}_2(\alpha^-)/dx_1 = PL^2/H_{33}^c \mu\bar{s}^2 = P\mu/K_{22} = -R/K_{22}$. Clearly, the slope discontinuity is a direct consequence of the shear force discontinuity at the same location.

Example 15.5. *Unit load method for beams including shear deformations*
The transverse deflections of beams including shear deformations can be calculated using the unit load method developed section 9.7.6. The unit load method is a direct application of the principle of complementary virtual work and states that $\Delta\delta D + \delta W_I' = 0$, where Δ is the prescribed displacement, δD the virtual driving force, and $\delta W_I'$ the complementary internal virtual work in the beam. For an Euler-Bernoulli beam, this latter quantity is given by eq. (9.79b). To include shear deformations in the unit load method, the complementary internal virtual work done by virtual shear forces undergoing actual shear strains must also be taken into account.

For beams presenting symmetry with respect to plane $(\bar{\imath}_1, \bar{\imath}_2)$ and subjected to bending moments M_3 only, the complementary internal virtual work reduces to $\delta W_I' = -\int_0^L \kappa_3 \delta M_3 dx_1$. With the addition of the complementary internal virtual work done by virtual shear forces undergoing actual shear strains, this becomes

$$\delta W_I' = - \int_0^L \left(\kappa_3 \delta M_3 + \gamma_{\text{ave}} \delta V_2 \right) dx_1, \tag{15.31}$$

where γ_{ave} is given by eq. (15.12).

Following the reasoning developed section 9.7.6, the unit load method applied to shear deformable beams leads to the following expression for the displacement at a point of the beam

$$\Delta = \int_0^L \left(\frac{M_3 \hat{M}_3}{H_{33}^c} + \frac{V_2 \hat{V}_2}{K_{22}} \right) dx_1. \tag{15.32}$$

A unit load is applied at the point and in the direction of the desired displacement component. The bending moment distribution, M_3, and shear force distribution, V_2, are those acting in the beam under the action of the externally applied loads. The bending moment distribution, $\hat{M}_3$, and shear force distribution, $\hat{V}_2$, are statically admissible bending moment and shear force distributions in equilibrium with the unit load. The displacement component, Δ, is then computed by eq. (15.32).

To illustrate the unit load method applied to shear deformable beams, consider a cantilevered beam of length L subjected to a tip load P. For this isostatic problem, the bending moment and shear force distributions are easily found from statics considerations as $M_3 = -Px_1$ and $V_2 = P$, respectively. The statically admissible bending moment and shear force distributions in equilibrium with a unit tip load are then $\hat{M}_3 = -x_1$ and $\hat{V}_2 = 1$, respectively. Equation (15.32) then yields the desired tip displacement as

$$\Delta = \int_0^L \left[\frac{(-Px_1)(-x_1)}{H_{33}^c} + \frac{(P)(1)}{K_{22}} \right] dx_1 = \frac{1}{3} \frac{PL^3}{H_{33}^c} + \frac{PL}{K_{22}}.$$

Introducing the shear flexibility parameter, $\bar{s}$, defined in eq. (15.24), leads to

$$\Delta = \frac{1}{3} \frac{PL^3}{H_{33}^c} \left(1 + 3\bar{s}^2\right),$$

which agrees with eq. (15.26).

15.2.2 Shearing effects on buckling

In chapter 14, the buckling load of a simply supported beam under axial compressive loads is found using Euler-Bernoulli beam theory. If the beam is shear deformable, the additional compliance of the system will lower the buckling load, implying that the predictions based on Euler-Bernoulli theory overestimate the actual buckling load.

To investigate the effect of shear deformation on buckling loads, an energy approach will be used. Consider a uniform, simply-supported beam subjected to an end compressive load of magnitude P. The total potential energy of the system is obtained by combining the strain energy for a shear deformable beam, eq. (15.18), and the potential of the axial load, eq. (14.37), to find

$$\Pi = \frac{1}{2} \int_0^L H_{33}^c \left(\frac{d\Phi_3}{dx_1}\right)^2 dx_1 + \frac{1}{2} \int_0^L K_{22} \left(\frac{d\bar{u}_2}{dx_1} - \Phi_3\right)^2 dx_1$$
$$- \frac{1}{2} \int_0^L P \left(\frac{d\bar{u}_2}{dx_1}\right)^2 dx_1,$$

where the first term represents the strain energy associated with the bending of the beam, the second term that associated with its shearing, and the last term the work done by the axial compressive load, P.

The following displacement shape functions will be assumed

$$\bar{u}_2(x_1) = q_1 \sin \frac{\pi x_1}{L}, \quad \Phi_3(x_1) = q_2 \cos \frac{\pi x_1}{L}. \tag{15.33}$$

The mode shape assumed for the transverse displacement corresponds to the exact solution of the problem using Euler-Bernoulli theory. The mode shape assumed for the sectional rotation is the derivative of the assumed transverse displacement. Because the sectional rotation is not equal to the slope of the beam, coefficients q_1 and q_2 are different for a shear deformable beam.

Introducing the assumed shape functions into the expression for the total potential energy and integrating over the beam span then yields

$$\Pi = \frac{1}{2}\frac{L}{2}\left[H_{33}^c \left(\frac{\pi}{L}\right)^2 q_2^2 + K_{22}\left(\left(\frac{\pi}{L}\right)^2 q_1^2 - \frac{2\pi}{L}q_1 q_2 + q_2^2 \right) - P\left(\frac{\pi}{L}\right)^2 q_2^2 \right].$$

This expression can be recast into a matrix format as two quadratic forms given by

$$\Pi = \frac{1}{2}\underline{q}^T \frac{L}{2}\begin{bmatrix} \left(\frac{\pi}{L}\right)^2 K_{22} & -\frac{\pi}{L}K_{22} \\ -\frac{\pi}{L}K_{22} & \left(\frac{\pi}{L}\right)^2 H_{33}^c + K_{22} \end{bmatrix}\underline{q} - \frac{1}{2}\underline{q}^T\frac{PL}{2}\begin{bmatrix} \left(\frac{\pi}{L}\right)^2 & 0 \\ 0 & 0 \end{bmatrix}\underline{q}$$

$$= \frac{1}{2}\underline{q}^T\underline{\underline{K}}\,\underline{q} - \frac{1}{2}\underline{q}^T P\underline{\underline{K}}_G\underline{q},$$

where $\underline{q} = \{q_1, q_2\}^T$ is the solution array, $\underline{\underline{K}}$ the stiffness matrix, and $\underline{\underline{K}}_G$ the geometric stiffness matrix. The buckling equation is now given by eq. (14.48), and the vanishing of the determinant leads to

$$\left(\frac{\pi}{L}\right)^2\left[\left(\frac{\pi}{L}\right)^2 H_{33}^c + K_{22}\right]P_{cr} - \left(\frac{\pi}{L}\right)^4 K_{22}H_{33}^c = 0.$$

Solving for the buckling load yields

$$P_{cr} = \frac{\pi^2 H_{33}^c/L^2}{1 + \pi^2 H_{33}^c/(K_{22}L^2)}.$$

The buckling load can be written in terms of the shear flexibility parameter defined by eq. (15.24) and the buckling load for an Euler-Bernoulli beam, $P_{Euler} = \pi^2 H_{33}^c/L^2$, eq. (14.25), as

$$P_{cr} = \frac{P_{Euler}}{1 + \pi^2 \bar{s}^2}.$$

For Euler-Bernoulli beams, $\bar{s}^2 = 0$, and $P_{cr} = P_{Euler}$, as expected. For shear deformable beams, $\bar{s}^2 > 0$, and the buckling load is always lower than P_{Euler} due to the additional flexibility of the system.

To quantify this effect, the beams with solid rectangular and sandwich sections described in section 15.2.1 will be examined again. Table 15.2 lists the ratios of the buckling loads to the Euler loads for the various cases. Shear deformation has a more pronounced effect on buckling loads than on static deflections. For the sandwich sections, the results indicate a 26% or 58% decrease in buckling load as compared to Euler-Bernoulli predictions for aspect ratios $h/L = 1/10$ and $1/5$, respectively. Clearly, the inclusion of shear deformation effects is critically important when computing the buckling loads of a system, even for long, slender sandwich beams.

Table 15.2. Effect of shear deformation on the buckling load of a simply supported beam.

	Rectangular section		Sandwich section	
h/L	1/10	1/5	1/10	1/5
$\bar{s}^2$	2.6×10^{-3}	10.4×10^{-3}	35.0×10^{-3}	140.0×10^{-3}
$P_{\text{cr}}/P_{\text{E}}$	0.98	0.91	0.74	0.42

15.2.3 Problems

Problem 15.4. Cantilevered beam under distributed bending moment

Consider a shear deformable, cantilevered beam subjected to a uniform, distributed bending moment $q_3(x_1) = q_0$. *(1)* Write the governing differential equations of the problem. *(2)* Write the boundary conditions of the problem. *(3)* Find the tip transverse displacement and sectional rotation distributions along the beam. *(4)* Is the tip deflection affected by shear deformations?

Problem 15.5. Four-point bending test

The four-point bending test set-up depicted in fig. 5.19 is routinely used to experimentally determine the bending stiffness of a beam. The load, P, applied by the testing machine is transmitted to the test sample through two rollers; the applied load is reacted underneath the test sample by two additional rollers. The deformation of the test sample is measured by two strain gauges, one located on top and the other on the bottom of the sample, as shown in fig. 5.19. Let ϵ_t and ϵ_b be the strain measurements at the top and bottom locations, respectively. *(1)* Using Euler-Bernoulli beam theory, describe the data reduction procedure that evaluates the test sample's bending stiffness given the measured load, P, and strain gauge readings, ϵ_t and ϵ_b. *(2)* What correction should be made to the data reduction procedure if the test sample is a shear deformable, sandwich structure.

Problem 15.6. Cantilevered beam with a uniform distributed load

Derive the governing equations and associated boundary conditions for the shear deformable cantilevered beam with uniform distributed load depicted in fig. 15.14. *(1)* Develop a solution for the bending deflection, $\bar{u}_2^b$, (assume $\bar{s}^2 = 0$). *(2)* Construct an approximate solution for the shear deformable beam. Use $\bar{u}_2^b$ as the shape function for the transverse displacement field and $d\bar{u}_2^b/dx_1$ as that for the rotation field. *(3)* Use the principle of minimum total potential energy to find the approximate solution. *(4)* Construct a table similar to table 15.1 and discuss the results.

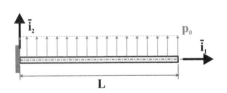

Fig. 15.14. Cantilevered beam with a uniform distributed load.

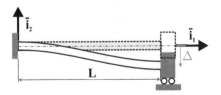

Fig. 15.15. Clamped-clamped beam with tip deflection.

Problem 15.7. Clamped-clamped beam with support misalignment

A uniform beam of length, L, is clamped at both ends, but the right hand support is misaligned by a vertical distance, Δ, as depicted in fig. 15.15. (1) Plot the transverse displacement field $\bar{u}_2(\eta)/\Delta$ over the span of the beam, $\eta = x_1/L$. (2) Plot the bending moment distribution, $L^2 M_3(\eta)/(H^c_{33}\Delta)$, over the span of the beam. (3) Plot the shear force distribution, $L^3 V_2(\eta)/(H^c_{33}\Delta)$, over the span of the beam. For each question, consider the following cases: (a) the beam has no shearing deformations (i.e., assume Euler-Bernoulli beam theory); (b) the beam is made of steel and has a rectangular cross-section with $E/G = 2.6$ and $h/L = 1/5$, see section A in fig. 15.10; (c) the beam has a sandwich cross-section with $E_f/G_c = 35$, $t/h = 1/10$, $h/L = 1/5$, see section B in fig. 15.10. Parameter $\bar{s}^2 = H^c_{33}/(K_{22}L^2)$ can be used to characterize the shearing deformations. For each of the three questions, plot the results for the three cases on on a single graph.

Problem 15.8. Measuring sectional shear stiffness

Figure 15.16 depicts the experimental set-up for the four-point and three-point bending tests. The four-point bending test is discussed in example 5.3. In the four-point bending test, the test section is subjected to bending only, and in the three-point bending test, the test section is subjected to combined bending and shearing. Assume that the test sample is of sandwich construction with the configuration shown in fig. 15.10. The beam shear stiffness is denoted K_{22}. (1) Compute the deflection of point $\mathbf{M}$, denoted Δ_4, for the four-point bending test. (2) Compute the deflection of point $\mathbf{M}$, denoted Δ_3, for the three-point bending test. (3) Use these results to develop an expression for the shear stiffness, K_{22}, in term of the deflections, Δ_4 and Δ_3, and the beam properties. (4) Using a simple first order analysis, determine the sensitivity of this expression to errors in the measurement of Δ_4 and Δ_3. This can be done by constructing a first order Taylor's series expansion for K_{22} in terms of Δ_4 and Δ_3. Comment on your results.

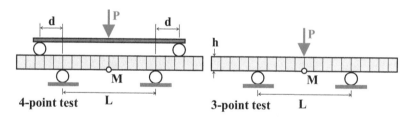

Fig. 15.16. Beam under 4-point and 3-point loading.

Problem 15.9. Cantilevered beam with tip rotational spring

Consider a cantilevered beam of length, L, with a tip rotational spring of stiffness, k, depicted in fig. 15.17. The beam is subjected to an end compressive load, P. Use an energy method to compute the buckling load of the system. The effects of shearing deformations should be included in your analysis. Assume the following displacement modes: $\bar{u}_2 = a\,(1 - \cos\pi x_1/2L)$, $\Phi_3 = b\,\sin\pi x_1/2L$. The following notation will be convenient to use: $P_E = \pi^2 H^c_{33}/(4L^2)$, $\bar{s}^2 = \pi^2 H^c_{33}/(4K_{22}L^2)$, $\bar{k} = 8kL/(\pi^2 H^c_{33})$, where P_{euler} is the buckling load of the system without the tip spring and ignoring shearing deformations, $\bar{s}^2$ is the shearing deformation parameter, and k^* is the non-dimensional stiffness of the torsional spring.

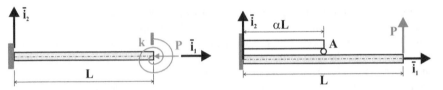

Fig. 15.17. Cantilevered beam with tip torsional spring.

Fig. 15.18. Cantilevered beam with intermediate support.

Problem 15.10. Simply supported beam on an elastic foundation

Consider a simply supported beam of length L resting on an elastic foundation of stiffness constant k, as shown in fig. 15.19. The beam is subjected to an axial compressive force, P. Shearing deformations should be taken into account. Use an energy approach with the following assumed modes: $\bar{u}_2 = \sum_{n=1}^{\infty} U_n \sin n\pi x_1/L$; $\Phi_3 = \sum_{n=1}^{\infty} \Phi_n \cos n\pi x_1/L$. The following notation will be convenient to use: $\bar{s}^2 = \pi^2 H_{33}^c/(K_{22}L^2)$, $\bar{k} = kL^4/(\pi^4 H_{33}^c)$. *(1)* Find the buckling load of the system as a function $\bar{k}$. *(2)* For a value of $\bar{k} = 12.0 \times 10^3$, find the buckling mode shape n and the buckling load of the system P/P_{euler} when shearing deformations are neglected. *(3)* Same questions when shearing deformation are taken into account. Use a beam of rectangular cross-section with $h/L = 1/10$ and $E/G = 2.6$. *(4)* Same questions for a beam of rectangular cross-section with $h/L = 1/10$ and $E/G = 28$. Note: In section 14.3, this problem is treated under the assumption of negligible shearing deformations.

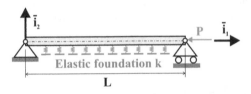

Fig. 15.19. Simply supported beam on an elastic foundation.

Problem 15.11. Cantilevered beam with intermediate support

The cantilevered beam depicted in fig. 15.18 is subjected to a tip load P. The tip of a second cantilevered beam contacts the first at point **A**. The lower and upper beams have a uniform bending stiffness, H_{33}^c, uniform shearing stiffnesses, K_{22}, and are of length L and αL, respectively. *(1)* Find the displacement fields for the two beams. *(2)* On one graph, plot the distribution of non-dimensional transverse displacement, $H_{33}^c \bar{u}_2/(PL^3)$, for both beams. *(3)* Plot the distribution of non-dimensional rotation, $H_{33}^c \Phi_3/(PL^2)$, for both beams. *(4)* Plot the distribution of non-dimensional bending moment, $M_3/(PL)$, for both beams. *(5)* Plot the distribution of non-dimensional transverse shear force, V_2/P, for both beams. Use $\alpha = 0.25$. *(6)* Study the behavior of the system as $\alpha \to 0$. *(7)* Plot the non-dimensional force in the intermediate support, F/P, as a function of $\alpha \in [0, 0.5]$. *(8)* Plot the distribution of non-dimensional transverse tip displacement, $H_{33}^c \bar{u}_2/(PL^3)$, as a function of $\alpha \in [0, 0.5]$. *(9)* Plot the distribution of non-dimensional root bending moment, $M_3/(PL)$, as a function of $\alpha \in [0, 0.5]$ for both beams. *(10)* Plot the distribution of non-dimensional root shear force, V_2/P, as a function of $\alpha \in [0, 0.5]$ for both beams. Comment on your results. For each question, consider the following three cases: *(a)* the beam has no shearing deformations (*i.e.*, assume Euler-Bernoulli beam theory); *(b)* the beam is made of steel and has a rectangular cross-section with $E/G = 2.6$ and $h/L = 1/5$, see section A in fig. 15.10; *(c)* the beam has a sandwich cross-section with $E_f/G_c = 35$, $t/h = 1/10$, $h/L = 1/5$, see section B in fig. 15.10.

16

Kirchhoff plate theory

Chapter 5 develops the analysis of beams, which are structures presenting one dimension that is much larger than the other two. The present chapter focuses on another type of structural component, plates, which are defined as structures possessing one dimension far smaller than the other two. The mid-plane of the plate lies along the two long dimensions of the plate, whereas the normal to the plate extends along the shorter dimension. The term "plate" is usually reserved for flat structures, while the term "shell" refers to a curved plate.

The long, slender wings of an aircraft can be analyzed, to a first approximation, as beams, but a more refined analysis will treat the upper and lower skins of the wing as thin plates supported by ribs and longerons or stiffeners. Aircraft wings with a small aspect ratio cannot be treated as beams because two of their dimensions are large compared to their thickness. Such structures, however, can often be treated as plates. Aircraft fuselages are also constructed of thin-walled structures stiffened with ribs and longerons, and the thin-walled portions between the stiffeners can be viewed as thin plates. Finally, portions of thin-walled beams such as those studied in chapter 8 can be modeled as plates when considering localized behavior induced by attachments or supports, for instance.

Solid mechanics theories describing plates, more commonly referred to simply as *plate theories*, play an important role in structural analysis because they provide tools for the analysis of structures that are commonly used. Although more sophisticated tools, such as fully three-dimensional finite element methods, are now widely available for the analysis of complex structures, plate and shell models are often used because they provide valuable insight into the behavior of these structures and are computationally simpler.

Beam theories reduce the analysis of complex, three-dimensional structures to one-dimensional problems. Indeed, the equations of Euler-Bernoulli and Timoshenko beam theories, as presented in chapters 5 and 15, respectively, lead to ordinary differential equations expressed in terms of a single span-wise variable along the axis of the beam. In contrast, plate theories reduce the analysis of three-dimensional structures to two-dimensional problems. The equations of plate theory are partial differential equations in the two dimensions defining the mid-plane of the plate.

One widely used theory for thin plates, *Kirchhoff plate theory*, is based on assumptions that are closely related to those of Euler-Bernoulli beam theory described in section 5.1. Whereas the assumptions of beam theory deal with the kinematics of the cross-section of the beam, the assumptions of Kirchhoff plate theory deal with the kinematics of a *normal material line*, *i.e.*, a set of material particles initially aligned in a direction normal to the mid-plane of the plate.

A fundamental assumption of Kirchhoff plate theory is that the normal material line is infinitely rigid along its length, *i.e.*, no deformations occur in the direction normal to plate's mid-plane. This assumption parallels the beam theory assumption that the cross-section of the beam is infinitely rigid in its own plane, *i.e.*, no deformations occur in the plane of the cross-section. Two additional assumptions deal with the displacements of the material line in the plate's mid-plane. During deformation, the normal material line is assumed to remain, *(1)* straight, and *(2)* normal to the deformed mid-plane of the plate. These assumptions parallel the beam theory assumptions stating that during deformation, the beam's cross-section remains, *(1)* plane, and *(2)* normal to the deformed axis of the beam.

16.1 Governing equations of Kirchhoff plate theory

The development of Kirchhoff plate theory closely parallels the development of Euler-Bernoulli beam theory, and a similar treatment will be followed. First, the Kirchhoff assumptions and their implications are discussed in detail in section 16.1.1. These assumptions are of a purely kinematic nature and allow the displacement and strain fields of the plate to be expressed in terms of a two-dimensional, mid-plane displacement field. Next, the stress resultants are defined in section 16.1.2 and the equilibrium equations of the plate are developed in section 16.1.3. The plate constitutive laws presented in section 16.1.4 complete the development of Kirchhoff plate theory.

16.1.1 Kirchhoff assumptions

A plate is a structure that possesses one dimension that is far smaller than the other two. Consider, for instance, the thin rectangular plate of width, a, length, b, and thickness, h, depicted in fig. 16.1. This structure is considered to be a plate when $h/a \ll 1$ and $h/b \ll 1$. The coordinate system used for the problem is selected in such a way that axis $\bar{\imath}_3$ is normal to the plane of the plate, whereas axes $\bar{\imath}_1$ and $\bar{\imath}_2$ define the mid-plane of the plate, as shown in fig. 16.1.

As discussed in the previous section, the assumption of Kirchhoff plate theory parallel those of Euler-Bernoulli beam theory presented in section 5.1. As is the case for Euler-Bernoulli beam theory, the Kirchhoff assumptions are of a purely kinematic nature, and can be summarized by the following statements.

Assumption 1: The normal material line is infinitely rigid along its own length.

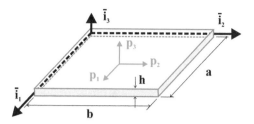

Fig. 16.1. Rectangular plate of thickness h.

Assumption 2: The normal material line of the plate remains a straight line after deformation.

Assumption 3: The straight normal material line remains normal to the deformed mid-plane of the plate.

These assumptions are known as the Kirchhoff assumptions for plates. Experimental measurements show that these assumptions are valid for thin plates made of homogeneous, isotropic materials. When one or more of theses conditions are not met, the predictions of Kirchhoff plate theory might become inaccurate. The mathematical and physical implications of these assumptions will be discussed in detail in this chapter.

Displacement field

Since the normal material line does not deform along its length, the displacement of any point along this line in the direction normal to the plate is the same, leading to

$$u_3(x_1, x_2, x_3) = \bar{u}_3(x_1, x_2), \tag{16.1}$$

where $u_3(x_1, x_2, x_3)$ is the transverse displacement of any point of the plate and $\bar{u}_3(x_1, x_2)$ that of the normal material line. The normal material line may rotate during this displacement, but these rotations are assumed to be small enough to have negligible effect on the transverse displacement field.

Next, because the normal material line remains a straight line after deformation, its displacement field in the plane of the plate, also called the in-plane displacement field, is at most a linear function of coordinate x_3, and can be written as

$$u_1(x_1, x_2, x_3) = \bar{u}_1(x_1, x_2) + x_3 \Phi_2(x_1, x_2), \tag{16.2a}$$

$$u_2(x_1, x_2, x_3) = \bar{u}_2(x_1, x_2) - x_3 \Phi_1(x_1, x_2), \tag{16.2b}$$

where $u_1(x_1, x_2, x_3)$ and $u_2(x_1, x_2, x_3)$ are the in-plane displacements of any point of the plate, $\bar{u}_1(x_1, x_2)$ and $\bar{u}_2(x_1, x_2)$ those of the mid-point of the normal material line, and $\Phi_1(x_1, x_2)$ and $\Phi_2(x_1, x_2)$ the rotations of the material line about axes $\bar{\imath}_1$ and $\bar{\imath}_2$, respectively.

Note the sign convention used here: displacements u_1, u_2, and u_3 or $\bar{u}_1$, $\bar{u}_2$, and $\bar{u}_3$ are positive along axes $\bar{\imath}_1$, $\bar{\imath}_2$, and $\bar{\imath}_3$, respectively. Similarly, Φ_1 and Φ_2 are the

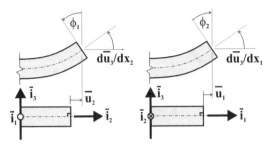

Fig. 16.2. Rotations of the normal line and slopes of the plate.

rotations of the normal material line, counted positive about axes $\bar{\imath}_1$ and $\bar{\imath}_2$, respectively.

Finally, the normal material line is assumed to remain normal to the deformed mid-plane of the plate. As depicted in fig. 16.2, this assumption implies that the slopes of the plate are equal to the rotations of the material line, and leading to

$$\Phi_1 = \frac{\partial \bar{u}_3}{\partial x_2}, \quad \Phi_2 = -\frac{\partial \bar{u}_3}{\partial x_1}. \tag{16.3}$$

The complete displacement field of the plate is found by combining eqs. (16.1), (16.2), and (16.3), to find

$$u_1(x_1, x_2, x_3) = \bar{u}_1(x_1, x_2) - x_3 \frac{\partial \bar{u}_3}{\partial x_1}, \tag{16.4a}$$

$$u_2(x_1, x_2, x_3) = \bar{u}_2(x_1, x_2) - x_3 \frac{\partial \bar{u}_3}{\partial x_2}, \tag{16.4b}$$

$$u_3(x_1, x_2, x_3) = \bar{u}_3(x_1, x_2). \tag{16.4c}$$

This three-dimensional displacement field is defined in terms of three displacements, $\bar{u}_1(x_1, x_2)$, $\bar{u}_2(x_1, x_2)$, and $\bar{u}_3(x_1, x_2)$, and the derivatives of the transverse displacement with respect to variables x_1 and x_2. The three displacement components, $\bar{u}_1(x_1, x_2)$, $\bar{u}_2(x_1, x_2)$, and $\bar{u}_3(x_1, x_2)$, can be interpreted as the mid-point displacements of the normal material line.

The structure of the displacement field implied by eqs. (16.4) is a direct consequence of the Kirchhoff assumptions and allows the development of a two-dimensional plate theory in which the unknown displacement field is a spatial function solely of the in-plane coordinates, x_1 and x_2.

Strain field

The strain field can be evaluated from the displacement field defined by eqs. (16.4) using eqs. (1.63) and (1.71) to find the transverse strain

$$\epsilon_3 = \frac{\partial u_3}{\partial x_3} = 0. \tag{16.5}$$

The vanishing of the strain component in the direction normal to the plate is a direct consequence of the first assumption of Kirchhoff plate theory: the normal material line is infinitely rigid along its own length.

The transverse shearing strains are found is a similar manner

$$\gamma_{13} = \frac{\partial u_1}{\partial x_3} + \frac{\partial u_3}{\partial x_1} = -\frac{\partial \bar{u}_3}{\partial x_1} + \frac{\partial \bar{u}_3}{\partial x_1} = 0, \; \gamma_{23} = \frac{\partial u_2}{\partial x_3} + \frac{\partial u_3}{\partial x_2} = -\frac{\partial \bar{u}_3}{\partial x_2} + \frac{\partial \bar{u}_3}{\partial x_2} = 0.$$
(16.6)

The vanishing of the transverse shear strain components is a direct consequence of the third assumption of Kirchhoff plate theory: the normal material line remains normal to the deformed mid-plane of the plate. The 90 degree angle between the normal material line and the plate's mid-plane remains a 90 degree angle, implying the vanishing of the corresponding shear strain components, see eq. (1.64).

The in-plane strains become

$$\epsilon_1 = \frac{\partial u_1}{\partial x_1} = \frac{\partial \bar{u}_1}{\partial x_1} - x_3 \frac{\partial^2 \bar{u}_3}{\partial x_1^2}, \quad \epsilon_2 = \frac{\partial u_2}{\partial x_2} = \frac{\partial \bar{u}_2}{\partial x_2} - x_3 \frac{\partial^2 \bar{u}_3}{\partial x_2^2},$$
(16.7)

and finally, the in-plane shear strain is

$$\gamma_{12} = \frac{\partial u_1}{\partial x_2} + \frac{\partial u_2}{\partial x_1} = \frac{\partial \bar{u}_1}{\partial x_2} + \frac{\partial \bar{u}_2}{\partial x_1} - 2x_3 \frac{\partial^2 \bar{u}_3}{\partial x_1 \partial x_2}.$$
(16.8)

The in-plane strains vary linearly through the thickness of the plate, as implied by their linear dependency on coordinate x_3.

It is convenient to introduce the array of *plate mid-plane strains*, defined as

$$\underline{\epsilon}_0 = \left\{ \frac{\partial \bar{u}_1}{\partial x_1}, \frac{\partial \bar{u}_2}{\partial x_2}, \frac{\partial \bar{u}_1}{\partial x_2} + \frac{\partial \bar{u}_2}{\partial x_1} \right\}^T = \left\{ \epsilon_1^0, \epsilon_2^0, \epsilon_{12}^0 \right\}^T,$$
(16.9)

where the notation $(\cdot)_0$ or $(\cdot)^0$ is a reminder that these quantities are strains at the mid-plane of the plate, *i.e.*, at $x_3 = 0$. Similarly, the array of *plate curvatures* is defined as

$$\underline{\kappa} = \left\{ \frac{\partial^2 \bar{u}_3}{\partial x_2^2}, -\frac{\partial^2 \bar{u}_3}{\partial x_1^2}, 2\frac{\partial^2 \bar{u}_3}{\partial x_1 \partial x_2} \right\}^T = \left\{ \kappa_1, \kappa_2, \kappa_{12} \right\}^T.$$
(16.10)

In summary, the strain component in the direction normal to the mid-plane of the plate and the transverse shearing strain components vanish. The only non-vanishing strain components are the in-plane strains that can be written as

$$\epsilon_1 = \epsilon_1^0 + x_3 \, \kappa_2, \quad \epsilon_2 = \epsilon_2^0 - x_3 \, \kappa_1, \quad \gamma_{12} = \epsilon_{12}^0 - x_3 \, \kappa_{12}.$$
(16.11)

In addition, a more compact matrix notation is defined,

$$\underline{\epsilon} = \underline{\epsilon}_0 + x_3 \, \underline{\underline{S}} \, \underline{\kappa},$$
(16.12)

where the array of in-plane strains is

$$\underline{\epsilon} = [\epsilon_1, \; \epsilon_2, \; \gamma_{12}]^T, \tag{16.13}$$

and matrix $\underline{\underline{S}}$ is defined as

$$\underline{\underline{S}} = \begin{bmatrix} 0 & 1 & 0 \\ -1 & 0 & 0 \\ 0 & 0 & -1 \end{bmatrix}. \tag{16.14}$$

The *permutation matrix*, $\underline{\underline{S}}$, is introduced solely to change signs and reorder the curvatures components; it enables the compact notation of eq. (16.12). Thus, the product $\underline{\underline{S}}\,\underline{\kappa} = \underline{\hat{\kappa}}$ simply defines a new array of curvatures, $\underline{\hat{\kappa}}$, which differ from those defined in eq. (16.10) by only a sign convention: $\hat{\kappa}_1 = \kappa_2$, $\hat{\kappa}_2 = -\kappa_1$, and $\hat{\kappa}_{12} = -\kappa_{12}$. Note that $\hat{\kappa}_1$ is the curvature of the plate about axis $\bar{\imath}_1$, and $\hat{\kappa}_2$ is the negative of the curvature about axis $\bar{\imath}_2$. Finally, the twisting curvature, $\hat{\kappa}_{12}$, is the negative of κ_{12}. Matrix $\underline{\underline{S}}$ simply performs a change in the sign convention for the curvatures, and it will be used frequently for this purpose throughout the remainder of this chapter. Finally, note that matrix $\underline{\underline{S}}$ is orthogonal, *i.e.*, $\underline{\underline{S}}^T \underline{\underline{S}} = \underline{\underline{I}}$, where $\underline{\underline{I}}$ is the identity matrix.

16.1.2 Stress resultants

The internal forces and moment in beam theory are expressed in terms of sectional stress resultants defined in section 5.3. Similar quantities will be defined for plates. Consider a differential element of the plate of infinitesimal dimensions dx_1 and dx_2 in the plane of the plate, but of finite thickness, h, equal to that of the plate, as depicted in the right portion of fig. 16.3.

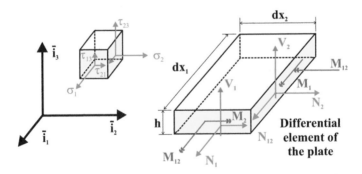

Fig. 16.3. Stress resultants acting on a differential element of the plate.

First, the in-plane stress components, σ_1, acting on the face normal to axis $\bar{\imath}_1$ is integrated over this face to find the axial force, N_1. A similar integral of the stress component σ_2 over the face normal to axis $\bar{\imath}_2$ yields the axial force N_2, and finally, the in-plane force, N_{12}, results from integration of the shear stress component, τ_{12}. The three *in-plane forces* are defined as

$$N_1(x_1, x_2) = \int_h \sigma_1 \, dx_3, \tag{16.15a}$$

$$N_2(x_1, x_2) = \int_h \sigma_2 \, dx_3, \tag{16.15b}$$

$$N_{12}(x_1, x_2) = \int_h \tau_{12} \, dx_3, \tag{16.15c}$$

where the notation $\int_h (\cdot) \, dx_3$ is used instead of the more cumbersome $\int_{-h/2}^{+h/2} (\cdot) \, dx_3$.

It is important to understand that the term "in-plane force" is incorrect, because the in-plane forces are, in fact, *in-plane forces per unit span*. Indeed, examination of eqs. (16.15) reveals that the units of these forces are force per unit length, N/m in the SI system. The in-plane shear forces acting on the faces normal to axes $\bar{\imath}_1$ and $\bar{\imath}_2$ are equal due the principle of reciprocity of shear stresses, see eq. (1.5). Equations (16.15) can be written in a more compact form as

$$\underline{N} = \int_h \underline{\sigma} \, dx_3, \tag{16.16}$$

where $\underline{N}$ is the array of in-plane forces defined as

$$\underline{N} = \{N_1, N_2, N_{12}\}^T, \tag{16.17}$$

and $\underline{\sigma}$ the array of in-plane stresses

$$\underline{\sigma} = \{\sigma_1, \sigma_2, \tau_{12}\}^T. \tag{16.18}$$

Next, the transverse shearing stress components, τ_{13} and τ_{23}, are integrated over the faces normal to axes $\bar{\imath}_1$ and $\bar{\imath}_2$, respectively, to find the corresponding *transverse shear forces*, Q_1 and Q_2, respectively, as

$$Q_1(x_1, x_2) = \int_h \tau_{13} \, dx_3, \tag{16.19a}$$

$$Q_2(x_1, x_2) = \int_h \tau_{23} \, dx_3. \tag{16.19b}$$

Here again, the term "transverse shear force" is misleading, because these forces are, in fact, *transverse shear forces per unit span*. Indeed, examination of eqs. (16.19) reveals that the units of these forces are force per unit length, N/m in the SI system.

In beams, the transverse shear forces are denoted V_2 and V_3. Shear force V_2 acts on the beam's cross-section, along axis $\bar{\imath}_2$ and shear force V_3 acts on the same cross-section, along axis $\bar{\imath}_3$. In plates, the transverse shear forces per unit span are denoted Q_1 and Q_2 but the subscript now refers to the orientation of the face on which they act. Shear force Q_1 acts on the face normal to axis $\bar{\imath}_1$, along axis $\bar{\imath}_3$ and shear force Q_2 acts on the face normal to axis $\bar{\imath}_2$, along axis $\bar{\imath}_3$.

Finally, the first moments of the in-plane stress components, σ_1 and σ_2, are determined by integration over the faces normal to axes $\bar{\imath}_1$ and $\bar{\imath}_2$, respectively, to yield the *bending moments*, M_2 and $-M_1$, respectively Similarly, the first moment of the in-plane shear stress component, τ_{12}, is called the *twisting moment*, M_{12}. These three stress resultants are defined as

$$M_1(x_1, x_2) = -\int_h x_3\sigma_2 \, \mathrm{d}x_3, \tag{16.20a}$$

$$M_2(x_1, x_2) = \int_h x_3\sigma_1 \, \mathrm{d}x_3, \tag{16.20b}$$

$$M_{12}(x_1, x_2) = -\int_h x_3\tau_{12} \, \mathrm{d}x_3. \tag{16.20c}$$

Note the sign convention used for the bending moments: M_1 and M_2 are positive about axes $\bar{\imath}_1$ and $\bar{\imath}_2$, respectively, but act on faces normal to axes $\bar{\imath}_2$ and $\bar{\imath}_1$, respectively[1]. In view of the principle of reciprocity of shearing stresses, see eq. (1.5), the twisting moments acting on mutually orthogonal faces are equal in magnitude. The sign convention implied by eq. (16.20c) is that twisting moment M_{12} is *positive* about axis $\bar{\imath}_1$ on the face normal to this axis, but *negative* about axis $\bar{\imath}_2$ on the face normal to this axis. It is important to understand that the terms "bending moment" or "twisting moment" are incorrect, because these moments are, in fact, *moment per unit span*. Indeed, examination of eqs. (16.20) reveals that the units of these moment are those of a force, N·m/m = N in the SI system.

Equations (16.20) can be recast in a more compact form as

$$\underline{M} = \underline{\underline{S}}^T \int_h \underline{\sigma}x_3 \, \mathrm{d}x_3, \tag{16.21}$$

where $\underline{M}$ is the array of bending moments defined as

$$\underline{M} = \{M_1, M_2, M_{12}\}^T. \tag{16.22}$$

Permutation matrix $\underline{\underline{S}}$ is defined by eq. (16.14). The product $\underline{\hat{M}} = \underline{\underline{S}}\,\underline{M}$ simply defines a new array of plate bending moments which differ from those defined in eq. (16.20) by only a sign convention: $\hat{M}_1 = M_2$, $\hat{M}_2 = -M_1$, and $\hat{M}_{12} = -M_{12}$.

16.1.3 Equilibrium equations

Figure 16.1 shows the loading applied to the plate; it consists of in-plane pressures, denoted $p_1(x_1, x_2)$ and $p_2(x_1, x_2)$, acting along axes $\bar{\imath}_1$ and $\bar{\imath}_2$, respectively. Furthermore, a transverse pressure, $p_3(x_1, x_2)$, also acts along axis $\bar{\imath}_3$. These externally applied forces have units of pressure or load per unit area of the plate, N/m² or Pascals in the SI system.

The left portion of fig. 16.4 shows the free body diagram of a differential element of the plate with all the in-plane forces and externally applied loads. Imposing

---

[1] In some texts, this notation is reversed and M_1 is defined as the moment due to σ_1 about axis $\bar{\imath}_2$, and *vice versa* for M_2.

the vanishing of the sum of the forces along axis $\bar{\imath}_1$ yields $-N_1 dx_2 - N_{12} dx_1 + (N_{12} + \partial N_{12}/\partial x_2\, dx_2) dx_1 + (N_1 + \partial N_1/\partial x_1\, dx_1) dx_2 + p_1\, dx_1 dx_2 = 0$. After simplification, this leads to

$$\frac{\partial N_1}{\partial x_1} + \frac{\partial N_{12}}{\partial x_2} = -p_1, \tag{16.23a}$$

$$\frac{\partial N_{12}}{\partial x_1} + \frac{\partial N_2}{\partial x_2} = -p_2, \tag{16.23b}$$

where the second equations results from imposing the vanishing of the forces along axis $\bar{\imath}_2$.

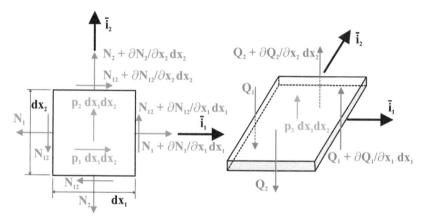

Fig. 16.4. Left figure: free body diagram for the equilibrium of in-plane forces. Right figure: free body diagram for the equilibrium of transverse shear forces.

The right portion of fig. 16.4 shows the free body diagram of a differential element of the plate with all the out-of-plane forces and externally applied loads. Imposing the vanishing of the sum of the forces along axis $\bar{\imath}_3$ yields $-Q_2 dx_1 - Q_1 dx_2 + (Q_2 + \partial Q_2/\partial x_2\, dx_2) dx_1 + (Q_1 + \partial Q_1/\partial x_1\, dx_1) dx_2 + p_3\, dx_1 dx_2 = 0$. After simplification, this becomes

$$\frac{\partial Q_1}{\partial x_1} + \frac{\partial Q_2}{\partial x_2} = -p_3. \tag{16.24}$$

Equilibrium of the differential element of the plate also implies the vanishing of the sum of the moments about axes $\bar{\imath}_1$, $\bar{\imath}_2$, and $\bar{\imath}_3$. Figure 16.5 shows the free body diagram of a differential element of the plate with all applied moments and shear forces. Imposing the vanishing of the sum of the moments about axis $\bar{\imath}_1$ yields $-M_1 dx_1 - M_{12} dx_2 + (M_1 + \partial M_1/\partial x_2\, dx_2) dx_1 + (M_{12} + \partial M_{12}/\partial x_1\, dx_1) dx_2 + Q_2\, dx_1 dx_2 = 0$. After simplification, this becomes

$$\frac{\partial M_2}{\partial x_1} - \frac{\partial M_{12}}{\partial x_2} - Q_1 = 0, \qquad (16.25a)$$

$$\frac{\partial M_{12}}{\partial x_1} + \frac{\partial M_1}{\partial x_2} + Q_2 = 0, \qquad (16.25b)$$

where the second equation results from imposing the vanishing of the sum of the moments about axis $\bar{\imath}_2$.

The free body diagram appearing in the left portion of fig. 16.4 is conveniently used to express the moment equilibrium condition about axis $\bar{\imath}_3$ and leads to $N_{12} - N_{12} = 0$, but this equations bring no new information about equilibrium. The reason for this apparent loss of information is clear: the in-plane shear forces acting on two normal faces are equal in magnitude because of the principle of reciprocity of shear stresses, see eq. (1.5). The principle of reciprocity of shear stresses is obtained from moment equilibrium conditions. Because this moment equilibrium condition is already used to infer the equality of in-plane shear forces, it is not unexpected that a second application of the same equilibrium condition brings no new information to light.

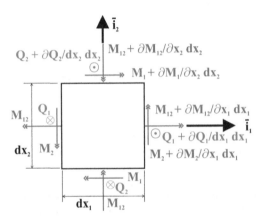

Fig. 16.5. Free body diagram for the equilibrium of bending moment and shear forces.

16.1.4 Constitutive laws

The plate is assumed to be made of a linearly elastic, isotropic material that obeys Hooke's law, see eqs. (2.4). The magnitudes of the stresses acting in the plane of the plate, σ_1 and σ_2, should be much larger than that of the transverse stress component, σ_3: $\sigma_1 \gg \sigma_3$ and $\sigma_2 \gg \sigma_3$. Consequently, the transverse stress component is assumed to be vanishingly small, $\sigma_3 \approx 0$. For this stress state, the generalized Hooke's law, eqs. (2.4), reduce to

$$\sigma_1 = \frac{E}{1-\nu^2}\left(\epsilon_1 + \nu\epsilon_2\right), \quad \sigma_2 = \frac{E}{1-\nu^2}\left(\nu\epsilon_1 + \epsilon_2\right), \quad \sigma_3 \approx 0. \qquad (16.26)$$

The constitutive laws for shear stress and shear strain components, see eqs. (2.9), re-main unchanged. The constitutive laws for the in-plane stress and strain components are written in a compact matrix form as

$$\underline{\sigma} = \underline{\underline{C}}\,\underline{\epsilon}, \tag{16.27}$$

where $\underline{\underline{C}}$ is the stiffness matrix for the plane stress state given by eq. (2.16), $\underline{\epsilon}$ the in-plane strain array defined by eq. (16.13) and $\underline{\sigma}$ the in-plane stress array defined by eq. (16.18).

When describing the plate's kinematics, it is assumed that the normal material line does not deform along its length, and therefore, the transverse strain vanishes, see eq. (16.5). When dealing with the plate's constitutive laws, the transverse stress component is assumed to vanish. This is an inconsistency of Kirchhoff plate theory that uses two contradictory assumptions: the vanishing of both transverse strain and stress components. In view of Hooke's law, these two quantities cannot vanish si-multaneously. Indeed, if $\sigma_3 = 0$, eq. (2.4c) results in $\epsilon_3 = -\nu(\sigma_1 + \sigma_2)/E$, which implies that the transverse strain does not vanish due to Poisson's effect. Because this effect is very small, assuming the vanishing of this strain component when describing the plate's kinematics does not cause significant errors for most problems.

The relationship between the in-plane force, N_1, and plate deformations is found by introducing the reduced Hooke's law into the definition of the in-plane force, eq. (16.15a), to find

$$
\begin{aligned}
N_1 &= \int_h \sigma_1 \, \mathrm{d}x_3 = \int_h \frac{E}{1-\nu^2} \left[(\epsilon_1^0 + x_3\kappa_2) + \nu(\epsilon_2^0 - x_3\kappa_1) \right] \, \mathrm{d}x_3 \\
&= \frac{hE}{1-\nu^2}(\epsilon_1^0 + \nu\epsilon_2^0),
\end{aligned}
$$

where the plate in-plane strain field is expressed in terms of plate mid-plane strains and curvatures using eq. (16.12). Similar developments will yield the other in-plane force components. It is convenient to recast the results in a matrix form as

$$\underline{N} = \underline{\underline{A}}\,\underline{\epsilon}_0, \tag{16.28}$$

where the *in-plane stiffness matrix* is

$$\underline{\underline{A}} = h\underline{\underline{C}} \tag{16.29}$$

and $\underline{\underline{C}}$ the material stiffness matrix defined in eq. (2.16).

Next, the relationship between the bending moment, M_1, and plate deformations is found by introducing the reduced Hooke's law into the definition of the bending moment, eq. (16.20a), to find

$$
\begin{aligned}
M_1 &= -\int_h x_3\sigma_2 \, \mathrm{d}x_3 = -\int_h \frac{Ex_3}{1-\nu^2} \left[\nu(\epsilon_1^0 + x_3\kappa_2) + (\epsilon_2^0 - x_3\kappa_1) \right] \, \mathrm{d}x_3 \\
&= \frac{Eh^3}{12(1-\nu^2)}(-\nu\kappa_2 + \kappa_1),
\end{aligned}
$$

where the plate in-plane strain field is expressed in terms of plate mid-plane strains and curvatures using eq. (16.12). Similar developments will yield the other bending moment component, M_1, and the twisting moment, M_{12}. It is convenient to recast the results in a matrix form as

$$\underline{M} = \widetilde{\underline{\underline{D}}}\,\underline{\kappa}, \tag{16.30}$$

where the *bending stiffness matrix* for homogeneous, isotropic plates is

$$\widetilde{\underline{\underline{D}}} = \frac{Eh^3}{12(1-\nu^2)} \begin{bmatrix} 1 & -\nu & 0 \\ -\nu & 1 & 0 \\ 0 & 0 & (1-\nu)/2 \end{bmatrix}. \tag{16.31}$$

16.1.5 Stresses due to in-plane forces and bending moments

Once a constitutive law is defined, the stresses in the plate due to bending and stretching can be determined. Using the constitutive laws, eqs. (16.27), the in-plane stress components can be expressed in terms of the in-plane strain components as

$$\underline{\sigma} = \underline{\underline{C}}\,\underline{\epsilon} = \underline{\underline{C}}(\underline{\epsilon_0} + x_3\,\underline{S}\,\underline{\kappa}),$$

where the second equality follows from eq. (16.12). Next, the in-plane strain and curvature components are expressed in terms of the in-plane forces and bending moments using eqs. (16.28) and (16.30), respectively, to find

$$\underline{\sigma} = \underline{\underline{C}} \left[\underline{\underline{A}}^{-1}\underline{N} + x_3\underline{\underline{S}}\,\widetilde{\underline{\underline{D}}}^{-1}\underline{M} \right]. \tag{16.32}$$

16.1.6 Summary of Kirchhoff plate theory

In summary, Kirchhoff plate theory is characterized by the following sets of equations.

1. *Six strain-displacement equations*: three equations define the mid-plane strains in terms of the plate in-plane displacements, see eqs. (16.9), and three equations define the plate curvatures in terms of the transverse displacement, see eqs. (16.10).
2. *Five equilibrium equations*: two equations express the equilibrium conditions for the in-plane forces, see eqs. (16.23), one equation expresses the vertical force equilibrium condition, see eq. (16.24), and finally, two equations express the moment equilibrium conditions, see eqs. (16.25).
3. *Six constitutive laws*: three equations state the relationship between the in-plane forces and mid-plane strains, see eqs. (16.28), and three equations state the relationship between the bending moments and plate curvatures, see eqs. (16.30).

Kirchhoff plate theory involves seventeen equations for the following seventeen unknowns: six components of strain, three mid-plane strains and three curvatures, eight stress resultants, three in-plane forces, three bending moments and two transverse shear forces, and three components of displacement. Given the proper boundary conditions, it therefore should be possible to solve this problem.

A cursory investigation of the governing equations indicates that the problem can be separated into two simpler problems.

1. *The in-plane problem.* This problem involves eight unknowns: the three in-plane forces, the three mid-plane strains, and the two in-plane displacement components. The eight governing equations are: three stain-displacement equations relating the mid-plane strains to the in-plane displacements, see eqs. (16.9), two equations of equilibrium involving the in-plane forces, see eqs. (16.23), and three constitutive laws relating the in-plane forces to the mid-plane strains, see eqs. (16.28).

2. *The bending problem.* This problem involves nine unknowns: the three bending moments, the two transverse shear forces, the three curvatures, and the transverse displacement. The nine governing equations are: three stain-displacement equations relating the curvatures to the transverse displacement, see eqs. (16.10), three equations of equilibrium involving the bending moments, see eqs. (16.25), and shear forces, see eq. (16.24), and three constitutive laws relating the bending moments to the curvatures, see eqs. (16.30).

For linearly elastic, homogeneous and isotropic materials, the in-plane and bending problems decouple from each other. The in-plane problem is, in fact, identical to the two-dimensional, plane stress elasticity problem described in section 3.3. The in-plane forces of plate theory are the through-the-thickness integral of the corresponding in-plane stress components which are assumed to be constant through the thickness of the plate. Because it is a two-dimensional elasticity problem of the kind treated in section 3.3, solutions of the in-plane problem will not be pursued here; the solutions methods developed in chapter 3 are directly applicable. Solution procedures for the bending problem will be developed in the following sections.

16.2 The bending problem

The bending problem can be reduced to a single partial differential equation for the plate transverse displacement, $\bar{u}_3$. First, the moment equilibrium equations, eqs. (16.25), are used to find the shear forces, which are then introduced into the vertical equilibrium equation, eq. (16.24), to find a single relationship among the three moment components,

$$\frac{\partial^2 M_2}{\partial x_1^2} - 2\frac{\partial^2 M_{12}}{\partial x_1 \partial x_2} - \frac{\partial^2 M_1}{\partial x_2^2} = -p_3. \tag{16.33}$$

Next, the strain-displacement equations, eqs. (16.10), are introduced into the constitutive laws, eqs. (16.30), to express the bending and twisting moments in terms of derivatives of the transverse displacements as

$$M_1 = \frac{Eh^3}{12(1-\nu^2)}(\kappa_1 - \nu\kappa_2) = \frac{Eh^3}{12(1-\nu^2)}\left(\frac{\partial^2 \bar{u}_3}{\partial x_2^2} + \nu\frac{\partial^2 \bar{u}_3}{\partial x_1^2}\right), \quad (16.34a)$$

$$M_2 = \frac{Eh^3}{12(1-\nu^2)}(-\nu\kappa_1 + \kappa_2) = \frac{Eh^3}{12(1-\nu^2)}\left(-\nu\frac{\partial^2 \bar{u}_3}{\partial x_2^2} - \frac{\partial^2 \bar{u}_3}{\partial x_1^2}\right), \quad (16.34b)$$

$$M_{12} = \frac{Eh^3}{24(1+\nu)}\kappa_{12} = \frac{Eh^3}{12(1+\nu)}\frac{\partial^2 \bar{u}_3}{\partial x_1 \partial x_2}. \quad (16.34c)$$

Finally, the bending and twisting moment components are introduced into the moment equilibrium equation, eq. (16.33), to find a single, partial differential equation for the transverse displacement component

$$\frac{\partial^4 \bar{u}_3}{\partial x_1^4} + 2\frac{\partial^4 \bar{u}_3}{\partial x_1^2 \partial x_2^2} + \frac{\partial^4 \bar{u}_3}{\partial x_2^4} = \frac{p_3}{D}, \quad (16.35)$$

where the *plate bending stiffness*, D, is defined as

$$D = \frac{Eh^3}{12(1-\nu^2)}. \quad (16.36)$$

The basic equation of Kirchhoff plate bending theory, eq. (16.35), is the *biharmonic partial differential equation* for the transverse displacement, which can be written in a more compact manner with the help of the Laplacian operator, ∇^2, defined by eq. (3.2),

$$\nabla^4 \bar{u}_3 = \frac{p_3}{D}. \quad (16.37)$$

16.2.1 Typical boundary conditions

The basic governing differential equation for plates is a fourth-order, partial differential equation for the transverse displacement, $\bar{u}_3$, given by eq. (16.37). The solution of this differential equation will require a proper set of boundary conditions. For a rectangular plate, two boundary conditions are required along each of the four edges of the plate. This section will focus on different types of boundary conditions applied to the edge of the plate located at $x_1 = a$; similar developments will yield the corresponding boundary conditions along the other three edges of the plate.

1. *Clamped edge.* Figure 16.6 illustrates the situation where the edge of the plate at $x_1 = a$ is clamped. Clearly, the transverse displacement must vanish along the edge, and because the rotation of the normal line is equal to the slope of the plate, see eq. (16.3), the slope of the plate also vanishes, leading to the following geometric boundary conditions,

$$\bar{u}_3 = 0; \quad \frac{\partial \bar{u}_3}{\partial x_1} = 0. \quad (16.38)$$

2. *Simply supported edge.* Figure 16.6 also illustrates the situation where the edge of the plate at $x_1 = a$ is simply supported. Clearly, the transverse displacement

must vanish along the edge, and because the pivot line cannot resist a bending moment, the bending moment, M_2, must also vanish along the edge. In view of eq. (16.34b), this latter condition implies $-\nu\, \partial^2 \bar{u}_3/\partial x_2^2 - \partial^2 \bar{u}_3/\partial x_1^2 = 0$. Because the transverse displacement vanishes along the edge, so does its derivative with respect to x_2, and this boundary condition reduces to $\partial^2 \bar{u}_3/\partial x_1^2 = 0$. The boundary conditions along a simply supported edge become

$$\bar{u}_3 = 0; \qquad \frac{\partial^2 \bar{u}_3}{\partial x_1^2} = 0. \tag{16.39}$$

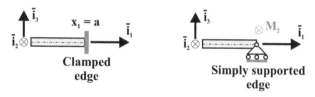

Fig. 16.6. Boundary conditions for clamped and simply supported edges.

3. *Free edge.* The free edge boundary condition is shown in fig 16.7 for an edge perpendicular to axis $\bar{\imath}_1$. No geometric conditions apply, and because the edge is free of any externally applied loads, the bending moment, twisting moment, and transverse shear force must all three vanish, *i.e.*, the boundary conditions are $M_2 = 0$, $M_{12} = 0$, and $Q_1 = 0$. Unfortunately, it is not possible to enforce these three boundary conditions because the fourth-order partial differential equations only requires two boundary conditions along each edge. To overcome this problem, Kirchhoff introduced the concept of *total vertical load*, illustrated on the right portion of fig. 16.7. The twisting moment, $M_{12}\mathrm{d}x_2$, acting on a differential element of length, $\mathrm{d}x_2$, along the free edge can be replaced by an equipollent system consisting of two vertical force of magnitude M_{12} separated by a moment arm, $\mathrm{d}x_2$. A similar representation of the twisting moment is depicted in the figure for the neighboring differential element. At each of the common boundaries between elements, the total vertical force, $V_1\mathrm{d}x_2$, is equal to $M_{12}\mathrm{d}x_2 - M_{12}\mathrm{d}x_2 - \partial M_{12}/\partial x_2\mathrm{d}x_2 + Q_1\mathrm{d}x_2$, or expressed in force per unit length

$$V_1 = Q_1 - \frac{\partial M_{12}}{\partial x_2} = \frac{\partial M_2}{\partial x_1} - 2\frac{\partial M_{12}}{\partial x_2}, \tag{16.40}$$

where the last equality follows from the moment equilibrium equation (16.25a). Note that the total vertical edge load, V_1, has the same units as the vertical shear force, Q_1, namely load per unit span, N/m. The proper boundary conditions to apply along the free edge perpendicular to axis $\bar{\imath}_1$ are now $M_2 = 0$ and $V_1 = 0$. Finally, the boundary conditions must be expressed in terms of the transverse displacement $\bar{u}_3$ and its derivatives with the help of eqs. (16.34), to find the moment and shear conditions,

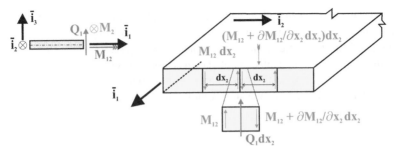

Fig. 16.7. Boundary conditions for free edge.

$$\frac{\partial^2 \bar{u}_3}{\partial x_1^2} + \nu \frac{\partial^2 \bar{u}_3}{\partial x_2^2} = 0; \quad \frac{\partial^3 \bar{u}_3}{\partial x_1^3} + (2 - \nu)\frac{\partial^3 \bar{u}_3}{\partial x_1 \partial x_2^2} = 0. \quad (16.41)$$

While a free edge is an easily understood concept, the associated boundary conditions are not easily expressed.

4. *Edge with a linear spring.* It is often the case that an edge of the plate is resting on an elastic foundation which is conveniently represented by a distributed linear spring as illustrated in fig. 16.8. Note that the spring representing the elastic foundation is a *distributed* spring with a stiffness, k, having units of force per length per length or, N/m^2. In addition, the bending moment, M_2, is assumed to vanish along the edge. Considering the free body diagram shown in the right portion of fig. 16.8, the total vertical load, V_1, must be equal to the load the distributed spring applies to the plate, so $V_1 = -k\bar{u}_3$. The minus sign in this relationship comes from the sign convention for the plate: the transverse displacement is positive along axis $\bar{\imath}_3$, but the total vertical load applied to the spring is positive in the opposite direction. These two boundary conditions can be expressed in terms of the transverse displacement, leading to the following moment and shear conditions,

$$\frac{\partial^2 \bar{u}_3}{\partial x_1^2} + \nu \frac{\partial^2 \bar{u}_3}{\partial x_2^2} = 0; \quad \frac{\partial^3 \bar{u}_3}{\partial x_1^3} + (2 - \nu)\frac{\partial^3 \bar{u}_3}{\partial x_1 \partial x_2^2} - \frac{12k(1 - \nu^2)}{Eh^3}\bar{u}_3 = 0. \quad (16.42)$$

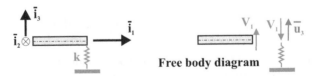

Fig. 16.8. Boundary conditions for an edge with a linear spring.

5. *Edge with a rotational spring.* It may often be the case that the rotation of a simply supported edge of the plate is restrained by an elastic connection which is conveniently represented by a distributed rotational spring as illustrated in

fig. 16.9. Note that the spring representing the elastic connection is a *distributed* spring with a stiffness, k, having units of moment per radian per length or, N/rad. In addition, the transverse displacement, $\bar{u}_3$, is assumed to vanish along the edge (similar to the simply supported boundary condition for a beam). Considering the free body diagram shown in the right portion of fig. 16.9, the distributed bending moment, M_2, must be equal to the moment the distributed spring applies to the plate, $M_2 = k \, \partial\bar{u}_3/\partial x_1$. Note that due to the sign conventions, the bending moment applied to the spring and the plate rotation are both negative about axis $\bar{i}_2$. These two boundary conditions can be expressed in terms of the transverse displacement, leading to

$$\bar{u}_3 = 0; \quad \frac{\partial^2 \bar{u}_3}{\partial x_1^2} + \nu \frac{\partial^2 \bar{u}_3}{\partial x_2^2} + \frac{12k(1-\nu^2)}{Eh^3}\frac{\partial \bar{u}_3}{\partial x_1} = 0. \tag{16.43}$$

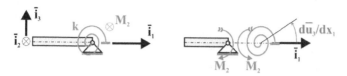

Fig. 16.9. Boundary conditions for an edge with a rotational spring.

The boundary conditions along the other three edges of the plate at $x_1 = 0$, $x_2 = 0$, or $x_2 = b$, are similar to the expressions given above. It is also possible to consider different combinations of geometric and force boundary conditions depending on the particular physical configuration under study. For example, an edge may be supported by both a rotational and linear spring, or by a linear spring with fixed rotation. In other cases, it may be necessary to permute the subscripts for other edges not considered above. For example, for a free edge condition along $x_2 = 0$, or $x_2 = b$, the total vertical load, V_2, must be introduced

$$V_2 = Q_2 - \frac{\partial M_{12}}{\partial x_1} = -\frac{\partial M_1}{\partial x_2} - 2\frac{\partial M_{12}}{\partial x_1}, \tag{16.44}$$

and recast in terms of the transverse displacement, $\bar{u}_3$.

Twisting moment

Twisting moments applied along one or more edges of the plate require further discussion. Consider the case of an edge, say at $x_1 = a$, subjected to a uniform twisting moment, $M_{12} = M_0$. The boundary conditions are similar to those of a free edge, $M_2 = 0$ and $V_1 = Q_1 - \partial M_{12}/\partial x_2 = 0 - 0 = 0$. It is clear that the edge does not "see" the applied constant twisting moment, which seems to indicate that an edge subjected to a constant twisting moment is, in fact, a free edge.

Next, consider the case where two adjacent edges of the plate are subjected to constant twisting moments, M_0. Note that using the same reasoning as employed for the free edge analysis, the total vertical load along the edge vanishes, $V_1 = Q_1 - \partial M_{12}/\partial x_2 = 0 - 0 = 0$. At the corner between the two edges, however, the same reasoning leads to an applied *corner force*, $R = 2M_{12}$.

As shown in fig. 16.10, if the plate is subjected to constant twisting moments, $M_{12} = M_0$, along its four edges, this loading is equivalent to free edges but four forces, each of magnitude $R = 2M_0$, are applied in alternating up and down directions at each of the four corners. This type of loading is called *corner force bending*. The corner force can be expressed in terms of the derivatives of the transverse displacement at the corner

$$R = 2M_{12} = 2D\frac{1-\nu}{2}\kappa_{12} = 2(1-\nu)D\frac{\partial^2\bar{u}_3}{\partial x_1\partial x_2}. \tag{16.45}$$

If a plate is subjected to a constant twisting moment, $M_{12} = M_0$, along one of its edges, two forces, each of magnitude $R = M_0$ and opposite direction, will arise at the boundaries of the edge. Corner forces will appear when the edges of the plate are subjected to twisting moments. For instance, a plate simply supported along its four edges will develop reaction twisting moments along the edges and reaction forces at the four corners.

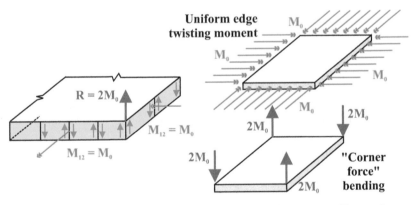

Fig. 16.10. Loading the edge of a plate with a uniform twisting moment. Plate under corner force bending

16.2.2 Simple plate bending solutions

Consider a plate subjected only to loading applied along its edges. The governing differential equation, see eq. (16.37), becomes the homogeneous biharmonic equation, $\nabla^4\bar{u}_3 = 0$. Consider now a solution of the type $\bar{u}_3 = A\,x_1^2 + B\,x_1x_2 + C\,x_2^2$, where A, B, and C, are three arbitrary constants. This solution satisfies the governing equation, and the bending and twisting moments are easily evaluated with the help of eqs. (16.34) to find

$$M_1 = D \left(\frac{\partial^2 \bar{u}_3}{\partial x_2^2} + \nu \frac{\partial^2 \bar{u}_3}{\partial x_1^2} \right) = D(2C + \nu 2A), \tag{16.46a}$$

$$M_2 = D \left(\nu \frac{\partial^2 \bar{u}_3}{\partial x_2^2} + \frac{\partial^2 \bar{u}_3}{\partial x_1^2} \right) = -D(2\nu C + 2A), \tag{16.46b}$$

$$M_{12} = D(1 - \nu) \frac{\partial^2 \bar{u}_3}{\partial x_1 \partial x_2} = D(1 - \nu)B, \tag{16.46c}$$

which reveals that the bending and twisting moments are uniform throughout the plate. From eqs. (16.25), it follows that the shear forces vanish throughout the plate: $Q_1 = \partial M_2/\partial x_1 - \partial M_{12}/\partial x_2 = 0$, and $Q_2 = -\partial M_{12}/\partial x_1 - \partial M_1/\partial x_2 = 0$. Finally, eqs. (16.40) and (16.44) show that the total vertical load also vanish, $V_1 = Q_1 - \partial M_{12}/\partial x_2 = 0$, and $V_2 = Q_2 - \partial M_{12}/\partial x_1 = 0$.

The unknown constants of the problem are readily evaluated in terms of the bending moments by inverting eqs. (16.46) to find

$$A = -\frac{M_2 + \nu M_1}{2D(1 - \nu^2)}, \quad C = \frac{M_1 + \nu M_2}{2D(1 - \nu^2)}, \quad B = \frac{M_{12}}{D(1 - \nu)}. \tag{16.47}$$

Three case are now considered that correspond to different edge loading of the plate.

Synclastic bending

In the first case, all four edges of the plate are subjected to bending moments of equal magnitude in such a way that $M_1 = M_0$ and $M_2 = -M_0$, as illustrated in fig. 16.11. The unknown constants corresponding to this loading case are obtained from eqs. (16.47) as $A = M_0/[2D(1 + \nu)]$, $C = M_0/[2D(1 + \nu)]$, and $B = 0$. Hence, the transverse displacement field in the plate is

$$\bar{u}_3 = \frac{M_0}{2D(1 + \nu)} (x_1^2 + x_2^2). \tag{16.48}$$

This solution is called *synclastic bending*, and the plate deforms into the shape of a sphere, as depicted in fig. 16.12.

Anticlastic bending

In the second case, all four edges of the plate are subjected to bending moments of equal magnitude in such a way that $M_1 = M_0$ and $M_2 = M_0$, as illustrated in fig. 16.13. The unknown constants corresponding to this loading case are obtained from eqs. (16.47) as $A = -M_0/[2D(1 - \nu)]$, $C = M_0/[2D(1 - \nu)]$, and $B = 0$. Hence the transverse displacement field in the plate is

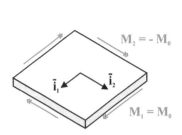

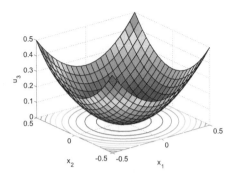

Fig. 16.11. Transverse displacement distribution for synclastic bending.

Fig. 16.12. Transverse displacement distribution for synclastic bending.

$$\bar{u}_3 = -\frac{M_0}{2D(1-\nu)}(x_1^2 - x_2^2). \tag{16.49}$$

This solution of the problem is called *anticlastic bending*, and the plate deforms into the shape of a hyperbolic paraboloid, depicted in fig. 16.14.

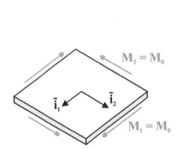

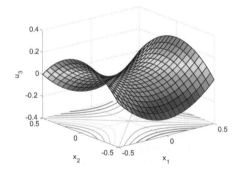

Fig. 16.13. Transverse displacement distribution for anticlastic bending.

Fig. 16.14. Transverse displacement distribution for anticlastic bending.

Corner force bending

In the last case, all four edges of the plate are subjected to twisting moments of equal magnitude, in such a way that $M_{12} = M_0$, as illustrated in fig. 16.15. The applied bending moments are zero, $M_1 = M_2 = 0$. The unknown constants corresponding to this loading case are obtained from eqs. (16.47) as $A = C = 0$, and $B = M_0/[D(1-\nu)]$. Hence, the transverse displacement field for the plate is

$$\bar{u}_3 = \frac{M_0}{D(1-\nu)}x_1 x_2. \tag{16.50}$$

This solution of the problem is called *corner force bending*, and the plate deforms into the shape of a hyperbolic paraboloid, depicted in fig. 16.16.

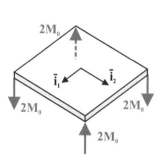

Fig. 16.15. Transverse displacement distribution for corner force bending.

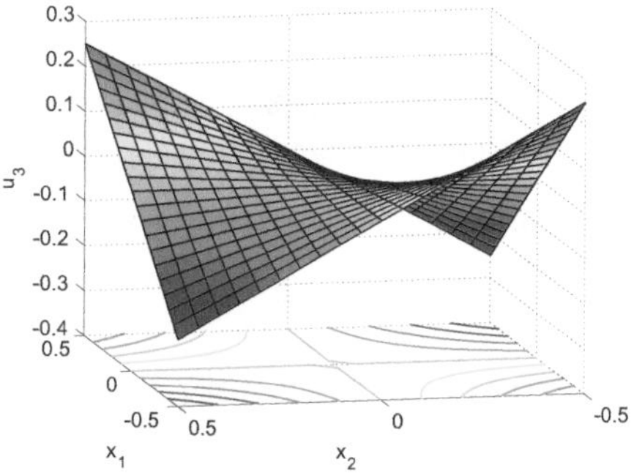

Fig. 16.16. Transverse displacement distribution for corner force bending.

Other plate bending solutions

The treatment of plates with pressure loading, p_3, or with concentrated transverse loads requires development of general solutions to the governing nonhomogeneous biharmonic partial differential equation, eq. (16.37). This is a rich area of mathematical theory and exact and approximate solutions will be developed in later sections.

16.2.3 Problems

Problem 16.1. Plate with various boundary conditions
State the governing differential equation and the boundary conditions for the plate bending problem shown in fig. 16.17.

Problem 16.2. Rectangular plate with edge moments
A rectangular plate is acted upon by edge moments M_0, as shown in fig. 16.18. Find the solution of this problem for the transverse displacement $\bar{u}_3(x_1, x_2)$, bending moments $M_1(x_1, x_2)$, $M_2(x_1, x_2)$ and $M_{12}(x_1, x_2)$, and transverse shear forces $Q_1(x_1, x_2)$ and $Q_2(x_1, x_2)$. Plot the deflected shape of the plate and the various stress resultants.

Problem 16.3. Plate with various boundary conditions
In section 16.2.1, five types of boundary conditions are given for the edge $x_1 = a$ of a rectangular plate. Write the corresponding five sets of boundary conditions for the edge $x_2 = 0$.

Problem 16.4. Plate with given deflection field
Consider the solution $U_3 = \alpha x_1^2 x_2$ for a rectangular plate of dimensions $x_1 \in [0, a]$, $x_2 \in [0, b]$. *(1)* Check that this solution satisfies the governing differential equations of plates. *(2)* State the boundary conditions of the problems it solves and sketch the problem statement. Give the boundary conditions along all four edges.

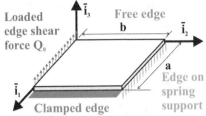

Fig. 16.17. Rectangular plate with various boundary conditions.

Fig. 16.18. Rectangular plate with edge moments.

16.3 Anisotropic plates

The last section treats plates made of homogeneous, linearly elastic, isotropic materials. However, many aerospace applications feature anisotropic plates. The anisotropy of these structures arises from the use of layered composite materials or from directional stiffening. Laminated composite plates and directionally stiffened plates will examined in this section.

16.3.1 Laminated composite plates

Section 2.6 describes the mechanical behavior of lamina consisting of fibers all aligned in a single direction and embedded in a matrix material. The mechanical properties, both stiffness and strength, are highly directional. Consequently, such lamina are not suitable for carrying stresses in several directions simultaneously. A possible solution to this problem is the use of *laminated plates* which consist of a number of lamina, often called "layers" or "plies," stacked on top of each other to form the plate.

Figure 16.19 depicts such a laminated plate consisting of N superposed plies. Axes $\bar{\imath}_1$ and $\bar{\imath}_2$ are located at the mid-plane of the plate, and axis $\bar{\imath}_3$ is perpendicular to that plane. These axes form the reference system for the plate. The ith lamina has a fiber orientation angle $\theta^{[i]}$, and is located at a distance $x_3^{[i]}$ from the plate's mid-plane. The fiber orientation angle is counted positive in the counter clockwise direction from axis $\bar{\imath}_1$. The thickness of each lamina is $t^{[i]} = x_3^{[i+1]} - x_3^{[i]}$. The various lamina often have the same thickness, although this is not necessary.

Laminates will be described by the ply angle characterizing each layer, starting from the bottom ply. For instance, $[\pm 45, 0_2, 0_2, \mp 45]$, is the notation used to describe an 8-ply laminate. The first layer, starting from the bottom of the laminate, has a fiber angle of +45 degrees, the next a -45 degree orientation. Next come two lamina with a 0 degree angles, and so on up to the last layer at the top of the laminate which has a +45 degree angle.

In many instances, laminates possess mid-plane symmetry, *i.e.*, for each lamina below the mid-plane, there is a lamina of identical thickness, orientation angle, and position above the mid-plane. In such case, the top half of the laminate is a mirror

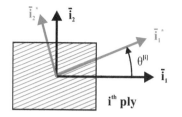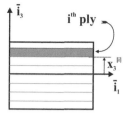

Fig. 16.19. Definition of a laminated plate.

image of the bottom half. This symmetry is reflected in the notation as well. It is not necessary to repeat the description of the top half of the laminate: the notation $[\pm 45, 0_2, 0_2, \mp 45]$ is equivalent to $[\pm 45, 0_2]_S$, where the notation $[.]_S$ indicates the mid-plane symmetry.

16.3.2 Constitutive laws for laminated composite plates

The constitutive laws for laminated plates relate the loads applied to the plate to the resulting deformations, as is done in section 16.1.4 for homogeneous, isotropic plates. These constitutive laws must be distinguished from the constitutive laws that characterize an individual lamina, as discussed in sections 2.6.1 and 2.6.2. The goal of the present analysis is to relate the behavior of the laminate to that of the individual lamina.

The in-plane forces, eq. (16.16), can be related to the plate deformations by introducing the constitutive laws for an individual lamina, eq. (2.78), and the in-plane strain field resulting from Kirchhoff assumptions, eqs. (16.13), to find

$$\underline{N} = \int_h \underline{\underline{C}} \, (\underline{\epsilon}^0 + x_3 \, \underline{\underline{S}} \, \underline{\kappa}) \, \mathrm{d}x_3 = \left[\int_h \underline{\underline{C}} \, \mathrm{d}x_3 \right] \underline{\epsilon}^0 + \left[\int_h \underline{\underline{C}} \, x_3 \, \mathrm{d}x_3 \right] \underline{\underline{S}} \, \underline{\kappa}.$$

A similar treatment for the bending moments, eq. (16.21), yields

$$\underline{\underline{S}} \, \underline{M} = \int_h \underline{\underline{C}} \, (\underline{\epsilon}^0 + x_3 \, \underline{\underline{S}} \, \underline{\kappa}) x_3 \, \mathrm{d}x_3 = \left[\int_h \underline{\underline{C}} \, x_3 \, \mathrm{d}x_3 \right] \underline{\epsilon}^0 + \left[\int_h \underline{\underline{C}} \, x_3^2 \, \mathrm{d}x_3 \right] \underline{\underline{S}} \, \underline{\kappa}.$$

The following matrices are defined

$$\underline{\underline{A}} = \int_h \underline{\underline{C}} \, \mathrm{d}x_3; \tag{16.51a}$$

$$\underline{\underline{D}} = \int_h \underline{\underline{C}} \, x_3^2 \, \mathrm{d}x_3; \tag{16.51b}$$

$$\underline{\underline{B}} = \int_h \underline{\underline{C}} \, x_3 \, \mathrm{d}x_3. \tag{16.51c}$$

With the help of these matrices, the laminate constitutive laws can be written in a compact matrix form as

$$\left\{ \begin{matrix} N \\ \underline{S}\,M \end{matrix} \right\} = \begin{bmatrix} A & B \\ B & D \end{bmatrix} \left\{ \begin{matrix} \epsilon^0 \\ \underline{S}\,\kappa \end{matrix} \right\}. \tag{16.52}$$

If the laminate presents mid-plane symmetry, matrix $\underline{B}$ vanishes. Indeed, the lamina stiffness matrix $\underline{C}$ is then a symmetric function through the plate thickness, but x_3 clearly is antisymmetric, so the product $\underline{C}\,x_3$ is antisymmetric as well and integrates to zero through the thickness of the plate. Consequently, the constitutive laws for mid-plane symmetric laminates reduce to

$$\underline{N} = \underline{A}\,\underline{\epsilon}^0, \tag{16.53}$$

and

$$\underline{M} = \underline{S}^T \underline{D}\,\underline{S}\,\underline{\kappa}. \tag{16.54}$$

The meaning of the three matrices defined in eqs. (16.51) is now clear. Matrix $\underline{A}$ relates the in-plane forces to the plate mid-plane strains; it is called the *in-plane stiffness* matrix. Matrix $\underline{D}$ relates the bending moments and the plate curvatures; it is called the *bending stiffness* matrix. Matrix $\underline{B}$, called the *coupling stiffness* matrix is nonzero for laminates that do not present mid-plane symmetry.

16.3.3 The in-plane stiffness matrix

Consider a laminate consisting of N plies each made of an identical material oriented at various angles, $\theta^{[i]}$. The following array stores the independent entries of the in-plane stiffness matrix, $\underline{A}$, arranged as follows

$$\underline{A} = \left\{ A_{11}, A_{22}, A_{12}, A_{66}, A_{16}, A_{26} \right\}^T. \tag{16.55}$$

In view of the definition of the in-plane stiffness matrix, eq. (16.51a), it follows that $\underline{A} = \int_h \underline{C}(\theta)\,dx_3$, where array $\underline{C}$ contains the corresponding components of the lamina stiffness matrix, $\underline{C}$, defined by eq. (2.84). The integration through the thickness of the plate can be seen as a summation of the integrals over the various plies,

$$\underline{A} = \sum_{i=1}^{N} \int_{x_3^{[i]}}^{x_3^{[i+1]}} \underline{C}(\theta^{[i]})\,dx_3.$$

where $\theta^{[i]}$ is the fiber orientation angle for the ith ply, see fig. 16.19. The ith lamina can be assumed to be a homogeneous, transversely isotropic material with constant stiffness properties through its thickness, hence,

$$\underline{A} = \sum_{i=1}^{N} \underline{C}(\theta^{[i]}) \int_{x_3^{[i]}}^{x_3^{[i+1]}} dx_3 = \left[\sum_{i=1}^{N} \chi(\theta^{[i]}) \int_{x_3^{[i]}}^{x_3^{[i+1]}} dx_3 \right] \underline{\alpha}$$

$$= h \left[\sum_{i=1}^{N} \frac{x_3^{[i+1]} - x_3^{[i]}}{h} \chi(\theta^{[i]}) \right] \underline{\alpha},$$

where the constitutive law for the lamina, eq. (2.86), is used, and h is the total laminate thickness.

The components of the in-plane stiffness array now become

$$\underline{A} = h \, \underline{\underline{\chi}}^A \underline{\alpha}, \tag{16.56}$$

where matrix $\underline{\underline{\chi}}^A$ is defined as

$$\underline{\underline{\chi}}^A = \begin{bmatrix} 1 & 1 & \chi_1^A & \chi_2^A \\ 1 & 1 & -\chi_1^A & \chi_2^A \\ 1 & -1 & 0 & -\chi_2^A \\ 0 & 1 & 0 & -\chi_2^A \\ 0 & 0 & \chi_3^A/2 & \chi_4^A \\ 0 & 0 & \chi_3^A/2 & -\chi_4^A \end{bmatrix}. \tag{16.57}$$

Coefficient h appearing in eq. (16.56), is introduced to render the following *in-plane stiffness lay-up parameters* as non-dimensional quantities,

$$\chi_1^A = \sum_{i=1}^{N} \frac{x_3^{[i+1]} - x_3^{[i]}}{h} \cos 2\theta^{[i]}, \tag{16.58a}$$

$$\chi_2^A = \sum_{i=1}^{N} \frac{x_3^{[i+1]} - x_3^{[i]}}{h} \cos 4\theta^{[i]}, \tag{16.58b}$$

$$\chi_3^A = \sum_{i=1}^{N} \frac{x_3^{[i+1]} - x_3^{[i]}}{h} \sin 2\theta^{[i]}, \tag{16.58c}$$

$$\chi_4^A = \sum_{i=1}^{N} \frac{x_3^{[i+1]} - x_3^{[i]}}{h} \sin 4\theta^{[i]}. \tag{16.58d}$$

Although the laminate may comprise a large number of plies, the in-plane stiffness matrix depends only on the four lay-up parameters defined in eqs. (16.58). Arrays (2.83) and (16.58) for a lamina, and laminate, respectively, present a striking similarity: the trigonometric function, $\cos 2\theta^{[i]}$, for the lamina, is replaced by the lay-up parameter, χ_1^A, for the laminate. In fact, the lay-up parameters are obtained from a rule of mixture, each lamina contribution, $\cos 2\theta^{[i]}$, is weighted by the lamina relative thickness, $t^{[i]}/h = (x_3^{[i+1]} - x_3^{[i]})/h$. The other lay-up parameters are weighted averages of the other trigonometric functions, $\cos 4\theta^{[i]}$, $\sin 2\theta^{[i]}$, and $\sin 4\theta^{[i]}$.

Because the weighting factor is simply the lamina thickness, the position of the lamina in the laminate is unimportant, and therefore the in-plane stiffness matrix is said to be *stacking sequence independent*. It is interesting to note that the lay-up parameters are bounded by -1 and +1, like the trigonometric functions they contain,

$$-1 \leq \chi_{1,2,3,4}^A \leq 1. \tag{16.59}$$

If the laminate possesses mid-plane symmetry, the laminate constitutive laws reduce to (16.53), which, in a reduced form, become

$$\frac{1}{h}\underline{N} = \bar{\underline{\sigma}} = \frac{1}{h}\underline{\underline{A}}\,\underline{\epsilon}^0, \tag{16.60}$$

where $\bar{\underline{\sigma}}$ is the *average stress through-the-thickness of the laminate*. The reduced constitutive laws express a proportionality between the laminate average stresses and the mid-plane strains.

Inverting eq. (16.60) yields the constitutive laws in compliance form

$$\underline{\epsilon}^0 = \left[\frac{1}{h}\underline{\underline{A}}\right]^{-1}\bar{\underline{\sigma}}, \tag{16.61}$$

where $\left[\underline{\underline{A}}/h\right]^{-1}$ is the laminate in-plane compliance matrix. This matrix can be experimentally measured using tests similar to those described in section 2.6.1.

Some laminates are built in such a way that for each lamina at an angle θ, there is a corresponding lamina of equal thickness at an angle $-\theta$. In that case, the last two lay-up parameters χ_3^A and χ_4^A vanish, because the sine functions are odd functions of the fiber orientation angle. For such laminates, called *balanced laminates*, the stiffness matrix components A_{16} and A_{26} are zero. This means that balanced laminates present no coupling between extension and shearing.

16.3.4 Problems

Problem 16.5. Shear moduli of tubes
The shear moduli G of thin-walled tubes have been measured experimentally for tubes made of an identical material with the various lay-ups listed in table 16.1. Determine how well this data correlates with classical lamination theory by plotting the shear modulus as a function of a lay-up parameter to be determined. From the experimental data, determine the material invariants α_2 and α_4.

Table 16.1. Tube lay-ups and corresponding shear moduli.

Lay-up	G [GPa]	Lay-up	G [GPa]	Lay-up	G [GPa]
$[\pm 45°]_3$	29.36	$[0°]_{12}$	4.64	$[\pm 30°]_3$	23.90
$[0_3°, \pm 45°]_s$	14.59	$[0_2°, \pm 45°]_s$	17.36	$[0_4°, (\pm 45°)_3]$	20.25
$[\pm 15°]_3$	11.80				

Problem 16.6. Uniaxial test on a cross-ply laminate
Experimental measurements have been made for the following material constants of a lamina: $E_1 = 134$ GPa, $E_2 = 8.6$ GPa, and $\nu_{12} = 0.29$. The shearing modulus for the lamina is to be measured from a uniaxial test on a cross-ply laminate made of that material with the lay-up $[\pm 45°]_s$. The laminate is subjected to a force F_1 and the corresponding strain ϵ_1 is measured, as depicted in fig. 16.20. The slope of the force F_1 versus strain ϵ_1 is found to be 9.5 MN/m. If the ply thickness of the material is $t_p = 125\ \mu$m, find the lamina shearing modulus G_{12}.

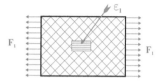

Fig. 16.20. Configuration of the uniaxial test on a cross-ply laminate.

16.3.5 The bending stiffness matrix

Consider once more a laminate consisting of N plies each made of an identical material oriented at various angles, $\theta^{[i]}$. The following array stores the independent entries of the bending stiffness matrix, $\underline{D}$, arranged as follows

$$\underline{D} = \left\{ D_{11}, D_{22}, D_{12}, D_{66}, D_{16}, D_{26} \right\}^T. \tag{16.62}$$

In view of the definition of bending stiffness matrix, eq. (16.51b), it follows that $\underline{D} = \int_h \underline{C}(\theta) x_3^2 \, dx_3$, where array $\underline{C}$ contains the corresponding components of the lamina stiffness matrix, $\underline{C}$, defined by eq. (2.84). The integration through the thickness of the plate can be seen as a summation of the integrals over the various plies

$$\underline{D} = \sum_{i=1}^{N} \int_{x_3^{[i]}}^{x_3^{[i+1]}} \underline{C}(\theta^{[i]}) x_3^2 \, dx_3.$$

where $\theta^{[i]}$ is the fiber orientation angle for the ith ply, see fig. 16.19. The ith lamina is assumed to be a homogeneous, transversely isotropic material with constant stiffness properties through its thickness, and hence,

$$\underline{D} = \sum_{i=1}^{N} \underline{C}(\theta^{[i]}) \int_{x_3^{[i]}}^{x_3^{[i+1]}} x_3^2 \, dx_3 = \left[\sum_{i=1}^{N} \underline{\chi}(\theta^{[i]}) \int_{x_3^{[i]}}^{x_3^{[i+1]}} x_3^2 \, dx_3 \right] \underline{\alpha}$$

$$= \left[\sum_{i=1}^{N} \frac{x_3^{[i+1]3} - x_3^{[i]3}}{3} \underline{\chi}(\theta^{[i]}) \right] \underline{\alpha},$$

where the constitutive law for the lamina, eq. (2.86), is used.

The components of the bending stiffness array now become

$$\underline{D} = \frac{h^3}{12} \underline{\underline{\chi}}^D \underline{\alpha}, \tag{16.63}$$

where matrix $\underline{\underline{\chi}}^D$ is defined as

$$\underline{\underline{\chi}}^D = \begin{bmatrix} 1 & 1 & \chi_1^D & \chi_2^D \\ 1 & 1 & -\chi_1^D & \chi_2^D \\ 1 & -1 & 0 & -\chi_2^D \\ 0 & 1 & 0 & -\chi_2^D \\ 0 & 0 & \chi_3^D/2 & \chi_4^D \\ 0 & 0 & \chi_3^D/2 & -\chi_4^D \end{bmatrix}. \tag{16.64}$$

The factor $h^3/12$ appearing in eq. (16.63) is introduced to render the following *bending stiffness lay-up parameters* as non-dimensional quantities,

$$\chi_1^D = \frac{12}{h^3} \sum_{i=1}^{N} \frac{x_3^{[i+1]3} - x_3^{[i]3}}{3} \cos 2\theta^{[i]}, \qquad (16.65a)$$

$$\chi_2^D = \frac{12}{h^3} \sum_{i=1}^{N} \frac{x_3^{[i+1]3} - x_3^{[i]3}}{3} \cos 4\theta^{[i]}, \qquad (16.65b)$$

$$\chi_3^D = \frac{12}{h^3} \sum_{i=1}^{N} \frac{x_3^{[i+1]3} - x_3^{[i]3}}{3} \sin 2\theta^{[i]}, \qquad (16.65c)$$

$$\chi_4^D = \frac{12}{h^3} \sum_{i=1}^{N} \frac{x_3^{[i+1]3} - x_3^{[i]3}}{3} \sin 4\theta^{[i]}. \qquad (16.65d)$$

Although the laminate may comprise a large number of plies, the bending stiffness matrix depends only on the four lay-up parameters defined in eqs. (16.65). Arrays (2.83) and (16.65) for a lamina, and laminate, respectively, present a striking similarity: the trigonometric function, $\cos 2\theta^{[i]}$, for the lamina, is replaced by the lay-up parameter, χ_1^D, for the laminate. Here again, the lay-up parameters are obtained from a rule of mixtures: each lamina contribution, $\cos 2\theta^{[i]}$, is weighted by a purely geometric factor, $(x_3^{[i+1]3} - x_3^{[i]3})/3$. This weighting factor explicitly depends on the position of the lamina in the stacking sequence, and therefore the bending stiffness matrix is said to be *stacking sequence dependent*. The weighting factor for the outermost ply is considerably larger than that of a lamina near the mid-plane where x_3 is nearly zero. Here again, the lay-up parameters are bounded,

$$-1 \leq \chi_{1,2,3,4}^D \leq 1. \qquad (16.66)$$

The stacking sequence effect can be illustrated by computing the lay-up parameter for two laminates $[0_2, \pm45]_S$ and $[\pm45, 0_2]_S$. All plies are assumed to have the same thickness t. The lay-up parameter χ_1^D of the first laminate is

$$\chi_1^D = \frac{8}{(8t)^3} \left\{ \left[(4t)^3 - (2t)^3\right] \cos(0) + \left[(2t)^3 - (t)^3\right] \cos(90) + \left[(t)^3\right] \cos(-90) \right\}$$

$$= \frac{1}{64} \left\{ 56 \times \cos(0) + 7 \times \cos(90) + 1 \times \cos(-90) \right\} = 0.875.$$

Similarly, for the second laminate:

$$\chi_1^D = \frac{8}{(8t)^3} \left\{ \left[(4t)^3 - (3t)^3\right] \cos(90) + \left[(3t)^3 - (2t)^3\right] \cos(-90) + (2t)^3 \cos(0) \right\}$$

$$= \frac{1}{64} \left\{ 37 \times \cos(90) + 19 \times \cos(-90) + 8 \times \cos(0) \right\} = 0.125.$$

The two laminates differ only by their stacking sequence, but their bending stiffness lay-up parameters are very different.

If the laminate possesses mid-plane symmetry, the laminate constitutive laws reduce to eq. (16.54). Assume the laminate to be made of an equivalent, homogeneous material. The linear distribution of in-plane strain through the thickness of the plate then implies a linear distribution of in-plane stress. The maximum stress, $\underline{\sigma}^{\text{max}}$ is found at the top and bottom of the laminate, and is given by

$$\underline{\sigma}^{\text{max}} = \frac{6}{h^2}\underline{M} = \frac{12}{h^3}\underline{\underline{D}}\frac{h}{2}\underline{\kappa}. \tag{16.67}$$

The assumed linearity of the in-plane strain distribution through the thickness of the plate further implies that the strains at the top and at the bottom of the laminate are $\underline{\epsilon}^{\text{top}} = \underline{\epsilon}^0 + h\underline{\kappa}/2$ and $\underline{\epsilon}^{\text{bottom}} = \underline{\epsilon}^0 - h\underline{\kappa}/2$, respectively. When the laminate is in pure bending $|\underline{\epsilon}^{\text{top}}| = |\underline{\epsilon}^{\text{bottom}}| = |h\underline{\kappa}/2|$, and eq. (16.67) becomes

$$\underline{\sigma}^{\text{max}} = \frac{12}{h^3}\underline{\underline{D}}\,\underline{\epsilon}^{\text{top}}. \tag{16.68}$$

These reduced constitutive laws express a proportionality between the stresses and strains at the top of the laminate if the plate is made of an *equivalent*, homogeneous material. Inverting eq. (16.67) yields the constitutive laws in compliance form as

$$\frac{h}{2}\underline{\kappa} = \left[\frac{12}{h^3}\underline{\underline{D}}\right]^{-1}\frac{6\underline{M}}{h^2}, \tag{16.69}$$

where $\left[12\underline{\underline{D}}/h^3\right]^{-1}$ is the laminate bending compliance matrix. This compliance matrix could be experimentally measured using the four-point bending test described in example 5.3.

Note that a *balanced laminate* construction does not imply the vanishing of the bending matrix components D_{16} and D_{26}. Indeed the lay-up parameters χ_3^D and χ_4^D depend not only on the orientation angle but also on the stacking sequence. In other words, the contributions of two adjacent plies at angles θ and $-\theta$, respectively, do not cancel because the two plies are at a different position in the stacking sequence. However, it often happens that the bending stiffness matrix components D_{16} and D_{26} are very small compared to the others. For such laminates, called *specially orthotropic laminates*, no coupling exists between bending and twisting.

As a final note, the bending stiffness matrix for plate made of a homogeneous, isotropic material is

$$\underline{\underline{D}} = \underline{\underline{S}}\,\underline{\underline{\tilde{D}}}\,\underline{\underline{S}}^T = \frac{Eh^3}{12(1-\nu^2)}\begin{bmatrix} 1 & \nu & 0 \\ \nu & 1 & 0 \\ 0 & 0 & (1-\nu)/2 \end{bmatrix}, \tag{16.70}$$

where $\underline{\underline{\tilde{D}}}$ is defined by eq. (16.31) and matrix $\underline{\underline{S}}$ by eq. (16.14).

16.3.6 The coupling stiffness matrix

Again, consider a laminate consisting of N plies each made of an identical material oriented at various angles, $\theta^{[i]}$. The following array stores the independent entries of the coupling stiffness matrix, $\underline{\underline{B}}$, arranged as follows

$$\underline{B} = \{B_{11}, B_{22}, B_{12}, B_{66}, B_{16}, B_{26}\}^T . \tag{16.71}$$

In view of the definition of coupling stiffness matrix, eq. (16.51c), it follows that $\underline{B} = \int_h \underline{C}(\theta)x_3 \, dx_3$, where array $\underline{C}$ contains the corresponding components of the lamina stiffness matrix, $\underline{C}$, defined by eq. (2.84). The integration through the thickness of the plate can be viewed as a summation of the integrals over the various plies

$$\underline{B} = \sum_{i=1}^{N} \int_{x_3^{[i]}}^{x_3^{[i+1]}} \underline{C}(\theta^{[i]})x_3 \, dx_3.$$

where $\theta^{[i]}$ is the fiber orientation angle for the ith ply, see fig. 16.19. The ith lamina can be assumed to be a homogeneous, transversely isotropic material with constant stiffness properties through its thickness. Hence,

$$\underline{B} = \sum_{i=1}^{N} \underline{C}(\theta^{[i]}) \int_{x_3^{[i]}}^{x_3^{[i+1]}} x_3 \, dx_3 = \left[\sum_{i=1}^{N} \chi(\theta^{[i]}) \int_{x_3^{[i]}}^{x_3^{[i+1]}} x_3 \, dx_3 \right] \underline{\alpha}$$

$$= \left[\sum_{i=1}^{N} \frac{x_3^{[i+1]2} - x_3^{[i]2}}{2} \chi(\theta^{[i]}) \right] \underline{\alpha},$$

where the constitutive law for the lamina, eq. (2.86), is used.

The components of the coupling stiffness matrix now become

$$\underline{B} = \frac{h^2}{2} \underline{\underline{\chi}}^B \underline{\alpha}, \tag{16.72}$$

where array $\underline{\underline{\chi}}^B$ is defined as

$$\underline{\underline{\chi}}^B = \begin{bmatrix} 0 & 0 & \chi_1^B & \chi_2^B \\ 0 & 0 & -\chi_1^B & \chi_2^B \\ 0 & 0 & 0 & -\chi_2^B \\ 0 & 0 & 0 & -\chi_2^B \\ 0 & 0 & \chi_3^B/2 & \chi_4^B \\ 0 & 0 & \chi_3^B/2 & -\chi_4^B \end{bmatrix}. \tag{16.73}$$

The factor $h^2/2$ appearing in eq. (16.72) is introduced to render the following *coupling stiffness lay-up parameters* as non-dimensional quantities,

$$\chi_1^B = \sum_{i=1}^N \frac{x_3^{[i+1]2} - x_3^{[i]2}}{h^2} \cos 2\theta^{[i]}, \tag{16.74a}$$

$$\chi_2^B = \sum_{i=1}^N \frac{x_3^{[i+1]2} - x_3^{[i]2}}{h^2} \cos 4\theta^{[i]}, \tag{16.74b}$$

$$\chi_3^B = \sum_{i=1}^N \frac{x_3^{[i+1]2} - x_3^{[i]2}}{h^2} \sin 2\theta^{[i]}, \tag{16.74c}$$

$$\chi_4^B = \sum_{i=1}^N \frac{x_3^{[i+1]2} - x_3^{[i]2}}{h^2} \sin 4\theta^{[i]}. \tag{16.74d}$$

Although the laminate may comprise a large number of plies, the coupling stiffness matrix depends only on the four lay-up parameters defined in eqs. (16.74). Arrays (2.83) and (16.74) for a lamina and laminate, respectively, present a striking similarity: the trigonometric function, $\cos 2\theta^{[i]}$, for the lamina, is replaced by the lay-up parameter, χ_1^B, for the laminate. Here again, the lay-up parameters are obtained from a rule of mixtures, each lamina contribution, $\cos 2\theta^{[i]}$, is weighted by a purely geometric factor, $(x_3^{[i+1]2} - x_3^{[i]2})/2$. This weighting factor explicitly depends on the position of the lamina in the stacking sequence, and therefore the coupling stiffness matrix is said to be *stacking sequence dependent*. The weighting factor for the outermost ply is considerably larger than that of a lamina near the mid-plane where x_3 is nearly zero. Here again, the lay-up parameters are bounded,

$$-1 \leq \chi_{1,2,3,4}^B \leq 1. \tag{16.75}$$

Note that if the laminate possesses mid-plane symmetry, the coupling bending stiffnesses vanishes.

16.3.7 Problems

Problem 16.7. Computing stresses in a laminate
The strain states at the upper and lower surfaces of a laminated composite plate have been measured by means of strain gauges. Figure 16.21 depicts the arrangement of the three gauges on the upper surface of the plate, noted ε_1^u, ε_2^u, and ε_3^u. The corresponding gauges on the lower surface of the plate are ε_1^ℓ, ε_2^ℓ, and ε_3^ℓ. The following strains are measured: $\varepsilon_1^u = 3561\ \mu$, $\varepsilon_2^u = 2804\ \mu$, $\varepsilon_3^u = 5011\ \mu$, $\varepsilon_1^\ell = 1069\ \mu$, $\varepsilon_2^\ell = -464.1\ \mu$, and $\varepsilon_3^\ell = -1603\ \mu$. The material properties are as follows: $E_1 = 140$ GPa, $E_2 = 9.0$ GPa, $G_{12} = 6.0$ GPa, $\nu_{12} = 0.3$. The ply thickness is $t = 125\ \mu m$ and the lay-up is $[0_2^\circ, \pm 45^\circ]_s$. *(1)* Compute the axial forces $\underline{N}$ and bending moments $\underline{M}$ in the laminate. *(2)* Compute the stresses in the 45° ply, in the axes aligned with the fiber.

16.3.8 Directionally stiffened plates

The previous sections treat laminated composite plates composed of a stack of layers made of highly anisotropic material. This type of plate offers the potential for strongly anisotropic behavior.

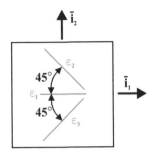

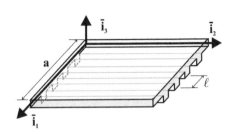

Fig. 16.21. Configuration of the three strain gauges at the upper surface of the plate.

Fig. 16.22. Rectangular plate with directional reinforcement along axis $\bar{i}_2$.

A second type of plate structure that also presents anisotropic behavior is the directionally reinforced plate. Figure 16.22 depicts such a plate presenting stiffeners along axis $\bar{i}_2$, for instance. Typically, such plates are realized by adhesively bonding or riveting stiffeners to the plate, or they can be manufactured by starting with a thick plate and machining or chemically etching away the material between the stiffeners, thereby creating a set of stiffeners in a desired direction.

If the plate features a large number of closely spaced stiffeners, *i.e.*, if $\ell/a \ll 1$, it is possible to smear the effect of the stiffeners to obtained *equivalent plate bending stiffnesses* by adding the bending stiffness of the stiffeners to those of the plate. On the other hand, if the plate features just a few stiffeners, the smearing process is clearly not justified, and each stiffener must be represented individually in the analysis.

As an example, if the number of stiffeners is sufficient to justify smearing their properties, the bending stiffnesses of the plate become

$$
\begin{aligned}
D_{11} &= \frac{Eh^3}{12(1-\nu^2)}, & D_{22} &= \frac{Eh^3}{12(1-\nu^2)} + \frac{H_{\text{bend}}^{\text{stif}}}{\ell}, \\
D_{12} &= \frac{\nu Eh^3}{12(1-\nu^2)}, & D_{66} &= \frac{Eh^3}{24(1+\nu)} + \frac{H_{\text{tor}}^{\text{stif}}}{\ell}.
\end{aligned}
\tag{16.76}
$$

In this approximation, the bending stiffness terms D_{11} and D_{12} are unaffected by the presence of the stiffeners. The term D_{22} is the sum of the base plate contribution, $Eh^3/[12(1-\nu^2)]$, and of the smeared stiffener bending stiffness per unit length, $H_{\text{bend}}^{\text{stif}}/\ell$. Similarly, the term D_{66} is the sum of the base plate contribution, $Eh^3/[24(1+\nu)]$, and of the smeared stiffener torsional stiffness per unit length, $H_{\text{tor}}^{\text{stif}}/\ell$.

Because the plate no longer presents mid-plane symmetry, the coupling stiffness matrix, $\underline{\underline{B}}$, no longer vanishes. These terms, however, can often be neglected. Of course, the plate might present stiffeners located in a symmetric manner on both sides of the plate; in such case, the coupling matrix vanishes exactly due to the symmetry condition.

16.3.9 Governing equations for anisotropic plates

The only difference between isotropic and anisotropic plates is the form of the constitutive laws. For isotropic plates, the constitutive laws split into two separate sets of equations: one set relates in-plane forces to mid-plane strains, $\underline{N} = \underline{\underline{A}}\,\underline{\epsilon}_0$, and the other relates bending moments to curvatures, $\underline{SM} = \underline{D}\,\underline{S}\,\underline{\kappa}$. Furthermore, the in-plane and out-of-plane responses of the plate are uncoupled.

In contrast, anisotropic, laminated plates have the potential for coupling in-plane and out-of-plane behavior due to the presence of the coupling stiffness matrix matrix, $\underline{\underline{B}}$, see eq. (16.51c).

Generally anisotropic plates

The governing equations for generally anisotropic plates can be reduced to three coupled partial differential equations for the three displacement components. First, the strain-displacement equations, eqs. (16.9) and (16.10), are introduced into the constitutive laws, eqs. (16.52), to express the in-plane forces and the bending and twisting moments in terms of derivatives of the three displacement components. Next, these in-plane forces and the bending and twisting moments are introduced into the in-plane and moment equilibrium equations, eqs. (16.23) and (16.33), respectively, to yield the three governing equations for generally anisotropic plates.

These three equations involve numerous mixed partial derivatives up to fourth order. They are presented below using a shorthand notation for derivatives in which a spatial derivative of a variable is indicated by adding to the variable's subscript a comma followed by the spatial variable's index. For instance, $\bar{u}_{3,1122}$, is interpreted as $\partial^4 \bar{u}_3/(\partial x_1^2 \partial x_2^2)$.

With this notation, the governing equations for generally anisotropic plates are

$$A_{11}\bar{u}_{1,11} + 2A_{16}\bar{u}_{1,12} + A_{66}\bar{u}_{1,22} + A_{16}\bar{u}_{2,11} + (A_{12} + A_{66})\bar{u}_{2,12} + A_{26}\bar{u}_{2,22}$$
$$- B_{11}\bar{u}_{3,111} - 3B_{16}\bar{u}_{3,112} - (B_{12} + 2B_{66})\bar{u}_{3,122} - B_{26}\bar{u}_{3,222} = -p_1, \quad (16.77a)$$
$$A_{16}\bar{u}_{1,11} + (A_{12} + A_{66})\bar{u}_{1,12} + A_{26}\bar{u}_{1,22} + A_{66}\bar{u}_{2,11} + 2A_{26}\bar{u}_{2,12} + A_{22}\bar{u}_{2,22}$$
$$- B_{16}\bar{u}_{3,111} - (B_{12} + 2B_{66})\bar{u}_{3,112} - 3B_{26}\bar{u}_{3,122} - B_{22}\bar{u}_{3,222} = -p_2, \quad (16.77b)$$
$$D_{11}\bar{u}_{3,1111} + 4D_{16}\bar{u}_{3,1112} + 2(D_{12} + 2D_{66})\bar{u}_{3,1122} + 4D_{26}\bar{u}_{3,1222} + D_{22}\bar{u}_{3,2222}$$
$$- B_{11}\bar{u}_{1,111} - 3B_{16}\bar{u}_{1,112} - (B_{12} + 2B_{66})\bar{u}_{1,122} - B_{26}\bar{u}_{1,222}$$
$$- B_{16}\bar{u}_{2,111} - (B_{12} + 2B_{66})\bar{u}_{2,112} - 3B_{26}\bar{u}_{2,122} - B_{22}\bar{u}_{2,222} = p_3. \quad (16.77c)$$

Clearly, the in-plane behavior, associated with the in-plane displacement components, $\bar{u}_1$ and $\bar{u}_2$, cannot be decoupled from the out-of-plane behavior associated with the transverse displacement component, $\bar{u}_3$. Analytical solutions of these three coupled partial differential equations cannot be obtained for the most general cases. Instead, numerical techniques are required, and these lead to approximate solutions.

Anisotropic plates with zero coupling stiffness matrix

An important class of plate problems is that for which the *coupling stiffness matrix vanishes*. For a plate possessing *mid-plane symmetry*, the coupling stiffness matrix vanishes. The governing equations for this class are easily obtained by setting all entries of the coupling stiffness matrix to zero in eqs. (16.77). The first two equations, eqs. (16.77a) and (16.77b), now define the in-plane problem

$$
\begin{aligned}
A_{11}\bar{u}_{1,11} + 2A_{16}\bar{u}_{1,12} + A_{66}\bar{u}_{1,22} & \\
+ A_{16}\bar{u}_{2,11} + (A_{12} + A_{66})\bar{u}_{2,12} + A_{26}\bar{u}_{2,22} &= -p_1, \quad (16.78a)
\end{aligned}
$$
$$
\begin{aligned}
A_{16}\bar{u}_{1,11} + (A_{12} + A_{66})\bar{u}_{1,12} + A_{26}\bar{u}_{1,22} & \\
+ A_{66}\bar{u}_{2,11} + 2A_{26}\bar{u}_{2,12} + A_{22}\bar{u}_{2,22} &= -p_2. \quad (16.78b)
\end{aligned}
$$

These two equations involve only the two in-plane displacement components, $\bar{u}_1$ and $\bar{u}_2$. These equations are, in fact, the general equations for two dimensional, anisotropic elasticity.

A subclass of problems of practical significance is that of *balanced laminates* for which $A_{16} = A_{26} = 0$. Equations (16.78) then reduced to

$$
A_{11}\bar{u}_{1,11} + A_{66}\bar{u}_{1,22} + (A_{12} + A_{66})\bar{u}_{2,12} = -p_1; \quad (16.79a)
$$
$$
(A_{12} + A_{66})\bar{u}_{1,12} + A_{66}\bar{u}_{2,11} + A_{22}\bar{u}_{2,22} = -p_2. \quad (16.79b)
$$

When the coupling stiffness matrix vanishes, the last governing equation is readily obtained from eq. (16.77c) as

$$
\begin{aligned}
D_{11}\bar{u}_{3,1111} + 4D_{16}\bar{u}_{1,1112} + 2(D_{12} + 2D_{66})\bar{u}_{3,1122} & \\
+ 4D_{26}\bar{u}_{3,1222} + D_{22}\bar{u}_{3,2222} &= p_3.
\end{aligned} \quad (16.80)
$$

This is the governing equation for bending of anisotropic plates when the coupling stiffness matrix vanishes. It should be compared to eq. (16.37) for bending of isotropic plates.

Specially orthotropic plates

In many practical applications, the D_{16} and D_{26} entries in the bending stiffness matrix are much smaller than the diagonal terms D_{11} and D_{22}. Setting these two terms to zero, $D_{16} \approx 0$ and $D_{26} \approx 0$, is, then, a reasonable approximation, and such plates are called *specially orthotropic plates*.

The governing equation for the bending of specially orthotropic plates is found by setting $D_{16} = D_{26} = 0$ in eq. (16.80), which results in

$$
D_{11}\bar{u}_{3,1111} + 2(D_{12} + 2D_{66})\bar{u}_{3,1122} + D_{22}\bar{u}_{3,2222} = p_3. \quad (16.81)
$$

Note that for a plate made of a homogeneous, isotropic, linearly elastic material, the bending stiffness matrix is given by eq. (16.70), and $D_{11} = D_{22} = D$, $D_{12} = \nu D$, $D_{66} = (1 - \nu)D/2$, and $D_{16} = D_{26} = 0$, where D is the plate bending stiffness given by eq. (16.36). The governing differential equation now reduces $\bar{u}_{3,1111} + 2\bar{u}_{3,1122} + \bar{u}_{3,2222} = p_3/D$, which matches the previously results, see eq. (16.37).

16.4 Solution techniques for rectangular plates

For rectangular plates, the bending equations for anisotropic plates can be solved using various techniques. The following sections present Navier's and Lévy's solution procedures.

16.4.1 Navier's solution for simply supported plates

The governing partial differential equation for a specially orthotropic plate, given by eq. (16.81), is repeated here

$$D_{11}\bar{u}_{3,1111} + 2(D_{12} + 2D_{66})\bar{u}_{3,1122} + D_{22}\bar{u}_{3,2222} = p_3. \tag{16.82}$$

The rectangular plate is simply supported along its four edges, leading to the following boundary conditions, see eqs. (16.39), $\bar{u}_3 = \bar{u}_{3,11} = 0$ at $x_1 = 0$ and a, and $\bar{u}_3 = \bar{u}_{3,22} = 0$ at $x_2 = 0$ and b.

The solution of the problem is assumed to be represented by a double infinite series expansion of sine functions

$$\bar{u}_3(x_1, x_2) = \sum_{m=1}^{\infty}\sum_{n=1}^{\infty} q_{mn} \sin\alpha_m x_1 \sin\beta_n x_2, \tag{16.83}$$

where the coefficients of the expansion, q_{mn}, are now the unknowns of the problem, and the following short hand notation is used

$$\alpha_m = \frac{m\pi}{a}, \quad \beta_n = \frac{n\pi}{b}. \tag{16.84}$$

Note that due to the periodic properties of trigonometric functions, each term of this expansion satisfies all boundary conditions, and hence, the complete solution automatically satisfies all boundary conditions.

Next, the applied transverse pressure is also represented by a double Fourier series as

$$p_3(x_1, x_2) = \sum_{m=1}^{\infty}\sum_{n=1}^{\infty} p_{mn} \sin\alpha_m x_1 \sin\beta_n x_2, \tag{16.85}$$

where the coefficients, p_{mn}, of the expansion are known, and their evaluation will be deferred until later.

The assumed transverse displacement field, eq. (16.83), and the transverse pressure expansion, eq. (16.85), are introduced into the governing partial differential equation of the problem, eq. (16.82), to find

$$\sum_{m=1}^{\infty}\sum_{n=1}^{\infty} \Big\{ q_{mn} \left[D_{11}\alpha_m^4 + 2(D_{12} + 2D_{66})\alpha_m^2\beta_n^2 + D_{22}\beta_n^4 \right]$$

$$-p_{mn} \Big\} \sin\alpha_m x_1 \sin\alpha_m x_2 = 0.$$

This equation must be satisfied at every point of the plate, *i.e.*, for any value of $0 \leq x_1 \leq a$ and $0 \leq x_2 \leq b$. This is only possible if the term in the braces exactly vanishes for all integers m and n, and hence,

$$q_{mn} = \frac{p_{mn}}{D_{11}\alpha_m^4 + 2(D_{12} + 2D_{66})\alpha_m^2\beta_n^2 + D_{22}\beta_n^4} = \frac{p_{mn}}{\Delta_{mn}}.$$

where $\Delta_{mn} = D_{11}\alpha_m^4 + 2(D_{12} + 2D_{66})\alpha_m^2\beta_n^2 + D_{22}\beta_n^4$ defines the plate's transverse stiffness to the applied pressure.

The solution of the problem then follows from eq. (16.83) as

$$\bar{u}_3 = \sum_{m=1}^{\infty} \sum_{n=1}^{\infty} \frac{p_{mn}}{\Delta_{mn}} \sin \alpha_m x_1 \sin \beta_n x_2. \tag{16.86}$$

Once the transverse displacements have been determined, it is possible to evaluate the plate curvatures from eq. (16.10), and the bending moment distribution in the plate follows from the constitutive laws, eqs. (16.54), as

$$M_1 = -\sum_{m=1}^{\infty} \sum_{n=1}^{\infty} \frac{(D_{12}\alpha_m^2 + D_{22}\beta_n^2)p_{mn}}{\Delta_{mn}} \sin \alpha_m x_1 \sin \beta_n x_2, \tag{16.87a}$$

$$M_2 = \sum_{m=1}^{\infty} \sum_{n=1}^{\infty} \frac{(D_{11}\alpha_m^2 + D_{12}\beta_n^2)p_{mn}}{\Delta_{mn}} \sin \alpha_m x_1 \sin \beta_n x_2, \tag{16.87b}$$

$$M_{12} = \sum_{m=1}^{\infty} \sum_{n=1}^{\infty} \frac{2D_{66}\alpha_m\beta_n p_{mn}}{\Delta_{mn}} \cos \alpha_m x_1 \cos \beta_n x_2. \tag{16.87c}$$

Finally, the transverse shear forces are evaluated with the help of the equilibrium equations, eqs. (16.25), to find

$$Q_1 = \sum_{m=1}^{\infty} \sum_{n=1}^{\infty} \frac{\alpha_m[D_{11}\alpha_m^2 + (D_{12} + 2D_{66})\beta_n^2]p_{mn}}{\Delta_{mn}} \cos \alpha_m x_1 \sin \beta_n x_2, \tag{16.88a}$$

$$Q_2 = \sum_{m=1}^{\infty} \sum_{n=1}^{\infty} \frac{\beta_n[D_{22}\beta_n^2 + (D_{12} + 2D_{66})\alpha_m^2]p_{mn}}{\Delta_{mn}} \sin \alpha_m x_1 \cos \beta_n x_2. \tag{16.88b}$$

Along the edges of the plate, the vertical reaction forces are given by the total vertical forces, defined by eqs. (16.40) and (16.44),

$$V_1 = \sum_{m=1}^{\infty} \sum_{n=1}^{\infty} \frac{\alpha_m[D_{11}\alpha_m^2 + (D_{12} + 4D_{66})\beta_n^2]p_{mn}}{\Delta_{mn}} \cos \alpha_m x_1 \sin \beta_n x_2, \tag{16.89a}$$

$$V_2 = \sum_{m=1}^{\infty} \sum_{n=1}^{\infty} \frac{\beta_n[D_{22}\beta_n^2 + (D_{12} + 4D_{66})\alpha_m^2]p_{mn}}{\Delta_{mn}} \sin \alpha_m x_1 \cos \beta_n x_2. \tag{16.89b}$$

To complete the solution process, the loading coefficients, p_{mn}, defined by eq. (16.85), must be evaluated for a given pressure distribution, $p_3(x_1, x_2)$. The coefficients, p_{mn}, are the Fourier coefficients of the applied pressure, $p_3(x_1, x_2)$. To evaluate these coefficients, both sides of eq. (16.85) are multiplied by $\sin \alpha_i x_1 \sin \beta_j x_2$, then integrated over the surface of the plate, leading to

$$\int_0^a \int_0^b p_3 \sin \alpha_i x_1 \sin \beta_j x_2 \, dx_1 dx_2$$

$$= \sum_{m=1}^\infty \sum_{n=1}^\infty p_{mn} \int_0^a \sin \alpha_m x_1 \sin \alpha_i x_1 \, dx_1 \int_0^b \sin \beta_n x_2 \sin \beta_j x_2 \, dx_2.$$

Using the orthogonality properties of sine functions, the two integrals on the right-hand side are easily evaluated with the help of eq. (A.45a). The only non-vanishing term in the double summation is that for which $m = i$ and $n = j$. Solving for Fourier coefficients of the pressure distribution, p_{ij}, and replacing i and j by m and n, respectively, for consistency with eq. (16.85), results in

$$p_{mn} = \frac{4}{ab} \int_0^a \int_0^b p_3(x_1, x_2) \sin \alpha_m x_1 \sin \beta_n x_2 \, dx_1 dx_2. \qquad (16.90)$$

Navier's solution is developed for a rectangular plate with four simply supported edges and subjected to an arbitrary transverse pressure distribution. If other boundary conditions such as clamped or free edges are present, the representation of the transverse displacement field given by eq. (16.83) no longer satisfies the boundary conditions, and an alternative solution procedure must be used.

Example 16.1. Rectangular plate subjected to a uniform transverse pressure

Consider a uniform, rectangular plate made of a homogeneous, isotropic material and subjected to a uniform transverse pressure, p_0. First, the Fourier coefficients of the transverse pressure distribution are evaluated, based on eq. (16.90), to find

$$p_{mn} = \frac{4p_0}{ab} \int_0^a \int_0^b \sin \alpha_m x_1 \sin \beta_n x_2 \, dx_1 dx_2$$

$$= \frac{4p_0}{ab} \int_0^a \sin \alpha_m x_1 \, dx_1 \int_0^b \sin \beta_n x_2 \, dx_2.$$

Integration of the sine functions is given by eq. (A.48), leading to

$$p_{mn} = \frac{16p_0}{\pi^2 mn} \begin{cases} 1, & \text{if } m \text{ and } n \text{ odd}, \\ 0, & \text{otherwise}. \end{cases} \qquad (16.91)$$

The solution of the problem then follows from eq. (16.86) with Δ_{mn} reduced to $D[\alpha_m^4 + \beta_n^4]$ for an isotropic plate where $D = Eh^3/[12(1 - \nu^2)]$. The transverse displacement field becomes

$$\bar{u}_3 = \frac{16p_0}{\pi^2 D} \sum_{\substack{m,n \ odd}}^{\infty} \frac{\sin \alpha_m x_1 \sin \beta_n x_2}{mn(\alpha_m^2 + \beta_n^2)^2}. \tag{16.92}$$

Consider a square plate of length and width a. The distribution of non-dimensional transverse displacement, $\bar{u}_3 D/(p_0 a^4)$, over the plate is shown in fig. 16.23. Next, the non-dimensional bending moment distributions, $M_1/(p_0 a^2)$ and $M_2/(p_0 a^2)$, and twisting moment distribution, $M_{12}/(p_0 a^2)$, in the plate can be computed from eqs. (16.87) and are depicted in figs. 16.25, 16.26 and 16.24, respectively. Finally, the non-dimensional shear force distributions, $Q_1/(p_0 a)$ and $Q_2/(p_0 a)$ in the plate are obtained from eqs. (16.88), and shown in figs. 16.27 and 16.28, respectively.

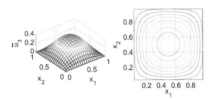

Fig. 16.23. Distribution of transverse displacement, $\bar{u}_3$, over the plate.

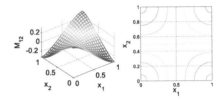

Fig. 16.24. Distribution of twisting moment, M_{12}, over the plate.

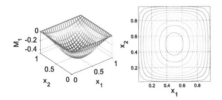

Fig. 16.25. Distribution of bending moment, M_1, over the plate.

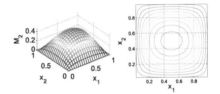

Fig. 16.26. Distribution of bending moment, M_2, over the plate.

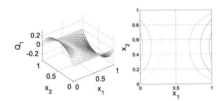

Fig. 16.27. Distribution of transverse shear force, Q_1, over the plate.

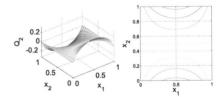

Fig. 16.28. Distribution of transverse shear force, Q_2, over the plate.

Although the solution is given as an infinite series, the convergence is rather fast for the transverse displacements. Keeping only the first term of the series in eq. (16.92), the displacement becomes

$$\bar{u}_3 \approx \frac{4}{\pi^6} \frac{p_0 a^4}{D} \sin \frac{\pi x_1}{a} \sin \frac{\pi x_2}{a} = 0.00416 \frac{p_0 a^4}{D} \sin \frac{\pi x_1}{a} \sin \frac{\pi x_2}{a}.$$

As expected, the maximum displacement is found at the middle of the plate $\bar{u}_3 \approx 0.00416 p_0 a^4/D$ while the converged solution gives $\bar{u}_3 = 0.00406 p_0 a^4/D$. Since the bending moments are proportional to second derivatives of the displacements and shear forces to third derivatives, much slower convergence is expected for these quantities. To ensure convergence, at least 100 terms are used in computing the series used to evaluate the displacement, moment, and force fields shown in figs. 16.23 to 16.28.

It is interesting to compute the overall equilibrium condition of the plate. The total transverse pressure applied to the plate, $p_0 ab$, must be equilibrated by the sum of the total vertical forces along the four edges of the plate plus the four corner forces

$$p_0 ab = - \int_0^b V_1(a, x_2)\, dx_2 + \int_0^b V_1(0, x_2)\, dx_2 - \int_0^a V_2(x_1, b)\, dx_2$$
$$+ \int_0^a V_2(x_1, 0)\, dx_2 - R(0,0) - R(a,0) - R(0,b) - R(a,b).$$

In view of the sign conventions, the corner forces are obtained as follows: $R(0,0) = 2M_{12}(0,0)$, $R(a,0) = -2M_{12}(a,0)$, $R(0,b) = -2M_{12}(0,0)$, and $R(a,b) = 2M_{12}(a,b)$. Performing the indicated integrations along the edges of the plate leads to

$$p_0 ab = 4 \sum_{m,n \text{ odd}}^{\infty} \frac{\alpha_m^2 [D_{11}\alpha_m^2 + (D_{12} + 4D_{66})\beta_n^2] p_{mn}}{\alpha_m \beta_n \Delta_{mn}}$$

$$+ 4 \sum_{m,n \text{ odd}}^{\infty} \frac{\beta_n^2 [D_{22}\beta_n^2 + (D_{12} + 4D_{66})\alpha_m^2] p_{mn}}{\alpha_m \beta_n \Delta_{mn}} - 16 \sum_{m,n \text{ odd}}^{\infty} \frac{D_{66}\alpha_m^2 \beta_n^2 p_{mn}}{\alpha_m \beta_n \Delta_{mn}}$$

$$= 4 \sum_{m,n \text{ odd}}^{\infty} \frac{p_{mn}}{\alpha_m \beta_n} = p_0 ab \frac{64}{\pi^2} \sum_{m,n \text{ odd}}^{\infty} \frac{1}{m^2 n^2}.$$

The double series expression reduces to $\sum_{m,n \text{ odd}}^{\infty} 1/(m^2 n^2) = \pi^4/64$, which confirms that equilibrium is achieved. The double infinite series converges slowly, and this implies that many terms must be included in the solution to accurately capture the shear force distribution in the plate and the corner forces.

Example 16.2. Rectangular plate subjected to a concentrated load

Consider a uniform, rectangular plate made of a homogeneous, isotropic material and subjected to a concentrated load, P, applied at point **A** with coordinates (x_{1a}, x_{2a}). This loading can be viewed as a special case of a pressure loading, $p_3 = P/\mathcal{A}$, acting over a small rectangle of area, $\mathcal{A}$, centered at point **A**. In the limit, $\mathcal{A} \to 0$, while the integral of the pressure remains equal to P. Such a loading can be represented as, $p_3(x_1, x_2) = P\delta(x_1 - x_{1a})\delta(x_2 - x_{2a})$ where $\delta(x - c)$ is the *Dirac delta function*, which is infinite when its argument vanishes and zero everywhere else. Its integral, however, is $\int \delta \mathrm{d}x = 1$.

Using this representation of the concentrated force, P, as a pressure, $P\delta(x_1 - x_{1a})\delta(x_2 - x_{2a})$, in eq. (16.90) leads to

$$
\begin{aligned}
p_{mn} &= \frac{4}{ab} \int_0^a \int_0^b P\delta(x_1 - x_{1a})\delta(x_2 - x_{2a}) \sin \alpha_m x_1 \sin \beta_n x_2 \, \mathrm{d}x_1 \mathrm{d}x_2 \\
&= \frac{4P}{ab} \sin \alpha_m x_{1a} \sin \beta_n x_{2a},
\end{aligned}
$$

and the solution of the problem then follows from eq. (16.86) as

$$
\bar{u}_3 = \frac{4P}{abD} \sum_{m,n}^{\infty} \frac{\sin \alpha_m x_{1a} \sin \beta_n x_{2a}}{(\alpha_m^2 + \beta_n^2)^2} \sin \alpha_m x_1 \sin \beta_n x_2. \tag{16.93}
$$

The solutions for the bending moment and shear force distributions are then readily obtained in the same manner as example 16.1 above, but they lead to very slowly converging series.

16.4.2 Problems

Problem 16.8. Rectangular plate on an elastic foundation
Consider a simply supported rectangular plate made of a homogeneous, isotropic material resting on an elastic foundation of stiffness constant k, expressed in N/m^3, and subjected to a transverse pressure $p_3(x_1, x_2)$. Derive a Navier type solution for this problem. What non-dimensional parameter governs the behavior of the elastic foundation?

Problem 16.9. Rectangular plate under uniform loading
Consider a simply supported, rectangular plate with an aspect ratio $b/a = 3$. Use Navier's solution to compute the transverse deflection of the plate under a uniform transverse pressure $p_3(x_1, x_2) = p_0$. The following seven cases are to be investigated: a homogeneous, aluminum plate of thickness $h = 1.5 \times 10^{-3}$ m, and six different laminated composite plates with lay-ups: $[0_{12}^\circ]$, $[90_{12}^\circ]$, $[0_4^\circ, \pm 45^\circ]_s$, $[90_4^\circ, \pm 45^\circ]_s$, $[\pm 45^\circ, 0_4^\circ]_s$, and $[\pm 45^\circ, 90_4^\circ]_s$. For simplicity, the laminated composite plates can be assumed to be specially orthotropic. (1) Use Navier's solution to compute the transverse deflection, bending moment, and shear force distributions in the plate. (2) For three cases plot the distribution of non-dimensional transverse deflection, $\bar{u}_3 \sqrt{D_{11}D_{22}}/(p_0 a^4)$, bending moments, $M_1/(p_0 a^2)$ and $M_2/(p_0 a^2)$, twisting moment, $M_{12}/(p_0 a^2)$, and shear forces, $Q_1/(p_0 a)$ and $Q_2/(p_0 a)$, over the plate in a contour and/or surface plot that preserves the geometric aspect ratio. (3) Determine the location of the maximum transverse deflection, bending moment, and shear force. (4) For each of the seven plates, determine the maximum loading it can sustain and the maximum deflection under this maximum load. (5) For one case, determine the total vertical load along the four edges of the plate and check the vertical equilibrium of the entire plate.

The plate stiffnesses and reserve factors for the various cases can be obtained from the CLT code. Material properties for all cases are given in the input file.

Problem 16.10. Rectangular plate under concentrated load

Consider a simply supported, rectangular plate with an aspect ratio $b/a = 3$. Use Navier's solution to compute the transverse deflection of the plate under a point load, P, applied at the center of the plate. The following seven cases will be investigated: a homogeneous, aluminum plate of thickness $h = 1.5 \times 10^{-3}$ m, and six different laminated composite plates with lay-up $[0^\circ_{12}]$, $[90^\circ_{12}]$, $[0^\circ_4, \pm 45^\circ]_s$, $[90^\circ_4, \pm 45^\circ]_s$, $[\pm 45^\circ, 0^\circ_4]_s$, and $[\pm 45^\circ, 90^\circ_4]_s$. For simplicity, the laminated composite plates can be assumed to be specially orthotropic. *(1)* Use Navier's solution to compute the transverse deflection, bending moment, and shear force distributions in the plate. *(2)* For three cases plot the distribution of non-dimensional transverse deflection, $U u_3 sqrt D_{11} D_{22}/(P a^2)$, bending moments, M_1/P and M_2/P, twisting moment, M_{12}/P, and shear forces, $Q_1 a/P, Q_2 a/P$ over the plate on a contour and/or surface plot that preserves the geometric aspect ratio. *(3)* Determine the location of the maximum transverse deflection, bending moment, and shear force. *(4)* For each of the seven plates, determine the maximum loading it can sustain and the maximum deflection under this maximum load. *(5)* For one case, determine the total vertical load along the four edges of the plate and check the vertical equilibrium of the entire plate.

The plate stiffnesses and reserve factors for the various cases can be obtained from the CLT code. Material properties for all cases are given in the input file.

16.4.3 Lévy's solution

Navier's solution presented in the previous section suffers two major drawbacks. First, it is limited to rectangular plates that are simply supported along all four edges, and second, it leads to slowly converging series, especially for bending moments and shear forces.

Lévy's solution remedies these two deficiencies by considering plates that are simply supported along two opposite edges but may present a variety of boundary conditions along the other two as depicted in fig. 16.29. The solution method proposed by Lévy leads to single series, in contrast with the double series of Navier's solution, and furthermore, these single series present much faster convergence rates.

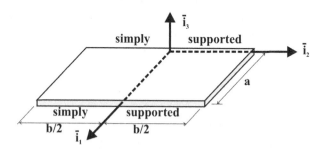

Fig. 16.29. Rectangular plate simply supported along two opposite edges.

Although Lévy's solution can be applied directly to the same homogeneous, isotropic plates analyzed using Navier's method, it is instructive to extend Lévy's solution to specially orthotropic plates. The governing partial differential equation for a specially orthotropic plate is given by eq. (16.81).

As a first step, the following *affine transformation* of the coordinates is performed

$$\hat{x}_1 = \frac{x_1}{\sqrt[4]{D_{11}}}, \quad \hat{x}_2 = \frac{x_2}{\sqrt[4]{D_{22}}}. \tag{16.94}$$

Introducing this change of variables into eq. (16.81) leads to

$$\frac{\partial^4 \bar{u}_3}{\partial \hat{x}_1^4} + 2\bar{D} \frac{\partial^4 \bar{u}_3}{\partial \hat{x}_1^2 \partial \hat{x}_2^2} + \frac{\partial^4 \bar{u}_3}{\partial \hat{x}_2^4} = p_3, \tag{16.95}$$

where $\bar{D}$ is defined as

$$\bar{D} = \frac{D_{12} + 2D_{66}}{\sqrt{D_{11}D_{22}}}. \tag{16.96}$$

Note that the affine transformation eliminates most bending stiffness coefficients from the governing equation, leaving only $\bar{D}$. Moreover, for homogeneous, isotropic plates, $\bar{D} = 1$, as can be verified by introducing into eq. (16.96) the bending stiffness coefficients given by eq. (16.70). It fact, it can be shown that $0 \leq \bar{D} \leq 1$ for all possible specially orthotropic plates. With the affine transformation defined by eq. (16.94), the orthotropy of the plate is represented by a single coefficient, $\bar{D}$, which varies from zero to unity.

The rectangular plate is simply supported along two opposite edges, leading to the following boundary conditions, see eqs. (16.39): $\bar{u}_3 = \partial^2 \bar{u}_3/\partial \hat{x}_1^2 = 0$ at $x_1 = 0$ and $x_1 = \hat{a}$, where $\hat{a} = a/\sqrt[4]{D_{11}}$. The solution of the problem is assumed to be represented by an infinite series expansion of sine functions,

$$\bar{u}_3(\hat{x}_1, \hat{x}_2) = \sum_{m=1}^{\infty} g_m(\hat{x}_2) \sin \frac{m\pi \hat{x}_1}{\hat{a}}, \tag{16.97}$$

where $g_m(\hat{x}_2)$ are unknown functions of the affine variable $\hat{x}_2$ only. Note that due to the properties of trigonometric functions, each term of this expansion satisfies the boundary conditions at $x_1 = 0$ and $x_1 = \hat{a}$, and hence, the complete solution automatically satisfies the same boundary conditions.

Next, the applied transverse pressure is also expanded in Fourier series as

$$p_3(\hat{x}_1, \hat{x}_2) = \sum_{m=1}^{\infty} p_m(\hat{x}_2) \sin \frac{m\pi \hat{x}_1}{\hat{a}}, \tag{16.98}$$

where the coefficients of the expansion, $p_m(\hat{x}_2)$, are known functions of the affine coordinate $\hat{x}_2$ only and can be obtained in the same manner as used in example 16.1.

The assumed transverse displacement field, eq. (16.97), and the transverse pressure expansion, eq. (16.98), are introduced into the governing partial differential equation of the problem, eq. (16.95), to find

$$\sum_{m=1}^{\infty} \left[\left(\frac{m\pi}{\hat{a}}\right)^4 g_m - 2\bar{D} \left(\frac{m\pi}{\hat{a}}\right)^2 \frac{d^2 g_m}{d\hat{x}_2^2} + \frac{d^4 g_m}{d\hat{x}_2^4} - p_m \right] \sin \frac{m\pi \hat{x}_1}{\hat{a}} = 0.$$

This equation must be satisfied at every point of the plate, *i.e.*, for any value of $0 \leq \hat{x}_1 \leq \hat{a}$. This is only possible if the bracketed term vanishes for all integers m, leading to

$$\frac{\mathrm{d}^4 g_m}{\mathrm{d}\hat{x}_2^4} - 2\bar{D} \left(\frac{m\pi}{\hat{a}}\right)^2 \frac{\mathrm{d}^2 g_m}{\mathrm{d}\hat{x}_2^2} + \left(\frac{m\pi}{\hat{a}}\right)^4 g_m = p_m.$$

This is a fourth order, ordinary differential equation for the unknown function $g_m(\hat{x}_2)$.

To further simplify the algebra, a non-dimensional spatial variable, η, is introduced: $\eta = m\pi\hat{x}_2/\hat{a}$. By defining another parameter, λ_m, variable η can be written as

$$\eta = \lambda_m \frac{2x_2}{b}, \qquad \lambda_m = \frac{m\pi}{2} \frac{b}{a} \sqrt[4]{\frac{D_{11}}{D_{22}}}. \tag{16.99}$$

With this definition, the two edges of the plate with unspecified boundary conditions are at $x_2 = \pm b/2$, or $\eta = \pm\lambda_m$. The ordinary differential equation governing the unknown function $g_m(\eta)$ now becomes

$$g_m'''' - 2\bar{D} g_m'' + g_m = p_m, \tag{16.100}$$

where the notation $(\cdot)'$ denotes a derivative with respect to η.

The solution of this equation requires specification of four remaining boundary conditions: two at each of the edges of the plate at $x_2 = \pm b/2$. For instance, if the plate is simply supported at $x_2 = \pm b/2$, the corresponding boundary conditions are $\bar{u}_3 = \mathrm{d}^2\bar{u}_3/\mathrm{d}\hat{x}_2^2 = 0$.

Introducing the assumed solution, eq. (16.97), leads to $g_m \sin(m\pi\hat{x}_1)/\hat{a} = 0$ and $\mathrm{d}^2 g_m/\mathrm{d}\hat{x}_2^2 \sin(m\pi\hat{x}_1)/\hat{a} = 0$. Because these two conditions must be satisfied all along the edges, *i.e.*, for any value of $0 \leq \hat{x}_1 \leq \hat{a}$, it follows that $g_m = g_m'' = 0$ at $\eta = \pm\lambda_m$.

On the other hand, if the plate is clamped along the same two edges, a similar reasoning leads to the following boundary conditions: $g_m = g_m' = 0$ at $\eta = \pm\lambda_m$.

To complete the solution process, the loading coefficients, $p_m(\hat{x}_2)$, defined by eq. (16.98), must be evaluated. Because of the orthogonality of the sine functions, see section A.4, this can be accomplished in the same manner used in Navier's solution in section 16.4.1 above. First, both members of eq. (16.98) are multiplied by $\sin(i\pi\hat{x}_1)/\hat{a}$, then integrated over variable $\hat{x}_1$, leading to

$$\int_0^{\hat{a}} p_3 \sin \frac{i\pi\hat{x}_1}{\hat{a}} \, \mathrm{d}\hat{x}_1 = \sum_{m=1}^{\infty} p_m(\hat{x}_2) \int_0^{\hat{a}} \sin \frac{m\pi\hat{x}_1}{\hat{a}} \sin \frac{i\pi\hat{x}_1}{\hat{a}} \, \mathrm{d}\hat{x}_1.$$

The integral on the right hand side is zero if $m \neq i$ and is $\hat{a}/2$ when $m = i$, and this eliminates all but the ith term in the summation. As a result,

$$p_m(\hat{x}_2) = \frac{2}{\hat{a}} \int_0^{\hat{a}} p_3(\hat{x}_1, \hat{x}_2) \sin \frac{m\pi\hat{x}_1}{\hat{a}} \, \mathrm{d}\hat{x}_1, \tag{16.101}$$

which can be evaluated for any given pressure distribution, $p_3(x_1, x_2)$.

The general solution of the governing ordinary differential, eq. (16.100), for the unknown function $g_m(\hat{x}_2)$ depends on the level of orthotropy of the plate, *i.e.*, on the value of $\bar{D}$. For homogeneous isotropic plates, $\bar{D} = 1$, whereas for orthotropic plates, $0 \leq \bar{D} < 1$. These two cases are now addressed separately.

Homogeneous isotropic plates

If the plate is homogeneous and isotropic, $\bar{D} = 1$, the characteristic equation of the governing ordinary differential equation, eq. (16.100), for $g_m(\eta)$ is $(z^2 - 1)^2 = 0$. The solutions to the characteristic equation imply that the general solution is

$$g_m = (C_1 + C_2\eta)\cosh\eta + (C_3 + C_4\eta)\sinh\eta + g_{mp}, \qquad (16.102)$$

where g_{mp} is the particular solution associated with the nonzero right hand side of eq. (16.100).

Specially orthotropic plates

If the plate is specially orthotropic, $0 \leq \bar{D} < 1$, the characteristic equation of the governing ordinary differential equation, eq. (16.100), for $g_m(\eta)$ is $z^4 - 2\bar{D}z^2 + 1 = 0$. The solutions to the characteristic equation are $z = \pm(\alpha \pm \mathbf{i}\beta)$, where $\mathbf{i} = \sqrt{-1}$, and

$$\alpha = \sqrt{\frac{1 + \bar{D}}{2}}, \quad \beta = \sqrt{\frac{1 - \bar{D}}{2}}. \qquad (16.103)$$

The general solution of the problem is then

$$g_m = C_1 a_1(\eta) + C_2 a_2(\eta) + C_3 a_3(\eta) + C_4 a_4(\eta) + g_m^p, \qquad (16.104)$$

where g_m^p is the particular solution associated with the nonzero right hand side of eq. (16.100), and $a_i(\eta)$ are the following transcendental functions

$$a_1(\eta) = \cosh\alpha\eta\,\cos\beta\eta, \qquad (16.105a)$$
$$a_2(\eta) = \cosh\alpha\eta\,\sin\beta\eta, \qquad (16.105b)$$
$$a_3(\eta) = \sinh\alpha\eta\,\cos\beta\eta, \qquad (16.105c)$$
$$a_4(\eta) = \sinh\alpha\eta\,\sin\beta\eta. \qquad (16.105d)$$

For future reference, the first and second derivatives of these functions are

$$a_1' = \alpha a_3 - \beta a_2, \quad a_1'' = \bar{D}a_1 - 2\alpha\beta a_4, \qquad (16.106a)$$
$$a_2' = \alpha a_4 + \beta a_1, \quad a_2'' = \bar{D}a_2 + 2\alpha\beta a_3, \qquad (16.106b)$$
$$a_3' = \alpha a_1 - \beta a_4, \quad a_3'' = \bar{D}a_3 - 2\alpha\beta a_2, \qquad (16.106c)$$
$$a_4' = \alpha a_2 + \beta a_3, \quad a_4'' = \bar{D}a_4 + 2\alpha\beta a_1, \qquad (16.106d)$$

whereas their third and fourth derivatives are

$$a_1''' = -\beta(2\alpha^2 + \bar{D})a_2 + \alpha(\bar{D} - 2\beta^2)a_3, \quad a_1'''' = -4\alpha\beta\bar{D}a_4 + (2\bar{D}^2 - 1)a_1,$$
$$(16.107a)$$

$$a_2''' = +\beta(2\alpha^2 + \bar{D})a_1 + \alpha(\bar{D} - 2\beta^2)a_4, \quad a_2'''' = +4\alpha\beta\bar{D}a_3 + (2\bar{D}^2 - 1)a_2,$$
$$(16.107b)$$

$$a_3''' = -\beta(2\alpha^2 + \bar{D})a_4 + \alpha(\bar{D} - 2\beta^2)a_1, \quad a_3'''' = -4\alpha\beta\bar{D}a_2 + (2\bar{D}^2 - 1)a_3,$$
$$(16.107c)$$

$$a_4''' = +\beta(2\alpha^2 + \bar{D})a_3 + \alpha(\bar{D} - 2\beta^2)a_2, \quad a_4'''' = +4\alpha\beta\bar{D}a_1 + (2\bar{D}^2 - 1)a_4.$$
$$(16.107d)$$

Example 16.3. Rectangular isotropic plate under a uniform transverse pressure
Consider a uniform, rectangular plate made of a homogeneous, isotropic material
and subjected to a uniform transverse pressure, p_0. All edges of the plate are simply
supported. This configuration is identical to that considered in example 16.1 using
Navier's solution.

First, the coefficients of the transverse pressure expansion are evaluated based on
eq. (16.101), to find

$$p_m(\hat{x}_2) = \frac{2}{\hat{a}} \int_0^{\hat{a}} p_3(\hat{x}_1, \hat{x}_2) \sin \frac{m\pi\hat{x}_1}{\hat{a}} \, d\hat{x}_1$$

$$= \frac{2p_0}{\hat{a}} \int_0^{\hat{a}} \sin \frac{m\pi\hat{x}_1}{\hat{a}} \, d\hat{x}_1 = \frac{4p_0}{\pi m} \begin{cases} 1, & \text{if } m \text{ odd,} \\ 0, & \text{otherwise.} \end{cases} \qquad (16.108)$$

The particular solution for eq. (16.100) is now $g_m^p = 4p_0\hat{a}^4/(\pi^5 m^5)$, and as
mentioned earlier, the boundary conditions for this problem are $g_m = g_m'' = 0$ at
$\eta = \pm\lambda_m$, allowing the determination of the integration constants as

$$C_1 = -\frac{2 + \lambda_m \tanh \lambda_m}{2 \cosh \lambda_m} \frac{4p_0\hat{a}^4}{\pi^5 m^5}, \quad C_2 = C_3 = 0, \quad C_4 = \frac{1}{2 \cosh \lambda_m} \frac{4p_0\hat{a}^4}{\pi^5 m^5}.$$

The solution of the problem then follows from eq. (16.97) as

$$\bar{u}_3 = \frac{4}{\pi^5} \frac{p_0 a^4}{D} \sum_{\substack{m \text{ odd}}}^{\infty} \frac{1}{m^5} \left[-\frac{2 + \lambda_m \tanh \lambda_m}{2 \cosh \lambda_m} \cosh \lambda_m \frac{2x_2}{b} \right.$$

$$\left. + \frac{\lambda_m}{2 \cosh \lambda_m} \frac{2x_2}{b} \sinh \lambda_m \frac{2x_2}{b} + 1 \right] \frac{\sin m\pi x_1}{a}. \qquad (16.109)$$

The distributions of bending moments, twisting moment, and shear forces in the plate
are then readily obtained. The results are identical to those presented in figs. 16.23
to 16.28. The convergence of Lévy's solution is much faster than that of Navier's
solution. Consequently, fewer terms are needed in the single series to achieve comparable accuracy.

Example 16.4. Rectangular orthotropic plate under a uniform transverse pressure
Consider a uniform, rectangular plate made of a homogeneous, orthotropic material
and subjected to a uniform transverse pressure, p_0. All edges of the plate are simply
supported.

The coefficients of the transverse pressure expansion are still given by eq. (16.108), leading to the same particular solution, $g_m^p = 4p_0\hat{a}^4/(\pi^5 m^5)$.

As described earlier, the boundary conditions for this problem are $g_m = g_m'' = 0$ at $\eta = \pm\lambda_m$. For eqs. (16.106), it is clear that $g_m'' = \bar{D}g_m + 2\alpha\beta(-C_1 a_4 + C_2 a_3 - C_3 a_2 + C_4 a_1)$, and the integration constants become

$$
C_1 = -\frac{4p_0\hat{a}^4}{\pi^5 m^5 \Delta_m} a_1(\lambda_m), \quad C_2 = C_3 = 0, \quad C_4 = -\frac{4p_0\hat{a}^4}{\pi^5 m^5 \Delta_m} a_4(\lambda_m),
$$

where $\Delta_m = a_1^2(\lambda_m) + a_4^2(\lambda_m)$. The solution of the problem then follows from eq. (16.97) as

$$
\bar{u}_3 = \frac{4}{\pi^5} \frac{p_0 a^4}{D_{11}} \sum_{m\ odd}^{\infty} \frac{1}{m^5}
$$

$$
\left[1 - \frac{a_1(\lambda_m)\, a_1(\lambda_m 2x_2/b) + a_4(\lambda_m)\, a_4(\lambda_m 2x_2/b)}{\Delta_m}\right] \frac{\sin m\pi x_1}{a}.
$$

(16.110)

The distributions of bending moments, twisting moment, and shear forces in the plate are then readily obtained. Again, due to the appearance of the m^5 term in the denominator, the convergence is much better for this single series compared to Navier's solution.

16.4.4 Problems

Problem 16.11. Rectangular plate under uniform loading
Use Lévy's solution to determine the transverse deflection of a rectangular, simply supported plate with an aspect ratio $b/a = 3$. A uniform transverse pressure $p_3(x_1, x_2) = p_0$ acts on the plate. The following seven cases are to be investigated: a homogeneous, aluminum plate of thickness $h = 1.5 \times 10^{-3}$ m, and six different laminated composite plates with lay-ups: $[0_{12}^\circ]$, $[90_{12}^\circ]$, $[0_4^\circ, \pm 45^\circ]_s$, $[90_4^\circ, \pm 45^\circ]_s$, $[\pm 45^\circ, 0_4^\circ]_s$, and $[\pm 45^\circ, 90_4^\circ]_s$. For simplicity, the laminated composite plates can be assumed to be specially orthotropic. (1) Use Lévy's solution to compute the transverse deflection, bending moment, and shear force distributions. (2) For three cases plot the distribution of non-dimensional transverse deflection, $\bar{u}_3\sqrt{D_{11}D_{22}}/(p_0 a^4)$, bending moments, $M_1/(p_0 a^2)$ and $M_2/(p_0 a^2)$, twisting moment, $M_{12}/(p_0 a^2)$, and shear forces, $Q_1/(p_0 a)$, $Q_2/(p_0 a)$, over the plate on a contour and/or surface plot that preserves the geometric aspect ratio. (3) Determine the location of the maximum transverse deflection, bending moment, and shear force. (4) For each of the seven plates, determine the maximum loading it can sustain and the maximum deflection under this maximum load. (5) For one case, determine the total vertical load along the four edges of the plate and check the vertical equilibrium of the entire plate.

Problem 16.12. Rectangular plate under uniform loading
Use Lévy's solution to determine the transverse deflection of a rectangular plate with an aspect ratio $b/a = 3$. Two opposite edges of the plate are simply supported and the other two are clamped. A uniform transverse pressure $p_3(x_1, x_2) = p_0$ acts on the plate. The following seven cases will be investigated: a homogeneous, aluminum plate of thickness $h = 1.5 \times 10^{-3}$ m, and six laminated composite plates with lay-ups: $[0_{12}^\circ]$, $[90_{12}^\circ]$, $[0_4^\circ, \pm 45^\circ]_s$, $[90_4^\circ, \pm 45^\circ]_s$,

$[\pm 45^\circ, 0^\circ_4]_s$, and $[\pm 45^\circ, 90^\circ_4]_s$. For simplicity, the laminated composite plates can be assumed to be specially orthotropic. *(1)* Use Lévy's solution to compute the transverse deflection, bending moment, and shear force distributions. *(2)* For three cases plot the distribution of non-dimensional transverse deflection, $\bar{u}_3 \sqrt{D_{11} D_{22}}/(p_0 a^4)$, bending moments, $M_1/(p_0 a^2)$ and $M_2/(p_0 a^2)$, twisting moment, $M_{12}/(p_0 a^2)$, and shear forces, $Q_1/(p_0 a)$, $Q_2/(p_0 a)$, over the plate on a contour and/or surface plot that preserves the geometric aspect ratio. *(3)* Determine the location of the maximum transverse deflection, bending moment, and shear force. *(4)* For each of the seven plates, determine the maximum loading it can sustain and the maximum deflection under this maximum load. *(5)* For one case, determine the total vertical load along the four edges of the plate and check the vertical equilibrium of the entire plate.

16.5 Circular plates

The previous sections focused on rectangular plates. A common configuration for thin plates is the circular plate depicted in fig. 16.30. Clearly, the circular symmetry of the problem calls for the use of polar instead of Cartesian coordinates. Kirchhoff plate theory will be extended here to deal with circular plates. Of course, basic kinematic assumptions underlying Kirchhoff plate theory and discussed in section 16.1.1 are directly applicable to circular plates. To develop a set of equations relevant to circular plates, the governing equations of Kirchhoff plate theory must be rewritten in the polar coordinate system. The results presented in this section will be limited to bending of homogeneous, isotropic plates only.

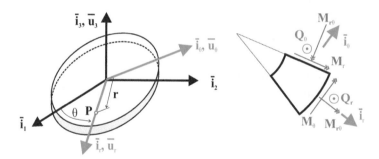

Fig. 16.30. Circular plate with Cartesian and polar coordinate systems.

16.5.1 Governing equations for the bending of circular plates

Kirchhoff's assumptions, as developed in section 16.1.1, imply the vanishing of the transverse strain, ϵ_3, see eq. (16.5), and the vanishing of the transverse shearing strains, γ_{r3} and $\gamma_{\theta3}$, see eq. (16.6). The in-plane strains are linear through-the-thickness of the plate as implied by eq. (16.12). Combining these results and replacing the Cartesian partial derivatives with partial derivatives in polar coordinates, see eqs. (3.33)-(3.34), the curvatures can be recast in polar form as

$$\underline{\kappa} = \left\{ \begin{matrix} \kappa_r \\ \kappa_\theta \\ \kappa_{r\theta} \end{matrix} \right\} = \left\{ \begin{matrix} \dfrac{\bar{u}_{3,r}}{r} + \dfrac{\bar{u}_{3,\theta\theta}}{r^2} \\ -\bar{u}_{3,rr} \\ \dfrac{2\bar{u}_{3,r\theta}}{r} - \dfrac{2\bar{u}_{3,\theta}}{r^2} \end{matrix} \right\}, \tag{16.111}$$

where the compact, indicial notation for partial derivatives introduced previously is used.

The equations of equilibrium of the problem are found by considering the forces and moments acting on a differential element of the plate shown on the right side of fig. 16.30. The equilibrium condition for forces acting in the vertical direction leads to

$$Q_{r,r} + \frac{Q_{\theta,\theta}}{r} + \frac{Q_r}{r} = -p_3, \tag{16.112}$$

and the equilibrium conditions for moments about axes $\bar{\imath}_r$ and $\bar{\imath}_\theta$ are

$$\frac{M_{r,\theta}}{r} + M_{r\theta,r} + \frac{2M_{r\theta}}{r} + Q_\theta = 0, \tag{16.113a}$$

$$M_{\theta,r} - \frac{M_{r\theta,\theta}}{r} + \frac{M_r + M_\theta}{r} - Q_r = 0. \tag{16.113b}$$

Finally, by analogy with eq. (16.31), the constitutive laws for a linearly elastic, homogeneous and isotropic plate are

$$\left\{ \begin{matrix} M_r \\ M_\theta \\ M_{r\theta} \end{matrix} \right\} = D \begin{bmatrix} 1 & -\nu & 0 \\ -\nu & 1 & 0 \\ 0 & 0 & (1-\nu)/2 \end{bmatrix} \left\{ \begin{matrix} \kappa_r \\ \kappa_\theta \\ \kappa_{r\theta} \end{matrix} \right\}, \tag{16.114}$$

where D is the isotropic plate bending stiffness given by eq. (16.36).

To develop the governing equations for bending of circular plates, the bending and twisting moments are first expressed in terms of derivatives of the transverse displacement by introducing the definition of curvatures, eqs. (16.111), into the constitutive laws, eqs. (16.114), to find

$$M_r = D\left[\frac{\bar{u}_{3,r}}{r} + \frac{\bar{u}_{3,\theta\theta}}{r^2} + \nu\bar{u}_{3,rr}\right], \tag{16.115a}$$

$$M_\theta = -D\left[\bar{u}_{3,rr} + \frac{\nu\bar{u}_{3,r}}{r} + \frac{\nu\bar{u}_{3,\theta\theta}}{r^2}\right], \tag{16.115b}$$

$$M_{r\theta} = D(1-\nu)\left[\frac{\bar{u}_{3,r\theta}}{r} - \frac{\bar{u}_{3,\theta}}{r^2}\right]. \tag{16.115c}$$

Next, the shear forces are expressed in terms of the derivatives of the transverse displacement by introducing the above expressions for the bending and twisting moments into the moment equilibrium equations, eqs. (16.113), leading to

$$Q_r = -D\frac{\partial}{\partial r}(\nabla^2\bar{u}_3), \tag{16.116a}$$

$$Q_\theta = -D\frac{1}{r}\frac{\partial}{\partial\theta}(\nabla^2\bar{u}_3), \tag{16.116b}$$

where Laplace's operator in polar coordinates is given by eq. (3.40).

Finally, the governing equation for the bending of circular plates is obtained by introducing the shear forces given by eqs. (16.116) into the vertical equilibrium equation, eq. (16.112), to find

$$DV^4\bar{u}_3 = p_3.$$
(16.117)

This equation describes the bending of circular plates and is a fourth-order, partial differential equation for the transverse displacement, $\bar{u}_3$.

The solution of this differential equation will require a proper set of boundary conditions. For a circular plate, two boundary conditions are required along the outer edge of the plate. Note that the theory developed here is also valid for a plate in the form of a circular annulus; in that case, two boundary conditions must also be applied along the inner edge of the annulus. Typical boundary conditions are similar to those described in section 16.2.1 for rectangular plates.

1. *Clamped edge.* If the outer edge of the plate is clamped, the transverse displacement must vanish along this edge, and because the rotation of the normal line is equal to the slope of the plate, see eq. (16.3), the slope of the plate also vanishes, leading to the following boundary conditions,

$$\bar{u}_3 = 0; \quad \bar{u}_{3,r} = 0.$$
(16.118)

2. *Simply supported edge.* If the outer edge of the plate is simply supported, the transverse displacement must vanish along the edge, and because the pivot line cannot resist a bending moment, the bending moment, M_θ, must also vanish along the edge. In view of eq. (16.115b), this latter condition implies $\bar{u}_{3,rr} + \nu\bar{u}_{3,r}/r + \nu\bar{u}_{3,\theta\theta}/r^2 = 0$, however, because the transverse displacement vanishes along the edge, so does its derivative with respect to θ, and this boundary condition reduces to $\bar{u}_{3,rr} + \nu\bar{u}_{3,r}/r = 0$. The boundary conditions along a simply supported edge thus become

$$\bar{u}_3 = 0; \quad \bar{u}_{3,rr} + \frac{\nu}{r}\bar{u}_{3,r} = 0.$$
(16.119)

3. *Free edge.* If the outer edge of the plate is free, the bending moment, M_θ, and the total vertical load, V_r, must both vanish. As for rectangular plates, this total vertical load is defined as

$$V_r = Q_r - \frac{M_{r\theta,\theta}}{r}.$$
(16.120)

The boundary conditions must be expressed in terms of the transverse displacement, $\bar{u}_3$, and its derivatives with the help of eqs. (16.34) to find

$$\bar{u}_{3,rr} + \frac{\nu\bar{u}_{3,r}}{r} + \frac{\nu\bar{u}_{3,\theta\theta}}{r^2} = 0,$$
$$\bar{u}_{3,rrr} + \frac{\bar{u}_{3,rr}}{r} - \frac{\bar{u}_{3,r}}{r^2} + \frac{2-\nu}{r^2}\bar{u}_{3,r\theta\theta} - \frac{3-\nu}{r^3}\bar{u}_{3,\theta\theta} = 0.$$
(16.121)

16.5.2 Circular plates subjected to loading presenting circular symmetry

Circular plates presenting circular symmetry lead to considerable simplification of the governing equations and boundary conditions and therefore will be examined first. Consider a circular plate subjected to a loading presenting circular symmetry, i.e., $p_3(r, \theta) = p_3(r)$. In view of the circular symmetry of the plate and applied loading, the solution is also expected to present circular symmetry, and hence all derivatives with respect to variable θ should vanish. This implies $\kappa_{r\theta} = 0$, $M_{r\theta} = 0$ and $Q_\theta = 0$, and Laplace's operator reduces to

$$\nabla^2 = \frac{\partial^2}{\partial r^2} + \frac{1}{r}\frac{\partial}{\partial r} + \frac{1}{r^2}\frac{\partial^2}{\partial \theta^2} = \frac{\partial^2}{\partial r^2} + \frac{1}{r}\frac{\partial}{\partial r} = \frac{1}{r}\frac{d}{dr}\left(r\frac{d}{dr}\right). \tag{16.122}$$

From eqs. (16.115), the two bending moments become

$$M_r = D\left[\frac{\bar{u}_{3,r}}{r} + \nu\bar{u}_{3,rr}\right], \tag{16.123a}$$

$$M_\theta = -D\left[\bar{u}_{3,rr} + \frac{\nu\bar{u}_{3,r}}{r}\right]. \tag{16.123b}$$

Considering eq. (16.120), the total vertical force reduces to

$$V_r = Q_r = -D\frac{d}{dr}\left[\frac{1}{r}\frac{d}{dr}\left(r\frac{d\bar{u}_3}{dr}\right)\right]. \tag{16.124}$$

Finally, the governing equation for the circular plate, eq. (16.117), becomes

$$D\frac{1}{r}\frac{d}{dr}\left\{r\frac{d}{dr}\left[\frac{1}{r}\frac{d}{dr}\left(r\frac{d\bar{u}_3}{dr}\right)\right]\right\} = p_3(r). \tag{16.125}$$

This is now an ordinary differential equations to be solved for the transverse deflection of the plate, $\bar{u}_r(r)$, under the pressure $p_3(r)$. Integrating this equation four times yields the general solution of the problem as

$$\bar{u}_3 = C_1 + C_2 r^2 + C_3 \ln\frac{r}{R} + C_4 r^2 \ln\frac{r}{R} + \frac{1}{D}\int\frac{1}{r}\int r\int\frac{1}{r}\int r\, p_3(r)\, dr^4, \tag{16.126}$$

where C_1, C_2, C_3, and C_4 are the four integration constants and R the radius of the circular plate.

Example 16.5. Simply supported circular plate under uniform pressure
Consider a simply supported circular plate of radius R under a uniform pressure p_0. The general solution of the problem is in the form of eq. (16.126). For a uniform applied pressure, the solution becomes

$$\bar{u}_3 = C_1 + C_2 r^2 + C_3 \ln\frac{r}{R} + C_4 r^2 \ln\frac{r}{R} + \frac{p_0 r^4}{64D}.$$

The two boundary conditions around the outer edge of the plate are given by eqs. (16.119). Clearly, these conditions are not sufficient to determine the four integration constants. The general solution involves two terms featuring the logarithmic function, $\ln r/R$. At the center of the plate, these terms would yield an infinite transverse deflection, $\ln r/R \rightarrow \infty$ for $r \rightarrow 0$. This solution must be discarded, and hence, $C_3 = C_4 = 0$, leading to $\bar{u}_3 = C_1 + C_2 r^2 + p_0 r^4/(64D)$. The remaining two integration constants are found with the help of the boundary conditions around the outer edge of the plate, eqs. (16.119), leading to

$$C_1 = \frac{5+\nu}{1+\nu} \frac{p_0 R^4}{64D}, \quad C_2 = -2\frac{3+\nu}{1+\nu} \frac{p_0 R^4}{64D}.$$

The transverse displacement of the plate then becomes

$$\bar{u}_3 = \frac{p_0 R^4}{64D} \left(\frac{5+\nu}{1+\nu} - 2\frac{3+\nu}{1+\nu} \frac{r^2}{R^2} + \frac{r^4}{R^4} \right) = \frac{p_0 R^4}{64D} \left(1 - \frac{r^2}{R^2} \right) \left(\frac{5+\nu}{1+\nu} - \frac{r^2}{R^2} \right).$$
$$(16.127)$$

The corresponding bending moment distribution then follow from eqs. (16.123) as

$$M_r = \frac{p_0 R^2}{16} \left[-(3+\nu) + (1+3\nu)\frac{r^2}{R^2} \right], \quad M_\theta = \frac{p_0 R^2}{16}(3+\nu) \left(1 - \frac{r^2}{R^2} \right).$$

Finally, the shear force is readily obtained from eq. (16.124) as $Q_r/(p_0 R) = 1/2\, r/R$. Figures 16.31 and 16.32 show the distributions of transverse displacement, bending moment, and shear force along the radial direction. Note that the maximum bending moments are found at the center of the plate, where $M_\theta/(p_0 R^2) = -M_r/(p_0 R^2) = (3+\nu)/16$.

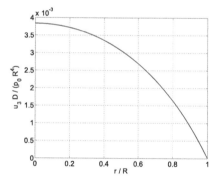

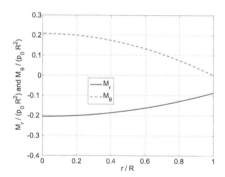

Fig. 16.31. Radial distribution of the non-dimensional displacement, $\bar{u}_3 D/(p_0 R^4)$.

Fig. 16.32. Radial distribution of the non-dimensional bending moments, $M_r/(p_0 R^2)$ and $M_\theta/(p_0 R^2)$.

Example 16.6. Simply supported annular plate under inner edge shear force
Consider an annular plate of internal and external radii, R_i and R_o, respectively. Along its inner edge, the plate is subjected to a uniformly distributed shear force, V_0, as shown in fig. 16.33. The total shear force is $P = 2\pi R_i V_0$. Find the deflection of the plate under this loading.

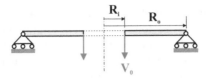

Fig. 16.33. Diametric cross-section through a simply supported annular plate subjected to a uniform shear force around its inner edge.

Since no distributed pressure is applied, the general solution given by eq. (16.126) reduces to

$$\bar{u}_3 = C_1 + C_2 r^2 + C_3 \ln \frac{r}{R} + C_4 r^2 \ln \frac{r}{R}.$$

The boundary conditions for the problem are as follows. Around the outer edge of the plate, $\bar{u}_3 = 0$ and $M_\theta = 0$. Around the inner edge of the plate, $M_\theta = 0$ and $V_r = Q_r = V_0$. Using eq. (16.124), this latter condition implies $C_4 = -P/(8\pi D)$. The vanishing of the transverse displacement around the outer edge implies $C_1 + C_2 R_o^2 = 0$. Finally, the vanishing of the bending moment M_θ around the two edges leads to

$$C_2 = \left(\frac{\bar{R}^2 \ln \bar{R}}{1 - \bar{R}^2} - \frac{1}{2}\frac{3 + \nu}{1 + \nu} \right) C_4, \quad C_3 = 2\frac{1 + \nu}{1 - \nu}\frac{\bar{R}^2 \ln \bar{R}}{1 - \bar{R}^2} R_o^2 C_4,$$

where $\bar{R} = R_i/R_o$. The transverse displacement of the plate becomes

$$\bar{u}_3 = \frac{1}{8\pi} \frac{PR_o^2}{D} \left[\left(\frac{\bar{R}^2 \ln \bar{R}}{1 - \bar{R}^2} - \frac{1}{2}\frac{3 + \nu}{1 + \nu} \right)(1 - \rho^2) - 2\frac{1 + \nu}{1 - \nu}\frac{\bar{R}^2 \ln \alpha}{1 - \bar{R}^2} \ln \rho - \rho^2 \ln \rho \right],$$

where $\rho = r/R_o$ is the non-dimensional radial variable. The radial distribution of the non-dimensional displacement is shown in fig. 16.34.

The bending moments are then computed with the help of eqs. (16.123) to find

$$\frac{M_\theta}{P} = \frac{1 + \nu}{4\pi} \left[\ln \rho - \frac{\alpha^2 \ln \alpha}{1 - \alpha^2} \left(\frac{1}{\rho^2} - 1 \right) \right],$$

$$\frac{M_r}{P} = \frac{1 + \nu}{4\pi} \left[\frac{1 - \nu}{1 + \nu} - \ln \rho - \frac{\alpha^2 \ln \alpha}{1 - \alpha^2} \left(\frac{1}{\rho^2} + 1 \right) \right].$$

Figure 16.35 shows the radial distribution of non-dimensional displacement and bending moments. Note that for annular plates with a very small internal radius, a boundary layer phenomenon is observed: the bending moment M_θ presents increasingly larger peak values, at a radial location that is increasingly close to the inner edge. The same bending moment, however, vanishes around the inner edge.

It is interesting to consider the limiting case where $R_i \to 0$ while P remains constant. Since $P = 2\pi R_i V_0$, this implies that $V_0 \to \infty$ as $R_i \to 0$, and their product remains finite. At the limit, a circular plate subjected to a concentrated load,

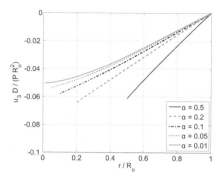

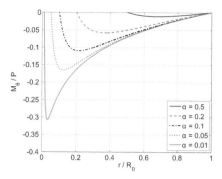

Fig. 16.34. Radial distribution of non-dimensional displacement, $\bar{u}_3 D/(PR_o^2)$.

Fig. 16.35. Radial distribution of non-dimensional bending moments, M_θ/P.

P, at its center is recovered. The solution of this problem is obtained by letting $\bar{R} \to 0$ in the above expressions to yield

$$\bar{u}_3 = -\frac{1}{8\pi}\frac{PR_o^2}{D}\left[\frac{1}{2}\frac{3+\nu}{1+\nu}(1-\rho^2)+\rho^2\ln\rho\right],$$

and

$$M_\theta = P\frac{1+\nu}{4\pi}\ln\rho, \quad \text{and} \quad M_r = P\frac{1+\nu}{4\pi}\left(\frac{1-\nu}{1+\nu}-\ln\rho\right).$$

The transverse displacement remains finite at the center of the plate, but the bending moment components grow without bound at the point of application of the concentrated load.

16.5.3 Problems

Problem 16.13. Clamped circular plate under uniform loading
Determine the transverse deflection of a clamped circular plate under a uniform transverse loading p_0.

Problem 16.14. Annular plate loaded by a center shaft
Consider the simply supported, annular plate of radius R_o with a rigid shaft of radius R_i clamped at its center, as depicted in fig. 16.36. Let $\bar{R} = R_i/R_o$. *(1)* Determine the solution of the problem corresponding to a vertical motion Δ of the rigid shaft. *(2)* Determine the equivalent spring rate of the plate, $k_\Delta(\alpha) = P/\Delta$. Plot k_Δ as a function of $\bar{R}$, use a log scale for k_Δ. *(3)* Plot the distributions of transverse displacement $\bar{u}_3 D/(PR_o^2)$ and bending moment M_r/P and M_θ/P for $\bar{R} = 0.2$.

Problem 16.15. Simply supported annular plate under uniform loading
Consider a simply supported annular plate of inner and outer radii R_i and R_o, respectively, subjected to a uniform transverse pressure, p_0, as depicted in fig. 16.37. Let $\bar{R} = R_i/R_o$. *(1)* Determine the transverse deflection of the plate and the bending moment distributions, M_r and M_θ. *(2)* Plot the distributions of transverse displacement, $\bar{u}_3 D/(p_0 R_o^4)$, and bending moment, $M_r/(p_0 R_o^2)$ and $M_\theta/(p_0 R_o^2)$, for $\bar{R} = 0.2$.

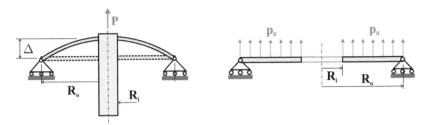

Fig. 16.36. Simply supported circular plate with center rigid shaft.

Fig. 16.37. Simply supported annular plate under uniform pressure.

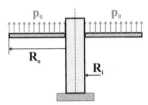

Fig. 16.38. Circular plate clamped to a center rigid shaft.

Problem 16.16. Clamped annular plate under uniform loading

Consider an annular plate of radius R_o cantilevered to a rigid shaft of radius R_i, as shown in fig. 16.38. The plate is subjected to a uniform pressure distribution p_0. Let $\bar{R} = R_i/R_o$. *(1)* Determine the transverse deflection of the plate and the bending moment distributions, M_r and M_θ. *(2)* Plot the distributions of transverse displacement, $\bar{u}_3 D/(p_0 R_o^4)$, and bending moment, $M_r/(p_0 R_o^2)$ and $M_\theta/(p_0 R_o^2)$, for $\bar{R} = 0.2$.

16.5.4 Circular plates subjected to arbitrary loading

In the previous section, the loading applied to the plate is assumed to present circular symmetry. In general, the applied loading will be a function of both radial and circumferential variables, *i.e.*, $p_3 = p_3(r, \theta)$. To deal with this problem, the applied loading is expanded in Fourier series in terms of angle θ,

$$p_3(r, \theta) = p_0(r) + \sum_{m=1}^{m=\infty} [p_m^c(r) \cos m\theta + p_m^s(r) \sin m\theta], \tag{16.128}$$

where $p_0(r)$ is the applied pressure component presenting circular symmetry, and $p_m^c(r)$ and $p_m^s(r)$ are the cosine and sine components, respectively, of the Fourier expansion.

The transverse deflection of the plate is also expanded in Fourier series as

$$\bar{u}_3(r, \theta) = \bar{u}_0(r) + \sum_{m=1}^{m=\infty} [\bar{u}_m^c(r) \cos m\theta + \bar{u}_m^s(r) \sin m\theta], \tag{16.129}$$

where $\bar{u}_0(r)$ is the transverse displacement component presenting circular symmetry, and $\bar{u}_m^c(r)$ and $\bar{u}_m^s(r)$ the cosine and sine components, respectively, of the Fourier expansion.

Next, the Fourier expansions of the applied pressure and transverse displacement, eqs. (16.128) and (16.129), respectively, are introduced into the governing differential equation for plate bending, eq. (16.117). Regrouping all terms for the various harmonics then leads to

$$
\sum_{m=1}^{m=\infty} \left[\left(\frac{\partial^2}{\partial r^2} + \frac{1}{r}\frac{\partial}{\partial r} - \frac{m^2}{r^2} \right) \left(\frac{\partial^2}{\partial r^2} + \frac{1}{r}\frac{\partial}{\partial r} - \frac{m^2}{r^2} \right) \bar{u}_m^c(r) - \frac{p_m^c(r)}{D} \right] \cos m\theta
$$

$$
+ \sum_{m=1}^{m=\infty} \left[\left(\frac{\partial^2}{\partial r^2} + \frac{1}{r}\frac{\partial}{\partial r} - \frac{m^2}{r^2} \right) \left(\frac{\partial^2}{\partial r^2} + \frac{1}{r}\frac{\partial}{\partial r} - \frac{m^2}{r^2} \right) \bar{u}_m^s(r) - \frac{p_m^s(r)}{D} \right] \sin m\theta
$$

$$
+ \left[\left(\frac{\partial^2}{\partial r^2} + \frac{1}{r}\frac{\partial}{\partial r} \right) \left(\frac{\partial^2}{\partial r^2} + \frac{1}{r}\frac{\partial}{\partial r} \right) \bar{u}_0(r) - \frac{p_0(r)}{D} \right] = 0.
$$

This equation must be satisfied for all arbitrary values of angle θ, and hence, the bracketed terms, corresponding to the coefficients of the various harmonics, must vanish, leading to the following uncoupled equations for the various harmonics of the transverse displacement

$$
\left(\frac{\partial^2}{\partial r^2} + \frac{1}{r}\frac{\partial}{\partial r} \right) \left(\frac{\partial^2}{\partial r^2} + \frac{1}{r}\frac{\partial}{\partial r} \right) \bar{u}_0 = \frac{p_0(r)}{D}, \qquad (16.130a)
$$

$$
\left(\frac{\partial^2}{\partial r^2} + \frac{1}{r}\frac{\partial}{\partial r} - \frac{m^2}{r^2} \right) \left(\frac{\partial^2}{\partial r^2} + \frac{1}{r}\frac{\partial}{\partial r} - \frac{m^2}{r^2} \right) \bar{u}_m^c(r) = \frac{p_m^c(r)}{D}, \qquad (16.130b)
$$

$$
\left(\frac{\partial^2}{\partial r^2} + \frac{1}{r}\frac{\partial}{\partial r} - \frac{m^2}{r^2} \right) \left(\frac{\partial^2}{\partial r^2} + \frac{1}{r}\frac{\partial}{\partial r} - \frac{m^2}{r^2} \right) \bar{u}_m^s(r) = \frac{p_m^s(r)}{D}. \qquad (16.130c)
$$

The first equation, eq. (16.130a), gives the response of the plate under a loading presenting circular symmetry which is the problem treated in section 16.5.2. For each harmonic, the sine and cosine components of the transverse displacement are governed by identical equations, eqs. (16.130b) or (16.130c), as expected. Indeed, because the plate presents circular symmetry, angle θ can be measured from an arbitrary origin, and hence the sine and cosine components of the Fourier series are interchangeable.

In summary, the response of the plate under an arbitrary loading is obtained by superposing the response of the loading presenting circular symmetry, as obtained from solving eq. (16.130a), and the responses to the various loading harmonics. For this latter problem, a single generic problem can be investigated

$$
\left(\frac{\partial^2}{\partial r^2} + \frac{1}{r}\frac{\partial}{\partial r} - \frac{m^2}{r^2} \right) \left(\frac{\partial^2}{\partial r^2} + \frac{1}{r}\frac{\partial}{\partial r} - \frac{m^2}{r^2} \right) \bar{u}_m(r) = \frac{p_m(r)}{D}, \qquad (16.131)
$$

If $m = 1$, the solution of this equation is

$$
\bar{u}_1 = A_1 r + B_1 r^3 + \frac{C_1}{r} + D_1 r \ln \frac{r}{R_o} + \bar{u}_{1p}, \qquad (16.132)
$$

and if $m > 1$, the solution becomes

$$\bar{u}_m = A_m r^m + B_m r^{m+2} + \frac{C_m}{r^m} + D_m r^{2-m} + \bar{u}_{mp}, \qquad (16.133)$$

where $\bar{u}_{1p}$ and $\bar{u}_{mp}$ are particular solutions for the non-vanishing right hand sides of the governing equation.

Example 16.7. Simply supported circular plate under a linear pressure distribution

Consider a simply supported plate of radius R_o subjected to a linear pressure distribution, as depicted in fig.16.39. The linear pressure distribution can written as $p_3(r, \theta) = p_0 + p_a r/R_o \cos \theta$, where p_0 is the uniform pressure distribution over the plate, and p_a the additional, linear component of pressure, which corresponds to a first harmonic of loading when expressed in the polar coordinate system. Note that the pressure loading is symmetric with respect to the axis defined by $\theta = 0$.

Using the principle of superposition, the response of the plate is that due to the uniform pressure, given by eq. (16.127), plus that due to the first harmonic of loading, $p_3(r, \theta) = p_a r/R_o \cos \theta$. Because the first part of the response is already known, only the response of the plate to the first harmonic of loading is addressed here.

The response of the plate to the first harmonic of loading is given by eq. (16.132) as

$$\bar{u}_1 = A_1 r + B_1 r^3 + \frac{C_1}{r} + D_1 r \ln \frac{r}{R_o} + \frac{p_a R_o^4}{192 D} \left(\frac{r}{R_o} \right)^5,$$

where the last term corresponds to the particular solution of the problem. The solution must remain finite at the center of the plate, $i.e.$, at $r = 0$, and this implies $C_1 = D_1 = 0$. The additional boundary conditions are $\bar{u}_1 = 0$ and $M_\theta = 0$ at $r = R_o$, leading to

$$A_1 = \left(1 - 2\frac{5 + \nu}{3 + \nu} \right) 2\frac{p_a R_o^4}{192 D}, \quad \text{and} \quad B_1 = -\frac{5 + \nu}{3 + \nu} 2\frac{p_a R_o^4}{192 D},$$

respectively.

The complete solution of the problem is now obtained by superposing the solution under the uniform pressure, eq. (16.127), with the above result to obtain

$$\bar{u}_3 = \frac{p_0 R_o^4}{64 D} \left(1 - \rho^2 \right) \left(\frac{5 + \nu}{1 + \nu} - \rho^2 \right) + \frac{p_a R_o^4}{192 D} \rho \left(1 - \rho^2 \right) \left(\frac{7 + \nu}{3 + \nu} - \rho^2 \right) \cos \theta,$$

where $\rho = r/R_o$. The bending moment and shear distributions in the plate can be obtained readily from this solution using eq. (16.115) and (16.116).

16.5.5 Problems

Problem 16.17. Annular plate pitched by a center shaft

Consider the simply supported, annular plate of radius R_o with a rigid shaft of radius R_i clamped at its center, as depicted in fig. 16.40. Let $\bar{R} = R_i/R_o$. (1) Determine the solution

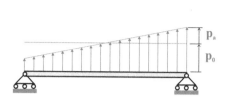

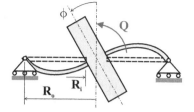

Fig. 16.39. Diametric cross-section at $\theta = 0$ of a simply-supported circular plate.

Fig. 16.40. Simply supported circular plate with center rigid shaft.

of the problem corresponding to a rotation ϕ of the rigid shaft. *(2)* Determine the equivalent spring rate of the plate, $k_\Phi(\bar{R}) = Q/\phi$. Plot k_Φ as a function of $\bar{R}$ using a log scale for k_Φ. *(3)* Plot the distributions of transverse displacement $\bar{u}_3 D/(PR_o^2)$ and bending moment M_r/P and M_θ/P for $\bar{R} = 0.2$.

16.6 Energy formulation of Kirchhoff plate theory

Kirchhoff plate theory can also be formulated using the work and energy principles developed in chapters 9 and 10. Given the kinematic assumptions of Kirchhoff plate theory, the principle of minimum total potential energy can be used to derive the governing equations and boundary conditions of the problem. This approach is especially helpful in understanding the concepts of total vertical edge load and corner force. In addition, both the work and energy principles lead directly to powerful approximate methods for realistic plate problems, such as those involving anisotropic plates or plates featuring complex shapes and boundary conditions.

Consider a thin plate of arbitrary shape with its mid-plane surface, denoted $\mathcal{S}_m$, lying in the plane defined by axes $\bar{\imath}_1$ and $\bar{\imath}_2$, as depicted in fig. 16.41. The upper and lower surfaces of the plate are parallel to its mid-plane, and the plate is bounded by an outer cylindrical surface, denoted $\mathcal{S}_b$. Let $\mathcal{C}$ be the curve that is at the intersection of the bounding cylindrical surface of the plate with its midplane; a curvilinear variable s measures length along curve $\mathcal{C}$. Unit vector $\bar{s}$ is tangent to $\mathcal{C}$ whereas $\bar{n}$ denotes the outer normal to the same curve. The plate is subjected to in-plane pressures, denoted p_1 and p_2, along axes $\bar{\imath}_1$ and $\bar{\imath}_2$, respectively; the applied transverse pressure, denoted p_3, acts along axis $\bar{\imath}_3$. Three surface tractions are applied at the bounding cylindrical surface, $\mathcal{S}_b$; in-plane surface tractions, denoted t_1 and t_2, act along axes $\bar{\imath}_1$ and $\bar{\imath}_2$, respectively, whereas the surface traction t_3 acts along axis $\bar{\imath}_3$.

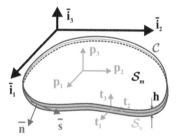

Fig. 16.41. Plate of arbitrary shape.

For this three-dimensional structure, the principle of virtual work states that

$$- \int_{\mathcal{V}} (\sigma_1 \delta\epsilon_1 + \sigma_2 \delta\epsilon_2 + \sigma_3 \delta\epsilon_3 + \tau_{23}\delta\gamma_{23} + \tau_{13}\delta\gamma_{13} + \tau_{12}\delta\gamma_{12}) \, \mathrm{d}\mathcal{V}$$

$$+ \int_{\mathcal{S}_m} (p_1 \delta\bar{u}_1 + p_2 \delta\bar{u}_2 + p_3 \delta\bar{u}_3) \, \mathrm{d}\mathcal{S}_m + \int_{\mathcal{S}_b} (t_1 \delta u_1 + t_2 \delta u_2 + t_3 \delta u_3) \, \mathrm{d}\mathcal{S}_b = 0,$$

for all arbitrary virtual displacements and associated compatible strains. The first integral represents the virtual work done by the internal stresses for a general three-dimensional solid, see eq. (9.77a). The last two integral represent the virtual work done by the externally applied loads.

The Kirchhoff kinematic assumptions discussed in section 16.1.1 are now used to simplify this expression. These assumptions imply the vanishing of the transverse strain, ϵ_3, see eq. (16.5), and the vanishing of the transverse shear strains, γ_{13} and γ_{23}, see eq. (16.6). Finally, the in-plane strains are linearly distributed through-the-thickness of the plate, see eq. (16.12). With these assumption, the principle of virtual work becomes

$$- \int_{\mathcal{V}} \left[\sigma_1(\delta\epsilon_1^0 + x_3\delta\kappa_2) + \sigma_2(\delta\epsilon_2^0 - x_3\delta\kappa_1) + \tau_{12}(\delta\epsilon_{12}^0 - x_3\delta\kappa_{12}) \right] \, \mathrm{d}\mathcal{V}$$

$$+ \int_{\mathcal{S}_m} (p_1 \delta\bar{u}_1 + p_2 \delta\bar{u}_2 + p_3 \delta\bar{u}_3) \, \mathrm{d}\mathcal{S}_m + \int_{\mathcal{S}_b} (t_1 \delta u_1 + t_2 \delta u_2 + t_3 \delta u_3) \, \mathrm{d}\mathcal{S}_b = 0.$$
$$(16.134)$$

The first line of this expression represents the virtual work done by the internal stresses; this term will be evaluated more precisely in section 16.6.1. The last two terms represents the virtual work done by the externally applied loads; these will be evaluated in section 16.6.2.

16.6.1 The virtual work done by the internal stresses

Since the strain distribution through-the-thickness of the plate is known, it is now possible to carry out the first integral in eq. (16.134) to find the internal virtual work, δW_I, in the plate as

$$\delta W_I = \delta W_I^s + \delta W_I^b = - \int_{\mathcal{S}_m} \left[N_1 \delta\epsilon_1^0 + N_2 \delta\epsilon_2^0 + N_{12}\delta\epsilon_{12}^0 \right] \, \mathrm{d}\mathcal{S}_m$$
$$- \int_{\mathcal{S}_m} \left[M_1 \delta\kappa_1 + M_2 \delta\kappa_2 + M_{12}\delta\kappa_{12} \right] \, \mathrm{d}\mathcal{S}_m. \qquad (16.135)$$

The first integral in this equation involves the in-plane forces and mid-plane strains, and represents the virtual work done in stretching the plate, denoted δW_I^s. The second integral involves the bending moments and curvatures and represents the virtual work done in bending the plate, denoted δW_I^b. These two expressions will be evaluated in more details in the two sections below.

The virtual work done by in-plane forces

First, the virtual work done by in-plane forces, denoted δW_I^s in eq. (16.135), is investigated in more detail. The mid-plane strains are expressed in terms of the in-plane displacements using the strain-displacement equations, eqs. (16.9), to find

$$\delta W_I^s = - \int_{\mathcal{S}_m} [N_1 \delta \bar{u}_{1,1} + N_2 \delta \bar{u}_{2,2} + N_{12}(\delta \bar{u}_{1,2} + \delta \bar{u}_{2,1})] \, d\mathcal{S}_m. \quad (16.136)$$

Green's theorem is used in each term to find

$$\delta W_I^s = \int_{\mathcal{S}_m} [(N_{1,1} + N_{12,2})\delta \bar{u}_1 + (N_{2,2} + N_{12,1})\delta \bar{u}_2] \, d\mathcal{S}_m$$
$$- \int_{\mathcal{C}} [(n_1 N_1 + n_2 N_{12})\delta \bar{u}_1 + (n_2 N_2 + n_1 N_{12})\delta \bar{u}_2] \, ds,$$

where n_1 and n_2 are the components of the unit outer normal to the plate in the global coordinate system, *i.e.*, $\bar{n} = n_1 \bar{\imath}_1 + n_2 \bar{\imath}_2$. The second integral along curve $\mathcal{C}$ that bounds the plate's mid-plate is a consequence of the application of Green's theorem. Along the outer edge of the plate, it is more natural to work with the local coordinate system, $(\bar{n}, \bar{s})$, rather than the global coordinate system, $(\bar{\imath}_1, \bar{\imath}_2)$, as illustrated in fig. 16.42.

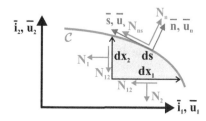

Fig. 16.42. The global coordinate system and the local coordinate system along the edge of the plate.

The displacement components, $\bar{u}_n$ and $\bar{u}_s$, measured along the local coordinate system are related to their counterparts, $\bar{u}_1$ and $\bar{u}_2$, measured along the global coordinate system, through the following transformations, see section A.3.3,

$$\bar{u}_n = n_1 \bar{u}_1 + n_2 \bar{u}_2, \quad \bar{u}_s = -n_2 \bar{u}_1 + n_1 \bar{u}_2, \quad (16.137a)$$
$$\bar{u}_1 = n_1 \bar{u}_n - n_2 \bar{u}_s, \quad \bar{u}_2 = n_2 \bar{u}_n + n_1 \bar{u}_s. \quad (16.137b)$$

The boundary term resulting from the application of Green's theorem now becomes

$$\int_{\mathcal{C}} [(n_1 N_1 + n_2 N_{12})\delta \bar{u}_1 + (n_2 N_2 + n_1 N_{12})\delta \bar{u}_2] \, ds =$$
$$\int_{\mathcal{C}} [(n_1 N_1 + n_2 N_{12})(n_1 \delta \bar{u}_n - n_2 \delta \bar{u}_s) + (n_2 N_2 + n_1 N_{12})(n_2 \delta \bar{u}_n + n_1 \delta \bar{u}_s)] \, ds$$
$$= \int_{\mathcal{C}} \left\{ [n_1^2 N_1 + n_2^2 N_2 + 2n_1 n_2 N_{12}] \, \delta \bar{u}_n \right.$$
$$\left. + [n_1 n_2 N_2 - n_1 n_2 N_1 + (n_1^2 - n_2^2)N_{12}] \, \delta \bar{u}_s \right\} \, ds.$$

The two bracketed terms in the last integral are the in-plane force components measured in the local coordinate system, $N_n = n_1^2 N_1 + n_2^2 N_2 + 2n_1 n_2 N_{12}$ and $N_{ns} = n_1 n_2 N_2 - n_1 n_2 N_1 + (n_1^2 - n_2^2)N_{12}$. The in-plane force components form a second order tensor, and the components of this tensor in the global and local coordinate systems are related by transformations laws identical to those for the components of the stress tensor, see eqs. (1.31) and (1.32). These transformation laws should be contrasted with those for first order tensors, such as in-plane displacement components, which are given by eq. (16.137). Finally, the virtual work done by the in-plane forces becomes

$$\delta W_I^s = \int_{\mathcal{S}_m} [(N_{1,1} + N_{12,2})\delta\bar{u}_1 + (N_{2,2} + N_{12,1})\delta\bar{u}_2] \, d\mathcal{S}_m$$
$$- \int_{\mathcal{C}} (N_n\delta\bar{u}_n + N_{ns}\delta\bar{u}_s) \, ds. \tag{16.138}$$

The virtual work done by the bending moments

Next, the virtual work done by the bending moments, defined by the second integral in eq. (16.135), is investigated in more detail. The curvatures are expressed in terms of the transverse displacement using the strain-displacement equations, eqs. (16.10), to find

$$\delta W_I^b = - \int_{\mathcal{S}_m} (M_1\delta\bar{u}_{3,22} - M_2\delta\bar{u}_{3,11} + 2M_{12}\delta\bar{u}_{3,12}) \, d\mathcal{S}_m. \tag{16.139}$$

Next, Green's theorem is used in each terms, leading to

$$\delta W_I^b = - \int_{\mathcal{S}_m} [(M_{2,1} - M_{12,2})\delta\bar{u}_{3,1} - (M_{12,1} + M_{1,2})\delta\bar{u}_{3,2}] \, d\mathcal{S}_m$$
$$- \int_{\mathcal{C}} [(-n_1M_2 + n_2M_{12})\delta\bar{u}_{3,1} + (n_1M_{12} + n_2M_1)\delta\bar{u}_{3,2}] \, ds.$$

Here again, the integral along curve $\mathcal{C}$ that bounds the plate's mid-plate is a consequence of the application of Green's theorem. The same theorem is applied once more to find

$$\delta W_I^b = - \int_{\mathcal{S}_m} (M_{1,22} + 2M_{12,12} - M_{2,11}) \, \delta\bar{u}_3 \, d\mathcal{S}_m$$
$$- \int_{\mathcal{C}} [(-n_1M_2 + n_2M_{12})\delta\bar{u}_{3,1} + (n_1M_{12} + n_2M_1)\delta\bar{u}_{3,2}] \, ds$$
$$- \int_{\mathcal{C}} [(M_{2,1} - M_{12,2})n_1 - (M_{1,2} + M_{12,1})n_2] \, \delta\bar{u}_3 \, ds.$$

Along the outer edge of the plate, it is more natural to work with the local coordinate system, $(\bar{n}, \bar{s})$, rather than the global coordinate system, $(\bar{\imath}_1, \bar{\imath}_2)$, as illustrated in fig. 16.43. Since $Q_1 = M_{2,1} - M_{12,2}$ and $Q_2 = -M_{1,2} - M_{12,1}$, see eqs. (16.25), the last integral becomes

$$\int_{\mathcal{C}} [(M_{2,1} - M_{12,2})n_1 - (M_{1,2} + M_{12,1})n_2] \, \delta\bar{u}_3 \, ds$$
$$= \int_{\mathcal{C}} (Q_1n_1 + Q_1n_2) \, \delta\bar{u}_3 \, ds = \int_{\mathcal{C}} Q_n\delta\bar{u}_3 \, ds,$$

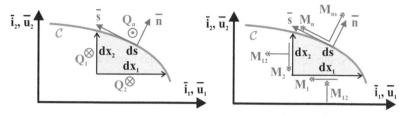

Fig. 16.43. The global coordinate system and the local coordinate system along the edge of the plate.

where the last equality follows from the vertical equilibrium of the differential element of the plate represented on the left portion of fig. 16.43, which implies $Q_n = Q_1 n_1 + Q_1 n_2$.

The other boundary term along curve $\mathcal{C}$ is treated as follows

$$\int_{\mathcal{C}} [(-n_1 M_2 + n_2 M_{12})\delta\bar{u}_{3,1} + (n_1 M_{12} + n_2 M_1)\delta\bar{u}_{3,2}]\ ds$$

$$= \int_{\mathcal{C}} [(-n_1 M_2 + n_2 M_{12})(n_1\delta\bar{u}_{3,n} - n_2\delta\bar{u}_{3,s})$$
$$+ (n_1 M_{12} + n_2 M_1)(n_2\delta\bar{u}_{3,n} + n_1\delta\bar{u}_{3,s})]\ ds$$

$$= \int_{\mathcal{C}} \left\{ - \left[n_1^2 M_2 - n_2^2 M_1 - 2n_1 n_2 M_{12} \right] \delta\bar{u}_{3,n} \right.$$
$$\left. - \left[-n_1 n_2 M_1 + n_1 n_2 M_2 - (n_1^2 - n_2^2)M_{12} \right] \delta\bar{u}_{3,s} \right\}\ ds$$

The two bracketed terms in the last integral are the bending moment components measured in the local coordinate system, $M_n = n_1^2 M_2 - n_2^2 M_1 - 2n_1 n_2 M_{12}$ and $M_{ns} = -n_1 n_2 M_1 + n_1 n_2 M_2 - (n_1^2 - n_2^2)M_{12}$. Here again, the transformation of the bending moment components follow the laws of transformation for second order tensors.

The virtual work done by the bending moments becomes

$$\delta W_I^b = -\int_{\mathcal{S}_m} (M_{1,22} + 2M_{12,12} - M_{2,11})\,\delta\bar{u}_3\ d\mathcal{S}_m$$
$$- \int_{\mathcal{C}} (Q_n \delta\bar{u}_3 - M_n \delta\bar{u}_{3,n} - M_{ns}\delta\bar{u}_{3,s})\ ds.$$

The final step in the evaluation of the virtual work done by the bending moments is to realize that variations $\delta\bar{u}_3$ and $\delta\bar{u}_{3,s}$ are not independent quantities along curve $\mathcal{C}$, because the knowledge of $\bar{u}_3(s)$ for all s implies the knowledge of $\bar{u}_{3,s}(s)$. Hence, the term $-M_{ns}\delta\bar{u}_{3,s}$ appearing in the last integral can be integrated by parts to yield

$$\delta W_I^b = -\int_{\mathcal{S}_m} (M_{1,22} + 2M_{12,12} - M_{2,11})\,\delta\bar{u}_3\ d\mathcal{S}_m$$
$$- \int_{\mathcal{C}} [-M_n \delta\bar{u}_{3,n} + (Q_n + M_{ns,s})\delta\bar{u}_3]\ ds - [M_{ns}\delta\bar{u}_3]_{\mathcal{C}}\,,$$

(16.140)

where the notation $[\cdot]_C$ indicates the boundaries of curve C. If the plate is square, the boundaries of C would be defined by its four corners, whereas if the plate is circular, C has no boundaries.

In summary, the virtual work done by the in-plane forces and bending moments is the sum of expressions (16.138) and (16.140).

16.6.2 The virtual work done by the applied loads

The virtual work done by the externally applied loads is given by the second integral in the statement of the principle of virtual work, eq. (16.134), as

$$
\delta W_E = \int_{S_m} (p_1 \delta \bar{u}_1 + p_2 \delta \bar{u}_2 + p_3 \delta \bar{u}_3) \, \mathrm{d}S_m + \int_{S_b} (t_1 \delta u_1 + t_2 \delta u_2 + t_3 \delta u_3) \, \mathrm{d}S_b,
$$

where the first integral represents the virtual work done by the pressure distribution applied over the mid-plane of the plate and the second integral represents the virtual work done by the surface tractions acting over edge surface S_b that bounds the plate.

First, Kirchhoff's kinematic assumptions for the displacement field, see eqs. (16.4), are introduced, leading to

$$
\begin{aligned}
\delta W_E = \ & \int_{S_m} (p_1 \delta \bar{u}_1 + p_2 \delta \bar{u}_2 + p_3 \delta \bar{u}_3) \, \mathrm{d}S_m \\
& + \int_{S_b} [t_1(\delta \bar{u}_1 - x_3 \delta \bar{u}_{3,1}) + t_2(\delta \bar{u}_2 - x_3 \delta \bar{u}_{3,2}) + t_3 \delta \bar{u}_3] \, \mathrm{d}S_b,
\end{aligned}
$$

Because the displacement distribution through-the-thickness of the plate is known, it is now possible to integrate the second integral to find

$$
\begin{aligned}
\delta W_E = \ & \int_{S_m} (p_1 \delta \bar{u}_1 + p_2 \delta \bar{u}_2 + p_3 \delta \bar{u}_3) \, \mathrm{d}S_m \\
& + \int_{C} \left[\widetilde{N}_1 \delta \bar{u}_1 + \widetilde{N}_2 \delta \bar{u}_2 + \widetilde{Q}_n \delta \bar{u}_3 - \widetilde{M}_2 \delta \bar{u}_{3,1} + \widetilde{M}_1 \delta \bar{u}_{3,2} \right] \, \mathrm{d}s,
\end{aligned}
$$

where $\widetilde{N}_1$ and $\widetilde{N}_2$ are the in-plane forces applied along the external boundary of the plate, $\widetilde{Q}_n$, are the vertical force, and finally, $\widetilde{M}_1$ and $\widetilde{M}_2$, are the bending moments applied along the same boundary. Here again, the last integral is more naturally expressed in the local coordinate system shown in figs. 16.42 and 16.43.

$$
\begin{aligned}
\delta W_E = \ & \int_{S_m} (p_1 \delta \bar{u}_1 + p_2 \delta \bar{u}_2 + p_3 \delta \bar{u}_3) \, \mathrm{d}S_m \\
& + \int_{C} \left[\widetilde{N}_n \delta \bar{u}_n + \widetilde{N}_{ns} \delta \bar{u}_s + \widetilde{Q}_n \delta \bar{u}_3 - \widetilde{M}_n \delta \bar{u}_{3,n} - \widetilde{M}_{ns} \delta \bar{u}_{3,s} \right] \, \mathrm{d}s.
\end{aligned}
$$

As is the case for bending strain energy, the final step in the evaluation of the work done by the external forces is to observe that variations $\delta \bar{u}_3$ and $\delta \bar{u}_{3,s}$ are not independent quantities along curve $\mathcal{C}$. Hence, the term $-\widetilde{M}_{ns}\delta\bar{u}_{3,s}$ appearing in the last integral can be integrated by parts to yield

$$
\begin{aligned}
\delta W_E = \int_{\mathcal{S}_m} & (p_1\delta\bar{u}_1 + p_2\delta\bar{u}_2 + p_3\delta\bar{u}_3)\, \mathrm{d}\mathcal{S}_m \\
+ \int_{\mathcal{C}} & \left[\tilde{N}_n\delta\bar{u}_n + \tilde{N}_{ns}\delta\bar{u}_s - \widetilde{M}_n\delta\bar{u}_{3,n} + (\tilde{Q}_n + \widetilde{M}_{ns,s})\delta\bar{u}_3 \right]\, \mathrm{d}s - \left[\widetilde{M}_{ns}\delta\bar{u}_3 \right]_{\mathcal{C}}.
\end{aligned}
$$
(16.141)

16.6.3 The principle of virtual work for Kirchhoff plates

The internal and external virtual work given in the last two sections by eqs. (16.138), (16.140), and (16.141), can be combined to yield the following expression for the principle of virtual work

$$
\begin{aligned}
\int_{\mathcal{S}_m} & \left\{ -\left[N_{1,1} + N_{12,2} + p_1 \right]\delta\bar{u}_1 - \left[N_{2,2} + N_{12,1} + p_2 \right]\delta\bar{u}_2 \right. \\
& \left. - \left[M_{2,11} - M_{1,22} - 2M_{12,12} + p_3 \right]\delta\bar{u}_3 \right\}\, \mathrm{d}\mathcal{S}_m \\
+ \int_{\mathcal{C}} & \left\{ \left[N_n - \tilde{N}_n \right]\delta\bar{u}_n + \left[N_{ns} - \tilde{N}_{ns} \right]\delta\bar{u}_s \right. \\
& \left. + \left[Q_n + M_{ns,s} - \tilde{Q}_n - \widetilde{M}_{ns,s} \right]\delta\bar{u}_3 - \left[M_n - \widetilde{M}_n \right]\delta\bar{u}_{3,n} \right\}\, \mathrm{d}s \\
- & \left\{ \left[M_{ns} - \widetilde{M}_{ns} \right]\delta\bar{u}_3 \right\}_{\mathcal{C}} = 0.
\end{aligned}
$$
(16.142)

Because all variations are arbitrary, the bracketed terms of this expression must all vanish. The vanishing of the coefficients of the variations $\delta\bar{u}_1$, $\delta\bar{u}_2$, and $\delta\bar{u}_3$ in the first integral leads to the equilibrium equations of the problem which are identical to those obtained earlier, eqs. (16.23a), (16.23b), and (16.33), respectively.

The vanishing of the bracketed terms in the integral along the outer boundary of the plate, $\mathcal{C}$, leads to the natural boundary conditions of the problem: $N_n = \tilde{N}_n$ and $N_{ns} = \tilde{N}_{ns}$, $Q_n + M_{ns,s} = \tilde{Q}_n + \widetilde{M}_{ns,s}$, and $M_n = \widetilde{M}_n$. Finally, the vanishing of the last term bracketed term implies $M_{ns} = \widetilde{M}_{ns}$ at the boundaries of curve $\mathcal{C}$.

As expected, the principle of virtual work yields all the equations of equilibrium of the problem and the natural boundary conditions. It is interesting to note that the concepts of total vertical force and corner force are a natural byproduct of the application of Green's theorem. Indeed, the natural boundary condition, $Q_n + M_{ns,s} = \tilde{Q}_n + \widetilde{M}_{ns,s}$, implies the equality of the total vertical force in the plate, $Q_n + M_{ns,s}$, with its externally applied counterpart. The equality of the corner forces, M_{ns}, with their externally applied counterparts is also identified. This contrasts with the classical approach to the development of the governing equations of Kirchhoff's plate theory, where these two concepts are introduced in an *ad hoc* manner to overcome the problems associated with the boundary conditions to be applied along the free edge of a plate, as discussed in section 16.2.1.

16.6.4 The principle of minimum total potential energy for Kirchhoff plates

As discussed in section 12.2.5 for the general elasticity case, the principle of minimum total potential energy can be derived from the principle of virtual work. Two assumptions are involved in this process: first, it is assumed that the constitutive laws for the material can be expressed in terms of a strain energy density function, see eq. (12.31), and second, it is assumed that the externally applied loads can be derived from a potential. These two steps are described in the next two sections.

The strain energy for Kirchhoff plates

The expression for the strain energy in a plate can be obtained from that for a general three-dimensional solid developed in section 10.5. Introducing the kinematic assumptions of Kirchhoff theory then leads to

$$
A = \frac{1}{2} \int_{\mathcal{S}_m} \left(\underline{N}^T \, \underline{\epsilon}_0 + \underline{M}^T \, \underline{\kappa} \right) \, \mathrm{d}\mathcal{S}_m,
$$

where the arrays of forces and moments are given by eqs. (16.17) and (16.22), respectively, and the arrays of mid-plane strains and curvatures by eqs. (16.9) and (16.10), respectively.

Next, the constitutive laws for the material, which are assumed to be linearly elastic but generally anisotropic as given by eqs. (16.52), are introduced to find the strain energy stored in the plate

$$
A = \int_{\mathcal{S}_m} \frac{1}{2} \left(\underline{\epsilon}_0^T \underline{\underline{A}} \, \underline{\epsilon}_0 + 2\underline{\epsilon}_0^T \underline{\underline{B}} \, \underline{\underline{S}} \, \underline{\kappa} + \underline{\kappa}^T \underline{\underline{S}}^T \underline{\underline{D}} \, \underline{\underline{S}} \, \underline{\kappa} \right) \, \mathrm{d}\mathcal{S}_m. \tag{16.143}
$$

An important class of plate problems is that for which the coupling stiffness matrix, $\underline{\underline{B}}$, vanishes. In such case, the strain energy in the plate becomes

$$
A = A_s + A_b = \frac{1}{2} \int_{\mathcal{S}_m} \underline{\epsilon}_0^T \underline{\underline{A}} \, \underline{\epsilon}_0 \, \mathrm{d}\mathcal{S}_m + \frac{1}{2} \int_{\mathcal{S}_m} \underline{\kappa}^T \underline{\underline{S}}^T \underline{\underline{D}} \, \underline{\underline{S}} \, \underline{\kappa} \, \mathrm{d}\mathcal{S}_m, \tag{16.144}
$$

where the first term represents the strain energy associated with stretching of the plate, denoted A_s, and the second term that associated with bending of the plate, denoted A_b. The strain energy associated with stretching depends solely on the in-plane displacements, whereas the strain energy associated with bending depends solely on the transverse displacement. This decoupling between in-plane and out-of-plane behavior is also observed in the classical approach described in section 16.3.9.

Using the strain-displacement equations for curvatures, see eq. (16.10), the bending strain energy can be written as

$$
A_b = \frac{1}{2} \int_{\mathcal{S}_m} \left[D_{11}\bar{u}_{3,11}^2 + 2D_{12}\bar{u}_{3,11}\bar{u}_{3,22} + D_{22}\bar{u}_{3,22}^2 \right.
$$
$$
\left. + 4D_{66}\bar{u}_{3,12}^2 + 4D_{16}\bar{u}_{3,11}\bar{u}_{3,12} + 4D_{26}\bar{u}_{3,22}\bar{u}_{3,12} \right] \, \mathrm{d}\mathcal{S}_m. \tag{16.145}
$$

For specially orthotropic plates, the bending strain energy reduces to

$$A_b = \frac{1}{2} \int_{S_m} \left[D_{11} \bar{u}_{3,11}^2 + 2D_{12} \bar{u}_{3,11} \bar{u}_{3,22} + D_{22} \bar{u}_{3,22}^2 + 4D_{66} \bar{u}_{3,12}^2 \right] \, \mathrm{d}S_m.$$

(16.146)

Finally, if the plate is isotropic, then $D_{11} = D_{22} = D$, $D_{12} = \nu D$ and $D_{66} = D(1 - \nu)/2$, where D is the plate bending stiffness defined by eq. (16.36). In this case, eq. (16.146) further reduces to

$$A_b = \frac{1}{2} \int_{S_m} D \left[\bar{u}_{3,11}^2 + \bar{u}_{3,22}^2 + 2\nu \bar{u}_{3,11} \bar{u}_{3,22} + 2(1 - \nu) \bar{u}_{3,12}^2 \right] \, \mathrm{d}S_m. \quad (16.147)$$

The potential of the externally applied loads

The externally applied loads consist of the pressures, p_1, p_2, and p_3 along with the edge loads $\tilde{N}_n$, $\tilde{Q}_n$, $\widetilde{M}_n$, and $\widetilde{M}_{12}$, and the external virtual work is given by eq. (16.141) developed previously. For simplicity, only the in-plane and transverse pressure components will be considered here, although it is a simple matter to add the contributions due to edge loads.

The virtual work done by the externally applied forces is then simply

$$\delta W_E = \delta W_E^s + \delta W_E^b = \int_{S_m} (p_1 \delta \bar{u}_1 + p_2 \delta \bar{u}_2) \, \mathrm{d}S_m + \int_{S_m} p_3 \delta \bar{u}_3 \, \mathrm{d}S_m,$$

where the expression is split into the work done by the in-plane pressures and that done by the transverse pressure, leading to the stretching and bending parts of the work done by the externally applied pressures, denoted δW_E^s and δW_E^b, respectively.

If these pressure can be derived from a potential, *i.e.*, if each pressure component is written as $p_i = -\partial \phi_i / \partial \bar{u}_i$, $i = 1, 2, 3$, the virtual work done by the externally applied pressure becomes

$$\delta W_s + \delta W_b = -\delta \int_{S_m} (\phi_1 + \phi_2) \, \mathrm{d}S_m - \delta \int_{S_m} \phi_3 \, \mathrm{d}S_m = -\delta \Phi_s - \delta \Phi_b,$$

where Φ_s and Φ_b are the total potentials of the in-plane and transverse pressures, respectively, and

$$\Phi_s = - \int_{S_m} (\phi_1 + \phi_2) \, \mathrm{d}S_m, \tag{16.148a}$$

$$\Phi_b = - \int_{S_m} \phi_3 \, \mathrm{d}S_m. \tag{16.148b}$$

Note that if the pressure applied on the plate is of constant magnitude and direction, as is typically the case, $\phi_i(\bar{u}_i) = -\bar{u}_i p_i$, because the pressure can then be obtained from this potential as $p_i = -\partial \phi_i / \partial \bar{u}_i$.

Total potential energy

The total potential energy for a general Kirchhoff plate is given by $\Pi = A_s + A_b + \Phi_s + \Phi_s$ where each term is given above. When the coupling stiffness matrix vanishes, the total potential energy can be separated into two separate problems: one involving stretching only, and one involving bending only. In this case, the total potential energy due to stretching of the plate is

$$\Pi_s = A_s + \Phi_s, \qquad (16.149)$$

and that due to bending is

$$\Pi_b = A_b + \Phi_b. \qquad (16.150)$$

The principle of minimum total potential energy can now be used to determine the equilibrium configuration of the plate. The variational methods from section 12.2.5 can be used to find the Euler-Lagrange governing equations for the plate along with the possible boundary conditions. The result will be the same equations obtained in eq. (16.142) using the principle of virtual work and relevant constitutive equations.

The principle of minimum total potential energy does provide an effective approach for constructing approximate solutions to Kirchhoff plate problems because of the integral formulation and the reduced (weaker) continuity requirements placed on the solution. The approach follows closely that developed in chapter 10 to solve beam bending and torsion problems.

16.6.5 Approximate solutions for Kirchhoff plates

Approximate solutions of beam problems can be obtained from the principle of virtual work or from the principle of minimum total potential energy, and the general solution procedures using these two principles are presented in sections 11.4.2 and 11.4.3, respectively. In this section, a similar approach is developed for Kirchhoff plate bending problems. The use of the principle of minimum total potential energy will be illustrated, although the principle of virtual work could be employed with equal ease.

Following the overall procedure described in section 11.4.3, The first step of the solution procedure is to assume a specific form of the transverse displacement field as

$$\bar{u}_3(x_1, x_2) = \sum_{i=1}^{N} h_i(x_1, x_2)\, q_i, \qquad (16.151)$$

where the q_i are unknown coefficients, often called *degrees of freedom*, that determine the solution of the problem, and the known functions, $h_i(x_1, x_2)$, form a set of functions, called *shape functions*, which each must satisfy the geometric boundary conditions of the problem.

As is the case for the beam problems, monomials or trigonometric functions can be selected as shape functions. Polynomial or transcendental functions can also be used as long as they form a set of linearly independent functions that satisfy the geometric boundary conditions.

Expression (16.151) represents an approximation to the problem because the solution is assumed to consists of a linear combination of a finite number, N, of shape functions where each coefficient, q_i, indicates how much the corresponding shape function contributes to the final solution. This explains the expression "modal participation factors" frequently used to denote the coefficients q_i.

It is convenient to recast the expressions for the assumed displacements, eq. (16.151), into a matrix form

$$\bar{u}_3(x_1, x_2) = \underline{H}^T(x_1, x_2)\,\underline{q}, \tag{16.152}$$

where $\underline{q} = \{q_1, q_2, \cdots, q_N\}^T$ is an array of size N that stores the N degrees of freedom of the problem. The *displacement interpolation array*, $\underline{H}(x_1, x_2)$, is defined as

$$\underline{H}(x_1, x_2) = \{h_1(x_1, x_2), h_2(x_1, x_2), h_3(x_1, x_2), \cdots, h_{N(x_1, x_2)}\}^T. \tag{16.153}$$

Next, the assumed displacements are introduced in the curvature-displacement relationship, eq. (16.10), to find the curvature distribution

$$\underline{S}\,\underline{\kappa}(x_1, x_2) = \left\{ \begin{array}{c} -\bar{u}_{3,11} \\ -\bar{u}_{3,22} \\ -2\bar{u}_{3,12} \end{array} \right\} = \underline{B}(x_1, x_2)\,\underline{q}, \tag{16.154}$$

where the *strain* or *curvature interpolation matrix* is

$$\underline{B}(x_1, x_2) = \begin{bmatrix} \cdots & -h_{i,11} & \cdots \\ \cdots & -h_{i,22} & \cdots \\ \cdots & -2h_{i,12} & \cdots \end{bmatrix} \tag{16.155}$$

The total strain energy in the plate for the approximate solution is now obtained by introducing the curvatures, eq. (16.154), into the strain energy expression, eq. (16.144), to obtain

$$A_b = \frac{1}{2}\int_{\mathcal{S}_m} \underline{\kappa}^T \underline{S}^T \underline{D}\,\underline{S}\,\underline{\kappa}\; \mathrm{d}\mathcal{S}_m = \frac{1}{2}\,\underline{q}^T\left[\int_{\mathcal{S}_m} \underline{B}^T \underline{D}\,\underline{B}\; \mathrm{d}\mathcal{S}_m\right]\underline{q} = \frac{1}{2}\,\underline{q}^T \underline{K}\,\underline{q}, \tag{16.156}$$

where $\underline{K}$ is the stiffness matrix for the plate

$$\underline{K} = \int_{\mathcal{S}_m} \underline{B}^T(x_1, x_2)\underline{D}\,\underline{B}(x_1, x_2)\; \mathrm{d}\mathcal{S}_m, \tag{16.157}$$

Note that this expression for the stiffness matrix of the plate is formally identical to those obtained for beam problems, see eqs. (11.67) and (11.73), for beams under axial and transverse loading, respectively.

Next, the total potential of the externally applied pressure is evaluated by introducing the assumed displacement field, eq. (16.152), into the total potential energy expression, eq. (16.148), to obtain

$$\Phi = \int_{\mathcal{S}_m} \phi_3 \, d\mathcal{S}_m = -\underline{q}^T \left[\int_{\mathcal{S}_m} \underline{H}(x_1, x_2) p_3(x_1, x_2) \, d\mathcal{S}_m \right] = -\underline{q}^T \underline{Q}, \quad (16.158)$$

where $\underline{Q}$ is the *load array* for the system

$$\underline{Q} = \int_{\mathcal{S}_m} \underline{H}(x_1, x_2) p_3(x_1, x_2) \, d\mathcal{S}_m. \quad (16.159)$$

This expression for the load array is formally identical to that found for beam problems, see eq. (11.74).

With the help of eqs. (16.156) and (16.159), the total potential energy for plate bending, eq. (16.150), now reduces to

$$\Pi_b(\underline{q}) = \frac{1}{2} \underline{q}^T \underline{K} \underline{q} - \underline{q}^T \underline{Q}. \quad (16.160)$$

The first term is a positive-definite quadratic form in the degrees of freedom, whereas the second term is a linear form of the same variables.

According to the principle of minimum total potential energy, the system is in equilibrium when this expression assumes a minimum with respect to all possible choices of the displacements, which are now limited to the choices of the degrees of freedom. Imposing the vanishing of the derivatives of the total potential energy with respect to $\underline{q}$ leads to the following equations

$$\underline{K} \underline{q} - \underline{Q} = 0. \quad (16.161)$$

where eqs. (A.27) and (A.29) are used to evaluate the derivatives of the potential of the externally applied loads and strain energy, respectively.

The general procedure for solving plate bending problems using the principle of minimum total potential energy can be summarized by the following steps, which mirror the corresponding steps used for the solution beam bending problems, see section 11.4.3.

1. Select a set of shapes functions that satisfy the geometric boundary conditions.
2. Construct the displacement interpolation array, eq. (16.153), and strain interpolation matrix, eq. (16.155).
3. Compute the total strain energy of the system, $A_b(\underline{q}) = 1/2 \, \underline{q}^T \underline{K} \, \underline{q}$, where the stiffness matrix is given by eq. (16.157).
4. Compute the total potential of the externally applied loads, $\Phi(\underline{q}) = -\underline{q}^T \underline{Q}$, where the load array $\underline{Q}$ is given by eq. (16.159).
5. Solve the set of linear equations, $\underline{K} \, \underline{q} = \underline{Q}$, for the solution vector, q.
6. Determine the strain distribution from eq. (16.154), and the internal forces from the constitutive law, eq. (16.54).

Choice of shape functions

When the above procedure is applied to rectangular plates, it is often convenient to write the assumed displacement field as a product of two sets of functions, $f_m(x_1)$ and $g_n(x_2)$, which depend solely on the variables x_1 and x_2, respectively,

$$\bar{u}_3(x_1, x_2) = \sum_{m=1}^{M} \sum_{n=1}^{N} f_m(x_1) g_n(x_2)\, q_{mn}. \tag{16.162}$$

Functions $f_m(x_1)$ must satisfy the geometric boundary conditions of the problem at $x_1 = 0$ and a, but are otherwise arbitrary. Similarly, functions $g_n(x_2)$ must satisfy the geometric boundary conditions of the problem at $x_2 = 0$ and b, but are otherwise arbitrary.

Consider, for instance, a plate that is simply supported at $x_1 = 0$ and a, for which the boundary conditions are given by eq. (16.39). An appropriate set of shape functions would be

$$f_m(x_1) = \sin \frac{m\pi x_1}{a}. \tag{16.163}$$

This choice clearly satisfies the geometric boundary condition, $\bar{u}_3 = 0$, at $x_1 = 0$ and a. A similar expression could be used for functions $g_n(x_2)$ if the plate is simply supported at $x_2 = 0$ and b.

Consider next a plate that is clamped at $x_1 = 0$ and a, for which the boundary conditions are given by eq. (16.38). An appropriate set of shape functions would be

$$f_m(x_1) = \frac{1}{2} \left[\cos \frac{(m-1)\pi x_1}{a} - \cos \frac{(m+1)\pi x_1}{a} \right]. \tag{16.164}$$

This choice clearly satisfies the geometric boundary conditions, $\bar{u}_3 = 0$ and $\bar{u}_{3,1} = 0$, at $x_1 = 0$ and a. A similar expression could be used for functions $g_n(x_2)$ if the plate is clamped at $x_2 = 0$ and b. Another set of functions that can be used for clamped plates is

$$f_m(x_1) = (x_1 - a)^2 x_1^{2+m}. \tag{16.165}$$

Finally, free edges are easily treated as well. Indeed, the associated boundary conditions correspond to the vanishing of the bending moment and total vertical force along the free edge, two natural boundary conditions. Hence, no requirements at all must be imposed on the shape functions along these edges.

Elastic supports

Energy methods are also very versatile when it comes to the modeling edges of a plate resting on elastic foundations. Consider the case of a rectangular plate with the edge located at $x_1 = a$ supported by a linear distributed spring of stiffness k, as depicted in fig. 16.8. The elastic system is defined as comprising the plate and the distributed elastic foundation along the edge, and hence, the total strain energy of the system now becomes

$$A = A_b + \frac{1}{2} \int_0^b k \bar{u}_3^2(a, x_2) \, \mathrm{d}x_2, \tag{16.166}$$

where A_b is the strain energy stored in the elastic plate due to bending, and the second term represents the strain energy stored in the deformation of the linear spring.

If, instead, a rotational spring of stiffness constant k is connected to the plate along the same edge as depicted in fig. 16.9, the total strain energy of the system becomes

$$A = A_b + \frac{1}{2} \int_0^b k \bar{u}_{3,1}^2(a, x_2) \, \mathrm{d}x_2, \tag{16.167}$$

where the second term represents the strain energy stored in the deformation of the rotational spring. If the plate rests on elastic foundations along several of its edges, the strain energies of the various springs are added to the strain energy of the plate to find the total strain energy of the system.

A similar approach can be used to treat directionally reinforced plates, such as those depicted in fig. 16.22. In section 16.3.8, the contributions of individual stiffeners is smeared to obtained equivalent plate bending stiffnesses, see eqs. (16.76). While such approximation is reasonable for a plate stiffened by a large number of stiffeners, *i.e.*, if $\ell/a \ll 1$, the smearing process is clearly not justified when the plate features just a few stiffeners. In such case, it is possible to take into account the contributions of each stiffener by adding their individual strain energies to that of the base plate. Consider the case of a plate reinforced by three stiffeners running along the direction parallel to axis $\bar{\imath}_2$ and located at $x_1 = \alpha_1, \alpha_2,$ and α_3, respectively.

$$A = A_b + \sum_{i=1}^3 \left[\frac{1}{2} \int_0^b H_i^b \bar{u}_{3,22}^2(\alpha_i, x_2) \, \mathrm{d}x_2 + \frac{1}{2} \int_0^b H_i^t \bar{u}_{3,12}^2(\alpha_i, x_2) \, \mathrm{d}x_2 \right],$$
$$\tag{16.168}$$

where H_i^b and H_i^t are the bending and torsional stiffnesses of the stiffeners, respectively.

Example 16.8. *Simply supported rectangular isotropic plate*
Consider a simply supported rectangular isotropic plate subjected to a uniform transverse pressure, p_0. An approximate solution of this problem can be constructed based on the general procedure described above. The shape functions for this problem with simply supported boundaries are assumed to take the following form

$$h_1 = \sin \frac{\pi x_1}{a} \sin \frac{\pi x_2}{b}, \quad h_2 = \sin \frac{\pi x_1}{a} \sin \frac{3\pi x_2}{b},$$
$$h_3 = \sin \frac{3\pi x_1}{a} \sin \frac{\pi x_2}{b}, \quad h_4 = \sin \frac{3\pi x_1}{a} \sin \frac{3\pi x_2}{b}.$$

Clearly, these shape functions satisfy the geometric boundary conditions along the four edges of the plate.

The displacement interpolation matrix is now given by eq. (16.153), and the curvature interpolation defined by eq. (16.155) becomes

$$B(x_1, x_2) = - \begin{bmatrix} h_{1,11} & h_{2,11} & h_{3,11} & h_{4,11} \\ h_{1,22} & h_{2,22} & h_{3,22} & h_{4,22} \\ 2h_{1,12} & 2h_{2,12} & 2h_{3,12} & 2h_{4,12} \end{bmatrix}.$$

Next, the plate stiffness matrix is evaluated using eq. (16.157). A typical entry of this matrix is

$$K_{ij} = \int_{S_m} D \left[h_{i,11} h_{j,11} + \nu (h_{i,11} h_{j,22} + h_{i,22} h_{j,11}) \right. \tag{16.169}$$
$$\left. + h_{i,22} h_{j,22} + 2(1 - \nu) h_{i,12} h_{j,12} \right] dS_m.$$

where the plate bending stiffness is given by eq. (16.36) for an isotropic plate. The entries of the stiffness matrix are found by introducing the shape functions into this equation and integrating over the area of the plate. Due to the orthogonality properties of sine functions (see section A.4), the stiffness matrix turns out to be a diagonal matrix,

$$\underline{\underline{K}} = \frac{abD}{4} \operatorname{diag} \left[\left(\left(\frac{\pi}{a} \right)^2 + \left(\frac{\pi}{b} \right)^2 \right)^2, \left(\left(\frac{\pi}{a} \right)^2 + \left(\frac{3\pi}{b} \right)^2 \right)^2, \right.$$
$$\left. \left(\left(\frac{3\pi}{a} \right)^2 + \left(\frac{\pi}{b} \right)^2 \right)^2, \left(\left(\frac{3\pi}{a} \right)^2 + \left(\frac{3\pi}{b} \right)^2 \right)^2 \right].$$

Finally, the load array is computed with the help of eq.(16.159). A typical entry of this array is

$$Q_i = \int_{S_m} h_i p_3 \, dS_m. \tag{16.170}$$

For a plate subjected to a uniform transverse pressure, $p_3 = p_0$, and introducing the selected shape function, the load vector becomes

$$\underline{Q} = \frac{4}{\pi^2} p_0 ab \left[1, \frac{1}{3}, \frac{1}{3}, \frac{1}{9} \right]^T.$$

In the solution phase of the procedure, the linear system, eq. (16.161), is solved for vector $\underline{q}$. This step is particularly simple in this case because the stiffness matrix is diagonal. The final solution for the transverse displacement is

$$\bar{u}_3 = \frac{16}{\pi^6} \frac{p_0}{D} \left[\frac{\sin(\pi x_1/a) \sin(\pi x_2/b)}{[1/a^2 + 1/b^2]^2} + \frac{\sin(\pi x_1/a) \sin(3\pi x_2/b)}{3 [1/a^2 + 9/b^2]^2} \right.$$
$$\left. + \frac{\sin(3\pi x_1/a) \sin(\pi x_2/b)}{3 [9/a^2 + 1/b^2]^2} + \frac{\sin(3\pi x_1/a) \sin(3\pi x_2/b)}{9 [9/a^2 + 9/b^2]^2} \right].$$

This solution corresponds to the first four terms of Navier's solution applied to the same problem, see example 16.1. This should not be unexpected because Navier's solution is based on an infinite, double sine wave expansion of the solution, whereas the present approach only takes the first four terms of this infinite expansion.

Example 16.9. Clamped rectangular anisotropic plate using numerical quadrature

Consider a clamped, rectangular plate made of an anisotropic material and subjected to a transverse pressure, p_3. For simplicity, the coupling stiffness matrix is assumed to vanish.

The shape functions are selected to be products of the functions given by eq. (16.164). The first step of the procedure is to compute the stiffness matrix, $\underline{\underline{K}}$, given by eq. (16.157). Clearly, this step will be very tedious given the selected shape functions. The Gauss-Legendre quadrature scheme, see section A.5, can be used to carry out the integrations numerically, rather than symbolically, leading to

$$
\begin{aligned}
\underline{\underline{K}} &= \int_0^a \int_0^b \underline{\underline{B}}^T(x_1, x_2)\underline{\underline{D}}(x_1, x_2)\underline{\underline{B}}(x_1, x_2)\ \mathrm{d}x_1\mathrm{d}x_2 \\
&= \frac{ab}{4}\int_{-1}^{+1}\int_{-1}^{+1}\underline{\underline{B}}^T(\eta, \zeta)\underline{\underline{D}}(\eta, \zeta)\underline{\underline{B}}(\eta, \zeta)\ \mathrm{d}\eta\mathrm{d}\zeta \qquad (16.171) \\
&\approx \frac{ab}{4}\sum_{i=1}^N\sum_{j=1}^M w_i w_j \underline{\underline{B}}^T(\eta_i, \zeta_j)\underline{\underline{D}}(\eta_i, \zeta_j)B(\eta_i, \zeta_j),
\end{aligned}
$$

where η_i and ζ_j are the locations of the points Gauss points and w_i and w_j the associated weights, as defined in section A.5.

Note that in the above expression, the bending stiffness matrix of the plate can vary over the surface of the plate, $\underline{\underline{D}} = \underline{\underline{D}}(x_1, x_2)$. This would happen, for instance, if the plate features reinforcements over specific portions of its surface. Clearly, realistic plate configurations are easily treated numerically within the framework of energy methods.

The computation of the load array is based on eq. (16.159) and is treated in similar manner

$$
\begin{aligned}
\underline{Q} &= \int_0^a \int_0^b \underline{H}(x_1, x_2)p_3(x_1, x_2)\ \mathrm{d}x_1\mathrm{d}x_2 = \frac{ab}{4}\int_{-1}^{+1}\int_{-1}^{+1}\underline{H}(\eta, \zeta)p_3(\eta, \zeta)\ \mathrm{d}\eta\mathrm{d}\zeta \\
&\approx \frac{ab}{4}\sum_{i=1}^N\sum_{j=1}^M w_i w_j \underline{H}(\eta_i, \zeta_j)p_3(\eta_i, \zeta_j). \qquad (16.172)
\end{aligned}
$$

Arbitrary pressure distributions over the surface of the plate are easily accommodated within the framework of this approach.

The solution of the problem now reduces to the solution of a linear system, eq. (16.161). The main advantage of energy approach to the solution of plate bending problems is that it is very systematic. For a very large class of problems, the stiffness matrix and load array are evaluated numerically based on eqs. (16.171) and (16.172), respectively. This phase involves the computation of the displacement and strain interpolation matrices only at the sampling points of the selected Gauss-Legendre quadrature scheme. The second phase of the process involves the solution of a linear system of equations. All operations are readily and systematically implemented in a computer algorithm.

16.6.6 Solutions based on partial approximation

Approximate solutions for complex plate bending problems can be found using a combination of known and unknown functions in much the same way that the Lévy solution is developed for solutions to the equilibrium equation in section 16.4.3. A transverse deflection function for the plate is chosen in the form

$$\bar{u}_3(x_1, x_2) = \sum_{i=1}^{n} f_i(x_1) g_i(x_2), \qquad (16.173)$$

where it is assumed that $f_i(x_1)$ are *unknown functions* and $g_i(x_2)$ are *known functions*. Of course, it is also possible to reverse the role of the two sets of functions, selecting $g_i(x_2)$ as unknown functions and $f_i(x_1)$ as known functions.

To illustrate this approach, consider the plate shown in fig. 16.44, which is used to model an aircraft wing with a simple planform configuration. Axis $\bar{\imath}_1$ is in the span-wise direction and $\bar{\imath}_2$ is in the chordwise direction. The span of the wing is L; both leading and trailing edges are linearly tapered. The location of the leading edge is given by $c_\ell(x_1)$ and that of the trailing edge by $c_t(x_1)$. The wing is subjected to a transverse pressure distribution, $p_3(x_1, x_2)$, and a tip distributed shear force, $V_t(x_2)$.

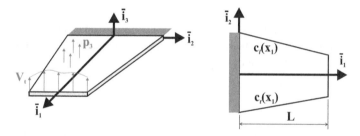

Fig. 16.44. Wing modeled as a thin flat plate.

The plate's transverse displacement field will be approximated in the following manner

$$\bar{u}_3(x_1, x_2) = \bar{u}(x_1) + x_2 \phi(x_1), \qquad (16.174)$$

where the unknown functions are $f_1(x_1) = \bar{u}(x_1)$ and $f_2(x_1) = \phi(x_1)$. The first unknown function represents the transverse deflection of the plate along axis $\bar{\imath}_1$ and the second defines the rotation of the chord line. The known functions are selected as $g_1(x_2) = 1$ and $g_2(x_2) = x_2$.

The transverse displacement field given by eq. (16.174) is expressed in terms of two unknown functions, $\bar{u}(x_1)$ and $\phi(x_1)$, that depend on variable x_1 only. The dependency of the solution on variable x_2 is explicitly assumed. Clearly, the procedure will lead to one-dimensional governing equations, which are the basis of a beam theory similar to that developed in chapter 5 based on the Euler-Bernoulli assumptions. Both bending and torsion are considered here since both displacements and rotations of the plate are considered.

The strain energy due to plate bending is given by substituting eq. (16.174) into eq. (16.147) to obtain

$$
A_b = \frac{1}{2} \int_0^L \int_{c_t}^{c_\ell} D \left[(\bar{u}'' + x_2 \phi'')^2 + 2(1 - \nu)\phi'^2 \right] \mathrm{d}x_1 \mathrm{d}x_2,
$$

where $(\cdot)'$ denotes a derivative with respect to x_1. Expanding the integrand leads to

$$
A_b = \frac{1}{2} \int_0^L \left\{ \left[\int_{c_t}^{c_\ell} D \mathrm{d}x_2 \right] \bar{u}''^2 + \left[\int_{c_t}^{c_\ell} D x_2^2 \mathrm{d}x_2 \right] \phi''^2 \right.
$$
$$
\left. + 2 \left[\int_{c_t}^{c_\ell} D x_2 \mathrm{d}x_2 \right] \bar{u}'' \phi'' + \left[2 \int_{c_t}^{c_\ell} D(1 - \nu) \mathrm{d}x_2 \right] \phi'^2 \right\} \mathrm{d}x_1.
$$

Each of the bracketed terms can be integrated over the wing's chord to find the following property distributions

$$
I(x_1) = \int_{c_t}^{c_\ell} D \, \mathrm{d}x_2, \qquad J(x_1) = 2 \int_{c_t}^{c_\ell} D(1 - \nu) \, \mathrm{d}x_2,
$$
$$
\alpha(x_1) = \int_{c_t}^{c_\ell} D x_2 \, \mathrm{d}x_2, \qquad \beta(x_1) = \int_{c_t}^{c_\ell} D x_2^2 \, \mathrm{d}x_2.
$$
(16.175)

$I(x_1)$ is the bending stiffness of beam, $J(x_1)$ the torsional stiffness, $\alpha(x_1)$ the bending-torsion coupling stiffness, and $\beta(x_1)$ the warping stiffness. With these definitions, the strain energy reduces to

$$
A_b = \frac{1}{2} \int_0^L \left(I \bar{u}''^2 + J \phi'^2 + 2\alpha \bar{u}'' \phi'' + \beta \phi''^2 \right) \mathrm{d}x_1.
$$
(16.176)

The potential of the externally applied loads consists of the negative work done by the transverse pressure loading and transverse tip shear,

$$
\Phi = - \int_0^L \int_{c_t}^{c_\ell} p_3 \left(\bar{u} + x_2 \phi \right) \mathrm{d}x_1 \mathrm{d}x_2 - \int_{c_t}^{c_\ell} V_t \left[\bar{u}(L) + x_2 \phi(L) \right] \mathrm{d}x_2.
$$

Here again, this expression can be integrated over the wing's chord and the potential of the externally applied loads reduces to

$$
\Phi = - \int_0^L \left(p \bar{u} + q \phi \right) \mathrm{d}x_1 - P_t \bar{u}(L) - Q_t \phi(L),
$$
(16.177)

where the following loading parameters are defined

$$
p(x_1) = \int_{c_t}^{c_\ell} p_3 \, \mathrm{d}x_2, \qquad q(x_1) = \int_{c_t}^{c_\ell} p_3 x_2 \, \mathrm{d}x_2,
$$
$$
P_t = \int_{c_t}^{c_\ell} V_t \, \mathrm{d}x_2, \qquad Q_t = \int_{c_t}^{c_\ell} V_t x_2 \, \mathrm{d}x_2.
$$
(16.178)

$p(x_1)$ is the applied load per unit span, $q(x_1)$ the applied torque per unit span, P_t the tip shear force, and Q_t the tip torque.

The principle of stationary total potential energy now implies the vanishing of the variations of the total potential energy, $\Pi = A_b + \Phi$, which leads to

$$\delta\Pi = \int_0^L [I\bar{u}''\delta\bar{u}'' + J\phi'\delta\phi' + \alpha\bar{u}''\delta\phi'' + \alpha\phi''\delta\bar{u}'' + \beta\phi''\delta\phi''] \, dx_1$$
$$- \int_0^L (p\delta\bar{u} + q\delta\phi) \, dx_1 - P_t\delta\bar{u}(L) - Q_t\delta\phi(L) = 0.$$

Next, integrations by parts are performed on all terms involving derivative of the variations to find

$$\delta\Pi = \int_0^L \left\{ \left[(I\bar{u}'' + \alpha\phi'')'' - p \right] \delta\bar{u} + \left[(\alpha\bar{u}'' + \beta\phi'')'' - (J\phi')' - q \right] \delta\phi \right\} dx_1$$
$$+ [(I\bar{u}'' + \alpha\phi'')\delta\bar{u}']_0^L + [(\alpha\bar{u}'' + \beta\phi'')\delta\phi']_0^L + [J\phi'\delta\phi]_0^L$$
$$- [(I\bar{u}'' + \alpha\phi'')'\delta\bar{u}]_0^L - [(\alpha\bar{u}'' + \beta\phi'')'\delta\phi]_0^L - P_t\delta\bar{u}(L) - Q_t\delta\phi(L) = 0.$$

Because all variations are arbitrary, the two bracketed terms in the first integral must vanish, revealing the governing equations of the problem,

$$(I\bar{u}'' + \alpha\phi'')' = p(x_1), \tag{16.179a}$$
$$(\alpha\bar{u}'' + \beta\phi'')'' - (J\phi')' = q(x_1). \tag{16.179b}$$

For the configuration shown in fig. 16.44, the wing is cantilevered at the root, and the appropriate geometric boundary conditions are $\bar{u}(0) = \phi(0) = \bar{u}'(0) = \phi'(0) = 0$; the corresponding variations vanish. At the tip of the wing, variations $\delta\bar{u}(L)$, $\delta\phi(L)$, $\delta\bar{u}'(L)$, and $\delta\phi'(L)$ are all arbitrary, leading to the following natural boundary conditions

$$-(I\bar{u}'' + \alpha\phi'')' = P_t, \quad -(\alpha\bar{u}'' + \beta\phi'')' + J\phi' = Q_t,$$
$$I\bar{u}'' + \alpha\phi'' = 0, \quad \alpha\bar{u}'' + \beta\phi'' = 0.$$

The approach described here can be applied to general configurations. Both isotropic or anisotropic plates can be treated, and the properties can vary over the surface of the plate. It is interesting to note that although very simple kinematic assumptions are made by selecting the displacement field in the form of eq. (16.174), the behavior of the plate under nonuniform torsion conditions is recovered. The governing equations derived here should be compared to those obtained in section 12.3.4.

Example 16.10. Flat plate model for a straight, uniform wing
A simple example of a partial approximation approach developed above will be treated here. Consider a straight, uniform wing similar to that shown in fig. 16.44. The wing has a constant chord, $c_t(x_1) = -c/2$ and $c_\ell(x_1) = c/2$, and is subjected to a tip load, P_t, and tip torque, Q_t.

The properties of the wing given by eqs. (16.175) now remain constant along its span,

$$I(x_1) = \int_{-c/2}^{+c/2} D\, dx_2 = cD = \frac{ch^3 E}{12(1-\nu^2)},$$

$$J(x_1) = 2\int_{-c/2}^{+c/2} D(1-\nu)\, dx_2 = 2cD(1-\nu) = \frac{ch^3 E}{6(1+\nu)} = \frac{1}{3}Gch^3,$$

$$\alpha(x_1) = \int_{-c/2}^{+c/2} Dx_2\, dx_2 = 0,$$

$$\beta(x_1) = \int_{-c/2}^{+c/2} Dx_2^2\, dx_2 = D\frac{c^3}{12} = \frac{c^3 h^3 E}{144(1-\nu^2)}.$$

Because $\alpha(x_1) = 0$ the equations of the problem decouple: eq. (16.179a) can be solved to find the transverse displacement field of the wing, and eq. (16.179b) yields the rotation field. For the bending problem, the geometric boundary conditions at the wing's root are $\bar{u} = \bar{u}' = 0$, and the natural conditions at the wing's tip are $-I\bar{u}''' = P_t$ and $I\bar{u}'' = 0$. The solution of the bending problem is

$$\bar{u}(x_1) = \frac{P_t L^3}{6I}\left[3\left(\frac{x_1}{L}\right)^2 - \left(\frac{x_1}{L}\right)^3\right],$$

which is simply the displacement field for a cantilevered beam subjected to a tip load, P_t.

The torsion problem is governed by eq. (16.179b). The geometric boundary conditions at the wing's root are $\phi = \phi' = 0$, and the natural conditions at the wing's tip are $-\beta\phi''' + J\phi' = Q_t$ and $\beta\phi'' = 0$. Integrating the governing equation yields $\beta\phi''' - J\phi' = C_1$ and the first boundary condition at the tip then leads to $C_1 = -Q_t$. Integrating this equation a second time and dividing by β results in $\phi'' - J\phi/\beta = (C_2 - Q_t)/\beta$. The solution to this second order nonhomogeneous ordinary differential equation is

$$\phi(x_1) = C_3 \cosh\lambda\frac{x_1}{L} + C_4 \sinh\lambda\frac{x_1}{L} + \frac{1}{J}(Qx_1 + C_2),$$

where $\lambda = L\sqrt{J/\beta} = \sqrt{24(1-\nu^2)}L/c$. The remaining three integration constants are determined from the remaining boundary conditions to find

$$\phi(x_1) = \frac{QL}{J}\frac{x_1}{L} - \frac{QL}{J}\left[\frac{\sinh\lambda - \sinh\lambda(1 - x_1/L)}{\lambda\cosh\lambda}\right].$$

The first term in this solution is Saint-Venant's solution for torsion of a beam with a thin rectangular cross-section. The method correctly identifies the torsional stiffness of the plate as $J = H_{11} = Gch^3/3$, see eq. (7.58). The second term represents the effect of the constrained warping of the cross-section at the root where the geometric boundary conditions are enforced. The present solution should be compared the the non-uniform torsion solution developed in section 12.3.6; $\beta = H_w$ is the warping stiffness of the thin rectangular strip.

16.6.7 Problems

Problem 16.18. Plate with two clamped and two free edges

The plate shown in fig. 16.45 has two adjacent edges clamped and the other two edges are free. Find an approximate solution of the problem using the following assumed deflection field: $\bar{u}_3(x_1, x_2) = \alpha_1 x_1^2 x_2^2 + \alpha_2 x_1^2 x_2^3 + \alpha_3 x_1^3 x_2^2$. Compare your solution with α_1 only, and with α_1, α_2, and α_3. Use $p_3(x_1, x_2) = p_0$, $b = 3a$, and $\nu = 0.3$.

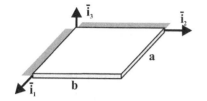

Fig. 16.45. Plate with two adjacent clamped edges and two free edges.

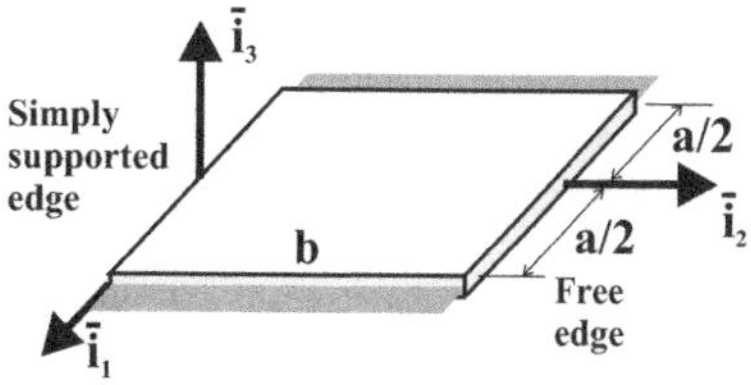

Fig. 16.46. Plate with two opposite edges clamped.

Problem 16.19. Plate with two opposite clamped edges

Consider a uniform plate with two opposite edges clamped, one simply supported edge, and a free edge, as depicted in fig. 16.46. Use an energy method to find an approximate solution of this problem with the following assumed deflection field: $\bar{u}_3(x_1, x_2) = a_1 \left[1 - (2x_1/a)^2\right]^2 (x_2/b)$. Use $p_3(x_1, x_2) = p_0$, $b = 3a$, and $\nu = 0.3$.

Problem 16.20. Plate with opposite edges clamped

Consider a uniform plate with two opposite edges clamped, one simply supported edge, and a free edge, as depicted in fig. 16.46. It is desired to use an energy method to find an approximate solution of this problem with the following assumed deflection field: $\bar{u}_3(x_1, x_2) = a_1 \left[1 - (2x_1/a)^2\right] (x_2/b)$. Is this a valid approach? Justify your answer.

Problem 16.21. Clamped plate under uniform loading

Consider a uniform rectangular plate of length a and width b, clamped along all four of its edges, and subjected to a uniform transverse pressure distribution p_0. The plate is made of a homogeneous, isotropic material with a bending stiffness, D. Use an energy method to solve this problem with the following assumed shape functions, $f_m(x_1) = 1/2 \left[\cos(m-1)\pi x_1/a - \cos(m+1)\pi x_1/a\right]$ and $g_n(x_2) = 1/2 \left[\cos(n-1)\pi x_2/b - \cos(n+1)\pi x_2/b\right]$. Use a Gauss-Legendre integration procedure to evaluate the stiffness matrix and load vector. (1) For $b/a = 3$, plot the distribution of non-dimensional transverse deflection, $D\bar{u}_3/(p_0 a^4)$, versus physical coordinates x_1 and x_2 on a plot that preserves the geometric aspect ratio.

Problem 16.22. Plate with three stiffeners

The simply supported, rectangular plate shown fig. 16.47 is of uniform thickness $h = a/10$, width a, and length $b = 3a$. It is made of aluminum (plate bending stiffness D) and subjected to a uniform transverse pressure p_0. The plate is reinforced by three equally spaced sets of back-to-back aluminum stiffeners of square cross-section $h \times h$. In *configuration 1*, the stiffeners run along the long direction of the plate, whereas in *configuration 2* they run along its short direction. *(1)* Use an energy method to investigate this problem with the following assumed displacement field: $\bar{u}_3(x_1, x_2) = \sum_{m,n} q_{mn} \sin(m\pi x_1/a) \sin(n\pi x_2/b)$. *(2)* For two configurations plot the distribution of non-dimensional transverse deflection, $D\bar{u}_3/(p_0 a^4)$, bending moments, $M_1/(p_0 a^2)$, $M_2/(p_0 a^2)$, and $M_{12}/(p_0 a^2)$, and shear force $Q_1/(p_0 a)$, $Q_2/(p_0 a)$, versus physical coordinates x_1 and x_2 on a plot that preserves the geometric aspect ratio. *(3)* For two configurations plot the distribution of bending moment, $M/(p_0 a^3)$ and torsional moment in the stiffeners, $Q/(p_0 a^3)$. *(4)* Compare the performance of *configurations 1* and 2 in terms of maximum load carrying capability and maximum deflection.

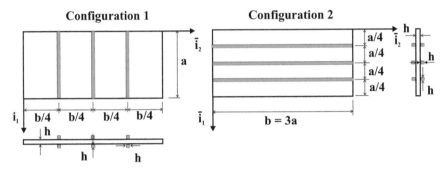

Fig. 16.47. Simply supported rectangular plate with three stiffeners.

Problem 16.23. Uniform clamped wing

Consider the uniform, anisotropic wing of constant chord length subjected to an aerodynamic pressure $p_3(x_1, x_2)$, as depicted in fig. 16.48. Use a partial approximate solution method to solve this problem with the following assumed deflection $\bar{u}_3(x_1, x_2) = \bar{u}(x_1) + \eta\phi(x_1)$, where $\eta = 2x_2/c$. The applied aerodynamic pressure is $p_3(x_1, x_2) = p_0 + \eta q_0$. The following two cases will be investigated: a homogeneous, aluminum plate of thickness $h = 1.5 \ 10^{-03}$ m, a laminated composite plate with lay-up $[0_{12}^\circ]$. Consider two aspect ratios $L/c = 5$ and 10. *(1)* Plot the distribution of transverse displacement $\bar{u}(x_1)$ for the four cases. *(2)* Plot the distribution of twist $\phi(x_1)$ for the four cases. NOTE: the bending stiffness matrix for the laminated composite plate with lay-up $[0_{12}^\circ]$ is

$$D = \begin{bmatrix} 39.0 & 0.76 & 0 \\ 0.76 & 2.53 & 0 \\ 0 & 0 & 2.00 \end{bmatrix}$$

16.7 Buckling of plates

Thin plates used in aerospace vehicles are often subjected to in-plane loads applied along their boundaries. These loads can arise from tensile, compressive, or shear forces applied along the edges of the plate and acting in its plane. For instance, upper and lower skins of an aircraft wing are thin plates subjected to in-plane loads. Similarly, fuselage panels are also subjected to in-plane loads.

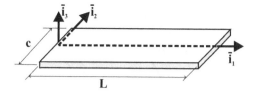

Fig. 16.48. Uniform clamped wing.

In the case of in-plane compressive or shear loads, the plate can buckle. This type of instability is similar to the buckling of beams that is addressed in chapter 14. For isotropic plates, it is shown in section 16.1.6 that the general Kirchhoff plate formulation decouples into two independent problems: an in-plane problem and a bending problem. Consequently, according to this theory, in-plane compressive or shear loads create in-plane displacements only. Clearly, the plate theory developed thus far is unable to predict the buckling phenomenon.

Similar observations are made in section 14.2 concerning Euler-Bernoulli beam theory. In its basic formulation, this theory is unable to predict the buckling of slender beams. In section 14.2.1, the equilibrium equations of a differential element of the beam are written for the deformed configuration of the element. The resulting transverse equilibrium equation, eq. (14.13), differs from its counterpart, eq. (5.38), written on the undeformed configuration of the differential element. The governing equation for beam bending resulting from this updated transverse equilibrium equation, eq. (14.15), depends on the axial force and is the basis for the analysis of the buckling phenomenon in beams.

A similar approach will be taken here to investigate plate buckling problems. It will be assumed that at the onset of buckling, large in-plane forces are present in the plate, but the transverse displacement and shear forces are still small. The transverse equilibrium equation of the plate will then be developed for the deformed configuration of a differential element of the plate. The resulting governing equations will then be applied to the analysis of plate buckling problems.

16.7.1 Equilibrium formulation

Equilibrium equations

Consider an infinitesimal element of the plate subjected to in-plane and transverse forces as well as bending moments. In contrast with the developments of section 16.1.3, a free body diagram of a *deformed* differential element of the plate will be analyzed. Figure 16.49 shows the deformed differential element of the plate subjected to the various loading components. At the onset of buckling, it is assumed that the in-plane forces are large, but the transverse displacement and shear forces are still small.

Because the differential element is in its deformed configuration, the in-plane forces N_1, N_2, and N_{12} all contribute to the transverse equilibrium equation. To simplify the development of the equilibrium equation, the two faces perpendicular to

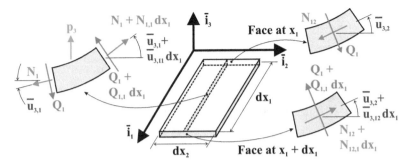

Fig. 16.49. Free body diagram of a deformed differential element of the plate.

axis $\bar{\imath}_1$ will be considered first, as shown in fig. 16.49. The force components acting on those two faces along axis $\bar{\imath}_3$ are

$$-Q_1 dx_2 \cos(\bar{u}_{3,1}) + (Q_1 + Q_{1,1} dx_1) dx_2 \cos(\bar{u}_{3,1} + \bar{u}_{3,11} dx_1)$$
$$-N_1 dx_2 \sin(\bar{u}_{3,1}) + (N_1 + N_{1,1} dx_1) dx_2 \sin(\bar{u}_{3,1} + \bar{u}_{3,11} dx_1)$$
$$-N_{12} dx_2 \sin(\bar{u}_{3,2}) + (N_{12} + N_{12,1} dx_1) dx_2 \sin(\bar{u}_{3,2} + \bar{u}_{3,12} dx_1).$$

At the onset of buckling, the transverse displacement is small, resulting in the following simplifications: $\cos(\bar{u}_{3,1}) \approx 1$ and $\sin(\bar{u}_{3,1}) \approx \bar{u}_{3,1}$. Neglecting higher order differential terms then leads to

$$\left[Q_{1,1} + N_1 \bar{u}_{3,11} + N_{1,1} \bar{u}_{3,1} + N_{12} \bar{u}_{3,12} + N_{12,1} \bar{u}_{3,2}\right] dx_1 dx_2.$$

Next, the two faces perpendicular to axis $\bar{\imath}_2$ will be considered and lead to a similar result. Finally, summing up all the forces acting in the vertical direction leads to the following equilibrium equation

$$\begin{aligned}
Q_{1,1} + N_1 \bar{u}_{3,11} + N_{12} \bar{u}_{3,12} + N_{1,1} \bar{u}_{3,1} + N_{12,1} \bar{u}_{3,2} \\
Q_{2,2} + N_2 \bar{u}_{3,22} + N_{12} \bar{u}_{3,12} + N_{2,2} \bar{u}_{3,2} + N_{12,2} \bar{u}_{3,1} + p_3 = 0.
\end{aligned} \tag{16.180}$$

It will be left to the reader to verify that at the onset of buckling, the in-plane force equilibrium equations, eqs. 16.23, are unchanged, even when written for a deformed element of the plate. Similarly, the moment equilibrium equations, eqs. (16.25), also remain unchanged at the onset of buckling.

The governing equation is obtained by introducing the shear forces, Q_1 and Q_2, from eqs. (16.25) into eq. (16.180) to find

$$\begin{aligned}
\frac{\partial^2 M_2}{\partial x_1^2} - 2\frac{\partial^2 M_{12}}{\partial x_1 \partial x_2} - \frac{\partial^2 M_1}{\partial x_2^2} = -p_3 \\
- \frac{\partial}{\partial x_1}\left(N_1 \frac{\partial \bar{u}_3}{\partial x_1} + N_{12}\frac{\partial \bar{u}_3}{\partial x_2}\right) - \frac{\partial}{\partial x_2}\left(N_{12}\frac{\partial \bar{u}_3}{\partial x_1} + N_2 \frac{\partial \bar{u}_3}{\partial x_2}\right).
\end{aligned} \tag{16.181}$$

This equation should be compared to the basic moment equilibrium equation of Kirchhoff plate theory, eq. (16.33). The terms appearing on the second line of eq. (16.181) give the effect of the large in-plane loads on the moment equilibrium equation of the plate.

Governing equations for plates with large in-plane loads

The governing equation for plate buckling is obtained by expressing the bending moment in terms of curvature using the constitutive laws, and the curvatures in terms of the transverse displacement using eq. (16.10). For linearly elastic, homogeneous isotropic plates, the constitutive laws are given by eqs. (16.30), and the governing equation of the problem becomes

$$D\nabla^4 \bar{u}_3 = p_3 + \frac{\partial}{\partial x_1}\left(N_1 \frac{\partial \bar{u}_3}{\partial x_1} + N_{12}\frac{\partial \bar{u}_3}{\partial x_2}\right) + \frac{\partial}{\partial x_2}\left(N_{12}\frac{\partial \bar{u}_3}{\partial x_1} + N_2 \frac{\partial \bar{u}_3}{\partial x_2}\right).$$
(16.182)

In the absence of in-plane loads, this equation reduces to the governing equation of Kirchhoff plates, eq. (16.37). The boundary conditions along a clamped or simply supported edges are unchanged. Along free edges, however, the natural boundary conditions must be expressed on a deformed configuration of the plate.

If the plate is made of a specially orthotropic material, the governing equation of the problem becomes

$$D_{11}\frac{\partial^4 \bar{u}_3}{\partial x_1^4} + 2(D_{12} + 2D_{66})\frac{\partial^4 \bar{u}_3}{\partial x_1^2 \partial x_2^2} + D_{22}\frac{\partial^4 \bar{u}_3}{\partial x_2^4} = p_3$$
$$+ \frac{\partial}{\partial x_1}\left(N_1 \frac{\partial \bar{u}_3}{\partial x_1} + N_{12}\frac{\partial \bar{u}_3}{\partial x_2}\right) + \frac{\partial}{\partial x_2}\left(N_{12}\frac{\partial \bar{u}_3}{\partial x_1} + N_2 \frac{\partial \bar{u}_3}{\partial x_2}\right).$$
(16.183)

It is convenient here again to introduce the affine transformation defined by eq. (16.94) to simplify the governing equation

$$\frac{\partial^4 \bar{u}_3}{\partial \hat{x}_1^4} + 2\bar{D}\frac{\partial^4 \bar{u}_3}{\partial \hat{x}_1^2 \partial \hat{x}_2^2} + \frac{\partial^4 \bar{u}_3}{\partial \hat{x}_2^4} = p_3 + \frac{\partial}{\partial \hat{x}_1}\left(\frac{N_1}{\sqrt{D_{11}}}\frac{\partial \bar{u}_3}{\partial \hat{x}_1} + \frac{N_{12}}{\sqrt[4]{D_{11}D_{22}}}\frac{\partial \bar{u}_3}{\partial \hat{x}_2}\right)$$
$$+ \frac{\partial}{\partial \hat{x}_2}\left(\frac{N_{12}}{\sqrt[4]{D_{11}D_{22}}}\frac{\partial \bar{u}_3}{\partial \hat{x}_1} + \frac{N_2}{\sqrt{D_{22}}}\frac{\partial \bar{u}_3}{\partial \hat{x}_2}\right).$$
(16.184)

where $\bar{D}$ is defined by eq. (16.96). In the affine space, the plate dimensions are $\hat{a} = a/\sqrt[4]{D_{11}}$ and $\hat{b} = b/\sqrt[4]{D_{22}}$.

Example 16.11. Buckling of a simply supported plate
Consider a rectangular plate of length a and width b simply supported along its four edges, as depicted in fig. 16.50. A uniform in-plane compressive load , $N_1 = -N_0$, is acting on two opposite edges of the plate. For this case, the governing equation, eq. (16.182), reduces to

$$D\nabla^4 \bar{u}_3 + N_0 \frac{\partial^2 \bar{u}_3}{\partial x_1^2} = 0.$$
(16.185)

The boundary conditions along the loaded edges at $x_1 = 0$ and $x_1 = b$ are $\bar{u}_3 = \bar{u}_{3,11} = 0$, while the boundary conditions along the unloaded edges at $x_2 = 0$ and $x_2 = b$ are $\bar{u}_3 = \bar{u}_{3,22} = 0$.

The solution of the problem is assumed to be in the form of a double infinite series identical to that used for Navier's solution, see eq. (16.83), where $\alpha_m = m\pi/a$ and

$\beta_n = n\pi/b$. The trigonometric functions satisfy all the boundary conditions of the problem. Substituting this expansion into the governing equation, eq. (16.185), leads to

$$\sum_{m,n=1}^{\infty} \left[D \left(\alpha_m^2 + \beta_n^2 \right)^2 - N_0 \alpha_m^2 \right] q_{mn} \sin \alpha_m x_1 \sin \beta_n x_2 = 0.$$

This equation must be satisfied for all values of x_1 and x_2. This is only possible if the bracketed term vanishes, leading to the critical in-plane load

$$N_0 = N_{cr} = k_c \frac{\pi^2 D}{b^2}, \qquad (16.186)$$

where the buckling parameter, k_c, is defined as

$$k_c = \left[\left(\frac{m}{a/b} \right) + n^2 \left(\frac{a/b}{m} \right) \right]^2. \qquad (16.187)$$

The critical load depends on the plate aspect ratio, a/b, but also on the for wave numbers, m and n. To find the buckling load, the lowest critical load must be found. As n increase, so does buckling parameter and the critical load. Hence, the lowest critical load is found for $n = 1$, which corresponds to a single sine wave across the width of the plate.

The buckling parameter still depends on the second wave number, m. Because it appears both in the numerator of the first term and the denominator of the second, the wave number m that yields the lowest value of the buckling parameter is not easily found. Figure 16.51 shows the variation of the buckling parameter as a function of the plate aspect ratio, a/b, for increasing values of the wave number.

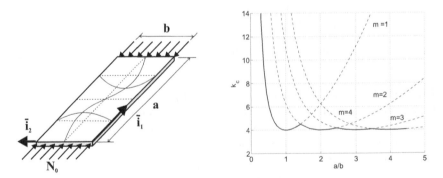

Fig. 16.50. Simply supported plate buckling under in-plane loading with $m = 2$ and $n = 1$ wave numbers.

Fig. 16.51. Bucking parameter versus plate's aspect ratio.

For aspects ratio $a/b \leq \sqrt{2}$, the lowest value of the buckling parameter is obtained for $m = 1$, which corresponds to a single wave along the loading direction.

For aspect ratios $\sqrt{2} < a/b \leq \sqrt{6}$, however, the lowest value of the buckling parameter is obtained for $m = 2$, which corresponds to two waves along the loading direction. The scalloped curve shown with a solid line in fig. 16.51 gives the lowest value of the buckling parameter as a function of the plate's aspect ratio. For $a/b > 1$, the buckling parameter remains nearly constant, $k_c \approx 4$, and the corresponding buckling load is

$$N_{\mathrm{cr}} = 4\frac{\pi^2 D}{b^2}. \tag{16.188}$$

Although the buckling load remains nearly constant, the buckling mode shape is a function of the plate's aspect ratio. m sine waves appear along the loading direction for plates with aspects ratio that satisfy the following inequality

$$\sqrt{m(m-1)} < a/b < \sqrt{m(m+1)}.$$

Figure 16.50 illustrates the case of $m = 2$. It is often convenient to express the buckling load as a buckling stress

$$\sigma_{\mathrm{cr}} = \frac{N_{\mathrm{cr}}}{h} = \frac{\pi^2}{3(1-\nu^2)} E \left(\frac{h}{b}\right)^2, \tag{16.189}$$

which may be easier to use in certain design calculations.

At small aspect ratios, the plate becomes a wide strip loaded across its long sides. In this case, $k_c \rightarrow 1/(a/b)^2$, and $N_{\mathrm{cr}} = \pi^2 D/a^2$. Introducing the plate bending stiffness given by eq. (16.36), leads to

$$P_{\mathrm{cr}} = bN_{\mathrm{cr}} = \frac{1}{1-\nu^2} \frac{\pi^2 H_{22}^c}{a^2} = \frac{P_{\mathrm{Euler}}}{1-\nu^2}.$$

If the plate is approximated by a simply supported beam, its buckling load is given by eq. (14.28) as $P_{\mathrm{Euler}} = \pi^2 H_{22}^c/a^2$, where $H_{22}^c = Ebh^3/12$ is the bending stiffness of the beam. This result is generally referred to as the "wide beam" formula; the corrective factor, $1/(1-\nu^2)$, is due to the fact that the plate is simply supported along its four edges, whereas the beam is only supported at its two end points.

Example 16.12. Simply supported plate under pressure and in-plane loading

Consider a rectangular plate of length a and width b simply supported along its four edges, as depicted in fig. 16.50. A uniform in-plane load, N_1, is acting on two opposite edges of the plate, which is also subjected to a uniform transverse pressure. The in-plane load could be tensile of compressive.

The solution of the problem is assumed to be in the form of a double infinite series identical to that used for Navier's solution, see eq. (16.83), where $\alpha_m = m\pi/a$ and $\beta_n = n\pi/b$. The trigonometric functions satisfy all the boundary conditions of the problem. The uniform transverse pressure distribution, $p_3 = p_0$, is expanded in Fourier series to find the loading coefficients, p_{mn}, given by eq. (16.91).

Substituting the assumed solution and the pressure expansion into the governing equation, eq. (16.182), leads to

$$\sum_{m,n=1}^{\infty} q_{mn} \left[D \left(\alpha_m^2 + \beta_n^2 \right)^2 + N_1 \alpha_m^2 \right] \sin \alpha_m x_1 \sin \beta_n x_2$$

$$= \sum_{m,n=\text{odd}}^{\infty} \frac{16 p_0}{\pi^2 mn} \sin \alpha_m x_1 \sin \beta_n x_2.$$

For even values of m and n, the right hand side is zero, and the corresponding solution is $q_{mn} = 0$. For odd values, the equation reduces to

$$\sum_{m,n=\text{odd}}^{\infty} \left[q_{mn} \left(D \left(\alpha_m^2 + \beta_n^2 \right)^2 + N_1 \alpha_m^2 \right) - \frac{16 p_0}{\pi^2 mn} \right] \sin \alpha_m x_1 \sin \beta_n x_2 = 0.$$

This equation must be satisfied for all values of x_1 and x_2 and hence, the bracketed term must vanish. It is then possible to solve for the unknown coefficients, q_{mn}, which can be substituted back into the assumed solution, eq. (16.83), to yield

$$\bar{u}_3(x_1, x_2) = \frac{16}{\pi^6} \frac{p_0 b^4}{D} \sum_{m,n=\text{odd}}^{\infty} \frac{1}{mn} \frac{\sin \alpha_m x_1 \sin \beta_n x_2}{\left[\left(\dfrac{m}{a/b} \right)^2 + n^2 \right]^2 + \dfrac{N_1 b^2}{\pi^2 D} \left(\dfrac{m}{a/b} \right)^2}$$

As expected, when $N_1 = 0$, this solution is identical to Navier's solution given by eq. (16.92). If an in-plane tensile load is applied to the plate, i.e., if $N_1 > 0$, the denominator increases and the plate's transverse displacement decreases. The in-plane tensile load effectively stiffens the plate. On the other hand, if an in-plane compressive load is applied to the plate, i.e., if $N_1 < 0$, the denominator decreases and the plate's transverse displacement increases. The in-plane compressive load effectively softens the plate. In fact, when the denominator vanishes, the transverse displacement increases without bound. This occurs when $N_1 = N_{\text{cr}}$, where N_{cr} is given by eq. (16.188).

Finally, it should be noted that the solution presented here is valid only for small transverse displacements, $\bar{u}_3 \ll h$, during which the in-plane force, N_1, remains uniform throughout the plate. For larger deflections, the fully nonlinear, coupled bending and stretching equations must be used to solve the problem.

Example 16.13. Buckling of a simply supported anisotropic plate

Consider a rectangular plate of length a and width b made of a specially orthotropic material and simply supported along its four edges, as depicted in fig. 16.50. A uniform in-plane compressive load , $N_1 = -N_0$, is acting along two opposite edges of the plate. For this case, the governing equation in the affine space, eq. (16.184), reduces to

$$\bar{u}_{3,1111} + 2\bar{D}\bar{u}_{3,1122} + \bar{u}_{3,2222} + \frac{N_0}{\sqrt{D_{11}}} \bar{u}_{3,11} = 0.$$

where the partial derivatives are with respect to the affine variables $\hat{x}_1$ and $\hat{x}_2$, defined by eq. (16.94). The affine dimensions of the plate are $\hat{a} = a/\sqrt[4]{D_{11}}$ and $\hat{b} = b/\sqrt[4]{D_{11}}$.

The transverse displacement field is assumed to be in the form of a double infinite series expansion of sine functions

$$\bar{u}_3(\hat{x}_1, \hat{x}_2) = \sum_{m,n=1}^{\infty} q_{mn} \sin \hat{\alpha}_m \hat{x}_1 \sin \hat{\beta}_n \hat{x}_2,$$

where $\hat{\alpha}_m = m\pi/\hat{a}$ and $\hat{\beta}_n = n\pi/\hat{b}$. Each term of this expansion satisfies all the boundary conditions of the problem. Introducing this solution into the governing equation leads to

$$\sum_{m,n=1}^{\infty} \left[\hat{\alpha}_m^4 + 2\bar{D}\hat{\alpha}_m^2 \hat{\beta}_n^2 + \hat{\beta}_n^4 - \frac{N_0}{\sqrt{D_{11}}} \hat{\alpha}_m^2 \right] q_{mn} \sin \hat{\alpha}_m \hat{x}_1 \sin \hat{\beta}_n \hat{x}_2 = 0.$$

Because this equation must vanish for all values of $\hat{x}_1$ and $\hat{x}_2$, the bracketed term must vanish, leading to the following critical load

$$N_0 = N_{\text{cr}} = \frac{\pi^2 \sqrt{D_{11} D_{22}}}{\hat{b}^2} \left[2\bar{D}n^2 + \left(\frac{m}{\hat{a}/\hat{b}} \right)^2 + \left(\frac{\hat{a}/\hat{b}}{m} \right)^2 n^4 \right].$$

The critical load is a function of the affine aspect ratio of the plate, $\hat{a}/\hat{b}$, and of the wave numbers, m and n. To determine the buckling load, the lowest critical load must be found. Because the critical load is an increasing function of the wave number n, the lowest critical load is obtained for $n = 1$,

$$N_{\text{cr}} = (2\bar{D} + k_c) \frac{\pi^2 \sqrt{D_{11} D_{22}}}{\hat{b}^2},$$

where the buckling parameter, k_c, is defined as

$$k_c = \left(\frac{m}{\hat{a}/\hat{b}} \right)^2 + \left(\frac{\hat{a}/\hat{b}}{m} \right)^2.$$

Figure 16.52 shows the variation of this buckling parameter as a function of the plate's affine aspect ratio, $\hat{a}/\hat{b}$. For $\hat{a}/\hat{b} > 2$, the buckling parameter remains nearly constant, $k_c \approx 2$, and the buckling load

$$N_{\text{cr}} = 2(\bar{D} + 1) \frac{\pi^2 \sqrt{D_{11} D_{22}}}{\hat{b}^2},$$

Although the buckling load remains nearly constant, the buckling mode shape is a function of the plate's aspect ratio. m sine waves appear along the loading direction for plates with aspects ratio that satisfy the following inequality $\sqrt{m(m-1)} < \hat{a}/\hat{b} < \sqrt{m(m+1)}$. It is left to the reader to verify that for an isotropic plate, the results of example 16.11 are recovered.

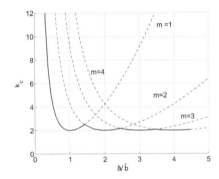

Fig. 16.52. Buckling parameter versus plate's affine aspect ratio.

16.7.2 Energy formulation

The examples presented in the previous section demonstrate the difficulty of determining the buckling loads of plates. For realistic problems, closed form solutions cannot be obtained, and energy based solution procedures providing approximations to the buckling load are desirable.

The principle of minimum total potential energy for plate bending problems is presented in section 16.6.4. The total strain energy in the plate is given by eqs. (16.145), (16.146), or (16.147), for anisotropic, specially orthotropic, or isotropic plates, respectively. The potential of the externally applied transverse pressure is given by eq. (16.148b).

Work done by in-plane forces

If the plate is subjected to large in-plane forces, the associated potential must be added to the potential of the externally applied loads. In section 14.2.5, the work done by large axial forces applied to beams is evaluated and is given by eq. (14.37). Following a similar reasoning for plates, the potential of large in-plane forces is found to be

$$
\begin{aligned}
\Phi &= \frac{1}{2} \int_{\mathcal{S}_m} \left[N_1 \bar{u}_{3,1}^2 + 2 N_{12} \bar{u}_{3,1} \bar{u}_{3,2} + N_2 \bar{u}_{3,2}^2 \right] \mathrm{d}\mathcal{S}_m \\
&= \frac{1}{2} \int_{\mathcal{S}_m} \underline{g}^T \underline{\underline{N}}\, \underline{g}\, \mathrm{d}\mathcal{S}_m,
\end{aligned}
\tag{16.190}
$$

where $\underline{g}$ is the *array of displacement gradients* defined as

$$
\underline{g} = \begin{Bmatrix} \bar{u}_{3,1} \\ \bar{u}_{3,2} \end{Bmatrix},
\tag{16.191}
$$

and $\underline{\underline{N}}$ the *matrix of in-plane loads*,

$$
\underline{\underline{N}} = \begin{bmatrix} N_1 & N_{12} \\ N_{12} & N_2 \end{bmatrix}.
\tag{16.192}
$$

Solution procedure

A general procedure for obtaining approximate solutions of plate problems is described in section 16.6.5 and starts with an assumed transverse displacement field such as that given by eq. (16.151). The stiffness matrix can then be written in a generic manner as eq. (16.157), and the load array as eq. (16.159).

To address buckling problems, it is also necessary to obtain an expression for the potential of the large in-plane forces. First, the array of displacement gradients is interpolated by introducing the assumed solution, eq. (16.151), into eq. (16.191) to find

$$\underline{g} = \underline{\underline{G}}\,\underline{q}, \tag{16.193}$$

where $\underline{\underline{G}}$ is the *displacement gradient interpolation matrix*,

$$\underline{\underline{G}} = \begin{bmatrix} \cdots & \dfrac{\partial h_i(x_1, x_2)}{\partial x_1} & \cdots \\ \cdots & \dfrac{\partial h_i(x_1, x_2)}{\partial x_2} & \cdots \end{bmatrix}. \tag{16.194}$$

Next, the potential of large in-plane forces, eq. (16.190), becomes

$$\Phi = \frac{1}{2} \int_{\mathcal{S}_m} (\underline{\underline{G}}\,\underline{q})^T \underline{\underline{N}}(\underline{\underline{G}}\,\underline{q})\, \mathrm{d}\mathcal{S}_m$$
$$= \frac{1}{2}\, \underline{q}^T \left[\int_{\mathcal{S}_m} \underline{\underline{G}}^T \underline{\underline{N}}\, \underline{\underline{G}}\, \mathrm{d}\mathcal{S}_m \right] \underline{q} = \frac{1}{2}\, \underline{q}^T \underline{\underline{K}}_G\, \underline{q}, \tag{16.195}$$

where the *geometric stiffness matrix* is defined as

$$\underline{\underline{K}}_G = \int_{\mathcal{S}_m} \underline{\underline{G}}^T \underline{\underline{N}}\, \underline{\underline{G}}\, \mathrm{d}\mathcal{S}_m. \tag{16.196}$$

The formulation developed here parallels that presented for the analysis of beam buckling problems in section 14.2.5. The geometric stiffness matrix for beam problems defined by eq. (14.45) should be compared to that for plate problems, eq. (16.196).

Example 16.14. Buckling of a plate with one free edge and the others simply supported

Consider an isotropic plate of length a a width b simply supported along three edges and free along the fourth (at $x_2 = b$), as depicted in fig. 16.53. A uniform compressive in-plane loading is applied along two opposite simply supported edges. Intuitively, the buckling load for this configuration should be much lower than that for a similar plate simply supported along all four edges.

This problem will be treated using the energy approach described in section 16.7.2. The transverse displacement field is assumed to be of the following form

$$\bar{u}_3(x_1, x_2) = \sum_{m=1}^{\infty} q_m x_2 \sin \alpha_m x_1, \tag{16.197}$$

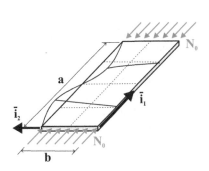

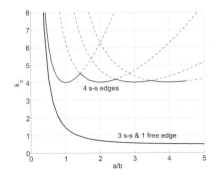

Fig. 16.53. Plate with three edges simply supported and the fourth side edge free.

Fig. 16.54. Buckling parameter versus the plate's aspect ratio ($\nu = 0.25$).

where $\alpha_m = m\pi/a$ and q_m are the degrees of freedom of the problem. Each of the shape functions satisfies the geometric boundary conditions of the problem. A single shape function, x_2, is selected to described the displacement field across the width of the plate; it corresponds to a rigid body rotation of the plate about the simply supported edge along $x_2 = 0$.

The total strain energy in the plate is given by eq. (16.147) and introducing the assumed displacement field, eq. (16.197), it becomes

$$A_b = \frac{D}{2} \sum_{m,n=1}^{\infty} q_m q_n \alpha_m^2 \alpha_n^2 \int_0^b x_2^2 \, dx_2 \int_0^a \sin \alpha_m x_1 \sin \alpha_n x_1 \, dx_1$$

$$+ D(1-\nu)b \sum_{m,n=1}^{\infty} q_m q_n \alpha_m \alpha_n \int_0^a \cos \alpha_m x_1 \cos \alpha_n x_1 \, dx_1.$$

The orthogonality properties of the sine and cosine functions given eqs. (A.45a) and (A.45b), respectively, readily enable the evaluation of the integrals, and the total strain energy in the plate reduces to

$$A_b = \frac{D}{2} \sum_{m=1}^{\infty} q_m^2 \left[\alpha_m^4 \frac{ab^3}{6} + 2(1-\nu)\alpha_m^2 \frac{ab}{2} \right].$$

The potential of the in-plane forces in the plate is given by eq. (16.190) with $N_1 = -N_0$. Introducing the assumed displacement field, eq. (16.197), and integrating over the plate gives

$$\Phi = -\frac{N_0}{2} \sum_{m=1}^{\infty} q_m^2 \alpha_m^2 \frac{ab^3}{6}.$$

The principle of stationary total potential energy implies the vanishing of the derivatives of the total potential energy, $\Pi = A_b + \Phi$, with respect to the degrees of freedom of the problem, leading to

$$\left\{ D \left[\alpha_m^4 \frac{ab^3}{6} + 2(1-\nu)\alpha_m^2 \frac{ab}{2} \right] - N_0 \alpha_m^2 \frac{ab^3}{6} \right\} q_i = 0, \quad \text{for } i = 1, 2, \ldots \infty,$$

This forms a set of homogeneous linear equations for the degrees of freedom. The trivial solution is $q_i = 0$. For a non-trivial solution to exist, the determinant of the system must vanish, leading to the following critical load

$$N_0 = N_{\text{cr}} = \frac{\pi^2 D}{b^2} \left[\left(\frac{m}{a/b} \right)^2 + \frac{6(1-\nu)}{\pi^2} \right].$$

Here again, the critical load is a function of the plate's aspect ratio, a/b, and of the wave number m. Because the critical load is an increasing of the wave number, the buckling load is found for $m = 1$, which corresponds to a single sine wave along the length of the plate,

$$N_{\text{cr}} = k_c \frac{\pi^2 D}{b^2},$$

where the buckling parameter, k_c, is defined as

$$k_c = \frac{6(1-\nu)}{\pi^2} + \left(\frac{1}{a/b} \right)^2.$$

Figure 16.54 shows the variation of the buckling parameter as a function of the plate's aspect ratio for $\nu = 0.25$. For reference, the figure also shows the buckling parameter for a comparable plate simply supported along all four edges, see example 16.11.

As expected, the presence of the free edge significantly reduces the buckling load of plates. For high aspect ratios, the buckling parameter becomes $k_c \approx 6(1-\nu)/\pi^2$. The buckling stress then becomes

$$\sigma_{\text{cr}} = \frac{N_{\text{cr}}}{h} = \frac{E}{2(1+\nu)} \left(\frac{h}{b} \right)^2 = G \left(\frac{h}{b} \right)^2,$$

where G is the shear modulus of the material.

To illustrate the importance of these results when analyzing the buckling of beams with free edges, consider a thin-walled beam of length L with the cross-section shown in fig. 8.52 on page 347, which combines a closed trapezoidal box and overhanging rectangular strips. The overhanging portions of the cross-section can be approximated as plates simply supported along three edges and free along the fourth. If the simply supported beam is subjected to compressive loads, two buckling modes are possible.

The first buckling mode is the buckling of the overall beam discussed in section 14.2.2. It is characterized by the Euler buckling load, $P_{\text{Euler}} = \pi^2 H_{22}^c / L^2$, where H_{22}^c is the bending stiffness of the cross-section; the buckling mode shape is a single sine wave over the entire length of the beam.

The second buckling mode is the buckling of the overhanging portions as discussed in this example. The corresponding buckling stress is $\sigma_{\text{cr}} = Gt^2/w^2$, where t and w are the thickness and width of the overhang, respectively. This buckling mode is called *local crippling* because it only involves the overhang, in contrast with the Euler buckling mode which involves the entire beam.

Depending on the design of the beam, either of the two buckling modes could have the lowest buckling load. The local crippling load can be increased by adding a lip at the free edge of the overhanging portion of the section to create a stiffener that provides some small measure of support at the free edge.

Local crippling may also arise from purely bending loads if they produce significant compressive stresses in portions of the cross-section with a free edge. To illustrate this, assume now that the cross-section shown in fig. 8.52 is subjected to bending and that the overhanging portion of the section is under compressive stress. Euler buckling is no longer a buckling mode because the overall beam is no longer subjected to a compressive axial force. However, local crippling, may still occur, depending on the level of compressive stress induced in the free edge portion due to bending.

Example 16.15. Buckling of a plate with clamped loaded edges and simply supported sides

In examples 16.11 and 16.13, the rectangular plate is assumed to be simply supported along its four edges. Consider now an isotropic plate of length a and width b subjected to a uniform compressive load, $N_1 = -N_0$. The plate is clamped along the two opposite edges where the compressive load is applied, and is simply supported along the other two edges.

The following transverse displacement field will be assumed for this problem,

$$\bar{u}_3(x_1, x_2) = \sum_{m=1}^{\infty} q_m \left[\cos \frac{(m-1)\pi x_1}{a} - \cos \frac{(m+1)\pi x_1}{a} \right] \sin \frac{\pi x_2}{b}. \quad (16.198)$$

A sinusoidal shape function is assumed across the width of the plate because it satisfies the geometric boundary conditions at the simply supported edges. It is shown in example 16.11 that for a plate simply supported along its four edges, the lowest critical load is found for a wave number $n = 1$ across the width of the plate. It seems reasonable to select that single mode shape for the problem at hand. Along the length of the plate, the shape functions must satisfy the geometric boundary conditions associated with clamped edges. The shape functions given by eq. (16.164) are selected here.

The series solution defined by eq. (16.198) will involve complex integrals of products of cosine functions and leads to a fully populated stiffness matrix. This approach would be practical is a Gauss-Legendre numerical procedure is used to evaluate the entries of the stiffness matrix, as discussed in example 16.9.

An easier approach is to select a single term of the sum indicated in eq. (16.198). The analysis can be perform for a generic wave number, m. This approach is expected to be less accurate than that using the entire summation, but should provide preliminary predictions for this problem.

The total strain energy for an isotropic plate is given by eq. (16.147), and the potential of the in-plane loads by eq. (16.190). Introducing a single term of the summation expressed by eq. (16.198) leads to the total potential energy of the system

$$\Pi = \frac{\pi^2 q_m^2}{4a^3 b^3} \left\{ \pi^2 D \left[a^4 + 2a^2 b^2 (1 + m^2) + (1 + 6m^2 + m^4) b^4 \right] \right.$$
$$\left. - N_0 a^2 b^4 (1 + m^2) \right\}$$

The principle of minimum total potential energy requires the total potential energy to be a minimum with respect to the choice of the degree of freedom. This is achieved by imposing the vanishing of the derivative of Π with respect to q_m, resulting in a single homogeneous equation for the single degree of freedom, q_m. A nontrivial solution exists only if the coefficient of this homogeneous equation vanishes, leading to the following critical load, $N_0 = N_{\mathrm{cr}} = k_c \pi^2 D / b^2$, where the buckling parameter, k_c, is defined as

$$k_c = \frac{m^4 + [1 + (a/b)^2]^2 + 2m^2 [3 + (a/b)^2]}{(1 + m^2)(a/b)^2}. \tag{16.199}$$

Figure 16.55 shows the buckling parameter as a function of the plate's aspect ratio for $m = 1$ to 4. The buckling load is determined by the lowest value of the buckling parameter. Here again, as the plate's aspect ratio increases, an increasing number of waves appears along the length of the plate.

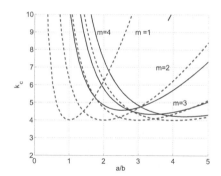

Fig. 16.55. Buckling factor, k_c, versus the plate's aspect ratio.

For reference, the buckling parameter for a comparable plate simply supported along its four edges is also shown in dash-dotted lines in the same figure. Clearly, the effect of the clamped edges is greatest for small aspect ratio plates, but becomes negligible for aspect ratios greater than about 4.

Example 16.16. Buckling of a plate under shear loading
Consider a specially orthotropic plate simply supported along all four edges and subjected to an in-plane shear load, N_{12}. As this shear load increases, a critical level is reached for which the plate buckles. This phenomenon will also occur in thin-walled beams subjected to large shear flows. For instance, the vertical web of a deep "I" beam carries most of the shear force applied to the beam. If the web thickness is reduced too much to lighten the structure, the web could buckle under the action of in-plane shear loading.

The transverse displacement field is assumed to be in the form of a double infinite series expansion of sine functions

$$\bar{u}_3(\hat{x}_1, \hat{x}_2) = \sum_{m,n=1}^{\infty} q_{mn} \sin \hat{\alpha}_m \hat{x}_1 \sin \hat{\beta}_n \hat{x}_2,$$

where $\hat{\alpha}_m = m\pi/\hat{a}$ and $\hat{\beta}_n = n\pi/\hat{b}$. Each term of the series satisfies the geometric boundary conditions along the four edges of the plate. The affine variables, $\hat{x}_1$ and $\hat{x}_2$, are given by eq. (16.94) and the affine dimension of the plate are $\hat{a} = a/\sqrt[4]{D_{11}}$ and $\hat{b} = b/\sqrt[4]{D_{22}}$.

The total strain energy stored in the plate is found by introducing the assumed displacement field into eq. (16.146). The orthogonality properties of the sine functions, see section A.4, are used to evaluate the integrals, leading to

$$A_b = \frac{1}{2} \frac{\hat{a}\hat{b}}{4} \sum_{m=1}^{\infty} \sum_{n=1}^{\infty} \left(\hat{\alpha}_m^4 + \hat{\beta}_n^4 + 2\bar{D}\hat{\alpha}_m^2 \hat{\beta}_n^2 \right) q_{mn}^2,$$

where $\bar{D}$ is defined in eq. (16.96).

Similarly, the potential of the in-plane shear loading is obtained by introducing assumed displacement field into eq. (16.190) to obtain

$$\Phi = \frac{1}{2} \sum_{m,n=1}^{\infty} \sum_{k\ell=1}^{\infty} \frac{2N_0}{\sqrt[4]{D_{11}D_{22}}} q_{mn} q_{k\ell}$$

$$\hat{\alpha}_m \hat{\beta}_n \int_0^{\hat{a}} \cos \hat{\alpha}_m \hat{x}_1 \sin \hat{\alpha}_k \hat{x}_1 \mathrm{d}\hat{x}_1 \int_0^{\hat{b}} \cos \hat{\beta}_m \hat{x}_2 \sin \hat{\beta}_\ell \hat{x}_2 \mathrm{d}\hat{x}_2.$$

Due to orthogonality between the sine and cosine functions, the first integral vanishes unless $m + k =$ odd and the second integral vanishes unless $n + \ell =$ odd, leading to the following expression for the total potential energy

$$\Pi = \frac{1}{2} \frac{ab}{4} \sum_{m,n=1}^{\infty} \left(\hat{\alpha}_m^4 + \hat{\beta}_n^4 + 2\bar{D}\hat{\alpha}_m^2 \hat{\beta}_n^2 \right) q_{mn}^2$$

$$+ \frac{1}{2} \frac{ab}{4} \sum_{m+k=\text{odd}} \sum_{n+\ell=\text{odd}} q_{mn} q_{k\ell} \frac{mnk\ell}{(k^2 - m^2)(n^2 - \ell^2)} \frac{32N_0}{\hat{a}\hat{b}\sqrt[4]{D_{11}D_{22}}}.$$

The principle of minimum total potential energy requires the vanishing of the derivatives of the total potential energy with respect to the degrees of freedom, $\partial \Pi/\partial q_{ij} = 0$. This results in

$$\frac{ab}{4} \left[\left(\hat{\alpha}_i^4 + \hat{\beta}_j^4 + 2\bar{D}\hat{\alpha}_i^2 \hat{\beta}_j^2 \right) q_{ij} \right.$$

$$\left. + \frac{32N_0}{ab} \sum_{m+i=\text{odd}} \sum_{n+j=\text{odd}} \frac{mnij}{(i^2 - m^2)(n^2 - j^2)} q_{mn} \right] = 0.$$

These equations form a set of linear, homogeneous equations for the unknown coefficients, q_{ij}. Since indices i and j take on values up to infinity, the system is of
infinite size. This systems admits a trivial solution, $q_{ij} = 0$ for all values of i and j.
For a nontrivial solution to exist, the determinant of the system must vanish. To ease
the evaluation of this determinant, the following non-dimensional form of the system
is developed first,

$$\left[i^4 + (\hat{\lambda}j)^4 + 2\bar{D}i^2(\hat{\lambda}j)^2 \right] q_{ij} + \bar{N}_0\hat{\lambda}^2 \sum_{\substack{m+i=\\ \text{odd}}}^{\infty} \sum_{\substack{n+j=\\ \text{odd}}}^{\infty} \frac{mnijq_{mn}}{(i^2 - m^2)(n^2 - j^2)} = 0,$$

where $\hat{\lambda} = \hat{a}/\hat{b}$ is the plate affine aspect ratio, and $\bar{N}_0 = 32N_0ab/(\pi^4\sqrt{D_{11}D_{22}})$
the non-dimensional load parameter.

In practice, it is not possible to evaluate the determinant of a matrix of infinite
size, and hence, a finite number of terms must be used in the expansion. It can be
shown, however, that the complete set of equations decouples into two independent
sets corresponding to $i + j =$ odd and $i + j =$ even. Furthermore, the lowest critical
loads for low aspect ratios are found for $i + j =$ even.

Keeping only two terms in the expansion, a 2×2 system of equations is obtained,

$$\begin{bmatrix} 1 + \hat{\lambda}^4 + 2\bar{D}\hat{\lambda}^2 & -4\bar{N}_0\hat{\lambda}^2/9 \\ -4\bar{N}_0\hat{\lambda}^2/9 & 16(1 + \hat{\lambda}^4 + 2\bar{D}\hat{\lambda}^2) \end{bmatrix} \begin{Bmatrix} q_{11} \\ q_{22} \end{Bmatrix} = 0,$$

which corresponds to $(i, j) = (1, 1)$ and $(2, 2)$. The critical load is found by imposing
the vanishing of the determinant, leading to $16(1 + \hat{\lambda}^4 + 2\bar{D}\hat{\lambda}^2)^2 = 16\bar{N}_0^2\hat{\lambda}^4/81$.
The critical load becomes $\bar{N}_0 = \pm9(1 + \hat{\lambda}^4 + 2\bar{D}\hat{\lambda}^2)/\hat{\lambda}^2$. The two solutions of
equal magnitude indicate that the plate behaves in the same manner for positive or
negative shear loading, as expected. When transformed back to dimensional form,
the buckling load becomes $N_{cr} = k_c\sqrt{D_{11}D_{22}}/(ab)$, where the bucking parameter,
k_c, is defined as

$$k_c = \frac{9\pi^4}{32}\left(2\bar{D} + \hat{\lambda}^2 + \frac{1}{\hat{\lambda}^2}\right). \tag{16.200}$$

Finally, if the plate is homogeneous and isotropic, $D_{11} = D_{22} = D$, $\bar{D} = 1$ and
$\hat{\lambda} = \lambda = a/b$. The buckling load then becomes $N_{cr} = k_c\pi^2D/b^2$, where the bucking
parameter, k_c, is defined as

$$k_c = \frac{9\pi^2(2 + \lambda^2 + 1/\lambda^2)}{32\lambda}. \tag{16.201}$$

Of course, the buckling loads predicted by eqs. (16.200) and (16.201) are approximate because only two terms are kept in the infinite series. For example, if the
plate is square, $a/b = 1$, the approximate solution gives $k_c = 11.1$, while the exact
solution is $k_c = 9.34$. A five term approximation corresponding to $(i, j) = (1, 1)$, (1,
3), (2, 2), (3, 1), and (3, 3) yields $k_c = 9.42$, which is only about 0.9% above the exact buckling load. For plates with higher aspect ratios, say $a/b > 1.5$, an increasing
number of terms of the series must be kept to obtain accurate predictions.

Example 16.17. Buckling of an isotropic plate under shear loading

Consider an isotropic plate simply supported along all four edges and subjected to an in-plane shear load, N_{12}. The solution developed in the previous example becomes increasingly laborious as the plate's aspect ratio increases because an increasing number of terms must be kept in the series expansion to obtain accurate predictions.

To overcome this problem, an energy approach will be used based the the following assumed transverse displacement field

$$\bar{u}_3(x_1, x_2) = q_1 \sin \alpha_m x_1 \sin \beta_1 x_2 + q_2 \cos \alpha_m x_1 \sin \beta_2 x_2,$$

where $\alpha_m = m\pi/a$ and $\beta_n = n\pi/b$. In this case, m defines the number of waves appearing along the length of the plate. The second shape function is depicted in fig. 16.56 for $m = 3$. Clearly, this shape function does not satisfy the geometric boundary conditions along the short edges of the plate. This should not significantly affect the predictions for plates with high aspect ratios.

Using the energy approach outlined in section 16.6.5, the curvature interpolation matrix defined by eq. (16.155) is evaluated first, and the stiffness matrix given by eq. (16.157) then follows as

$$\underline{\underline{K}} = \frac{\pi^4 D}{4a^3 b^3} \begin{bmatrix} (a^2 + b^2 m^2)^2 & 0 \\ 0 & (4a^2 + b^2 m^2)^2 \end{bmatrix}.$$

Next, the geometric stiffness matrix is obtained by following the procedure described in section 16.7.2. The displacement gradient interpolation matrix, $\underline{\underline{G}}$, is evaluated with the help of eq. (16.194), and the matrix of in-plane loads, $\underline{\underline{N}}$, defined in eq. (16.192) reduces to its off diagonal terms, $N_{12} = N_0$. Finally, the geometric stiffness matrix defined by eq. (16.196) becomes

$$\underline{\underline{K}}_G = \frac{4\pi}{3} m N_0 \begin{bmatrix} 0 & -1 \\ -1 & 0 \end{bmatrix}.$$

Here again, the governing equations form a set of homogeneous linear equations, $(\underline{\underline{K}} + \underline{\underline{K}}_G)\underline{q} = 0$. A nontrivial solution of the problem exists only when $\det(\underline{\underline{K}} + \underline{\underline{K}}_G) = 0$, leading to a critical load $N_{cr} = k_c \pi^2 D/b^2$ where the buckling parameter, k_c, is defined as

$$k_c = \frac{3\pi}{16m} \sqrt{16\lambda^2 + \frac{33m^4}{\lambda^2} + \frac{10m^6}{\lambda^4} + \frac{m^8}{\lambda^6} + 40m^2}, \qquad (16.202)$$

and $\lambda = a/b$ is aspect ratio of the plate.

Given the assumptions made at the onset of this analysis, the present solution is only valid for plates with long aspect ration. On the other hand, the solution developed in the previous example is reasonably accurate for plates with a small aspect ratio. Figure 16.57 collects the results of the two analyses. The dashed curves are the results for buckling parameter, k_c, obtained in example 16.16 with two and five term expansions. The solid curves give the results of the present analysis for $m = 2, 3$, and 5. For high aspect ratio plates, the present analysis predicts a buckling parameter $k_c = 5.60$, while the exact solution is $k_c = 5.34$.

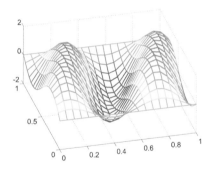
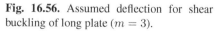

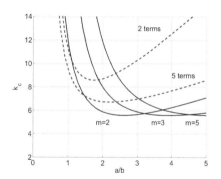

Fig. 16.56. Assumed deflection for shear buckling of long plate ($m = 3$).

Fig. 16.57. Buckling parameter, k_c, versus plate aspect ratio.

16.7.3 Problems

Problem 16.24. Buckling of simply supported plate

Consider a plate of length a and width b simply supported along its four edges. An in-plane compressive loading is applied along two opposite edges. Prove that the following statements made earlier are correct. *(1)* Show that the aspect ratio, a/b, at which the wave number in the axial direction changes from m to $m + 1$ is $a/b = \sqrt{m(m + 1)}$. *(2)* Show that the minimum value of the buckling parameter, k_c, occurs at integer values of the plate's aspect ratio, a/b, and $k_c = 4$ for those integer values.

Problem 16.25. Buckling of a plate with simply supported side edges and clamped loaded edges

Consider the same basic plate buckling problem treated in example 16.11, but now assume that the edges through which the in-plane loading is applied are clamped instead of simply supported. *(1)* Explain why a solution of the form $\bar{u}_3 = g_{mn}(\cos(m - 1)\pi/a - \cos(m + 1)\pi/a) \sin n\pi/b$ is possible. *(2)* Using this solution, determine the buckling equation for the plate. *(3)* Construct a plot of k_c versus a/b similar to fig. 16.51. *(4)* Discuss the trend of k_c as $a/b \to \infty$ compared to the behavior of k_c for the simply supported plate. What is the effect on the boundary conditions at $x_1 = 0, a$ as $a \to \infty$?

Problem 16.26. Cantilevered plate under tip shear load

Consider the thin, specially orthotropic, cantilevered plate depicted in fig. 16.58. A shear load P acts along the edge of the plate. Find the critical value of the load P for which the plate will buckle in the transverse direction. Use an energy method with the following assumed mode $\bar{u}_3(x_1, x_2) = x_1^2\, q_1 + x_1 x_2\, q_2$. The in-plane loading due to the tip shear force can be approximated as $N_1 = (12P/b^3)(a - x_1)x_2, N_2 \approx 0, N_{12} \approx 0$.

Problem 16.27. Rectangular plate on elastic foundation

Consider a specially orthotropic plate of length a and width b simply supported along its four edges. The plate rests on an elastic foundation of stiffness constant, k, and an in-plane compressive loading, N_0, is applied along two opposite edges, as shown in fig. 16.59. Use an energy method to compute the buckling load of the system. Assume the transverse displacement field to be in the following form

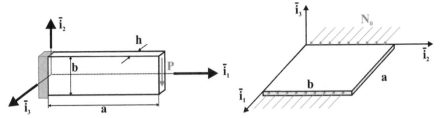

Fig. 16.58. Cantilevered plate under tip shear load.

Fig. 16.59. Simply supported plate on an elastic foundation.

$$\bar{u}_3(\hat{x}_1, \hat{x}_2) = \sum_{m,n=1}^{\infty} q_{mn} \sin \hat{\alpha}_m \hat{x}_1 \sin \hat{\beta}_n \hat{x}_2,$$

where $\hat{\alpha}_m = m\pi/\hat{a}$ and $\hat{\beta}_n = n\pi/\hat{b}$. Note: the strain energy stored in the elastic foundation is $A_{ef} = 1/2 \int_{S_m} k \, \bar{u}_3^2 \, dS_m$.

Problem 16.28. Buckling equations derived from energy principle
In section 14.2.5, the differential equation of beams subjected to large axial forces is derived from the total potential energy of the system using calculus of variations. Starting form the expression for the total potential energy of plates, use calculus of variations of derive the equations of plates subjected to large in-plane forces, eqs. (16.182) or (16.183), for isotropic or specially orthotropic plates, respectively.

Problem 16.29. Simply supported plate with various lay-up configurations
Consider two laminates: laminate A, $[0_2, \pm 45]_S$, and laminate B, $[\pm 45, 0_2]_S$. Consider next two simply supported rectangular plates, denoted plates A and B, of dimensions a and b along axes $\bar{\imath}_1$ and $\bar{\imath}_2$, respectively, such that $b = 3a$. Plates A and B are constructed with laminates A and B, respectively. The 0 degree fibers of the laminates run along axis $\bar{\imath}_1$, as shown in fig. 16.60. *(1)* If the plates are subjected to a uniform transverse pressure, which of the two plates will carry the largest load at failure? *(2)* If the plates are subjected to uniform in-plane compressive loads, N_0, applied along the edges and acting in a direction parallel to axis $\bar{\imath}_2$, which of the two plates will feature the largest load at buckling? *(3)* If the plates are subjected to uniform in-plane shear loads, N_0, applied along all four edges, which of the two plates will feature the largest load at buckling?

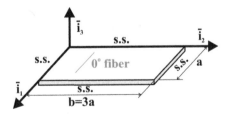

Fig. 16.60. Simply-supported rectangular plate

Erratum to:

Structural Analysis

O. A. Bauchau and J. I. Craig, School of Aerospace Engineering, Georgia Institute of Technology
April 12, 2012

ISBN 978-90-481-2515-9

Chapter 1: Basic equations of linear elasticity

p.12; Last paragraph: The equation for $\bar{s}$ should be $\bar{s} = s_1 \bar{\imath}_1 + s_2 \bar{\imath}_2 + s_3 \bar{\imath}_3$.
p.16; Example 1.1: The equation for $\underline{S}$ is missing 0 in the (3,1) position.
p.20; Problem 1.3: Replace $\tau_{12} = V_3/(bh)$ with $\tau_{12} = V_2/(bh)$.
p.35; Eq. (1.59): In second line replace $\bar{u}(x_1)$ with $\underline{u}(x_1)$.

Chapter 2: Constitutive behavior of materials

p.73; Problem 2.7: Replace last sentence with "Use a range of Poisson's ratios $0 \leq \nu \leq 0.5$."
p.95; Table 2.9: In column #5, replace σ_{1c}^{*f} with σ_{2c}^{*f}.

Chapter 3: Linear elasticity solutions

There are no errata.

Chapter 4: Engineering structural analysis

p.155; Fifth paragraph, last sentence: Capitalize "Introducing..."
p.160; Third paragraph, sentence 2: Delete second "a" at start.
p.161; First paragraph: Replace equation with: $\frac{d_B}{L} = \frac{e_B}{L} = \frac{\alpha \Delta T}{1 + 2k_A \cos^3 \theta}$.

Chapter 5: Euler-Bernoulli beam theory

p.207; Example 5.12: Last equation replace -1 by +1.

p.212; Problem 5.8: Add the following sentence: "All plots are to be constructed for values of $\bar{k} = 0, 10, 1000$."

p213; Problem 5.9: Sentence #4: change "fall" to "falls". Item (3): delete "that".

p.213; Problem 5.10: Title: change "Cantilever" to "Cantilevered". Sentence #2: change "on" to "to".

p.213; Problem 5.11: Title: change "Cantilever" to "Cantilevered". Sentence #2: delete "concentrated".

p.214; Problem 5.12: Item *(4)*, replace $V_3(\eta)/P$ with $V_2(\eta)/P$.

Chapter 6: Three-dimensional beam theory

p.243; Fig. 6.11: Swap the point labels on the two axes.

p.247; Problem 6.3: Title: replace "bema" with "beam". Item (6): replace "defined" with "define".

p.257; Fig. 6.25: The figure should show the $\bar{\imath}_2 - \bar{\imath}_3$ axes located at the midpoint of the vertical web with axis $\bar{\imath}_3$ pointing upwards.

Chapter 7: Torsion

p.286; Third paragraph: Delete second sentence which reads: "This forms a set of equations for the unknown coefficients, C_{ij}."

p.292; Fig. 7.31: The shear flow arrows in the small circled blow-up in the upper right of the figure should be reversed.

Chapter 8: Thin-walled beams

p.305; Problem 8.7: Delete "at point C." Add as second sentence: "Define $b = a/2$ and $\alpha = \arcsin(3/5)$."

p.306; Problem 8.10: Add as second sentence: "Define $a = \alpha R$." Add to item (4) the following sentence: " Assume that $\alpha = 1/2$." Add as NEW item the following: "*(5)* Find the critical value of $\alpha = \alpha_{cr}$ such that the maximum bending stress σ_1 assumes equal positive and negative values on the section."

p.306; Problem 8.11: Add as second sentence: "Define $\alpha = b/h$." Add to item (4) the following sentence: "Assume that $\alpha = 1/2$ and $\beta = 2$."

p.317; Problem 8.16: Add to the second sentence: "..., and assume that $b = \beta h$."

p.318; Problem 8.20: Add as second sentence: "Define $a = \alpha R$ and assume $\alpha = 1$." Add NEW item: "*(5)* What is the effect of α on the maximum value of the shear flow?"

p.318; Problem 8.22: Add as a second sentence: "Define $b = \alpha h$ and assume that $\alpha = 1/2$ with $\beta = 3$."

p.333; Problem 8.38: Replace the last sentence with: "Use $b = a$ and $c = 2a$ and $t_1 = t_2 = t_w = t$."

p.337; Last 2 lines: Replace "is C is" with "is C" in two places.

p.338; Last equation: Replace "360t" with "360" immediately following the second equal sign.

p.341; Problem 8.39: Change the second *(3)* to *(4)*.

p.341; Problem 8.41: Add "on page 295" following "Figure 7.34..."

p.341; Problem 8.43: Add a third sentence: "Assume $t_1 = t_2 = t_w = t$, $b = a$ and $c = 2a$."

p.342; Problem 8.44: In item (2) change "$d/b \in [0, 1.5]$" to "$0 \leq d/b \leq 1.5$."

p.344; Third paragraph: Replace "10that" with "10 that".

p.350; Fig. 8.56: Replace f_1 and f_2 with $f^{[1]}$ and $f^{[2]}$.

p.353; Problem 8.54: Switch item *(1)* with item *(2)*, *i.e.*, reverse their order.

p.353; Problem 8.55: Switch item *(1)* with item *(2)*, *i.e.*, reverse their order.

p.353; Problem 8.56: Switch item *(1)* with item *(2)*, *i.e.*, reverse their order. Also, add as last sentence: "Assume $b = a$, $c = 2a$ and $t_1 = t_2 = t_w = t$."

p.361; Problem 8.60: As specified the beam develops unrealistically large twisting. Change the beam properties to: $h = 0.2$ m, $b = 0.1$ m and $t = 10$ mm.

p.362; Eqs. (8.80a,b): Change x_2 to $x_2(s)$ and x_3 to $x_3(s)$.

p.364; Second paragraph: In first line change "8.5.1" to "7.5".

p.376; Problem 8.64: Title: change "Cantilever" to "Cantilevered".

p.386; Fig. 8.84: Change "countour" to "contour" in the figure.

p.390; Problem 8.67: Add to end of first sentence: "... and specified in problem 8.66."

p.390; Problem 8.68: Add as second sentence: "Assume $b = a$, $c = 2a$ and $t_1 = t_2 = t_w = t$."

Chapter 9: Virtual work principles

p.403; Fig. 9.3: Change "$F_2 = -3$" to "$F_2 = 3$" in figure.

p.403; Fig. 9.4: Change "to" to "by" in caption.

p.403; Example 9.2: In second paragraph, change first "s_2" to "s_1."

p.429; Fig. 9.29: Swap Δ_1 and Δ_2 in figure.

p.440; First paragraph: Change second u_1 to u_2.

p.448; Last paragraph: Change "eq. (9.59)" to "eq. (9.61)".

p.453; Eq. (9.68): Reverse subscripts "A" and "C" in numerator of final result.

p.456; Fig. 9.47: Change both vertical dimensions from L to $L/2$. Note: this affects only problem 9.17.

p.457; Fig. 9.49: Reverse the direction of M_1 acting on rear face.

p.460; Eq. (9.75): Change V to $\mathcal{V}$ at integral symbol.

p.461; Third equation: Change V to $\mathcal{V}$ at integral symbol.

p.464; First paragraph: In next to last sentence change "his" to "this": "For this expression..."

p.471; Problem 9.23: Change last item *(2)* to *(3)*.

p.472; Problem 9.27: Change second item *(1)* to *(2)*.

p.482; Fig. 9.71 caption: Replace caption's second sentence with: "The isostatic system is obtained by cutting the moment restraint at the left end."

p.482; Example 9.29: Equation in second paragraph should read: $M_3(\eta) = -p_0 L^2 (1 - \eta)^2/2$.

p.485; Example 9.31: In first paragraph, second sentence, add after "infinite stiffness" the following: "$(EA \to \infty)$"

p.486, Next to last paragraph: Add to end of sentence beginning with "Finally," the following: "...for $0 \le \eta \le 1$ (symmetric)."

p.487; Third paragraph, second sentence: Replace "possible" with "possibly".

p.490; Fig. 9.79: Replace the upward distributed load with a single downward concentrated load P applied at the mid-span point D. (Otherwise, problem 9.32 is identical to example 9.31.)

p.490; Problem 9.34: Item (1): replace "displacement" with "vertical displacement".

Energy methods

p.521; Last paragraph: Swap "$\sin \theta$" and "$\cos \theta$" in equations for "e_1" and "e_3".

p.522; First equation: Change "$u_1 \cos \theta$" to "$u_1 \sin \theta$" in 2 places and change "$u_2 \sin \theta$" to "$u_2 \cos \theta$" in 2 places. Also, replace the last (3rd) line in equation with: $= \frac{1}{2} \frac{EA}{L} \left[2u_1^2 \sin^2 \theta \cos \theta + u_2^2 (1 + 2 \cos^3 \theta) \right]$.

p.522; Second equation set: Replace second pair of equations with the following:

$$\frac{\partial \Pi}{\partial u_1} = \frac{EA}{L} 2u_1 \sin^2 \theta \cos \theta - P_1 = 0,$$

$$\frac{\partial \Pi}{\partial u_2} = \frac{EA}{L} u_2 (1 + 2 \cos^3 \theta) = 0.$$

p.522; Third equation set: Replace matrix equation with the following:

$$\begin{bmatrix} 2 \sin^2 \theta \cos \theta & 0 \\ 0 & 1 + 2 \cos^3 \theta \end{bmatrix} \begin{Bmatrix} u_1 \\ u_2 \end{Bmatrix} = \frac{L}{EA} \begin{Bmatrix} P_1 \\ 0 \end{Bmatrix}.$$

p.522; Sentence following third equation set: Replace sentence with new sentence: "Solving these equations then yields $u_1 = P_1 L/(2EA \sin^2 \theta \cos \theta)$ and $u_2 = 0$."

p.522; Fourth equation set: Replace the fourth equation set with the following:

$$\frac{e_1}{L} = \frac{1}{2 \sin \theta \cos \theta} \frac{P_1}{EA}, \quad e_2 = 0, \quad \frac{e_3}{L} = -\frac{1}{2 \sin \theta \cos \theta} \frac{P_1}{EA}.$$

p.522; Last equation set: Replace the last equation set with the following:

$$\frac{F_1}{P_1} = \frac{1}{2 \sin \theta}, \quad F_2 = 0, \quad \frac{F_3}{P_1} = -\frac{1}{2 \sin \theta}.$$

p.526; Problem 10.7: Replace "generalized coordinates" with "nodal displacements".

p.526; Problem 10.8: Replace "generalized coordinates" with "nodal displacements".

p.526; Problem 10.9: In sentence 2, replace "bar 3" with "bar 2." Also, replace "generalized coordinates" with "nodal displacements".

p.534; Equations #1-3: Replace all 6 instances of $\underline{b}$ with $\underline{\hat{b}}$.

p.567; Example 10.17: Replace title, "Ring under internal forces" with "Ring under inplane and out of plane loads".

p.568; Third equation: Change H_{33}^c to H_{22}^c and change M_3 to M_2 in two places, and in following line also change M_3 to M_2.

p.568; Fourth equation: Change H_{33}^c to H_{22}^c and H_{33} to H_{22}^c.

p.576; Fig. 10.50: Change "half-span" to "triangular" in figure caption.

p.576; Problem 10.21: In title, replace "Simply supported" with "Cantilevered".

p.576; Problem 10.22: In last line on page, replace "turn to" with "consider".

p.577; Problem 10.23: Replace "move in the only in the vertical" with "move in only the vertical". Swap items *(1)* and *(2)*, and replace "reaction Q" with "reaction B".

p.577; Problem 10.24: In second sentence, replace "while that bars" with "while that of bars". Also, change the second item *(1)* to *(2)*.

p.577; Problem 10.25: In title, change "simple" to "mid-span".

Chapter 11: Variational and approximate solutions

p.589; Last equation: Change l^4 to L^4 in numerator.

p.596; Last paragraph: First sentence, last word: change to "ends."

p.608; First equation: Change third integral to read: $\int_0^L wp_1 \mathrm{d}x_1$.

p.608; Eq. (11.16): Change second integral to read: $\int_0^L wp_1 \mathrm{d}x_1$.

p.616: Problem 11.6: In Item (1), replace $\bar{u}(x_1$ with $\bar{u}_1(x_1)$.

p.625; Problem 11.10: Add to end of problem statement the following sentence: "Assume $\alpha = 1/2$ for all plots."

p.646; Third equation: Replace $-\pi^2$ with $-\frac{\pi^2}{L^2}$.

p.650; Problem 11.12: Replace last word in first sentence with: "and a uniform load p_0 is acting upwards."

p.651; Problem 11.13: In second sentence in item *(2)*, replace "Construct 5" with "Construct 3".

p.651; Problem 11.14: In title, replace "two" with "end-point and".

p.652; Problem 11.17: In last line on page, replace x_1^{2+i} with x_1^{1+i}.

p.653; Problem 11.17: In first sentence on page, replace "3 cases" with "4 cases".

p.653; Problem 11.18: In second sentence of item *(2)*, replace x_1^{2+i} with x_1^{1+i}.

p.654; Problem 11.19: In last sentence of item (2), replace $8\ 10^3$ with 8×10^3.

p.660; Last equation: Change limits of first integral to $-\ell/2$ and $\ell/2$.

p.661; Third equation: Change limits of first integral to $-\ell/2$ and $\ell/2$, and change $\hat{p}_2$ to p_2.

p.671; Fig. 11.50: A pinned (simple) support is missing at the right end of the beam.

Chapter 12: Variational and energy principles

p.681; Last sentence on page: Change punctuation to "matrix, $\underline{\underline{S}}$, are" in sentence.

p.686; Third equation: Remove leading minus sign.

p.689; Last sentence on page: Insert space after first comma.

p.699; Fourth equation: Second line should include leading minus sign: $-2tL \int_0^1 \tau_0 (g + \frac{2h}{3}) \, d\eta$,

p.701; Fig. 12.10: Change label on abscissa from "η" to "ζ".

p.703; Paragraph #3: Change $\tau_{13} = \partial \phi / \partial x_2$ with $\tau_{13} = -\partial \phi / \partial x_2$

p.713; Problem 12.1: Replace all 6 occurrences of u_2 with $\bar{u}_2$.

p.713; Problem 12.2: In item (1), replace $H_{11}/(4abG)$ with $H_{11}/(16ab^3 G)$.

p.715; Problem 12.11: In third sentence, replace γ_{12}^2 with γ_{ave}^2, replace ϵ_1 with $\bar{\epsilon}_1$, du_1 with $d\bar{u}_1$, γ_{12} with γ_{ave}, and du_2 with $d\bar{u}_2$. In the next sentence, replace F_1 with N_1 in 2 places and F_2 with V_2 in 3 places. Replace items *(1)* and *(2)* with the following: "*(1)* Develop the principle of virtual work from the equilibrium equations and boundary conditions. *(2)* Develop the governing differential equations and boundary conditions using the principle of minimum total potential energy."

p.716; Problem 12.11: In item *(5)*, replace u_2 with $\bar{u}_2$. In item *(6)*, replace F_2 with V_2 in 1 place and F_0 with V_0 in 2 places; also replace u_2 with $\bar{u}_2$. In item *(7)*, replace $(H_{33}u_2)$ with $(H_{33}\bar{u}_2)$, replace $s^2 = 2.0 \, 10^{-3}$ with $\bar{s}^2 = 2.0 \times 10^{-3}$, and replace $s^2 = H_{33}/K_{22}l^2$ with $\bar{s}^2 = H_{33}/(K_{22}L^2)$. In item *(8)*, replace F_2 with V_2 and γ_{12} with γ_{ave}. In item *(11)*, replace s^2 with $\bar{s}^2$ in 2 places.

p.716; Problem 12.12: In 1st sentence, replace "of axial stiffness S." with "with cross-sectional area $\mathcal{A}_s$ and axial stiffness S." In 4th sentence, replace "in to" with "into". Throughout entire problem, replace subscript $(.)_f$ with subscript $(.)_0$ on variables σ and P (a total of 10 places). In the last sentence in item *(2)*, replace $(du_1/dx_1)_{x_2=h}^2$ with $(du_1/dx_1)_{x_2=b}^2$. In item *(8)*, replace $h\tau^{12}/P$ with $\mathcal{A}_s\tau_{12}/P$. In item *(9)*, replace $h\tau_{12}/P(\eta, \zeta = 1)$ with $\mathcal{A}_s\tau_{12}/P(\eta, \zeta = 1)$

p.717; Problem 12.12: In first line on page, replace "same question" with "repeat this in another plot..." In item (12), replace "same question" with "repeat this in another plot". In item (13), replace "$k \in [0,1]$" with "β for $0 \leq \beta \leq 1$". In the last sentence of the last paragraph, replace "$\mathcal{A}$" with "$\mathcal{A}_s$" in 3 places; replace "k" with "β"; replace "; " with ", and".

Chapter 13: Introduction to plasticity and thermal stresses

p.731; Problem 13.3: In item (2), delete the end of the sentence, "on a non-dimensional scale P^y vs. Δ/Δ^y". Add item (3): "*(3)* Find an expression for P/P^y as a function of Δ/Δ^y in the elasto-plastic region, $P^y < P < P^p$." Add the following sentence to the end of item *(5)*: "Your plot can either be generic or you can choose specific values for θ and $\bar{k}$."

p.731; Problem 13.4: Change "cylinder" to "tube" in the problem title and in the first sentence. In item (1), change p_i/p_i^E to p_i/p_i^y.

p.741; First paragraph: change κ_3^y to κ_1^y in three instances.

p.745; First equation: Change second plus (+) sign to a minus (-) sign so first equation reads: $\frac{d^2 u_r}{dr^2} + \frac{1}{r}\frac{du_r}{dr} - \frac{u_r}{r^2} - \frac{(1+\nu)}{(1-\nu)}\alpha\frac{dT}{dr} = 0$.

p.746; Fig. 13.16 caption: Add to end of last sentence: "and $\nu = 0.3$."

p.746; Problem 13.7: Add to the end of the problem this sentence: "Assume $\bar{R}_i = 0.2$ and $\nu = 0.3$."

p.758; Second paragraph: Change $d\bar{u}_1/dx_1 = P_1^e/S = E\alpha T_0/3$ to $d\bar{u}_1/dx_1 = P_1^e/S = \alpha T_0/3$.

p.758; Second equation: Change L^3 to L^2.

p.760; Problem 13.12: In first sentence change "cress-section" to "cross-section". In last 2 sentences do the following. Change $\hat{h}_1$ to $\bar{h}_1$ and change $\hat{h}_2$ to $\bar{h}_2$. Change η_1 to λ_1 in 2 places and change η_2 to λ_2 in 2 places.

p.761; Problem 13.13: In first sentence change "cress-section" to "cross-section". In last 2 sentences do the following. Change $\hat{h}_1$ to $\bar{h}_1$ and change $\hat{h}_2$ to $\bar{h}_2$. Change η_1 to λ_1 in 2 places and change η_2 to λ_2 in 2 places.

p.761; Problem 13.14: In item (5), insert "for" after "cross-section".

p.762, Problem 13.15: In the second paragraph replace "example 6.6." with "example 6.6 on page 249."

Chapter 14: Buckling of beams

p.767; First sentence: Insert after "beam" the following: "with a symmetrical cross-section".

p.767; Fig. 14.5: Remove rollers from left support.

p.769; Eq. (14.15): Replace H_{33}^{*c} with H_{33}^c.

p.770; Eqs. (14.17 & 14.19): Replace H_{33}^{*c} with H_{33}^c.

p.771; Eq. (14.24-25): Replace H_{33}^{*c} with H_{33}^c.

p.771; Eq. (14.27): Replace H_{22}^{*c} with H_{22}^c.

p.772; Second paragraph: Replace H_{33}^{*c} with H_{33}^c in 2 places.

p.772; Eq. (14.30): Replace H_{33}^{*c} with H_{33}^c.

p.773; Eq. (14.33): Replace H_{33}^{*c} with H_{33}^c.

p.774; Eq. (14.34-35) and text between: Replace H_{33}^{*c} with H_{33}^c in 4 places.

p.775; Full page: Page must be reformatted at fig. 14.11.

p.775; Eq. (14.38) and remaining equations on page: Replace H_{33}^{*c} with H_{33}^c in a total of 6 places.

p.776; Second paragraph & first equation: Replace H_{33}^{*c} with H_{33}^c in 3 places.

p.778; Second equation from bottom: Change $[\underline{K} + P\underline{K}_G]$ to $[\underline{K} - P\underline{K}_G]$.

p.778; Eq. (14.47): Change $[\underline{K} + P\underline{K}_G]$ to $[\underline{K} - P\underline{K}_G]$.

p.779; Eq. (14.49) & two following lines: Replace H_{33}^{*c} with H_{33}^c in 5 places.

p.781; Second paragraph: Replace H_{33}^{*c} with H_{33}^c in 2 places.

p.781; Second equation & following paragraph: Replace H_{33}^{*c} with H_{33}^c in 5 places.

p.782; Entire page: Replace H_{33}^{*c} with H_{33}^c in 5 places.

p.784; First paragraph: Replace H^{*c} with H_{33}^c.

p.784; Equation above Table 14.1: Replace H_{33}^{*c} with H_{33}^{c}.

p.784; Problem 14.1: Add to end of second sentence: "by solving the governing differential equation."

p.784; Problem 14.2: Replace H_{33}^{*c} with H_{33}^{c} in 2 places.

p.785; Figs. 14.16-17: In figs. 14.16-17 add rollers to right supports.

p.785; Problem 14.3: In second sentence, change a_3 to q_3. Also, replace H_{33}^{*c} with H_{33}^{c} in 2 places.

p.785; Problem 14.5: In second sentence replace "on" with "for".

p.786; Problem 14.6: In item (1) change italicized text to: "*subjected solely to the transverse load p_0.*" Also, change σ_{ult} to σ_{allow} in 4 places.

p.786; Problem 14.7: In the inline equation in item (1) replace a with q_1.

p.788; Problem 14.11: In the last sentence, replace "$\bar{\imath}_3$," with "$\bar{\imath}_3$ directions, respectively."

p.788; Problem 14.13: In item (4) replace $hM_3^{\text{mid}}/H_{33}^{c}$ with $M_3^{\text{mid}}L^2/(hH_{33}^{c})$.

p.790; Sentence before last equation: Replace "apendix" with "appendix".

Chapter 15: Shear deformation in beams

p.801; Problem 15.1: Replace "fig. 15.4" with "fig. 15.3." Also, in item (2) replace "improved deformation mode" with "parabolic stress distribution".

p.801; Problem 15.2: In item (5) replace "cantilevered" with " clamped".

p.802; Fig. 15.8: Change angle Φ_3 to $-\Phi_3$.

p.807; Second paragraph, last sentence: Replace "fast" with "faster" and delete the following "to".

p.815; Problem 15.9: In the fifth sentence replace a with q_1 and b with q_2. In the following sentence, add a ")" at the end of the equation for $\bar{k}$. Finally, in the last sentence, change k^* to $\bar{k}$.

p.816; Problem 15.10: In the fourth sentence replace Φ_n with Z_n.

p.816; Problem 15.11: At the end of items (2), (3) and (4), add "for $\alpha = 0.25$." Also, change item (5) to read the same way. In the remainder of the problem statement, replace $\alpha \in [0, 0.5]$ with $0 \le \alpha \le 0.5$ in 4 places.

Chapter 16: Kirchhoff plate theory

p.822; Second paragraph: Delete "and" in line before eq. (16.3).

p.823; Eq. (16.9): Replace ϵ_{12}^{0} with γ_{12}^{0}.

p.824; Fig. 16.3: Change V_1 to Q_1 and V_2 to Q_2.

p.824; Last paragraph: Change "components" to "component" in first sentence.

p.830; First paragraph: Change M_1 to M_2 in 2nd sentence.

p.834; Item 4: Change "linear" to "rectilinear" in title.

p.836; Fig. 16.10: Reverse direction of moment arrows for M_0 in upper-right, lower-left direction (*i.e.*, axis $\bar{\imath}_1$).

p.837; Eq. (16.46b): Change first D to $-D$.

p.839; Problem 16.4: Change U_3 to $\bar{u}_3$ and replace $x_1 \in [0,a]$, $x_2 \in [0,b]$ with $0 \le x_1 \le a$ and $0 \le x_2 \le b$.

p.840; Fig. 16.18: Reverse direction of M_0 arrow on left edge of figure.

p.844; Problem 16.6: Third sentence: Replace "The laminate" with "A laminate of width b."

p.851; First paragraph: Replace $\underline{S}\underline{M} = \underline{D}\,\underline{S}\,\underline{\kappa}$ with $\underline{S}\underline{M} = \underline{D}\,\underline{S}\,\underline{\kappa}$.

p.852; Eq. (16.80): Change second term to $4D_{16}\bar{u}_{3,1112}$.

p.853; Paragraph following eq. (16.82): Replace a with $x_1 = a$ and b with $x_2 = b$.

p.855; Last paragraph: Replace $D[\alpha_m^4 + \beta_m^4]$ with $D(\alpha^2 + \beta^2)^2$.

p.858; Problem 16.8: Change N/m^3 to N/m^3.

p.858; Problem 16.9: Add to end of 4th sentence: "with material properties for T300/5208 graphite-epoxy from Table 2.7 on page 87." Insert as last sentence of item (2): "Comment on the effect of fiber direction on these plots." Insert as last sentence of item (4): "Assume $\sigma_a = 420$MPa for aluminum and use the failure stress for T300/5208 graphite-epoxy from Table 2.9 on page 95." Delete the last paragraph entirely.

p.859; Problem 16.10; Add to end of 4th sentence: "with material properties for T300/5208 graphite-epoxy from Table 2.7 on page 87." In item (2) replace $U u_3 sqrt D_{11} D_{22}/(Pa^2)$ with $\bar{u}_3\sqrt{D_{11}D_{22}}/(Pa^2)$. Insert as last sentence of item (2): "Comment on the effect of fiber direction on these plots." Insert as last sentence of item (4): "Assume $\sigma_a = 420$MPa for aluminum and use the failure stress for T300/5208graphite-epoxy from Table 2.9 on page 95." Delete the last paragraph entirely.

p.860; Third paragraph: Replace x_1 with $\hat{x}_1$ in all 4 places.

p.860; Fourth paragraph: Replace "Fourier" with "a Fourier".

p.861; Eq. (16.100): Replace p_m with $\left(\frac{\hat{a}}{m\pi}\right)^4 p_m$.

p.861; Last line: Replace $p_3(x_1, x_2)$ with $p_3(\hat{x}_1, \hat{x}_2)$.

p.862; Eq. (16.102): Change g_{mp} to g_m^p in equation and in first line of text immediately following.

p.863; Eq. (16.109): Change $\frac{\sin m\pi x_1}{a}$ with $\sin\frac{m\pi x_1}{a}$.

p.864; Second paragraph, sentence 2: Change "For" to "Using".

p.864; Eq. (16.110): Change $\frac{\sin m\pi x_1}{a}$ with $\sin\frac{m\pi x_1}{a}$.

p.864; Problem 16.11: Add to end of 4th sentence: "with material properties for T300/5208 graphite-epoxy from Table 2.7 on page 87." Insert as last sentence of item (2): "Comment on the effect of fiber direction on these plots." Insert as last sentence of item (4): "Assume $\sigma_a = 420$MPa for aluminum and use the failure stress for T300/5208 graphite-epoxy from Table 2.9 on page 95." Delete the last paragraph entirely.

p.864; Problem 16.12: In 2nd sentence, replace "Two" with "The two long", and replace "other" with "short edges". Add to end of 4th sentence: "with material properties for T300/5208 graphite-epoxy from Table 2.7 on page 87." Insert as last sentence of item (2): "Comment on the effect of fiber direction on these plots." Insert as last sentence of item (4): "Assume $\sigma_a = 420$MPa for aluminum and use the failure stress for T300/5208 graphite-epoxy from Table 2.9 on page 95." Delete the last paragraph entirely.

p.867; Line above eq. (16.121): Change "eqs. (16.34)" to "eqs. (16.115)".

p.868; Sentence following eq. (16.125): Change "equations" to "equation".

p.870; Third equation: Replace α with $\bar{R}$.

p.870; Last pair of equations: Replace α with $\bar{R}$ in 6 places.

p.871; Problem 16.14: Add to 2nd sentence: "and $\rho = r/R_0$." In item (2) change α to $\bar{R}$, change k_Δ to $k_\Delta R_0^2/D$, replace "use" with "using" and replace the last occurrence of k_Δ with "the stiffness and assuming $\nu = 0.3$." Add a last sentence: "Assume $\nu = 0.3$."

p.871; Problem 16.15: Add to 2nd sentence: "and $\rho = r/R_0$." Add a last sentence: "Assume $\nu = 0.3$."

p.872; Problem 16.16: Add to last sentence: "and assume $\nu = 0.3$."

p.873; all equations: Replace partial derivative symbol ∂ with ordinary derivative symbol d (56 instances).

p.875; Problem 16.17: Change all instances of ϕ to Φ (2 instances). In the second sentence in item (2), change k_ϕ to k_Φ/D in 2 places. Add as the last sentence: "For all plots assume $\nu = 0.3$."

p.875; Fig. 16.40: Change ϕ to Φ.

p.885; Eqs. (16.152-16.153): Change last term in eq. (16.153) to $h_N(x_1, x_2)$.

p.885; Sentence before eq. (16.154): Add to sentence, "needed in eq. (16.144)" and add to continuing sentence before eq. (16.155), "defined as".

p.889; First equation: Change B to $\underline{B}$.

p.895; Problem 16.18: Change $\alpha_1, \alpha_2, \alpha_3$ to q_1, q_2, q_3 in 2 places.

p.895; Problem 16.21: Insert (1) at the start of the third sentence to read: "(1) Use..." Also, change "(1)" to "(2)" at beginning of the last sentence.

p.895; Problem 16.22: In second sentence, insert after "D" and within the parentheses: "and $\nu = 0.3$".

p.896; Problem 16.22: Insert after the first sentence and before (1): "Because of the wide spacing of the stiffeners, they must be treated individually and cannot be smeared into an anisotropic model." Also, change (2) to read: "(2) For the two..."

p.896; Problem 16.23: In the fourth sentence change $h = 1.5\ 10^{-03}$ to $h = 1.5 \times 10^{-3}$. Also add to the end of the last sentence: "(in N.m units)".

p.901; Sentence including eq. (16.189): Move this sentence to the end of the example (just before **Example 16.12**).

p.901; Example 16.12: In the last sentence of the first paragraph, replace "could" with "can" and replace "of" with "or".

p.906; Last paragraph: Change "stationary" to "minimum".

p.908; Third paragraph from bottom: Replace "complex" with "complicated". Replace "is a Gauss-Legendre" with "if a Gauss-Legendre".

p.909; Fig. 16.55: The solid curve for m=1 did not print in this figure.

p.909; Example 16.16: Change the title to: "**Example 16.16 Buckling of an anisotropic plate under shear loading**".

p.910; Third equation: In second integral in second line of the equation, change $\hat{\beta}_m$ to $\hat{\beta}_n$.

p.910; Last 2 equations: In the last two equations on the page, change the variables a and b to $\hat{a}$ and $\hat{b}$ in a total of 3 places each.

p.912; Last paragraph: In first sentence, change "ration." to "ratios." Also, in the fourth sentence, insert "for $m = 1$" after "example 16.16".

p.913; Fig. 16.57: Add "m=1" to the labels "2 terms" and "5 terms" in the figure.

p.913; Problem 16.25: Replace the first sentence with: "Consider the same plate buckling problem treated in example 16.15. In item (2) insert x_1 after π in 3 places. Also in item (2) replace "Using this solution..." with " Using a 2-term solution..."

p.913; Problem 16.26: Add as the last sentence: "Assume $\nu = 0.3$."

Appendix: Mathematical tools

p.927; eq. (A.36): Change the second row of the rotation matrix from $\ell_2 \; n_2 \; m_2$ to $\ell_2 \; m_2 \; n_2$.

A

Appendix: mathematical tools

A.1 Notation

It is traditional to use a bold typeface to represent vectors, arrays, and matrices. While this typographical convention is elegant in print, it is difficult to reproduce in handwriting or on a white board in a lecture hall. Students are then faced with the confusing dilemma of using a notation in handwriting that does not match that used in textbooks. The notation used in this book is selected to eliminate this problem. The printed notation uses single and double underlines to indicate arrays and matrices, respectively, and these are easily reproduced in handwriting.

Vectors and arrays are denoted using an underline, *i.e.*, $\underline{u}$ or $\underline{F}$. A vector is first order tensors such as a position, displacement, or force vector. An array is a container used to store a collection of scalars. When defining the components of an array, the scalars it consists of are listed in a column delimited by curly braces, see eq. (A.2).

Unit vectors are vectors of unit magnitude and are denoted with an overbar; for instance, $\bar{\imath}_1$ indicates the first unit vector of a triad, or $\bar{n}$ denotes a unit vector in a particular direction in Euclidean space. The overbar is also used to denote non-dimensional scalar quantities; for instance, $\bar{k}$ denotes a non-dimensional stiffness coefficient.

Matrices are indicated using a double underline. For instance, $\underline{\underline{C}}$ indicates a matrix with M rows and N columns, see eq. (A.5). Matrices are used the store the components of second order tensors such the stress and strain tensors. They are also used to store the coefficients of linear systems of equations.

The indicial notation is used throughout the book. The traditional notation, $\bar{\imath}, \bar{\jmath}, \bar{k}$, for a Cartesian axis system is replaced by $\mathcal{I} = (\bar{\imath}_1, \bar{\imath}_2, \bar{\imath}_3)$ and the corresponding coordinates x, y, and z, become x_1, x_2, and x_3, respectively. Similarly, force components commonly denoted F_x, F_y, and F_z become F_1, F_2, and F_3, respectively.

A.2 Vectors, arrays, matrices and linear algebra

Vectors are a fundamental part of mechanics and provide a powerful abstraction for manipulating forces, moments, and displacements in statics and mechanics of deformable solids. Arrays and matrices are also very useful constructs in mechanics, especially when dealing with vectors and linear algebra concepts. Vectors will frequently be represented by arrays, and array operations can be used to carry out vector operations. Arrays and matrices provide powerful tools to represent sets of simultaneous linear algebraic equations and express their solutions. Many numerical procedures for approximating the solution of complex mechanics problems will be described in terms of arrays and matrices. Coordinate transformations can also be represented in a compact manner using rotation matrices.

This section provides a summary of some of the key properties of vectors, arrays and matrices that are used in this book. The presentation is by no means complete or rigorous; in most cases, useful results will be presented without proof. Introductory texts such as that of Strang [10] provide in-depth coverage of linear algebra and its applications.

A.2.1 Vectors, arrays and matrices

Vectors

Vectors describe quantities that have both a magnitude and a direction, whereas *scalars* have only a signed magnitude. In this book, the term vector will be used to describe a quantity with a magnitude and a direction in Euclidean space, that is, a quantity with three independent directional components. Typically, these three components are defined in a Cartesian coordinate system, but cylindrical and spherical coordinate systems may also be used. Cartesian coordinates are defined by a triad,

$$\mathcal{I} = (\bar{\imath}_1, \bar{\imath}_2, \bar{\imath}_3), \tag{A.1}$$

where $\bar{\imath}_1$, $\bar{\imath}_2$, and $\bar{\imath}_3$ are three mutually orthogonal unit vectors.

Vector quantities in this book include forces, moments, positions, and displacements. An underscore is used to indicate a vector, $\underline{v}$.

Arrays

An *array* is a container used to store a collection of scalars. An underscore[1] is used to indicate an array, $\underline{a}$. The N elements of an array are arranged into a column of size N. In this book, an array is always defined as a column,

---

[1] Many texts use a bold font to indicate arrays or vectors but this is not adopted here because of the difficulty of creating a bold symbol in handwriting.

$$\underline{a} = \left\{ \begin{array}{c} a_1 \\ a_2 \\ a_3 \\ \vdots \\ a_N \end{array} \right\}. \tag{A.2}$$

The ith element of the array, a_i, is identified by a subscript that indicates its position in the array. Curly braces will be used to denote an array.

The *transpose* of an array of size N is a row of N elements, and a superscript $(\cdot)^T$ is used to indicate a transpose. Thus, the transpose of array $\underline{a}$ defined in eq. (A.2) is written as

$$\underline{a}^T = \left\{ a_1, a_2, a_3, \cdots, a_N \right\}. \tag{A.3}$$

Frequently, the following notation will also be used

$$\underline{a} = \left\{ a_1, a_2, a_3, \cdots, a_N \right\}^T. \tag{A.4}$$

Equations (A.2), (A.3), and (A.4), all define the same column array, $\underline{a}$.

The components of a vector can be stored in an array with three elements. For instance, array $\underline{f} = \left\{ f_1, f_2, f_3 \right\}^T$ could represent a force vector with components f_1, f_2, and f_3 in a Cartesian system.

Matrices

A matrix is a container used to store a collection of N arrays all of the same size. Each array is of size M and forms a column of the matrix, which is of size $M \times N$. When specifying the size of a matrix, the notation $M \times N$ will be used: the matrix consists of M rows and N columns. A matrix of size 2×3 consists of 2 rows and 3 columns. A double underscore is used to indicate a matrix,

$$\underline{\underline{A}} = [\underline{a}_1 \, \underline{a}_2 \, \cdots \, \underline{a}_N] = \begin{bmatrix} a_{11} & a_{12} & \cdots & a_{1N} \\ a_{21} & a_{22} & \cdots & a_{2N} \\ \cdots & \cdots & \cdots & \cdots \\ a_{M1} & a_{M2} & \cdots & a_{MN} \end{bmatrix}. \tag{A.5}$$

The elements of a matrix with the same indices (or subscripts) define the *diagonal of the matrix*. When the number of rows is equal to the number of columns, the matrix is said to be a *square matrix*.

The *transpose of a matrix* is represented as $\underline{\underline{A}}^T$ and is defined by switching the rows and columns in the original matrix,

$$\underline{\underline{A}}^T = \begin{bmatrix} a_{11} & a_{21} & \cdots & a_{M1} \\ a_{12} & a_{22} & \cdots & a_{M2} \\ \cdots & \cdots & \cdots & \cdots \\ a_{1N} & a_{2N} & \cdots & a_{NM} \end{bmatrix}. \tag{A.6}$$

The matrix transpose can also be defined by simply reversing the subscripts of all the individual elements. The transpose of a square matrix is also a square matrix, but the transpose of a matrix of size $M \times N$ is a matrix of size $N \times M$.

A *symmetric matrix* is a square matrix that is identical to its transpose, that is, a matrix for which $\underline{\underline{A}} = \underline{\underline{A}}^T$. A *skew-symmetric matrix* is a square matrix whose transpose is also its negative, that is, a matrix for which $\underline{\underline{A}}^T = -\underline{\underline{A}}$. Any square matrix can be expressed aa the sum of its symmetric and skew-symmetric parts,

$$\underline{\underline{A}} = \frac{1}{2}\left(\underline{\underline{A}} + \underline{\underline{A}}^T\right) + \frac{1}{2}\left(\underline{\underline{A}} - \underline{\underline{A}}^T\right) = \underline{\underline{A}}_s + \underline{\underline{A}}_a, \tag{A.7}$$

where $\underline{\underline{A}}_s$ is symmetric because $\underline{\underline{A}}_s^T = (\underline{\underline{A}} + \underline{\underline{A}}^T)^T = \underline{\underline{A}}_s$ and $\underline{\underline{A}}_a$ is skew-symmetric because $\underline{\underline{A}}_a^T = (\underline{\underline{A}} - \underline{\underline{A}}^T)^T = -\underline{\underline{A}}_a$.

A *diagonal matrix* is a matrix whose only non-zero elements lie along its diagonal. The *identity matrix* is a square diagonal matrix whose diagonal elements are all unity.

A.2.2 Vector, array and matrix operations

Basic operations

Vectors, arrays and matrices can be added or subtracted only if all quantities are of the same dimensions. Consequently, only vectors can be added to vectors, arrays to arrays and matrices to matrices. The resulting vector, array or matrix is a new quantity of the same type and dimension as the those being added or subtracted.

Vectors, arrays, or matrices can also be multiplied by a signed constant, resulting in a new vector, array, or matrix whose elements are each multiplied by the same signed constant. These operations follow the associative, distributive and commutative rules of scalar algebra.

Scalar product

Let a_1, a_2, and a_3 be the component of vector $\underline{a}$ in a given triad, and b_1, b_2, and b_3 those of vector $\underline{b}$ in the same triad. The *scalar product* of the vectors, denoted $\underline{a} \cdot \underline{b}$, is defined as

$$\underline{a} \cdot \underline{b} = a_1 b_1 + a_2 b_2 + a_3 b_3 = \|\underline{a}\| \, \|\underline{b}\| \cos(\widehat{ab}), \tag{A.8}$$

where $\|\underline{a}\|$ and $\|\underline{b}\|$ are the magnitudes of vectors $\underline{a}$ and $\underline{b}$, respectively, and $\widehat{ab}$ denotes the angle between vectors $\underline{a}$ and $\underline{b}$. The scalar product is so named because this operation involving two vectors results in a scalar quantity. The scalar product is also referred to as the *dot product*, because of the notation used to represent this operation.

If a_i and b_i are the components of arrays $\underline{a}$ and $\underline{b}$, both of size N, the scalar product of the two arrays, denoted $\underline{a} \cdot \underline{b}$, is defined as

$$\underline{a} \cdot \underline{b} = a_1 b_1 + a_2 b_2 + \cdots + a_N b_N = \sum_{i=1}^{N} a_i b_i. \tag{A.9}$$

The scalar product is also expressed with the following notation

$$\underline{a} \cdot \underline{b} = \underline{a}^T \underline{b} = \underline{b}^T \underline{a} = \sum_{i=1}^{N} a_i b_i. \tag{A.10}$$

Norm of an array

The *norm* of array, $\underline{a}$, denoted $\|\underline{a}\|$, is defined as

$$\|\underline{a}\| = \sqrt{\underline{a} \cdot \underline{a}}. \tag{A.11}$$

This norm is also called the magnitude or length of the array. The norm is always a non-negative scalar quantity.

A vector whose norm is unity is called a *unit vector* and is denoted with an overscore. Thus, $\bar{a}$ is a vector with unit magnitude or a unit vector. The definition of a triad, see eq. (A.1), involves three unit vectors, denoted $\bar{\imath}_1$, $\bar{\imath}_2$, and $\bar{\imath}_3$. The scalar product of two unit vectors is the cosine of the angle between them as can be seen from eq. (A.8). Also, the scalar product of a unit vector and another vector is the projection of that vector in the unit vector direction.

Matrix determinant

The determinant of a matrix is a scalar quantity, denoted $\det(\underline{\underline{A}})$, that plays an important role in linear algebra. The determinant of a matrix is defined as the sum of the entries of any row (or column) times its co-factors: $\det(\underline{\underline{A}}) = a_{i1} C_{i1} + a_{i2} C_{i2} + \cdots + a_{iN} C_{iN}$, where the co-factor is defined as $C_{ij} = (-1)^{i+j} \det(\underline{\underline{M}}_{ij})$; $\underline{\underline{M}}_{ij}$ is the sub-matrix obtained by deleting the ith row and j^{th} column of matrix $\underline{\underline{A}}$.

This formal recursive definition is not necessarily the most computationally efficient manner to compute the determinant of a matrix. Many efficient numerical algorithms to perform this task are available in most numerical analysis software packages.

The determinant of a 2×2 matrix is easy to evaluate: $\det(\underline{\underline{A}}) = a_{11} a_{22} - a_{21} a_{12}$. The determinant of a 3×3 matrix is: $\det(\underline{\underline{A}}) = a_{11} a_{22} a_{33} + a_{12} a_{23} a_{31} + a_{13} a_{21} a_{32} - a_{31} a_{22} a_{13} - a_{32} a_{23} a_{11} - a_{33} a_{21} a_{12}$. For matrices of larger size, it is preferable to rely on computer software.

Several properties of the determinant are important to note:

1. The determinant of a product of matrices is the product of the determinants: $\det(\underline{\underline{A}}\,\underline{\underline{B}}) = \det(\underline{\underline{A}}) \det(\underline{\underline{B}})$.
2. The determinant of the transpose is the same as the determinant of the matrix: $\det(\underline{\underline{A}}^T) = \det(\underline{\underline{A}})$.
3. The determinant of a diagonal matrix is the product of the diagonal elements.

4. Interchanging two rows or columns changes the sign of the determinant.
5. Adding or subtracting a multiple of one row (or column) to another row (or column) does not change the determinant.
6. If two rows (or columns) are the same or multiples of each other, the determinant is zero.
7. If a row (or column) is zero, the determinant is zero.
8. A matrix whose determinant is zero is called a *singular matrix*.

Vector product

Let a_1, a_2, and a_3 be the component of vector $\underline{a}$, and b_1, b_2, and b_3 those of vector $\underline{b}$, both resolved in the same triad $\mathcal{I} = (\bar{\imath}_1, \bar{\imath}_2, \bar{\imath}_3)$. The *vector product* of two vectors, denoted $\underline{a} \times \underline{b}$ yields a vector quantity defined as

$$\underline{a} \times \underline{b} = (a_2 b_3 - a_3 b_2)\bar{\imath}_1 + (a_3 b_1 - a_1 b_3)\bar{\imath}_2 + (a_1 b_2 - a_2 b_1)\bar{\imath}_3. \tag{A.12}$$

The vector product can also be defined in a more geometric fashion as $\underline{a} \times \underline{b} = \|\underline{a}\| \|\underline{b}\| \sin(\widehat{\underline{a}\underline{b}}) \, \bar{n}$, where $\bar{n}$ is a unit vector perpendicular to both $\underline{a}$ and $\underline{b}$ and whose direction is determined by the right-hand rule. It follows that $\underline{a} \times \underline{a} = 0$, $\bar{\imath}_1 \times \bar{\imath}_2 = \bar{\imath}_3$, and $\underline{a} \times \underline{b} = -\underline{b} \times \underline{a}$.

Matrix multiplication

Let matrix $\underline{\underline{A}}$ be of size $M \times K$ and matrix $\underline{\underline{B}}$ of size $K \times N$. The product of these two matrices, simply denoted $\underline{\underline{A}}\,\underline{\underline{B}}$, results in a third matrix, $\underline{\underline{C}}$, of size $M \times N$, whose components are

$$c_{ij} = \sum_{k=1}^{K} a_{ik} b_{kj}, \quad i = 1, 2, \ldots, M, \quad j = 1, 2, \ldots, N, \tag{A.13}$$

Multiplication of two matrices is only possible if the number of columns of the first matrix matches the number of rows of the second.

Let matrix $\underline{\underline{A}}$ be of size $M \times K$ and array $\underline{b}$ of size K. The product of the matrix by the array, simply denoted $\underline{\underline{A}}\,\underline{b}$, results in an array, $\underline{c}$, of size M, whose components are

$$c_i = \sum_{k=1}^{K} a_{ik} b_k, \quad i = 1, 2, \ldots, M. \tag{A.14}$$

From these definitions, the following properties can easily be proved.

1. $\underline{\underline{A}}\,\underline{\underline{B}} \neq \underline{\underline{B}}\,\underline{\underline{A}}$: matrix multiplication is not a commutative operation.
2. $(\underline{\underline{A}}\,\underline{\underline{B}})^T = \underline{\underline{B}}^T \underline{\underline{A}}^T$.
3. Operation $\underline{b}\,\underline{\underline{A}}$ is not defined because of dimension mismatch between the array and matrix.
4. Operation $\underline{b}^T \underline{\underline{A}}$ is defined if array $\underline{b}$ is of size equal to the number of rows in $\underline{\underline{A}}$.
5. $(\underline{\underline{A}}\,\underline{\underline{B}})\underline{\underline{C}} = \underline{\underline{A}}(\underline{\underline{B}}\,\underline{\underline{C}})$ (associative rule).
6. $\underline{\underline{A}}(\underline{\underline{B}} + \underline{\underline{C}}) = \underline{\underline{A}}\,\underline{\underline{B}} + \underline{\underline{A}}\,\underline{\underline{C}}$ (distributive rule).
7. Product of a matrix by the identity matrix gives the matrix itself: $\underline{\underline{A}}\,\underline{\underline{I}} = \underline{\underline{I}}\,\underline{\underline{A}} = \underline{\underline{A}}$.

Matrix inverse

Matrix division is not defined, but the *inverse of a square matrix* is defined. By definition, the multiplication of a square matrix, $\underline{\underline{A}}$, of size $N \times N$ by its inverse, denoted $\underline{\underline{A}}^{-1}$, produces the identity matrix of the same size,

$$\underline{\underline{A}}\,\underline{\underline{A}}^{-1} = \underline{\underline{A}}^{-1}\underline{\underline{A}} = \underline{\underline{I}}. \tag{A.15}$$

If the determinant of a matrix vanishes, its inverse does not exist. Note that the product of a matrix by its inverse is commutative.

Calculation of the inverse of a matrix is difficult when its dimensions exceed 3. For a matrix of size 2×2, the inverse is

$$\underline{\underline{A}}^{-1} = \frac{1}{\det(\underline{\underline{A}})} \begin{bmatrix} a_{22} & -a_{12} \\ -a_{21} & a_{11} \end{bmatrix}. \tag{A.16}$$

For square matrices of size larger than 3, numerical software packages should be used to compute the inverse. Detailed descriptions of the numerical procedures can be found in Strang [10].

A.2.3 Solutions of simultaneous linear algebraic equations

Matrix and array notations provide powerful abstractions for dealing with sets of simultaneous linear algebraic equations. A set of N linear equations in N unknowns, q_i, can be written explicitly as

$$a_{11}q_1 + a_{12}q_2 + \cdots + a_{1N}q_N = b_1$$
$$a_{21}q_1 + a_{22}q_2 + \cdots + a_{2N}q_N = b_2$$
$$\vdots$$
$$a_{N1}q_1 + a_{N2}q_2 + \cdots + a_{NN}q_N = b_N.$$

When expressed in matrix notation, this problem takes a much more compact form,

$$\underline{\underline{A}}\,\underline{q} = \underline{b}, \tag{A.17}$$

where $\underline{\underline{A}}$ is the square matrix of size $N \times N$ storing the coefficients of the system, $\underline{q}$ the array of size N storing the unknowns of the problem, and $\underline{b}$ the array of size N storing the right-hand side coefficients. Expanding the matrix-array product using eq. (A.14) will show that the compact notation of eq. (A.17) is equivalent to the more explicit form given above.

To find the solution of the system of equation, the inverse of matrix $\underline{\underline{A}}$, denoted $\underline{\underline{A}}^{-1}$, is computed first. Equation (A.17) is then pre-multiplied by this inverse to find $\underline{\underline{A}}^{-1}\underline{\underline{A}}\,\underline{q} = \underline{\underline{A}}^{-1}\underline{b}$. In view of the definition of the inverse, eq. (A.15), this implies $\underline{\underline{I}}\,\underline{q} = \underline{\underline{A}}^{-1}\underline{b}$, and finally

$$q = \underline{\underline{A}}^{-1}\underline{b}. \qquad (A.18)$$

If matrix $\underline{\underline{A}}$ is singular, its inverse does not exist and the linear system cannot be solved.

When the right-hand side coefficients vanish, the system of linear equations of called a *homogeneous system*,

$$\underline{\underline{A}}\,\underline{q} = 0. \qquad (A.19)$$

The *trivial solution*, $\underline{q} = 0$, is clearly a solution of the system of equations. If matrix $\underline{\underline{A}}$ is not singular, it is the only possible solution. When matrix $\underline{\underline{A}}$ is singular, however, an infinite number of solutions exist.

A.2.4 Eigenvalues and eigenvectors

The following matrix equation

$$\underline{\underline{A}}\,\underline{q} = \lambda \underline{q}, \qquad (A.20)$$

is called an *eigenvalue problem*. Matrix $\underline{\underline{A}}$ is a known square matrix of size $N \times N$, $\underline{q}$ an unknown array of size N, and λ an unknown scalar.

To solve the problem, it is recast in the homogeneous form as

$$(\underline{\underline{A}} - \lambda \underline{\underline{I}})\underline{q} = 0. \qquad (A.21)$$

If matrix $(\underline{\underline{A}} - \lambda \underline{\underline{I}})$ is non singular, *i.e.*, if $\det(\underline{\underline{A}} - \lambda \underline{\underline{I}}) \neq 0$, the only possible solution is the trivial solution, $\underline{q} = 0$.

If matrix $(\underline{\underline{A}} - \lambda \underline{\underline{I}})$ is singular, *i.e.*, if $\det(\underline{\underline{A}} - \lambda \underline{\underline{I}}) = 0$, non-trivial solutions becomes possible. For the determinant of matrix $(\underline{\underline{A}} - \lambda \underline{\underline{I}})$ to vanish, scalar λ must satify the following equation

$$\det \begin{bmatrix} a_{11} - \lambda & a_{12} & \cdots & a_{1N} \\ a_{21} & a_{22} - \lambda & \cdots & a_{2N} \\ \vdots & \vdots & \ddots & \vdots \\ a_{N1} & a_{N2} & \cdots & a_{NN} - \lambda \end{bmatrix} = 0.$$

When the determinant is expanded, the following N^{th} order polynomial equation for λ results

$$\lambda^N + c_1 \lambda^{N-1} + c_{N-2}\lambda^{N-2} + \cdots + c_N = 0, \qquad (A.22)$$

where the coefficients, c_i, are determined for the evaluation of the determinant. This equation is called the *characteristic equation*. The N values λ which satisfy this characteristic equation are called *eigenvalues*, and for each eigenvalue, a non-trivial solution can be found, called an *eigenvector*.

If all the entries of matrix $\underline{\underline{A}}$ are real, the N eigenvalues could be real or complex number. It can be shown, however, that for symmetric matrices, the eigenvalues and associated eigenvectors are always real [10]. For each distinct eigenvalue, λ_i, eq. (A.21) can be solved for the corresponding eigenvector, $\underline{q}_i$. Repeated eigenvalues require different treatment, see [10]. If one or more eigenvalues are zero, matrix $\underline{\underline{A}}$ is singular. Because eq. (A.21) is a homogeneous equation, the solutions for the eigen-vectors are not unique. Each eigenvector is defined within an arbitrary constant. A unique definition of the eigenvectors is obtained by requiring their norm to be unity.

The eigenvectors can be used to diagonalize matrix $\underline{\underline{A}}$. First, eq. (A.21) can be written for each eigenvectors and the results are collected in a matrix form as

$$\underline{\underline{A}} \left[\underline{q}_1, \underline{q}_2, \cdots, \underline{q}_N\right] = \left[\lambda_1 \underline{q}_1, \lambda_2 \underline{q}_2, \cdots, \lambda_N \underline{q}_N\right] = \left[\underline{q}_1, \underline{q}_2, \cdots, \underline{q}_N\right] \begin{bmatrix} \lambda_1 & & \\ & \ddots & \\ & & \lambda_N \end{bmatrix}.$$

Next, the *eigenvector matrix*, $\underline{\underline{S}}$, is constructed; each column of this matrix stores one of the eigenvectors of $\underline{\underline{A}}$, *i.e.*, $\underline{\underline{S}} = [\underline{q}_1 \ \underline{q}_2 \ \cdots \underline{q}_N]$. With this notation, the above equation can be written in a compact form as

$$\underline{\underline{A}} \, \underline{\underline{S}} = \underline{\underline{S}} \, \underline{\underline{\Lambda}}, \tag{A.23}$$

where $\underline{\underline{\Lambda}}$ is the diagonal matrix whose diagonal elements contain the eigenvalues. Pre-multiplying eq. (A.23) by the inverse of the eigenvector matrix then leads to

$$\underline{\underline{S}}^{-1} \underline{\underline{A}} \, \underline{\underline{S}} = \underline{\underline{\Lambda}}, \tag{A.24}$$

This transformation is called the diagonalization of matrix $\underline{\underline{A}}$.

A.2.5 Positive-definite and quadratic forms

Matrix-array products of the form $\Phi = \underline{q}^T \underline{\underline{A}} \, \underline{q}$, where Φ is a scalar, are frequently encountered in structural analysis. Expanding the array and matrix product leads to

$$\Phi = \{q_1, q_2, \cdots, q_N\} \begin{bmatrix} a_{11} & a_{12} & \cdots & a_{1N} \\ a_{21} & a_{22} & \cdots & a_{2N} \\ \vdots & \vdots & \ddots & \cdots \\ a_{N1} & a_{N2} & \cdots & a_{NN} \end{bmatrix} \begin{Bmatrix} q_1 \\ q_2 \\ \vdots \\ q_N \end{Bmatrix} = \sum_{i=1}^{N} \sum_{j=1}^{N} a_{ij} q_i q_j.$$

Scalar Φ is clearly a quadratic expression in q_i. Consequently, Φ is referred to as a *quadratic form*.

A symmetric matrix, $\underline{\underline{A}}$, is a *positive-definite matrix* if the quadratic form

$$\underline{q}^T \underline{\underline{A}} \, \underline{q} \geq 0, \tag{A.25}$$

for any non-zero, real valued $\underline{q}$ and it is equal to zero only when $\underline{q} = 0$. It is possible to show that the eigenvalues of a positive-definite matrix are all only positive, and non-zero. It is also possible to show that a symmetric matrix with positive, non-zero eigenvalues is positive-definite.

A.2.6 Partial derivatives of a linear form

Consider a scalar, Φ, that is a linear function of three variables, q_1, q_2, and q_3,

$$\Phi = 13q_1 + 8q_2 - 19q_3.$$

The derivatives of this scalar with respect to the three variables are obtained by using elementary rules of calculus to find

$$\frac{\partial \Phi}{\partial q_1} = 13, \quad \frac{\partial \Phi}{\partial q_2} = 8, \quad \frac{\partial \Phi}{\partial q_3} = -19.$$

It is convenient to express this scalar in a more compact form using the following matrix notation

$$\Phi = \{q_1, q_2, q_3\} \left\{ \begin{array}{c} 13 \\ 8 \\ -19 \end{array} \right\} = \underline{q}^T \underline{Q}. \tag{A.26}$$

The derivatives of scalar Φ also can be represented using the compact notation. First, the array of partial derivatives is defined as

$$\frac{\partial \Phi}{\partial \underline{q}} = \left\{ \frac{\partial \Phi}{\partial q_1}, \frac{\partial \Phi}{\partial q_2}, \frac{\partial \Phi}{\partial q_3} \right\}^T,$$

from which it follows that $\partial \Phi / \partial \underline{q} = \{13, 8, -19\}^T = \underline{Q}$. The desired partial derivatives are then readily obtained as

$$\frac{\partial \Phi}{\partial \underline{q}} = \frac{\partial}{\partial \underline{q}} \left(\underline{q}^T \underline{Q} \right) = \underline{Q}. \tag{A.27}$$

These derivatives are identical to those obtained from elementary rules of calculus.

A.2.7 Partial derivatives of a quadratic form

Consider a quadratic form defined by symmetric matrix, $\underline{A}$, and array $\underline{q}$,

$$\Phi = \frac{1}{2} \underline{q}^T \underline{\underline{A}} \, \underline{q} = \frac{1}{2} \{q_1, q_2, q_3\}^T \begin{bmatrix} a_{11} & a_{12} & a_{13} \\ a_{12} & a_{22} & a_{23} \\ a_{13} & a_{23} & a_{33} \end{bmatrix} \left\{ \begin{array}{c} q_1 \\ q_2 \\ q_3 \end{array} \right\}$$

$$= \frac{a_{11}}{2} q_1^2 + \frac{a_{22}}{2} q_2^2 + \frac{a_{33}}{2} q_3^2 + a_{12} q_1 q_2 + a_{23} q_2 q_3 + a_{13} q_1 q_3.$$

The derivatives of this scalar with respect to the three variables are obtained by using elementary rules of calculus to find

$$\frac{\partial \Phi}{\partial q_1} = a_{11} q_1 + a_{12} q_2 + a_{13} q_3,$$

$$\frac{\partial \Phi}{\partial q_2} = a_{12} q_1 + a_{22} q_2 + a_{23} q_3, \tag{A.28}$$

$$\frac{\partial \Phi}{\partial q_3} = a_{13} q_1 + a_{23} q_2 + a_{33} q_3.$$

First, the array of partial derivatives is defined as

$$\frac{\partial \Phi}{\partial \underline{q}} = \left\{ \frac{\partial \Phi}{\partial q_1}, \frac{\partial \Phi}{\partial q_2}, \frac{\partial \Phi}{\partial q_3} \right\}^T,$$

and using this with eq. (A.28) results in the desired partial derivative

$$\frac{\partial \Phi}{\partial \underline{q}} = \frac{\partial}{\partial \underline{q}} \left(\frac{1}{2} \underline{q}^T \underline{\underline{A}} \, \underline{q} \right) = \underline{\underline{A}} \, \underline{q}. \tag{A.29}$$

These derivatives are identical to those obtained from elementary rules of calculus.

A.2.8 Stationarity and quadratic forms

Consider a scalar, Π defined as

$$\Pi(\underline{q}) = \frac{1}{2} \underline{q}^T \underline{\underline{A}} \, \underline{q} - \underline{q}^T \underline{b}, \tag{A.30}$$

where $\underline{q}$ is an array of size N that stores N independent variables, $\underline{\underline{A}}$ is a known symmetric, positive-definite matrix of size $N \times N$, and $\underline{b}$ an array of size N storing known coefficients. Determine the value of the independent variables that make Π stationary, *i.e.*, for which the partial derivatives of Π all vanish $\partial \Pi / \partial \underline{q} = \underline{0}$.

The derivatives are found using eqs. (A.27) and (A.29) as

$$\frac{\partial \Pi}{\partial \underline{q}} = \frac{\partial}{\partial \underline{q}} \left(\frac{1}{2} \underline{q}^T \underline{\underline{A}} \, \underline{q} - \underline{q}^T \underline{b} \right) = \underline{\underline{A}} \, \underline{q} - \underline{b} = \underline{0}, \tag{A.31}$$

The vanishing of the partial derivatives of Π leads to a system of linear equations, $\underline{\underline{A}} \, \underline{q} = \underline{b}$. which can be solved with the help of eq. (A.18) to find $\underline{q} = \underline{\underline{A}}^{-1} \underline{b}$.

A.2.9 Minimization and quadratic forms

Consider once again the scalar Π defined by eq. (A.30). Determine the value of the independent variables that make Π minimum. An elegant solution of this problem is given by Strang [10].

Define a second array, $\underline{p}$, that is the same size as $\underline{q}$ but otherwise arbitrary and construct the following scalar function

$$\Phi(\underline{p}) - \Phi(\underline{q}) = \frac{1}{2} \underline{p}^T \underline{\underline{A}} \, \underline{p} - \underline{p}^T \underline{b} - \frac{1}{2} \underline{q}^T \underline{\underline{A}} \, \underline{q} + \underline{q}^T \underline{b}.$$

Next, let $\underline{b} = \underline{\underline{A}} \, \underline{q}$ to find

$$\Phi(\underline{p}) - \Phi(\underline{q}) = \frac{1}{2} \underline{p}^T \underline{\underline{A}} \, \underline{p} - \underline{p}^T \underline{\underline{A}} \, \underline{q} + \frac{1}{2} \underline{q}^T \underline{\underline{A}} \, \underline{q} = \frac{1}{2} (\underline{p} - \underline{q})^T \underline{\underline{A}} \, (\underline{p} - \underline{q}). \tag{A.32}$$

Because matrix $\underline{\underline{A}}$ is symmetric, $\underline{p}^T \underline{\underline{A}} \, \underline{q} = (\underline{p}^T \underline{\underline{A}} \, \underline{q})^T = \underline{q}^T \underline{\underline{A}} \, \underline{p}$. This identity is used to obtain the last equality. By definition of a positive-definite matrix, eq. (A.25), the last expression of eq. (A.32) must be positive, resulting in $\Phi(\underline{p}) > \Phi(\underline{q})$. Because $\underline{p}$ is arbitrary, this means that the minimum value of Π is achieved for $\underline{b} = \underline{\underline{A}}\,\underline{q}$, which can be solved to find $\underline{q} = \underline{\underline{A}}^{-1}\underline{b}$

Comparing the results obtained here with those of the previous section, it is concluded that if $\underline{\underline{A}}$ is a symmetric, positive-definite matrix, the stationary point of Π is a minimum.

A.2.10 Least-square solution of linear systems with redundant equations

Consider a system of N linear equations, $\underline{\underline{A}}\,\underline{x} = \underline{b}$, where $\underline{\underline{A}}$ is an $N \times N$ square matrix, $\underline{x}$ the array storing the N unknowns of the problem, and $\underline{b}$ the known right-hand side array. If $\underline{\underline{A}}$ is not singular, the solution of this system is simply $\underline{x} = \underline{\underline{A}}^{-1}\underline{b}$, see eq. (A.18). Consider next a system of N linear equations, $\underline{\underline{A}}\,\underline{x} = \underline{b}$, where $\underline{\underline{A}}$ is a rectangular matrix of size $N \times M$, $N > M$, $\underline{x}$ the array of size M storing the unknowns of the problem, and $\underline{b}$ the known right-hand side array. This problem features more equations than unknowns. Such a system is known as an *over-determined system of equations*, and in general, no solution exists.

To obtain an approximate solution of the problem, it is assumed that each of the N equations is not exactly satisfied, but rather, presents an error, hopefully small. This is written as $\underline{\underline{A}}\,\underline{x} - \underline{b} = \underline{e}$, where $\underline{e}$ is the array of errors. A solution is now sought that minimizes the square of the norm of the error array, and this can be stated as

$$\min_{\underline{x}} \|\underline{e}\|^2 = \min_{\underline{x}} \|(\underline{\underline{A}}\,\underline{x} - \underline{b})\|^2 = \min_{\underline{x}} \left[(\underline{\underline{A}}\,\underline{x} - \underline{b})^T (\underline{\underline{A}}\,\underline{x} - \underline{b}) \right].$$

The minimum of this quadratic expression is obtained by requiring the vanishing of its derivatives with respect to $\underline{x}$, *i.e.* $\partial\|\underline{e}\|^2/\partial\underline{x} = 0$. This results in the following equation

$$\frac{\partial\|\underline{e}\|^2}{\partial\underline{x}} = 2\underline{x}^T\underline{\underline{A}}^T\underline{\underline{A}} - 2\underline{b}^T\underline{\underline{A}} = 0,$$

which, after taking the transpose, can be recast as $(\underline{\underline{A}}^T\underline{\underline{A}})\underline{x} = \underline{\underline{A}}^T\underline{b}$. Note that $(\underline{\underline{A}}^T\underline{\underline{A}})$ is now a square matrix of size $M \times M$, and hence, this linear system of equations is readily solved as

$$\underline{x} = (\underline{\underline{A}}^T\underline{\underline{A}})^{-1}\underline{\underline{A}}^T\underline{b}, \tag{A.33}$$

provided that matrix $(\underline{\underline{A}}^T\underline{\underline{A}})$ is not singular. Equation (A.33) provides a *least-squares solution* of the over-determined system of equations.

A.2.11 Problems

Problem A.1. Evaluation of a quadratic form
Evaluate the quadratic form $\Phi = \underline{q}^T\underline{\underline{A}}_a\,\underline{q}$, where $\underline{\underline{A}}_a$ is a skew symmetric form.

A.3 Coordinate systems and transformations

A.3.1 The rotation matrix

Consider two orthonormal bases, $\mathcal{I} = (\bar{\imath}_1, \bar{\imath}_2, \bar{\imath}_3)$ and $\mathcal{I}^* = (\bar{\imath}_1^*, \bar{\imath}_2^*, \bar{\imath}_3^*)$, as shown in fig. A.1. The relative orientation of these two bases is arbitrary.

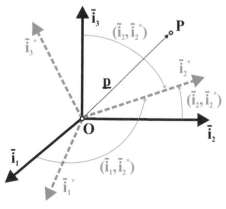

Let ℓ_1, ℓ_2 and ℓ_3 be the direction cosines of unit vector $\bar{\imath}_1^*$ with respect to axes $\bar{\imath}_1$, $\bar{\imath}_2$, and $\bar{\imath}_3$, respectively, *i.e.*, $\ell_1 = \bar{\imath}_1^* \cdot \bar{\imath}_1 = \cos(\bar{\imath}_1^*, \bar{\imath}_1)$, $\ell_2 = \bar{\imath}_1^* \cdot \bar{\imath}_2 = \cos(\bar{\imath}_1^*, \bar{\imath}_2)$, and $\ell_3 = \bar{\imath}_1^* \cdot \bar{\imath}_3 = \cos(\bar{\imath}_1^*, \bar{\imath}_3)$. The direction cosines of unit vector $\bar{\imath}_2^*$ are defined in a similar manner as $m_1 = \bar{\imath}_2^* \cdot \bar{\imath}_1 = \cos(\bar{\imath}_2^*, \bar{\imath}_1)$, $m_2 = \bar{\imath}_2^* \cdot \bar{\imath}_2 = \cos(\bar{\imath}_2^*, \bar{\imath}_2)$, and $m_3 = \bar{\imath}_2^* \cdot \bar{\imath}_3 = \cos(\bar{\imath}_2^*, \bar{\imath}_3)$; these quantities are highlighted in fig. A.1. Finally, the direction cosines of unit vector $\bar{\imath}_3^*$ are $n_1 = \bar{\imath}_3^* \cdot \bar{\imath}_1 = \cos(\bar{\imath}_3^*, \bar{\imath}_1)$, $n_2 = \bar{\imath}_3^* \cdot \bar{\imath}_2 = \cos(\bar{\imath}_3^*, \bar{\imath}_2)$, and $n_3 = \bar{\imath}_3^* \cdot \bar{\imath}_3 = \cos(\bar{\imath}_3^*, \bar{\imath}_3)$. With these definitions, it becomes possible to express the unit vectors of basis $\mathcal{I}^*$ as linear combinations of those of basis $\mathcal{I}$,

Fig. A.1. Position of a point in two coordinate systems.

$$\bar{\imath}_1^* = \ell_1 \bar{\imath}_1 + \ell_2 \bar{\imath}_2 + \ell_3 \bar{\imath}_3, \tag{A.34a}$$

$$\bar{\imath}_2^* = m_1 \bar{\imath}_1 + m_2 \bar{\imath}_2 + m_3 \bar{\imath}_3, \tag{A.34b}$$

$$\bar{\imath}_3^* = n_1 \bar{\imath}_1 + n_2 \bar{\imath}_2 + n_3 \bar{\imath}_3. \tag{A.34c}$$

Similarly, the unit vectors of basis $\mathcal{I}$ can be expressed as linear combinations of those of basis $\mathcal{I}^*$,

$$\bar{\imath}_1 = \ell_1 \bar{\imath}_1^* + m_1 \bar{\imath}_2^* + n_1 \bar{\imath}_3^*, \tag{A.35a}$$

$$\bar{\imath}_2 = \ell_2 \bar{\imath}_1^* + m_2 \bar{\imath}_2^* + n_2 \bar{\imath}_3^*, \tag{A.35b}$$

$$\bar{\imath}_3 = \ell_3 \bar{\imath}_1^* + m_3 \bar{\imath}_2^* + n_3 \bar{\imath}_3^*. \tag{A.35c}$$

It is convenient to defined the *direction cosine matrix* or *rotation matrix*,

$$\underline{\underline{R}} = \begin{bmatrix} \cos(\bar{\imath}_1^*, \bar{\imath}_1) & \cos(\bar{\imath}_2^*, \bar{\imath}_1) & \cos(\bar{\imath}_3^*, \bar{\imath}_1) \\ \cos(\bar{\imath}_1^*, \bar{\imath}_2) & \cos(\bar{\imath}_2^*, \bar{\imath}_2) & \cos(\bar{\imath}_3^*, \bar{\imath}_2) \\ \cos(\bar{\imath}_1^*, \bar{\imath}_3) & \cos(\bar{\imath}_2^*, \bar{\imath}_3) & \cos(\bar{\imath}_3^*, \bar{\imath}_3) \end{bmatrix} = \begin{bmatrix} \ell_1 & m_1 & n_1 \\ \ell_2 & n_2 & m_2 \\ \ell_3 & m_3 & n_3 \end{bmatrix}. \tag{A.36}$$

This matrix fully defines the orientation of basis $\mathcal{I}^*$ with respect to basis $\mathcal{I}$ because it stores the direction cosines of each of the three unit vectors defining $\mathcal{I}^*$ with respect to $\mathcal{I}$. The direction cosines must satisfy the following relationships: $\ell_1^2 + \ell_2^2 + \ell_3^2 = 1$, $m_1^2 + m_2^2 + m_3^2 = 1$, and $n_1^2 + n_2^2 + n_3^2 = 1$. It follows that the rotation matrix has the following property

$$\underline{\underline{R}}^T \underline{\underline{R}} = \begin{bmatrix} \ell_1 & \ell_2 & \ell_3 \\ m_1 & m_2 & m_3 \\ n_1 & n_2 & n_3 \end{bmatrix} \begin{bmatrix} \ell_1 & m_1 & n_1 \\ \ell_2 & m_2 & n_2 \\ \ell_3 & m_3 & n_3 \end{bmatrix} = \begin{bmatrix} 1 & 0 & 0 \\ 0 & 1 & 0 \\ 0 & 0 & 1 \end{bmatrix} = \underline{\underline{I}}, \tag{A.37}$$

where the vanishing of the off-diagonal terms arises from of the orthogonality among the unit vectors themselves: $\bar{\imath}_1^* \cdot \bar{\imath}_2^* = (\ell_1\bar{\imath}_1 + \ell_2\bar{\imath}_2 + \ell_3\bar{\imath}_3) \cdot (m_1\bar{\imath}_1 + m_2\bar{\imath}_2 + m_3\bar{\imath}_3) = \ell_1 m_1 + \ell_2 m_2 + \ell_3 m_3 = 0$. Since the product $\underline{\underline{R}}^T \underline{\underline{R}}$ is equal to the identity matrix, $\underline{\underline{R}}^T$ must be the inverse of the rotation matrix $\underline{\underline{R}}$, or $\underline{\underline{R}}^{-1} = \underline{\underline{R}}^T$. Such matrices are said to be *orthogonal matrices,* and therefore, the *matrix of direction cosines is an orthogonal matrix.*

A.3.2 Rotation of vector components

Consider now the position vector of point **P**, denoted $\underline{p}$ in fig. A.1, and its components expressed in the two bases, $\mathcal{I}$ and $\mathcal{I}^*$,

$$\underline{p} = p_1 \bar{\imath}_1 + p_2 \bar{\imath}_2 + p_3 \bar{\imath}_3 = p_1^* \bar{\imath}_1^* + p_2^* \bar{\imath}_2^* + p_3^* \bar{\imath}_3^*. \tag{A.38}$$

The components of vector $\underline{p}$ in basis $\mathcal{I}$ are denoted p_1, p_2, and p_3, whereas those in basis $\mathcal{I}^*$ are denoted p_1^*, p_2^*, and p_3^*. On the right-hand side of this equation, the unit vectors of basis $\mathcal{I}^*$ will now be expressed in term of their counterparts in basis $\mathcal{I}$ using eq. (A.34) to find

$$p_1 \bar{\imath}_1 + p_2 \bar{\imath}_2 + p_3 \bar{\imath}_3 = p_1^*(\ell_1\bar{\imath}_1 + \ell_2\bar{\imath}_2 + \ell_3\bar{\imath}_3) + p_2^*(m_1\bar{\imath}_1 + m_2\bar{\imath}_2 + m_3\bar{\imath}_3)$$
$$+ p_3^*(n_1\bar{\imath}_1 + n_2\bar{\imath}_2 + n_3\bar{\imath}_3).$$

A scalar product of this result by $\bar{\imath}_1$, $\bar{\imath}_2$, and $\bar{\imath}_3$ yields three equations, $p_1 = p_1^*\ell_1 + p_2^*m_1 + p_3^*n_1$, $p_2 = p_1^*\ell_2 + p_2^*m_2 + p_3^*n_2$, and $p_3 = p_1^*\ell_3 + p_2^*m_3 + p_3^*n_3$, respectively. These equations relate the components of vector $\underline{p}$ in bases $\mathcal{I}$ and $\mathcal{I}^*$, and can be summarized in a compact matrix form as

$$\begin{Bmatrix} p_1 \\ p_2 \\ p_3 \end{Bmatrix} = \begin{bmatrix} \ell_1 & m_1 & n_1 \\ \ell_2 & m_2 & n_2 \\ \ell_3 & m_3 & n_3 \end{bmatrix} \begin{Bmatrix} p_1^* \\ p_2^* \\ p_3^* \end{Bmatrix} = \underline{\underline{R}} \begin{Bmatrix} p_1^* \\ p_2^* \\ p_3^* \end{Bmatrix},$$

where the rotation matrix, $\underline{\underline{R}}$, is defined in eq. (A.36). The rotation matrix expresses the linear relationship between the components of vector $\underline{p}$ in bases $\mathcal{I}$ and $\mathcal{I}^*$. Of course, the inverse relationship is easy to find

$$\begin{Bmatrix} p_1 \\ p_2 \\ p_3 \end{Bmatrix} = \underline{\underline{R}} \begin{Bmatrix} p_1^* \\ p_2^* \\ p_3^* \end{Bmatrix} \Longleftrightarrow \begin{Bmatrix} p_1^* \\ p_2^* \\ p_3^* \end{Bmatrix} = \underline{\underline{R}}^T \begin{Bmatrix} p_1 \\ p_2 \\ p_3 \end{Bmatrix}, \tag{A.39}$$

because the rotation matrix is orthogonal, see eq. (A.37). The rotation matrix "rotates the components of vector $\underline{p}$" from one coordinate system to the other. A vector is a mathematical entity characterized by a magnitude and orientation in space. For practical reasons, however, it is often easier to represent a vector by its components in a specific basis. For instance, the three components, p_1, p_2, and p_3, represent vector $\underline{p}$ in basis $\mathcal{I}$. Had a different basis been selected, say $\mathcal{I}^*$, the same vector $\underline{p}$ would have been represent by a different set of components, p_1^*, p_2^*, and p_3^*. When vector $\underline{p}$ is represented in two different bases, $\mathcal{I}$ and $\mathcal{I}^*$, the corresponding components, p_1, p_2, p_3, and p_1^*, p_2^*, p_3^*, respectively, must be related by eqs. (A.39).

While the above development has focused on the position vector of an arbitrary point **P**, similar arguments could have been used for other vectors, such as displacement vectors or force vectors. Equations (A.39) are very general and express the relationship between the components of any vector in two different bases. In fact, eqs. (A.39) can be taken as the definition of a vector quantity: *a vector is a mathematical entity whose components in two different bases are related by eqs. (A.39).*

A.3.3 The rotation matrix in two dimensions

The previous section has focused on coordinate transformations in three dimensions. In many cases, however, a simpler, two-dimensional transformation is sufficient. Consider, for instance, the plane stress or plane strain problems investigated in sections 1.3 or 1.6, respectively. Two unit vectors, say $\bar{\imath}_1$ and $\bar{\imath}_2$, define the plane of the problem, whereas $\bar{\imath}_3$ is normal to this plane, as depicted in fig. A.2. A second set of unit vectors, $\bar{\imath}_1^*$ and $\bar{\imath}_2^*$, is now selected such that the angle between axes $\bar{\imath}_1$ and $\bar{\imath}_1^*$ is θ. Note that since $(\bar{\imath}_1, \bar{\imath}_2)$ and $(\bar{\imath}_1^*, \bar{\imath}_2^*)$ define the same plane, $\bar{\imath}_3 = \bar{\imath}_3^*$ are both normal to this plane.

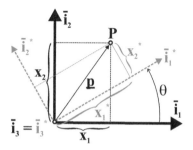

Fig. A.2. Position of a point in two coordinate systems.

Two orthonormal bases, $\mathcal{I} = (\bar{\imath}_1, \bar{\imath}_2, \bar{\imath}_3)$ and $\mathcal{I}^* = (\bar{\imath}_1^*, \bar{\imath}_2^*, \bar{\imath}_3^*)$, have now been defined, a situation similar to that of section A.3.1. Many of the direction cosines of the present problem, however, have special values, because $\bar{\imath}_3 = \bar{\imath}_3^*$. For instance, $\ell_1 = \bar{\imath}_1^* \cdot \bar{\imath}_1 = \cos\theta$, $\ell_2 = \bar{\imath}_1^* \cdot \bar{\imath}_2 = \sin\theta$, and $\ell_3 = \bar{\imath}_1^* \cdot \bar{\imath}_3 = 0$. Similarly, $m_1 = \bar{\imath}_2^* \cdot \bar{\imath}_1 = -\sin\theta$, $m_2 = \bar{\imath}_2^* \cdot \bar{\imath}_2 = \cos\theta$, and $m_3 = \bar{\imath}_2^* \cdot \bar{\imath}_3 = 0$; finally, $n_1 = \bar{\imath}_3^* \cdot \bar{\imath}_1 = 0$, $n_2 = \bar{\imath}_3^* \cdot \bar{\imath}_2 = 0$, and $n_3 = \bar{\imath}_3^* \cdot \bar{\imath}_3 = 1$. The rotation matrix defined by eq. (A.36) now simplifies to

$$\underline{\underline{R}} = \begin{bmatrix} \cos\theta & -\sin\theta & 0 \\ \sin\theta & \cos\theta & 0 \\ 0 & 0 & 1 \end{bmatrix}.$$

The entries in the last line and column simply imply that the component of a vector along axis $\bar{\imath}_3 = \bar{\imath}_3^*$ is unaffected by the change of basis. In many cases, it is not necessary to use a 3×3 rotation matrix; the use of the smaller size, 2×2 rotation matrix is then preferable,

$$\underline{\underline{R}} = \begin{bmatrix} \cos\theta & -\sin\theta \\ \sin\theta & \cos\theta \end{bmatrix}. \tag{A.40}$$

The fact that the direction cosine matrix is an orthogonal matrix, as show in eq. (A.37), is now a straightforward consequence of trigonometric identities,

$$\underline{\underline{R}}\,\underline{\underline{R}}^T = \begin{bmatrix} \cos\theta & -\sin\theta \\ \sin\theta & \cos\theta \end{bmatrix} \begin{bmatrix} \cos\theta & \sin\theta \\ -\sin\theta & \cos\theta \end{bmatrix} = \begin{bmatrix} 1 & 0 \\ 0 & 1 \end{bmatrix} = \underline{\underline{I}}. \tag{A.41}$$

A.3.4 Rotation of vector components in two dimensions

Consider now the position vector of point **P**, denoted $\underline{p}$ in fig. A.2, and its components expressed in the two bases, $\mathcal{I}$ and $\mathcal{I}^*$,

$$\underline{p} = p_1\,\bar{\imath}_1 + p_2\,\bar{\imath}_2 = p_1^*\vec{\imath}_1^* + p_2^*\vec{\imath}_2^*. \tag{A.42}$$

Vector $\underline{p}$ is assumed to be in plane $(\bar{\imath}_1, \bar{\imath}_2)$, and hence its component along axis $\bar{\imath}_3 = \bar{\imath}_3^*$ vanishes. The components of vector $\underline{p}$ in basis $\mathcal{I}$ are denoted p_1 and p_2, whereas those in basis $\mathcal{I}^*$ are denoted p_1^* and p_2^*. Following the procedure developed in section A.3.2, the following relationship is found between these two sets of components $p_1\,\bar{\imath}_1 + p_2\,\bar{\imath}_2 = p_1^*(\cos\theta\,\bar{\imath}_1 + \sin\theta\,\bar{\imath}_2) + p_2^*(-\sin\theta\,\bar{\imath}_1 + \cos\theta\,\bar{\imath}_2)$. A scalar product of this result by $\bar{\imath}_1$ and $\bar{\imath}_2$ the following equations

$$\begin{Bmatrix} p_1 \\ p_2 \end{Bmatrix} = \begin{bmatrix} \cos\theta & -\sin\theta \\ \sin\theta & \cos\theta \end{bmatrix} \begin{Bmatrix} p_1^* \\ p_2^* \end{Bmatrix} = \underline{\underline{R}} \begin{Bmatrix} p_1^* \\ p_2^* \end{Bmatrix}.$$

The rotation matrix expresses the linear relationship between the components of vector $\underline{p}$ in bases $\mathcal{I}$ and $\mathcal{I}^*$. Of course, the inverse relationship is easy to find

$$\begin{Bmatrix} p_1 \\ p_2 \end{Bmatrix} = \underline{\underline{R}} \begin{Bmatrix} p_1^* \\ p_2^* \end{Bmatrix} \Longleftrightarrow \begin{Bmatrix} p_1^* \\ p_2^* \end{Bmatrix} = \underline{\underline{R}}^T \begin{Bmatrix} p_1 \\ p_2 \end{Bmatrix}, \tag{A.43}$$

because the rotation matrix is orthogonal, see eq. (A.41). These equations should be compared to their three-dimensional counterparts, eq. (A.39).

A.4 Orthogonality properties of trigonometric functions

Trigonometric functions enjoy remarkable orthogonality properties, which are often used to obtain series solution of various problems. The Kronecker delta symbol will be used to express these properties in a compact manner and is defined as

$$\delta_{ij} = \begin{cases} 1 & i = j, \\ 0 & i \neq j. \end{cases} \tag{A.44}$$

Consider now the product of sine or cosine functions with different wave numbers, m an n. The integration of these products leads to the following results

$$\int_0^1 \sin m\pi\eta \sin n\pi\eta \, d\eta = \frac{\delta_{mn}}{2}, \tag{A.45a}$$

$$\int_0^1 \cos m\pi\eta \cos n\pi\eta \, d\eta = \frac{\delta_{mn}}{2}. \tag{A.45b}$$

The integration of the product of two sine functions with different wave numbers vanishes, except when the two wave numbers are identical; the same is true for the cosine function. Equation (A.45a) expresses the orthogonality of the sine functions: the set of functions, $\sin m\pi\eta$, $m = 1, 2, \ldots, \infty$, is said to be orthogonal over the range $\eta \in [0, 1]$ because eq. (A.45a) holds. The cosine functions, $\cos m\pi\eta$, $m = 1, 2, \ldots, \infty$, are also orthogonal over the same range. Similarly, the following results can be verified,

$$\int_{-1/2}^{+1/2} \sin m\pi\eta \sin n\pi\eta \, d\eta = \frac{\delta_{mn}}{2}, \tag{A.46a}$$

$$\int_{-1/2}^{+1/2} \cos m\pi\eta \cos n\pi\eta \, d\eta = \frac{\delta_{mn}}{2}. \tag{A.46b}$$

The sine functions, $\sin m\pi\eta$, $m = 1, 2, \ldots, \infty$, are also orthogonal over the range $\eta \in [-1/2, +1/2]$.

The following definite integral are also useful

$$\int_{-1/2}^{+1/2} \cos m\pi\eta \, d\eta = \frac{2}{m\pi} \begin{cases} 0, & m \text{ even,} \\ (-1)^{(m-1)/2}, & m \text{ odd.} \end{cases} \tag{A.47}$$

$$\int_{0}^{1} \sin m\pi\eta \, d\eta = \frac{2}{m\pi} \begin{cases} 0, & m \text{ even,} \\ 1, & m \text{ odd.} \end{cases} \tag{A.48}$$

A.5 Gauss-Legendre quadrature

When applying energy methods, the computation of the stiffness matrix and load array involves integrations of the product of the shape functions by the stiffness properties of the structure. As the number of assumed shape function increases, it becomes increasingly cumbersome to perform all these integration in closed form, especially when the expression for the shape functions becomes complex.

To circumvent this problem, numerical integration can be used. A very powerful tool for numerical integration is the Gauss-Legendre quadrature scheme. In its simplest form [11], this scheme approximately evaluates an integral by the following sum

$$\int_{-1}^{+1} f(\eta) \, d\eta \approx \sum_{i=1}^{N} w_i f(\eta_i), \tag{A.49}$$

where η_i, $i = 1, 2, \ldots N$ are the *Gauss-Legendre quadrature points*, and w_i the associated *weights*. The Gauss-Legendre quadrature points are often called *sampling points*, because the integral is evaluated by sampling the value of the integrand at these points. Table A.1 lists the Gauss-Legendre quadrature points and associated weights for $N = 2$, 3, and 4. The fundamental property of the N point Gauss-Legendre quadrature scheme is that it exactly integrates a polynomial of degree $2N - 1$.

Table A.1. Gauss points and associated weights for $N = 2, 3,$ and 4.

N	η_i	w_i
2	$\pm\sqrt{1/3}$	1
3	0	8/9
	$\pm\sqrt{3/5}$	5/9
4	$\pm\sqrt{(3 - 2\sqrt{6/5})/7}$	$(18 + \sqrt{30})/36$
	$\pm\sqrt{(3 + 2\sqrt{6/5})/7}$	$(18 - \sqrt{30})/36$

To illustrate the application of the Gauss-Legendre quadrature scheme, consider the following integral

$$I = \int_{-1}^{+1} \left[x^4 - 5x^3 + 3x^2 + 5x \right] \, \mathrm{d}x = 2.4.$$

At first, the 2-point quadrature formula is used to find

$$I \approx \left[\left(\frac{1}{3} \right)^2 + 5 \left(\frac{1}{3} \right)^{3/2} + 3\frac{1}{3} - 5 \left(\frac{1}{3} \right)^{1/2} \right]$$

$$+ \left[\left(\frac{1}{3} \right)^2 - 5 \left(\frac{1}{3} \right)^{3/2} + 3\frac{1}{3} + 5 \left(\frac{1}{3} \right)^{1/2} \right] = \frac{20}{9} = 2.22.$$

This 2-point formula exactly integrates a polynomial of degree $2 \times 2 - 1 = 3$; hence, an approximate answer is expected for this integral involving a polynomial of degree four. The approximate answer only incurs a 7.4% error. Next, the 3-point quadrature formula is used, leading to

$$I \approx \frac{5}{9} \left[\left(\frac{3}{5} \right)^2 + 5 \left(\frac{3}{5} \right)^{3/2} + 3\frac{3}{5} - 5 \left(\frac{3}{5} \right)^{1/2} \right]$$

$$+ \frac{5}{9} \left[\left(\frac{3}{5} \right)^2 - 5 \left(\frac{3}{5} \right)^{3/2} + 3\frac{3}{5} + 5 \left(\frac{3}{5} \right)^{1/2} \right] = \frac{60}{25} = 2.4.$$

This 3-point formula exactly integrates a polynomial of degree $3 \times 2 - 1 = 5$; hence, the exact solution is recovered.

Next, consider the following integral involving transcendental function

$$I = \int_1^5 \frac{1}{x} \, \mathrm{d}x = [\ln x]_1^5 = \ln 5 = 1.609.$$

To recast the problem in the standard form, a change of variable, $x = 2\eta + 3$, is first performed. The Jacobian of the coordinate transformation is readily evaluated, $\mathrm{d}x/\mathrm{d}\eta = 2$. The 2-point quadrature formula then yields a first approximation of the integral

$$I = \int_{-1}^{+1} \frac{1}{2\eta + 3} \frac{dx}{d\eta} \, d\eta \approx 2 \left[\frac{1}{-2\sqrt{1/3} + 3} + \frac{1}{2\sqrt{1/3} + 3} \right] = \frac{36}{23} = 1.565,$$

which only involves a 2.75% error. To improve the approximation, the 3-point quadrature formula is used, leading to

$$I \approx \frac{2}{9} \left[\frac{5}{-2\sqrt{3/5} + 3} + \frac{8}{3} + \frac{5}{2\sqrt{3/5} + 3} \right] = \frac{476}{297} = 1.603.$$

The error is now reduced to about 0.42%. Higher order Gauss-Legendre quadrature scheme can be derived that involve an increasing number of sampling points and associated weights. This data have been tabulated, see Abramowitz and Stegun [12], or can be readily calculated [11].

For integration over a rectangular domain, the basic Gauss-Legendre quadrature scheme of eq. (A.49) is generalized as

$$\int_{-1}^{+1} \int_{-1}^{+1} f(\eta, \zeta) \, d\eta d\zeta \approx \sum_{i=1}^{N} \sum_{j=1}^{M} w_i w_j f(\eta_i, \zeta_j), \tag{A.50}$$

where the sampling points, η_i and ζ_j, and associated weights, w_i and w_j, respectively, are those listed in table A.1.

References

1. B.K. Donaldson. *Analysis of Aircraft Structures*. Cambridge University Press, Cambridge, New York, second edition, 2008.
2. T.H.G. Megson. *Aircraft Structures for Enginnering Students*. Elsevier Aerospace Engineering Series, Oxford, fourth edition, 2007.
3. D.J. McGill and W.W. King. *An Introduction to Dynamics*. PWS-KENT Publishing Company, Englewood Cliffs, New Jersey, third edition, 1995.
4. J.H. Ginsberg. *Advanced Engineering Dynamics*. Cambridge University Press, Cambridge, second edition, 1998.
5. R. Ewing. *Calculus of Variations with Applications*. Dover Publications, Inc., New York, 1985.
6. R. Weinstock. *Calculus of Variations with Applications to Physics and Engineering*. Dover Publications, Inc., New York, 1974.
7. F.B. Hildebrand. *Advanced Calculus for Applications*. Prentice Hall, Inc., Englewood Cliffs, New Jersey, second edition, 1976.
8. T.J.R. Hughes. *The Finite Element Method*. Prentice Hall, Inc., Englewood Cliffs, New Jersey, 1992.
9. K.J. Bathe. *Finite Element Procedures*. Prentice Hall, Inc., Englewood Cliffs, New Jersey, 1996.
10. G. Strang. *Linear Algebra and its Applications*. Thomson-Brooks/Cole, Toronto, 2006.
11. W.H. Press, B.P. Flannery, S.A. Teutolsky, and W.T. Vetterling. *Numerical Recipes. The Art of Scientific Computing*. Cambridge University Press, Cambridge, 1990.
12. M. Abramowitz and I.A. Stegun. *Handbook of Mathematical Functions*. Dover Publications, Inc., New York, 1964.

Index

CPSIA information can be obtained at www.ICGtesting.com
Printed in the USA
LVOW01*2007110115

422388LV00001B/1/P